T0133229

Historical Biogeography, Plate Tectonics, and the Changing Environment

Historical Biogeography, Plate Tectonics, and the Changing Environment

Proceedings of the
Thirty-seventh Annual Biology Colloquium
and Selected Papers

Edited by Jane Gray and Arthur J. Boucot

Corvallis:
OREGON STATE UNIVERSITY PRESS

Library of Congress Cataloging in Publication Data

Biology Colloquium, 37th, Oregon State University, 1976.
Historical biogeography, plate tectonics, and the changing environment.
1. Geographical distribution of animals and plants—History—Congresses. 2. Plate tectonics —Congresses. 3. Marine fauna—Geographical distribution—History—Congresses. 4. Marine flora—Geographical distribution—History—Congresses. I. Gray, Jane, 1931- II. Boucot, Arthur James, 1924-

QH84.B56 1976 574.9 78-31376

ISBN 0-87071-176-8

Printed in the United States of America

Preface

The 37th Annual Biology Colloquium of Oregon State University, by its choice of topic—"Historical biogeography, plate tectonics, and the changing environment"—represents a radical departure from preceding Colloquia. For the first time, it has addressed a topic whose practitioners are primarily geologists, most with limited or no biological backgrounds, and many of whom are concerned with neither the present nor the immediate past, but with the past of 100 to 600 m. y. ago. Biogeography as a topic is not new to the Colloquium series, but the slim volume of 1947 (47 pages) related entirely to the biogeography of terrestrial organisms, including the human animal, of the Pleistocene and Tertiary. Plate tectonics as such had not been heard of at that time, and continental drift, although it had its adherents, particularly in Europe, was hardly the voguey topic of today. Wegener and continental drift do not go unmentioned in 1947, but the references are perfunctory. E. L. Packard remarks (1947, p. 11) that the Wegenerian hypothesis "is not yet fully accepted" and Ernst Antevs (1947, p. 8) refers briefly to "junctions and separations of continents and oceans" as "among past geographical changes." Continental drift is obviously perceived as not being fundamental to a serious discussion of biogeography.

Between 1947 and 1976, less than two generations, the world has witnessed a scientific revolution that rightly or wrongly, has had as far-reaching consequences for many disciplines as the revolution precipitated by Darwin and Wallace in the mid-1800's. In today's intellectual climate most discussions of biogeography and historical biogeography are predicated on the blanket acceptance of continental drift and plate tectonics, even though general agreement is lacking on the geometry of the world at most times in the past, with the consequences that continental masses are fragmented and reassembled with abandon, independently for each group of organisms and for each time span. Few concepts have enjoyed more of a bandwagon among scientists with biogeographic interests than continental drift, which is now being mindlessly touted as the "philosophers stone," the solution to all problems of distribution past and present. The collected essays that comprise this volume provide no exception to this general viewpoint. Despite other differences—the diversity of disparate organisms some of whose biological affinities are obscure, the adherence to "natural" or "artifical" taxonomies, and different ground rules in collecting and plotting their data—essentially all are united by an unquestioned acceptance of continental drift and plate tectonics.

Until the present decade, there have been little more than scattered forays into the biogeographic thicket posed by the fossil record, despite early syntheses for modern organisms in pre-Darwinian and penecontemporaneous Darwinian times by Alexander von Humboldt and P. L. Sclater, and the tremendous impetus given to biogeography by Alfred Wallace in 1879 who freed the field of biogeography for the first time from the "thus far and no further" distributional philosophy of the Creationists. There are, of course, outstanding exceptions to this generalization—wide-ranging attempts to resurrect biogeographic information from fossil organisms and make use of it constructively—the studies of M. Neumayr on Jurassic ammonites, those of Eric Florin on Mesozoic conifers, and the work of such notables as E. W. Berry and R. W. Chaney, whose biogeographic contributions dominated Cenozoic paleobotany during the first half of the 20th Century. But the renewed interest in historical biogeography (and in biogeography as a whole) marked by the current systematic effort on the part of many workers to accumulate relevant information, may be said to largely coincide with renewed interest in, and general acceptance of, continental drift.

The assembled essays represent a further departure from the Biogeography Colloquium of 1947 in that they are overwhelmingly devoted to marine organisms. The great bulk of biogeographic effort relating to living organisms, and to organisms of the most immediate past, deals with nonmarine biotas. For life of the more distant geologic past, shallow-water marine organisms have received the lion's share of the attention. The reasons for this dichotomy are simple enough. Fresh-water and terrestrial organisms are near at hand, more familiar than marine organisms and therefore perceived as being "easier to study" in many respects. Wide-ranging taxonomic studies on a global basis have also made available much information on the distribution of terrestrial organisms. The bulk of the readily available fossils, on the other hand, are remains of shallow-water marine invertebrates, chiefly low trophic level benthos, and these have been studied over the years for biostratigraphic purposes by those with "potential" biogeographic interests. What may be perceived as the most abundant of all fossil remains on a global scale, and as perhaps potentially the most useful—the organic microfossils—both plant and animal—have received little biogeographic attention in lieu of the initial necessity for taxonomic and biostratigraphic studies by students of these groups. Certain of these, in particular wind dispersed plant microfossils, present some sampling problems uniquely related to their mode of dispersal not encountered with other groups. They especially have attracted limited biogeographic attention.

The bias toward marine organisms and particularly marine invertebrates is readily apparent not only in the papers of the present volume, but in compilations of similar data presented in recent biogeographic summaries. Thus in Middlemiss and others (1971) there are 9 titles devoted to marine organisms, 1 to nonmarine. In Hallam (1973), 36 titles are devoted to marine organisms, 1 to nonmarine. In Hughes (1973), 10 titles are devoted to marine organisms, 5 to nonmarine. In Ross (1974), 7 titles relate to marine organisms, 2 to nonmarine. In the present volume, marine organisms again make up the bulk of the papers with 28 titles; these range over a wide spectrum of topics, from chitinozoans to walruses to more conventional organisms for biogeographic study, the brachiopods and corals. Nonmarine organisms are represented by only 7 titles. In each of these volumes, including the present one, there is interestingly, an inverse relation between the number of titles and the relative age of the fossils being considered, a relationship that appears as well to reflect the energy with which data is being collected relative to organisms in geologic time.

Despite omnipresent imperfections and omissions, we cannot but be impressed with the fact that historical biogeography has finally come of age as a lively, valuable adjunct to a number of disciplines, both physical and biological. It has come of age in the sense that paleontologists, most of whom are practising taxonomists, have finally come to view global paleobiogeographic syntheses as a useful, routine method for evaluating and integrating their data. We confidently predict that the next generation of taxonomic paleontologists will publish many more biogeographic syntheses that, in turn, will begin to influence the attitudes of practising taxonomists concerned with the classification and evolution of their special groups.

Jane Gray
A. J. Boucot

REFERENCES

Hallam, A., ed., 1973, Atlas of palaeobiogeography: Amsterdam, Elsevier, 531 p.

Hughes, N. F., ed., 1973, Organisms and continents through time: Spec. Papers Palaeontology no. 12, 334 p.

Middlemiss, F. A., Rawson, P. F., and Newall, G., eds., 1971, Faunal provinces in space and time: Liverpool, Seel House Press, 236 p.

Ross, C. A., ed., 1974, Paleogeographic provinces and provinciality: Soc. Econ. Paleontologists and Mineralogists Spec. Pub. 21, 233 p.

Editors' Disclaimer

Jane Gray, *Department of Biology, University of Oregon, Eugene, Oregon 97405*
A. J. Boucot, *Department of Geology, Oregon State University, Corvallis, Oregon 97331*

In what will be viewed as an unusual move, we wish to avail ourselves of this opportunity to recognize in a very general way, some of the deficiencies in the collected essays that will surely be noted and commented on by reviewers of this volume. This is in lieu of a formal introduction.

Editors of volumes of essays which they have specifically solicited, not infrequently find themselves in the paradoxical position of being in considerable disagreement with methodology and conclusions, but with less editorial latitude than editors to whom unsolicited papers have been submitted which are subject to peer review. Our task as editors has been largely confined to imposing stylist uniformity—which in some cases has itself been a formidable task. We have otherwise maintained a largely *laissez-faire* attitude toward the contents of the papers.

By specifically refraining from editorial comments to the authors, we have encouraged wide latitude in interpretation and variability in conclusions. Nevertheless, we believe that there is much that can be questioned, and much that must be subject to careful scrutiny by the scientific community at large. While we eschew *specific* comments here, since the authors will have been provided with no opportunity for rebuttal, we disclaim agreement with many of the viewpoints expressed, and many of the conclusions reached by the authors. All in all, as reviewers of such collections are often fond of noting, the assembled essays can be best described as "uneven."

We have also refrained from imposing uniformity with regard to certain of the ground rules used by the contributors to this volume. For example, the base maps on which data have been plotted and on which boundaries have been drawn epitomize a variety of divergent usages. One group prefers to plot its data and boundaries on the geography of the present. A second group tends to employ various continental drift plate tectonic reconstructions which move the present continents, as well as Peninsular India, Madagascar, and sometimes New Zealand, in a variety of ways like chessmen. A third group tends to fragment the present continents into geographic fractions corresponding to biogeographic units which are then sorted out so as to bring similar biogeographic units together or at least into proximity. The last procedure is most commonly employed in the Paleozoic where the information of marine geology and geophysics is no longer available for guidance (see van Andel, this volume). An entirely new and different approach to this problem for the Paleozoic has been attempted by us as a form of epilogue.

As any reader of this volume will quickly ascertain, the assembled essays represent a variety of levels of information. Some are culminations of years—even a lifetime—of work. Others can be viewed as merely progress reports along the road to ultimate knowledge of the organisms in question. The latter are unquestioningly pioneering studies, and the information they provide can only be regarded as preliminary. The potential for such information remaining unaltered for any length of time is relatively low.

Some are based on extensive raw data collected over a period of time; others are primarily exercises in logic, based on little or no actual fossil information. As such, they are largely theoretical or speculative. Of special interest is the disagreement, for example, in the number of biogeographic units arrived at by those applying logic as opposed to those who consider themselves practicing taxonomists. Among taxonomists dealing with various fossil groups the agreement about the number of units is reassuring enough to suggest that model-building which relies too heavily on unsupported assumptions may have limited value for actual problem solving.

In a slightly different context, it was pointed out during the course of the Colloquium, using the Devonian as an example, that certain groups provide far more biogeographic resolution than do others. This is not an artifact induced by disparate levels of attention given to various groups. Rather it reflects varied dispersal mechanisms combined with varying levels of eurytopy. In the Early Devonian, graptolites show far less biogeographic differentiation than do conodonts, which in turn show less than either brachiopods or corals, while trilobites among the groups considered appear to show the highest levels of biogeographic differentiation. Some of the Early Devonian work is done at the species level but most has been done at the generic and subgeneric levels. During the later Devonian, House (1973) has made the point that ammonites are very cosmopolitan, and it is clear that they co-occur with shelly benthos in the Middle Devonian that are far from cosmopolitan. This makes it obvious that the groups with cosmopolitan tendencies, particularly the more rapidly evolving taxa, will be of great value for purposes of global stratigraphic correlation, whereas the very provincial groups such as the trilobites of the Early and Middle Devonian will be of greatest value in recognizing biogeographic units. There is no reason to believe that the type of conclusions arrived at with Devonian marine invertebrates will not apply equally well to other time intervals characterized by other groups of organisms.

Some of the authors deal with data they realized to be biased either because of the limited availability of raw data on a global scale, or because the taxonomy of the group is in a state of flux that hampers understanding of the distribution patterns. Few of the contributors, however, specifically address the question of how and whether this data limits their conclusions perhaps tacitly assuming that data is always "biased" in paleontology.

Most of our contributors have failed to heed Hedgpeth's admonition (1957, and repeated here as well) that the sampling problems faced by biogeographers necessitates their ascertaining which communities are present and which have been sampled in each region *before* making conclusions about presences or absences of taxa on either a local or global scale. For example, the absence of demersal fishes from an area sampled only with a midwater trawl is not significant; neither is the absence of a suite of reef organisms until it is certain that reef community complexes in that area are not merely unstudied or uncollected. Obviously, the question of ascertaining which communities are present in any area will present a significantly different problem depending on whether one deals with benthic communities, pelagic micro-organisms, or wind-borne microfossils, such as pollen and spores whose source communities may be very disparate.

The problem of sample size, especially as it relates to the small sample, is one of which some paleontologists are only now becoming aware. It is well to recognize in this connection that in poorly sampled areas the small sample invariably overemphasizes the most common taxa in the assemblage; hence it provides limited information except about what are the dominant taxa. The small sample invariably overemphasizes as well, the cosmopolitan aspect of the biota—Figure 1 illustrates the fact that as number of taxa (number of genera and subgenera) known from a biogeographic unit decreases, the percentage of cosmopolitan taxa increases—purely a sampling artifact. The chances are thus great that where only a few specimens are available, the veritable traveller's cigarbox collection, that they will be of cosmopolitan taxa.

A different kind of sampling problem relates to the dilemma of trying to decide whether unlike biotas represent ecologic or biogeographic units. This problem is ever present and crucial. It requires the exercise of common sense, a large measure of experience, and willingness to retain a healthy scepticism until the evidence appears to be overwhelming. No formula or computer program will replace these attributes. There are already published examples that strongly suggest confusion of ecologic units with biogeographic units, as well as other examples in which the two have been thoroughly scrambled. It will not do to adopt a defeatist attitude of treating every distinctive biofacies as a biogeographic unit. The reverse approach, of assuming that because biofacies differences involve ecologic units there is no such thing as a biogeographic unit, is equally defeatist.

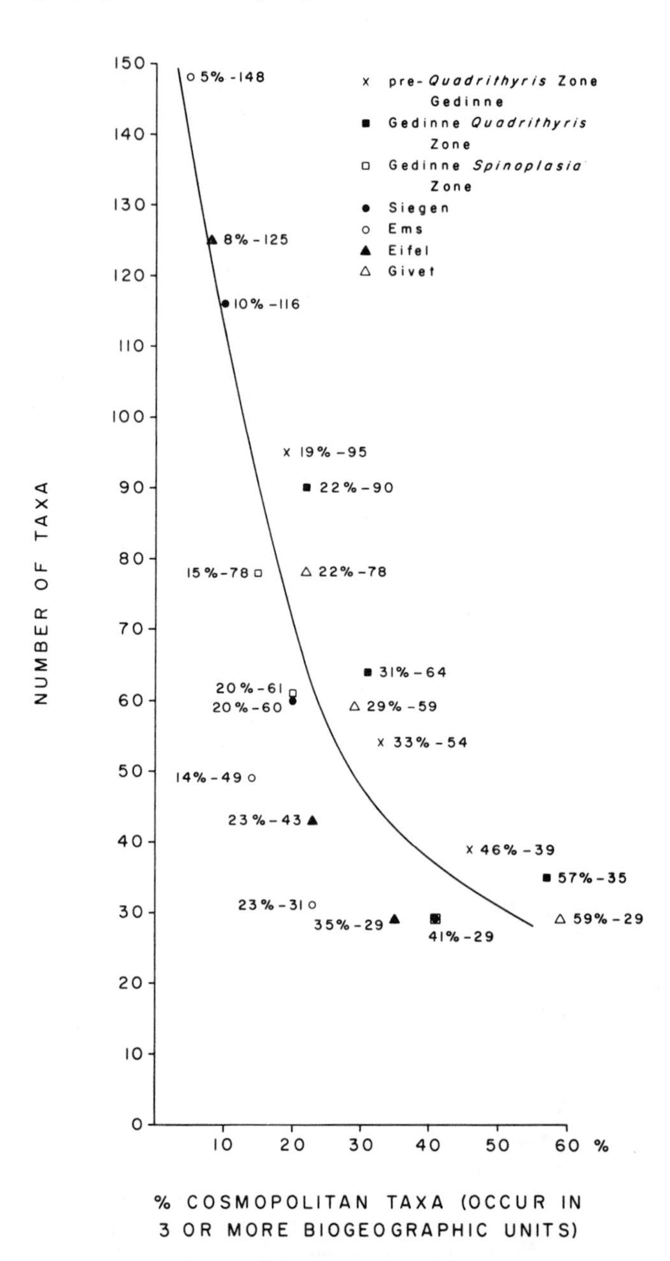

Figure 1. Number of taxa (genera and subgenera) plotted against percentage of cosmopolitan taxa of Devonian brachiopods (from Boucot, 1975, Fig. 28). Each set of figures represents roughly equivalent biogeographic units for Devonian time, Gedinne through Givet. In poorly sampled biogeographic units, the apparent percentage of cosmopolitan taxa is always greater than in units where collecting has been more adequate, since endemic taxa are invariably more rarely represented. The small sample thus introduces a significant sampling artifact.

Few of the assembled papers in the present collection are concerned with what *should* be the ultimate long range goals of historical biogeography—the integration of all available information into an internally consistent whole, consistent with all available biological information, as well as with all available physical evidence from geology and geophysics. For most practitioners, the goals are less ambitious and more immediate. It is sufficient merely to work out geographic distributions of organisms past and present. In so doing, as is amply borne out by the papers of the Colloquium volume, there is often a cavalier disregard for other than one's own data and one's own causal factors in establishing biogeographic provinces and biogeographic boundaries. The failure to be concerned with an internally consistent whole, a time integrated biogeography, appears to be due to the simple reason that most efforts in historical biogeography are carried out by paleontologists or biostratigraphers who have specialized in one time interval, commonly a geologic period or a fraction thereof. Few specialists are seriously interested in or involved with the biogeographic picture which either procedes or follows the interval of specific interest to them. Although we recognize that this state of affairs is part of the normal growth of the discipline, we must look ahead to, and encourage the possibilities inherent in, synthesizing data for time intervals longer than a geologic period or a portion of a geologic period. An internally consistent picture demands this type of attention, for we commonly find that the data available for any limited time duration are inadequate for ruling out most alternate interpretations. Provinces and boundaries necessary to explain one set of data may prove deficient or impossible with another set, while biogeographic barriers are clearly an embarassment when co-occurring taxa do not share the same boundary patterns.

Finally, we would like to relieve ourselves of a few personal comments engendered by our editorial responsibilities. As rank neophytes in this game, we were amazed to note the devil-may-care attitude often assumed toward bibliographical data. While attempting to rectify missing information, we uncovered errors running on the order of 25 percent in reference citations—including lack of correspondence between references cited in the text and those listed in the bibliography, wrong titles associated with journal citations, wrong journals, volumes and years of publications, to say nothing of that usual, less serious annoyance, wrong page citations. In one case, only about one-third, of a very long, imposing bibliography was cited in the body of the text, although the author asked that the rest of the references remain "for purposes of documentation"! All of us would clearly benefit our readers by the routine checking of such data.

We would also like to make a few comments about the current use of the term "Editor" in this day of camera-ready copy, and of ever more colloquia and symposia. We wonder if it is not time to restrict the title to those cases where some in-depth editing of manuscripts has taken place? Many of our current "Editors" have clearly been responsible for organizing symposia, but have done little more afterwards than collect the papers. While this last may be no mean chore, differences in responsibilities are involved and should be recognized, rather than denegrating the responsibilities of those who, in fact, assume old-fashioned editorial duties. Publishers tend to feel that use of the title "Editor" on the book spine relieves them of responsibilities toward the intellectual content of papers, but it may be well to recognize that the hapless "Editor" does not share that responsibility, nor responsibility for editorial deficiencies present in many of the papers.

REFERENCES

Boucot, A. J., 1975, Evolution and extinction rate controls: Amsterdam, Elsevier, 427 p.

Hedgpeth, Joel W., 1957, Concepts of marine ecology, *in* Hedgpeth, Joel W., ed., Treatise on marine ecology and paleoecology: Geol. Soc. Amer. Mem. 67, v. 1, p. 29-52.

House, M. R., 1971, Devonian faunal distributions, *in* Middlemiss, F. A., Rawson, P. F., and Newall, G., eds., Faunal provinces in space and time: Liverpool, Seel House Press, p. 77-94.

Contents

1. Prologue: At Sea with Provinces and Plates 1
Joel W. Hedgpeth

2. An Eclectic Overview of Plate Tectonics, Paleogeography, and Paleoceanography 9
Tjeerd H. van Andel

3. Geographic Distribution of Cambrian and Ordovician Rostroconch Mollusks 27
John Pojeta, Jr.

4. Gastropod Opercula as Objects for Paleobiogeographic Study 37
Ellis L. Yochelson

5. Geographic Distribution of the Ordovician Gastropod *Maclurites* 45
David M. Rohr

6. *Monorakos* in the Ordovician of Alaska and its Zoogeographic Significance 53
Allen R. Ormiston and Reuben J. Ross, Jr.

7. Swedish Late Ordovician Marine Benthic Assemblages and Their Bearing on Brachiopod Zoogeography 61
Peter M. Sheehan

8. Biogeography of Ordovician, Silurian, and Devonian Chitinozoans 75
Sven Laufeld

9. Biogeography of the Stylophoran Carpoids (Echinodermata) 91
Kraig Derstler

10. Graptolite Biogeography: A Biogeography of Some Lower Paleozoic Plankton 105
William B. N. Berry

11. Biogeography of the Silurian-Lower Devonian Echinoderms 117
Brian J. Witzke, Terrence J. Frest, and Harrell L. Strimple

12. Biogeography of Late Silurian and Devonian Rugose Corals in North America 131
W. A. Oliver, Jr. and A. E. H. Pedder

13. Biogeography of Silurian and Devonian Trilobites of the Malvinokaffric Realm 147
Niles Eldredge and Allen R. Ormiston

14. A Quantitative Analysis of Lower Devonian Brachiopod Distribution 169
Norman M. Savage, David G. Perry, and Arthur J. Boucot

15. Devonian Conodont Distribution—Provinces or Communities? 201
Peter G. Telford

16. Evolution of Fusulinacea (Protozoa) in Late Paleozoic Space and Time 215
Charles A. Ross

17. Biological and Physical Factors in the Dispersal of Permo-Carboniferous Terrestrial Vertebrates 227
Everett C. Olson

18. Permian Positions of the Northern Hemisphere Continents as Determined from Marine Biotic Provinces 239
T. E. Yancey

19. The Role of Fossil Communities in the Biostratigraphic Record and in Evolution 249
J. B. Waterhouse

20. Permian Ectoprocts in Space and Time 259
June R. P. Ross

21. Biogeographic Significance of Land Snails, Paleozoic to Recent 277
Alan Solem

22. Paleobiogeography of the Middle Jurassic Corals 289
Louise Beauvais

23. African Cretaceous Ostracodes and Their Relations to Surrounding Continents 305
Karl Krömmelbein

24. Pre-Tertiary Phytogeography and Continental Drift—Some Apparent Discrepancies 311
Charles J. Smiley

25. Fossil Birds of Old Gondwanaland: A Comment on Drifting Continents and Their Passengers 321
Pat Vickers Rich

26. A Biogeographical Problem Involving Comparisons of Later Eocene Terrestrial Vertebrate Faunas of Western North America 333
Jason A. Lillegraven

27. Biogeographic Significance of the Late Mesozoic and Early Tertiary Molluscan Faunas of Seymour Island (Antarctic Peninsula) to the Final Breakup of Gondwanaland 349
William J. Zinsmeister

28. Pinniped Biogeography 357
Charles A. Repenning, Clayton E. Ray, and Dan Grigorescu

29. Fossil Beetles and the Late Cenozoic History of the Tundra Environment 371
John V. Matthews

30. In Search of Lost Oceans: A Paradox in Discovery 379
Richard H. Benson

31. Dispersal of Pelagic Larvae and the Zoogeography of Tertiary Marine Benthic Gastropods 391
Rudolf S. Scheltema

32. Californian Transition Zone: Significance of Short-Range Endemics 399
William A. Newman

33. The Role of Circulation in the Parcelling and Dispersal of North Atlantic Planktonic Foraminifera 417
Richard Cifelli

34. The Architectural Geography of Some Gastropods 427
Geerat J. Vermeij

35. The Roles of Plate Tectonics in Angiosperm History 435
Daniel I. Axelrod

36. The Role of Biogeographic Provinces in Regulating Marine Faunal Diversity through Geologic Time 449
Thomas J. M. Schopf

37. Crustose Coralline Algae as Microenvironmental Indicators for the Tertiary 459
Walter H. Adey

38. Epilogue 465
A. J. Boucot and Jane Gray

Appendix 483

Taxonomic Index 485

Author Index 493

Biogeographic Index 499

Prologue: At Sea with Provinces and Plates

JOEL W. HEDGPETH, *5560 Montecito Avenue, Santa Rosa, California, 95404*

Soon after his appointment to the Chair of Natural History at Edinburgh in 1854, the Manx naturalist Edward Forbes began to write a book that was obviously intended to summarize his knowledge of the life of the seas. It would also reflect, to some extent, the growing body of knowledge about the occurrence of such attractive elements of the fauna of other seas as mollusks (often known in those days only as shells from far-off shores in the cabinets of collectors), crustacea, and bits of coral. Two decades later this fascination with these attractive collector's items was vividly evoked by passages in Jules Verne's ***Twenty Thousand Leagues Under the Seas,*** and undoubtedly stimulated the interest of many young men to become naturalists of the sea.

Forbes' book was to be a professional account of these wonders as expressed in the absorbing phenomena of their distribution in the various seas best known to him, of northern Europe and the Mediterranean. His tragically early death in 1855 at the age of 40 ended this project almost before it began, and of *The Natural History of European Seas,* continued and published in 1859 by his colleague Robert Godwin-Austen, only the first 126 pages are in the words of Edward Forbes. They are remarkably fresh reading, even in this jaded day. Forbes was both marine zoologist and paleontologist, and it is to him that we owe the field of study now known as paleoecology (Hedgpeth, 1957). Like many of the investigators of his time, Forbes found the subject of the origin and distribution of life in the seas of the utmost fascination and significance: "there is a deeper interest in the march of a periwinkle, and the progress of a limpet" than in the historian's concern with battles and conquests.

Even in those days, a hundred and twenty years ago, books seemed to be getting numerous enough to require an apology for adding still another (would that more of our contemporaries thought so!), and we find Forbes remarking, on page 2 of *The Natural History of European Seas,* "In this age of volumes, a man needs offer a good excuse before adding a new book, even though it be a small one, to the heap already accumulated. He should either have something fresh to say, or be able to tell that which is old in a new and pleasanter way."

As W. A. Herdman has pointed out in *Founders of Oceanography and their Work* (1923), Forbes was a worker in the borderlands of science, the meeting ground of biology and geology, and he was concerned about the evidence of the past for the distributions of the present time. This was apparent in his first treatment of the problems of present-day distributions as related to possible past states of landmasses in Europe (Forbes, 1846). As such, Forbes was the first great marine biogeographer. Despite certain usages expected of every cultured man of his times, such as "centers of Creation," for what we would now call centers of evolution, Forbes brought an enlightened intelligence to his first works in biogeography. He recognized Pleistocene relicts and maintained that an understanding of the distributions of the past is necessary to understanding the present.

In all, Forbes' ideas and speculations were tremendously influential in his day. A charismatic personality on the lecture platform, whose like as a marine biologist we have seen in our own times in the person of Gunnar Thorson (1906-1971), he often carried the day even when he was wrong, as in his famous generalization that there was no life in the sea below 300 fathoms. This did stimulate a great deal of dredging and was soon proved wrong; doubtless, as did Thorson, Forbes would have admitted his error as soon as he had proof. Forbes thrived on symposia, as well as those in the more literal meaning of the word, held in less arid facilities after the meetings. He enjoyed the give and take of scientific discussion which he likened to pagan activities before the days of Beowulf: "The old Scandanavian gods amused themselves all day in their Valhalla hacking each other to small pieces, but when the time of feasting came, sat down together whole and harmonious, all their wounds healed and forgotten. Our modern Thors, the hammer-wielders of Science, enjoy similar rough sport with like pleasant ending." (Forbes, 1854).

The evaluation of what and who is significant in the development of biogeography will depend not only on the systematic bias of the appraiser, but on the fashions of the day. This is amply demonstrated by the summary review by Karl P. Schmidt, a herpetologist and terrestrially oriented biologist of great competence (Schmidt, 1955). He begins his discussions with the work of Ludwig Schmarda (1853), and calls attention to Schmarda's abandonment of the concept of a single center of creation in favor of several such centers. Schmidt suggests this was a bold enough step in 1853, but it is obvious from the context of *The Natural History of European Seas* that Forbes was also considering distribution patterns in terms of several centers of creation at the same time. Also, according to Schmidt, Schmarda was aware of the concept of vicariation, "of the replacement of one species of animal by an obviously related one in adjacent areas or different regions." The significance of allopatric speciation or vicariance is discussed at some length by Rosen (1976) with respect to Caribbean biogeography and it may be argued that Forbes also understood this phenomenon (see p. 8-9 of *The Natural History of European Seas,* written in 1854).

At the time Schmidt wrote his review (in the early 1950's) isostasy was in flower and from this it followed that the continents had not changed very much through time except to tilt and readjust with the waxing and waning of the ice load during glacial epochs: "The conviction that the continental platforms are indeed permanent in broad outline, and that the ocean floor is of very different composition, became more and more an axiom of modern geology as the extension of gravity measurements failed to find exceptions to the lightness of the continental masses relative to the ocean floors." In Europe, however, continental drift still explained more of the puzzles of distribution than it created, in contrast to isostasy, but Schmidt thought little of it and cites Matthew (1915) in support of the permanence of the continents. Much of Schmidt's paper is an effort to revive Matthew's ideas and to keep the continents in their appointed places:

> It is quite evident that there are still numerous believers in the former existence of continents where the great oceans now are; of movement of the continents to their present positions from an original single continent; and of back and forth movement of "accordion-type" continents. There is a strong opposing school of conservatives, who hold to the belief that the continental platforms, though obviously often flooded by epicontinental seas, have been stable throughout the geological ages in which life has existed on land. Much of this controversy is primarily geological, and only secondarily zoogeographic. My concern in this matter has been lest the geologists base arguments on those of zoogeographers and that these then complete an argument in a circle by triumphantly pointing to the fact that the geologists support them. If the geological theories involved were restricted to pre-Paleozoic or even to Paleozoic times, zoogeographers could have little to say regarding them. (Schmidt, 1955, p. 782)

One wonders, in view of this, what will be said a quarter century hence of our obvious faith in the concept of plate tectonics as the modern *Deus ex machina* to explain all our major biogeographic puzzles of the present as well as all past epochs.

In a recent discussion of the nature and formulation of biogeographic hypotheses, Ball (1976) suggests that perhaps the first major impetus to biogeography as a science came from Sclater's review (1858) of the geographical distribution of birds. Perhaps so, for those interested in birds and terrestrial vertebrates, but with respect to marine biogeography the first impetus was that of Forbes, and this in turn depended on such studies as those of Audouin and Milne-Edwards (1832) in France and Carl Vogt (1848) in the Mediterranean. Ball, a terrestrial biologist, does not mention either Schmarda or Forbes, nor the significant contribution of Ortmann (1896) to marine zoogeography.

In all of this the primary effort has been to describe and elucidate the distribution of faunas, arranged in realms and provinces, and it was in the tradition of Forbes that Ekman (1935, 1953) stated that the final aim of biogeography is the elucidation of the history of the faunas, "an opinion with which it is difficult to disagree" according to Ball. In my review of the development of marine biogeography (Hedgpeth, 1957), I remarked that the biogeography of marine communities was a pioneer subject with much yet to be done; twenty years has not changed the state of affairs although there is much lively argument concerning the diversity and stability of communities in the sea. Some workers (e.g., Dunbar, 1960) have discussed community or ecosystem development in evolutionary terms, as if there were a "natural selection" of ecosystems. While it seems probable that there has been evolution in the broad meaning of development from simple to complex groupings (Hedgpeth, 1964), the implication of a sort of superorganism in which the component species might act like a pool of genes is a vitalistic overstatement. The tendency to regard the community or biocoenosis as a coherent, closed system was already evident at the outset in the original concept of the community as stated by Moebius (1883), and was so recognized by Bashford Dean (1893) in his criticism of the original concept of Moebius:

> Möbius maintains that the size of the banks in a given region can not be materially augmented—a matter which is of great interest even from the standpoint of pure biology. Not that it is at all to be questioned that a natural bank would under normal conditions remain more or less uniform in size and in the proportion of its component organisms—but it is the theory involved in this question that seems to the writer susceptible of broader interpretation than has been assigned it. *Biocoenosis* is the term applied by Möbius to express the mutual interdependence of species existing in a colony—a condition of happy-family existence in a natural cage whose limited food supply locks up the chances of permanent numerical increase. In accordance with this keenly poised life-balance Möbius infers that the banks of the Wattenmeer can not be permanently added to, even by artificial means (Auster u. Austernwirtschaft, p. 78). He notes, for example, that a season favorable to oysters, will, *per se* cause the oysters during the following seasons to fall back to their normal, inasmuch as food material has thus been prepared for the enemies (crab and starfish) whose increased progeny will restore the balance of life.
>
> The important inferences drawn from this doctrine of life-balance, do not, however, seem to be entirely warranted by the premises. We are led, for example, to infer that individuals are dependent upon the colony, and that the colony holds the curb, checking the permanent increase of one form at the expense of another. On the other hand, struggle for survival is undoubtedly the democracy of animal living, and in these days it has been pretty clearly established that the colony is but an incident more or less transient in the survival of the fittest. So the biocoenose, as we

must accordingly admit, becomes but an episode in colonial life, whose duration depends upon the enduring force of its component species, where quickly moving predatory forms have the right of might, where stationary and defenseless forms have become mimics to escape their enemies, or have developed a surprising fecundity to survive the dangers of a compressed living-area or unfavorable environment. It can not at present be doubted that the scale of the struggle may readily be turned in favor of but a single type or species. (Dean, 1893, p. 374)

Dean also disagreed with Moebius' idea that the food supply was finite and limits the size of the biocoenosis, in this case the oyster bank of the North Sea:

The boundaries of a natural bank are certainly not fixed by food quantum. The food stuff may, it is true, vary in quantity in different regions during the same season, or in the one locality at different stations, seasons, or even tides. But there seems in general to exist a food normal which is recognized as characteristic of a locality.

Aside, however, from this question of local variation, the amount of food that is actually brought to an oyster colony seems to the writer to be in direct proportion to the volume of water passing over it. If this volume be infinite, as it is in the Wattenmeer, exhaustion of food supply would seem an impossible condition. Barrenness and sterility of water could not occur; general transfusion of floating or of free-swimming microorganisms is very clearly one of the characters of the open sea. The lower water layers that may have been screened out by a thrifty oyster bank would not remain without organisms, but would immediately be replenished from above by the currents that exhausted the lower layers. (Dean, 1893, p. 375)

Nevertheless, the general concept of the biocoenosis has survived and constitutes one of the fundamental generalizations of ecology, however we may interpret the development or survival of communities. Paleontology is especially rich in inferences about past environments based on the implicit assumption of community or biocoenotic similarities. In this context the work of C. G. J. Petersen (1914) is fundamental. In setting out to ascertain the available food supply for bottom feeding fishes in Danish waters, Petersen did not anticipate that he would find that the organisms of the food supply were apparently arranged in persistently recurring combinations, but he immediately recognized that he was dealing with communities, and extended his consideration of them to the North Atlantic, providing the first biogeographic map of communities in the sea (Petersen, 1915, Chart I). Admittedly preliminary, Petersen intended his chart as "a graphical working hypothesis" for investigations in other parts of the world. With characteristic caution, he advised future workers:

Lists of species without information as to the quantitative occurrence of the various species are not sufficient for this purpose; nor are collections of material in which all, or at least several communities are mixed together, despite the fact that they exist separately in the sea. In making a comparison between the animal life of various portions of the globe, it is indispensable, if any thorough knowledge of the subject is to be obtained, to compare each community with the same or corresponding community elsewhere, and not with one entirely different. It is necessary, therefore, to know these various communities, to keep them separate, and gradually to note them in a detailed survey. (Petersen, 1915, p. 3-4)

Petersen's work was taken up in many other countries, including Russia, where it has become a major feature of marine research, and his methodology is now recommended for base line and environmental impact studies. In Denmark the work was carried on by Gunnar Thorson, who attempted a biogeographical synthesis of bottom communities for the world, especially the arctic and temperate shallow bottoms of the Northern Hemisphere (Thorson, 1957). From these observations and the somewhat atypical bottom community he observed in the Persian Gulf, Thorson developed a concept of parallel infaunal communities. These were communities or a relatively small number of macroscopic species of the same genera in the same types of bottom at comparable depths, "replacing each other in accordance with the geographical regions" (Thorson, 1957, p. 521). He thought that this generalization could be applied to tropical infaunal communities also, but toward the end of his life he had a chance to study more tropical infaunal communities and realized that the concept of parallel bottom communities in shallow seas was not universally applicable. He freely admitted his error and planned a revision of his theory of bottom communities, but died before he could write it.

In his later work, Thorson (1961) considered some of the problems of recruitment and dispersal related to the length of pelagic life of larvae of benthic invertebrates and their transport by ocean currents, and the control or establishment of benthic communities by the activity of predators, especially upon larval and juvenile stocks (Thorson, 1966). These studies were an outgrowth of his observation of the scarcity of planktotrophic larvae in Arctic seas. In the Antarctic, where pelagic or swimming larvae are also reduced in number and the duration of their free stages, the distribution of the benthos seems to be more uniform than in the Arctic, or at least the benthic species are more widely distributed throughout the Antarctic, in response at least in part to the prevailing current system of the Antarctic (Hedgpeth, 1969, 1971). The neatly timed interactions between reproduction, larval stages, and predator activity envisioned by Thorson has not been as enthusiastically postulated in recent years, as studies suggest there is often no clear separation between predators, cannibals, and scavengers in the sea (Isaacs, 1972).

In recent years there has been increased effort to ascertain the nature of the communities of the deep sea, which turn out to be much more complex and diverse (in terms of species composition) than previously thought (see especially Hessler, 1974; Hessler and Jumars, 1973; Sanders and Hessler, 1969; and Sokolova, 1972, as examples of recent studies of the deep sea benthos). As yet the information is not adequate even for the preliminary sort of synthesis attempted by Thorson for the shallow benthos. Nevertheless, it is unfortunate that Briggs (1974) did not

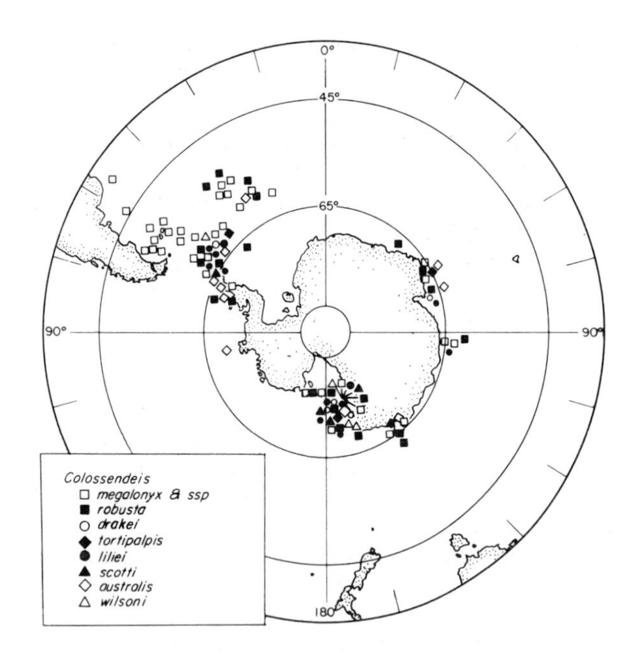

Figure 1. Distribution of Antarctic pycnogonida of the genus *Colossendeis* (after Hedgpeth, 1969b).

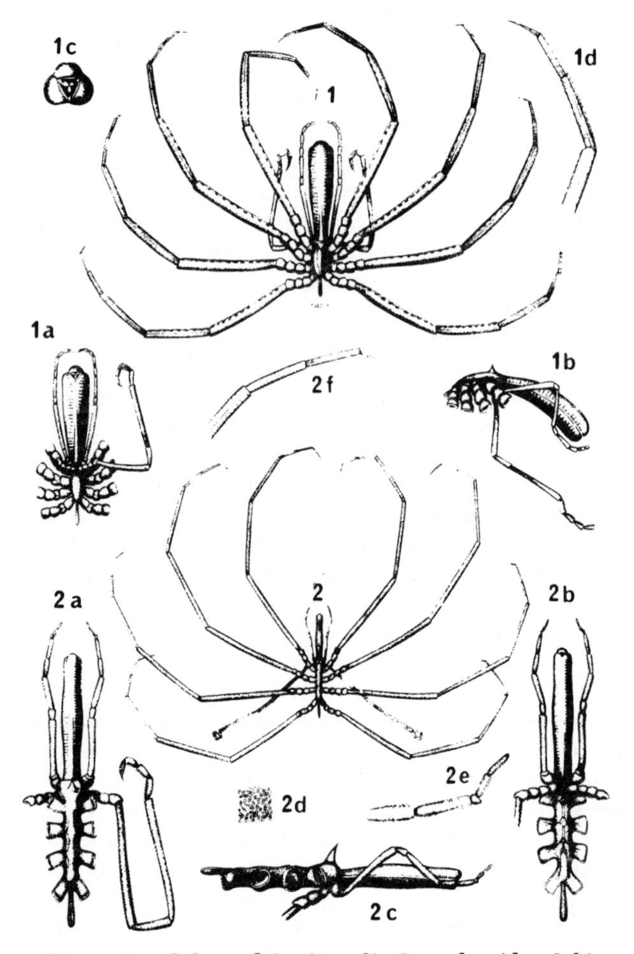

Figure 2. *Colossendeis:* (1 a-d) *C. proboscidea* Sabine; (2 a-f) *C. angusta* G. O. Sars (after Sars, 1891).

consider the development of biogeography of marine benthos in his effort to bring marine zoogeography up to date since Ekman's revision of 1953.

One of the fascinating aspects of biogeography is that despite all the sweeping generalizations or painstaking marshalling of details to produce a synthesis, despite the military hierarchy of realms, provinces, subprovinces, districts and all, and now, conveniently movable tectonic

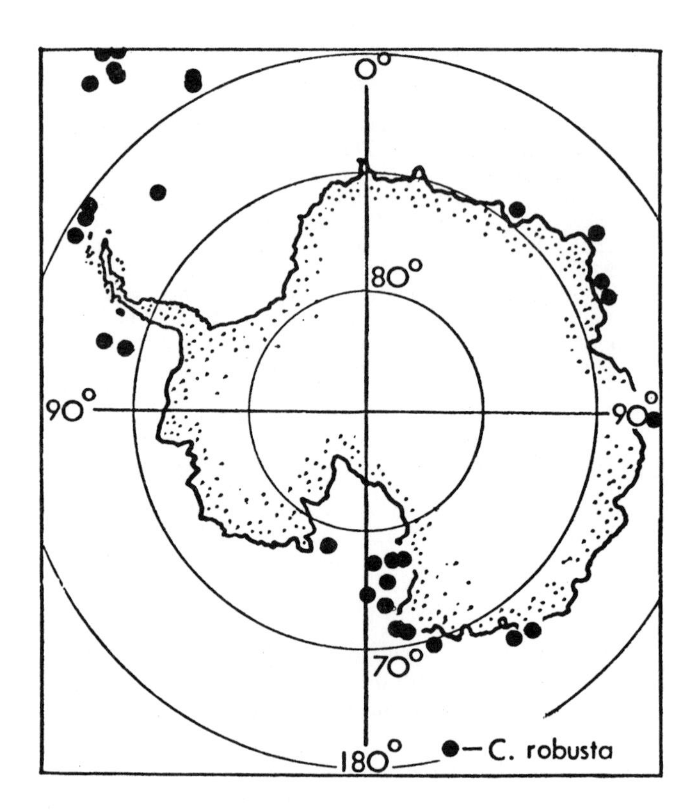

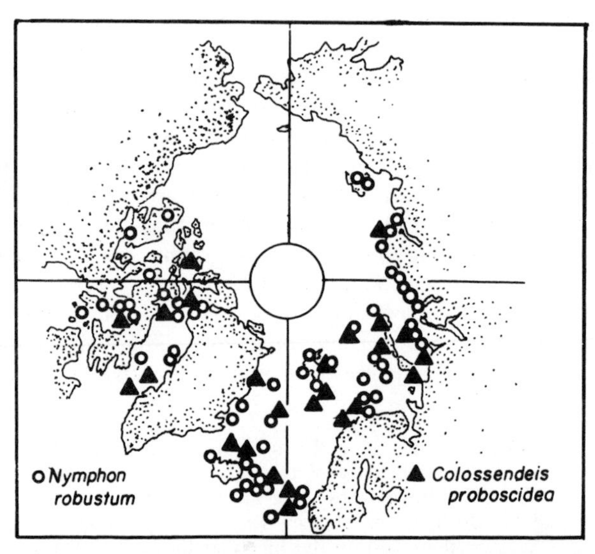

Figure 3. Distribution of pycnogonida. Top: *Colossendeis robusta* (from Fry and Hedgpeth, 1969); Bottom: *Nymphon robustum* and *C. proboscidea* (from Hedgpeth, 1971).

plates, there will always be puzzles that do not easily fit any hypothesis. Of course, this may be in part because we do not have all the information. But it will be a sad world when we know everything and there is nothing left to surprise or puzzle us. There will be no need or reason to have an idea; such an age would be very dull indeed. It is unlikely, however, that we will ever know so much about the life in the sea as to attain such a state of mindless Nirvana.

Perhaps each of us has his own private puzzles that stimulate and bewilder. Mine is the strange case of the predominantly deep sea pycnogonids of the genus *Colossendeis,* those pantographlike animals that appear to be dipping their enormous proboscides into the upper surface of the bottom ooze for meiofauna. They appear neither to breed nor reproduce; many of the largest are sexed with difficulty because they are not dimorphic and unless the gonopores are obvious (and they may not be when the animal is sexually quiescent) their sex cannot always be easily guessed. In most pycnogonids, the eggs are carried about by the male in clumps, clusters, or neat balls until hatching; no colossendeid, including those with 10 and 12 pairs of legs, has ever been found with an egg mass, and as yet no larval stages have been found in any possible or plausible host. The pycnogonids are the only other marine arthropods besides the Crustacea that have a free larval stage. This protonymphon in many species invades coelenterates, especially hydroids, to form galls. Unlike the crustacean nauplius, however, it is not free-swimming.

The center of origin for this family of pycnogonids seems to be the Antarctic regions. Of the more than 30 known species, 11 occur south of 45°S, and most of these are strictly Antarctic (Fig. 1). In addition, there are two polymerous forms, the 10-legged *Decalopoda australis,* and the 12-legged *Dodecolopoda mawsoni,* restricted to Antarctic waters. There are two basic forms or morphogroups, in Dunbar's (1968) sense, in the genus *Colossendeis s. s.:* the long slender form exemplified by the world-wide *Colossendeis colossea* (often illustrated in popular books as well as reference works) and the Antarctic *Colossendeis megalonyx,* and a heavier set group of species with shorter, thicker legs and a thick, batlike or bulbous proboscis, exemplified by such Antarctic species as *C. australis, C. scotti,* and *C. wilsoni* and the only Arctic species, *Colossendeis proboscidea.* These two kinds, groups, or morphs are exemplified in an old plate by G. O. Sars (Fig. 2). The Antarctic *C. robusta* is intermediate between these two types (Fig. 3). The two Antarctic polymerous forms also belong to this morphogroup, but the other polymerous species, *Pentacolossendeis reticulata,* from the Florida Strait, is more obviously related to the *C. colossea - C. angusta* complex.

This may possibly be the older group of colossendeids; a Jurassic fossil (Solenhofen) not yet officially named, appears to be of this type. It also is evidently a polymerous species, having 10 legs. On Figure 4 the distribution of some of the common species of this group, which will be provisionally referred to as Morphogroup A, is indicated together with that of the heavier set species, as Morphogroup B. What is most interesting about this distribution is that the Arctic *Colossendeis proboscidea* (see Fig. 4) is separated from the other members of Morphogroup B by most of the Atlantic Ocean, and no member of this group has been found in the Pacific or Indian Oceans. Another complex, of which several species were originally attrib-

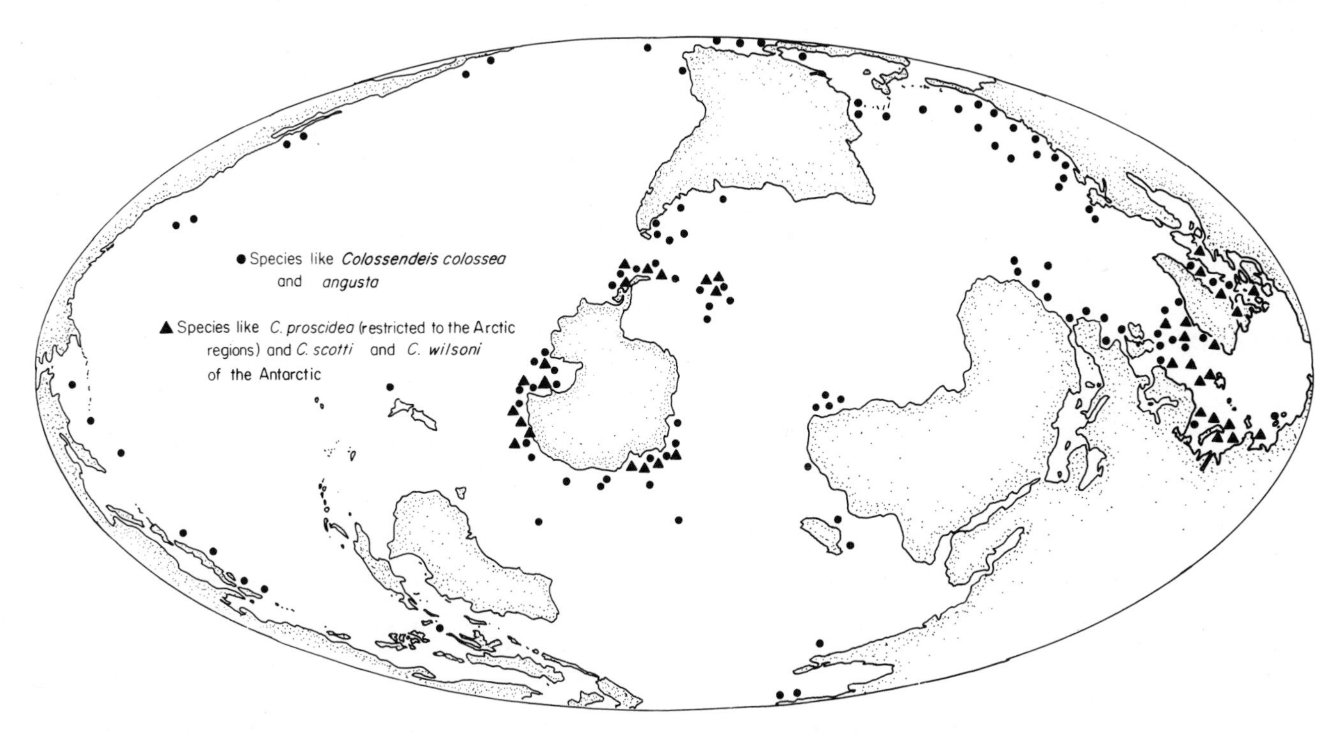

Figure 4. Distribution of the two *Colossendeis* "morphs"(from Hedgpeth, 1969a, 1969b, 1971).

uted to *Colossendeis* but are now considered a separate genus, are characteristically North Pacific and have no clear Antarctic relatives.

One could obviously interpret this distribution of the two morphogroups as a pattern suggesting two successive waves of speciation emanating from the Antarctic regions, since representatives of Morphogroup A are found only occasionally in Antarctic waters in contrast to their distribution in the deeper waters of the world ocean. However, the occurrence of *C. proboscidea* in the Arctic makes it difficult to accept it as a later emigrant from the Antarctic unless somehow it was transported northward through the Atlantic to be captured in the Arctic waters, which may have acted as a sort of biogeographic trap. It is interesting to note, however, that *C. proboscidea* does not occur generally in the Arctic basin, and is certainly absent from shallow water of the Beaufort and Chukchi Seas. It may be found in the deeper Arctic basins, but no collections have been reported yet. Can this species have arrived so far from its center of origin by some unknown means of dispersal? Does it have a larval stage in some bathypelagic medusa, or what? There is a world-wide bathypelagic species of pycnogonid, *Pallenopsis calcanea,* which may possibly be associated with some medusa, but there has been no confirmation of such an association (Hedgpeth, 1962).

The lack of known larval stages and apparent nonreproductive activity of all these species gives us little evidence to carry such speculations beyond the stage of questions. *Colossendeis colossea* has been observed to tread water near the bottom, or perhaps float with the deep currents, but the only reliable method of dispersal these animals seem to have is walking. One is tempted to suggest, albeit facetiously, that these animals have been unable to reproduce because they are searching for some lost hosts for their larval stages, but in that case, why not even eggs?

In any event, it does not seem to me that either isostasy and the perturbations of glacial epochs, or continents drifting about on convenient tectonic plates can help to elucidate this distribution pattern although it seems most plausible that the evolution and dispersal of these species has taken place since the separation of the Antarctic landmass and the formation of the Antarctic Ocean.

REFERENCES

Audouin, J. V. and Milne-Edwards, H., 1832-1834, Récherches pour servir à l'histoire naturelle de la France, ou Recueil de mémoires sur l'anatomie, la physiologie, la classification et les moeurs des animaux de nos côtes; ouvrage accompagné de planches faites d'apres nature: Paris, Crochard, 2 vols., 696 p.

Ball, Ian R., 1976, Nature and formation of biogeographical hypotheses: Systematic Zoology, v. 24, p. 407-430.

Briggs, John C., 1974, Marine zoogeography: New York, McGraw-Hill, 475 p.

Dean, Bashford, 1893, Report on the European methods of oyster-culture: U. S. Fish Commission Bull., v. 11, p. 357-406 (1891).

Dunbar, Max J., 1960, The evolution of stability in marine environments: Natural selection at the level of the ecosystem: Am. Naturalist, v. 94, p. 129-136.

——— 1968, Ecological development in polar regions. A study in evolution: Englewood Cliffs, N.J., Prentice-Hall, 119 p.

Ekman, Sven, 1935, Tiergeographie des Meeres: Leipzig, Akad. Verlag, 542 p.

——— 1953, Zoogeography of the sea: London, Sidgwick and Jackson, 417 p.

Forbes, Edward, 1846, On the connexion between the distribution of the existing fauna and flora of the British Isles, and the geological changes which have affected their area, especially during the epoch of the Northern Drift: Geol. Survey Great Britain Mem., v. 1, p. 336-432.

——— 1854, Murchison's Siluria (review): Literary Gazette, June 24, 1854.

Forbes, Edward and Godwin-Austen, Robert, 1859, The natural history of the European seas: London, John Van Voorst, 306 p. (p. 1-126 by Forbes).

Fry, W. G. and Hedgpeth, Joel W., 1969, Pycnogonida. I. Colossendeidae, Pycnogonidae, Endeidae, Ammotheidae, *in* The fauna of the Ross Sea, Pt. 7, New Zealand Oceanogr. Inst. Mem. 49, p. 1-139.

Hedgpeth, Joel W., 1957a, Concepts of marine ecology, *in* Hedgpeth, Joel W., ed., Treatise on marine ecology and paleoecology: Geol. Soc. Amer. Mem. 67, v. 1, p. 29-52.

——— 1957b, Marine biogeography, *in* Hedgpeth, Joel W., ed., Treatise on marine ecology and paleoecology: Geol. Soc. Amer. Mem. 67, v. 1, p. 359-382.

——— 1962, A bathypelagic pycnogonid: Deep-Sea Research, v. 9, p. 487-491.

——— 1964, Evolution of community structure, *in* Imbrie, John, and Newell, Norman, eds., Approaches to paleoecology: New York, John Wiley, 432 p.

——— 1969a, Introduction to Antarctic zoogeography, *in* Antarctic Map Folio Series, Folio 11: New York, American Geographical Society, p. 1-9.

——— 1969b, Pycnogonida, *in* Antarctic Map Folio Series, Folio 11: New York, American Geographical Society, p. 26-28.

——— 1971, Perspectives of benthic ecology in Antarctica, *in* Quam, Louis O., ed., Research in the Antarctic: Am. Assoc. Adv. Sci., Pub. 93, p. 93-136.

Herdman, William A., 1923, Founders of oceanography and their work. An introduction to the science of the sea: London, Edward Arnold, 340 p.

Hessler, Robert R., 1974, The structure of deep benthic communities from central oceanic waters, *in* Miller, Charles, ed., The biology of the oceanic Pacific: 33rd Ann. Biol. Colloquium Proc.: Corvallis, Oregon State Univ. Press, p. 79-93.

Hessler, Robert R., and Jumars, P. A., 1974, Abyssal community analysis from replicate box cores in the central North Pacific: Deep-Sea Research, v. 21, p. 185-209.

Isaacs, John D., 1972, Unstructured marine food webs and "pollutant analogues": U.S. Fishery Bull., v. 70, p. 1053-1059.

Matthew, W. D., 1915, Climate and evolution: New York Acad. Sci. Ann., v. 24, p. 171-318.

Moebius, Karl, 1883, The oyster and oyster-culture: U. S. Comm. Fish and Fisheries Rept., 1880, p. 683-751 (translation of German original of 1877).

Ortmann, A. E., 1896, Grundzüge der marinen Tiergeographie: Jena, Gustav Fischer, 96 p.

Petersen, C. G. Joh., 1914, Valuation of the sea. II. The animal communities of the sea-bottom and their importance for marine zoogeography: Danish Biol. Station Rept. no. 21, p. 1-68 (1913).

——— 1915, Notes to Charts I and II: Danish Biol. Station, Appendix to Rept. no. 21, 7 p. (1914).

Rosen, Donn E., 1976, A vicariance model of Caribbean biogeography: Systematic Zoology, v. 24, p. 431-464.

Sanders, H. L., and Hessler, R. R., 1969, Ecology of the deep-sea benthos: Science, v. 163, p. 1419-1424.

Sars, G. O., 1891, Pycnogonidea: Norwegian North Atlantic Expedition 1876-8, Zool., v. 20, p. 1-163.

Schmarda, Ludwig K., 1853, Die geographische Verbreitung der Thiere: Wien, C. Gerold u. Sohn, 755 p.

Schmidt, Karl P., 1955, Animal geography, *in* Kessel, E. L., ed., A century of progress in the natural sciences, 1853-1953: San Francisco, California Acad. Sci., p. 767-794.

Sclater, P. L., 1858, On the general geographical distribution of the members of the class Aves: Jour. Linn. Soc., Zool., v. 2, p. 130-145.

Sokolova, M. N., 1972, Trophic structure of deep-sea macrobenthos: Marine Biol., v. 16, p. 1-12.

Thorson, Gunnar, 1957, Bottom communities (sublittoral or shallow shelf), *in* Hedgpeth, Joel W., ed., Treatise on marine ecology and paleoecology: Geol. Soc. Amer. Mem. 67, v. 1, p. 461-534.

——— 1961, Length of pelagic life in marine bottom invertebrates as related to larval transport by ocean currents, *in* Sears, Mary, ed., Oceanography: Am. Assoc. Adv. Sci., Pub. 67, p. 455-474.

——— 1966, Some factors influencing the recruitment and establishment of marine benthic communities: Netherlands Jour. Sea Research, v. 3, p. 267-293.

Vogt, Carl, 1848, Ozean und Mittelmeer. Reisebriefe: Frankfurt am Main, J. Rütten, 2 vols.

An Eclectic Overview of Plate Tectonics, Paleogeography, and Paleoceanography

TJEERD H. VAN ANDEL, *Department of Geology, Stanford University, Stanford, California 94305*

ABSTRACT

This paper examines the consequences of plate tectonics and continental drift for the evolution of the global physical environment, with emphasis on the changing distribution of land and sea and the history of the oceanic circulation. A broad outline of the change in surface circulation since the breakup of Pangaea is supported by a more detailed synopsis of the response of the equatorial Pacific to the closure and opening of circum-global seaways and by the deteriorating climate of the last 50 m.y. Plate tectonic concepts are used in the reconstruction of the topography of the ocean basins of the past. Large eustatic sea-level changes in late Mesozoic and early Cenozoic times may have resulted from changes in the volume of mid-ocean ridges, and, in turn, had an effect on the availability of calcium carbonate and nutrients in the pelagic realm. Changing rates of formation of new oceanic crust may also have influenced the chemical composition of salt water. For the last 200 m.y. changes in the physical environment, climate, and paleogeography can be reconstructed from plate tectonics principles without recourse to paleobiological data. On the other hand, the loss of virtually all pre-Mesozoic ocean floor renders the reconstruction of patterns of continental drift impossible without the aid derived from paleomagnetic data and from fossil assemblages. At the present time, a major controversy exists between those who accept Pangaea as the primordial continent and those who perceive the large scale mobility of the continents as a permanent feature of the dynamics of the Earth.

PROLOGUE AND APOLOGY

This paper began in response to a request for a review of plate tectonics. Now, the principles of plate tectonics are quite simple, while their application can be infinitely complex. Thus, this might have been a short and perhaps superfluous paper or a long and likely boring one. Seeking a way out of this dilemma, I have chosen to discuss the use of plate tectonic concepts in the reconstruction of the physical paleoenvironment and its changes with time rather than the tectonic processes themselves. In doing so, I have emphasized the oceanic environment because this is my principal interest and because some of the most interesting recent results have been obtained in this area. I have further permitted myself the privilege and pleasure of a sweeping and speculative overview with minimal detail and documentation. This, then, is an essay and a rather personal one, and not a carefully constructed and comprehensive review supported by a large bibliography. However, although the references may at times appear sparse, and perhaps occasionally capricious, I have endeavored to make them complete enough to permit more substantive reading.

It seems probable to me that the impact of plate tectonics on historical geology, which is just beginning to be felt, will ultimately lead to a "new historical geology" analogous to the "new global tectonics" of the geophysicists. It is this exciting perspective that I wish to illustrate here with selected reasonable reconstructions without implying that others have no merit.

The price to be paid for potential deeper insight into the world's evolving environments, climates, and biota is the addition of spatial instability to the eternal change with time. This new degree of freedom has contributed enormously to the difficulty of understanding the world of the past, in part because motions on a sphere are not easily grasped and the quantitative methods that have evolved require considerable effort. Yet, the basic concepts of plate tectonics are simple. The motion of a set of points, or a surface on a sphere, can always be described as a rotation around a unique pole and is fully determined by the position of this pole and the rate of rotation in degrees per unit time. The rotation does not necessarily yield the actual travel path but it does give the net result. Similarly, the relative movements of two surfaces with respect to each other can be described by a rotation around a joint pole at a given rate (Fig. 1). If expressed in degrees per unit time, the rate of rotation for a given system remains constant wherever the surfaces are located, but if it is given in units of length it will increase toward the equator of the pole of rotation and decrease beyond. Various surface features of the Earth that mark the boundaries between moving plates, such as earthquake zones, magnetic anomalies (crustal growth lines parallel to plate boundaries of separation), and fracture zones can be used to determine poles and rates of rotation. It is also possible to determine the motion of a surface

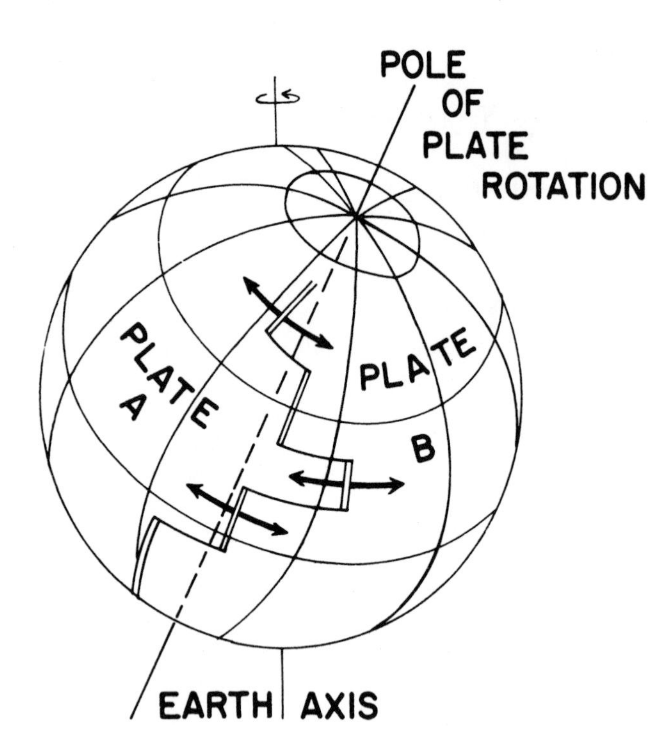

Figure 1. Rotation on a sphere of two plates which are separated by accreting edges or spreading axes (double lines) and transform faults (heavy single lines). Note that the pole of rotation for the two plates is not the same as the rotation axis of the sphere itself.

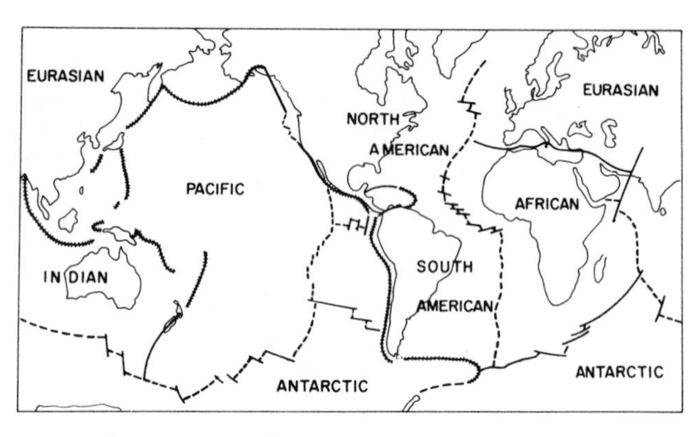

Figure 2. Present plate boundaries of the world. Hatched lines are divergent plate boundaries (subduction zones), dashed lines accreting (convergent or spreading) plate boundaries. Short solid lines are transform faults. Note that individual plates may contain both oceanic and continental areas except for the Pacific plate which is only oceanic.

element (plate) with respect to a fixed reference frame such as the grid of latitudes and longitudes. Many long and linear volcanic chains increase systematically in age from one end to the other. They are assumed to result from drift of the plate across a hot spot deep in the Earth and thus provide a record of motion with respect to this hot spot. If the spot is fixed in space, the plate motions with respect to a fixed reference frame can be determined. With a combination of data on all known two-plate motions and absolute drift, a reasonably satisfactory scheme of global plate rotations has been achieved for approximately the last 10 to 15 m.y. (Minster and others, 1974); with additional data on past configurations, this scheme can be extended backwards some 100 m. y.

Plates are not synonymous with continents and oceans; in fact, most plates contain both, so that the configuration of plates is not the same as the distribution of continents and oceans (Fig. 2). Present evidence indicates that the plates (lithosphere) are 75 to 100 km thick, cold, and of relatively low density, and rest on a deeper, hotter, and denser layer (asthenosphere) which behaves in some respects like a fluid and allows the lithosphere to slide. The driving forces are still unknown; pull by sinking cold and heavy slabs is one preferred mode. Three types of boundaries between plates exist (Fig. 3). When plates move apart (diverge), molten material wells up from below and accretes to both edges, thus forming a spreading boundary, usually a mid-ocean ridge. Where plates converge or collide, the excess lithosphere formed at the accreting edges is disposed of. Either one of the two plates overrides the other which sinks deep into the Earth, or the leading edges are compressed and build mountain ranges. The first is called subduction, and occurs between two oceanic edges or one oceanic and one continental edge; the second is where two continental edges, both of low density and unable to sink, collide. Oceanic trenches, deep earthquake zones, volcanic island arcs, or marginal mountain belts, mark colliding edges. Along the third boundary neither convergence nor divergence occurs; the plates slide along each other and form transform faults (strike-slip faults) and their fossil equivalents, the fracture zones. To achieve this, the edges must be exactly parallel to the direction of rotation and hence to a circle around the pole of rotation. Fracture zones can thus be used to locate this pole.

Because continents have low density they tend to be preserved in collisions, while oceanic crust, continuously formed at accreting edges, is also continuously destroyed in subduction zones. As a result, the present ocean floor records only a relatively short piece of Earth history; the oldest oceanic crust, no older than Early Jurassic or perhaps latest Triassic, occurs in the western Pacific. This limits to about 150 m. y. the reconstruction of paleogeography and paleoceanography from oceanic data. Prior to that, the much less complete and commonly intensely deformed continental record must be relied upon.

A more extensive discussion of the principles and processes of plate tectonics and the evidence that supports this theory is given in Cox (1973) and Gass and others (1971, p. 213-250, 263-300).

CONTINENTAL POSITIONS: THE PROBLEM OF PALEOLATITUDE AND PALEOLONGITUDE

The actual migration path of a pair of drifting continents is not necessarily identical with the simple rotation that moves them from their present position to an

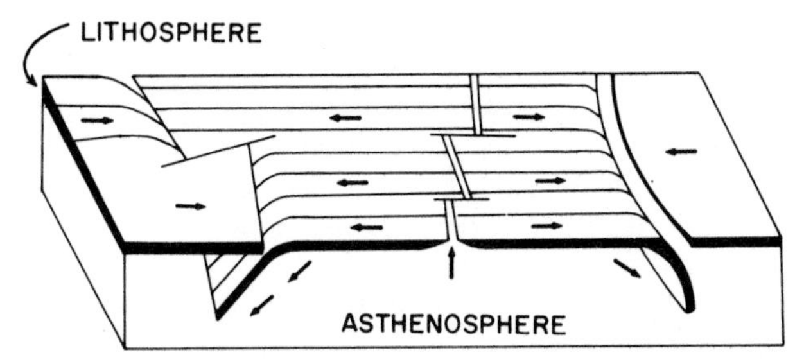

Figure 3. The basic structural components of plate tectonics. Accreting or spreading axis in center is offset by transform faults; subduction zones occur at left and right where lithosphere is being destroyed. Left subduction zone reverses direction of underthrusting across a transform fault. Lithosphere is black.

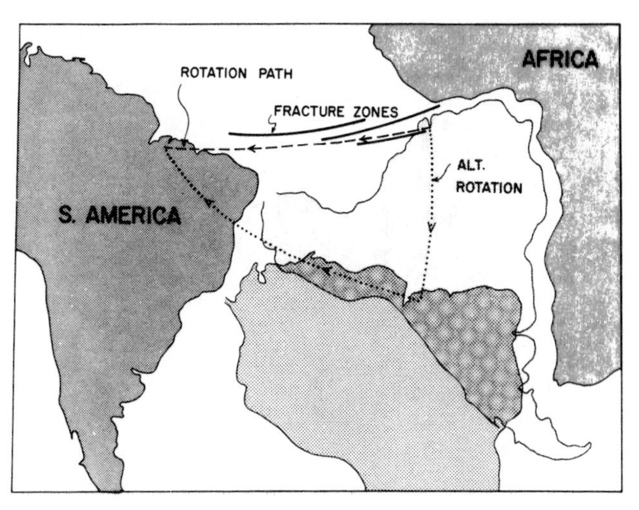

Figure 4. Rotation of South America with respect to Africa. Initial fit after Bullard and others (1965) on right (light shading of South America), and present position on left. Heavy curved arrow indicates most probable migration path [after Sclater and others, in press (b)] and is supported by large equatorial fracture zones. Arbitrary possible path that will lead to same final rotation shown with broken arrows through an intermediate position of South America.

initial best fit. Because paleolatitudes have a bearing on climatic and paleoceanographic inferences, and because paleolongitudes help in determining the size and orientation of ocean basins, we need to look at this issue in more detail. Bullard and others (1965), in a now classic paper, showed that South America can be rotated back in a single move to a near-perfect fit with Africa. Theirs, however, was a purely geometric operation and much more complex paths are possible that would lead to the same end result (Fig. 4). Path tracers are needed to fill in the details. Fracture zones, the fossil traces of plate boundaries parallel to circles around the pole of rotation, provide such tracers and changes in their azimuths record changes in pole position. Magnetic anomaly bands, which are parallel to accreting plate edges and can be dated, record the speed of rotation and its changes with time. With this kind of data the relative motions of one continent with respect to another can be charted rather precisely. The two plates, however, could have moved as a unit with respect to the latitude and longitude of the Earth so that we need independent indicators to establish these. For this we can either use the combined motions of all plate pairs, indicators of absolute motion such as volcanic chains marking hot-spot trails, or paleomagnetic data that give us paleolatitude. The migration path shown in Figure 4 results from this kind of analysis [Sclater and others, in press (b)].

If we start with the present location of all plates we can use the combined effect of their relative motions to determine their positions in the past. As far back as perhaps 10 to 15 m. y. this gives reasonably reliable results but for longer time spans we must consider the possibility that the entire complex of plates may have shifted with respect to the spin axis and equator of the Earth. Indicators of such shifts are therefore important and they do exist. The one that has recently received most attention is the trail of volcanic islands that is presumed to be produced when a plate passes over a hot spot fixed deep in the Earth's mantle (Morgan, 1972). Plate motions calculated for the last 10 m. y. on the basis of such absolute motion indicators have generally agreed well with each other, and with the data on relative motions of plate pairs (Minster and others, 1974). In the Pacific (Fig. 5), two separate trends of such

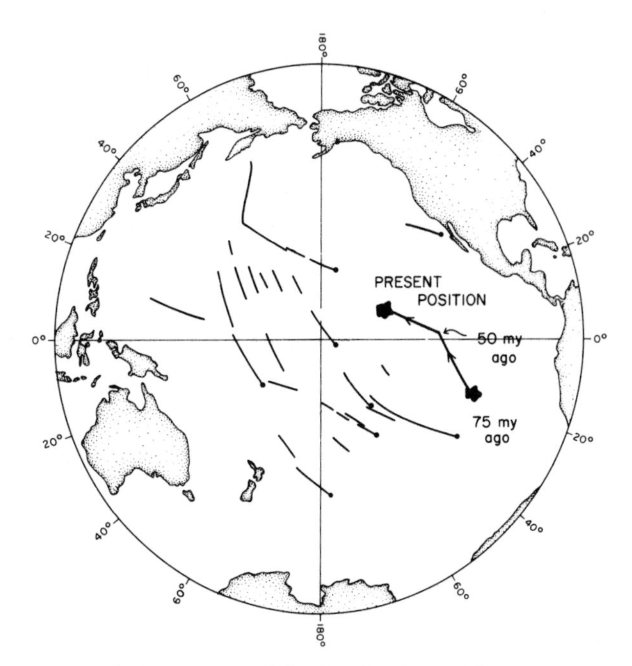

Figure 5. Movement of the Pacific plate with respect to a fixed reference frame is shown by linear chains of volcanic islands (solid lines) that are due to hot-spot activity. An imagined microcontinent initially located in the Southern Hemisphere would have drifted at high speed toward the equator, changing both speed and direction 45 m. y. ago to reach present position at 15°N and far west of its origin.

linear island chains exist. The younger one of these indicates a west-northwestward motion at a low angle to the equator for the past 45 to 50 m. y. Before that time, the motion occurred at a much higher angle to the equator and resulted in a fast transfer of the Pacific plate towards the north (Winterer, 1973). Unfortunately, this rather well-defined migration path of the Pacific plate does not tell us anything about the paths of other plates, because the Pacific is bordered by subduction zones and the movements of the surrounding plates can be independent.

As is usually the case, this use of linear island chains is not undisputed. There is still a reasonable possibility for doubt regarding the validity of the entire hot-spot concept and some small but growing body of evidence that suggests that, if hot spots exist at all, they may not be as fixed as one would like to assume (for example, Molnar and Atwater, 1973). The suggested amount of hot-spot drift has been small, however, and the reconstructions of plate migration based on this concept have thus far been in good general agreement with those achieved by other means. Nevertheless, independent and precise confirmation would be very welcome. Such confirmation is indeed available, but its precision often leaves a good deal to be desired.

Paleomagnetic measurements on rocks (Irving, 1964; McElhinny, 1973) can yield a paleolatitude from the magnetic inclination and the azimuth to the magnetic pole from the declination. Rock alteration, metamorphism, the difficulty of obtaining accurate ages, and many other pitfalls have combined to keep the number of really excellent measurements smaller than one would wish. Moreover, the azimuth yields a meridian to the pole but does not say which meridian (Fig. 6) so that the longitude of the rock at the time of magnetization remains unknown. Because the magnetic field of the Earth is in the habit of reversing frequently (Cox, 1973), it is necessary to know whether we are dealing with a normal or reversed magnetic field before the latitude tells us whether it was in the Northern or the Southern Hemisphere. The issue is further complicated by the fact that it is the magnetic, not the rotational pole of the Earth that is the point of reference. The two do not necessarily coincide and there is considerable evidence for a slow migration of the former over long distances (McElhinny, 1973), independent of the migration of plates. With increasing age, the frame of reference furnished by paleomagnetic data thus becomes less and less satisfactory.

Paleoclimatic indicators for latitude and sometimes longitude based on sediment and fossil evidence have been widely used both for (Wegener, 1915; Du Toit, 1937) and against (Meyerhoff, 1970) continental drift and plate tectonics. For example, in the equatorial Pacific the high biological productivity of the equatorial current system produces a narrow zone of rapidly depositing calcareous sediment on and just north of the equator. Similar zones of rapid sedimentation can be identified as far back as 45 m. y. ago, but with increasing age their positions gradually move northward (Fig. 7). Because the equatorial current system is the product of planetary winds generated by the rotation of the Earth, it is fair to assume that it was not the equatorial current that moved south but the Pacific plate that migrated northward, carrying the sediment marker with it. Consequently, the position of the pole and the rate of rotation of the Pacific plate can be computed by migrating all axes of maximum sedimentation rate back to the present equator (van Andel, 1974). The absolute pole of rotation of the Pacific plate and the rates of rotation determined in this manner fall within the confidence limits of poles obtained from linear volcanic chains, and from relative motions of plate pairs, and lend independent support to the assumed drift scheme.

RECONSTRUCTION OF OCEAN PALEOBATHYMETRY

Land bridges and island chains have long played an important part in the interpretation of biogeographic patterns. In the first flush of plate tectonic excitement the

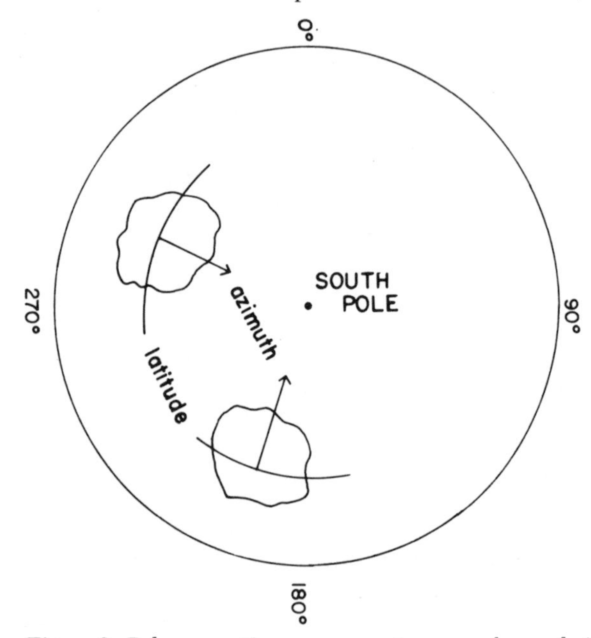

Figure 6. Paleomagnetic measurements on a rock sample indicate the latitude of the site and the direction towards the magnetic pole but do not establish the longitude. Polar projection.

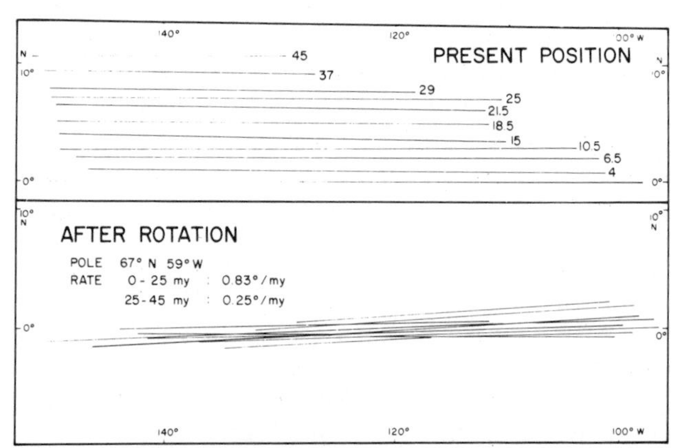

Figure 7. Using the equatorial zone of maximum sedimentation to derive the rotation parameters of the Pacific plate. Top: present positions of zones of maximum sedimentation since the middle Eocene (ages to right of axes in m. y. B. P.). Bottom: after rotation according to scheme on figure, axes are brought together on the equator (after van Andel, 1974).

need for such devices, often an easy way out of problems rather than a well-documented hypothesis, appeared to vanish as many patterns could be explained by continental drift paths. In particular, the recognition that drifting continents sometimes left behind trains of small fragments in the form of banks and islands [for example, the Seychelles-Saya de Malha Ridge in the western Indian Ocean: Fisher and others, 1967; see also Sclater and others, in press (a)], and thus provided shallow-water pathways well after the moment of initial separation, was regarded as a useful substitute. Recent deep-sea drilling in the northern North Atlantic has also provided evidence of the existence of land or shallow-water bridges between Europe and North America-Greenland until fairly late in the Cenozoic. As our knowledge grows and continental drift patterns and the positions of old island arcs and shallow ridges become better known, a firm physical basis will be established for inferences regarding intercontinental faunal and floral migrations.

A new element has recently been introduced with the emerging possibility of reconstructing the bathymetric evolution of ocean basins with time. This possibility rests on the recognition that oceanic crust is formed at shallow (2.5 to 3 km deep) mid-ocean ridge crests and subsides systematically with increasing age as a result of the cooling of the lithosphere. Several theoretical models have been developed (Sclater and Francheteau, 1970; Parker and Oldenburg, 1973) and supported by field observations (Sclater and others, 1971) and deep-sea drilling data (Fig. 8). A firm relationship has emerged that can be used to reconstruct the subsidence of any parcel of the ocean floor by means of its crustal age and present depth (Berger, 1972), provided it can be shown that it formed as normal oceanic crust. From this data we can then construct paleobathymetric charts for given time intervals [Sclater and others, in press (a), for the Indian Ocean]. Large parts of the ocean floor, however, are significantly shallower than one would infer from their ages; for example, the numerous rises studded with atolls of the western Pacific, or the oceanic banks of the Walvis Ridge and Rio Grande Rise in the South Atlantic. Some of these anomalously shallow regions are interpreted as hot-spot trails; others may represent large and complex fracture zones belonging to earlier spreading phases (such as the 90° East Ridge in the Indian Ocean); yet others are still unexplained. These shallow zones are of special interest because they form barriers to the flow of deep water and hence have a large influence on oceanic circulation and water-mass distribution. Some of them are near the sea surface or even break it and thus form special shallow habitats in the pelagic realm.

Our lack of understanding of the processes of their formation has prevented a theoretical treatment of their subsidence. Drilling and dredging evidence suggest that many of them started early at very shallow depth and subsided gradually at about the same rate as the adjacent oceanic crust formed normally at accreting plate edges. This hypothesis is currently being used in paleobathymetric studies, but the resultant precision (about ± 400 to 600 m) is two to three times less than that of the subsidence of normal oceanic crust (about ± 200 m).

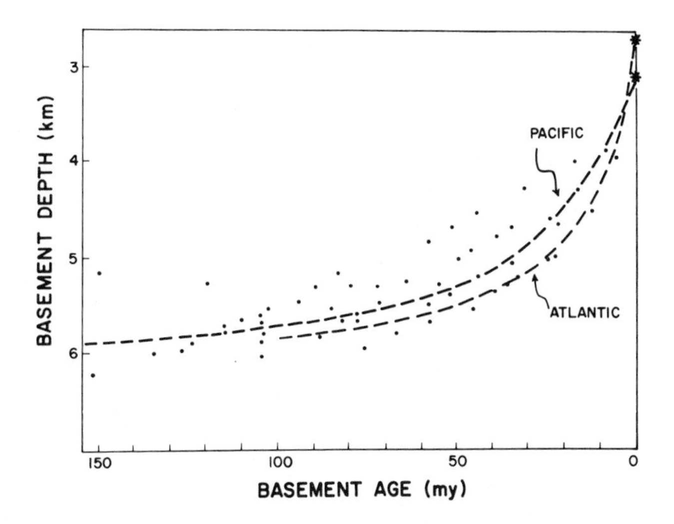

Figure 8. Relation between age of oceanic crust and depth below sea level. Dots are data from sites of the Deep Sea Drilling Project in the Atlantic, Pacific, and Indian Oceans. Dashed lines indicate subsidence curves fitted to the Pacific and Atlantic data which can be used to determine the subsidence history of the sea floor when age and present depth are known (modified after van Andel, 1975).

As an example of the construction and use of paleobathymetric charts, Figure 9 shows four stages in the evolution of the South Atlantic [modified after Sclater and others, in press (b)]. The North Atlantic opened first and 125 m. y. ago that ocean was about 1,500 km wide. Shortly afterwards, the South Atlantic began to open with a rotation of the tip of South America away from Africa and a hinge point in the Gulf of Guinea. By 95 m. y. ago (Fig. 9), a wedge of open ocean, a little more than 1,000 km wide, had formed in the south but only a narrow seaway, if any, existed in the north. The maximum depth in the flanking basins was just below 4,000 m, but a shallow transverse ridge, perhaps in part above sea level (Thiede, in press), divided the ocean into a southern and a northern half and prevented north-south circulation. In the early Cenozoic (65 m. y. ago), the flanking basins had widened and deepened considerably and the transverse ridge was beginning to break up into the present Rio Grande Rise and Walvis Ridge. As spreading continued, the two ridges moved farther away from the Mid-Atlantic Ridge, but 40 m. y. ago the passages through them were still less than 4,000 m and mostly blocked flow of Antarctic deep water into the northern South and the North Atlantic. Only early in the Miocene (20 m. y. ago) broad passages below 4 km were finally open and the way was clear for the development of the present south-to-north movement of Antarctic deep water and southward flow of intermediate Atlantic water.

Reconstructed paleobathymetries of this kind [Sclater and others, in press (a); van Andel, 1974] resemble those

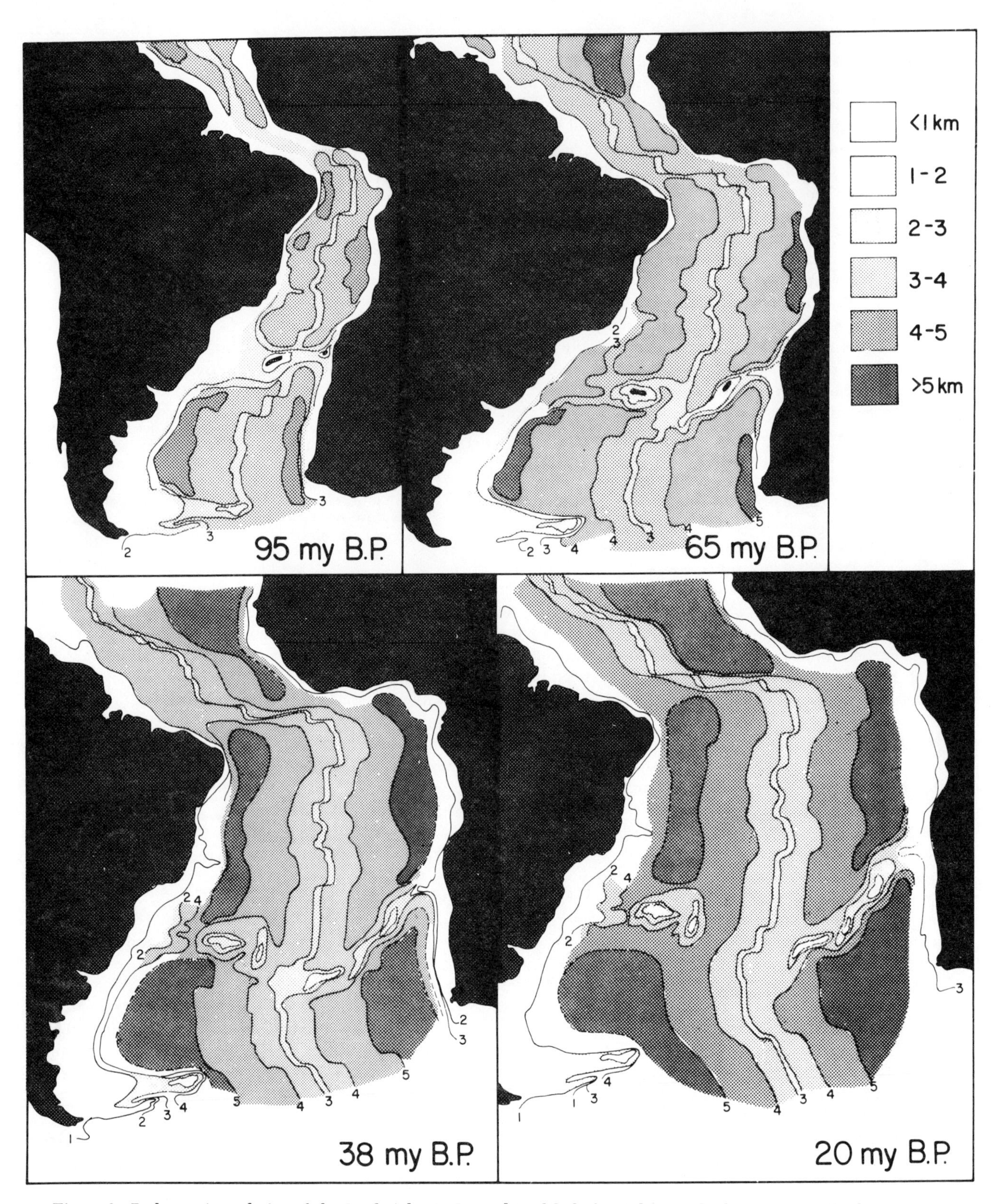

Figure 9. Bathymetric evolution of the South Atlantic Ocean [simplified after Sclater and others, in press (b)]. Reconstruction used depth-age relations as in Figure 8 and plate rotations from magnetic anomaly patterns and fracture zones as in Figure 4. Note transverse barrier in center formed by ancestral Rio Grande Rise and Walvis Ridge.

of present oceans because they have been constructed from them. Morphological features not present in today's oceans cannot be restored. This limits the resurrection of the bathymetry of the earliest phase of new oceans because we are not sure which rules we must apply in view of the lack or special nature of modern cases (Red Sea, Gulf of California). It also limits the reconstruction of the Mesozoic Pacific because over most of its circumference nothing older than the early Cenozoic has been spared by subduction. The principal remaining pre-Cenozoic floor occurs in the western Pacific and even a cursory glance at a map shows how strikingly different the pre-Cenozoic morphology west of the Line Islands is from that of the Cenozoic east of them (Fig. 10). Ridges, plateaus, and volcanic island chains occur in far greater abundance than elsewhere in the world where Cenozoic crust was formed by normal spreading from a mid-ocean ridge. This rugged topography has given rise to much speculation (Winterer, 1976) and hot spots, unusually rapid spreading rates, or anomalously closely spaced and elevated fracture-zone ridges have been invoked, together with the very complex spreading history, to explain it. One might also speculate, however, that this topography was the normal one for Panthalassa, that it resulted from plate tectonic processes subtly different from those of the Cenozoic (there is some independent evidence for this), and that the present open and relatively smooth ocean basins are a Cenozoic phenomenon. If that is true, then the Mesozoic and pre-Mesozoic ocean floor was a different world from what it is today.

BREAKUP OF PANGAEA AND THE EVOLUTION OF THE OCEANIC SURFACE CIRCULATION

For the last 100 to 125 m. y., the changing paleogeography can be reconstructed reasonably well with plate tectonic game rules working backwards from the present, although there is still disagreement regarding rotation parameters, the timing of shifts in direction, and many details. By Late Jurassic time, however, we begin to run out of preserved ocean floor, the Atlantic and Indian Oceans have essentially closed up, and too little is known of the floor of the Pacific to be of much help. The Pacific is a difficult ocean because it is nearly surrounded by subduction zones and contains no continents so that, although its internal configuration can be restored, it aids little in determining the position of the surrounding continents in the distant past.

Thus, the method of working backwards from the present fails and we must turn to other means. There is fair agreement on the late Paleozoic and early Mesozoic arrangement of the continents based on geometric and geologic fitting (Dietz and Holden, 1973; Smith and Hallam, 1970; Tarling and Runcorn, 1973), with some controversy regarding a precise fit of the southern blocks. The main disagreement centers on whether one (Pangaea) or two (Gondwana and Laurasia) supercontinents existed (Fig. 11). Paleomagnetic data, however, do not leave room for a wide intervening ocean, so that this problem does not materially affect subsequent drift patterns and the reconstruction of the oceanic circulation. From this supercontinent, knowing the shape and ultimate destination of the fragments, we can then work forward in time, using paleomagnetic data as a partial guide, until around 100 m. y. ago, we encounter the arrangement resulting from the backward extrapolation of plate rotations.

With the aid of such reconstructions of the drift of continents, together with some basic knowledge regarding the driving forces of the oceanic surface circulation and a vast mass of paleontologic and sedimentologic data (Berggren and Hollister, 1974, 1977), we can trace the evolution of the ocean surface circulation at least in a general manner (Fig. 12). During much of Mesozoic time, a circum-global seaway existed in low latitudes, well-known as the Tethys in the Middle and Near East and the Mediterranean. This Tethys Seaway connected a deep embayment in the eastern margin of Pangaea with the ocean beginning to develop in the Atlantic and completed its circum-global circuit through Central America. It began to close in the southeast Asian region in the early Tertiary as a result of northward movement of Australia (Moberly, 1972; Veevers, 1969); became choked in the Mediterranean and Near East in Oligocene and early Miocene time (Dewey and others, 1973; Phillips and Forsyth, 1972; Pitman and Talwani, 1972); and finally disappeared in the Pliocene when the Isthmus of Panama was uplifted (Bandy, 1970; Malfait and Dinkelman, 1972; Weyl, 1973). On the other hand, no circum-global seaway existed in high southern latitudes during most of the Mesozoic. The present circum-Antarctic passage began in the Cretaceous in the Indian Ocean; was extended eastward by the separation of Australia and Antarctica in the late Eocene and early Oligocene (Kennett and others, 1975); and was completed with the opening of a gap south of South America probably in the Miocene. As a result of the opening of the Atlantic and Indian Oceans, the Pacific lost its stature as the world ocean (Panthalassa) some 100 to 75 m. y. ago. The Atlantic first opened in the north in Early Cretaceous time, but about 75 m. y. ago the South Atlantic was also wide enough to provide a major north-south seaway. The history of the Indian Ocean is not yet fully resolved [Sclater and others, in press (a)], but it remained essentially a widening moat around a central continent (India) until the middle Tertiary and was strongly segmented for some time afterward.

This gradual change in distribution of continents, oceans, and seaways resulted in large modifications of the oceanic surface circulation. These have been sketched in Figure 12 with the aid of data from Berggren and Hollister (1977) and the laboratory modeling of Luyendyk and others (1972) who used a rotating water-filled pan with various arrangements of the continents. During the Jurassic, the world ocean probably possessed a complete circum-equatorial current system with two mid-latitude gyres in the north and south. As the North Atlantic opened, a northern return flow began to develop there also, while

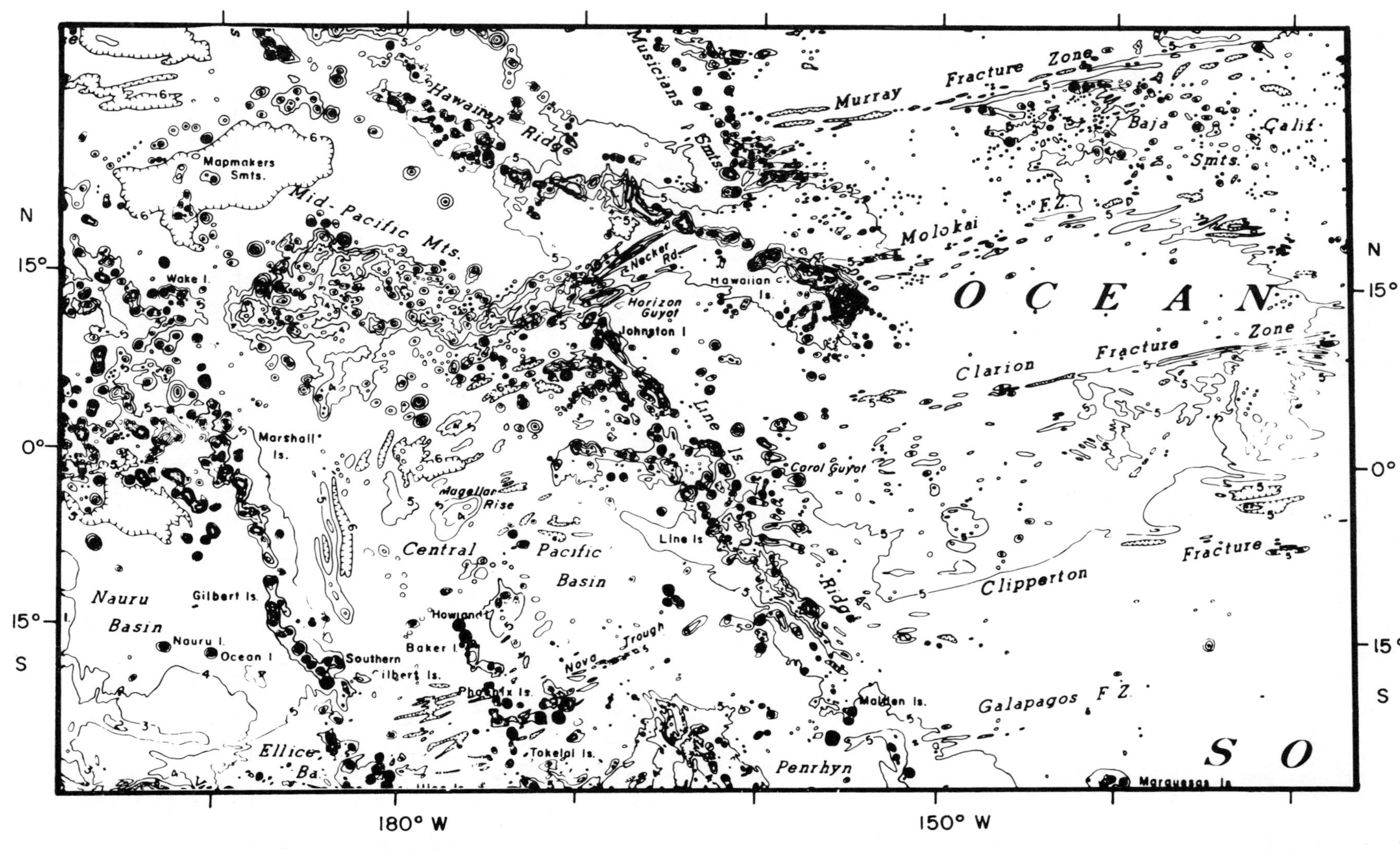

Figure 10. Bathymetry of the western Pacific (after Chase and others, 1975), contrasting rugged Mesozoic floor of the western Pacific with smoother Cenozoic floor in the east which was formed by normal spreading from the East Pacific Rise.

in the Indian Ocean southward flow between Africa and India and a return flow to the Pacific became possible (125 m. y. ago). At 65 m. y., the South Atlantic had opened enough to contribute water to the equatorial current system that still passed through the Tethys. Southward flowing currents existed along the east coast of South America; westward flow around the top of Africa supplied Indian Ocean water; and a current may have flowed east along the north edge of Antarctica. In the Indian Ocean, clockwise flow around India continued with an eastward branch into the Pacific north of Australia. An equatorial current system and two broad gyres with high latitude return flow persisted in the Pacific. At 35 m. y., finally, the Tethys was closed except for the Panamanian Isthmus which may have carried some water from the Pacific into the Caribbean. The equatorial current systems and the Atlantic flow patterns were much like those of today, and in the North Pacific the deteriorating climate had shifted the northern branch of the gyre southward. Circum-global high-latitude flow around Antarctica was established.

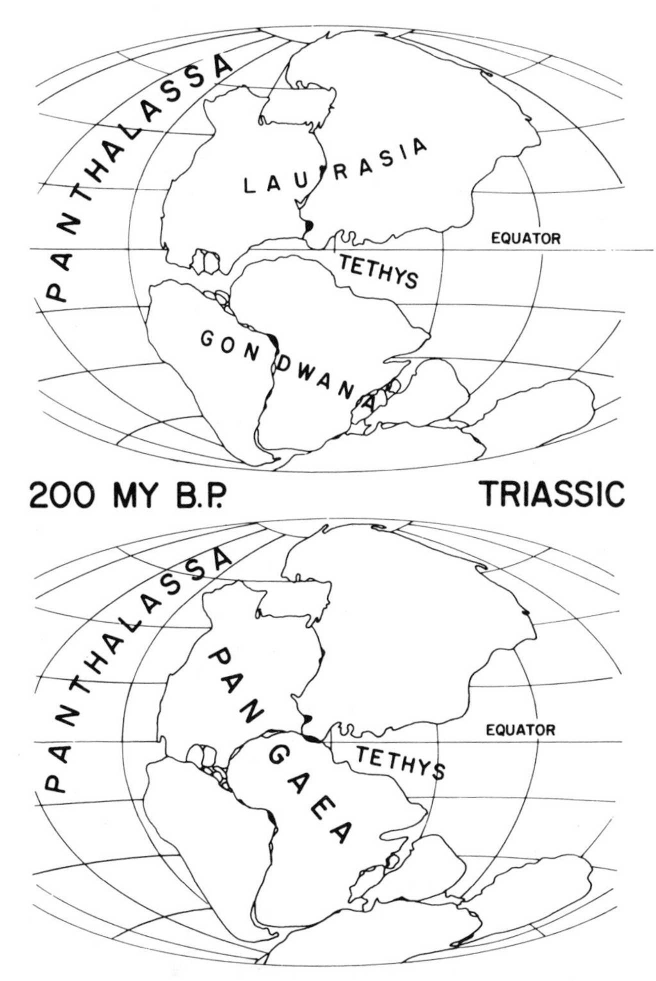

Figure 11. The possible configurations of Pangaea during the Triassic. Top: an open oceanic Tethys separates Gondwana and Laurasia. Bottom: a single supercontinent with an intracontinental Tethys. Gondwana fit after Dietz and Holden (1973).

The patterns of Figure 12 are generalized and in detail may be significantly in error. A large uncertainty stems from the fact that the plate tectonic reconstructions do not afford us an insight into the distribution of shallow epicontinental seas which, if extensive, may have had a major influence on flow patterns. Much more detailed studies of the regional effects of transgressions and regressions must be awaited before these influences can be assessed.

PALEOCEANOGRAPHY: RECONSTRUCTION OF OCEAN CIRCULATION IN THE EQUATORIAL PACIFIC DURING THE LAST 50 MILLION YEARS

The sediments and fossils of the deep ocean record the changing oceanic environment in considerable detail. Combined with the tectonic and paleobathymetric history of an ocean, this record, mainly available in the collections of the Deep Sea Drilling Project, can be interpreted in terms of its oceanographic evolution. So far there are few such studies and the available information is still fragmentary, partly because of an incomplete data synthesis, partly because some critical areas have not yet been drilled, and partly because we are still learning how to extract the information. The potential value of this kind of study, however, can be made clear with an example from the equatorial Pacific.

The study is based on three separate lines of evidence: (1) the tectonic changes in the configuration of the circum-Pacific continents and the rotation of the Pacific plate across the equator (Figs. 5, 12); (2) the paleoclimatic and glacial history of the circum-Antarctic region (Kennett and others, 1975); and (3) the sedimentary record of equatorial drill sites (van Andel and others, 1975). The principal features of the changing tectonic setting are the closure of the circum-global equatorial seaway and the opening of a circum-Antarctic one. The open equatorial seaway still existed 50 m. y. ago (Kennett and others, 1972) but became obstructed in the southeast Asian region in the Oligocene and in the Mediterranean and Near East in the late Oligocene or early Miocene. The last element, the Isthmus of Panama, did not close until the Pliocene. In the circum-Antarctic region, a shallow passage south of Australia opened in the early Oligocene and a deep one in the late Oligocene. The time of opening of the other critical passage, south of South America, has not yet been fully determined.

Broad circulation patterns can be inferred from these changes (Fig. 13) using oceanographic principles and the model experiments of Luyendyk and others (1972). A broad and rather fast equatorial current flowed westward in the open circum-global seaway, though convergence and hence fertility may have been less than at present. North of this equatorial current system existed a North Pacific gyre which broadened when the southeast Asian region became blocked and the equatorial current slowed and became narrower. As the western barrier became more impassable, the equatorial counter- and undercurrents may have developed which increased the up-

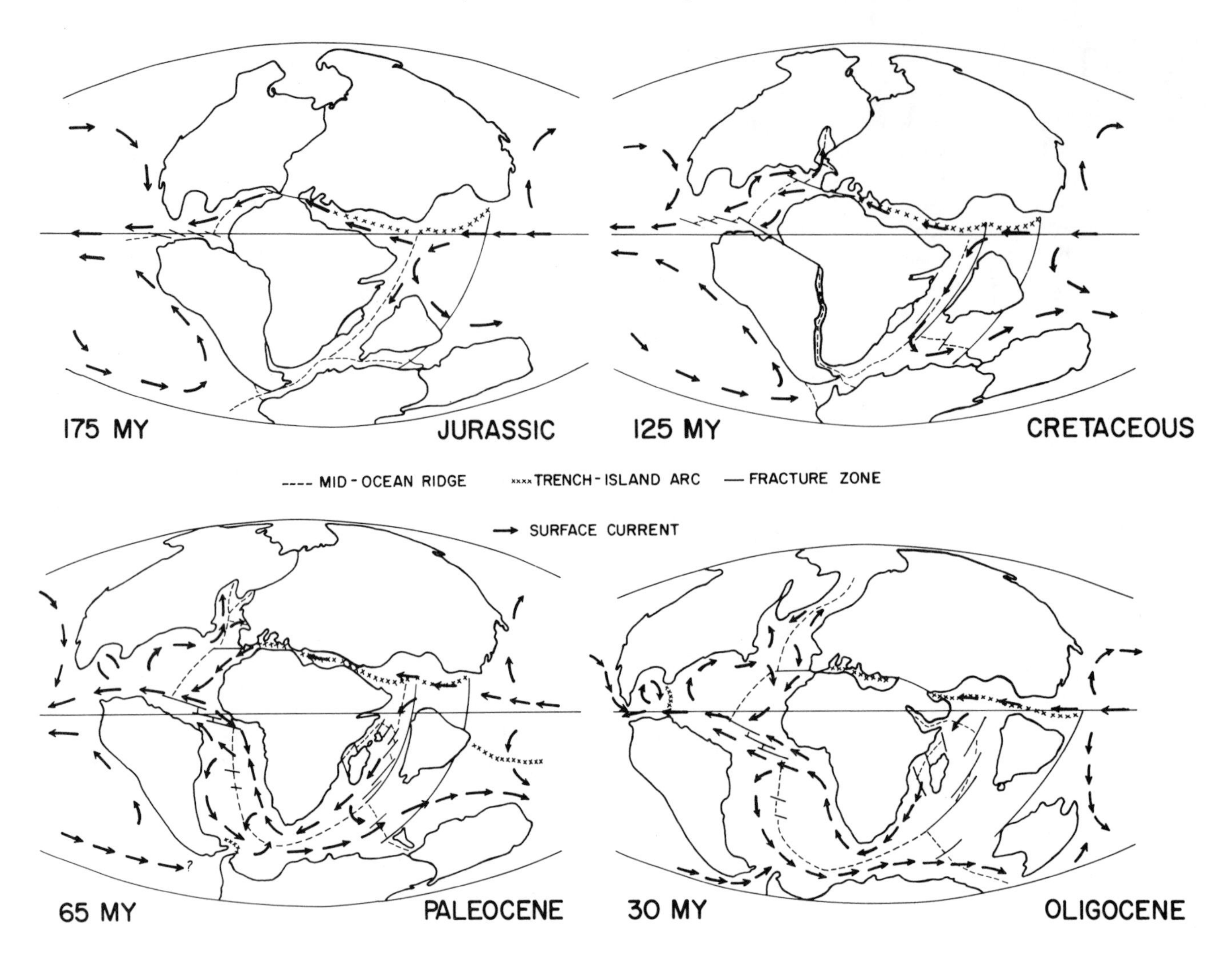

Figure 12. Paleogeographic evolution of the continents since the Jurassic and the surface circulation of the oceans. Continental positions after Dietz and Holden (1973); circulation patterns based on various sources including Berggren and Hollister (1974, in press) and Luyendyk and others (1972). Arrows indicate generalized current directions.

welling and hence the fertility, while the return flow of the North Pacific gyre shifted northward. For some time during the Miocene, Pacific water may have flowed eastward into the Caribbean through the Isthmus area.

The Antarctic climate was warm at the beginning of the period (Fig. 14) as shown by high ocean surface and bottom water temperatures (Shackleton and Kennett, 1975), but the temperature dropped sharply about 38 m. y. ago. While during the first part of the Cenozoic only the highest parts of Antarctica were glaciated, there is ample evidence that at least some glaciers reached sea level at this time and that sea ice was formed. There is no indication of an Antarctic ice cap, however, until about 15 m. y. ago when a major phase of Antarctic glaciation developed abruptly. Slightly earlier, progressive cooling and the continued widening of the circum-Antarctic seaway had resulted in the development first of the circum-Antarctic current system and of the Antarctic Convergence a little later. The maximal phase of Antarctic glaciation came 5 m. y. ago, and Arctic glaciation followed shortly afterwards.

These global changes drove the evolution of the ocean circulation in the equatorial Pacific which can be unraveled from several lines of evidence (Fig. 15). These include changes in the dissolution of carbonate in deep water inferred from variations in the rate of carbonate accumulation; a changing productivity of the surface waters deduced from accumulation rates of opal and carbonate; shifts in the depth boundary between calcareous and noncalcareous sediments (the Carbonate Compensation Depth, CCD); and variations in the intensity of sediment erosion measured by the frequency of depositional hiatuses in each interval. Until about 40 m. y. ago, opal accumulation rates indicate a fairly high equatorial fertility, but the dissolution rate of $CaCO_3$ is high and the CCD is shallow. This suggests an old bottom water rich in oxidative CO_2 and perhaps a low availability of $CaCO_3$ due to deposition elsewhere than in the deep ocean. At the Eocene-Oligocene

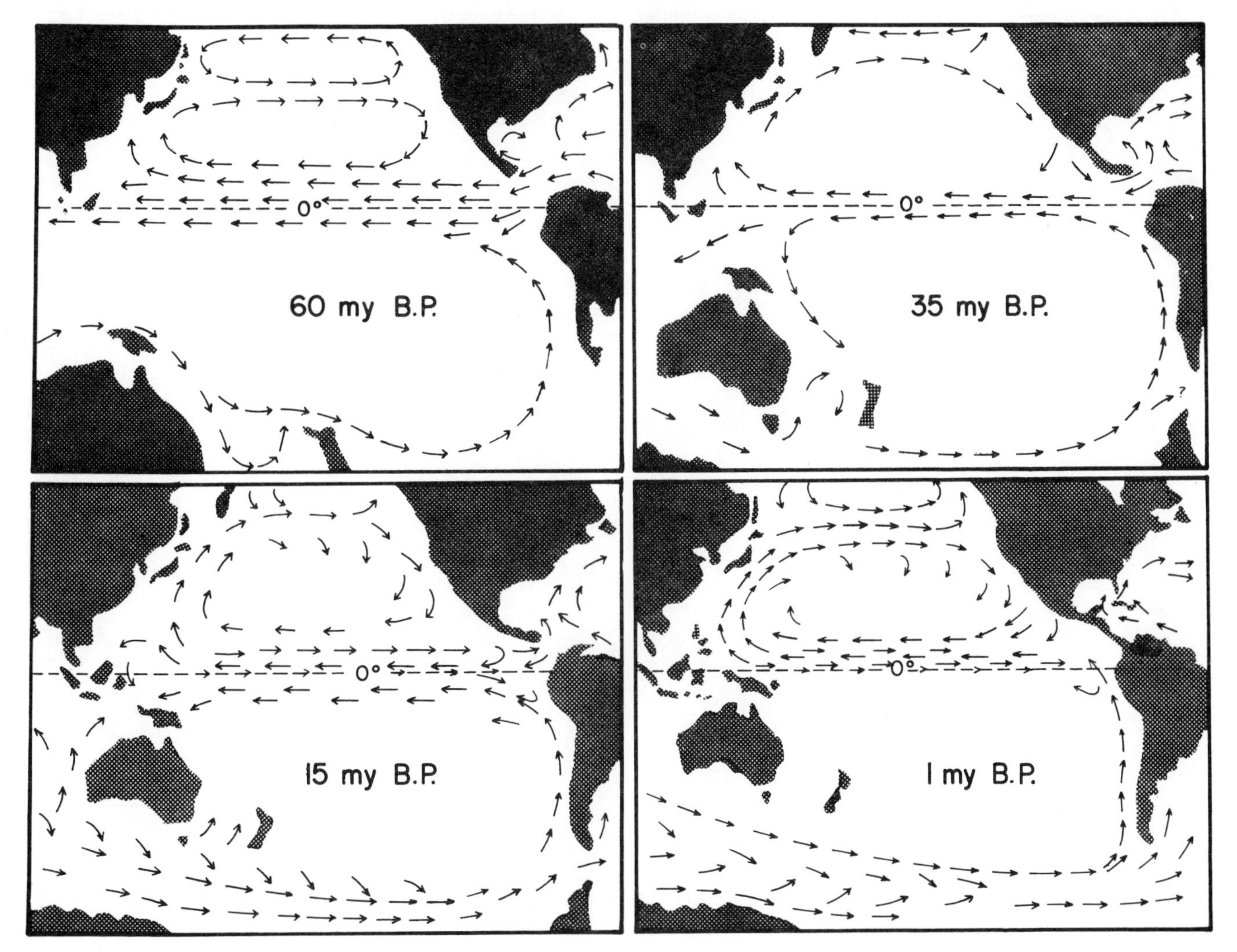

Figure 13. Changing configuration of continents in the circum-Pacific region during the last 60 m. y. (after Moore and others, 1976) and patterns of surface currents. Discussion in text.

boundary, conditions changed drastically. The fertility began to decrease, possibly as a result of the slowing of the current system due to blocking in southeast Asia. Carbonate dissolution and hiatuses diminished and the CCD dropped, perhaps because more vigorous flow of young bottom water was produced by cooling in the Antarctic. As the climate continued to deteriorate, carbonate dissolution and hiatuses gradually increased until 10 to 15 m. y. ago both reached a maximum accompanied by a steep rise of the CCD. This maximum falls at the time of a major Antarctic glaciation and of the first full development of the circum-Antarctic circulation. Equatorial fertility rose to a maximum about 5 m. y. ago that may have resulted from the full development of the equatorial countercurrent-undercurrent system (Leinen, in press). The subsequent decline of the opal productivity is somewhat mystifying; perhaps the massive extraction of opal that now characterizes the Antarctic Convergence developed at that time and reduced the availability of this limiting nutrient at the equator.

Obviously, all this is speculative and leaves many questions unanswered. It shows, however, that the fossil record, hitherto used primarily as a biostratigraphic tool, must be examined for its ecological information content. It is also now clear that, because of the many inter-ocean relations, only a true global approach will solve many of the problems. The next few years will undoubtedly see a rapid expansion of this kind of research leading to an improved understanding of the phenomena and eventually of the processes.

PLATE TECTONICS AND THE CHEMISTRY OF SEA WATER

Massive faunal and floral extinctions at the end of the Paleozoic and the Mesozoic have long been noticed. Bramlette (1965), in a thought-provoking paper, showed that a particularly massive extinction of planktonic taxa occurred at the Cretaceous-Tertiary boundary (Fig. 16) and suggested that this might have been due to a large

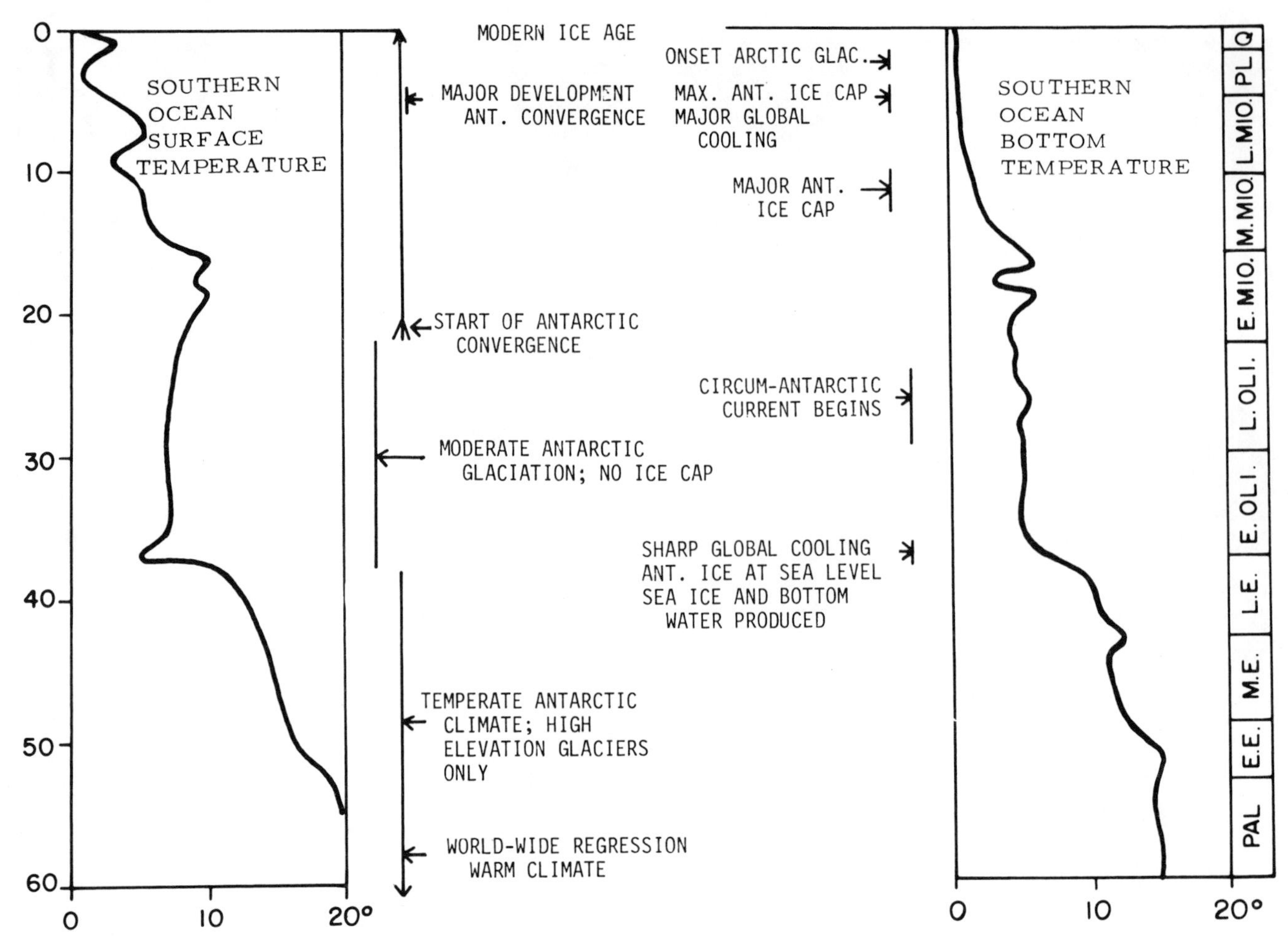

Figure 14. Evolution of Antarctic climate and oceanic conditions since the middle Cenozoic (after Kennett and others, 1975). Left and right: temporal variation of surface and bottom water temperatures based on oxygen isotope measurements of Shackleton and Kennett (1975).

reduction in the availability of nutrients brought about by deep weathering, low relief, and extensive transgression over the continents. This hypothesis is exceedingly difficult to prove because the principal nutrients, nitrogen, phosphorus, silica (limiting for diatoms), and trace elements and organic matter, are poorly preserved and extensively modified by later diagenesis. There is, however, growing evidence that the chemistry of sea water has changed, possibly substantially, over time.

Chemical interactions between plate processes and sea-water chemistry can be demonstrated. During the extrusion, cooling, and subsequent alteration of new oceanic crust at accreting plate edges, there is considerable chemical exchange between the basalt and sea water (Corliss, 1971). Hart (1973) has shown that the uptake of potassium, magnesium, and sodium by basalt and the release of calcium, silica, and iron are approximately equivalent to the amounts supplied from the continents by streams. Since the surface area of exposed basalt is a function of the spreading rate, changes in the rates of plate movement might significantly affect the composition of sea water and the residence times of some constituents. Unfortunately, processes active during the deposition and postdepositional alteration of oceanic sediments tend to obliterate the traces of temporal changes in sea-water element abundances. A few potential indicators of the chemical history of sea water, such as strontium and sulphur isotope ratios, support a correlation between changes in sea-water chemistry and tectonic processes. It has also been suggested that the extensive formation of evaporite deposits in newly-rifted ocean basins (Atlantic: Pautot and others, 1973) and in the Pliocene Mediterranean (Hsu and others, 1973) may have produced a significant reduction in the salinity of the world ocean. Even very optimistic estimates of the total amount of salt subtracted, however, suggest a reduction of only 10 to 20 percent, probably not enough for a global ecologic impact.

The relation between ocean-water chemistry and tectonics can be approached also from a quite different angle. Below a supersaturated near-surface layer, the dissolution of $CaCO_3$ in the ocean increases very slowly to a level called the lysocline where the gradient of the dissolution rate abruptly changes. At some point below the lysocline the dissolution rate matches the rate of carbonate supply; this level is called the Carbonate Compensation Depth

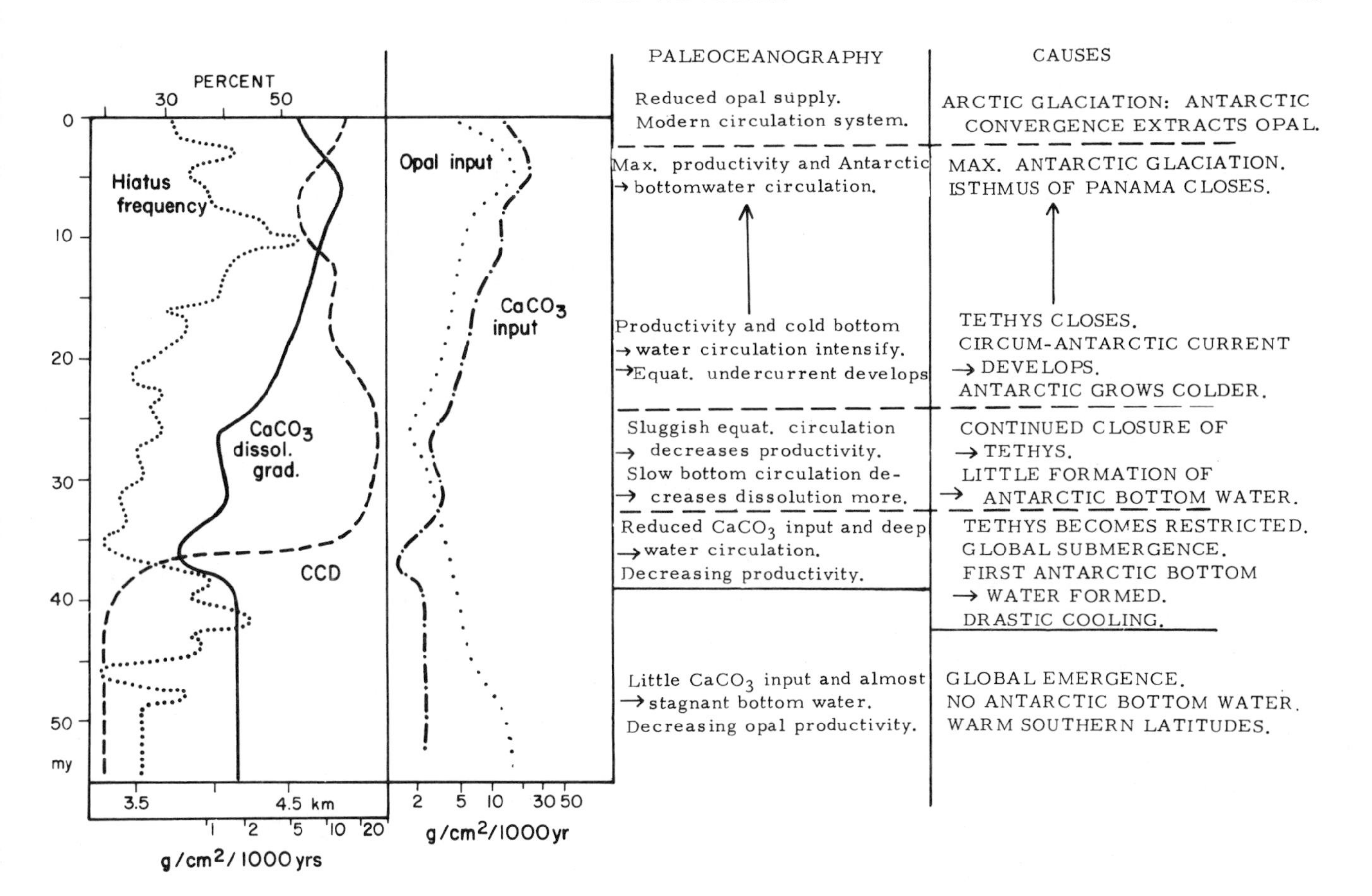

Figure 15. Summary of the paleoceanographic history of the central equatorial Pacific (after van Andel and others, 1975 and Leinen, in press). Discussion in text.

(CCD) and defines the boundary on the bottom of calcareous sediments above and noncalcareous sediments below. Its depth varies regionally and also with time (Berger and Winterer, 1974) indicating variations in the solution mechanism which are not yet well understood. For the last 10 to 15 m. y., the dissolving agent is widely assumed to be cold bottom water flowing northward from a surface source in the circum-Antarctic region.

Over the past 150 m. y. the CCD has varied in all oceans in an approximately similar manner (Fig. 17) (van Andel, 1975). As we have seen before, the later Cenozoic variations can be explained to some extent by the deteriorating climate of the Antarctic region, but this explanation is not fully satisfactory for the large drop at the Eocene-Oligocene boundary and for the high level during the late Mesozoic and early Cenozoic.

The CCD is a function of three variables: the input of carbonate as a result of biological processes near the sea surface; the depth of the lysocline; and the dissolution gradient below the lysocline (Heath and Culberson, 1970). Each of these can vary at least partly independently, but changes of the CCD do not indicate which variable is responsible. The explanation given above for late Cenozoic variation calls mainly on changes in the dissolution gradient but there is no evidence that such a climatic cause can be invoked for the shallow level prior to 40 m. y. ago. The parallelism of the curves of the three oceans (Fig. 17) suggests that a global cause must be responsible rather than more local events such as shifts in fertility patterns or bottom-water flow. One such cause (Berger and Winterer, 1974; van Andel, 1975) might be the partitioning of carbonate between shallow and deep seas. During the period 150 to 50 m. y., the continents were largely flooded and extensive neritic zones separated the deep sea from the continental carbonate sources. Many of these epeiric seas were, moreover, located at lower latitudes than today. Carbonate extraction and deposition rates in shallow water are higher than those in the deep sea (Chave and others, 1972) and little or no dissolution takes place at shallow depth. Thus, the shallow Mesozoic and early Cenozoic CCD might be a response to a lower availability of carbonate in pelagic areas and reflect the increased dissolution needed to maintain the same biological extraction rate (Broecker, 1974). The parallelism between the CCD and the best available curve for the eustatic change in sea level lends support to this assumption (P. R. Vail, personal commun. 1976). If this is correct, corresponding changes in other components might be expected since nutrients, trace elements, and needed organics would also be trapped in the epeiric seas. Bramlette's (1965) hypothesis of a reduced nutrient level as the cause for massive extinctions would thus receive support. Marine carbonate budgets for various time levels and from the shallow and deep environments are necessary to investigate this hypothesis further. There is also an intriguing temporal variation in the number of planktonic species that approximately parallels the CCD and sea-level change curves (compare also with Fig. 18).

Control of eustatic sea-level changes by plate motions is not intuitively obvious because the latter are mainly horizontal but many authors have speculated on the subject (Hallam, 1963, 1971; Valentine and Moores, 1970) and attributed them to changes in the volume of ocean basins. Such volume changes could result from the breakup or reassembly of continents, from changes in the number and length of oceanic accreting plate edges, or from variations in the spreading rate. Although an early view that changes in spreading rate produce variations in the crestal height of mid-ocean ridges was discredited, the fact that ridge elevations are a function of age alone implies that their volume must vary with the spreading rate. Hays and Pitman (1973), using all available information regarding temporal changes in spreading rate and in the number and length of mid-ocean ridges, calculated the resultant changes in ocean-basin volume. From this they derived, after correcting for isostatic effects on the continents and ocean floors, a curve of changing sea level which corresponds well with data on transgressions and regressions from independent sources (Fig. 18). The data are still rather sparse because our knowledge of Mesozoic mid-ocean ridges and spreading rates is inadequate, but Figure 18 shows that this is certainly a plausible cause for global sea-level changes.

The paleoceanographic and climatic effects of major global transgressions and regressions are important although at the moment quite speculative. During the height of the Mesozoic and early Cenozoic flooding, the much decreased area of land and increased area of shallow seas must have had a moderating influence on climate, because of the heat capacity of the epicontinental seas and because of the added pathways for the transfer of heat from low to high latitudes. With the onset of the regression, the continentality of the climate must have increased and the cross-latitudinal transfer of heat by sea water decreased. This process might have contributed significantly to a larger latitudinal air- and sea-temperature gradient, to a greater seasonal differentiation, and to the onset of glaciation in Antarctica (Fairbridge, 1973).

Other paleoceanographic effects can be postulated although not yet demonstrated. Hays and Pitman (1973) speculate that extensive shallow seas in the Late Cretaceous might have been sources of high-density water feeding the deep circulation. In low latitudes, the bottom water thus formed would have been supersaline as a result of evaporation; in high latitudes, it would have been cold. The Late Cretaceous deep water masses would thus vary markedly in nature from region to region and the resultant circulation could have been quite different from the circulation of deep water today. Hays and Pitman further suggest that, if the post-Cretaceous regression set in rather suddenly, the epicontinental sources of deep water might well have disappeared equally abruptly. Thus, a period of sluggish, old, and CO_2-rich bottom waters may have occurred before the onset of the Antarctic glaciation, which occurred considerably later, began to feed a new and vigorous south-to-north bottom-water flow. Such a sluggish, semistagnant phase is in good accord with data from

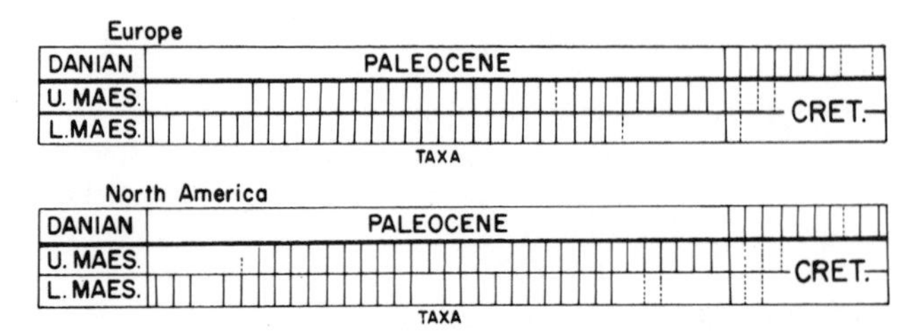

Figure 16. Massive extinction of pelagic taxa (calcareous nannofossils) at the Cretaceous-Tertiary boundary. Vertical lines indicate species ranges. This diagram, after Bramlette (1965), was the first to be published; since that time many striking examples have appeared in appropriate range charts published by the Deep Sea Drilling Project.

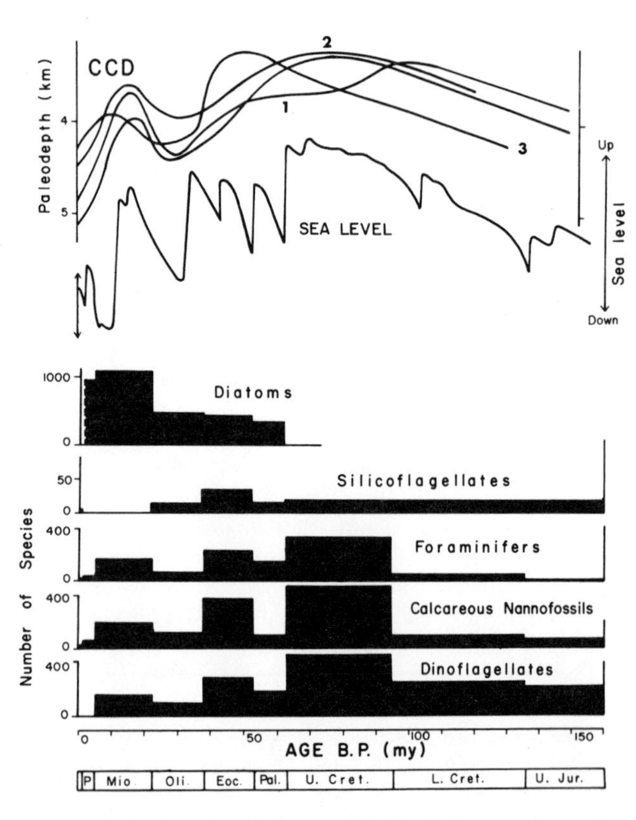

Figure 17. Temporal changes in the Carbonate Compensation Depth (CCD), the boundary between calcareous and noncalcareous sediments, for the Indian (1), Atlantic (2), and Pacific Oceans (3) (after van Andel, 1975) and eustatic sea-level changes (after P. R. Vail, personal commun. 1976). Underneath, variation of species number with time for several major planktonic groups (after Tappan and Loeblich, 1973).

the Pacific (Fig. 15). Moreover, a rapid withdrawal of the sea from the continents, perhaps accelerated by filling of the shallow seas with sediments and by isostatic compensation, would have caused drastic changes in the climate and the surface circulation of the oceans. All-in-all it is likely that late Mesozoic to early Cenozoic times imposed severe stresses on oceanic communities adapted to a long period of stability.

EPILOGUE: THE WORLD BEFORE AND AFTER PANGAEA

The discussion so far has been limited almost entirely to the world after the early Mesozoic breakup of Pangaea. I have attempted to show how far the reconstruction of paleogeography and oceanic paleoenvironment can be carried based only on physical data and without recourse to paleobiogeographic evidence. This, it seems to me, is very important. With a clear separation of the two lines of evidence, the resurrected physical world of the past can be used to interpret the biological one without the danger of circular reasoning that arises when paleobiogeography is first needed to define paleoenvironment. Much of what has been said is preliminary, speculative, or even a mere promise of good things to come. It should be evident, however, that in the near future a rather detailed picture will emerge of the ever-changing configuration of land and sea and its consequences for the evolution of ocean and atmosphere. This, in turn, will have a large influence on our future views of the history of life, the evolution of faunal and floral communities, and their effect on the environment.

For the world during and before Pangaea the situation is totally different. No longer can we work backwards from the present to reconstruct the past, and no ocean floor much older than 175 m. y. is likely to ever be found. Nothing remains of the floor of Panthalassa; very little remains of the Tethys except what was grafted onto the bordering continents in a complex and distorted manner (Bernouilli and Jenkyns, 1974). Prior to that, only the continents remain with their paleomagnetic records, their sutures (intensely deformed orogens where blocks were welded together while destroying an intervening ocean), and the paleobiological data. The paleomagnetic record, although in principle substantial, is much reduced by later overprints and is often ambiguous. Paleobiogeographic data are critical and always part of the argument but their use raises the spectre of circular reasoning.

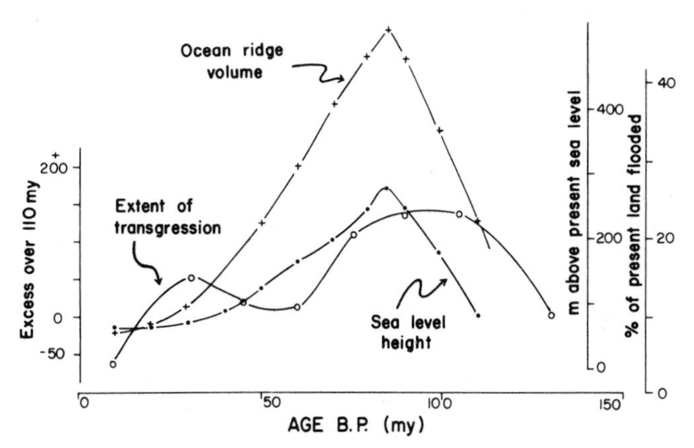

Figure 18. Variation with time of ocean ridge volume in km³ (expressed as the excess over the volume at 110 m. y. B. P.) and computed sea level after compensation for isostatic effects, compared with extent of transgressions and regressions over the continents (in percent of the present land area) (simplified after Hays and Pitman, 1973).

Even since Wilson (1966) proposed an open proto-Atlantic in the Paleozoic, the question has been hotly debated whether Pangaea was the original continent or merely a stage in a continuing sequence of separating, drifting, and colliding blocks. On one side are those, such as Engel and Kelm (1972), who accept only a single continent until the beginning of the Mesozoic although it may have shifted with respect to the Earth's axis. They recognize the presence of many ancient nuclei separated by zones of deformation, but perceive continuity of older structures across them (Hurley, 1973), and hence regard them as intracontinental deformation. On the other side are the defenders of large-scale plate motions as a permanent feature of Earth dynamics. Some very complex drift patterns have been postulated for pre-Mesozoic time, for instance, the reconstruction of the Asian plate from numerous smaller blocks once scattered widely (Burrett, 1974), or Badham and Halls' (1975) view of the Paleozoic North Atlantic as an ocean strewn with microcontinents. The paleomagnetic record has also been widely discussed in support of and against these notions (McElhinny and Briden, 1971; McElhinny, 1973; Creer, 1968), but remains inclusive and lacking in the detail that these complex reconstructions require. A fair sampling of the more recent views can be found in Tarling and Runcorn (1973).

It seems to me, at this time, that the evidence favors a drift history that was essentially as adventurous and complex during the last 2 b. y. as it has been shown to be during the last 200 m. y. The possibility of reconstructing the early history, however, would appear to be somewhat slim, with the full record of the oceanic floor or some 75 percent of the Earth's surface being irretrievably lost. Massive paleomagnetic work may eventually succeed in placing the continental masses of each time where they belonged on the globe. The filling of the intervening blanks will have to come from extrapolation of data from the very deformed old continental edges, from what little can be learned from ocean floor caught up in the final collisions, and from the fertile imagination of investigators.

ACKNOWLEDGMENTS

This paper was made possible by grant OCE75-21833 from the National Science Foundation.

REFERENCES

Badham, J. P. N., and Halls, C., 1975, Microplate tectonics, oblique collisions, and evolution of the Hercynian orogenic systems: Geology, v. 3, p. 373-392.

Bandy, O. L., 1970, Upper Cretaceous-Cenozoic paleobathymetric cycles, eastern Panama and northern Colombia: Gulf Coast Assoc. Geol. Soc. Trans., v. 20, p. 181-193.

Berger, W. H., 1972, Deep-sea carbonates: Dissolution facies and age-depth constancy: Nature, v. 236, p. 392-395.

Berger, W. H., and Winterer, E. L., 1974, Plate stratigraphy and the fluctuating carbonate line, *in* Hsü, K. J., and Jenkyns, H. C., eds., Pelagic sediments on land and in the sea: Internat. Assoc. Sedimentologists Spec. Pub. 1, p. 11-48.

Berggren, W. A., and Hollister, C. D., 1974, Paleogeography, paleobiogeography, and the history of circulation in the Atlantic Ocean, *in* Hay, W. W., ed., Studies in paleo-oceanography: Soc. Econ. Paleontologists and Mineralogists Spec. Pub. 20, p. 126-186.

——— 1977, Plate tectonics and paleocirculation: Commotion in the ocean: Tectonophysics, v. 38, p. 11-48.

Bernouilli, Daniel, and Jenkyns, H. C., 1974, Alpine, Mediterranean and central Atlantic Mesozoic facies in relation to the evolution of the Tethys, *in* Dott, R. H., and Shaver, R. H., eds., Modern and ancient geosynclinal sedimentation: Soc. Econ. Paleontologists and Mineralogists Spec. Pub. 19, p. 129-160.

Bramlette, M. N., 1965, Massive extinctions in biota at the end of Mesozoic time: Science, v. 148, p. 1696-1699.

Broecker, W. S., 1974, Chemical oceanography: New York, Harcourt, Brace, Jovanovich, 214 p.

Bullard, Edward, Everett, J. S., and Smith, A. G., 1965, The fit of the continents around the Atlantic: Royal Soc. London, Philos. Trans., ser. A, Math. Phys. Sci., no. 258, p. 41-51.

Burrett, C. F., 1974, Plate tectonics and the fusion of Asia: Earth and Planetary Sci. Letters, v. 21, p. 181-189.

Chase, T. E., Menard, H. W., and Mammerickx, J., 1975, Bathymetry of the North Pacific: La Jolla, California, Scripps Inst. Oceanog., 12 sheets.

Chave, K. E., Smith, S. V., and Roy, K. J., 1972, Carbonate production by coral reefs: Marine Geology, v. 12, p. 123-140.

Corliss, J. B., 1971, The origin of metal-bearing submarine hydrothermal solutions: Jour. Geophys. Research, v. 76, p. 8128-8138.

Cox, Allan, 1973, Plate tectonics and geomagnetic reversals: Readings with introductions: San Francisco, W. H. Freeman, 702 p.

Creer, K. M., 1968, Arrangement of continents during the Paleozoic Era: Nature, v. 219, p. 41-44.

Dewey, J. F., Pitman, W. C., III, Ryan, W. B. F., and Bonnin, Jean, 1973, Plate tectonics and the evolution of the alpine system: Geol. Soc. America Bull., v. 84, p. 3137-3180.

Dietz, R. S., and Holden, J. C., 1973, The breakup of Pangaea, *in* Continents adrift. Readings from Scientific American: San Francisco, W. H. Freeman, p. 102-113.

Du Toit, A. L., 1937, Our wandering continents: Edinburgh, Oliver and Boyd, 366 p.

Engel, A. E. J., and Kelm, D. L., 1972, Pre-Permian global tectonics: A tectonic test: Geol. Soc. America Bull., v. 83, p. 2325-2340.

Fairbridge, R. W., 1973, Glaciation and plate migration, *in* Tarling, D. H., and Runcorn, S. K., eds., Implications of continental drift for the Earth sciences, Vol. 1: London, Academic Press, p. 501-512.

Fisher, R. L., Johnson, G. L., and Heezen, B. C., 1967, Mascarene Plateau, western Indian Ocean: Geol. Soc. America Bull., v. 78, p. 1247-1266.

Gass, I. G., Smith, P. J., and Wilson, R. C. L., 1971, Understanding the Earth: A reader in the Earth sciences: Cambridge, Massachusetts Inst. Technology Press, 355 p.

Hallam, Anthony, 1963, Major epeirogenic and eustatic changes since the Cretaceous and their possible relationship to crustal structure: Am. Jour. Sci., v. 261, p. 397-423.

——— 1971, Re-evaluation of the paleogeographic argument for an expanding Earth: Nature, v. 232, p. 180-182.

Hart, R. A., 1973, Geochemical and geophysical implications of the reaction between seawater and the oceanic crust: Nature, v. 243, p. 76-79.

Hays, J. D., and Pitman, W. C., III, 1973, Lithospheric plate motion, sea level changes and climatic and ecological consequences: Nature, v. 246, p. 18-22.

Heath, G. R., and Culberson, Charles, 1970, Calcite: Degree of saturation, rate of dissolution and the compensation depth in the deep oceans: Geol. Soc. America Bull., v. 81, p. 3157-3160.

Hsü, K. J., Ryan, W. B. F., and Cita, M. B., 1973, Late Miocene desiccation of the Mediterranean: Nature, v. 242, p. 240-244.

Hurley, P. M., 1973, On the origin of the 450 $\pm$ 200 m. y. orogenic belts, *in* Tarling, D. H., and Runcorn, S. K., eds., Implications of continental drift for the Earth sciences, Vol. 2: London, Academic Press, p. 1083-1090.

Irving, E., 1964, Paleomagnetism and its application to geological and geophysical problems: New York, Wiley, 399 p.

Kennett, J. P. and others, 1972, Australian-Antarctic continental drift, paleocirculation changes and Oligocene deep-sea erosion: Nature, v. 239, p. 51-55.

Kennett, J. P. and others, 1975, Cenozoic paleoceanography in the southwest Pacific Ocean, Antarctic glaciation and the development of the Circum-Antarctic Current, *in* White, Stan M., ed., Initial Reports of the Deep Sea Drilling Project, Vol. 29: Washington, D. C., U. S. Govt. Printing Office, p. 1155-1169.

Leinen, Margaret, 1977, Biogenic silica sedimentation in the equatorial Pacific during the Cenozoic: Geol. Soc. America Bull. (in press).

Luyendyk, B. P., Forsyth, Donald, and Phillips, J. D., 1972, Experimental approach to the paleocirculation of the oceanic surface waters: Geol. Soc. America Bull., v. 83, p. 2649-2664.

Malfait, B. T., and Dinkelman, M. G., 1972, Circum-Caribbean tectonic and igneous activity and the evolution of the Caribbean plate: Geol. Soc. America Bull., v. 83, p. 251-272.

McElhinny, M. W., 1973, Paleomagnetism and plate tectonics: Cambridge, Cambridge Univ. Press, 358 p.

McElhinny, M. W., and Briden, J. C., 1971, Continental drift during the Phanerozoic: Earth and Planetary Sci. Letters, v. 10, p. 402-416.

Meyerhoff, A. A., 1970, Continental drift, implications of paleomagnetic studies, meteorology, physical oceanography and climatology: Jour. Geology, v. 18, p. 77-93.

Minster, J. B., Jordan, T. H., Molnar, Peter, and Haines, Eldon, 1974, Numerical modeling of instantaneous plate tectonics: Royal Astron. Soc. Geophys. Jour., v. 36, p. 541-576.

Moberly, Ralph, 1972, Origin of lithosphere behind island arcs, with reference to the western Pacific, *in* Shagam, R., ed., Studies in Earth and Space science: Geol. Soc. America Mem. 132, p. 35-56.

Molnar, Peter, and Atwater, Tanya, 1973, Relative motion of hot spots in the mantle: Nature, v. 246, p. 288-291.

Moore, T. C., Jr., van Andel, Tj. H., Sancetta, Constance, and Pisias, Nicholas, 1978, Cenozoic hiatuses in pelagic sediments, *in* Riedel, W. R., and Saito, Tsunemasa, eds., Marine plankton and sediments: New York, Micropaleontology Press (in press).

Morgan, W. J., 1972, Plate motions and deep mantle convection, *in* Shagam, R., ed., Studies in Earth and Space science, Geol. Soc. America Mem. 132, p. 7-22.

Parker, R. L., and Oldenburg, D. W., 1973, Thermal model of mid-ocean ridges: Nature, v. 242, p. 137-139.

Pautot, Guy, Renard, Vincent, Daniel, Jean, and Dupont, Jacques, 1973, Morphology, limits, origin and age of salt layer along South Atlantic African margin: Am. Assoc. Petroleum Geologists Bull., v. 57, p. 1658-1671.

Phillips, J. D., and Forsyth, Donald, 1972, Plate tectonics, paleomagnetism and the opening of the Atlantic: Geol. Soc. America Bull., v. 83, p. 1579-1600.

Pitman, W. C., III, and Talwani, Manik, 1972, Sea floor spreading in the North Atlantic: Geol. Soc. America Bull., v. 83, p. 619-646.

Sclater, J. G., Abbott, Dallas, and Thiede, Jörn, 1977a, Paleobathymetry and sediments of the Indian Ocean, *in* Heirtzler, J. R., and Davies, T. A., eds., Syntheses of deep-sea drilling in the Indian Ocean: Am. Geophys. Union Mon. (in press).

Sclater, J. G., Anderson, R. N., and Bell, M. L., 1971, Elevation of ridges and evolution of the central eastern Pacific: Jour. Geophys. Research, v. 76, p. 7888-7915.

Sclater, J. G., and Francheteau, Jean, 1970, The implications of terrestrial heat flow observations on current tectonic and geochemical models of the crust and upper mantle of the Earth: Royal Astron. Soc. Geophys. Jour., v. 20, p. 509-542.

Sclater, J. G., Hellmyer, S., and Tapscott, Paul, 1977b The paleobathymetry of the Atlantic Ocean: Jour. Geol. (in press).

Shackleton, N. J., and Kennett, J. P., 1975, Paleotemperature history of the Cenozoic and the initiation of Antarctic glaciation: Oxygen and carbon isotope analyses at DSDP sites 77, 279, 281, *in* White, Stan M., ed., Initial Reports of the Deep Sea Drilling Project, Vol. 29: Washington, D. C., U. S. Govt. Printing Office, p. 743-755.

Smith, A. G., and Hallam, Anthony, 1970, The fit of the southern continents: Nature, v. 225, p. 139-144.

Tappan, Helen, and Loeblich, A. R., Jr., 1973, Evolution of oceanic plankton: Earth-Sci. Rev. v. 9, p. 207-240.

Tarling, D. H., and Runcorn, S. K., 1973, Implications of continental drift for the Earth sciences: London, Academic Press, Vol. 1, 622 p.; Vol. 2, 559 p.

Thiede, Jörn, 1977, The subsidence of aseismic ridges: Evidence from sediments on the Rio Grande Rise (Atlantic Ocean): Am. Assoc. Petroleum Geologists Bull. (in press).

Valentine, J. W., and Moores, E. M., 1970, Plate tectonic regulation of biotic diversity and sea level, a model: Nature, v. 228, p. 657-659.

van Andel, Tj. H., 1974, Cenozoic migration of the Pacific plate, northward shift of the axis of deposition and paleobathymetry of the central equatorial Pacific: Geology, v. 2, p. 507-510.

——— 1975, Mesozoic-Cenozoic calcite compensation depth and the global distribution of carbonate sediments: Earth and Planetary Sci. Letters, v. 26, p. 187-194.

van Andel, Tj. H., Heath, G. R., and Moore, T. C., Jr., 1975, Cenozoic history and paleoceanography of the central equatorial Pacific: Geol. Soc. America Mem. 143, 134 p.

Veevers, J. J., 1969, Paleogeography of the Timor Sea region: Palaeogeography, Palaeoclimatology, Palaeoecology, v. 6, p. 125-140.

Wegener, Alfred, 1915, Die Entstehung der Kontinente und Ozeane: Braunschweig, Friedr. Vieweg und Sohn, 539 p.

Weyl, Richard, 1973, Die paläogeographische Entwicklung Mittelamerikas: Zentralbl. Geol. Paläont., v. 1, p. 433-466.

Wilson, J. T., 1966, Did the Atlantic close and then re-open?: Nature, v. 211, p. 676-681.

Winterer, E. L., 1973, Sedimentary facies and plate tectonics of the equatorial Pacific: Am. Assoc. Petroleum Geologists Bull., v. 57, p. 265-282.

——— 1976, Anomalies in the tectonic evolution of the Pacific, *in* Sutton, G. H., ed., The geophysics of the Pacific Ocean Basin and its margins: Am. Geophys. Union Mon. 19, 269-278.

Geographic Distribution of Cambrian and Ordovician Rostroconch Mollusks

JOHN POJETA, JR., *U. S. Geological Survey, E-501 U. S. National Museum, Washington, D. C. 20560*

ABSTRACT

Rostroconchs form a class of bivalved mollusks distinct from the class Pelecypoda. Cambrian and Ordovician rostroconchs were primarily tropical and subtropical in their distribution and are found mostly in carbonate rocks within 30° of the Cambrian and Ordovician equators. A marked decrease in the taxonomic diversity of rostroconchs after the Early Ordovician is attributed to competition of that group with the Pelecypoda. Pelecypods are probably far more efficient burrowers than rostroconchs were because the latter lack a ligament and adductor muscles.

INTRODUCTION

This paper documents the known geographic distribution of the molluscan class Rostroconchia in the lower Paleozoic. The concept of the Rostroconchia is relatively new (Pojeta and others, 1972), and much of the primary data used herein has been documented by Pojeta and Runnegar (1976), Pojeta and others (in press), and Runnegar and Jell (1976). The orders Ribeirioida and Conocardioida have been assigned to the Rostroconchia (Pojeta and Runnegar, 1976), and the class is defined as follows: mollusks with an uncoiled and untorted univalved larval shell which straddles the dorsal midline, and a bivalved adult shell with one or more shell layers continuous across the dorsal margin so that a dorsal commissure is lacking. Previously, the ribeirioids had been regarded as unusual arthropods, and the conocardioids, as aberrant pelecypods.

The Rostroconchia is an entirely Paleozoic group ranging in age from Early Cambrian (Tommotian: Matthews and Missarzhevsky, 1975) to the Late Permian (Makarewan: Waterhouse, 1967). Figure 1 shows the abundance of the class throughout the Paleozoic on the basis of the number of described species for each system. There are 29 known genera and about 125 named species of rostroconchs in the Cambrian and Ordovician. Figure 2 shows the stratigraphic distribution of the known genera in the lower Paleozoic.

Although rostroconchs are known from all modern continents except Antarctica, large well-documented lower Paleozoic faunas have been described only from Manchuria (Kobayashi, 1933; Pojeta and Runnegar, 1976), northern Australia (Pojeta and others, in press), and North America (Pojeta and Runnegar, 1976). Thus there are not yet enough data points to use rostroconchs paleogeographically in the way in which lower Paleozoic brachiopods and trilobites have been used. However, the data on rostroconchs can be presented and the known distributions of genera can be plotted on various paleogeographic maps which have been proposed for the Cambrian and Ordovician. I have done this on maps published for the Cambrian and Ordovician by Whittington and Hughes (1972), Smith and others (1973), and Ross (1975). All these maps present world-wide paleogeographic reconstructions for the time intervals concerned. Comparisons of these maps may be used to draw paleontological deductions on such matters as the climatic zones in which Cambrian and Ordovician rostroconchs may have lived.

Table 1 summarizes the known geographic, stratigraphic, and lithologic occurrences of Cambrian and Ordovician rostroconchs. Because most of the lithologic information was gathered from the literature or from museum specimens, only the broad subdivisions carbonate rocks and clastic rocks are used.

DISTRIBUTION OF CAMBRIAN ROSTROCONCHS

The 16 known species of Cambrian rostroconchs have been placed in 10 genera. Seven of the genera are known only from Cambrian rocks; 3 genera are known from both Cambrian and Ordovician rocks (Fig. 2).

Seven of the Cambrian genera are at present each known only from one outcrop area and thus may be endemics. *Cymatopegma* Pojeta, Gilbert-Tomlinson, and Shergold; *Pinnocaris* Etheridge; *Kimopegma* Pojeta, Gilbert-Tomlinson, and Shergold; *Oepikila* Runnegar and Pojeta; and *Pleuropegma* Pojeta, Gilbert-Tomlinson, and Shergold are known only from north-central Australia. In the Ordovician, *Pinnocaris* becomes more widespread. The genus *Watsonella* Grabau is known only from eastern Massachusetts. *Wanwania* Kobayashi is known only from southern Manchuria in the Cambrian, although it becomes more widespread in the Ordovician. The remaining 3 Cambrian genera are each known from two widely separated outcrop areas: *Heraultipegma* Pojeta and Runnegar occurs in southern France and the River Lena area of Siberia; *Myona* Kobayashi is known from South Korea and north-

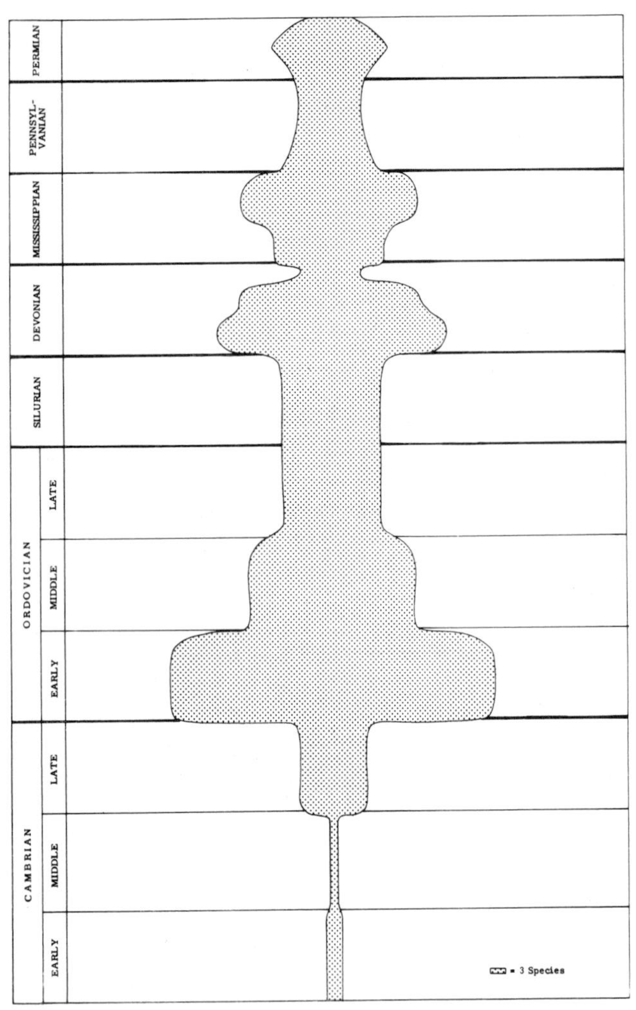

Figure 1. Abundance of rostroconchs throughout the Paleozoic; diagram based on the number of species for each time interval.

central Australia; and *Ribeiria* Sharpe is found in north-central Australia and eastern New York.

The lithologic occurrences of 14 of the 16 Cambrian species are known (Table 1). Six species belonging to the genera *Cymatopegma, Kimopegma, Pinnocaris,* and *Riberia* occur in shallow-marine quartzose sandstones in north-central Australia. Eight species belonging to the genera *Heraultipegma, Myona, Oepikila, Pleuropegma, Ribeiria* and *Watsonella* are known from carbonate rocks from various parts of the world.

On the Smith and others (1973) paleogeographic map of the Cambrian and Lower Ordovician (Fig. 3), all known Cambrian occurrences of rostroconchs plot between 40°S paleolatitude and 30°N paleolatitude; only four occurrences are outside the 20° interval between 10° and 30°N paleolatitude. On the Ross (1975) paleogeographic map for the Cambrian, the rostroconch occurrences plot between 60°S paleolatitude and 35°N paleolatitude; all but three occurrences are between 10° and 35°N paleolatitude.

The two paleogeographic reconstructions used show substantial areas of agreement about the likely climatic zones in which Cambrian rostroconchs lived, although they differ somewhat in the total life zone indicated for the group at that time. The data indicate that most Cambrian rostroconchs were tropical to subtropical forms, although a few taxa ranged into temperate and perhaps low polar latitudes. This information also suggests that the greatest amount of new data about Cambrian rostroconchs is likely to come from examining outcrops that were within 30° of the Cambrian equator.

DISTRIBUTION OF EARLY ORDOVICIAN ROSTROCONCHS

Five genera of rostroconchs are known from the latest Cambrian; 3 of these are also known from Ordovician rocks (Fig. 2). *Pinnocaris* is not yet known from the Early Ordovician, but presumably it will be found in this time interval because it occurs in both the latest Cambrian and Middle Ordovician. In addition to the genera that cross the Cambrian boundary, there are 16 other known genera of Early Ordovician rostroconchs and about 65 known Early Ordovician species. The Early Ordovician represents the greatest known generic-level diversity of the Rostroconchia; 12 of the 18 genera represented in the Early Ordovician are limited to this part of the column.

Twelve of the Early Ordovician genera are each known from only one outcrop area and may be endemics. *Apoptopegma* Pojeta, Gilbert-Tomlinson, and Shergold; *Ptychopegma* Pojeta, Gilbert-Tomlinson, and Shergold; *Pauropegma* Pojeta, Gilbert-Tomlinson, and Shergold; and *Bransonia* Pojeta and Runnegar are known only from north-central Australia; in younger Ordovician rocks, *Bransonia* is more widely distributed. *Ribeirina* Billings is known only from Ontario. *Anisotechnophorus* Pojeta and Runnegar is known only from New York. *Eoischyrina, Pseudotechnophorus, Wanwanella, Wanwanoidea, Euchasmella,* and *Pseudoeuchasma,* all named by Kobayashi, are known only from southern Manchuria.

Of the remaining genera, *Wanwania* occurs in the Lower Ordovician rocks of southern Manchuria and north-central Australia; *Tolmachovia* Howell and Kobayashi is known from north-central Australia and Tasmania, although it increases its geographic range in the Middle Ordovician; *Technophorus* Miller is known from Korea and north-central Australia, and becomes more widespread in younger Ordovician rocks; and *Ribeiria, Eopteria* Billings, and *Euchasma* Billings have world-wide Early Ordovician distributions, occurring in North America, southern Manchuria, north-central Australia, and Malaysia.

The lithologic associations of 35 Early Ordovician species of rostroconchs are known; 29 species occur in carbonate rocks and 6 species in clastic rocks (Table 1). All known species of the following genera occur in carbonate rocks in the Early Ordovician: *Apoptopegma, Ptychopegma, Anisotechnophorus, Pauropegma, Pseudotechnophorus, Eopteria, Euchasma,* and *Bransonia.* Eleven

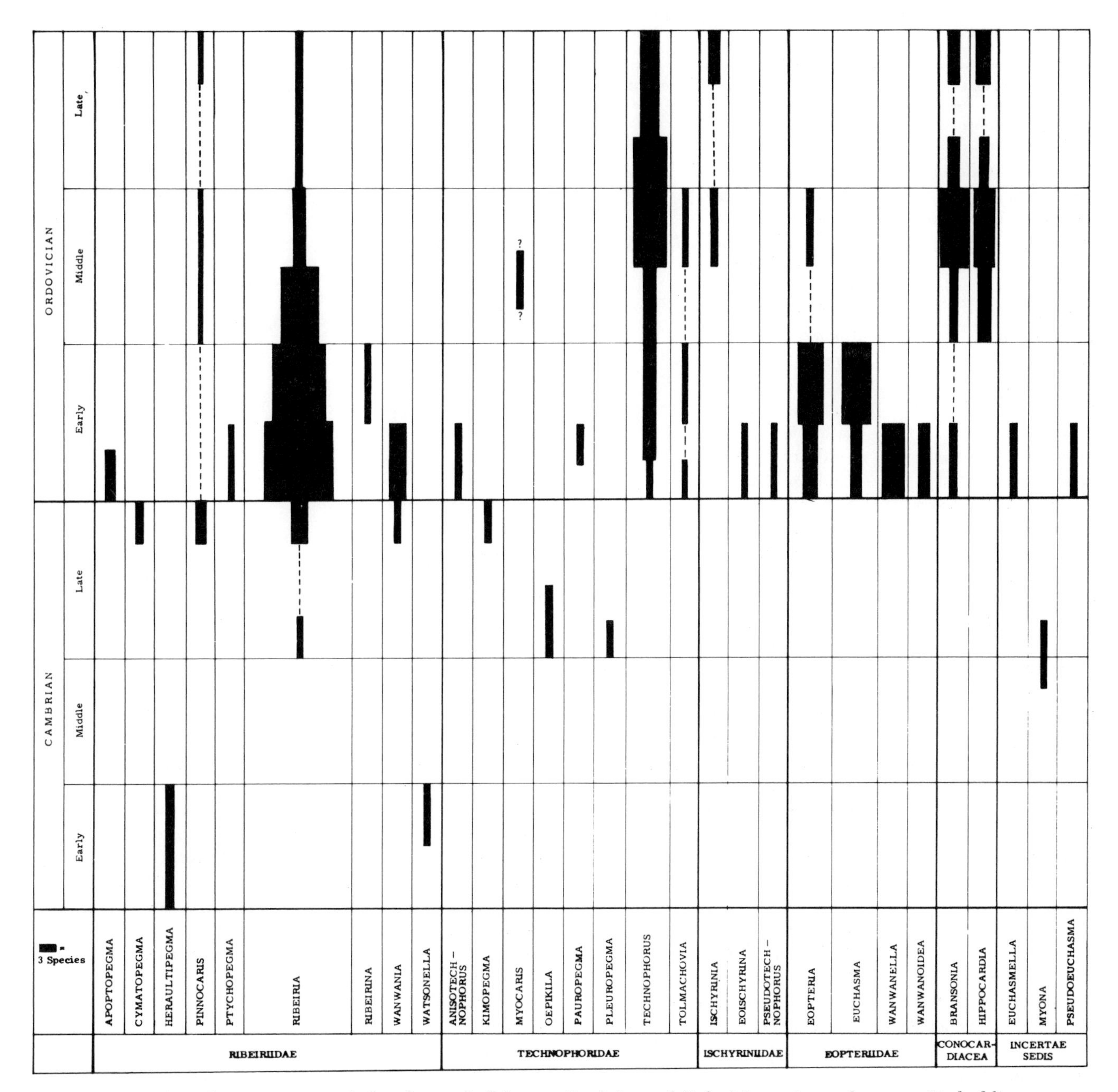

Figure 2. Chart showing range and abundance of all known Cambrian and Ordovician rostroconch genera. Dashed lines, genera presumed present.

of 13 species of *Ribeiria* occur in carbonate rocks, 1 of 4 species of *Technophorus* occurs in carbonate rocks, and 1 species of *Tolmachovia* occurs in clastic rocks.

On the Smith and others (1973) paleogeographic map for the Cambrian and Lower Ordovician (Fig. 3), all Early Ordovician occurrences of rostroconchs plot between 40°S paleolatitude and 40°N paleolatitude and are concentrated within 30° of the Ordovician equator. On the Whittington and Hughes (1972) paleogeographic map for Arenigian-Llanvirnian time, Early Ordovician occurrences of rostroconchs plot between 40°S paleolatitude and 25°N paleolatitude, mostly within 20° of the Ordovician equator.

The two paleogeographic maps used show substantial areas of agreement about the likely climatic zones in which Early Ordovician rostroconchs lived. There is no evidence of an increase in the total life zone of rostroconchs between the Cambrian and the Early Ordovician, and the class remains largely a tropical and subtropical group.

Table 1

Species	Geographic Occurrences	Stratigraphic Occurrences	Gross Lithology
Anisotechnophorus nuculitiformis	New York	Canadian	Carbonate
Apoptopegma dickinsi	North-central Australia	Datsonian	Carbonate
Apoptopegma sp. A	North-central Australia	Datsonian	Carbonate
Bransonia alabamensis	Alabama	Porterfieldian	
B. beecheri	New York, Vermont, Tennessee, Pennsylvania	Marmorian-Ashbyan	Carbonate
B. cressmani	Kentucky	Shermanian-Edenian	Carbonate
B. chapronierei	North-central Australia	Datsonian	Carbonate
B. immatura	Ontario	Wildernessian	
B. isbergi	Sweden	Ashgillian	
B. lindstromi	Sweden	Ashgillian	
B. paquettensis	Ontario, Virginia	Porterfieldian-Wildernessian	
B. townleyi	Tasmania	Trentonian	
Cymatapegma semiplicatum	North-central Australia	Payntonian	Clastic
Eoischyrina billingsi	Southern Manchuria	Wanwanian	
Eopteria conocardiformis	Alabama, Kentucky	Porterfieldian-Wildernessian	Carbonate
E. crassa	Southern France	Arenigian	
E. flora	Southern Manchuria	Wanwanian	
E. obsolata	Southern Manchuria	Wanwanian	
E. ornata	Quebec	Ozarkian?	
E. richardsoni	Quebec, Arkansas, Missouri, Nevada, Texas	Canadian-Whiterockian	Carbonate
E. typica	Newfoundland	Canadian	
E. ventricosa	Vermont	Canadian	Carbonate
E. struszi	North-central Australia	Datsonian	Carbonate
Eopteria sp. A	North-central Australia	Datsonian	Carbonate
Euchasma blumenbachii	Newfoundland, Quebec, Texas, Virginia	Canadian	Carbonate
E. caseyi	North-central Australia	Datsonian	Carbonate
E. jonesi	Malaysia	Canadian	Carbonate
E. mytiliforme	Malaysia	Canadian	Carbonate
E. shorinense	Korea	Lower Ordovician	
E. skwarkoi	North-central Australia	Arenigian	Carbonate
E. wanwanense	Southern Manchuria	Wanwanian	
Euchasma sp. A	North-central Australia	Arenigian	Carbonate
"E." eopteriforme	Southern Manchuria	Wanwanian	
Euchasmella multistriata	Southern Manchuria	Wanwanian	
Heraultipegma varensalense	Southern France	Georgien	Carbonate
Heraultipegma sp.	River Lena, Siberia	Tommotian	Carbonate
Hippocardia antiqua	Manitoba	Maysvillian	Carbonate
H. cooperi	Virginia	Porterfieldian	Carbonate
H. calcis	Southern Ireland	Llandeilian	
H. diptera	Scotland	Llandeilian	
H. limatula	Illinois and Missouri	Kimmswick Limestone	Carbonate
H. pygmaea	Estonia	Ashgillian	Carbonate
H. richmondensis	Ohio, Indiana	Richmondian	Carbonate
Hippocardia spp.	Illinois, Virginia	Porterfieldian-Wildernessian	Carbonate
Ischyrinia elongata	Northern Germany	Caradocian	
I. norvegica	Norway	Middle Caradocian	Carbonate
I. schmidti	Estonia	Ashgillian	
I. winchelli	Anticosti Island	Richmondian	Carbonate
Kimopegma pinnatum	North-central Australia	Payntonian	Clastic
Myocaris lutraria	England	Middle Ordovician	
Myona flabelliformis	Korea	Dresbachian	

Table 1. (Cont.)

Species	Geographic Occurrences	Stratigraphic Occurrences	Gross Lithology
Myona? queenslandica	North-central Australia	Post-Templetonian Middle Cambrian	Carbonate
Oepikila cambrica	North-central Australia	Mindyallan-Idamean	Carbonate
Pauropegma jelli	North-central Australia	Warendian-Arenigian	Carbonate
Pinnocaris americana	Iowa	Trentonian	Carbonate
P. curvata	Scotland	Ashgillian	Clastic
P. lapworthi	Scotland	Lower Caradocian	Clastic
P. robusta	North-central Australia	Payntonian	Clastic
P. wellsi	North-central Australia	Payntonian	Clastic
Pinnocaris sp. A	North-central Australia	Llanvirnian-Llandeilian	Clastic
Pleuropegma plicatum	North-central Australia	Mindyallan	Carbonate
Ptychopegma burgeri	North-central Australia	Datsonian-Warendian	Carbonate
Pseudoeuchasma typica	Southern Manchuria	Wanwanian	
Pseudotechnophorus typicalis	Southern Manchuria	Wanwanian	Carbonate
Ribeiria apusoides	Bohemia	Llanvirnian-Caradocian	Carbonate nodules
R. australiensis	North-central Australia	Mindyallan	Carbonate
R. bassleri	Southern Manchuria	Wanwanian	
R. calcifera	Ontario, Texas	Canadian	Carbonate
R. complanata	Northern Wales	Llandeilian	Clastic
R. compressa	New York, Vermont	Canadian	Carbonate
R. conformis	England	Middle Ordovician	
R. crassa	Southern France	Upper Tremadocian-Lower Arenigian	
R. csiro	North-central Australia	Llanvirnian-Llandeilian	Clastic
R. huckitta	North-central Australia	Payntonian	Clastic
R. jonesi	North-central Australia	Payntonian	Clastic
R. lucan	Alberta	Canadian	Carbonate
R. magnifica	England	Middle Ordovician	
R. manchurica	Southern Manchuria	Wanwanian	Carbonate
R. manchurica pennata	Southern Manchuria	Wanwanian	
R. parva	Pennsylvania	Canadian	Carbonate
R. personata	Southern France	Upper Tremadocian-Lower Arenigian	
R. personata lata	Southern France	Arenigian	
R. personata obsoleta	Southern France	Upper Tremadocian-Lower Arenigian	
R. pholadiformis	Normandy, Portugal	Arenigian-Llandeilian	Clastic
R. runnegari	North-central Australia	Datsonian	Clastic
R. soleaeformis	Southern France	Upper Tremadocian-Lower Arenigian	
R. taylori	New York	Trempealeauan	Carbonate
R. turgida	New York	Canadian	Carbonate
Ribeiria spp.	Bohemia, Bolivia, Morocco, north-central Australia, Maryland, Utah	Warendian, Canadian, Llanvirnian, Llandeilian, Ashgillian	Carbonate
Ribeirina longiuscula	Ontario	Canadian	Carbonate
Technophorus bellistriatus	Missouri	Trentonian	
T. cancellatus	New York, Quebec	Trentonian-Maysvillian	Clastic
T. cincinnatiensis	Ohio, Kentucky	Edenian-Maysvillian	Carbonate
T. coreanica	Korea	Wolungian	
T. divaricatus	Minnesota, Indiana	Trentonian-Richmondian	
T. extenuatus	Minnesota	Trentonian	
T. faberi	Ohio, Kentucky	Edenian-Maysvillian	Carbonate and clastic
T. filistriatus	Minnesota	Trentonian	Carbonate
T. kempae	North-central Australia	Datsonian	Clastic
T. marija	Khatango-Anabar region, northern Siberia	Middle Ordovician	
T. milleri	Ohio	Richmondian	Carbonate

Table 1. (Cont.)

Species	Geographic Occurrences	Stratigraphic Occurrences	Gross Lithology
T. nicolli	North-central Australia	Arenigian	Clastic
T. otaviensis	Bolivia	Middle Ordovician	Clastic
T. planei	North-central Australia	Warendian	Carbonate
T. plicatus	Anticosti Island	Richmondian	
T. sharpei	Bohemia	Llanvirnian-Ashgillian	Clastic
T. stoermeri	Norway	Middle Caradocian	Carbonate
T. subacutus	Minnesota	Wildernessian	Carbonate
T. walteri	North-central Australia	Arenigian	Clastic
Technophorus sp.	Kentucky	Edenian	Carbonate
Tolmachovia belfordi	North-central Australia	Datsonian	Clastic
T. concentrica	Khatango-Anabar region, northern Siberia	Middle Ordovician	
T. corbetti	Tasmania	Arenigian	
Wanwania ambonychiformis	Southern Manchuria	Wanwanian	
W. cambrica	Southern Manchuria	Upper Cambrian	
W. compressa	Southern Manchuria	Wanwanian	
W. drucei	North-central Australia	Datsonian	Carbonate
Wanwanella alta	Southern Manchuria	Wanwanian	
W. asiatica	Southern Manchuria	Wanwanian	
W. striata	Southern Manchuria	Wanwanian	
W. striata auriculata	Southern Manchuria	Wanwanian	
W. tumida	Southern Manchuria	Wanwanian	
Wanwanoidea trigonalis	Southern Manchuria	Wanwanian	
W. trigonalis delicata	Southern Manchuria	Wanwanian	
Watsonella crosbyi	Eastern Massachusetts	Lower Cambrian	Carbonate

DISTRIBUTION OF MIDDLE ORDOVICIAN ROSTROCONCHS

Nine genera of rostroconchs occur in the Middle Ordovician (Fig. 2), 5 of which are also known in Lower Ordovician rocks. *Pinnocaris,* known in the latest Cambrian but not in the Early Ordovician, recurs in the Middle Ordovician. *Myocaris* Salter, *Ischyrinia* Billings, and *Hippocardia* Brown first appear in the Middle Ordovician. The approximately 35 known species of Middle Ordovician rostroconchs constitute about half as many species as are known from the Early Ordovician. Thus, on the basis of the known number of genera and species, the diversity of rostroconchs in the Middle Ordovician shows a marked decrease.

On the other hand, rostroconch endemism decreased in the Middle Ordovician, most genera having widespread distribution. The genus *Pinnocaris* is known from north-central Australia, Scotland, and Iowa; *Ribeiria* is found in Bohemia, Morocco, Wales, England, north-central Australia, and Bolivia; *Technophorus* became widespread, occurring in Missouri, New York, Minnesota, Bolivia, Bohemia, Norway, and Siberia; *Bransonia* is known from Alabama, New York, Vermont, Tennessee, Pennsylvania, Kentucky, Virginia, Ontario, and Tasmania; *Hippocardia* occurs in Illinois, Missouri, Virginia, southern Ireland, and Scotland; *Ischyrinia* is known from Norway and northern Germany; *Myocaris* is known only from England; and *Tolmachovia* is found only in northern Siberia. *Eopteria* is known only from 1 Middle Ordovician species in Alabama and Kentucky; this may be a relict distribution, as the genus had a world-wide distribution in the Early Ordovician.

The lithologic associations of 19 species of Middle Ordovician rostroconchs are known, 12 species occurring in carbonate rocks and 7 in clastic rocks. Of 3 species of *Pinnocaris,* 1 is found in carbonate rocks, and 2 in clastic rocks. *Ribieria* is represented by 1 species known to occur in carbonate rocks and 2 in clastic rocks. *Technophorus* is represented by 3 species found in carbonate rocks and 3 in clastic rocks. The Middle Ordovician species of the genera *Eopteria, Bransonia,* and *Hippocardia,* for which lithologic associations are known, all occur in carbonate rocks.

On the Whittington and Hughes (1972) paleogeographic map for Llandeilian time, Middle Ordovician occurrences of rostroconchs plot between 70°S paleolatitude and 15°N paleolatitude. On the Ross (1975) paleogeographic map for the Middle Ordovician, the extremes of rostroconch distribution are almost identical with the paleolatitudes obtained from the Whittington and Hughes map (Fig. 4). Although in the Middle Ordovician rostroconch diversity decreased, a significant increase in the total life zone occupied by the class seems to have taken place, several species living in high polar paleolatitudes. Most Middle Ordovician species of rostroconchs, however,

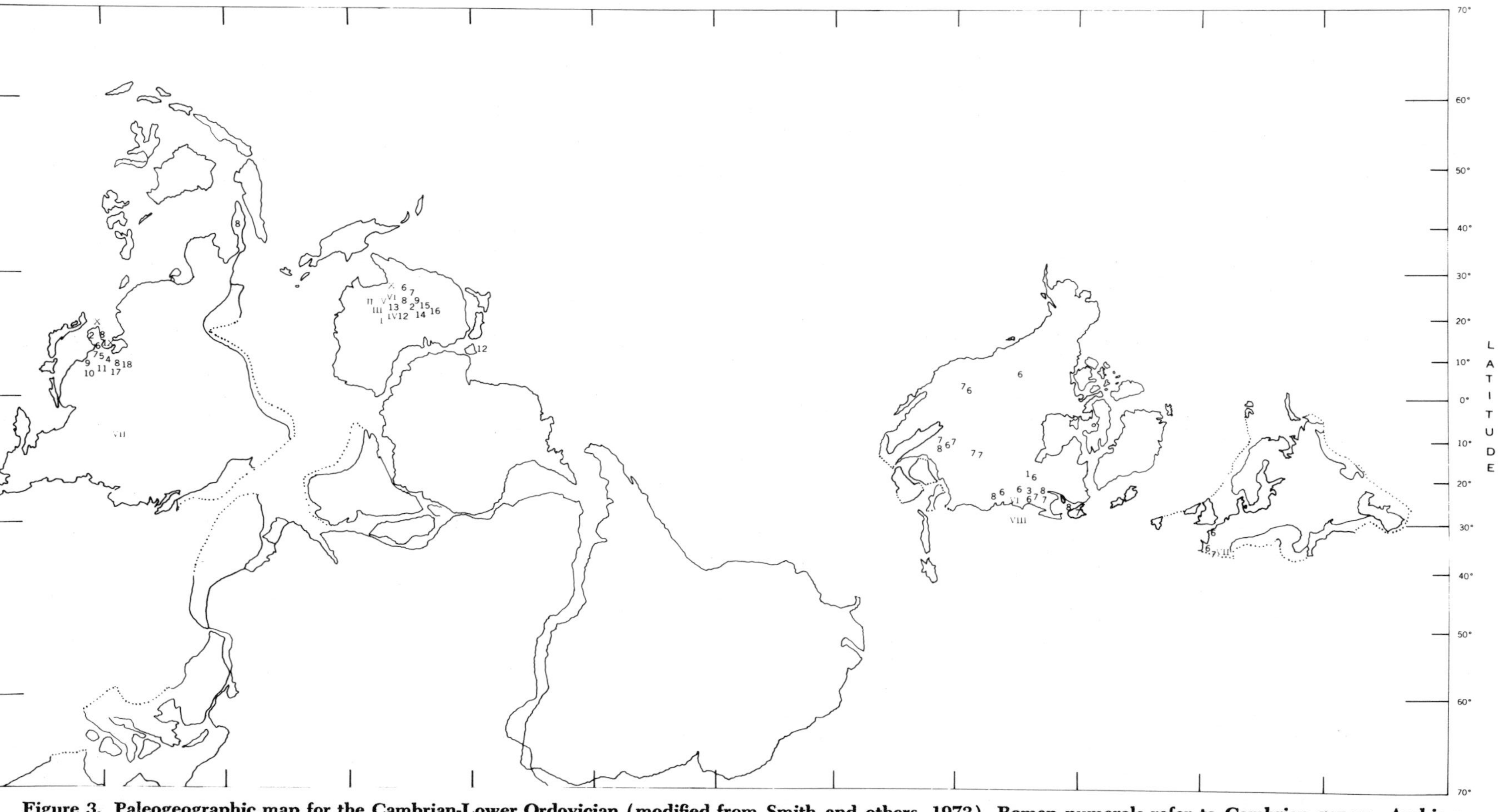

Figure 3. Paleogeographic map for the Cambrian-Lower Ordovician (modified from Smith and others, 1973). Roman numerals refer to Cambrian genera; Arabic numerals, to Lower Ordovician genera.

(I) *Cymatopegma;* (II) *Kimopegma;* (III) *Oepikila;* (IV) *Pinnocaris;* (V) *Pleuropegma;* (VI) *Ribeiria;* (VII) *Heraultipegma;* (VIII) *Watsonella;* (IX) *Wanwania;* (X) *Myona.*

(1) *Ribeirina;* (2) *Technophorus;* (3) *Anisotechnophorus;* (4) *Eoischyrina;* (5) *Pseudotechnophorus;* (6) *Ribeiria;* (7) *Eopteria;* (8) *Euchasma;* (9) *Wanwania;* (10) *Wanwanella;* (11) *Wanwanoidea;* (12) *Tolmachovia;* (13) *Apoptopegma;* (14) *Bransonia;* (15) *Pauropegma;* (16) *Ptychopegma;* (17) *Euchasmella;* (18) *Pseudoeuchasma.*

remained tropical and subtropical in their distribution (Fig. 4).

DISTRIBUTION OF LATE ORDOVICIAN ROSTROCONCHS

All of the 6 known genera of Late Ordovician rostroconchs (Fig. 2) are also known from the Middle Ordovician. About 21 species of Late Ordovician rostroconchs are known; the decline in diversity of the class, which began in the Middle Ordovician, continued through the Late Ordovician. By the end of Ordovician time, all rostroconchs except the genera *Bransonia* and *Hippocardia,* which belong to the superfamily Conocardiacea, had become extinct. All post-Ordovician rostroconchs are placed in the Conocardiacea, and the class never regained the morphological or generic diversity seen in the Cambrian and Ordovician.

In the Late Ordovician, *Pinnocaris* is represented by a single species known only from Scotland; *Ribeiria* is known only from Bohemia; *Ischyrinia* occurs in Anticosti Island (Canada) and Estonia; *Bransonia* is known from Kentucky and Sweden; *Hippocardia* occurs in Manitoba, Ohio, Indiana, and Estonia; and *Technophorus* is known from Ohio, Kentucky, Indiana, Quebec, Anticosti Island, and Bohemia. The single occurrences of *Pinnocaris* and *Ribeiria* may represent relict distributions, as the genera were much more widely distributed in older Ordovician rocks.

The lithologic associations of 13 species of Late Ordovician rostroconchs are known; 9 species occur in carbonate rocks and 4 in clastic rocks (Table 1). The single known Late Ordovician species of *Pinnocaris* is found in clastic rocks; 1 species of *Ischyrinia* is known from carbonate rocks; 1 species of *Bransonia* and 3 species of *Hippocardia* occur in carbonate rocks; and the genus *Technophorus* is represented by 4 species found in carbonate rocks and 3 species in clastic rocks.

On the Whittington and Hughes (1972) paleogeographic map for Ashgillian time, the Late Ordovician occurrences of rostroconchs plot between 20°S paleolatitude and 10°N paleolatitude. On the Ross (1975) paleogeographic map for the Late Ordovician, rostroconchs range from the equator to 40°S paleolatitude (Fig. 5), although there are only two occurrences south of 20°S paleolatitude. Thus, as throughout their lower Paleozoic history, Late Ordovician rostroconchs are primarily tropical and subtropical in their distribution.

DECLINE OF ORDOVICIAN ROSTROCONCHS

From a high point in the Early Ordovician, the diversity of genera and species of rostroconchs in the Middle and Late Ordovician showed a marked decline, which resulted in only 2 genera surviving into the Early Silurian. After a low point in the Late Ordovician and Silurian, some expansion of genera and species of rostroconchs took place in the Devonian; about 48 species in the Early Devonian are assigned to 5 genera (Fig. 1). However, in the middle and late Paleozoic, rostroconchs are not as morphologically diverse as in the Ordovician.

Because the Middle and Late Ordovician decline in diversity is on a world-wide scale, tectonic events were probably not responsible for it, although they could have been contributing factors locally or regionally in some basins. Late Ordovician glaciation was probably not responsible for the decline either, because the decline began in the early Middle Ordovician (Fig. 2). When glaciation became an important factor in Ordovician history, it may have accelerated the decline of an already waning group.

The decline of the rostroconchs can be explained as a biologic interaction of this group with the other major group of bivalved mollusks, the Pelecypoda. Both rostroconchs and pelecypods evolved in the Early Cambrian, and rostroconchs were the first to diversify in the Late Cambrian and Early Ordovician. At about the time that rostroconchs began to decline in the early Middle Ordovician, pelecypods began a spectacular diversification, and in the Middle and Late Ordovician, they are represented by about 140 genera and 1,400 species. The fit of the empirical evidence of one bivalved mollusk group almost entirely replacing another is extremely good (compare Fig. 2 herein with Pojeta, 1971, p. 27); the problem is what competitive advantage the pelecypods may have had over the rostroconchs.

The major conchological difference between rostroconchs and pelecypods is that in the former group, shell growth proceeds from a single center of calcification, whereas in the latter group, shell growth proceeds from two centers of calcification, the uncalcified part between the two valves becoming the ligament. The layers of the ligament in pelecypods are continuous with comparable shell layers of the two valves. The ligament of pelecypods serves to open the valves and make them gape, and muscular force must be applied to close the valves. The rostroconch shell is calcified across the dorsum, and there is no ligament; to add growth increments rostroconchs may have opened their valves by means of their foot or by hydrostatic means (Pojeta and Runnegar, 1976). Feeding and excretion in rostroconchs was accomplished by means of permanent shell gapes at the anterior and posterior ends of the shell.

Most lower Paleozoic rostroconchs and pelecypods were infaunal forms living at various depths in burrows in the sediment (Pojeta, 1971; Pojeta and Runnegar, 1976); the effiiciency with which burrowing could be accomplished and new burrows made when the animals were exposed would have high adaptive significance for survival.

When living pelecypods burrow, their valves are pressed against the sides of the burrow cavity by the opening thrust of the ligament. Subsequently, as the foot forms the pedal anchor, the sediment around the shell is loosened by ejection of water from the mantle cavity produced by contraction of the adductor muscles of the shell. Contraction of the pedal retractor muscles then pulls the animal into the loosened sediment. The sequence of events must be repeated several times for the animal to become entirely buried (Trueman, 1968). Pelecypods are thus highly efficient burrowers. It seems likely that rostroconchs were

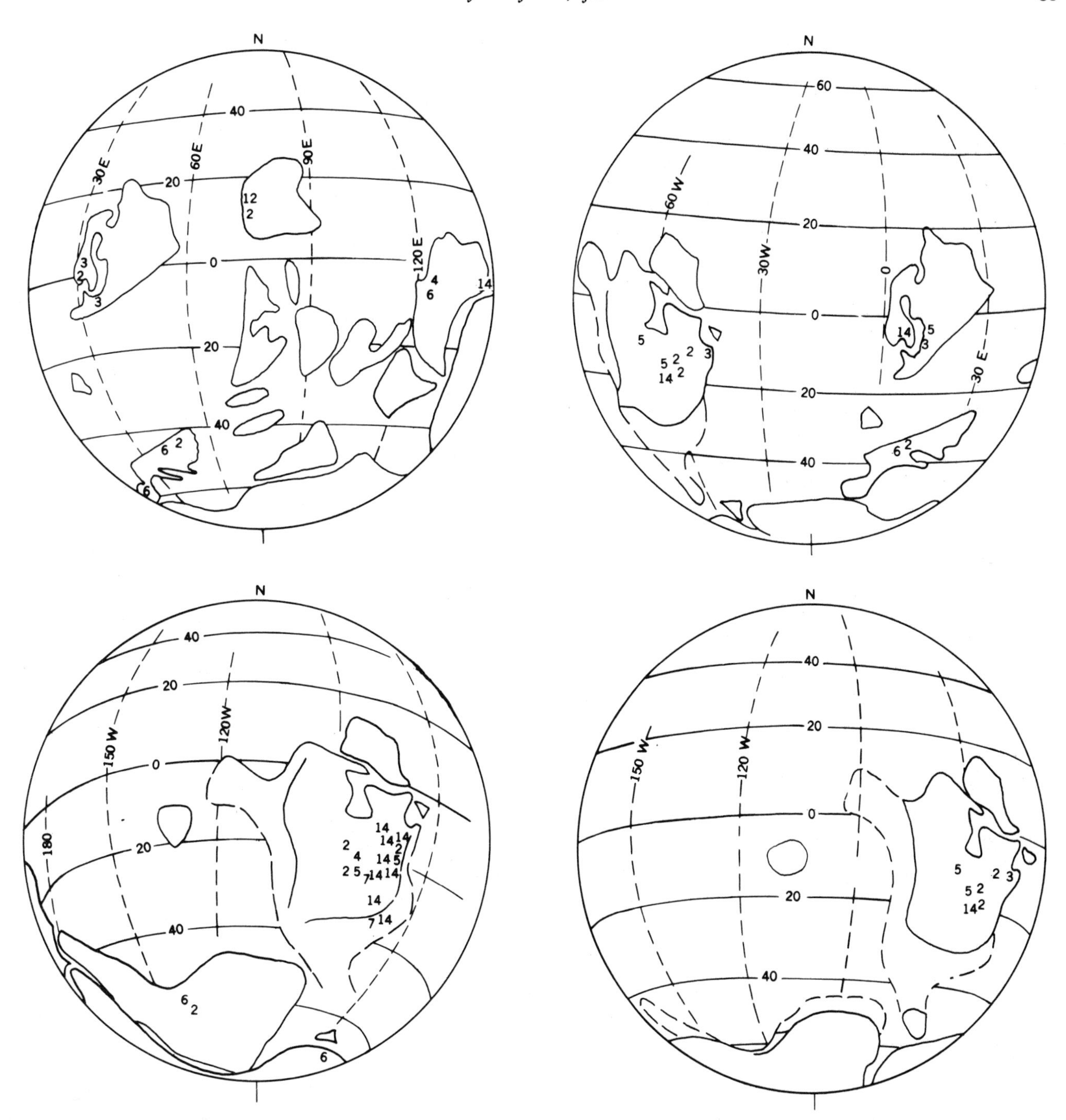

Figure 4. Paleogeographic map for the Middle Ordovician (modified from Ross, 1975). (2) *Technophorus;* (3) *Ischyrinia;* (4) *Pinnocaris;* (5) *Hippocardia;* (6) *Ribeiria;* (7) *Eopteria;* (12) *Tolmachovia;* (14) *Bransonia.* Genera that occur in both the Early Ordovician and Middle Ordovician are indicated by the same Arabic number as on Figure 3. Various taxa occurring in the British Isles are not plotted because the British Isles are not shown on this reconstruction. These taxa are indicated in Table 1.

Figure 5. Paleogeographic map for the Late Ordovician (modified from Ross, 1975). (2) *Technophorus;* (3) *Ischyrinia;* (5) *Hippocardia;* (6) *Ribeiria;* (14) *Bransonia.* Genera that occur in the Early, Middle, and Late Ordovician are indicated by the same Arabic number as on Figures 3 and 4. *Pinnocaris* occurs in the Late Ordovician of Scotland, but this reconstruction does not show the position of Scotland.

less efficient burrowers than pelecypods for the following reasons. Although rostroconchs had large pedal retractor muscles which could be used to draw the shell into the sediment against the pedal anchor, they lacked adductor muscles and could not soften the sediment into which they were burrowing by ejecting water from the mantle cavity. This would mean that they could only burrow into relatively soft sediment. Rostroconchs also lacked a ligament and could not press the valves against the sides of the burrow in preparation for extending the burrow downward. The fit of the burrow would depend entirely upon the initial excavation.

Thus, it is postulated that most Ordovician rostroconchs became extinct because of the rise of burrowing pelecypods, which could better establish and maintain their position in the sediment, and which could burrow into a wider variety of substrates than could rostroconchs. Throughout the Paleozoic, rostroconchs remain relatively minor elements of the fauna, being represented by a few tens of species in each system, whereas pelecypods are represented by hundreds of genera and thousands of species.

ACKNOWLEDGMENTS

I thank Elinor Stromberg and Marija Balanc, U.S. Geological Survey, for preparing the illustrations. To E. G. Kauffman, Smithsonian Institution, I extend sincere thanks for stimulating discussions as to how pelecypods and rostroconchs may have interacted biologically.

REFERENCES

Kobayashi, Teiichi, 1933, Faunal studies of the Wanwanian (basal Ordovician) series with special notes on the Ribeiridae and the ellesmeroceroids: Tokyo Imperial Univ. Fac. Sci. Jour., sec. 2, v. 3, pt. 7, p. 249-328.

Matthews, S. C., and Missarzhevsky, V. V., 1975, Small shelly fossils of late Precambrian and Early Cambrian age: A review of recent work: Geol. Soc. London Jour., v. 131, p. 289-304.

Pojeta, John, Jr., 1971, Review of Ordovician pelecypods: U. S. Geol. Survey Prof. Paper 695, 46 p.

Pojeta, John, Jr., Gilbert-Tomlinson, Joyce, and Shergold, J. H., 1977, Cambrian and Ordovician rostroconch molluscs from northern Australia: Australia Bur. Mineral Resources Bull. 171 (in press).

Pojeta, John, Jr., and Runnegar, Bruce, 1976, The paleontology of rostroconch mollusks and the early history of the phylum Mollusca: U. S. Geol. Survey Prof. Paper 968, 88 p.

Pojeta, John, Jr., Runnegar, Bruce, Morris, N. J., and Newell, N. D., 1972, Rostroconchia: A new class of bivalved mollusks: Science, v. 177, p. 264-267.

Ross, R. J., Jr., 1975, Early Paleozoic trilobites, sedimentary facies, lithospheric plates, and ocean currents, *in* Martinsson, A., ed., Evolution and morphology of the Trilobita, Trilobitoidea, and Merostomata: Fossils and Strata, no. 4, p. 307-329.

Runnegar, Bruce, and Jell, P. A., 1976, Australian Middle Cambrian molluscs: Alcheringa, v. 1, p. 109-138.

Smith, A. G., Briden, J. C., and Drewry, G. E., 1973, Phanerozoic world maps, *in* Hughes, N. F., ed., Organisms and continents through time: Spec. Papers Palaeontology no. 12, p. 1-42.

Trueman, E. R., 1968, The burrowing activities of bivalves: Symposia Zoological Soc. London, no. 22, p. 167-186.

Waterhouse, J. B., 1967, Proposal of series and stages for the Permian in New Zealand: Royal Soc. New Zealand Trans., v. 5, p. 161-180.

Whittington, H. B., and Hughes, C. P., 1972, Ordovician geography and faunal provinces deduced from trilobite distribution: Royal Soc. London Philos. Trans., ser. B, Biol. Sci., v. 263, p. 235-278.

Gastropod Opercula as Objects for Paleobiogeographic Study

ELLIS L. YOCHELSON, *U. S. Geological Survey, Washington, D. C. 20560*

ABSTRACT

Most primitive marine gastropods have an operculum, a plate that closes the aperture when soft parts are retracted into the shell. Opercula of some modern taxa are consistently calcified, whereas others are not; calcified opercula are far more likely to be preserved as fossils. The oldest known gastropod opercula are late Early Ordovician. The record of opercula is generally poor, but at least 3 generic-level morphologies occur in the late Early Ordovician, and distributional data permit some speculation concerning biogeography.

Opercula of the pleurotomariacean *Ceratopea* are abundant, diverse, and widespread in one continental mass. *Ceratopea* might have been an inhabitant of shallow, subtropical to tropical waters. A model of distribution using continental drift and rotation fits this interpretation. Opercula of the macluritacean *Teiichispira* occur through a shorter time interval but are as abundant and diverse as *Ceratopea* within that interval. They are widespread, but distribution is exceedingly patchy; no models using continental drift satisfactorily explain this distribution, though rotation of continents reduces longitudinal spread. Another macluritacean operculum, "Billings' operculum," is less well-known and has limited distribution and diversity; its distribution is satisfactorily explained by invoking continental drift.

INTRODUCTION

Much professional interest and even more amateur interest has been taken in present-day sea shells; snails have been looked at and written about for centuries. Fossil gastropod shells have also been collected and described in the literature. Conchologists also collect the opercula associated with snail shells, and some literature exists on present-day opercula. Little has been written about fossil opercula, and few paleontologists study them, for they are uncommon and show great irregularity in distribution and relative abundance. Nevertheless, they are interesting objects for paleobiogeographic speculation.

HOLOCENE OPERCULA

One characteristic of many living gastropods is an operculum, a plate grown on the dorsal, mid-posterior of the foot. This operculum is the last part of the foot retracted inward when soft parts are withdrawn into the shell; the operculum effectively seals the aperture. Once retracted, the animal may expand slightly and open just a tiny space between shell and operculum to peer out; it can close this gape rapidly if danger still threatens.

Many prosobranch gastropods, generally considered to be the most primitive subclass, have an operculum; this structure is reduced in size or even absent among "advanced" families of the Mesogastropoda and Neogastropoda. In less advanced snails, the only types without an operculum among the Archaeogastropoda are those that attach themselves to a hard surface. This "limpet" habit has evolved in several lines of prosobranchs, but in every instance the larval forms develop an operculum, only to lose it later. Thus, an operculum or at least a vestigial opercular gland is characteristic of most present-day marine snails and probably was a feature of fossil prosobranchs. All opisthobranchs that have lost their shell and a few that have a shell reduced in size lack an operculum. Pulmonate gastropods also lacks an operculum.

The operculum is commonly composed of a tough organic substance, but in a few modern prosobranchs the operculum is calcified. All living Neritacea (except the shell-less Titiscaniidae) have a calcified operculum; neritaceans are the most abundant members of the living gastropod fauna having such an operculum. The consistent calcification of the operculum is one conchological reason for suggesting that this group, currently placed in the Archaeogastropoda, may form a separate order of prosobranchs.

Thick calcified opercula, such as the "cat's eye" of *Turbo,* are strongly resistant to erosion; in Fiji, some beaches are formed to a significant extent by gastropod opercula. Opercula have different shapes and weights from the shells with which they are associated in life. With few exceptions, shell and operculum are separated after the death of the snail; each may be deposited in different sediments. If the operculum is not calcified, there is little chance that it will be preserved.

The mechanism for calcification of opercula is not clear; it is even less clear what advantage there is to calcification. For example, *Cittarium* and *Astraea* today live by scraping algae off rocks in virtually the same habitat, on rock crevices bathed by warm waters. The first does not have a calcified operculum; the second does.

Some species of the carnivorous naticids that move through sand have calcified opercula. The older literature has accounts of some individuals in several species having calcified opercula and some not. Recent revision of living Naticidae has indicated that this is a question of incorrect identification (L. Marincovich, oral commun. 1976). Members of the Naticinae have a calcified operculum, whereas members of the Policinae do not. One species of *Amauropsis* in the Ampullospirinae has a calcareous layer, or coating, over the central part of the corneous operculum. In the Sininae, the fourth subfamily, again 1 species of the genus *Eunaticina* shows partial calcification; calcareous beads that touch and coalesce cover the central part of the operculum.

Even though no reason for the calcification is readily apparent, one gains the impression from the study of modern gastropods that possession of a calcified operculum is a constant characteristic. On the generic and subfamily level, this characteristic seems to have high systematic value in distinguishing taxa.

An organic operculum may be retracted a considerable distance into the aperture. Because the material of this operculum is flexible, it may be bent and modified when it is withdrawn, so that the shape is not a perfect replica of the aperture. In contrast, the inflexibility of the calcified operculum limits the degree to which it can be retracted. In living *Turbo* and *Astraea,* the operculum is withdrawn only 1 to 2 mm into the aperture; in Devonian *Omphalocirrus,* the operculum could be retracted a short distance into the shell. The outline of the calcified operculum conforms to the outline of the apertural opening so that a good seal is formed. If the gastropod shell is simple, the outline of the operculum is a close approximation of the outline of the outer surface of the whorl, but if the shell is quite thick and modified by angulations and ridges, this approximation may not be very close.

Opercula may have essentially the form of a plate or partial coil that grows at the edges, as in the neritaceans. Most opercula are coiled, either through a few whorls (paucispiral), as in *Turbo* and *Astraea,* or many whorls (multispiral), as in *Cittarium.* No living gastropod species are known to possess calcified multispiral opercula, such as that of *Omphalocirrus.*

FOSSIL OPERCULA

Literature reports of fossil opercula indicate spotty distribution through geologic time. Deposits of many ages that have produced abundant gastropod shells have yet to yield any opercula. Where opercula do occur, more often than not they are among the rarest of fossils.

Rarely is a fossil operculum found in place within the shell aperture. Assignment of a loose operculum to a particular species or genus includes an element of subjectivity, depending on the variety of shells in the deposit yielding the operculum, the apparent amount of movement of shells after death, and a host of other factors. Such factors combine to make collection and examination of fossil gastropod opercula difficult.

The concept of an operculum—a plate to close a tube—is a sound notion in bioengineering. The operculum is not confined to the gastropods or to the mollusks. It was used in some fossil cephalopods and in hyoliths; this last group is judged by some to be an extinct class of mollusks (Marek and Yochelson, 1976) and by others to be a separate phylum (Runnegar and others, 1975). Many living and fossil worms have an operculum, and several fossil corals had them, to name only a few diverse examples. Some pelecypods and brachiopods modified one valve so that it functioned effectively as an operculum, although the hinge line did not allow the flexibility of movement characteristic of true opercula. Some terrestrial pulmonates secrete an epiphragm that may be calcified, but this is a seal across the aperture during hibernation or estivation and is not functionally comparable with the movable operculum.

A pertinent question is the age of the oldest gastropod operculum. This cannot be answered with certainty; among other factors, it naturally depends on the age of the earliest authentic gastropod. The oldest operculum found is Early Cambrian; in my view and in the view of many others, it is the operculum of some worm tube (Bengtson, 1968, and references therein). The class Gastropoda is defined on features of the soft parts. Assignment of fossils to the class usually is not difficult, though the farther one goes back in time from the Holocene, the greater becomes the uncertainty. Small coiled shells occur in the Early Cambrian, but in my view these are not gastropods (Yochelson, 1975b). I consider the earliest authentic gastropods to be Late Cambrian. The earliest gastropod opercula currently known occur in rocks of the upper half of the Early Ordovician. What gastropod opercula, if any, there may have been between the Late Cambrian and late Early Ordovician is conjecture.

Compared with later occurrences, gastropod opercula are well-known and occur in disproportionate numbers in the Early Ordovician; most of the detail in this paper is devoted to these opercula. The literature on Middle and Late Ordovician opercula is scattered. There is at least one other shape operculum in the Middle Ordovician (Yochelson, 1966), in addition to that of the common *Maclurites* mentioned below.

In the mid-Paleozoic, multispiral calcified opercula of several kinds, as well as sparce paucispiral opercula are known (Yochelson and Linsley, 1972, and references therein). Some Silurian and Devonian multispiral forms appear widespread; it is of interest to note for biogeographic purposes that in the Early Devonian identical multispiral opercula occur in Victoria, Australia, and in the Carnic Alps of Austria (Spitz, 1907, p. 125). Multispiral opercula in several gastropod groups appear related to a reef-dwelling habitat.

In the upper Paleozoic, neritacean opercula first occur; although sparse, they are by far the most common opercula found. The Triassic and Jurassic record of opercula is no better than that of the mid-Paleozoic. The Cretaceous and Cenozoic gastropod fauna is essentially like that of today, and the record of opercula holds no surprises, only the complications posed by taphonomy.

The operculum of 1 Pliocene nonmarine gastropod, *Scalez,* is abundant and has considerable utility as a local zone marker in southern California (Woodring and others, 1932, p. 32-38). However, except in the Early Ordovician, far too few opercula are known to permit any sort of speculation concerning biogeography.

If the known record of fossil opercula is a fair approximation of reality, it suggests that calcification first took place moderately early in the history of the class. It further suggests that such calcification occurred irregularly but repeatedly in several different stocks.

EARLY ORDOVICIAN OPERCULA

Three different opercula, at least, are found in the late Early Ordovician of North America. For many years they were considered to be closely related. It is clear now that one stock is associated with a pleurotomariacean gastropod and that the other two are associated with macluritaceans. The pleurotomariacean *Ceratopea* died out at the end of the Early Ordovician. *Liospira* and other similar shells, possibly in the same family, occur in Middle Ordovician and younger beds, but no calcified opercula are known. One of the macluritacean groups that has a calcified operculum, *Teiichispira,* is also confined to the Early Ordovician and no shell having a profile close to that of *Teiichispira* is known from the Middle Ordovician. The shell of the third form is not well-known, beyond the fact that it is a macluritacean with strongly sloping sides. Similar shells are found in the Early, Middle, and Late Ordovician. Indeed, *Maclurites* opercula are the most common Middle Ordovician forms (see paper by Rohr in this volume). However, there may be

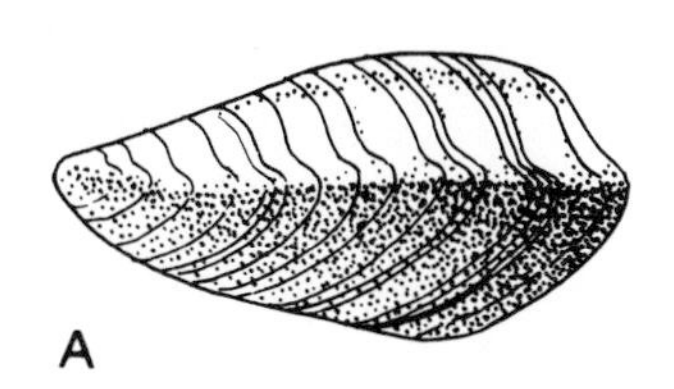

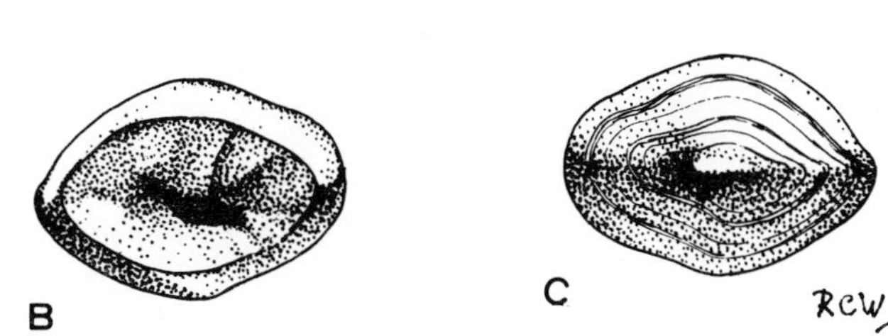

Figure 1. The operculum of *Ceratopea keithi* Ulrich, at natural size: (A) side view; (B) interior view of attachment surface; (C) apical view.

a fundamental difference between the Early Ordovician and Middle Ordovician opercula in the mode of muscle attachment, suggesting that the shells may not be closely related but might be convergent within the superfamily. This puzzle remains to be solved.

Ceratopea

The first Early Ordovician operculum recognized was that of the genus *Ceratopea* (Ulrich, 1911, p. 665-666). Characteristically, the operculum is found dissociated from the shell. For many years only the operculum was known and there was argument as to the shape of the shell. One life association of a *Ceratopea* operculum has been described (Yochelson and Wise, 1972), and another has been found (M. Fix, unpub. data). This association clarifies some points, but it is not the shell of the type species, and many minor details of shell morphology remain to be worked out when other life associations or plausible associations of shells and opercula are found.

Commonly, opercula are isolated, and commonly, they are secondarily silicified. I have speculated that these gastropods lived in exceedingly shallow water and grazed on algae, and that in some places they were killed by a change in water chemistry (Yochelson, 1975a); after death, when the flesh had rotted, the shell floated away. At rare localities *Ceratopea* shells are found without opercula (Yochelson and Copeland, 1974).

Silicification also took place in other environments. It is a poorly understood phenomenon but is exceedingly important. At many localities in the southeastern United States, Tertiary weathering has removed much Paleozoic limestone. Silicified opercula remain in the residuum and may be used to map Ordovician units that cannot otherwise be distinguished (Wise and others, 1975).

The ceratopean operculum is difficult to interpret. Most opercula of living snails are flat plates that may be pauci- to multi-spiral. The operculum of *Ceratopea* shows no spiral, and for many species it is difficult to detect any twist of this sort. Rather than being a flat plate, it is greatly elongated (Fig. 1). The type species is best described as a twisted wedge with a point at one end and a slight depression at the other. This depression, the site of attachment of the operculum to the gastropod foot, may be a simple cup, or there may be a groove and one or two deep pits in the area. The operculum grew by adding layers at the cup end, each slightly wider than the preceeding one, so that as age increased and the shell expanded, the opercular width also increased to provide a close fit. The operculum is so elongate that one wonders how the animal could have moved during life. Specimens more than 6 cm long are known.

Because of the accentuation of growth in the third dimension, it has been assumed that all *Ceratopea* opercula belong to 1 genus. Differences in the profile and in the curvature allow one to distinguish readily a variety of discrete forms. Possibly, the presence of a calcified operculum should be treated as a subfamily characteristic, and several genera or subgenera should be recognized.

Figure 2. Known occurrences of *Ceratopea* opercula, plotted on a reconstruction of Early Ordovician geography (modified slightly from Ross, 1976).

Be that as it may, distinct species of *Ceratopea* are recognized from the shape of the operculum. These species have limited stratigraphic ranges and are exceedingly useful in distinguishing zones in an otherwise monotonous pile of Early Ordovician limestone (Yochelson and Bridge, 1957; Yochelson, 1973). An important test of the utility of the opercula is that in place after place they are found in the same stratigraphic succession and do not conflict with other biologic evidence of age of the rocks or geologic evidence of local structures. They are excellent old-fashioned guide fossils, and they are widespread.

Ceratopea opercula are known from Colorado, Texas, Oklahoma, Kansas (in a subsurface core), Arkansas, Missouri, Alabama, Georgia, Tennessee, Virginia, Maryland, Pennsylvania, and New York (Yochelson and Barnett, 1972) in the United States. They occur in Newfoundland, Baffin Island (Trettin, 1975, p. 18), northern Greenland (Yochelson and Peel, 1975), eastern Greenland (Yochelson, 1964), northern Scotland, Vestspitsbergen (Major and Wisnes, 1955, p. 20-21) and Bjornøya (Bear Island) (Holtedahl, 1919). An occurrence in the Mackenzie Mountains, Northwest Territories, Canada, is known (James Derby, unpub. data).

One species occurs from New York to Oklahoma (Yochelson and Barnett, 1972), another species is found in Arkansas and northern Greenland (Yochelson and Peel, 1975), and yet another species, in Georgia and Bear Island (Yochelson, unpub. data). These are extremely long lateral ranges for species. Plotted on the present-day globe, they show a range of about one-sixth the circumference of the Earth. The distribution of *Ceratopea* opercula is summarized (Fig. 2) on a model of the globe that embodies one possible reconstruction of Early Ordovician continental drift. This reconstruction reduces some of the range, but species still are widely distributed.

Teiichispira

A second kind of calcified operculum also grows mainly in the third dimension (Fig. 3). Most specimens of *Teiichispira* are more elongate and larger than the average operculum of *Ceratopea,* from which they also differ in a rod-like inner construction, so that in cross section the operculum resembles a miniature honeycomb. The basal surface in *Teiichispira* has a prong for muscle attachment, unlike the broad cup or pit and groove that characterize the basal surface of opercula of *Ceratopea.* Because of the prong, this operculum was originally assigned to *Maclurea,* now a synonym of *Maclurites.*

The first species, *M. odenvillensis* Butts, was collected from the Odenville Limestone of Alabama, a poorly exposed stratigraphic unit. Opercula also are found in residuum a few miles from the type locality, but the species is not known from other States. The Odenville is dated as late Early Ordovician. Judging from collections, specimens were fairly common.

A second species was found in Malaysia and a third in Utah; these occurrences are also dated as very late Early Ordovician. Specimens are abundant. These opercula when combined with that of *M. odenvillensis* were distinctive enough from *Maclurites* to warrant another generic name, and *Teiichispira* was chosen (Yochelson and Jones, 1968). In one of the Malaysian occurrences, shells and opercula are in a highly plausible association. The shell is also distinct

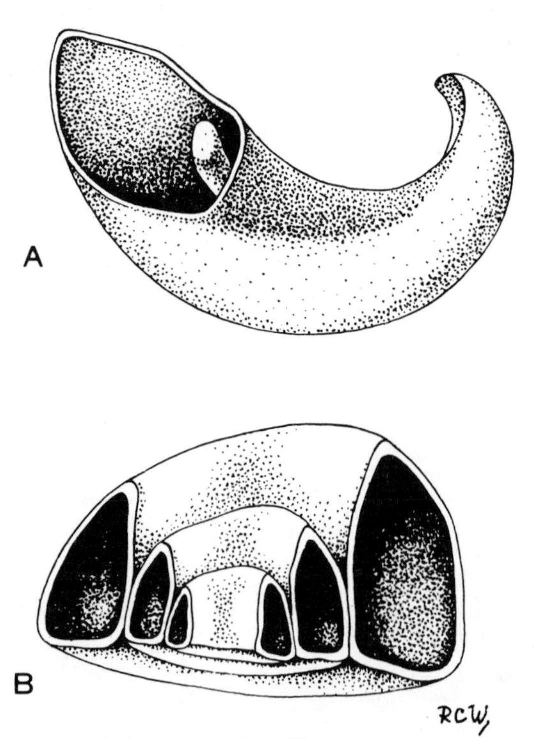

Figure 3. *Teiichispira odenvillensis* (Butts), at natural size: (A) the operculum; (B) reconstruction of the poorly known shell, based in part on the cross section of the operculum.

from that of typical *Maclurites* in being higher and narrower. Gastropoda taxa can be named on shells or opercula, but it is comforting when both indicate that a taxon should be differentiated.

After 1968, *Teiichispira* was reported from several localities in Queensland and western Australia (Gilbert-Tomlinson, 1973). So far as I know, no additional species have been described. However, R. H. Flower, New Mexico Bureau of Mines, has found specifically indeterminate specimens of *Teiichispira* opercula associated with opercula of a species of *Ceratopea* in a collection made in the upper part of the St. George Formation, near St. George, Newfoundland. These scraps show the characteristic honeycomb fibrous construction and, as both kinds of opercula are silicified, they rule out the possibility that the peculiar internal structure of *Teiichispira* opercula is a result of diagenesis.

When plotted on a present-day configuration of the globe (Fig. 4), the straight-line distance between occurrences of *Teiichispira* in western Australia and Newfoundland is about 13,000 miles (20,800 km). The first occurrence is at about lat 20°S and the second is at lat 50°N. When these points are plotted on a reconstruction of the globe embodying a model of Ordovician continental drift (Fig. 5), this latitudinal difference is reduced, but the straight-line distance is not changed materially.

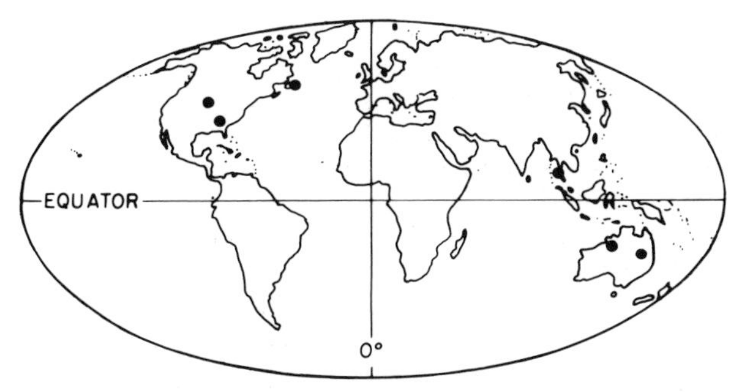

Figure 4. Known occurrences of *Teiichispira* opercula, plotted on Mollweide's homolographic projection.

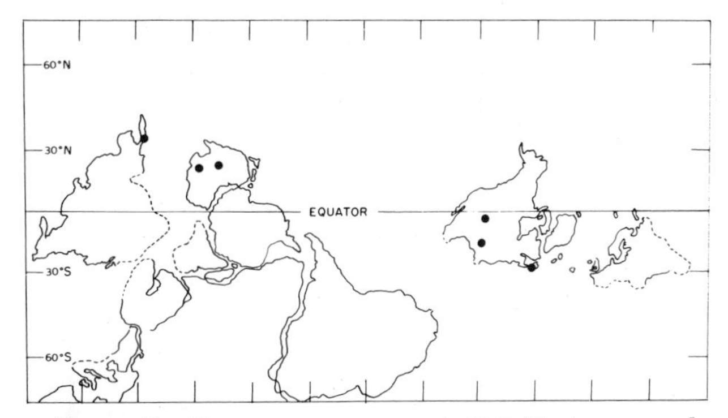

Figure 5. Known occurrences of *Teiichispira* opercula, plotted on a reconstruction of Early Ordovician geography (modified from Whittington and Hughes, 1973).

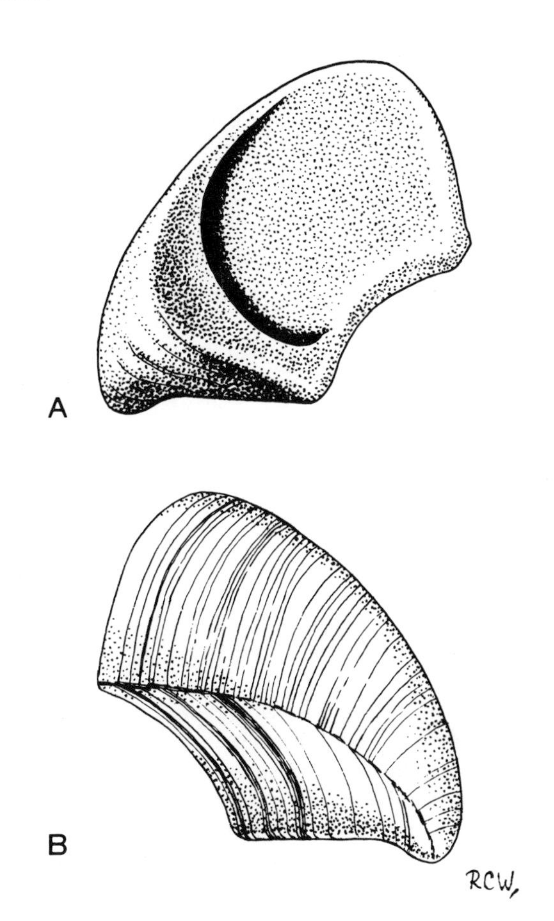

Figure 6. "Billings' operculum," at natural size: (A) interior view; (B) exterior view.

"Billings' Operculum"

Report of a third late Early Ordovician operculum has been in the literature for more than a century. Billings (1865, p. 243) illustrated and described, but did not name, this form and examples of 2 species of *Ceratopea* from what is now the upper part of the St. George Formation at Cape Norman, Newfoundland.

In contrast to the opercula of *Ceratopea* and *Teiichispira,* this form is low and platelike, but greatly thickened at the apex (Fig. 6). The interior surface lacks the prong present on 1 species of *Teiichispira* and several species of Middle Ordovician *Maclurites.* The margin of the operculum toward the apex is excavated, forming a "pocket" which may have functioned as an area of muscle attachment. The species has not been carefully studied, and too much should not be made of the seeming absence of this prong. In any event, the external morphology is distinctive in having a strongly arched side and a gently sloping face. Specimens are not common, possibly in part because the plate shape was more subject to post-mortem destruction.

"Billings' operculum" has only been reported in the literature once, from northern Newfoundland, but R. H. Flower (unpub. data) has found it in southwest Newfoundland. It occurs also in the Durness Limestone on Skye,

Scotland (University of Glasgow collections and, observed in the field in 1974), and on Ellesmere Island, Canada (collected by Robert Christie, Geological Survey of Canada). I have not seen it in Early Ordovician collections from the United States.

The type of the poorly known *Pondia* from the Knox Group at Knoxville, Tennessee (Oder, 1932), might be a badly worn fragment of the same general form of operculum. Except for the 2 specimens constituting the type lot of *Pondia,* only opercula of *Ceratopea* are known from the Knox Group. Thus, until additional and better specimens of *Pondia* are reported, it is better to keep it distinct from "Billings' operculum."

The occurrences of "Billings' operculum" are plotted (Fig. 7) on a map showing part of Earth's current geography (compare with Figs. 2, 5).

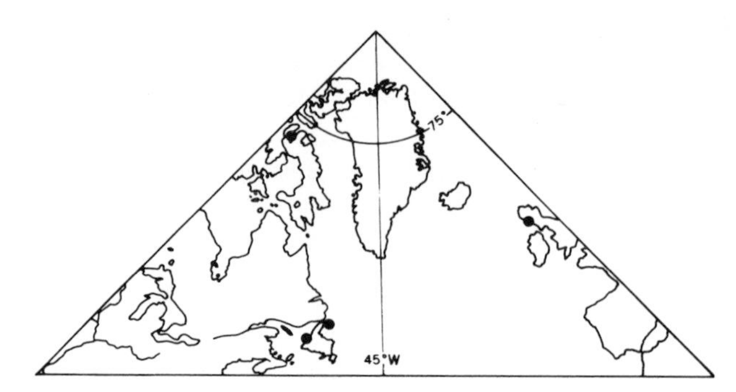

Figure 7. Known occurrences of "Billings' operculum," plotted on stereographic polar projection.

CONCLUSIONS

One excuse for writing a small essay is to summarize data and call attention to a particular field. However, the framework into which this essay should fit is that of paleobiogeography. Early Ordovician opercula show three different patterns of geographic distribution, temporal abundance, and morphologic diversity, which will illustrate how these factors may interact in paleobiogeographic studies.

The opercula of *Ceratopea* are common, highly diverse and widespread, with few gaps in the lateral distribution of species.

Teiichispira opercula are locally abundant, have limited diversity (though probably within the length of geologic time through which this genus was distributed, it achieved more diversity than *Ceratopea*), and though widespread, have a markedly disjunct distribution.

"Billings' operculum" is relatively rare, shows no taxonomic diversity, and has a narrowly restricted geographic distribution.

Because opercula of *Teiichispira, Ceratopea,* and Billings' unnamed form have all been found at Port-au-Port, southwestern Newfoundland, one may deduce that this island was the center of calcification of opercula during late Early Ordovician times. I know that this is a nonsense statement, but I have rendered it in extreme form for emphasis. There is a danger in drawing too much inference from the distribution of fossils, particularly if one deals with forms that are rare, or at least little collected and studied.

Biogeographic data from fossils is always changing. Some change comes from taxonomic refinement and some comes from new finds. My outrageous hypothesis put forth above is based on a new find; 10 years ago, the data to propose such a ridiculous idea would not have been present. One should look at the distribution data for fossils with the view that it is a progress report, not a final statement.

As to calcification, to wander a bit from the subject of biogeography, I suspect that it is fortuitous that all 3 kinds of opercula occur in the late Early Ordovician. The Early Ordovician was a time of experimentation, and the elongate ceratopean operculum may certainly be considered an experiment! Succeeding opercula are lighter and less awkward in form. *Ceratopea* was a good candidate for extinction when other similar pleurotomariaceans evolved that lacked this type of operculum. Perhaps the same might be said for the aberrant *Teiichispira;* however, other macluritacean opercula, including that figured by Billings, are one of the adaptations of a highly successful group of organisms that persisted through the Ordovician.

In the example of Early Ordovician opercula discussed here, continental drift may be of some assistance in refining and interpreting the current distribution of *Ceratopea* and of "Billings' operculum" in the present Northern Hemisphere. Principal major faults are one of the prime lines of geologic evidence for assuming early continental drift in the North Atlantic area. Reversal of presumed movement bring areas of similar lithology together. The presence of similar faunas, as exemplified by these 2 opercula, reinforces this similarity and does add a modicum of support to the notion of movement. The lateral distribution of "Billings' operculum" in the North Atlantic would be reduced nearly 50 percent.

Likewise, there is considerable discussion not only of continents moving laterally but also of their changing position with reference to the poles and the equator. The data from the distribution of *Ceratopea,* when placed on one reconstruction of the early Paleozoic, indicate distribution in a belt parallel to the equator. The occurrence in the Mackenzie Mountains of Canada fits well as being within the equatorial belt, though north of the equator, like Greenland. Pushing the continents together minimizes some of the distances in the present North Atlantic, but the lateral distribution of *Ceratopea* is still enormous, approaching one-fourth of the Earth's circumference. As a critic kindly pointed out, there is nothing wrong with such a broad distribution; some animals today are as widely distributed. Still, this wide a distribution is not common in the geography of any time.

The known distribution of *Teiichispira* is comparable with that of living horseshoe crabs, that is, within one time plane at three or four isolated spots. Placing those spots on a globe showing early Paleozoic geography does reduce latitudinal differences and position them all as subequatorial. One consequence of continental drift closing the Atlantic would be that the Pacific would be wider. If this reasoning is correct, the distances between the Newfoundland and Australian occurrences on the Pacific side of the globe would be increased to more than half the diameter of the globe; on the other hand, on the Atlantic side of the globe, Newfoundland and Australia would be closer together. One cannot tell where *Teiichispira* began and in which direction it traveled. One can only say that it covered enormous distances in essentially a geologic instant and that in the course of travel it had to cross at least one large ocean basin, wherever that may have been.

This was not a unique event. Middle Ordovician *Maclurites* from Tasmania (Banks and Johnson, 1957) have a different specific name from those in Tennessee (Knight, 1941, p. 184-185), but both shell and operculum show inconsequential differences, and the Tasmanian species may be a synonym. A fairly large number of living gastropods cross the Atlantic during their larval stage.

I conclude that continental drift helps to refine our understanding of some features of the distribution of Early Ordovician opercula, but it does not shed great light on them all. To help *Teiichispira,* one might try moving Australia close to the western margin of America or scooting it around so that it was adjacent to Newfoundland, but no one would like to see that much wandering of continents just to accommodate a dead snaïl. Continental drift is not a panacea to solve all aspects of past distribution of animals.

ACKNOWLEDGMENTS

This manuscript has profited greatly from criticism by Carole Hickman, N. F. Sohl, and R. J. Ross, who was the first to consider *Ceratopea* as an animal distributed subequatorially. R. H. Flower and James Derby contributed unpublished data. Colleagues in field and laboratory, including John Peel, O. A. Wise, Jr., and Charles Cressler, as well as Flower and Ross, have contributed to my understanding of Early Ordovician opercula. Roberta Widger and Frances Smith contributed drawings and maps, respectively.

REFERENCES

Banks, M. R., and Johnson, J. H., 1957, *Maclurites* and *Girvanella* in the Gordon River Limestone (Ordovician) of Tasmania: Jour. Paleontology, v. 31, p. 632-640.

Bengtson, S. 1968, The problematic genus *Mobergella:* Lethaia, v. 1, p. 325-351.

Billings, E., 1865, Palaeozoic fossils, containing descriptions and figures of new or little known organic remains from the Silurian rocks 1861-1865: Canada Geol. Survey, Montreal, Dawson Brothers, 426 p.

Gilbert-Tomlinson, J., 1973, The Lower Ordovician gastropod *Teiichispira* in northern Australia: Australia Bur. Mineral Resources, Geology and Geophysics Bull. 126, p. 65-88.

Holtedahl, O., 1919, On the Paleozoic series of Bear Island, especially on the Heclahook system: Norsk Geol. Tidsskr., v. 5, 121-148.

Knight, J. B. 1941, Paleozoic gastropod genotypes: Geol. Soc. America Spec. Paper 32, 510 p.

Major, H., and Wisnes, T. S., 1955, Cambrian and Ordovician fossils from Sørkapp Land, Spitsbergen: Norsk Polarinst., Skr. 106, 47 p.

Marek, L., and Yochelson, E. L., 1976, Aspects of the biology of Hyolitha (Mollusca): Lethaia, v. 9, p. 165-182.

Oder, C. R. L., 1932, Fossil opercula from the Knox dolomite: Am. Midland Naturalist, v. 13, p. 133-153.

Ross, R. J., Jr., 1976, Ordovician sedimentation in the western United States, *in* Bassett, M. G., ed., The Ordovician system: Palaeontological Assoc. Symposium, Birmingham 1974, Proc.: Cardiff, Univ. Wales Press and Natl. Mus. Wales, p. 73-105.

Runnegar, Bruce and others, 1975, Biology of the Hyolitha: Lethaia, v. 8, p. 181-191.

Spitz, A., 1907, Die Gastropodan des Karnischen Unterdevon: Beiträge Paläont. Geol. Österreich-Ungarns und des Orients, v. 20, p. 115-190.

Trettin, H. P., 1975, Investigations of lower Paleozoic, Foxe Basin, northeastern Melville Peninsula, and parts of northwestern and central Baffin Island: Canada Geol. Survey Bull. 251, 177 p.

Ulrich, E. O., 1911, Revision of the Paleozoic systems: Geol. Soc. America Bull., v. 22, p. 281-680.

Whittington, H. B., and Hughes, C. P., 1973, Ordovician trilobite distribution and geography, *in* Hughes N. F., ed., Organisms and continents through time: Spec. Papers Palaeontology no. 12, p. 235-240.

Wise, O. A., Jr., Yochelson, E. L., and Clardy, B. F. 1975, Lower Ordovician stratigraphic relationships at Smithville, Arkansas and adjacent areas, *in* Contributions to geology of the Arkansas Ozarks: Little Rock, Arkansas Geological Commission, p. 38-60.

Woodring, W. P., Roundy, P. V., and Farnsworth, M. R., 1932, Geology and oil resources of the Elk Hills, California: U. S. Geol. Survey Bull. 835, 82 p.

Yochelson, Ellis L., 1964, The Early Ordovician gastropod *Ceratopea* from East Greenland: Medd. om Grønland, v. 164, 12 p.

——— 1966, An operculum associated with the Ordovician gastropod *Helicotoma:* Jour. Paleontology, v. 40, p. 748-749.

——— 1973, The late Early Ordovician gastropod *Ceratopea* in the Arbuckle Mountains, Oklahoma: Oklahoma Geology Notes, v. 33, p. 67-78.

——— 1975a, Early Ordovician gastropod opercula and epicontinental seas: U. S. Geol. Survey Jour. Research, v. 3, p. 447-450.

——— 1975b, Discussion of Early Cambrian "molluscs": Geol. Soc. London Jour., v. 131, p. 661-662.

Yochelson, E. L., and Barnett, S. G., 1972, The Early Ordovician gastropod *Ceratopea* in the Plattsburg, New York, area: Jour. Paleontology, v. 46, p. 685-687.

Yochelson, E. L., and Bridge, Josiah, 1957, The Lower Ordovician gastropod *Ceratopea:* U. S. Geol. Survey Prof. Paper 294-H, p. 281-304 (1958).

Yochelson, E. L., and Copeland, M. J., 1974, Taphonomy and and taxonomy of the Early Ordovician gastropod *Ceratopea canadensis* (Billings), 1865: Canadian Jour. Earth Sci., v. 11, p. 189-207.

Yochelson, E. L., and Jones, C. R., 1968, *Teiichispira,* a new Early Ordovician gastropod genus: U. S. Geol. Survey Prof. Paper 613-B, p. B1-B15.

Yochelson, E. L., and Linsley, R. M., 1972, Opercula of two gastropods from the Lilydale Limestone (Early Devonian) of Victoria, Australia: Natl. Mus. Victoria. Mem. v. 33, 14 p.

Yochelson, E. L., and Peel, J. S., 1975, *Ceratopea* and the correlation of the Wandel Valley Formation, eastern North Greenland: Grønlands Geol. Undersøgelse, v. 75, p. 28-31.

Yochelson, E. L., and Wise, O. A. Jr., 1972, A life association of shell and operculum in the Early Ordovician gastropod *Ceratopea unguis:* Jour. Paleontology, v. 46, p. 681-684.

Geographic Distribution of the Ordovician Gastropod *Maclurites*

DAVID M. ROHR, *Department of Geology, Oregon State University, Corvallis, Oregon 97331*

ABSTRACT

The operculate gastropod, *Maclurites*, is a cosmopolitan, relatively common fossil in carbonate rocks of the Middle and Upper Ordovician. Because of its massive shell and distinctive shape, it is easily recognized, but the commonly poor preservation makes specific assignment difficult. A set of diagnostic features (i.e., whorl profile, rate of whorl expansion, shape of base, height) discernable from cross sections of poorly preserved specimens should allow more accurate determination of species, which in turn, might result in biogeographic groupings at the specific level. *Maclurites* lived in shallow, subtidal (about Benthic Assemblage 3), warm marine waters, and is found almost everywhere that Middle and Upper Ordovician carbonate rocks occur, with the central part of the Siberian Platform and the Urals being the only major exceptions. The genus occurs on all continents except Africa and Antarctica (where there are no carbonate rocks in the Middle and Upper Ordovician) which contrasts with many other Middle and Upper Ordovician fossil groups that show provincialism during that interval.

INTRODUCTION

The gastropod *Maclurites*, the first new Paleozoic gastropod genus to be described in North America (Lesueur, 1818), has since been recognized in Ordovician rocks throughout the world. The genus is commonly used as a guide fossil for the Ordovician because of its massive shell, distinctive morphology, abundance, and wide geographic distribution. It is most abundant in the Middle Ordovician, but it also occurs in the Upper Ordovician and uppermost Lower Ordovician.

The genus *Maclurites* includes *Maclurea* (an emendation of the generic name by Emmons in 1842 which was used extensively for many years) and *Maclurina* Ulrich and Scofield, 1897. *Maclurina* differs from *Maclurites* in the lack of projections for attachment of muscles on the inner side of the operculum. Since the operculum is rarely found in place, this method of determination is somewhat impractical, and *Maclurina* has since been synonymized with *Maclurites*. *Paramaclurites* Vostokova, 1955, is a very similar form, but has been designated as a separate genus by Vostokova.

In spite of its stratigraphic significance and numerous citations, the genus has not been studied in any great detail. Over 50 species of *Maclurites* have been established since 1818, but a large number of them are probably synonyms. Much of the confusion is due to the poor preservation of specimens which are rarely silicified, and lack of agreement on what constitute diagnostic specific features.

A reliable summary of the geographic distribution of *Maclurites* may be compiled from the literature even when the specimens are not illustrated. Assuming even a mediocre state of preservation, the genus is difficult to mistake for any other gastropod. Identification to the species level is, however, an entirely different matter.

Maclurites is by far the most common operculate gastropod of the Middle and Late Ordovician. Yochelson (this volume) discusses the operculate gastropods of the Lower Ordovician—*Ceratopea, Teiichispira,* and the "Billings' operculum." *Ceratopea* is a pleurotomariacean and the other two belong to the same superfamily as *Maclurites* (Macluritacea).

DESCRIPTION AND DIAGNOSTIC SPECIFIC FEATURES

Specimens of *Maclurites* (see Fig. 1) may be as large as 30 cm across the base; most are commonly 5 to 10 cm. The genus exhibits pseudosinistral (hyperstrophic) coiling. Counterclockwise ornamentation on the aperture indicates, however, that the shell is actually dextral with the aperture to the right when the axis of the shell is oriented vertically. Thus, the flat side of the shell is the base, and the convex side with the umbilicus or apical cavity is the top; the shell has a depressed spire. A thick calcareous operculum closes the aperture and although commonly preserved, it is rarely found in place. The operculum is concave on the inner side and in most species is characterized by a roughened, knoblike process to which the retractor muscle of the gastropod was attached (Fig. 1).

The geographic distribution of the species of *Maclurites* is poorly known, in part, because the species themselves are inadequately defined. If a method of specific determination were agreed on, it should be possible to combine many of the approximately 50 species of *Maclurites*, perhaps into subgenera, and patterns of provincialism

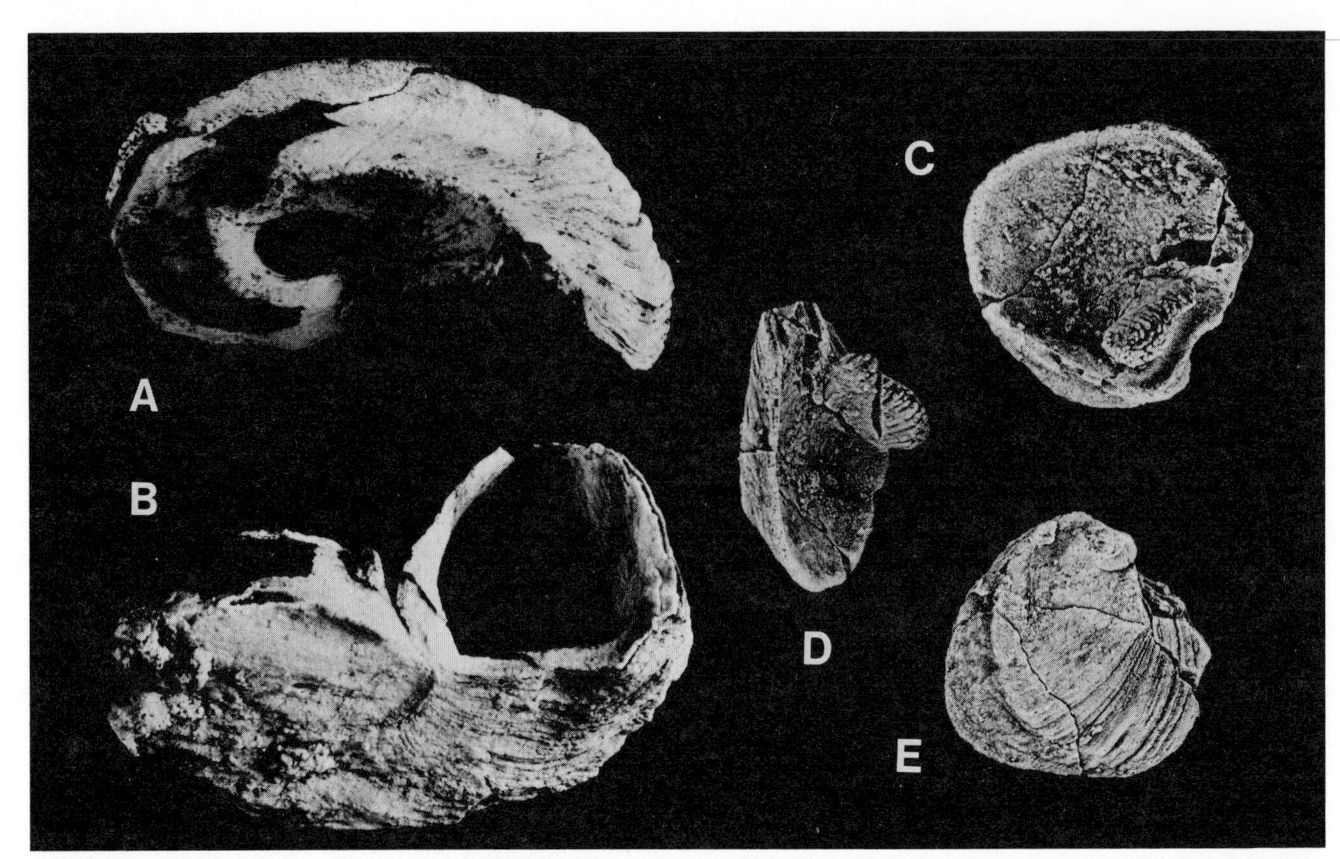

Figure 1. *Maclurites* sp. from the Klamath Mountains, California. (A) View of the top of shell showing apical cavity, X1.5; (B) Oblique view of flat base and aperture of shell, X1.5; (C), (D) Interior views of the operculum with knoblike process for attachment of retractor muscle, X1.7; (E) Exterior of operculum. The counterclockwise ornamentation is characteristic of dextral shells, X1.7.

might appear. In view of the relationship between taxonomy and biogeography, a short review and evaluation of morphologic features used to diagnose species is appropriate.

Most specific descriptions of *Maclurites* stress the relative height, whorl profile, width of the apical cavity, ornamentation, shape of the base, rate of expansion of the shell, size of the shell, and nature of the operculum. Some of these features are more significant than others. Since the opercula are rarely found in place, and many times not in the same bed, they are of little use in specifically assigning the shells. An independent classification of opercula alone would be useful, however.

The size of the shell may be diagnostic in some cases. Only certain species seem to attain very large size, but since all shells, even the large ones, were presumably small at one time, "small size" is not diagnostic.

The surface ornamentation of most species of *Maclurites* is limited to growth lines, but a few species like *M. logani* Salter and *M. bigsbyi* Hall have revolving striae which are diagnostic.

The width of the apical cavity ("umbilicus") is often used to distinguish species. Wilson (1931, p. 300) noted that, "the wideness of the umbilicus of *M. manitobensis* really only appears in the final whorl." Thus, if the width of the apical cavity is to be used as a diagnostic character, it should be taken from a number of specimens of different sizes to determine how much variation, if any, is present.

In many cases, the height of the shell is used as a diagnostic character. For example, *M. profunda* Butts and *M. magnus* Lesueur seem to be identical except for the height of the shell. Since Butts (1926) did not say how many specimens he studied, it is not possible to tell if the difference in height is variation within a species or a specific difference. As with the width of the apical cavity, it is necessary to measure enough specimens to determine what is variation.

Knight and others (1944, 1960) list a flat base as one of the diagnostic features of *Maclurites*. Actually, the bottom may vary by a few degrees from 180° making it slightly convex or concave. *M. depressa* Ulrich and Scofield is distinguished by the concave surface of the bottom of the individual whorls.

The shape of the whorl section is almost always given with descriptions and may vary from oval to triangular to trapezoidal. The whorl profile is diagnostic but must usually be employed in conjunction with other features.

Another useful taxonomic feature of *Maclurites* is the rate of expansion of the whorls. Banks and Johnson (1957,

p. 638) noted that many authors refer to the rate of increase as a specific character but rarely quote figures. Terms such as "rapidly enlarging whorls" are used in most descriptions, but Banks and Johnson (1957) give a detailed description of increase in terms of successive whorl-width ratios. They find that while some species have about the same rate of increase (*M. magnus, M. florentinensis, M. profunda*), others (*M. manitobensis* and *M. bigsbyi*) differ. Wilson (1931, p. 305) used the difference in rate of expansion to distinguish among *M. borealis, M. alta,* and *M. manitobensis*.

All the characteristics mentioned above are obvious on well-preserved specimens, but most published descriptions are based on unsilicified specimens on which the aperture is rarely preserved and the surface ornamentation is totally or partially lacking. What characteristics of poorly preserved specimens can be used to attempt specific determination? When cut in half longitudinally (as many specimens are illustrated) it is possible to determine size, whorl profile, whorl-width ratios, height, and in some cases, shape of base. If these few characteristics are observed, it should be possible to at least assign specimens to subgeneric groups which might differentiate biogeographic units.

ECOLOGY

The mode of life of *Maclurites* is slightly better known than that of most other Ordovician gastropods. Salter (1859, p. 10) concluded that *Maclurites*, with its heavy shell, was "probably stationary or nearly so on the bottom, seeing that its upper or convex side is constantly overgrown with sponge . . . while the flat or lower side preserves the sharp lines of growth, which would have been abraded had the animal been endowed with much locomotion."

Banks and Johnson (1957) reported the association of *Maclurites, Girvanella* (an alga), and bellerophontid gastropods in the Gordon River Formation of Tasmania. They observed that *Maclurites* occurred in dolomitic limestone and in some cases was encrusted by stromotoporoids. These observations and associations indicate a shallow-water environment within the photic zone. The algal masses (oncolites) indicate a moderately turbulent to quiet environment, and the encrusted shells are consistent with shallow water at about Benthic Assemblage 3 of the depth scheme suggested by Boucot (1975).

Yochelson (personal commun. 1976) has noted that the opercula and shells of *Maclurites* are, in many cases, not found in the same beds and he speculated (Yochelson, 1975) that since *Maclurites* is closely related to *Ceratopea* it might live in a similar habitat. He hypothesized that *Ceratopea* grazed on algal mats at depths of less than 20 ft (6 m) in marine environments having a greater than normal salinity. Occasional influxes of hypersaline water killed the animals, after which the operculum fell off, the soft parts decayed, and the shell was partially or completely buoyed up and transported away.

Opercula and shells of *Maclurites* are found together in the Klamath Mountains, California, indicating that this hypothesis does not always hold true, although in the widespread epicontinental platform environments it may be reasonable. In the Klamath Mountains, *Maclurites* occurs with brachiopods that suggest Benthic Assemblage 3 to 4, or slightly deeper than most other Ordovician gastropods from the area. Calcareous algae have not been observed with the Klamath Mountain *Maclurites*.

DISTRIBUTION OF *MACLURITES*

Figure 2 shows the world-wide distribution of *Maclurites* as compiled from the literature. The sources for this compilation listed in the Appendix are not intended to be comprehensive for each locality; they are intended to provide a reasonable documentation for each point on the map. In some areas such as Nevada and the eastern United States, *Maclurites* is widespread and only representative occurrences are noted.

The publications consulted for this study reported *Maclurites* only from carbonate rocks and most commonly from limestone. *Maclurites* is found in virtually every area that has Middle or Upper Ordovician carbonate rocks. None were recovered from shale or sandstone, and areas that have only noncarbonate, clastic Ordovician rocks (Africa, central Europe, and most of South America) do not have *Maclurites*. The genus occurs in geosynclinal areas (Klamath Mountains of California, central Norway) as well as platform sequences. The absence of *Maclurites* from the main part of the Siberian Platform (between the Yenissei and Lena Rivers) may be due to lack of study, because it is found around the margins of this Platform.

From this summary of occurrences, *Maclurites* appears to be a shallow, warm-water, cosmopolitan genus present on all the continents except Africa and Antarctica (which have no Ordovician carbonate rocks). Even allowing for plate tectonic reconstructions, the Ordovician distribution of the genus is still widespread.

COMPARISON TO OTHER MIDDLE AND UPPER ORDOVICIAN TAXA

A preliminary compilation from *The Treatise on Invertebrate Paleontology*, Part I (Knight and others, 1960), indicates that most gastropods confined to the Early Ordovician to Late Silurian are relatively cosmopolitan. With few exceptions, all genera occur in at least North America and Europe (usually the Baltic area). Most of the "Treatise" information on Ordovician gastropods is from North America and the Baltic area, however, with little data for the rest of the world. Thus, some genera may appear to be less widespread than they actually are, and others may appear to be more cosmopolitan than they actually are due to inclusion of species that do not belong to them (Boucot, personal commun. 1976).

Other Middle and Upper Ordovician fossil groups show more provincialism than is apparent from gastropod

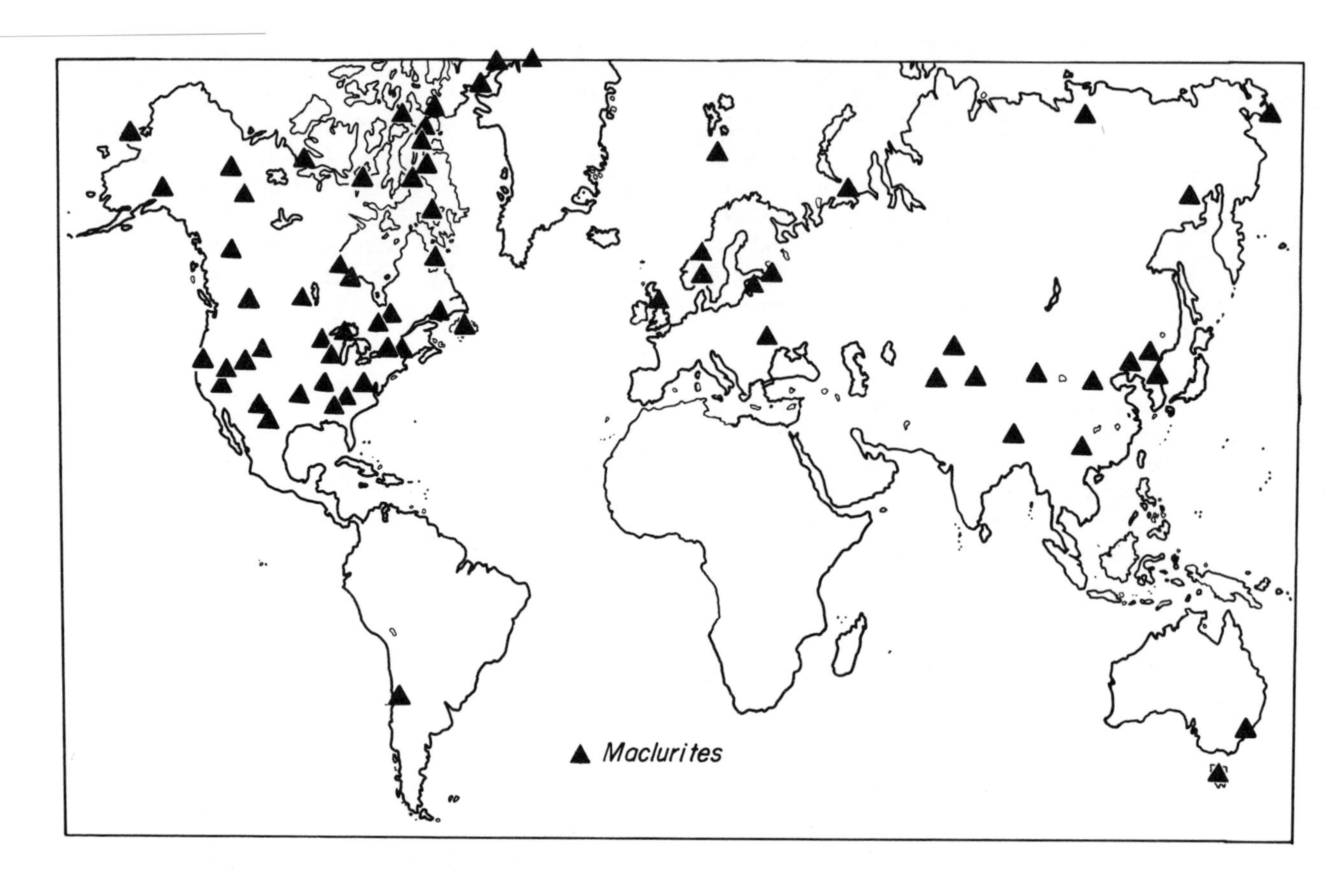

Figure 2. Geographic distribution of *Maclurites* in the Middle and Upper Ordovician of the world. Note added in proof: Additional localities not shown on Figure 2 are in the Kosva River region of the central Ural Mountains, U.S.S.R. (Ivanov, A. N. and Miyagkova, E. I., 1950, Stratigrafiya nizhnego i srednego Paleozoya zapadnogo sklona srednego Urala: Akad. Nauk SSSR, Ural. Filial, Gorno-Geol. Inst. Trudy, v. 17, Sbornik Voprosam stratigrafi no. 1, p. 7) and in the Middle Ordovican Børglum River Formation of Peary Land, eastern North Greenland (Christie, R. L., and Peel, J. S., 1977, Cambrian-Silurian stratigraphy of Børglum Elv, Peary Land, eastern North Greenland: Greenland Geol. Survey Rept. 82, p. 22).

biogeography. Bergström (1973) was able to distinguish three conodont subprovinces in the Midcontinent of North America. Provincialism is also apparent in Middle and Upper Ordovician trilobites (Whittington, 1973), and in articulate brachiopods (Jaanusson, 1973).

The Middle and Upper Ordovician distribution of the graptolites (Berry, this volume) is very similar to that of *Maclurites* in that neither occur in the Malvinokaffric Realm.

ACKNOWLEDGMENTS

I gratefully acknowledge the encouragement and assistance of A. J. Boucot during this project. The following people kindly provided information on the distribution of *Maclurites:* Dr. B. S. Norford, Geological Survey of Canada; Dr. Valdar Jaanusson, Naturhistoriska Riksmuseet, Stockholm; and Dr. Ellis Yochelson, U.S. Geological Survey.

REFERENCES

Banks, M. R., and Johnson, J. H., 1957, *Maclurites* and *Girvanella* in the Gordon River Limestone (Ordovician) of Tasmania: Jour. Paleontology, v. 31, p. 632-640.

Bergström, Stig, 1973, Ordovician conodonts, *in* Hallam, A., ed., Atlas of palaeobiogeography: Amsterdam, Elsevier, p. 47-58.

Boucot, A. J., 1975, Evolution and extinction rate controls: Amsterdam, Elsevier, 427 p.

Butts, Charles, 1926, The Paleozoic rocks, *in* Geology of Alabama: Alabama Geol. Survey Spec. Rept. 14, p. 40-230.

Jaanusson, V., 1973, Ordovician articulate brachiopods, *in* Hallam, A., ed., Atlas of palaeobiogeography: Amsterdam, Elsevier, p. 19-26.

Knight, J. B., Bridge, Josiah, Shimer, H. W., and Shrock, R. R., 1944, Paleozoic Gastropoda, *in* Shimer, H. W., and Shrock, R. R., eds., Index fossils of North America: Cambridge, Technology Press, Massachusetts Inst. Technology, p. 433-479.

Knight, J. B. and others, 1960, Systematic descriptions, *in* Moore, R. C., ed., Treatise on invertebrate paleontology. Pt. I. Mollusca 1: Geol. Soc. America and Univ. Kansas, p. I169-I309.

Lesueur, C. A., 1818, Observations on a new genus of fossil shells: Acad. Nat. Sci. Philadelphia Jour., v. 1, p. 310.

Salter, J. W., 1859, Canadian organic remains: Geol. Survey Canada, decade 1, Montreal, Dawson Brothers, 47 p.

Ulrich, E. O., and Scofield, W. H., 1897, The Lower Silurian Gastropoda of Minnesota, *in* The geology of Minnesota, Vol. 3, Pt. 2, Paleontology: Minnesota Geol. and Nat. History Survey, p. 813-1081.

Whittington, H. B., 1973, Ordovician trilobites, *in* Hallam, A., ed., Atlas of palaeobiogeography: Amsterdam, Elsevier, p. 13-18.

Wilson, A. E., 1931, Notes on the Baffinland fossils collected by J. Dewey Soper during 1925 and 1929: Royal Soc. Canada Trans., v. 25, p. 285-306.

Yochelson, E. L., 1975, Early Ordovician gastropod opercula and epicontinental seas: U. S. Geol. Survey Jour. Research, v. 3, p. 447-450.

APPENDIX

World-wide Distribution of *Maclurites* by County, Location, and Geologic Age Used to Compile Figure 2

Argentina

Talacasta: limestone (Ordovician)

Kayser, E., 1925, Contribuciones a la paleontologia de la Republica Argentina: Sobre fossiles primordiales y infrasilurianos: Cordoba, Acad. Nac. Ciências Act., v. 8, p. 297-333.

Australia

Orange, New South Wales: unnamed limestone (Middle Ordovician?)

Stevens, N. C., 1952, Ordovician stratigraphy of Cliefden Cliffs, near Mandurama, N. S. W.: Linnean Soc. New South Wales Proc., v. 77, p. 114-120.

Tasmania: Gordon River Limestone (Chazy?)

Banks, M. R., and Johnson, J. H., 1957, *Maclurites* and *Girvanella* in the Gordon River Limestone (Ordovician) of Tasmania: Jour. Paleontology, v. 31, p. 632-640.

Canada

Alberta-British Columbia, Palliser Pass: Owen Creek Formation (Middle Ordovician) and Skoki Formation (Whiterock, Middle Ordovician)

Norford, B.S., 1969, Ordovician and Silurian stratigraphy of the southern Rocky Mountains: Canada Geol. Survey Bull. 176, p. 24, 37-38.

Arctic Islands and vicinity:

Akpatok Island: unnamed limestone (Richmond, Upper Ordovician)

Wilson, A. E., 1938, Gastropods from Akpatok Island, Hudson Strait: Royal Soc. Canada Trans., v. 32, p. 33.

Baffin Island: Baillarge Formation at Admiralty Inlet (upper Middle or lower Upper Ordovician)

Trettin, H. P., 1965, Lower Paleozoic sediments of northwestern Baffin Island, District of Franklin: Canada Geol. Survey Paper 64-47, p. 17.

Lemon, R. R. H., and Blackadar, R. G., 1963, Admiralty Inlet area Baffin Island, District of Franklin: Canada Geol. Survey Mem. 328, p. 23.

Baffin Island: unnamed limestone at Putnam Highlands, Koukjauk River (Richmond, Upper Ordovician)

Wilson, A. E., 1931, Notes on the Baffinland fossils collected by J. Dewey Soper during 1925 and 1929: Royal Soc. Canada Trans., v. 25, p. 301-305.

Baffin Island: Baillarge Formation at Brodeur Peninsula (upper Middle or lower Upper Ordovician)

Trettin, H. P., 1965, Middle Ordovician to Middle Silurian carbonate cycle, Brodeur Peninsula, northwest Baffin Island: Canadian Petroleum Geology Bull., v. 13, p. 165.

Cornwallis and Little Cornwallis Islands: Cornwallis Formation (Middle Ordovician)

Thorsteinsson, Raymond, 1958, Cornwallis and Little Cornwallis Islands, District of Franklin, N.W.T.: Canada Geol. Survey Mem. 294, p. 36, 39.

Devon Island: Eleanor River Formation (Chazy, Middle Ordovician)

Kerr, J. W., 1967, New nomenclature for Ordovician rock units of the eastern and southern Queen Elizabeth Islands, Arctic Canada; Canadian Petroleum Geology Bull., v. 15, p. 91-113.

Ellsmere Island: Eleanor River Formation (Chazy, Middle Ordovician)

Kerr, J. W., 1967, New nomenclature for Ordovician rock units of the eastern and southern Queen Elizabeth Islands, Arctic Canada: Canadian Petroleum Geology Bull., v. 15, p. 91-113.

Ellsmere Island: Cornwallis Formation (Upper Ordovician)

Kerr, J. W., 1970, *in* Berry, W. B. N., and Boucot A. J., eds., Correlation of the North American Silurian rocks: Geol. Soc. America Spec. Paper 102, p. 132.

Iglulik Island: unnamed limestone (Middle Ordovician)

Teichert, Curt, 1937, Ordovician and Silurian faunas from Arctic Canada: 5th Thule Expedition, 1921-24, Rept., v. 1, p. 71-75.

King William Island: unnamed limestone (Ordovician)

Teichert, Curt, 1937, Ordovician and Silurian faunas from Arctic Canada: 5th Thule Expedition, 1921-24, Rept., v. 1, p. 24.

Read and Sutton Islands: unnamed dolomite (Upper Ordovician)

Miller, A. K., and Youngquist, W., 1947, Ordovician fossils from the southwestern part of the Canadian Arctic Archipelago: Jour. Paleontology, v. 21, p. 2.

District of Mackenzie: Sunblood Formation (Middle Ordovician)

Gabrielse, H., Blusson, S. L., and Roddick, J. A., 1973, Geology of Flat River, Glacier Lake, and Wrigley Lake map-areas; District of Mackenzie and Yukon Territories: Canada Geol. Survey Mem. 266, p. 54-57.

Manitoba: Bad Cache Rapids Formation (Caradoc-Ashgill, Upper Ordovician)

Norford, B. S., 1970, Ordovician and Silurian biostratigraphy of the Sogepet-Aquitaine Kaskattama Province No. 1 well, northern Manitoba: Canada Geol. Survey Paper 69-8, 36 p.

Lake Winnipeg: Doghead Horizon of the Red River Formation (Upper? Ordovician)

Wilson, A. E., 1938, Gastropods from Akpatok Island, Hudson Strait: Royal Soc. Canada Trans., v. 32, p. 33.

Manitoba-Ontario, Hudson Bay Lowlands: Portage Chute Formation (Eden, Middle to Upper Ordovician); Churchill Group (Upper Ordovician); Chasm Creek Formation (Richmond, Upper Ordovician); Bad Cache Rapids Group ("post-Chazy," post early Middle Ordovician)

Cumming, L. M., 1975, Ordovician strata of the Hudson Bay Lowlands: Canada Geol. Survey Paper 74-28, 93 p.

Nelson, S. J., 1963, Ordovician paleontology of northern Hudson Bay Lowland: Geol. Soc. America Mem. 90, p. 66-67.

Newfoundland, Table Head and Point Rich: "Quebec Group" (Ordovician)

Billings, E., 1865, Paleozoic fossils, Vol. 1: Canada Geol. Survey, Montreal, Dawson Brothers, p. 237-245.

Newfoundland: Cutwell Group (Middle Ordovician)

Williams, H., 1962, Botwood (west half) map-area, Newfoundland: Canada Geol. Survey Paper 62-9, p. 6.

Ontario-Quebec, Lake Timiskaming area: limestone beds of Liskeard Formation (Middle or Upper Ordovician).

Ollerenshaw, N. C., and MacQueen, R. W., 1960, Ordovician and Silurian of the Lake Timiskaming area: Canada Geol. Assoc. Proc., v. 12, p. 106.

Quebec, Waswanipi Lake: Liskeard Limestone (Richmond, Upper Ordovician)

Clark, T. H., and Blake, D. A. W., 1952, Ordovician fossils from Waswanipi Lake, Quebec: Canadian Field-Naturalist, v. 66, p. 119.

Quebec, Paquette Rapids: Rockland Beds of the Ottawa Formation (Middle Ordovician)

Salter, J. W., 1859, Canadian organic remains: Canada Geol. Survey decade 1, Montreal, Dawson Brothers, p. 7-10.

Wilson, A. E., 1951, Gastropoda and Conularida of the Ottawa Formation of the Ottawa-St. Lawrence Lowland: Canada Geol. Survey Bull. 17, p. 61.

China

South Setchouan-North Koueitcheou: Altsziachan Series, limestone (Middle Ordovician)

Central Liaoning: Matsziagoou Series limestone (Middle Ordovician)

North Chansi: Matsziagoou Series limestone (Middle Ordovician)

Kansu: Tchjotszychan Limestone (Middle and Upper Ordovician)

Dubertret, L., ed., 1964, Lexique Stratigr. Internat., Vol. 3, Asie, Fasc. 3, Republique Populaire Chinoise I, p. 213, 509, 511, 649.

Tibet: Mount Everest region ("0_1")

Mu En-zhi and others, 1973, Stratigraphy of the Mt. Jolmo Lungma region in southern Tibet: Scienta Geologica Sinica no. 1, pl. 1 (in Chinese).

Yü Wen, 1975, The gastropod fossils from the Qomolangma Feng region, *in* A report of scientific investigations in the Qomolangma Feng region (Palaeontology, Fasc. 1): Academia Sinica, Nanking Inst. Geology and Palaeontology, pl. 2-3 (in Chinese).

Manchuria: Toufangkou Limestone (Middle Ordovician)

Kobayashi, Teiichi, 1930, Ordovician fossils from Korea and South Manchuria. Pt. II. On the Bantatsu Bed of the Ordovician age: Japanese Jour. Geology and Geography, v. 7, p. 96-99.

T'ien Shan area: unnamed limestone (dated Middle Ordovician on the presence of *Maclurites*)

Norin, Erik, 1941, Geologic reconnaissance of the Chinese T'ien Shan: Stockholm, Rept. Scientific Expedition to N.W. Province of China, Sino-Swedish Expedition Pub. 16, Pt. 3, Geology, p. 42, 57.

Greenland

Cape Calhoun: Cape Calhoun Beds (Middle or Upper Ordovician)

Troedsson, Gustaf, 1928, On the Middle and Upper Ordovician faunas of northern Greenland. Pt. II. Jubilaeumsekkspeditionen Nord om Grønland, 1920-1923, no. 7, p. 18-22.

Great Britain

Girvan: Stinchar Limestone, Balclatchie Group (Upper Ordovician)

Longstaff, Jane, 1924, Descriptions of gastropods chiefly in Mrs. Robert Gray's collections from the Ordovician and Lower Silurian of Girvan: Geol. Soc. London Quart. Jour., v. 80, p. 431-433.

Korea

"Unkaku bed of Mt. Bantatsu, North Korea" (Middle Ordovician)

Kobayashi, Teiichi, 1930, Ordovician fossils from Korea and South Manchuria, Pt. II. On the Bantatsu Bed of the Ordovician age: Japanese Jour. Geology and Geography, v. 7, p. 96-99.

Norway

Oslo: "Gastropod Limestone" (Ashgill, Upper Ordovician)

Jaanusson, Valdar, personal commun. 1976.

Central Norway: Otta Conglomerate (early Middle Ordovician)

Yochelson, E. L., 1963, Gastropods from the Otta Conglomerate: Norsk Geol. Tidsskr. 43, p. 76.

Smølen Island: Smølen Limestone ("Lower Ordovician")

Holtedahl, Olaf, 1924, On the rock formations of Novaya Zemlya, *in* Rept. Sci. Results Norwegian Expedition to Novaya Zemlya 1921, no. 22: Videnskaps. i Kristiania: Oslo, A. W. Brøggers Boktrykkeri, p. 110-112.

Bear Island: unnamed dolomite possibly correlative of Smølen Limestone (Lower Ordovician)

Holtedahl, Olaf, 1924, p. 111-112.

Soviet Union

Soviet Union: northeast

Taskan River (Llanvirn, Middle Ordovician)

Uelenskoe Uplift: (Upper Ordovician)

Tas-Khayakhtakh Mountains: (Upper Ordovician)

Oradovskaya, M. N., 1970, Ordovician System, *in* Geology of the USSR, 30, northeastern USSR, geological description, Book 1, p. 80-104.

Soviet Union: arctic

Ostrova Vaigach (Upper Ordovician)

Vostokova, V. A., 1961, Paleozoic gastropods from Novaya Zemlya and Ostrova Vaigach: Naucho-Issledovatelskii Inst. Geologi Arktika, Sbornik Statei Po Paleontologia i Biostratigrafi, Vipusk 25, p. 8 (in Russian).

Estonia: (Ashgillian Pirgu Stage, Upper Ordovician)

Jaanusson, Valdar, personal commun. 1976.

Leningrad District: "Orthoceratite Limestone" (lower Middle Ordovician)

Jaanusson, Valdar, personal commun., 1976.

T'ien Shan region

Yochelson, Ellis, oral commun. 1976.

Kazakhstan: Chu-Ili Mountains (Upper Ordovician)

Keller, B. M. and others, 1956, The Ordovician of Kazakhstan. II. Ordovician stratigraphy of Chu-Ili Mountains: Akad. Nauk SSSR, Trudy Geol. Inst., Vipusk 1, p. 171 (in Russian).

Podolia (Middle Dnestr River): Molodovo Horizon (Upper Ordovician)

Nikiforova, O. I., and Predtechenskij, N. N., 1968, A guide to the geological excursion on Silurian and Lower Devonian deposits of Podolia: Internat. Symposium on Silurian-Devonian boundary and Lower and Middle Devonian stratigraphy, 3rd, Leningrad 1968, VSEGEI, p. 9.

Mironova, M. G., 1971, Gastropods of the Molodovo horizon in Podolia: Leningrad Univ. Vestnik Geology Geography, v. 18, p. 168.

United States

Alabama, Odenville: Odenville Limestone (lower Middle Ordovician)

Butts, Charles, 1926, The Paleozoic rocks, *in* Geology of Alabama: Alabama Geol. Survey Spec. Rept. 14, p. 106-108.

Alaska, Seward Peninsula: unnamed limestone (Upper Ordovician)

Sainsbury, C. L., Dutro, J. T., and Churkin, M., Jr., 1972, The Ordovician-Silurian boundary in the York Mountains, western Seward Peninsula, Alaska: U. S. Geol. Survey Prof. Paper 750-C, p. C55.

Alaska: Medfra D-2 quadrangle, unnamed limestone (Middle or Upper Ordovician)

Dutro, J. T., unpubl. data.

California: Klamath Mountains (Llandeilo, Middle Ordovician)

Rohr, unpubl. data.

California-Nevada order area, Nye and Inyo Counties: limestone of the Pogonip Group (Middle Ordovician)

Cornwall, H. R., and Kleinhampl, F. J., 1964, Geology of Bullfrog quadrangle and ore deposits related to Bullfrogs Hills caldera, Nye Country, Nevada and Inyo County, California: U. S. Geol. Survey Prof. Paper 454J, p. J5.

Illinois, Dixon: "Black River" (Middle Ordovician)

Ulrich, E. O., and Scofield, W. H., 1897, The Lower Silurian Gastropoda of Minnesota, *in* The geology of Minnesota, Vol. 3, Pt. 2: Minneapolis, Harrison and Smith State Printers, p. 1038-1043.

Iowa, Dubuque: "Galena Dolomite" (Middle Ordovician)

Ulrich and Scofield, 1897, p. 1038-1043.

Michigan, Houghton County: "Stewartsville or Upper Galena" Dolomite (Middle Ordovician)

Case, E. C., and W. I. Robinson, 1914, The geology of Limestone Mountain and Sherman Hill in Houghton County, Michigan: Ann. Rept. Board Geol. and Biological Survey, p. 258.

Minnesota:

Minneapolis: "Black River (Platteville)" (Middle Ordovician)

Stewartville: Stewartville Dolomite (Middle Ordovician)

Ulrich and Scofield, 1897, p. 1038-1043.

Nevada:

Eureka and Nye Counties: Ninemile Formation "Early Ordovician); Antelope Valley Formation, (Chazy, Middle Ordovician)

Eureka: Pogonip Group (Middle Ordovician)

Merriam, C. W., 1963, Paleozoic rocks of Antelope Valley, Eureka and Nye Counties, Nevada: U. S. Geol. Survey Prof. Paper 423, p. 23.

Walcott, C. D., 1884, Paleontology of the Eureka district: U. S. Geol. Survey Mon., v. 8, p. 81-83.

New Mexico, southern: El Paso Limestone (Lower or Middle Ordovician)

Darton, N. H., 1917, A comparison of Paleozoic sections in southern New Mexico: U. S. Geol. Survey Prof. Paper 108-C, p. 36.

New York, Lake Champlain area: Chazy Limestone (Middle Ordovician)

Raymond, Percy E., 1908, Gastropods of the Chazy Formation: Carnegie Mus. Ann., v. 4, p. 199-202.

Oklahoma, Adair County: lower Tyner Formation (Chazy, lower Middle Ordovician)

Huffman, G. G., and Starke, John M., Jr., 1960, A Chazyan faunule from the lower Tyner, northeastern Oklahoma: Oklahoma Geology Notes, v. 20, p. 268-271.

Pennsylvania, south-central: Stones River Limestone (Middle Ordovician)

Willard, B., and Cleaves, A. B., 1938, A Paleozoic section in south-central Pennsylvania: Penn. Geol. Survey, 4th ser., Bull. G8, p. 4.

Tennessee Knoxville: Lenoir Limestone (Middle Ordovician)

Ulrich and Scofield, 1897, p. 1038-1043.

Texas, Marathon region: limestone of the Alsate Formation (lower Middle Ordovician)

King, P. B., 1937, Geology of the Marathon Region, Texas: U. S. Geol. Survey Prof. Paper 187, p. 32.

Utah, Ute Peak: Pogonip (Middle Ordovician)

Hall, James, and Whitfield, R. P., 1877, *in* King, Clarence, Prof. Paper Engineer Dept., U. S. Army. No. 18. Rept. Geol. Exploration of the 40th Parallel, Vol. 4, Pt. 2, Paleontology, p. 235-236.

Virginia, Smyth, Scott, and Giles Counties: Lenoir Limestone (lower Middle Ordovician)

Butts, Charles, 1941, Geology of the Appalachian Valley in Virginia, Pt. 2, Fossil plates and explanations: Virginia Geol. Survey Bull. 52, p. 58-59.

Wyoming, Lander: Bighorn Dolomite (Upper Ordovician)

Yochelson, E. L., unpub. data.

ADDENDUM

Australia

New South Wales: Cliefden Caves Limestone (Caradoc, Middle Ordovician)

Dr. Barry Webby kindly loaned the following specimens of *Maclurites* sp. from the University of Sydney collections: SUP No. 18770-18775.

Canada

British Columbia: Sunblood Formation? (early Middle Ordovician)

Gabrielse, H., 1975, Geology of Fort Grahame E. ½ map area, British Columbia (94CE½): Canada Geol. Surv. Paper 75-33, p. 25.

Ellesmere Island, South Cape: "brown limestone" (Ordovician).

Holtedahl, Olaf, 1914. On the fossil faunas from Per Schei's Series B in south western Ellesmereland, *in* Rept. Second Norwegian Arctic Expedition in the "Fram" 1898-1902, no 32: Videnskaps-Selskabet i Kristiania, p. 4.

Newfoundland: Table Head Series, Long Point Series, Cow Head Limestone Breccia (Middle Ordovician)

Schuchert, Charles and Dunbar, C.O., 1934, Stratigraphy of western Newfoundland: Geol. Soc. America Mem. 1, p. 64, 68-71, 79.

United States

Missouri, Ste. Genevieve Co.: Plattin Limestone (Black River, Middle Ordovician)

Fenton, C. L., 1928, The statigraphy and larger fossils of the Plattin Formation in Ste. Genevieve County, Missouri: Am. Midland Naturalist, v. 1, p. 137-138.

Monorakos in the Ordovician of Alaska and Its Zoogeographic Significance

ALLEN R. ORMISTON, *Amoco Production Company, Tulsa, Oklahoma 74103*
REUBEN J. ROSS, JR., *United States Geological Survey, Mail Shop 919 Denver Federal Center, Denver, Colorado 80255*

ABSTRACT

The trilobite *Monorakos* Schmidt, 1886, which is widely distributed in rocks of late Middle and(or) Late Ordovician age in the northern Soviet Union, is recorded also from the Seward Peninsula, Alaska. The unusual distribution of *Monorakos* is used to diagnose a new zoogeographic subdivision, the Monorakid Subprovince. This newly defined subprovince suggests to us that the distribution of continents and plates in late Middle Ordovician time should be modified to place Alaska and Kolyma in more northerly latitudes than previously suggested by Ross (1975a). The Seward Peninsula in this reconstruction was not a part of Alaska in the Ordovician.

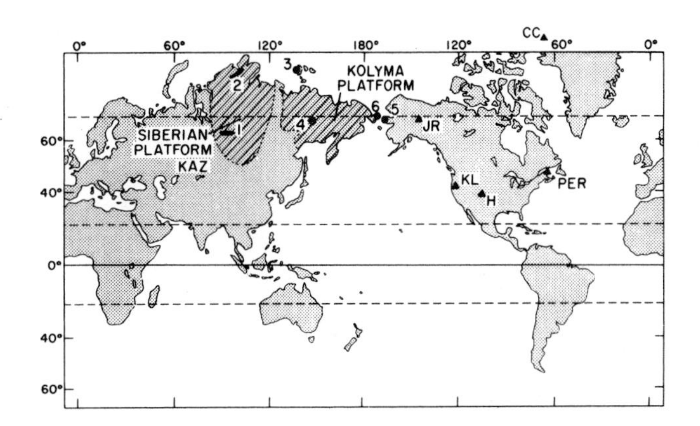

Figure 1. Areas from which monorakid trilobites have been reported in upper Middle or Upper Ordovician strata are indicated by numbers; other localities mentioned in the text are designated by letters. (1) Stony Tunguska River; (2) Taimyr Peninsula; (3) Kotelny Island; (4) Inani River Valley and Omulev Mountains; (5) Seward Peninsula (Don River); (6) Chukotsk Massif; (CC) Cape Calhoun Formation, northwestern Greenland; (H) Harding Sandstone, Colorado; (KL) Klamath Mountains, California; (KAZ) Kazakhstan; (JR) Jones Ridge, Alaska; and (PER) Perce, Quebec.

INTRODUCTION

Monorakid trilobites of Middle or Late Ordovician age are restricted to the Siberian and Kolyma Platforms, USSR, and the Seward Peninsula, Alaska (Fig. 1). Despite earlier descriptions, monorakids probably are not present in northern Greenland or Colorado. Although somewhat different from the distribution envisioned by Whittington (1966, p. 726), and by Whittington and Hughes (1972, p. 225), occurrences of monorakids may be significant in suggesting the Ordovician paleopositions of the Seward Peninsula relative to the rest of Alaska and to the Kolyma and Siberian Platforms, and the paleolatitudes of these Asian plates.

Monorakids in the Siberian Platform

Monorakos was originally described by Schmidt (1886) based on 2 species collected from Ordovician strata exposed along the Stony Tunguska River (Fig. 1). Several other species and another monorakid genus, *Evenkaspis*, were described in this same area by Kramarenko (1952, 1957), and *Monorakos* was reported in Taimyr to the north by Balashova (1960).

Monorakids in the Kolyma Platform

To the east of the Siberian Platform, *Evenkaspis* was recorded (Maksimova, 1962, p. 117) on Kotelny Island in the New Siberian Islands (Fig. 1). Subsequently, Chugaeva and others (1964) identified *Monorakos mutabilis* in the Middle Ordovician strata of the Inani River valley, south of the Omulev Mountains (Fig. 1). Balashov and others (1968, p. 116-117, Pl. 42, Fig. 5) reported the same species in this area and listed the stratigraphic position as the Llandeilian Darpirski horizon, which we suspect might be Caradocian. Bogdanov (1963, p. 8, Figs. 1, 2) considered this terrain to be part of the western margin of the Kolyma Platform. The most easterly occurrence of *Monorakos* in the USSR is in the Chukotsk folded belt, directly across the Bering Straits from the Seward Peninsula (Oradovskaia, 1970, p. 100 and correlation chart 2).

Monorakos in the Seward Peninsula

Monorakos was first recognized in Alaska by Ross (cited in Sainsbury and others, 1971, p. 55) from one [68-ADu-18(USGS Coll. D2036CO)] of a series of collections made by Dutro and Churkin from Ordovician and Silurian rocks exposed near the Don River, Seward Peninsula (Figs. 1, 2). Ross identified a new species of *Mono-*

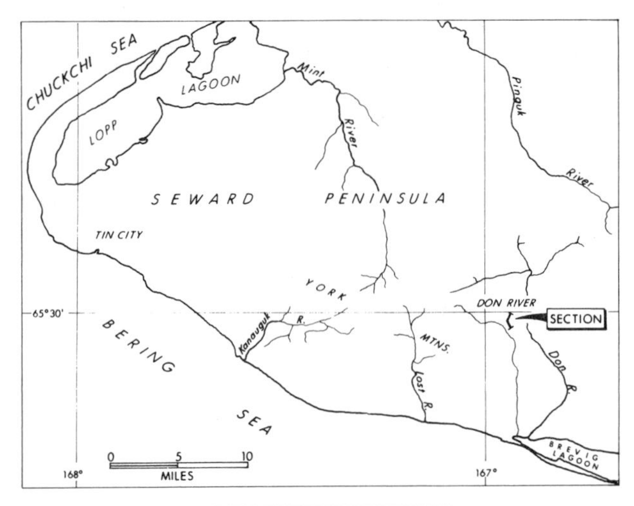

Figure 2. Map of part of Seward Peninsula, Alaska, showing location of Don River Section where *Monorakos* occurs and Lost River from which Early Ordovician trilobites were described by Ross (1965).

rakos in the collection, and also the trilobite *Remipyga* Whittington, 1954. Working independently, Ormiston in 1969 also recognized the presence of *Monorakos* in two samples collected 50 ft (15 m) apart, stratigraphically, from the Don River area. Apparently, this collection was made at essentially the same interval used by Dutro and Churkin. The higher sample contains both *Monorakos* and *Remipyga* (Fig. 3), the lower only *Monorakos*. The higher sample (Sample 11, Fig. 3) is considered to represent essentially the same horizon as collection 68-ADu-18 (USGS Coll. D2036CO) of Dutro and Churkin, from which Ross identified his trilobites.

The trilobites here assigned to *Remipyga* are represented only by cranidia. No pygidia are associated which can be assigned to any ceraurinid genus. As a result we cannot be certain that *Remipyga* is present. The shape of glabella, arrangement of glabellar furrows, and form of cephalic border are those of *Remipyga* (Whittington, 1954, p. 127-128). In our Alaskan specimens the eye is so positioned that the palpebral lobe is almost entirely between lateral glabellar furrows 2p and 3p. The center of the lobe is marked by a large pit and is very close to the dorsal furrow. The lateral glabellar lobes 1p are nearly isolated by furrows that curve inward and backward, a feature which is not demonstrated from *Ceraurinus marginatus* Barton, the type species of *Ceraurinus*, but which is present in *R. glabra* Whittington and *R. daedalus* Cox. In fact, the anterior position of the eye suggests a closer affinity to *R. daedalus* than to *R. glabra*. Species currently assigned to *Ceraurinus* in association with *Monorakos* should be reexamined to assess their possible assignment to *Remipyga*.

Location and Stratigraphy

The presence of fossiliferous Ordovician and Silurian strata in the vicinity of the Don River, Alaska (Fig. 2), has long been known, as indicated by faunal lists in Steidtmann and Cathcart (1922, p. 25-26). Sainsbury and others (1971, Figs. 2, 3) provided both a geologic map and a measured section for Ordovician and Silurian rocks near the Don River. Their measured section is from the same location as the one in this paper (Fig. 3) and covers an interval beginning and ending higher stratigraphically than our section, although the two partly overlap, particularly in the *Monorakos*-bearing part of the stratigraphic sequence. The top of our measured section is interpreted to equate with a point about 500 ft (150 m) above the base of their section where *Pentamerus* occurs with abundant favositid corals in dolomite, and the interval 425 to 450 ft (130 to 137 m) below the top of our measured section is equated with the base of their section. We equate the *Monorakos* and *Remipyga*-bearing Sample 11 (Fig. 3) with the faunally and lithologically similar Sample 68-ADu-18 (USGS Coll. D2036CO) which occurs 300 ft (91 m) above the base of their section (Sainsbury and others, 1971, Fig. 3). The carbonate rock from both collections can be described as an ostracode pelmicrite. J. M. Berdan (cited *in* Sainsbury and others, 1971, p. 55) provided a faunal list for the abundant ostracods from this horizon. Sample 10 (Fig. 3), which also yields *Monorakos*, is an ostracod-molluscan pelmicrite in which cephalopods, gastropods, and bivalves are conspicuous faunal elements.

Biostratigraphy

The association of *Remipyga* and *Monorakos* in our Sample 11 suggests a late Middle or Late Ordovician age. *Remipyga* is known from beds of late Middle or Late Ordovician age in Baffin Island (Whittington, 1954), of Middle or Late Ordovician age in the Northwest Territories (Ludvigsen, 1975, p. 682), and of Late Ordovician age in Siberia (Maksimova, 1962). *Monorakos* is known from the Late Ordovician Mangazeysky Stage (Caradocian), and possibly also from slightly younger Ordovician beds (Maksimova, 1962, p. 108) in the Soviet Union. A large part of Sample 11 was processed for conodonts, but only simple cones were recovered not diagnostic beyond a general Ordovician age (H. R. Lane, personal commun. 1974).

The stratigraphic position of Sample 68-ADu-18 (USGS Coll. D2036CO) supports a Late Ordovician age inasmuch as it lies above samples 68-ADu-16 and 17 which contain corals regarded by W. A. Oliver (cited *in* Sainsbury and others, 1971, p. 56) as Late Ordovician, and both it and Sample 11 (Fig. 3) lie below *Pentamerus*-bearing Silurian rocks. In fact, Sample 11 is only 75 ft (23 m) below Silurian dolomites (see Fig. 3).

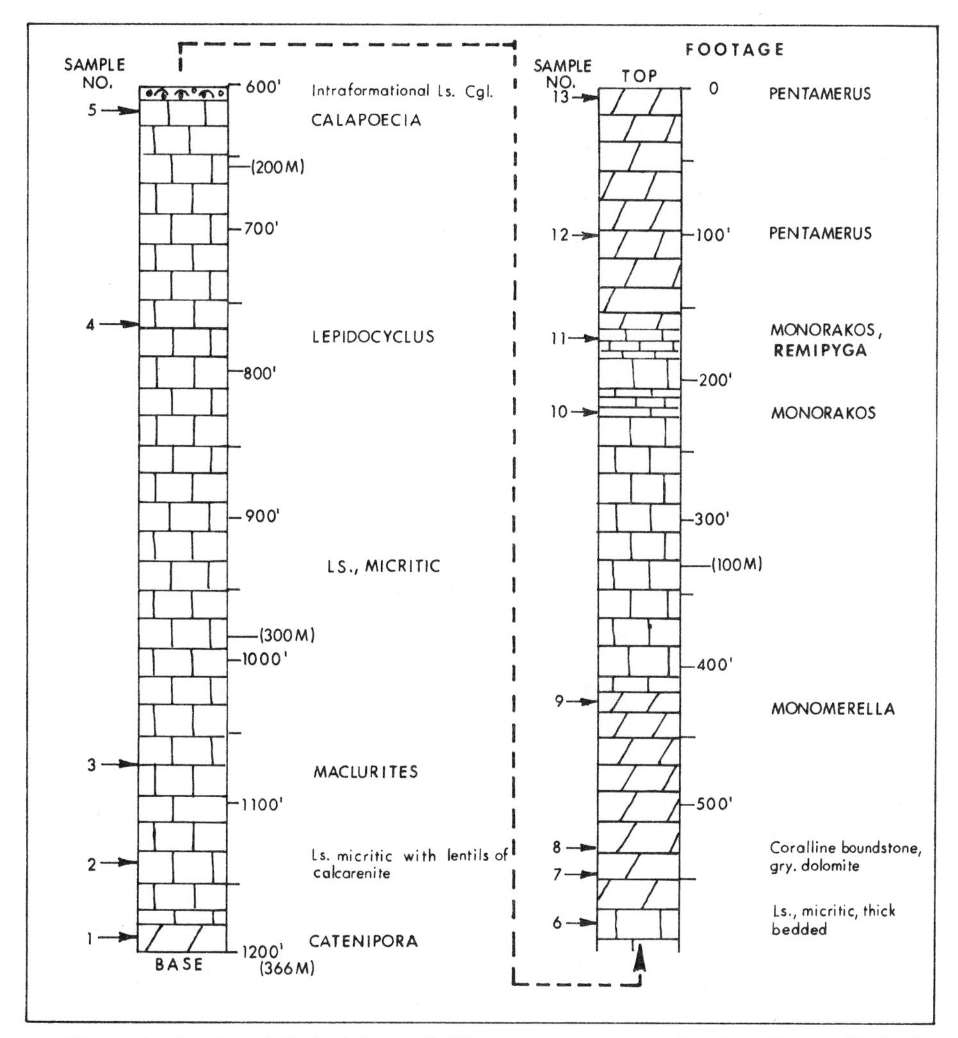

Figure 3. Section of Ordovician and Silurian strata measured at Don River, Alaska, by Ormiston in July, 1969 showing position of *Monorakos*-bearing samples. The right column vertically succeeds that on the left.

Older Ordovician trilobites were earlier described by Ross (1965) from an area just to the west of the Don River in the vicinity of Lost and Kanauguk Rivers (see Fig. 2). These trilobites are clearly Early Ordovician and considerably predate the *Monorakos* beds.

Revised Monorakid Assignments

Troedsson (1928, p. 79) assigned two trilobites from the Cape Calhoun Formation of northwestern Greenland to *Monorakos*. However, neither *M. wimani* Troedsson nor *M. schucherti* Troedsson exhibits three discrete pairs of lateral glabellar furrows expressed as deepened areas or pits within the prominent longitudinal furrow, and both have a well-developed transverse S2 glabellar furrow dividing the lateral lobe. We agree with Maksimova (1962, p. 108) that the Greenland species do not belong to *Monorakos* and should be considered pterygometopids. The genus *Isalaux*, originally established by Frederickson and Pollack (1952, p. 641) for material from the Harding Sandstone of Colorado, has been regarded as a monorakid by Wolfgang Struve (*in* Harrington and others, 1959, p. 494) and by Whittington and Hughes (1972, p. 255). This genus occurs in the same general region of east-central Siberia, as does *Monorakos*. Morphologically, however, it differs from the monorakids in lacking the three discrete pairs of pits in the longitudinal furrows, in having an interrupted longitudinal furrow, and in the presence of a prominent transverse S2 furrow. Maksimova (1962, p. 101) placed *Isalaux* in the family Pterygometopidae; its distribution in Siberia and Colorado may have special zoogeographic significance.

ZOOGEOGRAPHY

The occurrence of *Monorakos* (Fig. 5) along the Don River in the Seward Peninsula indicates a faunal tie between Alaska and the Chukotsk massif, the western Kolyma Platform, the New Siberian Islands, the Taimyr region, and the Stony Tunguska River valley of Siberia in the Middle and, possibly, Late Ordovician. Although the absence

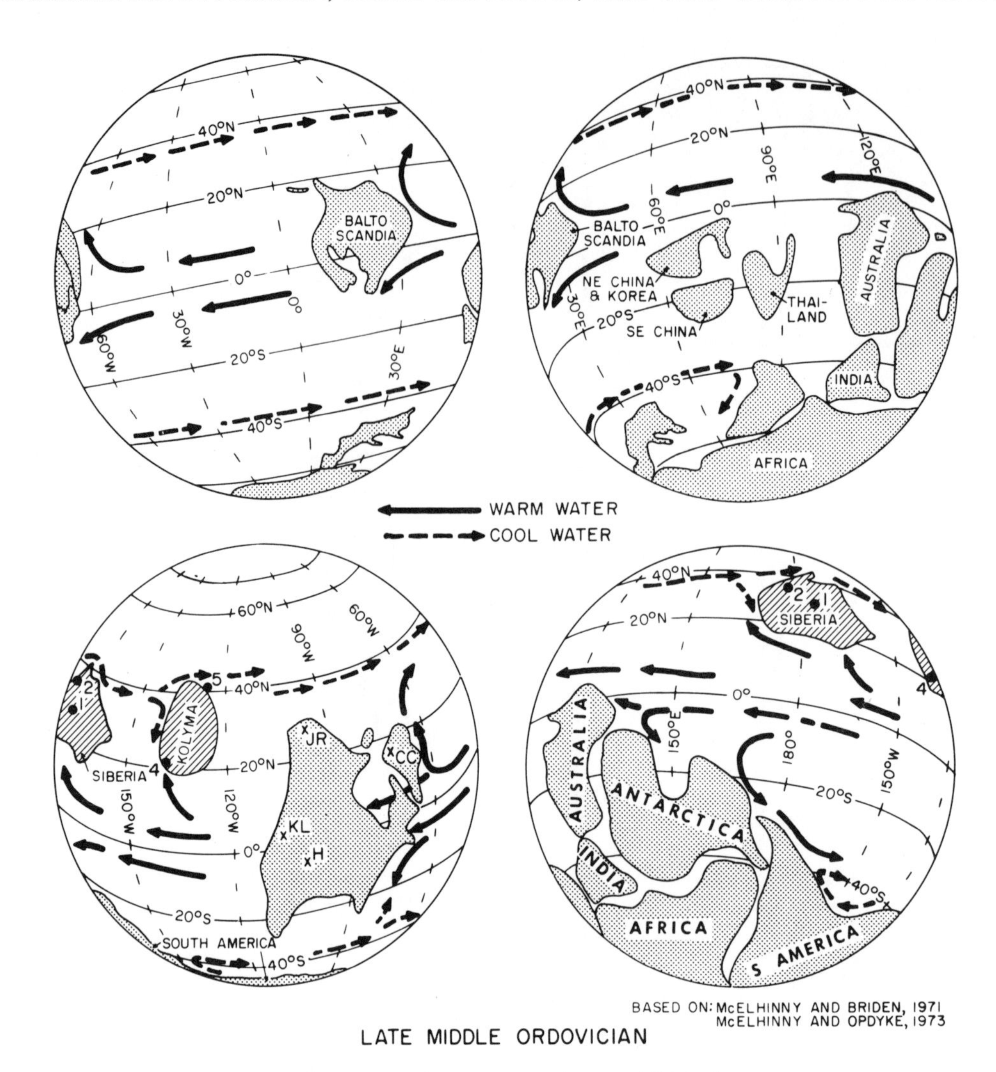

Figure 4. Hypothetical positions of continents in late Middle Ordovician time. Occurrences of monorakid trilobites and other localities as indicated in Figure 1.

of monorakids from elsewhere in Alaska may be more apparent than real, it strongly suggests to us that the Seward Peninsula was not a part of Alaska in Ordovician time. Similarly, we conclude that the tectonic plates with these monorakid-bearing strata were located in a very similar, if not the same, Ordovician environment, despite the late Paleozoic and Mesozoic age of sutures that now join them (Hamilton, 1970).

Whittington (1966, Fig. 16) showed the distribution of Caradocian-Ashgillian trilobite faunas and indicated monorakid genera to be restricted to most of the areas we have noted and (based on *Isalaux*) to the western United States (Colorado). From this distribution he inferred faunal interchange between the Soviet Union and the United States through western Alaska. The exclusion of *Isalaux* from the monorakids suggests that this inference was drawn for the wrong reasons, but the reported occurrence of *Isalaux* in the central Siberian Platform still poses a distributional problem.

Whittington and Hughes (1972, p. 255, Fig. 8) proposed that the distribution of monorakids (again with *Isalaux* in mind) might signify a subdivision of their Remopleuridid Province. That such a Monorakid Subprovince did exist, including the Seward Peninsula and excluding Colorado, seems obvious. Its paleogeographic significance may not be so clear.

Monorakid distribution is not in accord with Rozman's (1970, Figs. 65, 66) scheme of Middle and Late Ordovician faunas, because it occurs in parts of both her Arctic Canada-Siberia (I) and Kolyma-Alaska (IV) Provinces, but it is absent from all of Arctic Canada and most of Alaska. In fact, the brachiopod assemblage from Jones Ridge on the Alaska-Yukon boundary (Ross and Dutro, 1966), on which the Alaskan part of Rozman's Province IV is based, is very similar to assemblages from the Klamath Mountains of northern California (Boucot and others, 1973) and from Percé, Quebec (Schuchert and Cooper, 1930).

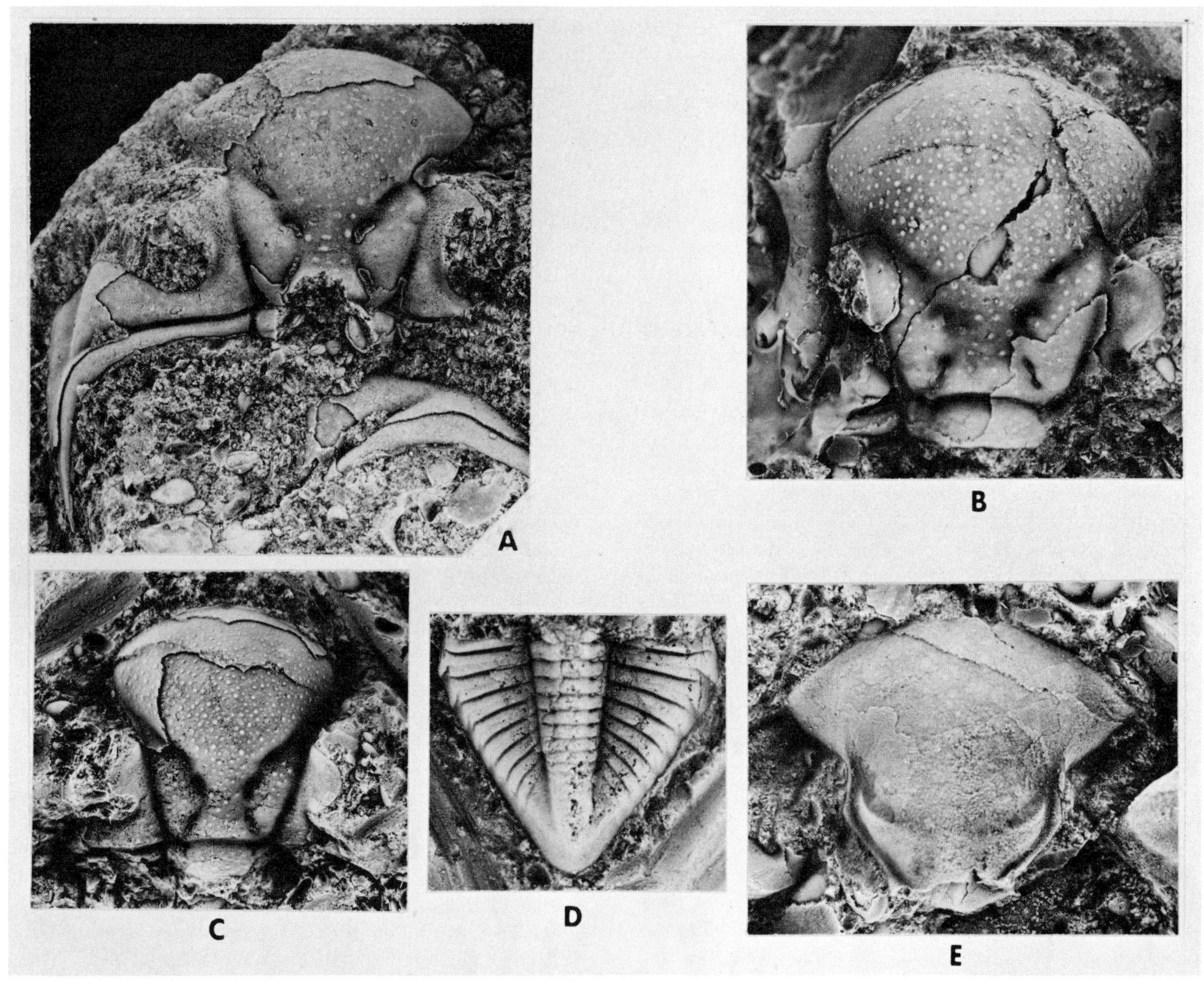

Figure 5. ***Monorakos*** **n. sp. All specimens are from the Don River section, Seward Peninsula, Alaska. (A) Dorsal view of cranidium, USNM 235773, X4; (B) Dorsal view of glabella showing crescentic 2S, USNM 235774, X5; (C) Dorsal view of cranidium, USNM 234782, X4; (D) Dorsal view of pygidum, USNM 235776, X5; (E) Exterior view of hypostome, USNM 235780, X5.**

As noted by Oradovskaia (1970, p. 103) and by Chugaeva (1972, p. 57), most of the trilobite genera associated with *Monorakos* are common to the Soviet Union and North America but *Monorakos* itself is not. Chugaeva (1973) suggested three trilobite provinces for the Lower Ordovician of the Northern Hemisphere, one of which is the American-Siberian Province. Her study emphasizes the close faunal ties between the combined Siberia-Kolyma Platforms and the North American Platform in the Early Ordovician. However, this generalization should be considered with caution when the Seward Peninsula is taken into account.

A small assemblage of Early Ordovician trilobites was described by Ross from the Lost River area (1965, Fig. 2) about 25 miles west of the Don River. This assemblage included *Triarthrus* sp., *Niobella, Nileus, Harpides,* and a species questionably referred to *Hystricurus*. This last species, *H.? sainsburyi* (Ross, 1965, p. 18, Pl. 8, Figs. 1-3, 5-7, 10, 11) may be congeneric with, or closely related to, *Nyaya novozemelica* Burskyi and *N. paichoica* Burskyi, both from Nova Zemlya (Burskyi, 1970, Pl. II, Figs. 1, 3, 4, 6). The trilobite fauna from Nova Zemlya includes all of these Lost River genera except *Triarthrus* and suggests a close faunal tie between Nova Zemlya, Alaska, and eastern Balto-Scandia in the Early Ordovician. However, the slightly younger fauna from Nova Zemlya also includes such North American genera as *Cybelurus* of the Toquima-Table Head "realm" of Ross and Ingham (1970). This evidence, like that derived from the distribution of the monorakids, suggests that the Seward Peninsula was not a part of Alaska until after Ordovician time, although there may have been faunal interchange between various parts of the Soviet Union and North America during the Ordovician.

The combined trilobite evidence from the Seward Peninsula and from the nearby Chukotsk massif might be cited in support of several theories (for instance, Hamilton, 1970, Fig. 6) on the movement of northern Alaska from a position somewhere near the Lomonosov Ridge, but that was a late Mesozoic or Cenozoic event that had little to do with the Ordovician whereabouts of a Monorakid Subprovince. The evidence reasonably supports Richards' theory (1974, p. 83, Figs. 3, 5) for the insertion of the Seward Peninsula into Alaska.

Using the benthic theory of trilobite migration, Whittington and Hughes (1972) concluded that a single large continental mass composed of the Kazakhstan area, and the Siberian and Kolyma Platforms lay adjacent to North America throughout the Ordovician. As noted by Hamilton (1970) and by Burrett (1973), impressive post-Ordovician suture zones lie between these components, making continental unity unlikely in the Ordovician. Nonetheless, all these components have marked faunal similarities with North America.

Excluding Kazakhstan, the north Asian area is characterized by occurrence of Middle to Late Ordovician monorakid trilobites not found anywhere else to date. It is possible that monorakids were endemic to a single Siberia-Kolyma continent that broke apart in the post-Ordovician time and was reconstituted in late Mesozoic or early Cenozoic time. If such a continent had been juxtaposed to North America astride the equator, the absence of monorakids in North America is unexplained.

In seeking a pelagic alternative to the benthic theory of trilobite migration of Whittington and Hughes (1972), Ross (1975a) relied heavily on paleomagnetic evidence to place Ordovician continents and on trilobite distributions to support the placements. In so doing he neglected the evidence of the monorakids, including his own Alaskan identifications (*in* Sainsbury and others, 1971, p. 55). Subsequently, paleomagnetic evidence for the Ordovician position of North America has much improved (McElhinny and Opdyke, 1973; Ross, 1975b, Fig. 1). It suggests that Alaska (exclusive of the Seward Peninsula and Brooks Range) should be moved from approximately 20°N in the Late Cambrian to 30°N in earliest Cincinnatian (Trenton) time and to 10°N in the Early Silurian as North America rotated counterclockwise. In other words, none of North America lay in high northern latitudes after the early Caradocian. As Hamilton (1970) noted, the subcontinent of the Siberian Platform seems to have straddled the equator in the Early Ordovician and arrived at 45°N in the Devonian. We, therefore, purpose a possible arrangement of continents in late Middle Ordovician time, as shown in Figure 4.

Warm, west-flowing equatorial currents could have spread North American faunal elements to the Kolyma, Siberian, and Balto-Scandic plates throughout the Early and early Middle Ordovician (Arenig-Llanvirn). We speculate that as the Siberian Platform (and presumably the Kolyma Platform and the Seward Peninsula) moved northward into the path of the cold, eastward-flowing currents at approximately 40°N, some species of the *Monorakos* fauna of Taimyr and the Stony Tunguska River Valley were carried to the Omulev Mountain and Don River areas in late Middle or Late Ordovician time (Caradocian). Perhaps no other shore lay far enough north in the path of the eastward-moving currents to be hospitable for these distinctive trilobites, which remained isolated on the only continent then present near lat 40°N. Corals occurring in the Don River section are cosmopolitan forms (W. A. Oliver, oral commun. 1975); although their presence does not necessitate a connection with North America, it does suggest that oceanic waters were not excessively cold. The presence of highly organic limestone at a paleolatitude of 40°N but not at 40°S may have been the result of contrasting continenal dispositions and patterns of ocean currents. Whereas an ice cap grew on a polar continent in the south, open oceanic circulation in the northern Arctic probably resulted in a far milder general climate.

CONCLUSIONS

The presence of *Monorakos*, previously unknown outside of the northern Soviet Union, in the Late Ordovician of the Seward Peninsula requires explanation. It is conceivable that only direct Ordovician faunal migration is indicated, but this explanation seems insufficient because of the absence of *Monorakos* from eastern Alaska where only North American trilobites are reported from Late Ordovician rocks (Jones Ridge).

A more likely explanation is that monorakids were highly endemic, occupying a circumscribed area, here termed the Siberia-Kolyma continent, of which Seward Peninsula was a part. Hamilton (1970) and Burrett (1973) have noted that post-Ordovician suture zones separate the proposed components of such a continent, apparently making their Ordovician unity unlikely. However, monorakid distributions suggest Ordovician unity of this mass (Siberian Platform, Kolyma, New Siberian Islands, Chukotsk, and the Seward Peninsula), followed by post-Ordovician breakup, and later (Mesozoic?) reunification of all components except for the Seward Peninsula which, as described in Richards' (1974) Asian Plate Theory, came to be fused with Alaska. The Ordovician separation between the North American continent and the Siberia-Kolyma continent (including the Seward Peninsula) was probably both latitudinal and oceanic.

REFERENCES

Balashov, Z. G. and others, 1968, Polevoi Atlas Ordovikskoi Fauni severo-vostoka SSSR (Field Atlas to Ordovician faunas of northeastern SSSR): Minister. Geol. RSFSR, 141 p.

Balashova, E. A., 1960, Sredne i Verkhneordovikskie i Nyzhnesiluriskie Trilobity Vostoschnogo Taimyra i ikh Stratigrafricheskoe Znastenie: Sbornik Statei po Paleontologii i Biostratigraffi, v. 17, p. 5-40.

Bogdanov, N. A., 1963, Tectonic development of Kolyma massif and eastern Arctic region during the Paleozoic: Akad. Nauk SSSR, Geol. Inst. Trudy, v. 99, 235 p. (in Russian).

Boucot, A. J. and others, 1973, Pre-late Middle Devonian biostratigraphy of the eastern Klamath belt, northern California: Geol. Soc. America, Abs. with Programs, v. 5, p. 15.

Burrett, Clive, 1973, Ordovician biogeography and continental drift: Palaeogeography, Palaeoclimatology, Palaeoecology, v. 13, p. 161-202.

Burskyi, A. Z., 1970, Early Ordovician trilobites of northern Pai-Khoi, *in* Opornyy razrez ordovika Pay-Khoya, Vaygacha i yuga Novoy Zemli: Leningrad, Nauch.-issled. Inst. Geol. Arkt., p. 96-131 (in Russian).

Chugaeva, M. N., 1973, Biogeography of the uppermost Lower Ordovician, *in* Biostratigraphy of the lower part of the Ordovician in the northeast USSR and biogeography of the uppermost Lower Ordovician: Akad. Nauk SSSR Trans., v. 213, p. 237-280 (in Russian).

Chugaeva, M. N., Rozman, K. S., and Ivanova, V. A., 1964, Comparative biostratigraphy of Ordovician deposits in the northeast of the USSR: Akad. Nauk SSSR, Geol. Inst. Trudy, v. 106, 236 p. (in Russian).

Frederickson, E. A., and Pollack, J. M., 1952, Two trilobite genera from the Harding Formation (Ordovician) of Colorado: Jour. Paleontology, v. 26, p. 641-644.

Hamilton, Warren, 1970, The Uralides and the motion of the Russian and Siberian Platforms: Geol. Soc. America Bull., v. 81, p. 2553-2576.

Harrington, H. J. and others, 1959, Trilobita, *in* Moore, R. C., ed., Treatise on invertebrate paleontology. Pt. O, Arthropoda 1: Geol. Soc. America and Univ. Kansas, p. O38-O539.

Kramarenko, N. N., 1952, Novye trilobity iz Silura Bassejna riki Podkamemnaja Tunguska: Akad. Nauk SSSR Doklady, v. 86, p. 401-404.

——— 1957, Novye predstaviteli ordoviskich trilobitov roda *Monorakos* Schmidt Sibirskoi Platformii: Akad. Nauk Paleont. Inst., "Osnovam Paleontologii," 1, p. 49-55.

Ludvigsen, Rolf, 1975, Ordovician formations and faunas, southern Mackenzie Mountains: Canadian Jour. Earth Sci., v. 12, p. 663-697.

Maksimova, Z. A., 1962, Trilobity Ordovika i Silura Sibirskoi Platformy: Vsesovizmyi Geol. Inst. (VSEGEI) Trudy, nov. ser., v. 76, 214 p.

McElhinny, M. W., and Briden, J. C., 1971, Continental drift during the Paleozoic: Earth and Planetary Sci. Letters, v. 10, p. 407-416.

McElhinny, M. W., and Opdyke, N. D., 1973, Remagnetization hypothesis discounted, a paleomagnetic study of the Trenton Limestone: Geol. Soc. America Bull., v. 84, p. 3697-3708.

Oradovskaia, M. N., 1970, Ordovician System, *in* Geology of the USSR, 30, northeastern USSR, geological description, Book 1, p. 80-104 (in Russian).

Richards, H. G., 1974, Tectonic evolution of Alaska: Am. Assoc. Petroleum Geologists Bull., v. 58, p. 79-105.

Ross, R. J., Jr., 1965, Early Ordovician trilobites from the Seward Peninsula, Alaska: Jour. Paleontology, v. 39, p. 17-20.

——— 1975a, Early Paleozoic trilobites, sedimentary facies, lithospheric plates and ocean currents, *in* Martinsson, A., ed., Evolution and morphology of the Trilobita, Trilobitoidea, and Merostomata: Fossils and Strata, no. 4, p. 307-329.

——— 1975b, Ordovician sedimentation in the western United States, *in* Bassett, M. G., ed., The Ordovician System: Palaeontological Assoc. Symposium, Birmingham 1974, Proc., p. 61-93.

Ross, R. J., Jr., and Dutro, J. T., 1966, Silicified Ordovician brachiopods from east-central Alaska: Smithsonian Misc. Colln., no. 149, 22 p.

Ross, R. J., Jr., and Ingham, J. K., 1970, Distribution of the Toquima-Table Head (Middle Ordovician Whiterock) faunal realm in the Northern Hemisphere: Geol. Soc. America Bull., v. 81, p. 393-408.

Rozman, K. S., 1970, Biostratigraphy and paleobiogeography of the Upper Ordovician of the northeast of the USSR, *in* Rozman, K. S., Ivanova, V. A., Krasilova, I. N., and Modzalevskaia, E. A., Upper Ordovician biostratigraphy of the northeastern USSR: Acad. Nauk SSSR Trans., v. 205, p. 212-270 (in Russian).

Sainsbury, C. L., Dutro, J. T., Jr., and Churkin, Michael, Jr., 1971, The Ordovician-Silurian boundary in the York Mountains, western Seward Peninsula, Alaska: U.S. Geol. Survey Prof. Paper 750-C, p. 52-57.

Steidtmann, Edward, and Cathcart, S. H., 1922, Geology of the York tin deposits, Alaska: U.S. Geol. Survey Bull. 733, 130 p.

Schmidt, F., 1886, Über einige neue ostsibirische Trilobiten und verwandte Thierformen: Acad. Imperial Sciences St. Petersbourg Bull. v. 30, p. 501-512.

Schuchert, Charles, and Cooper, G. A., 1930, Upper Ordovician and Lower Devonian stratigraphy and paleontology of Percé, Quebec Pts. I, II: Am. Jour. Sci., ser. 5, v. 20, p. 161-176, 265-288, 365-392.

Troedsson, G. T., 1928, On the Middle and Upper Ordovician faunas of northern Greenland: Medd. om Grønland, v. 72, 197 p.

Whittington, H. B., 1954, Ordovician trilobites from Silliman's Fossil Mount, *in* Miller, A. K., Youngquist, W. L., and Collinson, C. W., eds., Ordovician cephalopods of Baffin Island: Geol. Soc. America Mem. 62, p. 119-149.

——— 1966, Phylogeny and distribution of Ordovician trilobites: Jour. Paleontology, v. 40, p. 696-737.

Whittington, H. B., and Hughes, C. P., 1972, Ordovician geography and faunal provinces deduced from trilobite distribution: Royal Soc. London Philos. Trans., ser. B, Biol. Sci., v. 263, p. 235-278.

Swedish Late Ordovician Marine Benthic Assemblages and Their Bearing on Brachiopod Zoogeography

PETER M. SHEEHAN, *Department of Geology, Milwaukee Public Museum, Milwaukee, Wisconsin 53233*

ABSTRACT

Two contemporary faunas may differ either because they represent different communities (they live in differing local environments) or because they belong to different faunal provinces (they were separated historically, having evolved in separate regions). In this study, Benthic Assemblage analysis of brachiopods from Sweden reveals that only North European Province communities were present in the Viruan and early Harjuan. In the latter part of the Harjuan, Mediterranean Province communities entered the region. Initially deep, then progressively shallower environments were occupied by Mediterranean Province communities. Mediterranean Province communities apparently evolved in cold, high-latitude environments associated with a developing glacial period. It is postulated that cold, deep water moved into the Swedish region, bringing with it Mediterranean Province communities; as glacial climates intensified, cold water moved into shallower environments accompanied by shallower communities. A thermocline may have separated communities of the two Provinces.

INTRODUCTION

"Provinces may be defined as regions in which communities maintain characteristic taxonomic compositions" (Valentine, 1973, p. 377). The Late Ordovician brachiopod fauna of Europe has been studied since the early 19th Century. Preliminary examinations of the paleozoogeographic relationships of these faunas have been undertaken by several workers, but to date, the community ecology of the fauna has been ignored. I intend to examine the Benthic Assemblage relationships of north European faunas from Sweden in the hope that they will facilitate paleozoogeographic studies. The interval considered is the Upper Viruan and Harjuan Series of the Baltic region which is approximately equivalent to the Caradoc and Ashgill Series of Britain and the Upper Beroun and Králův Dvůr Series of Bohemia. Primary emphasis is on the latter part of the interval.

Spjeldnaes (1961, 1967), Dean (1967), Williams (1969, 1973), Jaanusson (1973), and Havlíček (1974) examined certain aspects of the zoogeographic distribution of European Late Ordovician brachiopods. Three provinces have been recognized: (1) the Mediterranean Province included southern Europe, North Africa, Bohemia, and possibly parts of Wales; (2) the North European Province included western Europe north from Belgium (Scandinavia, the Baltic countries, and the Russian Platform), and rimmed North America; characteristic faunas are known from open ocean regions marginal to the continent in Maine (Neuman, 1968), Gaspé (Schuchert and Cooper, 1930; Cooper, 1930; Cooper and Kindle, 1936), Alaska (Ross and Dutro, 1966), and California (Potter and Boucot, 1971); and (3) the North American Province which was centered in epicontinental seas on the North American plate but included the Trondheim region of Norway. For a brief interval, faunas characteristic of the North American Province invaded the Baltic region (Jaanusson, 1973), but that Province will not be considered in this study.

Paleogeography

At first sight, the geographic distribution of provinces in Europe seems bewildering. However, the paleogeography of Europe during the Paleozoic has been interpreted recently within the plate tectonic paradigm, and the Ordovician positions of landmasses which now comprise Europe go far in explaining the provincial patterns. The case for a mid-European ocean separating the plates of North Europe and South Europe-Africa has been made by a number of workers (Whittington and Hughes, 1972; Smith and others, 1973; Williams, 1973; McKerrow and Ziegler, 1972; Burrett, 1973). The proto-Atlantic Ocean, which separated the North European and North American plates in the Early Ordovician, was closing in the Late Ordovician (Wilson, 1966; Bird and Dewey, 1970; McKerrow and Ziegler, 1972). Burrett and Griffiths (in press) summarize this work and their analysis of the paleogeography of Europe during the Ordovician is accepted in this study. A polar projection map (Fig. 1) has been prepared to illustrate possible relative positions of landmasses considered in this study (Asia is omitted). The North European Province included the North European plate and extended around the nearby open ocean margins of North America. The Trondheim region was part of the North American plate. The Mediterranean Province was centered in high-latitude Gondwanaland.

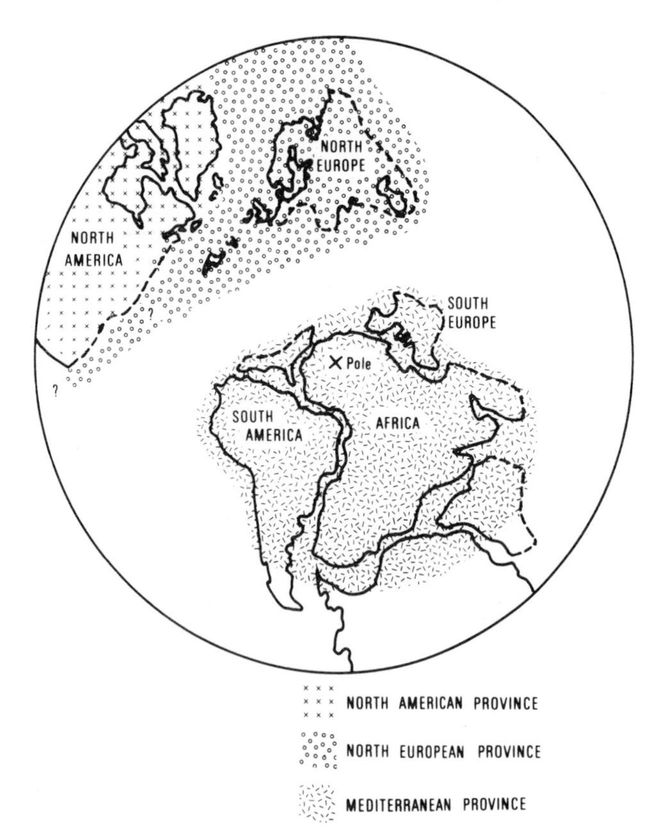

Figure 1. Polar view of Late Ordovician paleogeography and brachiopod provinces. Asia is omitted from the stereographic projection.

Mediterranean Province

Spjeldnaes (1961, 1967) and Dean (1967) discussed the Mediterranean Province and ascertained some of its characteristic brachiopods. Spjeldnaes (1961, 1967) attributed faunal patterns to a temperature gradient from warm in North Europe to cold in North Africa. He believed the Mediterranean Province was developed in high latitudes around a pole, located by paleomagnetic data in central Africa.

Since Spjeldnaes' original work, the presence of a Late Ordovician glacial episode has been established in North Africa (Beuf and others, 1971). More recent paleomagnetic data (Smith and others, 1973) places the pole in northwest Africa. The knowledge of brachiopod faunas of the Mediterranean Province has increased tremendously through the work of Havlíček (1970, 1971, 1974) in Bohemia, Morocco, and Libya. Mediterranean Province brachiopods were strongly endemic during the Caradoc and early Ashgill. Toward the end of the Ashgill some Mediterranean Province faunas were present in north Europe.

North European Province

Williams (1969, 1973) studied the world-wide brachiopod zoogeography in the Ordovician. His study is the most detailed account of the North European Province during the Late Ordovician. His approach was to select from earlier systematic studies only those which detailed brachiopod occurrences of two specific ages (late Caradoc and middle Ashgill). The scope of his work necessitated that some data on small faunas be ignored, and his study of two specific intervals naturally omitted a great deal of data on faunas of other ages in the Late Ordovician. Jaanusson (1973) presented brief summaries of the distribution of certain brachiopods, thereby extending the work of Spjeldnaes (1961, 1967) and Williams.

Provincial Ecology

Paleozoogeographic province analysis shares with modern zoogeographic studies the problem of determining whether the differences between taxa at two localities are due to provincial or merely community relationships. Paleozoogeographic analysis has an additional factor with which students of modern distributions do not contend —changes of taxonomic composition through geologic time. The time factor permits the paleozoogeographer to examine aspects of faunal changes that cannot be dealt with by examining modern faunas. However, the complexity added by the time factor makes it important to consider what Valentine (1973, p. 365) terms provincial ecology.

Williams (1973) noted the need to consider community relationships in zoogeographic analysis. By studying assemblages from a wide variety of rocks he attempted to include a representative suite of communities in each province. This methodology is unlikely to be truly representative of the community spectrum—witness, for example, the mid-Ashgill Bohemian Province of Williams (1969, 1973) which is based on collections from only one relatively deep-water community (see discussion below). To be certain that representative samples of the various communities of a province are included in a zoogeographic study, at least some preliminary analysis of communities is necessary.

The term Benthic Assemblage was proposed by Boucot (1975) to replace the marine benthic life zone of Berry and Boucot (1972; see also Watkins and others, 1973). Benthic Assemblages are groupings of benthic communities into parallel bands which generally are aligned perpendicular to the depth gradient. Though the Benthic Assemblages are depth-related, water depth *per se* is only a minor factor controlling distribution. Rather, the primary controls of community distribution, such as temperature, available light (for autotrophs), sediment type, and strength of water currents also vary with depth and it is this linkage which makes Benthic Assemblages depth-related.

Boucot (1975) has shown that six Benthic Assemblages can commonly be recognized in the middle Paleozoic. Benthic Assemblages 1 through 5 generally correspond to the *Lingula* through *Clorinda* "Community" positions of Ziegler (1965). Benthic Assemblage 6 is deeper than the *Clorinda* "Community," and marginal to the pelagic graptolite lithotope.

BENTHIC ASSEMBLAGES IN SWEDEN

The following section presents a Benthic Assemblage analysis of the Late Ordovician faunas of Sweden. Emphasis is on the brachiopod elements of the communities. This is the first synecologic analysis of this kind to be applied to Late Ordovician North European Brachiopod Province faunas. It must be emphasized that this is not an attempt to recognize specific communities but only to sort the faunas into parallel, depth-related bands. This synthesis covers an interval of six graptolite zones and it is to be expected that in any of the Benthic Assemblages numerous communities evolved and replaced prior communities. In addition, at any particular time numerous communities may have existed lateral to each other within each Benthic Assemblage.

The southern half of Sweden is one of the best-suited regions for synecologic studies in the North European Province. The Viruan and Harjuan rocks include a broad spectrum of ecologic habitats, from reefs and shallow-water carbonates to mudstones and graptolitic shales. The stratigraphic succession in the area of outcrop (Fig. 2) is well established, as is the correlation of the units (Fig. 3). The depositional patterns and bathymetric gradients of the Baltic Basin during the Viruan and Harjuan have been summarized by Männil (1966) in a series of twelve lithofacies maps. Männil's study provides an excellent model of the physical environment to use as a basis for synecologic analysis.

Principal areas of outcrop (Fig. 2) are in the Siljan region (Dalarna), Västergötland, Östergötland, and Skåne. The lower part of the sequence is preserved on Öland.

The time interval under study is divided into six graptolite zones in the Viruan and Harjuan Stages. Figure 3 presents a correlation of rock units that is taken largely from Männil (1966, Fig. 1), but certain changes have been made. The most important change is the extension of the *Dalmanitina* Beds in Östergötland upwards in the *Dicellograptus anceps* Zone (cf. Männil, 1966, Fig. 68); this is done to coincide with the most common usage in Sweden. The *Dalmanitina* Beds are missing in some sections in Östergötland but probably are present in other sections; current knowledge of the faunas does not permit accurate correlation of these beds. Following the suggestion of Dr. Jan Bergström, the Nittsjo Mudstones in Västergötland are included in the *Dalmanitina* Beds.

Many of the Viruan and Harjuan brachiopods from Sweden are poorly known. Much of the descriptive work was done in the 19th Century and is out-of-date. With few exceptions (e.g., Bergström, 1968a), recent work has involved only the compilation of species lists. During the preparation of this study, I examined the substantial brachiopod collections from Sweden housed in the Paleontological Institute, Lund University. From this examination it is clear that detailed synecologic descriptions (for instance, community relationships) must await taxonomic revison of the faunas. However, since it will be several years before there are any major additions to the knowledge of Viruan and Harjuan brachiopods, it seems reasonable to prepare a review of broad Benthic Assemblage relationships of the Swedish articulate brachiopods.

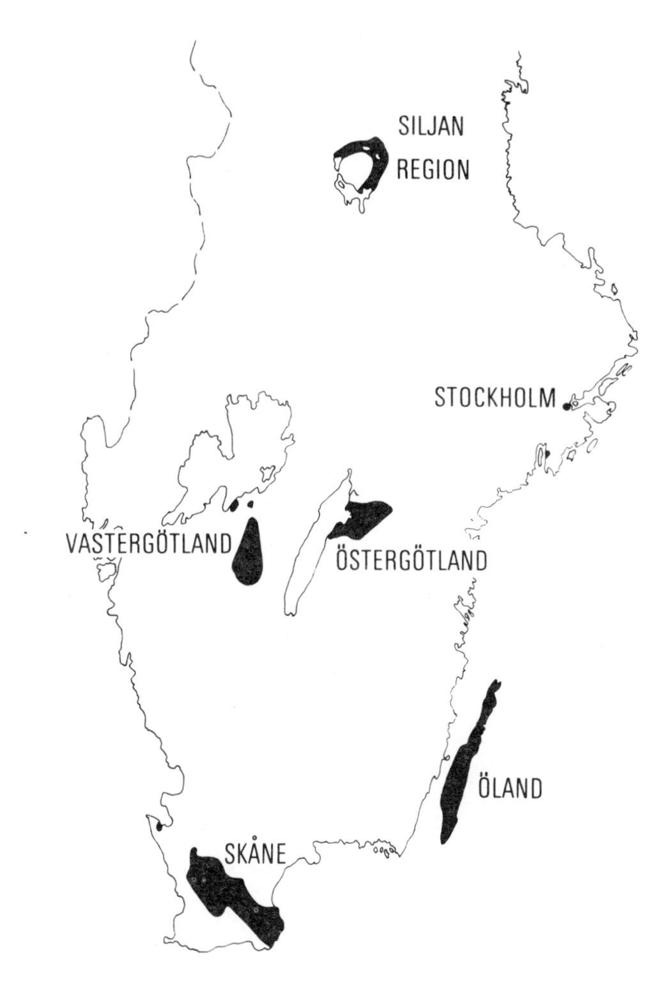

Figure 2. Areas in which Ordovician rocks outcrop in southern Sweden.

Männil (1966) recognized thirteen distinct sediment types in this area during the Viruan and Harjuan. For the purpose of this study, these sediment types are grouped into five major lithotopes: reef, detrital carbonate, noncarbonate siltstone, mud-clay, and graptolitic shale. Männil's work indicates that the reef, detrital carbonate, and siltstone lithotopes were deposited in relatively shallow water, the mudstone-clay in water of intermediate depth, and the graptolitic shale in relatively deep water. Each of the five lithotopes contains a distinctive assemblage of organisms. The lithotopes are generally arrayed in parallel bands that lie perpendicular to the bathymetric gradient.

The reef lithotope has an extremely diverse assemblage of algae, trilobites, articulate brachiopods, gastropods, cephalopods, tabulate corals, and others. Articulate brachiopods are very diverse and locally abundant. The detrital carbonate lithotope, deposited in relatively shallow water, has a diverse assemblage of trilobites, ostracodes, brachiopods, and other groups. Brachiopods are

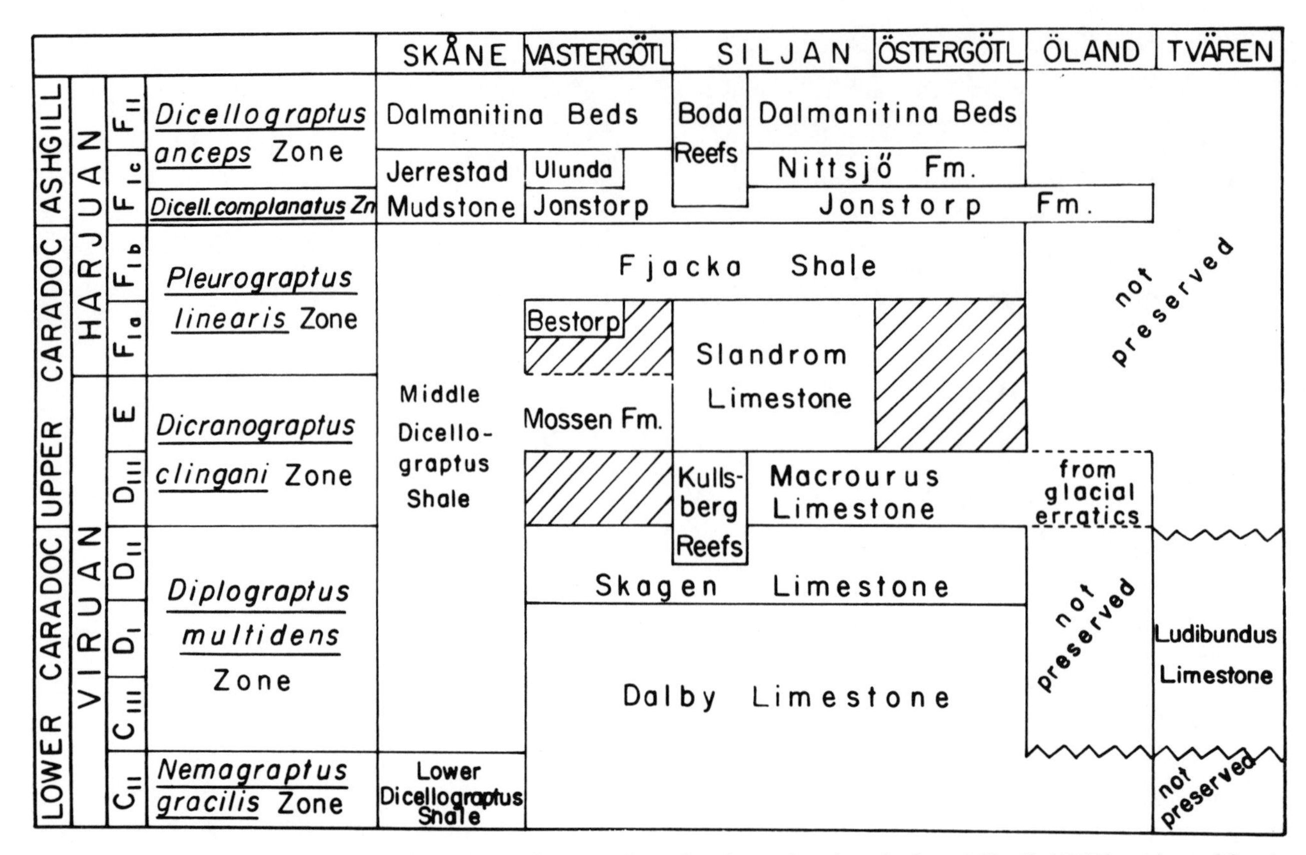

Figure 3. Correlation of Viruan and Harjuan rocks in southern Sweden, taken largely from Männil (1966) with modifications described in the text.

diverse, including many large species. The noncarbonate siltstone lithotope, deposited in moderately shallow water, has a fauna strongly dominated by diverse brachiopods, including many large species. The mud-clay lithotope, deposited in relatively deep water, has a fauna dominated by trilobites; ostracodes and graptolites may be abundant. Sedentary organisms are relatively uncommon. Brachiopods are uncommon, but may be moderately diverse, and commonly are very small size. The graptolitic shale lithotope, deposited in relatively deep water, has a fauna dominated by diverse graptolites which fell to the sea floor after death. Inarticulate brachiopods are locally abundant. Trilobites and ostracodes are occasionally found. Articulate brachiopods are scarce, represented by few species, and these are very small. Some of the brachiopods may have been epiplanktic, living attached to seaweed.

Assigning the faunas of the lithotopes to the numbered Benthic Assemblages of Boucot (1975) is straightforward. The specialized, extremely shallow-water communities of Benthic Assemblage 1 have not been recognized. On the other end of the depth spectrum, the graptolitic shale or pelagic community is well-represented. The mud-clay lithotope was developed adjacent to the graptolitic shale lithotope, and since its fauna is dominantly vagile it fits well into Benthic Assemblage 6. The detrital carbonate lithotope is assigned a Benthic Assemblage 2 to 5 position. The noncarbonate siltstone lithotope occurs only in the latest Ordovician and contains the well-known *Hirnantia* Community (see Cocks and Price, 1975). The *Hirnantia* Community grades laterally into deeper trilobite-dominated communities of Benthic Assemblage 5 to Benthic Assemblage 6 position (Lespérance and Sheehan, 1976). In other regions, the *Hirnantia* Community lies lateral to the Benthic Assemblage 3 *Holorhynchus* Community (Boucot, 1975). The range of the *Hirnantia* Community is thus limited to Benthic Assemblage 4 or possibly 5.

A series of maps (Fig. 4A-F) illustrates changes of Benthic Assemblage distribution through time. In the following section, representative examples of the fauna are given for each of the Benthic Assemblages. Comparison of the Benthic Assemblage maps with Männil's (1966) lithofacies maps shows the close relationship of Benthic Assemblages and sediment types.

Representative Benthic Assemblage Faunas

Reefs. Reefs are biologically constructed, rough-water structures which are best treated separately from the Benthic Assemblage bands.

Reef-building occurred in the Siljan region during two intervals (Fig. 3). The Kullsberg reefs were formed during the *D. multidens* and *D. clingani* Zones (Fig. 4B) and the Boda reefs were formed during the *D. complanatus* and *D. anceps* Zones (Fig. 4E-F). The reefs range from a few

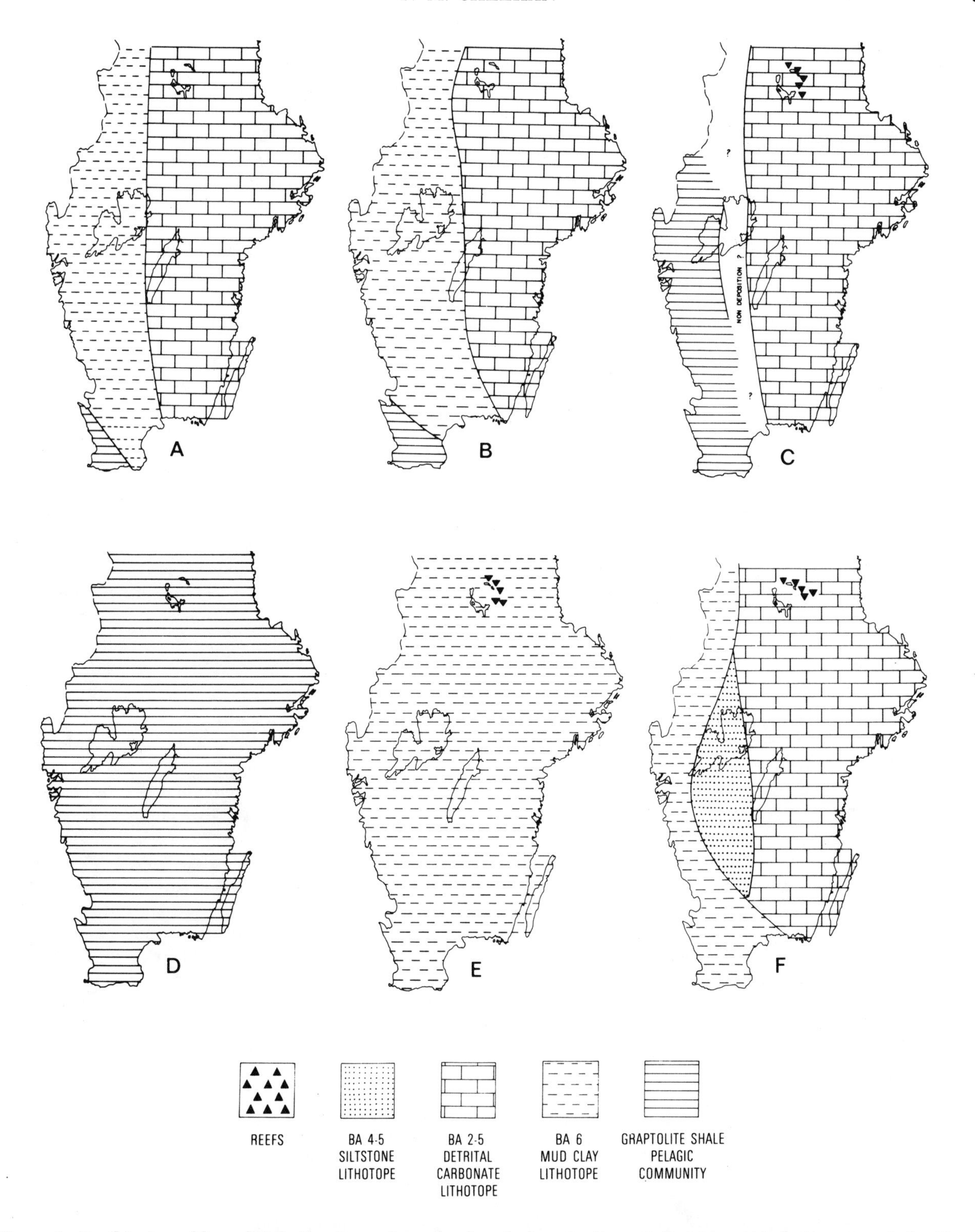

Figure 4. Benthic Assemblage distribution in southern Sweden during the Late Ordovician: (A) during part of the *Nemagraptus gracilis* Zone (compare with lithofacies patterns of Stage C_{II} in Männil, 1966 Fig. 57); (B) during part of the *Diplograptus multidens* Zone (compare with lithofacies patterns of Stage C_{III} in Männil, Fig. 58); (C) during part of the *Dicranograptus clingani* Zone (compare with lithofacies patterns of Stage D_{III} in Männil, Fig. 61); (D) during part of the *Pleurograptus linearis* Zone (compare with lithofacies patterns of Stage F_{Ib} in Männil, Fig. 65); (E) during part of the *Dicellograptus complanatus* Zone (compare with lithofacies patterns of Stage $F_{Ic}1$ in Männil, Fig. 66); (F) during part of the *Dicellograptus anceps* Zone (compare with lithofacies patterns of Stage $F_{Ic}2$ in Männil, Fig. 67). The entire fauna previously reported in the literature together with additional data from museum collections and collections made by me has been considered in preparing the maps. The data used in assembling the maps are available from me on request.

tens of meters in thickness to the largest reef (a Boda reef at Ösmundsberg) which is 1 km long and 100 m thick (Thorslund, 1960b). The reefs are broadly divisable into unbedded reef cores, and bedded, clastic reef-flank deposits. In the reef core the role of sediment binding is dominated by encrusting "coolinea" algae (Hadding, 1941).

Dominant elements of the Kullsberg reefs include dasyclad algae, trilobites, and articulate brachiopods; gastropods and cephalopods are less common (Hadding, 1941; Thorslund, 1960b). Brachiopods from the reef cores have been little studied.

Dominant elements of the reef flanks include articulate brachiopods, cystoids, bryozoans, and tabulate corals (Thorslund, 1960b). Thorslund (1936), Thorslund and Jaanusson (1960, p. 29), and Jaanusson (1958, p. 182-183) report the following brachiopods: *Nicolella* n. sp., *Dolerorthis* n. sp., *Platystrophia lynx* (Eichwald), *"Orthis"* cf. *lyckholmiensis* (Wys.), *Eoplectodonta* n. sp., *Actinomena* cf. *A. asmusi* (Vernueil), *Actinomena* n. sp., *Ptychoglyptus* n. sp. (Thorslund, 1936, P1. 1, Fig. 10), and *Leptaena* sp. (Thorslund, 1936, Pl. 1, Fig. 11).

The Boda reefs contain a fauna similar in overall composition to the earlier Kullsberg reefs, but it is almost entirely different at the genus-species level. The reefs are divisible into dense, nonbedded limestone of the reef core and varicolored, bedded, clastic limestones of the reef flanks (Thorslund, 1960a, 1960b).

Abundant organisms are sediment-binding algae, trilobites (see Warburg, 1925), articulate brachiopods, and dasyclad algae; gastropods, cephalopods, and tabulate corals are locally abundant (Hadding, 1941; Thorslund, 1960a, 1960b). As in the Kullsberg reefs, groups which are not common may be very diverse in numbers of species as exemplified by the pelecypods (Isberg, 1934) and cystoids (Regnéll, 1945). Thorslund (1960b) noted that fossils commonly occur in coquina "nests" of 1 species. Articulate brachiopods from the reef core are: *"Eoplectodonta" schmidti*, *"Sowerbyella"* n. sp., *Aphanomena luna, Cryptothyrella terebratulina, Cliftonia psittacina*, and *Athyris? portlockiana*. (The original reports and subsequent revisions of these species are given in Appendix A).

The reef-flank deposits are dominated by articulate brachiopods, cystoids, bryozoans, and tabulate corals (Thorslund, 1960a). Trilobites (see Warburg, 1925), pelecypods (see Isberg, 1934), crinoids, and streptelasmatid corals (Neuman, 1969) are also present. More plicate and costate brachiopods are found in the reef flanks than the reef cores.

Articulate brachiopods from the reef flanks are: ?*Spirigerina* (*Eospirigerina*) n. sp., *Nicolella* cf. *oswaldi*, *"Orthis"* cf. *lyckholmiensis*, *Glyptorthis* sp., *Ptychopleurella* n. sp., *Platystrophia lynx*, *Eoplectodonta* n. sp., *Hyattidina* sp. *Christiania* sp., *"Eoplectodonta" schmidti, Dicoelosia indenta, Dicoelosia transversa*, and *Epitomyonia glypha*. (The original reports and subsequent revisions of these species are given in Appendix B).

Additional brachiopods reported from the reefs but not assigned to the reef-core or reef-flank deposits are: *Camerella salteri, Leangella* cf. *scissa, Leptaenopoma trifidum, Parastrophina angulosa, Camerella rapa, "Orthis" concinna, "Orthis" conferta, "Orthis" umbo, Ptychoglyptus corrugatella, Strophomena imbrex, Spirigerina* (*Eospirigerina*) cf. *gaspeensis, Holorhynchus* cf. *giganteus, Dictyonella* sp., *Cyclospira*? sp., *Parastrophina lindstroemi, Oxoplecia* sp., and *Parastrophinella* sp. (The original reports and subsequent revisions of these species are given in Appendix C).

Benthic Assemblage 2 to 5: Detrital Carbonate Lithotope. The fauna of the detrital carbonate lithotope was developed in Benthic Assemblages 2 to 5 position in a variety of fine- to coarse-grained limestones, and was present at some time in all areas except Skåne. Calcarenites predominate, but calcilutites are also present. The assemblage includes a diverse group of both sessile and vagile organisms. Common bottom-dwelling groups are articulate brachiopods, bryozoa, locally abundant cystoids, and less common pelecypods and inarticulate brachiopods. Common vagile organisms are trilobites and ostracodes; less common ones are gastropods and orthoceratites. Articulate brachiopods are diverse and include both large and small species. Dominant groups are many plicate and coarsely ribbed orthacids, large and small dalmanellids, clitambonitacids, many large plectambonitacids, and strophomeniacids.

This lithotope supported many diverse communities during the Late Ordovician. The faunas can be segregated into separate communities for each Benthic Assemblage. However, grouping the communities together is adequate as a basis for examining communities as an aid to paleozoogeographic analysis. All faunas known from this lithotope were part of the North European Province. Representative examples from the *Nemagraptus gracilis* and *Dicellograptus anceps* Zones are given below. During the *N. gracilis* Zone, the detrital carbonate lithotope includes the lower part of the Dalby Limestone in the Siljan region, in Östergötland, and on Öland (Fig. 4A). Dominant faunal elements are brachiopods, trilobites, bryozoa, ostracodes, and, less commonly, cystoids. The brachiopods have not been adequately described, but Jaanusson (1960, 1962a, 1962b, 1963) made extensive collections from this lithotope and presented check lists of brachiopods.

In the Siljan region, the type section of the Dalby Limestone has been extensively studied by Jaanusson and Martna (1948) and Jaanusson (1963). Sediments are dominantly gray, fine-to-coarse calcarenities with some calcilutites; oöidlike grains are present in the lower part (Jaanusson, 1963). Jaanusson (1963, p. 36) noted that the grain size is considerably smaller than that of the Dalby Limestone in Östergötland, but the effect of this lithologic difference on the biota is uncertain because of the limited knowledge of Östergötland fossils. The brachiopods from the Siljan region listed by Jaanusson (1963) are presented in Table 1. An indication of the high brachiopod diversity is given, even though Jaanusson (1963, p. 36) noted that his faunal lists are provisional.

In Östergötland, the stratigraphic interval of the *N. gracilis* Zone has been described in the Smedsby Gård boring by Jaanusson (1962a), who noted that the dominant lithology is gray, coarse calcarenite, regularly bedded, and somewhat nodular. The fauna from the lower 10 m, which is taken to be included in the *N. gracilis* Zone, contains articulate brachiopods (Table 1), trilobites, ostracodes, and locally abundant cystoids (Jaanusson, 1962a).

On Öland, the youngest bedrock is the Lower Dalby Limestone of *N. gracilis* Zone age (Jaanusson, 1960). The rocks are coarse gray calcarenites with abundant sedentary organisms (Jaanusson, 1960, p. 277). From Böda Hamn Jaanusson (1960) listed a fauna dominated by articulate brachiopods (Table 1), trilobites, bryozoa, and cystoids. An additional brachiopod (*Delthyris subsulcata* Dalman, 1828, of unknown generic affinity) is known from this horizon.

Table 1. Articulate Brachiopods of the Detrital Carbonate Lithotope: *Nemagraptus gracilis* Zone

	Öland	Östergotland	Siljan Region
Nicolella demissa (Dalman, 1828, Pl. 2, Fig. 7)	X		X
Nicolella sp.		X	
Hesperorthis inostrancefi (Wysogorski)	X		
Schizoramma? sp.	X		
Platystrophia dentata (Pander)	X		(cf.)
Platystrophia sublimis Öpik	X		
Platystrophia n. sp.	X		
Onniella cf. *navis* (Öpik)	X		
Onniella sp.		X	X
Vellamo n. sp.	X		
Kullervo sp.	X		
Oxoplecia dorsata (Hisinger, 1837, Pl. 21, Fig. 14a-c)	X		X
Leptestia n. sp.	X		
Tetradontella biseptata Jaanusson, 1962	X	X	X
Sowerbyella (*Viruella*) cf. *minima* Röömusoks	X		
Bilobia musca (Öpik)	X		
Bilobia sp.			X
Eoplectodonta? sp.		X	
Anisopleurella? n. sp.	X		
Actinomena sp.	X		
Christiania cf. *holtedahli* Spjeldnaes	X		

Faunal lists are from the following publications: Öland-Böda Hamn (Jaanusson, 1960, p. 23-24); Östergötland-Smedsby Gard boring (Jaanusson, 1962a, p. 13-16); Siljan region—type locality of the Dalby Limestone (Jaanusson, 1963, p. 28-29). *O. dorsata* is also illustrated by Lindström (1880, Pl. 12, Figs. 24-25, Pl. 14, Figs. 33-37).

The uppermost Ordovician deposits in southern Sweden (*Dicellograptus anceps* Zone) are a lithologically diverse set of rocks referred to everywhere as the *Dalmanitina* Beds. In the Siljan region, the *Dalmanitina* Beds yield a fauna probably belonging to the detrital carbonate lithotope (Fig. 4F). The *Dalmanitina* Beds are described by Thorslund (1960b) as gray calcareous shales with a basal band of fine sandy limestone (clink-limestone). They contain a poorly known fauna of trilobites, brachiopods, and less common rugose corals (Neuman, 1969). Other groups recorded by Thorslund (1935) are orthoceratites, inarticulate brachiopods, ostracodes, *Hyolithus* sp., *Tentaculites*? sp., and conodonts. Brachiopods reported by Thorslund (1935) are: *Meristella "crassa," Leptaena* sp., *Sowerbyella* sp., *Dicoelosia* (longer type) [= *D. indenta* (Cooper) according to Wright, 1968b, p. 27], *Vellamo* sp., *Sowerbyella* cf. *5-costata* M'Coy, *Orthis* sp., *Platystrophia* sp., and *Dalmanella*? sp. These identifications are tenuous and can only be taken as an indication of the diversity of the brachiopod fauna.

Thus, a suite of communities was developed in the detrital carbonate lithotope with diverse sessile and vagile benthic elements. The brachiopods are poorly known, but seemingly diverse. Jaanusson (1972) noted that echinoderms are often underrepresented in faunal lists, and they may have been more abundant in some communities than indicated here.

Benthic Assemblage 4: Noncarbonate Siltstone Lithotope. The noncarbonate siltstone lithology was found only in the *D. anceps* Zone in Västergötland (Fig. 4F). Articulate brachiopods strongly dominate the assemblage. Trilobites and inarticulate brachiopods are found in small numbers.

Common articulate brachiopods are several large and small dalmanellacids, several large strophomenacids (commonly flat, but some resupinate), a davidsoniacid, a triplesiacid, a large, coarsely ribbed rhynchonelliacid, and a large, smooth athyridacid.

The articulate brachiopod fauna was described and illustrated by Bergström (1968a) who identified as abundant: *Hirnantia sagittifera, Kinnella kielanae, Cliftonia psittacina, Eostropheodonta hirnantensis, Coolinia dalmani,* and *Plectothyrella crassicosta;* as common: *Dalmanella testudinaria, Draborthis caelebs, Aphanomena schmalenseei,* and "*Cryptothyrella*" n. sp. (= *Hindella crassa* according to Sheehan, in press); as fairly common: *Horderleyella fragilis, Drabovia westrogothica, Leptaena rugosa,* and *Leptaenopoma trifidum* (? = *Leptaena* n. sp. according to Jan Bergström, personal commun.); and as rare: *Titanomena grandis, Kjerulfina*? sp., *Leangella* cf. *scissa, Giraldiella bella,* and an undescribed rhynchonellid (Bergström, 1968a, p. 6).

This fauna contains many genera derived from the Mediterranean Province to which it was assigned by Jaanusson (1973). Brachiopod-dominated faunas with many of these genera occur widely in Europe, the British Isles, and the northeastern margin of North America in the latest Ordovician. This assemblage has been referred to as the *Hirnantia* Community by Cocks and Price (1975) and Lespérance and Sheehan (1976). The deeper water (Benthic Assemblages 5 to 6) trilobite-dominated *Mucronaspis* Community (Lespérance and Sheehan, 1976) is associated with the *Hirnantia* Community in other regions. The essentially single taxon *Holorhynchus* Community (Benthic Assemblage 3) discussed by Boucot (1975) is a shallower water contemporary of the *Hirnantia* Community. However, the zoogeographic origin of *Holorhynchus* is not known.

Though both the detrital carbonate lithotope and the noncarbonate siltstone lithotope were developed during the *Dicellograptus anceps* Zone, it is not certain that both were deposited at the same time. It is possible that the *Hirnantia* Community is somewhat younger. Only a few of the *Hirnantia* Community species have been found in southern Sweden outside of Västergötland. Bergström (1968a) recorded *Cliftonia psittacina* and *Leangella* cf. *scissa* in the Boda reefs and *Dalmanella testudinaria* and *Leptaena rugosa* from the detrital carbonate lithotope at Borenshult in Östergötland. Bergström (1968a) also recorded *Leptaenopoma trifidum* from these areas, but he now believes these specimens belong to a new species (Bergström, personal commun.). The record of a single specimen of *Holorhynchus* in the Boda reefs (Isberg, 1934) provides evidence that some reef faunas were contemporaries of the *Hirnantia* Community.

It is interesting that the *Hirnantia* Community is a part of the Mediterranean Province whereas possibly contemporaneous communities in the detrital carbonate lithotope were part of the North European Province. In addition, there are only 4 species in common between the communities of the two Provinces and of these, 3 belong to the cosmopolitan genera *Leptaena, Leangella,* and *Cliftonia;* only *Dalmanella testudinaria* is an endemic of the Mediterranean Province.

Benthic Assemblage 6: Mud-Clay Lithotope. The mud-clay lithotope may or may not have a high carbonate content. The lithotope was widespread in all areas with the exception of Öland where the Late Ordovician sequence is only partially preserved.

The assemblage is characteristically dominated by vagile organisms of which trilobites predominate and ostracodes are locally abundant. Sessile organisms are scarce; articulate and inarticulate brachiopods are the most common forms. Frequently, pelagic graptolites are preserved.

Articulate brachiopods are commonly very small forms. The small plectambonitacid *Sericoidea* is ubiquitous. Very small dalmanellids, strophomenacids, and dayiacids are also commonly present. Overall, the articulate brachiopods are scarce, but locally they are quite abundant and may even dominate assemblages.

Communities in this life zone have been relatively little studied. In the Silurian, Ziegler and others (1968, p. 23) noted a trilobite-dominated fauna (Benthic Assemblage 6) between their deepest water shelly community (*Clorinda* Community) and the pelagic graptolite-dominated community; the Silurian fauna is in a position analogous to the mud-clay life zone with its trilobite-dominated communities.

Representative Benthic Assemblage 6 faunas from the *Nemagraptus gracilis, Diplograptus multidens*, and *Dicellograptus complanatus* Zones are discussed below.

During the *N. gracilus* Zone, the mud-clay lithotope was present in Västergötland (Fig. 4A) where it includes the Lower Member of the Dalby Limestone, which is a calcilutite (Jaanusson, 1964). The fauna, though poorly known, is summarized by Jaanusson (1964, p. 68-70) who noted that it is characterized by few large sedentary organisms with the exception of a cystoid. Trilobites dominate the assemblage; ostracodes are diverse, graptolites and cystoids relatively common. Articulate brachiopods and bryozoa are relatively uncommon. Jaanusson (1964) recorded the following brachiopods: *Skenidioides*? sp., *Onniella* sp., *Palaeostrophomena*? sp., and *Sericoidea restricta* (Hadding).

In Skåne, graptolite shale deposition was briefly interrupted during the *Diplograptus multidens* Zone by deposition of the Sularp Shale. The Sularp Shale is composed of beds of dark siliceous shale with interbeds of probable metabentonite (Lindström, 1953). Faunal lists and descriptions by Lindström (1953) and Olin (1906) reveal a fauna dominated by the brachiopod *Dalmanella bancrofti* (Lindström), with *Sericoidea restricta* (Hadding) and inarticulate brachiopods common, and with graptolites, ostracodes, trilobites, and hyolithids also present. This is an interesting assemblage because other occurrences of the mud-clay life zone commonly have *Dalmanella bancrofti* as a minor element, with trilobites and ostracodes dominating. Jaanusson (1964, p. 53) notes that many of the trilobites and ostracodes from the Fågelsång region of Skåne are common species in the Upper Member of the Dalby Limestone on Mt. Kinnekulle.

In the Siljan region, the Fjäcka Shale (*Pleurograptus linearis* Zone) is a dark brown to black, thin-bedded shale from which Jaanusson (1958, p. 41) recorded 7 species of trilobites, 4 species of graptolites, 2 species of brachiopods [*Resserella argentea* Hisinger (? = *Dalmanella bancrofti*) Lindström, 1880, Table 14, Figs. 12-15; and *Actinomena arachnoidea* (Lindström, 1880, Pl. 14, Figs. 41-42)]. Fossils in the Lund Paleontological Institution include *Glyptorthis* sp., *Sericoidea* sp., and *Drabovia* sp. The fauna is representative of the mud-clay lithotope, but graptolites are common, and it may be regarded as transitional with the graptolitic shale lithotope.

In Östergötland, graptolites are reported from the Fjäcka Shale (Skoglund, 1963); its fauna is otherwise unstudied. In the collections at Lund, brachiopods from the Fjäcka shale are: *Sericoidea* sp., *Draborthis*? sp., and *Drabovia* sp. The records of *Drabovia* sp. in the Fjäcka Shale are the earliest occurrence of a Mediterranean Province brachiopod in Sweden.

During the *Dicellograptus complanatus* Zone, the mud-clay lithotope spread throughout southern Sweden (Fig. 4E). In Skåne, the Jerrestad Formation (*Dicellograptus complanatus* Zone) is composed of gray mudstones. It contains a trilobite-dominated fauna. In the Lindergård boring in the Fågelsång district, Glimberg (1961) reported a fauna with trilobites, graptolites, inarticulate and articulate brachiopods, and ostracodes. The fauna is similar in the Koängen boring also in the Fågelsång district. Abundant articulate brachiopods described by Sheehan (1973a) from the Koängen boring are *Foliomena folium* and *Cyclospira? scanica;* much less common species are *Glyptorthis* sp., a dolerorthid, *Heterorthina?* sp., *Dedzetina* sp., *Anoptambonites* sp., *Leptestiina prantli, Eoplectodonta (Kozlowskites) ragnari, Sericoidea* sp., an aegiromeninid, and *Christiania nilssoni.*

Both the articulate brachiopods (Sheehan, 1973a) and trilobites (Ragnar Nilsson, oral commun.) are more closely allied with Bohemian faunas than with contemporary faunas from the North European Province. The only genera that are not known from the Mediterranean Province are *Glyptorthis* and *Christiania,* both of which are common in the North European Province.

In Västergötland, the Jonstorp Formation (*Dicellograptus complanatus* Zone) is (from bottom to top) a grey, and then a red mudstone. Henningsmoen (1948) recorded the following brachiopods in the Kullatorp core: *Sowerbyella?* cf. *restricta* (? = *Sericoidea restricta*), *Ptychoglyptus rosettana* (Henningsmoen, 1948, p. 396-399, P1, 24, Figs. 9-12), *Sowerbyella?* sp., and *Christiania* sp. Fossils from Västergötland in the collections of the Paleontological Institute at Lund University and in my collections include species of *Cyclospira?* and *Leangella.*

Valdar Jaanusson (personal commun.) kindly informed me that he has identified many species of the *Foliomena* Community in the Jonstorp Mudstones in Västergötland including species of *Foliomena, Leptestiina,* and *Christiania.*

Pelagic Community: Graptolitic Shale Lithotope. Pelagic graptolites are the most abundant fossils in the graptolitic shale. Sessile organisms are scarce; inarticulate brachiopods are the most common. A few articulate brachiopods may be present. Trilobites and ostracodes are the most abundant vagile organisms.

The articulate brachiopods are very small, and include species also found in the mud-clay lithotope. But their diversity is lower than in the mud-clay lithotope and they are less abundant numerically, compared to other organisms. Usually only 1 or 2 species are present. The most common form is *Sericoidea.* Small dalmanellids are also found.

Representative Pelagic Community faunas from the *Nemagraptus gracilis, Diplograptus multidens,* and *Pleurograptus linearis* Zones are discussed below.

The upper part of the Lower *Dicellograptus* Shale in central Skåne is of *N. gracilis* Zone age (Fig. 4A). It is largely a dark, thinly bedded graptolitic shale (Hadding, 1913; Hede, 1951). The fauna is dominated by diverse pelagic graptolites. Vagile organisms recorded by Hadding (1913) are a rather diverse trilobite assemblage, with several ostracodes, and 3 species of ?phyllocarids; sessile organisms are a few, small, commonly abundant inarticulate brachiopod species, and 1 articulate brachiopod, *Sericoidea restricta* (Hadding, 1913). Bergström (1968b) suggested that *Sericoidea restricta* may have been epipelagic, living attached to seaweed.

During the *Diplograptus multidens* Zone, the lithotope is represented in Skåne by the lower part of the Middle *Dicellograptus* Shale (Fig. 4B). The fauna is dominated by pelagic graptolites. Possible sessile organisms are common inarticulate brachiopods, and rare articulate brachiopods; vagile organisms (trilobites and ostracodes) are rare (Hadding, 1915). According to Hadding (1915), articulate brachiopods are *Leptaena sericea* var. *restricta* Hadding (? = *Sericoidea restricta*) and *"Orthis" argentia* (? = *Dalmanella bancrofti*).

During the *Pleurograptus linearis* Zone, the black shale interbeds of the Fjäcka Shale in Västergötland (Fig. 4D) contain a graptolite-dominated fauna. In the Kullstorp boring, the associated fauna recorded by Henningsmoen (1948, p. 377) has abundant inarticulate brachiopods, common ostracodes, 2 articulate brachiopods, *Sowerbyella?* cf. *restricta* (? = *Sericoidea restricta*) and *"Orthis" argentea* (? = *Dalmanella bancrofti*), rare trilobites, and indeterminate gastropods.

Mediterranean Province Communities in Sweden

No community analysis of Mediterranean Province brachiopods has been attempted to date. Havlíček (1974) draws a distinction between platform and deeper water miogeosynclinal areas but has not yet attempted to segregate the faunal assemblages into communities.

One distinctive community is important to recognize for comparison with other regions. The *Foliomena* Community in Sweden (see above) is closely comparable to a community preserved in the Králův Dvůr Shales (Sheehan, 1973a). It is this assemblage upon which Williams (1969, 1973) based his Bohemian Province. In Sweden, the Community was developed in Benthic Assemblage 6 position, between the graptolitic shale and detrital carbonate lithotopes. In Bohemia, the Community also occurs closely associated with graptolitic shales (Havlíček and Vanek, 1966). The Community has not been certainly identified elsewhere, but one of the characteristic and most easily identifiable genera, *Foliomena,* has been identified by me in Benthic Assemblage 6 position in Belgium (Sheehan, 1975b) and Gaspé, Québec. Shallow-water faunas in North Africa contain the earliest occurrences of many of the *Hirnantia* Community genera (Havlíček, 1970, 1971; Havlíček and Massa, 1973).

The preceding discussion of Benthic Assemblages reveals a trend in the changing zoogeographic patterns of Swedish brachiopods. With time, Mediterranean Province genera increased in abundance, first in deep then in shallower Benthic Assemblages. It is postulated here that the Mediterranean Province developed in high latitudes which became progressively colder during the Ordovician and that Mediterranean Province communities moved into Sweden following cold water masses that were at first deep then became progressively shallower as glaciation intensified toward the end of the Ordovician.

Throughout the Viruan and most of the Harjuan, Swedish brachiopods from all Benthic Assemblages were clearly allied with the North European Province. However, toward the end of the Harjuan, progressively more Mediterranean Province genera are recorded in Sweden. The first such genus is *Drabovia* which was found in the *Pleurograptus linearis* Zone, Benthic Assemblage 6, in both the Siljan region and Östergötland; it occurred with a North European Province fauna. During the *Dicellograptus complanatus* Zone, the Benthic Assemblage 6 *Foliomena* Community in Skåne (and probably Västergötland) was dominated by Mediterranean Province genera. By the *Dicellograptus anceps* Zone, the Benthic Assemblage 4 (possibly 5) *Hirnantia* Community was composed largely of Mediterranean Province genera.

The incursion of Mediterranean Province communities into Sweden was probably due to climatic changes associated with a developing glacial regime. Spjeldnaes (1961, 1967) postulated a climatic gradient which produced faunal temperature zonation during the Ordovician. Subsequent to Spjeldnaes' work, evidence has accumulated for Late Ordovician glaciation in North Africa (summarized by Beuf and others, 1971). Glacial maximum occurred in the latest Ordovician. Sheehan (1973b, 1975a) and Berry and Boucot (1973) tied in eustatic sea-level decline (associated with the glaciation) with drastic changes in North American faunas. Temperature changes were probably relatively unimportant to the North American faunal turnover. The equator passed through North America (Fig. 1), and, by analogy with the Pleistocene, tropical faunas were probably little effected by actual temperature decline, though the width of the tropical and subtropical bands may have narrowed (Newell, 1971). Northern Europe and the Mediterranean Province were in higher latitudes than North America (Fig. 1) and temperature gradients should have had a significant effect on faunas.

During the Cenozoic, warm, deep waters persisted until about the late Eocene-early Oligocene at which time the first cold, bottom water mass originated (Benson, 1975; Menzies and others, 1973; Savin and others, 1975). Modern cold abyssal faunas first developed at this time. Depth estimates of Benthic Assemblage 6 range from about 150 to 200 m (Boucot, 1975, p. 50) to as much as 1,500 m (Hancock and others, 1974). Even the maximum Benthic Assemblage 6 depth estimate (which I do not believe is as well substantiated as the estimate of Boucot) places Benthic Assemblage 6 much shallower than the modern abyssal environment. On the other hand, open ocean water at 200 m can be significantly colder than the surface water (Tait, 1968; Savin and others, 1975) and such a temperature gradient could have been the basis of Ordovician faunal differentiation in Sweden.

Minimum temperature at depth in an ocean is no less that the minimum surface temperature in the polar regions (Menzies and others, 1973). Therefore, as the high latitude Late Ordovician glaciation developed, it can be assumed that temperature stratification of the oceans intensified. Surface waters were progressively warmer toward the equator and temperature depth gradients must have been established. In the present low- to mid-latitude oceans a thermocline separates warm, shallow water from cold, deep water (Tait, 1968; Kinne, 1970). The thermocline commonly lies between 100 and 500 m (Tait, 1968). Faunas change markedly across the thermocline (Tait, 1968; Kinne, 1970; Menzies and others, 1973). Given the high-latitude glaciation a thermocline should have been present in the Ordovician oceans. The Mediterranean Province comunities may have lived below the thermocline and North European communities may have lived in warmer water above the thermocline (see Cook and Taylor, 1975, for a Cambrian example).

If present, the thermocline could have become shallower as glaciation intensified. The incursion of first deep then shallower cold-water communities into Sweden may reflect the shallowing of the thermocline as high-latitude marine surface temperatures declined.

Spjeldnaes (1961) drew attention to the absence of carbonate deposits in the Mediterranean Province and suggested this was due to the presence of cold water. Mediterranean Province faunas in both Sweden and Québec occur exclusively in carbonate-poor sediments.

Supportive evidence for the proposed zoogeographic development of Swedish faunas is provided by faunas from Gaspé, Québec, which have a similar history. In the Percé region of Gaspé, the zoogeographic affinities of faunas from the Whitehead Formation with the North European Province have been shown by Schuchert and Cooper (1930), Cooper (1930), and Cooper and Kindle (1936). The faunas are largely from limestones but diverse North European faunas have been identified in noncarbonate siltstones by me. At one locality early or middle Ashgill mudstones in Benthic Assemblage 6 position have yielded *Foliomena folium*. At the top of the Ordovician section, noncalcareous siltstones contain the *Hirnantia* Community (Lespérance, 1974; Lespérance and Sheehan, 1976). Mediterranean Province faunas have been found in North America only in Gaspé and Maine (Neuman, 1968). These were the highest latitude regions of North America and they faced the Mediterranean Province (Fig. 1).

If the *Hirnantia* Community is from cold, shallow-water Benthic Assemblage 4, its distribution might be restricted to one hemisphere. If the thermocline was in progressively deeper water toward the equator as it is today (Kinne, 1970), a depth-temperature barrier may have prevented movement of the *Hirnantia* Community into the tropical region, thus excluding the community from the opposite hemisphere. Notable here is that during the Late Ordovician, faunas of the North European Province occur in a band that crossed the equator in marginal areas of North America in Maine, Gaspé, Alaska, and California (see above) which suggests that the North European Province included warm regions. Therefore, it is reasonable to consider temperature as a barrier which might separate the two Provinces.

Patterns of graptolite distribution also indicate that cold waters moved progressively toward the equator during the Late Ordovician. Skevington (1974) proposed that graptolites retreated to lower and lower latitudes as cold water moved toward the equator, eventually restricting graptolites to the tropics.

ACKNOWLEDGMENTS

I would like to thank J. Bergström, W. B. N. Berry, A. J. Boucot, P. W. Bretsky, V. Jaanusson and J. G. Johnson for comments on an early version of this manuscript. R. Nilson kindly offered information on faunas from Skåne. The study began during a U.S. National Science Foundation Postdoctoral Fellowship at Lund University, and the work was completed with the aid of grants from the National Research Council of Canada. I would also like to thank Professor G. Regnéll of Lund University for making the facilities of the Paleontological Institution available to me.

REFERENCES

Benson, R. H., 1975, The origin of the psychrosphere as recorded in changes of deep sea ostracode assemblages: Lethaia, v. 8, p. 69-83.

Bergström, J., 1968a, Upper Ordovician brachiopods from Västergötland, Sweden: Geologica et Palaeontologica, v. 2, p. 1-35.

——— 1968b, Some Ordovician and Silurian brachiopod assemblages: Lethaia, v. 1, p. 230-237.

Berry, W. B. N., and A. J. Boucot, 1973, Glacio-eustatic control of Late Ordovician-Early Silurian platform sedimentation and faunal changes: Geol. Soc. America Bull., v. 84, p. 275-284.

Beuf, S., Biju-Duval, B., Decharpal, O., Rognon, P., Gariel, O., and Bennacef, A., 1971, Les grès du Paléozoique inférieur au Sahara—sédimentation et discontinuités, évolution structurale d'un craton: Inst. Français Pétrole—Science et Technique du Pétrol., no. 18, 464 p.

Bird, J. M., and Dewey, J. F., 1970, Lithospheric plate-continental margin tectonics and the evolution of the Appalachian orogen: Geol. Soc. America Bull., v. 81, p. 1031-1059.

Boucot, A. J., 1975, Evolution and extinction rate controls: Amsterdam, Elsevier, 427 p.

Boucot, A. J., and Johnson, J. G. 1967, Silurian and Upper Ordovician atrypids of the genera *Plectatrypa* and *Spirigerina:* Norsk Geol. Tidsskr., v. 47, p. 79-101.

Burrett, C., 1973, Ordovician biogeography and continental drift: Palaeogeography, Palaeoclimatology, Palaeoecology, v. 13, p. 161-201.

Burrett, C., and Griffiths, J., 1977, A case for a mid-European ocean, *in* Cogne, J., ed., La chaine varisque d'Europe Moyenne et Occidentale: Bur. Recherches Géol. Minières Mém. (in press).

Cocks, L. R. M., and Price, D., 1975, The biostratigraphy of the Upper Ordovician and Lower Silurian of southwest Dyfed, with comments on the *Hirnantia* fauna: Palaeontology, v. 18, p. 703-724.

Cook, H. E., and Taylor, M. E., 1975, Early Paleozoic continental margin sedimentation, trilobite biofacies, and the thermocline, western United States: Geology, v. 3, p. 559-562.

Cooper, G. A., 1930, New species from the Upper Ordovician of Percé: Am. Jour. Sci., ser. 5, v. 20, p. 265-288, 365-392.

Cooper, G. A., and Kindle, C. H., 1936, New brachiopods and trilobites from the Upper Ordovician of Percé, Québec: Jour. Paleontology, v. 10, p. 348-372.

Dalman, J. W., 1828, Uppstallning och Beskrifning af de i Sverige funne Terebratuliter: Kgl. Svenska Vetenskapsakad. Handl. 1827, p. 85-155.

Dean, W. I., 1967. The distribution of Ordovician shelly faunas in the Tethyan region, *in* Adams, C. G., and Ager, D. V., eds., Aspects of Tethyan biogeography: Systematics Assoc. Pub. 7, p. 11-44.

Glimberg, C. F., 1961, Middle and Upper Ordovician strata at Lindegård in the Fågelsång District, Scania S. Sweden: Geol. Fören. Stockholm Förh., v. 83, p. 79-85.

Hadding, A., 1913, Undre dicellograptusskiffern i Skåne jämte några därmed ekvivalenta bildningar: Lunds Univ. Årsskr., n.f., v. 9, p. 1-93.

——— 1915, Undre och mellersta dicellograptusskiffern i Skåne och å Bornholm: Dansk Geol. Foren. Medd., v. 4, p. 361-382.

——— 1941, The pre-Quaternary sedimentary rocks of Sweden: Lunds Univ. Årsskr., n.f., v. 37, p. 1-137.

Hancock, N. J., Hurst, J. M., and Fürsich, F. T., 1974, The depths inhabited by Silurian brachiopod communities: Geol. Soc. London Jour., v. 130, p. 151-156.

Havlíček, V., 1970, Heterorthidae (Brachiopoda) in the Mediterranean Province: Sbornik Geol. Ved. Paleont., v. 12, p. 7-39.

——— 1971, Brachiopodes de l'Ordovicien du Maroc: Notes et Mém. Serv. Géol. Maroc, no. 230, p. 1-135.

——— 1974, Some problems of the Ordovician in the Mediterranean region: Vestnik Ustredni Ustavu Geol., v. 49, p. 343-348.

Havlíček, V., and Massa, D., 1973, Brachiopodes de l'Ordovicien supérieur de Libye occidentale implications stratigraphiques régionales: Geobios, v. 6, p. 267-290.

Havlíček, V., and Vaněk, J., 1966, The biostratigraphy of the Ordovician of Bohemia: Sbornik Geol. Ved. Paleont., v. 8, p. 7-69.

Hede, J. E., 1951, Boring through Middle Ordovician-Upper Cambrian strata in the Fågelsång District, Scania (Sweden): Lunds Univ. Årsskr., n.f., v. 46, p. 1-84.

Henningsmoen, G., 1948, The Tretaspis Series of the Kullatorp core, *in* Waern, B., Thorslund, P., and Henningsmoen, G., Deep boring through Ordovician and Silurian strata at Kinnekulle, Västergötland: Uppsala Univ. Geol. Inst. Bull., v. 32, p. 374-432.

Isberg, O., 1934, Studien über Lamellibranchiaten des Leptaenakalkes in Dalarna: Lund, Hakan Ohlssons Buchdruckerei, 492 p.

Jaanusson, V., 1956, Untersuchungen über den Oberordovizischen Lyckholm-Stufenkomplex in Estland: Uppsala Univ. Geol. Inst. Bull., v. 36, p. 369-399.

——— 1958, Black Tretaspis Shale; Boda Limestone; Kullsberg Limestone; Leptaena Limestone; Macrourus Limestone, *in* Lexique Stratigr. Internat., Vol. 1, Europe, Fasc. 2c, Sweden, p. 41, 42-44, 182-183, 189-191, 219-221.

——— 1960. The Viruan (Middle Ordovician) of Öland: Uppsala Univ. Geol. Inst. Bull., v. 38, p. 207-288.

——— 1962a, The Lower and middle Viruan sequence in two borings in Östergötland, central Sweden: Uppsala Univ. Geol. Inst. Bull., v. 39, p. 1-30.

——— 1962b, Two plectambonitacean brachiopods from the Dalby Limestone (Ordovician) of Sweden: Uppsala Univ. Geol. Inst. Bull., v. 39, pt. 10, p. 1-8.

——— 1963, Lower and middle Viruan (Middle Ordovician) of the Siljan District: Uppsala Univ. Geol. Inst. Bull., v. 42, p. 1-40.

——— 1964, The Viruan (Middle Ordovician) of Kinnekulle and northern Billingen, Västergötland: Uppsala Univ. Geol. Inst. Bull., v. 43, p. 1-73.

——— 1972, Constituent analysis of an Ordovician limestone from Sweden: Lethaia, v. 5, p. 217-237.

——— 1973, Ordovician articulate brachiopods, *in* Hallam, A., ed., Atlas of palaeobiogeography: Amsterdam, Elsevier, p. 20-25.

Jaanusson, V., and Martna, J., 1948, A section from the Upper Chasmops Series to the Lower Tretaspis Series at Fjäcka Riverlet in the Siljan area, Dalarne: Uppsala Univ. Geol. Inst. Bull., v. 32, p. 183-193.

Kinne, O., 1970, Temperature, general introduction, *in* Kinne, O., ed. Marine ecology, Vol. 1: London, Wiley-Interscience, p. 321-346.

Lespérance, P. J., 1974, The Hirnantian fauna of the Percé area (Quebec) and the Ordovician-Silurian boundary: Am. Jour. Sci., v. 274, p. 10-30.

Lespérance, P. J., and Sheehan, P. M., 1976, Brachiopods from the Hirnantian Stage (Ordovician-Silurian) at Percé, Quebec: Palaeontology, v. 19, p. 719-731.

Lindström, G., 1880, *in* Angelin, N. P., and Lindström, G., Fragmenta Silurica e dono Caroli Henrici Wegelin. Holmiae: Stockholm, Samson and Wallin, p. 1-60.

Lindström, M., 1953, On the lower Chasmops beds in the Fågelsång District (Scania): Geol. Fören. Stockholm Förh. v. 75, p. 125-148.

Männil, R. M., 1966, Evolution of the Baltic Basin during the Ordovician: Tallinn, EESTI NSV Tead. Akad. Geol. Inst., p. 1-199.

McKerrow, W. S., and Ziegler, A. M., 1972, Palaeozoic oceans: Nature, v. 240, p. 92-94.

Menzies, R. J., George, R. Y., and Rowe, G. T., 1973, Abyssal environment and ecology of the world oceans: New York, Wiley-Interscience, 488 p.

Neuman, R. B., 1968, Paleogeographic implications of Ordovician shelly fossils in the Magog belt of the northern Appalachian region, *in* Zen, E-An, White, W. S. Hadley, J. B., and Thompson, J. B., Jr., eds., Studies of Appalachian geology: Northern and maritime: New York, Wiley-Interscience, p. 35-48.

Neuman, B., 1969, Upper Ordovician streptelasmatid corals from Scandinavia: Uppsala Univ. Geol. Inst. Bull., n.s., v. 1, p. 1-73.

Newell, N. D., 1971, An outline of tropical organic reefs: Am. Mus. Novitates, no. 2465, p. 1-37.

Olin, E., 1906, Om de chasmopskalken och trinucleusskiffern motsvarande bildningarna i Skåna: Lunds Univ. Årsskr., n.f., v. 2, p. 1-79.

Potter, A. W., and Boucot, A. J., 1971, Ashgillian, Late Ordovician brachiopods from the eastern Klamath Mountains, California: Geol. Soc. America Abs. with Programs, v. 3, p. 180-181.

Regnéll, G., 1945, Non-crinoid Pelmatazoa from the Paleozoic of Sweden: Lunds Geol.-Mineral. Inst. Medd., no. 108, 255 p.

Ross, R. J., and Dutro, J. T., 1966, Silicified Ordovician brachiopods from east-central Alaska: Smithsonian Misc. Colln., v. 149, p. 1-22.

Savin, S. M., Douglas, R. G., and Stehli, F. G., 1975, Tertiary marine paleotemperatures: Geol. Soc. America Bull., v. 86, p. 1499-1510.

Schuchert, C., and Cooper, G. A., 1930, Upper Ordovician and Lower Devonian stratigraphy and paleontology of Percé, Quebec, Pt. 1. Stratigraphy and faunas: Am. Jour. Sci., ser. 5, v. 20, p. 161-176.

Sheehan, P. M., 1973a, Brachiopods from the Jerrestad Mudstone (early Ashgillian, Ordovician) from a boring in southern Sweden: Geologica et Palaeontologica, v. 7, p. 59-76.

——— 1973b, The relation of Late Ordovician glaciation to the Ordovician-Silurian changeover in North American brachiopod faunas: Lethaia, v. 6, p. 147-154.

——— 1975a, Brachiopod synecology in a time of crisis (Late Ordovician-Early Silurian): Paleobiology, v. 1, p. 205-212.

——— 1975b, Late Ordovician brachiopods from Belgium: Geol. Soc. America Abs. with Programs, v. 7, p. 1267.

——— 1977, Late Ordovician and earliest Silurian meristellid brachiopods from Scandinavia: Jour. Paleontology, v. 51 (in press).

Skevington, D., 1974, Controls influencing the composition and distribution of Ordovician graptolite faunal provinces, *in* Rickards, R. B., Jackson, D. E., and Hughes, C. P., eds., Graptolite studies in honour of O. M. Bulman: Spec. Papers Palaeontology no. 13, p. 59-73.

Skoglund, R., 1963, Uppermost Viruan and lower Harjuan (Ordovician) stratigraphy of Västergötland and lower Harjuan graptolite faunas of central Sweden: Uppsala Univ. Geol. Inst. Bull., v. 42, p. 1-55.

Smith, A. G., Briden, J. C., and Drewry, G. E., 1973, Phanerozoic world maps, *in* Hughes, N. F., ed., Organisms and continents through time: Spec. Papers Palaeontology no. 12, p. 1-42.

Spjeldnaes, N., 1961, Ordovician climatic zones: Norsk Geol. Tidssk., v. 41, p. 45-77.

——— 1967, The palaeogeography of the Tethys region during the Ordovician, *in* Adams, C. G., and Ager, D. V., eds., Aspects of Tethyan biogeography: Systematics Assoc. Pub. 7, p. 45-57.

Tait, R. V., 1968, Elements of marine ecology: London, Butterworths, 272 p.

Thorslund, P., 1935, Über den Brachiopodenschiefer und den Jungeren Riffkalk in Dalarna: Nova acta Regiae soc. scientiarum upsaliensis, ser. 4, v. 9, no. 9, p. 1-48.

——— 1936, Silijansområdets Brännkalkstenar och Kalkindustri: Sveriges Geol. Undersökning Årsb., ser. C, no. 398, p. 1-64.

——— 1960a, The Cambro-Silurian, *in* Magnusson, H. H., Thorslund, P., Brotzen, F., Asklund, B., and Kulling, O., Description to accompany the map of the pre-Quaternary rocks of Sweden: Sveriges Geol. Undersökning Årsb., ser. Ba, no. 16, p. 69-110.

——— 1960b, Notes on the geology and stratigraphy of Dalarna, *in* Thorslund, P., and Jaanusson, V., The Cambrian, Ordovician, and Silurian in Västergötland, Närke, Dalarna, and Jämtland, central Sweden: Internat. Geol. Cong., 21st, Norden 1960, Guide to excursions nos. A23 and C18 (Guidebook e), p. 23-27.

Thorslund, P., and Jaanusson, V., 1960, The Cambrian, Ordovician, and Silurian in Västergötland, Närke, Dalarna, and Jämtland, central Sweden: Internat. Geol. Cong., 21st, Norden 1960, Guide to excursions nos. A23 and C18 (Guidebook e), 51 p.

Valentine, J. W., 1973, Evolutionary paleoecology of the marine biosphere: Englewood Cliffs, N. J., Prentice Hall, 511 p.

Warburg, E., 1926, The trilobites of the Leptaena Limestone in Dalarne: Uppsala Univ. Geol. Inst. Bull., v. 17, p. 1-446.

Watkins, R., Berry, W. B. N., and Boucot, A. J., 1973, Why "Communities"?: Geology, v. 1, p. 55-58.

Whittington, H. B., and Hughes, C. P., 1972, Ordovician geography and faunal provinces deduced from trilobite distribution: Royal Soc. London Philos. Trans, ser. B., Biol. Sci., v. 263, p. 235-278.

Williams, A., 1969, Ordovician faunal provinces with reference to brachiopod distribution, *in* Wood, A., ed., The Pre-Cambrian and Lower Paleozoic rocks of Wales: Cardiff, Univ. Wales Press, p. 117-154.
——— 1973, Distribution of brachiopod assemblages in relation to Ordovician palaeogeography, *in* Hughes, N. F., ed., Organisms and continents through time: Spec. Papers Palaeontology no. 12, p. 241-269.
Wilson, J. T., 1966, Did the Atlantic close and then re-open?: Nature, v. 211, p. 676-681.
Wiman, C., 1907, Über die Fauna des Westbaltischen Leptaenakalks: Arkiv Zoologi, v. 3, p. 1-20.
Wright, A. D., 1968a, A new genus of dicoelosiid brachiopod from Dalarna: Arkiv Zoologi, ser. 2, v. 22, p. 127-138.
——— 1968b, The brachiopod *Dicoelosia biloba* (Linnaeus) and related species: Arkiv Zoologi, ser. 2, v. 20, p. 261-319.
——— 1974, A revision of the Upper Ordovician brachiopod *"Pentamerus angulosus* Törnquist": Geol. Fören. Stockholm Förh., v. 96, p. 237-246.
Ziegler, A. M., 1965, Silurian marine communities and their environmental significance: Nature, v. 207, p. 270-272.
Ziegler, A. M., Cocks, L. R. M., and Bambach, R. K., 1968, The composition and structure of Lower Silurian marine communities: Lethaia, v. 1, p. 1-27.

APPENDIX

The original citations together with revised identifications of the brachiopods from the Boda reefs are discussed more fully below.

A. Articulate brachiopods from the reef cores include ones cited by Jaanusson (1958, p. 42-44): *"Sowerbyella" schmidti* (placed in *"Eoplectodonta"* by Jaanusson, 1960), *"Sowerbyella"* n. sp., *Actinomena luna* (Lindström, 1880, Pl. 14, Figs. 25-26; see Thorslund, 1936, Pl. 2, Fig. 14; placed in *Aphanomena* by Bergström, 1968a, p. 13), and *Meristella? terebratulina* (see Jaanusson, 1956, Pl. 1, Figs. 1-6; placed in *Cryptothyrella* by Sheehan, in press).

In addition, *Cliftonia psittacina* (Wahlenberg) (see Bergström, 1968a, p. 11-12) occurs in the reef core. This species was called *Atrypa altijugata* by Lindström (1880, p. 23, Pl. 13, Figs. 9-13) (see Bergström, 1968a, p. 11-12). It was incorrectly recorded by Boucot and Johnson (1967, p. 88) as being found on Gotland and incorrectly assigned to *?Spirigerina.*

Athyris? portlockiana Lindström (1880, p. 22, Pl. 13, Figs. 20-22) (non Davidson) is present in reef-core deposits in the Anderarvet reef (personal observation).

B. Brachiopods from the reef-flank deposits include those listed by Jaanusson (1958, p. 42-44): *Plectatrypa* n. sp. [? = *Spirigerina* (*Eospirigerina*)], *Nicolella* cf. *oswaldi* (Buch.), *"Orthis"* cf. *lyckholmiensis* Wys., *Glyptorthis* sp., and *Ptychopleurella* n. sp.

Additional articulate brachiopods listed by Thorslund and Jaanusson (1960, p. 31) are: *Platystrophia lynx, Eoplectodonta* n. sp., *Hyattidina* sp., *Christiania* sp., and *"Eoplectodonta" schmidti* (Lindström).

Wright (1968a, 1968b) recorded: *Dicoelosia indentata* (Cooper) (Wright, 1968b, Pl. 4, Figs. 3-13), *Dicoelosia transversa* Wright (1968b, Pl. 5, Figs. 1-5), and *Epitomyonia glypha* Wright (1968a, p. 128, Pl. 1, Figs. 1-16).

C. Several additional authors have reported brachiopods from reef-core or reef-flank deposits.

Wiman (1907, p. 6-7, Pl. 2, Figs. 4, 5) recorded *Camerella salteri* Davidson.

Bergström (1968a) recorded *Leangella* cf. *scissa* (Davidson) and *Leptaenopoma trifidum* Marek and Havlíček (Bergström, 1968a, p. 16). The latter may = *Leptaena* sp. of Thorslund (1936, Pl. 2, Fig. 12) which is probably a new species of *Leptaena* (Jan Bergström, personal commun.).

Lindström (1880) recorded *Camerella angulosa* Lindström (1880, p. 23, Pl. 13, Figs. 14-19) (= *Parastrophina angulosa,* see Wright, 1974), *Camerella rapa* Lindström (1880, p. 23-24, Pl. 12, Figs. 1-5), *Orthis concinna* Lindström (1880, p. 26, Pl. 13, Figs. 26-31), *O. conferta* Lindström (1880, p. 26, Pl. 13, Figs. 1-3), *O. umbo* Lindström (1880, p. 27, Pl. 14, Figs. 18-22), *Strophomena corrugatella* Davidson (Lindström, 1880, p. 28-29, Pl. 14, Fig. 24), *Ptychoglyptus corrugatella,* and *Strophomena imbrex* Pander (Lindström, 1880, p. 29, Pl. 14, Figs. 27-32).

Boucot and Johnson (1967, p. 88) reported *Spirigerina* (*Eospirigerina*) cf. *gaspeensis* from the Boda reef at Ösmundsberg.

Isberg (1934) figured *Holorhynchus* cf. *giganteus* Kiaer from the Kallholn reef.

Wright (1974) recorded *Dictyonella* and *Cyclospira* and illustrated *Parastrophina angulosa* (Fig. 1 A-J, 2 A-E, 3 A-1), *Parastrophina lindstroemi* n. sp. (Fig. 1 K-O), *Oxoplecia* sp. (Fig. 2 K-O), and *Parastrophinella* sp. (Fig. 1 P-Q, Fig. 2 F-J, P-R).

Biogeography of Ordovician, Silurian, and Devonian Chitinozoans

SVEN LAUFELD, *Geological Survey of Sweden, S-104 05 Stockholm, Sweden*

ABSTRACT

Thirteen paleobiogeographic charts are presented for Ordovician, Silurian, and Devonian chitinozoans. The compilation of these has been difficult for several reasons: little is known of the chitinozoans from some geographic regions; stratigraphic information about the occurrence of these microfossils is often imprecise; and there is little uniformity in chitinozoan taxonomy. Paleobiogeographic interpretations must also take into consideration the paleoecology of chitinozoans, since some of these animals had a planktic, others a benthic mode of life. All chitinozoans were bathymetrically controlled but as shown in a graph of abundance, the benthic forms were more common than the planktic. Despite the possible sources of error, it is evident that Chitinozoa display some provincialism. Ordovician chitinozoans are conspicuously provincial. Silurian Chitinozoa are predominantly cosmopolitan. Devonian taxa are halfway between these extremes.

INTRODUCTION

Chitinozoa are a group of extinct, organic-walled microfossils that are common in several kinds of marine rocks of Ordovician through Devonian age. The fossils are club-, vase-, or flask-shaped and are radially symmetrical with an opening at the so-called oral end. The vesicles of some Chitinozoa have a smooth surface. Others show a more-or-less conspicuous ornamentation that covers parts or all of the vesicle. Some taxa are characterized by long branching appendices located at the aboral edge of the vesicle or by transversally or longitudinally arranged series of spines on other parts of the vesicle. A few forms are shorter than 100 μm or longer than 1 mm, but the majority are 100 to 400 μm in length. Due to the chemical resistance of the body wall, chitinozoans are readily extracted from rocks by the use of inorganic acids.

Chitinozoa have proved to be very useful tools for intra- and inter-continental correlation despite the fact that their mode of life and systematic affinities are controversial and raise heated arguments among most paleontologists working with them. Some specialists maintain that Chitinozoa are protozoans; a few advocate an algal relationship; but most workers now argue that these fossils represent an early embryonic stage of metazoans. There is no unanimous view, however, on which group or groups of metazoans are involved. "Chitinous" hydroids, annelid worms, gastropods, cephalopods, and graptolites are among the groups suggested, but no definitive evidence for a specific systematic assignment has been presented. It is also theoretically possible that Chitinozoa belong to a separate phylum of unknown affinity. Although I do not wish to discuss further the systematic affinities of Chitinozoa, it is pertinent to note that if the group should prove to be polyphyletic, it will alter the interpretation of the distribution maps presented in this paper.

Chitinozoans were discovered in 1929. Between 1930 and 1976 fewer than 300 publications were devoted to the group. Fewer than 100 of these publications are serious scientific studies. At the present time, less than fifty people work regularly with Chitinozoa.

It is difficult to synthesize present knowledge about chitinozoans to provide global maps of their biogeographic distribution in Ordovician, Silurian, and Devonian time. In large part, this is due to the incompleteness and inconsistency of the data. Although the general bias of the fossil record should be kept in mind, four major pitfalls pertaining specifically to the distribution of chitinozoans have complicated the task: poor geographic coverage, poor stratigraphic data, inconsistencies in taxonomic usage, and inattention to paleoecology.

Even a quick glance at the maps of chitinozoan distribution (Figs. 2 to 14) shows the poor global coverage of the studies made so far. This is due primarily to the limited number of students working with these fossils and not to the absence of chitinozoans in unmetamorphosed rocks of the proper age. But, if nothing else, the maps indicate suitable geographic areas for work. There is no published record of Chitinozoa from Antarctica or Greenland.[1] A handful of papers record Chitinozoa from western North America, and South American chitinozoans have been reported in about ten publications which deal only with a small part of the continent. There is a single report of chitinozoans from the African continent south of the Sahara and two papers briefly mention the occurrence of Chitinozoa in Australia. The greatest obstacle to a synthesis of chitinozoan paleobiogeography, however, is the

[1] Recently, John Peel discovered chitinozoans in the Silurian of northernmost Greenland and I have recovered many specimens of a single species in rocks of "Precambrian" age in eastern Greenland.

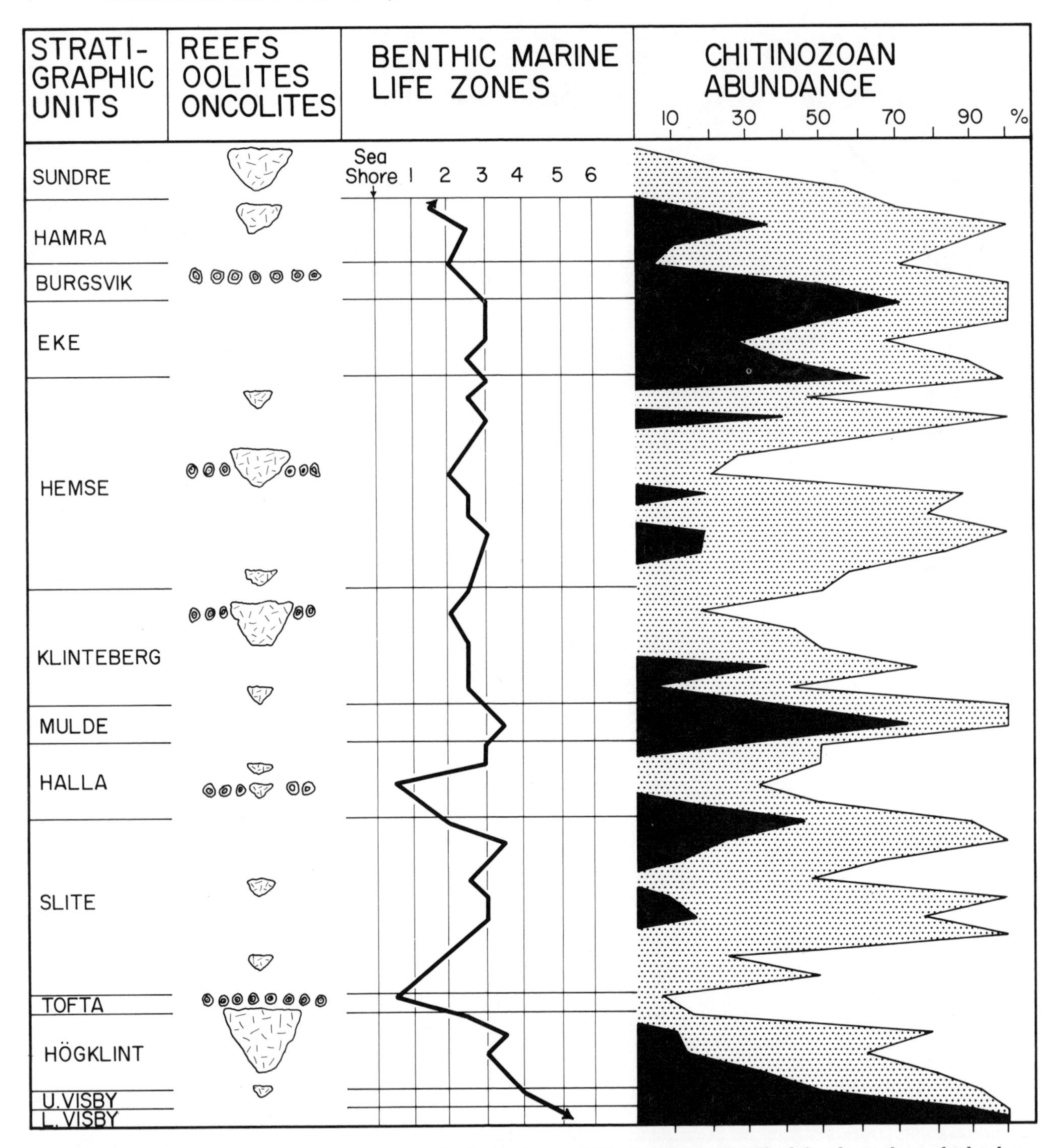

Figure 1. Diagram showing bathymetric control of abundance of Silurian Chitinozoa. The left column shows the local stratigraphic units of the Silurian of Gotland, Sweden. The sequence of strata is about 500 m thick and embraces about two-thirds of Silurian time. The Llandovery-Wenlock boundary is located immediately above the boundary between the Lower and Upper Visby Beds; the Wenlock-Ludlow boundary is at the transition between the middle and upper part of the Klinteberg Beds. For a detailed description of these rock units, see Laufeld (1974, p. 7-13) and for an explanation of the time- and rock-stratigraphic framework used, see Laufeld and Jeppsson (1976, Fig. 4).

The second column shows the stratigraphic occurrence of reef bodies, oolites, and oncolites. The reefs were formed at a water depth probably not exceeding 75 to 80 m. Most of them grew at a depth of 2 to 50 m. The oncolites and oolites on Gotland were formed in water representing low intertidal or shallower positions.

The "bathymetric" graph is based on brachiopod Benthic Marine Life Zones (=Benthic Assemblages of Boucot, 1975, p. 49-51). Most of the Benthic Marine Life Zones are based on the data in Gray and others (1974, Table 1); for those units and subunits not covered by Gray and coworkers, I constructed the curve by using key brachiopod species listed in Hede's explanations almost complete lack of data from eastern Asia and the European part of the Soviet Union.

The poor geographical coverage becomes even more evident on a world map where the Atlantic Ocean has been eliminated by fitting South America along the continental margin of Africa, and North America with Europe. However, the generic distributions were plotted on a base map of Van der Grinten's Projection where the continents are shown in their present positions since replotting of data from distorted continents with vague coastlines is almost impossible and always inaccurate. The skewed geographical distribution makes it probable that the chitinozoans heretofore reported in the literature represent only a limited latitudinal or longitudinal area of the Ordovician, Silurian, and Devonian continental shelves and slopes, depending on the direction of the paleoequator in each Period.

A full understanding of the stratigraphical usefulness and paleobiogeography of Chitinozoa is hampered by the lack of accurate and detailed stratigraphic information in most publications. Nowadays, it is not meaningful to describe and propose new taxa of chitinozoans and other fossils from erratic boulders and float. Unfortunately, a great number of publications on Chitinozoa are concerned with series of strata whose age is not established or is obscure even with reference to geologic system. Thus, a lot of valuable morphological and paleobiological work cannot be properly evaluated and used in a biogeographic synthesis until some kind of stratigraphic framework for the areas concerned has been established. Hence, I had to exclude from this study a number of otherwise interesting publications and to disregard parts of several of the papers used. This screening does not effect the geographical coverage, however.

The absence of a consistent chitinozoan taxonomy has been another source of difficulty in this synthesis. No taxonomic monographs or broad systematic summaries of *Treatise* type have been published for the Chitinozoa and recent inquiry indicates that several specialists would consider such an undertaking premature at this time. My judgment about the taxonomy used in this paper differs in several cases with that used by other authors. To cite one extreme example, I have assigned to 5 genera and 10 to 12 species, chitinozoans from South America that had originally been assigned to 16 genera, 53 species (9 new) and 4 subspecies (2 new). This is presently the only way, though not necessarily the best, to reduce a false "world-wide" distribution of Chitinozoa. The taxonomy expressed in this paper thus represents the conservative view of a taxonomic "lumper," and has a certain internal consistency.

Taxonomic revisions of the Chitinozoa are severely hampered by several factors: the poor quality of a substantial number of the published illustrations; the loss of much type material; the deposition of type material in private or company collections where it is inaccessible to most workers. In addition, information on type localities and type strata are poor in many cases, or lacking entirely, making it impossible to recollect topotype material. Furthermore, a great number of taxa in need of revision, are based on specimens from exploration wells in inacccessible areas. In sum, major taxonomic revisions, though badly needed, are difficult to undertake and beyond the scope of this review.

Due to the fact that described suprageneric taxa of Chitinozoa are limited in number (and in my opinion biologically unwarranted at the present time), and because the taxonomy at species level is unstable and confused, the genus will be the unit used in this synthesis.

Finally, it should be emphasized that an interpretation of the distribution of chitinozoans must take into consideration their paleoecology. It has been suggested (Laufeld, 1967, p. 282-283; Chaiffetz, 1972, p. 501) that chitinozoans with long hollow spines or appendices, e.g., forms belonging to the genera *Ancyrochitina* and *Plectochitina*, were planktic, whereas others were benthic and strictly facies-controlled (Staplin, 1961, p. 399-400; Urban and Kline, 1970, p. 75). Staplin pointed out that Middle Devonian chitinozoans are most abundant in shallow-water environments favorable to reef development (1961, p. 402) and Laufeld (1973a, p. 138; 1974, p. 119-123) argued for a bathymetric control of both abundance and diversity of Silurian Chitinozoa and singled out *Sphaerochitina* as showing preference for very shallow, agitated water and a fairly coarse-grained sediment substrate. Miller (1975, 1976) shows that Ordovician Chitinozoa also were bathymetrically controlled. Recent (Laufeld, 1975) and

for the geologic map sheets of Gotland. The "bathymetric" graph is tentative although the major trends will not change. Zone 1, High Intertidal; zone 2, Low Intertidal; zone 3, depths between 15 and 75 m; zones 4 and 5, depths between 75 m and 175 to 200 m.

The right column shows abundance of Chitinozoa expressed as the percentage of samples that yielded chitinozoans (stippled), compared to the total number of samples analysed per stratigraphic subunit. For example, 6 percent of the Tofta samples yield Chitinozoa, 94 percent are barren. Black areas represent percentage of samples that yielded more than 10 specimens per gram of rock. For example, all samples from the Lower Visby Beds contain more than 10 chitinozoan specimens per gram of rock. These black peaks, although possibly biased by slightly different rates of deposition, represent optimum living conditions for chitinozoans. The graph of abundance represents nearly 1000 analysed samples from 63 stratigraphic subunits and 13 units.

Reefs were formed during regressive phases of the Silurian sea, which were also coupled with a decrease in abundance of Chitinozoa. The absence of chitinozoans in a large proportion of the analysed samples can be correlated with increasingly shallower Benthic Marine Life Zones and reef growth, commonly associated with oolitic and oncolitic limestones. Peaks of chitinozoan abundance are related to transgressive phases of the Silurian sea. The highest abundances occur in Benthic Marine Life Zone 3 and deeper, with an optimum at, say 150 to 200 m in zone 5, where all samples processed yielded more than 10 specimens of Chitinozoa per gram of rock.

ongoing studies further substantiate a bathymetric control of chitinozoan abundance (Fig. 1) and diversity.

It can be concluded that all chitinozoans were bathymetrically controlled and that some were benthic, others planktic, and some possibly epiplanktic. This being so, there is risk of interpreting different but more-or-less synchronous faunas from geographically widespread localities as belonging to different faunal realms or provinces when in fact the faunal differences can be explained by different paleobathymetric positions. Hence, paleobiogeographical interpretations based on Chitinozoa distribution will have to be cautious until we know more about the mode of life of individual genera and species.

CHITINOZOAN PALEOBIOGEOGRAPHY

Ordovician

The distribution of genera of Ordovician Chitinozoa is plotted in Figures 2 to 6. Some of these genera were cosmopolitan; others show a geographically restricted distribution. *Cyathochitina* (Fig. 2) is a typical cosmopolitan genus whose species were unaffected by water temperature. Species have been reported from Alaska and widely in North America from Newfoundland to Texas, from South and North Africa, from Australia, and from several localities in western Europe. The three occurrences of *Cyathochitina* in North Africa are located fairly close to an Ordovician pole position and the Russian localities are not very far from the probable contemporary equator. Even though present concepts of Ordovician pole and equator positions may change, the distribution of *Cyathochitina* is so wide that almost all climatic zones were occupied by species of this genus. Another cosmopolitan genus with an even wider distribution than *Cyathochitina* is *Conochitina* (Fig. 5), a genus that occurs also in Asia. *Ancyrochitina* and *Spinachitina* (Fig. 3), and probably *Desmochitina* (Fig. 5) were also cosmopolitan. The wide distribution of these 3 genera indicates that they, like *Cyathochitina,* were independent of water temperature.

The Ordovician species of *Lagenochitina* (Fig. 3), *Angochitina* (Fig. 4), and *Acanthochitina* (Fig. 2), and possibly also *Tanuchitina* (Fig. 5) show a northern distribution in the present geography that is very difficult to explain without taking continental drift into account. Species of these 4 genera occur mainly in the "North Atlantic Area" and are among the best examples of Ordovician chitinozoan provincialism.[2] Most of them were probably restricted to temperate and warm water.

It is of interest to note that chitinozoans of the above-mentioned genera can be used for very detailed correlations (fractions of a graptolite zone) within the North Atlantic Area, but it is very difficult to correlate rock in the North Atlantic Area with those of, for instance, Spain or North Africa by means of Chitinozoa.

In Ordovician time, *Eremochitina* (Fig. 3) and *Pseudoclathrochitina* (Fig. 4) are characteristic elements of a southern area or province that includes France, the Iberian Peninsula, North Africa, and interestingly enough, Florida.

It is possible that *Hercochitina-Kalochitina* (Fig. 2) are characteristic of a western area in Ordovician time. The data on *Rhabdochitina* (Fig. 4) are not conclusive, but indicate a wide, possibly cosmopolitan distribution. *Pterochitina-Hoegisphaera* and *Siphonochitina* (Fig. 6) are not abundant but it seems probable that the former two are cosmopolitan and the latter more restricted in occurrence.

My conclusions based on the data presented in the maps (Figs. 2 to 6) support Jenkins' (1967, p. 485) opinion that there was fairly free communication between the Chitinozoa of Shropshire and Baltoscandia in the Ordovician, especially in Caradocian time, whereas the communication between North European and North African chitinozoans was considerably more limited. Laufeld (*in* Carter and Laufeld, 1975, p. 462) pointed out that the Ordovician Chitinozoa from northernmost Alaska "belong to the same chitinozoan faunal province" as the Middle and Late Ordovician Chitinozoa of the Baltic area, a conclusion that is further substantiated here.

Silurian

Silurian chitinozoans are less provincial than those of the Ordovician. This is evident from the distribution of genera shown on the maps (Figs. 7 to 11). In Silurian time, the following genera can be classified as cosmopolitan: *Ancyrochitina* (Fig. 7), *Angochitina* (Fig. 10), *Conochitina* (Fig. 9), probably *Cyathochitina* (Fig. 9), *Desmochitina* (Fig. 11), *Eisenackitina* (including *Bursachitina*) (Fig. 8), *Linochitina* (Fig. 8), *Margachitina* (Fig. 10), probably *Pterochitina* (Fig. 10), and *Sphaerochitina* (Fig. 7).

Angochitina is represented by a fairly small number of species restricted to the North Atlantic Area in Ordovician time, but by the Silurian the genus is represented by a great number of more-or-less cosmopolitan species. However, there are several Silurian species of *Angochitina* that show restricted geographic occurrence. This can be explained, in part, by a depth-controlled distribution.

At the present time, no Silurian chitinozoan genera are known to be confined exclusively to northern areas (in the present geography). Two genera, however, are restricted to southern areas, e.g. *Urochitina* (Fig. 9) and *Plectochitina* (Fig. 8). Species of these genera are characterized by long basal appendices. If these appendices were buoyancy structures they might indicate something specific, for instance, about water temperature, salinity, paleocurrents, and so on. To test this idea, I plotted the occurrence of 2 species which possibly belong to the cosmopolitan genera *Sphaerochitina* and *Ancyrochitina*. These 2 species were selected because of their deviating gross morphology. "*Ancyrochitina fragilis*" (it has no close relationship with *Ancyrochitina fragilis* Eisenack, 1955) is provided with very long, spongy basal appendices; *Sphaerochitina longicollis* has an enormously elongated neck. Both features might be interpreted as buoyancy structures. As is evident

[2] Although it is premature to introduce formal chitinozoan provinces at the present stage of knowledge, the "North Atlantic Area" is as close to a province as we can come.

from the maps, *Urochitina* (Fig. 9), *Plectochitina* (Fig. 8), *"Ancyrochitina fragilis"* (Fig. 11), and *Sphaerochitina longicollis* (Fig. 7) have an almost identical distribution and are confined to a restricted, southern area (a province or part of a province?). Species from other genera with very conspicuous basal appendices show a similar distribution. Lange (1970 *in* Cramer 1971a, p. 4756) pointed out the similarity between the Silurian chitinozoans of Florida and Brazil, a part of this southern area or province.

Cramer (1971c, p. 62) advocated two Silurian chitinozoan provinces related to paleolatitude, the "Baltic Province and the Sahara Province," or three Silurian "chitinozoan-facies": "(1) The Baltic chitinozoan-facies (Arctic Canada, Sweden); (2) The Appalachian chitinozoan-facies (Kentucky, Tennessee, Belgium); (3) The Iberian-Sahara chitinozoan-facies." (Cramer 1971b, p. 235). The following year, Cramer and Diez de Cramer (1972c, p. 116) discussed Silurian Chitinozoa and concluded that "there are faint indications of a paleolatitude-related chitinozoan provinciation. But provinciation, if it exists, is not at all as pronounced as that of acritarchs."

It will be difficult to delineate specific Silurian chitinozoan provinces until the paleoecology of Chitinozoa is reasonably well understood.

Devonian

Devonian Chitinozoa seem to be intermediate in provincialism compared to Ordovician and Silurian taxa. The cosmopolitan genera are *Ancyrochitina* (Fig. 12), *Angochitina* (Fig. 13), *Eisenackitina-Bursachitina* (Fig. 12), probably *Linochitina* (Fig. 13), and *Sphaerochitina* (Fig. 13).

Urochitina (Fig. 12) is geographically restricted to the same southern area or province that it was in Silurian time. *"Sphaerochitina longicolla"* (Fig. 14) shows the same pattern of distribution.

The exclusively Devonian genus *Alpenachitina* (Fig. 14) is restricted to the New World, whereas *Margachitina* (Fig. 14) is restricted to the Old World.

Unfortunately, the data on Devonian Chitinozoa are very sparse and since chitinozoans became extinct during this period, the decreasingly diversified and decreasingly abundant chitinozoan assemblages reveal little about the paleobiogeography of Chitinozoa.

DISCUSSION

It is premature to define the geographic extent of chitinozoan provinces in Ordovician, Silurian, and Devonian time and to attach formal names to such paleogeographic entities, even though provincialism has been shown to exist among Chitinozoa.

It seems evident that the Ordovician Period was one of fairly pronounced provincialism for Chitinozoa. Of the 13 Ordovician genera discussed here, only 5 were cosmopolitan. This stands in strong contrast to the cosmopolitanism displayed by Silurian Chitinozoa. Ten of the 12 Silurian genera plotted on the maps were cosmopolitan. Chitinozoans were on the decline in the Devonian and became extinct by the end of that Period; 6 of 8 genera were cosmopolitan.

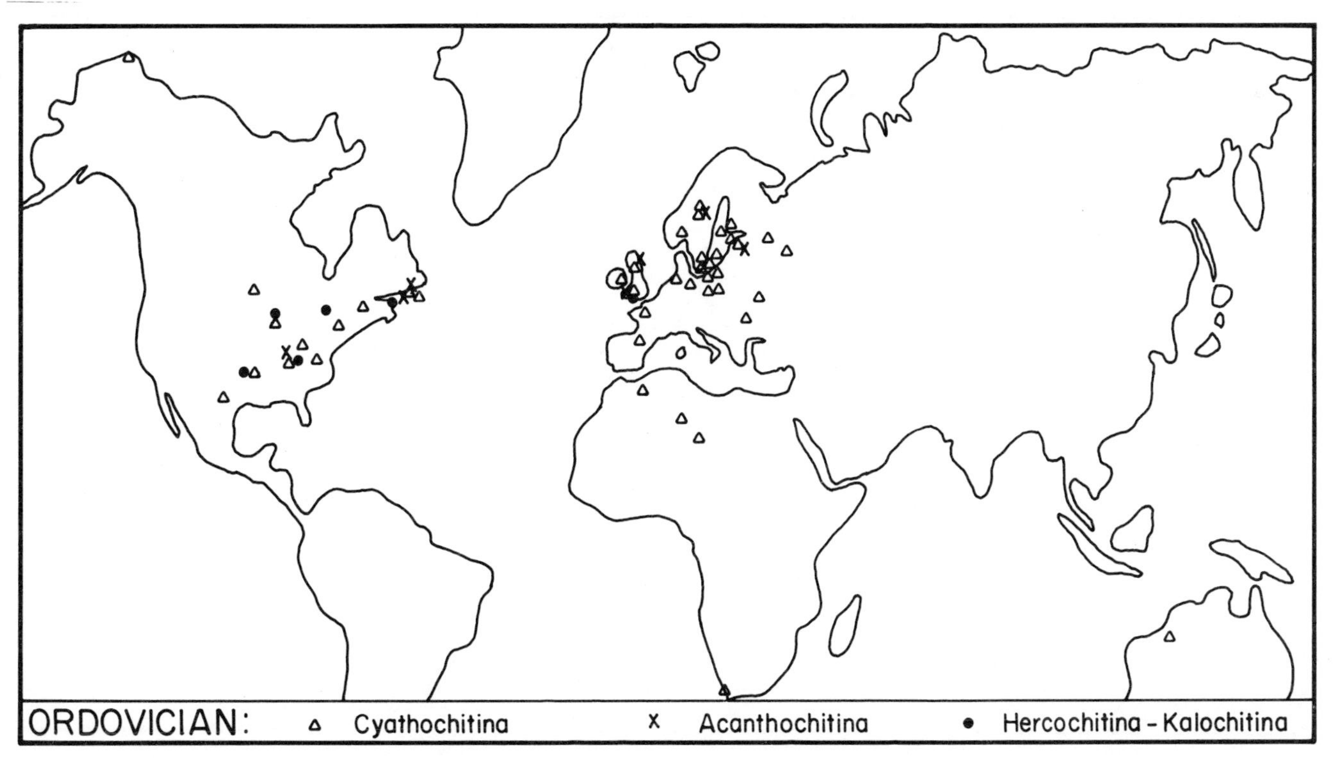

Figure 2. ***Cyathochitina*** **is cosmopolitan;** ***Hercochitina-Kalochitina*** **includes species with a westerly distribution;** ***Acanthochitina*** **is restricted to a fairly narrow zone from the Baltic area across the British Isles and along the Appalachians.**

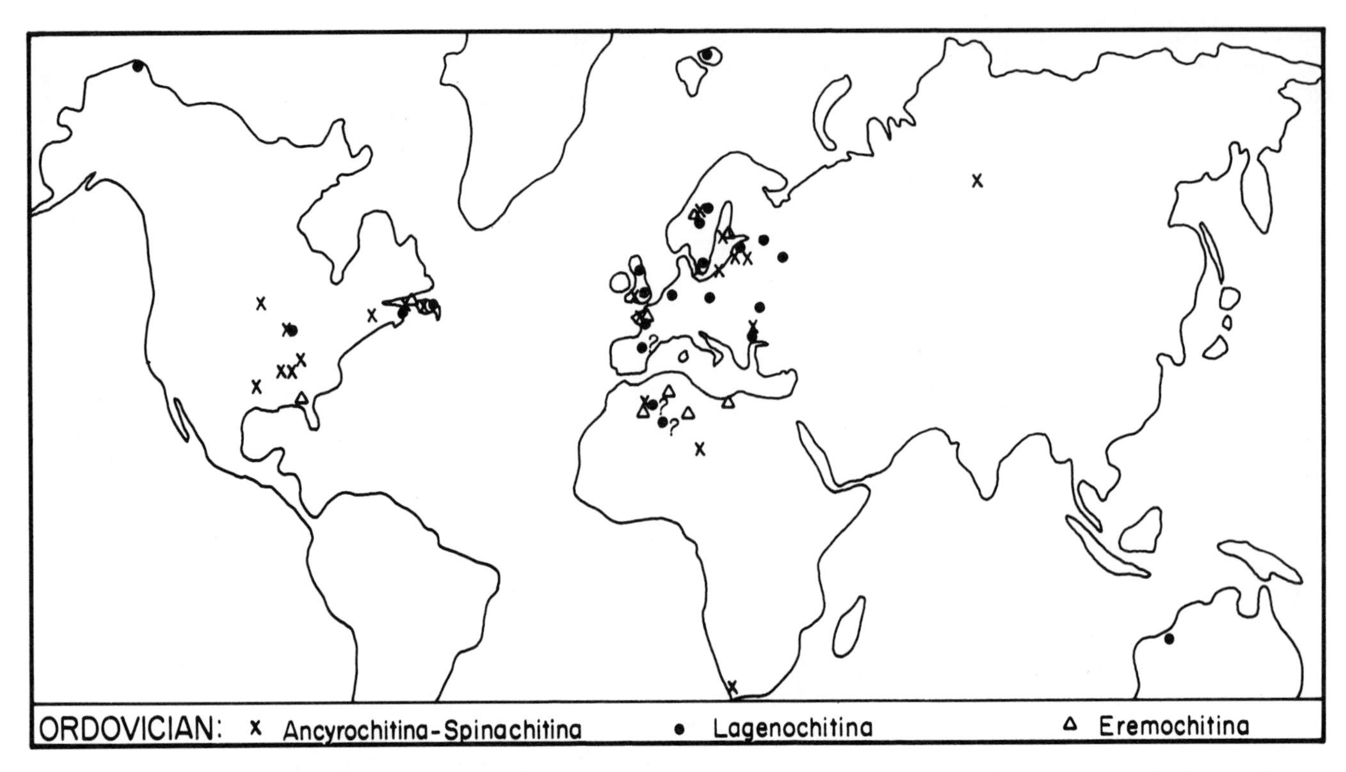

Figure 3. *Ancyrochitina-Spinachitina* embraces planktic and cosmopolitan species; *Lagenochitina* has a northerly distribution (the Spanish and North African occurrences are doubtful); *Eremochitina* is more common in southern areas.

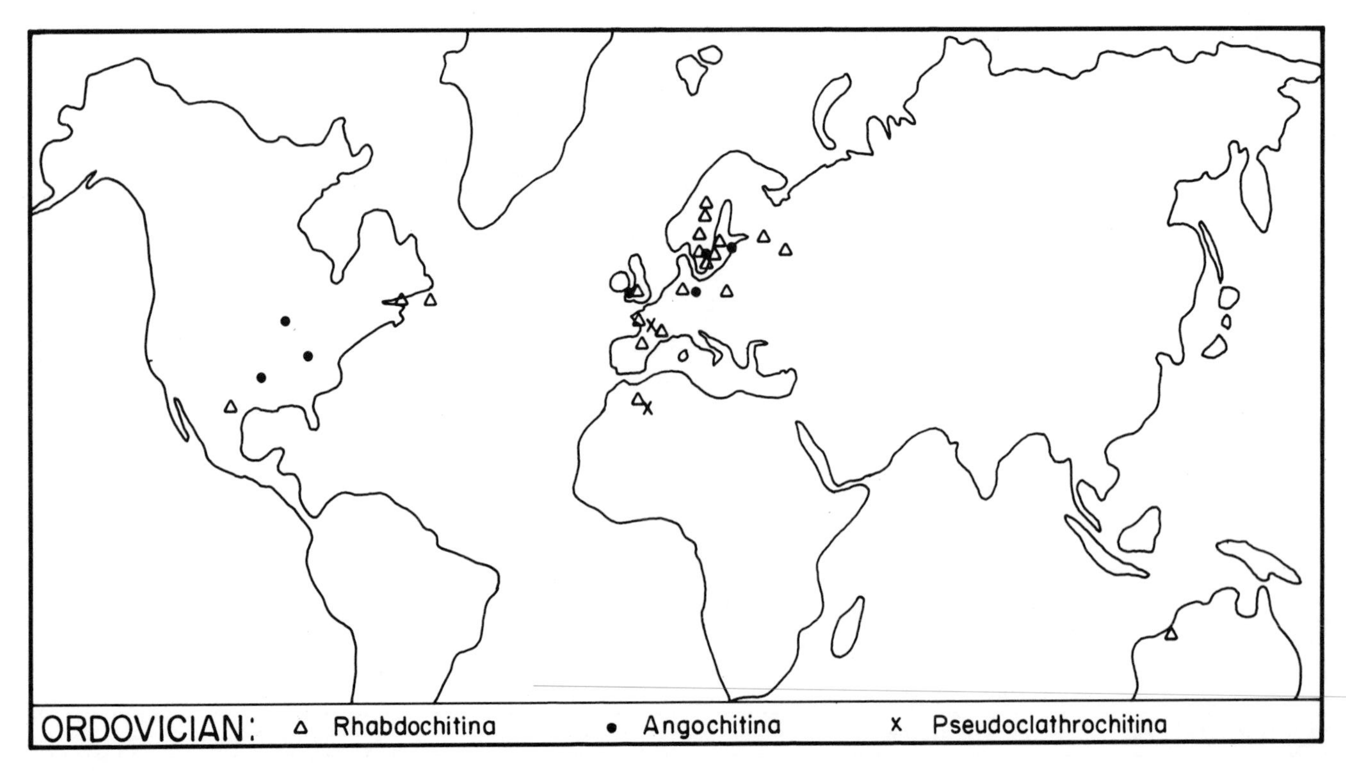

Figure 4. *Pseudoclathrochitina* has a southerly distribution; *Angochitina* is restricted to northerly areas; *Rhabdochitina* is composed of large, benthic, cosmopolitan species.

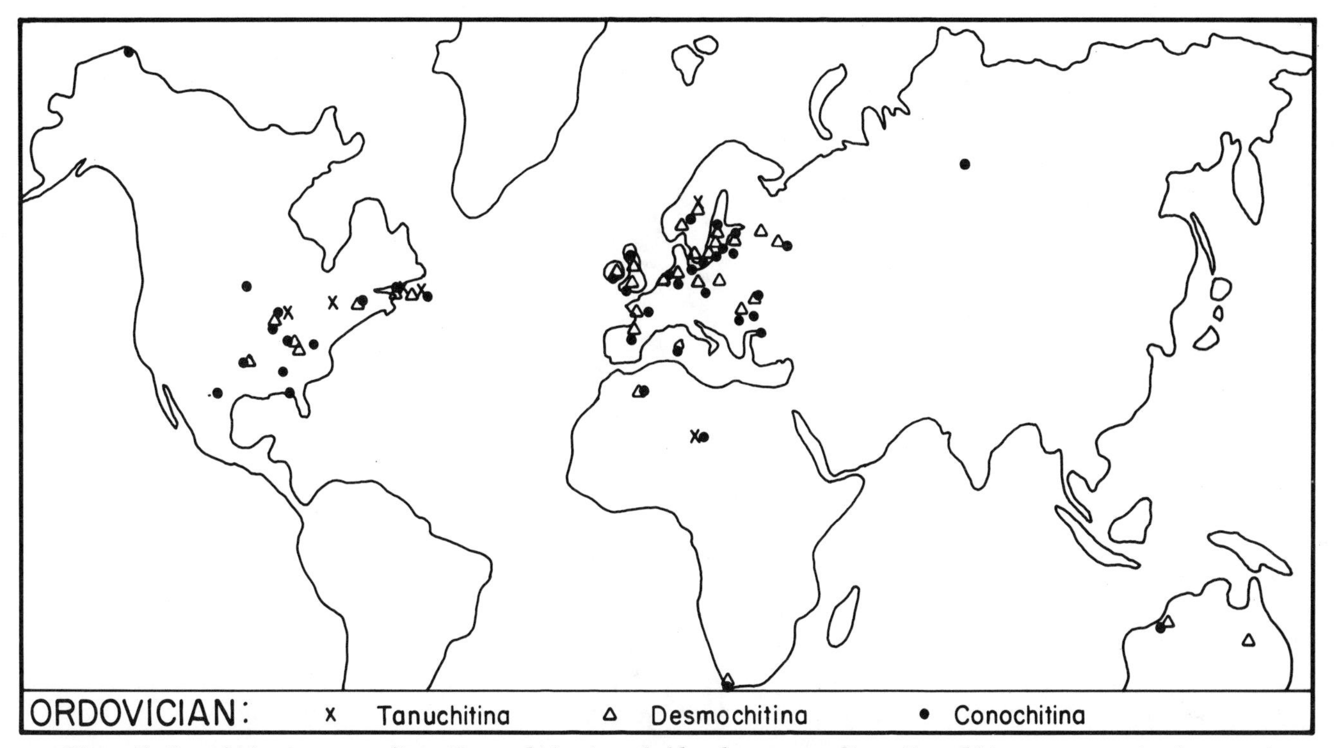

Figure 5. *Conochitina* is cosmopolitan; *Desmochitina* is probably also cosmopolitan; *Tanuchitina* seems restricted to northern areas except for a single occurrence in the southern part of North Africa.

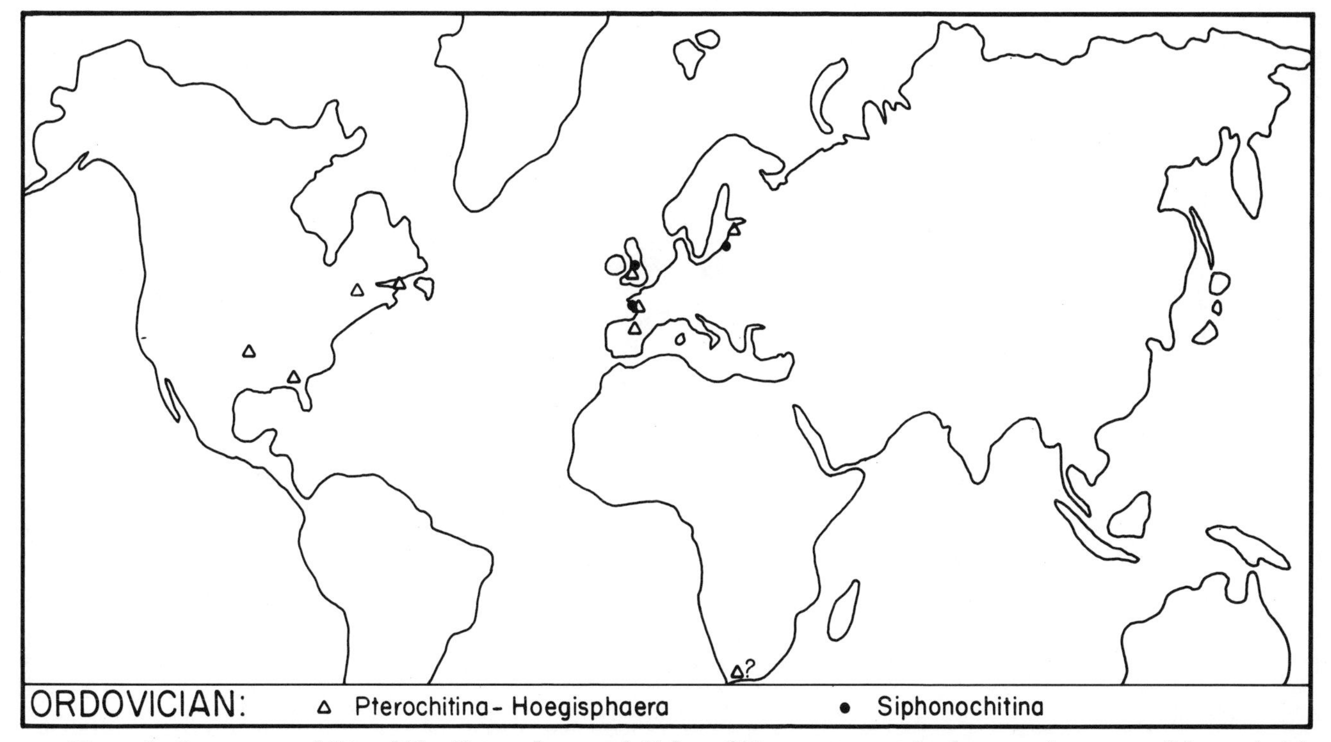

Figure 6. Occurrences of *Pterochitina-Hoegosphaera* and *Siphonochitina* are too restricted to speculate on possible provincialism.

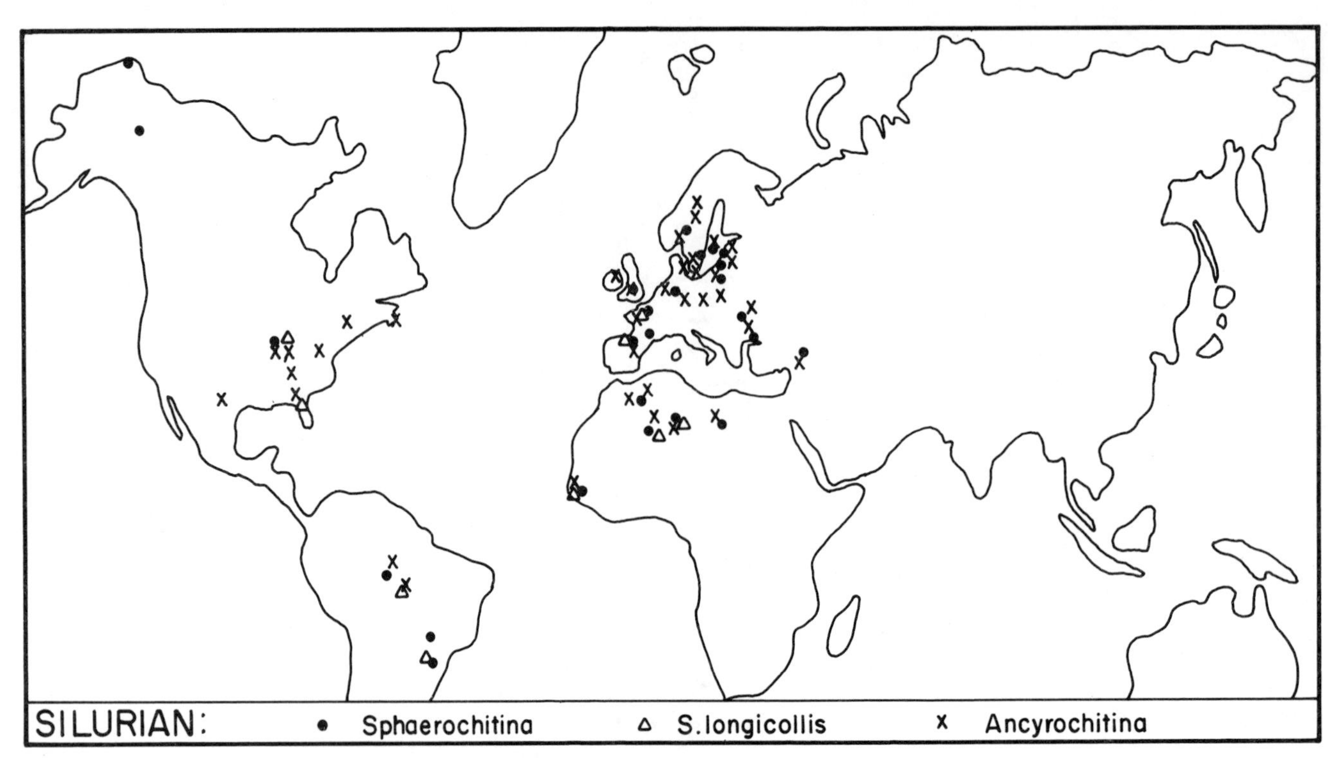

Figure 7. Species of *Sphaerochitina* are benthic and some are restricted to fairly shallow water; *S. longicollis* may have been planktic with preference for southerly areas.

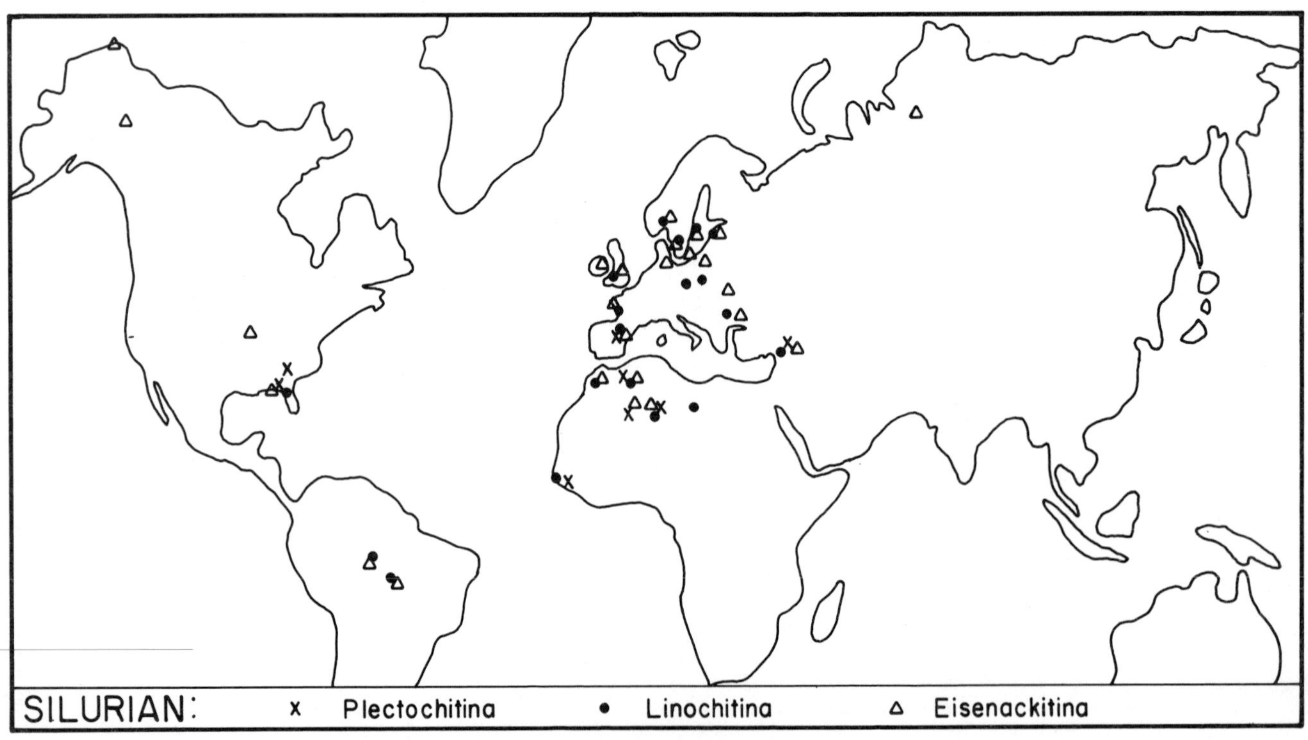

Figure 8. *Eisenackitina* and *Linochitina* are probably cosmopolitan but occur only in fairly shallow water; *Plectochitina* species are planktic with a southerly distribution.

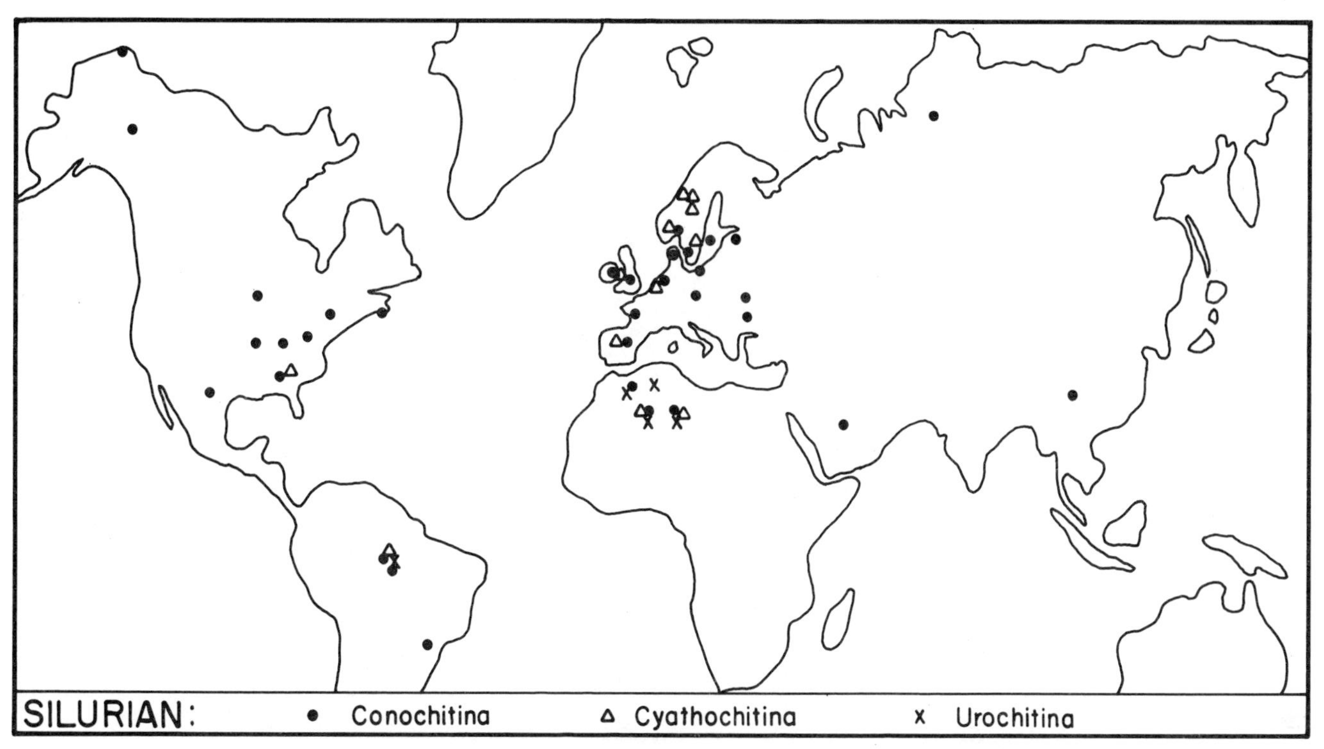

Figure 9. *Conochitina* embraces cosmopolitan taxa; *Cyathochitina* species are cosmopolitan but with a narrowing area of occurrence before their extinction in latest Early Silurian time; *Urochitina* is restricted to southerly areas.

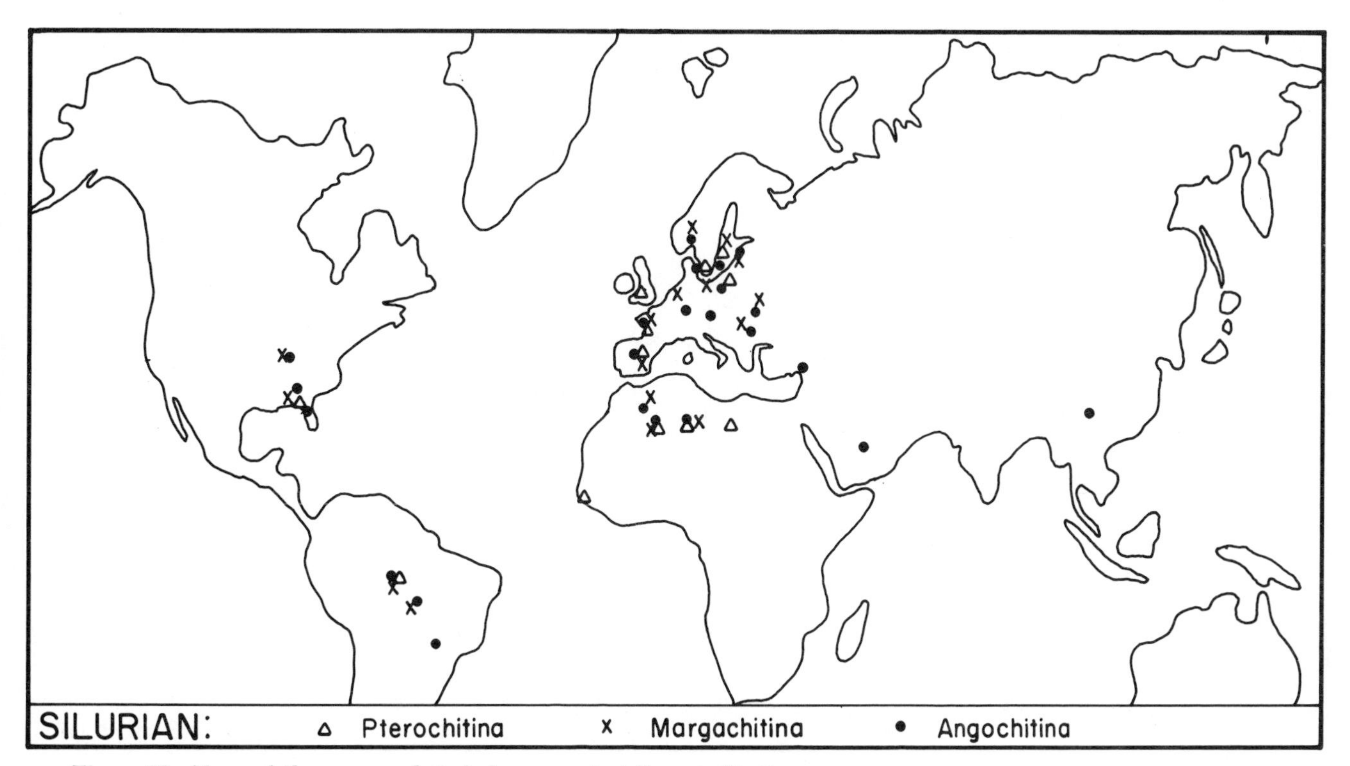

Figure 10. None of the genera plotted shows provincialism at this time.

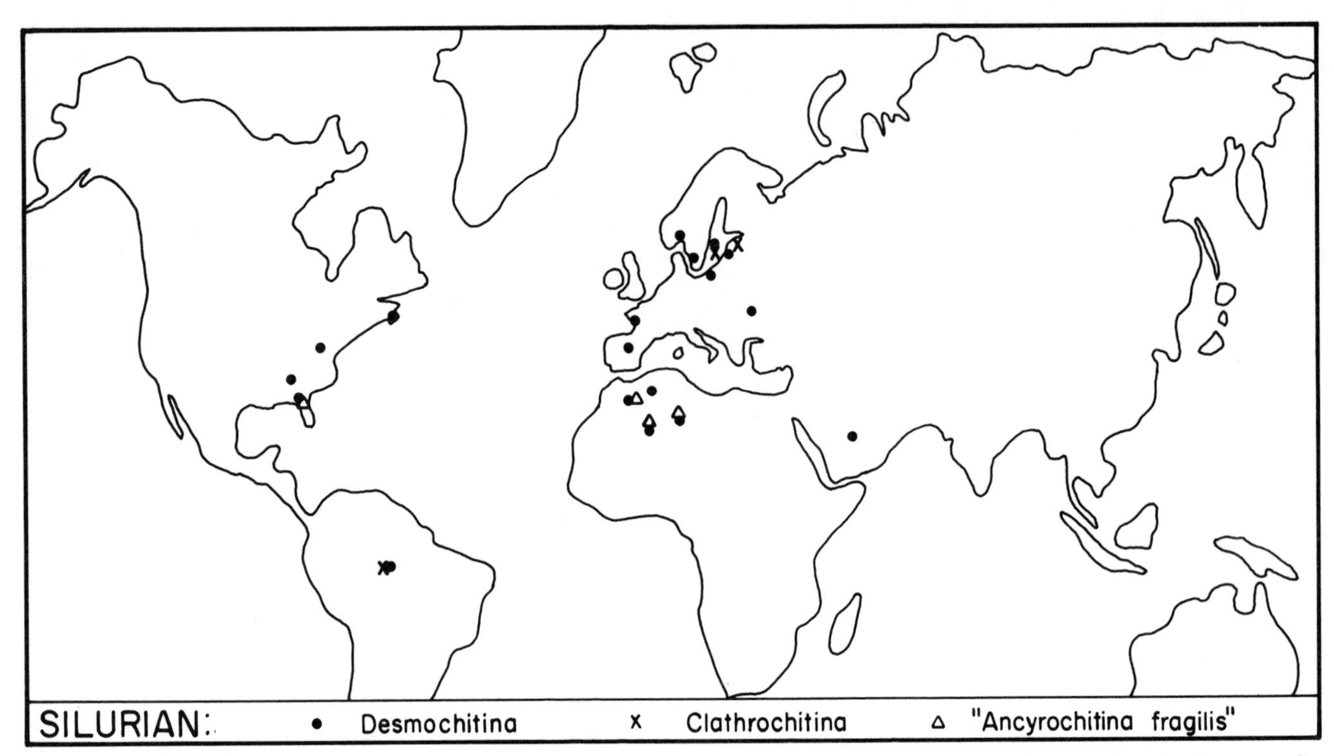

Figure 11. The planktic "*Ancyrochitina fragilis*" seems to have a very restricted occurrence but no provincialism can be discerned at the generic level.

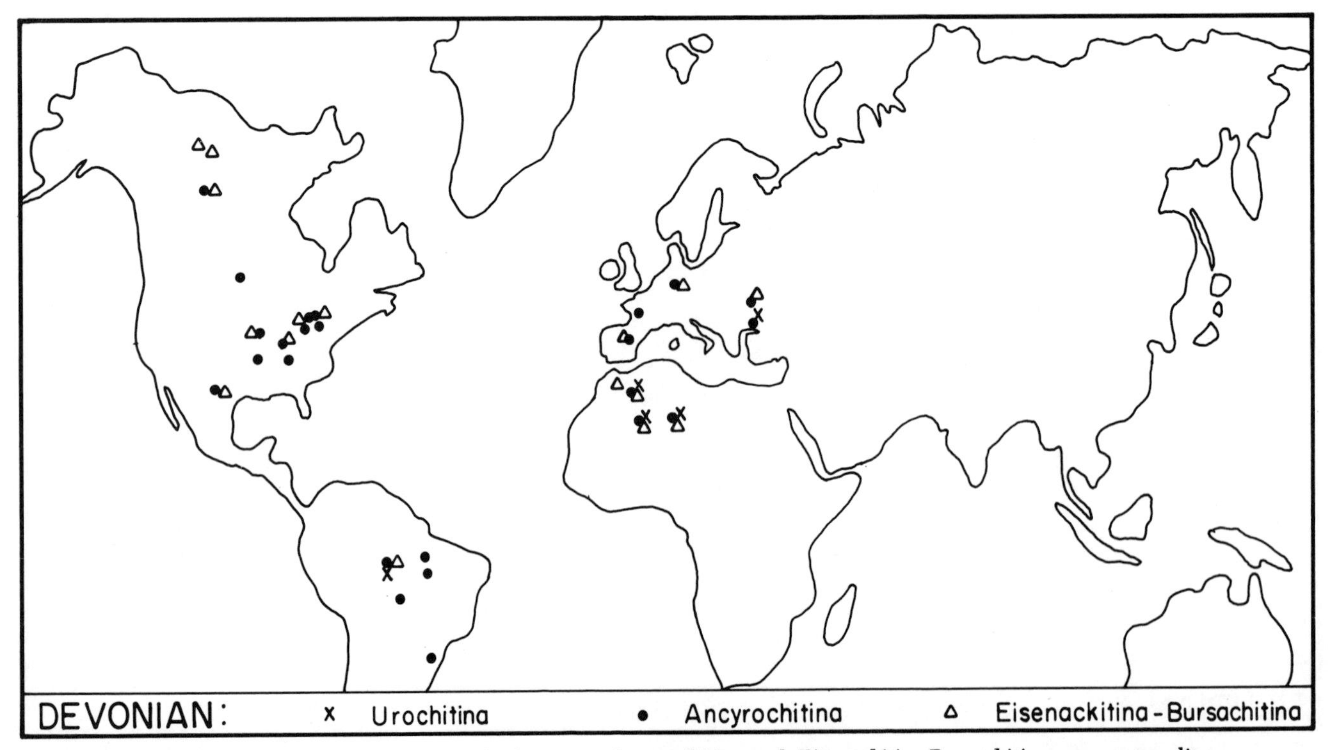

Figure 12. *Urochitina* is restricted to southerly areas; *Ancyrochitina* and *Eisenackitina-Bursachitina* are cosmopolitan.

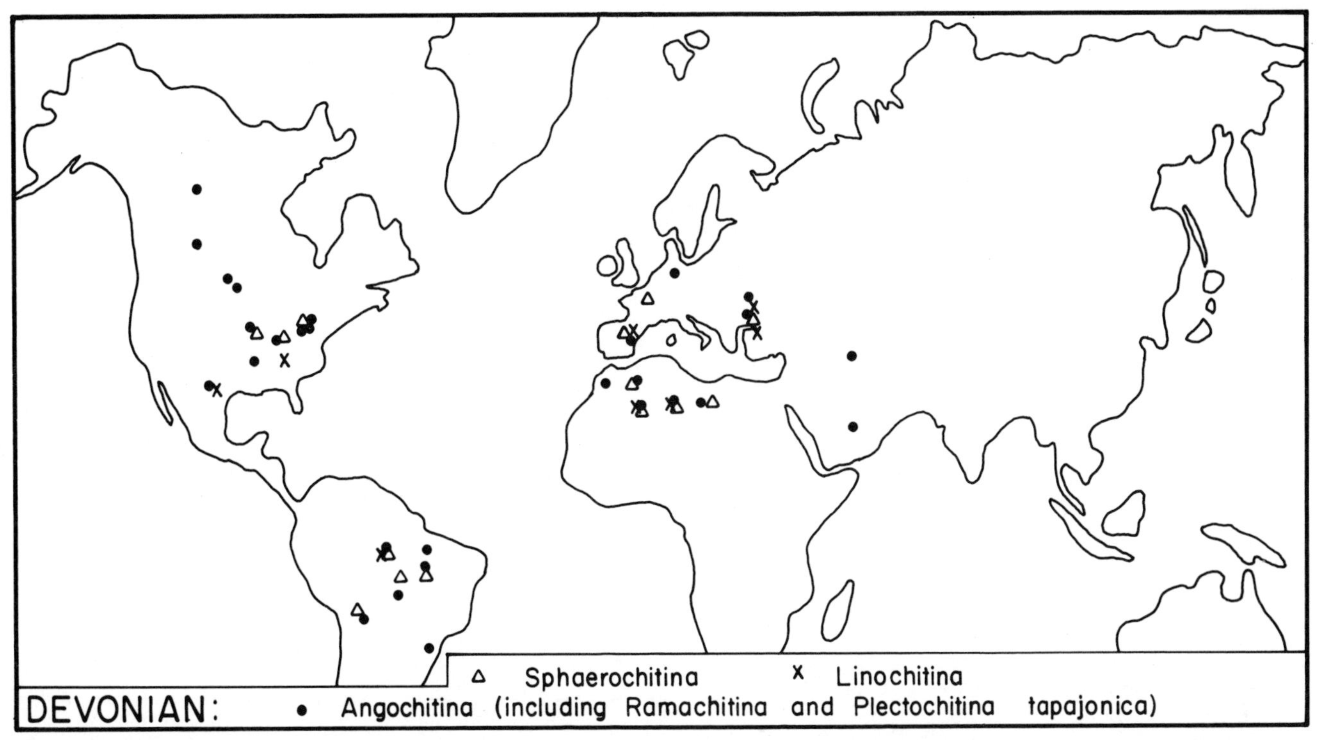

Figure 13. The genera plotted are probably cosmopolitan; there is an occurrence of *Ancyrochitina* on the Falkland Islands, outside the map area of South America.

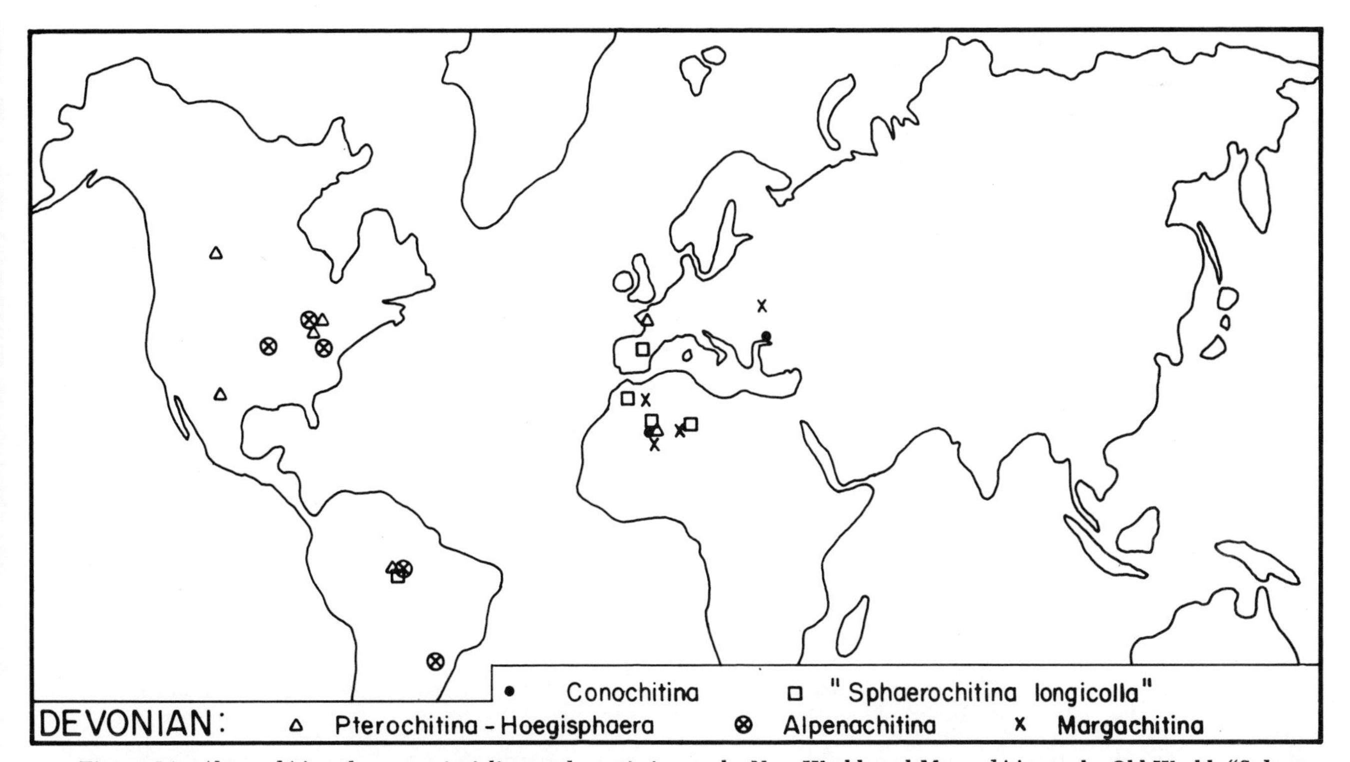

Figure 14. *Alpenochitina* show provincialism and restriction to the New World, and *Margachitina* to the Old World; "*Sphaerochitina longicolla*" occurs in the same southerly areas as it did in Silurian time; there is an occurrence of *Angochitina* on the Falkland Islands, outside the map area of South America.

REFERENCES

Achab, Aïcha, 1977, Les Chitinozoaires de la zone à *Dicellograptus complanatus*, Ordovicien supérieur, Ile d'Anticosti-Québec: Canadian Jour. Earth Sci., v. 14, p. 413-425.

Atkinson, K., and Moy, R. Lesley, 1971, Lower Caradocian (Upper Ordovician) Chitinozoa from North Wales: Rev. Palaeobotany and Palynology, v. 11, p. 239-250.

Bachmann, A., and Schmid, M. E., 1964, Mikrofossilien aus dem österreichischen Silur: Geol. Bundesanst. Verhandl. 1964, p. 53-64.

Beju, D., 1961, Zonare şi corelare a paleozoicului din platforma moesică pe baza asociaţiilor palino-protistologice. I: Petrol şi Gaze, v. 12, p. 714-722.

Beju, D., and Dănet, N., 1962, Chitinozoare siluriene din platforma moldovenească şi platforma moezică: Petrol şi Gaze, v. 13, p. 527-536.

Benoit, A., and Taugourdeau, P., 1961, Sur quelques Chitinozoaires de l'Ordovicien du Sahara: Rev. Inst. Français Pétrole, v. 16, p. 1403-1421.

Bergström, J., Bergström, S. M., and Laufeld, S., 1968, En ny skärning genom överkambrium och mellanordovicium i Rävatofta-området, Skåne: Geol. Fören. Stockholm Förh., v. 89, p. 160-165.

Bergström, S. M., and Nilsson, R., 1974, Age and correlation of the Middle Ordovician bentonites on Bornholm: Geol. Soc. Denmark Bull., v. 23, p. 27-48.

Boekel, Norma M., van, 1967, New Devonian chitinozoans from the Tapajós River, Pará: Acad. Brasil. Ciências An., v. 39, p. 273-278.

Boneham, R. F., 1967, Hamilton (Middle Devonian) Chitinozoa from Rock Glen, Arkona, Ontario: Am. Midland Naturalist, v. 78, p. 121-125.

Boneham, R. F., and Masters, W. R., 1971, Silurian Chitinozoa from Indiana. I. The Mississinewa Shale Member of north-central Indiana: Indiana Acad. Sci. Proc., v. 80, p. 320-329.

——— 1973, Silurian Chitinozoa of Indiana. II. The Waldron Shale and Osgood Member of the Salamonie Dolomite: Am. Midland Naturalist, v. 90, p. 87-96.

Bouché, P. M., 1965, Chitinozoaires du Silurien s.l. du Djado (Sahara Nigérien): Rév. Micropaléontologie, v. 8, p. 151-164.

Boucot, A. J., 1975, Evolution and extinction rate controls: Amsterdam, Elsevier, 427 p.

Brenchley, P. J., Harper, J. C., Romano, M., and Skevington, D., 1967, New Ordovician faunas from Grangegeeth, Co. Meath: Royal Irish Acad. Proc., v. 65, sec. B, p. 297-304.

Brück, P. M., and Downie, C., 1974, Silurian microfossils from west of the Leinster Granite: Jour. Geol. Soc. (London), v. 130, p. 383-386.

Brück, P. M., Potter, T. L., and Downie, C., 1974, The Lower Palaeozoic stratigraphy of the northern part of the Leinster Massif: Royal Irish Acad. Proc., v. 74, sec. B, p. 75-84.

Carter, Claire, and Laufeld, S., 1975, Ordovician and Silurian fossils in well cores from North Slope of Alaska: Am. Assoc. Petroleum Geologists Bull., v. 59, p. 457-464.

Chaiffetz, M. S., 1972, Functional interpretation of the sacs of *Ancyrochitina fragilis* Eisenack, and the paleobiology of the ancyrochitinids: Jour. Paleontology, v. 46, p. 499-502.

Chauvel, J.-J., Deunff, J., and Le Corre, C., 1970, Découverte d'une association minerai de fer-microplancton dans l'Ordovicien du flanc nord du Bassin de Laval (Mayenne): étude petrographique et micropaléontologique: Acad. Sci. Comptes Rendus, sér. D, v. 270, p. 1219-1222.

Chlebowski, R., and Szaniawski, H., 1974, Chitinozoa from the Ordovician conglomerates at Miedzygorz in the Holy Cross Mts: Acta Geol. Polonica, v. 24, p. 221-228.

Collinson, C., and Schwalb, H., 1955, North American Paleozoic Chitinozoa: Illinois State Geol. Survey Rept. Invest. no. 186, 33 p.

Collinson, C., and Scott, A. J., 1958, Chitinozoan faunule of the Devonian Cedar Valley Formation: Illinois State Geol. Survey Circ. 247, 34 p.

Combaz, A., 1965, Un microbios a Chitinozoaires dans le Paléozoique du Queensland (Australie): Acad. Sci. Comptes Rendus, sér. D, v. 260, p. 3449-3451.

——— 1967, Un microbios de Trémadocien dans un sondage d'Hassi-Messaoud: Actes Soc. Linné. Bordeaux, v. 104, sér. B, no. 29, 26 p.

Combaz, A., and Poumot, C., 1962, Observations sur la structure des Chitinozoaires: Rév. Micropaléontologie, v. 5, p. 147-160.

Correia, M., 1964, Présence de Chitinozoaires dans le Gotlandien des environs de Rabat (Maroc): Soc. Géol. France, Compte Rendu Somm. Séances, no. 3, p. 105.

Costa, Norma M. van Boekel, da, 1966, Quitinozoáres de Riberão do Monte, Goiás: Div. Geol. Min., Notas Prel. Estud., no. 132, 25 p.

——— 1967, Quitinozoários Silurianos e Devonianos de Bacia Amazônica e sua correlação estratigráfica: Simpósio Biota Amazônica Atas, v. 1, p. 87-119.

——— 1968, Microfósseis Devonianos do Rio Tapajós, Pará. II-Chitinozoa: Div. Geol. Min., Notas Prel. Estud., no. 146, 19 p.

——— 1971a, Quitinozoários Brasileiros e sua importância estratigráfica: Acad. Brasil. Ciências An., v. 43 Supl., p. 209-270.

——— 1971b, Quitinozoários Silurianos do Igarapé da Rainha, Estudo do Pará: Div. Geol. Mineral. Bol., no. 255, 96 p.

Cousminer, H. L., 1964, Devonian Chitinozoa and other palynomorphs of medial South America and their biostratigraphic value: Ph.D. dissertation, New York Univ., New York, 266 p.

Cramer, F. H., 1964, Microplankton from three Palaeozoic formations in the province of León, NW-Spain: Leidse Geologische Meded., v. 30, p. 254-361.

——— 1966, Hoegispheres and other microfossils *incertae sedis* of the San Pedro Formation (Siluro-Devonian boundary) near Valporquero, León, NW Spain: Inst. Geol. Minero España Notas Comuns., v. 86, p. 75-84.

——— 1967, Chitinozoans of a composite section of upper Llandoverian to basal Gedinnian sediments in northern León, Spain. A preliminary report: Soc. Belge Géol. Bull., v. 75, p. 69-129.

——— 1968a, Palynologic microfossils of the Middle Silurian Maplewood Shale in New York: Rév. Micropaléontologie, v. 11, p. 61-70.

——— 1968b, Considérations paléogéographiques a propos d'une association de microplanctontes de la série gothlandienne de Birmingham (Alabama, U.S.A.): Soc. Géol. France Bull., sér. 7, v. 10, p. 126-131.

——— 1969, Possible implications for Silurian paleogeography from phytoplancton assemblages of the Rose Hill and Tuscarora Formations of Pennsylvania: Jour. Paleontology, v. 43, p. 485-491.

——— 1970a, Acritarchs and chitinozoans from the Silurian Ross Brook Formation, Nova Scotia: Jour. Geology, v. 78, p. 745-749.

——— 1970b, *Angochitina sinica*, a new Siluro-Devonian chitinozoan from Yunnan Province, China: Jour. Paleontology, v. 44, p. 1122-1124.

——— 1971a, Position of the north Florida lower Paleozoic block in Silurian time. Phytoplankton evidence: Jour. Geophys. Research, v. 76, p. 4754-4757.

——— 1971b, A palynostratigraphic model for Atlantic Pangea during Silurian time: Bur. Recherches Géol. Minières Mém., no. 73, p. 220-235.

——— 1971c, Implications from middle Paleozoic palynofacies transgressions for the rate of crustal movement, especially during the Wenlockian: Acad. Brasil. Ciências An., v. 43 Supl., p. 51-66.

——— 1973, Middle and Upper Silurian chitinozoan succession in Florida subsurface: Jour. Paleontology, v. 47, p. 279-288.

Cramer, F. H., and Díez de Cramer, M.d. Carmen, 1972a, Subsurface section from Portugese Guinea dated by palynomorphs as middle Silurian: Am. Assoc. Petroleum Geologists Bull., v. 56, p. 2271-2272.

——— 1972b, Exclusive occurrence of chitinozoans and miospores in a shale of Devonian age from the Malvinas Islands: Ameghiniana, v. 9, p. 220-222.

——— 1972c, North American Silurian palynofacies and their spatial arrangement: Acritarchs: Palaeontographica, v. 138, Abt. B, p. 107-180.

Cramer, F. H., Julivert, M., and Díez, Carmen, 1972, Llandeilian chitinozoans from Rioseco, Asturias, Spain. Preliminary note: Breviora Geol. Astúrica, v. 16, p. 23-25.

Cramer, F. H., Rust, I. C., and Díez de Cramer, M.d. Carmen, 1974, Upper Ordovician chitinozoans from the Cedarberg Formation of South Africa. Preliminary Note: Geol. Rundschau, v. 63, p. 340-345.

Deflandre, G., 1942, Sur les microfossiles des calcaires siluriens de la Montagne Noire, les Chitinozoaires (Eisenack): Acad. Sci. Comptes Rendus, sér. D, v. 215, p. 286-288.

——— 1945, Microfossiles des calcaires siluriens de la Montagne Noire: Ann. Paléont., v. 31, p. 41-75.

Deflandre, G., and Ters, Mireille, 1970, Présence de microplancton silurien fixant l'âge des ampélites associées aux phtanites de Brétignolles (Vendée): Acad. Sci. Comptes Rendus, sér. D, v. 270, p. 2162-2166.

Deunff, J., 1959, Microorganismes planctoniques du Primaire armoricain. I. Ordovicien du Veryhac'h (Presqu'île de Crozon): Soc. Géol. Minéral. Bretagne Bull., n.sér., v. 2, p. 1-41.

Deunff, J., and Chauris, L., 1974, Découverte d'un microplancton à Acritarches, Chitinozoaires et Spores du Silurien supérieur près de Landerneau (Nord-Finistère): Acad. Sci. Comptes Rendus, sér. D, v. 278, p. 2091-2093.

Deunff, J., and Chauvel, J.-J., 1970, Un microplancton à Chitinozoaires et Acritarches dans des niveaux schisteux du grès armoricain (Mayenne et Sud de Rennes): Soc. Géol. France, Compte Rendu Somm. Séances, no. 6, p. 196-198.

Deunff, J., Lefort, J.-P., and Paris, F., 1971, Le microplancton ludlovien des formations immergees des Minquiers (Manche) et sa place dans la distribution du paléoplancton silurien: Soc. Géol. Minéral. Bretagne Bull., sér. C, v. 3, p. 9-28.

Deunff, J., and Paris, F., 1972, Présence d'un paleoplancton a Acritarches, Chitinozoaires, Spores, Scolecodontes et Radiolaires dans les formations Siluro-Devoniennes de la region de Plourach (Côtes-du-Nord): Soc. Géol. Minéral. Bretagne Bull., sér. C, v. 3, p. 83-88.

Dicevitchius, E., 1970, Novye vidy khitinozoyev iz ordovikskikh i siluriyskikh otlozheniy yuzhnoy Pribaltiki i severozapodnoy Belorussii. I. *Acanthochitina* i nekotoryye *Conochitina*, *in* Paleontologiya i stratigrafiya Pribaltiki i Belorussii, sbornik II [VII]: USSR Min. Geol. Inst. Geol. (Vil'nyus), Vilnius, Publishing House "Mintis," p. 5-15.

Díez, Maria d. C. R., and Cramer, F. H., 1974, Morphology of *Pseudoclathrochitina carmenchui* (Cramer 1964), a chitinozoan species from the Ludlovian of Spain: Breviora Geol. Astúrica, v. 18, p. 9-16.

Doubinger, Jeanne, 1963, Etude palyno-planctologique de quelques échantillons du Dévonien inférieur (Siegénien) du Cotentin: Serv. Carte Géol. Alsace Lorraine Bull., v. 16, p. 261-273.

Dunn, D. L., 1959, Devonian chitinozoans from the Cedar Valley Formation in Iowa: Jour. Paleontology, v. 33, p. 1001-1017.

Dunn, D. L., and Miller, T. H., 1964, A distinctive chitinozoan from the Alpena Limestone (Middle Devonian) of Michigan: Jour. Paleontology, v. 38, p. 725-728.

Echols, Dorothy Jung, and Levin, H. L., 1966, Ordovician Chitinozoa from Missouri: Oklahoma Geol. Notes, v. 26, p. 134-139.

Eisenack, A., 1939, Chitinozoen und Hystrichosphaerideen im Ordovizium des Rheinischen Schiefergebirges: Senckenbergiana, v. 21, p. 135-152.

——— 1948, Mikrofossilien aus Kieselknollen des böhmischen Ordoviziums: Senckenbergiana, v. 28, p. 105-117.

——— 1955a, Chitinozoen, Hystrichosphären und andere Mikrofossilien aus dem *Beyrichia*-Kalk: Senckenbergiana Lethaea, v. 36, p. 157-188.

——— 1955b, Neue Chitinozoen aus dem Silur des Baltikums und dem Devon der Eifel: Senckenbergiana Lethaea, v. 36, p. 311-319.

——— 1958, Mikrofossilien aus dem Ordovizium des Baltikums. 1. Markasitschicht, *Dictyonema*-Schiefer, Glaukonitsand, Glaukonitkalk: Senckenbergiana Lethaea, v. 39, p. 389-405.

——— 1959, Neotypen baltischer Silur-Chitinozoen und neue Arten: Neues Jahrb. Geologie Paläontologie Abh., v. 108, p. 1-20.

——— 1962a, Neotypen baltischer Silur-Chitinozoen und neue Arten: Neues Jahrb. Geologie Paläontologie Abh., v. 114, p. 291-316.

——— 1962b, Mikrofossilien aus dem Ordovizium des Baltikums. 2. Vaginatenkalk bis Lyckholmer Stufe: Senckenbergiana Lethaea, v. 43, p. 349-366.

——— 1964, Mikrofossilien aus dem Silur Gotlands. Chitinozoen: Neues Jahrb. Geologie Paläontologie Abh., v. 120, p. 308-342.

——— 1968, Über Chitinozoen des Baltischen Gebietes: Palaeontographica, v. 131, Abt. A, p. 137-198.

——— 1970, Mikrofossilien aus dem Silur Estlands und der Insel Ösel: Geol. Fören. Stockholm Förh., v. 92, p. 302-322.

——— 1972, Chitinozoen und andere Mikrofossilien aus der Bohrung Leba, Pommern: Palaeontographica, v. 139, Abt. A, p. 64-87.

Fink, R. P., 1968, Chitinozoa, Paleozoic problematic from the Middle Devonian of Ohio: M.A. thesis, Washington Univ., Saint Louis, Missouri, 86 p.

Goldstein, R. F., 1970, Comparison of Silurian chitinozoans from Florida well samples with those from the Red Mountain Formation in Alabama and Georgia: M.Sc. thesis, Florida State Univ., Tallahassee, 90 p.

Goldstein, R. F., Cramer, F. H., and Andress, N. E., 1969, Silurian chitinozoans from Florida well samples: Gulf Coast Assoc. Geol. Soc. Trans., v. 19, p. 377-384.

Graindor, M.-J., Robardet, M., and Taugourdeau, P., 1966, Chitinozoaires du Siluro-dévonien dans le nord du Massif Armoricain: Soc. Géol. Nord Ann., v. 85, p. 337-343.

Gray, Jane, Laufeld, S., and Boucot, A. J., 1974, Silurian trilete spores and spore tetrads from Gotland: Their implication for land plant evolution: Science, v. 185, p. 260-263.

Grignani, D., and Mantovani, Maria P., 1964, Les chitinozoaires du sondage Oum Doul 1 (Maroc): Rév. Micropaléontologie, v. 6, p. 243-258.

Henry, J.-L., 1969, Micro-organismes *incertae sedis* (Acritarches et Chitinozoaires) de l'ordovicien de la presqu'île de Crozon (Finistère): Gisements de Mort-Anglaise et de Kerglintin: Soc. Géol. Mineral. Bretagne Bull., n. sér. (1968), p. 61-100.

Jansonius, J., 1964, Morphology and classification of some Chitinozoa: Canadian Petroleum Geology Bull., v. 12, 901-918.

——— 1967, Systematics of the Chitinozoa: Rev. Palaeobotany and Palynology, v. 1, p. 345-360.

——— 1970, Classification and stratigraphic application of Chitinozoa, *in* Ultramicroplankton: North American Paleont. Conv., 1st, Chicago 1969, Proc., Pt. G., p. 789-808.

Jardiné, S., and Yapaudjian, L., 1968, Lithostratigraphie et palynologie du Dévonien-Gothlandien gréseux du bassin de Polignac (Sahara): Rév. Inst. Français Petrole, v. 23, p. 439-469.

Jenkins, W. A. M., 1967, Ordovician Chitinozoa from Shropshire: Palaeontology, v. 10, p. 436-488.

——— 1969, Chitinozoa from the Ordovician Viola and Fernvale Limestones of the Arbuckle Mountains, Oklahoma: Spec. Papers Palaeontology no. 5, 44 p.

——— 1970a, Chitinozoa, *in* Perkins, B. F., ed., Geoscience and Man, v. 1, p. 1-21.

——— 1970b, Chitinozoa from the Ordovician Sylvan Shale of the Arbuckle Mountains, Oklahoma: Palaeontology, v. 13, p. 261-288.

Jodru, R. L., and Campau, D. E., 1961, Small pseudochitinous and resinous microfossils: New tools for the subsurface geologist: Am. Assoc. Petroleum Geologists Bull., v. 45, p. 1378-1391.

Kaljo, D. L., ed., 1970, Silur Estonii: Tallinn, Izdatelstvo "Valgus", 343 p.

Kauffman, A. E., 1971, Chitinozoans in the subsurface lower Paleozoic of west Texas: Kansas Univ. Paleont. Contr., Paper 54, 12 p.

Lange, F. W., 1949, Novos microfósseis devonianos do Paraná: Mus. Paranaense Arquiv., v. 7, p. 287-298.

——— 1952, Chitinozoários do Folhelho Barreirinha, Devoniano do Pará: Dusenia, v. 3, p. 373-385.

——— 1967a, Subdivisão bioestratigráfica e revisão da coluna Siluro-Devoniana da Bacia do Baixo Amazonas: Simpósio Biota Amazônica Atas, v. 1, p. 215-326.

——— 1967b, Biostratigraphic subdivision and correlation of the Devonian in the Paraná basin, *in* Bigarella, J. J., ed., Problems in Brazilian Devonian geology: Paranaense de Geosciências Bol. 21/22, p. 62-98.

Lange, F. W., and Petri, S., 1967, The Devonian of the Paraná basin, *in* Bigarella, J. J., ed., Problems in Brazilian Devonian geology: Paranaense de Geosciências Bol. 21/22, p. 5-56.

Laufeld, S., 1967, Caradocian Chitinozoa from Dalarna, Sweden: Geol. Fören. Stockholm Förh., v. 89, p. 275-349.

——— 1968, Finds of Chitinozoa in the overthrust nappes of Jemtland: Geol. Fören. Stockholm Förh., v. 90, p. 463.

——— 1971, Chitinozoa and correlation of the Molodova and Restevo Beds of Podolia, U.S.S.R.: Bur. Recherche Géol. Minières Mém., no. 73, p. 291-300.

——— 1973a, Chitinozoa—en dåligt känd mikrofossilgrupp: Fauna och Flora, v. 68, p. 135-141.

——— 1973b, Ordovician chitinozoans from Portixeddu, Sardinia: Soc. Paleont. Italiana Boll., v. 12, p. 3-7.

——— 1974, Silurian Chitinozoa from Gotland: Fossils and Strata, no. 5, 130 p.

——— 1975, Paleoecology of Silurian polychaetes and chitinozoans in a reef-controlled sedimentary regime: Geol. Soc. America Abs. with Programs, v. 7, p. 804-805.

Laufeld, S., Bergström, J., and Warren, P. T., 1975, The boundary between the Silurian *Cyrtograptus* and *Colonus* Shales in Skåne, southern Sweden: Geol. Fören. Stockholm Förh., v. 97, p. 207-222.

Laufeld, S., and Jeppsson, L., 1976, Silicification and bentonites in the Silurian of Gotland: Geol. Fören. Stockholm Förh., v. 98, p. 31-44.

Lefort, J.-P., and Deunff, J., 1971, Esquisse géologique de la partie méridionale du golfe normano-breton (Manche): Acad. Sci. Comptes Rendus, sér. D, v. 272, p. 16-19.

——— 1974, Etude du socle antémésozoïque de la partie septentrionale du golfe normano-breton: Bur. Recherches Géol. Minières Bull., sér. 2, section 4, no. 2, p. 73-83.

Legault, Jocelyne A., 1973a, Mode of aggregation of *Hoegisphaera* (Chitinozoa): Canadian Jour. Earth Sci., v. 10, p. 793-797.

——— 1973b, Chitinozoa and Acritarcha of the Hamilton Formation (Middle Devonian), southwestern Ontario: Canada Geol. Surv. Bull., v. 221, 103 p.

Lindström, M., 1953, On the Lower Chasmops Beds in the Fågelsång District (Scania): Geol. Fören. Stockholm Förh., v. 75, p. 125-148.

Lister, T. R., Cocks, L. R. M., and Rushton, A. W. A., 1969, The basement beds in the Bobbing borehole, Kent: Geol. Mag., v. 106, p. 601-603.

Lister, T. R., and Downie, C., 1967, New evidence for the age of the primitive echinoid *Myriastiches gigas:* Palaeontology, v. 10, p. 171-174.

Magloire, Lily, 1967, Etude stratigraphique, par la Palynologie, des depots argilo-greseux du Silurien et du Devonien inferieur dans la Region du Grand Erg Occidental (Sahara Algerien), *in* Oswald, D. H., ed., Internat. Symposium on the Devonian System, Calgary 1967: Calgary, Alberta Soc. Petroleum Geologists, v. 2, p. 473-491.

Männil, R., 1972a, The zonal distribution of chitinozoans in the Ordovician of the East Baltic area: Internat. Geol. Cong., 24th, Montreal 1972, Proc., Section 7, p. 569-571.

——— 1972b, Korrelyatsya verkhneviruskikh otlozheniy (Sredniy Ordovik) Shvetsii i Pribaltiki po Chitinozoyam [Chitinozoan correlation of the Upper Viruan (Middle Ordovician) rocks of Sweden and the East Baltic area]: Eesti NSV Tead. Akad., Toim., Keemia-Geol., v. 21, p. 137-142.

Männil, R. M., Polma, L. J., and Hints, L. M., 1968, Stratigrafiya Viruskikh i Harjuskikh otlozheniy (Ordovik) sredney Pribaltiki [Stratigraphy of the Viru and Harju Series (Ordovician) of the central east Baltic area], *in* Stratigrafiya nizhnego Paleozoya Pribaltiki i korrelatsiya s drugimi regionam: USSR Min. Geol. Inst. Geol. (Vil'nyus) Vilnius, Publishing House "Mintis," p. 81-110.

Martin, Francine, 1969, Chitinozoaires de l'Arenig supérieur-Llanvirn inférieur en Condroz (Belgique): Rév. Micropaléontologie, v. 12, p. 99-106.

——— 1974, Ordovicien supérieur et Silurien inférieur à Deerlijk (Belgique). Palynofacies et microfacies: Inst. Royal Sci. Natur. Belg. Mém., v. 174, 71 p.

——— 1975, Sur quelques Chitinozoaires ordoviciens du Quebec et de l'Ontario, Canada: Canadian Jour. Earth Sci., v. 12, p. 1006-1018.

Martin, Francine, Michot, P., and Vanguestaine, M., 1970, Le Flysch Caradocien d'Ombret: Soc. Géol. Belg. Ann., v. 93, p. 337-362.

Miller, T. H., 1967, Techniques for processing and photographing chitinozoans: Kansas Univ. Paleont. Contr., Paper 21, 10 p.

Miller, M. A., 1975, Chitinozoa from the Upper Ordovician Maysvillian age strata, Kentucky: Geol. Soc. America Abs. with Programs, v. 7, p. 823-824.

——— 1976, Biostratigraphy and paleoecology of Maysvillian (Middle Ordovician) Chitinozoans in the Cincinnati region: M.Sc. thesis, Ohio State Univ. Columbus. 199 p.

Morgan, D. H., 1964, Chitinozoa from the Williston Basin in North Dakota: Compass (Gamma Sigma Epsilon), v. 41, p. 156-161.

Neville, R. W., 1974, Ordovician Chitinozoa from western Newfoundland: Rev. Palaeobotany and Palynology, v. 18, p. 187-221.

Nilsson, R., 1960, A preliminary report on a boring through Middle Ordovician strata in western Scania (Sweden): Geol. Fören. Stockholm Förh., v. 82, p. 218-226.

Obut, A. M., 1973, O geograficheskom rasprostranenii, sravintel'-noy morfologii, ekologii, filogenii i sistematicheskom polozhenii Chitinozoa, *in* Zhuravleva, I. I., ed., Morfologiya i ekologiya vodnykh organismov: Novosibirsk, Nauka Publishers, Siberian Branch, p. 72-84, 147-152, 165-174.

Paris, F., 1971, L'Ordovicien du Synclinorium du Ménez-Bélair (Synclinorium médian armoricain). Ses caractères et sa place dans la paléogeographie centre-armoricaine: Soc. Géol. Nord Ann., v. 91, p. 241-251.

Paris, F., and Deunff, J., 1970, Le Paléoplancton Llanvirnien de la Roche-au-Merle (Commune de Vieux-Vy-sur-Couesnon, Ille-et-Vilaine): Soc. Géol. Mineral. Bretagne Bull., v. 2, p. 25-43.

Patrulius, D., Iordan, Magdalena, and Mirauta, Elena, 1967, Devonian of Romania, *in* Oswald, D. H., ed., Internat. Symposium on the Devonian System, Calgary 1967: Calgary, Alberta Soc. Petroleum Geologists, v. 1, p. 127-134.

Pichler, R., 1971, Mikrofossilien aus dem Devon der südlichen Eifeler Kalkmulden: Senckenbergiana Lethaea, v. 52, p. 315-357.

Poumot, C., 1968, *Amphorachitina, Ollachitina, Velatachitina;* trois nouveaux genres de chitinozoaires de l'Erg oriental (Algérie-Tunisie): Centr. Recherche Pau - SNPA Bull., v. 2, p. 45-53.

Răileanu, G., Iordan, Magdalena, Mehmed-Dăneţ, N. A., and Beju, D., 1965, Studiul Devonianului din forajul de la Mangalia: Dări Seamă Şedinţelor, v. 52, p. 323-339.

Răileanu, G., Semaka, A., Iordan, Magdalena, and Mehmed-Dăneţ, N. A., 1965, Le Devonien de la Dobrogea meridionale: Carpatho-Balkan Geol. Assoc. Cong., 7th, Sofia, Rept. II, v. 1, p. 11-15.

Rauscher, R., 1968, Chitinozoaires de l'Arenig de la Montagne Noire (France): Rév. Micropaléontologie, v. 11, p. 51-60.

——— 1971, Les chitinozoaires de l'Ordovicien du Synclinal de May-súr-Orne (Calvados): Soc. Linné. Normandie Bull., v. 101, p. 117-127.

Rauscher, R., and Doubinger, Jeanne, 1967, Associations de Chitinozoaires de Normandie et comparaisons avec les faunes déjà décrites: Serv. Carte Géol. Alsace Lorraine Bull., v. 20, p. 307-328.

——— 1970, Les Chitinozoaires des schistes à Calymène (Llanvirnien) de Normandie: Cong. Nat. Soc. Savantes, 92e, Strasbourg et Colmar 1967, v. 2, p. 471-484.

Rhodes, F. H. T., 1961, Chitinozoa from the Ordovician Nod Glas Formation of Merioneth: Nature, v. 192, p. 275-276.

Robardet, M. and others, 1972, La Formation du Pont-de-Caen (Caradocien) dans les synclinaux de Domfront et de Sées (Normandie): Soc. Géol. Nord Ann., v. 92, p. 117-137.

Saidji, M., 1968, Chitinozoans from the Upper Ordovician Maquoketa Formation: M.Sc. thesis, Wisconsin Univ., Madison, 57 p.

Schultz, G., 1967, Mikrofossilien des oberen Llandovery von Dalarna: Univ. Cologne, Geol. Inst., Sonderveröff, no. 13, p. 175-185.

Sommer, F. W., and Boekel, Norma M. van, 1964, Quitinozoários do Devoniano de Goiás: Acad. Brasil. Ciências An., v. 36, p. 423-431.

Sommer, F. W., and Costa, Norma M. van Boekel da, 1965, Novas espécies de Quitinozoários do furo 56, de Bom Jardin, Itaituba, Pará: Div. Geol. Min., Notas Prel. Estud., no. 130, 20 p.

Spielman, J. R., 1974, Chitinozoa from the Ordovician Point Pleasant and Kope Formations of southwestern Ohio: Senior thesis, Ohio State Univ., Columbus, 26 p.

Staplin, F. L., 1961, Reef-controlled distribution of Devonian microplankton in Alberta: Palaeontology, v. 4, p. 392-424.

Stauffer, C. R., 1933, Middle Ordovician Polychaeta from Minnesota: Geol. Soc. America Bull., v. 44, p. 1209-1215.

Taugourdeau, P., 1961, Chitinozoaires du Silurien d'Aquitaine: Rév. Micropaléontologie, v. 4, p. 135-154.

——— 1962a, Associations de Chitinozoaires dans quelques sondages de la région d'Edjelé (Sahara): Rév. Micropaléontologie, v. 4, p. 229-236.

——— 1962b, Association de Chitinozoaires sahariens du Gothlandien supérieur (Ludlowien): Soc. Géol. France Bull., sér. 7, v. 4, p. 806-808.

——— 1963, Etude de quelques espèces critique de Chitinozoaires de la région d'Edjelé et compléments a la faune locale: Rév. Micropaléontologie, v. 6, p. 130-144.

——— 1965, Trois petites associations de Chitinozoaires du Frasnian du Boulonnais: Rév. Micropaléontologie, v. 8, p. 64-70.

——— 1966a, Les Chitinozoaires. Techniques d'études, morphologie et classification: Soc. Géol. France Mém., n. sér., v. 45 (Mém. 104), 62 p.

——— 1967a, Néotypes de Chitinozoaires: Rév. Micropaléontologie, v. 9, p. 258-264.

Taugourdeau, P., and Abdusselamoglu, S., 1962, Présence de Chitinozoaires dans le Siluro-Dévonien turc des environs d'Istambul: Soc. Géol. France, Compte Rendu Somm. Séances, no. 8, p. 238-239.

Taugourdeau, P., and Jekhowsky, B. de, 1960, Répartition et déscription des Chitinozoaires Siluro-dévoniens de quelques sondages de la C.R.E.P.S., de la C.F.P.A. et de la S.N. REPAL au Sahara: Rév. Inst. Français Pétrole, v. 15, p. 1199-1260.

——— 1964, Chitinozoaires Siluriens de Gotland: Comparaison avec les formes sahariennes: Rév. Inst. Français Pétrole, v. 19, p. 845-871.

Taugourdeau, P., and Magloire, Lily, 1965, Développement interne et croissance chez quelques Chitinozoaires: Grana Palynologica, v. 6, p. 128-146.

Tynni, R., 1975, Ordovician hystrichospheres and chitinozoans in limestone from the Bothnian Sea: Finland Geol. Survey Bull. 279, 59 p.

Umnova, Nina I., 1969: Rasprostraneniye Chitinozoa v ordovike Russkoy platformy: Paleont. Zhurn., no. 3, p. 45-62.

Urban, J. B., 1972, A reexamination of Chitinozoa from the Cedar Valley Formation of Iowa with observations on their morphology and distribution: Bull. Am. Paleontology, v. 63, no. 275, 43 p.

Urban, J. B., and Kline, Judith K., 1970, Chitinozoa of the Cedar City Formation, Middle Devonian of Missouri: Jour. Paleontology, v. 44, p. 69-76.

Urban, J. B., and Newport, R. L., 1973, Chitinozoa from the Wapsipinicon Formation (Middle Devonian) of Iowa: Micropaleontology, v. 19, p. 239-246.

Wood, G. D., 1974, Chitinozoa of the Silica Formation (Middle Devonian, Ohio): Vesicle ornamentation and paleoecology: Michigan State Univ., Mus. Pub., Paleontology, ser. 1, p. 127-162.

Wright, R. P. 1976, Occurrence, stratigraphic distribution, and abundance of Chitinozoa from the Middle Devonian Columbus Limestone of Ohio: Ohio Jour. Sci., v. 76, p. 214-224.

APPENDIX

Sources for Plots of Distribution of Chitinozoa

Ordovician

Achab, 1977; Atkinson and Moy, 1971; Beju 1972; Benoit and Taugourdeau, 1961; Bergström and others, 1968; Bergström and Nilsson, 1974; Bouché, 1965; Brenchley and others, 1967; Brück and others, 1974; Carter and Laufeld, 1975; Chauvel and others, 1970; Chlebowski and Szaniawski, 1974; Combaz, 1965, 1968; Cramer-Díez and others, 1972; Cramer and others, 1974; Deunff, 1959; Deunff and Chauvel, 1970; Dicevicius, 1970; Echols and Levin, 1966; Eisenack, 1939, 1948, 1955b, 1958, 1959, 1962a, 1962b, 1968; Goldstein, 1970; Henry, 1969; Jansonius, 1964, 1967; Jenkins, 1967, 1969, 1970b; Kauffman, 1971; Laufeld, 1967, 1968, 1971, 1973b; Lindström, 1953; Lister and others, 1969; Männil, 1972a, 1972b; Männil and others, 1968; Martin, 1969, 1974, 1975; Martin and others, 1970; Miller, T. H., 1967; Miller, M. A., 1975, 1976; Morgan, 1964; Neville, 1974; Nilsson, 1960; Obut, 1973; Paris, 1971; Paris and Deunff, 1970; Poumot, 1968; Rauscher, 1968, 1971; Rauscher and Doubinger, 1967, 1970; Rhodes, 1961; Robardet and others, 1972; Saidji, 1968; Spielman, 1974; Stauffer, 1933; Taugourdeau, 1961, 1967a; Taugourdeau and Jekhowsky, 1960; Tynni, 1975; Umnova, 1969.

Silurian

Bachmann and Schmid, 1964; Beju, 1972; Beju and Danet, 1962; Brück and Downie, 1974; da Costa, 1966, 1967b, 1971b; Boneham and Masters, 1971, 1973; Carter and Laufeld, 1975; Chaiffetz, 1972; Collinson and Schwalb, 1955; Combaz and Poumot, 1962; Correia, 1964; Cramer, 1964, 1966, 1967, 1968a, 1968b, 1969, 1970a, 1970b, 1973; Cramer and Díez, 1972a; Deflandre, 1942, 1946; Deflandre and Ters, 1970; Deunff and Chauris, 1974; Deunff and others, 1971; Deunff and Paris, 1972; Dicevicius, 1970; Díez and Cramer, 1974; Eisenack, 1955a, 1955b, 1959, 1962a, 1964, 1968, 1970, 1972; Goldstein, 1970; Goldstein and others, 1969; Jardiné and Yapaudjian, 1968; Kaljo, 1970; Kauffman, 1971; Lange, 1967a; Laufeld, 1971, 1973a, 1974; Laufeld and others, 1975; Lefort and Deunff, 1971; Lister and Downie, 1967; Magloire, 1967; Martin, 1974; Obut, 1973; Rauscher and Doubinger, 1967; Schultz, 1967; Sommer and van Boekel, 1965; Taugourdeau, 1962a, 1962b, 1963, 1966a, 1967a; Taugourdeau and Abdusselamoglu, 1962; Taugourdeau and Jekhowsky, 1960, 1964; Taugourdeau and Magloire, 1965.

Devonian

Beju, 1972; van Boekel, 1967a; Boneham, 1967; Collinson and Schwalb, 1955; Collinson and Scott, 1958; Combaz and Poumot, 1962; da Costa, 1967; Cousminer 1964; Cramer, 1964; Cramer and Díez, 1972b; Doubinger, 1964; Dunn, 1959; Dunn and Miller, 1964; Eisenack, 1955b; Fink, 1968; Jansonius, 1964, 1967, 1970; Jardiné and Yapaudjian, 1968; Jodru and Campau, 1961; Kauffman, 1971; Lange, 1949, 1952, 1967a, 1967b; Lange and Petri, 1967; Lefort and Deunff, 1974; Legault, 1973a, 1973b; Magloire, 1967; Miller, T. H., 1967; Morgan, 1964; Obut, 1973; Patrulius and others, 1967; Pichler, 1971; Raileanu and others, 1965; Raileanu and others, 1965; Sommer and van Boekel, 1964; Staplin, 1961; Taugourdeau, 1962a, 1963, 1965, 1966a, 1967a; Taugourdeau and Jekhowsky, 1960; Urban, 1972; Urban and Kline, 1970; Urban and Newport, 1973; Wood, 1974; Wright, 1976.

Biogeography of the Stylophoran Carpoids (Echinodermata)

KRAIG DERSTLER, *Department of Geological Sciences, University of Rochester, Rochester, New York 14627*

ABSTRACT

Present information supports the hypothesis that stylophoran carpoids migrated to Gondwana from North America during the Silurian. Cambrian and Ordovician stylophorans are known only from North America and Europe, while Silurian and Devonian representatives are found on these continents as well as in South America, Africa, Australia, and New Zealand. The Gondwana stylophorans were derived from and are closely related to North American and European forms. Stylophoran distribution, limited to certain continents but widespread in these areas, suggests that ocean basins formed a barrier to their dispersal. Paleomagnetic and regional stratigraphic data indicate North America and Gondwana were in close proximity during the Late Silurian through Early Devonian. If the presently known distribution of the stylophoran carpoids is representative of their actual biogeographic range, the stylophorans may have migrated to Gondwana from North America when these two continents approached each other in the Early Silurian.

INTRODUCTION

Caster (1954) noted that the carpoid echinoderms known from the Northern and Southern Hemispheres are distinctly different. He suggested that the Devonian "austral" carpoids were derived from Ordovician "boreal" forms and that the southern forms had been isolated from the northern stock for a sufficient length of time to have evolved into several uniquely austral suborders (Caster, 1954, p. 124). Although new material from North America shows that the austral stylophorans are not as distinctive as originally envisioned, Caster was correct in surmising that these southern carpoids were derived from European or North American (boreal) parent lineages (Derstler, 1975b).

Caster (1954) implied that stylophorans migrated to the southern continents from the Northern Hemisphere sometime during or after the Ordovician and prior to the Early Devonian. These apparent migrations become evident when all known occurrences of the stylophorans are plotted on Paleozoic continental reconstructions (Fig. 1). Carpoid echinoderms were limited to North America and Balto-Europe in the Cambrian and Ordovician. Three suborders survived beyond the Ordovician and are known from North America and Balto-Europe. All 3 appear on Gondwana as well.

In this paper, I examine the possibility that several different groups of stylophoran carpoids migrated to Gondwana from North America during the Siluro-Devonian. Both paleomagnetic and regional stratigraphic information support the interpretation that the two continents came into close proximity during this time interval. If the known data are representative of the actual stylophoran distribution, such a migration event is likely to have taken place.

Appendix 1 contains the stylophoran occurrence data used in this paper. Appendix 2 consists of a taxonomic hierarchy for the stylophoran genera. Anglicized versions of the suprageneric taxa are used throughout the text. Morphologic terminology comes from Ubaghs (1968).

BASE MAPS

Three maps (Fig. 1) are used to show reconstructed continental configurations during the early Middle Cambrian, Cambro-Ordovician, and Early Devonian. These time periods correspond to the first appearance, adaptive radiation, and final proliferation of the stylophoran carpoids.

The early Middle Cambrian map is based upon the continental reconstruction of Jell (1974). It is the only map used herein which is based on biogeographic data, in this case from trilobites. The map does not contradict any geophysical data, but these data are too incomplete to allow an independent global reconstruction (see Klootwijk, 1976). This limitation does not seriously interfere with the interpretations presented in this paper because the few pre-Upper Cambrian stylophorans known are found in areas that were clearly part of the Middle Cambrian North American and European cratons. The early Middle Cambrian reconstruction was projected onto a Lambert Equal Area base so that the map would conform with the other two maps used.

The Cambro-Ordovician and Lower Devonian maps were taken from Briden and others (1974). These maps were modified to show the movement of the Kolyma plate (Ross and Ingham, 1970), the original position of northern Scotland on North America (Ross and Ingham,

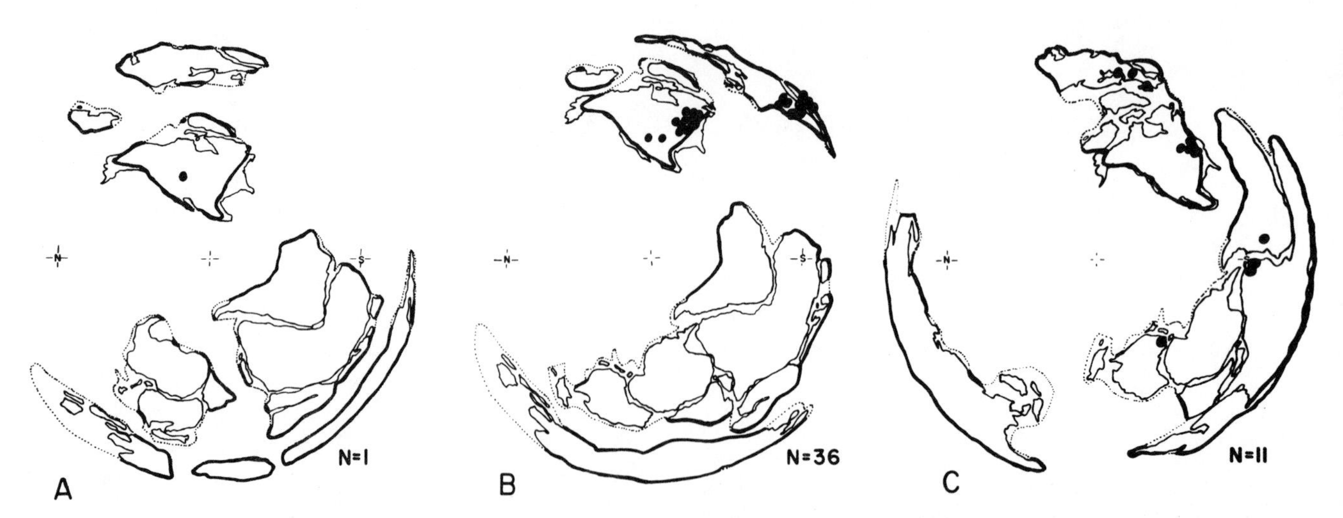

Figure 1. Distribution of the phyllocystid, mitrocystid, and anomalocystid stylophoran carpoids, using Lambert Equal Area projections of paleogeographic reconstructions modified from Jell (1974) and Smith and others (1973). (A) Lower Middle Cambrian; (B) Cambro-Ordovician; (C) Silurian and Lower Devonian.

Dotted lines represent probable Paleozoic cratonic margins; lighter solid lines are present-day coastlines; heavier solid lines are places where Paleozoic cratonic margins lie within present-day coastlines. (N) total occurrences for each time interval.

1970), and the original position of Morocco on the Balto-European craton (Jell, 1974). The maps were rotated about their poles to show North America, Balto-Europe, and Gondwana to best advantage. Schematic renditions of the three continental reconstructions are presented in Figure 4.

Middle Cambrian stylophorans were plotted on the first map, Upper Cambrian through Upper Ordovician stylophorans on the second, and Lower Silurian through Lower Devonian stylophorans on the third. When several different species were found at a single locality, this locality was plotted as a single point, but each species was counted as a separate occurrence. In several instances, a carpoid fauna is known from several localities that are separated by a few kilometers. In these cases, the localities are grouped together as a single collection. Included in this category are the Šárka (Czechoslovakia), the Montagne Noire (France), the Martinsburg (Pennsylvania), the Ponta Grossa (Brazil), the Cincinnatian (Ohio, Kentucky, and Indiana), and the Hunsrückschiefer (West Germany) faunas.

DERIVATION OF POST-ORDOVICIAN STYLOPHORANS

The class Stylophora consists of 3 orders (Derstler, 1975b). The order Cornuta forms a diverse, fairly widespread group that includes the earliest member of the class, an undescribed form from the Middle Cambrian of Utah (Sprinkle, 1976). A second order, the Ceratocystida n. ord., contains only a single species, *Ceratocystis perneri,* from the Middle Cambrian of Czechoslovakia (Ubaghs, 1967). The Cornuta and Ceratocystida have a series of shared, derived (synapomorphic) characters and also many primitive (synplesiomorphic) characters. They are here judged to be sister groups. One family of cornutes, the Amygdalothecidae or advanced phyllocystids, and the third order, Mitrata, have synapomorphic characters and are also believed to be sister groups. Thus, the Ceratocystida and the Cornuta had a common ancestry and the Mitrata evolved from an advanced group of cornutes (Fig. 2).

The Mitrata is a polyphyletic order (see Appendix 3). The mitrate suborder Paranacystida and the advanced phyllocystid-mitrocystid-anomalocystid clade have one set of synapomorphic characters, as well as many plesiomorphic characters that also appear in the cornutes. The advanced phyllocystids, mitrocystids, and anomalocystids have an additional set of synapomorphic features not found in the paranacystids. The Paranacystida appear to have evolved from a cornute ancestor, independent of the advanced phyllocystids and the other mitrates.

The earliest paranacystid is a new species of *Paranacystis* from the "Lower" Eden (Upper Ordovician), giving the minimum age for origination of the suborder. The earliest advanced phyllocystids and mitrates (Mitrocystida and Lagynocystida) come from the Arenig (Lower Ordovician) of the Montagne Noire in France. Thus, both mitrate clades evolved well before the end of the Ordovician.

The Lagynocystida occupy an uncertain position within the Mitrata. The species assigned to this suborder have many unique derived characters. Some of these characters, such as the multibladed stylocone, are convergent upon analogous features in other mitrates. *Chinianocarpos thorali* and the mitrocystids have a series of shared, derived characters, such as paripores and a large dibladed stylocone. Thus, *C. thorali* is judged to be a member of the Mitrocystida. The only uniquely shared character between the mitrocystids and lagynocystids is the subanal plate. This structure is known only in *C. thorali* but is characteristic of the lagynocystids. Considering the lack of other observed synapomorphic characters, the relationship between these two stylophoran groups must remain undecided. If the subanal plate of *C. thorali* and the lagynocystids is homologous, then perhaps these two are sisters. I favor this view but cannot further support it. On the other hand, if subanal plates were independently derived in the two groups, there is no basis for claiming that the two are closely related.

Some advanced mitrocystids and the final mitrate suborder, Anomalocystida, have a series of synapomorphic characters. Only two of these characters are mentioned in the text below (see Appendix 3). One relatively advanced genus of mitrocystid, *Mitrocystella,* attains a notable degree of bilateral symmetry in its theca. This is one character that typifies the Anomalocystida. *M. miloni* has pseudoimbricate ornament, another feature characteristic of the Anomalocystida. This ornament is believed to have served an important food-gathering function (Derstler 1975a, 1975b). The development of this new food-gathering structure may have provided an adaptive breakthrough allowing the stylophorans to occupy new habitat areas. I believe that an advanced mitrocystid lineage evolved a character complex which gave a significant adaptive advantage to the group and allowed for the rapid diversification of the lineage. This lineage gave direct rise to the Anomalocystida.

There is a well-defined morphocline beginning with the mitrocystid *Mitrocystites mitra,* and continuing with *M. dobrotivaensis, Mitrocystella incipens, M. miloni,* an undescribed genus from the Caradoc of North America, and an undoubted anomalocystid, a Caradoc enoplourid (Derstler, 1975b). Perhaps arbitrarily, the anomalocystids are distinguished from the mitrocystids by the presence of two well-developed, laterally-situated anal platelets called posterior spines. The undescribed genus from the Caradoc of North America has posterior spines and here is considered to be the oldest true anomalocystid. There is little difference between the oldest members of the suborder and subsequent species.

Three groups of stylophorans are presently known from post-Ordovician rocks. They include the Anomalocystida, Mitrocystida, and Paranacystida. All are mitrates. These three groups evolved prior to the end of the Ordovician, as demonstrated by the earliest appearance of members of each group. The morphology within each group is fairly consistent although few of the post-Ordovician forms can be assigned to Ordovician genera (Derstler, 1975b). Cambrian and Ordovician stylophorans are known only from rocks that were deposited on the Paleozoic North American and Balto-European cratons (see below). If the occurrence data are representative of the actual biogeographic and temporal ranges of the Stylophora, it would seem that all post-Ordovician forms were derived from North American and Balto-European stocks.

Reconstructions of the major stylophoran groups are presented in Figure 3.

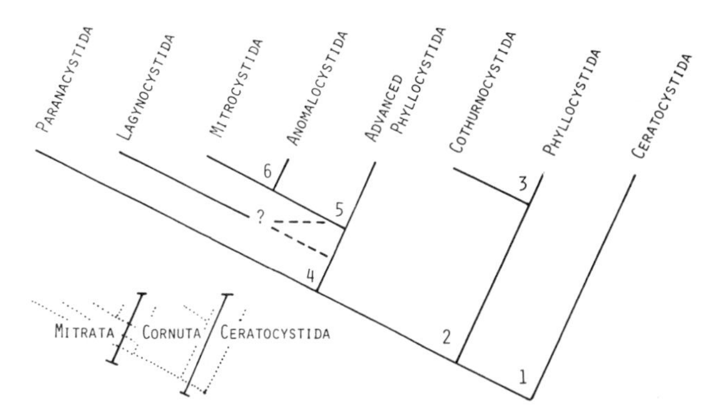

Figure 2. Cladogram of the stylophoran carpoids at the family and subordinal level. Inset shows the ordinal assignment of each of the stylophoran groups. Derived characters at the nodal points and for each end member are given in Appendix 3.

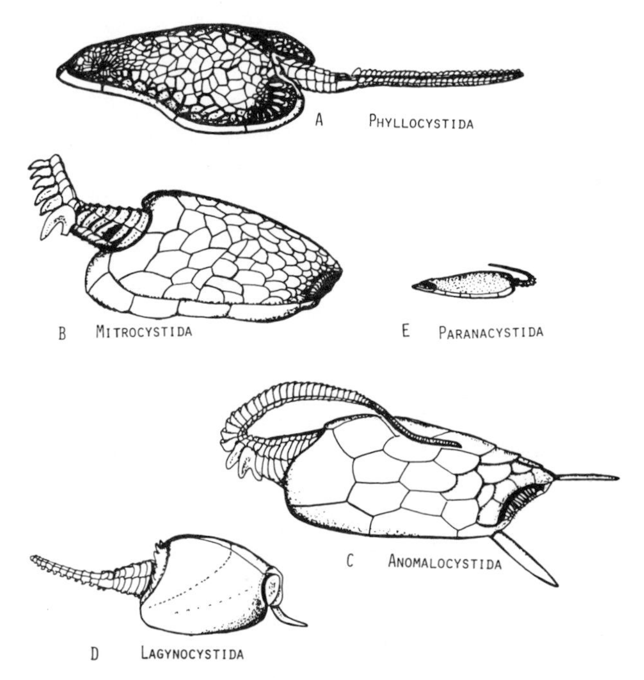

Figure 3. Reconstructions of the stylophoran carpoids discussed in the text. Specimens viewed from the upper posterior lateral position; all figures X1. (A) Phyllocystid cornute, *Phyllocystis blayaci*, Lower Ordovician, France; (B) Mitrocystid mitrate, *Mitrocystites mitra*, Middle Ordovician, Czechoslovakia; (C) Anomalocystid mitrate, *Ateleocystites huxleyi*, Middle and Upper Ordovician, Quebec and Pennsylvania; (D) Lagynocystid mitrate, *Balanocystites* n. sp., Middle Ordovician, Pennsylvania; (E) Paranacystid mitrate, *Paranacystis* n. sp., Upper Ordovician, Pennsylvania.

BIOGEOGRAPHIC DISTRIBUTION

Cornuta and Ceratocystida

The first stylophorans encountered in the stratigraphic record belong to the Cornuta and Ceratocystida (Fig. 4A). The oldest form presently known is an undescribed cornute from the Middle Cambrian Spence Shale of northern Utah

(Sprinkle, 1976). Another species, the only known representative of the Ceratocystida, *Ceratocystis perneri*, is found in the slightly younger Jince Formation from the Middle Cambrian of Czechoslovakia (Ubaghs, 1967). The only other Cambrian stylophorans known are undescribed cornutes from the lower Franconian Davis Formation of southeastern Missouri (Sprinkle, personal commun. 1975) and two cornutes from the uppermost Trempeleau of Nevada (Ubaghs, 1963).

Post-Cambrian cornutes have been described from Europe and Africa. Arenig faunas in Morocco and southern France have produced diverse assemblages (Chauvel, 1971; Ubaghs, 1969). Less diverse cornute assemblages are known from the Llanvirn and Llandeilo of the Šárka and Sv. Dobrotivá Formations of Czechoslovakia (Ubaghs, 1968). Finally, 2 species of specialized cothurnocystids come from the celebrated Starfish Beds in the uppermost Ordovician (Ashgill) of Scotland (Bather, 1913). In general, phyllocystids and cothurnocystids are found in the same faunas. As a result, the distribution patterns for the two groups are nearly coincident.

Advanced phyllocystids occur in the Arenig of France (Ubaghs, 1969) and possibly Morocco (Chauvel, 1971). One additional species has been described from the Llanvirn of Czechoslovakia (Jefferies and Prokop, 1972).

Mitrata

There are 4 suborders of the Mitrata (see Appendix 2). Each suborder is considered separately in the discussion that follows.

MITROCYSTIDA The initial record of this suborder (Fig. 4B) is from the Arenig faunas of Herault, France, where *Chinianocarpos* is found (Ubaghs, 1961). Mitrocystids are found scattered throughout the Llanvirn and Llandeilo rocks of Czechoslovakia, France, and Morocco. One Ashgill species, *Mitrocystites riadensis*, has been reported from Brittany (Chauvel, 1941). *Placocystella capensis* from the Lower Devonian Bokkeveld Beds of South Africa is probably a mitrocystid, based upon the published figures of this form (Caster, 1955; Rennie, 1936). Another Lower Devonian form, "*Mitrocystites? styloideus*," is poorly known but does not appear to be a mitrocystid (Ubaghs, 1968).

ANOMALOCYSTIDA The oldest anomalocystid fossils (Fig. 4C) encountered are 2 undescribed middle Caradoc forms from Pennsylvania. An advanced mitrocystid quite similar to these earliest anomalocystids, *Mitrocystella miloni*, is described from the Llandeilo Schistes à Calymènes of Brittany. No Llandeilo or lower Caradoc anomalocystids have been described to date. Various anomalocystids are found scattered through the Caradoc of Pennsylvania, Kentucky, Quebec, Ontario, Virginia, and possibly Great Britain. Upper Ordovician anomalocystids have been found at a number of localities in North America. If the Ashgillian "*Placocystites bohemicus*" from Czechoslovakia (Barrande, 1887) and Morocco (Chauvel, 1971) are anomalocystids, the Ordovician representatives of the suborder occur outside of North America.

Silurian anomalocystids are uncommon. Single species are known from New York, England, and possibly Gotland. *Victoriacystis wilkinsi* occurs in the Ludlow *Monograptus nilssoni* Zone of Victoria, Australia (Gill and Caster, 1960). Although poorly preserved, these fossils show unmistakably close affinities with typical North American anomalocystids.

Another Australian anomalocystid was reported by Gill and Caster (1960) from rocks they tentatively assigned to the Lower Silurian. Fossils were reported to be extremely rare at this locality. The age was based upon a fauna found in rocks of similar lithology several kilometers away. While these fossils are most assuredly anomalocystids, the correlation of the rocks bearing them is open to question. Until a more diagnostic fauna is found associated with these carpoids, the Ludlow *Victoriacystis wilkinsi* must be considered the oldest anomalocystid known outside of North America, Europe, or Morocco.

Anomalocystid carpoids are moderately widespread in the Lower Devonian. They are known from Australia, North America, West Germany, South Africa, and Brazil. No anomalocystids or other stylophorans have been found in post-Lower Devonian rocks.

LAGYNOCYSTIDA The earliest representatives of the third mitrate suborder, the Lagynocystida (Fig. 4D), occur in the Arenig of southern France, where they are associated with a variety of cornutes and a mitrocystid (Ubaghs, 1969). Lagynocystids also are found in the Middle Ordovician rocks of France and Czechoslovakia. There are 2 Caradoc lagynocystids from North America (Bassler, 1943; Derstler, 1975a). Two Upper Ordovician forms comprise the latest record of this suborder. Both are Upper Ordovician, one from the Ashgill of Scotland (Bather, 1913), and the other from similar age rocks in Morocco (Chauvel, 1971).

Species of *Antifopsis* are not considered in this paper. Pope (1975) briefly mentioned that this genus was probably a lagynocystid, but did not offer any evidence in support of this suggestion, nor were any species or occurrences explicitly mentioned.

PARANACYSTIDA Only 3 species of the Paranacystida are known (Fig. 4E). These include an undescribed species from the Martinsburg Formation of Pennsylvania (Derstler, 1975a), *Paranacystis petrii* from the Lower Devonian of Paraná, Brazil (Caster, 1954), and *Allanicytidium flemingi* from the Lower Devonian of New Zealand (Caster and Gill *in* Ubaghs, 1968).

Occurrence data for the Phyllocystidae, Amygdalothecidae, Mitrocystida, and Anomalocystida are given in Figure 1. Members of these four groups loosely define a "central array" in the Stylophora, perhaps approximating the lineage which began with the earliest stylophoran and ended with the anomalocystids. The other carpoids not represented in this central array are considered, in an equally loose sense, less successful lineages. The members of this central array are more widespread and occur more

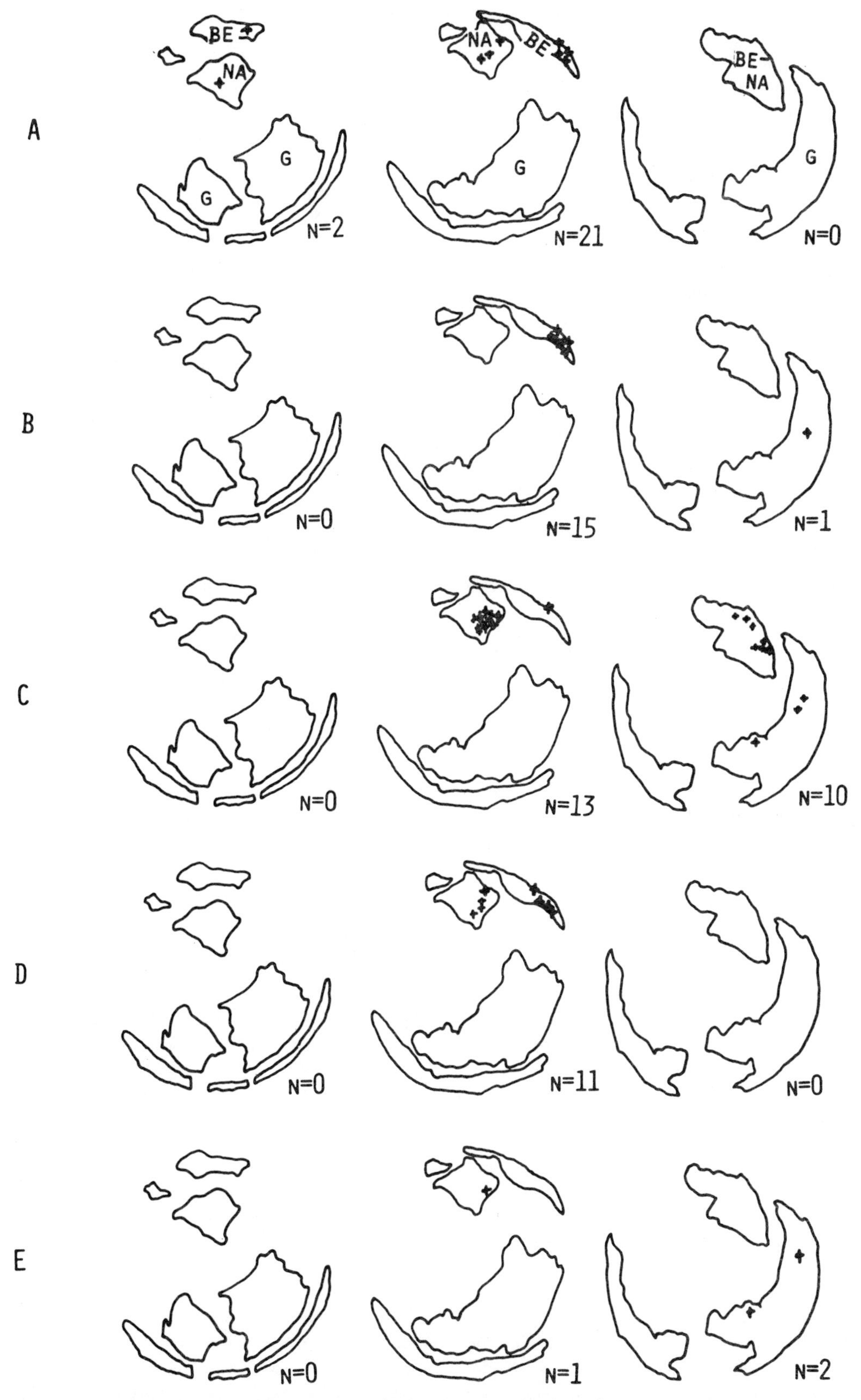

Figure 4. **Biogeographic distribution of the various stylophoran carpoids through time. Map sequence as in Figure 1. (A) Cornuta and Ceratocystida; (B) Mitrocystida; (C) Anomalocystida; (D) Lagynocystida; (E) Paranacystida. (NA) North America; (BE) Balto-Europe; (G) Gondwana; (N) total number of occurrences within each time interval.**

frequently (63 percent of all occurrences) than the other stylophorans. Precarious though it may be to define such a grouping of animals, their biogeographic distribution illustrates the fundamental pattern I wish to discuss. This same pattern is shown in more rigorous fashion in Figure 4.

DISCUSSION

Paleogeography

In this paper, biogeographic inferences are based upon paleogeographic interpretation. The inferences can be valid only to the degree that the paleogeography is based upon sound, nonbiogeographic sources. So it is important to examine the strengths and possible weak points of the continental reconstructions used.

The crux of my argument involves the change in relative positions of the continents between the Ordovician and Early Devonian, as shown in Figures 1B and 1C. The reconstructions are modified from the Cambro-Ordovician and Lower Devonian continental reassemblies of Smith and others (1973). In large part, the original maps prepared by those authors relied upon paleomagnetic information. At best, paleomagnetic data allow ambiguous continental reconstruction. Usually, only paleolatitude can be obtained and this only within the limits of original magnetic declination and experimental error. Such data do not yield any direct information about paleolongitude. However, geometric considerations of the size of the continents can sometimes provide indirect measures of paleolongitude. Possible magnetic polar wandering introduces further uncertainty into the reconstruction. Magnetic polarity reversals complicate paleolatitude determination since there is no way to tell whether magnetic south was coincident or opposite true south. Finally, uncertainty and error in dating the paleomagnetic pole position contribute another limitation to continental reassembly.

Original magnetic declination and possible polar wandering are assumed to be negligible in the data used by Smith and others (1973). Such possible complications cannot be evaluated and must be assumed to contribute no significant error, although they could contribute as much as 20° deviation in paleolatitude determinations. Smith and others minimized experimental error by averaging the available data on each continent for each time interval. The problem of pole reversals was effectively eliminated by implicitly assuming generally slow rates for plate movement. I cannot critically evaluate this assumption, but can only hope that it has led paleogeographers toward the truth. There are unavoidable problems in dating and correlating paleomagnetic data on different continents. These were discussed by Smith and others. The most serious example is in their Cambro-Ordovician reconstruction, where their data spanned roughly 100 m. y. As a result of these limitations and the lack of data in some areas, Smith and others (1973, p. 36) made arbitrary decisions in some cases. In other cases, nonpaleomagnetic data were used to help select between various possible reconstructions.

Regional and historical geology are two prime sources for additional paleogeographic information. Stable cratonic masses, fossil subduction zones, continental plate sutures, and other major features can often be identified, as can the locale, and occasionally the timing of major tectonic events. Smith and others (1973) used such geological interpretations throughout their paper.

Although Smith and others (1973, p. 18) did not directly use sedimentologic data, unique sources of distinctive sediments can provide further resolution in continental reassembly. This information can be used without introducing circularity in biogeographic studies. Such data from Isaacson (1975) are used below.

Continental reassemblies, such as those used herein, imply acceptance of the basic tenets of global tectonic theory. Assumptions about the rates of plate movement, as in Klootwijk (1976), and the constant volume of the globe through time are implied in these continental reconstructions. In the current intellectual climate, these assumptions seem reasonable. Arguments contrary to these assumptions have been offered. For example, H. G. Owen (*in* Smith and others, 1973, p. 40) argued that the Earth may have increased substantially in volume during the Phanerozoic. More recently, other workers have convincingly argued the same point (Carey, 1975). With some evidence to back up their claims, a handful of geologists have urged caution in accepting the entire body of global tectonic theory. So it is with a certain degree of caution that the continental reconstructions of Smith and others (1973) are used in this paper. These reconstructions serve as a working model, albeit a very likely one, to explain the biogeographic distribution of the Stylophora.

Stylophoran carpoids are known from North America, Europe, Africa, Australia, New Zealand, and South America. Smith and others (1973) show these to belong to three major cratons during the lower Paleozoic, here called North America, Balto-Europe, and Gondwana. The main parts of Africa, Australia, New Zealand, and South America formed part of Gondwana (Fig. 1). North America and Europe formed large parts of the Paleozoic continents bearing their names.

Two carpoid-bearing assemblages occur in areas that were originally part of other continents. These relocated continental fragments were left behind as Paleozoic continents became sutured and then split, the rift not perfectly coinciding with the suture. The part of Scotland that includes the Girvan Inlier is believed to have originally been attached to North America (Ross and Ingham, 1970). The western and northern part of Morocco is similarly thought to have been part of the Balto-European continent (Jell, 1974). Unfortunately, the restored positions of both these areas are based on biogeographic data. The faunas are remarkably similar to those found in the hypothesized original continent. The restored positions of Scotland and Morocco are plausible and they are compatible with the histories of plate movement, as I understand these histories, but there are no confirmatory geophysical data.

Biogeographic Distribution Patterns

The stylophoran orders Cornuta and Ceratocystida are limited to North America and Balto-Europe (Figs. 1, 4). Middle Cambrian stylophorans exist on both continents. These are the earliest known stylophorans. Cornutes on both continents have many primitive morphologic features and they have synapomorphic characters as well, suggesting there was some contact between the faunas at some time in the Cambrian or Ordovician. The available data do not offer any insight into where the Stylophora first evolved, nor can anything be said about the extent of the communication between the Balto-European and North American cornute faunas.

The advanced phyllocystids and pre-Caradoc mitrates (suborders Mitrocystida and Lagynocystida) are known only from Balto-Europe. Not until the Caradoc do mitrates appear in North America. Stylophoran fossils generally occur in pelitic rocks (Ubaghs, 1968), and they are usually associated with abundant remains of other echinoderms. *Most* pre-Caradoc Ordovician echinoderm faunas in North America occur in carbonate lithologies, so it is possible that advanced phyllocystids and pre-Caradoc mitrates have been overlooked in this area. Alternatively, there are few pre-Caradoc Ordovician carbonates in Europe to determine if stylophorans lived in carbonate environment in that area. Cambrian and post-Caradoc carbonates of both continents do contain stylophorans, suggesting that these carpoids should be expected in the North American sediments if they were originally there, regardless of the lithology. Relatively numerous and widespread occurrences of Caradoc mitrates, in both carbonate and pelitic sediments, support this view. If the available data are representative of the actual ranges, perhaps the advanced phyllocystids, mitrocystids, and lagynocystids evolved in Balto-Europe and were then introduced into North America. Such a migration would have taken place during or immediately prior to the Caradoc.

Smith and others (1973) show Balto-Europe and North America relatively near each other in the Cambro-Ordovician. Their paleomagnetic data place both cratons on the same approximate latitude. They separated these cratons from each other and from other cratons by an arbitrary longitudinal degree (1973, p. 12). Dewey and Bird (1970) have suggested that the ocean between North America and Europe reached its maximum width in the Late Cambrian and Early Ordovician. After this time, the ocean began to contract, bringing these continents closer together. As the two continents approached each other, they would come within the dispersal range of organisms living on both sides of the ocean. This situation would fit with the data such that mitrates, which had evolved in Balto-Europe, would be able to migrate to North America during the latter half of the Middle Ordovician.

The Paranacystida evolved from advanced phyllocystids. Since the occurrences of advanced phyllocystids seems to indicate that this group was limited to Balto-Europe, the paranacystids would have had to evolve in this area as well. They too would have migrated to North America, in this case sometime prior to their first appearance in North America in the Eden. However, this suggestion is based on very limited data.

Anomalocystids are the most abundant and widespread stylophorans. They first appear in the Caradoc. Advanced mitrocystids, from which the anomalocystids apparently evolved, occur in the slightly older Llandeilo. These ancestral mitrocystids are limited to Balto-Europe during the Ordovician, while the earliest known members of the daughter group are found in North America. An undocumented report of a Caradoc mitrate from Great Britain, possibly an anomalocystid, may indicate that the daughter group was also in Balto-Europe at this time. Perhaps the anomalocystids evolved in Balto-Europe and migrated to North America at the same time as the other mitrates, but the available biogeographic data allow no more than speculation on this point.

One piece of information, not discussed in the section on phylogeny, offers further insight into anomalocystid biogeography. There are two groups of anomalocystids (Derstler, 1975b). *Australocystis langei* is certainly an anomalocystid, but its reported morphology (Caster, 1954) does not fit with either group. Only one-half of the animal is known and the few described fossils are crushed and otherwise poorly preserved. Until more diagnostic material is studied, this form cannot be classified, other than to say it belongs within the suborder Anomalocystida. The one group of anomalocystids, called enoplourids, and the mitrocystids have a suite of relatively synapomorphic characters. The most significant of these is a *pair* of plates in the position of the placocystid plate (Ubaghs, 1968). The second group of anomalocystids, assigned to the family Anomalocystitidae Bassler, have a *single plate* in this same position. Specimens of *Ateleocystites huxleyi,* the earliest member of the family Anomalocystitidae, show a single placocystid plate with two growth centers. Early in the adult ontogeny these two growth centers fused together into a single plate, indicating that the placocystid plate in this species was originally two plates. Later members of the Anomalocystitidae show a single growth center for the adult placocystid plate. All other morphologic characters are consistent with the interpretation that the Anomalocystitidae were derived from enoplourids, rather than being independently derived from the mitrocystids (Derstler, 1975b). Enoplourids and pre-Silurian Anomalocystitidae are known only from North America. All Silurian and Early Devonian anomalocystids have a single plate and are placed within the family Anomalocystitidae. It would seem that the Anomalocystitidae evolved in North America and then spread into Balto-Europe. This point could be tested by documenting the morphology of the Caradoc and Upper Ordovician mitrates reported from Europe.

The first stylophorans encountered outside North America or Balto-Europe are Silurian anomalocystids from Gondwana (Figs. 1, 4). Additional Early Devonian stylophorans are known from Gondwana in widely separated areas. These Early Devonian stylophorans belong to 3 different suborders of the Mitrata and were derived from groups already well-established in North America and Balto-Europe. The appearance of several different lineages on a new continent, previously isolated by oceans, requires more than chance dispersal of 1 or 2 species of stylophorans.

Geological evidence supports the hypothesis that the mitrate stylophorans migrated to Gondwana from North America. Smith and others (1973) show North America and Gondwana separated by approximately 30° in their Lower Devonian map. At this time, Balto-Europe was either sutured to North America or very close to it and in a position far from other continents. Using the map of Smith and others (1973, Fig. 12) and assuming for the moment that the world had equal volume then as now, the separation between the Devonian east coast of North America and the west coast of Gondwana (map direction after Smith and others) could have been as great as 80°, along 50°S latitude, with western North America abutting eastern Gondwana. At the opposite extreme, eastern North America could have been in contact with the west coast of Gondwana. Paleomagnetic information does not allow anything further to be said about the relative separation of the two continents in the Early Devonian.

Regional geologic studies show that North America may have been close to the western coast of Gondwana in the mid-Paleozoic. Several workers have proposed the presence of a western extracratonic sediment source for the Silurian and Devonian rocks in the central Andes, South America. Placing this in the context of the paleogeographic reconstruction, the proposed source of the Andean sediment would have been off the western coast of Gondwana. Berry and Boucot (1972) suggested this western sediment source for the Andean Silurian, and Isaacson (1975) offered compelling evidence for the same source in the Early Devonian. Isaacson and others (1976) briefly reviewed this earlier work. The hypothesis for such a sediment source is based upon a westward-thickening 3,000 m sequence of quartz-rich, shallow-marine, and possibly terrestrial clastic rocks that outcrop in the central Andes. The sequence contains Silurian and Lower Devonian fossils. Isaacson (1975) believed that there were not sufficient rocks exposed in South America during the Devonian to provide the volume of sediment found in the central Andes. Further, his paleoenvironmental reconstructions were in total disagreement with an eastern or northern source for these sediments and strongly suggested a western source. Paleomagnetic data show North America to be the only potential landmass in the proper position to shed large amounts of quartz-rich sediment onto Gondwana. If Berry and Boucot (1972) and Isaacson (1975) are correct, their hypothesis would fit the geophysical data of Smith and others (1973) such that North America could have been the source of the Andean sediment. The Andean sediments were deposited in a shallow-marine environment (Isaacson, 1975). If these sediments were derived from a western source, as they seem to have been, a continuous shelf must have connected Gondwana and the western source area. The paleomagnetic and sedimentologic data indicate to my satisfaction that North America was in contact with the western edge of Gondwana during the Silurian and Early Devonian. At that time, the North American stylophorans could have spread into Gondwana without having to cross an ocean.

The earliest Gondwana stylophorans are Silurian. All three groups on Gondwana were derived from lineages already well-established on North America and Balto-Europe. The paleogeography indicates North America and Gondwana were not in close contact until the Silurian. The available data on stylophoran biogeography and phylogeny, on paleolatitudes of the continents, and on the actual proximity of these continents in the Silurian and Early Devonian are in agreement with the hypothesis that several groups of North American and Balto-European stylophorans migrated to Gondwana during the Silurian.

Paleoecology

Little is known about the paleoecology of the stylophoran carpoids. Many occur in assemblages that could be assigned to Benthic Assemblage 6, the Starfish Community, of Boucot (1975), or its equivalent in Ordovician (and Cambrian?) rocks. These carpoids are generally found associated with relatively abundant articulated remains of starfish and other echinoderms. In regressive sequences, carpoids sometimes are found in the fossiliferous rocks immediately above thick packages of graptolitic shales, as in the Martinsburg Formation in the Appalachians. No systematic survey has been carried out to determine whether stylophorans actually lived in deeper water on the outer shelf thought to be indicated by Benthic Assemblage 6. The most satisfying survey of this sort should rely upon nonpaleontologic evidence of bathymetry. Unfortunately, it is rare to find means of determining paleobathymetry without the use of fossil assemblages.

I have assumed that carpoids did not live in abyssal depths (off the continental shelf), and that their larvae were not able to disperse across wide expanses of ocean. Stylophoran distribution actually may have been limited by some combination of temperature, food availability, competitive exclusion, the high pressure of deep water itself or other environmental parameters, and larval development time. Regardless of the barrier to their dispersal, stylophoran carpoids are limited to a few Paleozoic continents and they were not able to migrate across ocean basins into new continental shelf areas.

Extinction

Worldwide, the latest known stylophorans occur in Lower Devonian rocks. Precise correlations are lacking for some of these faunas, but all are clearly pre-Middle Devonian. Nine Lower Devonian stylophorans belonging to 3 suborders are presently known. Some of these are represented by sizable collections of fossils (Caster, 1955; Derstler, 1975b; Stürmer, personal commun. 1976), indicating moderately large population size. The diversity in the collections is relatively high. From this vantage point of relatively high diversity and abundance, the Stylophora quickly became extinct. For some unknown reason these carpoids died out within a relatively short span of time in the Early Devonian.

SUMMARY AND CONCLUSIONS

Four major points can be made concerning the biogeography of the Stylophora:

(1) Primitive stylophorans, belonging to the orders Cornuta and Ceratocystida, were endemic in the lower Paleozoic to the North American and Balto-European cratons. These echinoderms first appeared in the Middle Cambrian and died out near the end of the Ordovician.

(2) Representatives of the third order, Mitrata, evolved during the earliest Ordovician, probably on the Balto-European craton. They subsequently spread to North America.

(3) Three suborders of the Mitrata survived past the Ordovician and all 3 migrated to Gondwana, probably during the Early Silurian. This migration took place as North America and Gondwana approached and the ocean previously separating the two continents ceased to be a barrier to stylophoran dispersal.

(4) For unknown reasons, all 3 post-Ordovician suborders of Stylophora became extinct during the Early Devonian. Within the resolution of available data, this extinction appears to have happened nearly simultaneously on all continents.

ACKNOWLEDGMENTS

Financial support for this project was provided by the Trustees of the Minnich Fund. E. C. Beutner and M. E. Kauffman, Franklin and Marshall College, Lancaster, Pennsylvania; E. L. Yochelson, U.S. Geological Survey, Washington, D.C.; B. M. Bell, New York State Geological Survey, Albany; and D. M. Raup, University of Rochester, New York, read the manuscript and offered useful suggestions, D. C. Fisher, University of Rochester, provided valuable advice on various intricacies of Hennigean phylogenetic concepts. Judy Tumosa and Lee Gray of the University of Rochester, proofread the manuscript. I am particularly grateful to P. E. Isaacson, Amherst College and the University of Massachusetts, for his discussion concerning this project. I thank these people and their institutions for their aid.

After the manuscript was submitted, several collections of new stylophorans were provided by Craig Clement, Grand Island, New York and Thomas J. DeVries, Amherst College, Amherst, Massachusetts. These are included in Appendix I. I thank both colleagues for their contributions.

REFERENCES

Barrande, J., 1887, Classe des Échinodermes. Ordre des Cystidées, *in* Système Silurien du Centre de la Bohême, Pt. 1: Recherches Paléontologiques, continuation éditée par le Musée Bohême, v. 7, pt. 1, 233 p.

Bassler, R. S., 1935, Description of Paleozoic fossils from the Central Basin of Tennessee: Washington Acad. Sci. Jour., v. 25, p. 403-409.

——— 1943, New Ordovician cystidean echinoderms from Oklahoma: Am. Jour. Sci., v. 241, p. 694-703.

——— 1950, New genera of American Middle Ordovician "Cystoidea": Washington Acad. Sci. Jour., v. 40, p. 273-277.

Bather, F. A., 1913, Caradocian Cystidea from Girvan: Royal Soc. Edinburgh Trans., v. 49, pt. 2, p. 359-529.

Berry, W. B. N., and Boucot, A. J., eds., 1972, Correlation of the South American Silurian rocks: Geol. Soc. Amer. Spec. Paper 133, 59 p.

Billings, E., 1858, On the Cystideae of the Lower Silurian rocks of Canada, *in* Figures and descriptions of Canadian organic remains, decade 3: Canadian Geol. Survey, Montreal, Dawson Brothers, p. 9-74.

Boucot, A. J., 1975, Evolution and extinction rate controls: Amsterdam, Elsevier, 427 p.

Briden, J. C., Drewry, G. E., and Smith, A. G., 1974, Phanerozoic equal-area world maps: Jour. Geology, v. 82, p. 555-574.

Carey, S. W., 1975, The expanding Earth—an essay review: Earth-Sci. Rev., v. 11, p. 105-143.

Caster, K. E., 1952, Concerning *Enoploura* of the Upper Ordovician and its relation to the other carpoid Echinodermata: Bull. Am. Paleontology, v. 34, 47 p.

——— 1954, A new carpoid echinoderm from the Paraná Devonian: Acad. Brasil. Ciências An., v. 26, p. 123-147.

——— 1955, A Devonian placocystoid echinoderm from Paraná, Brasil: Paleontologia do Paraná, Comemorativo Vol., v. 1, p. 137-148.

Chauvel, J., 1941, Recherches sur les Cystoïdes et les Carpoïdes Armoricains: Soc. Géol. Minéral. Bretagne Mém., v. 5, 286 p.

——— 1966, Échinodermes de l'Ordovicien du Maroc: Fr. C.N.R.S. (Cahiers de Paléontologie), 120 p.

——— 1971, Les echinodermes carpoides des Paleozoïque inférieur Marocain: Notes et Mem. Serv. Géol. Maroc, v. 31, no. 237, p. 49-60.

Dehm, R., 1932, Cystoideen aus dem rheinischen Unterdevon: Neues Jahrb. Mineral. Geol. Paläont., v. 69, Abt. B, p. 63-93.

——— 1934, Untersuchungen an Cystoideen des rheinischen Unterdevons: Bayer. Akad. Wiss., mat.-nat. Abt., Sitzungsb., 1934, p. 19-43.

de Koninck, L. G., 1870, On some new and remarkable echinoderms from the British Paleozoic rocks: Geol. Mag., v. 7, p. 258-263.

Derstler, K., 1975a, Carpoid echinoderms from Pennsylvania: Geol. Soc. America Abs. with Programs, v. 7, p. 48.

——— 1975b, Homoiostelean carpoids: B.A. thesis, Franklin and Marshall College, Lancaster, 186 p.

Derstler, K., and Price, J. W., Sr., 1975, *Anomalocystites* in Pennsylvania: Pennsylvania Geology, v. 6, p. 7-9.

Dewey, J. F., and Bird, J. M., 1970, Mountain belts and new global tectonics: Jour. Geophys. Research, v. 75, p. 2625-2647.

Gigout, M., 1954, Sur un hétérostélé de l'Ordovicien marocain: Soc. Sci. Nat. Maroc Bull., v. 34, p. 3-7.

Gill, E. D., and Caster, K. E., 1960, Carpoid echinoderms from the Silurian and Devonian of Australia: Bull. Am. Paleontology, v. 41, 71 p.

Hall, J., 1861, Crinoidea and Cystoidea of the Lower Helderberg Limestones and Oriskany Sandstone: New York Geol. Survey, Paleont., v. 3, p. 99-152.

Isaacson, P. E., 1975, Evidence for a western extracontinental land source during the Devonian Period in the central Andes: Geol. Soc. America Bull., v. 86, p. 39-46.

Isaacson, P. E., Antelo, B., and Boucot, A. J., 1976, Implications of a Llandovery (Early Silurian) brachiopod fauna from Salta Province, Argentina: Jour. Paleontology, v. 50, p. 1103-1112.

Jaekel, O., 1901, Über Carpoideen, eine neue Klasse von Pelmatozoen: Deutsch. Geol. Gesell. Zeitschr., 1900, v. 52, p. 661-677.

——— 1921, Phylogenie und System der Pelmatozoen: Paläont. Zeitschr., v. 3, p. 1-128.

Jefferies, R.P.S., 1968, The subphylum Calcichordata (Jefferies, 1967)—primitive fossil chordates with echinoderm affinities: British Mus. (Nat. History) Bull., Geology, v. 16, p. 243-339.

——— 1973, The Ordovician fossil *Lagynocystis pyramidalis* (Barrande) and the ancestry of *Amphioxus:* Royal Soc. London Philos. Trans., ser. B, Biol. Sci., v. 265, p. 409-469.

——— 1975, Fossil evidence concerning the origin of the chordates: Symposia Zool. Soc. London, no. 36, p. 253-318.

Jefferies, R. P. S., and Prokop, R. J., 1972, A new calcichordate from the Ordovician of Bohemia and its anatomy, adaptations and relationships: Biol. Jour. Linnean Soc. London, v. 4, p. 69-115.

Jell, P. A., 1974, Faunal provinces and possible planetary reconstructions of the Middle Cambrian: Jour. Geology, v. 82, p. 319-350.

Klootwijk, C. T., 1976, The drift of the Indian subcontinent; An interpretation of recent palaeomagnetic data: Geol. Rundschau, v. 65, p. 885-909.

Marr, J. E., 1913, The Lower Paleozoic rocks of the Cautley District (Yorkshire): Geol. Soc. London Quart. Jour., v. 69, p. 1-13.

Meek, F. B., 1872, Description of new species of fossils from the Cincinnati Group of Ohio: Am. Jour. Sci., 3rd ser., v. 3, p. 423-428.

Parsley, R. L., 1969, Studies in Middle Ordovician primitive echinoderms: Ph.D. thesis, Univ. of Cincinnati, Cincinnati, 322 p.

Pope, J. K., 1975, Evidence for relating the Lepidocoleidae, machaeridian echinoderms, to the mitrate carpoids: Bull. Am. Paleontology, v. 67, p. 385-406.

Prokop, R. J., 1963, *Dalejocystis,* n. gen., the first representative of the Carpoidea in the Devonian of Bohemia: Jour. Paleontology, v. 37, p. 648-650.

Reed, F. R. C., 1925, Revision of the fauna of the Bokkeveld Beds: South African Mus. Ann., v. 22, p. 27-226.

Regnéll, G., 1945, Non-crinoid Pelmatozoa from the Paleozoic of Sweden: Lunds Geol.-Mineral. Inst. Medd., no. 108, 255 p.

——— 1960, The Lower Paleozoic echinoderm faunas of the British Isles and Balto-Scandia: Palaeontology, v. 2, p. 161-179.

Rennie, J. V. L., 1936, On *Placocystella,* a new genus of cystids from the Lower Devonian of South Africa: South African Mus. Ann., v. 31, p. 269-275.

Ross, R. J., Jr., and Ingham, J. K., 1970, Distribution of the Toquima-Table Head (Middle Ordovician, Whiterock) faunal realm in the Northern Hemisphere: Geol. Soc. America Bull., v. 81, p. 393-408.

Schuchert, C., 1904, On Siluric and Devonic Cystidea and *Camerocrinus:* Smithsonian Misc. Colln., v. 47, p. 201-272.

Smith, A. G., Briden, J. C., and Drewry, G. E., 1973, Phanerozoic world maps, *in* Hughes, N. F., ed., Organisms and continents through time: Spec. Papers Paleontology no. 12, p. 1-42.

Sprinkle, J., 1973, Morphology and evolution of blastozoan echinoderms: Harvard Univ. Mus. Comparative Zoology, Spec. Pub., 284 p.

——— 1976, Biostratigraphy and paleoecology of Cambrian echinoderms from the Rocky Mountains: Brigham Young Univ. Research Studies Geology Series, v. 23, pt. 2, p. 61-73.

Thoral, M., 1935, Contribution à l'étude paléontologique de l'Ordovicien Inférieur de la Montagne Noire et revision sommaire de la faune Cambrienne de la Montagne Noire: Montpellier, 363 p.

Ubaghs, G., 1961, Un échinoderms nouveau de la classe des carpoïdes dans l'Ordovicien inférieur du département de l'Hérault (France): Acad. Sci. Comptes Rendus, sér. D, v. 253, p. 2565-2567.

——— 1963, *Cothurnocystis* Bather, *Phyllocystis* Thoral and an undetermined member of the order Soluta (Echinodermata, Carpoidea) in the uppermost Cambrian of Nevada: Jour. Paleontology, v. 37, p. 1133-1142.

——— 1967, Le Genre *Ceratocystis* Jaekel (Echinodermata, Stylophora): Kansas Univ. Paleont. Contr., Paper 22, p. 1-16.

——— 1968, Stylophora, *in* Moore, R. C., ed., Treatise on invertebrate paleontology. Pt. S, Echinodermata I: Geol. Soc. America and Univ. Kansas, p. S495-S565.

——— 1969, Les échinodermes carpoïdes de l'Ordovicien Inférieur de la Montagne Noire (France): Fr. C.N.R.S. (Cahiers de Paleontologie), 112 p.

Wetherby, A. G., 1879, Description of a new family and genus of Lower Silurian Crustacea: Cincinnati Soc. Nat. History Jour., v. 1, p. 162-166.

Wilson, A. E., 1946, Echinodermata of the Ottawa Formation of the Ottawa-St. Lawrence Lowland: Canada Geol. Survey Bull., v. 4, 46 p.

Woodward, H., 1880, Notes on the Anomalocystidae, a remarkable family of Cystoidea, found in the Silurian rocks of North America and Britain: Geol. Mag., n. ser., decade 2, v. 7, p. 193-201.

APPENDIX 1

Known Occurrences of Stylophoran Carpoids; Specific Occurrence Data Available in the References Cited After Each Entry; Taxonomy and Unpublished Data from Derstler (1975b)

Middle Cambrian

Ceratocystida n. ord.

Ceratocystis perneri Jaekel: Jince Formation, Skryje and Tejrovic, Czechoslovakia (Ubaghs, 1967).

Cornuta: Phyllocystidae

Undescribed genus and species: Spence Shale, Langston Formation, Calls Fort, Utah (Sprinkle, 1976).

Upper Cambrian-Upper Ordovician

Cornuta: Phyllocystidae

Phyllocystis blayaci Thoral:

(a) Lower Arenig, Herault, France (Ubaghs, (1969).

(b) Upper Fezouata Series, Imfout, Morocco (Gigout, 1954).

Phyllocystis crassimarginata Thoral: Lower Arenig, Herault, France (Ubaghs, 1969).

Phyllocystis sp.: Whipple Cave Formation, Lund, Nevada (Ubaghs, 1963).

Undetermined form: Davis Formation, southeastern Missouri (J. Sprinkle, personal commun. 1975).

Cornuta: Amygdalothecidae ("Advanced Phyllocystids")

Galliaecystis ligniersi Ubaghs: Lower Arenig, Herault, France (Ubaghs, 1969).

Amygdalotheca griffei Ubaghs: Lower Arenig, Herault, France (Ubaghs, 1969).
Reticulocarpos hanusi Jefferies and Prokop: Šárka Formation, Praha, Czechoslovakia (Jefferies and Prokop, 1972).

Cornuta: Cothurnocystidae

Cothurnocystis elizae Bather: Drummuck Group, Girvan, Scotland (Bather, 1913).
Cothurnocystis fellinensis Ubaghs: Lower Arenig, Herault, France (Ubaghs, 1969).
Cothurnocystis courtessolei Ubaghs: Lower Arenig, Audi, France (Ubaghs, 1969).
Cothurnocystis primaeva Thoral: Lower Arenig, Herault, France (Ubaghs, 1969).
Nevadacystis americana (Ubaghs): Whipple Cave Formation, Lund, Nevada (Ubaghs, 1963).
Chauvelicystis ubaghsi (Chauvel): Lower Arenig, Jbel el-Khanutra, Morocco (Chauvel, 1966).
Chauvelicystis spinosa Ubaghs: Lower Arenig, Herault, France (Ubaghs, 1969).
Scotiacystis curvata (Bather): Drummuck Group, Girvan, Scotland (Jefferies, 1968).
Thoralicystis griffei (Ubaghs): Lower Arenig, Herault and Audi, France (Ubaghs, 1969).
Thoralicystis zagoraensis Chauvel: Arenig, Zagora, Morocco (Chauvel, 1971).
Bohemiaecystis bouceki Caster:
(a) Šárka Formation, Barrandium, Czechoslovakia (Caster *in* Ubaghs, 1968).
(b) Sv. Dobrotivá Formation, Barrandium, Czechoslovakia (Caster *in* Ubaghs, 1968).
Undetermined forms: Llandeilo, Jbel Bou-Isidane, Morocco (Chauvel, 1971).

Mitrata: Mitrocystida

Mitrocystites mitra Barrande:
(a) Šárka Formation, Praha, Czechoslovakia (Chauvel, 1941).
(b) Sv. Dobrotivá Formation, Praha, Czechoslovakia (Chauvel, 1941).
Mitrocystites lata Jaekel: Šárka Formation, Praha, Czechoslovakia (Chauvel, 1941).
Mitrocystites dobrotivaensis Chauvel:
(a) Šárka Formation, Praha, Czechoslovakia (Chauvel, 1941).
(b) Sv. Dobrotivá Formation, Praha, Czechoslovakia (Chauvel, 1941).
Mitrocystites riadensis Chauvel: Schistes de Riadan, Poligné, France (Chauvel, 1941).
Mitrocystites? sp.: Middle Ordovician, Andouillé, France (Chauvel, 1941).
Mitrocystites? sp.: Llandeilo, Jbel Bou-Isidane, Morocco (Chauvel, 1971).
Mitrocystella incipens (Barrande): Sv. Dobrotivá Formation, Praha, Czechoslovakia (Chauvel, 1941).
Mitrocystella barrandei Jaekel: Šárka Formation, Rokycany and Praha, Czechoslovakia (Chauvel, 1941).
"*Mitrocystella barrandei*": Llandeilo, Zagora, Morocco (Chauvel, 1971).
Mitrocystella miloni Chauvel:
(a) Schistes à Calymènes, Guichen-Traveusot, France (Chauvel, 1941).
(b) Middle Ordovician, Andouillé, France (Chauvel, 1941).
Mitrocystella? *bohemicus* (Barrande): Zdice Group (Králův Dvůr Shale?), Chadouň, Czechoslovakia (Chauvel, 1941).
Chinianocarpos thorali Ubaghs: Lower Arenig, Herault, France (Ubaghs, 1969).

Mitrata: Anomalocystida

Ateleocystites huxleyi Billings:
(a) Hull Formation, Hull, Quebec (Wilson, 1946).
(b) Martinsburg Formation, Lebanon County, Pennsylvania (Derstler, 1975a).
(c) Kirkfield Limestone, Brechin, Ontario (Derstler, unpub. data).
Enoploura balanoides (Meek):
(a) Maysville, Cincinnati Arch, Ohio, Kentucky, and Indiana (Caster, 1952).
(b) Richmond, Cincinnati Arch, Ohio and Indiana (Caster, 1952).
Enoploura punctata Bassler: Cannon Formation, Pulaski, Tennessee (Bassler, 1935).
Undescribed genus, species 1: Salona Formation, Salona, Pennsylvania (Derstler, 1975a).
Undescribed genus, species 2: Trenton Limestone, Kirkfield, Ontario (Parsley, 1969).
Undescribed genus, species 3: Curdsville Limestone, Lexington Formation, Curdsville, Kentucky (Parsley, 1969).
Undetermined form: Trenton Limestone, Trenton Falls, New York (Derstler, 1975b).
Undetermined form: Salona Formation, Salona, Pennsylvania (Derstler, 1975a).
Undetermined form: Martinsburg? Formation, Salem, Virginia (Derstler, unpub. data).
Undetermined form: Decorah Formation, St. Paul, Minnesota (Derstler, unpub. data).
Undetermined form ("?*Placocystites bohemicus*"): Ashgill, Tazarine, Morocco (Chauvel, 1971).

Mitrata: Lagynocystida

Lagynocystis pyramidalis (Barrande): Šárka Formation, Osek and Praha, Czechoslovakia (Jefferies, 1973).
Peltocystis cornutus Thoral: Lower Arenig, Herault, France (Ubaghs, 1969).
Balanocystites lagenula Barrande:
(a) Šárka Formation, Praha, Czechoslovakia (Chauvel, 1941).
(b) Schistes à Calymènes, Guichen Traveusot, France (Chauvel, 1941).
Anitiferocystis barrandei Chauvel: Šárka Formation, Praha, Czechoslovakia (Chauvel, 1941).
Anitiferocystis minuta Chauvel: Schistes à Calymènes, Guichen Traveusot, France (Chauvel, 1941).

Kirkocystis papillata Bassler: Poolville Member, Bromide Formation, Carter County, Oklahoma (Bassler, 1943).

Balanocystites n. sp.:
(a) Shippensburg Formation, Marion, Pennsylvania (Derstler, 1975a).
(b) Salona Formation, Plainsfield, Pennsylvania (Derstler, unpub. data).

Undetermined form: Drummuck Group, Girvan, Scotland (Bather 1913, Pl. 4, Fig. 47a, b).

Undetermined form: Ashgill, Tazarine, Morocco (Chauvel, 1971).

Mitrata: Paranacystida

Paranacystis n. sp.: Martinsburg Formation, Lebanon County, Pennsylvania (Derstler, 1975a).

Lower Silurian-Lower Devonian

Mitrata: Mitrocystida

Placocystella capensis Rennie: Bokkeveld Series, Gamka Poort, South Africa (Rennie, 1936).

Mitrata: Anomalocystida

Anomalocystites disparilis Hall:
(a) Ridgely Formation, Cumberland, Maryland and Ridgely, West Virginia (Schuchert, 1904).
(b) Shriver Formation, Curtin, Pennsylvania (Derstler and Price, 1975).

Anomalocystites cornutus Hall: Olney Member, Manlius Formation, Dayville, New York (Schuchert, 1904).

Victoriacystis wilkinsi Gill and Caster: Dargile Formation, Heathcote, Victoria, Australia (Gill and Caster, 1960).

"*Victoriacystis* aff. *wilkinsi*": Lower Devonian, Kingslake West, Victoria, Australia (Gill and Caster, 1960, p. 54).

Placocystites forbesianus de Koninck: Dudley Limestone, Dudley, England (Woodward, 1880).

Rhenocystis latipedunculata Dehm: Hunsrückschiefer, Bundenbach, West Germany (Dehm, 1932).

Australocystis langei Caster: Ponta Grossa Shale, Santa Cruz, Paraná, Brazil (Caster, 1955).

"*Placocystis africanus*" Reed: Bokkeveld Series, Buffelskraal, South Africa (Reed, 1925).

Undescribed genus: Keyser Formation, Woodmont, Maryland (Derstler and DeVries, unpub. data).

Undetermined form: Rochester Shale, Clinton Group, Rochester, New York (Derstler, unpub. data).

Undetermined form: Power Glen Sandstone, Medina Group, Jordan, Ontario (Derstler and Clement, unpub. data).

Mitrata: Paranacystida

Paranacystis petrii Caster: Ponta Grossa Shale, Paraná, Brazil (Caster, 1954).

Allanicytidium flemingi Caster and Gill: Reefton Mudstone, Waitahn, South Island, New Zealand (Caster and Gill *in* Ubaghs, 1968).

The following stylophoran carpoids have been reported in the literature but their occurrence and morphology have not yet been documented.

"*Placocystites* n. sp.": Silurian, Gotland (Regnéll, 1945).

"*Ateleocystites* n. sp." and "*Ateleocystites oblongatus* ms.": Shoalshook Limestone?, Shoalshook, England (Regnéll, 1960).

"*Ateleocystites* (?)": lower Ashgill, Cautley, England (Marr, 1913).

"mitrate": Caradoc?, Shropshire, England (Jefferies, 1968).

Undetermined cornute: Šárka or Sv. Dobrotivá Formation, Barrandium, Czechoslovakia (Ubaghs, 1968).

The following forms have been described and their occurrence documented but their affinities or stratigraphic position have not been sufficiently well established for utilization in the present work.

Mitrocystites? *styloideus* Dehm: Hunsrückschiefer, Bundenbach, West Germany (Dehm, 1934).

Spermacystis ensifer (Barrande): Letná Formation, Trubin, Czechoslovakia (Barrande, 1887).

"*Victoriacystis* aff. *wilkinsi*": Melbourne Series, Melbourne and Parkville, Victoria, Australia (Gill and Caster, 1960, p. 52).

Undetermined form (amygdalothecoid?, paranacystid?): Llandeilo, Izgouren, Morocco (Chauvel, 1971).

Pope (1975) has suggested that *Anitifopsis* Barrande was a lagynocystid stylophoran carpoid. The various species assigned to this genus have not been critically evaluated in light of this suggestion. If any, or all, of these species are stylophorans, their geographic distribution would not significantly alter the interpretations put forward in the present work. As such, *Anitifopsis* is not included herein.

Cigara dusli Barrande from the Middle Cambrian of Czechoslovakia may be a primitive stylophoran (Sprinkle, 1973; Derstler, 1975b). Since this fossil is found in the same stratigraphic interval and geographic area as *Ceratocystis perneri*, the conclusions presented in this paper would not be significantly changed. For this reason, and because the form is too poorly known to accurately determine its affinities, it is not incorporated in the data used herein.

Sprinkle (1973, p. 195, 1976) mentioned an undescribed ?carpoid from the Middle Cambrian at Secret Canyon, Nevada. This fossil may also be a stylophoran but determination of its affinities must await future study.

Dalejocystis casteri Prokop was described as a mitrate stylophoran (Prokop, 1963). The original text figures show a plate arrangement and overall morphology similar to the iowacystid solutan carpoids. This form is included within the Soluta in another paper (Derstler, in prep.).

APPENDIX 2

Taxonomic Hierarchy of the Stylophoran Echinoderms (Modified from Derstler, 1975b)

Class Stylophora Gill and Caster, 1960

Order Ceratocystida n. ord.

Family Ceratocystidae Jaekel, 1901: *Ceratocystis* Jaekel, 1901

Order Cornuta Jaekel, 1901

Family Amygdalothecidae Ubaghs, 1969: *Amygdalotheca* Ubaghs, 1969; *Reticulocarpos* Jefferies and Prokop, 1972; *Galliacystis* Ubaghs, 1969

Family Phyllocystidae n. fam.: *Phyllocystis* Thoral, 1935; undescribed genus Sprinkle, 1976

Family Cothurnocystidae Bather, 1913: *Cothurnocystis* Bather, 1913; *Nevadacystis* Ubaghs, 1968; *Chauvelicystis* Ubaghs, 1969; *Thoralicystis* Chauvel, 1971; *Scotiacystis* Caster and Ubaghs *in* Ubaghs, 1968; *Bohemiaecystis* Caster *in* Ubaghs, 1968

Order Mitrata Jaekel, 1921

Suborder Mitrocystida Caster, 1952

Family Mitrocystitidae Ubaghs, 1968: *Mitrocystites* Barrande, 1887; *Mitrocystella* Jaekel, 1901; *Placocystella* Rennie, 1936; *Chinianocarpos* Ubaghs, 1961

Suborder Anomalocystidae Caster, 1952

Family Anomalocystitidae Meek, 1872: *Ateleocystites* Billings, 1858; *Placocystites* de Koninck, 1869; *Rhenocystis* Dehm, 1932; *Anomalocystites* Hall, 1861; *Victoriacystis* Gill and Caster, 1960; *Enoploura* Wetherby, 1879; undescribed genus Derstler, 1975a; *Australocystis* Caster, 1955

Suborder Paranacystida Caster, 1954

Family Paranacystidae Caster, 1954; *Paranacystis* Caster, 1954; *Allanicytidium* Caster and Gill *in* Ubaghs, 1968

Suborder Lagynocystida Caster, 1952

Family Lagynocystidae Jaekel, 1921: *Lagynocystis* Jaekel, 1921

Family Kirkocystidae Caster, 1952: *Peltocystis* Thoral, 1935; *Balanocystites* Barrande, 1887; *Anitiferocystis* Chauvel, 1941; *Kirkocystis* Bassler, 1950

The taxonomic hierarchy used in this paper is designed to conform to *Treatise* convention (Ubaghs, 1968) where possible. New higher taxa are defined only where such action clarifies my interpretations. Resulting from this policy, the major subgroupings of the Mitrata are listed at the subordinal level, while equivalent groups of the Cornuta are considered at the family level.

APPENDIX 3

Derived characters at the nodal points and for each end member of the cladogram shown as Figure 3 in this paper; morphologic terminology after Ubaghs (1968).

(1) dorso-ventrally flattened, asymmetrical theca; right and left thecal lobes, with the right lobe quadrate, having a series of sutural pores on the superior side of the right anterior corner; left lobe rectangular, being the same width as the right lobe but roughly twice as long; periproct on the posterior margin of the left lobe; short glossal, digital and spinal processes; aulacophore insertion at the anterior lateral junction of the two thecal lobes; three adoral plates and two anterior marginals form aulacophore insertion; notch on the anterior margin of the median adoral plate; small pore on the right side of the aulacophore insertion; aulacophore bilaterally symmetrical, extending anteriorly from the theca; proximal aulacophore covered by many small plates; simple cone-shaped stylocone with a few pair of podial basins on the oral surface; distal aulacophore composed of a uniseries of brachial plates, each with a central longitudinal groove and a lateral pair of podial basins on the oral (superior) surface; distally imbricate, subovoid, planar coverplates weakly articulated to either lateral margin of the oral surface of the brachial plates, one pair of coverplates per each pair of podial basins; proximal aulacophore with lateral flexure only; distal aulacophore and stylocone with extremely limited range of flexure.

(2) marginal framework of plates, each plate having a rounded cross section; marginal frame surrounds anterior and lateral portions of the theca; adorals and anterior marginals form a framework around the aulacophore insertion; superior and inferior surfaces covered by many small plates; zygal separates right and left thecal lobes; internal ridge on zygal; anus surrounded by a pyramid of elongate plates.

(3) posterior margin of the theca enclosed by marginal plates, resulting in a complete marginal framework; periproct on the posterior extremity of the superior surface; bilaterally symmetrical tectals and inferiolaterals cover the proximal aulacophore.

(4) thecal outline rounded and somewhat more bilaterally symmetrical than before; no lateral projections; no discrete pores on the right anterior corner of the superior surface; superiocentrals few and with carious texture; distal aulacophore with the ability of flexure in a vertical plane; periproct wide, occupying most of the posterior margin of the theca; tectals and inferiolaterals crudely developed and are analogous with those at (3) but with tectals wider and many small, irregularly arranged platelets between each of these larger elements of the proximal aulacophore; stylocone with a single large aboral spine; anal respiration and stylocone-aided locomotion acquired here (Derstler, 1975a, 1975b).

(5) aboral spines on the distal aulacophore; large, carious inferiocentrals; marginal plates develop marginal flange; anterior lateral corners of theca with serrate margin.

(6) inferiocentral and superiocentrals imperforate and relatively large; posterior superiocentrals imbricate; adoral plates fuse to adjacent anterior superiocentrals; coverplates on the distal aulacophore highly arched over the oral surface; anal platelets modified into small vanes which articulate in small sockets on the interior edge of the posterior "marginals"; a solid ring of plates surrounds the large anal opening; all superiocentrals except the anterior range have a central node on the internal surface; septal system of internal ridge is well developed, being an elaboration upon the primitive internal zygal ridge; marginal plates of the theca have a strong angular cross section, the inferior and superior sides of these plates continuous with the inferior and superior plate surfaces, respectively; inferior surface nearly flat; stylocone with plow-shaped aboral blade and a second, similar, blade extending from the proximal aboral portion of the plate, where it articulates with the proximal aulacophore; groove along both articulatory surfaces of each brachial plate, forming a sutural canal that connects the oral and aboral sides of the distal aulacophore.

Ceratocystida n. ord.: large theca covered by relatively few large plates; accessory thecal pores arranged around the aulacophore insertion; some variability in the number of podial basins on each brachial plate, with occasional fusion of two, three, or four individual plates.

Phyllocystidae n. fam.: theca with a rounded outline that lacks lateral projections and spines; 1 species retains primitive sutural pores in the right anterior corner on the superior surface, all other species with cothurnopores; zygal more symmetrically disposed than in other cornutes, with right and left thecal lobes roughly equal in size. Type genus: *Phyllocystis* Thoral, 1935.

Cothurnocystidae: glossal, digital and spinal highly developed; cothurnopores present (lamellipores derived from cothurnopores within the family); median and right adoral plates fuse into a single plate in all except 1 species (*Cothurnocystis fellenisis*) where primitive condition of three adorals is retained.

Advanced Phyllocystids: right and median adoral plates fuse together; free zygal plate.

Paranacystida: inferior surface of theca with exactly two broad, flat inferiocentrals; stylocone spines quite prominent; anal platelets lost.

Mitrocystida: paripores on the inferior thecal surface; thecal outline not perfectly symmetrical; some relatively advanced members of this group developed pseudoimbricate ornament on the superiocentrals and anterior marginals.

Anomalocystida: pseudoimbricate ornament on superior surface of theca; median plate of the second range of superiocentrals very large; two median plates from the third range of superiocentrals also quite large and with prominent internal nodes (these two plates fuse into a single plate in all subgroups except the most primitive one, the enoplourids); posterior spines, one at either posterior lateral corner of the theca.

Lagynocystida: subanal plate, two or three large adoral plates composed of fused homologs of cornute adorals and superio-centrals, a subanal spine and a stylocone that has several aboral spines. Present observations do not provide sufficient phylogenetic resolution to determine if the mitrocystid-anomalocystid or the mitrocystid-anomalcystid —advanced phyllocystid clade is the sister group of the Lagynocystida.

The mitrate suborder Paranacystida and the advanced phyllocystid-mitrocystid-anomalocystid clade have many synapomorphic characters. The advanced phyllocystids, mitrocystids, and anomalocystids have other synapomorphic characters that are not shared with the Paranacystida. These relationships indicate that the paranacystids have a common ancestry with the other Mitrata and the advanced phyllocystids, but that the Mitrata is a polyphyletic order.

From a strict cladistic point of view, the definition of the Mitrata should be expanded to include the advanced phyllocystids. Alternatively, the suprageneric taxonomy of the Mitrata would require extensive revision to accommodate the present polyphyly within the order. In the interests of stabilizing the higher taxonomy of the Stylophora, neither tack is taken. The overall similarity of appearance within the Cornuta and Mitrata as presently defined (Ubaghs, 1969), and the general acceptance of the distinctness of these 2 higher taxa (e.g., Jefferies, 1975) argue that this taxonomic concept is firmly entrenched. I thus feel justified in defining and accepting the Mitrata as a polyphyletic order.

Cladistic analysis is based upon detailed comparative morphologic analysis of several hundred specimens of stylophoran carpoids and functional interpretations of these forms (see Derstler, 1975b for a list of the specimens studied). That work includes systematic descriptions of the Stylophora and discussion of the functional morphology of these forms. A detailed cladistic analysis of the Stylophora is not included in that work.

Graptolite Biogeography: A Biogeography of Some Lower Paleozoic Plankton

William B. N. Berry, *Department of Paleontology, University of California, Berkeley, California 94720*

ABSTRACT

Graptolite biogeographic regions and provinces were well-developed in the early part of the Ordovician (the latter part of the Tremadoc, the Arenig, and the early part of the Llanvirn). Atlantic and Pacific Faunal Regions may be delineated, and faunal provinces appear to have been present within each Region. The Atlantic Region included most of Europe and North Africa, and northern Argentina, Bolivia, and Peru in South America. Pacific Region faunas are known from North America, western Argentina, Australia, and Asia. Temperature appears to have been a primary control on the Regions. The Pacific Region was tropical and subtropical and the Atlantic Region was situated in relatively cooler waters. The barrier between the Regions began to break down in about mid-Ordovician time (about late Llanvirn-Llandeilo), followed by a general trend toward cosmopolitanism. In addition to the breakup of marked faunal provincialism and regionalism during the mid-Ordovician, many extinctions took place, particularly among dichograptid graptolites. Increased current circulation and a narrowing of the warm-water areas of the seas apparently occurred at about the same time. Most graptolite taxa became extinct in the latest Ordovician, a consequence of widespread lowered sea levels and climatic cooling during glaciation. Silurian-Early Devonian faunas were essentially cosmopolitan. Early Silurian faunas were rich, but extinctions took place at several intervals commencing early in the Wenlock. The reductions were followed by modest radiations. Ecologic selection during the Silurian was for graptolites that lived in near-surface waters and were widely distributed. A consideration of Ordovician graptolite biogeography in conjunction with major lithofacies patterns, indicates that an Ordovician pole lay in, or near, North Africa and that the equator lay within the Pacific Region. These conclusions are consistent with remnant magnetism data for the Early Ordovician.

INTRODUCTION

Patterns in the distribution of animals and plants have both ecological and historical facets. As any adventuresome traveler is aware, similar climatic conditions may be experienced, for example, in desert areas of southern California, Australia, and northern Africa. Despite essentially similar climatic and ecologic situations in these widely separated areas, animals and plants living in them differ. Certain morphological similarities exist among some organisms, but these organisms have, in general, little, if any phyletic relationship. The reasons for the differences lie both in the history of the geographic areas in which the organisms live and in the phyletic history of the organisms themselves. The histories of the areas differ through geologic time and the animals and plants now living in each of them came from different places at different times. The organisms, too, had different histories. Those distributions of organisms that reflect the interplay of ecology, phyletic histories of organisms, and the histories of the areas in which they are found, may be termed *historical biogeographic distributions.* Simpson and others (1957) and George (1962), among others, described several factors pertaining to present-day historical biogeographic distributions of organisms.

Biogeographers (see Darlington, 1957; George, 1962; Simpson and others, 1957) recognize an hierarchical order among the floras and faunas distinguished in historical biogeographic distributions. The hierarchial order stems from analyses of such 19th Century naturalists as A. R. Wallace (1876) and S. P. Woodward (1856), who drew attention to the broad, regional distribution patterns of certain large faunas. Relatively widespread or broad areas with essentially similar faunas came to be distinguished as *faunal regions.* Somewhat smaller areas within each region that are inhabited by a characteristic fauna may be recognized as *faunal provinces.* Commonly, no hard and fast line exists between regions and provinces; because of faunal intergradation, each region and province has a certain degree of faunal resemblance throughout its extent; each also has certain organisms that give it its unique character.

As Simpson and others (1957) stated, ecologic biogeographic distribution of organisms must be kept distinct from historic biogeographic distribution in order to understand why animals and plants are where they are. Simpson and others (1957, p. 712) noted: "Ecological explanations of the distributions of plants and animals . . . explain why a certain kind of community lives in one place, and another kind a mile away, why one lives at the foot of a mountain and another on top, or one in northern Canada and another in the United States." They (Simpson and others, 1957, p. 712) also noted that although "ecological and historical aspects interact and overlap," when broad, regional distribution patterns are examined, it is the historical aspects that predominate. Historical biogeographic regions and provinces provide clues, taken with other evidence, to past positions of continents and oceans and to possible movements between the continents and oceans. Changes in the development of faunal regions and provinces, viewed with the perspective of geologic time, provide insight into changes in ocean currents, land barriers and land connections, climatic barriers, marine connections, and other environmental and climatic developments through time. The changes or developments are, indeed, a fundamental reason for examining faunal regions and provinces at different intervals within geologic time. They form a goal in historical biogeography more basic to geologic history than the simple delineation of historical biogeographic regions and provinces.

GRAPTOLITE HISTORICAL BIOGEOGRAPHY

Graptoloid graptolites, in general, appear to have been marine plankton of relatively large size (when compared with most marine plankton) that lived from early in the Ordovician through the early part of the Devonian. Graptoloid graptolites (hereafter, simply, graptolites) have been recorded from many parts of all continents except Antarctica. They are so abundant in certain rock suites of Ordovician and Silurian age that those rocks have commonly been termed "the graptolitic facies" or "graptolitic biofacies."

Historical biogeography of Ordovician graptolites has been discussed by many authors (Berry, 1959, 1960a, 1967, 1972; Bulman, 1971; Boucek, 1972, 1973; Erdtmann, 1972; Jackson, 1964, 1969; Mu, 1963; Skevington, 1968, 1969, 1973, 1974). Skevington's (1974) recent survey of Ordovician graptolite biogeography went beyond documenting faunal provinces to suggest that latitudinal differences in Ordovician oceanic surface-water temperatures exerted a primary influence on the distribution patterns of Ordovician graptolite faunas. Skevington (1974) and Boucek (1973) clearly indicate in their discussions that the several aspects of graptolite phylogeny and ecologic history that lie at the roots of the historical biogeographic provinces remain to be clarified. As Boucek and Pribyl (1954) and Boucek (1973) have outlined, Ordovician graptolite phylogenies are little understood. For example, Cooper's (1973) analysis of isograptids of the *victoriae* and *caduceus* stocks suggests that these lineages may be closely allied, but what of certain other taxa currently considered isograptids such as *I. gibberulus*? Is *I. gibberulus* allied phyletically with either the *caduceus* or *victoriae* stocks, or is it an example of convergence? How closely allied are all the pendant or so-called "tuning fork" graptolites? These and parallel questions may be raised; answers are needed before a more precise analysis of the historical biogeographic distribution of Ordovician graptolites can be achieved. Siluro-Devonian graptolite phylogeny has been examined far more closely (Hutt and others, 1972; Rickards and others, 1977), and it has been considered in an analysis of Siluro-Devonian graptolite historical biogeography (Berry, 1973).

Ecologic factors have yet to be explored that may have influenced graptolite distribution into possible communities (Warren, 1971) and into possibly coeval yet dissimilar faunas that reflect control by the physical and chemical characteristics of particular water masses. Berry (1962), Berry and Boucot (1972), Boucek (1973), and Lenz (1972) have shown that at least certain graptolites may have been depth-distributed in Ordovician and Silurian oceanic water columns. Whether all graptolites were depth-distributed remains to be tested.

A second ecologic factor that has received but slight attention is the relationship between plankton faunas and water masses. Fager and McGowan (1963) and Russell (1952) described modern plankton faunas that are restricted to oceanic water masses with specific hydrographic characteristics. Russell (1939, 1952) drew attention to the use of certain plankton faunas in establishing the presence or absence of particular water masses near the British Isles. These documented relationships between water masses and plankton faunas in modern seas open the possibility of recognizing coeval but dissimilar graptolite faunas and then using the distribution of specific faunas to establish the extent and influence of possible marine water masses in the past. Ecologic biogeographic distributions must be documented and kept separate from those indicative of historical biogeographic regions and provinces.

These introductory remarks provide a context for the following summary review of historical biogeographic regions and provinces exhibited by the graptolites and for some implications of the patterns that can be delineated. When phyletic histories have been more clearly documented and ecologic controls on distribution of graptolite faunas more precisely understood, the patterns recorded here and the suggested implications of them will require revision.

The time increments used herein are: for the Ordovician—Tremadoc, Arenig-Llanvirn, Llandeilo, Caradoc-Ashgill (or latter part of the Ordovician); for the Silurian—Llandovery, Wenlock, Ludlow, and Pridoli. The characteristic elements of the faunal regions and provinces have been discussed by Berry (1960a, 1967, 1973), Bulman (1971), Boucek (1972, 1973), and Skevington (1969, 1973, 1974). Berry (1972), Boucek (1972), Erdtmann (1972), and Skevington (1974) drew attention to aspects of plate tectonics in relation to Ordovician graptolite provincialism and changes in provincialism. Skevington (1974) depicted Ordovician graptolite provinces in relation to a possible Early Ordovician paleoequator.

Tremadoc

Tremadoc graptolites, particularly those from the upper part of the Tremadoc, are relatively little known. Occurrences of the apparently early Tremadoc *Dictyonema flabelliforme* stock have been cited from many parts of the world, but little can be indicated regarding possible provincialism other than to note that many areas appear to have locally endemic subspecies of this widely cited stock. Other, possibly endemic, taxa may occur locally with members of the *D. flabelliforme* stock.

Successions of graptolites that seem to represent all of the Tremadoc and thus include faunas indicative of the latter part of the Tremadoc, have been recorded from northwestern Canada (Jackson, 1974), the western United States (Berry, 1960b, Ross and Berry, 1963), Quebec (Bulman, 1950a; Osborne and Berry, 1966), Argentina (Harrington and Leanza, 1957), Victoria, Australia (Harris and Keble, 1932), and the Taimyr Peninsula in the USSR (Obut and Sobolevskaya, 1964). Monsen (1925) and Spjeldnaes (1936) described probable late Tremadoc graptolites from the Oslo region, Norway.

Jackson (1974) reviewed certain aspects of Tremadoc graptolite provincialism. His comments and the general similarity of Tremadoc faunas in North America, western Argentina, Victoria, Australia, and Taimyr, USSR, suggest that these parts of the present-day world lay within a Pacific Faunal Region. Tremadoc faunas from Britain, Scandinavia, and North Africa suggest that, collectively, these faunas comprised an Atlantic Faunal Region. Anisograptids, adelograptids, and certain clonograptids characterized the Pacific Region. The Matane fauna of Quebec (Bulman, 1950a) is perhaps the richest known Tremadoc Pacific Region fauna. Atlantic Region Tremadoc faunas are few in number and markedly poorer in species diversity than coeval Pacific Region faunas. Certain species of *Clonograptus* (*C. tenellus*, for example) and *Bryograptus* characterized the Atlantic Region.

Arenig-Llanvirn

Atlantic Faunal Region. Although graptolite faunal regionalism barely had begun to blossom during Tremadoc time, it reached full flower during the Arenig-Llanvirn interval. At no other time throughout the history of the graptolites were faunas so highly provincial. Analyses of Arenig-Llanvirn graptolites (Berry, 1967; Skevington, 1973, 1974) indicate that two faunal regions were in existence, a Pacific and an Atlantic (Skevington, 1974, described these regions as provinces and suggested the existence of subprovinces in them). In terms of present-day areas, the Atlantic Region included Britain (except western Ireland and northwest Scotland), the Baltic, and the remainder of continental Europe, North Africa, and northern Argentina, Bolivia, and Peru in South America. The characteristic elements of the Atlantic Region fauna are the *Didymograptus murchisoni-D. geminus-D. clavulus* group of pendent didymograptids, certain pseudoclimacograptids (*P. paradoxus*, for example), *Aulograptus, Azygograptus*, and certain species in the form-genus *Didymograptus*, such as *D. retroflexus* (*Corymbograptus* of Boucek, 1973) and *D. vacillans*.

The Atlantic Region appears to include at least three provinces, each of which is characterized by its own distinctive association of taxa, certain of which are endemic to each province. Britain represents one of the faunal provinces (excluding western Ireland and northern Scotland). Many phyllograptids, didymograptids of several stocks, and tetragraptids are associated with the earliest known graptolites with biserial scandent rhabdosome form, such as members of the genera *Climacograptus* and *Glyptograptus*. British Arenig and Llanvirn graptolites have been discussed by Jackson (1962) and Skevington (1970), respectively.

Bohemia appears to represent a second faunal province within the Atlantic Region. Boucek (1972, 1973) described Bohemian Arenig-Llanvirn graptolites and cited their stratigraphic ranges. Bohemian Arenig age faunas include few taxa with biserial scandent rhabdosome form. They do include many endemic species (the didymograptids *v-similis*, and *minutus*, for example) as well as *Schizograptus tardibrachiatus*. Bohemian Llanvirn faunas are typified by pseudoclimacograptids (*P. jaroslavi* and *P. paradoxus*, for example) and pendent didymograptids (*D. clavulus* and *D. pseudogeminus*).

Baltic area Arenig-Llanvirn graptolites include endemic species such as *D. balticus s. s., D. vacillans, Tetragraptus phyllograptoides*, and *Phyllograptus angustifolius elongatus*. Monsen (1937), Tjernvik (1956, 1960), and Kaljo (1974) described Baltic Arenig graptolites and summarized Baltic Arenig graptolite biostratigraphy. Jaanusson (1960) recorded Llanvirn taxa that appear to be endemic to the Baltic (*Didymograptus pakrianus*, for example), and Berry (1964) noted their presence in Llanvirn faunas of the Olso region, Norway. The large number of apparently endemic species, such as those cited, appears to characterize a third faunal province of the Atlantic Region.

Only scattered elements of the Atlantic Region faunas have been recorded from Spain (Tamain, 1971), Portugal, Belgium (Bulman, 1950b), Germany (Jaeger, 1967), and northern Africa (Destombes, 1960, 1970; Legrand, 1966).

Bulman (1931) discussed relatively rich Atlantic Region Llanvirn age faunas from Bolivia and Peru. Harrington and Leanza (1957) and Turner (1960) recorded Atlantic Region graptolites from localities in the Jujuy and Salta areas of northern Argentina. The Baltic and South American faunas appear to be the most closely similar of all the faunas from areas within the Atlantic Region, and they are the most highly diversified of the Atlantic Region faunas. The Baltic and South American areas may have lain in the same province.

Pacific Faunal Region. Characteristic Pacific Region graptolites include pseudisograptids, isograptids of the *caduceus* and *victoriae* groups, *Paraglossograptus, Oncograptus, Cardiograptus*, and *Skiagraptus*. Pacific Region faunas are markedly richer in number of different genera and species than any Atlantic Region fauna (Skevington, 1974). Pacific Region faunas have been recorded from many places, particularly those situated in areas now located about the present-day Pacific Ocean. Faunas from Victoria, Australia perhaps are the most typical of the

Pacific Region. Although many taxa remain to be studied closely, the lists provided by Harris and Thomas (1938) and Thomas (1960) suggest the richness of Pacific Region graptolite faunas. Essentially similar faunas have been recorded from North America (Ruedemann, 1947; Berry, 1960b; Jackson, 1964, 1966; Jackson and others, 1965; Ross and Berry, 1963), China (Mu, 1963; Mu and others, 1960; Lee, 1961, 1963; Board of Editors of the Geological Society of China and Institute of Geology of the Chinese National Academy of Science, 1956, 1958), Taimyr, USSR, (Obut and Sobolevskaya, 1964), and western Argentina (Harrington and Leanza, 1957; Turner, 1960). Cooper's (1973) re-evaluation of isograptids described as diverse subspecies of "*I. gibberulus*" by Turner (1960) from western Argentinan localities clarifies the relationship of western Argentina faunas with others in the Pacific Region.

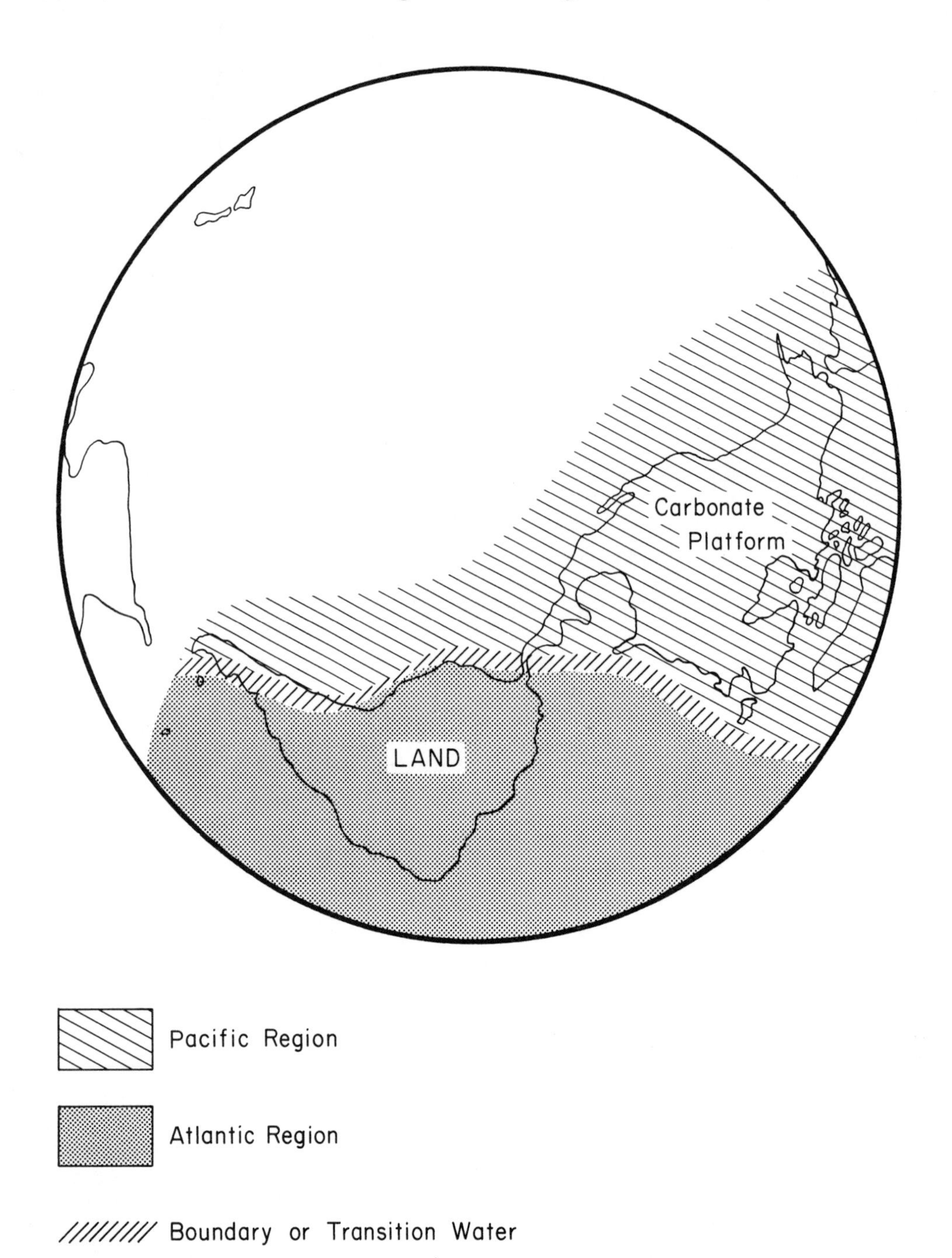

Figure 1. Hemispherical projection of the New World showing positions of the Pacific and Atlantic Faunal Regions during the Arenig-Llanvirn. Transition water between the two Regions, the North American carbonate platform, and an indication that much of South America may have been land are also depicted.

Although most taxa that typify the Pacific Region have been recorded from many localities in the Region, some provinces appear to have been in existence. Faunas from the following areas may be suggested as potential provincial faunas: North America and western Argentina, Australia, southeast Asia (particularly south China), and Asiatic USSR.

Elements of the Atlantic Region fauna appear to intermingle with characteristic Pacific Region faunas in Kazachstan (Nikitin, 1972; Mikhailova, 1974; Tzaj, 1974a), Kirghizia (Zima, 1974), and in the Chu-llyisk Mountains (Tzaj, 1974b). These areas may have lain relatively close to the boundary between the Atlantic and Pacific Regions, and during slight shifts in currents or oceanic climatic

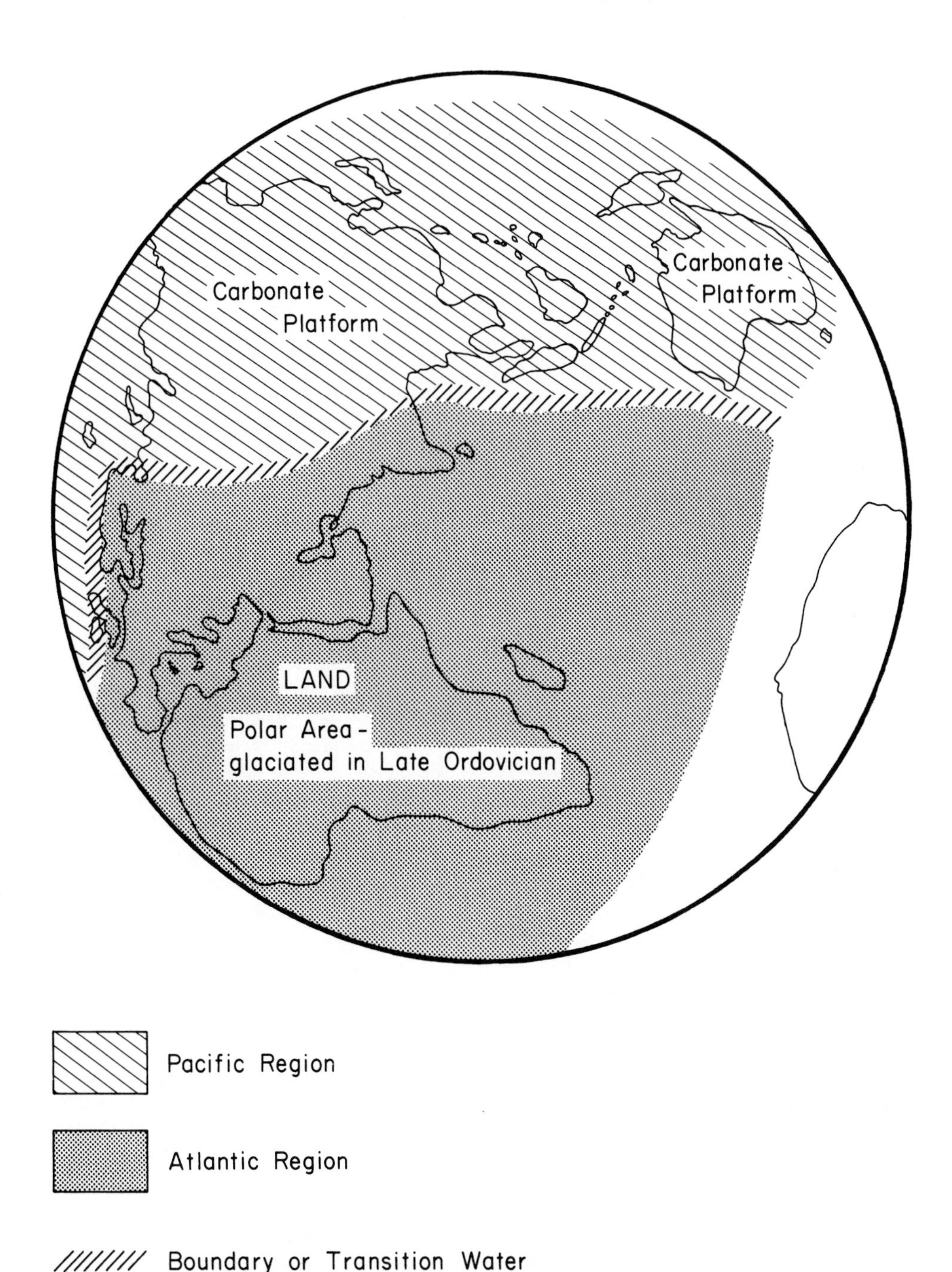

Figure 2. Hemispherical projection of the Old World showing positions of the Pacific and Atlantic Faunal Regions during the Arenig-Llanvirn. Transition water between the two Regions, Asiatic and Australian carbonate platforms, and an indication that much of Africa was land are also depicted.

conditions, faunas from the Atlantic Region may have drifted into the Pacific Region.

In addition to the mingling of Atlantic and Pacific Region taxa in Kazachstan and nearby areas, certain taxa of what may have been primarily Atlantic Region lineages appear to have floated into the Pacific Region during the Llanvirn, when the boundary between the two faunal regions was beginning to break down. Pendent didymograptids of the *bifidus* and *geminus* types are found in certain localities in the Pacific Region. They may occur as "floods" of one or two taxa on certain bedding planes, or they may occur as a few individuals on bedding surfaces rich in numbers of Pacific Region taxa. Kobayashi (1969, p. 254) drew attention to the records of *D. murchisoni* and *D. geminus* in collections made in western Yunnan. *Didymograptus bifidus* and related taxa also occur at many localities in North America from the eastern United States and adjacent parts of Canada westward to the eastern Great Basin (Levis, Quebec; eastern New York; Ozark area, Arkansas; Marathon region, Texas; western Utah). These pendent didymograptids have not been recorded from localities in central or western Nevada, Idaho, or western Canada. They appear to have floated in near-surface waters, because they occur in both relatively near-shore and offshore environments of deposition in association with both relatively shallow-water as well as relatively deep-water benthic organisms. The presence of these near-surface floating didymograptids in that part of North America closest to the boundary between the Atlantic and Pacific Regions at about the time when the boundary was beginning to break down, suggests that these organisms could have drifted relatively readily from one Region to the other; their presence in the Pacific Region suggests, in fact, that currents were beginning to increase after a long interval of relatively minimal activity.

These near-surface dwellers may be seen as harbingers of the more marked incursion of Atlantic Region taxa into the Pacific Region that was soon to follow. The general trend of graptolite development from the late Llanvirn into the Llandeilo was marked by a significant increase in the Pacific Region of stocks that had developed in the Atlantic Region. Berry (1967) illustrated certain graptolite stocks with biserial scandent rhabdosome form that appear to have been a part of this trend during the latter part of the Llanvirn.

Llandeilo

Boucek (1973) drew attention to the marked reduction in number of graptolite species in both Atlantic and Pacific Region faunas during the Llandeilo. Most pendent didymograptids became extinct, as did most phyllograptids, isograptids, pseudisograptids, and tetragraptids. Marked reduction in dichograptid lineages characterized the Llandeilan extinctions. Lineages with biserial scandent rhabdosome form appear to go through this extinction interval virtually unscathed, although some species disappeared from among the climacograptids, glyptograptids, amplexograptids, and pseudoclimacograptids. The general trend in morphologic change during the Llandeilo is toward those rhabdosomes with biserial scandent form, regardless of their phyletic origins. Selection was demonstrably against dichograptids.

The Llandeilo was a time of sharp change in graptolite phyletic history; it was also a time when the pronounced regionalism of the Arenig-Llanvirn was markedly diminished. Never again did graptolite faunas become as fractionated into faunal provinces and regions as they had been in the early part of the Ordovician. The change toward reduced provincialism among graptolites was paralleled by a similar and coeval change in trilobite provinciality (Whittington, 1966).

In addition to the biological changes already noted, certain physical changes also took place. Much of the North American carbonate platform became emergent and a karst topography developed upon it (Berry, 1974). Lands, perhaps islandlike, formed on the eastern and northern margins (in terms of present-day orientation) of the North American platform (Berry, 1974). Beuf and others (1971) indicated that land developed in North Africa across much of the terrain that had been shallow shelf sea during the early part of the Ordovician. Harrington and Leanza (1957) suggested that western and northern Argentina and Bolivia included areas of uplift and marked change in sedimentation in the late Llanvirn-Llandeilo interval.

Late Ordovician (Caradoc-Ashgill)

Late Ordovician graptolite faunas are neither so widely recorded nor so well-known as those from the early part of the Ordovician. Late Ordovician graptolites are not present in North Africa (Destombes, 1971; Skevington, 1974). A few early and possibly middle Caradoc graptolites are known from South America (Turner, 1960), but demonstrable late Caradoc and Ashgill graptolites have not been found in South America. Glaciomarine beds are present in both North African and northern Argentinian-Bolivian sequences of Late Ordovician age; they stratigraphically overlie graptolite-bearing Arenig-Llanvirn age beds in those areas (Beuf and others, 1971; Harrington and Leanza, 1957).

Breakdown of the faunal Region boundary in the late Llanvirn-Llandeilo interval resulted in invasion of the Pacific Region by many stocks with biserial scandent rhabdosome form, as well as by dicellograptids, dicranograptids, nemagraptids, and leptograptids. Most, if not all, of these lineages seem to have originated in the Atlantic Region and to have spread relatively rapidly since their members distinguish the graptolite-bearing beds at the base of the Caradoc in nearly all parts of the world. Their relative abundance and wide geographic spread has led to the suggestion that graptolites were essentially cosmopolitan during the Late Ordovician. Skevington (1969, 1974) suggested, however, that some provinces may have existed during the Caradoc-Ashgill. Certain local centers of endemism appear to have existed because a number of species, as well as combinations of species, known from eastern Australia differ from those found elsewhere. Similarly, south China, North America, western Europe, and Asiatic USSR may have been provincially distinct from each other. Even eastern and western North American faunas differ somewhat, but ecological restriction of certain taxa to eastern areas where they lived in waters close to a landmass, needs to be evaluated.

The richest Late Ordovician faunas appear to be those of North America, Britain, Asiatic USSR, and eastern Australia. These areas may have lain, therefore, in essentially warm, perhaps tropical-subtropical, waters of the Late Ordovician.

Graptolite faunas were reduced markedly at the close of the Ordovician. The extinction of many lineages, including dicellograptids, dicranograptids, leptograptids, amplexograptids, and orthograptids of the *calcaratus* and *quadrimucronatus* groups, characterizes the latest Ordovician. Graptolites with the most morphologically generalized rhabdosome form appear to have been the survivors, whereas extinctions occurred in many other lineages.

Silurian

Silurian graptolites were essentially cosmopolitan (Berry, 1973). Relatively few taxa were endemic to any area, although numbers of local subspecies have been recorded from many geographically small areas. Such geographic subspecies appear to be the result of a certain degree of isolation of local populations. Not enough combinations of species and subspecies are known from any area to recognize distinct provinces.

The major phyletic developments among Siluro-Devonian graptolites were a relatively rapid radiation in the early part of the Llandovery (Early Silurian) followed by reduction in a number of stocks in the latter part of the Llandovery and earliest Wenlock. Graptolites almost became extinct during a short interval in the early Wenlock, about the time of the British *Monograptus riccartonensis* Zone. Only the morphologically simplest stocks with large populations spread widely about the world, and that floated highest in the water column, appear to have survived this short but intense extinction interval. No significant change from essentially cosmopolitan faunas appears to have taken place during the Wenlock, despite the obvious restructuring of planktic graptolite communities.

A moderate radiation followed the early Wenlock extinction, with lineages now grouped in the cyrtograptids, the pristiograptids, and the retiolitids developing significantly. The essential ecological trend appears to have been toward those stocks that lived relatively high in the water column. In general, those colonies that lived in relatively deep waters (based on their occurrences with relatively deep benthic faunas) became fewer in number. A group of extinctions occurred late in the Wenlock. Again, those lineages with relatively simple rhabdosome form that lived in the uppermost water layers survived, whereas taxa with relatively complex structures died out. The latest Wenlock-earliest Ludlow interval is characterized by few graptolite stocks. Those that did exist during the interval are primarily pristiograptids or descendants from that stock, and retiolitids, both of which lived in surface or near-surface waters.

A relatively modest expansion of lineages ensued in the early Ludlow. Neodiversograptids and several stocks that developed from the pristiograptids characterize the early Ludlow radiation. Nearly all stocks appear to have lived in the upper layers of the seas. Graptolites apparently survived only in the near-surface layers. The latter part of the Ludlow, the Pridoli, and the Early Devonian are characterized by a pattern in graptolite radiation that is similar to that of the early Wenlock into early Ludlow. Graptolite lineages were reduced by extinctions in the latter part of the Ludlow and in the latter part of the Pridoli. Final extinction came near the end of the Early Devonian. The extinction intervals are punctuated by slight radiations in the early Pridoli and the Early Devonian. Indeed, the Early Devonian is characterized by a surprising number of new stocks (the *microdon* group, *Abiesgraptus*, the *uniformis* group, the *aequabilis* group, and the *hemiodon* group). All of these stocks appear to have been near-surface dwellers and to have been widespread. Distinct communities can be recognized, but provincial distinctions are not clear, if they were present at all.

In general, Siluro-Devonian graptolites appear to have developed under conditions of at least moderate marine circulation; most taxa were widely distributed and essential cosmopolitanism was the rule. The general ecological trend was one from graptolites spread at different depths in the water column toward stocks that were surface or near-surface dwellers. Community distinctions have been suggested by Warren (1971) and Watkins (in press) for certain Silurian faunas. The physical characteristics of the water in which these faunas lived may have been at the roots of the community distinctions.

POTENTIAL SIGNIFICANCES OF GRAPTOLITE DISTRIBUTION PATTERNS

The general patterns of phyletics, provincialism, and ecology indicated in the foregoing summary can be used to suggest something about marine climates and circulation, and even basic nutrient supply, during the Ordovician-Early Devonian interval. Skevington (1974) summarized data that indicate that the Pacific Region fauna was a warm-water fauna and that the fauna of the Atlantic Region lived in cool waters. He noted that diversity increased toward the waters or areas he included in the Pacific Faunal Region; he documented the distribution of greatest numbers of individuals and peak species diversity in warm waters as contrasted with low species diversity and numbers of individuals in cool-water areas. The data indicate, as Skevington (1974) implied, that when waters become cold, as they must have in present-day North Africa late in the Ordovician, graptolites could no longer survive. Absence of late Caradoc and Ashgill graptolites from South America, and the presence there of latest

Ordovician glaciomarine beds, is consistent with the evidence in North Africa where glacial beds have been documented (see Beuf and others, 1971). The distribution pattern of graptolite faunas in relation to water temperature is closely similar to the pattern exhibited by modern planktic foraminifer faunas in the Pacific Ocean described by Bradshaw (1959). The distribution of widespread carbonate platforms (see Berry, 1974 and Figs. 1, 2) is consistent with Skevington's (1974) suggestion that the Pacific Region lay in a warm, perhaps tropical and subtropical area.

The relatively widespread Pacific Region faunas (Figs. 1, 2) during the Arenig-Llanvirn and the richness of those faunas suggests that warm waters covered a major part of the Earth's surface at that time. Wide distribution of waters with relatively similar temperature indicates that marine currents were probably relatively weak, a factor that could have enhanced the integrity of the provinces. Restructuring of the land-sea relationships in the late Llanvirn-Llandeilo and widespread cooling, at least over much of what is today Africa and South America, probably led to more pronounced marine circulation followed by a general break down of the barrier between the two faunal Regions. The mid-Ordovician invasions of Atlantic Region stocks into the Pacific Region suggests that a general trend in circulation was from colder to warmer water areas.

Many of the phyletic lineages that dominated most Late Ordovician graptolite faunas appear to have had their origins in the Atlantic Region. Faunal provinces in the Early Ordovician Atlantic Region were relatively more limited in area than those in the Pacific, and each of them was characterized by a unique fauna with relatively many endemic taxa. These factors suggest that many populations in the Atlantic Region could have been small. The possibilities for genetic recombination are greater in small populations than in large, thus the relatively small population sizes of stocks in the Atlantic Region faunas could have been an important aspect of the origin of many late Ordovician lineages in the Atlantic Region.

Cool water temperatures and perhaps different types of food resources in the Atlantic Region (many more land areas were present there than in the Pacific Region during the Early Ordovician) appear to have been factors in selection for rhabdosomes with biserial scandent rhabdosome form during marked changes that took place in the late Llanvirn-Llandeilo interval. As water temperatures cooled in the middle part of the Ordovician and as more lands developed in the area of the Pacific Region, ecological conditions of the Pacific Region came increasingly to resemble those in the Atlantic Region. Two consequences of this gradual ecologic change in the Pacific Region were: (1) that graptolites with biserial scandent rhabdosome form were selected for worldwide, and (2) that the dichograptids slowly became extinct.

Circulation apparently improved significantly during the latter part of the Ordovician, with the consequence that faunas became increasingly cosmopolitan and centers of endemism were limited both in number and geographic extent. Late Ordovician extinctions probably reflect both widespread cold waters and marked reduction of those habitats inhabited by graptolites. Possibly, even food resources were sharply diminished.

Deglaciation and general ocean warming appears to have resulted in rich Early Silurian graptolite faunas. Circulation apparently was good because faunas were cosmopolitan.

Locally, large nutrient supplies may have become available in shelf seas during the early part of the Silurian because, as the ice melted, nutrient-rich runoff would have been carried from lands that had been under ice to nearby shelf seas. Increased abundance of nutrients should have resulted in richer and probably more diverse foods (in the form of small planktic organisms, presumably) upon which graptolites could have fed. An abundant and diverse food reservoir may have been a factor in the richness, as well as diversity, of Early Silurian graptolite faunas. In addition, occurrences of certain graptolite taxa with benthic shelly faunal communities indicative of different depths in the shelf seas led to the suggestion that Llandovery graptolites were depth-distributed (Berry and Boucot, 1972). Early Silurian graptolite diversity may thus have been enhanced by different taxa feeding in different segments of the water column.

As oceanic climates continued to warm, circulation probably diminished and graptolite food resources doubtless stabilized in the late Llandovery-Wenlock at a level that was less than that during the early part of the Llandovery. The phyletic history of the graptolites during the late Llandovery is consistent with the speculation that individuals in different lineages came into competition with each other for diminished food resources and extinctions began. Those taxa that lived relatively near the surface would have an advantage in the competition. The fossil record suggests that it was primarily those stocks that survived the middle and late Wenlock extinctions.

Extinction of the graptolites at about the close of the Early Devonian appears to have coincided approximately with the maximum development of a relatively distinctive fauna in the brachiopod-based Malvinokaffric Faunal Province (see Boucot and others, 1969). That fauna has some characteristics of faunas known today from relatively cool waters. Oliver (1976) described Devonian coral distribution and concluded that the Malvinokaffric Province was an area of cold water. The presence of cool to cold waters in the Malvinokaffric Province suggests that perhaps circulation patterns were changing in the early part of the Devonian. Certain aspects of early Middle Devonian history are similar to developments during the late Llanvirn-Llandeilo interval in the Middle Ordovician. These include a general tendency for tectonism in many parts of the world, a change from widespread relatively warm seas to overall cooler oceanic temperatures, and a change toward increased strength in circulation. Significant graptolite extinctions occurred, probably as a result of these and related changes.

In summary, graptolite biogeographic, ecologic, and phyletic relationships may be used to suggest certain possible developments in oceans of the past. At least in terms of an Ordovician world, the Pacific Faunal Region appears to have lain within the tropical-subtropical areas of the time. A pole (Figs. 1, 2) probably was situated in present-day northern Africa, or close to it. The tropical-subtropical belts probably diminished during the Ordovician, to become markedly limited for a time in the latest Ordovician. Warm areas expanded during the early part of the Silurian, to include about the same area that they did early in the Ordovician. That expansion ended during the Wenlock. Warm-water areas may have begun to be reduced during the latter part of the Silurian, during the early development of faunas characteristic of the cool-water Malvinokaffric Province. Latest Early Devonian-early Middle Devonian may have been a relatively cold interval, worldwide. Much of Africa and South America apparently continued to be near a pole from the Ordovician and Silurian into the Middle Devonian, judged from the position and extent of the Malvinokaffric Faunal Province (see Boucot and others, 1969; Oliver, 1976).

These conclusions, considered together with the lithofacies patterns for the Ordovician (Berry, 1974), have led to the position for the Ordovician pole suggested on the accompanying maps (Figs. 1, 2). Smith and others (1973) reviewed certain remnant magnetism data for the Ordovician and suggested that an Early Ordovician South Pole lay in about the same position in northern Africa as suggested herein. The Early Ordovician equator depicted by Smith and others (1973) passed through western North America and Asiatic USSR. The data discussed herein are in close agreement with that suggested position for an Early Ordovician equator.

The change from these Ordovician polar and equatorial positions to those of the present day may be the result of drifting continental plates, or they may be the consequence of a wholesale shift of the Earth's crust over the mantle, or of major shifts in the position of the poles. Shifts are implied from the Earth's crustal configuration indicated in the accompanying maps (Figs. 1, 2).

Although drift episodes may have taken place during the Ordovician-Early Devonian interval, distribution patterns exhibited by the graptolites do not require them. Indeed, Ordovician-Early Devonian graptolite distributions may be explained readily by changes in climates, circulation, and ecology rather than by appealing to drift episodes in that interval.

REFERENCES

Berry, W. B. N., 1959, Distribution of Ordovician graptolites, *in* Sears, Mary, ed., Internat. Oceanogr. Cong. Preprints: Am. Assoc. Adv. Sci., p. 273-274.

——— 1960a, Correlation of Ordovician graptolite-bearing sequences: Internat. Geol. Cong., 21st, Norden 1960, Rept. Pt. 7, p. 97-108.

——— 1960b, Graptolite faunas of the Marathon region, west Texas: Texas Univ. Pub. 6005, 179 p.

——— 1962, Graptolite occurrence and ecology: Jour. Paleontology, v. 36, p. 285-293.

——— 1964, The Middle Ordovician of the Oslo region, Norway. No. 16. Graptolites of the Ogygiocaris Series: Norsk Geol. Tidsskr., v. 44, p. 61-169.

——— 1967, Comments on correlation of the North American and British Lower Ordovician: Geol. Soc. America Bull., v. 78, p. 419-428.

——— 1972, Early Ordovician bathyurid province lithofacies, biofacies, and correlations—their relationship to a proto-Atlantic Ocean: Lethaia, v. 5, p. 69-83.

——— 1973, Silurian-Early Devonian graptolites, *in* Hallam, A., ed., Atlas of palaeobiogeography: Amsterdam, Elsevier, p. 81-87.

——— 1974, Facies distribution patterns of some marine benthic faunas in early Paleozoic platform environments: Palaeogeography, Palaeoclimatology, Palaeoecology, v. 15, p. 158-168.

Berry, W. B. N., and Boucot, A. J., 1972, Silurian graptolite depth zonation: Internat. Geol. Cong., 24th, Montreal 1972, Proc., Section 7, p. 59-65.

Beuf, S. and others, 1971, Les grès du Paléozoique inférieur au Sahara—sedimentation et discontinuites, évolution structurale d'un craton: Inst. Français Pétrole—Science et Technique du Petrole, no. 18, 464 p.

Board of Editors of the Geological Society of China and Institute of Geology of the Chinese National Academy of Science, 1956, Tables of the regional stratigraphy of China: Peking, Science Press, 693 p. (in Chinese).

——— 1958, Tables of the regional stratigraphy of China, Supplement: Peking, Science Press, 190 p. (in Chinese).

Boucek, B., 1972, The palaeogeography of Lower Ordovician graptolite faunas: A possible evidence of continental drift: Internat. Geol. Cong., 24th, Montreal 1972, Proc., Section 7, p. 266-272.

——— 1973, Lower Ordovician graptolites of Bohemia: Prague, Czechoslovak Academy of Sciences, 185 p.

Boucek, B., and Pribyl, A., 1954, Phylogeny and the taxonomy of the graptolites; proposal for a revision of their system: Internat. Geol. Cong., 19th, Algiers 1952, Comptes Rendus, fasc. 19, p. 115-119.

Boucot, A. J., Johnson, J. G., and Talent, J. A., 1969, Early Devonian brachiopod zoogeography: Geol. Soc. America Spec. Paper 119, 106 p.

Bradshaw, J. S., 1959, Ecology of living planktonic Foraminifera in the North and Equatorial Pacific Ocean: Cushman Found. Foram. Research Contr., v. 10, p. 25-64.

Bulman, O. M. B., 1931, South American graptolites with special reference to the Nordenskiöld Collection: Arkiv Zoologi, v. 22A, p. 1-111.

——— 1950a, Graptolites from the *Dictyonema* Shales of Quebec: Geol. Soc. London Quart. Jour., v. 106, p. 63-99.

——— 1950b, On some Ordovician graptolite assemblages of Belgium: Inst. Royal Sci. Natur. Belg. Bull., v. 26, p. 1-8.

——— 1971, Graptolite faunal distribution, *in* Middlemiss, F. A., Rawson, P. F., and Newall, G., eds., Faunal provinces in space and time: Liverpool, Seel House Press, p. 47-60.

Cooper, R. A., 1973, Taxonomy and evolution of *Isograptus* Moberg in Australasia: Palaeontology, v. 16, p. 45-115.

Darlington, P. J., 1957, Zoogeography: The geographical distribution of animals: New York, Wiley, 675 p.

Destombes, J., 1960, Stratigraphie de l'Ordovicien de la partie occidentale du Jbel Bani et du Jbel Zini, Anti-Atlas occidental (Maroc): Soc. Géol. France Bull., sér. 7, v. 7, p. 747-751.

——— 1970, Cambrien moyen et Ordovicien, *in* Colloque internat. sur les corrélation du Précambrien, Agadir-Rabat, Mai 1970, Livret-guide de l'excursion: Anti-Atlas occidental et central: Notes et Mém. Serv. Géol. Maroc, no. 229, p. 161-170.

——— 1971, L'Ordovicien au Maroc: Essai de synthèse stratigraphique: Bur. Recherches Géol. Minières Mém. 73, p. 237-263.

Erdtmann, B.-D., 1972, Ordovician graptolite provincialism: Evidence for continental drift?: Internat. Geol. Cong., 24th, Montreal 1972, Proc., Section 7, Abs., p. 274.

Fager, E. W., and McGowan, J. A., 1963, Zooplankton species groups in the North Pacific: Science, v. 140, p. 453-460.

George, Wilma, 1962, Animal geography: London, Heinemann, 142 p.

Harrington, H. J., and Leanza, A. F., 1957, Ordovician trilobites of Argentina: Kansas Univ. Dept. Geol. Spec. Pub. 1, 276 p.

Harris, W. J., and Keble, R. A., 1932, Victorian graptolite zones with correlations and descriptions of species: Royal Soc. Victoria Proc., v. 44, p. 25-48.

Harris, W. J., and Thomas, D. E., 1938, A revised classification and correlation of the Ordovician graptolite beds of Victoria: Victoria Mining and Geol. Jour., v. 1, p. 62-73.

Hutt, J. E., Rickards, R. B., and Berry, W. B. N., 1972, Some major elements in the evolution of Silurian and Devonian graptoloids: Internat. Geol. Cong., 24th, Montreal 1972, Proc., Section 7, p. 163-173.

Jaanusson, V., 1960, Graptoloids from the Ontikan and Viruan (Ordovician) Limestones of Estonia and Sweden: Uppsala Univ. Geol. Inst. Bull., v. 38, p. 289-366.

Jackson, D. E., 1962, Graptolite zones in the Skiddaw Group in Cumberland, England: Jour. Paleontology, v. 36, p. 300-313.

——— 1964, Observations on the sequence and correlation of Lower and Middle Ordovician graptolite faunas of North America: Geol. Soc. America Bull., v. 75, p. 523-534.

——— 1966, Graptolitic facies of the Canadian Cordillera and Arctic Archipelago: A review: Canadian Petroleum Geology Bull., v. 14, p. 469-485.

——— 1969, Ordovician graptolite faunas in lands bordering North Atlantic and Arctic Oceans: Am. Assoc. Petroleum Geologists Mem. 12, p. 504-512.

——— 1974, Tremadoc graptolites from Yukon Territory Canada, *in* Rickards, R. B., Jackson, D. E., and Hughes, C. P., eds., Graptolite studies in honour of O. M. B. Bulman: Spec. Papers Palaeontology no. 13, p. 35-58.

Jackson, D. E., Steen, G., and Sykes, D., 1965, Stratigraphy and graptolite zonations of the Kechika and Sandpile Groups in northwestern British Columbia: Canadian Petroleum Geology Bull., v. 13, p. 139-154.

Jaeger, H., 1967, Ordoviz auf Rugen, Datierung und Vergleich mit andern Gebieten: Deutsch. Gesell. Geol. Wiss., Ber., Reihe A, Geol. und Paläont., v. 12, no. 1/2, p. 165-176.

Kaljo, D. L., 1974. On graptolite zones of Tremadoc and Arenig of Pribaltic and Moscow synclises, *in* Obut, A. M., ed., Graptolites of the USSR: Novosibirsk, Publishing House "Nauka" Siberian Branch, p. 31-36 (in Russian).

Kobayashi, T., 1969, The Cambro-Ordovician formations and faunas of South Korea. Part X. Stratigraphy of the Chosen Group in Korea and South Manchuria and its relation to the Cambro-Ordovician formations of other areas. Section D. The Ordovician of Eastern Asia and other parts of the continent: Tokyo Imperial Univ. Fac. Sci. Jour., Sec. 11, v. 17, pt. 2, p. 163-316.

Lee, C. K., 1961, Graptolites from the Dawan Formation (Lower Ordovician) of W. Hupeh and S. Kueichou: Acta Palaeontologica Sinica, v. 9, p. 64-74.

——— 1963, Some Middle Ordovician graptolites from Gueizhou: Acta Palaeontologica Sinica, v. 11, p. 570-578.

Legrand, P., 1966, Précisions biostratigraphiques sur l'Ordovicien inferieur et le Silurien des chaines d'Ougarta (Sahara algérien): Soc. Géol. France, Compte Rendu Somm. Séances, no. 7, p. 243-244.

Lenz, A., 1972, Silurian graptolites from eastern Gaspé, Quebec: Canadian Jour. Earth Sci., v. 12, p. 77-89.

Mikhailova, N. F., 1974, New data on Late Ordovician-Early Silurian biostratigraphy of Kazahkstan, *in* Obut, A. M., ed., Graptolites of the USSR: Novosibirsk, Publishing House "Nauka" Siberian Branch, p. 72-82 (in Russian).

Monsen, A., 1925, Über eine neue Ordovizische Graptolithen fauna: Norsk Geol. Tidsskr., v. 8, p. 147-187.

——— 1937, Die Graptolithenfauna im Unteren Didymograptusschiefer (Phyllograptus-schiefer) Norwegens: Norsk Geol. Tidsskr., v. 16, p. 57-266.

Mu, A. T., 1963, Research in graptolite faunas of Chilianshan: Scientia Sinica, v. 12, p. 347-371.

Mu, A. T., Lee, C. K., and Geh, M. Y., 1960, Ordovician graptolites from Xinjiang (Sinkiang): Acta Palaeontologica Sinica, v. 8, p. 27-39.

Nikitin, I. F., 1972, Ordovician of Kazahkstan. Part 1. Stratigraphy: Alma-Alta, Publishing House "Nauka" Kazahkstan, USSR, 240 p. (in Russian).

Obut, A. M., and Sobolevskaya, R. F., 1964, Ordovician graptolites of Taimyr: Akad. Nauk SSSR, Sibir. Otdel., Inst. Geol. i Geofiz.-Nauch.-issled. Inst. Geol. Arkt. 86 p. (in Russian).

Oliver, W. A., 1976, Biogeography of rugose corals: Jour. Paleontology, v. 50, p. 365-373.

Osborne, F. F., and Berry, W. B. N., 1966, Tremadoc rocks at Levis and Lauzon: Naturaliste Canadien, v. 93, p. 133-143.

Rickards, R. B., Hutt, J. E., and Berry, W. B. N., 1977, Evolution of the Silurian and Devonian graptoloids: British Mus. (Nat. Hist.) Bull., Geology, v. 28, no. 1, 120 p.

Ross, R. J., Jr., and Berry, W. B. N., 1963, Ordovician graptolites of the Basin Ranges in California, Nevada, Utah, and Idaho: U. S. Geol. Survey Bull. 1134, 177 p.

Ruedemann, R., 1947, Graptolites of North America: Geol. Soc. America Mem. 19, 652 p.

Russell, F. S., 1939, Hydrographical and biological conditions in the North Sea as indicated by plankton organisms: Jour. Réun. Cons. Internat. Explor. Mer, v. 14, p. 171-192.

——— 1952, The relation of plankton research to fisheries hydrography: Rapp. P.-v. Réun. Cons. Internat. Explor. Mer, v. 131, p. 28-34.

Simpson, G. G., Pittendrigh, C. S., and Tiffany, L. H., 1957, Life: An introduction to biology: New York, Harcourt, Brace, 845 p.

Skevington, D., 1968, British and North American Lower Ordovician correlation: Discussion: Geol. Soc. America Bull., v. 79, p. 1259-1264.

——— 1969, Graptolite faunal provinces in Ordovician of northwest Europe, *in* Kay, Marshall, ed., North Atlantic—geology and continental drift: Am. Assoc. Petroleum Geologists Mem. 12, p. 557-562.

——— 1970, A lower Llanvirn graptolite fauna from the Skiddaw Slates, Westmoreland: Yorkshire Geol. Soc. Proc., v. 37, p. 395-444.

——— 1973, Ordovician graptolites, *in* Hallam, A., ed., Atlas of palaeobiogeography: Amsterdam, Elsevier, p. 27-35.

——— 1974, Controls influencing the composition and distribution of Ordovician graptolite faunal provinces, *in* Rickards, R. B., Jackson, D. E., and Hughes, C. P., eds., Graptolite studies in honour of O. M. B. Bulman: Spec. Papers Palaeontology no. 13, p. 59-73.

Smith, A. G., Briden, J. C., and Drewry, G. E., 1973, Phanerozoic world maps, *in* Hughes, N. F., ed., Organisms and continents through time: Spec. Papers Palaeontology no. 12, p. 1-42.

Spjeldnaes, N., 1963, Some upper Tremadoc graptolites from Norway: Palaeontology, v. 6, p. 121-131.

Tamain, G., 1971, L'Ordovicien est-marianique (Espagne): Sa place dans la province mediterranéenne: Bur. Recherches Géol. Minières, Mém. 73, p. 403-416.

Thomas, D. E., 1960, The zonal distribution of Australian graptolites: Royal Soc. New South Wales Jour. Proc., v. 94, p. 1-58.

Tjernvik, T., 1956, On the early Ordovician of Sweden: Stratigraphy and fauna: Uppsala Univ. Geol. Inst. Bull., v. 36, p. 107-284.

——— 1960, The Lower *Didymograptus* Shales of the Flagabro drilling core: Geol. Fören. Stockholm Förh., v. 82, p. 203-217.

Turner, J. C. M., 1960, Faunas graptoliticas de America del Sur: Asoc. Geol. Argentina Rev., v. 14, p. 5-180.

Tzaj, D. T., 1974a, Lower Ordovician graptolites of Kazahkstan: Moscow Akad. Nauk USSR, Publishing House "Nauka," 127 p. (in Russian).

——— 1974b, Middle Ordovician graptolites of the Chullyisk Mountains, *in* Obut, A. M., ed., Graptolites of the USSR: Novosibirsk, Publishing House "Nauka" Siberian Branch, p. 40-63 (in Russian).

Wallace, A. R., 1876, The geographical distribution of animals: New York, Harper & Bros., Vol. 1, 503 p.; Vol. 2, 607 p.

Warren, P. T., 1971, The sequence and correlation of graptolite faunas from the Wenlock-Ludlow rocks of North Wales: Bur. Recherches Géol. Minières, Mém. 73, p. 451-460.

Watkins, R. M., 1977, British Ludlow palaeoecology and its bearing on the Silurian marine ecosystem: British Mus. (Nat. Hist.) Bull., Geology (in press).

Whittington, H. B., 1966, The phylogeny and distribution of Ordovician trilobites: Jour. Paleontology, v. 40, p. 696-737.

Woodward, S. P., 1856, A manual of the Mollusca, Vol. 3: London, John Weale, p. 331-486.

Zima, M. B., 1974, Ordovician graptolite complexes of northern Kirghizia, *in* Obut, A. M., ed., Graptolites of the USSR: Novosibirsk, Publishing House "Nauka" Siberian Branch, p. 36-40 (in Russian).

Biogeography of the Silurian-Lower Devonian Echinoderms

BRIAN J. WITZKE, *University of Iowa, Department of Geology, Iowa City, Iowa 52242*
TERRENCE J. FREST, *University of Iowa, Department of Geology, Iowa City, Iowa 52242*
HARRELL L. STRIMPLE, *University of Iowa, Department of Geology, Iowa City, Iowa 52242*

ABSTRACT

Review of the distributional record of the better-known Silurian and Lower Devonian echinoderms (mostly crinoids and cystoids) indicates that some groups were markedly cosmopolitan or endemic. The development of widespread, platform-carbonate environments in the Silurian led to a rapid increase in diversity in some echinoderm groups. From a North American Llandoverian center of origin, the echinoderms reached their lower Paleozoic zenith in the Wenlockian. Many Wenlockian families and genera are cosmopolitan but some are endemic to either Europe or North America. Ludlovian faunas show a general decline in diversity, probably related to reduction in area of carbonate platforms, that culminates in extinction of many groups at the close of the Silurian. Lower Devonian echinoderms are strongly provincial; the echinoderm provinces are largely coextensive with brachiopod provinces in the same area. Emsian echinoderm diversity is extremely limited in North America but had dramatically increased in the Rhenish-Bohemian Region. Favorable environments returned to North America in the Middle Devonian and these were populated largely by Rhenish-Bohemian immigrants.

Occurrences of *Scyphocrinites* and *Petalocrinus*, 2 crinoid genera widely used for stratigraphic correlation, are discussed. *Scyphocrinites* is known to range from the Llandoverian into the Gedinnian, and its utility as a Pridolian marker is questioned. *Petalocrinus* appears to be a useful guide fossil for the late Llandoverian through the middle Wenlockian.

INTRODUCTION

The distributional record of the Silurian and Lower Devonian echinoderms was examined with two objectives in view: first, to delineate cosmopolitan or endemic groups; second, to determine the presence or absence of paleobiogeographic provinces among the echinoderms of that time. It soon became apparent that some echinoderm classes exhibited both endemism and cosmopolitanism in the Silurian-Lower Devonian. However, difficulties, individually not unique to the Echinodermata but affecting the phylum more than is typical of other groups, have hindered analysis; some of these are discussed below. As a result, statistical comparisons of the echinoderm faunas of different geographical regions were not attempted, and with few exceptions the family is the taxonomic unit utilized. Nonetheless, enough information is available to enable construction of a tentative model of Silurian-Lower Devonian echinoderm biogeography and to allow comparison of this model with results obtained from the study of better-known coexistent invertebrate phyla such as the Brachiopoda.

Preservational Problems

Any attempt at an interpretation of the biogeography of the Silurian and Lower Devonian echinoderms is fraught with preservational problems generally not encountered in the study of other contemporary benthic invertebrate groups. The classification of most echinoderm taxa is based almost exclusively upon articulated skeletal remains, yet this mode of preservation is not characteristic for the phylum in most depositional environments. The tissue support of the individual calcitic plates and ossicles readily and rapidly decays after death; most echinoderms are subsequently rendered into a jumbled disarticulated mass of skeletal debris by currents or scavengers.

Echinoderms are abundant sediment contributors in many Paleozoic communities, but their identities are often indeterminate using current taxonomic criteria. The majority of described echinoderm faunas have been found in rock units where the bulk of the skeletal debris makes the preservation of at least some articulated remains probable, often in "pockets" where a number of individuals have been rapidly buried before disarticulation could occur. The occurrence of articulated echinoderm remains usually indicates that post-mortem transportation was an insignificant or nonexistent factor. The preservational problem becomes an important consideration in interpreting any echinoderm taxon in a paleobiogeographic context.

Many crinoid column types have been given Linnaean binomens, particularly by a number of Russian and American workers. Although a classification of crinoid column types may have "considerable importance for stratigraphic correlation" (Dubatolova and Yeltysheva, 1967, p. 537), such a classification may not reflect real phylogenetic relationships since it is established on arbitrary morphological criteria (Stukalina, 1967). The placement of those "genera" based exclusively on columns within the classification scheme devised for articulated crinoid calices is often uncertain or unspecified. The occurrence of major column types in several unrelated crinoid groups or the presence of xenomorphic columns in many crinoids may cause false phylogenetic and biogeographic conclusions to be drawn. Ideally, columns should not be classified in the Linnaean system until the relationships between the column and calyx are known, and "column-genera" are not considered in this report unless such relationships are known.

Western and arctic North America, South America, Asia, Africa, Antarctica, and Australia lack large described Silurian-Devonian echinoderm faunas. The majority of Silurian and Lower Devonian echinoderm taxa are known from Europe and eastern and central North America, and comparisons between these two regions should have a greater degree of confidence than those between other continents. Much of the material presented herein will undoubtedly be modified and corrected as new echinoderm faunas are collected and described.

In the Silurian and Lower Devonian, representatives of 13 echinoderm classes have been noted: Crinoidea, "Cystoidea" (Rhombifera and Diploporita), Blastoidea, Stylophora, Homoiostelea, Asteroidea, Ophiuroidea, Edrioasteroidea, Cyclocystoidea, Ophiocistioidea, Echinoidea, and Holothuroidea. The isolated nature of occurrences and extreme rarity of representatives of many of these classes make paleobiogeographic considerations of most echinoderm classes difficult at present. Therefore, the echinoderm distribution section is concerned primarily with the two echinoderm groups most abundantly represented in the Silurian and Lower Devonian, the crinoids and cystoids.

Dispersal of Benthic Invertebrates

The dispersal of benthic echinoderms, as in most other benthic marine groups, is accomplished by a pelagic larval stage. Physical (environmental) barriers control and restrict the dispersal. Ideally, paleobiogeographic reconstructions should reveal the presence or absence of physical barriers, of any kind, between adjacent regions. Assuming that areas with a favorable environment are contemporaneous but spatially separated, the migration of benthic communities to new and suitable areas will be limited by the nature and extent of any physical barrier as well as the capabilities of larval dispersal exhibited by each species in the community. A deep ocean basin may, for example, allow the larval dispersal of many shallow-water benthic invertebrates across it, but the same ocean basin may act as an effective barrier to those benthic species with low larval-dispersal potentials, that is, those with shorter lived or less environmentally tolerant larvae. Larval-dispersal potential and independent physical dispersal agents, such as current direction and velocity, define a limiting distance of dispersal for each species between suitable environments. Benthic species may be geographically isolated if the dispersal mechanisms are insufficiently rapid to transport the larvae to another environment suitable for spatfall.

Approximately equivalent benthic communities in two geographically separated regions can exhibit varying degrees of endemism due to (1) the nature of any selective physical barrier to larval transport between the two regions and (2) the actual distance between the two regions. Both may act to inhibit the migration of those species with limited larval-dispersal potentials. Marked provincialism of benthic marine invertebrate communities may indicate that other regions with potentially suitable environments either are isolated by physical barriers restricting the larval dispersal, or are nonexistent. The most obvious and effective hindrance to the spread of marine communities is a land barrier, although other barriers that are less obvious to the paleobiogeographer may also act to restrict migration to varying degrees. Water depth, temperature, circulation, salinity, or turbidity barriers between two spatially separated but essentially equivalent environments can also act to restrict larval exchange. With any given barrier, the larvae of some taxa will be able to cross the barrier, but others will not. Cosmopolitan taxa are not useful in delineating the more subtle of the physical barriers since their distribution may be an artifact of a greater larval-dispersal potential. Better suited are benthic taxa endemic to, and abundantly represented in, a certain region also inhabited by cosmopolitan benthic taxa.

The environmental potentials of certain taxa may change as evolution occurs, and new regions may become available for migration as a result. Likewise, any shift in the environment can produce migration events of taxa from other regions or marked evolutionary changes of endemic groups during periods of increased competition and stress. The distribution of benthic communities is controlled by the presence or absence of physical barriers, by the changing environmental potentials of evolving taxa, and by the availability of suitable environments through time.

Maximum taxonomic differentiation of the crinoids occurred at two times in the Paleozoic: the "Middle" Silurian (Wenlockian-Ludlovian) and Lower Mississippian (Osage) (Moore, 1952). Maximum abundance and diversity of the Silurian crinoids seems to have occurred primarily in carbonate-platform environments. The widespread expansion of carbonate-platform environments in the Silurian caused an "enormous expansion" in numbers of crinoid taxa and their distributions; "nearly four-fifths of all Middle Silurian species belong to cosmopolitan genera" (Moore, 1952, p. 341). The maximum proliferation of the Silurian crinoids seems to correspond closely to the maximum spread of the epicontinental seas at a time when suitable habitats were numerous and widespread (Moore, 1952, p. 352; Brower, 1973, p. 330). In more unfavorable times, i.e., during regressions, habitats are restricted or eliminated and taxonomic diversity is reduced (Brower, 1973, p. 330). On a broad scale, "provinciality increased when seaways became more restricted" and cosmopolitanism increased "as broad continental areas were inundated by shallow seas" (Johnson and Boucot, 1973, p. 95).

The Silurian and Lower Devonian Setting

The Llandoverian deglaciation produced a marked onlap of marine waters onto several widespread continental platforms (Berry and Boucot, 1973). Uniformity of depositional environments in the Silurian is evident on at least three large-scale carbonate platforms currently termed the North American, the Siberian, and the Russian (Baltoscandian) Platforms. Throughout the early Paleozoic, these three Platforms are believed to have remained in a single latitudinal zone with stable, warm climates (Ross, 1975, p. 322). The Russian and North American Platforms apparently were separated by a deeper ocean barrier, the proto-Atlantic, during most of the Silurian (McKerrow and Ziegler, 1972). Platform mudstones extended over portions of South America, Africa, southern Europe, and eastern Asia. Slope and deeper water deposits are found in Australia and flanking the North American, Russian, and Siberian Platforms.

The extensive slope, mudstone, and carbonate-platform environments of the Silurian supported a variety of cosmopolitan benthic taxa, the most studied of which are the brachiopods. Certain brachiopod taxa have been used to define a series of depth-controlled communities that recur in a wide range of depositional environments. These and other widespread taxa have been used to delineate the North Silurian Realm (Boucot, 1975, p. 307) which today includes most land north of the equator as well as eastern Australia. Limited brachiopod endemism is noted in the Mongolo-Okhotsk Region (eastern Asia), the Uralian-Cordilleran Region (southern Europe, Asia, western North America), the North Atlantic Region (northern Europe, North America), and the Yangzte Valley Subprovince (China, Japan) (Boucot, 1975, p. 307-311). Within the North American Platform, local brachiopod endemism is noted among the triplesioids in Wenlockian deposits of Arkansas, Oklahoma, and Texas (Boucot, 1975, p. 310). An apparent colder water endemic brachiopod fauna defines the Malvinokaffric Realm in the Silurian, identifiable today in South America and South Africa.

The high degree of cosmopolitanism exhibited by many benthic groups in the Silurian indicates that any physical barriers present were easily crossed by these organisms during their larval stages. Other benthic groups could conceivably remain endemic to certain platform or slope environments if their dispersal potentials were comparatively low. The Silurian fauna, although primarily cosmopolitan, also contains a number of endemic benthic forms and Silurian echinoderms provide examples of both cosmopolitanism and endemism. Detailed examination of their occurrences, especially that of the endemic forms, should provide useful information on Silurian evolutionary patterns, paleoecology, and paleotectonics from a previously neglected source.

The epicontinental seas that typified the Silurian had generally withdrawn from the continental interior of North America by the start of the Devonian. Land barriers emerged between restricted shelf seas, and the resulting geographic isolation produced strongly endemic faunas. Brachiopod distribution patterns have delineated a series of Lower Devonian biogeographic entities (Boucot and others, 1969; Johnson and Boucot, 1973; Boucot, 1975). The Eastern Americas Realm includes the Amazon-Colombian Subprovince (northern South America) and the Appohimchi Subprovince (eastern North America). The Old World Realm reveals increasing provinciality through the Lower Devonian, enabling discrimination of the Rhenish-Bohemian Region (Europe, North Africa, and Nova Scotia), the Tasman Region (Australia), the New Zealand Region, the Uralian Region (Asia, eastern Europe), the Mongolo-Okhotsk Region (eastern Asia), and the Cordilleran Region (western North America). Nevada oscillates in the Lower Devonian between Eastern Americas Realm and Old World Realm brachiopod faunas (Boucot, 1975, p. 319). The Malvinokaffric Realm includes portions of South America, South Africa, and Antarctica.

ECHINODERM DISTRIBUTIONS

Information on echinoderm distributions has been derived largely from references listed in Bassler and Moodey (1943), Webster (1973), Regnéll (1975), and from our own investigations and correspondence. The classification used is primarily that of published and unpublished parts of the Treatise on Invertebrate Paleontology. Time-stratigraphic information is based largely on Berry and Boucot (1970) for North America, the other Geological Society of America Special Papers on Silurian correlation for the remaining continents, and personal correspondence. Subsequent studies should modify and enlarge the distributional record of the Silurian and Lower Devonian echinoderms.

Latest Ordovician-Early Llandoverian

Echinoderms from the earliest Silurian are poorly known; a few forms have been described from the Mid-continent of North America. The latest Ordovician Girardeau Limestone of Missouri (Brower, 1973) has yielded a crinoid fauna dominated by a number of genera also common during the Silurian (*Alisocrinus, Macrostylocrinus, Dendrocrinus, Protaxocrinus, Clidochirus,* and a hapalocrinid). The Noix and Cyrene Limestones of the Edgewood Group (Missouri) produced a meager crinoid fauna (Rowley, 1904) that includes *Calceocrinus,* a patelliocrinid, and a periechocrinid (Brower, 1973, p. 332). The Edgewood faunas may be latest Ordovician or earliest Silurian. The Manitoulin Dolomite (early Llandoverian) of the Lake Huron region contains *Calceocrinus* and *Brockocystis,* a callocystitid rhombiferan. Early Llandoverian "*Scyphocrinites*" are known from India and China. No middle Llandoverian echinoderms have yet been described.

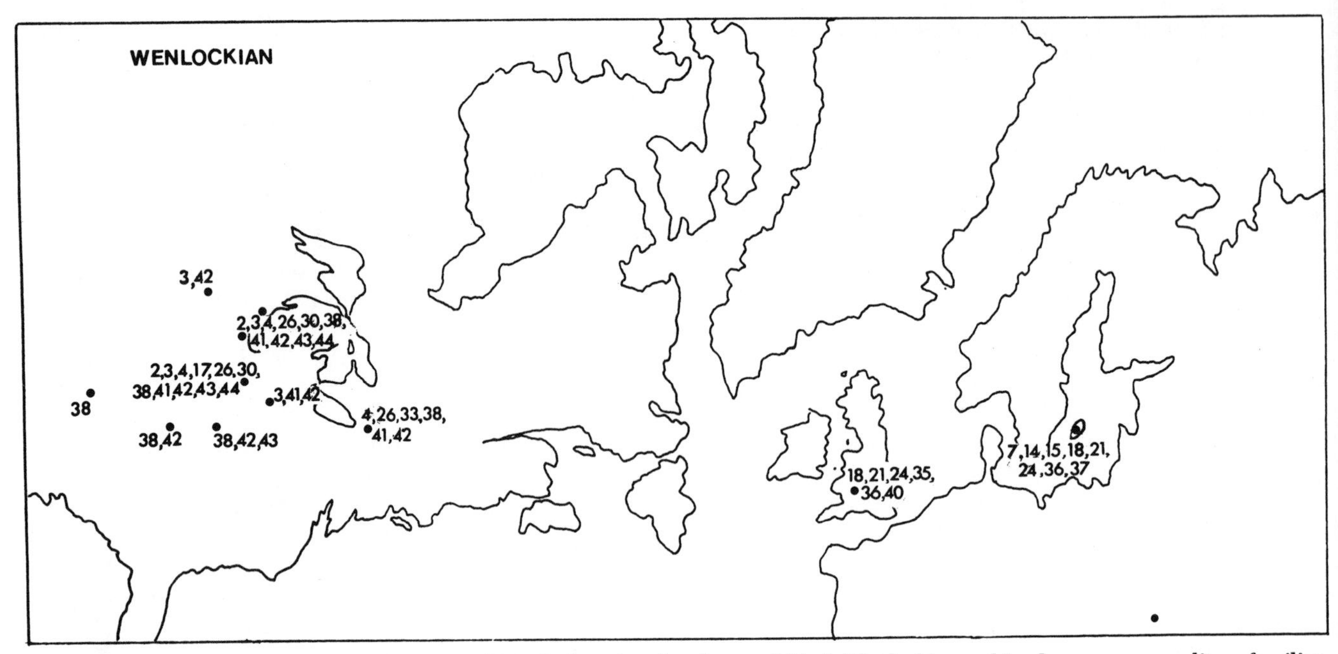

Figure 1. Localities (dots) in Europe and North America that have yielded Wenlockian echinoderms; cosmopolitan families not shown. The numbers correspond to endemic Wenlockian echinoderm groups listed in the Appendix. The basemap used here and in Figure 3 is not an accurate representation of continental positions in the Wenlockian and Ludlovian.

Late Llandoverian and Late Llandoverian-Early Wenlockian Transition

Abundant and diverse Silurian echinoderm faunas are first recognized in several late Llandoverian carbonate formations in eastern and central North America. The echinoderms of the Jupiter Formation (Anticosti Island), although not completely described as yet, include common Silurian genera such as *Dimerocrinites, Crotalocrinites, Eucalyptocrinites,* and *Caryocrinites.* The Brassfield Limestone (Ohio) contains a large echinoderm fauna including *Myelodactylus, Calceocrinus, Clidochirus, Icthyocrinus, ?Lyriocrinus, Dimerocrinites, Eucalytocrinites,* a hapalocrinid, and *Brockocystis.* The Sandpile Group (British Columbia) has produced *Petalocrinus* and *Pisocrinus* (Norford, 1962). Another fauna currently under investigation, from the Hopkinton Dolomite (Iowa) includes forms of about mid-late Llandoverian to about C_6 in age (M. Johnson, 1975). Included are undescribed genera as well as many well-known Silurian taxa such as *Eucalyptocrinites, Calliocrinus, Dimerocrinites, Siphonocrinus, Marsupiocrinus,* hapalocrinids, *Macrostylocrinus, Icthyocrinus, Stephanocrinus, Myelodactylus, Petalocrinus, Crotalocrinites, Thalamocrinus, Pisocrinus, Lysocystites, Caryocrinites,* and a callocystitinid.

The placement of the Llandoverian-Wenlockian boundary in eastern North America is often difficult. The Irondequoit Limestone (New York) has yielded *Caryocrinites, Closterocrinus, Pisocrinus, Stephanocrinus,* and *Icthyocrinus* from approximately this time boundary. The Euphemia Dolomite (Ohio) contains *Lampterocrinus.* The Chicotte Formation (Anticosti Island) has yielded *Periechocrinus.*

Outside of North America a few late Llandoverian echinoderms have been identified. *Osculocystis,* a scoliocystinid rhombiferan cystoid, and *Petalocrinus* are recorded from England. *Petalocrinus* is noted from the Visby Marl of Gotland (Sweden). *Myelodactylus* and a variety of crinoid stem types have been found in Podolia, USSR (Nikiforova and Predtechenskij, 1968). Since the described echinoderm faunas of this time period are almost exclusively North American, no regional comparisons can be made.

Wenlockian

Diverse and abundant Wenlockian echinoderm faunas are found primarily in eastern North America, England, and Gotland, but smaller collections are also known from the USSR, China, and Australia (Fig. 1). In North America, the Osgood (Indiana), Laurel (Indiana), Waldron (Indiana, Tennessee), Rochester (New York), Lockport (New York, Ohio), Cedarville (Ohio), Louisville (Indiana, Kentucky), St. Clair (Arkansas), Gower (Iowa), and Racine (Wisconsin, Illinois) Formations have yielded the bulk of Wenlockian echinoderms. The Racine also contains Ludlovian echinoderms. The Wenlock Limestone of England and the Högklint and Slite of Gotland (Sweden) are the primary Wenlockian echinoderm-bearing rocks of Europe. Wenlockian faunal comparisons between eastern and central North America and Europe should have some degree of confidence, particularly for those taxa that are abundant community elements in one or both regions.

A variety of Wenlockian crinoid genera are found in both Europe and North America and are probably cosmo-

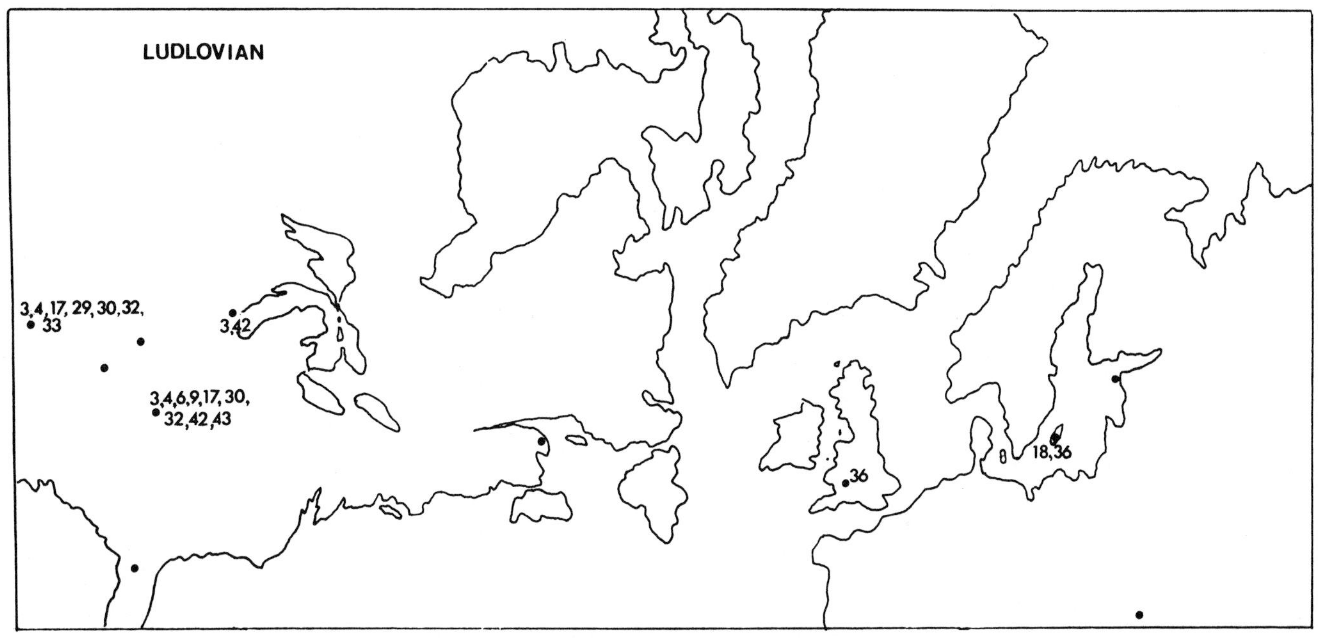

Figure 2. Localities (dots) in Europe and North America that have yielded Ludlovian echinoderms; cosmopolitan families not shown. Numbers correspond to endemic Ludlovian echinoderm groups listed in the Appendix.

politan. These include such common forms as *Dimerocrinites, Desmidocrinus, Carpocrinus, Periechocrinus, Hapalocrinus, Eucalyptocrinites, Macrostylocrinus, Lecanocrinus, Icthyocrinus, Calceocrinus, Pisocrinus, Myelodactylus, Cyathocrinites, Crotalocrinites, Gissocrinus, Botryocrinus,* and others. Two families of camerate crinoids, the Gazacrinidae and Lampterocrinidae, include 3 common genera (*Gazacrinus, Lampterocrinus, Siphonocrinus*) that are apparently endemic to North America and numerically dominate some faunas, e.g., that of the Racine. Other North American endemic groups include the camerate family Archaeocrinidae (*s.l.*) (*Wilsonicrinus, Paulocrinus, "Archaeocrinus"*), the inadunate crinoid families Homocrinidae (*Homocrinus*), Zophocrinidae (*Zophocrinus*), Dendrocrinidae (*Dendrocrinus*), and Sphaerocrinidae (*Thalamocrinus*), the rhombiferan subfamily Callocystitinae (*Hallicystis, Coelocystis, Callocystites*), the enigmatic *Stephanocrinus,* the spiraculate blastoid *Troosticrinus,* and *Lysocystites,* a blastozoan of uncertain affinities. The rhombiferan cystoid *Caryocrinites,* noted from the Ordovician of the Old World, is one of the most abundant of all Wenlockian echinoderms in eastern and central North America, yet it is curiously lacking from all European Silurian faunas with the exception of one formation in France. Ten-armed *Marsupiocrinus* are North American, those with twenty arms are European.

Other Wenlockian echinoderm groups appear to be endemic to Europe and include the camerate crinoid families Abacocrinidae (*Abacocrinus*), Polypeltidae (*Polypeltes*), and Stelidocrinidae (*Stelidocrinus*), the flexible crinoid families Dactylocrinidae (*Calpiocrinus, Lithocrinus, Temnocrinus*) and Taxocrinidae (*Protaxocrinus*), the inadunate crinoid families Thenarocrinidae (*Thenarocrinus*), Metabolocrinidae (*Cyliocrinus*), and Mastigocrinidae (*Mastigocrinus, Bathericrinus, Dictenocrinus, Streptocrinus*), and the rhombiferan subfamily Scoliocystinae (*Glansicystis, Schizocystis*).

Additionally, Wenlockian deposits in southeastern Australia have produced *Botryocrinus,* a hapalocrinid, *Lecanocrinus,* and *Pisocrinus. Crotalocrinites, Syndetocrinus,* and *Myelodactylus* have come from the Wenlockian of the Urals (USSR). Podolian (USSR) Wenlockian forms include *Desmidocrinus* and *Myelodactylus.* Chinese representatives, probably Wenlockian, include *Petalocrinus, Pisocrinus,* and 2 genera apparently unique to China, *Spirocrinus* and *Dazhucrinus* (Mu and Wu, 1974).

The majority of Wenlockian echinoderm families are cosmopolitan in their distributions. However, a significant number of genera and families are endemic to either North America or Europe, and two biogeographic entities that include the mentioned Wenlockian localities may be indicated. If more echinoderm collections are made, particularly in Asia, the validity of these two apparent provincial areas can be tested.

Ludlovian

The Ludlovian echinoderm faunas are primarily known from the eastern United States, Gotland (Hemse-Klinteberg, Eke, Burgsvik Formations), England, and Bohemia (Fig. 2). Limited occurrences have also been noted in India (Kashmir), the USSR (Podolia, Oesel, Urals, Tien Shan, Baschkiria), Morocco, and Australia. In North America, Ludlovian echinoderms are most diversely and abundantly represented in the Brownsport (Tennessee) and Henryhouse (Oklahoma) Formations (Strimple, 1963). The Read Bay (Canadian Arctic), West Point

(Quebec), Dixon (Tennessee), Moccasin Springs (Missouri), Florida subsurface, Lafferty (Arkansas), upper Racine (Illinois), and possibly Decatur (Tennessee) Formations have also yielded Ludlovian echinoderms.

Several abundantly represented cosmopolitan genera such as *Lecanocrinus, Pisocrinus, Periechocrinus,* and *Crotalocrinites* typify many of the Ludlovian echinoderm faunas. In North America the same crinoid families that appeared to be endemic in the Wenlockian remain endemic in the Ludlovian. These include the camerate crinoid families Lampterocrinidae (*Lampterocrinus, Siphonocrinus*) and Gazacrinidae (*Gazacrinus*), the inadunate crinoid families Dendrocrinidae (*Bactrocrinites*), Sphaerocrinidae (*Thalamocrinus*), and Zophocrinidae (*Zophocrinus*), and spiraculate blastoids (*Troosticrinus*). Additionally, 3 other endemic crinoid families make their appearance in North America during the Ludlovian: the camerate Nyctocrinidae (*Nyctocrinus*) and Coelocrinidae (*Aorocrinus*), and the inadunate Synbathocrinidae (*Abyssocrinus, Stylocrinus*). The flexible family Taxocrinidae, unknown from the abundant Wenlockian faunas of North America, apparently migrated to North America from Europe in the Ludlovian. *Caryocrinites* remains typically North American. Ten-armed *Marsupiocrinus* are still apparently restricted to North America; twenty-armed *Marsupiocrinus* remain solely European. The Diploporita apparently disappeared from North America in the Ludlovian, although these cystoids survived in Europe through the Lower Devonian.

In Europe the 2 prominent Wenlockian endemic crinoid families, the Dactylocrinidae and Mastigocrinidae, remain endemic to that region. A family that makes its appearance in the Ludlovian, the Methabocrinidae, is also a European endemic. All European echinoderm groups show a reduction in generic diversity from the Wenlockian to the Ludlovian. *Stephanocrinus,* which throughout the Wenlockian does not seem to appear outside of North America, is noted from Kashmir, India, in the Ludlovian (Gupta and Webster, 1971). Russian Ludlovian echinoderms include *Crotalocrinites* (northern Baschkiria, Oesel, Podolia), *Eucalyptocrinites* (Oesel), *Cicerocrinus* (Oesel), *Syndetocrinus* (Urals, Tien Shan), *Prohexacrinites* (Urals), and a callocystitid cystoid (Podolia). Australian Ludlow-age echinoderms include *Pisocrinus, Thylacocrinus,* and *Eucalyptocrinites.*

A major offlap of the Silurian epeiric seas probably began during the Ludlovian. The echinoderms, apparently best adapted to life in large carbonate-platform environments, were restricted by the loss of available habitat space and a drop in generic diversity resulted. In Iowa, for example, the Wenlockian bioherms supported a large and diverse echinoderm fauna, but during the Ludlovian shallower and harsher environments not suitable for echinoderm growth appeared, and the echinoderms disappeared from that section of the North American Midcontinent. Likewise the English Ludlovian crinoids, when compared with the prolific Wenlockian crinoid faunas found there, represent "an impoverished fauna, probably due to the onset of unfavorable conditions" (Ramsbottom, 1958, p. 106). During the Ludlovian many of the major echinoderm groups in both Europe and North America became extinct, although some Silurian forms survived through the Pridolian into the Early Devonian.

Pridolian

The most widespread and common of the Pridolian echinoderms is *Scyphocrinites* (discussed in a later section). The remainder of the identifiable Pridolian echinoderms are known only from a few rock units, all of which have also yielded *Scyphocrinites:* West Virginia (Keyser Limestone), Tennessee (Decatur Limestone), Bohemia, Podolia (Dzwinograd), and Kazakhstan.

The Bohemian genera include *Bohemicocrinus, Calpiocrinus, Pisocrinus,* and *Ctenocrinus.* The Dzwinograd horizon in Podolia has yielded *Crotalocrinites, Hexacrinus,* and *Pisocrinus.* Kazakhstanian deposits, probably Pridolian, have produced *Crotalocrinites* and *Herpetocrinus.* The Keyser Limestone contains a sizable echinoderm fauna, although the stratigraphic position of the crinoids is not completely known. *Scyphocrinites* and 5 genera of callocystitid rhombiferan cystoids definitely occur in the *Eccentricosta jerseyensis* Zone (late Pridolian), and it is likely that the remaining crinoids also came from the same horizon (Schuchert and others, 1913). The Keyser crinoids include *Clidochirus, Pycnosaccus, Synchirocrinus, Myelodactylus, Protaxocrinus, Hapalocrinus,* and *Sphaerotocrinus.* The Decatur Limestone, in part, may be Pridolian based on the presence of the *Spathognathodus remscheidenis* group (Rexroad and Nicoll, 1971). The upper levels of the Decatur have yielded *Aorocrinus, Abathocrinus, Clonocrinus, Gazacrinus, Desmidocrinus, Eucalytocrinites, Marsupiocrinus, Lecanocrinus, Pisocrinus,* and *Caryocrinites* (Springer, 1917, p. 25; 1926); these forms may be Pridolian or late Ludlovian.

The Gazacrinidae, endemic to North America since at least the Wenlockian, became extinct during the Pridolian. The cystoid family Caryocrinitidae remained endemic to North America during the Pridolian. The Callocystitidae, the last surviving family of rhombiferan cystoids, became entirely restricted to North America during the Pridolian and remained so until their extinction in the Upper Devonian. The Dactylocrinidae, known only from Europe since at least the Wenlockian, remained endemic to that region in the Pridolian. The Hexacrinitacea apparently disappeared from eastern North America during the Pridolian and Lower Devonian, and the superfamily went on to become one of the most abundant crinoid groups in the Lower Devonian of Australia, Asia, Europe, and North Africa. Cosmopolitan or extremely widespread crinoid families in the Pridolian include the Scyphocrinitidae, Desmidocrinidae, Myelodactylidae, and Pisocrinidae.

Gedinnian-Early Siegenian

Gedinnian-early Siegenian (Devonian) echinoderms in the Eastern Hemisphere are known from North Africa (Spanish Sahara, Algeria, Morocco), the USSR (Urals, Kazakhstan, Tadzhikistan, Kuznetsk Basin, and Podolia),

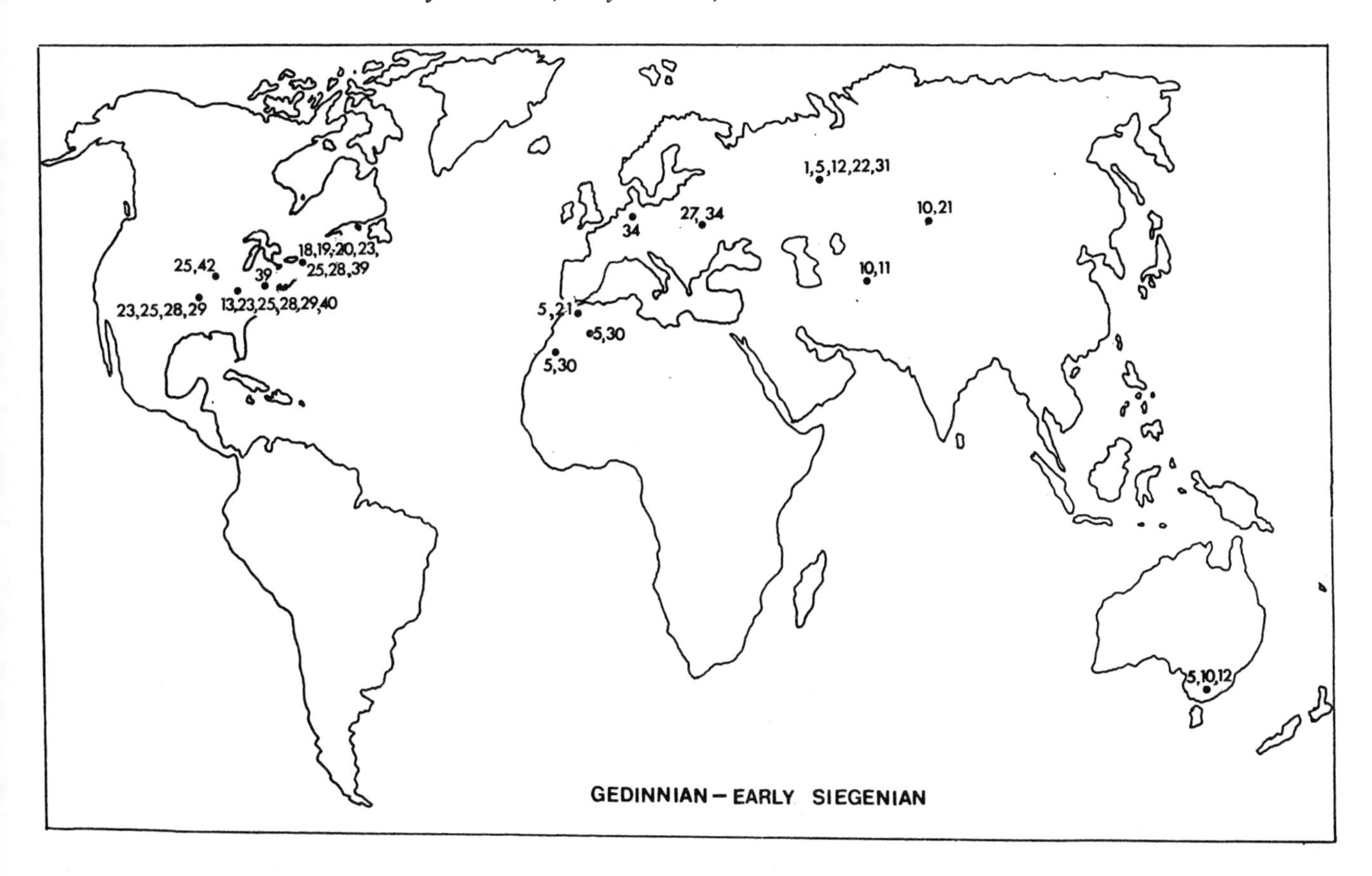

Figure 3. Gedinnian-early Siegenian echinoderm distributions (cosmopolitan families not shown). Numbers correspond to endemic Gedinnian-early Siegenian echinoderm groups as listed in the Appendix.

and southeastern Australia (Fig. 3). Three genera of echinoderms have been reported from the type Siegenian in Germany and are questionably included in this section. In the United States, early Gedinnian echinoderms are known from the Rockhouse Shale (Tennessee) and Coeymans Limestone (New York); late Gedinnian echinoderms are noted in the Haragan Formation (Oklahoma), New Scotland Limestone (New York), Bailey Limestone (Missouri), Ross Limestone and Birdsong Shale (Tennessee); early Siegenian forms are meagerly represented in the Becraft Limestone (New York) and equivalents (Virginia), and Bois D'Arc Limestone (Oklahoma).

Ten of 13 crinoid families present are endemic to part or all of the region that includes North Africa, Russia, and Australia during the Gedinnian-early Siegenian. These include the camerate families Rhodocrinitidae (*Acanthocrinus, Condylocrinus, Thylacocrinus*), Hexacrinitidae Parahexacrinidae, and Spyridiocrinidae (*Spyridiocrinus*), and the inadunate families Zophocrinidae and Botryocrinidae (*Botryocrinus, Gastrocrinus*). The inadunate family Pisocrinidae is represented in Podolia by a Silurian holdover, *Pisocrinus.* Four other families, endemic to the Old World Realm in the Gedinnian, are represented by Silurian holdover genera in the Urals, Kazakhstan, or the Kuznetsk Basin: Eucalyptocrinitidae (*Eucalyptocrinites, Calliocrinus*), Nipterocrinidae (*Pycnosaccus*), Taxocrinidae (*Eutaxocrinus*), and Cyathocrinitidae (*Gissocrinus*). *Eucalyptocrinites* has also been found in Australia, and taxocrinid stems have been noted in Morocco for this time interval. The inadunate family Dendrocrinidae is known from Gedinnian-age rocks in North America (*Alsopocrinus*) and Russia (*Bactrocrinites*). *Ctenocrinus* and *Lecanocrinus* (a Silurian holdover), are found in the Gedinnian of both North America and the Old World.

Four families found in the Gedinnian of the eastern United States are not noted in the Old World until the Emsian. These include the Hapalocrinidae (cosmopolitan in the Silurian), Mastigocrinidae (endemic to Europe in the Silurian), and the Synbathocrinidae (endemic to North America in the Silurian). The Silurian holdover *Icthyocrinus* (Icthyocrinidae) is noted from the Gedinnian of the United States and the Emsian of Europe. These 4 families either have not been identified as yet in the Gedinnian of the Old World, or they may actually be endemic to North America in the Gedinnian. If the latter choice is correct, then faunal interchange between Europe and North America must have taken place sometime in the Siegenian or early Emsian. Eastern and central North American endemic families include the camerate Dolatocrinidae (*Dolatocrinus*), the rhombiferan Callocystitidae and Caryocrinitidae (*Caryocrinites*): endemic genera include 2 Silurian holdovers, *Marsupiocrinus* (Marsupiocrinidae) and *Myelodactylus* (Myelodactylidae), and the common stemless flexible crinoid *Edriocrinus* (Flexibilia, *incertae sedis*). The incrusting root *Aspidocrinus*, the camerate crinoid *Coronocrinus*, and the Silurian holdover *Clidochirus* (Ichthyocrinidae) are also endemic to eastern North America.

The distribution of the Gedinnian-early Siegenian echinoderms is similar to that noted for the brachiopods (Boucot and others, 1969) and lends further support for the recognition of two major marine realms, the Eastern Americas and Old World Realms. Most of the Silurian holdover genera of the Old World Realm (especially the Eucalyptocrinitidae) are apparently restricted to the Urals, Asia, and Australia, and the establishment of a Uralian-Tasman Region in the Gedinnian is indicated.

Late Siegenian-Emsian

Echinoderms from the late Siegenian-Emsian interval have been described from all continents except Antarctica. European localities have produced the greatest quantity and diversity of echinoderm material: notable are Spain (Upper La Vid, Lower Santa Lucia), Germany (Coblenz, Hunsrück, Kahleberg), Bohemia (Koněprusy, Prokop, Zlichov), Austria (Carnic Alps), France, and Belgium. The Hunsrück black shale, one of the classic Paleozoic echinoderm-bearing units, has produced in excess of 60 genera of echinoderms divided almost equally between the crinoids and asteroids. There are several late Siegenian-Emsian localities in North Africa (Algeria, Morocco, and the Spanish Sahara). Occurrences in the USSR include the Urals, Kuznetsk Basin, Tadzhikistan, Far East, and the Soviet Arctic. Emsian rocks in southeastern Australia have produced crinoids and other echinoderms, and stylophoran carpoids are known from the Reefton Beds of New Zealand. South African and Brazilian Emsian stylophoran carpoids are also known, a single Emsian crinoid has been reported from the Falkland Islands, and crinoids and blastoids are noted from Bolivia.

No Emsian echinoderms have yet been described from eastern or central North America, although two small late Siegenian faunas have been noted in the Glenerie Limestone (New York) and the Ridgeley Sandstone (West Virginia, Maryland). The apparent lack of Emsian echinoderms in eastern and central North America seriously handicaps paleobiogeographic interpretations. Two crinoids, both apparently Emsian in age, are noted from Nevada (Johnson and Lane, 1969).

Edriocrinus, a stemless flexible crinoid, and *Macrostylocrinus*, a Silurian holdover, are endemic to eastern North America in the late Siegenian. Crinoid families found in both the late Siegenian of eastern North America and the Emsian of Europe include the camerate Dimerocrinitidae and Clonocrinidae, and the inadunate Calceocrinidae and Botryocrinidae (*Ancyrocrinus*). The Mastigocrinidae or Dendrocinidae (*"Homocrinus"*) may also be present in the late Siegenian of eastern North America. The stylophoran Anomalocystitidae are noted from both eastern North America and Europe.

The late Siegenian-Emsian Old World echinoderm Realm is characterized by several widespread crinoid families including the camerate Hexacrinitidae, Periechocrinidae, Polypeltidae, Melocrinitidae, and Rhodocrinidae, the inadunate Crotalocrinitidae and Pisocrinidae, and the flexible Taxocrinidae and Lecanocrinidae. Localities within the Rhenish-Bohemian brachiopod Region have produced the greatest diversity of Emsian echinoderms and include numerous apparently endemic genera and families. The most diversified crinoid families in the Emsian of Europe include the Periechocrinidae, Rhodocrinidae, Mastigocrinidae, Hapalocrinidae, and Botryocrinidae. The European Emsian Mastigocrinidae, a family restricted to Europe throughout the Silurian, includes 3 Silurian holdovers (*Antihomocrinus, Bathericrinus, Dictenocrinus*), a possible migrant from the Gedinnian of North America (*Lasiocrinus*), and endemic genera. With the exception of an *Ancyrocrinus* from New York and *Botryocrinus* from the Falkland Islands, the Botryocrinidae are known mostly from the Rhenish-Bohemian brachiopod Region. Emsian genera of the Hapalocrinidae are all noted from the Rhenish-Bohemian Region and include *Cyttarocrinus*, an apparent migrant from eastern North America. Most genera of the Periechocrinidae are uniquely European, although *Megistocrinus* is noted from Australia and *Gennaeocrinus* is noted from Germany and Nevada. Probable pre-Emsian migrants to Europe include the North American Gedinnian *Icthyocrinus* (Icthyocrinidae) and *Phimocrinus* (Synbathocrinidae). Other Emsian crinoid families endemic to the Rhenish-Bohemian Region include the Spyridiocrinidae, Orthocrinidae, Platycrinidae, Cupressocrinitidae, Pygmaeocrinidae, Zophocrinidae, Rhenocrinidae, Codiacrinidae, Poteriocrinidae, and Cyathocrinitidae (2 Silurian holdovers: *Cyathocrinites, Gissocrinus*). The rhombiferan cystoid family Pleurocystitidae, unknown in the geologic record during the entire Silurian, makes an unexpected reappearance in the Hunsrück Shale of Germany and the Meadfoot Beds of England (Paul, 1974) and is the last known rhombiferan family noted from the Old World. The last respresentatives of the Diploporita, all included in the Sphaeronitidae, are known from Emsian rocks of Europe and North Africa.

Late Siegenian-Emsian representatives of the Eucalyptocrinitidae include the Silurian holdovers *Eucalyptocrinites* from the Urals and Australia and *Calliocrinus* from the Urals. The Parahexacrinidae are described only from Tadzhikistan, and the Gasterocomidae are known only from the Kuznetsk Basin and Nevada for this time interval. These occurrences suggest the presence of a biogeographic boundary separating the Urals, Australia, and the Cordillera from the Rhenish-Bohemian Region in the latter half of the Early Devonian. The stylophoran family Allanicytidiidae is known only from the Emsian of New Zealand, but it is not known if this isolated occurrence has any biogeographic significance.

The stylophoran family Australocystidae is known only from Emsian rocks in Brazil and South Africa. The apparent lack of the Australocystidae in other late Siegenian-

Emsian homalozoan-bearing beds in North America, Europe, Australia, and New Zealand suggests that the Australocystidae are endemic to the Malvinokaffric Realm. The occurrence of *Botryocrinus* in the Falkland Islands (Malvinokaffric Realm) and in the Rhenish-Bohemian Region suggests that some communication was established between these two areas in the Emsian. The Emsian of Bolivia has revealed a variety of undescribed echinoderms including crinoids and blastoids (Branisa, 1965).

Late Siegenian-Emsian echinoderm distributions and the apparent endemism of certain groups can be used to delineate biogeographic boundaries that parallel those established using brachiopod distributions. These include the Eastern Americas, Old World, and Malvinokaffric Realms. Within the Old World Realm certain echinoderm elements of the Uralian-Tasman-Cordilleran Regions are lacking from the abundant faunas of the Rhenish-Bohemian Region (Europe, North Africa). The Appohimchi Subprovince (eastern and central North America) underwent a serious reduction in echinoderm diversity from the Gedinnian through Emsian, whereas the Rhenish-Bohemian Region underwent a marked increase in echinoderm diversity for the same time interval. Several Gedinnian Appalachian echinoderm elements made their appearance in the Rhenish-Bohemian Region in the Emsian, and this is probably indicative of the establishment of limited communication between these two areas. The Emsian echinoderm faunas of the Rhenish-Bohemian Region share close affinities with the Middle Devonian faunas of North America (Breimer, 1962, p. 181). The re-establishment of luxuriant echinoderm populations did not occur in North America until after the Emsian. Carbonate environments again became widespread in the continental interior of North America during the Middle Devonian, and echinoderm families that were endemic to the Old World Realm in the Emsian migrated to eastern and central North America during the Middle Devonian. The most prominent of these migrant Middle Devonian North American families are the Hexacrinitidae, Rhodocrinitidae, Periechocrinidae, Taxocrinidae, Gasterocomidae, and Crotalocrinitidae.

Petalocrinus and *Scyphocrinites*

Most crinoids can be identified only if the dorsal cup remains articulated during burial. However, even when the rigid arm-fans (or "petals") of *Petalocrinus*, an inadunate crinoid, become separated from the calyx they are easily identifiable (Fig. 4). The arm-fans are solidly fused rays with branching ambulacra on the ventral surface.

Petalocrinus was first described from the Hopkinton Dolomite of Iowa and has subsequently been noted in the Cedarville Dolomite of Ohio and Illinois (Lowenstam *in* Berry and Boucot, 1970, p. 201), strata below the Racine Dolomite of Wisconsin (D. Mikulic, personal commun.

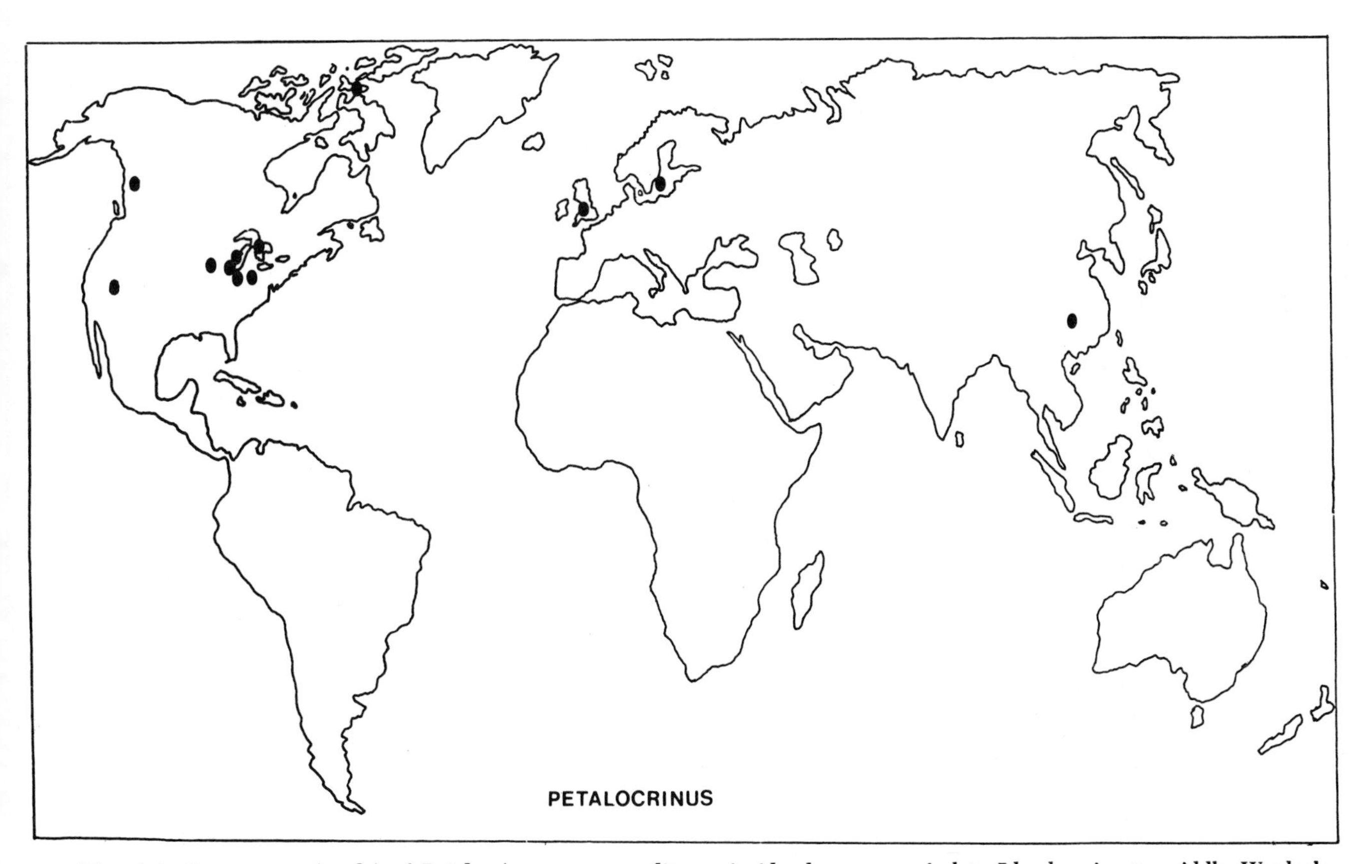

Figure 4. Occurrences (ovals) of *Petalocrinus*, a cosmopolitan crinoid whose range is late Llandoverian to middle Wenlock.

1975), the Fossil Hill Formation of Manitoulin Island (*ibid.*), the Laurel Limestone of Indiana (Springer, 1926), Silurian strata of Nevada (Boucot, personal commun. 1976), the Sandpile Group of British Columbia (Norford, 1962), on Baillie-Hamilton Island, Arctic Canada (Boucot, personal commun. 1976); the base of the Woolhope Limestone in England (Ziegler and others, 1968, p. 757), the Visby and possibly Högklint Formations of Gotland, and in southeast China (Mu and Wu, 1974, p. 209). In all cases the occurrences of *Petalocrinus* are apparently restricted to the late Llandoverian through the middle Wenlockian. *Petalocrinus* has the potential of serving as a guide fossil for the time interval including part or all of the *celloni*, *amorphognathoides*, and *patula* conodont zones.

Scyphocrinites is among the most widespread of the Late Silurian echinoderms (Fig. 5). This camerate crinoid genus, which commonly has an exceptionally large crown, has a column ending in a small encrusting root or in a large bulbous chambered structure (Ubaghs, in press). The chambered structure, or lobolith, of *Scyphocrinites* has been given the synonomous generic name, *Camarocrinus*. It has been interpreted either as a functional float (i.e., *Scyphocrinites* was planktic) or as a bulbous anchor (i.e., it was benthic). *Scyphocrinites* is found in deposits with only crowns preserved, with loboliths only, or with both associated together (Springer, 1917).

Large loboliths, possibly of *Scyphocrinites* or a related genus, have been reported from Burma and China (Yuennan) beneath beds bearing early Llandoverian (Zone 18) graptolites (Berry and Boucot, 1972, p. 20). In France, *Scyphocrinites* has been identified in beds associated with early Wenlockian graptolites (*Monograptus riccartonensis* and others) (LeMaitre and Heddebaut, 1963). Early Ludlovian graptolites have been found in southern Morocco above beds yielding *Scyphocrinites* (Hollard, 1962). Occurrences of *Scyphocrinites* in the upper Henryhouse Formation of Oklahoma and in Cornwall, England, may be either Ludlovian or Pridolian in age.

The majority of reports of undoubted *Scyphocrinites* are from Pridolian sequences in North America, Europe, USSR, and North Africa. North American Pridolian *Scyphocrinites* are known from West Virginia (Keyser Limestone-*Eccentricosta jerseyensis* Zone), Tennessee (upper Decatur Limestone, Rockhouse Formation), Missouri (Moccasin Springs Formation), Newfoundland (Clam Bank Series), and the Gaspe Peninsula (Griffin Cove River, Sirois, and St. Leon Formations). European and Russian Pridolian occurrences are known from France (Pyrenees, Cabrieres, Artois), Spain (Pyrenees), Germany (Rhineland, Kellerwald, Thuringia), Bulgaria, Poland, Czechoslovakia (Bohemia), and the USSR (Podolia, western Siberia, Kazakhstan). In North Africa, Pridolian *Scypho-*

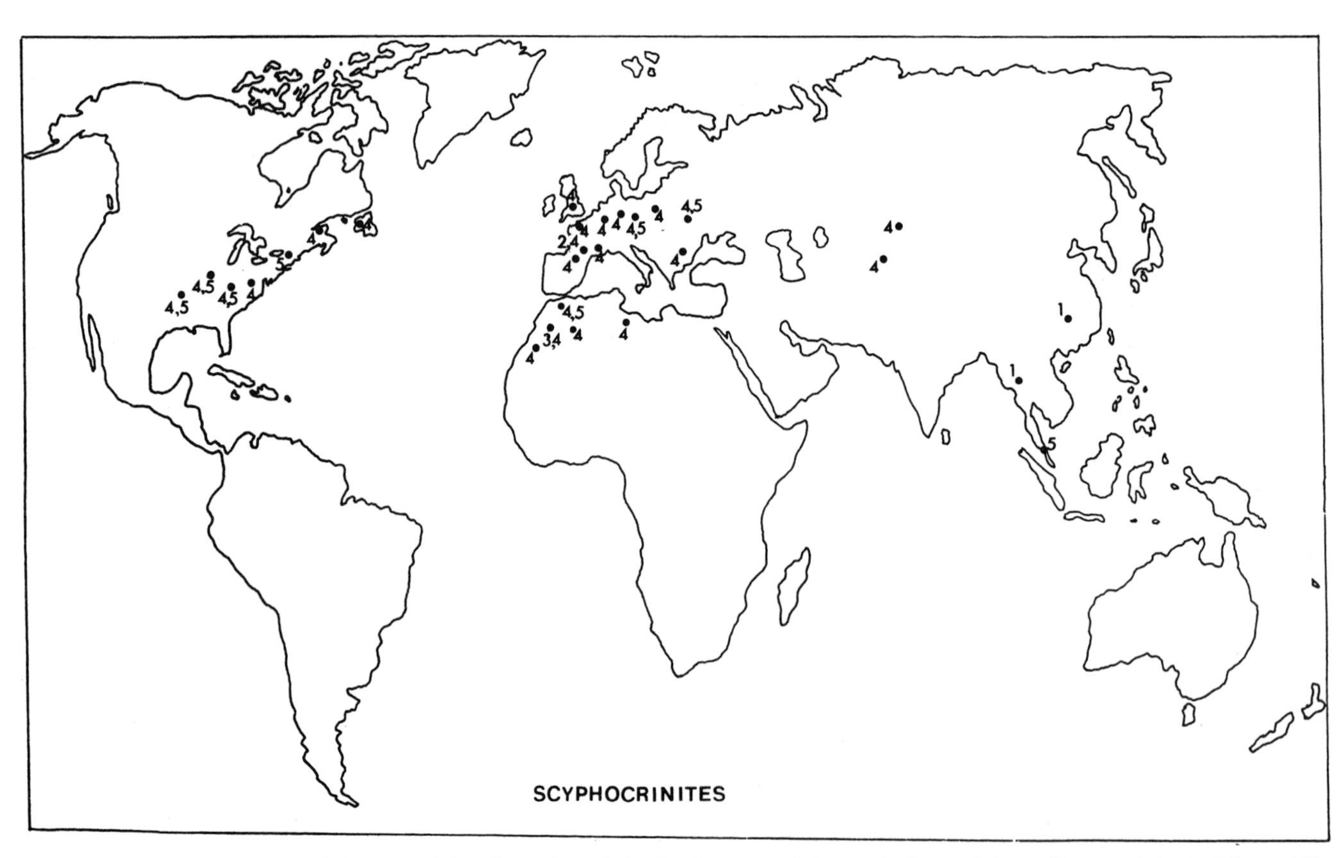

Figure 5. Distribution of the crinoid *Scyphocrinites:* (1) Llandoverian, (2) Wenlockian, (3) Ludlovian, (4) Pridolian, (5) Gedinnian.

crinites have been identified from Algeria, Tunisia, Morocco, and Spanish Sahara. Both in North Africa and Europe *Scyphocrinites* is most abundant in the late Pridolian [e.g., *M. transgrediens* Zone of Bohemia (Jaeger, 1962, p. 110)], although the genus is known from as low as the *Monograptus ultimus* Zone (early Pridolian) in Bohemia and possibly North Africa.

Scyphocrinites ranges upward into the basal Devonian (*Monograptus uniformis* Zone) in Bohemia (Lochkov), Podolia (Borschov), Morocco, and Malaysia. The last occurrence has not been previously recorded (T. E. Yancey, personal commun. 1975). North American *Scyphocrinites* are known from approximate early Gedinnian-age rocks in New York (Coeymans Limestone) and from New Scotland-age rocks (approximate late Gedinnian) in Oklahoma (Haragan Formation), Tennessee (Ross Limestone, Birdsong Shale), and Missouri (Bailey Limestone).

Although *Scyphocrinites* is apparently most abundant in Pridolian rocks, it should not be used as a Pridoli index fossil except locally, as its documented geologic range is from early Llandoverian through the late Gedinnian. *Scyphocrinites* appears to have originated in Asia in the Late Ordovician or early Llandoverian and migrated to Europe by at least the Wenlockian. It appeared in North America during the late Ludlovian or Pridolian where it remained until the late Gedinnian. In almost all cases *Scyphocrinites* is found in calcareous shales and limestones that are interbedded with shales. The crinoidal limestones may contain shelly faunas of bivalves, ostracodes, nautiloids, or brachiopods; the shales above or below the *Scyphocrinites*-bearing zones generally lack benthic faunas but often yield abundant graptolites. In Europe, North Africa, and North America many of the carbonate rocks found associated with graptolitic shales have yielded abundant *Scyphocrinites* in the Pridolian and Gedinnian. However, the well-known Upper Silurian-Lower Devonian limestone-graptolitic shale sequences of Australia have yet to yield a specimen of this bizarre crinoid; its apparent absence in Australia indicates that it was not a completely cosmopolitan genus.

SUMMARY AND CONCLUSIONS

The Silurian expansion of platform-carbonate environments, particularly in North America and Europe, led to accelerated diversification of certain echinoderm groups, particularly the crinoids. Some Late Ordovician crinoid faunas of North America are dominated by taxa that became widely distributed in the Silurian. Unlike the replacement of North American endemic brachiopods by Old World groups at the beginning of the Silurian (Sheehan, 1975, p. 206), the Silurian echinoderm radiation has one of its major focal points in the Ordovician faunas of North America. Crinoids and cystoids were already diversified and abundant in the late Llandoverian carbonate environments of North America. The late Llandoverian increase in echinoderm diversity was characteristically North American at a time when other groups were becoming cosmopolitan.

Wenlockian echinoderm faunas became among the most diversified of the Paleozoic, and cosmopolitan distributions of many echinoderm families and genera accompanied this diversification. However, some abundantly represented crinoid and cystoid families and genera can be demonstrated to have remained endemic to either Europe or North America during the Silurian. Especially significant is the endemism of some dominant and highly successful crinoid families such as the Lampterocrinidae in North America and Mastigocrinidae in Europe. No control other than restrictive factors inherent to the larval-dispersal potentials of some taxa is invoked to explain the endemism of certain abundant benthic echinoderm groups during a time of general cosmopolitanism. The Silurian crinoid and cystoid faunas apparently included some forms with low larval-dispersal potentials and other forms with potentials more like those of the other contemporary cosmopolitan groups such as the brachiopods. The proto-Atlantic Ocean of the Silurian probably served as a barrier to the larval dispersal of those echinoderm families that remained endemic to the carbonate environments on either side of it.

Ludlovian echinoderm faunas show a general decline in taxonomic diversity, although diverse faunas are still known from a few areas (Hemse-Klinteberg, Brownsport, Henryhouse Formations). Changes restricting the availability of favored environments (e.g., carbonate platforms) toward the close of the Silurian instigated a crisis among many echinoderm groups. The Silurian offlap from the continental interior of North America resulted in increased stress within the marine carbonate communities as the organisms began to compete for habitat space on the continental margins. Extinction of many common and widely distributed Silurian echinoderm genera had taken place by the end of the Period. Surviving groups in the Old World and eastern North America became isolated from each other during the Early Devonian, and marked provincialism occurred among the echinoderms in these two regions. Sparse echinoderm evidence also indicates the presence of the Malvinokaffric Realm during the Emsian. During the final phases of the Tippecanoe Sequence (Sloss, 1963) in the Emsian, an extreme reduction in diversity is noted among the North American echinoderms. In the Rhenish-Bohemian Region, on the other hand, echinoderm diversity dramatically increased during the Emsian. In the Middle Devonian the Rhenish-Bohemian echinoderms migrated into North America and populated newly available environments created during the onlap of the Kaskaskia Sequence.

Scyphocrinites, commonly used as a Pridolian marker, is known to range from the Llandoverian into the Gedinnian of eastern and central North America. It seems to occur most commonly in Pridolian and Gedinnian strata. *Petalocrinus,* widely distributed and easily recognized by its distinctive and rigid arm-fans, may be a useful guide fossil for the late Llandoverian through the middle Wenlockian.

ACKNOWLEDGMENTS

We gratefully acknowledge important distributional and stratigraphic occurrences of certain taxa supplied by A. J. Boucot and D. Mikulic, Oregon State University; T. Bolton, Geological Survey of Canada; and T. E. Yancey, Idaho State University. Helpful suggestions during manuscript preparation and revision were received from B. F. Glenister, T. W. Broadhead, and G. Klapper, University of Iowa. Much appreciated input was also provided by R. L. Lewis, University of Texas at Austin.

REFERENCES

Bassler, R. S., and Moodey, M. W., 1943, Bibliographic and faunal index of Paleozoic pelmatozoan echinoderms: Geol. Soc. America Spec. Paper 45, 734 p.

Berry, W. B. N., and Boucot, A. J., 1970, Correlation of the North American Silurian rocks: Geol. Soc. America Spec. Paper 102, 289 p.

——— 1972, Correlation of the southeast Asian and Near Eastern Silurian rocks: Geol. Soc. America Spec. Paper 137, 65 p.

——— 1973, Glacio-eustatic control of Late Ordovician-Early Silurian platform sedimentation and faunal changes: Geol. Soc. America Bull., v. 84, p. 275-284.

Boucot, A. J., 1975, Evolution and extinction rate controls: Amsterdam, Elsevier, 427 p.

Boucot, A. J., Johnson, J. G., and Talent, J. A., 1969, Early Devonian brachiopod zoogeography: Geol. Soc. America Spec. Paper 119, 113 p.

Branisa, L., 1965, Index fossils of Bolivia. I. Paleozoic: Servicio Geol. Bolivia Bol. 6, 282 p.

Breimer, A., 1962, A monograph on Spanish Paleozoic Crinoidea: Leidse Geologische Medel., v. 27, p. 1-189.

Brower, J. C., 1973, Crinoids from the Girardeau Limestone (Ordovician): Palaeontographica Americana 7, p. 263-499.

Dubatolova, J. A., and Yeltysheva, R. S., 1967, Stratigraphic importance of the Devonian crinoids of Siberia, *in* Oswald, D. H., ed., Intern. Symposium on the Devonian System, Calgary 1967: Calgary, Alberta Soc. of Petroleum Geologists, v. 2, p. 537-542.

Gupta, V. J., and Webster, G. D., 1971, *Stephanocrinus angulatus* Conrad (Crinoidea) from the Silurian of Kashmir: Palaeontology, v. 14, p. 262-265.

Hollard, H., 1962, État des recherches sur la limite Siluro-Devonienne dans le Sud du Maroc, *in* Erben, H. K., ed., Symposiums-Band Internationalen Arbeitstagung über die Silur/Devon-Grenze und die Stratigraphie von Silur und Devon, 2nd, Bonn-Bruxelles 1960, E. Schweizerbart'sche Verl., p. 95-97.

Jaeger, H., 1962, Das Silur (Gotlandium) in Thuringen und am Ostrand des Rheinischen Schiefergebirges (Kellerwald, Marburg, Giessen), *in* Erben, H. K., ed., Symposiums-Band Internationalem Arbeitstagung über die Silur/Devon-Grenze und die Stratigraphie von Silur und Devon, 2nd, Bonn-Bruxelles 1960, E. Schweizerbart'sche Verl., p. 108-135.

Johnson, J. G., and Boucot, A. J., 1973, Devonian brachiopods, *in* Hallam, A., ed., Atlas of palaeobiogeography: Amsterdam, Elsevier, p. 89-96.

Johnson, J. G., and Lane, N. G., 1969, Two new Devonian crinoids from central Nevada: Jour. Paleontology, v. 43, p. 69-73.

Johnson, M. E., 1975, Recurrent community patterns in epeiric seas: The Lower Silurian of eastern Iowa: Iowa Acad. Sci. Proc., v. 82, p. 130-139.

LeMaître, D., and Heddabaut, C., 1963, Présence de gisements à *Scyphocrinites* dan les Pyrénees basques: Soc. Géol. France Compte Rendu Somm. Seances, 1963, p. 273-274.

McKerrow, W. S., and Ziegler, A. M., 1972, Silurian paleogeographic development of the proto-Atlantic Ocean: Internat. Geol. Cong., 24th, Montreal 1972, Proc., Section 6, p. 4-10.

Moore, R. C., 1952, Evolution rates among crinoids: Jour. Paleontology, v. 26, p. 338-352.

Mu, En-Chih, and Wu, Yung-Jung, 1974, Silurian Crinoidea, *in* Chao, Kingkoo and Yuan, Chen Yun, eds., A handbook of the stratigraphy and paleontology in southwest China: Nanking Inst. Geol. and Palaeontology, Academia Sinica, p. 208-211 (in Chinese).

Nikiforova, O. I., and Predtechenskij, N. N., 1968, A guide to the geological excursion on Silurian and Lower Devonian deposits of Podolia: Internat. Symposium on Silurian-Devonian boundary and Lower and Middle Devonian stratigraphy, 3rd, Leningrad 1968, Ministry Geol. USSR, 58 p.

Norford, B. S., 1962, The Silurian fauna of the Sandpile Group of northern British Columbia: Canada Geol. Survey Bull. 78, 51 p.

Paul, C. R. C., 1974, *Regulaecystis devonica*, a new Devonian pleurocystitid cystoid from Devon: Geol. Mag., v. 3, p. 349-352.

Ramsbottom, W. H. C., 1958, British Upper Silurian crinoids from the Ludlovian: Palaeontology, v. 1, p. 106-115.

Regnéll, G., 1975, Review of recent research on "pelmatozoans": Palaont. Zhurn., v. 49, p. 530-564.

Rexroad, C. B., and Nicoll, R. S., 1971, Summary of conodont biostratigraphy of the Silurian System of North America: Geol. Soc. America Mem. 127, p. 207-225.

Ross, R. J., Jr., 1975, Early Paleozoic trilobites, sedimentary facies, lithospheric plates, and ocean currents, *in* Martinsson, A., ed., Evolution and morphology of the Trilobita, Trilobitoidea, and Merostomata: Fossils and Strata, no. 4, p. 307-329.

Rowley, R. R., 1904, The Echinodermata of the Missouri Silurian and a new brachiopod: Am. Geologist, v. 34, p. 269-282.

Schuchert, C., Swartz, C. K., Maynard, T. P., and Row, R. B., 1913. The Lower Devonian deposits of Maryland, *in* Swartz C. K. and others, Lower Devonian [of Maryland]: Maryland Geol. Survey, p. 67-132.

Sheehan, P. M., 1975, Brachiopod synecology in a time of crisis (Late Ordovician-Early Silurian): Paleobiology, v. 1, p. 205-212.

Sloss, L. L., 1963, Sequences in the cratonic interior of North America: Geol. Soc. America Bull., v. 74, p. 93-114.

Springer, F., 1917, On the crinoid genus *Scyphocrinus* and its bulbous root *Camarocrinus:* Smithsonian Inst. Pub. 2440, 74 p.

——— 1926, American Silurian crinoids: Smithsonian Inst. Pub. 2871, 239 p.

Strimple, H. L., 1963, Crinoids of the Hunton Group: Oklahoma Geol. Survey Bull. 100, 169 p.

Stukalina, G. A., 1967, Stratigraphic significance of the stems of crinoids in solving the Silurian-Devonian boundary problem, *in* Oswald, D. H., ed., Internat. Symposium on the Devonian System, Calgary 1967: Calgary, Alberta Soc. Petroleum Geologists, v. 2, p. 893-896.

Ubaghs, G., 1976, Camerata, *in* Moore, R. C., ed., Treatise on invertebrate paleontology, Pt. T, Crinoidea: Geol. Soc. America and Univ. Kansas (in press).

Webster, G. D., 1973, Bibliography and index of Paleozoic crinoids, 1942-1968: Geol. Soc. America Mem. 137, 341 p.

Ziegler, A. M., Cocks, L. R. M., and McKerrow, W. S., 1968, The Llandovery transgression of the Welsh Borderland: Palaeontology, v. 11, p. 736-782.

APPENDIX

Endemic echinoderm taxa: numbers at left index families and genera appearing on Figures 1 to 3

Crinoidea: Camerata

1. Spyridiocrinidae
2. Archaeocrinidae
3. Lampterocrinidae
4. Gazacrinidae
5. Rhodocrinidae
6. Nyctocrinidae
7. Abacocrinidae
8. Metnapocrinidae
9. Coelocrinidae
10. Hexacrinitidae
11. Parahexacrinidae
12. Eucalyptocrinitidae
13. Dolatocrinidae
14. Polypeltidae
15. Stelidocrinidae
16. Marsupiocrinidae
17. *Marsupiocrinus* (*Amarsupiocrinus*)—10 armed
18. *M.* (*Marsupiocrinus*)—20-armed
19. Hapalocrinidae
20. Family uncertain—*Coronocrinus*

Crinoidea: Flexibilia

21. Taxocrinidae
22. Nipterocrinidae
23. Icthyocrinidae
24. Dactylocrinidae
25. Family uncertain—*Edriocrinus*

Crinoidea: Inadunata

26. Homocrinidae
27. Pisocrinidae
28. Myelodactylidae
29. Synbathocrinidae
30. Zophocrinidae
31. Cyathocrinidae
32. Sphaerocrinidae
33. Dendrocrinidae
34. Botryocrinidae
35. Thenarocrinidae
36. Mastigocrinidae
37. Metabolocrinidae

Crinoidea: *Incertae Sedis*

38. *Stephanocrinus*
39. *Aspidocrinus*

Cystoidea: Rhombifera

40. Scoliocystinae
41. Callocystitinae
42. Caryocrinitidae

Crinoidea: Miscellaneous

43. *Troosticrinus*
44. *Lysocystites*

Biogeography of Late Silurian and Devonian Rugose Corals in North America

W. A. Oliver, Jr., *U.S. Geological Survey, Washington, D.C. 20244*
A. E. H. Pedder, *Geological Survey of Canada, Calgary, Alberta*

ABSTRACT

A stage-by-stage analysis of the distribution of rugose corals in the Late Silurian and Devonian shows the following:

(1) Lower and Middle Devonian faunas in Eastern and Western North America are markedly different. In the East, numbers of endemic genera increased through the Early Devonian to a high of 92 percent in the late Emsian; endemism then decreased through the Middle Devonian so that faunas were cosmopolitan by Late Devonian time. In Western North America, generic-level endemism did not exceed 24 percent in any stage. This east-west contrast permits the assignment of Western American areas to the Old World Biogeographic Realm, while Eastern American areas formed a distinct realm of their own, the Eastern Americas Realm.

(2) Within Western North America, during the Early Devonian, endemism generally increased and faunal similarity coefficients decreased from arctic Canada, to western Canada-Alaska to the Great Basin. Furthermore, relict Silurian genera persisted longer into the Early Devonian in the Great Basin than in the other Western American areas. This suggests that migration routes to and from the rest of the Old World Realm were along the northern side of a joined North America-European continent or that Old World coral faunas in the Great Basin were diluted by emigration of corals from the southwestern part of the Eastern Americas Realm. Middle Devonian coral assemblages in the West contained few endemics. This suggests that water movements at this time were adequate to homogenize the faunas.

(3) Coral faunas from Pacific Coast areas, possibly eugeosynclinal in Devonian time, are not significantly different from the "carbonate shelf" faunas farther inland, even though they may have been separated by deep water.

(4) Eastern North American corals were isolated through much of the Devonian. Movement of Old World corals into Eastern America, apparently from the Williston Basin to the Michigan Basin, began in Eifelian and increased in Givetian time. Simultaneously, Eastern American genera moved into an apparently adjacent Africa. The only known Devonian corals from northern South America are Eifelian; they are distinctly Eastern North American in affinities.

INTRODUCTION

Three first-order marine biogeographic divisions of the Devonian world are recognized (see Oliver, 1976a, 1977b, for review). These are: (1) the relatively small Eastern Americas Realm, including Eastern North America and northern South America; (2) the Old World Realm, including Western and arctic North America, northern Africa, Eurasia, and Australia; and (3) the Malvinokaffric Realm, including southern South America, southern Africa, and Antarctica.

North America is unique among present-day continents in including richly coralliferous parts of both the Eastern Americas and the Old World Realms. Figure 1 shows the areas occupied by the Realms in North America (separated by the Transcontinental Arch) and the principal geographic units discussed in this paper. We capitalize East and West (and Eastern and Western) when we are making specific reference to the two parts of North America. During most of the Devonian, the Transcontinental Arch was a positive area that may have been mostly emergent during much of the time.

In this paper we consider the evolution of this two-part division of North America from its apparent beginning in Late Silurian time to its termination in the Late Devonian. Several questions seem pertinent:

(1) What were the relationships between rugose coral faunas in the two areas or Realms?

(2) How did each realm-fauna evolve in its own area? And, are major biogeographic subdivisions recognizable?

(3) What were the intercontinental relationships of each half of North America? Specifically:

(4) Was the Pacific Coast area an island arc, a detached platelet, or part of another continent, as has been suggested by some geologists?

(5) Do Western American faunas show any gradation from arctic Canada through western Canada to the Great Basin as would be suggested by contemporary restorations of Devonian geography?

Table 1. Comparison of rugose coral assemblages in Western and Eastern North America for the Late Silurian and Devonian. Data are the total number of genera known to occur in each area (N); the number (n) and percentage of these genera that are endemic; the number (*N*) of genera common to both areas; and the Otsuka Similarity Coefficient.

			West			*N* in common	Otsuka Coeff.	East		
			N	Endemic				N	Endemic	
				n	%				n	%
Devonian	Late	Frasnian	24	0	0	9	51	13	0	0
Devonian	Middle	Givetian	34	2	6	13	38	35	11	31
Devonian	Middle	Eifelian	33	6	18	5	16	28	18	64
Devonian	Early	Late Emsian	29	5	17	2	7	25	23	92
Devonian	Early	Pragian	29	7	24	3	19	9	4	44
Devonian	Early	Lochkovian	26	3	12	2	10	14	8	57
Late Silurian			52	8	15	14	32	36	11	31

(6) How did Eastern American coral faunas relate to coral faunas in South America, Africa, and western Europe, all areas presumed to be adjacent to Eastern North America during the Devonian?

This discussion is based on our studies of rugose corals and includes their history in North America and what they can indicate of physical conditions and events during the Late Silurian and Devonian. The entire discussion is based on genera, including numbers of genera in well-defined geographic areas, the percentage of these that were endemic, and a comparison of generic compositions made by using the Otsuka Similarity Coefficient (see Cheetham and Hazel, 1969, for discussion of this coefficient). This is primarily a qualitative analysis rather than a quantitative one; the numbers are included to show the data base and, insofar as is practical, to show the reasoning behind the interpretations. Our generic identifications are attached (Tables 2 to 36). We know that many North American faunas are unstudied by modern techniques and that future studies will undoubtedly modify some of our generic determinations but we expect that this will not alter our conclusions significantly.

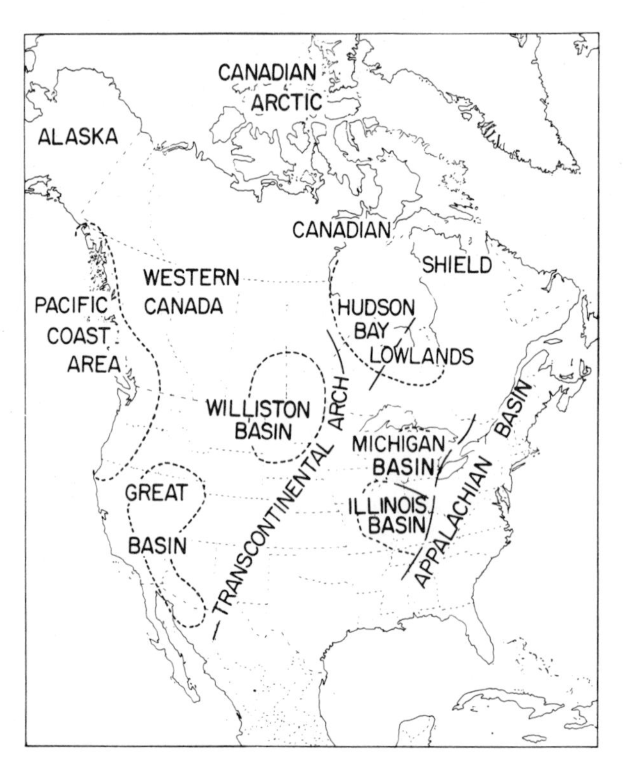

Figure 1. Index map showing North American areas discussed in text.

The following discussion proceeds in a stage-by-stage sequence. Table 1 summarizes the relationships of corals in the two halves of North America. The number of genera known from each stage and the percentage of these that are endemic are shown for both West and East. We define endemic as not being known to occur anywhere else in the world in rocks of the same or any earlier stage.

Table 1 emphasizes two points:

(1) Endemism in the East increased through the Early Devonian to a high of 92 percent in the late Emsian, and then decreased gradually to zero in the early Late Devonian (Frasnian Stage). In contrast, levels of endemism never exceeded 24 percent in the West.

(2) Few genera are common to both East and West in most stages. Similarities are at a medium level in the Late Silurian, but are very low during the Early Devonian, start to build up in the early Middle Devonian, and become high in the Late Devonian Frasnian Stage.

LATE SILURIAN

The generalized distribution of Late Silurian corals is shown in Figure 2. Two stages are represented because

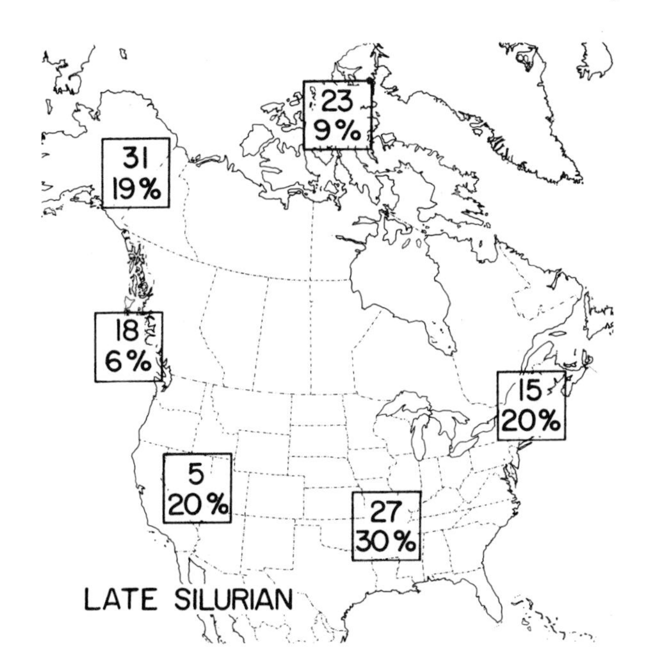

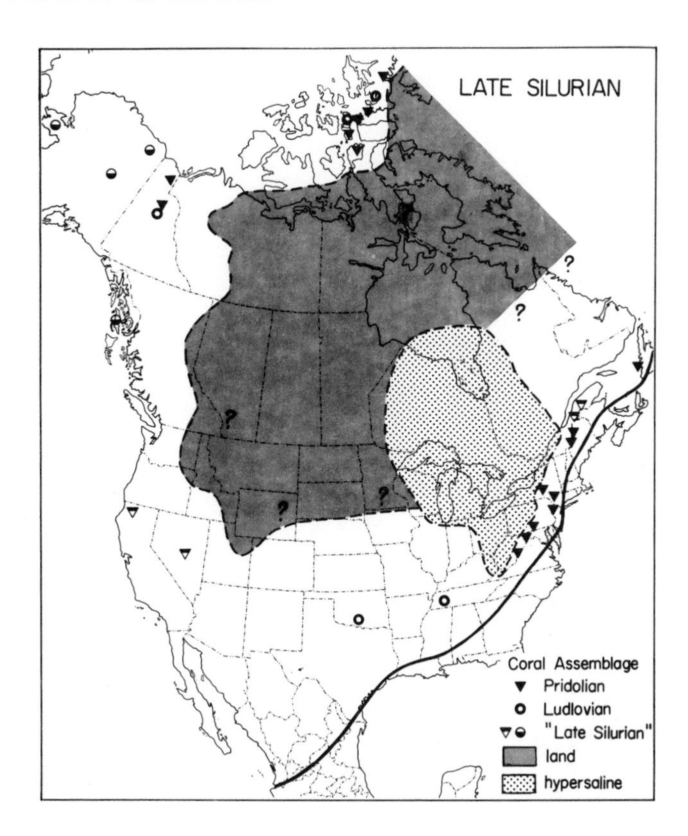

Figure 2. Generalized distribution of Late Silurian corals. Left (a): Late Silurian data. Squares enclose number of genera known from each area and the percentage of these that are endemic to their half of North America (see Tables 2 to 7). Right (b): Late Silurian paleogeography. The heavy line from Newfoundland to Mexico represents the plate boundary used by Oliver (1976a, 1977b).

in some of the areas we cannot now separate Ludlovian corals from Pridolian corals and because some of the samples are too small. Present stratigraphic control is poor in Alaska, the Pacific Coast area (SE Alaska and NW California), and the Great Basin. Numbers of genera are small in the Great Basin in any case. Both stratigraphic control and numbers are good in the other areas, but similarities are low for reasons suggested by the paleogeographic map (Fig. 2b).

The Late Silurian arctic corals are quite unlike the Eastern ones. Ideally they should be compared with the Yukon samples (too small), or with those from Alaska where control is poor, but few genera have been found in common between any of the areas, possibly because of inadequate sampling and control.

The Midcontinent corals (Oklahoma and Tennessee) are Ludlovian, whereas the Appalachian ones are mostly Pridolian; therefore, these cannot be compared either.

The Ludlovian corals in the Midwest show a medium level of endemism (30 percent). They may have been isolated by the beginnings of hypersaline conditions (Fig. 2b) and(or) by an early Transcontinental Arch (Fig. 1), but they do not seem to represent an early stage of the Eastern Americas Realm coral fauna.

The beginning of realm differentiation is traceable to Pridolian time. During the Late Silurian in general, Eastern North America is thought to have been a carbonate shelf extending along the south side of a North American-European continent (Oliver, 1976a, Fig. 3). To the south lay the closing proto-Atlantic Ocean, which by this time was very narrow but possibly quite deep. North of the carbonate shelf in the late Ludlovian and Pridolian (Fig. 2b), were extensive areas of hypersaline water and probably broad, lowland areas that are generalized on the map. A distinct gradient is shown by the shelf coral faunas in that the most eastern faunas (New England and maritime Canada) are very European in aspect, whereas those farther from Europe (New York-Virginia) have more endemic genera and species. There was probably no barrier to migration along the belt from the Appalachians to the Russian Platform, other than distance.

EARLY DEVONIAN

During the Lochkovian Stage (Table 1), the East-West separation was well-defined. Figure 3a shows the corals grouped into three broad areas (compare Fig. 3b). From each area we know 14 to 16 genera, but 57 percent of these are endemic in the Appalachians, whereas only 6 percent and 13 percent are endemic in western Canada and the Great Basin, respectively. The number of genera in common between each combination of areas and the corresponding Otsuka Similarity Coefficients are also shown. Only 2 of the Eastern genera occur in either Western area, and the similarities are low. Six genera are common to both Western areas, and the Otsuka Coefficient is 38. This is not a very high degree of similarity and we think that this may reflect the position of the Great Basin at the end of the Old World Realm (see world maps in

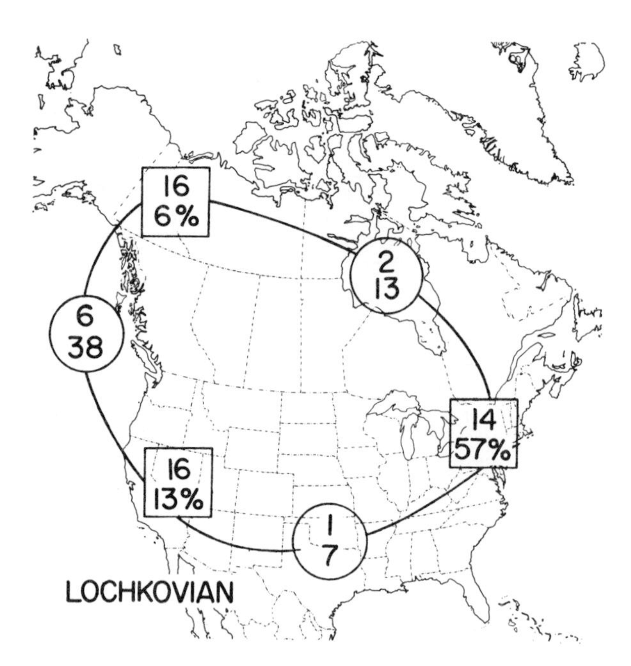

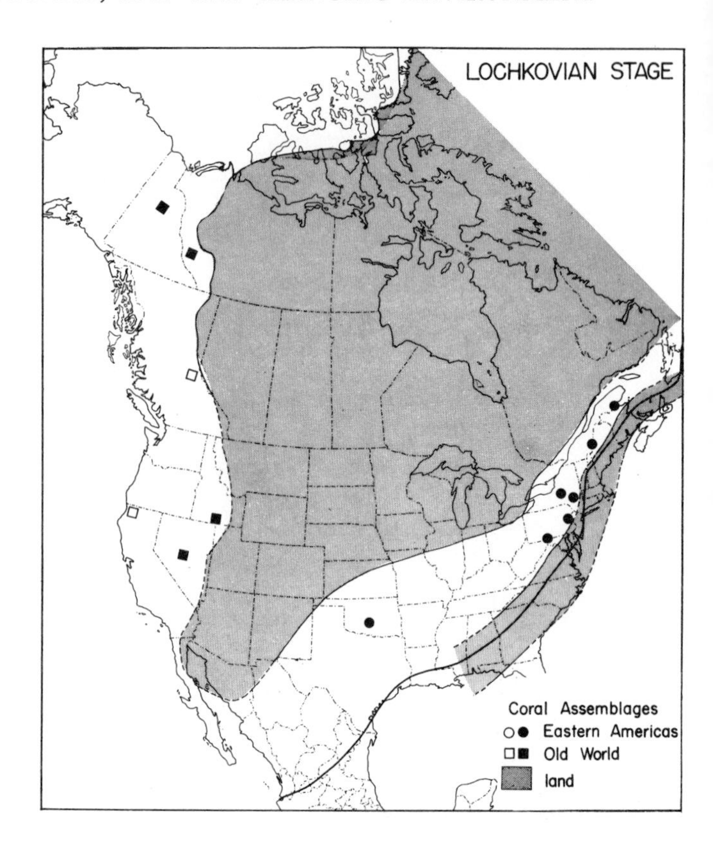

Figure 3. Generalized distribution of Lochkovian corals. Left (a): Lochkovian data. Squares enclose number of genera known from each area and the percentage that are endemic to their half of North America. Circles enclose the number of genera that occur in both of the two areas connected by the lines and the Otsuka Similarity Coefficient for the two areas (see discussion in text and Tables 8 to 10). Right (b): Lochkovian paleogeography. Symbols mark areas from which rugose corals have been studied: circles, Eastern Americas Realm corals; squares, Old World Realm corals; solid circles or squares represent major data points; open circles or squares represent one or few genera. Distribution of land (shaded) and sea is generalized.

Oliver 1976a, 1977b). In addition, the Great Basin coral faunas contain a number of relict Silurian genera that are not known in western Canada; these decrease the apparent similarity of the two assemblages. Convincing evidence may come from larger numbers of genera, and data from other Old World areas.

The most significant fact for the Early Devonian is that every coral genus now known from the Lochkovian of Eastern North America is either endemic (57 percent) or was in the same area during the Late Silurian. By this time, there was no need for a connection with any part of the Old World Realm as far as the Eastern America rugose corals were concerned.

The Western American corals are distinctly Old World. Apparently they were separated from the East by an emergent Transcontinental Arch (Fig. 3b), but some evidence indicates that some of the associated brachiopods migrated around the southern end of this land barrier.

By the Early Devonian, a land area was emerging along the southeastern edge of North America or the northern edge of Gondwanaland. This is shown as a continuous belt on Figure 3b, but it was more likely discontinuous (Oliver, 1976a, Fig. 4). Terrestrial fossils are found locally within the belt, and several areas north of the belt have coarse clastic deposits with sedimentary features that indicate a southeastern source.

Sometime during the Early Devonian, the eastern end of the Appalachian seaway was closed off by the developing Old Red Continent, but it may have been still open during the Lochkovian. In Europe, the nearest known Lochkovian coral faunas are in the Alps or possibly on the north coast of Spain. Eastern America may have been still isolated only by distance.

Pragian Stage coral assemblages have been found in the same three areas (Figs. 4a, b) as for the Lochkovian but the assemblages differ in size. Only 5 genera of this age are known in the Appalachians (Table 11). The Great Basin corals had a relatively high level of endemism and a low similarity with those of western Canada. Western Canada had the same endemic level that both Western areas had in the previous stage.

These data can be interpreted in the framework suggested by Boucot and Johnson (1967, 1968) on the basis of brachiopods. Boucot and Johnson showed that there were major faunal movements between the Great Basin and Eastern North America during Pragian time (arrows on Fig. 4b), and they considered the Great Basin to be temporarily a part of the Eastern Americas Realm rather than of the Old World Realm. If this interpretation is accepted and the Great Basin and Eastern data are combined, we find that Eastern Realm corals are 47 percent endemic and that the Otsuka Similarity between the North American parts of the two Realms is only 16 (Fig. 4a).

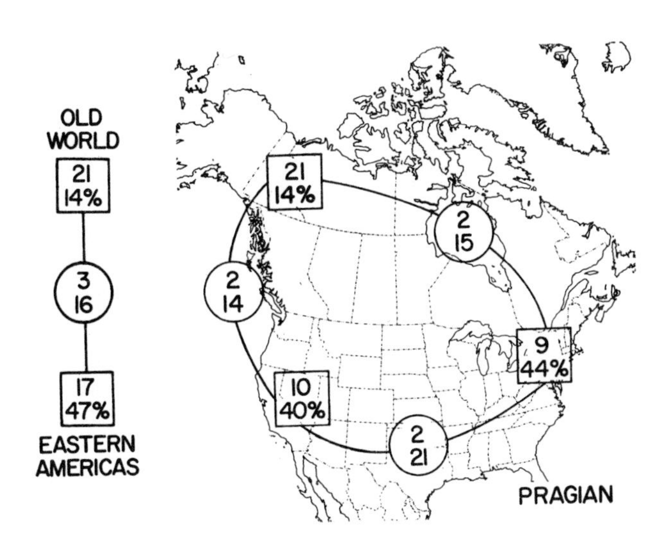

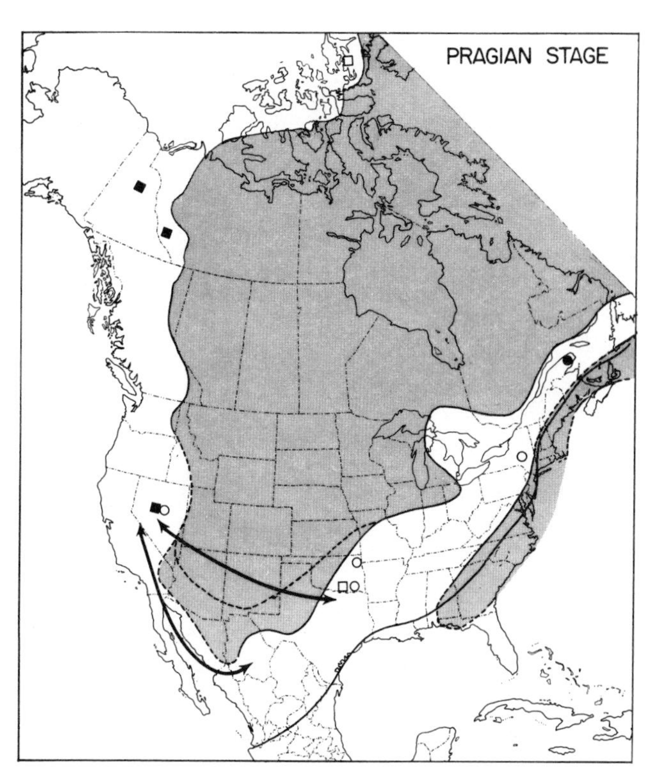

Figure 4. Generalized distribution of Pragian corals. Left (a): Pragian data. Squares enclose number of genera known from each area and the percentage that are endemic to their half of North America. Circles enclose the number of genera that occur in both of the two areas connected by the lines and the Otsuka Similarity Coefficient for the two areas. On the left: comparison of North American parts of Old World and Eastern Americas Realms (see discussion in text and Tables 11 to 13). Right (b): Pragian paleogeography (see Figure 3b explanation for key).

To this point it has been most convenient to discuss the North American Devonian coral faunas in terms of the Czechoslovakian stages. Younger Devonian faunas are better considered in terms of the German stages. As the Pragian overlaps the Emsian (in time), our next (and last) unit of Early Devonian time is the late Emsian.

The late Emsian marks the high point of endemism in the East (92 percent, Fig. 5a). Twenty-three Eastern genera are endemic. In the West, three areas have enough genera to analyze (Fig. 5a). Numbers are low, but endemism is highest and similarities lowest toward the south. This fits our hypothesis that the Great Basin corals ought to be more different from those of other parts of the Old World Realm than are corals from other Old World parts of North America.

The paleogeographic map (Fig. 5b) shows a more isolated East and many more data points in both East and West than earlier maps.

MIDDLE DEVONIAN

The Middle Devonian was the time of declining endemism in the East and of increasing East-West similarities. This began in the Eifelian Stage (Fig. 6).

During the Eifelian, the Eastern Americas Realm included northern South America (see Oliver, 1976a, 1977b, for discussion of this fauna and questions regarding its age). Twelve rugose coral genera are known from Venezuela (Fig. 6a); all of them occur in Eastern North America as well (although 4 of them a stage earlier or later), and 10 of the 12 are Eastern Americas Realm endemics. There is no basis for the division of the Eastern Realm into provinces; the fauna is one and the same.

Similarities between East and West are low (Fig. 6a), but within the West similarities are high. Only 7 arctic genera are known, but all of them also occur in western Canada. The Great Basin coral assemblage contains few endemics and has a high similarity with the combined list of western and arctic Canadian genera (Fig. 6a).

Eifelian paleogeography (Fig. 6b) is not very different from that of the Emsian but more data are available from more widely scattered areas. The marine embayment in northern Alberta and Saskatchewan appears to have been the beginning of a principal connection between East and West.

During the Givetian Stage, two provinces can be recognized within the Eastern Americas Realm (Oliver, 1977b). The Michigan Basin-Hudson Bay Lowlands Province has more genera (27) and lower endemism (26 percent) than the Appohimchi Province (22 genera, 36 percent endemic) (Fig. 7a). Fourteen genera are common to both Provinces, and the coefficient of similarity is high (57). Part of the faunal difference is the result of 3 new genera endemic to the Michigan Basin-Hudson Bay Province and 4 new endemics in the Appohimchi Province, but the principal difference is the presence of 9 Old World genera in the Michigan-Hudson Bay area. These are not known farther south or east in the rest of Eastern North America, and we can only conclude that they moved in from the west or north, most likely the west (Fig. 7b).

Logical East-West connections would seem to have been from the Williston Basin (Fig. 1) to Michigan or

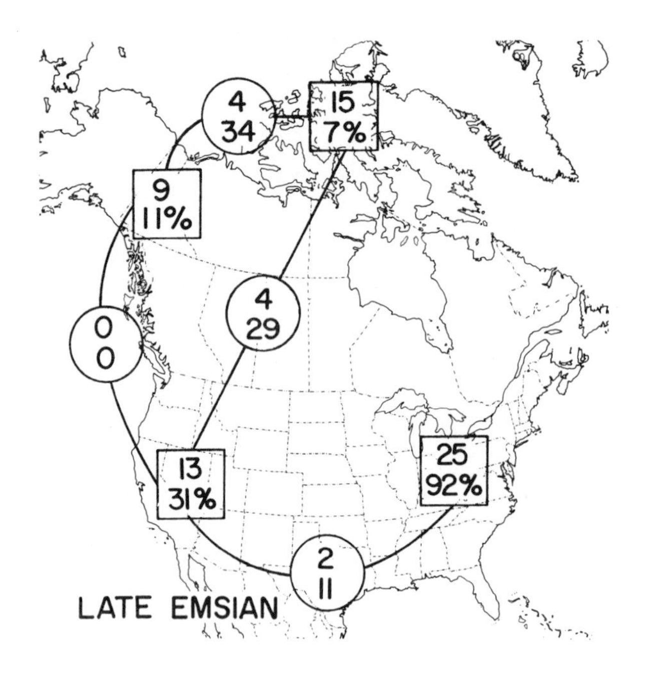

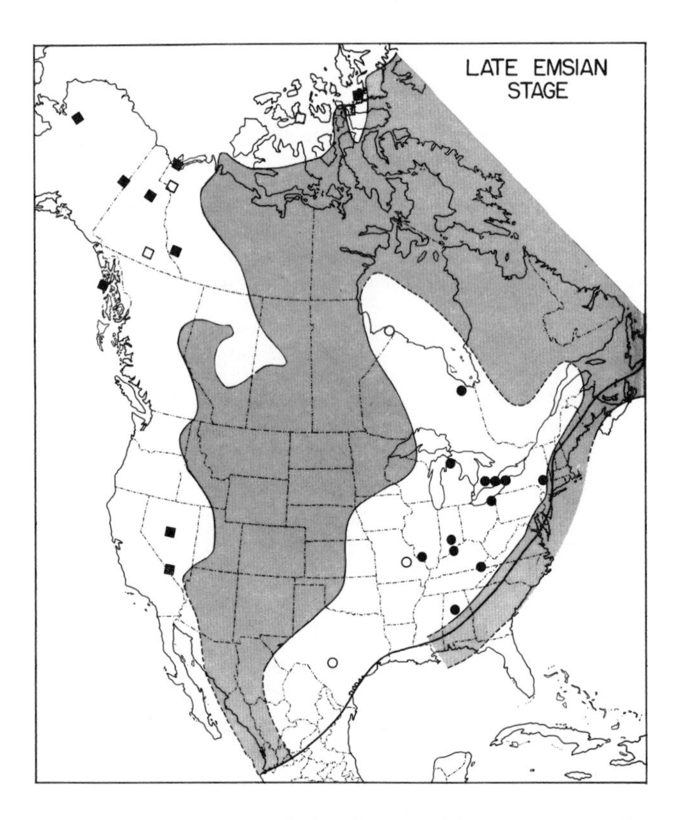

Figure 5. Generalized distribution of late Emsian corals. Top (a): Late Emsian data. Squares enclose number of genera known from each area and the percentage that are endemic to their half of North America. Circles enclose the number of genera that occur in both of the two areas connected by the lines and the Otsuka Similarity Coefficient for the two areas (see discussion in text and Tables 14 to 17). Bottom (b): Late Emsian paleogeography (see Figure 3b explanation for key).

Ontario, but Williston Basin samples are apparently too small to show this (Fig. 7a). We do get the expected result by combining all the western Canada data. The Michigan-Hudson Bay corals show high similarity to those of western Canada whereas the corals of other Eastern areas show much less similarity (Fig. 7a).

Within the West, we compare several assemblages (Fig. 7a). Endemism is very low and similarities are mostly high. There is no apparent basis for subdividing this area and no indication that the Great Basin was in any way set off from the other areas.

Middle Devonian coral assemblages from Alaska and the Pacific Coast area (SE Alaska and central Oregon) are more similar to Western Givetian assemblages than they are to Eifelian ones and may be principally of Givetian age. In Figure 7a, these Middle Devonian assemblages are compared with Canadian and Great Basin Givetian assemblages. It may be significant that the Pacific Coast corals are more similar to those in Alaska than they are to those in western Canada and the Great Basin; however, the Alaskan corals are even less similar to those of western Canada, and it is most likely that these results are due to the dilution of the two assemblages by an unknown number of Eifelian corals.

Givetian paleogeography is outlined in Figure 7b (compare world maps in Oliver, 1976a, 1977a, 1977b). The Old World genera in the Michigan Basin-Hudson Bay Province are located on the map by black squares in northern Indiana, northern Michigan, and near James Bay. The occurrence of the Old World genera in the East raises some questions that are, as yet, unanswered. The northern Indiana occurrence is in the Miami Bend Formation, the *Stringocephalus*-bearing limestone described by Cooper and Phelan (1966). From this unit we have identified 6 Old World genera and only 2 Eastern Americas endemic genera; the brachiopods and molluscs are dominantly Old World also (Cooper and Phelan, 1966). The Miami Bend is immediately overlain by another limestone (Logansport) containing a pure Eastern fauna and none of the Old World elements. The corals are better known in Michigan (in the Traverse Group) but there the details of stratigraphic occurrence are unclear. It is important to find out whether the Old World elements are more-or-less segregated in certain beds (as in the Miami Bend), or whether there was clear intermixing of the Old World and Eastern Americas corals.

The major movement of corals was from West to East, but a few Eastern genera did move west. These are located on Figure 7b by the open circles. Each circle in the West represents 1 Eastern genus except for the one in Nevada which represents 2 genera.

Oliver (1976a, 1977b) reviewed the relationship of the Eastern Americas corals to those in Africa. Nine genera, approximately 25 percent of those known from North African Middle Devonian rocks, were Eastern Realm endemics before appearing in Africa. It is logical to assume that these American genera migrated to Africa around, or

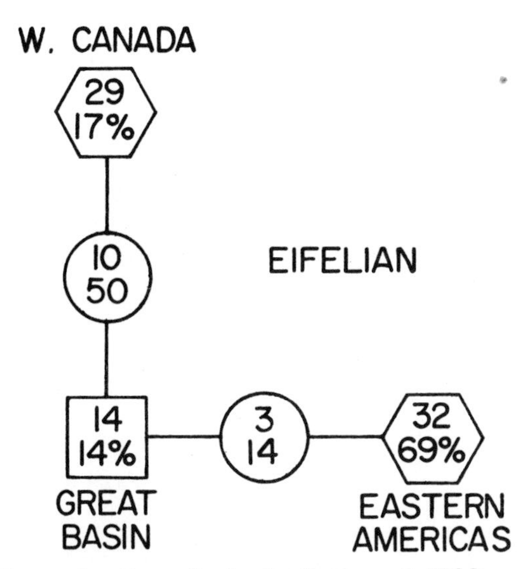

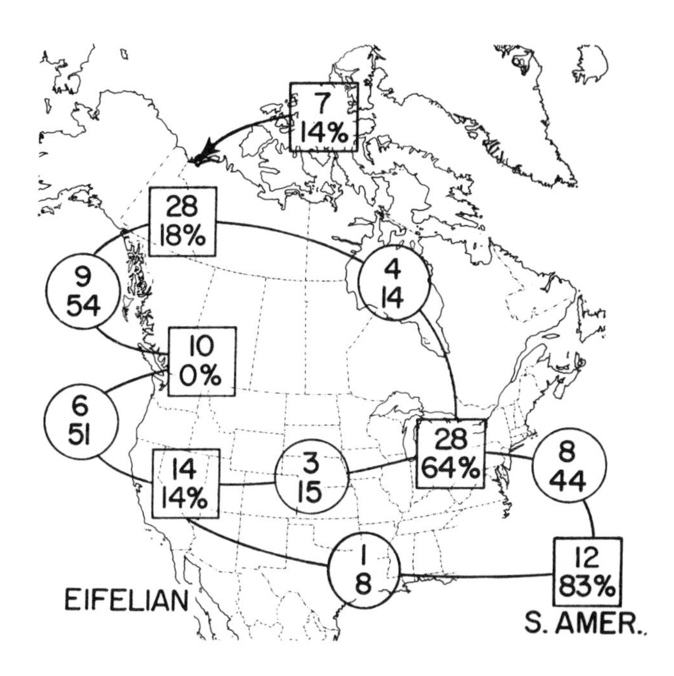

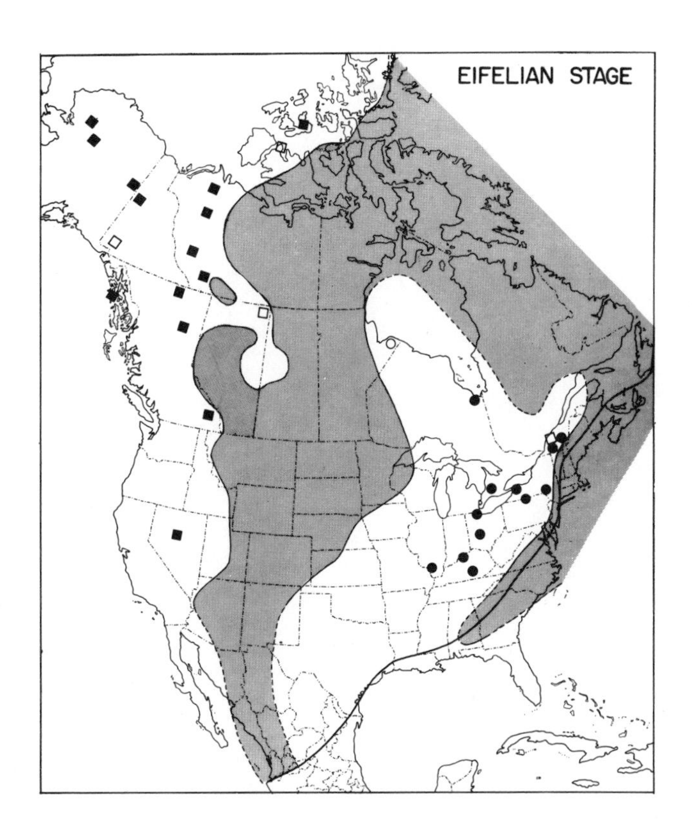

Figure 6. Generalized distribution of Eifelian corals. Top (a): Eifelian data. Squares enclose number of genera known from each area and the percentage that are endemic to their half of North America. Circles enclose the number of genera that occur in both of the two areas connected by the lines and the Otsuka Similarity Coefficient for the two areas. On the left: hexagons enclose the combined data for western Canada and for the Eastern Americas Realm for comparison with the Great Basin (see discussion in text and Tables 19 to 24). Bottom (b): Eifelian paleogeography (see Figure 3b explanation for key).

through, breaks in the peninsular extension of the Old Red Continent, shown on Figure 7b.

We noted that movement from Western North America into the Michigan-Hudson Bay Province was dominantly one way. A few genera moved west, but those that did are uncommon, whereas many of the Western genera are very common in the East. Similarly, movement into Africa was one way—and the same way, west to east. Presumably, surface-water movements carrying coral larvae are responsible for this pattern, but we have not attempted to plot currents on a map.

In general, the waxing of endemism in the Eastern Americas Realm coincided with the buildup of the land area on the southeast side of the Appalachian belt. This can logically be related to the collision of the North American-European continent with Gondwanaland and is consistent with the appearance of Eastern Americas genera in Africa. The waning of endemism was due to the breakdown of the barriers and especially to the Givetian influx of Western American corals into the East.

THE END OF ENDEMISM

Eastern endemism and the Eastern Americas Realm ended with the Middle Devonian. Early Late Devonian corals of East and West were notably cosmopolitan. Fairly large faunas are known from Iowa, in what had been the Eastern Americas Realm, and smaller faunas are known as far east as New York (Fig. 8). None of the known genera were endemic, and none are even thought to have evolved from former endemic genera. The whole endemic coral fauna seems to have become extinct.

Instead, the Late Devonian corals in the Eastern area are cosmopolitan genera, known from all Old World continents where marine Upper Devonian rocks have been studied. They or their ancestors are generally well-known in Middle Devonian rocks of the Old World Realm. This would appear to be an excellent example of a fauna

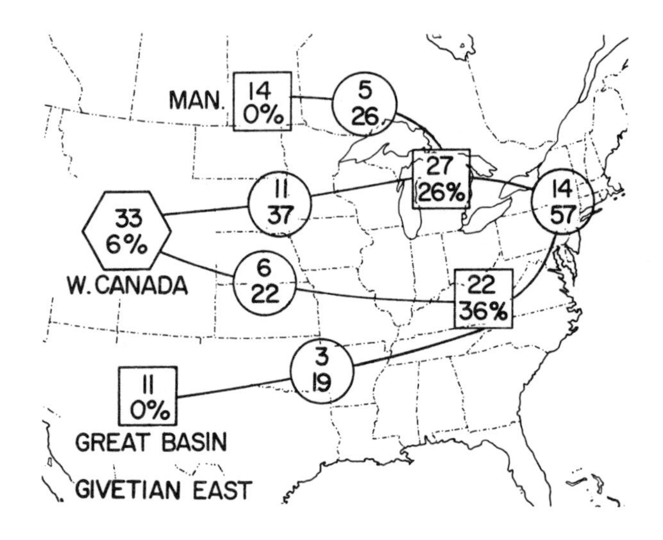

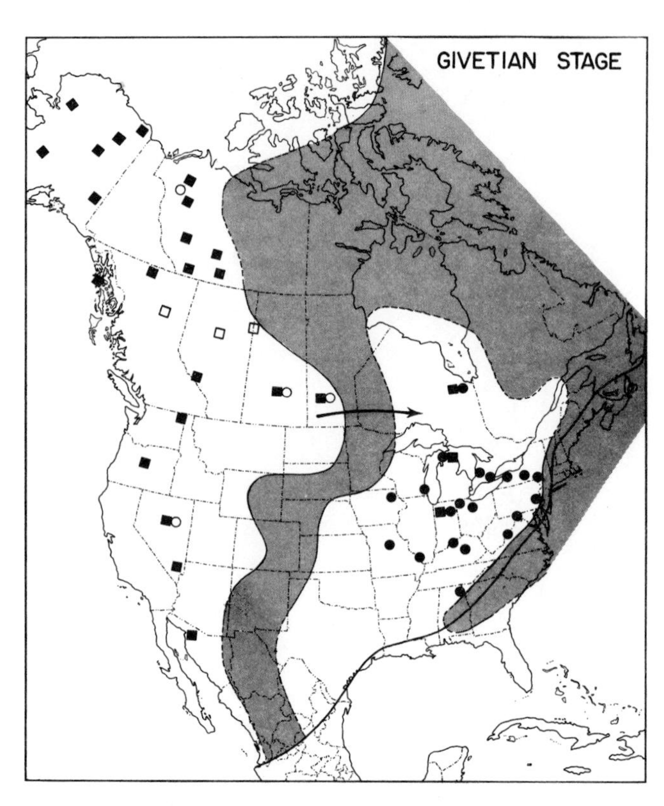

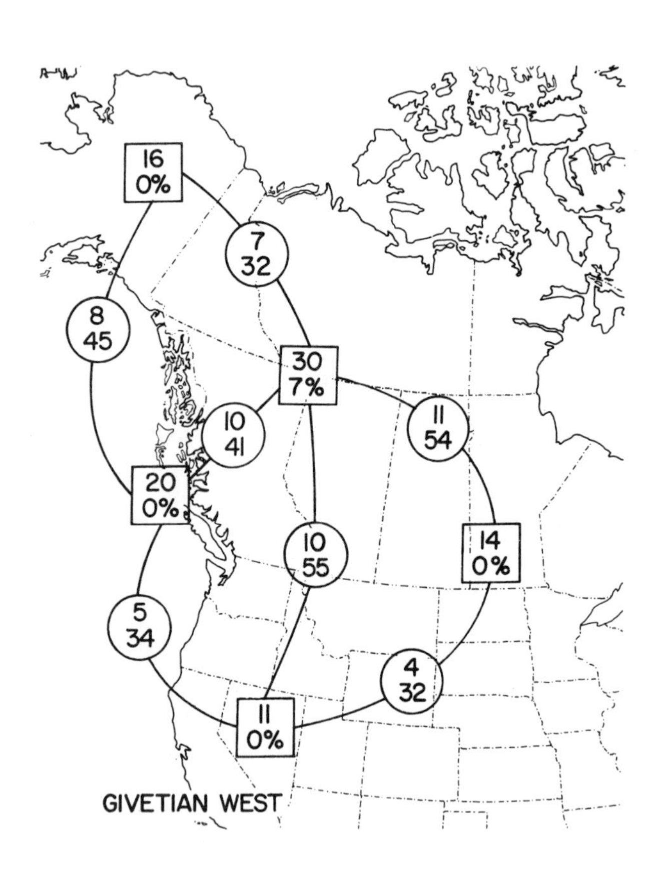

Figure 7. Generalized distribution of Givetian corals. Left (a): Givetian data. Squares enclose number of genera known from each area and the percentage that are endemic to their half of North America. Circles enclose the number of genera that occur in both of the two areas connected by the lines and the Otsuka Similarity Coefficient for the two areas. Hexagon encloses combined data as mentioned in text. Top: Eastern North America. Bottom: Western North America (note that Alaska and Pacific Coast data include all Middle Devonian) (see discussion in text and Tables 25 to 31). Right (b): Givetian paleogeography (see Figure 3b explanation for key).

evolving in isolation for a long period of time, and then being unable to compete when its isolation was ended.

There are several possible reasons for this inability to compete. The Eastern Americas corals thrived in the Eastern Realm, but the environment was probably not quite "normal" open marine. Given an epicontinental sea, bordered on three sides by land, and having only limited access to the open ocean (Figs. 3 to 7), it seems safe to assume that the water was warmer and that salinity was slightly higher than in comparable Old World areas.

In addition, the Old World Realm was much larger, by a factor of 10 or more. Most parts of the Realm bordered the ocean, and water circulation must have been better than in the Eastern Realm. The area was ecologically more diverse, as is indicated by its size, latitudinal extent, and the presence of a continental slope (indicated by the presence of a surrounding ocean). Numbers of genera were not in proportion to area, but there were at least two to three times as many Old World genera as there were Eastern Americas ones.

The Old World corals were more diverse and were living in a Realm with more diverse and more "normal" environments. The Eastern Americas corals were less diverse and were living in more limited, probably somewhat specialized environments. When the Transcontinental Arch ceased to separate the Realms in North America, the less adaptable Eastern Americas corals were less able to cope with the changes that resulted, and the Old World corals took over.

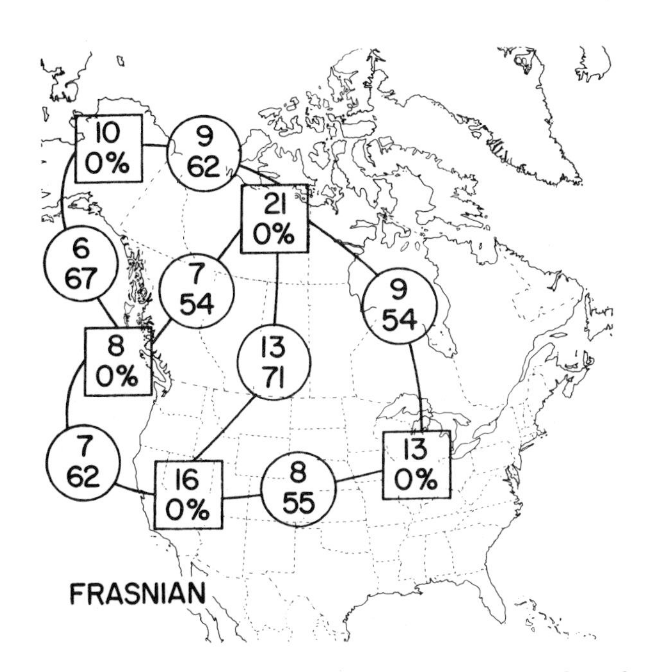

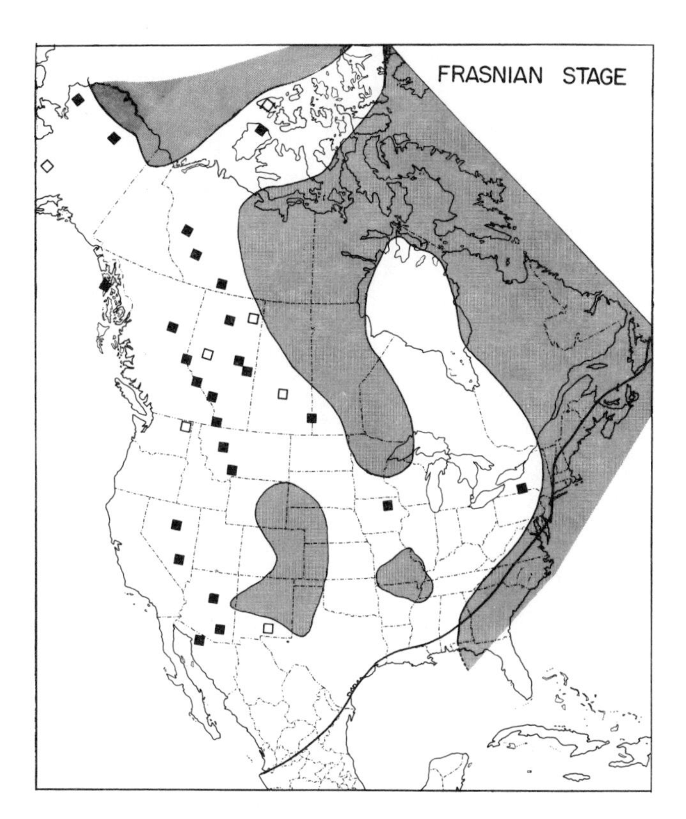

Figure 8. Generalized distribution of Frasnian corals. Left (a): Frasnian data. Squares enclose number of genera known from each area and the percentage that are endemic to their half of North America. Circles enclose the number of genera that occur in both of the two areas connected by the lines and the Otsuka Similarity Coefficient for the two areas (see discussion in text and Tables 32 to 36). Right (b): Frasnian paleogeography (see Figure 3b explanation for key).

CONCLUSIONS

We have given at least generalized answers to four of the six questions posed in our introduction. Questions (4) and (5), about Western American geography, require elaboration.

(4) Current interpretations of Pacific Coast tectonics and stratigraphy suggest that the Pacific Coast area (Fig. 1) was a volcanic island arc, separated from the continent by a deep trench (Churkin, 1974; Stewart and Poole, 1974; Davis, 1974). It has also been suggested that the area was not part of the North American plate during the middle Paleozoic (Wilson, 1968; Hughes, 1975). In either case, we would expect that Pacific area corals might be different from the continental-shelf assemblages known from central and northern Alaska, western Canada, and the Great Basin. They are not significantly different, and we conclude that either the migration barriers were no greater between island arc and carbonate shelf than they were along the shelf, or that our data are inadequate to show the differences.

(5) The Great Basin is thought to have been one end of the Old World Realm which extended all the way around the Devonian world from Australia (Oliver, 1976a, 1977b). If this is approximately correct, the positions of the Western American areas should be reflected in the generic makeup of their coral faunas. For example, the corals of Alaska and western Canada should be more European in aspect than those of the Great Basin, but less so than those of arctic Canada. The Early Devonian distributions do fit this pattern in that apparent endemism was higher and similarity coefficients were lower toward the south. However, the Middle Devonian data do not fit this at all; instead, endemic levels were uniformly low and similarity coefficients high throughout the West. Apparently, Middle Devonian water movements were adequate to homogenize the coral fauna.

ACKNOWLEDGMENTS

The Silurian paleogeographic map (Fig. 2b) is based partly on data in Berry and Boucot (1970); Shell Oil Co. (1975); and unpublished information. The Devonian maps ("b" parts of Figs. 3 to 8) are extensively revised from Oliver (1977b); references to source materials are included in that work. Revisions, primarily in the western half of North America and especially in Canada, are based on Ziegler (1969), and unpublished data of various geologists of the Geological Survey of Canada.

REFERENCES

Amsden, T. W., 1949, Stratigraphy and paleontology of the Brownsport Formation (Silurian) of western Tennessee: Yale Univ. Peabody Mus. Nat. History Bull. 5, 134 p.

Berry, W. B. N., and Boucot, A. J., eds., 1970, Correlation of the North American Silurian rocks: Geol. Soc. America Spec. Paper 102, 289 p.

Boucot, A. J., 1975, Evolution and extinction rate controls: Amsterdam, Elsevier, 427 p.

Boucot, A. J., and Johnson, J. G., 1967, Paleogeography and correlation of Appalachian Province Lower Devonian sedimentary rocks: Tulsa Geol. Soc. Digest, v. 35, p. 35-87.

——— 1968, Appalachian Province Early Devonian paleogeography and brachiopod zonation, *in* Oswald, D. H., ed., Internat. Symposium on the Devonian System, Calgary 1967: Calgary, Alberta Soc. Petroleum Geologists, v. 2, p. 1255-1267.

Cheetham, A. H., and Hazel, J. E., 1969, Binary (presence-absence) similarity coefficients: Jour. Paleontology, v. 43, p. 1130-1136.

Churkin, Michael, Jr., 1974, Paleozoic marginal ocean basin-volcanic arc systems in the Cordilleran fold belt: Soc. Econ. Paleontologists and Mineralogists Spec. Pub. 19, p. 174-192.

Cooper, G. A., and Phelan, Thomas, 1966, *Stringocephalus* in the Devonian of Indiana: Smithsonian Misc. Colln., v. 151, p. 1-20.

Crickmay, C. H., 1960, The older Devonian faunas of the Northwest Territories: Calgary, Evelyn de Mille Books, 21 p.

——— 1962, New Devonian fossils from western Canada: Calgary, Evelyn de Mille Books, 16 p.

——— 1968, Lower Devonian and other coral species in northwestern Canada: Calgary, Evelyn de Mille Books, 9 p.

Davis, G. A., 1974, Pre-Mesozoic history of California and the west: San Joaquin Geol. Soc. Short Course (Bakersfield, Feb., 1974), paper 10, 6 p.

Ehlers, G. M., and Stumm, E. C., 1949a, Corals of the Devonian Traverse Group of Michigan. Pt. I. *Spongophyllum:* Michigan Univ. Mus. Paleontology Contr., v. 7, p. 123-130.

——— 1949b, Corals of the Devonian Traverse Group of Michigan. Pt. II. *Cylindrophyllum, Depasophyllum, Disphyllum, Eridophyllum* and *Synaptophyllum:* Michigan Univ. Mus. Paleontology Contr., v. 8, p. 21-41.

——— 1951, Corals of the Devonian Traverse Group of Michigan. Pt. IV. *Billingsastraea:* Michigan Univ. Mus. Paleontology Contr., v. 9, p. 83-92.

Fenton, C. L., and Fenton, M. A., 1924, The stratigraphy and fauna of the Hackberry Stage of the Upper Devonian: Michigan Univ. Mus. Paleontology Contr., v. 1, 260 p.

Hughes, T., 1975, The case for creation of the north Pacific ocean during the Mesozoic era: Palaeogeography, Palaeoclimatology, Palaeoecology, v. 18, p. 1-43.

Johnson, J. G., and Oliver, W. A., Jr., 1977, Silurian and Devonian coral zones in the Great Basin: Geol. Soc. America Bull. (in press).

Lambe, L. M., 1901 (1900), A revision of the genera and species of Canadian Palaeozoic corals. The Madreporaria Aporasa and the Madreporaria Rugosa: Canada Geol. Survey Contr. Canadian Palaeontology, v. 4, pt. 2, p. 97-197.

MacKenzie, W. S., Pedder, A. E. H., and Uyeno, T. T., 1975, A Middle Devonian sandstone unit, Grandview Hills area, District of Mackenzie: Canada Geol. Survey Paper 75-1, pt. A, p. 547-552.

McCammon, Helen, 1960, Fauna of the Manitoba Group in Manitoba: Manitoba Dept. Mines and Nat. Resources Pub. 59-6, 109 p.

McLaren, D. J., 1959, A revision of the Devonian coral genus *Synaptophyllum* Simpson: Canada Geol. Survey Bull. 48, p. 15-33.

McLaren, D. J., Norris, A. W., and McGregor, D. C., 1962, Illustrations of Canadian fossils. Devonian of western Canada: Canada Geol. Survey Paper 62-4, 34 p.

Merriam, C. W., 1972, Silurian rugose corals of the Klamath Mountains region, California: U. S. Geol. Survey Prof. Paper 738, 50 p.

——— 1974a (1973), Silurian rugose corals of the central and southwest Great Basin: U. S. Geol. Survey Prof. Paper 777, 66 p.

——— 1974b (1973), Middle Devonian rugose corals of the central Great Basin: U. S. Geol. Survey Prof. Paper 799, 53 p.

——— 1974c, Lower and Lower Middle Devonian rugose corals of the central Great Basin: U. S. Geol. Survey Prof. Paper 805, 83 p.

Oliver, W. A., Jr., 1958, Significance of external form in some Onondagan rugose corals: Jour. Paleontology, v. 32, p. 815-837.

——— 1960a, Rugose corals from reef limestone in the Lower Devonian of New York: Jour. Paleontology, v. 34, p. 59-100.

——— 1960b, Devonian rugose corals from northern Maine; a description of two new faunules from rocks of Helderberg and Schoharie age: U. S. Geol. Survey Bull. 1111-A, 22 p.

——— 1962, Silurian rugose corals from the Lake Témiscouata area Quebec: U.S. Geol. Survey Prof. Paper 430-B, p. 11-19.

——— 1964a, The Devonian colonial coral genus *Billingsastraea* and its earliest known species: U. S. Geol. Survey Prof. Paper 483-B, p. 1-5.

——— 1964b, New occurrences of the rugose coral *Rhizophyllum* in North America: U. S. Geol. Survey Prof. Paper 475-D, p. 149-157.

——— 1971, The coral fauna and age of the Famine Limestone in Quebec: Smithsonian Contr. Paleobiology no. 3, p. 193-201.

——— 1974, Classification and new genera of noncystimorph colonial rugose corals from the Onesquethaw Stage in New York and adjacent areas: U. S. Geol. Survey Jour. Research, v. 2, p. 165-174.

——— 1976, Noncystimorph colonial rugose corals from the Onesquethaw and lower Cazenovia Stages (Early and Middle Devonian) in New York and adjacent areas: U.S. Geol. Survey, Prof. Paper 869, 156 p.

——— 1976a, Biogeography of Devonian rugose corals: Jour. Paleontology, v. 50, p. 365-373.

——— 1977a, Devonian rugose coral assemblages in the United States: Internat. Symposium Fossil Corals, 2nd, Paris 1975: Bur. Recherches Géol. Minières Mém. 89, p. 167-174.

——— 1977b, Biogeography of Late Silurian and Devonian rugose corals in the Eastern Americas Realm: Palaeogeography, Palaeoclimatology, Palaeoecology v. 22, p. 85-135.

Oliver, W. A., Jr., Merriam, C. W., and Churkin, Michael, 1975, Ordovician, Silurian, and Devonian corals of Alaska: U. S. Geol. Survey Prof. Paper 823-B, 44 p.

Pedder, A. E. H., 1964, Correlation of the Canadian Middle Devonian Hume and Nahanni Formations by tetracorals: Palaeontology, v. 7, pt. 3, p. 430-451.

——— 1965, Some North American species of the Devonian tetracoral *Smithiphyllum:* Palaeontology, v. 8, pt. 4, p. 618-628.

——— 1971a, Two new aphroid corals from the Middle Devonian Hume Formation of western Canada: Canada Geol. Survey Bull. 192, p. 45-61.

——— 1971b, Middle Devonian coelenterates from the Nahanni Formation of H. B., Amerada Camsell A-37 well, District of Mackenzie: Canada Geol. Survey Bull. 192, p. 63-81.

——— 1971c, An Upper Silurian (Pridolian) coral faunule from northern Yukon Territory: Canada Geol. Survey Bull. 197, p. 13-21.

——— 1971d, *Dohmophyllum* and a new related genus of corals from the Middle Devonian of northwestern Canada: Canada Geol. Survey Bull. 197, p. 37-77.

——— 1972, Species of the tetracoral genus *Temnophyllum* from Givetian/Frasnian boundary beds of the District of Mackenzie, Canada: Jour. Paleontology, v. 46, p. 696-710.

——— 1973, Description and biostratigraphical significance of the Devonian coral genera *Alaiophyllum* and *Grypophyllum* in western Canada: Canada Geol. Survey Bull. 222, p. 93-127.

——— 1975, Sequence and relationships of three Lower Devonian coral faunas from Yukon Territory: Canada Geol. Survey Paper 75-1, pt. B, p. 285-295.

——— 1976a, Initial records of two unusual Late Silurian rugose coral genera from Yukon Territory: Canada Geol. Survey Paper 76-1, pt. B, p. 285, 286.

——— 1976b, First records of five rugose coral genera from Upper Silurian rocks of the Canadian Arctic Islands: Canada Geol. Survey Paper 76-1, pt. B, p. 287-293.

Pedder, A. E. H., and McLean, R. A., 1976, New records and range extensions of seven rugose coral genera in Silurian strata of northwestern and arctic Canada: Canada Geol. Survey Paper 76-1, pt. C, p. 131-141.

Shell Oil Company Exploration Department, 1975, Stratigraphic atlas North & Central America: Houston, Shell Oil Company.

Scrutton, C. T., 1973, Palaeozoic coral faunas from Venezuela. II. Devonian and Carboniferous corals from the Sierra de Perija: British Mus. (Nat. Hist.) Bull., Geol., v. 23, p. 221-281.

Smith, Stanley, 1945, Upper Devonian corals of the Mackenzie River region, Canada: Geol. Soc. America Spec. Paper 59, 126 p.

Sorauf, J. E., 1972, Middle Devonian coral faunas (Rugosa) from Washington and Oregon: Jour. Paleontology, v. 46, p. 426-439.

Stewart, J. H., and Poole, F. G., 1974, Lower Paleozoic and uppermost Precambrian Cordilleran miogeocline, Great Basin, western United States: Soc. Econ. Paleontologists and Mineralogists Spec. Pub. 22, p. 28-57.

Stumm, E. C., 1940, Upper Devonian rugose corals of the Nevada Limestone: Jour. Paleontology, v. 14, p. 57-67.

——— 1948, Upper Devonian compound tetracorals from the Martin Limestone: Jour. Paleontology, v. 22, p. 40-47.

——— 1960, New rugose corals from the Middle and Upper Devonian of New York: Jour. Paleontology, v. 34, p. 161-163.

——— 1962a, Corals of the Traverse Group of Michigan. Pt. VII. The Digonophyllidae: Michigan Univ. Mus. Paleontology Contr., v. 17, p. 215-231.

——— 1962b, Silurian corals from the Moose River Synclinorium, Maine: U. S. Geol. Survey Prof. Paper 430-A, p. 1-9.

——— 1962c, Corals of the Traverse Group of Michigan. Pt. VIII. *Stereolasma* and *Heterophrentis:* Michigan Univ. Mus. Paleontology Contr., v. 17, p. 233-240.

——— 1962d, Corals of the Traverse Group of Michigan. Pt. X. *Tabulophyllum:* Michigan Univ. Mus. Paleontology Contr., v. 17, p. 291-297.

——— 1963, Corals of the Traverse Group of Michigan. Pt. XI. *Tortophyllum, Bethanyphyllum, Aulacophyllum,* and *Hallia:* Michigan Univ. Mus. Paleontology Contr., v. 18, p. 135-155.

——— 1965 (1964), Silurian and Devonian corals of the Falls of the Ohio: Geol. Soc. America Mem. 93, 184 p.

——— 1968a, The corals of the Middle Devonian Tenmile Creek Dolomite of northwestern Ohio: Michigan Univ. Mus. Paleontology Contr., v. 22, p. 37-44.

——— 1968b, Rugose corals of the Silica Formation (Middle Devonian) of northwestern Ohio and southeastern Michigan: Michigan Univ. Mus. Paleontology Contr., v. 22, p. 61-70.

——— 1970, Corals of the Traverse Group of Michigan. Pt. XIII. *Hexagonaria:* Michigan Univ. Mus. Paleontology Contr., v. 23, p. 81-91.

Stumm, E. C., and Tyler, J. H., 1962, Corals of the Traverse Group of Michigan. Pt. IX. *Heliophyllum:* Michigan Univ. Mus. Paleontology Contr., v. 17, p. 265-276.

Sutherland, P. K., 1965, Henryhouse rugose corals: Oklahoma Geol. Survey Bull. 109, 92 p.

Swartz, C. K., 1913, Coelenterata, *in* Swartz, C. K. and others, Lower Devonian [of Maryland]: Maryland Geol. Survey, p. 195-227.

Warren, P. S., and Stelck, C. R., 1956, Reference fossils of Canada. Pt. 1. Devonian faunas of western Canada: Geol. Assoc. Canada Spec. Paper 1, 15 p.

Wells, J. W., 1943, The Rio Cachiri section in the Sierra de Perijà, Venezuela. Pt. II. Paleontology, B: Anthozoa: Bull. Am. Paleontology, v. 27, p. 95-100.

Wilson, J. T., 1968, Static or mobile Earth; the current scientific revolution: Am. Philos. Soc. Proc., v. 112, p. 309-320.

Ziegler, P. A., 1969, The development of sedimentary basins in western and arctic Canada: Calgary, Alberta Soc. Petroleum Geologists, Spec. Pub., 89 p.

APPENDIX

Table 2. Late Silurian genera from Eastern North America, Appalachian area (Swartz, 1913; Stumm, 1962b; Oliver, 1962, 1964b, unpub. data). (N) Newfoundland; (M) Maine and Quebec; (V) Virginia and New York; (1) Late Silurian, undifferentiated; (2) Pridolian; (E) endemic.

Genus		Genus	
Acanthophyllum	N2	*Phaulactis*	M1
Acmophyllum?	M1, V2	*Ptychophyllum?*	M1
Cystiphyllum	M1, V2	*Pycnactis*	M1
Embolophyllum	M1, V2, E	*Rhizophyllum*	M1
Entelophylloides	M1, V2, E	*Spongophylloides*	M1
Entelophyllum	M1	*Tryplasma* (solitary)	M1, V2
Holmophyllum	M1	*Zelophyllia?*	V2, E
Microplasma	V2		

Table 3. Ludlovian genera from Eastern North America, Midcontinent area: Henryhouse Formation of Tennessee (Amsden, 1949) and Brownsport Formation of Oklahoma (Sutherland, 1965. (E) endemic.

Genus		Genus	
Allotropiophyllum?	E	*Oliveria*	E
Amsdenoides	E	*Palaeocyathus*	
Anisophyllum		*Petraia?*	
Arachnophyllum		*Phaulactis?*	
Capnophyllum	E	*Pilophyllum?*	
Craterophyllum		*Pseudocryptophyllum?*	E
Cyathactis?		*Rhizophyllum*	
Cystiphyllum		*Spongophylloides*	
Dentilasma		*Stereoxylodes*	
Ditoecholasma	E	*Sutherlandinia*	
Duncanella	E	*Syringaxon*	
Entelophyllum		*Tryplasma* (solitary)	
Lamprophyllum		*Zelophyllum?*	
Oligophyllum?	E		

Table 4. Late Silurian genera from Great Basin: Roberts Mountains Formation (part) and equivalents (Merriam, 1974c, zone C fauna; Johnson and Oliver, in press). (1) Late Silurian, undifferentiated; (2) Pridolian; (E) endemic to Western North America.

Genus		Genus	
Cystiphyllum	1	*Prohexagonaria*	1
Denayphyllum	2, E	*Tryplasma*	1, 2
Microplasma?	1		

Table 5. Late Silurian genera from Pacific Coast areas (Merriam, 1972, unpub. data; Oliver and others, 1976). (A) SE Alaska; (C) NW California; (E) endemic.

Aphyllum	A	*Rhizophyllum*	C
Cyathactis	A, C	*Scyphophyllum*	A
Cystiphyllum	A, C	*Spongophylloides*	A, C
Entelophyllum	A	*Syringaxon*	C
Hedstroemophyllum	A	*Tryplasma*	A
Kodonophyllum	C	*Wintunastraea*	C, E
Lamprophyllum	A	*Xystriphyllum*	C
Microplasma	A	*Yassia*	C
Phaulactis	A	*Zelophyllum*	A

Table 6. Late Silurian genera from Alaska (exclusive of SE Alaska) and Yukon Territory (Oliver and others, 1975; Pedder, 1971c, 1976a, unpub. data; Pedder and McLean, 1976). (A) Alaska; (Y) Yukon Territory; (1) Late Silurian, undifferentiated; (2) Ludlovian; (3) Pridolian; (E) endemic.

Arachnophyllinid genus	Y3, E	*Pseudomphyma*	Y3
		Pseudomicroplasma	Y3
Cyathactis	A1, Y3	*Spinolasma*	A1, Y3
Cystiphyllum	A1, Y3	*Stathmoelasma?*	Y3, E
Denayphyllum	Y3, E	*Stauria*	A1
Entelophyllum	A1	*Stereoxylodes*	Y3
Holmophyllum	Y3	Streptelasmatid genus	Y3
Kozlowiaphyllum	Y3		
Lamprophyllum	A1	*Stylopleura?*	Y3
Lycokystiphyllum?	A1	*Tryplasma*	A1, Y2, Y3
Maikottia	A1	Tryplasmatid genus	Y3
Microplasma?	A1	*Xystriphyllum*	Y3
Migmatophyllum	Y3, E	*Yassia*	Y3
Mucophyllum	Y2, Y3	*Zelophyllum*	Y1
Neomphyma	Y3	New genus ("*Ketophyllum*")	A1, E
Niajuphyllum	Y2		
Prohexagonaria	Y3	New genus ("undet. rugose coral")	A1, E

Table 7. Late Silurian genera from Arctic Islands (Pedder, 1976b; Pedder and McLean, 1976). (1) Late Silurian, undifferentiated; (2) Ludlovian (Douro Formation and equivalents); (3) Pridolian (Read Bay Formation, restricted and equivalents); (E) endemic.

Arachnophyllinid genus	3	*Plasmophyllum*	3
		Prohexagonaria	3
Camurophyllum	3	*Ptychophyllum*	2
Cyathactis?	2	*Pycnostylus?*	3
Cystiphylloides	3	*Radiastraea*	2, 3, E
Cystiphyllum	2, 3	*Rhizophylloides*	3
Dinophyllum	2	*Stylopleura*	3
Entelophyllum	1, 3	*Tryplasma*	3
Kozlowiaphyllum	2	*Wintunastraea?*	3, E
Kymocystis	3	*Xyphelasma*	3
Mazaphyllum	3	*Yassia*	3
Nanshanophyllum	3		
Neobrachyelasma	2		

Table 8. Lochkovian (early and middle Helderbergian) genera from Eastern North America (Oliver, 1960a, 1960b, unpub. data). (E) endemic.

Aknisophyllum	E	*Palaeocyathus*	
Breviphrentis	E	*Pseudoblothrophyllum*	E
Briantelasma	E	*Spongophylloides*	
Embolophyllum		*Syringaxon*	
Fletcherina	E	*Tryplasma* (colonial)	
Heterophrentis	E	*Zelophyllia?*	
"*Nalivkinella*"	E	New genus	E

Table 9. Lochkovian genera from the Great Basin (Oliver, 1964b; Merriam, 1974a, "Silurian" D and E faunas; Pedder, 1975b, unpub. data; Johnson and Oliver, in press). (E) endemic.

Aphroidophyllum	E	Ptenophyllid genus	
"*Chonophyllum*"		*Rhizophyllum*	
Cystiphylloides		"*Salairophyllum?*"	
Kodonophyllum		*Spongophyllum*	
Kozlowiaphyllum		Streptelasmatid genus	
Mucophyllum		*Stylopleura*	
Neomphyma		*Tonkinaria*	E
Neostringophyllum		*Tryplasma*	

Table 10. Lochkovian genera from western Canada (Pedder, 1975, unpub. data). (1) Royal Creek and Knorr Range, Yukon Territory; (2) SW District of Mackenzie; (3) Quesnel Lake area, British Columbia; (E) endemic.

Chlamydophyllum	1, 2	*Pseudamplexus*	1
Cystiphylloides	2	*Pseudogrypophyllum*	1
Dohmophyllum	1, 2	*Spongophylloides*	1
Kodonophyllum	1	*Spongophyllum*	1, 3
Kyphophyllid genus	1, E	*Stylopleura*	1
Loboplasma	1	*Tryplasma*	2
Lyrielasma	1	*Xystriphyllum*	1
Neomphyma	1	*Zelophyllum*	1

Table 11. Pragian (late Helderbergian, Deerparkian and early Sawkillian) genera from Eastern North America (Oliver, 1964a, unpub. data). (G) Gaspe; (NY) New York; (O) Oklahoma; (1) late Helderbergian and Deerparkian; (2) early Sawkillian; (E) endemic to Realm.

Asterobillingsa	G2, E	*Siphonophrentis*	G2, E
Breviphrentis	G2, E	*Syringaxon*	O2, NY1
Duncanella?	O1, E	Undetermined genus	O2
Palaeocyathus	O1?, G2	Undetermined genus	O2
cf. *Petraia*	O2		

Table 12. Pragian genera from the Great Basin (Merriam, 1974c, Devonian A-D1 faunas; Johnson and Oliver, in press). (E) endemic.

Aulacophyllum		"*Odontophyllum*"	E
Breviphrentis		*Papiliophyllum*	E
Cystiphylloides		"*Sinospongophyllum*"	E
Kobeha	E	*Syringaxon*	
Kozlowiaphyllum		*Zonophyllum*	

Table 13. Pragian genera from NW Canada (Pedder, 1975b, unpub. data). (M) SW District of Mackenzie; (Y) Yukon Territory; (E) endemic.

Arachnophyllinid genus	Y, E	*Metrionaxon*	M, E
		Metriophyllum	M
Barrandeophyllum	M, Y	*Palaeocyathus*	M
Cystiphylloides	Y	*Pseudamplexus*	M, Y
Cystiphyllid genus	Y, E	Streptelasmatid genus	Y
Diplochone	M	*Syringaxon*	M
Dohmophyllum	M, Y	*Taimyrophyllum*	Y
Fasciphyllum	Y	*Tryplasma*	Y
Gurievskiella	Y	*Vepresiphyllum*	Y
Lyrielasma	Y	*Xystriphyllum*	Y
Neostringophyllum	M, Y		
Martinophyllum	Y		

Table 14. Late Emsian (late Sawkillian) genera from Eastern North America (Lambe, 1901; Oliver, 1958, 1974, 1976, unpub. data; Stumm, 1965). (E) endemic.

Acinophyllum	E	*Grewgiphyllum*	E
Acrophyllum	E	*Heliophyllum*	E
Aemulophyllum	E	*Heterophrentis*	E
Asterobillingsa	E	*Homalophyllum*	E
Aulacophyllum	E	*Kionelasma*	E
Blothrophyllum	E	*Prismatophyllum*	E
Bowenelasma	E	*Scenophyllum*	E
Briantelasma	E	*Skoliophyllum*	E
Cladionophyllum	E	*Stereolasma*	E
Compressiphyllum	E	*Syringaxon*	
Cylindrophyllum	E	New genus ("*Disphyllum*")	E
Cystiphylloides		New genus (cf.	
Edaphophyllum	E	*Palaeocyathus*)	E

Table 15. Late Emsian genera from the Great Basin (Merriam, 1974c, Devonian D2, 3 faunas; Johnson and Oliver, in press). (E) endemic.

Breviphrentis		cf. *Peneckiella*	
Cyathophyllum	E	*Pinyonastraea*	E
Cystiphylloides		*Radiastraea*	
Lekanophyllum		*Sinospongophyllum*	
Mesophyllum		*Syringaxon*	
Moravophyllum	E	*Zonophyllum*	
Nevadaphyllum	E		

Table 16. Late Emsian genera of NW Canada (Crickmay, 1968; Pedder, unpub. data). (Y) Yukon Territory; (M) W District of Mackenzie; (E) endemic to western North America.

Dohmophyllum	Y	*Spongonaria*	Y, M
Embolophyllum	Y	*Taimyrophyllum*	Y
Exilifrons	Y, M, E	*Vepresiphyllum*	M
Loomberaphyllum	Y	*Xystriphyllum*	M
Neostringophyllum	Y, M		

Table 17. Late Emsian genera from Ellesmere Island (Pedder, unpub. data). (E) endemic to Western North America.

Cavanophyllum		*Neostringophyllum*	
Digonophyllum		*Radiastraea*	
Dohmophyllum		*Rhizophyllum*	
Exilifrons	E	*Salairophyllum*	
Glossophyllum		*Spongonaria*	
Lekanophyllum		cf. *Temnophyllum*	
Mesophyllum		*Zonophyllum*	
Minussiella			

Table 18. Early Devonian genera from Alaska (Oliver and others, 1975).

Acanthophyllum	*Spongonaria*
Fasciphyllum	*Spongophyllum*
Hexagonaria	*Syringaxon*
Martinophyllum	*Utaratuia*
"*Peneckiella*" (cerioid)	*Xystriphyllum*
Peripaedium	

Table 19. Eifelian genera from Eastern North America (Lambe, 1901; McLaren, 1959; Stumm, 1965; Oliver, 1971, 1974, 1976, unpub. data). (E) endemic to Realm.

Acinophyllum	E	*Hallia*	E
Asterobillingsa	E	*Heliophyllum*	E
Aulacophyllum		*Heterophrentis*	E
Blothrophyllum	E	*Microplasma?*	
Breviphrentis	E	"*Nalivkinella*"	
Bucanophyllum	E	*Phymatophyllum*	E
Cyathocylindrium	E	*Prismatophyllum*	E
Cylindrophyllum	E	*Siphonophrentis*	
Cystiphylloides		*Skoliophyllum*	
Dendrostella		*Stereolasma*	E
Diplochone		*Synaptophyllum*	E
Disphyllum		*Syringaxon*	
Eridophyllum	E	*Zaphrentis*	E
Grewgiphyllum	E	New genus	E

Table 20. Eifelian genera from Venezuela (Wells, 1943; Scrutton, 1973; Oliver, unpub. data). (E) genera endemic to Eastern North America and northern South America.

Acinophyllum	E	*Hadrophyllum*	E
Bowenelasma	E	*Heliophyllum*	E
Breviphrentis	E	*Heterophrentis*	E
Briantelasma	E	*Stereolasma*	E
Cylindrophyllum	E	"*Stewartophyllum*"	E
Cystiphylloides		*Syringaxon*	

Table 21. Eifelian genera from the Great Basin (Merriam, 1973, Devonian E and F faunas; Johnson and Oliver, in press). (E) endemic.

Acanthophyllum		*Moravophyllum*	
Cystiphylloides		*Orthocyathus*	E
Digonophyllum		"*Siphonophrentis*"	
cf. *Disphyllum*		*Sociophyllum*	
"*Grypophyllum*"		*Tabulophyllum*	
Hexagonaria		*Taimyrophyllum*	
Lekanophyllum		*Utaratuia*	E

Table 22. Eifelian genera from the Harrogate Formation of SE British Columbia (Pedder, unpub. data). None is endemic.

Ceratophyllum	*Kozlowiaphyllum*
Cystiphylloides?	*Lekanophyllum*
Digonophyllum	*Mansuyphyllum*
Disphyllum	*Moravophyllum?*
Hexagonaria?	*Xystriphyllum*

Table 23. Eifelian genera from western Canada (Crickmay, 1960, 1962; Pedder, 1964, 1971a, 1971b, 1971d, unpub. data). (A) Ernestina Lake Formation of NE Alberta and NW Saskatchewan; (B) Headless, Nahanni and equivalent Formations of central and NE British Columbia, SE Yukon Territory and SW District of Mackenzie; (C) Hume Formation of N Yukon Territory and NW District of Mackenzie; (E) endemic.

Aphoidophyllum	B, C	*Dohmophyllum*	B, C
Ceratophyllum	C	*Exilifrons*	B, C, E
Cystiphylloides	B, C	*Grypophyllum*	B
Dendrostella	B, C	*Hexagonaria*	B, C
Digonophyllum	B, C	*Kozlowiaphyllum*	B, C
Disphyllum	B, C	*Kunthia*	B, C

Table 23. (Cont.)

Lekanophyllum	B, C	*Radiastraea*	B, C
Mackenziephyllum	B, C	*Redstonea*	B, C, E
"Microcyclus"	B, C	*Sociophyllum*	B, C
Microplasma	B, C	*Stringophyllum*	C
Minussiella	C	*Taimyrophyllum*	B, C
Moravophyllum	B, C	*Utaratuia*	B, C, E
Planetophyllum	A, E	*Xystriphyllum*	B, C
Psydracophyllum	B, E	*Zonophyllum*	C

Table 24. Eifelian genera from Canadian Arctic Islands (Pedder, unpub. data). (E) endemic.

Aphroidophyllum	*Taimyrophyllum*	
Dendrostella	*Utaratuia*	E
Digonophyllum	*Zonophyllum*	
Minussiella		

Table 25. Givetian genera from Michigan Basin and Hudson Bay Lowlands (Ehlers and Stumm, 1949a, 1949b, 1951; Stumm, 1962a, 1962c, 1962d, 1963, 1970; Stumm and Tyler, 1962; Oliver, unpub. data). (M) Miami Bend Formation; (T) Traverse Group; (E) endemic to Realm.

Alaiophyllum	M	*Hexagonaria*	M, T
Asterobillingsa	T, E	*Iowaphyllum*	T
Aulacophyllum	T	*Kozlowiaphyllum*	M
Bethanyphyllum	M, T, E	*Lekanophyllum*	M, T
cf. *Chlamydophyllum*	M	*Microplasma*	T
Cylindrophyllum	T	*Microcyclus*	T
Cystiphylloides	T	*Prismatophyllum*	T
Depasophyllum	T, E	*"Spongophyllum"*	T
Disphyllum	T	*Stereolasma*	T, E
Diversophyllum	T, E	*Stringophyllum?*	M
Eridophyllum	T	*Tabulophyllum*	T
Hallia	T	*Tortophyllum*	T, E
Heliophyllum	M, T	*Xenocyathellus*	T, E
Heterophyrentis	T		

Table 26. Givetian genera from the Appohimchi Province of the Eastern Americas Realm (Stumm, 1968a, 1968b; Oliver, unpub. data). (E) endemic to Realm.

Asterobillingsa	E	*Metriophyllum*	
Aulacophyllum		*Microcyclus*	
Bethanyphyllum	E	*"Nalivkinella"*	
Cylindrophyllum		*Odontophyllum*	E
Cystiphylloides		*Phymatophyllum*	E
Depasophyllum	E	*Prismatophyllum*	
Eridophyllum		*Siphonophrentis*	
Hadrophyllum	E	*Stereolasma*	E
Hallia		*Stewartophyllum*	E
Heliophyllum		*Syringaxon*	
Heterophrentis		*Tabulophyllum*	

Table 27. Givetian genera from the Great Basin (Merriam, 1973, Devonian G fauna; Johnson and Oliver, in press). None is endemic.

Columnaria	*Moravophyllum*
Cystiphylloides	*Neostringophyllum?*
Grypophyllum	*Prismatophyllum*
Heliophyllum	*Spongophyllum*
Hexagonaria	*Temnophyllum*
Lekanophyllum	

Table 28. Givetian genera from western Canada, exclusive of Manitoba and southern Saskatchewan (Lambe, 1901; Smith, 1945; Warren and Stelck, 1956; Crickmay, 1960; McLaren and others, 1962; Pedder, 1972, 1973, unpub. data; Mackenzie and others, 1975). (M) Methy Formation, NE Alberta and NW Saskatchewan; (B) unnamed limestones, McDame and Fort Grahame areas of British Columbia; (H) Horn Plateau Formation, SW District of Mackenzie; (P) Pine Point Formation, Great Slave Lake; (S) Sulphur Point and Slave Point Formations, SW District of Mackenzie; (R) Ramparts and Kee Scarp Formations and equivalents, W District of Mackenzie; (E) endemic to Western North America.

Alaiophyllum	S, R	*Lekanophyllum*	H, P
Argutastraea	R	*Moravophyllum*	B, P, S, R
"Columnaria"	R	*"Moravophyllum"*	R, E
Cylindrophyllum	H	*Neocolumnaria*	R
Cyathophyllid genus	H	*Neostringophyllum*	B, H
Cystiphylloides	B, R	*"Peripaedium*	H
Dendrostella	B, S	*Prismatophyllum*	M
Digonophyllum	P	*Sinospongophyllum*	H
Disphyllum	M, B, H, S, R	*Sociophyllum*	H
Dohmophyllum?	S	Streptelasmatid genus	H
Grypophyllum	S, R	*Stringophyllum*	S, R
Heliophyllum	H, R	*Tabulophyllum*	R
Hexagonaria	M, B, R	*Temnophyllum*	S, R
Kunthia	P	*Utaratuia*	H, E
Kozlowiaphyllum	H	*Zonophyllum*	H

Table 29. Givetian genera from southern Saskatchewan and Manitoba (Lambe, 1901; Warren and Stelck, 1956; McCammon, 1960; Pedder, unpub. data). (S) Upper Elm Point Formation, Steep Rock area; (W) Winnipegosis Formation, Lakes Winnipegosis and Manitoba; (D) Dawson Bay Formation, Lake Winnipegosis and subsurface. None is known to be endemic.

Argutastrea	D	*Moravophyllum*	D
"Buschophyllum"	D	*Nalivkinella?*	D
Cylindrophyllum	D	*Neocolumnaria*	D
Dendrostella	W	*Prismatophyllum*	S
Disphyllum	D	*Tabulophyllum*	?W, D
Kunthia?	D	*Temnophyllum*	D
Lekanophyllum	D, ?W	*Xystriphyllum*	W

Table 30. Middle Devonian genera from Alaska, exclusive of SE Alaska (Oliver and others, 1975; Oliver, unpub. data). None is endemic.

Acanthophyllum	cf. *Neostringophyllum*
Cystiphylloides?	cf. *Pseudamplexus*
Dendrostella	"*Pseudomicroplasma*"
Dohmophyllum	*Pseudotryplasma*
Diplochone	cf. *Siphonophrentis*
Disphyllum	*Sociophyllum*
Grypophyllum	*Taimyrophyllum*
Hexagonaria	*Tryplasma*

Table 31. Middle Devonian genera from Pacific Coast areas (Sorauf, 1972; Oliver and others, 1975; Merriam and Oliver, unpub. data). (A) SE Alaska; (O) Oregon. None is endemic.

Acanthophyllum	A	*Mesophyllum*	A
"*Breviphyllum*"	A	*Microplasma*	A
Cystiphylloides	A	*Moravophyllum*	A
Dendrostella	A	*Pseudamplexus*	A
Digonophyllum	A	*Sociophyllum*	A
Dohmophyllum	O	*Spongophyllum*	A
Fasciphyllum	O	*Taimyrophyllum*	A
Grypophyllum	A	*Temnophyllum*	A
Kozlowiaphyllum	A	*Xystriphyllum*	A
Loyalophyllum	A	"*Zonophyllum*"	A

Table 32. Frasnian genera from Eastern North America (Fenton and Fenton, 1924; Stumm, 1960).

"*Amplexus*"	*Metriophyllum*
Charactophyllum	"*Mictophyllum*"
"*Chonophyllum*"	*Pachyphyllum*
Disphyllum	*Smithiphyllum*
Hexagonaria	*Syringaxon*
Iowaphyllum	*Tabulophyllum*
Macgeea	

Table 33. Frasnian genera from the Great Basin (Stumm, 1940, 1948; Oliver, unpub. data).

"*Breviphyllum*"	*Phacellophyllum*
Charactophyllum	*Phillipsastraea*
"*Chonophyllum*"	"*Ptychophyllum*"
Disphyllum	*Smithiphyllum*
Hexagonaria	"*Smithiphyllum*" (cerioid)
Macgeea	*Tabulophyllum*
Pachyphyllum	*Temnophyllum*
Peneckiella	*Trapezophyllum*

Table 34. Frasnian genera from Pacific Coast areas (SE Alaska) (Oliver and others, 1975; Merriam, unpub. data).

Disphyllum?	*Phacellophyllum*
Macgeea	*Phillipsastraea*
Pachyphyllum	*Pseudamplexus*
Peneckiella	*Tabulophyllum*

Table 35. Frasnian genera from Alaska (exclusive of SE Alaska) (Oliver and others, 1975; Oliver, unpub. data).

"*Breviphyllum*"	*Peneckiella*
Charactophyllum	*Phacellophyllum*
Disphyllum	*Phillipsastraea*
Hexagonaria	*Smithiphyllum*
Macgeea	*Tabulophyllum*

Table 36. Frasnian genera of rugose corals from arctic and western Canada (Smith, 1945; McLaren and others, 1962; Pedder, 1965, unpub. data). (1) Arctic Islands; (2) SW District of Mackenzie, N Alberta; (3) Central and S Alberta and adjacent British Columbia; (4) S Saskatchewan and Manitoba.

Argutastraea	3	*Phacellophyllum*	1, 2, 3, 4
Charactophyllum	2, 3, 4	*Phillipsastraea*	1, 2, 3, 4
Chonophyllum	2	*Ptychophyllum*	3
Disphyllum	1, 2, 3, 4	"*Ptychophyllum*" (sensu Smith, 1945)	2
Frechastraea	2	*Smithiphyllum*	2, 3, 4
Hexagonaria	1, 2, 3	"*Smithiphyllum*" (cerioid)	2
Hunanophrentis	2, 3	*Sudatea*	1, 2, 3
Macgeea	2, 3, 4	*Tabulophyllum*	2, 3, 4
Mictophyllum	2, 3	*Temnophyllum*	2
Neocolumnaria	3, 4		
Pachyphyllum	2, 3		
Peneckiella	2, 3		

Biogeography of Silurian and Devonian Trilobites of the Malvinokaffric Realm

Niles Eldredge, *American Museum of Natural History, New York, New York 10024*
Allen R. Ormiston, *Research Center, Amoco Production Company, Tulsa, Oklahoma 74101*

ABSTRACT

The level of endemism of Malvinokaffric trilobites is great enough to indicate the existence of a Malvinokaffric Realm during the Devonian. Faunal similarity values suggest subdivision into three Provinces: Andean, Brazilian, and South African-Malvinan. Low diversity Malvinokaffric communities such as the homalonotid are not amenable to zoogeographic analysis; comparisons are better confined to more diverse trilobite faunas.

Colombia and Venezuela do not belong to the Malvinokaffric Realm. Appalachian immigrants were not an important source for Malvinokaffric endemics which had already appeared at least by Late Silurian before an Appalachian trilobite province existed. Cosmopolitan stocks present in the Silurian of South America (such as dalmanitids, acastids, proetids, otarionids, calymenids) are an adequate source for the endemics.

The separateness of the Malvinokaffric Realm is readily understandable from a Gondwana reconstruction combined with a paleocurrent analysis. The absence of Malvinokaffric elements from Australia is explained by paleocurrents; an emergent area may have to be invoked to explain their absence from northwest Africa.

INTRODUCTION

The name Malvinokaffric Province was coined by Rudolph and Emma Richter (1942) as a designation for the strongly endemic trilobite fauna which they knew from several austral localities. The endemic level of this fauna is truly high, high enough to indicate the existence of a Realm following the definition of Kauffman (1973). At the subgeneric level, fully 80 percent of the Bolivian Lower Devonian trilobites are endemic, 17 percent are cosmopolitan, and 3 percent can be described as quasi-cosmopolitan (known from scattered localities, but without true world-wide distribution). Three quarters of these endemics are calmoniids or plesiomorphic forms considered calmoniids herein (e.g., ?*Acastoides*). These calmoniids are considered to be derived from a cosmopolitan acastid stock and a marked radiation among them characterizes all subdivisions of the Malvinokaffric Realm. No calmoniids are known from beyond this Realm, true *Acastoides* being excluded from this family. *Neocalmonia* Pillet from the Frasnian of Afghanistan, originally assigned to the calmoniids, is an asteropyginid.

Nine other trilobite families are represented in the Malvinokaffric Devonian, many of them by endemic genera. The radiation of endemic dalmanitids is scarcely less spectacular than that of the calmoniids.

Recent taxonomic work on Malvinokaffric trilobites has been concentrated on Andean forms from South America with those elsewhere receiving scarcely any attention.

PROCEDURE

Inasmuch as no zoogeographic assessment is possible without a modern and consistent taxonomic framework, it has been necessary to re-evaluate many Malvinokaffric collections in the course of this study. The study has also profited from a taxonomic revision of the trilobites of the Bolivian *Scaphiocoelia* Zone now in preparation by Eldredge and Branisa. As a consequence, much of the taxonomy in this paper has had to be cited in open nomenclature, an unwieldy but necessary procedure. In addition to the Bolivian study mentioned, a re-examination and reevaluation has been made of Caster's trilobites from the Floresta fauna of Colombia, Weisbord's trilobites from Venezuela, all of J. M. Clarke's types and casts from South America and the Falkland Islands, the extensive Branisa collections from Bolivia housed in the American Museum of Natural History and the National Museum of Natural History, many South African specimens, and stratigraphically useful collections from Bolivia made available to us by Peter Isaacson. Homalonotids have received scant attention from us because their morphologic conservatism limits their use in zoogeographic discrimination.

MALVINOKAFFRIC TRILOBITES

Among trilobites, the calmoniids are the hallmark of the Malvinokaffric Realm. Fully 34 genera and subgenera of this family are recognized within it and none (according to our taxonomic revisions) outside of it. To further gild the lily, there are at least 6 endemic dalmanitids, and 1 each of endemic lichads, proetids, calymenids, and brachymetopids.

Calmoniids

Four calmoniid subgroups are recognized in this paper. The first of these is designated the plesiomorphic group, consisting of the genera *Andinacaste* n. gen., *Phacopina*, *?Acastoides*, and probably including *Vogesina*, and a new genus related to *Vogesina*. Of these, *?Acastoides*, *Phacopina*, and *Andinacaste* (misidentified as *Eophacops* and *Scotiella* in Wolfart, 1961) are the only calmoniid genera which have supposedly been recognized outside of the Malvinokaffric Realm. This is of interest and has two possible interpretations: (a) these genera *do* occur as quasi-cosmopolitans; and (b) their plesiomorphic retentions make them more prone to be confused with other plesiomorphic (i.e., true) acastids elsewhere. The second is the hypothesis favored. *Phacopina* is herein restricted to the Malvinokaffric Realm and the designation *?Acastoides* is used for Malvinokaffric forms (see section on systematic paleontology). Regardless of taxonomy, these genera are not particularly valuable for correlation. They are also among the oldest of the Malvinokaffric calmoniids, although *?Acastoides* especially, but also *Phacopina*, do persist higher in the section in Peru and Bolivia. The expression of the pattern of auxilliary impressions (big tubercles on the internal mold) is an especially compelling plesiomorphy, shared with the Acastinae and Asteropyginae.

The *Calmonia* group appears to be a monophyletic one. *Schizostylus* and *Bainella* are slightly doubtful members, and there is a real problem in distinguishing between *Calmonia* and *Pennaia*. This should prove to be the critical group for correlations within the Malvinokaffric Realm. This group has no affinities outside of the Realm.

The *Probolops* group appears to be an excellent monophyletic assemblage. It is perhaps most closely related to the *Metacryphaeus* group and is perhaps not found outside of the Andean portion of the Malvinokaffric Realm.

The *Metacryphaeus* group also appears to be monophyletic and includes some taxa restricted to the Andean region (*Bouleia* and *Paraboulеia*), some cosmopolitans within the entire Realm (e.g., *Metacryphaeus* itself), and some interesting groups such as *Malvinella* with strong ties to the Amazon fauna, and also *Typhloniscus* from South Africa, as well as its sister taxon from Bolivia.

Homalonotids

While widespread in the Realm, homalonotids are not considered particularly sensitive zoogeographic indicators. They appear to be more affected by their common, shallow-water community position, which reduces their utility for zoogeographic assessment. Thus, the homalonotid-bearing strata at Accra, Ghana, and in the Horlick Mountains of Antarctica, contain no other trilobites. They can reasonably be inferred to belong to the Malvinokaffric Realm, but no real comparisons are possible.

DIVERSITY

Boucot (1974, p. 166) has characterized the brachiopod faunas of the Malvinokaffric Realm as being "very low in diversity," a condition he considers consistent with Caster's (1952) thesis of a cold-water setting. While low diversity appears to characterize Malvinokaffric Silurian trilobites (Eldredge, 1975), those from the Devonian do not conform to this pattern. For example, the Belen Beds of northern Bolivia contain a fauna of at least 22 genera and subgenera assignable to 8 families, implying reasonable diversity. The majority of our samples, being museum collections, are not suitable for rigorous diversity calculations, but we fortunately have some collections made by Peter Isaacson from carefully measured sections which qualify. An analysis of three such closely spaced collections, USNM 17952-17954, from the Belen Beds shows the following diversity. As calculated by the Shannon formula (see Pielou, 1975, p. 8), $H = 1.92$ for a total of 75 specimens. This is at least a moderate diversity which does not seem entirely compatible with the cold-water theory of maintenance of the Malvinokaffric Realm. Since that theory seems so comfortably to explain so many other attributes of the Malvinokaffric Devonian (e.g., lack of carbonates, low diversity in other groups), one is not inclined to abandon it easily. A heightened diversity in the Lower Devonian of Bolivia suggests a favorable environment, cold or not. The low diversity of the Malvinokaffric Silurian seems more a reflection of monotonous community composition (mostly homalonotid) than of any intrinsic biotic character of the region.

THE APPALACHIAN INFLUENCE IN SOUTH AMERICA

The trilobites of the Floresta fauna of Colombia have been examined by us and involve but 4 taxa: *Phacops* (*Viaphacops*) sp., *Dechenella boteroi* Caster and Richter (*in* Richter and Richter, 1950, p. 161), indeterminate odontopleurid (= *?Cyphaspis* of Caster, 1939), and homalonotid indeterminate. The presence of *Dechenella* is a probable link with the Appalachians and the *Viaphacops* also appears to be. There is nothing identifiably Malvinokaffric about this assemblage which is probably best termed impoverished Appalachian. The *Phacops* reported by Weisbord (1926) from Venezuela apparently also belongs to *Viaphacops* and, again, occurs without endemic Malvinokaffric trilobites. The most northerly occurrences of endemic Malvinokaffric forms are to be found in the Amazon Basin of Brazil and in the Peruvian Andes. In neither case is there greater adulteration by Appalachian forms than elsewhere in South America. Thus, the boundary between impoverished Appalachian and basically Malvinokaffric trilobites can be sharply drawn. The lack of a gradational zone could be adduced as evidence of the relatively minor impact of Appalachian immigrants on the composition of true Malvinokaffric Realm. The trilobites may not entirely coincide with the distribution pattern for other shelled benthos in particular in the region of the Amazon Basin which Harrington (1968, p. 664) contends shows mixing of Appalachian and Malvinokaffric elements. Copper's (1977, p. 183) exclusion of the Amazonian Devonian from the Malvinokaffric Realm is opposed by the clear Malvinokaffric affinity of the trilobites of the Maecuru Formation of the Amazon Basin: *Malvinella goeldi, Phacopina braziliensis,* n. gen. aff. *Vogesina, ?Acastoides menurus,* n. gen. aff. *Fenestraspis,* and other new genera having affinities with *Malvinella* (p. 160), among others.

TRILOBITE COMMUNITIES

Sorting out the changes in faunal composition caused by changes in community composition from those larger-scale changes which reflect geographic distribution should be a major concern in any zoogeographic study. In particular, it is important to be able to identify those taxa whose occurrence can be suspected to be more closely related to changes in local environmental conditions than to any regionally significant changes in distribution. Unfortunately, our ability to do this on a regional basis with the Malvinokaffric trilobites is severely limited by the nature of the available samples. Study of the Isaacson collections from Bolivia does, however, provide at least a window into the composition of Malvinokaffric trilobite communities useful in evaluating assemblages elsewhere in the Realm. Analysis of the vertical succession of trilobites in sections in central Bolivia leads us to a conclusion compatible with that arrived at by Boucot (1971) on the basis of the study of brachiopods, namely, that an onlap-offlap sequence is present (see Fig. 1). The trilobite fauna of the Silurian Catavi Formation and its correlatives can be assigned to the Homalonotid Community of shallow-water setting. *Leonaspis* also occurs abundantly in the Catavi Formation. It is of interest to note that its failure to persist upward into the Devonian does not seem to have been a function of community change although one would intuitively expect a taxon capable of surviving the rigors of the shallow

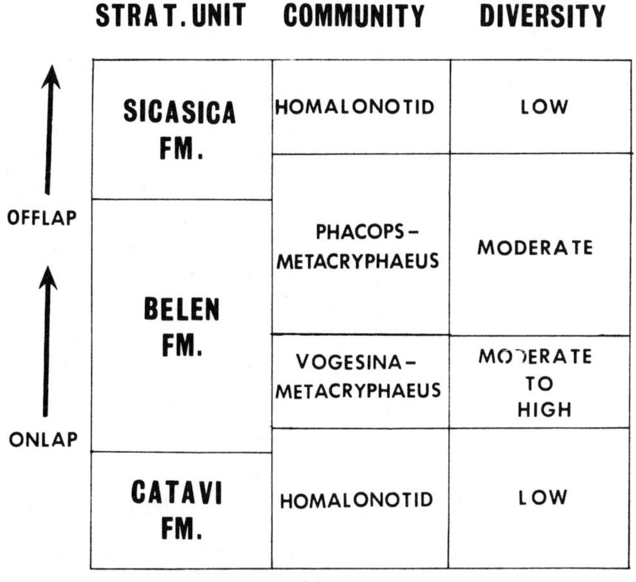

Figure 1. Trilobite communities of the Siluro-Devonian of Bolivia.

end of the spectrum to persist into deepening conditions. Elsewhere, *Leonaspis* is not found in community associations thought to be markedly shallow.

The lower part of the overlying Devonian continues to bear homalonotids (Wolfart, 1968, p. 61). It is only somewhat higher in the Belen Beds that one encounters a new community consisting of abundant *Vogesina* and *Metacryphaeus* and characterized, as noted elsewhere, by at least a moderate diversity. A deeper setting as compared with the Homalonotid Community is suggested, not only by the increased diversity, but also by the association of these trilobites with brachiopods such as *Notiochonetes* and *Metaplasia.* In the uppermost part of the Belen Beds and the lower part of the Sicasica Formation a slightly different assemblage appears, this one dominated by *Phacops* (*Viaphacops*) and *Metacryphaeus.* The exact environmental parameters controlling this assemblage are not clear, but changes in lithology and in the rest of the megafauna suggest that this reflects the beginning of the offlap sequence. The highest available trilobite collections from the upper part of the Sicasica Formation represent, again, the Homalonotid Community (USNM 17928) and strongly suggest a continuance of the offlap and a return to very shallow-water settings.

The tendency for homalonotids to recur in very shallow-water settings was noted previously. By contrast, one might conclude that that part of the section represented by the middle Belen Beds represents the most marine conditions (*Vogesina* and *Metacryphaeus* Community) and is the assemblage on which regional zoogeographic comparisons ought to be made. That is the procedure we have followed in this paper.

COMPARISONS WITH OTHER AREAS

Because of its conspicuous uniqueness, it is of interest to explore quantitative comparisons between the Malvinokaffric assemblage and those of other areas. We have made the appropriate calculations using the Simpson index and the provinciality index proposed by Johnson (1971, p. 259). The formula for the provinciality index is: $PI = C/2E_1$, where C is the number of genera common to the two areas being compared, and E_1 is the smaller of the two values for the number of endemic genera present. In general, PI values above 1 indicate cosmopolitan conditions, those below 1 provincialism. The smallest taxonomic unit employed for the calculations shown on Table 1 is the subgenus.

As shown on Table 1, a comparison between Bolivia and Australia (data for Australia from Campbell and Davoren, 1972), the only austral continent not previously considered part of the Malvinokaffric Realm, reveals a PI of 0.14, and a Simpson index of 22. This is a very low degree of faunal similarity, and substantiates the earlier conclusion of Campbell and Davoren that Australia was not Malvinokaffric.

A comparison of Bolivia with the Appalachian Province reveals, if possible, an even lower provinciality index of

	1RE	2RE	C	N	SI	PI
BOLIVIA - SOUTH AFRICA	32	7	8	47	53	0.57
BRAZIL - SOUTH AFRICA	12	10	6	28	37	0.30
BOLIVIA - BRAZIL	33	10	7	50	41	0.35
BOLIVIA-APPALACHIAN	35	38	6	79	14	0.08
BOLIVIA-ARGENTINA	33	4	7	44	64	0.78
BOLIVIA-AUSTRALIA	35	21	6	62	22	0.14
BOLIVIA-ALASKA+YUKON	38	17	3	56	15	0.09

1RE = endemics in region 1
2RE = endemics in region 2
C = genera common to both regions
N = total number of genera in both regions
SI = Simpson index
PI = provinciality index

Table 1. Comparison of Devonian trilobite faunas of the Malvinokaffric Realm.

0.08 and a Simpson index of 14. These values not only show the zoogeographic distinctness of these regions, but also raise the question of how Appalachian immigrants could be supposed to have been the source of nearly half the Bolivian trilobite fauna (Wolfart, 1968) and yet leave so little trace of their former presence. As stated elsewhere, we find cosmopolitan Silurian trilobites to be an adequate source for South American Devonian trilobites, a view for which these low, extra-Malvinokaffric faunal similarity indices provide indirect support.

For completeness, we may compare the Malvinokaffric Realm with one which could be expected to be very different. A comparison of Siegen-Ems trilobites of Alaska and the adjacent Yukon (Ormiston, 1972, 1975, unpub. data) with those of Bolivia shows a provinciality index of 0.09 and a Simpson index of 15. These values are essentially identical with those for the Bolivia-Appalachian comparison. The conclusion with regard to Appalachian-Bolivian relationships is obvious.

SUBDIVISION OF THE MALVINOKAFFRIC REALM

The apparent coherence of the Malvinokaffric trilobite fauna has been noted by Kobayashi and Hamada (1975, p. 1) and by Campbell and Davoren (1972, p. 92), the last mentioned describing it as homogeneous. Homogeneity is more apparent than real as first noted by Clarke (1913). Wolfart (1968, p. 7) subdivided the Malvinokaffric Realm into a South African-Falklands-Antarctic Province and a South American Province. Our quantitative data suggest even further subdivisions are warranted. There are really fewer genera in common between Bolivia and Brazil than between Bolivia and South Africa, and all of those regions have relatively low provinciality indices with one another (Table 1).

It is the uniqueness of the Malvinokaffric endemics more than their shared number which has caused investigators more familiar with northern faunas to somewhat uncritically lump the Malvinokaffric Realm. Wolfart's suggestion (1968) that the South American Province differs from the South African-Malvinan Province because of differing sources of immigrants to those two regions, is not supported by our faunal analysis, which suggests, instead, that these were each somewhat isolated centers of endemism with imperfect exchange with one another, and further, that there seems to have been better exchange from Bolivia to South Africa than from Bolivia to Brazil. The greater differences between the trilobites of Bolivia and Brazil are interpreted by us to reflect community changes attendant on migration from the Andean trough shelfward into the Brazilian region. By contrast, the greater degree of similarity between Bolivia and South Africa may have involved exploitation of the more continuously marine Andean seaway as a migration route (see Fig. 4). That the Falkland Islands and South Africa were provincially related is reflected by the fact that all trilobite genera known from the Devonian of the Falklands are also present in South Africa. Within the Andean belt, there is even a measure of low-level provincialism within Bolivia. Northern and southern parts of Bolivia have mostly different trilobite species in the Devonian and even some different genera. As subdivisions of the Malvinokaffric Realm, we would then recognize an Andean Province including Peru, Bolivia, Paraguay, and Argentina; a Brazilian Province including Brazil and possibly Uruguay; and a South African-Malvinan Province possibly including Antarctica. In addition to its low level of faunal similarity with both Bolivia and South Africa (Table 1), the Brazilian Province is diagnosed by the presence of such endemic genera as *Paracalmonia*, new genus cf. *Malvinella*, and new genus aff. *Vogesina*.

Within the Andean region, trilobites do not appear to show as much adulteration by "Appalachian" forms as do the Devonian brachiopod faunas in northern South America. Also, contrary to the brachiopod situation, no marked Appalachian influence is reflected in the trilobite fauna of the Amazon Basin. However, with a better biostratigraphic framework, it may become clear that many forms (e.g., *Tropidoleptus*) once casually accepted as Appalachian are actually of Malvinokaffric origin and have only subsequently reached the Appalachians as has already been suggested by Isaacson (1974).

ORIGINS OF THE MALVINOKAFFRIC TRILOBITES

Wolfart (1968, p. 7) has proposed that the South American Devonian fauna immigrated from the Appalachian Province, whereas, the African Subprovince fauna

Table 2. Genera and subgenera of Devonian trilobites present in Bolivia, Brazil, and South Africa. Data for existence of three subdivisions of the Malvinokaffric Realm.

Bolivia	Brazil	South Africa
n. gen. aff. *Malvinella*	n. gen. cf. *Malvinella* (= *Dalmanites gemellus* Clarke, 1890) and other new genera related to *Malvinella*	
Bouleia		
Parabouleia		
n. gen. aff. *Typhloniscus*		*Typhloniscus*
Probolops		
Cryphaeoides		
n. gen. aff. *Cryphaeoides*		
Tarijactinoides	?*Tarijactinoides*	
n. gen. aff. *Tarijactinoides*		
Chiarumanipyge		
Phacops (*Viaphacops*)		
Francovichia		*Francovichia*
Fenestraspis	n. gen. aff. *Fenestraspis*	?*Fenestraspis*
Chacomurus		
"*Dalmanites*"		*Dalmanites lunatus*
dalmantid n. gen.		
Gamonedaspis	?*Gamonedaspis*	
Dipleura		*Dipleura*
Burmeisteria	*Burmeisteria*	*Burmeisteria*
Homalonotus		
calymenid n. gen.		
Boliviproetus n. gen.	indet. proetid	*Dechenella*? *malacus*
Acanthopyge? n. subgen.		
Otarion (*Otarion*)		
Otarion (*Maurotarion*)		
"*Australosutura*"		
Phacopina	*Phacopina*	?*Phacopina*
Andinacaste		
Vogesina (*Vogesina*)	n. gen. aff. *Vogesina*	
Vogesina n. subgen.		
?*Acastoides*	possibly present	
	Calmonia	*Calmonia*
	Paracalmonia	
	Pennaia	*Pennaia*
Deltacephalaspis (*D.*)		n. gen. cf. *Pennaia*
Deltacephalaspis (*Prestalia*)		n. gen. *pseudoconvexus*
Bainella (*Belenops*)		*Bainella* (*Belenops*)
		Bainella (*B.*)
Schizostylus (*Curuyella*)		*Schizostylus* (*Curuyella*)
Schizostylus (*Schizostylus*)		
Metacryphaeus	*Metacryphaeus*	*Metacryphaeus*
Kozlowskiaspis (*K.*)		
Kozlowskiaspis (*Romaniella*)		
Malvinella	*Malvinella*	

came partly from the eastern part of a cosmopolitan Devonian fauna (Boreal Province) via New Zealand and Antarctica. We cannot concur with either part of this thesis. Rather, we agree with Clarke (1913, p. 9), who concluded that "the special traits of this austral fauna appear, with present knowledge, to have been derived by inheritance from a Silurian stock which served as ancestors to both the northern and southern faunas." While it is true that some Appalachian elements invaded the Malvinokaffric Realm, as has been known from the work of Caster (1939) and more recently through documentation by Boucot (1975) and Oliver (1975), these Appalachian intruders are largely confined to northern South America (Colombia and Venezuela), and those which penetrated further south appeared after the appearance of distinctly Malvinokaffric elements there. The division of the Malvinokaffric fauna (especially that of Bolivia) into Boreal elements (*Otarion, Phacops, Acanthopyge,* calymenids and endemic calmoniids) which Wolfart (1968) would derive from a European Boreal acastid stock, is artificial. Those elements which have been considered Boreal by Wolfart are in reality cosmopolitan stocks, and otarionids, calymenids, proetids, and acastids are known to have been present in the Silurian of South America. The presence of such families and genera in South America is not *prima facie* evidence of migration of Boreal forms but simply an indication of their Silurian presence in South America as well as in most other places of the world. To borrow an analogy from systematics, their presence is most likely "primitive" and, therefore, not useful in analyzing provincial biogeographic relationships. By the same token, there is no need to derive the calmoniids by migration from a northern primitive Tethyan acastid stock. Silurian acastids of Bolivia are generalized types which provide as good a morphotype as any other for deriving all the so-called calmoniids. If we are to look in the Silurian for the origins of the Malvino-

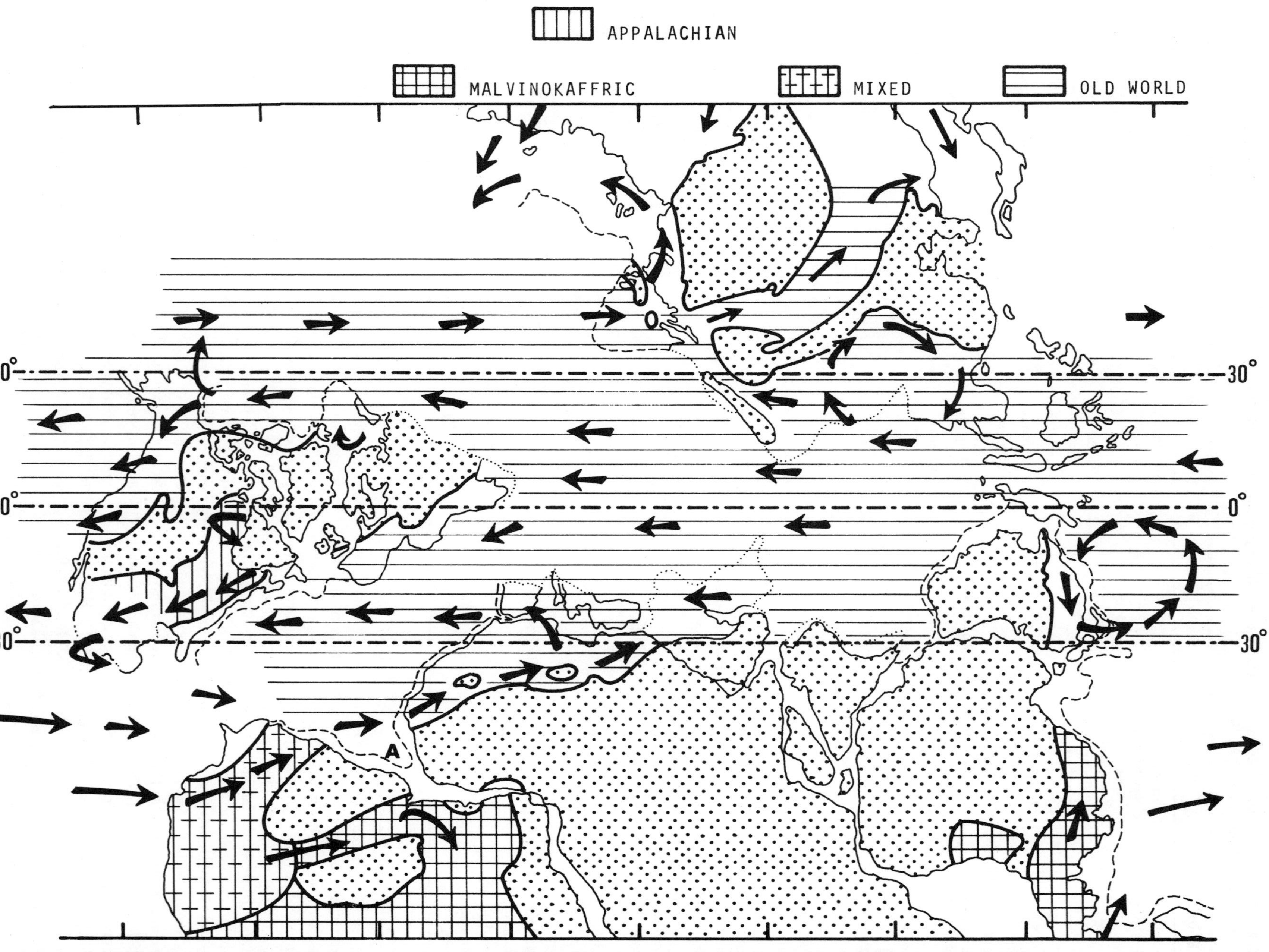

Figure 2. Middle Devonian paleogeography, paleocurrents, and trilobite biogeography. Continental positions modified from Smith and others (1973) to incorporate North American and Soviet Devonian paleomagnetic data. Emergent areas indicated by coarse stipple are modified from Boucot (1975).

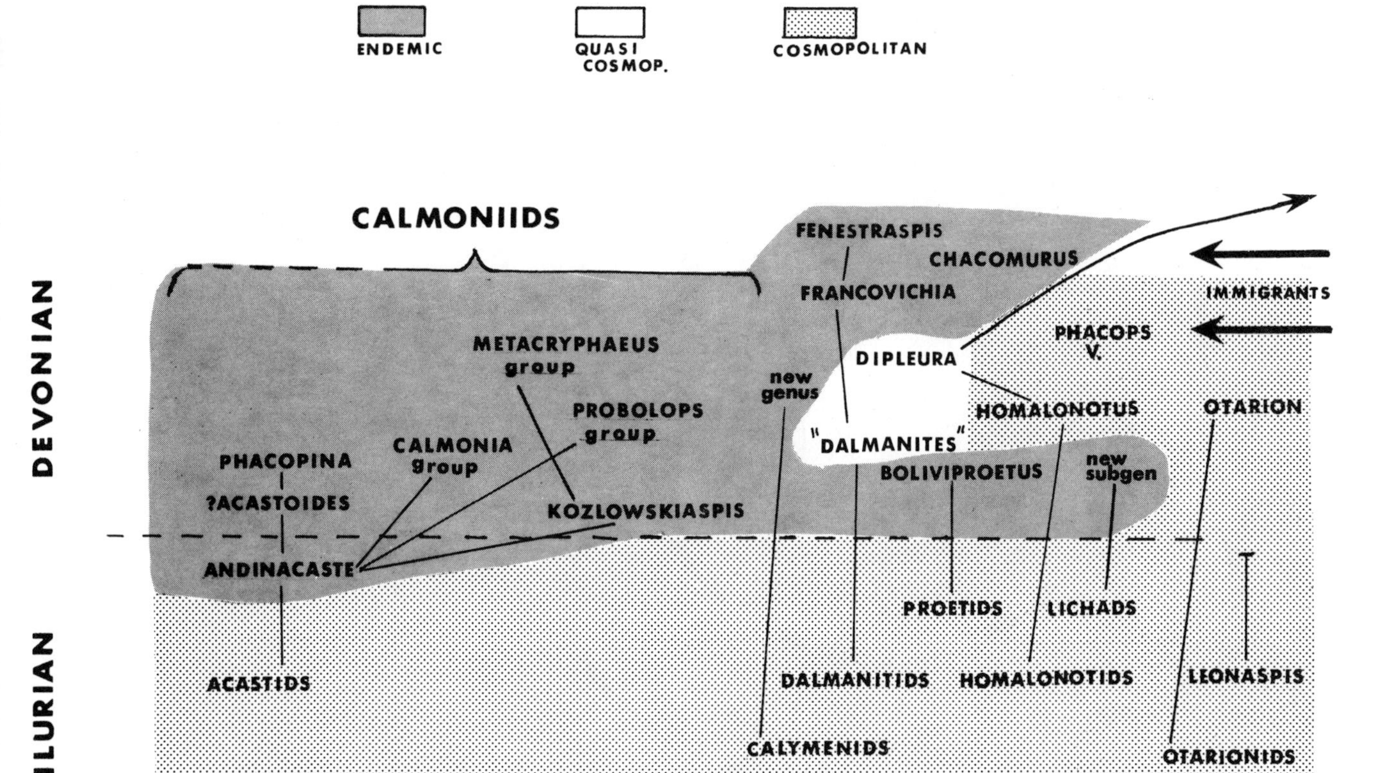

Figure 3. The origin of the Malvinokaffric trilobite fauna of the Andean Province.

kaffric trilobites, we must automatically discount Appalachian elements as having been a source inasmuch as a discrete Appalachian trilobite fauna did not exist in Silurian time, and immigrants therefrom could not have been the raw material for the Malvinokaffric Realm trilobites. Silurian acastids are rare in the area of the Appalachian Province, for example, except for the Maritimes which are zoogeographically related to the Lower Devonian Rhenish Communities complex of the Old World. Other cosmopolitan taxa present in the Late Silurian and Early Devonian of Bolivia include *Leonaspis,* dalmanitids, proetids, and even lichads. On the other hand, the presence of *Phacops* (*Viaphacops*) relatively low in the Bolivian Devonian section (Belen Beds) coupled with the consideration of the geographic distribution of that subgenus outside of South America (Appalachians, Australia, and Kazakstan) and inside South America (Colombia and Venezuela where it occurs without Malvinokaffric elements), and only as far south as the Andean part of the continent, suggests that this was truly an immigrant from the Appalachian Province. Another trilobite species, the homalonotid *Dipleura dekayi,* which is common to both Bolivia and the Hamilton faunas of the Appalachians, has long been cited as evidence of an Appalachian immigrant into the Malvinokaffric Realm. However, its occurrence in Bolivia must predate that in the Appalachians as indicated by its presence with *Viaphacops,* which in the Appalachians occurs no later than Eifelian age rocks and, thus, must predate the Hamilton. The only logical explanation is that *Dipleura dekayi* reached the Appalachians (Fig. 2) from South America (probably via northern Africa), not vice versa. The question of the presence of synphoriids in the Devonian of Bolivia hinges on a series of yet unsolved morphologic interpretations. There is at least a possibility, however, that this Appalachian stock did immigrate into the Malvinokaffric Realm to reach Bolivia. The genus *Chacomurus* Branisa and Vanek, 1973, in particular, may have synphoriid affinities. The rapidity of the development of the Malvinokaffric trilobite fauna (a true radiation) from the Silurian cosmopolitan stocks (not all of which survived to participate as indicated by the absence of *Leonaspis* above the Catavi Formation) is shown by the presence of at least 14 endemic genera in the *Scaphiocoelia* Zone of the Icla Formation and Belen Beds. Figure 3 illustrates our tentative interpretation of the general course of development of the Malvinokaffric trilobite fauna of Bolivia and indicates the minor direct role which we regard immigrants to have had.

THE MALVINOKAFFRIC REALM AND PLATE TECTONICS

The recognition that the Malvinokaffric fauna is restricted to the austral continents, with the exception of Australia, has long been ascribed to the existence of a Gondwana supercontinent. However, those reconstructions (e.g., Dietz and Holden, 1970) which include Australia

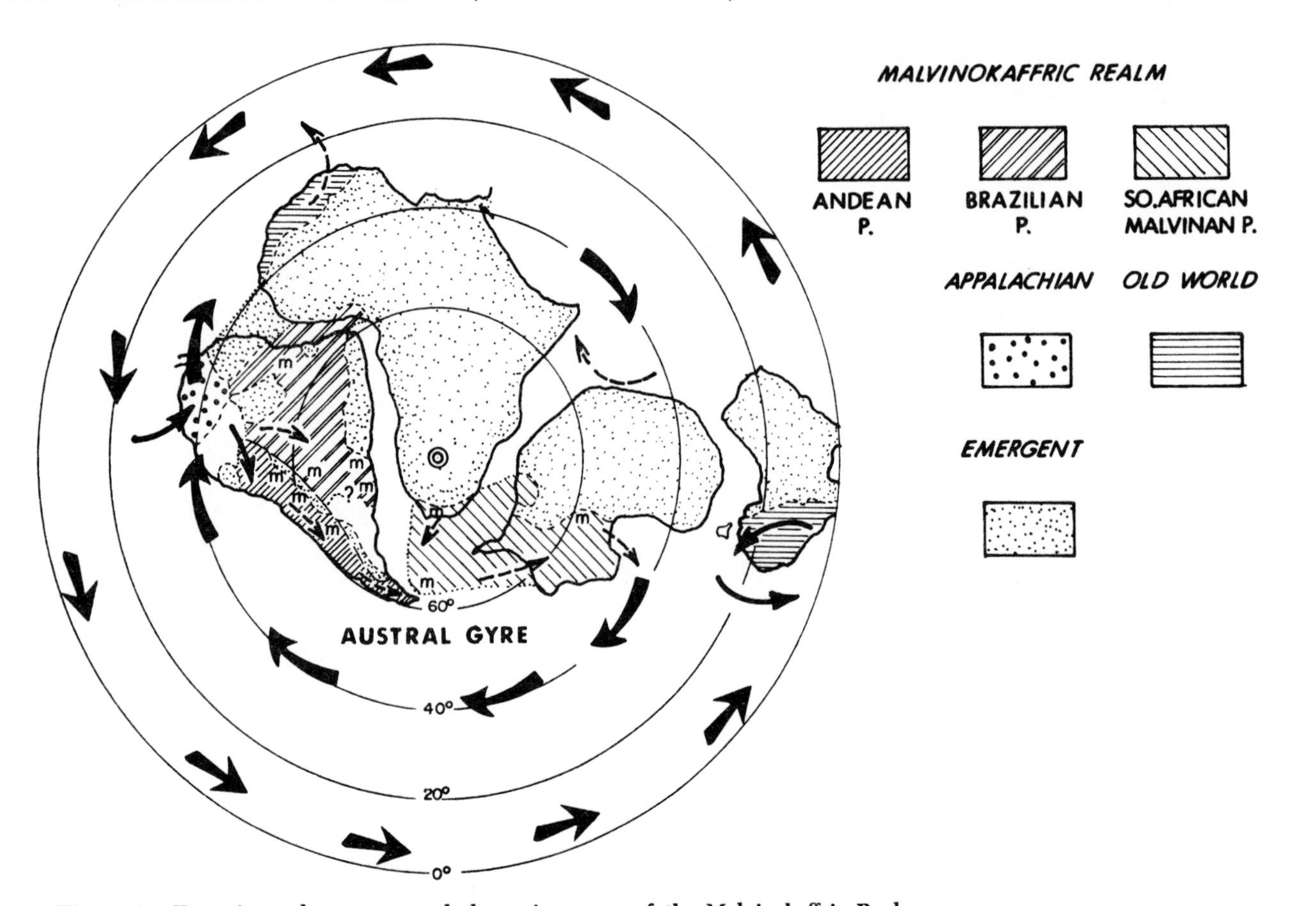

Figure 4. Devonian paleocurrents and the maintenance of the Malvinokaffric Realm.

in the Gondwana supercontinent do not seem compatible with what we know about the distribution of Malvinokaffric trilobites. Figure 2, which is a slightly modified version of the Devonian map published by Smith and others (1973) (modifications primarily involve a clockwise rotation of North America to conform with the North American and Soviet paleomagnetic data) seems to provide a much better explanation for where we find the Malvinokaffric forms. On this reconstruction, Australia is positioned above 30° S paleolatitude, whereas, the Malvinokaffric forms are positioned to the south of 40° S paleolatitude. Moreover, if we employ uniformitarian principles to the reconstruction of global current patterns, we can infer that the belt between the equator and 30° N and 30° S was one of predominant westwardly directed currents, whereas, that at 40° S latitude was one of predominant and strong easterly directed (the roaring 40's) currents. Using these principles, a paleocurrent pattern has been portrayed on Figure 4 which would conveniently explain, through the intervention of the roaring 40's current, why Malvinokaffric elements did not migrate northward to the Australian continent. It would not necessarily explain the current pattern that we surmise to have existed through the Andean part of South America and around the Horn to South Africa on the basis of levels of faunal similarity shown on Table 1, but the Austral Gyre, shown on Figure 4, would. The presence of a few Appalachian immigrants in northern South America does seem explicable on the basis of the deflection of the southwesterly directed Appalachian flow where it passes south of 30° S latitude to an easterly directed flow which would carry those forms to northern South America. The absence of Malvinokaffric forms from northwestern Africa (Alberti, 1969, 1970) is not readily explained by the current pattern shown on Figure 2 unless we invoke the existence of a land barrier connecting west Africa and northern South America at the position labeled "A" on the Figure.

ACKNOWLEDGMENTS

We thank Peter Isaacson of the University of Massachusetts for making available his carefully assembled trilobite collections from Bolivia, and Leonardo Branisa for his contributions toward the taxonomic revision of Bolivian trilobites. LeGrand Smith provided numerous trilobite specimens for which we are grateful.

REFERENCES

Ahlfeld, F., and Branisa, Leonardo, 1960, Geologia de Bolivia: La Paz, Inst. Boliviano del Petróleo Pub., 245 p.

Alberti, G. K. B., 1969, Trilobiten des jüngeren Siluriums sowie des Unter- und Mitteldevons. I. Beiträgen zur Silur-Devon Stratigraphie einiger Gebiete Marokkos und Oberfrankens: Senckenbergische Naturforsch. Gesell. Abh., v. 520, p. 1-692.

——— 1970, Trilobiten des jüngeren Siluriums sowie des Unter- und Mitteldevons. II: Senckenbergische Naturforsch. Gesell. Abh., v. 525, p. 1-233.

Amos, A. J., Campbell, K. S. W., and Goldring, R., 1960, *Australosutura* gen. nov. (Trilobita) from the Carboniferous of Australia and Argentina: Palaeontology, v. 3, p. 227-336.

Baldis, B. A. J., 1968, Some Devonian trilobites of the Argentine Precordillera, *in* Oswald, D. H., ed., Internat. Symposium on the Devonian System, Calgary 1967: Calgary, Alberta Soc. Petroleum Geologists, v. 2, p. 789-796.

——— 1972, Trilobites devonicos de la Sierra de Santa Barbara (provincia De Jujuy): Ameghiniana, v. 9, p. 35-44.

Boucot, A. J., 1971, Malvinokaffric Devonian marine community distribution and implications for Gondwana: Acad. Brasil. Ciências An., v. 43 Supl., p. 23-49.

——— 1974, Silurian and Devonian biogeography, *in* Ross, C. A., ed., Paleogeographic provinces and provinciality: Soc. Econ. Mineralogists and Paleontologists Spec. Pub. 21, p. 165-176.

——— 1975, Evolution and extinction rate controls: New York, Elsevier, 427 p.

Branisa, Leonardo, 1965, Los Fosiles Guias de Bolivia. I. Paleozoico: Serv. Geol. Bolivia Bol. 6, 282 p.

Branisa, Leonardo, and Vanek Jiri, 1973, Several new trilobite genera of the superfamily Dalmanitacea Vodges, 1890, *in* The Devonian of Bolivia: Ústřed Ústavu Geol. Vêstn., v. 48, 97-101.

Brink, A. S., 1951, On the compound eye of an unusually large trilobite from the Bokkeveld Beds south of Steytlerville, Cape Province: South African Jour. Sci., v. 47, p. 162-164.

Campbell, K. S. W., and Davoren, P. J., 1972, Biogeography of Australian Early to early Middle Devonian trilobites, *in* Talent, J., ed., Provincialism and Australian Early Devonian faunas: Geol. Soc. Australia Jour., v. 19, p. 88-93.

Caster, Kenneth E., 1939, A Devonian fauna from Colombia: Bull. Am. Paleontology, v. 24, p. 101-318.

——— 1952, Stratigraphic and paleontologic data relevant to the problem of Afro-American ligation during the Paleozoic and Mesozoic: Am. Mus. Nat. History Bull., v. 99, p. 105-152.

Clarke, John M., 1890, As Trilobitas do Grez de Ereré e Maecuru Estada do Pará, Brazil: Archivos Mus. Nac., v. 9, p. 1-58.

——— 1913, Fosseis Devonianos do Paraná: Serv. Geol. Mineral. Brasil Mon., v. 1, p. 1-353.

Dietz, R. S., and Holden, J. C., 1970, Reconstruction of Pangaea: Breakup and dispersion of the continents, Permian to Present: Jour. Geophys. Research, v. 75, p. 4939-4956.

Delo, David M., 1935, A revision of the Phacopid trilobites: Jour. Paleontology, v. 9, p. 402-420.

Copper, P. A., 1977, Paleolatitudes in the Devonian of Brazil and the Frasnian-Famennian mass extinction: Palaeogeography, Palaeoclimatology, Palaeoecology, v. 21, p. 165-207.

Eldredge, Niles, 1972, Morphology and relationships of *Bouleia* Kozlowski, 1923, (Trilobita, Calmoniidae): Jour. Paleontology, v. 46, p. 140-151.

——— 1974, Revision of the suborder Synziphosurina (Chelicerata, Merostomata), with remarks on merostome phylogeny: Am. Mus. Novitates, no. 2543, p. 1-41.

——— 1975, Evolution and extinction rate controls, by Arthur J. Boucot, a review: Systematic Zoology, v. 24, p. 389-391.

Eldredge, Niles, and Branisa, Leonardo, (in prep.), Calmoniid trilobites of the *Scaphiocoelia* Zone of Bolivia.

Fagerstrom, J. A., 1961, The fauna of the Middle Devonian Formosa Reef Limestone of southwestern Ontario: Jour. Paleontology, v. 35, p. 1-48.

Ferruglio, Egidio, 1930, Fossili devonici del Quemado (Sän Pedro de Jujuy) nella regione subandina dell' Argentina settentrionale: Giornale di Geologia, v. 5, p. 71-96.

Groth, J. 1912, Sur Quelques Trilobites du Devonien de Bolivie: Soc. Géol. France Bull., v. 12, p. 605-608.

Haas, W., 1968, Trilobiten aus dem Silur und Devon von Bithynien (NW Türkei): Palaeontographica, v. 130, Abt. A., 207 p.

Harrington, H. J., 1950, Geologia del Paraguay Oriental: Fac. Ciências Exactas, Contr. Ciências, Sec. E Geol. (Buenos Aires), v. 1, p. 1-82.

——— 1968, Devonian of South America, *in* Oswald, D. H., ed., Internat. Symposium on the Devonian System, Calgary 1967: Calgary, Alberta Soc. Petroleum Geologists, v. 1, p. 651-671.

Hartt, C. F., and Rathbun, R., 1875, Morgan expeditions, 1870-71: On the Devonian trilobites and mollusks of Erere, Province of Para, Brazil: Lyceum Nat. History New York Ann., v. 11, p. 110-127.

Isaacson, P., 1974, First South American occurrence of *Globithyris:* Its ecological and age significance in the Malvinokaffric Realm: Jour. Paleontology, v. 48, p. 778-784.

Johnson, J. G., 1971, A quantitative approach to faunal province analysis: Am. Jour. Sci., v. 270, p. 257-280.

Katzer, Friedrich, 1903, Grundzuge der Geologie des unteren Amazonasgebietes: Leipzig, Verlag Max Weg, 298 p.

Kauffman, E., 1973, Cretaceous Bivalvia, *in* Hallam, A., ed., Atlas of palaeobiogeography: New York, Elsevier p. 353-383.

Kayser, E., 1897, Beiträge zur Kenntniss einiger paläozoischer Faunen Südamerikas: Deutsche Geol. Gesell. Zeitschr., v. 49, p. 274-317.

Knod, Reinhold, 1908, Beiträge zur Geologie und Paläontologie von Südamerika. XIV. Devonische Faunen boliviens: Neues Jahrb. Mineral. Geol. Paläont., v. 25, p. 493-600.

Kobayashi, T., and Hamada, T., 1975, Devonian trilobite provinces: Japan Acad. Proc., v. 51, p. 447-451.

Kozlowski, Roman, 1913, Fossiles Devoniens de L'État de Parana (Brésil): Ann. Paléont., v. 8, p. 105-123.

——— 1923, La faune dévonienne de Bolivie: Ann. Paléont., v. 12, p. 1-112.

Lake, Philip, 1904, The trilobites of the Bokkeveld Beds: South African Mus. Ann., v. 4, p. 201-220.

Maksimova, Z. A., 1968, Middle Paleozoic trilobites of central Kazahkstan: Vse. Nauch.-issled. Geol.-razved. Inst. Trudy, no. 165, 208 p. (in Russian).

Méndez-Alzola, R., 1934, Contributión al conocimiento de la fauna devónica de Rincón de Alonso: Inst. Geol. Perforaciones Uruguay Bol. no. 21, p. 21-54.

——— 1938, Fósiles devónicos del Uruguay: Inst. Geol. Uruguay Bol. no. 24, p. 3-115.

Müller, H., 1964, Zur Altersfrage der Eisenerzlagerstätte Sierra Grande/Rio Negro, Nordpatagonien Aufgrund neuer Fossilfunde: Geol. Rundschau, v. 54, p. 715-732.

Oliver, W. A., 1975, Biogeography of Devonian rugose corals: Jour. Paleontology, v. 50, p. 365-373.

Ormiston, A. R., 1972, Lower and Middle Devonian trilobite zoogeography in northern North America: Internat. Geol. Cong., 24th, Montreal 1972, Proc. Section 7, p. 594-604.

——— 1975, Siegenian trilobite zoogeography in Arctic North America, *in* Martinsson, A., ed., Evolution and morphology of the Trilobita, Trilobitoidea, and Merostomata: Fossils and Strata, no. 4, p. 391-398.

Petri, Setembrino, 1949, Nota sôbre fosseis devonianos do oriente boliviano: Mineração e Metal., v. 13, p. 279-281.

Pielou, E. C., 1975, Ecological diversity: New York, Wiley, 165 p.

Reed, F. R. Cowper, 1925, Revision of the fauna of the Bokkeveld Beds: South African Mus. Ann., v. 22, p. 27-225.

——— 1927, Recent work on the Phacopidae: Geol. Mag., v. 64, p. 308-322, 337-353.

Rennie, John V. L., 1930, Some Phacopidae from the Bokkeveld Series: Royal Soc. South Africa Trans., v. 18, p. 327-360.
Richter, R., and Richter, E., 1942, Die Trilobiten der Weismes-Schichten am Hohen Venn, mit Bemerkungen über die Malvinocaffrische Provinz: Senckenbergische Naturforsch. Gesell. Abh., v. 25, 156-179.
——— 1950, Arten der Dechenellinae (Tril.): Senckenbergische Naturforsch. Gesell. Abh., v. 31, p. 151-184.
Salfeld, H., 1911, Versteinerungen aus dem Devon von Bolivien, dem Jura und der Kreide von Peru: Wissenschaftliche Veröffentlichungen der Gesell. für Erdkunde zu Leipzig, v. 7, p. 205-220.
Salter, J. W., 1856, Description of Palaeozoic Crustacea and Radiata from South Africa: Geol. Soc. London Trans., ser. 2, v. 7, p. 215-224.
Schwarz, E. H. L., 1906, South African Palaeozoic fossils: Albany Mus. Recs., v. 1, p. 347-404.
Shand, S. J., 1914, On the occurrence of the Brazilian trilobite *Pennaia* in the Bokkeveld Beds: Geol. Soc. South Africa Trans., v. 17, p. 24-28.
Smith, G. A., Briden, J. C., and Drewry, G. E., 1973, Phanerozoic world maps: *in* Hughes, N. F., ed., Organisms and continents through time: Spec. Papers Palaeontology 12, p. 1-42.
Steinmann, G., and Hoek, H., 1912, Das Silur und Kambrium des Hochlandes von Bolivia und ihre Fauna: Neues Jahrb. Mineral. Geol. Paläont., v. 34, p. 176-252.
Struve, W., 1958, Beiträge zur Kenntnis der Phacopacea (Trilobita). I. Die Zeliszkellinae: Senckenbergiana Lethaea, v. 39, p. 165-219.
——— 1959, Family Calmoniidae, *in* Harrington, H. J., ed., Treatise on invertebrate paleontology. Pt. O, Arthropoda 1: Geol. Soc. America and University of Kansas, p. 483-489.
Suarez-Soruco, R., 1971, *Tarijactinoides jarcasensis* n. gen. n. sp. del Devonico Inferior de Tarija: Serv. Geol. Bolivia Bol. 15, p. 53-56.
Swartz, Frank M., 1925, The Devonian fauna of Bolivia: Johns Hopkins Univ. Studies in Geol., no. 6, p. 29-68.
Thomas, Ivor, 1905, Neue Beiträge zur Kenntnis der devonische Fauna argentiniens: Deutsche Geol. Gesell. Zeitschr., v. 57, p. 233-290.
Tripp, R. P., 1958, Stratigraphic and geographic distribution of the named species of the trilobite superfamily Lichacea: Jour. Paleontology, v. 32, p. 574-582.
Ulrich, Arnold, 1892, Palaeozoische Versteinerungen aus Bolivien: Neues Jahrb. Mineral. Geol. Paläont., v. 8, p. 5-116.
Weisbord, N. E., 1926, Venezuelan Devonian fossils: Bull. Am. Paleontology, v. 11, p. 220-272.
Wolfart R., 1961, Stratigraphie und Fauna des älteren Paläzoikums (Silur, Devon), in Paraguay: Geol. Jahrb., v. 78, p. 29-102.
——— 1968, Die Trilobiten aus dem Devon Boliviens und ihre Bedeutung für Stratigraphie und Tiergeographie, *in* Wolfart, R., and Voges, A., Beiträge zur Kenntnis des Devons von Bolivien: Geol. Jahrb. Beih., v. 74, 241 p.

APPENDIX 1

Master List by Phylogenetic Affinity

Calmoniidae, 34 Genera and Subgenera

Calmoniidae plesiomorphs
- *Andinacaste*
- *Phacopina*
- *Vogesina* (prob.)
- ?*Acastoides*
- n. gen. prob. aff. *Vogesina* (Maecuru Sandstone)

Calmonia Group
- *Calmonia*
- *Pennaia*
- *Paracalmonia*
- *Deltacephalaspis* (*Deltacephalaspis*)
- *Deltacephalaspis* (*Prestalia*)
- *Bainella* (*Bainella*)
- *Bainella* (*Belenops*)
- *Schizostylus* (*Curuyella*)
- *Schizostylus* (*Schizostylus*)
- n. gen. (S Africa)
- cf. *Pennaia* or n. gen.
- n. gen. (Uruguay)
- gen. indet. (various discrete taxa)

Metacryphaeus Group
- *Metacryphaeus*
- *Kozlowskiaspis* (*Kozlowskiaspis*)
- *Kozlowskiaspis* (*Romaniella*)
- *Malvinella*
- n. gen. aff. *Malvinella* (Bolivia)
- cf. *Malvinella*, n. gen. (1 or 2 taxa Maecuru Sandstone)
- *Bouleia*
- *Parabouleia*
- *Typhloniscus*
- n. gen. aff. *Typhloniscus* (Bolivia)

Probolops Group
- *Probolops*
- *Cryphaeoides*
- n. gen. aff. *Cryphaeoides*
- *Tarijactinoides*
- n. gen. aff. *Tarijactinoides*

Calmoniidae aff. indet.
- *Chiarumanipyge*
- n. gen. (Bolivia)

Phacopidae
- *Phacops* (*Viaphacops*)

Dalmanitidae (possible affinity with Synphoriidae)
- *Francovichia*
- *Fenestraspis*
- *Chacomurus*
- n. gen. aff. *Fenestraspis* (Maecuru Sandstone)
- "*Dalmanites*"
- *Gamonedaspis*
- *Dalmanitoides*
- n. gen. indet. cf. *Odontochile* or *Gamonedaspis* (several discrete taxa)

Dechenellidae
- *Dechenella*

Proetidae
- *Boliviproetus* n. gen.

Otarionidae
- *Otarion* (*Otarion*)
- *Otarion* (*Maurotarion*)

Lichadae
- ?*Acanthopyge* n. subgen.

Brachymetopidae
- "*Australosutura*"

Calymenidae
- n. gen.

Odontopleuridae
- *Leonaspis* (Silurian)

Homalonotidae
- *Trimerus*
- *Digonus*
- *Dipleura*
- *Homalonotus*

APPENDIX 2

Systematic Paleontology

Calmoniidae Plesiomorphs

Andinacaste Eldredge and Branisa, in prep.

Type species: *Phacopina braziliensis* var. *chojñacotensis* Swartz, 1925. Silurian, Catavi Formation, northern Bolivia.

Other species: *Andinacaste legrandi* Eldredge and Branisa, in prep., Silurian, Cordillera Real Formation, northern Bolivia (Branisa informs us [personal commun. 1977] that the Cordillera Real Formation clearly underlies the Catavi Formation in the Quime Syncline, Bolivia): *Scotiella obsoleta perroana* Wolfart, 1961, Silurian, Paraguay; *Eophacops* n.sp. A, Wolfart, 1961, Silurian, Paraguay.

Notes: Plesiomorph closely resembles *Acaste longicaudata* Shergold, Lower Devonian, Australia; = *Cyphaspis* sp. Kozlowski, 1923; = *Proetus* sp. Salfeld, 1911 (*Chacaltaya*); = *Phacopina* (*Phacopina*) *chojñacotensis* Wolfart, 1958, *chojñacotensis* Zone, Pampa Beds, Bolivia.

Occurrence: Silurian: Bolivia, Catavi, Chojñacota (Swartz, 1925); Argentina, Arroyo Moralito, Jujuy (Ferruglio, 1930); Paraguay.

References: Salfeld, 1911; Swartz, 1925; Ferruglio, 1930; Ahlfeld and Branisa, 1960; Wolfart, 1961, 1968; Eldredge and Branisa, in prep.

Specimens: USNM, AMNH Branisa Collections; Swartz types of *A. chojñacotensis;* Eldredge and Branisa type and other specimens *A. legrandi* n. gen. n. sp.

Calmoniidae Plesiomorphs
Phacopina Clarke, 1913

Type species: *Phacops braziliensis* Clarke, 1890, also Katzer, 1903. Devonian, Maecuru Sandstone, Para, Brazil.

Other species: *Phacopina convexa* Eldredge and Branisa, in prep., *Scaphiocoelia* Zone, southern Bolivia, ?northern Bolivia; *P. padilla* Eldredge and Branisa, in prep., Lower Icla Shales, Padilla, Bolivia.

Notes: See Eldredge and Branisa, in prep., for extensive reasons why all Silurian and Devonian taxa from North America referred to *Phacopina* by various authors are rejected. Affinities of this taxon lie *within* the Malvinokaffric Realm, rather than with Appalachian forms.

Occurrence: Devonian: Bolivia, *Scaphiocoelia* Zone and basal Icla Formation; Brazil, Maecuru Sandstone, Para.

References: Clarke, 1890; Katzer, 1903; Delo, 1935; Struve, 1959; Suarez-Soruco, 1971; Eldredge and Branisa, in prep.

Specimens: USNM, Clarke's types of *P. braziliensis;* AMNH, NMNH Collections, types and otherwise of *P. padilla* and *P. convexa.*

Calmoniidae Plesiomorphs

Vogesina (*Vogesina*) Wolfart, 1968 (=*Phacopina* [*Vogesina*] Wolfart, 1968)

Type species: *Acaste devonica* Ulrich, 1892. Devonian, Conularid Beds, Chahuarani, Bolivia.

Other species: *V. lacunafera* Wolfart, 1968, *lacunafera, giganteus,* and *branisi* Zones of Wolfart, Bolivia; *V. aspera* Wolfart, 1968, *cornutus* and *aspera* Zones of Wolfart, Bolivia.

Notes: Restricted to Bolivia and Peru, but note related form from Maecuru Sandstone, Brazil (n. gen. aff. *Vogesina,* p. 159). Autapomorphic to a large degree and affinities in doubt; placed with question in the plesiomorphic group.

Occurrence: Devonian: Bolivia; Peru.

References: Kozlowski, 1923; Branisa, 1965 (*Platyceras, Francovichia, A. unispina* Zones); Wolfart, 1968.

Specimens: NMNH, AMNH Collections, Peru and Bolivia.

Calmoniidae Plesiomorphs
Vogesina new subgenus

Type species: Two undescribed specimens of a new species; one from Chiarumani, the other from Chacoma, Bolivia.

Notes: Pygidium shows some affinity with *Kozlowskiaspis* but is plesiomorphic with regard to *Vogesina* (*Vogesina*); doubtful if it should remain in plesiomorphic group.

Calmoniidae Plesiomorphs
New genus aff. *Vogesina*

Type species: *Dalmanites galeus* Clarke, 1890. Devonian, Maecuru Sandstone, Para, Brazil.

Notes: Appears to be a morphological link between *Vogesina* and *?Acastoides.*

Occurrence: Devonian: Brazil, Maecuru Sandstone, Para.

References: Clarke, 1890; Katzer, 1903.

Specimens: NYSM, cast of *Dalmanites galeus.*

Calmoniidae Plesiomorphs
?*Acastoides* Delo, 1935

Type species: *Acaste henni* Richter, 1916. Lower and ?Middle Devonian, Wetteldorf Sandstone, Kondel Group, Laubach Group, and Heisdorf Beds, Rheinische Schiefergebirge.

Other species: ?*Acastoides verneuili* (d'Orbigny, 1842), Devonian, *giganteus, cornutus, aspera* Zones of Wolfart, Bolivia; ?*A. acutilobata* Knod, 1908, Conulariid Beds, ?Chahuarani, Bolivia; ?*A. gamonedensis* Eldredge and Branisa, in prep. ?*A. menurus* Clarke, 1890, also Katzer, 1903, Maecuru Sandstone, Para, Brazil; ?*A. pro Acaste* (*Pennaia*) *pasiliana* Mendez-Alzola, 1938, Arroyo del Cordobes, Uruguay.

Notes: According to Eldredge and Branisa in prep., this genus is different from boreal species and is only with doubt referred to *Acastoides* as ?*Acastoides.* It is definitely known only from Bolivia and Peru within the Malvinokaffric Realm, but its resemblance with *Pennaia* is a source of confusion. *Phacops impressus* Reed, 1925, herein referred to as *Pennaia,* shares cephalic characters with ?*Acastoides* based on a single NMNH cranidium from Gamka Poort, South Africa.

Occurrence: Devonian: Bolivia; Peru; Brazil?, Maecuru Sandstone; ?Devonian, Uruguay.

References: Clarke, 1890; Katzer, 1903; Knod, 1908; Kozlowski, 1923; Delo, 1935; Mendez-Alzola, 1938; Ahlfeld and Branisa, 1960; Wolfart, 1968; Eldredge and Branisa, in prep.

Specimens: NMNH, AMNH, Bolivia, Peru; cast of *Phacops menurus* Clarke, 1890, Struve 1959.

Calmoniidae Plesiomorphs

Indet. but close to ?*Acastoides* or ?*Phacopina*

Type species: *Phacops* (*Bouleia*) *sharpei* Reed, 1925. Devonian, Keurbooms River, Upper Hex River Valley, South Africa.

Notes: Reed (1925) has figured 2 pygidia, one of which looks like *Phacopina*, the other like ?*Acastoides*.

Calmonia Group

Calmonia Clarke, 1913

Type species: *Calmonia signifer* Clarke, 1913. Devonian, Ponta Grossa Shale, Parana, Brazil.

Other species: *Calmonia sensu* Clarke, 1913 = *C. ocellus* Lake (*pars*, rest is *Metacryphaeus*), Pebble Island, Falkland Islands; *C. signifer* var. *micrischia*, Ponta Grossa Shale, Parana, Brazil; *C. subseciva*, Ponta Grossa Shale, Tybagy, Jaguariahyva, Parana, Brazil; ?*Acaste* (*Calmonia*) *callistris*, Bokkeveld Beds, South Africa (Rennie, 1930; Schwarz, 1906); *Pennaia africana* Shand, 1914, Bokkeveld Beds, Hex River Valley, South Africa; *non Calmonia curvioculata* Wolfart, 1968, Bolivia (see *Metacryphaeus*).

Occurrence: Devonian: Falkland Islands, Pebble Island; Brazil, Ponta Grossa Shale, Parana; South Africa, Bokkeveld Beds.

References: Schwarz, 1906; Clarke, 1913; Shand, 1914; Rennie, 1930; Struve, 1959.

Specimens: Types of *C. ocellus* seen; most decisions from literature.

Calmonia Group

Paracalmonia Struve, 1959

Type species: *Proboloides cuspidatus* Clarke, 1913. Lower Devonian, Ponta Grossa Shale, Parana, Brazil.

Other species: *Proboloides pessulus* Clarke, 1913, Devonian, Jaguariahyva, Sao Paulo, Brazil.

Notes: Very much like *Deltacephalaspis*.

Occurrence: Lower Devonian: Brazil, Ponta Grossa Shale, Parana, São Paulo.

References: Clarke, 1913; Struve, 1959.

Specimens: None seen.

Calmonia Group

Pennaia Clarke, 1913

Type species: *Pennaia pauliana* Clarke, 1913. Lower Devonian, Ponta Grossa Shale, Parana, Brazil.

Other species: ?*Phacops* (*Pennaia*) *gydowi* Reed, 1925, Devonian, Bokkeveld Beds, South Africa.

References: Clarke, 1913; Reed, 1925; Struve, 1959.

Specimens: None seen; all decisions from literature.

Calmonia Group

Deltacephalaspis (*Deltacephalaspis*) Eldredge and Branisa, in prep.

Types species: *D.* (*D.*) *comis* Eldredge and Branisa, in prep. Devonian, *Scaphiocelia* Zone, Gamoneda Formation, southern Bolivia.

Other species: *D.* (*D.*) *magister* Eldredge and Branisa in prep., Devonian, *Scaphiocoelia* Zone, Lower Belen Beds, northern Bolivia; *D.* (*D.*) *retrospina* Eldredge and Branisa, *Scaphiocoelia* Zone, Gamoneda Formation, southern Brazil.

Notes: This is the typical vicariant version of *Paracalmonia*, *Calmonia*, and other genera of the Andes.

Occurrence: Devonian: northern and southern Bolivia, *Scaphiocoelia* Zone.

References: Eldredge and Branisa, in prep.

Specimens: NMNH, AMNH Collections, types and other material.

Calmonia Group

D. (*Prestalia*) Eldredge and Branisa, in prep.

Type species: *D.* (*P.*) *tumida* Eldredge and Branisa, in prep. Devonian, *Scaphiocoelia* Zone, Icla Formation, east-central and southern Bolivia.

Other species: *D.* (*P.*) species A, *Scaphiocoelia* Zone, Lower Belen Beds, northern Bolivia.

Occurrence: Devonian, northern and southern Bolivia, *Scaphiocoelia* Zone.

Reference: Eldredge and Branisa, in prep.

Specimens: AMNH, NMNH Collections, including types.

Calmonia Group

Bainella (*Bainella*) Rennie, 1930

Type species: *B. bokkeveldensis* Rennie, 1930. Devonian, Gamka Poort and elsewhere, Bokkeveld Beds, South Africa.

Other species: *B.* (*B.*) *sanjuanina*? Baldis, 1968, Devonian, Talacasto Formation, San Juan, Argentina; *B.* (*B.*) *"acacia"* (*sensu* Clarke, 1913), Devonian, Falkland Islands; (?) *B. falklandicus* (*pars*, Clarke, 1913), Devonian, Falkland Islands; (?) *Phacops africana* (*pars*), Devonian, Bokkeveld Beds, Gamka Poort, South Africa (Lake, 1904); cf. *Phacops arbuteus*, Devonian, Bokkeveld Beds, Gamka Poort, South Africa (Lake, 1904); *Phacops crista-galli*, Devonian, Bokkeveld Beds, Gamka Poort, South Africa (Lake, 1904); ?*Dalmanites* (*Anchiopella*) *baini* Reed, 1925, Devonian, Bokkeveld Beds, South Africa; *Dalmanites* (*Anchiopella*) *africana*, Devonian, Cedarberg, Winterhoek Mts., South Africa (Reed, 1925); ?*Phacops* (*Phacopina*) *hiemalis*, Devonian, Winterhoek Mts., South Africa; *Bainella acacia*, Devonian, Bokkeveld Beds, South Africa (Rennie, 1930).

Occurrence: Devonian: Argentina, Talacasto Formation, San Juan; Falkland Islands, Fox Bay, Pebble Island; South Africa, Bokkeveld Beds, Gamka Poort and Cedarberg, Winterhoek Mts.

References: Lake, 1904; Clarke, 1913; Reed, 1925, Rennie, 1930; Struve, 1959; Baldis, 1968; Eldredge and Branisa, in prep.

Specimens: NYSM, Falkland Islands, including types of *"acacia"*; 1 cephalon and thorax-pygidium, Gamka Poort, South Africa.

Calmonia Group
B. (*Belenops*) Eldredge and Branisa, in prep.

Type species: *Acastoides insolitus* Wolfart, 1968. Devonian, *Scaphiocoelia* Zone, northern, east-central, and southern Bolivia.

Other species: *B.* (*B.*) *gamkaensis* Rennie, 1930, Basal Shales, Bokkeveld Beds, Gamka Poort, South Africa.

Notes: The 2 species appear nearly identical.

Occurrence: Devonian: throughout Bolivia, *Scaphiocoelia* Zone; South Africa, Basal Shales, Bokkeveld Beds, Gamka Poort.

References: Rennie, 1930; Wolfart, 1968; Eldredge and Branisa, in prep.

Specimens: USNM, AMNH Collections, type and others of *B.* (*B.*) *insolita.*

Calmonia Group
Schizostylus (*Curuyella*) Eldredge and Branisa, in prep.

Type species: *S.* (*C.*) *granulata* Eldredge and Branisa, in prep. Devonian, *Scaphiocoelia* Zone, Gamoneda Formation, southern Bolivia; ?Lower Belen Beds, northern Bolivia.

Other species: ?*Schizostylus* (*Curuyella*) *ensifer* (Reed, 1925), locality unknown, Bokkeveld Beds, South Africa; ?*Cryphaeus allardyceae,* Pebble Island, Falkland Islands (head only; assume tail and thorax = *Metacryphaeus*).

Notes: The plesiomorphic subgenus best showing links between this genus and other members of the *Calmonia* group.

Occurrence: Devonian: southern, northern Bolivia, *Scaphiocoelia* Zone; South Africa, ?Bokkeveld Beds; Falkland Islands, ?Pebble Island.

References: Clarke, 1913; Reed, 1925, Eldredge and Branisa, in prep.

Specimens: USNM, AMNH, types and others of *granulata;* NYSM, Clarke's single specimen of *allardyceae.*

Calmonia Group
S. (*Schizostylus*) Delo, 1935

Type species: *Dalmanites brevicaudatus* Kozlowski. Devonian, ?*giganteus, branisi, cornutus* Zones of Wolfart, Bolivia.

Other species: *S. cottreaui* (Kozlowski), Devonian, Bolivia.

Notes: Wolfart and others have synonymized the above species; they may be distinct. The subgenus occurs in the basal portion of the Icla Shales at Padilla, Bolivia. There may be more than 1 or 2 species involved in the entire section.

Occurrence: Devonian: Bolivia, Icla Formation (above *Scaphiocoelia* Zone) and basal part of Upper Belen Beds (including *M. cornutus Zone*).

References: Kozlowski, 1923; Delo, 1935; Struve, 1959; Wolfart, 1968; Branisa and Vanek, 1973; Eldredge and Branisa, in prep.

Specimens: AMNH, NMNH Collections.

Calmonia Group
New genus

Type species: *Acaste* (*Calmonia*) *signifer* var. *brevicaudata* Mendez-Alzola, 1938. Devonian, Arroyo del Cordobes, Uruguay.

Notes: Most like *Calmonia,* except pygidium; also recalls *Pennaia* and *Deltacephalaspis.*

Occurrence: Devonian: Uruguay, Arroyo del Cordobes.

Specimens: None seen.

Calmonia Group
cf. *Pennaia* or new genus

Type species: *Phacops* (*Calmonia*) *impressus* var. *vicina* Reed, 1925. Devonian, Gamka Poort, South Africa.

Notes: This is Lake's species, also listed as *P.* (*C.*) *impressus* by Reed. Lake's species may also be questionably referred to *Metacryphaeus. P.* (*C.*) *impressus* (*sensu* Reed) listed as indet. *P.* (*C.*) *impressus* var. *vicina* appears to be new, and closest to *Pennaia.*

References: Reed, 1925; Rennie, 1930.

Specimens: USNM, a cephalon from Gamka Poort, South Africa.

Calmonia Group
New genus

Type species: *Dalmanites* (*Acastella*?) *pseudoconvexus* Reed, 1925. Devonian, Bokkeveld Beds, Hoenderfontein, South Africa.

Calmonia Group
Genus indet. (various discrete taxa)

?*Dalmanites gonzaganus* Clarke, 1890, strata unknown, Parana, Jaguariahyva, Brazil (Clarke, 1913); *Phacops ocellus* Lake, 1904, Cedar Mts., Bokkeveld Beds, South Africa (closest to *Calmonia* and *Pennaia*); *Acaste* (*Calmonia*) *terrarocenai* Mendez-Alzola, 1934, 1938, Rincon de Alonso (type and only identified specimen), ?Arroyo del Cordobes, Uruguay (closest to *Calmonia* or *Pennaia*); *A.* (*C.*) *subseciva* Mendez-Alzola, 1938, Arroyo del Cordobes, Uruguay; *A.* (*C.*) *signifer* Mendez-Alzola, 1938, Arroyo del Cordobes, Uruguay; *Phacops* (*Calmonia*) *lakei* Reed, 1925, Touwes River Road, South Africa (indet. *Calmonia* group); *Dalmanites* (*Metacryphaeus*) *ceres* Rennie, 1930, Bokkeveld Beds, South Africa; *Phacops* (*Cryphaeus*) *ceres* Schwarz, 1906?, Bokkeveld Beds, South Africa.

Metacryphaeus Group
Metacryphaeus

Type species: *Phacops caffer* Salter, 1856. Devonian, Ezelfontein and other places near Ceres, South Africa (Lake, 1904).

Other species: *Calmonia ocellus* (*pars, sensu* Clarke, 1913), Mt. Robinson, Chartres River, Falkland Islands (Clarke, 1913); *Cryphaeus australis* Clarke, 1913, Parana, Brazil (Clarke, 1913); *C.* sp. nov.?, Ponta Grossa Shale, Parana, Brazil (Clarke, 1913); *C.* (?) *allardyceae* (*pars*) Clarke, 1913, Devonian, Falkland Islands; *C.* (*Acaste*) *convexa* and *C. dereimsi* Groth, 1912, Icla Formation, Icla, Bolivia; *C.* sp. α + β Kozlowski, 1913, locality unknown, Parana, Brazil; *C. boulei* Kozlowski, 1923, Letanias, Bolivia; C. *australis* var. *tuberculatus* Petri, 1949, Devonian, Bolivia; *C. concavus* Swartz, 1925, Devonian, Tihuanacu, Bolivia; *C. convexus, C. giganteus* Ulrich, 1892, Devonian, Bolivia; ?*Homalonotus* (*Schizopyge*) *parana* (=*Tibagya*), Tibagy, Parana, Brazil (Clarke, 1913); ?*Dalmanites* (*Mesembria*) sp., Lake Titicaca, Bolivia (Clarke, 1913); ?*D.* (*Corycephalus*) *capensis* Reed, 1925, Devonian, Bokkeveld Beds, South Africa; *Dalmanites falklandicus* (*pars*) Clarke, 1913, Fox Bay, Falkland Islands; *D.* (*Cryphaeus*) *paituna* Clarke, 1890, also Hartt and Rathbun, 1875, also Katzer, 1903, Erere Sandstone, Para, Brazil; *D. ulrichi* Katzer, 1903, Erere Sandstone, Para, Brazil; *D.* (*Cryphaeus*) cf. *pentlandi* Reed, 1925, Devonian, Bokkeveld Beds, South Africa; *D.* (*C.*) *caffer* Reed, 1925; *D.* (*C.*) *caffer* var. *albana*, Devonian, Bokkeveld Beds, South Africa; *D.* (*Metacryphaeus*) *caffer* Rennie, 1930, Devonian, Bokkeveld Beds, South Africa; *Acaste convexa* Knod, 1908, *Cryphaeus giganteus*?, Icla Formation, Icla and Chiarumani, Bolivia; ?*Phacops pupillus* Lake, 1904, Bokkeveld Beds, Gamka Poort, South Africa; ?*P. impressus* Lake, 1904 (see later treatment as ?*Pennaia* or new genus), Bokkeveld Beds, Gamka Poort, South Africa; *Phacops caffer* (*pars*) Salter, 1856, Devonian, Bokkeveld Beds, South Africa; *Calmonia curvioculata, Metacryphaeus praecursor, M. giganteus, M. cornutus, M. venustus, M. tuberculatus, M. boulei boulei, M. boulei boulei*?, ?*M. boulei pujravii, M. convexus* Wolfart, 1968, Devonian, Bolivia.

Notes: The *caffer* and *giganteus* groups of Wolfart, 1968 do not appear to have merit as monophyletic groups.

Occurrence: Devonian: Bolivia, whole section above *Scaphiocoelia* Zone through *dekayi* Zone; Argentina, Chavelas Formation, San Juan; Brazil, Erere Sandstone, Para, Ponta Grossa Shale, Parana; Falkland Islands, Fox Bay, Pebble Island; South Africa, Bokkeveld Beds, Gamka Poort, Ezelfontein, Ceres, Winterhoek.

References to *Metacryphaeus*: Ahlfeld and Branisa, 1960; Branisa, 1965; Baldis, 1968. *Asteropyge*, Branisa, 1965. *Andinopyge*, Branisa, 1965. Clarke, 1890 (*Dalmanites* [*Cryphaeus*] *paituna*); Clarke, 1913 (*pars Calmonia ocellus*); Clarke, 1913 (*Cryphaeus australis*); Clarke, 1913 *Homalonotus* (*Schizopyge*) *parana;* Clarke 1913 (?)*Dalmanites* (*Mesembria*) sp.; Clarke, 1913 (*pars Dalmanites falklandensis*); Clarke, 1913 [*pars Cryphaeus* (?) *allardyceae*]; Groth, 1912 (*Cryphaeus* [*Acaste*] *convexa—C. dereimsi*); Hartt and Rathbun, 1875 (*Dalmania paituna*); Katzer, 1903 (*Dalmanites ulrichi, D.* [*Cryphaeus*] *paituna*); Knod, 1908 (*Acaste convexa, Cryphaeus giganteus*); Kozlowski, 1913 (*Cryphaeus* sp. α and β); Kozlowski, 1923 (*C. boulei, C. australis, C. pentlandi, C. australis* var. *tuberculatus*); Lake, 1904 (?*P. pupillus, P.* ?*impressus, P. caffer*); Petri, 1949; Reed, 1925; Rennie, 1930; Swartz, 1925; Ulrich, 1892; Struve, 1959; Wolfart, 1968.

Specimens: USNM, AMNH, including Clarke's types from the Falkland Islands and Para, Brazil.

Metacryphaeus Group
Kozlowskiaspis (*Kozlowskiaspis*) Branisa and Vanek, 1973

Type species: *K.* (*K.*) *superna* Branisa and Vanek, 1973. Devonian, Icla Formation, Padilla region, Bolivia.

Notes: Monotypic and the plesiomorphic taxon of *Metacryphaeus* group with retentions reminiscent of *Calmonia* group.

Occurrence: Devonian: Bolivia, Icla Formation, Padilla region.

References: Branisa and Vanek, 1973; Eldredge and Branisa, in prep.

Specimens: USNM, AMNH Collections, topotypes and other material.

Metacryphaeus Group
K. (*Romaniella*) Eldredge and Branisa, in prep.

Type species: *K.* (*R.*) *borealis* Eldredge and Branisa, in prep. Devonian, *Scaphiocoelia* Zone, Lower Belen Beds, northern Bolivia.

Other species: *K.* (*R.*) *australis* Eldredge and Branisa, in prep., Devonian, *Scaphiocoelia* Zone, Gamoneda Formation, southern Bolivia.

Notes: In some ways this subgenus is more primitive than *Metacryphaeus* but more like *Metacryphaeus* in others.

Occurrence: Devonian: northern and southern Bolivia, *Scaphiocoelia* Zone.

References: Eldredge and Branisa, in prep.

Specimens: NMNH, AMNH, types of both species, plus other material.

Metacryphaeus Group
Malvinella Wolfart, 1968

Type species: *Anchiopella haugi* Kozlowski, 1923. Devonian, *branisi* and *cornutus* Zones of Wolfart, Bolivia.

Other species: There is apparently a rather large, plesiomorphic species, as yet undescribed from the lower Icla Formation at Padilla (USNM, AMNH Branisa Collections), which seems to connect *Metacryphaeus* with *Malvinella. Phacops goeldi* Katzer, 1903 from the Devonian of Brazil belongs to *Malvinella.*

Notes: A very distinctive genus, especially in terms of reduction and depression of 1p lobes and ridges and facial suture morphology around front of head. Not a dominant element of Bolivian fauna, but has affinities with other taxa within and especially outside, Bolivia. Restudy of Katzer's types (NYSM) of *Phacops goeldi* show that this taxon definitely is a species of *Malvinella.* It differs from Bolivian species by reduction of 3p and 2p furrows and greater curvature (rotundity) of glabella, but all other characteristic features of *Malvinella* are present.

Occurrence: Devonian: Bolivia, throughout Icla section (counting plesiomorphic species) and lower part of the Upper Belen Beds.

References: Katzer, 1903; Clarke, 1913; Kozlowski, 1923; Wolfart, 1968; Branisa and Vanek, 1973.

Specimens: AMNH, NMNH Collections.

Metacryphaeus Group
New genus aff. *Malvinella*

Type species: Undescribed species from Chacoma and Cahuanota, Bolivia; precise horizon and age unknown.

Notes: An undoubted relative of *Malvinella.*

Occurrence: Devonian: northern Bolivia, zone indeterminate, Chacoma, Cahuanota.

Specimens: AMNH Collection, 2 specimens.

Metacryphaeus Group
New genera or aff. *Malvinella*

Dalmanites gemellus (like calmoniid indet. A. of Eldredge and Branisa, in prep.,), Clarke, 1890, also Katzer, 1903, Maecuru Sandstone, Para, Brazil; *D. tumilobus* Clarke, 1890, also Katzer, 1903, Maecuru Sandstone, Para, Brazil; *D. australis* Clarke, 1890, also Katzer, 1903, Maecuru Sandstone, Para, Brazil; *Phacops scirpeus* Clarke, 1890, also Katzer, 1903, Maecuru Sandstone, Para, Brazil; *Acaste perplana* Knod, 1908, locality not known, Devonian, Bolivia.

Specimens: NYSM, casts seen of some of Clarke's (1890) taxa.

Metacryphaeus Group
Bouleia Kozlowski, 1923

Type species: *Phacops dagincourti* Ulrich, 1892. Devonian, *giganteus* and *cornutus* Zones of Wolfart, Bolivia; not in "*insolitus*" (= *Scaphiocoelia*) Zone.

Other species: *Bouleia sphaericeps* (Kozlowski, 1923), Devonian, Icla Formation, Bolivia.

Notes: *Dereimsia* Kozlowski, 1923 is a synonym of *Bouleia* (Eldredge, 1972).

Occurrence: Devonian: Bolivia, Icla Formation and lower Upper Belen Beds; Argentina, Chavelas Formation, San Juan.

References: Ulrich, 1892; Kozlowski, 1923; Struve, 1959; Ahlfeld and Branisa, 1960; Branisa, 1965; Baldis, 1968; Wolfart, 1968 (*pars*); Eldredge, 1972.

Specimens: AMNH, NMNH Collections, excluding types.

Metacryphaeus Group
Parabouleia Eldredge, 1972

Type species: *P. calmonensis* Eldredge, 1972. Devonian, *Scaphiocoelia* Zone, northern Bolivia.

Notes: Possible interacting (competition) with *Tarijactinoides* affecting morphology and biogeography (Eldredge and Branisa, in prep.).

Occurrence: Devonian: northern Bolivia, *Scaphiocoelia* Zone.

References: Branisa, 1965 (*pars Bouleia*); Wolfart, 1968 (*pars* under "*Bouleia dagincourti*"); Eldredge, 1972, Eldredge and Branisa, in prep.

Specimens: USNM, AMNH Collections, holotype and many others.

Metacryphaeus Group
Typhloniscus Salter, 1856

Type species: *Typhloniscus bainii* Salter, 1856. Devonian, Basal Shales of Bokkeveld Beds, Gamka Poort (*fide* Rennie) and presumably elsewhere, Bokkeveld Beds, South Africa.

Notes: Pygidium is slightly reminiscent of *Tarijactinoides.*

Occurrence: Devonian: South Africa, Basal Shales of Bokkeveld Beds, Gamka Poort (Rennie, 1930).

References: Salter, 1856; Lake, 1904; Reed, 1927; Rennie, 1930.

Specimens: BMNH, photographs and casts.

Metacryphaeus Group
New genus aff. *Typhloniscus*

Type species: An undescribed species based on a single cephalon from Chacoma, Bolivia, horizon not known.

Notes: Glabella more like *Metacryphaeus* than *Typhloniscus;* small eyes flush on surface of gena.

Occurrence: Devonian: Bolivia, Belen Beds, Chacoma.

Specimens: AMNH Collections, 1 specimen.

Probolops Group
Probolops Delo, 1935

Type species: *Proboloides glabellirostris* Kozlowski, 1923. Devonian, Icla Formation, Padilla region, Bolivia.

Occurrence: Devonian: Bolivia, above *Scaphiocoelia* Zone, Lower Icla Formation, Padilla.

References: Kozlowski, 1923; Delo, 1935; Struve, 1959; Eldredge and Branisa, in prep.

Specimens: NMNH (245619) Branisa Collection; 1 other specimen, now presumed to be lost, was studied by Kozlowski.

Probolops Group
Cryphaeoides Delo, 1935

Type species: *Cryphaeus rostratus* Kozlowski, 1923. Devonian, Sicasica Beds, *dekayi* Zone, Bolivia.

Notes: Probably a member of the *Probolops* Group.

Occurrence: Devonian: northern Bolivia, Sicasica Beds (only with *dekayi*); Argentina, Chavelas Formation, San Juan.

References: Kozlowski, 1923; Delo, 1935; Ahlfeld and Branisa, 1960; Branisa, 1965; Baldis, 1968; Wolfart, 1968.

Specimens: NMNH, AMNH Collections, only Bolivian specimens excluding types.

Probolops Group
New genus aff. *Cryphaeoides*

Type species: A single cephalon of an undescribed species from the Patacamaya region, Bolivia, horizon unknown; collected by L. Smith and now in the AMNH Collections.

Probolops Group
Tarijactinoides Suarez-Soruco, 1971

Type species: *T. jarcasensis* Suarez-Soruco, 1971, (=*Bolivianaspis scrutator*) Branisa and Vanek, 1973. Devonian, *Scaphiocoelia* Zone, Gamoneda Formation, Jarcas and Curuyo, Bolivia.

Other species: *T. tikanensis* Eldredge and Branisa, in prep., *Scaphiocoelia* Zone, Lower Belen Beds, northern Bolivia; ?=*Homalonotus* (*Calymene*) *acanthurus pars* Clarke, 1890, also Katzer, 1903 (doubtful), Maecuru Sandstone, Para, Brazil.

Occurrence: Devonian: northern and southern Bolivia, *Scaphiocoelia* Zone; Brazil, ?Maecuru Sandstone, Para.

References: Clarke, 1890; Katzer, 1903; Suarez-Soruco, 1971; Branisa and Vanek, 1973; Eldredge and Branisa, in prep.

Specimens: AMNH, NMNH Collections, topotypes and types.

Probolops Group
New genus aff. *Tarijactinoides*

Type species: A single undescribed cephalon of an undescribed species, Belen region, Bolivia, horizon unknown.

Specimens: NMNH Branisa Collection.

Calmoniidae aff. indet.
Chiarumanipyge Branisa and Vanek, 1973

Type species: *C. profligata* Branisa and Vanek, 1973. Devonian, Lower Belen Beds, Chiarumani-Machacoyo, Bolivia.

Notes: This is the sole reference to this unusual taxon.

Specimen: AMNH Collections, possibly present.

Calmoniidae aff. indet.
New genus

Type species: A single pygidium with some posterior thoracic segments, of an undescribed species from Chacoma, Bolivia; Devonian, horizon unknown.

Notes: Posteriorly rounded, but might be aff. *Vogesina*.

Specimen: AMNH Branisa Collections.

Calmoniidae aff. indet.

Australops Baldis, 1968 (=*Metacryphaeus*), Devonian, Cachipunco Formation, Jujuy, Argentina; *Cryphaeus* sp. Kayser, 1897, Devonian, Conularid Beds, west of Jachal, Argentina; *Acaste cordobesa* Mendez-Alzola, 1938, Devonian, Arroyo del Cordobes, Uruguay; New genus? *Bainella* sp. Rennie, 1930, Basal Shales of Bokkeveld Beds, Gamka Poort, South Africa.

Specimens: None examined.

Phacopinae
Phacops (*Viaphacops*) Maximova, 1972

Type species: *Phacops cristata* var. *pipa* Hall and Clarke, 1888. Eifelian, Onondaga Formation, eastern North America (=*P.* [*V.*] *bombifrons*).

Other austral species: *P. salteri* Kozlowski, *giganteus, cornutus, dekayi* Zones of Wolfart; *P.* (?*V.*) *argentinus* Thomas, 1905.

Notes: All specimens of the taxon from Peru and Bolivia examined by Eldredge are *Viaphacops* but do not appear particularly close to *P.* (*V.*) *cristata*. Wolfart's illustrated specimens from the *dekayi* Zone are definitely *Viaphacops*, implying that *Dipleura dekayi* is older here than in the Appalachians. Taxon restricted to northern South America and the Andean portion of the Malvinokaffric Realm.

Occurrence: Devonian: Peru; Colombia (Caster, 1939); Bolivia, Icla, Belen, Sisasica Beds; Argentina, Cerro del Fuerte (Kayser, 1897), Cerro del Fuerte (Thomas, 1905).

References: Kayser, 1897; Thomas, 1905; Groth, 1912; Kozlowski, 1923; Caster, 1939; Ahlfeld and Branisa, 1960; Branisa, 1965; Wolfart, 1968.

Specimens: AMNH (Peru, Bolivia), NMNH (Bolivia) Collections; Caster's specimens, Colombia.

Dalmanitidae *s.l.* (incl. Synphoriidae?)
Francovichia Branisa and Vanek, 1973

Type species: *F. branisi* (Wolfart, 1968).

Other species: ?*Dalmanitoides* Branisa, 1965; ?*Dalmanites maecurua*, Icla Formation, Bolivia (Groth, 1912); *Dalmanites maecurua* Knod, 1908; ?*D. boehmi* Knod, 1908, no locality data; *D. boehmi* var. *boliviensis* Kozlowski, 1923 (may be a *Gamonedaspis*), Tarabuco, Bolivia; *D.* sp. Lake, 1904, Gamka Poort, South Africa; *D.* (*Hausmannia*) *dunni* Reed, 1925, Gamka Poort, South Africa; ?*D.* Salfeld, 1911; *D. clarkei* Ulrich, 1892, Chahuarani, Bolivia; *Odontochile branisi* Wolfart, 1968.

Occurrence: Devonian: Bolivia, *Francovichia* Zone (Branisa, 1965), below Condoriquina Quartzite, Belen, Lower Belen Beds, Limabamba (Branisa, 1965; Wolfart, 1968); South Africa, Gamka Poort (Lake, 1904; Salfeld, 1911; Reed, 1925).

Specimens: AMNH, NMNH Collections.

Dalmanitidae *s.l.*
Fenestraspis Branisa and Vanek, 1973

Type species: *F. amauta* Branisa and Vanek, 1973.

Occurrence: Devonian: Bolivia, Lower Belen Beds, Chacoma (Branisa and Vanek, 1973); South Africa, ?Bokkeveld Beds, Bavianskloof (as *Phacops*) (Brink, 1951).

References: Branisa and Vanek, 1973.

Specimens: AMNH, NMNH Collections, Bolivia.

Dalmanitidae *s.l.*
Chacomurus Branisa and Vanek, 1973

Type species: *C. confragosus* Branisa and Vanek, 1973.

Notes: May have synphoriid affinities.

Occurrence: Devonian: northern Bolivia, Lower Belen Beds, Chacoma region.

References: Branisa and Vanek, 1973.

Specimens: AMNH Collections, Bolivia.

Dalmanitidae *s.l.*
New genus aff. *Fenestraspis*

Type species: *Dalmanites maecurua* Clarke, 1890; Katzer, 1903. Devonian, Maecuru Sandstone, Para, Brazil.

References: Clarke, 1890; Katzer, 1903.

Specimens: Cast of some types seen.

Dalmanitidae *s.l.*
"Dalmanites" Barrande, 1852

Type species: *D. caudatus* (Brünnich).

Other austral species: *"D." andii*, Cordillera Real Formation, northern Bolivia (=*Odontochile andii* Wolfart, 1968), Cordillera Real Formation, northern Bolivia.

Notes: Material as yet inadequately studied, hence quotation marks.

Occurrence: Silurian: Bolivia, Catavi and Cordillera Real Formations.

References: Steinmann and Hoek, 1912; Kozlowski, 1923; Swartz, 1925; Ahlfeld and Branisa, 1960; Branisa, 1965.

Specimens: AMNH, NMNH Collections.

Dalmanitidae *s.l.*
Gamonedaspis Branisa and Vanek, 1973

Type species: *G. scutata* Branisa and Vanek, 1973. Devonian, *Scaphiocoelia* Zone, Jarcas, southern Bolivia.

Other species: ?*Dalmanites accola* Clarke, 1913 (close, may be *Odontochile*), Devonian, Ponta Grossa Shale, Parana, Brazil.

Notes: A plesiomorphic, *Odontochile*-like dalmanitid with an upturned posterior region of pygidium = ? *Synphoria* Branisa, 1965, *Dalmanites litchfieldensis* Branisa, 1965, *D.* α and β Branisa, 1965.

Occurrence: Devonian: southern Bolivia, *Scaphiocoelia* Zone; Brazil, ?Ponta Grossa Shale, Parana (Clarke, 1913).

Specimens: AMNH, NMNH Bolivian Collections.

Dalmanitidae *s.l.*
Dalmanitoides Delo, 1935

Type species: *Dalmanites drevermanni* Thomas, 1905. Upper Lower Devonian, Jachal River, Cerro del Agua Negra, Argentina (Thomas, 1905).

References: Thomas, 1905; Delo, 1925.

Specimens: None seen.

Dalmanitidae *s.l.*
New genus Dalmanitidae

Type species: AMNH Bolivian Collection, an undescribed pygidium, horizon unknown.

Dalmanitidae *s.l.*

Dalmanitidae genus indet., cf. *Odontochile* or *Gamonedaspis* (several discrete taxa).

"Argentopyge" Baldis, 1972, Devonian, Cachipunco Formation, Jujuy, Argentina; *Dalmanites infractus* Clarke, 1890, also Katzer, 1903, Devonian, Maecuru Sandstone, Para, Brazil (may be a true dalmanitid); *D. patacamayensis*, Devonian, horizon unknown, Patacamaya, northern Bolivia; *D. lunatus*, Bokkeveld Beds, Gamka Poort, South Africa (Lake, 1904) (true dalmanitid); *D.* species 1 Thomas, 1905, Devonian, Cerro del Fuerte, Argentina; *D.* sp.? Kozlowski, 1923, Devonian, Bolivia (a true dalmanitid?).

Specimens: None seen.

Proetidae
Boliviproetus n. gen.

Diagnosis: Proetid with strongly inflated cephalon, cephalic height greater than glabellar length, glabella triangular in transverse profile, 3 pairs of glabellar furrows, lateral border uninflated; pygidium with axial furrow effaced.

Occurrence: Devonian: Bolivia, Chacoma.

Specimens: AMNH (34615-34617), 1 cranidium, 1 partial cephalon with 3 anteriormost thoracic segments preserved, 1 partial cephalon and partial thorax and pygidium; all material collected by Branisa.

Boliviproetus branisai n. gen. n. sp.
Plate 1, Figures 1 - 7

Description: Cephalon near semicircular, glabella as wide as long, available material with cranidial length up to 14.7 mm, high point of glabella a sagittal prominence centered posteriorly, bounded by adaxial ends of 2S, 1S, and posteriorly by the occipital furrow. Glabella drops off near vertically to narrow (sag.) occipital ring which lies at the level of palpebral lobes. With palpebral lobes oriented horizontally, preglabellar field and broad weakly separated, uninflated anterior border are steeply inclined at about 60° to horizontal. In transverse profile, glabella triangular, sides sloping at 35° to 40°; longitudinal profile forwardly inclined with a beaklike prominence above the occipital furrow. Three pairs of glabellar furrows: 1S angulate starts opposite midpoint of palpebral lobe, does not reach occipital furrow but bounds glabellar prominence; 2S runs backward and inward from opposite anterior edge of palpebral lobe; 3S weak and short not reaching axial furrow. Axial furrow near vertical against glabellar slope but not strongly impressed. Occipital ring short, lobes faintly indicated by slightly inflated triangular area. Palpebral lobe short and wide as long, eye low with surface not preserved. Genal fields plain, markedly sloping, border uninflated, weakly separated. Anterior branch of facial suture appears to diverge from midline at about 35°. Posterior branch obscured. Cephalic doublure strongly convex downward, semicircular in cross section. A poorly preserved pygidium is as wide as long, has a low, weakly segmented (3 rings visible) axis with narrow, wide articulating half-ring on first segment. Pleural fields and border forming one plane sloping at 25° to horizontal. Axial furrow effaced, first interpleural furrow present, subsequent ones effaced, pleural furrows are shallow, 3 are visible. Thorax consists of at least 5 segments.

Comparison: *Proetus*? *problematicus* Swartz (1925, Plate 1, Fig. 3) from the Devonian of Bolivia is not a proetid to judge from the illustrated sutural course. *Boliviproetus* shows the greatest similarity to *Proetus perinsignis* Chlupac and Vanek, 1965 (Plate 4, Figs. 1-3) from the Pragian of Bohemia. That species has a similar glabellar configuration with maximum glabellar height near its posterior end but a short preglabellar field, a weak and not angulate 1S, and more distinctly differentiated lateral border which is slightly inflated. *Unguliproetus gibbosus* Alberti, 1969 (p. 115, Plate 4, Figs. 9-11) from the late Pragian of Morocco has a vaguely similar longitudinal profile to that of *Boliviproetus* but lacks discrete glabellar furrows and has the glabellar prominence subcentrally located. The *Proetus* sp. illustrated by Müller (1965, Plate 5, Fig. 2) from the Silurian (*Clarkeia* Beds) of Patagonia differs in having an inflated lateral border.

Notes: In the style of glabellar inflation, *Boliviproetus* is homeomorphic with *Vogesina* suggesting that they share common morphologic adaptation to some environmental parameter.

Occurrence: Devonian: Bolivia, Chacoma.

Specimens: AMNH 34615, Collection no. B 3250, holotype, partial cephalon, partial thorax, pygidium; AMNH 34616, paratype, partial cephalon with anterior three thoracic segments; AMNH 34617, paratype, cranidium; all collected by Branisa.

Family Dechenellidae Pribyl, 1946

Dechenella boteroi Caster and Richter *in* Richter and Richter, 1950

?*Dalmanites* cf. *patacamayensis* Caster, 1939, p. 181, Plate 14, Figs. 3-6. *Dechenella* (*Basidechenella*?) *boteroi* Caster and Richter *in* Richter and Richter, 1950, p. 161.

Description: (Re-examination of Caster's original types permits the following remarks on this material which has never been thoroughly described.) Pygidium semi-elliptical with broad, somewhat flattened border, axis having at least 12 axial rings, somewhat crushed on holotype but notably deflected backward along midline, each axial ring with sagittal tubercle. Pleural fields strongly inflated, nine "ribs," interpleural furrow developed only on first rib. Each rib showing a row of tubercles along the anterior edge of the posterior pleural band and the posterior edge of the anterior pleural band. Pygidial doublure strongly convex downward.

Comparison: The pygidial prosopon is somewhat reminiscent of that of *Dechenella* (*Monodechenella*) *macrocephala* (Hall) from the Hamilton of New York State from which the Colombian species is distinguished in having a distinctly wider, flatter border and in having a relatively more elongate pygidium. Some similarity is also shown to the genus *Schizoproetoides* Ormiston, 1967, which, however, differs in having even more axial segments and a well-developed ridge on the inner edge of the pygidial border. No taxa are known elsewhere in South America that can be compared with *Dechenella boteroi.* The discovery of further material, especially including a cranidium, is necessary to help decide zoogeographic affinities of this apparently unique Colombian form.

Occurrence: Devonian: Colombia, Floresta Beds.

Specimens: PRI 5477A, holotype (external mold of pygidium); PRI 5477, internal mold of pygidium; PRI 5457, hypotype, pygidium.

?*Dechenella malaca* Lake, 1904

Proetus malacus Lake, 1904, p. 213-214, Plate 25, Fig. 10.

Notes: This is the only species from the true Malvinokaffric Realm that can be compared to *Dechenella,* but the assignment to that genus is doubtful. Although the pygidium is somewhat reminiscent of *Dechenella,* the head does not permit a definite decision. More and better material must be examined.

Occurrence: Devonian: South Africa, Bokkeveld Beds.

Specimens: No material examined.

Family Brachymetopidae Prantl and Pribyl, 1951

Genus *Australosutura* Campbell and Goldring, 1960

Type species: *Cordania gardneri* Mitchell, 1922.

"*Australosutura*" n. sp.

Plate 1, Figures 8, 9

Description: Cephalon semicircular, glabella trapezoidal, widest across base, gently rounded anteriorly, of modest transverse convexity, steeply declining anteriorly. Lobe 1L entirely isolated by furrow of length equaling 2/5 that of glabella, 2S faintly indicated as an inflection of the glabellar outline. Preglabellar field broad, concave laterally, anterior and lateral borders narrow with sharp crest and flat anterior face. Occipital furrow broad and shallow, ring short (sag.). Anterior branch of facial suture diverges at nearly 60° from midline, posterior branch transverse in terminal portion. Genal field broad (trans.), genal spine sulcate extending to fourth thoracic segment. Lateral doublure concave with near vertical inner slope. Glabella finely pitted and with large pustules, few large pustules on fixed cheek adjacent to axial furrow. Large pits are concentrated in trough of preglabellar field.

Comparison: The Chacoma species is morphologically intermediate between *Mystrocephala* Whittington, 1960, a North American genus of Eifelian and Givetian age and *Australosutura,* a genus of Carboniferous age. It resembles the former in glabellar outline, size of 1L, and degree of sutural divergence, and the latter in the faintness of 2S, and the style of cranidial prosopon.

It is distinguished from the type species of *Australosutura, A. gardneri* Mitchell, 1922, in having a more divergent facial suture, omega situated more distally, a broader (trans.) cheek, and cranidium less convex longitudinally. The Chacoma species may represent an early *Australosutura* or a bridge between *Mystrocephala* and *Australosutura.*

Australosutura is known from Visean strata elsewhere in South America in the province of Chubut, Argentina (Amos and others, 1960, p. 229).

Occurrence: Devonian: Bolivia, Belen Beds, Chacoma.

Specimens: AMNH 36971, 1 cephalon and attached partial thorax; AMNH 36972, 1 cranidium.

Family Otarionidae R. and E. Richter, 1926
Genus *Otarion* Zenker, 1833

Otarion (*Maurotarion*) *dereimsi* (Kozlowski, 1923)

Notes: The subquadrate glabella, which is strongly inflated both longitudinally and transversely, the semielliptical cephalic outline, which is slightly pointed on the sagittal line, the long genal spine reaching back to the eighth thoracic segment, and the very narrow occipital ring are all characteristics distinguishing this species. As Kozlowski (1923, p. 62) noted, this Bolivian species shows considerable similarity to *Otarion minuscula* (Hall, 1867) from the Bois Blanc, Schoharie, and Onondaga of Ontario and New York from which it is distinguished by its slightly larger size and stouter genal spine.

Occurrence: Devonian: Bolivia, Upper Belen Beds, Gamoneda, Chacoma.

References: Kozlowski, 1923; Swartz, 1925; Reed, 1925; Ahlfeld and Branisa, 1960: Branisa, 1965; Wolfart, 1968.

Specimens: One specimen lacking pygidium from Gamoneda, Bolivia, *M. giganteus* Zone, collected by LeGrand Smith; 1 cephalon and 1 cranidium from Chacoma, Bolivia, and 1 cephalon from Agua Castilla, near Zudanez, Bolivia, *Scaphiocoelia* Zone, all collected by Branisa.

Otarion n. sp.

Notes: A second otarionid present in the Belen Beds where it is associated in some samples with *Otarion dereimsi* (Kozlowski, 1923) is easily distinguished from that species by its semicircular cephalon and slender genal spines. The glabella is also somewhat rounded anteriorly, 1L slightly less than half of glabellar length, occipital ring low with low median node, palpebral lobe with small median pit, anterior and lateral borders narrow and semicircular in cross sections set off by distinct furrow. Cephalon bearing uniformly developed low tubercles. There are 14 thoracic segments, axis wider than 1 pleural field and low-lying in transverse view. Pygidium with broad axis showing five axial rings, only three pleural furrows perceptible on pleural field.

This species does not appear to be related to that described by Kozlowski (1923, p. 62, Plate 4, Fig. 18) from the Silurian of Chacaltaya, Bolivia, which consists of a pygidium and partial thorax of 9 segments. The pygidium shows eight axial rings on the axis and is assigned to *Andinacaste* (Eldredge and Branisa, in prep.). The Chacoma species shows considerable similarity to *Otarion novellum* (Barrande, 1852) from the Kopanina beds of Bohemia, especially in the presence of a distinct transverse eye ridge and a pit on the palpebral lobe. The Bohemian species differs, however, in having a finer prosopon on the glabella and fine pits on the preglabellar field.

Occurrence: Devonian: Bolivia, Upper Belen Beds, Chacoma.

Specimens: One nearly complete specimen, 1 cephalon with four thoracic segments, 1 external mold of a nearly complete specimen with half of cranidium missing, all collected by Branisa.

Family Odontopleuridae Burmeister, 1843
Genus *Leonaspis* R. and E. Richter, 1917

Type species: *Odontopleura leonhardi* Barrande, 1846.

Leonaspis aracana (Steinmann, 1912)
Leonaspis chacaltayana (Kozlowski, 1923)
Leonaspis berryi (Swartz, 1925)

Notes: All these taxa come from the Silurian Catavi Formation and seem to form a closely related plexus, differences among which may involve matters of dimorphism. Their closest relationship seems to be with the Bohemian Silurian species *Leonaspis minuta* Barrande. Alleged relationship to the Appalachian Lower Devonian species *tuberculatus* has been exaggerated. The lack of an occipital spine and the presence of a broad inner triangular area of the fixed cheek in the Bolivian taxon readily separates it from known Lower Devonian Appalachian species. No odontopleurids have been reported from younger strata in Bolivia.

Occurrence: Silurian: Bolivia, Catavi Formation.

Specimens: None seen.

Family Lichidae Hawle and Corda, 1857
Acanthopyge? (n. subgen.) *balliviani*

Type species: *Lichas balliviani* Kozlowski, 1923.

Notes: This species has certain similarities to *Lobopyge*, particularly in the form of the anterior and posterior pleural bands which are much more like those of *branikensis* than those of *haueri*. Unlike *Acanthopyge* H. and C., 1847, or any of the other ceratarginids, the hypostome of *balliviani* lacks large posterior middle lobes and has a posterior re-entrant like the Lichinae of Tripp. Wolfart describes the posterior border as rounded (1968, p. 133), but this is incorrect. *Lobopyge erinacea* Haas (1968, p. 177, Plate 37, Figs. 1-5) from the Emsian of Turkey has a much inflated posterior part of the median glabellar lobe, hypostome trapezoidal. Tripp (1958, p. 578) assigned *balliviani* to *Acanthopyge*. Wolfart states (1968, p. 133) that the tail has one to two distinct rings. Baldis (1968, p. 794) claims seven and figures a drawing of a tail (Plate 1, Fig. 6) with three pairs of lateral spines, plus the posterior pair. This must be wrong. *A. ?balliviani* shows considerable similarity with *L. longiaxis* Maksimova (1968, p. 39, Plate 6, Fig. 3; Plate 7, Figs. 2, 3) which lacks a postaxial ridge and has paired postaxial tubercles instead, but that species has a short third pair of spines, only one pronounced axial ring, and a more laterally persistent anterior band on the first pleural segment. *Lobopyge longiaxis* comes from the Lower Devonian (Coblenzian) of Kazachstan. *Lobopyge contusa* from the Onondaga Limestone (Hall and Clarke, 1888, Plate 19b, Figs. 3-6; Fagerstrom, 1961, Plate 14, Figs. 13-17) [=*Echinolichas parallelobatus* n. sp. of Fagerstrom, 1961], differs in having a *Lobopyge*-type pygidium with a pronounced postaxial ridge, a shorter depressed area behind the median glabellar lobe and longer (exs.) bicomposite lobes (bullar lobes of Chatterton). *Lobopyge docekali* Vanek, 1959, from the Lower Devonian of Bohemia differs in having less reniform bullar lobes. Since *Acanthopyge* and *Lobopyge* are distinguished only on their pygidia, and *A. parvula* and *L. docekali* are so similar in this pattern of segmentation, this generic distinction is not workable.

Occurrence: Devonian: Bolivia-Argentina.

References: Swartz, 1925; Branisa (*in* Ahlfeld and Branisa, 1960); Branisa, 1965; Baldis, 1968.

Specimens: Two cranidia, 2 hypostomes from Chacoma; 2 cephala, 1 pygidium from Pujravi all collected by Branisa.

Family Calymenidae Burmeister, 1843

"Calymene" n. sp. Kozlowski, 1923

Notes: The material on hand is an external mold which has a distinctly weaker preglabellar furrow than does the internal mold from Chacaltaya, Bolivia, illustrated by Kozlowski (1923, Plate 1, Fig. 16). As such differences are common between internal molds and external molds in calymenids and other features seem comparable, the two specimens are considered to represent the same species.

A latex cast of the external mold shows that the entire cephalon is evenly granulose except for the posterior part of the anterior border on which granules become somewhat sparser. The anterior border rises at about 30° from the horizontal and is relatively flat, preglabellar furrow not deeply incised. Glabella is trapezoidal and has three lateral lobes, 3L is weakly developed. There is no evidence of papillae or buttresses. A strong eye ridge runs transversely from a position opposite 3L, eye not preserved.

The generic assignment of this Bolivian calymenid is uncertain. The relative shallowness of the preglabellar furrow suggests that it does not belong to *Calymene*. In the form of its anterior border, it somewhat resembles *Thelecalymene* Whittington, 1970, from the Upper Ordovician of Iowa, but lacks other morphological characteristics of that genus. The Bolivian calymenid may represent a new genus but better material is required for its definition.

Calymene boettneri Harrington, 1950, from the Silurian of Paraguay has a similar preglabellar field but better differentiated glabellar lobes and no strong eye ridge.

Occurrence: Devonian: Bolivia, Huamampampa Formation, Icla-Cha-Kjeri, with *Phacops (Viaphacops) salteri.*

References: Kozlowski, 1923.

Material studied: One slightly distorted, largely complete specimen, collected by P. Isaacson.

Class Merostomata

Suborder Synziphosurina

Type species: *Legrandella lombardii* Eldredge, 1974.

Notes: This is the sister taxon of *Weinbergina optizi* from the Hunsruckschiefer, Germany.

Occurrence: Devonian: Bolivia, Icla Formation, with *Francovichia, Metacryphaeus giganteus,* and others, Aiquile region, Cochabamba Department.

References: Eldredge, 1974.

Specimens: AMNH, NMNH Collections, 3 specimens.

Plate 1. (1-3) *Boliviproetus branisai* n. gen. n. sp., AMNH 34615, holotype. Dorsal, anterior, and right lateral views showing bosslike inflated area on postero-median part of glabella, X3.4. AMNH collection B2350, Chacoma, Bolivia. (4-7) *Boliviproetus branisai* n. gen. n. sp. AMNH 34616, paratype (4-5), right lateral and dorsal views of incomplete cephalon, X2.8. AMNH 34617, paratype (6-7), dorsal and right lateral views of badly eroded cranidium showing steeply sloping preglabellar field and depressed border, X3.7 and X2.8. Both specimens from Chacoma, Bolivia. (8-10) "*Australosutura*" n. sp., AMNH 36971 and 36971, hypotype. Lateral and dorsal views of cephalon and partial thorax, X2.8 and X2.8. (9) "*Australosutura*" n. sp., AMNH 36972. Dorsal view of slightly flattened cranidium showing shallow occipital furrow, X2.6.

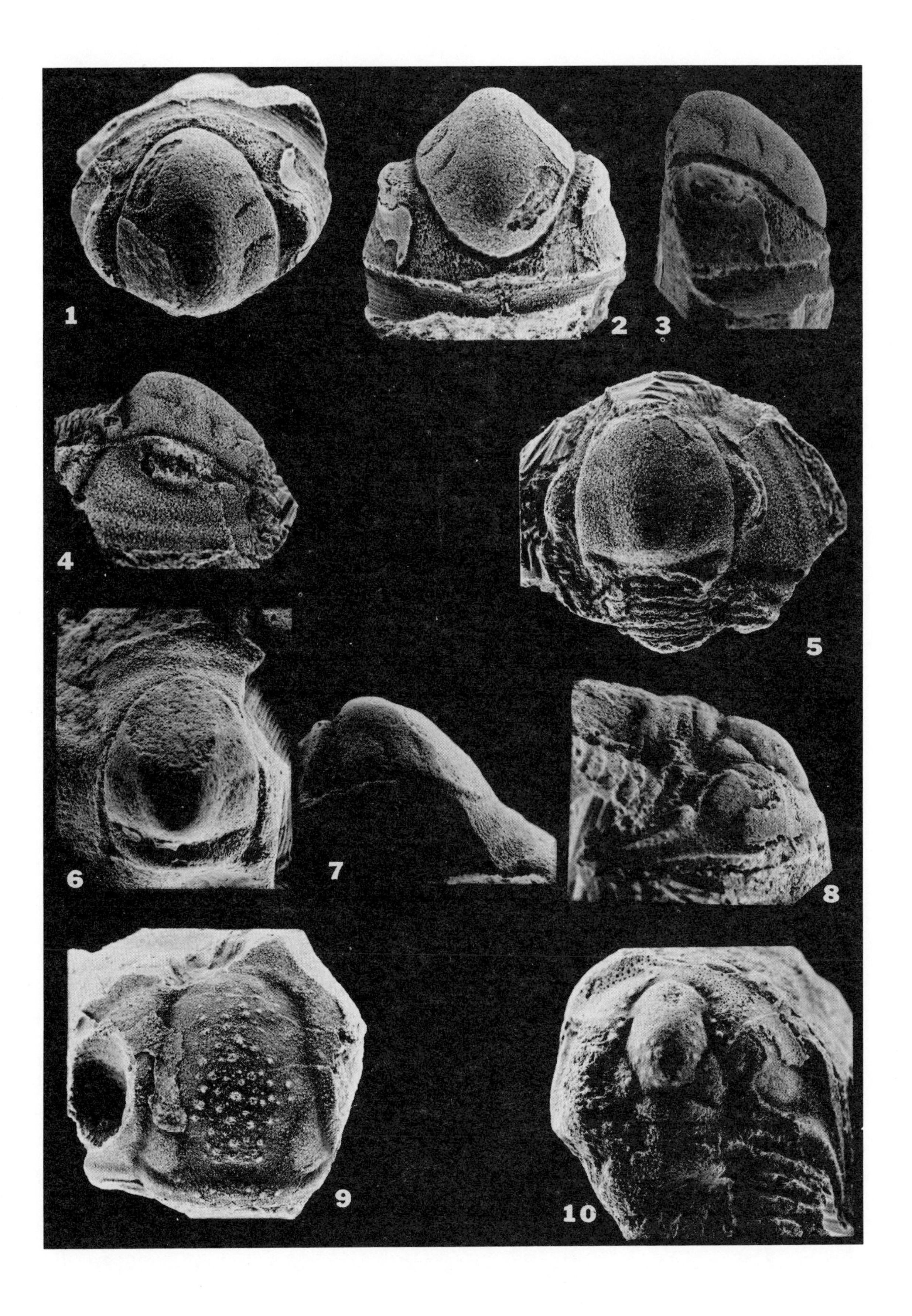
1
2
3
4
5
6
7
8
9
10

A Quantitative Analysis of Lower Devonian Brachiopod Distribution

NORMAN M. SAVAGE, *Department of Geology, University of Oregon, Eugene, Oregon 97403*
DAVID G. PERRY, *Department of Geology, University of British Columbia, Vancouver, B.C., Canada*
ARTHUR J. BOUCOT, *Department of Geology, Oregon State University, Corvallis, Oregon 97331*

ABSTRACT

In any given area, cosmopolitan taxa of marine benthos are generally more abundant as individuals than provincial taxa. Most formulae used to measure faunal affinities fail to allow for the disproportionately large number of cosmopolitan taxa in collections from less thoroughly sampled regions. We have attempted to obviate this difficulty by identifying the cosmopolitan taxa and excluding them from the comparisons.

On the basis of conodont faunas, we have divided Lower Devonian time into four units comprising early Lochkovian, late Lochkovian, early Pragian, and Emsian time. We have listed brachiopod genera from six geographic, not biogeographic, units comprising the Franklinian, Appalachian, Nevadan, Rhenish-Bohemian, Tasman, and Cordilleran regions.

The early Lochkovian brachiopod faunas were more cosmopolitan than in subsequent Lower Devonian time. The late Lochkovian brachiopod faunas are slightly less cosmopolitan than during early Lochkovian time but do not show the marked provinciality characteristic of subsequent Lower Devonian time. The early Pragian brachiopod faunas were much more provincial than those of early or late Lochkovian time. The Emsian brachiopod faunas showed a high degree of provinciality compared with early or late Lochkovian faunas but were slightly less provincial than the preceding early Pragian faunas.

We feel that the results of our quantitative analysis largely support the biogeographic units used previously by Boucot and others. The only modification we suggest is that the Nevadan, Cordilleran, and Franklinian geographic units should form the composite Cordilleran biogeographic unit, except during the early Pragian interval when the Nevadan geographic unit forms part of the Eastern Americas Realm.

The indices of faunal affinity produced by this study do not appear to provide compelling data which might help determine plate tectonic models although they should prove of permissive value as supplemental evidence. Environmental considerations and the spectrum of larval dispersal capabilities are probably as important as geographic proximity in determining benthic marine faunal affinities.

INTRODUCTION

The accumulation of systematic data on Lower Devonian brachiopods has continued for well over 100 years and yet new material is still being described and older species and genera subdivided in an attempt to reflect true taxonomic relationships. Frequently, researchers have endeavored to recognize the affinities between faunas in different parts of the world with the result that more obvious affinities, such as those between the Gondwana brachiopod faunas during Emsian time, have been known for many years. More recently, Boucot and others (1969), Boucot and Johnson (1973), and Johnson and Boucot (1973) presented descriptive summaries of global Lower Devonian brachiopod biogeography and these were later refined and expanded by Boucot (1975). Attempts to use quantitative methods to measure brachiopod faunal affinities are hindered by the subjectiveness and incompleteness of the taxonomic data, and by the constant modification of those data as understanding of Lower Devonian brachiopods moves forward from its present primitive condition. A further difficulty is that, in most areas, some of the brachiopod communities have not been preserved or have not been found. Because of these limitations, the analysis presented herein should be viewed as a preliminary, experimental analysis, rather than as a solidly based permanent statement of Lower Devonian brachiopod affinities.

METHODS OF ANALYSIS

For regional faunal comparisons and determinations of the level of provinciality, it is desirable to use a numerical measure of affinity to rank similarities between the qualitatively defined geographic areas. Calculations of this type compare faunas of the areas concerned at the taxonomic levels deemed reflective of true similarities and differences in the faunas. For these Lower Devonian brachiopod faunal comparisons, the generic-subgeneric level is employed as it is the only level at which there is sufficient data on the distribution of taxa and it is a level where there is general agreement on taxonomic identity among most brachiopod workers.

A method occasionally used in the past, and which may at first sight appear both rational and simple, is to compare the number of genera common to two areas with the total number of elements in those areas. This can be expressed by the formula

$$AI = \frac{C}{N_1 + N_2 - C}$$

where N_1 is the total number of genera in the smaller sample, N_2 is the total number of genera in the larger sample, C is the number of elements common to both samples, and AI is the affinity index. Simpson (1947) noted that, where there is a considerable difference between N_1 and N_2, the number of genera common to both samples (C) will be determined more by the total number of genera in the smaller sample (N_1) than by the total number of genera in the combined samples ($N_1 + N_2 - C$). He therefore modified the formula by using the relationship

$$\frac{C}{N_1}$$

as an index of taxonomic affinity and expressed this as a percentage using the formula

$$AI = \frac{100C}{N_1}$$

This formula has been widely used by other paleontologists (Jackson, 1969; Williams, 1969; Campbell and Davoren, 1972) in the discussion of affinities of Ordovician graptolite, brachiopod, and Devonian trilobite faunas, respectively. A bias is introduced into the affinity indices calculated using this formula by the disproportionately large number of cosmopolitan genera in samples from poorly studied regions. This is the result of the close correlation between local abundance and widespread geographic distribution of cosmopolitan genera. Cosmopolitan genera are defined here as genera occurring in five or six of the geographic areas discussed. Provincial genera are defined as occurring in only one geographic area. Initial collecting in an area will usually result in a sample composed largely of cosmopolitan genera and it is only after exhaustive collecting that representative numbers of the less common provincial genera begin to show up. This relationship is clearly shown in Figure 1 in which the total numbers of taxa known from different biogeographic units during successive Lower Devonian intervals are plotted against percentage of cosmopolitan taxa in those collections. Cheetham and Hazel (1969) present a useful comparison of various similarity coefficients showing the general effects of sample size.

In Figure 2, affinity indices calculated using the formula

$$AI = \frac{C}{N_1 + N_2 - C}$$

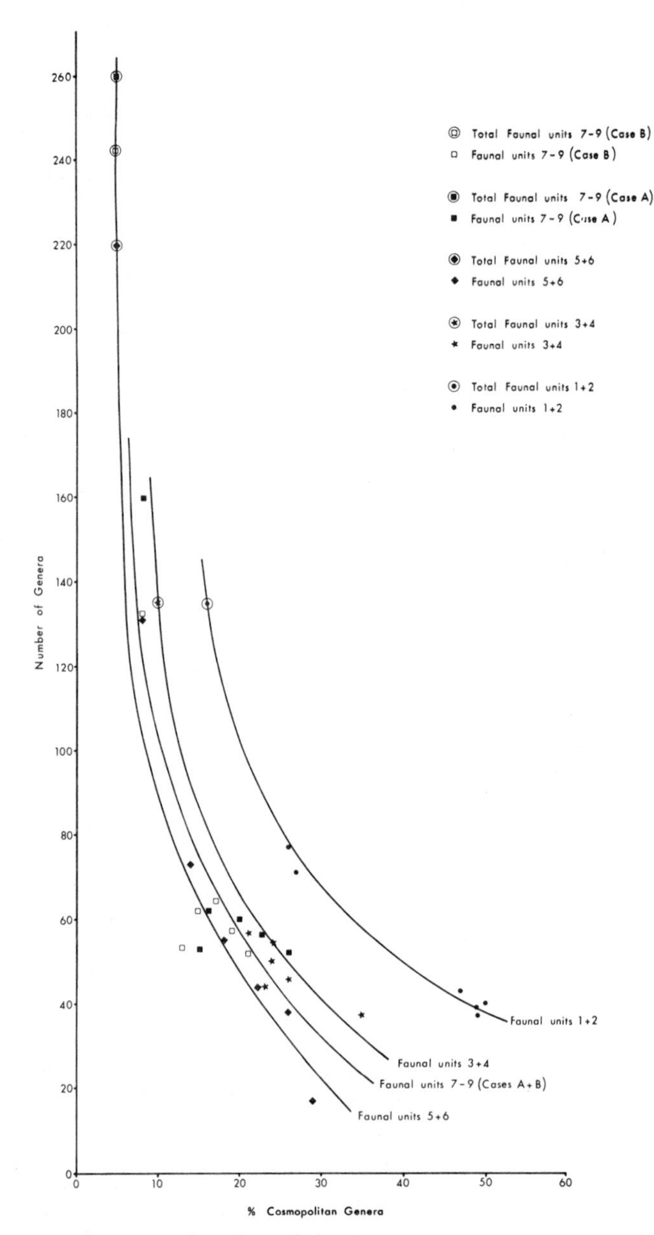

Figure 1. Plot of number of genera against percentage of cosmopolitan genera in successive Early Devonian stages using data from Table 1.

for the data from Table 1 are plotted to show the bias introduced into the affinity index by small sample size. In Figure 3, affinity indices calculated using Simpson's formula

$$AI = \frac{100C}{N_1}$$

for the same data are plotted to show a greatly reduced, but still evident, bias in which a small sample will tend to give an affinity index which is too high. Even with this bias, Simpson's formula will give a useful index of affinity, particularly where the samples in the two areas being com-

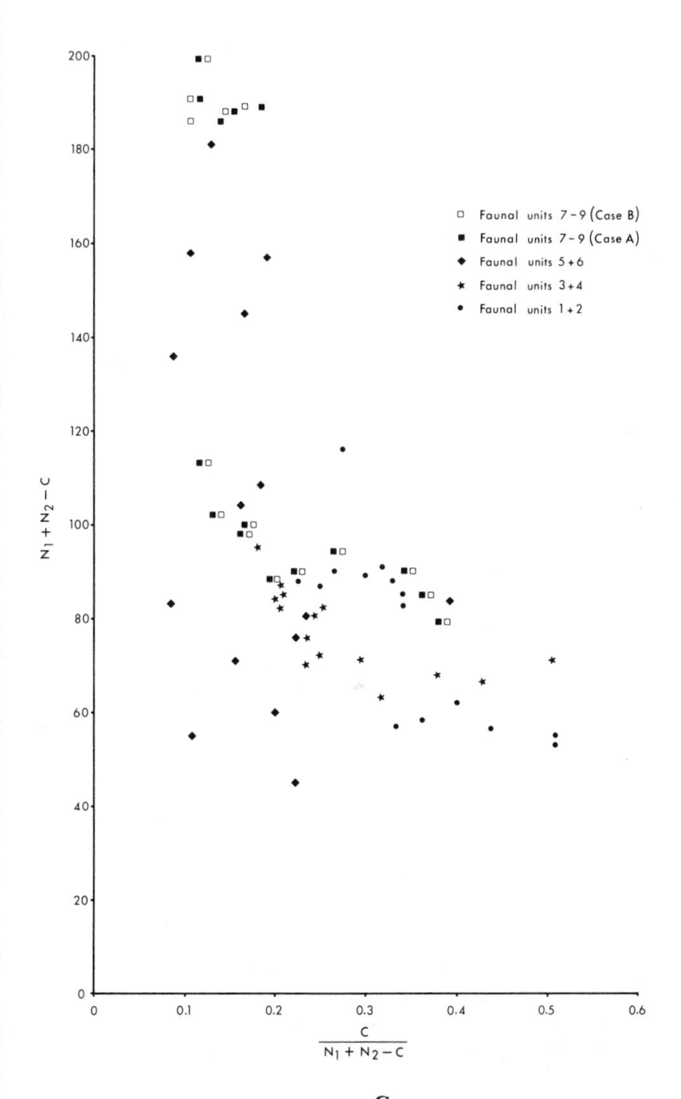

Figure 2. Plot of AI $= \dfrac{C}{N_1+N_2-C}$ against sample size (N_1+N_2-C) using data from Table 1, where AI is the affinity index, N_1 the total number of taxa in the smaller sample, N_2 the total number of taxa in the larger sample, and C the number of taxa common to both samples.

pared are of large size and truly represent the faunas present.

Johnson (1971a) has attempted to measure the affinities of brachiopod faunas by using the formula

$$\mathrm{PI}\ (=\mathrm{AI}) = \frac{C}{2E_1}$$

where C is the number of taxa common to both samples and E_1 is the smaller number of provincial[1] taxa in the two samples being compared. This formula removes cosmopolitan taxa from the denominator yet allows them to remain in the numerator. Because less exhaustively collected areas

[1] The word provincial is preferred over the word endemic because of the conflicting meanings of the latter word.

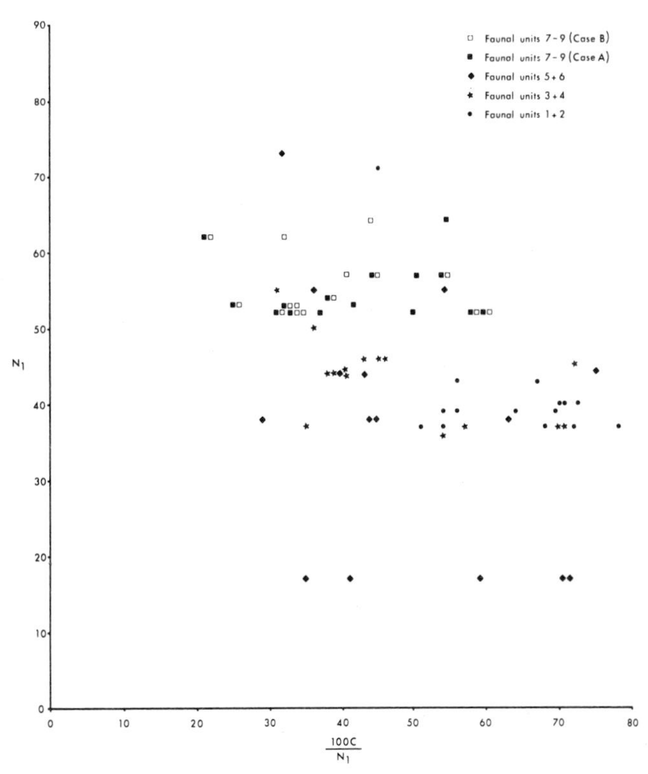

Figure 3. Plot of AI $= \dfrac{100C}{N_1}$ (Simpson's formula) against smaller sample size (N_1) using data from Table 1, where AI is the affinity index and C the number of taxa common to both samples.

will tend to show too few provincial taxa, the denominator in these cases will tend to be misleadingly small and the samples will appear to have a greater affinity than they should. This is clear from Figure 4 in which the data from Table 1 is plotted using the formula

$$\mathrm{PI}(=\mathrm{AI}) = \frac{C}{2E_1}$$

Sando and others (1975) used the Otsuka similarity index in the investigation of North American Mississippian coral zoogeography. This coefficient

$$\frac{C}{\sqrt{N_1N_2}}$$

is merely another variation of the basic Simpson formula except that there is an attempt to reduce the effect of sample size disparities by taking the square root of the product of the number of genera in the samples compared. This formula, like Simpson's, will still be susceptible to the disproportionately large number of cosmopolitan genera in smaller samples. Sando and others also introduce an endemism index which is merely the percentage of provincial genera compared to the total fauna of an area.

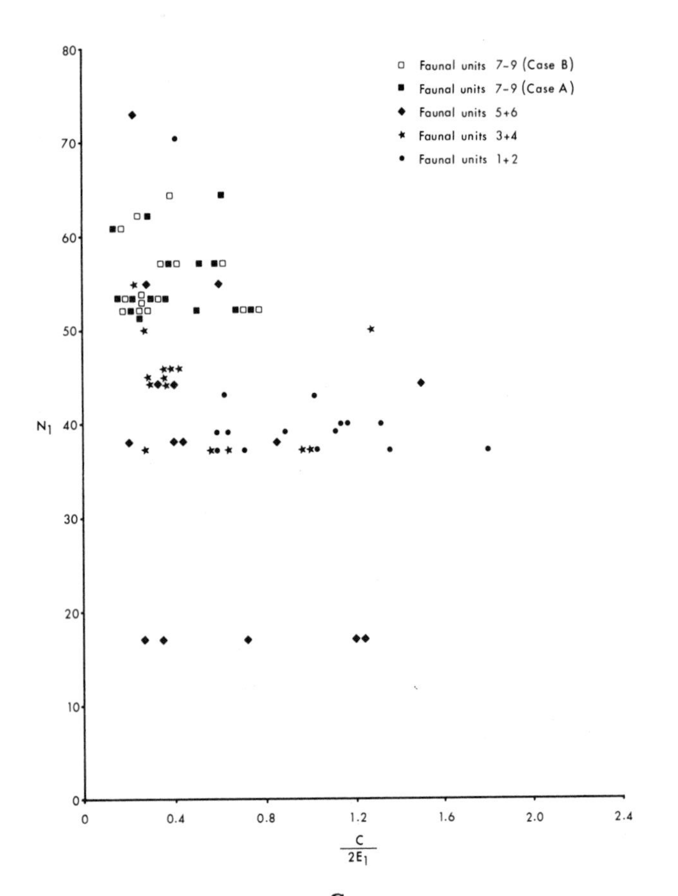

Figure 4. Plot of $AI = \frac{C}{2E_1}$ (Johnson's formula) against smaller sample size (N_1) using data from Table 1, where AI is the affinity index, C is the number of taxa common to both samples, and E_1 the smaller number of provincial taxa in the two samples.

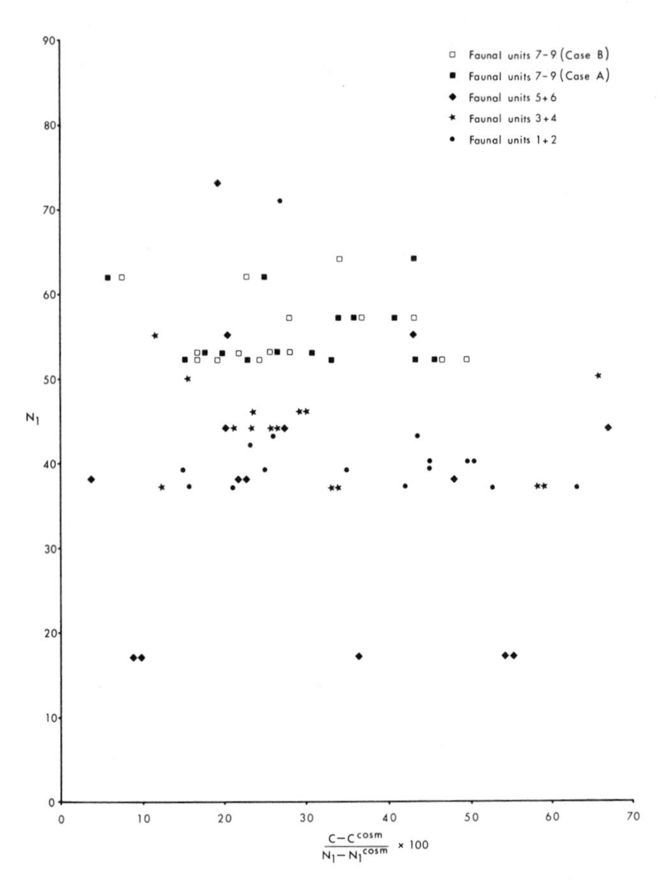

Figure 5. Plot of $AI = \frac{C - C^{cosm}}{N_1 - N_1^{cosm}} \times 100$ (this paper) against smaller sample size (N_1) using data from Table 1, where AI is the affinity index, C is the number of taxa common to both samples, C^{cosm} is the number of cosmopolitan taxa common to both samples, and N^{cosm} is the number of cosmopolitan taxa in the smaller sample.

Another possible method of faunal comparison is to calculate affinity values using the proportion of the shared genera which are biprovincial (known only from the two areas being compared). Taxa common only to certain areas of the world have long been considered significant indicators of biogeographic affinity (e.g., *Mesosaurus, Glossopteris, Eurydesma*) and it seems likely that a large proportion of biprovincial genera is an indication of general faunal affinity. However, when the number of known biprovincial genera is very low the affinity index may be doubled or reduced to nothing by the addition or subtraction of one or two genera. This method may be very useful for large samples but is clearly too clumsy where small samples are being compared.

The method adopted herein involves a modification of Simpson's formula. All cosmopolitan genera are removed from the numerator and the denominator, resulting in the formula

$$AI = \frac{C - C^{cosm}}{N_1 - N_1^{cosm}} \times 100$$

where C is the number of genera common to the two samples, C^{cosm} is the number of cosmopolitan genera common to the two samples, N_1 is the smaller sample, and N_1^{cosm} is the number of cosmopolitan genera in the smaller sample. This formula eliminates the effect of a higher number of cosmopolitan genera in the smaller sample. In essence this formula is removing genera such as *Atrypa, Cyrtina, Howellella,* and *Nucleospira* from the brachiopod affinity comparisons because these elements are found almost everywhere, often in abundance, and are commonly collected first. The method was tested for bias by plotting the data from Table 1. The resulting spread of points appears to indicate that this method is not biased by the smaller sample sizes (Fig. 5).

Many problems exist in analyses of this type because of the nature of the data available. The first problem is to decide which taxonomic level will make the biogeographic similarities and differences most evident. Use of the family level commonly overemphasizes similarities and results in a very low level of biogeographic differentiation on a global scale. Use of the specific level commonly leads to such a

Table 1. Generic list of Lower Devonian brachiopods showing occurrences in six geographic regions and four time intervals. F (Franklinian) Canadian Arctic Archipelago; A (Appalachian) Chihuahua to Gaspé exclusive of Nova Scotia; N (Nevada) Great Basin; R (Rhenish-Bohemian) Europe and North Africa exclusive of Carnic Alps and Urals; T (Tasman) eastern Australia; C (Cordilleran) northern British Columbia, District of Mackenzie and northern Yukon. Time intervals refer to informal conodont faunas of Klapper and others (1971). Superscripts refer to sources of information on brachiopod occurrences cited in references. Only a single reference is given for each entry although many genera are widely reported within one geographic region. Generic categories cited in quotation marks may include more than one generic-level element; where this is done, it should be noted that the effect may be to reduce the level of apparent provincialism. Notes to Table 1 begin on p. 195.

Conodont faunas of Klapper and others (1971)	1 + 2	3 + 4	5 + 6	7 + 8 + 9
Rhenish Stages	GEDINNIAN		SIEGENIAN	EMSIAN
Bohemian Stages	LOCHKOVIAN		PRAGIAN	ZLICHOVIAN
"Acanthospirifer"	F[125]			
Acrospirifer			A[16] N[77] R[1] T[42]	A[16] N[77] R[1] T[121]
Adrenia				T[29]
Aesopomum	A[81] N[84] C[100]	N[72] C[100]	R[59] C[100]	R[59] C[123]
Alatiformia			R[112]	R[152]
Ambocoelia	F[125] R[94] C[100]	F[97] A[16] N[72] T[139] C[100]	A[16] R[55] C[100]	F[81] A[16] N[72] *R[55] C[100]
Ambothyris				T[29]
Amissopecten				R[57]
Amoenospirifer				R[57]
Amphigenia				A[20] R[96]
"Amphistrophia"	A[53]			
Amsdenina	A[161]			
Anastrophia	A[16]	A[16]	A[16] N[107]	
Anatrypa-Desquamatia	C[100]	C[100]	F[118] C[100]	F[90] T[29] C[101]
Anathyris			R[1]	R[55]
Ancillotoechia	F[21] A[16] N[84] R[94] C[100]	F[79] A[16] C[123]	A[16] N[72] C[123]	F[80] N[84]
Ancylostrophia				R[54]
Anoplia	A[23]	A[16]	A[16] N[77]	A[20] N[77]
"Anoplia"				R[106]
Anoplotheca			R[1]	R[111]
Antispirifer			A[16]	
Arduspirifer			R[110]	R[110]
Areostrophia	R[66]	R[32]	R[59]	R[59]
Astutorhyncha	R[32]	R[32]	R[57]	N[72] R[57]
"Astutorhyncha"		F[73]		
Athyrhynchus			C[123]	F[75] N[75] C[75]
Athyris	R[133]	R[133]	R[112] T[42]	N[85] R[111] T[29]
Atrypa	F[125] A[16] N[84] R[32] T[136] C[100]	F[97] A[16] N[72] R[32] T[140] C[100]	F[118] A[16] N[72] R[1] T[42] C[100]	F[117] A[16] N[72] R[111] T[29] C[100]
Atrypina	A[16] N[84] T[136]	A[16] N[72] T[140] C[123]	A[16] C[100]	C[123]
Australirhynchia		T[138]		
Australocoelia		T[159]		
Barbaestrophia	F[125] R[32]	F[125] R[66]		
Bathyrhyncha	R[43]	R[43]		
Baturria	A[23] R[43]	T[11]		

Table 1. (Cont.)

Conodont faunas of Klapper and others (1971)	1 + 2	3 + 4	5 + 6	7 + 8 + 9
Rhenish Stages	GEDINNIAN		SIEGENIAN	EMSIAN
Bohemian Stages	LOCHKOVIAN		PRAGIAN	ZLICHOVIAN
Beachia			A[16]	
Biconostrophia			R[59]	R[59] C[123]
Bifida			R[55]	F[91] N[85] R[55] C[127]
Bisinocoelia			R[55]	R[55]
Bojodouvillina			R[59]	R[59]
Bojothyris			R[56]	*R[56]
Boucotia		T[168]	T[50]	
"Boucotia"	A[12]			
Brachyprion	F[125] N[76] R[32]	F[97] N[72] C[98]	A[33] C[123]	C[123]
Brachyspirifer			R[146]	N[87] R[146]
Brachyzyga	R[94]			
Branikia			R[56]	*R[56]
Browneella				T[29]
Buchanathyris				T[154]
Callicalyptella	N[17]			
Camarium	A[23]			
"Camarella"		F[79]		
Candispirifer			R[61]	*R[61]
Carinagypa		F[79] C[123]		F[80] N[86] C[86]
Carinatina			R[66]	F[80] *R[66] T[104] C[101]
Centronella			A[16]	A[20]
Charionella				A[16]
Charionoides				A[20]
Chonetes			R[5]	N[77] R[110]
"Chonetes"	F[125] A[16] N[76] R[32] T[144] C[99]	F[97] A[16] R[32] T[129]	F[73] T[42]	F[80] A[16] N[77] T[154]
new smooth chonetid				C[123]
Chonostrophia				A[16]
Chonostrophiella	A[20]	A[20]	A[20]	A[20]
Cimicinella				R[95]
Cingulodermis	R[61]	R[61]	R[61]	R[61]
Clorinda	R[32]	R[66] C[100]	R[66]	R[55]
Clorindina			C[123]	
Cloudella		A[16]	A[16]	
Cloudothyris				A[9]
Coelospira	F[15] A[16] N[76] C[100]	A[16] N[72]	A[16] N[72]	A[16] N[72] T[29]
Coelospirina			R[66]	R[55]
Concinnispirifer		A[16]		
Cordatomyonia	A[16]	A[16]		

Table 1. (Cont.)

Conodont faunas of Klapper and others (1971)	1 + 2						3 + 4						5 + 6						7 + 8 + 9					
Rhenish Stages	GEDINNIAN												SIEGENIAN						EMSIAN					
Bohemian Stages	LOCHKOVIAN												PRAGIAN						ZLICHOVIAN					
Cortezorthis							F[79]		N[79]			C[123]						C[123]	F[88]		N[72]			C[101]
Corvinopugnax																					N[72]			
Costellirostra		A[161]						A[16]						A[16]						A[16]				
Costellispirifer																				A[9]				
Costisorthis																R[64]						*R[64]		
Costispirifer														A[16]	N[72]					A[16]				
Crinistrophia																R[66]						R[28]		
Crurithyris																			F[80]					
Cryptatrypa			N[84]	R[81]		C[123]					T[141]	C[123]						C[123]						C[123]
Cryptonella																				A[16]	N[87]			
"Cryptonella"																R[112]						R[111]		
Cumberlandina														A[10]										
"Cupularostrum"		A[24]						A[16]										C[100]	F[81]	A[16]				
Cycladigera																R[62]						R[62]		
Cydimia																							T[29]	
Cymostrophia						C[100]	F[125]				T[141]	C[100]				R[59]	T[42]	C[100]	F[117]		N[72]	*R[59]	T[29]	
Cyrtina	F[21]	A[16]	N[84]	R[94]	T[144]	C[100]	F[97]	A[16]	N[72]		T[139]	C[100]	F[118]	A[16]	N[72]	R[110]	T[42]	C[100]	F[90]	A[16]	N[72]	R[111]	T[29]	C[100]
Cyrtinaella							F[125]		N[72]			C[123]						C[123]		A[16]				
Cyrtoniscus																				A[20]				
Dalejina	F[125]	A[16]	N[84]	R[94]	T[136]	C[100]	F[97]		N[72]			C[100]			N[72]	R[55]		C[100]	F[90]		N[72]	R[55]	T[29]	C[100]
Dalejodiscus																R[59]						R[59]		
Davidsoniatrypa																		C[100]						
Dawsonelloides														A[15]										
"Decoropugnax"												C[98]												
Delthyris		A[16]																						
"Delthyris"																							T[29]	
Dicoelosia		A[16]	N[84]	R[94]	T[136]	C[100]		A[16]								R[55]						*R[55]		
new dicoelosiid																								C[123]
Dichozygopleura				R[133]						R[133]														
Dictyonella																R[55]						*R[55]		
Dinapophysia																R[105]						R[37]		
Discomyorthis		A[16]						A[16]						A[16]	N[72]					A[16]				
Dolerorthis		A[16]		R[6]	T[136]	C[100]	F[79]		N[72]		T[141]	C[100]					T[170]	C[100]					T[121]	
Dubaria			N[84]	R[67]			F[97]		N[72]							R[66]						R[66]		
Duryeella														A[10]						A[10]				
Dyticospirifer															N[72]									
Eatonia		A[16]						A[16]						A[16]										
Elytha														A[16]	N[77]					A[16]	N[77]			

Table 1. (Cont.)

Conodont faunas of Klapper and others (1971)	1 + 2						3 + 4						5 + 6						7 + 8 + 9					
Rhenish Stages	GEDINNIAN												SIEGENIAN						EMSIAN					
Bohemian Stages	LOCHKOVIAN												PRAGIAN						ZLICHOVIAN					
Elythyna																			F[80]		N[85]		T[104]	C[80]
Eodevonaria																R[110]				A[20]		R[110]		
Eoglossinotoechia			N[84]	R[32]						R[32]	T[136]					R[57]	T[42]					R[57]	T[29]	
Eoreticularia																R[56]						R[56]		
Eoschuchertella	F[21]	A[16]	N[84]	R[94]	T[144]	C[100]	F[97]	A[16]	N[72]	R[133]	T[141]	C[100]		A[16]	N[72]	R[110]	T[42]	C[100]	F[80]	A[16]	N[72]	R[110]	T[29]	C[100]
Eospirifer				R[32]	T[136]					R[32]	T[156]					R[56]	T[129]					R[56]		
Etymothyris																				A[20]				
Eucharitina																R[57]						R[166]		
Eurekaspirifer																					N[72]			
Euryspirifer																						R[112]		
Eurythyris														A[16]						A[16]				
Falsatrypa																						R[60]		
Fascicostella				R[94]												R[65]						R[163]		
Felinotoechia																R[57]						*R[57]		
Fibulistrophia																						R[45]		
Fimbrispirifer																				A[16]				
"*Fimbrispirifer*"																			F[81]					
"*Franklinella*"							F[97]					C[123]						C[123]						
Fulcriphoria				R[27]												R[27]						R[27]		
Gladiostrophia																R[59]						R[59]		
Globithyris														A[16]						A[20]			T[40]	
Glossinotoechia																R[57]						R[57]		
Glossinulus																						R[55]		
Glossinulina																			F[80]					
Glossoleptaena				R[32]						R[32]														
Gorgostrophia				R[66]												R[59]						*R[59]		C[101]
Grayina	F[125]		N[84]	R[32]	T[145]	C[123]	F[79]		N[72]		T[141]	C[123]						C[100]						
Gypidula	F[125]	A[16]	N[84]	R[66]	T[136]	C[100]	F[97]	A[16]	N[72]	R[66]	T[141]	C[100]	F[118]	A[16]		R[66]	T[42]	C[100]	F[80]		N[72]	*R[66]		C[123]
Gypidulina																R[66]						R[89]		
Hanusatrypa																						R[60]		
Hebetoechia	F[125]		N[84]	R[32]		C[123]				R[32]		C[98]	F[81]			R[57]	T[42]					*R[57]		
Hedeina		A[16]						A[16]			T[168]			A[16]							N[85]			
(*Macropleura*)																								
Hipparionyx														A[16]										
"*Hipparionyx*"																R[5]	T[42]					R[1]		
Howellella	F[21]	A[16]	N[84]	R[94]	T[129]	C[100]	F[125]	A[16]	N[72]	R[32]	T[139]	C[100]	F[118]	A[16]	N[72]	R[56]	T[156]	C[123]	F 90	A[16]	N[72]	*R[56]	T[29]	C[123]
Howittia															N[77]							R[162]	T[29]	C[123]
Hysterolites																R[110]								

Table 1. (Cont.)

Conodont faunas of Klapper and others (1971)	1 + 2	3 + 4	5 + 6	7 + 8 + 9
Rhenish Stages	GEDINNIAN		SIEGENIAN	EMSIAN
Bohemian Stages	LOCHKOVIAN		PRAGIAN	ZLICHOVIAN
"Hysterolites"				F[81] N[85] T[29] C[123]
Iberirhynchia				R[39]
Innuitella				N[75] C[36]
Iridistrophia	F[125] A[16] N[84] R[32]	F[79] A[16] R[32] T[156]	F[81] A[16] R[59] C[123]	R[70]
Isopoma		C[123]	R[57]	*R[57]
Isorthis (Isorthis)	F[125] R[94]	R[66]	T[42]	T[29]
Isorthis (Protocortezorthis)	F[125] R[32] T[136] C[100]	F[97] R[32] T[141] C[100]	R[164] T[164] C[100]	R[164] C[123]
Isorthis (Arcualla)	A[164]	A[164]	A[164]	
Isorthis (Tyersella)	F[125] A[164] N[84] T[164] C[98]	A[164] T[164]	A[164] T[164]	F[81] R[164]
Ivanothyris	A[23] R[32]		T[170]	T[151]
"Janius"				C[98]
Katunia		F[79] N[72] C[100]	C[100]	
Kayserella			R[65] C[100]	*R[65] C[123]
Kransia				R[166]
Kymatothyris				R[153]
Lanceomyonia	R[32]	R[32]		F[80]
Latonotoechia	R[66]	R[66]	R[57] C[100]	R[57]
"Leiorhynchus"			F[118]	
Lepidoleptaena	N[84] C[123]		R[59]	R[59]
Leptaena-Leptagonia	F[125] A[16] N[84] R[32] T[156] C[100]	F[79] A[16] N[72] R[59] T[141] C[100]	A[16] N[72] R[59] T[121] C[100]	F[91] A[16] N[72] R[55] T[121] C[100]
Leptaenisca	A[16] N[84]	A[16] N[72]	A[16] N[72] R[59] T[14] C[123]	A[16] R[59] C[98]
Leptaenopyxis		F[90]	R[59]	R[28] T[151]
Leptathyris				F[80]
Leptocoelia			A[16]	A[16]
Leptocoelina			N[78]	
Leptodonta			R[132]	R[132]
Leptospira	A[16]			
"Leptostrophia"	A[53] N[84] R[32] T[144] C[100]	A[53] N[72] R[59] T[156]	A[53] N[72] R[53] T[168] C[123]	N[72] R[70] T[29] C[123]
Levenea	A[164]	F[81] A[164] N[72]	A[164] N[72]	A[164] N[72]
Lievinella	R[10]	R[10]		
Linguopugnoides	F[81] R[32] T[136]	F[97] R[32] T[141] C[100]	R[57] C[123]	R[57] C[127]
Lissatrypa	A[23] R[94] T[144] C[98]	N[72] T[156]	T[168]	R[55] C[123]
Lissopleura	A[16]			
Llanoella		A[16] N[72]	N[72]	
Machaeraria	F[125] A[16] R[141] T[136] C[100]	F[79] A[16] N[72] T[141] C[100]	A[16] T[42]	A[16] T[150]
Malurostrophia				T[29]

Table 1. (Cont.)

Conodont faunas of Klapper and others (1971)	1 + 2						3 + 4						5 + 6						7 + 8 + 9					
Rhenish Stages	GEDINNIAN												SIEGENIAN						EMSIAN					
Bohemian Stages	LOCHKOVIAN												PRAGIAN						ZLICHOVIAN					
Maoristrophia					T[144]						T[49]						T[129]						T[20]	
Markitoechia																						R[57]		
"McLearnites"				R[53]						R[53]											N[72]		T[151]	
Megakozlowskiella		A[16]	N[84]	R[32]	T[144]			A[16]	N[72]	R[32]				A[16]	N[72]					A[16]	N[72]		T[151]	
Meganterella																				A[16]				
Meganteris																R[110]						R[110]		
Megasalopina														A[16]										
Megastrophia												C[123]			N[72]			C[123]	F[91]	A[53]	N[72]			C[123]
Mendathyris														A[16]										
Merista		A[16]		R[94]						R[31]						R[55]						R[55]		
Meristella-Meristina	F[125]	A[16]	N[84]	R[94]		C[100]		A[16]	N[72]	R[31]	T[141]	C[123]		A[16]	N[72]	R[55]	T[42]	C[123]	F[73]	A[16]	N[72]	*R[55]		
"Mesodouvillina"	F[21]	A[53]	N[84]	R[32]	T[144]	C[100]	F[97]	A[53]	N[72]	R[32]	T[156]	C[123]		A[53]		R[59]	T[121]	C[100]				R[53]		C[123]
"Mesopholidostrophia"	F[125]			R[94]																				
Metaplasia		A[23]	N[84]											A[16]	N[72]					A[16]	N[72]			
Micidus																							T[29]	
Molongella				R[65]	T[144]																			
Molongia					T[156]						T[156]													
Monadotoechia																R[57]						*R[57]		
Mucrospirifer																				A[16]		R[149]		
Multispirifer																R[10]								
Muriferella					T[144]	C[123]	F[97]					C[100]	F[118]			R[66]	T[42]	C[100]	F[90]		N[88]	R[66]	T[29]	C[100]
Mutationella				R[94]						R[133]				A[16]		R[1]					N[72]	R[1]		
Mystrophora																								C[123]
Nadiastrophia																	T[42]						T[151]	
Najadospirifer																R[61]						R[8]		
Nanothyris		A[16]						A[16]						A[16]						A[16]				
Notanoplia					T[144]						T[20]						T[20]						T[20]	
Notoconchidium					T[20]						T[20]													
Notoleptaena					T[144]						T[156]						T[48]							
Notoparmella	F[21]		N[76]			C[123]	F[125]		N[123]			C[123]												
Nucleospira	F[125]	A[16]	N[84]	R[94]	T[156]	C[123]	F[97]	A[16]	N[72]	R[133]	T[141]	C[100]	F[118]	A[16]	N[72]	R[110]	T[155]	C[100]	F[91]	A[16]	N[72]	R[55]		C[123]
Nymphorhynchia				R[32]						R[66]		C[100]				R[57]		C[100]			N[87]	R[57]		C[100]
Obturamentella		A[16]						A[16]																
Ogilviella					T[144]	C[100]	F[79]				T[140]	C[100]												
Oligoptycherhynchus																R[110]						R[110]		
Oriskania														A[16]	N[78]									
Orthostrophonella		A[23]													N[77]									
Orthostrophia		A[16]						A[16]						A[16]										

Table 1. (Cont.)

Conodont faunas of Klapper and others (1971)	1 + 2					3 + 4					5 + 6					7 + 8 + 9					
Rhenish Stages	GEDINNIAN										SIEGENIAN					EMSIAN					
Bohemian Stages	LOCHKOVIAN										PRAGIAN					ZLICHOVIAN					
Pacificocoelia				T[144]			A[22]				A[22]	N[72]					A[16]	N[72]		T[169]	
Papillostrophia													R[59]						R[59]		
Parachonetes													R[72]			F[71]		N[72]	*R[72]	T[29]	C[100]
Parapholidostrophia																F[74]					C[123]
Paraspirifer																			R[112]		
Pegmarhynchia											A[16]	N[72]									
Peleicostella													R[62]						R[62]		
Pentagonia																	A[16]				
Pentamerella																	A[16]	N[85]			C[124]
"*Phoenicitoechia*"										C[123]			R[57]						*R[57]		
"*Pholidostrophia*"		A[53]								C[98]	A[20]	N[72]	R[53]				A[20]	N[77]	R[53]		
Pinguispirifer						F[79]							R[61]						*R[61]		
"*Phragmophora*"																					C[123]
Phragmostrophia										C[100]					C[100]	F[54]		N[72]			C[100]
Planicardinia									T[137]												
Platyorthis		A[16]	R[94]				A[16]	R[38]			A[16]		R[38]				A[16]		R[110]		
Plebejochonetes													R[110]						R[110]		
Plectodonta		A[23]	R[32]	T[144]			A[16]	R[32]	T[151]					T[155]					R[95]		
Plectorhynchella																			R[55]		
Plectospira			R[66]										R[66]						R[55]		
Pleiopleurina											A[16]	N[72]									
Plethorhyncha			R[32]			F[79]		R[32]			A[16]					F[81]					
Plicanoplia											A[16]						A[16]		R[28]		
Plicanoplites													R[63]						*R[63]		
Plicocyrtina													R[66]		C[100]				R[55]	T[29]	
Plicodevonaria																			R[15]		
Plicoplasia		A[81]								C[100]	A[20]	N[72]			C[100]	F[81]	A[16]				
Plicostropheodonta																			R[112]		
Podolella		A[16]	R[94]										R[4]								
Pradoia																			R[34]		
Praegnantenia													R[57]						*R[57]		
Prionothyris											A[16]						A[20]				
Procerulina						F[79]							R[2]						*R[2]		
Productella																	A[16]		R[72]		
Prokopia													R[55]						R[55]	T[158]	C[123]
Prorensselaeria											A[16]										
Proreticularia									T[139]				R[61]			F[98]			*R[61]		
Proschizophoria			R[27]					R[27]					R[112]								
Protathyris	F[21]		R[94]		C[100]	F[97]				C[100]					C[100]						C[100]

Table 1. (Cont.)

Conodont faunas of Klapper and others (1971)	1 + 2						3 + 4						5 + 6						7 + 8 + 9					
Rhenish Stages	GEDINNIAN												SIEGENIAN						EMSIAN					
Bohemian Stages	LOCHKOVIAN												PRAGIAN						ZLICHOVIAN					
"Protochonetes"																			F[80]				T[29]	
Protoleptostrophia														A[20]		R[59]				A[20]		R[59]		
Pseudoparazyga		A[72]						A[72]						A[72]	N[72]									
Ptychopleurella																R[66]						R[55]		
"Pugnax"																							T[29]	
Punctatrypa																R[66]			F[80]			R[55]		
Quadrifarius				R[6]																				
Quadrikentron														A[16]						A[16]				
Quadrithyrina																						R[66]	T[29]	
Quadrithyris					T[144]				N[72]		T[139]					R[56]			F[90]			R[56]		
Quasimartinia																R[56]						*R[56]		
Radiomena																			F[54]					C[101]
"Reeftonia"													F[118]										T[20]	
Rensselaeria								A[16]						A[16]	N[72]						N[72]			
Rensselaerina		A[16]						A[16]							N[72]									
Resserella		A[16]	N[84]	R[32]	T[144]	C[98]				R[65]		C[123]				R[66]	T[163]					R[38]	T[29]	
Reticulatrypa						C[123]					T[140]	C[100]												C[126]
Rhenorensselaeria																R[112]				A[13]		R[110]		
Rhenostrophia																R[7]						R[7]		
Rhenothyris																						R[153]		C[127]
"Rhynchotreta"		A[24]																						
"Rhynchospirina"		A[16]	N[84]	R[94]		C[123]		A[16]				C[123]		A[16]		R[66]						R[55]		
Rugoleptaena																R[59]						R[28]		
Salopina		A[165]		R[165]										A[16]						A[16]	N[72]			
Salopina (*crassiformis* type)	F[21]		N[84]	R[94]		C[100]	F[125]					C[100]						C[100]					T[29]	C[123]
Schizophoria	F[21]	A[16]	N[84]	R[94]	T[136]	C[123]	F[97]	A[16]	N[72]	R[133]		C[123]	F[118]	A[16]	N[72]	R[1]	T[42]	C[123]	F[91]	A[16]	N[72]	R[112]		C[127]
Septachonetes																							T[29]	
"Septalaria"																R[66]						R[66]		
Septathyris																R[20]						R[20]		
Septatrypa				R[94]																				
Shaleria				R[6]																				
"Sibirispira"	F[125]					C[123]																		C[123]
Sicorhyncha																R[57]						R[57]		
Sieberella		A[9]	N[84]					A[16]	N[72]			C[100]			N[72]	R[55]		C[100]				R[55]		C[100]
Skenidioides	F[125]		N[84]	R[94]	T[144]	C[100]	F[97]			R[66]		C[100]	F[118]			R[55]		C[100]	F[90]			R[55]		C[100]
Skenidium		A[161]						A[16]																
"Sphaerirhynchia"	F[73]	A[16]	N[84]	R[113]				A[16]						A[16]			T[155]							

Table 1. (Cont.)

Conodont faunas of Klapper and others (1971)	1 + 2						3 + 4						5 + 6						7 + 8 + 9					
Rhenish Stages	GEDINNIAN												SIEGENIAN						EMSIAN					
Bohemian Stages	LOCHKOVIAN												PRAGIAN						ZLICHOVIAN					
Spinatrypa				R[32]	T[144]	C[98]	F[79]		N[72]	R[66]	T[156]		F[118]			R[66]	T[155]	C[100]	F[80]	A[16]		R[95]	T[155]	
Spinatrypina												C[98]						C[100]						C[123]
Spinella																					N[72]		T[29]	
Spinoplasia		A[16]						A[16]						A[16]	N[72]									
Spinulicosta																			F[80]		N[72]		T[29]	C[100]
new spiriferid ("*Lenzia*")																								C[126]
Spirigerina	F[125]		N[84]		T[144]	C[100]	F[97]		N[72]		T[140]	C[100]												
Spirinella																			F[80]					
Spurispirifer				R[32]															F[80]					
Stegerhynchus		A[161]					F[81]					C[123]												
Straelenia																R[1]						R[1]		
Striispirifer				R[66]																				
Strixella		A[16]																						
"*Strophochonetes*"		A[16]		R[94]			F[97]	A[16]		R[47]		C[123]			N[72]	R[47]		C[100]	F[81]		N[72]	R[47]		C[101]
"*Strophodonta*"														A[53]	N[72]	R[34]		C[123]	F[90]	A[53]	N[72]	R[53]		C[123]
"*Strophonella*"	F[81]	A[53]	N[84]	R[32]	T[129]		F[81]	A[53]	N[72]	R[32]	T[156]			A[53]	N[72]	R[59]		C[123]	F[90]	A[53]	N[72]	R[59]		C[123]
Struveina																R[10]						R[10]		
Sturtella											T[141]													
Subcuspidella																R[109]						R[111]		
Taemostrophia																							T[29]	
Tastaria																R[59]						R[59]		
Teichertina																						R[66]		
Teichostrophia																R[53]						R[53]		
Tenellodermis				R[32]																				
Tetratomia																R[66]						R[57]		
Tenuicostella																R[110]						R[110]		
Thliborhynchia																		C[100]						
"*Thliborhynchia*"							F[79]					C[123]												
Tomheganella																				A[16]				
Toquimaella							F[79]		N[72]			C[100]												
Trematospira								A[16]			T[156]			A[16]	N[72]									
Triathyris																						R[34]		
Trigonirhynchia								A[16]					F[118]			R[166]					N[72]	R[111]		C[127]
Tropidoleptus																R[5]						R[110]		
Tubulostrophia																R[59]						*R[59]		
"*Uncinulus*"	F[81]	A[16]					F[81]	A[16]						A[16]		R[57]	T[42]					R[112]	T[41]	

Table 1. (Cont.)

Conodont faunas of Klapper and others (1971)	1 + 2			3 + 4			5 + 6				7 + 8 + 9		
Rhenish Stages			**GEDINNIAN**					**SIEGENIAN**				**EMSIAN**	
Bohemian Stages			**LOCHKOVIAN**					**PRAGIAN**				**ZLICHOVIAN**	
Vagrania									C[100]	F[19]			C[127]
Vandercammenina								R[56]				R[10]	
Velostrophia								R[59]				*R[59]	
Warrenella-Reticulariopsis	F[125]	C[123]	F[97]	N[72]	C[123]	F[118]	N[72]		C[100]	F[81]	N[72]	R[114]	C[102]
Werneckeella					C[100]				C[100]	F[90]	N[77]		
Xana								R[44]				R[44]	
Xenomartinia								R[55]				R[55]	
Zdimir												R[34]	
Zlichorhynchus								R[66]				R[58]	

Table 2. Comparative brachiopod distribution data for combinations of the six geographic regions and four time intervals. Affinity indices are calculated using Simpson's formula, Johnson's formula, and $\frac{C-C^{cosm}}{N_1-N_1^{cosm}} \times 100$. Superscripts indicate ranking of affinity by each method. Interval 7 to 9^A includes the case where genera known only from the Bohemian Pragian part of the Rhenish-Bohemian unit are listed in both intervals 5+6 and 7 to 9. Interval 7 to 9^B includes the case where these genera are listed in only interval 5+6.

COMBINED GEOGRAPHIC REGIONS	BIPROVINCIALS	COMMON GENERA	COSMOPOLITANS	SIMPSON INDEX	JOHNSON INDEX	$\frac{C-C^{cosm}}{N_1-N_1^{cosm}} \times 100$	BIPROVINCIALS	COMMON GENERA	COSMOPOLITANS	SIMPSON INDEX	JOHNSON INDEX	$\frac{C-C^{cosm}}{N_1-N_1^{cosm}} \times 100$
	INTERVAL 1 + 2						INTERVAL 3 + 4					
FA	1	20	16	54.1^{12}	0.59^{12}	21.1^{13}	1	18	12	36.0^{13}	0.28^{13}	15.8^{13}
FN	0	25	17	67.6^{7}	1.04^{7}	42.1^{8}	1	26	12	70.3^{3}	1.18^{3}	58.3^{3}
FR	3	29	17	78.4^{1}	1.81^{1}	63.2^{1}	2	18	9	40.9^{9}	0.35^{9}	26.5^{9}
FT	0	19	16	51.3^{14}	0.53^{14}	15.8^{14}	0	21	11	45.7^{7}	0.36^{8}	29.4^{7}
FC	2	27	17	73.0^{2}	1.35^{2}	52.6^{2}	7	36	11	72.0^{1}	1.28^{1}	65.8^{1}
AN	3	28	18	70.0^{4}	1.16^{4}	50.0^{4}	3	21	13	56.8^{4}	0.66^{4}	33.3^{4}
AR	7	32	18	45.1^{15}	0.41^{15}	26.9^{10}	1	17	10	38.6^{12}	0.31^{12}	20.6^{12}
AT	0	22	17	56.4^{10}	0.65^{10}	25.0^{12}	3	20	12	43.5^{8}	0.38^{7}	23.5^{10}
AC	1	24	18	53.8^{13}	0.58^{13}	15.0^{15}	1	17	12	30.9^{15}	0.22^{15}	11.9^{15}
NR	2	29	19	72.5^{3}	1.31^{3}	50.0^{3}	0	13	10	35.1^{14}	0.27^{14}	12.5^{14}
NT	0	21	18	53.8^{13}	0.58^{13}	15.0^{15}	2	20	12	54.1^{5}	0.59^{5}	33.3^{5}
NC	1	28	19	70.0^{5}	1.16^{5}	45.0^{5}	1	26	12	70.3^{2}	1.18^{2}	58.3^{2}
RT	2	27	18	69.2^{6}	1.12^{6}	45.0^{6}	2	18	9	40.9^{10}	0.35^{10}	26.5^{8}
RC	0	29	19	67.4^{8}	1.03^{8}	43.4^{7}	4	17	9	38.6^{11}	0.31^{11}	23.5^{11}
TC	2	25	18	64.1^{9}	0.89^{9}	35.0^{9}	2	21	11	45.7^{6}	0.42^{6}	29.4^{6}

COMBINED GEOGRAPHIC REGIONS	BIPROVINCIALS	COMMON GENERA	COSMOPOLITANS	SIMPSON INDEX	JOHNSON INDEX	$\frac{C-C^{cosm}}{N_1-N_1^{cosm}} \times 100$	BIPROVINCIALS	COMMON GENERA	COSMOPOLITANS	SIMPSON INDEX	JOHNSON INDEX	$\frac{C-C^{cosm}}{N_1-N_1^{cosm}} \times 100$
	INTERVAL 5 + 6						INTERVAL 7 + 8 + 9^A					
FA	0	7	6	41.2^{10}	0.35^{10}	9.1^{14}	2	16	10	30.8^{13}	0.22^{13}	15.4^{14}
FN	0	6	5	35.3^{13}	0.27^{13}	9.1^{15}	2	30	13	57.7^{2}	0.68^{2}	43.6^{2}
FR	1	12	6	70.6^{3}	1.20^{3}	54.5^{3}	4	26	13	50.0^{6}	0.50^{6}	33.3^{7}
FT	1	10	6	58.8^{5}	0.71^{5}	36.4^{6}	1	17	8	32.7^{11}	0.24^{11}	23.1^{11}
FC	1	12	6	70.6^{2}	1.20^{2}	54.5^{2}	3	31	13	59.6^{1}	0.74^{1}	46.2^{1}
AN	17	33	10	75.0^{1}	1.50^{1}	67.6^{1}	6	25	10	43.9^{7}	0.39^{7}	34.1^{6}
AR	4	23	11	31.5^{14}	0.23^{14}	19.4^{13}	8	23	10	37.1^{10}	0.29^{10}	25.0^{10}
AT	3	17	11	44.7^{8}	0.40^{8}	22.2^{9}	2	13	5	24.5^{14}	0.16^{14}	17.8^{13}
AC	2	20	11	36.4^{12}	0.29^{12}	20.5^{11}	0	13	10	21.0^{15}	0.13^{15}	5.8^{15}
NR	0	17	10	38.6^{11}	0.31^{11}	20.6^{10}	4	29	13	50.8^{5}	0.52^{5}	36.4^{5}
NT	0	11	10	28.9^{15}	0.20^{15}	3.7^{15}	1	20	8	37.7^{9}	0.30^{9}	26.7^{9}
NC	1	19	10	43.2^{9}	0.38^{9}	26.5^{7}	2	31	13	54.4^{4}	0.60^{4}	40.9^{4}
RT	5	24	11	63.2^{4}	0.86^{4}	48.2^{4}	6	22	8	41.5^{8}	0.35^{8}	31.1^{8}
RC	6	30	11	54.6^{6}	0.60^{6}	43.2^{5}	10	35	13	54.6^{3}	0.60^{3}	43.1^{3}
TC	1	17	11	44.7^{7}	0.40^{7}	22.2^{8}	1	17	8	32.1^{12}	0.24^{12}	20.0^{12}

Table 2. (Cont.)

COMBINED GEOGRAPHIC REGIONS	BIPROVINCIALS	COMMON GENERA	COSMOPOLITANS	SIMPSON INDEX	JOHNSON INDEX	$\frac{C-C^{cosm}}{N_1-N_1^{cosm}} \times 100$
	INTERVAL $7 + 8 + 9^B$					
FA	2	16	9	30.7^{13}	0.22^{13}	17.1^{14}
FN	2	30	11	57.7^{2}	0.68^{2}	46.3^{2}
FR	3	18	10	34.6^{8}	0.26^{8}	19.5^{12}
FT	1	17	7	32.7^{10}	0.24^{10}	24.4^{9}
FC	3	31	11	59.6^{1}	0.74^{1}	48.8^{1}
AN	6	25	9	43.8^{4}	0.39^{4}	34.8^{4}
AR	8	20	8	32.3^{11}	0.24^{11}	22.6^{10}
AT	2	13	5	24.5^{14}	0.16^{14}	17.4^{13}
AC	0	13	9	21.0^{15}	0.13^{15}	7.5^{15}
NR	4	23	10	40.4^{6}	0.34^{6}	28.3^{6}
NT	1	20	7	37.7^{7}	0.30^{7}	28.3^{7}
NC	2	31	11	54.4^{3}	0.60^{3}	43.5^{3}
RT	6	18	6	34.0^{9}	0.26^{9}	26.1^{8}
RC	8	28	10	43.8^{5}	0.39^{5}	34.0^{5}
TC	1	17	7	32.1^{12}	0.24^{12}	21.7^{11}

Table 3. Affinity indices produced by combining the Cordilleran and Franklinian units for times of relative cosmopolitanism (3+4) and relative provincialism (7 to 9^A) (see text page 191 for explanation).

	Six Areas $\frac{C-C^{cosm}}{N_1-N_1^{cosm}} \times 100$	Five Areas $\frac{C-C^{cosm}}{N_1-N_1^{cosm}} \times 100$	Six Areas $\frac{C-C^{cosm}}{N_1-N_1^{cosm}} \times 100$	Five Areas $\frac{C-C^{cosm}}{N_1-N_1^{cosm}} \times 100$
	Interval 3 + 4		Interval 7 to 9^A	
AN	33.3	28.5	34.1	30.8
AR	20.6	13.3	25.0	20.8
AT	23.5	13.8	17.8	12.5
NR	12.5	4.8	36.4	30.8
NT	33.3	23.8	26.7	20.0
RT	26.5	16.7	31.1	25.0
AC	11.9	15.8	5.8	10.4
FA	15.8	—	15.4	—
NC	58.3	71.4	40.9	48.7
FN	58.3	—	43.6	—
RC	23.5	33.3	43.1	37.3
FR	26.5	—	33.3	—
TC	29.4	31.0	20.0	22.5
FT	29.4	—	23.1	—
FC	65.8	—	46.2	—

high degree of biogeographic splitting as to result in serious loss of information regarding the affinities existing between most faunas. Furthermore, the specific level is less reliable for biogeographic purposes owing to the greater difficulty of distinguishing fossil species from mere phenotypic variants. It has been found from experience that genera and subgenera provide the best taxonomic level for investigating faunal affinities on a global scale. Because some fossil brachiopod genera are poorly understood at present, and distinction between them is uncertain, they have been hyphenated and treated as a single taxon in our data.

The first step in this compilation was the preparation of lists of genera and subgenera from previously published monographs and other sources. Identifications and correlations of other workers were carefully evaluated in an attempt to maintain a uniform standard in Table 1. Faunal lists alone are commonly unsatisfactory as data unless one is familiar with the faunas listed or has knowledge of the taxonomic habits of the author.

Of particular importance is the choice of geographic units for which faunal lists are to be prepared. This selection is done by qualitatively defining areas which appear to have distinctive faunas. Such areas may often be separated from each other by broad regions from which no data are available. It could be argued that this step is strongly biased because qualitative interpretations are used to select areas for quantitative study. However, this approach has the sanction of past practice and the experience of the past 130 years. Judgments, qualitative but incisive, made since the initial pioneering studies of Sclater and Wallace support this practice.

Northwestern Canada (Cordilleran unit of the paper) and Nevada are treated as distinct areas because of significant faunal differences at certain Lower Devonian time intervals, wide geographic separation, and the virtual absence of data for the intervening area of western North America. A useful Lower Devonian shelly fauna record in western Alberta-eastern British Columbia is unlikely

Table 4. Summary chart of total number of genera present and number and percentage of cosmopolitan (cosm) and provincial (prov) genera in each time interval for the six geographic regions. The second listing of data (with asterisk) in interval 7 to 9 shows modified results caused by restricting genera known only from the Bohemian Pragian part of the Rhenish-Bohemian unit to interval 5+6 (case B of text).

ZONES	All Units (FANRTC)	F	A	N	R	T	C
	260 Total	52 Total	62 Total	57 Total	160 Total	53 Total	64 Total
	*242 Total	*52 Total	*62 Total	*57 Total	*132 Total	*53 Total	*64 Total
7 + 8 + 9	13 Cosm= 5%	13 Cosm= 25%	10 Cosm= 16%	13 Cosm= 23%	13 Cosm= 8%	8 Cosm= 15%	13 Cosm= 20%
	*11 Cosm= 5%	*11 Cosm= 21%	*9 Cosm= 15%	*11 Cosm= 19%	*10 Cosm= 8%	*7 Cosm= 13%	*11 Cosm= 17%
	163 Prov= 63%	8 Prov= 15%	23 Prov= 37%	4 Prov= 7%	97 Prov= 61%	18 Prov= 34%	13 Prov= 20%
	*148 Prov= 61%	*9 Prov= 17%	*23 Prov= 37%	*4 Prov= 7%	*79 Prov= 60%	*18 Prov= 34%	*15 Prov= 23%
	220 Total	17 Total	73 Total	44 Total	131 Total	38 Total	55 Total
5+6	11 Cosm= 5%	5 Cosm= 29%	10 Cosm= 14%	10 Cosm= 23%	11 Cosm= 8%	10 Cosm= 26%	10 Cosm= 18%
	147 Prov= 67%	2 Prov= 12%	26 Prov= 36%	6 Prov= 14%	87 Prov= 66%	9 Prov= 24%	17 Prov= 31%
	135 Total	50 Total	55 Total	37 Total	44 Total	46 Total	57 Total
3+4	13 Cosm= 10%	12 Cosm= 24%	13 Cosm= 24%	13 Cosm= 35%	10 Cosm= 23%	12 Cosm= 26%	12 Cosm= 21%
	65 Prov= 48%	5 Prov= 10%	23 Prov= 42%	0 Prov= 0%	15 Prov= 34%	12 Prov= 26%	10 Prov= 18%
	135 Total	37 Total	71 Total	40 Total	77 Total	39 Total	43 Total
1+2	21 Cosm= 16%	18 Cosm= 49%	19 Cosm= 27%	20 Cosm= 50%	20 Cosm= 26%	19 Cosm= 49%	20 Cosm= 47%
	69 Prov= 51%	1 Prov= 3%	30 Prov= 42%	1 Prov= 2.5%	27 Prov= 35%	7 Prov= 18%	3 Prov= 7%

to emerge. Familiarity with the North Atlantic region Lower Devonian geology suggests inclusion of the Nova Scotia Devonian with that of western Europe. Although the Nova Scotian faunas were included with the Rhenish-Bohemian unit faunas in our compilation, there are no genera restricted to Nova Scotia and, therefore, the inclusion of Nova Scotia faunas in this way has not affected the results. The Carnic Alps and Uralian areas are not included in the Rhenish-Bohemian geographic unit because of the distinctiveness of their faunas. Inclusion of the faunas from the last two areas with the other European faunas would serve to increase provincialism for Europe as a whole in our analysis. On the other hand, the North African faunas appear similar to those of Europe, except for the Carnic Alps and the Urals. Arctic Canada is geographically distinct from northwestern Canada and its faunas have been treated separately.

Both quantitatively and qualitatively, it is apparent that the Cordilleran and Franklinian areas show considerable similarity during all four intervals discussed (Tables 1, 2). The question arises of the effect on the affinity indices if the two areas were grouped together and considered as one unit. Combining two areas will naturally increase the size of the sample (number of genera present) because no two areas have identical faunas. The definition of cosmopolitan genera must be suitably altered, e.g., to include genera occurring in at least four of the five geographic areas. This modification will tend to increase the number of cosmopolitan genera. Affinity indices relating to the combined unit will tend to increase because of the greater number of shared genera, whereas affinity indices not relating to the combined unit will tend to decrease because of the increased number of cosmopolitan genera. Affinity indices were calculated for the case where the Cordilleran and Franklinian areas are treated as a single unit during a time of relative cosmopolitanism (late Lochkovian 3 + 4 interval) and during a time of relative provincialism (Emsian 7 to 9^A interval). The results are presented in Table 3. Most comparisons involving the combined Cordilleran and Franklinian unit show higher affinity indices than comparisons where those units are kept separate. These data suggest that the combined Cordilleran and Franklinian unit has greatest affinity with the Nevadan unit in late Lochkovian and Emsian times. The affinity is sufficiently high in interval 3 + 4 to consider grouping the Cordilleran, Franklinian, and Nevadan units as one biogeographic unit during the late Lochkovian. This exercise of combining samples from various areas is a convenient method of testing the validity of biogeographic units and of ranking them in a biogeographic hierarchy.

One of the most serious sampling problems is the difficulty in obtaining a strictly comparable set of communities from the various Benthic Assemblages in each geographic area. This sampling effect is real but difficult to remove quantitatively from the available data. However, the eurytopic behavior of some genera helps to moderate the problem. For example, it is clear that more of the Early Devonian faunas from central Nevada represent shallower water communities (chiefly Benthic Assemblage 3 of Boucot, 1975) than do those from the Cordilleran area (northwestern Canada) which chiefly represent Benthic Assemblages 4 and 5. When community and Benthic Assemblage sampling are highly unequal some genera may appear as provincials (confined to one area) merely because the host Benthic Assemblage is either poorly sampled or absent in the other region. In addition, the density and intensity of sampling itself is unequal from area to area and from time interval to time interval. Poorly sampled regions commonly show an excess of cosmopolitan elements because the cosmopolitans tend to be far more abundant as individuals and, therefore, are more likely to be obtained in small samples. In northwestern Canada, fossiliferous platform carbonates of Emsian age are geographically much more widespread than older units because of the pronounced Emsian transgression (Perry and Lenz, 1977). Older Early Devonian strata are more restricted and hence have been sampled in fewer areas. The European Emsian is more geographically widespread than the European pre-Emsian and has been intensely studied for the past 150 years. This results in much better sampling and probably accounts for the greater number of European provincial genera.

The basic correlation problem revolves around the attempt to try and find a suitable means of tying together in a time framework areas with dissimilar faunas. The faunal lists for western Europe, especially those from the Pragian of Bohemia pose particular problems. The fourfold time division of the Early Devonian used here is in some cases more demanding than the data warrant. For example, in Bohemia no easy separation of early and late Pragian brachiopod faunas can be made. Because the Pragian-Zlichovian boundary is thought to lie at about the top of faunal unit 7, the Pragian Bohemian genera will be found in both the 5 + 6 and 7 to 9 time units. As some of these genera (18 of 260) are not known from the Zlichovian, this will bias the results of our analysis. In view of this, we have calculated and listed (Tables 1 to 4) two sets of results for Emsian time, the first with all of the Bohemian Pragian brachiopods included in unit 7 to 9, and the second with the Bohemian Pragian brachiopods (18 genera) excluded from the unit unless they are known to range into the Zlichovian or its equivalent. In western North America the informal conodont units of Klapper and others (1971) can be used to make meaningful correlations. To some extent the Australian faunas can be tied to this informal conodont scheme (Philip and Pedder, 1967; Savage, 1973a, 1973b). Appalachian North America (essentially the region between Chihuahua and Gaspé) presents considerable difficulty as only the two older Early Devonian conodont-based time units can be recognized with any degree of certainty. In the European sequences, the conodont faunas, combined with existing brachiopod correlations, can be used locally although the age of many of the Rhenish units remains somewhat uncertain.

Affinity tests only evaluate similarities between faunas from one area to another in a single time unit. Qualitative

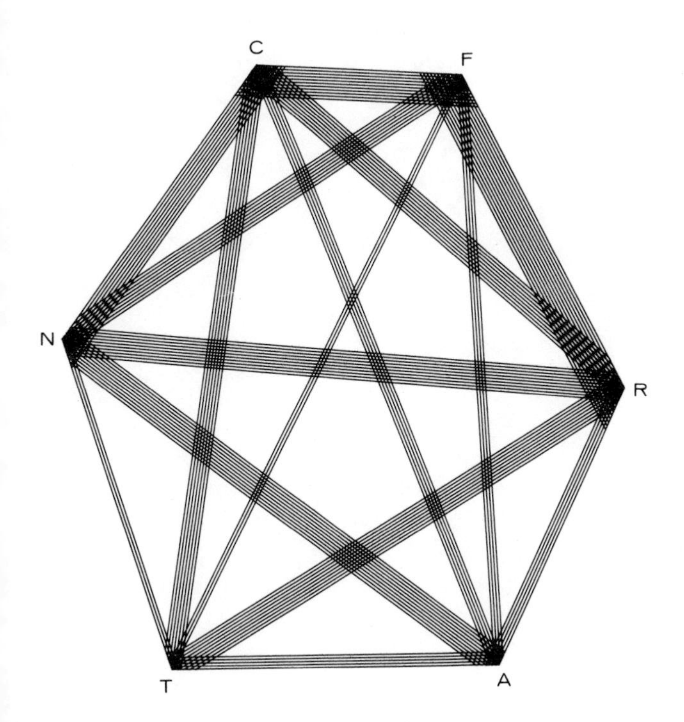

Figure 6. Graphical representation of affinity indices calculated for early Lochkovian time using the formula

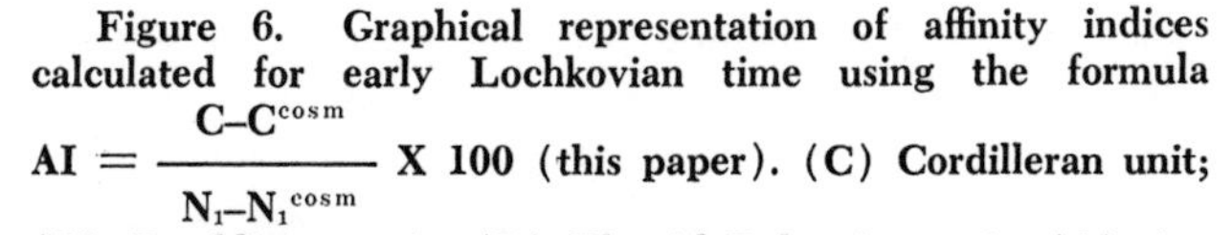

$$AI = \frac{C - C^{cosm}}{N_1 - N_1^{cosm}} \times 100$$

(this paper). (C) Cordilleran unit; (F) Franklinian unit; (R) Rhenish-Bohemian unit; (A) Appalachian unit; (T) Tasman unit; (N) Nevadan unit (see Table 1 caption for further details). The bundles of bars indicate relative affinity; a single bar represents affinity indices closest to 5 (the 2.6 to 7.5 interval), two bars represent affinity indices closest to 10 (the 7.6 to 12.5 interval), and so on.

observations of faunal lists show that some genera occur at earlier or later times in one area than in another. These shifts of genera in time may be facies controlled or may result from delayed migration. *Mystrophora*, for example, is known from Emsian beds in northwestern Canada (Perry, 1974) although it is not recognized until Middle Devonian times in western Europe. Examples of this type are not abundant although they can be used qualitatively to reinforce ideas of affinity. Alternatively such apparent anomalies may reflect sampling limitations rather than migration.

LOWER DEVONIAN BRACHIOPOD FAUNAL AFFINITIES

During early Lochkovian time (conodont faunas 1 + 2), the brachiopod faunas are more cosmopolitan than in subsequent Lower Devonian time. In the six geographic units (Table 4) a total of 135 genera are known, of which 21 (16 percent) are cosmopolitan (common to five or all six of the geographic units) and 69 (51 percent) are provincial (limited to a single geographic unit). Of the 21 cosmopolitan genera, 11 are common to all six geographic

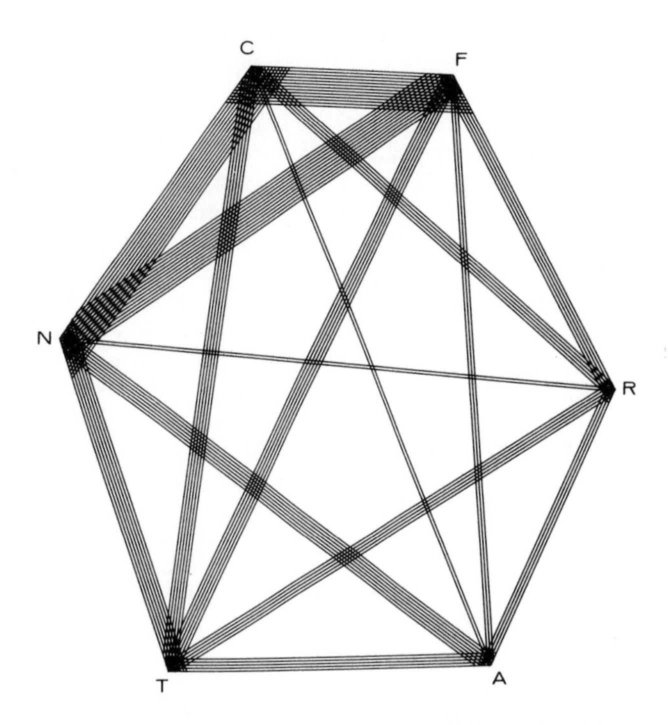

Figure 7. Graphical representation of affinity indices calculated for late Lochkovian time (see Fig. 6 caption for explanation).

units. These are *Atrypa*, *"Chonetes"*, *Cyrtina*, *Dalejina*, *Eoschuchertella*, *Gypidula*, *Howellella*, *Leptaena-Leptagonia*, *"Mesodouvillina"*, *Nucleospira*, and *Schizophoria*.

From the calculated affinity indices (shown here in parentheses), the brachiopods of the Franklinian geographic unit appear to have most affinity (Fig. 6) with those of the Rhenish-Bohemian (63.2) unit, less with the Cordilleran (52.6) and Nevadan (42.1) units, and least with the Appalachian (21.1) and Tasman (15.8) units; those of the Appalachian unit appear to have most affinity with the Nevadan unit (50.0), less with the Rhenish-Bohemian (26.9), Cordilleran (26.1), and Tasman (25.0) units, and least with the Franklinian (21.1) unit; those of the Nevadan unit appear to have most affinity with the Rhenish-Bohemian (50.0) and Appalachian (50.0) units, slightly less with the Cordilleran (45.0) and Franklinian (41.1) units, and least with the Tasman (15.0) unit; those of the Rhenish-Bohemian unit appear to have most affinity with the Franklinian (63.2) unit, somewhat less with the Nevadan (50.0), Tasman (45.0), and Cordilleran (43.5) units, and least with the Appalachian (26.9) unit; those of the Tasman unit appear to have most affinity with the Rhenish-Bohemian (45.0) unit, rather less with the Cordilleran (35.0) unit, considerably less with the Appalachian (25.0) unit, and least with the Franklinian (15.8) and Nevadan (15.0) units; and those of the Cordilleran unit appear to have most affinity with the Franklinian (52.6) unit, slightly less with the Nevadan (45.0) unit, slightly less again with the Rhenish-Bohemian (43.5) unit, considerably less with the Tasman (35.0) unit, and least with the Appalachian (26.1) unit.

The statistics are undoubtedly biased by unequal collecting and disproportionate representation of different communities but nevertheless appear to show that during early Lochkovian time the Rhenish-Bohemian unit had strong or fairly strong faunal affinities with all the other units except the Appalachian unit, where the faunal affinities were only moderate (Fig. 6). The mutual affinities between the Franklinian, Cordilleran, Rhenish-Bohemian, and Nevadan units are all strong at this time. The Tasman and Appalachian units appear to be more faunally isolated; the Tasman unit has fairly strong affinities with the Rhenish-Bohemian unit, moderate affinities with the Cordilleran unit, and weak affinities elsewhere; the Appalachian unit has strong affinities with the Nevadan unit but weak affinities with all the other units. Most of the faunal affinities apparent from these comparisons presumably represent active faunal migratory links but some may simply represent relict links inherited from more cosmopolitan times in the Late Silurian. A trend towards less cosmopolitan and more provincial brachiopod faunas is very evident during subsequent Lower Devonian time.

During late Lochkovian time (conodont faunas 3 + 4), the brachiopod faunas are slightly less cosmopolitan than during early Lochkovian time but do not show the marked provinciality characteristic of subsequent Lower Devonian time. In the six geographic units a total of 135 genera are known, of which 13 (10 percent) are cosmopolitan and 65 (48 percent) are provincial (Table 4). Of the 13 cosmopolitan genera, 7 are common to all six geographic units. These are *Atrypa, Eoschuchertella, Gypidula, Howellella, Leptaena-Leptagonia, "Mesodouvillina"*, and *Nucleospira.*

From the calculated affinity indices (Table 2), the brachiopods of the Franklinian geographic unit appear to have most affinity (Fig. 7) with those of the Cordilleran (65.8) unit, slightly less with the Nevadan (58.3) unit, much less with the Tasman (29.4) and Rhenish-Bohemian (26.5) units, and least with the Appalachian (15.8) unit; those of the Appalachian unit appear to have most affinity with the Nevadan (33.3) unit, less with the Tasman (23.5) and Franklinian (15.8) units, and least with the Cordilleran (11.9) unit; those of the Nevadan unit appear to have most affinity with the Franklinian (58.3) and Cordilleran (58.3) units, much less with the Tasman (33.3) and Appalachian (33.3) units, and least with the Rhenish-Bohemian (12.5) unit; those of the Rhenish-Bohemian unit appear to have most affinity with the Tasman (26.5) and Franklinian (26.5) units, slightly less with the Cordilleran (23.5) and Appalachian (20.6) units, and least with the Nevadan (12.5) unit; those of the Tasman geographic unit appear to have most affinity with the Nevadan (33.3) unit, slightly less with the Cordilleran (29.4) and Franklinian (29.4) units, and least with the Rhenish-Bohemian (26.5) and Appalachian (23.5) units; and those of the Cordilleran unit appear to have most affinity with the Franklinian (65.8) unit, slightly less with the Nevadan (58.3) unit, very much less with the Tasman (29.4) and Rhenish-Bohemian (23.5) units, and least with the Appalachian (11.9) unit.

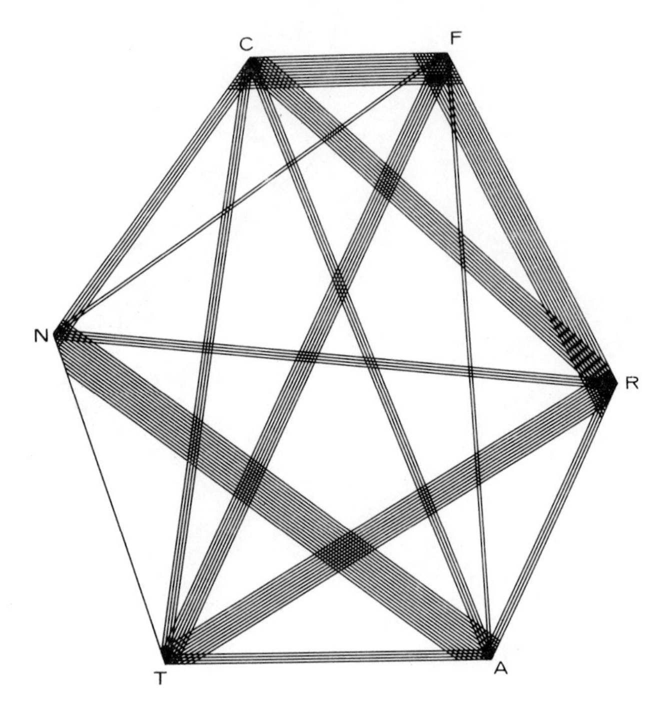

Figure 8. Graphical representation of affinity indices calculated for early Pragian time (see Fig. 6 caption for explanation).

The results indicate very strong mutual affinities of the faunas of the Nevadan, Cordilleran, and Franklinian units. Faunal links of these three units with the Rhenish-Bohemian, Tasman, and Appalachian units are moderate or weak, as are the mutual affinities between these latter units (Fig. 7).

During early Pragian time (conodont faunas 5 + 6) the brachiopod faunas are less cosmopolitan than during early or late Lochkovian time and are very much more provincial. In the six geographic units a total of 220 genera are known, of which 11 (5 percent) are cosmopolitan and 147 (67 percent) are provincial. Of the 11 cosmopolitan genera, 5 are common to all six geographic units. These are *Atrypa, Cyrtina, Howellella, Nucleospira,* and *Schizophoria.*

From the calculated affinity indices (Table 2), the brachiopods of the Franklinian geographic unit appear to have most affinity (Fig. 8) with those of the Cordilleran (54.5) and Rhenish-Bohemian (54.5) units, considerably less with the Tasman (36.4) unit, and very much less with the Appalachian (9.1) and Nevadan (9.1) units; those of the Appalachian unit appear to have most affinity with the Nevadan (67.6) unit, much less with the Tasman (22.2), Cordilleran (20.5), and Rhenish-Bohemian (19.4) units, and least with the Franklinian (9.1) unit; those of the Nevadan unit appear to have most affinity with the Appalachian (67.6) unit, much less with the Cordilleran (26.5) and Rhenish-Bohemian (20.6) units, and very little affinity with the Franklinian (9.1) and Tasman (3.7) units; those of the Rhenish-Bohemian unit appear to have most affinity with the Franklinian (54.5) unit, slightly less

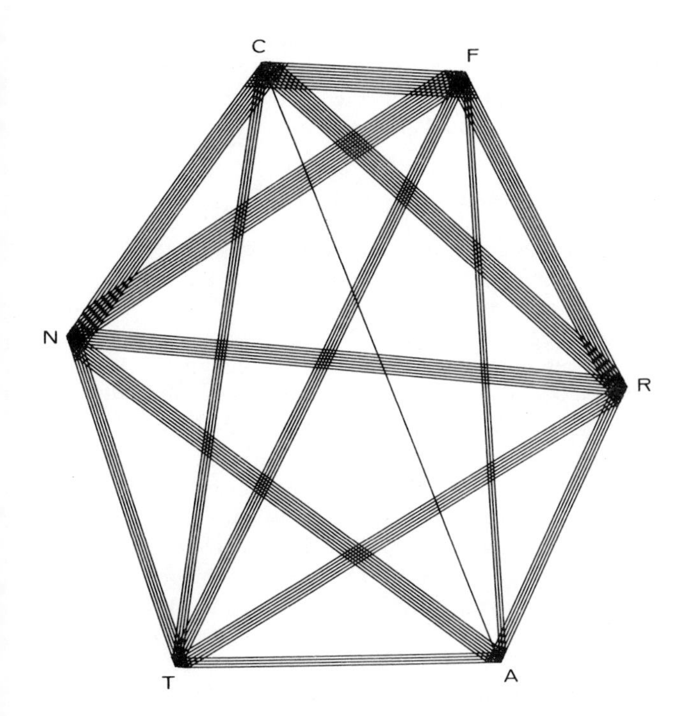

Figure 9. Graphical representation of affinity indices calculated for Emsian time (Case A, see p. 190) (see Fig. 6 caption for explanation).

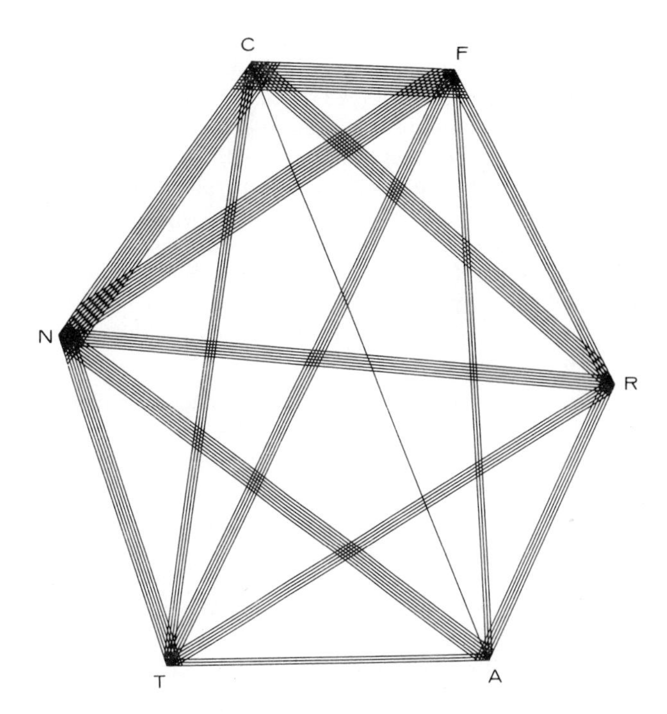

Figure 10. Graphical representation of affinity indices calculated for Emsian time (Case B, see p. 190) (see Fig. 6 caption for explanation).

with the Tasman (48.1) and Cordilleran (43.2) units, and much less with the Nevadan (20.6) and Appalachian (19.4) units; those of the Tasman unit appear to have most affinity with the Rhenish-Bohemian (48.1) unit, slightly less with the Franklinian (36.4) unit, much less with the Appalachian (22.2) and Cordilleran (22.2) units, and least with the Nevadan (3.7) unit; and those of the Cordilleran unit appear to have most affinity with the Franklinian (54.5) unit, rather less with the Rhenish-Bohemian (43.2) unit, and considerably less with each of the Nevadan (26.5), Tasman (22.2), and Appalachian (20.5) units.

A very high affinity of the faunas of the Nevadan and Appalachian units is characteristic of this time. However, the faunas of these two units have low affinity with faunas elsewhere. The Cordilleran, Franklinian, Rhenish-Bohemian, and Tasman units all have fairly high mutual affinities apart from a rather low affinity between the Cordilleran and Tasman units (Fig. 8).

During Emsian (late Pragian-Zlichovian) time (conodont faunas 7 to 9), the brachiopod faunas show a high degree of provinciality compared with faunas of early and late Lochkovian times but are slightly less provincial than the preceding early Pragian faunas (Table 4). When all the Bohemian Pragian brachiopods are included in the Rhenish-Bohemian totals for faunal unit 7 to 9 (Case A), 260 brachiopods are known in the six geographic units, and of these 5 are common to all six geographic units. These are ***Atrypa, Cyrtina, Eoschuchertella, Howellella,*** and *Leptaena-Leptagonia*. When only those Bohemian brachiopods known to range into the Zlichovian are included in the Rhenish-Bohemian unit (Case B), 242 brachiopods are known in the six geographic units, and of these 11 (5 percent) are cosmopolitan and 148 (61 percent) are provincial. Of the 11 cosmopolitan genera, 4 are common to all six geographic units (as Case A except for *Howellella*).

From the calculated affinity indices (Table 2), the brachiopods of the Franklinian geographic unit (Case A) appear to have most affinity (Fig. 9) with those of the Cordilleran (46.2) and Nevadan (43.6) units, less with those of the Rhenish-Bohemian (33.3) unit, and much less with those of the Tasman (23.1) and Appalachian (15.4) units. The Franklinian affinity indices are all affected by the Case B changes (Fig. 10) but only that with the Rhenish-Bohemian index is greatly affected, being reduced from 33.3 to 19.5. The other Case B Franklinian unit affinity indices are 17.1 with the Appalachian unit, 46.3 with the Nevadan unit, 24.4 with the Tasman unit, and 48.8 with the Cordilleran unit.

In Case A, the brachiopods of the Appalachian geographic unit have low affinities with other areas. Most affinity is with the Nevadan (34.1) unit, less with the Rhenish-Bohemian (25.0) unit, still less affinity with the Tasman (17.8) and Franklinian (15.4) units, and least with the Cordilleran (5.8) unit. In Case B, the Appalachian affinity indices (Fig. 10) are 17.1 with the Franklinian unit, 34.8 with the Nevadan unit, 22.6 with the Rhenish-Bohemian unit, 17.4 with the Tasman unit, and 7.5 with the Cordilleran unit.

In Case A, the brachiopod faunas of the Nevadan geographic unit have a moderately high affinity (Fig. 9) with those of the Franklinian (43.6) and Cordilleran (40.9) units, less with those of the Rhenish-Bohemian (36.4) and Appalachian (34.1) units, and least with those of the Tasman (26.7) unit. The affinity indices are all affected by the Case B changes (Fig. 10) but only the Rhenish-Bohemian index is significantly affected, being reduced from 36.4 to 28.3. The other Case B Nevadan unit affinity indices are 46.3 with the Franklinian unit, 34.8 with the Appalachian unit, 28.3 with the Tasman unit, and 43.5 with the Cordilleran unit.

In Case A, the brachiopods of the Rhenish-Bohemian geographic unit have moderate affinities with other areas. The strongest affinity is with the Cordilleran (43.1) unit, followed closely by the Nevadan (36.4), Franklinian (33.3), and Tasman (31.1) units, and the least affinity is with the Appalachian (25.0) unit. It is of interest to note that all but one (*Rhenothyris*) of the Rhenish-Bohemian and Cordilleran biprovincials are commonly found in carbonate faunas of the Bohemian region. The high number of Rhenish-Bohemian Emsian provincial genera is in part a result of the fact that Pragian time spans the conodont fauna 6 to 7 boundary so that, in Case A, an additional 18 genera are listed. The Case B affinities rank differently from the Case A affinities (Table 2). The affinities of the Rhenish-Bohemian unit are 34.0 with the Cordilleran unit, 28.3 with the Nevadan unit, 26.1 with the Tasman unit, 22.6 with the Appalachian unit, and 19.5 with the Franklinian unit.

In Case A, the brachiopod faunas of the Tasman geographic unit appear to have only moderate affinity with the Rhenish-Bohemian (31.1) unit, less affinity with the Nevadan (26.7) unit, and progressively weaker affinity with the Franklinian (23.1), Cordilleran (20.0), and Appalachian (17.8) units. In Case B, the Tasman unit affinity indices all change slightly compared with Case A. These indices are 28.3 with the Nevadan unit, 26.1 with the Rhenish-Bohemian unit, 24.4 with the Franklinian unit, 21.7 with the Cordilleran unit, and 17.4 with the Appalachian unit.

In Case A, the brachiopods of the Cordilleran geographic unit appear to have most affinity with the Franklinian (46.2) unit, slightly less with the Rhenish-Bohemian (43.1) and Nevadan (40.9) units, very much less with the Tasman (20.0) unit, and least with the Appalachian (5.8) unit. The affinity indices are all slightly raised by the Case B changes except for the Rhenish-Bohemian index which is lowered from 43.1 to 34.0. The other Cordilleran Case B indices are 48.8 for the Franklinian unit, 43.5 for the Nevadan unit, 21.7 for the Tasman unit, and 7.5 for the Appalachian unit.

The statistics show that during Emsian time the Rhenish-Bohemian, Franklinian, Cordilleran, and Nevadan units had fairly strong mutual faunal affinities (Figs. 9, 10). Of the remaining two units, the Tasman unit had fairly strong faunal affinities with the Rhenish-Bohemian unit, moderate affinities with the Nevadan unit, and weak affinities with the other units, and the Appalachian unit had fairly strong faunal affinities with the Nevadan unit, and weak affinities with all other units.

BIOGEOGRAPHIC IMPLICATIONS

Until now we have avoided employing a biogeographic nomenclature and instead have used geographic units as "samples" without any intent to imply that they are also biogeographic units. We consider this to have been a relatively objective approach to the analysis of the data. Now, however, it is time to consider how well the purely geographic units correspond to biogeographic units.

Lower Devonian brachiopod biogeography was discussed by Boucot (1960a) who recognized the similarity between Early Devonian brachiopod faunas of Nova Scotia and western European Rhenish faunas. Descriptive summaries of Late Silurian and Devonian brachiopod zoogeography were presented by Boucot and others (1969), Johnson and Boucot (1973), Boucot and Johnson (1973), and Boucot (1975) (see Table 5). These authors recognized Old World, Appalachian, and Malvinokaffric units as the major biogeographic entities. Old World faunas were observed to include numerous genera which "carryover" from the Late Silurian. In the context of finer subdivisions within the Lower Devonian, Johnson and Boucot (1973) recognized distinct Old World (Europe, western and Arctic North America) and Appalachian (eastern North America exclusive of Nova Scotia) biogeographic units during the Lochkovian. Within the European Old World faunas, nearshore terrigenous Rhenish Communities and offshore carbonate Bohemian Communities were identified. The Tasman area was included within the Old World unit although the numerous Tasman provincial genera were thought to make it a distinct biogeographic entity. Savage (1969, 1970, 1971) observed the strong affinity of Australian Lochkovian brachiopods with those of the Old World Bohemian area. During subsequent early Pragian time, Appalachian brachiopod genera were thought to have migrated into the western North America Nevada area although the remainder of northwestern and Arctic North America maintained Old World affinities (Boucot and Johnson, 1973). The early Pragian has been considered a time of regression in western and Arctic North America (Johnson, 1974a; Perry and Lenz, 1977), in eastern North America (Boucot, 1975), in South America, southern Africa, and Antarctica (Johnson and Boucot, 1973; Boucot, 1975), and in Europe (Johnson and Boucot, 1973; Boucot, 1975). A major Emsian transgression then resulted in Emsian marine deposits being more widespread than those of any other interval of Lower Devonian time (Boucot, 1975). Johnson and Boucot (1973) recognized the New Zealand, Tasman, Rhenish-Bohemian, Uralian, and Cordilleran biogeographic divisions of the Old World Emsian. Their Cordilleran unit is centered in Nevada whereas the northwestern Canadian Cordillera and the Franklinian areas were included in the Uralian unit. The Tasman faunas were in-

Table 5. Summary of Lower Devonian biogeographic divisions used by Boucot and others (1969), Telford (1972), Boucot (1975), Oliver (1975), and Savage and others (herein). The summary does not encompass the South American units recognized by Boucot (1975).

	EASTERN NORTH AMERICA	GREAT BASIN	
EMSIAN	**Appalachian Province**	**Old World Province** Cordilleran Subprovince	Boucot and others (1969)
	E North America "Province"	**W North America "Province"**	Telford (1972)
	E North America Province	**Old World Province**	Oliver (1975)
	Eastern Americas Realm Appohimchi Subprovince	**Old World Realm** Cordilleran Region	Boucot (1975)
	Eastern Americas Realm Appohimchi Subprovince	**Old World Realm** Cordilleran Region	Savage and others (herein)
EARLY PRAGIAN	**Appalachian Province**	**Old World Province**	Boucot and others (1969)
	E North America "Province"	**W North America "Province"**	Telford (1972)
	E North America Province	**Old World Province**	Oliver (1975)
	Eastern Americas Realm Appohimchi Subprovince	**Eastern Americas Realm** Nevadan Subprovince	Boucot (1975)
	Eastern America Realm Appohimchi Subprovince	**Eastern Americas Realm** Nevadan Subprovince	Savage and others (herein)
LATE LOCHKOVIAN	**Appalachian Province**	**Old World Province**	Boucot and others (1969)
	E North America "Province"	**W North America "Province"**	Telford (1972)
	E North America Province	**Old World Province**	Oliver (1975)
	Eastern Americas Realm Appohimchi Subprovince	**Eastern Americas Realm** Nevadan Subprovince	Boucot (1975)
	Eastern Americas Realm Appohimchi Subprovince	**Old World Realm** Cordilleran Region	Savage and others (herein)
EARLY LOCHKOVIAN	**Appalachian Province**	**Old World Province**	Boucot and others (1969)
	E North America "Province"	**W North America "Province"**	Telford (1972)
	E North America Province	**Old World Province**	Oliver (1975)
	Eastern Americas Realm Appohimchi Subprovince	**Old World Realm**	Boucot (1975)
	Eastern Americas Realm Appohimchi Subprovince	**Old World Realm** Cordilleran Region	Savage and others (herein)

cluded in the Old World unit as the Tasman Subprovince. Talent (1972) pointed out that the Emsian Tasman brachiopod faunas were characterized by a number of provincial genera and were very different from faunas in the Malvinokaffric unit.

Our results suggest to us that, throughout the entire Lower Devonian, the Appalachian unit is distinct from the other units except from the Nevadan unit during early Lochkovian and early Pragian times when numerous Appalachian genera moved into the central Nevada area. Even with these Nevada incursions, the Appalachian unit is clearly a distinct biogeographic unit as well as a geographic unit. Examination of our calculated affinity indices, and their graphical summary in Figures 6 to 10, shows strong mutual affinities of the brachiopod faunas of the Nevadan, Cordilleran, and Franklinian geographic units throughout the Early Devonian, except for the marked drop in affinity of the Nevadan unit with the other two units during early Pragian time when the greatest incursion of Appalachian genera into Nevada occurred. We do not feel that the available data warrants distinguishing between the Nevadan, Cordilleran, and Franklinian geographic units for biogeographic purposes but that a composite Cordilleran biogeographic unit, forming part of the Old World Realm, is indicated. However, the Nevadan geographic unit during early Pragian time has strong enough faunal affinities with the Appalachian geographic unit to form part of the Eastern Americas Realm at that time. The Rhenish-Bohemian geographic unit has faunas which during early Lochkovian time have strong affinities with all other geographic units except the Appalachian unit, and during early Pragian time have strong affinities with the Tasman, Franklinian, and Cordilleran geographic units. During other Early Devonian times, the affinities between the Rhenish-Bohemian unit faunas and those of other geographic units are moderate to weak so that a distinct Rhenish-Bohemian biogeographic unit is warranted. The Tasman geographic unit has strong faunal affinities with the Rhenish-Bohemian unit during early Lochkovian and early Pragian times but has only moderate or weak affinities with the other geographic units during the Early Devonian. This evidence supports the concept of a distinct Tasman biogeographic unit within the larger Old World Realm.

Table 5. (Cont.)

	CANADIAN CORDILLERA	ARCTIC ISLANDS	
EMSIAN	**Old World Province** Cordilleran-Uralian Subprovince	**Old World Province** Cordilleran-Uralian Subprovince	Boucot and others (1969)
	W North America "Province"	**W North America "Province"**	Telford (1972)
	Old World Province	**Old World Province**	Oliver (1975)
	Old World Realm Cordilleran Region	**Old World Realm** Cordilleran Region	Boucot (1975)
	Old World Realm Cordilleran Region	**Old World Realm** Cordilleran Region	Savage and others (herein)
EARLY PRAGIAN	**Old World Province**	**Old World Province**	Boucot and others (1969)
	W North America "Province"	**W North America "Province"**	Telford (1972)
	Old World Province	**Old World Province**	Oliver (1975)
	Old World Realm Cordilleran Region	**Old World Realm** Cordilleran Region	Boucot (1975)
	Old World Realm Cordilleran Region	**Old World Realm** Cordilleran Region	Savage and others (herein)
LATE LOCHKOVIAN	**Old World Province**	**Old World Province**	Boucot and others (1969)
	W North America "Province"	**W North America "Province"**	Telford (1972)
	Old World Province	**Old World Province**	Oliver (1975)
	Old World Realm Cordilleran Region	**Old World Realm** Cordilleran Region	Boucot (1975)
	Old World Realm Cordilleran Region	**Old World Realm** Cordilleran Region	Savage and others (herein)
EARLY LOCHKOVIAN	**Old World Province**	**Old World Province**	Boucot and others (1969)
	W North America "Province"	**W North America "Province"**	Telford (1972)
	Old World Province	**Old World Province**	Oliver (1975)
	Old World Realm	**Old World Realm**	Boucot (1975)
	Old World Realm Cordilleran Region	**Old World Realm** Cordilleran Region	Savage and others (herein)

We feel that the results of our quantitative analysis largely support the biogeographic units used previously by Boucot and others (1969), Johnson and Boucot (1973), Boucot and Johnson (1973), and Boucot (1975). The only modification we suggest is that the Nevadan, Cordilleran, and Franklinian geographic units should form the composite Cordilleran biogeographic unit, except during the early Pragian interval when the Nevadan geographic area forms part of the Eastern Americas Realm, as explained above. Our modified biogeographic divisions are summarized in Table 5. During early and late Lochkovian time, we recognize an Old World Realm consisting of a Rhenish-Bohemian Region, a Tasman Region, and a Cordilleran Region. The question of whether a Uralian Region and a Mongolo-Okhotsk Region can be recognized and included in the Old World Realm has not been considered in our faunal analysis. We also recognize, during early and late Lochkovian time, an Eastern Americas Realm which includes the Appalachian faunas and other adjacent eastern American faunas grouped together as the Appohimchi Subprovince. During early Pragian time, we recognize an Old World Realm consisting of a Rhenish-Bohemian Region, a Tasman Region, and a Cordilleran Region, but the Cordilleran Region at this time excludes the Nevadan geographic unit. We have not investigated the data relating to whether a distinct Uralian Region and a distinct Mongolo-Okhotsk Region are present. An early Pragian Eastern Americas Realm includes an Appohimchi Subprovince and a Nevadan Subprovince. During Emsian time, we recognize an Old World Realm consisting of a Rhenish-Bohemian Region, a Tasman Region, and a Cordilleran Region (which now includes the Nevadan faunas again, in addition to the Canadian Cordilleran faunas and the Franklinian faunas). The Emsian Old World Realm may also include a Uralian Region and a Mongolo-Okhotsk Region but these data were not examined in our analysis. The Emsian Eastern Americas Realm consists of an Appohimchi Subprovince and a possible Amazon-Colombian Subprovince but the data of the latter area were not investigated here. A third Emsian realm, the Malvinokaffric Realm, consisting of the Gondwana Emsian brachiopod faunas, is recognized by us but the data have not been considered herein. The advantage of using the term Malvinokaffric for those Silurian and Devonian faunas rather than the more familiar term Gondwana is not apparent to us.

Having examined the Early Devonian brachiopod data and attempted to relate these to biogeographic units, we will now briefly compare our conclusions with those of

Table 5. (Cont.)

	EUROPE	ASIA	
EMSIAN	**Old World Province** Rhenish-Bohemian Subprovince	**Old World Province** Uralian Subprovince	Boucot and others (1969)
	Europe "Province"		Telford (1972)
	Old World Province	**Old World Province**	Oliver (1975)
	Old World Realm Rhenish-Bohemian & Uralian Regs.	**Old World Realm** Uralian & Mongolo-Okhotsk Regs.	Boucot (1975)
	Old World Realm Rhenish-Bohemian Region		Savage and others (herein)
EARLY PRAGIAN	**Old World Province** Rhenish-Bohemian Subprovince	**Old World Province** Uralian Subprovince	Boucot and others (1969)
	Europe "Province"		Telford (1972)
	Old World Province	**Old World Province**	Oliver (1975)
	Old World Realm Rhenish-Bohemian & Uralian Regs.	**Old World Realm** Uralian Region	Boucot (1975)
	Old World Realm Rhenish-Bohemian Region		Savage and others (herein)
LATE LOCHKOVIAN	**Old World Province** Rhenish-Bohemian Subprovince		Boucot and others (1969)
	Europe "Province"		Telford (1972)
	Old World Province	**Old World Province**	Oliver (1975)
	Old World Realm Rhenish-Bohemian & Uralian Regs.	**Old World Realm** Uralian Region	Boucot (1975)
	Old World Realm Rhenish-Bohemian Region		Savage and others (herein)
EARLY LOCHKOVIAN	**Old World Province** Rhenish-Bohemian Subprovince		Boucot and others (1969)
	Europe "Province"		Telford (1972)
	Old World Province	**Old World Province**	Oliver (1975)
	Old World Realm Rhenish-Bohemian & Uralian Regs.	**Old World Realm**	Boucot (1975)
	Old World Realm Rhenish-Bohemian Region		Savage and others (herein)

workers investigating other fossil groups. Ormiston (1972, 1975) noted biogeographic affinities of western and Arctic North American Devonian trilobite faunas. Because of the smaller sample size (fewer Lower Devonian trilobite genera are known), quantitative comparisons involving trilobites suffer more from sampling inadequacies than do comparisons involving the brachiopod faunas. Appalachian Lochkovian trilobite faunas are recognized by Ormiston as distinct from those of western and Arctic North America which have strong Old World ties. Ormiston (1975) reported *Acastella* aff. *heberti elsana,* a Rhenish taxon, from the Nova Scotia Gedinnian, which is in keeping with our assignment of Nova Scotian brachiopod faunas to the Rhenish-Bohemian biogeographic Region. Although we do not agree with the Siegenian correlation (our late Lochkovian and early Pragian) of Ormiston (1975), his conclusions regarding trilobite zoogeography at that time agree in large part with our views on brachiopod zoogeography. Arctic North American trilobite faunas of this time still show some Old World influences although provincial genera appear and begin to diversify in the latter part of the interval. Ormiston's Siegenian (our early Pragian) trilobites from central Nevada include a mixture of Appalachian and Old World genera, as does the brachiopod fauna. He interprets the migration route of these Appalachian genera to be by way of northern Mexico, i.e., around the southern limit of the landmass separating the two geographical regions. Ormiston (1972, 1975) concluded that the Emsian and Eifelian were times of greatest provincialism for North American Devonian trilobites but we feel that this conclusion results mainly from the larger sample, and hence lower proportion of cosmopolitan genera, of Emsian and Eifelian trilobites. Much of his data is from Arctic Canada where early Pragian age beds are of limited distribution because of regressive conditions. Ormiston recognized four Emsian-Eifelian biogeographic units which he named Appalachian, Cordilleran (Nevada), Canadian-Siberian (northwestern Canada, Alaska, and Siberia), and Uralian (Arctic Islands and the Urals). These cannot be compared with the results of our brachiopod analysis herein as we did not include Uralian and Siberian faunas in our study. Campbell and Davoren (1972) showed that

Table 5. (Cont.)

	EASTERN AUSTRALIA	GONDWANA	
EMSIAN	**Old World Province** Tasman Subprovince	**Malvinokaffric Province**	Boucot and others (1969)
	E Australian "Province"		Telford (1972)
	Old World Province		Oliver (1975)
	Old World Realm Tasman Region	**Malvinokaffric Realm**	Boucot (1975)
	Old World Realm Tasman Region	**Malvinokaffric Realm**	Savage and others (herein)
EARLY PRAGIAN	**Old World Province** Tasman Subprovince		Boucot and others (1969)
	E Australian "Province"		Telford (1972)
	Old World Province		Oliver (1975)
	Old World Realm Tasman Region		Boucot (1975)
	Old World Realm Tasman Region		Savage and others (herein)
LATE LOCHKOVIAN	**Old World Province** Tasman Subprovince		Boucot and others (1969)
	E Australian "Province"		Telford (1972)
	Old World Province		Oliver (1975)
	Old World Realm Tasman Region		Boucot (1975)
	Old World Realm Tasman Region		Savage and others (herein)
EARLY LOCHKOVIAN	**Old World Province** Tasman Subprovince		Boucot and others (1969)
	E Australian "Province"		Telford (1972)
	Old World Province		Oliver (1975)
	Old World Realm Tasman Region		Boucot (1975)
	Old World Realm Tasman Region		Savage and others (herein)

Early Devonian Australian trilobites have greatest affinity with those of western Europe and North Africa, much less with those of Kazakhstan and the Appalachians, and least with those of the other southern continents. Insufficient data were available from western and Arctic North America for a comparison between Australian and western and Arctic North American Lochkovian trilobite faunas.

From the data available, the biogeographic differentiation of Lower Devonian conodont faunas is less well developed than for the brachiopod, trilobite, and coral benthos. Telford (1972) tentatively recognized four Early Devonian conodont biogeographic units based on the distribution and diversity of representatives of the major Early Devonian platform genera. He noted that each conodont biogeographic unit shared many more species with other biogeographic units than do brachiopods from the same units. Klapper and Johnson (1975) noted anomalies in Emsian conodont distributions at the generic level and suggested these anomalies might be better explained in terms of biofacies. If the conodonts are interpreted to be pelagic and show depth stratification (Seddon and Sweet, 1971), then they might be expected to show variance from biogeographic units based upon benthos whose provinciality would be more strongly controlled by regressive and transgressive phases.

Oliver (1973, 1975) reported the highly provincial nature and marked dissimilarity of Appalachian rugose corals from the Old World Lower Devonian coral faunas. Emsian coral faunas appear to show a greater degree of provincialism than older Lower Devonian faunas although Appalachian Siegenian (early Pragian) corals are few and are virtually unstudied. Pedder (1975) commented that late Lochkovian coral faunas from Royal Creek, Yukon, are very different from similar age (*Pedavis pesavis* conodont unit) corals from the Windmill Limestone of Nevada. Unlike other Old World late Lochkovian faunas, the Nevada coral fauna consists of a large number of Silurian "holdovers." The Lochkovian faunas associated with *P. pesavis* in Yukon Territory bear affinity with Old World rugose coral faunas of Europe, Asia, and Australia. Early Pragian corals from the Yukon have affinities with Old World faunas but differ even at the family level from taxa known from central Nevada. Pedder noted the Old World affinity of Yukon Emsian corals and the marked differences with similar age rugose corals from Nevada.

As more work has been done on the biogeography of Lower Devonian brachiopods than on other fossil groups of that age, it is not surprising that the biogeographic subdivisions are finer and better established. If it is found, after more research on all these groups, that the biogeographic areas for the different fossil groups largely coincide, it is hoped that a standard terminology can be agreed upon for describing and ranking these various biogeographic units. Table 5 summarizes some of the subdivisions recently proposed for different fossil groups.

The hope has sometimes been expressed that detailed quantitative analyses of the distribution of successive benthic marine faunas will prove of major significance in providing compelling data which would influence plate tectonic models, particularly for Paleozoic times. The indices of faunal affinity resulting from our study do not appear to provide such data although they should be of permissive value as supplemental evidence. One of the major problems in any attempt to employ biogeographic information derived from benthic marine invertebrates has to do with their dispersal mechanisms. Information from the Recent shows that some species characterized by teleplanic larvae have great dispersal capabilities permitting transoceanic distribution whereas for others there is a complete transition down to forms with very limited larval distribution possibilities. The spectrum of larval dispersal capabilities and restrictions relating to environmental requirements are probably as important as geographic proximity in determining benthic marine faunal affinities.

ACKNOWLEDGMENTS

Several colleagues helped in the compilation of the faunal lists by providing data from their own collections. Peter Carls, Würzburg, Germany, kindly commented on the correlation of Gedinnian Rhenish facies strata. J. G. Johnson, Oregon State University, discussed aspects of faunal correlations and allowed Perry access to his many unpublished brachiopod faunas and faunal lists. Vladimir Havlíček, Prague, Czechoslovakia, kindly made revisions and additions to our Bohemian faunal lists. A. C. Lenz, University of Western Ontario, London, provided unpublished data from Royal Creek Yukon and Lowther Island, Arctic Archipelago. R. E. Smith, Oregon State University, conveyed additional information regarding the distribution of some Arctic Lochkovian brachiopods discussed in his thesis. A. J. Wright, Wollongong University, New South Wales, made useful comments regarding faunal lists in his thesis dealing with faunas from Mudgee, New South Wales.

Notes to Table 1

These notes serve to explain the various generic listings where identifications have been changed or synonymized and where correlation of the host strata have been changed.

Open Nomenclature Citations: Many stropheodontid taxa are currently under revision by Harper and Boucot (1977). Their manuscript proposes many new genera and subgenera which serve to increase provincialism because of restricted known occurrences of new taxa. These new names have not been used here. Taxa in question are shown in quotation marks in this and other instances where genera need revision so that the reader is aware of possible future changes.

Grès de Gdoumont: This stratigraphic unit is thought to span the Silurian-Devonian boundary (Carls, personal commun. 1975) and taxa from it have been included in the Rhenish-Bohemian list for the early Lochkovian interval.

Zlichovian *v.* Emsian: It is realized that the top of the Zlichovian is considerably older than the top of the Emsian and a correlation very similar to that of Pedder (1975) is used for the Rhenish and Bohemian sequences along with the informal conodont units of Klapper and others (1971).

Appalachian rhynchonellids: These shells have not been carefully studied and consequently the estimate of number of provincial and cosmopolitan genera for this group probably errs in favor of cosmopolitanism.

Drake Bay Beds, Prince of Wales Island: Ormiston (1967) assigned these beds to the Siegenian (our faunas 5 + 6). Subsequently *Eognathodus sulcatus* Philip (late forms) were found in these beds confirming this age assignment. However, forms of *E. sulcatus* are reported to range as high as Emsian on Bathurst Island (Uyeno *in* McGregor and Uyeno, 1972).

Sutherland River Formation: Boucot and others (1960) assigned the Formation to the Late Silurian, however, it is now known to belong within the early Lochkovian interval based on both conodonts and shelly faunas.

"Acanthospirifer": A shell very similar to Llandovery *Acanthospirifer* has been discovered within conodont faunas 1 + 2 by R. E. Smith in Arctic Archipelago.

Anatrypa - Desquamatia: These "genera" are combined here because distinction between them is uncertain.

Amphigenia: The European representatives are the pedicle valves illustrated by LeMaitre (1934) which appear to have a more terebratuloid than gypiduloid septalium.

"Amphistrophia": This group of shells is under revision by Harper and Boucot (1977).

"Anoplia": *Anoplia* has been reported from several areas of the Rhenish-Bohemian late Lower Devonian. In all cases, internal structures of these European shells are distinct from those of *Anoplia s.s.* as is particularly evident in *"Anoplia" theorassensis* (Maillieux).

"Astutorhynchia": These are shells studied by J. G. Johnson which are apparently related to the Bohemian *Astutorhynchia,* but probably morphologically distinct at the generic level.

Baturria: Boucot has recovered *Baturria* from Kinglake, Victoria, Australia in association with *Hedeina (Macropleura) densilineata* (Chapman).

"Boucotia": These are Gaspé notanopliids which are distinct from other notanopliids but resemble *Boucotia.*

"Camerella": These are the shells referred to *"Camerella"* from Arctic Canada (J. G. Johnson, personal commun. 1975) but do not include the shells assigned to *"Camerella"* by Johnson (1970) from Nevada for which Boucot (1975) proposed the genus *Llanoella.*

"Chonetes": This category includes many Lower Devonian chonetids which are poorly preserved or poorly known and may not strictly belong to the same genus.

Cordatomyonia: This unit does not include material referred to *Cordatomyonia* (Bourque, 1973) which is closer to *Baturria* where it is included in our compilation.

"Cryptonella": These are various species assigned to *Cryptonella,* such as *C. rhenana,* from the Rhenish-Bohemian area which have internal structures different from *Cryptonella s.s.* and probably belong to a new genus.

"Cupularostrum": This is a poorly known Lower Devonian genus and the occurrences listed may include more than one genus.

"Decoropugnax": These are weakly ribbed rhynchonellids which are close to *Decoropugnax* or possibly a new genus.

Dalejina: Chatterton (1973) has assigned shells to *Aulacella* which have flat-topped and medially grooved (in cross section) peripheral crenulations and we have included this material in *Dalejina. Aulacella* has crenulations of rounded cross section and a brachial sulcus which is relatively angular in cross section compared to that of *Dalejina.*

"Delthyris": This is Chatterton's (1973) *Delthyris hudsoni* which has brachial cardinalia reminiscent of the Hysterolitinae and plications unlike those present in any other species of *Delthyris s.s.* Strong medial and lateral plications with v-shaped cross sections in *"D." hudsoni* are unlike those of the Gedinnian and older species of *Delthyris.*

Dubaria: Included here are shells commonly referred to the junior synonym *Atrypopsis* in the Bohemian literature.

Eoschuchertella: This name was proposed by Gratsianova (1974) for impunctate Lower Devonian schuchertellids.

Fascicostella: Fascizetina Havlíček (1975a) is regarded of subgeneric rank and included here.

"Fimbrispirifer": This includes *"Spirifer" scheii* Meyer of Arctic Canada which appears to be a paraspiriferid because of its shape and fine ornament of relatively flat-topped, medially grooved anterior costellae.

"Franklinella": The name *Franklinella* is already used for an ostracode taxon and the invalid brachiopod name is therefore used here in quotation marks.

Hedeina: Macropleura is here grouped with *Hedeina.* Although Brunton and others (1967) showed that *Macropleura* Boucot is an objective synonym of *Hedeina,* Boucot (1975) resurrected *Macropleura* as a subgenus of *Hedeina* for the very large Devonian species.

"Hipparionyx": This includes *"Hipparionyx" hipponyx* (Schnur) which is an orthotetacid characteristic of the Rhenish-Bohemian area and does not have the outline or robust form which would permit generic assignment to *Hipparionyx s.s.*

Howellella: Included in this unit is material from the Lower Devonian of Nevada described as *Undispirifer* by Johnson and others (1973b). This material is silicified and definite fine reticulariid ornament is not preserved. In view of this the material is best regarded as *Howellella.*

"Hysterolites:" These are the shells from Nevada, Australia, and the North American Cordillera which lack the typical hysterolitinid cardinalia, and may represent more than one genus related to *Howellella s.l.*

Iridistrophia: The Australian representatives are shells referred to *Schellwienella* by Talent (1965).

Isorthis (Isorthis): The genus *Zlichopyramis* Havlíček (1975a) is here assigned to *Isorthis (Isorthis) sensu* Walmsley and Boucot (1975).

"Janius": Included here are distinctive, yet poorly preserved shells from Royal Creek, Yukon with bifurcating lateral costae and radial striae.

Katunia: The Nevada representatives are shells referred to *Camarotoechia modica* by Johnson (1970) (Johnson, personal commun. 1975).

Kayserella: The genus *Biernatium* Havlíček (1975) whose type species is *Skenidium fallax* Gürich is here regarded as congeneric with *Kayserella* as suggested by Cooper (1955).

"Leiorhynchus": An undescribed leiorhynchid genus from Arctic Canada (Johnson, personal commun. 1975).

Leptaena - Leptagonia: These "genera" are combined here because distinction between them is uncertain.

Leptaenisca: The European example is *Taleoleptaena* of Havlíček (1967),

"Leptostrophia": This group is under revision by Harper and Boucot (1977).

"Mclearnites": This group is under revision by Harper and Boucot (1977).

Megakozlowskiella: This genus includes pre-Eifelian age European shells formerly assigned to *Cyrtinopsis.*

Meristella - Meristina: Meristina and *Meristella* are stated to be separable on the basis of their jugal structures. In common practice the form of the pedicle valve muscle field in large specimens (obsolescent dental lamellae, deeply impressed, trapezoidal muscle field in *Meristella)* is employed for distinction. Previously most Appalachian meristellids were thought to have a *Meristella*-like pedicle muscle field and most Old World genera were thought to be of the *Meristina* type. However, the muscle field alone has not been found to be consistent with jugal structures; thus the genera are grouped here.

"Mesodouvillina": This group is under revision by Harper and Boucot (1977).

"Mesopholidostrophia": This group is under revision by Harper and Boucot (1977).

Pacificocoelia: True *Leptocoelia* are known only from the Oriskany through Schoharie age beds of the Appalachians (Boucot and Rehmer, 1977); other leptocoeliids, with the exception of the Malvinokaffric Realm *Australocoelia,* belong to *Pacificocoelia* which also occurs in the Appalachian area.

"Phoenicitoechia": A poorly known group of Lower Devonian rhynchonellids which here include true *Phoenicitoechia* from Bohemia together with Cordilleran forms which may not be congeneric.

"Pholidostrophia": This group is under revision by Harper and Boucot (1977).

"Phragmophora": A probable new genus of septate dalmanellid closely related to the true Middle Devonian *Phragmophora.*

Platyorthis: The two valves assigned to *Platyorthis* from southeast Australia by Savage (1971) are of doubtful affinity and therefore not included in this compilation.

Plicocyrtina: Chatterton's (1973) *Cyrtinopsis* has brachial cardinalia, fine ornament, and medially conjunct dental lamellae adherent to a median septum, all of which are characteristics of *Plicocyrtina.* Furthermore it lacks the fine ornament and cardinalia characteristics of *Cyrtinopsis.* Therefore we include Chatterton's *Cyrtinopsis* in *Plicocyrtina.*

"Protochonetes": This unit comprises material described by Chatterton (1973) as *Protochonetes* and by Johnson (1975b) as *"Chonetes"* sp. B.

Prokopia: Miniprokopia Havlíček is included here with *Prokopia.*

"Pugnax": This is the material of Chatterton (1973) which shows some affinity to both *Pugnax* and *Parapugnax.*

"Reeftonia": This includes poorly known material from Arctic Canada and eastern Australia. True *Reeftonia* are known only with certainty from New Zealand. The assignment of some Australian shells to *Reeftonia* is questioned by Walmsley and Boucot (1975).

Resserella: The genus *Parmorthina* Havlíček (1975a) is included here with *Resserella.*

"Rhynchotreta": This comprises a poorly known group of Appalachian rhynchonellids which are distinct from true *Rhynchotreta.*

"Rhynchospirina": This includes material previously referred to *Retzia, Homeospira,* and *Rhynchospirina,* all poorly known taxa which may or may not be congeneric.

Salopina: It is important to realize that the nonseptate salopinids of the Silurian, occurring in Benthic Assemblage 2 in abundance, have similar descendants in the Appalachian and Rhenish-Bohemian Lower Devonian. These are referred herein to *Salopina.* Incipiently septate salopinids of the *crassiformis* and *submurifer* type persist from the Silurian into the Lower Devonian in a Benthic Assemblage 4-5 position. These are referred herein to *Salopina crassiformis* type and treated as a distinct unit.

"Septalaria": Septalaria and *Pseudocamarophoria* are poorly known, morphologically similar, rhynchonellid genera and grouped here as *"Septalaria." Amissopecten leidholdi* (Havlíček, 1961) is included in *"Septalaria".*

"Sibirispira": The northeastern Siberia genus *Sibirispira* Alekseeva is poorly known internally and North American shells of somewhat similar external morphology are for the time being referred to *"Sibirispira"* although they may well be a distinct genus.

Skenidioides: Some European workers have assigned shells with internal structures showing the characters of *Skenidioides* to *Skenidium.* True *Skenidium* is unknown to us outside the Appalachian area.

"Sphaerirhynchia": This unit includes forms with and without conjunct inner hinge plates which have been assigned previously to *Sphaerirhynchia* but some of which may belong to *Tadschikia* or other undescribed genera.

Straelina: The shells assigned to *Straelina* by Talent (1965) appear closer to *Trematospira* and are therefore included with *Trematospira* in this compilation.

Stegerhynchus: The material assigned to indet. rhynchonellid by Tillman (1967) appears to be a *Stegerhynchus.*

"Strophochonetes": This is a broadly conceived group which may include more than one genus.

"Strophodonta': This group is under revision by Harper and Boucot (1977). The types of the genus *Arbizustrophia* Garcia-Alcalde (1972c) may include more than one taxon and they all appear very close to *Strophodonta. Arbizustrophia* is here grouped with *"Strophodonta."*

"Thliborhynchia": This unit comprises extremely finely costellate rhynchonellids of *Thliborhynchia* shape which belongs to a new unnamed genus.

"Uncinulus": This unit includes the numerous Devonian uncinuloids of widely differing character which have been referred to *Uncinulus* and probably need revision.

Vagrania: A broad concept of the genus is used here. The recent revision of *Vagrania* and the erection of the related genus *Totia* by Rzhonsnitskaya and Mizens (1975) is not available to us.

Vandercammenina: Spirifer charybdis Barrande is included here in *Vandercammenina* whereas Havlíček (1959) assigned the species to *Fimbrispirifer.*

Warrenella - Reticulariopsis: This unit includes smooth Lower Devonian reticulariids which have been assigned previously to *Reticulariopsis* or *Warrenella.*

REFERENCES

Asselberghs, Etienne, 1946, L'Eodevonien de l'Ardenne et des Regions voisines: Louvain Univ. Inst. Geol. Mem., v. 14, 598 p. (1)

Barrande, Joachim, 1847, Über die brachiopoden der silurischen Schichten von Böhmen: Naturw. Abh. Wien, v. 1, p. 357-475. (2)

Binnekamp, J. G., 1965, Lower Devonian brachiopods and stratigraphy of North Palencia (Cantabrian Mountains, Spain): Leidse Geol. Meded., v. 33, p. 1-62. (4)

Boucot, A. J., 1960a, Implications of Rhenish Lower Devonian brachiopods from Nova Scotia: Internat. Geol. Cong., 21st, Norden 1960, Proc., Section 12, p. 129-137. (5)

——— 1960b, Lower Gedinnian brachiopods of Belgium: Louvain Univ. Inst. Geol. Mem., v. 21, p. 283-324. (6)

——— 1960c, A new Lower Devonian stropheodontid brachiopod: Jour. Paleontology, v. 34, p. 483-485. (7)

——— 1962, Observations regarding Silurian and Devonian spiriferoid genera: Senckenbergiana Lethaea, v. 43, p. 411-432. (8)

——— 1973, Early Paleozoic brachiopods of the Moose River Synclinorium, Maine: U.S. Geol. Survey Prof. Paper 784, 81 p. (9)

——— 1975, Evolution and extinction rate controls: New York, Elsevier, 427 p. (10)

——— Unpubl. collections from Kinglake District, Victoria, Australia. (11)

——— P. A. Bourque's unpubl. Gaspé materials reviewed with Boucot. (12)

Boucot, A. J., Cumming, L. M., and Jaeger, H., 1967, Contributions to the age of the Gaspé Sandstone and Gaspé Limestone: Canada Geol. Survey Paper 67-25, 27 p. (13)

Boucot, A. J., and Harper, C. W., 1968, Silurian to lower Middle Devonian Chonetacea: Jour. Paleontology, v. 42, p. 143-176. (15)

Boucot, A. J., and Johnson, J. G., 1967, Paleogeography and correlation of Appalachian Province Lower Devonian sedimentary rocks: Tulsa Geol. Soc. Digest, v. 35, p. 35-87. (16)

——— 1972, *Callicalpytella*, a new genus of notanopliid brachiopod from the Devonian of Nevada: Jour. Paleontology, v. 46, p. 299-302. (17)

——— 1973, Silurian brachiopods, *in* Hallam, A., ed., Atlas of palaeobiogeography: New York, Elsevier, p. 59-65. (18)

Boucot, A. J., Johnson, J. G., and Staton, R. D., 1964, On some atrypoid, retzioid, and athyridoid Brachiopoda: Jour. Paleontology, v. 38, p. 805-822. (19)

Boucot, A. J., Johnson, J. G., and Talent, J. A., 1969, Early Devonian brachiopod zoogeography: Geol. Soc. America Spec. Paper 119, 113 p. (20)

Boucot, A. J. and others, 1960, A late Silurian fauna from the Sutherland River Formation, Devon Island, Canadian Arctic Archipelago: Canada Geol. Survey Bull. 65, 51 p. (21)

Boucot, A. J. and others, 1966, *Skenidioides* and *Leptaenisca* in the Lower Devonian of Australia (Victoria, Tasmania) and New Zealand, with notes on other Devonian occurrences of *Skenidioides:* Royal Soc. Victoria Proc., v. 79, p. 363-369. (14)

Boucot, A. J., and Rehmer, Judith, 1977, *Pacificicocoelia acutiplicata* (Conrad, 1841) (Brachiopoda) from the Esopus Shale (Lower Devonian) of eastern New York: Jour. Paleontology (in press). (22)

Bourque, P. A., 1973, Stratigraphie du Silurien et du Dévonien basal du nord-est de la Gaspésie avec une illustration de la fauna à Brachiopodes: Ph.D. thesis, Univ. Montréal, Canada, 291 p. (23)

Bowen, Z. P., 1967, Brachiopoda of the Keyser Limestone (Silurian-Devonian) of Maryland and adjacent areas: Geol. Soc. America Mem. 102, 103 p. (24)

Brunton, C. H. C., Cocks, L. R. M., and Dance, S. P., 1967, Brachiopods in the Linnaean collection: Linnean Soc. London Proc., v. 178, p. 161-183. (25)

Campbell, K. S. W., and Davoren, P. J., 1972, Biogeography of Australian Early-early Middle Devonian trilobites, *in* Talent, J. A., Provincialism and Australian Early Devonian faunas: Australian Geol. Soc. Jour., v. 19, p. 88-93. (26)

Carls, Peter, 1974, Die Proschizophoriinae (Brachiopoda; Silurium-Devon) der Ostlichen Iberischen Ketten (Spanien): Senckenbergiana Lethaea, v. 55, p. 153-227. (27)

Carls, Peter and others, 1972, Neue Daten zur Grenze Unter-Mittel-Devon: Newsl. Stratigr., v. 2, p. 115-147. (28)

Chatterton, B. D. E., 1973, Brachiopods of the Murrumbidgee Group, Taemas, New South Wales: Australia Bur. Mineral Resources, Geology and Geophysics Bull. no. 137, 146 p. (29)

Cheetham, A. H., and Hazel, J. E., 1969, Binary (presence-absence) similarity coefficients: Jour. Paleontology, v. 43, p. 1130-1136. (30)

Chlupáč, Ivo, 1953, Stratigrafická studie o hraničních vrstvach mezi silurem a devonem ve strednich Čechách (Stratigraphical investigations of the border strata of the Silurian and the Devonian in central Bohemia): Ústřed. Úst. Geol. Sborn. v. 20, p. 277-380. (31)

——— 1972, The Silurian-Devonian boundary in the Barrandian (with contributions by Hermann Jaeger and Jana Zikmundova): Canadian Petroleum Geology Bull., v. 20, p. 104-174. (32)

Clarke, J. M., 1908, Early Devonic history of New York and eastern North America: New York State Mus. Mem. 9, pt. 1, 366 p. (33)

Comte, Pierre, 1938, Brachiopodes Dévoniens des Gisements de Ferroñes (Asturies) et de Sabero (Léon): Ann. Paléont., Invertébrès, v. 27, p. 41-87. (34)

Cooper, G. A., 1955, New genera of middle Paleozoic brachiopods: Jour. Paleontology, v. 29, p. 45-63. (35)

Crickmay, C. H., 1968, Discoveries in the Devonian of western Canada: Calgary, Evelyn de Mille Books, 12 p. (36)

Drot, Jeannine, 1964, Rhychonelloidea et Spiriferoidea Siluro-Devoniens du Maroc pre-Saharien: Notes et Mém. Serv. Géol. Maroc, no. 178, 286 p. (37)

——— 1975, Orthida (Brachiopodes) du Maroc Présaharien. I. Orthidina. II. Dalmanellidina de Dévonien Inferieur a l'Exclusion du Genre *Schizophoria:* Ann. Paléont., Invertébrès, v. 61, p. 4-99. (38)

Drot, Jeannine, and Westbroek, Peter, 1966, *Iberirhynchia santaluciensis,* nouveau rhynchonellacea du Dévonien de Leon (Espagne): Leidse Geol. Meded., v. 38, p. 165-172. (39)

Flood, P. G., 1969, Lower Devonian conodonts from the Lick Hole Limestone, southern New South Wales: Royal Soc. New South Wales Jour. Proc., v. 102, p. 5-9. (40)

——— 1973, *Uncinulus australia,* a new rhynchonellid species from the Lower Devonian of southern New South Wales: Australia Bur. Mineral Resources, Geology and Geophysics Bull. no. 126, p. 1-6. (41)

——— 1974, Lower Devonian brachiopods from the Point Hibbs Limestone of western Tasmania: Royal Soc. Tasmania Proc., v. 108, p. 113-136. (42)

Fuchs, Alexander, 1923, Über die Beziehungen der sauerlandischen Faciesgebietes zur belgischen Nord-und Sudfacies und ihre Bedeutung für das Alter der Verseschichten: Preuss. Geol. Landesanst. Jahrb., v. 42, p. 839-859. (43)

Garcia-Alcalde, J. L., 1972a, Braquiopodos Devonicos de la Cordillera Cantabrica. 2. Genero *Xana* Garcia-Alcalde, n. gen. (Terebratulida, Stringocephalacea): Breviora Geol. Asturica, v. 16, p. 4-12. (44)

——— 1972b, Braquiopodos Devonicos de la Cordillera Cantabrica. 3. *Fibulistrophia* n. gen. (Strophomenida, Strophodontacea): Breviora Geol. Asturica, v. 16, p. 42-48. (45)

——— 1972c, Braquiopodos Devonicos de la Cordillera Cantabrica. 4. *Arbizustrophia* n. gen. (Strophomenida, Strophodontacea): Breviora Geol. Asturica, v. 16, p. 56-64. (46)

Garcia-Alcalde, J. L., and Racheboeuf, P. R., 1975, Données paléobiologiques et paléobiogeographiques sur quelques Strophochonetinae du Dévonien d'Espagne et du Massif Armoricain: Lethaia, v. 8, p. 329-338. (47)

Gill, E. D., 1951, Two new brachiopod genera from Devonian rocks of Victoria: Natl. Mus. Victoria Mem. 17, p. 187-205. (48)

——— 1952, Palaeogeography of the Australian-New Zealand region in Lower Devonian time: Royal Soc. New Zealand Trans. Proc., v. 80, p. 171-185. (49)

——— 1969, Notanopliidae, a new family of Palaeozoic Brachiopoda from Australia: Jour. Paleontology, v. 43, p. 1222-1231. (50)

Gratsianova, R. T., 1974, "*Schuchertella*" of the Early and Middle Devonian of southwestern Siberia: Systematic relationships, elements of ecology, stratigraphic importance: Sreda i Zhizn v Geologicheskom Proshlom (Paleoekologicheskie, Problemy), Izdatelstvo "Nauka" Sibirskoe Otdelenie, Novosibirsk, p. 77-87 (in Russian). (51)

Harper, C. W., Jr., and Boucot, A. J., 1977, The Stropheodontacea: Palaeontographica, Abt. A (in press). (53)

Harper, C. W., Johnson, J. G., and Boucot, A. J., 1967, The Pholidostrophiinae (Brachiopoda; Ordovician, Silurian, Devonian): Senckenbergiana Lethaea, v. 48, p. 403-461. (54)

Havlíček, Vladimir, 1956, The brachiopods of the Braník and Hlubočepy Limestones in the immediate vicinity of Prague: Ústřed. Ûst. Geol. Sborn., v. 22, p. 535-665. (55)

——— 1959, Spiriferidae v českém siluru a devonu (Brachiopoda): Ústřed. Ûst. Geol. Rozpravy, v. 25, 275 p. (56)

——— 1961, Rhynchonelloidea des böhmischen älteren Paläozoikum (Brachiopoda): Ústřed. Ûst. Geol. Rozpravy, v. 27, 211 p. (57)

——— 1963, *Zlichorhynchus hiatus* n. g. et n. sp., neuer Brachiopode, von Unterdevon Böhmens: Ústřed Ûst. Geol. Věstn., v. 38, p. 403-404. (58)

——— 1967, Brachiopoda of the suborder Strophomenidina in Czechoslovakia: Ústřed. Ûst. Geol. Rozpravy, v. 33, 235 p. (59)

——— 1967, *Hanusatrypa* and *Falsatrypa* (Atrypacea, Brachiopoda) in the Lower Devonian of Bohemia: Ústřed. Ûst. Geol. Věstn., v. 42, p. 443-444. (60)

——— 1971a, Non-costate and weakly costate Spiriferidina (Brachiopoda) in the Silurian and Lower Devonian of Bohemia: Geol. Věd. Paleont. Sborn., Rada P, no. 14, p. 7-34. (61)

——— 1971b, New genera of enteletacean brachiopods in the Devonian of Bohemia: Ústřed. Ûst. Geol. Věstn., v. 46, p. 229-232. (62)

——— 1973, New brachiopod genera in the Devonian of Bohemia: Ústřed Ûst. Geol. Věstn., v. 48, p. 337-340. (63)

——— 1974, New genera of Orthidina (Brachiopoda) in the Lower Paleozoic of Bohemia: Ústřed. Ûst. Geol. Věstn., v. 49, p. 167-170. (64)

——— 1975a, New genera and species of Orthida (Brachiopoda): Ústřed. Ûst. Geol. Věstn., v. 50, p. 231-235. (65)

——— 1975b, Personal communication with regard to the Barrandian brachiopod distributions. (66)

Havlíček, Vladimir, and Plodowski, Gerhard, 1974, Über die systematische stellung der Gattung *Hircinisca* (Brachiopoda) aus dem bohmischen Ober-Silurium: Senckenbergiana Lethaea, v. 55, p. 229-249. (67)

Jackson, D. E., 1969, Ordovician graptolite faunas in lands bordering North Atlantic and Arctic Oceans, *in* Kay, Marshall, ed., North Atlantic—geology and continental drift: Am. Assoc. Petroleum Geologists Mem. 12, p. 504-512. (69)

Jahnke, Hans, 1971, Fauna und Alter der Erbslochgrauwacke (Brachiopoden und Trilobiten, Unter-Devon, Rheinisches Schiefergebirge und Harz): Göttinger Arbeiten Geol. Paläont. no. 9, 105 p. (70)

Johnson, J. G., 1966, *Parachonetes*, a new Lower and Middle Devonian brachiopod genus: Palaeontology, v. 9, p. 365-370. (71)

——— 1970, Great Basin Lower Devonian Brachiopoda: Geol. Soc. America Mem. 121, 421 p. (72)

——— 1971a, A quantitative approach to faunal province analysis: Am. Jour. Sci., v. 270, p. 257-280. (73)

——— 1971b, Lower Givetian brachiopods from central Nevada: Jour. Paleontology, v. 45, p. 301-326. (74)

——— 1973a, Late Early Devonian rhynchonellid genera from Arctic and western North America: Jour. Paleontology, v. 47, p. 465-472. (75)

——— 1973b, Mid-Lochkovian brachiopods from the Windmill Limestone of central Nevada: Jour. Paleontology, v. 47, p. 1013-1030. (76)

——— 1974a, Early Devonian brachiopod biofacies of western and Arctic North America: Jour. Paleontology, v. 48, p. 809-819. (77)

——— 1974b, *Oriskania* (terebratulid brachiopod) in the Lower Devonian of central Nevada: Jour. Paleontology, v. 48, p. 1207-1212. (78)

——— 1975a, Devonian brachiopods from the *Quadrithyris* Zone (Upper Lochkovian), Canadian Arctic Archipelago: Canada Geol. Survey Bull. 235, p. 5-57. (79)

——— 1975b, Late Early Devonian brachiopods from the Disappointment Bay Formation, Lowther Island, Arctic Canada: Jour. Paleontology, v. 49, p. 947-978. (80)

——— 1975c, Personal communication on brachiopod distribution in Arctic Archipelago, Bohemia, and New York State (Coeymans Limestone). (81)

Johnston, J. G., and Boucot, A. J., 1968, External morphology of *Anatrypa* (Devonian, Brachiopoda): Jour. Paleontology, v. 42, p. 1205-1207. (82)

——— 1973, Devonian brachiopods, *in* Hallam, A., ed., Atlas of palaeobiogeography: New York, Elsevier, p. 89-96. (83)

Johnson, J. G., Boucot, A. J., and Murphy, M. A., 1973, Pridolian and early Gedinnian age brachiopods from the Roberts Mountains Formation of central Nevada: Univ. California Pubs. Geol. Sci., v. 100, 75 p. (84)

Johnson, J. G., and Kendall, G. W., 1977, Late Early Devonian brachiopods and biofacies from central Nevada: Jour. Paleontology (in press). (85)

Johnson, J. G., and Ludvigsen, Rolf, 1972, *Carinagypa*, a new genus of pentameracean brachiopod from the Devonian of western North America: Jour. Paleontology, v. 46, p. 125-129. (86)

Johnson, J. G., and Niebuhr, W. W., II, 1977, Anatomy of an assemblage zone: Geol. Soc. America Bull. (in press). (87)

Johnson, J. G., and Talent, J. A., 1967, *Muriferella*, a new genus of Lower Devonian septate dalmanellid: Royal Soc. Victoria Proc., v. 80, p. 43-50. (88)

Kegel, Wilhelm, 1926, Unterdevon von böhmischer Facies (Steinberger Kalk) in der Lindener Mark bei Giessen: Preuss. Geol. Landesanst. Abh., n. f., no. 100, p. 1-77. (89)

Kerr, J. W., 1974, Geology of Bathurst Island Group and Byam Martin Island, Arctic Canada: Canada Geol. Survey Mem. 378, 152 p. (90)

Klapper, Gilbert, 1969, Lower Devonian conodont sequence, Royal Creek, Yukon Territory, and Devon Island, Canada (with a section on Devon Island stratigraphy by A. R. Ormiston): Jour. Paleontology, v. 43, p. 1-27. (91)

Klapper, Gilbert, and Johnson, D. G., 1975, Sequence in conodont genus *Polygnathus* in Lower Devonian at Lone Mountain, Nevada: Geologica et Palaeontologica, v. 9, p. 65-83. (92)

Klapper, Gilbert and others, 1971, North American Devonian conodont biostratigraphy: Geol. Soc. America Mem. 127, p. 285-316. (93)

Kozlowski, Roman, 1929, Les Brachiopodes Gothlandiens de la Podolie Polonaise: Palaeontologica Polonica, v. 1, 254 p. (94)

Langenstrassen, Frank, 1972, Fazies und Stratigraphie der Eifel-Stufe im östlichen Sauerland (Rheinisches Schiefergebirge, Bl. Schmallenberg und Girkhausen): Göttinger Arbeiten Geol. Paläont. no. 12, 106 p. (95)

LeMaître, Dorothée, 1934, Etudes sur la faune des calcaires dévoniens du Bassin d'Ancenis; Calcaire de Chaudefonds et calcaire de Chalonnes (Maine-et-Loire): Soc. Géol. Nord Mém., v. 12, 267 p. (96)

Lenz, A. C., 1973, *Quadrithyris* Zone (Lower Devonian) near-reef brachiopods from Bathurst Island, Arctic Canada, with a description of a new rhynchonellid brachiopod *Franklinella:* Canadian Jour. Earth Sci., v. 10, p. 1403-1409. (97)

——— 1975, Personal communication of brachiopod distributions from environs of Royal Creek, Yukon and Lowther Island, Arctic Archipelago. (98)

Lenz, A. C., and Jackson, D. E., 1964, New occurrences of graptolites from the South Nahanni region, Northwest Territories and Yukon: Canadian Petroleum Geology Bull., v. 12, p. 892-900. (99)

Lenz, A. C., and Pedder, A. E. H., 1972, Lower and Middle Paleozoic sediments and paleontology of Royal Creek and Peel River, Yukon, and Powell Creek, N.W.T.: Internat. Geol. Cong., 24th, Montreal 1972, Guidebook, Field Excursion A14, 43 p. (100)

Ludvigsen, Rolf, 1970, Age and fauna of the Michelle Formation, northern Yukon Territory: Canadian Petroleum Geology Bull., v. 18, p. 407-429. (101)

Ludvigsen, Rolf, and Perry, D. G., 1975, The brachiopod *Warrenella* in the Lower and Middle Devonian formations of northwestern Canada: Canada Geol. Survey Bull. 235, p. 59-107. (102)

McGregor, D. C., and Uyeno, T. T., 1972, Devonian spores and conodonts of Melville and Bathurst Islands, District of Franklin: Canada Geol. Survey Paper 71-13, 37 p. (103)

Malone, E. J., 1967, Devonian of the Anakie High area, Queensland Australia, *in* Oswald, D. H., ed., Internat. Symposium on the Devonian System, Calgary 1967: Calgary, Alberta Soc. Petroleum Geologists, v. 2, p. 93-97. (104)

Maillieux, E., 1935, Brachiopodes et pélécypodes Dévoniens: Musée Royal d'Histoire Nat. Belgique Mém., no. 70, 42 p. (105)

——— 1941, Les brachiopods de l'Emsien de l'Ardenne: Musée Royal d'Histoire Nat. Belgique Mém., no. 96, 74 p. (106)

Merriam, C. W., 1973, Paleontology and stratigraphy of the Rabbit Hill Limestone and Lone Mountain Dolomite of central Nevada: U.S. Geol. Survey Prof. Paper 808, 50 p. (107)

Middlemiss, F. A., Rawson, P. F., and Newall, G., eds., 1971, Faunal provinces in space and time: Liverpool, Seel House Press, 236 p. (108)

Mittmeyer, H. G., 1965, Die Bornischer Schichten in Gebiet zwischen Mittelrhein und Idsteiner Senke (Taunus, Rheinisches Schiefergebirge): Hess. Landesamt Bodenforsch. Notizbl., v. 93, p. 73-98. (109)

——— 1973a, Grenze Siegen/Unterems bei Bornhofen (Unter-Devon, Mittelrhein): Mainzer Geowiss. Mitteil., v. 2, p. 71-103. (110)

——— 1973b, Die Hunsrückschiefer-Fauna des Wisper-Gebietes im Taunus: Ulmen-Gruppe tiefes Unter-Ems, Rheinisches Schiefergebirge: Hess. Landesamt Bodenforsch. Notizbl., v. 101, p. 16-45. (111)

——— 1974, Zur Neufassung der Rheinischen Unterdevon-Stufen: Mainzer Geowiss. Mitteil., v. 3, p. 69-79. (112)

Nikiforova, O. I., 1954, Stratigraphy and brachiopods of the Silurian deposits of Podolia: All-Union Scientific Research Geological Institute (VSEGEI) Trudy, 218 p. (113)

Oehlert, D. P., 1901, Fossiles Dévoniens de Santa-Lucia (Province de Léon, Espagne): Soc. Géol. France Bull., ser. 4, v. 1, p. 233-250. (114)

Oliver, W. A., Jr., 1973, Devonian coral endemism in eastern North America and its bearing on paleogeography, *in* Hughes, N. F., ed., Organisms and continents through time: Spec. Papers Palaeontology no. 12, p. 318-319. (115)

——— 1975, Endemism and evolution of Late Silurian to Middle Devonian rugose corals in eastern North America, *in* Sokolov, B. S., ed., Ancient Cnidaria, Vol. 2: Novosibirsk, p. 148-150. (116)

Ormiston, A. R., 1967, Lower and Middle Devonian trilobites of the Canadian Arctic Islands: Canada Geol. Survey Bull. 153, 148 p. (117)

——— 1969, A new Lower Devonian rock unit in the Canadian Arctic Islands: Canadian Jour. Earth Sci., v. 6, p. 1105-1111. (118)

——— 1972, Lower and Middle Devonian trilobite zoogeography in northern North America: Internat. Geol. Congress, 24th, Montreal 1972, Proc., Section 7, p. 594-604. (119)

——— 1975, Siegenian trilobite zoogeography in Arctic North America, *in* Martinsson, A., ed., Evolution and morphology of the Trilobita, Trilobitoidea and Merostomata: Fossils and Strata, no. 4, p. 391-398.(120)

Packham, G. H., ed., 1969, The geology of New South Wales: Geol. Soc. Australia Jour., v. 16, 654 p. (121)

Pedder, A. E. H., 1975, Sequence and relationships of three Lower Devonian coral faunas from Yukon Territory: Canada Geol. Survey Paper 75-1, pt. B, p. 285-295. (122)

Perry, D. G., 1974, Paleontology and biostratigraphy of Delorme Formation (Siluro-Devonian), Northwest Territories: Ph.D. thesis, Univ. Western Ontario, London, Canada, 682 p. (123)

——— Personal observation of *Pentamerella* in Prongs Creek Formation reefs, Knorr Range, northern Yukon. (124)

——— Personal observation of brachiopod faunas from Arctic Archipelago collections of R. Thorsteinsson, H. Trettin, J. W. Kerr, and R. E. Smith at Oregon State University. (125)

Perry, D. G., and Boucot, A. J., 1978, Late Early Devonian brachiopods from Mount Lloyd George area north-central British Columbia: Canada Geol. Survey Paper (in press). (126)

Perry, D. G., Klapper, Gilbert, and Lenz, A. C., 1974, Age of the Ogilvie Formation (Devonian), northern Yukon: Based primarily on the occurrence of brachiopods and conodonts: Canadian Jour. Earth Sci., v. 11, p. 1055-1097. (127)

Perry, D. G., and Lenz, A. C., 1977, Emsian paleogeography and shelly fauna biostratigraphy of Arctic Canada: Geol. Assoc. Canada (in press). (128)

Philip, G. M., 1962, The paleontology and stratigraphy of the Siluro-Devonian sediments of the Tyers area, Gippsland, Victoria: Royal Soc. Victoria Proc., v. 75, p. 123-246. (129)

Philip, G. M., and Pedder, A. E. H., 1967, A correlation of some Devonian limestones of New South Wales and Victoria: Geol Mag., v. 104, p. 232-239. (130)

Renaud, Alzine, 1942, Le Dévonien du synclinorium médian Brest-Laval: Soc. Géol. Min. Bretagne Mém. 7, pt. 2, Paléontologie, 439 p. (132)

Renouf, J. T., 1972, Brachiopods from the Grès a *Orthis monnieri* Formation of northwestern France and their significance in Gedinnian/Siegenian stratigraphy of Europe: Palaeontographica, Abt. A, v. 139, p. 89-133. (133)

Rzhonsnitskaya, M. A., and Mizens, L. I., 1975, Novi Diagnoz Roda *Vagrania* i Novi Rod *Totia* (Brachiopoda): Ezhegod. Vsesouz Paleont., v. la, p. 1-5. (134)

Sando, J. W., Bamber, E. W., and Armstrong, A. K., 1975, Endemism and similarity indices: Clues to the zoogeography of North American Mississippian corals: Geology, v. 3, p. 661-664. (135)

Savage, N. M., 1968a, The geology of the Manildra district, New South Wales: Royal Soc. New South Wales Jour. Proc., v. 101, p. 159-169. (136)

——— 1968b, *Planicardinia,* a new septate dalmanellid brachiopod from the Lower Devonian of New South Wales: Palaeontology, v. 11, p. 627-632. (137)

——— 1968c, *Australirhynchia,* a new rhynchonellid brachiopod from the Lower Devonian of New South Wales: Palaeontology, v. 11, p. 731-735. (138)

——— 1969, New spiriferid brachiopods from the Lower Devonian of New South Wales: Palaeontology, v. 12, p. 472-487. (139)

——— 1970, New atrypid brachiopods from the Lower Devonian of New South Wales: Jour. Paleontology, v. 44, p. 665-668. (140)

——— 1971, Brachiopods from the Lower Devonian Mandagery Park Formation, New South Wales: Palaeontology, v. 14, p. 387-422. (141)

——— 1973a, Lower Devonian conodonts from New South Wales: Palaeontology, v. 16, p. 307-333. (142)

——— 1973b, Lower Devonian biostratigraphic correlation in eastern Australia and western North America: Lethaia, v. 5, p. 341-348. (143)

——— 1974, The brachiopods of the Lower Devonian Maradana Shale, New South Wales: Palaeontographica, Abt. A, v. 146, p. 1-51. (144)

——— Personal observations of brachiopod faunas in southeast Australia. (145)

Scupin, Hans, 1900, Die Spiriferen Deutschlands: Paläontogische Abh., n.f., v. 4, pt. 3, p. 207-344. (146)

Seddon, G., and Sweet, W. C., 1971, An ecological model for conodonts: Jour. Paleontology, v. 45, p. 869-880. (147)

Simpson, G. G., 1947, Holarctic mammalian faunas and continental relationships during the Cenozoic: Geol. Soc. America Bull., v. 58, p. 613-687. (148)

Solle, Gerhard, 1942, Die Kondel-Gruppe (Oberkoblenz) im Südlichen Rheinischen Schiefergebirge. VI-X: Senckenbergische Naturforsch. Gesell. Abh. 467, p. 157-240. (149)

Strusz, D. L., 1970, A new species of rhynchonellid brachiopod from the Devonian of New South Wales: Australia Bur. Mineral Resources, Geology and Geophysics Bull. no. 108, p. 305-323. (150)

Strusz, D. L. and others, 1972, Correlation of the Lower Devonian rocks of Australasia: Geol. Soc. Australia Jour., v. 18, p. 427-455. (151)

Struve, Wolfgang, 1964, Über *Alatiformia* - Arten und andere, äusserlich ähnliche Spiriferacea: Senckenbergiana Lethaea, v. 45, p. 325-346. (152)

——— 1970, "Curvate Spiriferen" der Gattung *Rhenothyris* und einige andere Reticulariidae aus dem Rheinischen Devon: Senckenbergiana Lethaea, v. 51, p. 449-577. (153)

Talent, J. A., 1956, Devonian brachiopods and pelecypods of the Buchan Caves Limestones, Victoria: Royal Soc. Victoria Proc., v. 68, p. 1-56. (154)

——— 1963, The Devonian of the Mitchell and Wentworth Rivers: Victoria Geol. Survey Mem. 24, 118 p. (155)

——— 1965, The Silurian and Early Devonian faunas of the Heathcote district, Victoria: Victoria Geol. Survey Mem. 26, 50 p. (156)

——— 1972, Brachiopods, *in* Talent, J. A., ed., Provincialism and Australian Early Devonian faunas: Geol. Soc. Australia Jour., v. 19, p. 81-83. (157)

——— Unpubl. communities manuscript dealing with Australian Lower Devonian brachiopod faunas. (158)

Talent, J. A., and Banks, M. R., 1967, Devonian of Victoria and Tasmania, *in* Oswald, D. H., ed., Internat. Symposium on the Devonian System, Calgary 1967: Calgary, Alberta Soc. Petroleum Geologists, v. 2, p. 147-163. (159)

Telford, P. G., 1972, Conodonts, *in* Talent, J. A., ed., Provincialism and Australian Early Devonian faunas: Geol. Soc. Australia Jour., v. 19, p. 83-88. (160)

Tillman, C. G., 1967, Silicified rhynchonellid brachiopods from beds of New Scotland age (Early Devonian) Virginia and West Virginia: Jour. Paleontology, v. 41, p. 1247-1255. (161)

Vandercammen, Antoine, and Krans, T. F., 1964, Révision de quelques types de Spiriferidae d'Espagne: Inst. Royal Sci. Nat. Belgique Bull., v. 40, 40 p. (162)

Walmsley, V. G., and Boucot, A. J., 1971, The Resserellinae—a new subfamily of Late Ordovician to Early Devonian dalmanellid brachiopods: Palaeontology, v. 14, p. 487-531. (163)

——— 1975, The phylogeny, taxonomy and biogeography of Silurian and Early to mid-Devonian Isorthinae (Brachiopoda): Palaeontographica, Abt. A, v. 148, p. 1-108. (164)

Walmsley, V. G., Boucot, A. J., and Harper, C. W., 1969, Silurian and Lower Devonian salopinid brachiopods: Jour. Paleontology, v. 43, p. 492-516. (165)

Westbroek, Peter, 1968, Morphological observations with systematic implications on some Paleozoic Rhynchonellida from Europe, with special emphasis on the Uncinulidae: Leidse Geol. Meded., v. 41, p. 1-82. (166)

Williams, A., 1969, Ordovician faunal provinces with reference to brachiopod distribution, *in* Wood, A., ed., The Pre-Cambrian and lower Palaeozoic Rocks of Wales: Cardiff, Univ. Wales Press, p. 117-154. (167)

Williams, G. E., 1964, The geology of the Kinglake district, central Victoria: Royal Soc. Victoria Proc., v. 77, p. 273-327. (168)

Wright, A. J., 1966, Studies in the Devonian of the Mudgee District, New South Wales: Ph.D. thesis, Univ. of Sydney, Sydney, Australia, 388 p. (169)

——— 1967, Devonian of the Capertee Geanticline, New South Wales, Australia, *in* Oswald, D. H., ed., Internat. Symposium on the Devonian System, Calgary 1967: Calgary, Alberta Soc. Petroleum Geologists, v. 2, p. 117-121. (170)

Devonian Conodont Distribution—Provinces or Communities?

PETER G. TELFORD, *Ontario Division of Mines, Queen's Park, Toronto, Ontario, Canada*

ABSTRACT

Spatial distribution of Devonian conodonts is controlled by environmental factors and various faunal communities have been recognized. However, major irregularities in occurrences of particular taxa can be interpreted as being the result of regional migrational barriers such as landmasses, deep-water basins, or temperature and salinity gradients which, during the Early Devonian, produced a system of conodont provinces. These can be defined using the known distribution of form element and multielement taxa of the families Icriodontidae and Polygnathidae. A European Province, characterized mainly by a diversity of icriodids and a *Spathognathodus steinhornensis* lineage, and encompassing present-day Europe, North Africa, and Turkey, developed during the earliest Devonian (Gedinnian) and persisted until the latest Early Devonian (Emsian). A Pacific Province, encompassing present-day eastern Australia and western North America, developed during the Siegenian and persisted till the late Emsian. It was typified by a *Spathognathodus expansus* lineage and several other endemic species of the form genera *Eognathodus*, *Spathognathodus*, and *Polygnathus*. Conodont faunas of the eastern North American region attained a separate identity during the Emsian, being of very low diversity and containing the endemic *Icriodus latericrescens robustus*. Conodont provinciality gradually waned during the Middle Devonian, coinciding with the breaching of the Taghanic Arch which allowed freer migration between eastern and western North America. By Late Devonian time, conodont faunas were uniformly distributed throughout the paleoequatorial regions.

INTRODUCTION

Conodonts are minute, phosphatic, toothlike or jawlike fossils of a primitive, marine, possibly vertebrate animal that existed from Cambrian to Early Triassic time. Much of the attention given to these microfossils has been directed by or to their usefulness in biostratigraphic correlation. However, as conodont research has become more detailed and comprehensive, and as conodont workers have become more discerning, it has become apparent that conodonts are not always the ideal index fossils that they were branded by earlier workers. It has become apparent that irregularities in the spatial distribution of certain conodont taxa are the result of provincial or community segregations.

The intensely studied Middle and Upper Ordovician conodont faunas of North America provide an example of this progression of attitudes, from those treating conodonts exclusively as stratigraphic tools to those promoting the evaluation of factors which may or may not discredit the stratigraphic value of conodonts (Sweet and others, 1971). Other parts of the geological record of conodonts have not been examined as fully. Telford (1972, 1975) provided brief, preliminary views on the possible provinciality of Lower Devonian "platform" conodonts. Klapper and Johnson (1975) disputed some of his generalizations, pointing out that community rather than provincial relationships may be the cause of certain inhomogeneities in Devonian conodont distribution. Fåhraeus (1976) has reached very similar conclusions to those expressed by Telford (1975, p. 86) for the Lower Devonian.

Therefore, the present synthesis of global distribution of Devonian conodonts was carried out with the aim of evaluating relationships among the occurrences of particular taxa, and determining whether provincial barriers, local environmental constraints, or a combination of these factors, was responsible. This study was possible in the light of much new data on conodont paleoecology (see following section) and taxonomy. Indeed, the latter aspect of conodont research has been revolutionized in recent years by several basic conceptual changes.

The most important of these changes has been the transfer from single element or form element taxonomy to a system of multielement taxonomy. The background and details of this change have been summarized by Lindstrom and Ziegler (1972) and most conodont workers are attempting to apply the new, supposedly more biologically natural taxonomic system. Of course, some problems have persisted (Telford, 1975, p. 8-10) and the taxonomy of several Devonian conodont genera is not defined agreeably. Klapper and Philip (1971, 1972) proposed that Devonian icriodid elements occurred in an apparatus with simple conelike (acodinid?) elements while Bultynck (1972) considers that icriodids are the only form elements represented in the *Icriodus* apparatus. Spathognathodid elements were assigned to 2 multielement genera, *Ozarkodina* and *Pandorinellina*, by Klapper and Philip (1971). For various reasons, Fåhraeus (1974) did not agree with homologizing spathognathodid elements in a *Pandorinellina* apparatus. Because of such problems and because many Devonian conodont form element taxa have not been brought to multielement nomenclature, the taxa cited in this paper include both form element and multielement representatives. All cited taxa are listed in the Appendix with explanatory notes or a reference to their most recent taxonomic revision.

There is no published record of Devonian conodont taxa from South America, central and southern Africa, Antarctica, and the Indian subcontinent.

CONODONT PALEOECOLOGY

At the beginning of this study it rapidly became obvious that before a meaningful interpretation of conodont biogeography was possible, the environmental factors involved in conodont distribution had to be assessed. Valentine (1973) defined "biotic provinces" as "regions inhabited by characteristic association of organisms bounded by barriers that prevent the spread of characterizing taxa into other regions and that also prevent the immigration of many foreign species." Boucot (1975, p. 26) described a community as "a specific set of organisms adapted to a specific set of environmental conditions, irrespective of their separation in time or space." Thus, until the environmental or ecological factors governing conodont distribution are understood, recurring associations of conodont taxa cannot be interpreted with absolute certainty as representing either provinces or communities.

In recent years there has been a great increase in numbers and quality of conodont ecological studies. Seddon and Sweet (1971), and Barnes and Fåhraeus (1975), summarized the historical background to these studies and the latter authors, in detailed analysis of Ordovician faunas, reviewed the conceptual framework of conodont paleoecology. They indicated that most conodont specialists now agree that the distribution of conodonts, in time and space, was controlled by environmental factors. However, they also raised the following basic and presently unanswered questions: (a) What degree of environmental dependency did the conodont animal possess? (b) What controls were imposed on these organisms? (c) How did these organisms adapt to environmental pressures?

A major barrier to resolution of these questions is the lack of knowledge of biological affinities and soft-tissue anatomy of the conodont animal. Paleoecological interpretations are dependent, therefore, on the spatial distribution of conodonts and the composition and depositional history of the rocks in which they are found.

Despite these limitations, various conodont ecological models have been hypothesized and they can be divided into two general types. One invokes depth stratification of conodont faunas (Seddon and Sweet, 1971); the other (Barnes and others, 1973) proposes a combination of depth stratification and, more importantly, lateral segregation of conodont faunas. The first model suggests that the conodont animal was free-swimming and pelagic with some taxa confined to deep-water layers and others restricted to shallow waters or the surface layers of deep-water basins (Seddon and Sweet, 1971; Druce, 1973). In the lateral segregation model (Barnes and others, 1973; revised by Barnes and Fåhraeus, 1975), it is suggested that most conodont animals were benthic or nektobenthic while a minority of taxa were pelagic. According to this model, conodont faunas or communities were governed by proximity to the shoreline so that certain taxa were characteristic of littoral conditions, or subtidal conditions, and so on. Some taxa would be ubiquitous due to their pelagic nature.

While studies of Ordovician (Barnes and Fåhraeus, 1975) and some Pennsylvanian faunas (Merrill, 1973) strongly favor the lateral segregation model, recent documentation of Devonian conodont distribution provides supporting evidence for both hypotheses. In a study of Gedinnian faunas from New York and New Jersey, Barnett (1972) found that *Spathognathodus remscheidensis* was most abundant in lagoonal (littoral and sublittoral) environments and the reef and shallow sublittoral parts of the basinal environment, while *Icriodus woschmidti* was most abundant in the shallow and intermediate sublittoral parts of the basinal environment. *S. remscheidensis* occurred in lesser abundance under the other basinal conditions but *I. woschmidti* was completely absent from lagoonal deposits. This situation is in line with the lateral segregation model.

Seddon (1970), Telford and von Bitter (1975), and Skipp and Sandberg (1975), also have implied that on carbonate platforms, the distribution of Devonian conodonts (particularly the form genera *Icriodus*, *Polygnathus*, and *Spathognathodus*) was controlled more strongly by sedimentary environments, e.g., reef and interreef conditions, than by water depth alone. Again, this follows the concept of lateral segregation. In addition, many Devonian studies have recorded the ubiquitous appearances of simple conelike conodonts. These taxa correspond to the pelagic forms included in the lateral segregation model (Barnes and Fåhraeus, 1975).

Davis (1975), in analysis of Middle Devonian conodonts from the Tully Limestone of New York, claimed to have confirmed Seddon and Sweet's (1971) depth-stratification model. However, some of his conclusions can also be considered to support the lateral-segregation model. He showed that species of *Icriodus* were most abundant in shallow-water, near-shore environments while species of *Polygnathus* were dominant in deeper water situations, and *Icriodus* decreased in abundance away from the shoreline. Davis' results indicated that *Icriodus* could tolerate changeable environments, especially varying salinities and temperatures, while *Polygnathus* preferred a more stable environment. Ferrigno (1971), in a study of Middle Devonian conodonts from the Dundee Formation in the Michigan Basin, also noted that *Icriodus* appeared to be tolerant of varying conditions but associated species of *Polygnathus* showed preferences for more specialized, constant environments.

Taken together, the above-mentioned Devonian studies suggest that a combination of depth stratification and lateral segregation (with the latter being dominant) very similar to that described by Barnes and Fåhraeus (1975) is the best possible ecological model to apply to interpretation of conodont distribution at this time.

Of even more direct importance to the present study is the actual definition of the different conodont faunas or communities. Druce (1973) instituted a system of conodont biofacies for the upper Paleozoic (Devonian-Permian) and Triassic. This system was based upon a depth-stratification model and the biofacies so defined (three or four were thought to occur) were considered to represent various water depths and to be characterized by the recurring presence or association of particular conodont taxa. Druce's (1973) speculations as to the environments represented by the various biofacies can be disputed. Nevertheless, his scheme shows that conodonts can be separated into distinct communities.

Specific compositions of Devonian conodont communities have yet to be determined with certainty, but the various paleoecological studies mentioned above give some indications. Simple conelike conodont genera, such as *Belodella* and *Panderodus*, may occur alone or may be associated with any other platform-type conodont elements (Telford and von Bitter, 1975). Species of *Icriodus* may occur to the exclusion of all other platform-type conodont elements (Davis, 1975) or, conversely, they may be entirely absent from faunas in which form species of *Polygnathus* and *Spathognathodus* are abundant (Skipp and Sandberg, 1975). The distribution of *Spathognathodus* is the least understood of all the platform-type form genera. Much further detailed research is required before these relationships are properly understood but they must not be ignored in attempting biogeographical syntheses.

DISTRIBUTION OF MIDDLE AND UPPER DEVONIAN CONODONTS

Upper Devonian

In their comprehensive studies and reviews of conodont biozonations in Europe, western Australia, and North America, Ziegler (1962, 1971), Glenister and Klapper (1966), and Klapper and others (1971), have demonstrated the virtually cosmopolitan distribution of Upper Devonian conodonts. Seddon (1970) described a Frasnian *Icriodus-Polygnathus* community occurring in forereef sediments and a *Palmatolepis* community that was restricted to interreef deposits of the Canning Basin, western Australia. However their occurrence does not hamper intercontinental correlation. The detailed Upper Devonian zonation scheme outlined by Ziegler (1962) has been extended worldwide and there is no evidence that conodont faunal provinces existed in Late Devonian time. The distribution of Upper Devonian conodonts mirrors the distribution of other fossil groups, particularly brachiopods (Johnson and Boucot, 1973; Boucot, 1975), and goniatites (House, 1973), although it must be noted that presently known conodont faunas are restricted to paleoequatorial regions.

Middle Devonian

Conodont faunas of this age are less well-known than those of the Upper Devonian. Limited studies of them suggest that they do not have as cosmopolitan a distribution as the Upper Devonian faunas (Telford, 1975, p. 84) but they also do not display marked provinciality. Irregularities in distribution of particular taxa may be related to community differences rather than major provincial barriers.

Middle Devonian species of *Polygnathus* are very widespread and useful for correlation purposes. Klapper (1971) identified the close relationship between the lower Middle Devonian polygnathid faunal sequence in New York and that of the Ardennes (Belgium) described by Bultynck (1970). Using this sequence of polygnathid faunas it has proven possible to correlate lower Middle Devonian faunas on a local and global scale.

For instance, conodonts were obtained recently from the Needmore Shale, directly underlying the Tioga Bentonite, in Pennsylvania and West Virginia. The fauna included *Polygnathus linguiformis linguiformis, P. costatus costatus,* and *P. angusticostatus; P. linguiformis cooperi* and *Icriodus latericrescens robustus* occurred in samples lower in the sections (measured and sampled by John Dennison, University of North Carolina); and *I.* cf. *corniger* was common throughout. The association of polygnathids could be correlated directly with Klapper's (1971) New York sequence and a late Onesquethaw age was postulated.

Intercontinental correlations cannot be performed with this precision. Indeed, Middle Devonian conodont zonations (Wittekindt, 1966; Orr, 1971) are much less detailed than Upper Devonian schemes (Ziegler, 1962). However, as is shown later, greater accuracy is possible with Middle Devonian faunas than with Lower Devonian ones. Sandberg and others (1975), Chatterton (1974), and Savage (personal commun. 1976) encountered lower Middle Devonian polygnathids in Idaho, Alberta, and Alaska, respectively, that can be correlated with the New York and Belgium faunas. Philip (1967), Pedder and others (1970), and Telford (1975) described similar lower Middle Devonian polygnathids from eastern Australia. Klapper and others (1970) described the worldwide distribution of upper Middle Devonian (upper Givetian) representatives of the *Polygnathus varcus* group.

The only major inhomogeneities in Middle Devonian conodont distribution, apparent at this time, concern species of *Icriodus. I. latericrescens robustus* is restricted to upper Lower and lower Middle Devonian sequences of eastern North America. *I. angustus* is common in lower Middle Devonian faunas of eastern North America but is not known definitely from Europe; Bultynck (1972)

Table 1. Distribution of Lower Devonian representatives of the conodont families Icriodontidae (*Icriodus, Pedavis, Pelekysgnathus*) and Polygnathidae (*Ancyrodelloides, Polygnathus, Eognathodus, Spathognathodus*). (X) definite occurrence; (?) questionable or unconfirmed occurrence. (A) Europe excluding the Spanish area; (B) Spain and North Africa; (C) eastern North America (Texas to Hudson Bay); (D) western North America including the Canadian Arctic Islands but excluding Nevada-Utah; (E) Nevada-Utah; (F) eastern Australia; (G) central Asia.

No.	Species	Age	Occurrence						
			A	B	C	D	E	F	G
1	*ANCYRODELLOIDES kutscheri*	upper Gedinnian?	X						
2	*A. trigonica*	upper Gedinnian	X	X		?			
3	*ICRIODUS angustoides*	upper lower Gedinnian—middle Siegenian	X	X		?			
4	*I. beckmanni*	upper lower—lower upper Emsian	X	X	?				
5	*I. bilatericrescens*	upper lower—lower upper Emsian	X	X					
6	*I. curvicauda*	lower Siegenian—lower Emsian	X	X		X			
7	*I. eolatericrescens*	lower Gedinnian	X	X		?			
8	*I. fusiformis*	upper Emsian	X	X					
9	*I. huddlei*	upper lower—upper Emsian	X	X	X	X	X		
10	*I. latericrescens robustus*	Emsian—lower Eifelian			X				
11	*I. latericrescens* n. subsp. B of Klapper 1969	middle Siegenian				X	X		
12	*I. postwoschmidti*	lower Gedinnian	X	X					
13	*I. rectangularis*	lower Gedinnian—middle Siegenian	X	X					
14	*I. sigmoidalis*	upper lower—lower upper Emsian	X	X					
15	*I. taimyricus*	lower Emsian				X			X
16	*I. woschmidti woschmidti*	lower Gedinnian	X		X	X		X	
17	*I. woschmidti hesperius*	lower Gedinnian				X	X	?	
18	*PEDAVIS pesavis pesavis*	lower Siegenian	X			X	X		X
19	*P. pesavis* n. subsp. A of Klapper and Philip 1972	upper Gedinnian	X	X	X		X	?	
20	*P.* n. sp. A of Klapper and Philip 1972	middle Siegenian				X			
21	*P.* n. sp. B of Klapper and Philip 1972	Siegenian						X	
22	*P.* n. sp. C of Klapper and Murphy 1974	upper Gedinnian					X		
23	*PELEKYSGNATHUS furnishi*	Emsian				X			
24	*P. glenisteri*	Emsian				X			
25	*P. serratus*	middle—upper Siegenian	X	X		X			
26	*P.* sp. of Telford 1975	Siegenian						X	
27	*POLYGNATHUS dehiscens*	lower Emsian	X	X		X	X	X	?
28	*P. gronbergi*	lower—middle Emsian	X				X		
29	*P. inversus*	upper Emsian	X			X	X	X	
30	*P. laticostatus*	middle—upper Emsian	X				X		
31	*P. perbonus*	middle Emsian				?	?	X	
32	*P. pirenae*	lower Emsian		X		X			
33	*P. serotinus*	upper Emsian				X	X	X	
34	*P.* sp. A of Klapper and Johnson 1975	upper Emsian					X		
35	*EOGNATHODUS sulcatus*	Siegenian				X	X	X	
36	*E. trilinearis*	upper Siegenian						X	
37	*SPATHOGNATHODUS asymmetricus*	upper Gedinnian	X	X					
38	*S. boucoti*	Siegenian				X			
39	*S. carinthiaca*	Emsian	X						
40	*S. eurekaensis*	upper Gedinnian					X		
41	*S. exiguus*	upper Siegenian—Emsian				X	X	X	
42	*S. expansus*	Emsian—Eifelian				X	X	X	
43	*S. inclinatus* (= *S. wurmi*)	middle Silurian—lower Emsian	X	X		X	X	X	
44	*S. johnsoni*	lower Siegenian				X	X		
45	*S. linearis*	Emsian						X	
46	*S. miae*	lower Emsian		X					
47	*S. optimus*	Siegenian—Emsian				X	X	X	X
48	*S. palethorpei*	upper Emsian						X	
49	*S. remscheidensis remscheidensis*	Gedinnian	X		X	X	X	?	X
50	*S. remscheidensis repetitor*	Gedinnian—lower Siegenian	X	X					
51	*S. steinhornensis*	Emsian	X	X		?			
52	*S. stygia*	upper Gedinnian—lower Siegenian	X			X	X		
53	*S. transitans*	upper Gedinnian	X	X	X		X		
54	*S.* n. sp. E of Klapper and Murphy 1974	lower Gedinnian				X	X		
55	*S.* n. sp. F of Klapper and Murphy 1974	Gedinnian					X		

pointed out that the North American holotype and paratypes described by Stewart and Sweet (1956) differ from European forms referred to *I. angustus*. *I. latericrescens latericrescens* has a very restricted stratigraphic range (late Givetian) in Europe (Ziegler, 1971) but is widespread throughout upper Middle Devonian strata of eastern North America (Klapper and Ziegler, 1967; Klapper and others, 1971). Thus, during the early Middle and part of the late Middle Devonian, icriodid conodonts possibly had difficulty in migrating between the European and eastern North American regions. As indicated by the stratigraphic and geographic distribution of *I. latericrescens latericrescens*, migration may have become easier after breaching of the Taghanic Arch (Johnson, 1970) in late Middle Devonian (late Givetian) time.

DISTRIBUTION OF LOWER DEVONIAN CONODONTS

Table 1 illustrates the stratigraphic range and geographic distribution of selected Lower Devonian conodonts. Only species of the families Icriodontidae and Polygnathidae are included as their Lower Devonian occurrences have been documented more thoroughly than other groups such as the Panderodontidae and Hibbardellidae. Also, the Icriodontidae and Polygnathidae contained many short-ranging, rapidly-evolving taxa that are useful for biostratigraphic and biogeographic interpretations. As explained previously, taxa listed in the tables or cited elsewhere in the paper may be multielement or form element species; their taxonomic status is clarified in the Appendix. For the purposes of this study it is not important, in most cases, whether a taxon represents multielement or form element nomenclature.

Gedinnian

During the early Gedinnian several important species had almost cosmopolitan distributions. *Icriodus woschmidti woschmidti* has been recorded from Europe, eastern and western North America (excluding the Nevada-Utah area), and eastern Australia (Table 1). A second subspecies, *I. woschmidti hesperius*, is known from western North America (Nevada, Yukon) and possibly occurred in eastern Australia (Klapper and Murphy, 1974). *Spathognathodus remscheidensis remscheidensis* is known definitely from Europe, eastern and western North America, and Pakistan (Barnett and others, 1966), and possibly occurred in eastern Australia (Table 1). *Spathognathodus remscheidensis repetitor* developed during the middle or late Gedinnian and is restricted to European faunas. Other possible Siegenian and Emsian descendants of *S. remscheidensis* are discussed later.

Icriodus woschmidti was the first member of a complex phylogeny which led to the evolution of all known species of *Icriodus* (Ziegler, 1971). A direct successor to *I. woschmidti* in European faunas was *I. postwoschmidti* and, in the late Gedinnian, this variable species gave rise to several lineages that produced species such as *I. angustoides*, *I. rectangularis*, and *I. eolatericrescens* (Ziegler, 1971, 1975). Upper Gedinnian European conodont faunas are characterized, therefore, by a diversity of endemic species of *Icriodus* (Tables 1, 2).

Spathognathodus asymmetricus was also endemic to upper Gedinnian faunas in the European region, and gave rise to the short-ranging *Ancyrodelloides* (Ziegler, 1971). This genus had a wide distribution in the European region but was rare or absent elsewhere.

There is considerable argument about the precise stratigraphic ranges of species of the unusual multielement genus *Pedavis* Klapper and Philip, 1971 (Carls, 1969; Ziegler, 1971; Klapper and others, 1971). Forms regarded herein (Table 1) as upper Gedinnian in age are *P. pesavis* n. subsp. A of Klapper and Philip (1972) and *P.* n. sp. C of Klapper and Murphy (1974). The former was widely distributed, being reported from Europe, Texas, and Nevada (Klapper and Philip, 1972), and possibly eastern Australia (Telford, 1975). *Pedavis* n. sp. C is known only from Nevada (Klapper and Murphy, 1974). However, the biogeographic significance of these species of *Pedavis* cannot be evaluated properly until the time-stratigraphic arguments are resolved.

As indicated by Tables 1 and 2 (and Fig. 1—see later biogeographic discussion), Gedinnian conodont faunas of eastern Australia, eastern North America, and western North America (excluding Nevada-Utah) are typified by a lack of endemic forms. The presence of several endemic species in Nevada (Table 2) may be an artifact of the more thorough stratigraphic and paleontological studies of this region (Klapper and Murphy, 1974).

Siegenian

European faunas continued to be characterized by a diversity of icriodids into Siegenian time (Table 1), although the rate of diversification of this group was not as great as during the late Gedinnian (Ziegler, 1971). However, the major differences that can be observed between European Siegenian faunas and those of other parts of the world involve the descendants of *Spathognathodus remscheidensis remscheidensis*. Bultynck (1971) outlined a phylogeny of European spathognathodids that began with *S. remscheidensis eosteinhornensis* in the Late Silurian and passed through *S. remscheidensis remscheidensis*, *S. remscheidensis repetitor*, and *S. miae*, and culminated in *S. steinhornensis* during the Emsian. Telford (1975) suggested that in eastern Australia and western North America a different lineage included *S. optimus*, *S. exiguus*, and *S. expansus* which are unknown from the European region. Fåhraeus (1974) organized these views into a cladogenetic scheme whereby the virtually cosmopolitan form *S. remscheidensis remscheidensis* (see Table 1) gave rise to the *S. steinhornensis* stock in Europe and the *S. expansus* stock in eastern Australia and western North America. During the Siegenian the only member of the *S. steinhornensis* stock present in European faunas was *S. remscheidensis repetitor* while, in eastern Australia and western North America, *S. optimus* and *S. exiguus* were abundant and

Table 2. Lower Devonian conodont faunal relationships among selected geographic regions (A to G as for Table 1). Numbers in brackets are figures for the European region including Spain and North Africa (A + B) and western North America including Nevada-Utah (D + E).

Region	Total Species		Endemic Species		Shared Species							
GEDINNIAN					**A**	**B**	**C**	**D**	**E**	**F**	**G**	
A	14		1			10	4	4	5	2	1	**A**
		(14)		(8)								
B	10		0				2	1	3	1	0	**B**
C	4		0					2	3	1	1	**C**
D	6		0						5	2	1	**D**
		(11)		(5)								
E	10		3							1	1	**E**
F	2		0								0	**F**
G	1		0									**G**
SIEGENIAN					**A**	**B**	**C**	**D**	**E**	**F**	**G**	
A	8		0			6	0	5	3	1	1	**A**
		(8)		(3)								
B	6		0				0	3	1	1	0	**B**
C	0		0					0	0	0	0	**C**
D	12		2						8	4	2	**D**
		(12)		(4)								
E	8		0							4	2	**E**
F	7		2								1	**F**
G	2		0									**G**
EMSIAN					**A**	**B**	**C**	**D**	**E**	**F**	**G**	
A	13		1			9	1	5	6	3	0	**A**
B		(15)		(7)								
	11		1				1	5	3	2	0	**B**
C	2		1					1	1	0	0	**C**
D	13		2						8	7	2	**D**
E		(16)		(3)								
	11		1							7	1	**E**
F	10		2								1	**F**
G	2		0									**G**

widespread; *S. optimus* has also been reported from central Asia (Table 1; Moskalenko, 1967).

A notable absentee from European Siegenian faunas is *Eognathodus sulcatus* which is widely distributed in eastern Australia and western North America and is very useful for biostratigraphic correlation (Klapper and others, 1971). Early and late variants of this species have been described from western North America (Klapper, 1969) and the same sequence has been reported from eastern Australia (Telford, 1975). A second eognathodid, *E. trilinearis,* has been reported only from southeastern Australia (Cooper, 1973).

As in the late Gedinnian, species of *Pedavis* had an irregular distribution. *P. pesavis pesavis* has been reported from lower Siegenian strata of Europe, western North America, and the same locality in central Asia that yielded *S. optimus* (Klapper and Philip, 1972). *Pedavis* n. sp. A and *P.* n. sp. B of Klapper and Philip (1972) were recorded from the Yukon and southeastern Australia, respectively.

Thus, global relationships among Siegenian conodont faunas differed from the situation during the Gedinnian. European faunas remained distinct and with the appearance of several endemic species and cladogenesis of *Spathognathodus,* faunas of eastern Australia and western North America also took on separate characters (Table 1). Table 2 shows that eastern Australian faunas had more in common with those of western North America than with European faunas. In fact, the only member of the Icriodontidae or Polygnathidae common to eastern Australia and Europe was the long-ranging *Spathognathodus inclinatus.*

There is no data available for Siegenian conodont faunas of eastern North America.

Emsian

In the European region species of *Icriodus* underwent a second burst of diversification during early Emsian time. *Icriodus huddlei* was abundant and widespread, giving rise to the endemic forms *I. beckmanni* and *I. bilatericrescens* (Ziegler, 1971). *Icriodus fusiformis* and *I. sigmoidalis* were also endemic to European Emsian faunas (Table 1). In western North America, species of *Icriodus* are rare in faunas of the northern Yukon and Canadian Arctic Islands but are abundant in Nevadan faunas (Klapper and Johnson, 1975). The unusual icriodid, *I. taimyricus,* has been recorded from Alaska, Yukon, and north-central Siberia (Lane, 1974). *Icriodus huddlei* is reported from eastern

North America, commonly in association with *I. latericrescens robustus* (Klapper and Ziegler, 1967). The latter species is not known outside eastern North America. Representatives of *Icriodus* are absent from Emsian faunas of eastern Australia.

Pelekysgnathus furnishi is known from the Yukon and *P. glenisteri* has been encountered in Alaska, Yukon, and the Canadian Arctic Islands (Klapper *in* Ziegler, 1975). These species, endemic to the western North American areas, may have taken the place of icriodids which, as noted above, are rare or absent from Emsian faunas of the region.

The two lineages of *Spathognathodus* that evolved during the Siegenian persisted through Emsian time. In the European region the *Spathognathodus steinhornensis* stock is represented by *S. miae,* known only from Spain (Bultynck, 1971), which supposedly gave rise to *S. steinhornensis.* In eastern Australia and western North America, *S. optimus* and *S. exiguus* possibly ranged into Emsian time and the latter gave rise to *S. expansus.* The exact stratigraphic range of *S. expansus* is not clear; it has been reported from lower and upper Emsian and lower Middle Devonian strata of eastern Australia (Telford, 1975; Pickett, personal commun. 1976) and upper Emsian to lower Middle Devonian strata in western North America (Uyeno and Mason, 1975). However, the species is a widespread and characteristic member of Emsian faunas of the circum-Pacific region.

Two other spathognathodids, *S. linearis* and *S. palethorpei,* are endemic to Emsian faunas of eastern Australia. The former is widespread (Philip and Pedder, 1967) but the latter has been found in only two areas, the Broken River Embayment in north Queensland (Telford, 1975) and central New South Wales (Pickett, personal commun. 1976). *S. palethorpei* may have evolved from *S. exiguus* during the middle or late Emsian.

One of the most important events in conodont history took place during the Emsian. This was the appearance of the genus *Polygnathus* whose species became increasingly useful in biostratigraphy for the remainder of the Lower Devonian (Klapper and Johnson, 1975) and Middle and Upper Devonian (Klapper, 1971; Ziegler, 1962). As indicated in Table 1, most Emsian species of *Polygnathus* had a wide distribution although they exhibit more diversity in eastern Australian and western North American faunas than in European faunas. Also, polygnathids are less diverse in Spain than in other parts of the European region. *Polygnathus perbonus* is common in eastern Australia but has only questionable occurrences in western North America (Klapper and Johnson, 1975). *Polygnathus serotinus* has not been reported from the European region but Snigireva (1975) noted a species called *P. totensis* (which appears conspecific with *P. serotinus*) from the northern Urals.

Generally, Emsian patterns of conodont distribution were similar to those of the Siegenian. Faunas of the European region maintained their distinctive character, typified by a diversity of icriodids and the *Spathognathodus steinhornensis* lineage. Faunas of western North America and eastern Australia were distinctive mainly due to a lack of icriodids (except in Nevada), a greater diversity of polygnathids, and the *S. expansus* lineage. Both eastern Australia and western North America had several endemic species but Table 2 illustrates the very close relationships between faunas of these circum-Pacific regions. For the first time, faunas of eastern North America (Texas to Hudson Bay) developed a separate identity, being characterized by endemic *Icriodus latericrescens robustus,* rare appearances of a single unnamed spathognathodid (Telford and von Bitter, 1975), and a total lack of Emsian species of *Polygnathus.*

PROVINCES OR COMMUNITIES?

Comments on Paleogeography

A far-ranging paleogeographic synthesis is not within the scope of this paper which is concerned primarily with comparisons of conodont provinces and communities. However, in making conclusions based on spatial distributions it is useful to have a background on which to arrange the patterns. Because of the tremendous increase in recent years of data available for Paleozoic continental reconstructions, development of a Devonian geographic model that can be used for testing interpretations of fossil distributions is possible.

Figures 1 to 3 are based on the Lower Devonian paleogeographic maps of Smith and others (1973). Shapes of landmasses or areas in which no marine rocks have been reported (after Johnson and Boucot, 1973) are superimposed on the maps. At this scale and with the present state of knowledge the shapes of the continental fragments are quite approximate. For example, Boucot (1975) noted that the supposed barrier between eastern North America and Europe may have been a series of islands rather than a continuous landmass. Also, data for various parts of the world, particularly in South America, Antarctica, and eastern Asia, is not complete.

Smith and others (1973) admitted several limitations to their Paleozoic reconstructions; the most important is that latitudinal positions of the continental fragments are considered to be fixed while their longitudinal positions are arbitrary. The ramifications of this problem are pursued in the later biogeographic discussions. Smith and others (1973) also noted that paleomagnetic data is lacking for eastern Asia, and positioning of an area such as China is very uncertain. This has some effect on interpretation of the conodont occurrences in south-central Asia (Barnett and others, 1966; Moskalenko, 1967).

Another possible weakness of the maps is that the southern Midcontinent area of North America appears to be in latitudes too high for the known carbonate deposition in the region (Amsden and others, 1967). This assumes, of course, that the climatic gradient was similar to the present. As suggested by Fåhraeus (1976), about 30° of clockwise rotation of the Euramerican fragment, to allow correspondence with the paleoequator related to

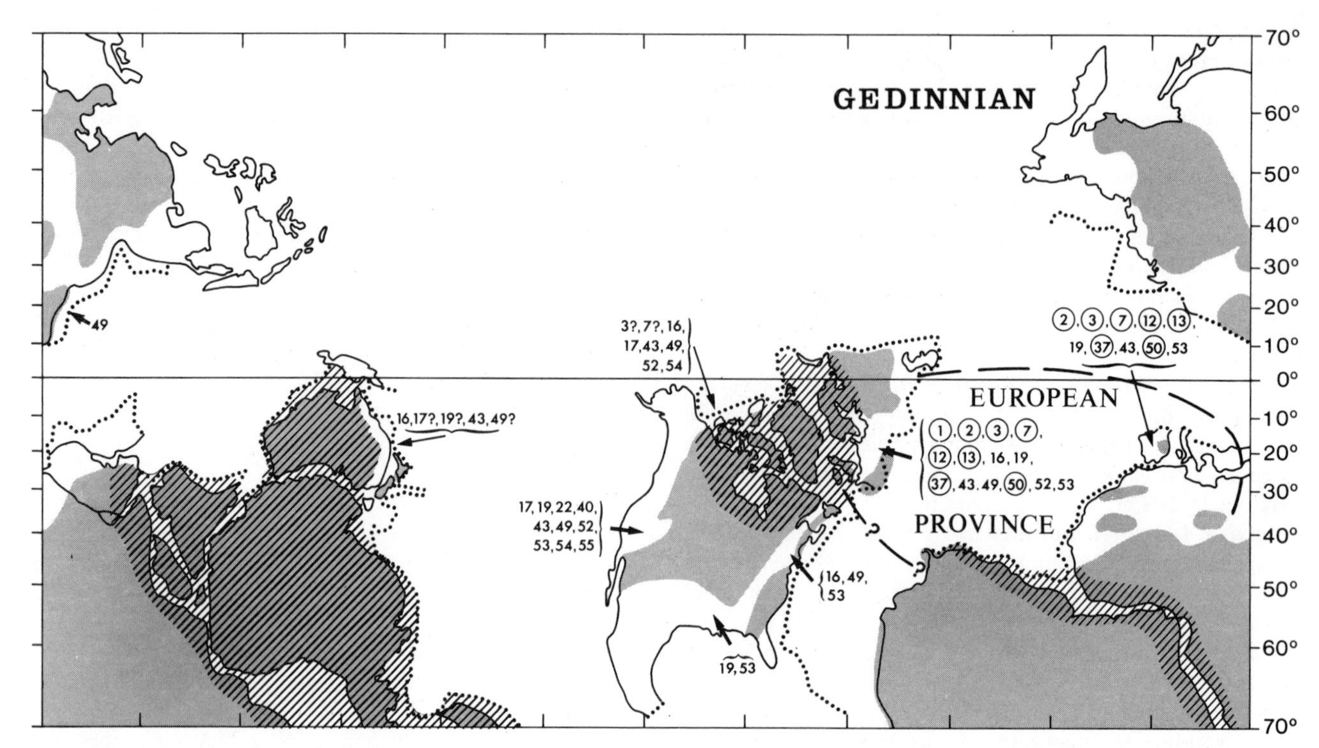

Figure 1. Provincial distribution of Gedinnian conodonts. Numbers correspond to species listed in Tables 1 and 2. Circled numbers represent endemic taxa. Paleogeographic reconstructions taken from Smith and others (1973) with additional data from Johnson and Boucot (1973).

North America defined by Irving (1964), would alleviate the carbonate problem.

Gedinnian Biogeography

During the early Gedinnian, species such as *Icriodus woschmidti* were distributed worldwide. However, by the middle or late Gedinnian, diversification of icriodids had established the European region as a separate faunal province. Except for the Nevada area (Klapper and Murphy, 1974), there is little information on upper Gedinnian faunas in regions outside Europe, but some species remained widely distributed, e.g., *Spathognathodus transitans* has been reported from Europe, eastern North America, and Nevada, although it has not been found in the Arctic or eastern Australian regions.

Isolation of the European icriodid fauna was possibly due to migrational limitations rather than absences of suitable environments in nearby regions. Depositional conditions similar to those of the European region seem to have been present elsewhere, e.g., southern Midcontinent of North America (Amsden and others, 1967), and the Canadian Arctic Islands (Kerr, 1974), and at least one icriodid, *I. woschmidti,* was widely distributed. The actual barriers to migration cannot be defined with any confidence, although judging from the configuration of the continental fragments (Fig. 1), it would appear that salinity and temperature gradients may have been more important than landmasses or deep-water basins. Contact between the European and other regions was probably by a southerly route via North Africa and Texas rather than a northerly route via the Urals and Arctic regions.

Thus, as indicated on Figure 1, the conodont biogeographic model for Gedinnian time includes a distinct European Province with a low diversity, almost cosmopolitan fauna inhabiting other regions.

Siegenian Biogeography

Two faunal provinces may be distinguished for the Siegenian. One is the European Province with boundaries similar to those of the Gedinnian (Fig. 1) and the other is termed herein the Pacific Province following Telford (1975). The latter encompasses eastern Australia and western North America including Nevada-Utah and the Canadian Arctic Islands (Fig. 2). The European Province was characterized by an abundance and diversity of icriodids. However, the best method of discriminating between the two Provinces is based on the *Spathognathodus steinhornensis* and *S. expansus* lineages together with the occurrences of *Eognathodus* (Table 1). During the Siegenian, *Spathognathodus exiguus, S. optimus,* and *Eognathodus sulcatus* were widely distributed in the Pacific Province but completely absent from faunas of the European Province which contained *S. remscheidensis repetitor* (an early member of the *S. steinhornensis* lineage).

Some arguments could still be raised indicating that the diversity of icriodids in the European region and general lack of them elsewhere is a "community effect." Never-

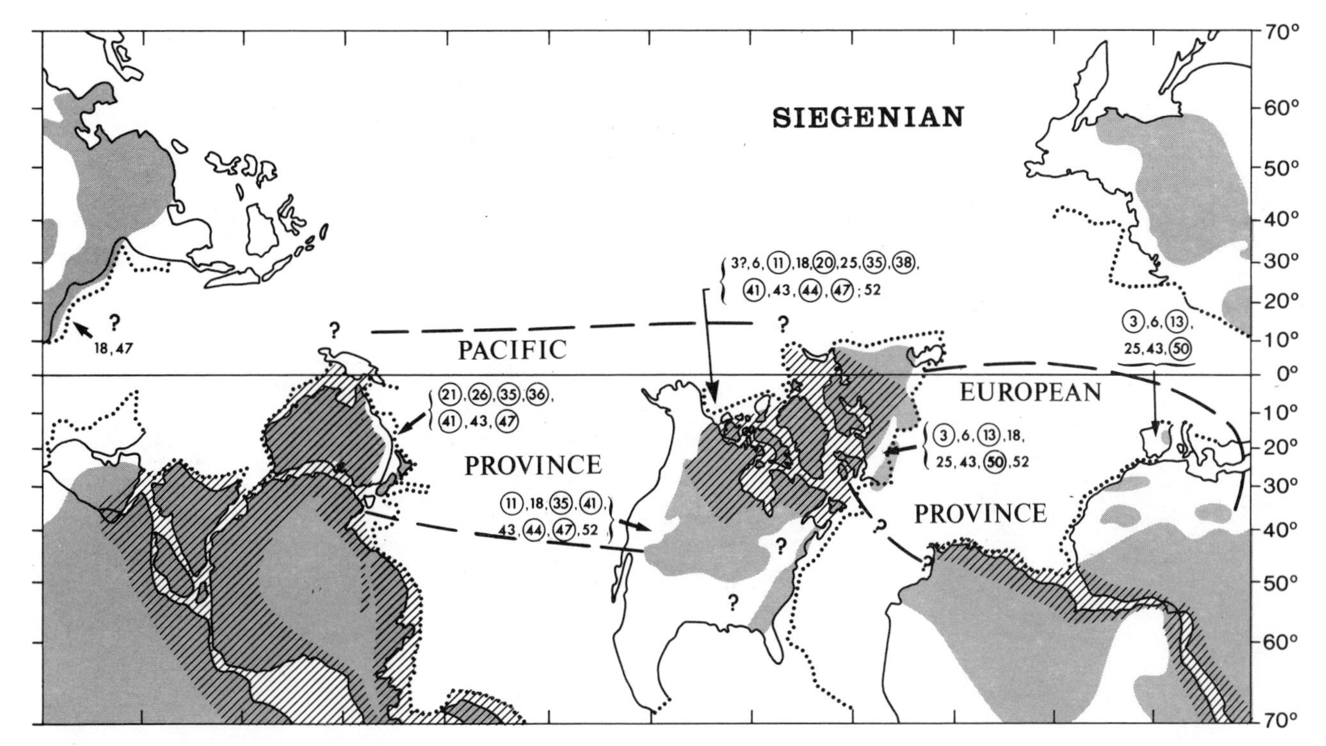

Figure 2. Provincial distribution of Siegenian conodonts (explanation as for Fig. 1).

theless, the spathognathodids occurring in the European and Pacific regions had basically similar morphologies (Fåhraeus, 1974; Telford, 1975; Klapper *in* Ziegler, 1973) and it may be assumed that their host animal had similar ecological requirements. Barriers to migration and to subsequent intermingling of taxa must have existed, thus producing a provincial distribution of faunas.

As suggested by the occurrence of *Spathognathodus optimus* (Fig. 2), central Asian conodont faunas (Moskalenko, 1967) may have constituted an extension of the Pacific Province, although there is too little data for firm speculation. If the entire Asian continental fragment were shifted westward to lie directly north of the Euramerican fragment (Cocks and McKerrow, 1973, Fig. 6) then Moskalenko's fauna would be in close proximity to very similar faunas in the northern Yukon and Canadian Arctic Islands (Klapper, 1969).

Again, it is difficult to predict the probable migration routes between the European and Pacific Provinces, especially as there is no faunal data for eastern North America. Table 2 shows that the northern Cordilleran faunas were slightly more closely related to the European faunas than were those of Nevada (note distribution of *Icriodus curvicauda* and *Pelekysgnathus serratus* in Table 1). Also, faunas of eastern Australia showed little relationship with the European ones (Table 2). It follows, therefore, that migration to or from the European region was perhaps by way of a northerly route, via the Urals and Arctic regions.

Emsian Biogeography

The problem of provinces versus communities is brought into focus by the distribution of Emsian conodonts. Icriodids were diverse in the European region and are reported to have been abundant in the Great Basin of Nevada (Klapper and Johnson, 1975). However, they are very scarce or absent from described northern Cordilleran (Alaska, Yukon) and eastern Australian faunas. Community (biofacies) differences are implied by the fact that the northern Cordilleran and many of the eastern Australian faunas are from intermediate-depth to deep-water carbonates while the European and Nevadan faunas may be from more shallow-water deposits. Unfortunately, adequate details of the Nevadan faunas have not been published so that it cannot be determined whether they were closely related taxonomically to the European icriodids or constituted a discrete fauna.

More uniformly distributed (which perhaps correlates with less environmentally restricted) spathognathodids and polygnathids are more appropriate taxa to use in Emsian biogeographic interpretations. As in the Siegenian, members of the *Spathognathodus expansus* lineage (*S. optimus*, *S. exiguus*, and *S. expansus*) are widespread in faunas of the circum-Pacific region while members of the *S. steinhornensis* lineage (*S. miae* and *S. steinhornensis*) inhabited the European region. Most Emsian polygnathids are distributed worldwide but *Polygnathus serotinus* is known only from faunas of eastern Australia, western North America, and possibly the northern Urals, and *P. perbonus* is not known definitely outside eastern Australia.

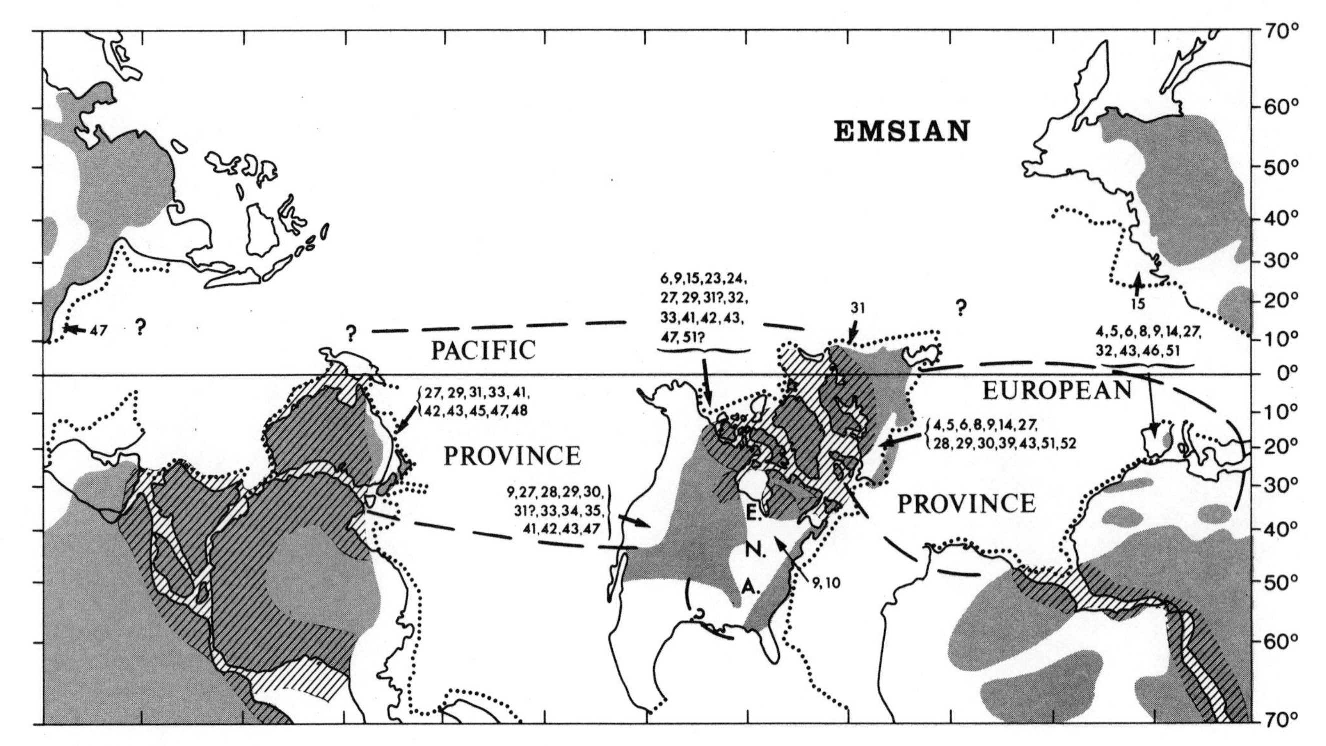

Figure 3. Provincial distribution of Emsian conodonts (explanation as for Fig. 1).

Therefore, during the Emsian, the European and Pacific Provinces remained distinct. Although eastern Australia and western North America had a number of endemic Emsian taxa (see Table 1), it is clear from Table 2 that they shared a considerable percentage of species, and separation of subprovinces does not seem feasible at this time.

The occurrences of *Icriodus taimyricus* (Table 1; Fig. 3) have interesting paleogeographic implications. If the Asian continental fragment is shifted westward (as postulated above following Cocks and McKerrow, 1973), the only known occurrences of *I. taimyricus* (Alaska, northern Yukon, and north-central Siberia) are in very close proximity.

As noted earlier, eastern North American Emsian faunas had an extremely low diversity and contained the endemic *Icriodus latericrescens robustus*. Rare specimens of *Spathognathodus* have been encountered (Telford and von Bitter, 1975) but polygnathids were completely absent. A combination of provincial and community factors, with the former considered to be most important, was possibly responsible for this situation. The majority of described faunas (Klapper and Ziegler, 1967; Klapper and others, 1971) are from near-shore or very shallow-water deposits which, according to the lateral segregation ecological model of Barnes and Fåhraeus (1975), could explain the total exclusion of polygnathids. However, spathognathodids, which also were almost totally absent, have been shown to inhabit very shallow waters (Barnett, 1972). Also, the diversity of icriodids remained relatively low compared to the geographically close (Fig. 3) European region.

Thus, the primary factor in development of the distinctive eastern North American fauna may have been a migrational barrier. In his comprehensive analysis of Devonian biogeography, Boucot (1975) proposed that the eastern North American region was isolated from neighboring regions (Nevada, Europe) by hypersaline waters over the continental backbone in the west (see Fig. 3) and by a thermal gradient kept in place by a marine current system hindering migration to and from the southeast and southwest. This model is effective in explaining the development of Emsian faunas in eastern North America.

Middle Devonian Biogeography

During the early Middle Devonian, polygnathid and spathognathodid taxa became widely distributed (Klapper, 1971; Telford, 1975) so that the boundaries of the Lower Devonian European and Pacific Provinces became indistinct. Some anomalies in distribution of icriodids have been noted (see p. 208) particularly with regard to the European and eastern North American regions. Perhaps the barriers to southerly migration between these regions, as described by Boucot (1975), remained partly in effect during early Middle Devonian time and could only be overcome by less environmentally restricted polygnathids and spathognathodids dwelling in deeper water. Only after breaching of the Taghanic Arch in late Middle Devonian (late Givetian) time could there have been free interplay of all conodont taxa between the eastern North American, western North American, and European regions on a northerly route through the Arctic and Ural regions.

SUMMARY

(1) Although data is poor or lacking for several critical areas, Devonian conodonts show significant inhomogeneities in distribution that are probably the result of both provincial and community factors.

(2) Community factors may be assessed using a conodont ecological model involving lateral segregation rather than simple depth stratification of faunas.

(3) During the early Early Devonian (Gedinnian), conodonts were either restricted to a European Province (encompassing present day Europe, North Africa, and Turkey) or formed part of a low diversity, almost cosmopolitan fauna which included species such as *Icriodus woschmidti* and *Spathognathodus remscheidensis*.

(4) Two conodont provinces can be defined for middle Early Devonian (Siegenian) time; these were the European Province and a Pacific Province that encompassed present day eastern Australia and western North America including Nevada-Utah and the Canadian Arctic Islands. Species of *Spathognathodus* and *Eognathodus* are most useful in separating the two Provinces.

(5) This pattern persisted into the late Early Devonian (Emsian) and, additionally, the eastern North American region (Texas to Hudson Bay) developed a particular faunal character (low diversity faunas dominated by *Icriodus latericrescens robustus*) due to a combination of provincial and community factors.

(6) Provinciality of conodonts gradually decreased during the Middle Devonian with only the eastern North American region maintaining a separate identity until breaching of the Taghanic Arch in late Givetian time. By the Late Devonian, conodont faunas were uniformly distributed through the paleoequatorial regions. The apparent absence of conodonts from the Devonian Malvinokaffric Realm may be a characteristic of that high paleolatitudinal region.

(7) With some relatively minor modification, paleogeographic maps of Smith and others (1973), combined with additional data from Johnson and Boucot (1973), agree fairly well with the presently known distribution of Lower Devonian conodonts.

ACKNOWLEDGEMENTS

John A. Talent gave initial impetus to these Devonian biogeographical studies. Norman M. Savage and John W. Pickett provided me with unpublished data on conodont faunas from northern California, Alaska, and New South Wales. Christopher R. Barnes reviewed the manuscript and offered valuable criticism. This paper is published with permission of the Director of the Geological Branch, Ontario Division of Mines.

REFERENCES

Amsden, T. W. and others, 1967, Devonian of the southern Midcontinent area, United States, *in* Oswald, D. H., ed., Internat. Symposium on the Devonian System, Calgary 1967: Calgary, Alberta Soc. Petroleum Geologists, v. 1, p. 913-932.

Barnes, C. R., and Fåhraeus, L. E., 1975, Provinces, communities and the proposed nektobenthic habit of Ordovician conodontophorids: Lethaia, v. 8, p. 133-150.

Barnes, C. R., Rexroad, C. B., and Miller, J. F., 1973, Lower Paleozoic conodont provincialism, *in* Rhodes, F. H. T., ed., Conodont paleozoology: Geol. Soc. America Spec. Paper 141, p. 156-190.

Barnett, S. G., 1972, The evolution of *Spathognathodus remscheidensis* in New York, New Jersey, Nevada, and Czechoslovakia: Jour. Paleontology, v. 46, p. 900-917.

Barnett, S. G., Kohut, J. J., Rust, C. C., and Sweet, W. C., 1966, Conodonts from Nowshera reef limestones (uppermost Silurian or lowermost Devonian), West Pakistan: Jour. Paleontology, v. 40, p. 435-438.

Bischoff, G., and Sannemann, D., 1958, Unterdevonische Conodonten aus dem Frankenwald: Hess. Landesamt Bodenforsch. Notizbl., v. 86, p. 87-110.

Boersma, K. T., 1974, Description of certain Lower Devonian platform conodonts of the Spanish central Pyrenees: Leidse Geol. Meded., v. 49, p. 285-301.

Boucot, A. J., 1975, Evolution and extinction rate controls: New York, Elsevier, 427 p.

Bultynck, P., 1970, Revision stratigraphique et paleontologique de la coupé type du Couvinien: Louvain Univ. Inst. Geol. Mem. 26, 152 p.

——— 1971, Le Silurien supérieur et le Devonien inférieur de la Sierra de Guadarrama (Espagne centrale). Deuxième partie. Assemblages de Conodontes à *Spathognathodus:* Inst. Royal Sci. Nat. Belgique Bull., v. 47, p. 1-43.

——— 1972, Middle Devonian *Icriodus* assemblages (Conodonta): Geologica et Paleontologica, v. 6, p. 71-86.

Carls, P. 1969, Die Conodonten des tieferen Unter-Devon der Guadarrama (Mittel-Spanien) und die Stellung des Grenzbereiches Lochkovium/Pragium nach der rheinischen Gliederung: Senckenbergiana Lethaea, v. 50, p. 303-355.

Carls, P., and Gandl, J., 1969, Stratigraphie und Conodonten des Unter-Devons der Ostlichen Iberischen Ketten (NE-Spanien): Neues Jahrb. Mineral. Geol. Paläont. Abh., v. 132, p. 155-218.

Chatterton, B. D. E., 1974, Middle Devonian conodonts from the Harrogate Formation, southeastern British Columbia: Canadian Jour. Earth Sci., v. 11, p. 1461-1484.

Cocks, L. R. M., and McKerrow, W. S., 1973, Brachiopod distributions and faunal provinces in the Silurian and Lower Devonian, *in* Hughes, N. F., ed., Organisms and continents through time: Spec. Papers Palaeontology no. 12, p. 291-304.

Cooper, B. J., 1973, Lower Devonian conodonts from Loyola, Victoria: Royal Soc. Victoria Proc., v. 86, p. 77-84.

Davis, W. E., Jr., 1975, Significance of conodont distribution in the Tully Limestone (Devonian), New York State: Jour. Paleontology, v. 49, p. 1097-1104.

Druce, E. C., 1973, Upper Paleozoic and Triassic conodont distribution and the recognition of biofacies, *in* Rhodes, F. H. T., ed., Conodont paleozoology: Geol. Soc. America Spec. Paper 141, p. 191-238.

Fåhraeus, L. E., 1974, Taxonomy and evolution of *Ozarkodina steinhornensis* and *Ozarkodina optima* (Conodontophorida): Geologica et Paleontologica, v. 8, p. 29-37.

——— 1976, Possible Early Devonian conodontophorid provinces: Palaeogeography, Palaeoclimatology, Palaeoecology, v. 19, p. 201-217.

Ferrigno, K. F., 1971, Environmental influences on the distribution and abundance of conodonts from the Dundee Limestone (Devonian), St. Mary's, Ontario: Canadian Jour. Earth Sci., v. 8, p. 378-386.

Glenister, B. F., and Klapper, G., 1966, Upper Devonian conodonts from the Canning Basin, western Australia: Jour. Paleontology, v. 40, p. 777-842.

House, M. R., 1973, Devonian goniatites, *in* Hallam, A., ed., Atlas of palaeobiogeography: New York, Elsevier, p. 97-104.

Irving, E., 1964, Paleomagnetism and its application to geological and geophysical problems: New York, Wiley, 399 p.

Johnson, J. G., 1970, Taghanic onlap and the end of North American Devonian provinciality: Geol. Soc. America Bull., v. 81, p. 2077-2106.

Johnson, J. G., and Boucot, A. J., 1973, Devonian brachiopods, *in* Hallam, A., ed., Atlas of palaeobiogeography: New York, Elsevier, p. 89-96.

Kerr, J. W., 1974, Geology of the Bathurst Island Group and Byam Martin Island, Arctic Canada: Canada Geol. Survey Mem. 378, 152 p.

Klapper, G., 1969, Lower Devonian conodont sequence, Royal Creek, Yukon Territory, and Devon Island, Canada: Jour. Paleontology, v. 43, p. 1-27.

——— 1971, Sequence within the conodont genus *Polygnathus* in the New York lower Middle Devonian: Geologica et Paleontologica, v. 5, p. 59-80.

Klapper, G., and Johnson, D. B., 1975, Sequence in conodont genus *Polygnathus* in Lower Devonian at Lone Mountain, Nevada: Geologica et Paleontologica, v. 9, p. 65-83.

Klapper, G., and Murphy, M. A., 1974, Silurian-Lower Devonian conodont sequence in the Roberts Mountains Formation of central Nevada: California Univ. Pubs. Geol. Sci., v. 111, p. 1-62.

Klapper, G., and Philip, G. M., 1971, Devonian conodont apparatuses and their vicarious skeletal elements: Lethaia, v. 4, p. 429-452.

——— 1972, Familial classification of reconstructed Devonian conodont apparatuses: Geologica et Paleontologica, SB 1, p. 97-114.

Klapper, G., Philip, G. M., and Jackson, J. H., 1970, Revision of the *Polygnathus varcus* Group (Conodonta, Middle Devonian): Neues Jahrb. Geol. Paläont. Monatsh., v. 11, p. 650-667.

Klapper, G. and others, 1971, North American Devonian conodont biostratigraphy: Geol. Soc. Amer. Mem. 127, p. 285-316.

Klapper, G., and Ziegler, W., 1967, Evolutionary development of the *Icriodus latericrescens* Group (Conodonta) in the Devonian of Europe and North America: Palaeontographica, Abt. A, v. 127, p. 68-83.

Lane, H. R., 1974, *Icriodus taimyricus* (Conodonta) from the Salmontrout Limestone (Lower Devonian), Alaska: Jour. Paleontology, v. 48, p. 721-726.

Lindstrom, M., and Ziegler, W., eds., 1972, Symposium on conodont taxonomy: Geologica et Paleontologica, SB 1, 158 p.

Merrill, G. K., 1973, Pennsylvanian conodont paleoecology, *in* Rhodes, F. H. T., ed., Conodont paleozoology: Geol. Soc. America Spec. Paper 141, p. 239-276.

Moskalenko, T. A., 1967, First find of Late Silurian conodonts in Zeravshan Range: Internat. Geol. Rev., v. 9, p. 195-204.

Orr, R. W., 1971, Conodonts from Middle Devonian strata of the Michigan Basin: Indiana Geol. Survey Bull. 45, 110 p.

Pedder, A. E. H., Jackson, J. H., and Ellenor, D. W., 1970, An interim account of the Middle Devonian Timor Limestone of north-eastern New South Wales: Linnean Soc. New South Wales Proc., v. 94, p. 242-272.

Philip, G. M., 1966, Lower Devonian conodonts from the Buchan Group, eastern Victoria: Micropaleontology, v. 12, p. 441-460.

——— 1967, Middle Devonian conodonts from the Moore Creek Limestone, northern New South Wales: Royal Soc. New South Wales Jour. Proc., v. 100, p. 151-161.

Philip, G. M., and Pedder, A. E. H., 1967, Stratigraphic correlation of the principal Devonian limestone sequences of eastern Australia, *in* Oswald, D. H., ed., Internat. Symposium on the Devonian System, Calgary 1967: Calgary, Alberta Soc. Petroleum Geologists, v. 2, p. 1025-1041.

Sandberg, C. A., Hall, W. E., Batchelder, J. N., and Axelsen, C., 1975, Stratigraphy, conodont dating, and paleotectonic interpretation of the type Milligen Formation (Devonian), Wood River area, Idaho: U. S. Geol. Survey Jour. Research, v. 3, p. 707-720.

Seddon, G., 1970, Frasnian conodonts from the Sadley Ridge-Bugle Gap area, Canning Basin, western Australia: Geol. Soc. Australia Jour., v. 16, p. 723-754.

Seddon, G., and Sweet, W. C., 1971, An ecologic model for conodonts: Jour. Paleontology, v. 45, p. 869-880.

Skipp, B., and Sandberg, C. A., 1975, Silurian and Devonian miogeosynclinal and transitional rocks of the Fish Creek Reservoir window, central Idaho: U. S. Geol. Survey Jour. Research, v. 3, p. 691-706.

Smith, A. G., Briden, J. C., and Drewry, G. E., 1973, Phanerozoic world maps, *in* Hughes, N. F., ed., Organisms and continents through time: Spec. Papers Palaeontology no. 12, p. 1-42.

Snigireva, M. P., 1975, Novyo konodonty iz srednedevonskikh otlozheniy severnogo urala: Paleont. Zhurn., v. 4, p. 24-31.

Stewart, G. A., and Sweet, W., 1956, Conodonts from the Middle Devonian bone beds of central and west-central Ohio: Jour. Paleontology, v. 30, p. 261-273.

Sweet, W. C., Ethington, R. L., and Barnes, C. R., 1971, North American Middle and Upper Ordovician conodont faunas: Geol. Soc. America Mem. 127, p. 163-193.

Telford, P. G., 1972, Conodonts, *in* Talent, J. A. and others, Provincialism and Australian Early Devonian faunas: Geol. Soc. Australia Jour., v. 19, p. 81-97.

——— 1975, Lower and Middle Devonian conodonts from the Broken River embayment, north Queensland, Australia: Spec. Papers Palaeontology no. 15, 96 p.

Telford, P. G., and von Bitter, P. H., 1975, Devonian conodont biostratigraphy and paleoecology, Niagara Peninsula, Ontario: Geol. Soc. America, Abs. with Programs, v. 7, p. 870-871.

Uyeno, T. T., and Mason, D., 1975, New Lower and Middle Devonian conodonts from northern Canada: Jour. Paleontology, v. 49, p. 710-723.

Valentine, J. W., 1973, Plates and provinciality, a theoretical history of environmental discontinuities, *in* Hughes, N. F., ed., Organisms and continents through time: Spec. Papers Palaeontology no. 12, p. 79-92.

Walliser, O. H., 1960, Scolecodonts, conodonts, and vertebrates, *in* Boucot, A. J. and others, A late Silurian fauna from the Sutherland River Formation, Devon Island, Canadian Arctic Archipelago: Canada Geol. Survey Bull., v. 65, p. 28-39.

Wittekindt, H., 1966, Zur Conodontenchronologie des Mitteldevons: Forschr. Geol. Rheinland Westfalens, v. 9, p. 621-646.

Ziegler, W., 1962, Taxonomie und Phylogenie Oberdevonischer Conodonten und ihre stratigraphische Bedeutung: Hess. Landesamt Bodenforsch. Abh., v. 38, p. 1-166.

——— 1971, Conodont stratigraphy of the European Devonian: Geol. Soc. America Mem. 127, p. 227-284.

——— ed., 1973, Catalogue of Conodonts, Vol. 1: Stuttgart, E. Schweizerbart'sche, 504 p.

——— ed., 1975, Catalogue of Conodonts, Vol. II: Stuttgart, E. Schweizerbart'sche, 404 p.

APPENDIX

Alphabetic List of Cited Taxa with References and Explanatory Notes. Supplementary References Indicate Recent Descriptions or Taxonomic Revisions

Ancyrodelloides kutscheri Bischoff and Sannemann, 1958

A. trigonica Bischoff and Sannemann, 1958

Eognathodus sulcatus Philip (Klapper and Philip, 1971)—multielement taxon

E. trilinearis (Cooper, 1973)

Icriodus angustoides Carls and Gandl (Ziegler, 1975)

I. angustus Stewart and Sweet (Ziegler, 1975)

I. beckmanni Ziegler (Ziegler, 1975)

I. bilatericrescens Ziegler (Ziegler, 1975)

I. curvicauda Carls and Gandl (Ziegler, 1975)

I. eolatericrescens Mashkova (Ziegler, 1975)

I. fusiformis Carls and Gandl (Ziegler, 1975)

I. huddlei Klapper and Ziegler (Ziegler, 1975)

I. latericrescens latericrescens Branson and Mehl (Ziegler, 1975)

I. latericrescens robustus Orr (Ziegler, 1975)

I. latericrescens n. subsp. B of Klapper, 1969

I. postwoschmidti Mashkova (Ziegler, 1975)

I. rectangularis Carls and Gandl, 1969

I. sigmoidalis Carls and Gandl (Ziegler, 1975)

I. taimyricus Kuzmin (Lane, 1974)

I. woschmidti hesperius Klapper and Murphy, 1974—multielement taxon

I. woschmidti woschmidti Ziegler (Ziegler, 1975; Klapper and Murphy, 1974)

I. sp. cf. *I. corniger* Wittekindt, 1966

Pedavis pesavis pesavis (Bischoff and Sannemann) (Klapper and Philip, 1972)—multielement taxon

P. pesavis n. subsp. A of Klapper and Philip, 1972—multielement taxon

P. n. sp. A of Klapper and Philip, 1972—multielement taxon

P. n. sp. B of Klapper and Philip, 1972—multielement taxon

P. n. sp. C of Klapper and Murphy, 1974—multielement taxon

Pelekysgnathus furnishi Klapper (Klapper and Philip, 1972)—multielement taxon

P. glenisteri Klapper (Klapper and Philip, 1972)—multielement taxon

P. serratus Jentzsch (Carls and Gandl, 1969)

P. sp. of Telford, 1975

Polygnathus angusticostatus Wittekindt (Klapper, 1971)

P. costatus costatus Klapper, 1971

P. dehiscens Philip and Jackson (Klapper and Johnson, 1975)

P. gronbergi Klapper and Johnson, 1975

P. inversus Klapper and Johnson, 1975

P. laticostatus Klapper and Johnson, 1975

P. linguiformis linguiformis Hinde (Bultynck, 1970)

P. perbonus (Philip) (Klapper and Philip, 1971)—multielement taxon

P. pireneae Boersma, 1974

P. serotinus Telford, 1975

P. totensis Snigireva, 1975 = *P. serotinus* Telford, 1975

P. varcus Stauffer (Klapper, Philip, and Jackson, 1970)

P. sp. A of Klapper and Johnson, 1975

Spathognathodus asymmetricus Bischoff and Sannemann, 1958

S. boucoti Klapper, 1969

S. carinthiaca Schulze (Klapper *in* Ziegler, 1973)

S. eurekaensis—P element of *Ozarkodina eurekaensis* Klapper and Murphy, 1974

S. exiguus Philip (includes *S. exiguus philipi* Klapper, 1969) (Telford, 1975)—P element of *Pandorinellina exigua* (Philip) (Klapper *in* Ziegler, 1973)

S. expansus—P element of *Pandorinellina expansa* Uyeno and Mason, 1975

S. inclinatus (Rhodes) = *S. wurmi* (Bischoff and Sannemann)—P element of *Ozarkodina excavata excavata* (Branson and Mehl) (Klapper *in* Ziegler, 1973)

S. johnsoni Klapper, 1969

S. linearis (Philip, 1966)

S. miae Bultynck, 1971

S. optimus Moskalenko (Telford, 1975) = *S. buchanensis* Philip, 1966—P element of *Pandorinellina optima* (Moskalenko) (Klapper and Philip, 1972)

S. palethorpei Telford, 1975

S. remscheidensis eosteinhornensis—P element of *Ozarkodina remscheidensis eosteinhornensis* (Walliser) (Klapper *in* Ziegler, 1973)

S. remscheidensis remscheidensis Ziegler = ? *S. canadensis* Walliser, 1960—P element of *Ozarkodina remscheidensis remscheidensis* (Ziegler) (Klapper *in* Ziegler, 1973)

S. steinhornensis Ziegler (Klapper *in* Ziegler, 1973)

S. stygia Flajs (Klapper *in* Ziegler, 1973)

S. transitans Bischoff and Sannemann (Klapper *in* Ziegler, 1973)

S. n. sp. E = *Ozarkodina* n. sp. E of Klapper and Murphy, 1974

S. n. sp. F = *Ozarkodina* n. sp. F of Klapper and Murphy, 1974

Evolution of Fusulinacea (Protozoa) in Late Paleozoic Space and Time

CHARLES A. ROSS, *Department of Geology, Western Washington State University, Bellingham, Washington 98225*

ABSTRACT

Late Paleozoic fusulinaceans had three times of rapid evolution: near the beginning of the Middle Carboniferous, in the early part of the Permian, and near the middle of the Permian. The first of these radiations gave rise to the Fusulinidae, Ozawainellidae, Schubertellidae, and Staffellidae which dominated the Middle Carboniferous fusulinacean faunas. Although the Schwagerinidae first appear in the Late Carboniferous, the family had a low generic diversity until the beginning of the Permian when it rapidly evolved many genera and several subfamilies. In the middle part of the Permian, the Staffellidae, Schubertellidae, and Ozawainellidae evolved a large number of new genera, several subfamilies, and 2 families, the Verbeekinidae and Neoschwagerinidae. The latest Permian fusulinaceans consist mostly of genera of the Schubertellidae and Ozawainellidae. The late Paleozoic fusulinacean extinction takes place by gradual attrition of genera and many previously dominant subfamilies are minor parts of, or are lacking in, latest Permian fusulinacean faunas.

Biogeographic provinciality was present in fusulinaceans during most of the late Paleozoic. Two regions were present in the Carboniferous and early part of Early Permian time. The largest region included the Tethyan, Uralian, Franklinian, and northern part of the Cordilleran Geosynclines and adjacent continental shelves. The smaller region was south of the transcontinental arch of North America, and included parts of northern and western South America in Late Carboniferous and Early Permian time. Irregular and incomplete dispersals occurred between these regions. In the later part of Early Permian time, fusulinaceans in the Tethyan portion were isolated from those in other areas and few or no dispersals took place. This resulted in three nearly independent biogeographical and evolutionary histories in fusulinaceans and in the establishment of a Tethyan Realm, a Uralian-Franklinian Realm, and a southwestern North American Realm. Youngest fusulinaceans are known only from the Tethyan Realm.

INTRODUCTION

The development and significance of high concentrations of endemic taxa in particular geographic regions during the geologic past has many ramifications when viewed from the perspectives of ocean-floor spreading and plate tectonic models. In general, the development in a region of a strongly endemic biota suggests that that biota did not have a free exchange with those in other regions. Because considerable interchange between marine faunas of adjacent regions is a common pattern, endemism implies strong ecological barriers to dispersal and population intermixing. Ecological barriers are of many kinds (Simpson, 1947), and may influence the distributions and movements of different species or groups within a biota in different ways. Most ecological barriers are related directly or indirectly to the physical environment, and include contrasts in temperature, orography, rainfall, and other factors that influence regional climates or are related to physical connections of land or sea areas. Circulation of the atmosphere and water currents are also of obvious importance to nonmobile or passively transported members of the biota. A cosmopolitan aspect to a biota with widely dispersed species and genera suggests ready dispersal and exchange between different regions.

The Recent terrestrial biota of the world shows a relatively high amount of endemism of terrestrial plants and animals which can be traced back through time to more cosmopolitan biotas of the Cretaceous and Jurassic Periods. Recent marine biota also have considerable endemism but these biota can be traced to more cosmopolitan early to middle Cenozoic distributions. Numerous attempts to quantify similarities and differences between different regions have resulted in placing emphasis on different features of the biota (Ross, 1974). The simplest of these comparisons uses taxonomic (or morphologic) similarities and differences in which the total number of taxa are considered. Some comparisons weight the hierarchal level of the taxon so that endemic or cosmopolitan genera carry more importance than species, and families have more importance than genera. Other comparisons of biota attempt to establish ecologically dominant parts of the biota, such as large plants or large herbivores, on which to base changes in the community and community structure. Still other types of comparative analysis stress phylogenetic continuity, centers of endemism, and centers of dispersal. Each of these approaches produces comparisons which greatly aid in understanding the history of biotas, and significant changes in world and regional biotas during the geologic past can be identified. Many of these changes can be directly or indirectly associated with geographic and climatic changes on the Earth's surface that relate to geological processes, such as marine transgression and regression across shallow epicontinental shelves or the phases of mountain building, such as those of the Himalayan Mountains.

Mountain-building activity also affects regional climates by altering climatic patterns because of orographic factors, such as height, width, and continuity, and by altering the place of origin and strength of ocean currents. In addition, periods of general climatic cooling or warming have influenced dispersals and colonization. Although many additional unanswered questions concern the detailed interaction of evolution and phylogeny, dispersals and colonization, climatic changes, community structure and biogeography, one general conclusion seems inescapable: changes that accumulate through time in any one of these aspects have a progressive effect on one or more of the other aspects.

The limitations of applying such analyses of Recent biota and biotic realms and provinces to the geologic past are obvious. The entire living biota, in all of its completeness, still leaves us with many unanswered questions about the interaction of these many dimensional aspects and a fossil biota represents only a small portion of the data that once was present. Also, as we delve farther back into the geological past the details of earlier paleogeography become less specific. Of particular interest is the apparent lack of ocean floor sediments or oceanic crust older than Jurassic. This suggests that the continental crust in pre-Jurassic time had associations with portions of oceanic crusts which have been returned to the mantle at Benioff zones and that much of the oceanic sediments on these portions of the oceanic crust also have been lost to the mantle or incorporated into the edges of continental crusts (Dietz and Holden, 1970). Thus, continental crustal positions for pre-Jurassic times cannot be cross referenced to oceanic floors for independent conformation, but must rely only on data from the continents themselves.

TETHYS

The name Tethys was proposed for a large east-west ocean or sea that stretched across southern Europe, Asia, and northern Africa during the Mesozoic Era. This ocean contained a distinctive marine fauna of probably tropical affinities in contrast to other marine faunas of northern Europe which appear to be temperate faunas. At times during the Mesozoic, particularly during the early part of the Jurassic and again in the middle and Late Cretaceous, faunas of Tethyan affinities became nearly cosmopolitan and widespread in shallow epicontinental seas. The Tethys as a seaway was gradually destroyed during the early and middle part of the Cenozoic by the Alpine-Himalayan orogeny and by the continued movement of North America away from Europe and Africa; the present Mediterranean and Caribbean Seas are but small isolated pieces of this once, much larger tropical ocean.

		URAL MTS. AND RUSSIAN PLATFORM	N.W. EUROPE	TETHYAN CENTRAL ASIA	SOUTHWEST AND CENTRAL N. AMERICA
PERMIAN	UPPER	TATARIAN	THURINGIAN	PAMIRIAN (=LOPINGIAN)	OCHOAN
		KAZANIAN		MURGABIAN	GUADALUPIAN
		UFIMIAN / KUNGURIAN	SAXONIAN		
	LOWER	ARTINSKIAN	AUTUNIAN	DARVASIAN	LEONARDIAN
		SAKMARIAN		KARACHATYRIAN	WOLFCAMPIAN
		ASSELIAN			
CARBONIFEROUS	UPPER	ORENBURGIAN	STEPHANIAN	GZHEL'IAN	VIRGILIAN
		GZHEL'IAN			
		KASIMOVIAN			MISSOURIAN
	MIDDLE	UPPER MOSCOVIAN	WESTPHALIAN	MOSCOVIAN	DESMOINESIAN
		LOWER MOSCOVIAN			ATOKAN
		BASHKIRIAN	U. M. L. NAMURIAN	BASHKIRIAN	MORROWAN
	LOWER	SERPUKOVIAN		SERPUKOVIAN	CHESTERIAN
		VISEAN	VISEAN	VISEAN	VALMEYERIAN
		TOURNAISIAN	TOURNAISIAN	TOURNAISIAN	KINDERHOOKIAN

Figure 1. Correlation chart of stages for different regions during the Carbonifeous and Permian. The boundaries of the series of the Carboniferous and Permian are used in the sense of usage in the Russian Platform, southern Urals, and Donetz Basin where shallow marine fossils are common in all except the Upper Permian Series.

The Tethys also has a pre-Mesozoic history, and this earlier history forms the subject for this investigation. Many marine invertebrate groups are known in considerable detail from late Paleozoic (Carboniferous and Permian) strata including foraminiferids, rugose corals, brachiopods, bryozoans, and cephalopods. This review concentrates on the foraminiferids, particularly the Fusulinacea, which have been more extensively documented both geographically and stratigraphically than the other four important groups.

Discussion of the subdivisions of the Carboniferous and Permian Systems will follow the general outline established for the Russian Platform, Donetz Basin, and southern Urals, and time equivalent stratigraphic intervals elsewhere (Fig. 1). The Carboniferous is divided into three series, Lower, Middle, and Upper; the Permian into two series, Lower and Upper. The total amount of time from the beginning of the Carboniferous (about 345 m.y.) to the end of the Permian (about 225 m.y.) is at least 120 m.y. or roughly twice as long as the whole of the Cenozoic Period. This is a major interval of geologic time and epochs within the Carboniferous and Permian are long time units so that the working units of correlation are the stages, many of which are given regional names.

The Lower Carboniferous stages are traceable throughout most of Eurasia with considerable certainty but with less certainty in the type region of the North American Mississippian System (Lower Carboniferous of this study). A general and persistent fall in sea level near the end of Early Carboniferous time resulted in the withdrawal of seas from epicontinental platforms followed by a gradual increase in sea level accompanying numerous marine transgressions and regressions during Middle Carboniferous time that again flooded these platforms. A second general lowering of sea level occurred near the end of Middle Carboniferous time and again sea level gradually rose in progressive steps related to marine transgressions and regressions during the Late Carboniferous. Nearly 90 cyclothems associated with repeated marine transgression and regression are recorded in Carboniferous epicontinental strata. The cause or causes of these short-term sea-level changes are difficult to establish beyond doubt, but they seem related to periods of glaciation in Gondwanan continents which suggest the same sort of relationship between sea-level changes and glaciation as in the Pleistocene during the last 1.5 m.y. In addition to sea-level fluctuations, the Middle and Late Carboniferous was a time of increasing orogenic activity along the Appalachian-Hercynian belt. This orogeny progressively modified the climates on either side of the mountain-building belt and caused increased provincial and regional differences in depositional environments. This is shown to some degree by the lateral extent of stage names which reflect the distribution of similar faunas (or floras) and also by the local increases in sedimentary rates and increases in clastic material in some regions.

The Lower Permian, or at least the first part of it, is in many aspects a continuation of Late Carboniferous depositional patterns. Marine transgressions and regressions continued but with decreasing magnitude and were possibly spaced farther apart in time. These correlate with a decrease in glacial deposits in Gondwanan continents during the Early Permian. The epicontinental shelves became increasingly evaporitic during the later part of the Early Permian. The last major phase of the Appalachian-Hercynian orogeny occurred near the end of Early Permian time and this late orogenic phase significantly altered the physical paleogeography. Afterwards, a Tethyan Faunal Realm is clearly identified from northern Africa, through the Alps, Yugoslavia, Greece, Turkey, Crimea, Iran, Iraq, Afghanistan, Pakistan, Kashmir, China, Japan, Indochina, New Zealand, the Maritime Provinces of USSR, and parts of Alaska, the Yukon, British Columbia, Washington, Oregon, and California—a much larger and more elongate realm than the Mesozoic Tethys. Marine fossils are known in the highest part of the Upper Permian only in the Eurasian part of the Tethys which more closely approximates the distribution of the Mesozoic Tethys. Many Paleozoic fossil groups, such as the fusulinaceans, become extinct at different times during the Late Permian and a few Paleozoic lineages in groups such as the brachiopods and bryozoans survived into the earliest part of the Triassic. For these reasons the Late Permian and Early Triassic are generally regarded as times of acute ecologic stress for shallow marine organisms.

FUSULINACEANS

The superfamily Fusulinacea (Fig. 2) is a group of large calcareous foraminiferids that evolved during the Carboniferous and Permian Periods. They have a planispiral shell with layered wall construction, or originate from such a wall construction, with chambers and an internal tunnel or tunnels (Fig. 3). Although the fusulinaceans range in size from a few millimeters in diameter and length to over 2 cm in diameter and 20 cm in length, most are 1 to 2 cm long. They are common limestone builders from the beginning of the Middle Carboniferous until very near the end of the Permian when they became extinct. Because of their physical size and abundance, fusulinaceans are usually easily seen in the field and are commonly collected for purposes of biostratigraphic age determinations.

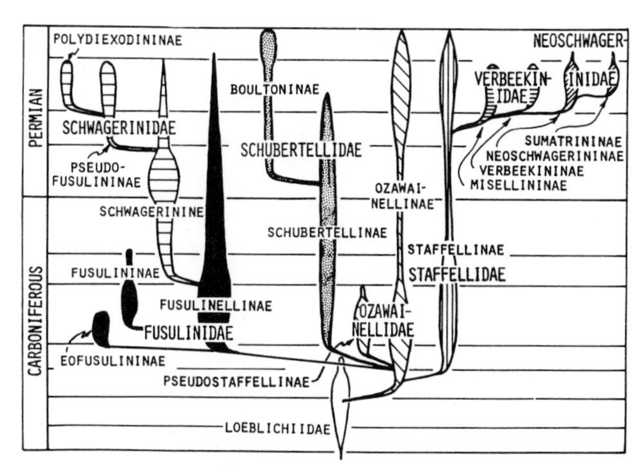

Figure 2. Generalized phylogenetic history of the families and subfamilies of Fusulinacea (modified after Rozovskaya, 1975).

The phylogeny and geologic history of fusulinaceans has recently been summarized by Rozovskaya (1975) and a slightly modified version of her phylogeny of 16 subfamilies of the Fusulinacea is shown in Figure 2. In the classification used in this paper, the Fusulinacea are retained as a superfamily rather than being placed as a suborder, and all of the families are assigned to the one superfamily. Seven families are recognized on the basis of differences in wall structure and other internal features (Fig. 3). The ancestral family, the Loeblichiidae, are small lenticular forms and are particularly important in Lower Carboniferous strata where they are widely used for stratigraphic zonation. During the Early Carboniferous, the Loeblichiidae gave rise to the first of the Ozawainellidae, which are also lenticular in shape (Fig. 2). Although the Ozawainellidae had a burst of evolution into at least 9 genera during the later part of the Early Carboniferous and earliest part of the Middle Carboniferous, and again into 3 or 4 genera in the Late Permian, their greater importance lies in the fact they gave rise to the Fusulinidae, Schubertellidae, and Staffellidae.

Fusulinidae are the first of the Fusulinacea to become markedly elongated along their axis of coiling (Figs. 3, 4). Of the 3 subfamilies (Fig. 2), the most primitive group, the Fusulinellinae, had the greatest geologic range, from the lower part of the Middle Carboniferous into the Upper Permian. Although this subfamily changes in some features, it retains basically simple unfolded septa and simple layered walls. The second subfamily, the Eofusulininae, is confined to the lower and middle parts of the Middle Carboniferous and has strongly folded septa and minor modification in the wall structure. The subfamily Fusulininae has well-developed septal folds and some genera have well-developed

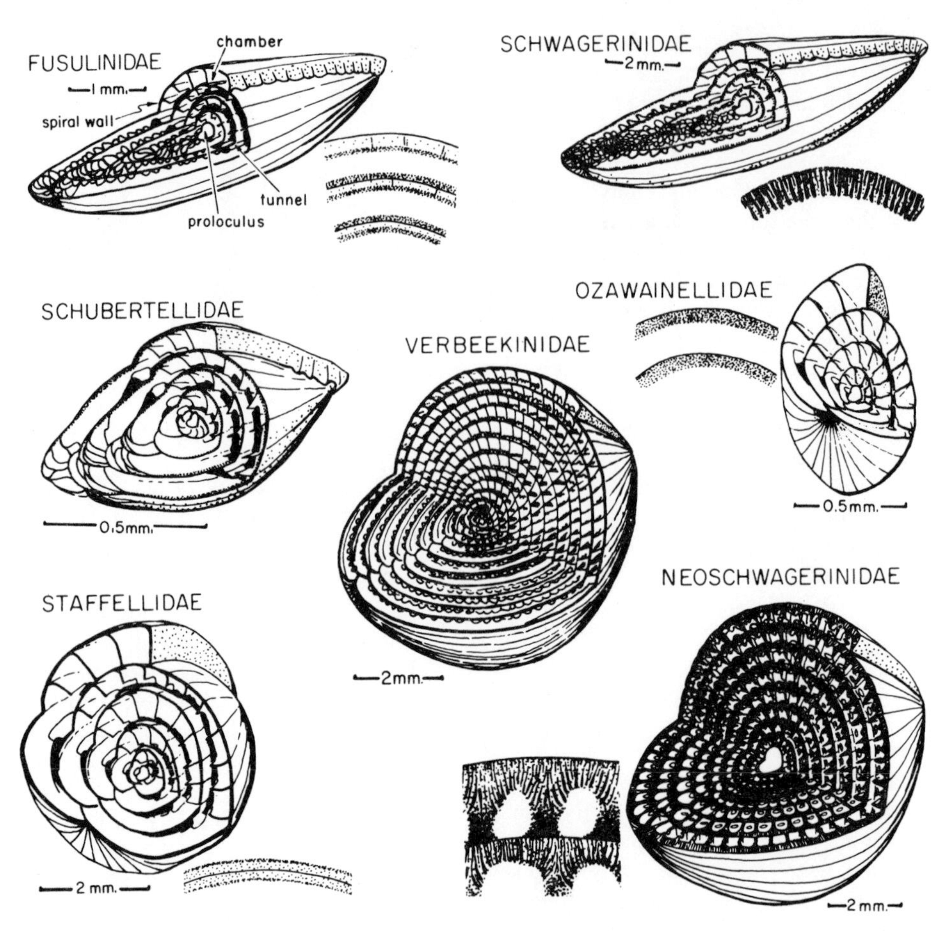

Figure 3. Representatives from the major families of Fusulinacea.

4-layered walls and massive chomata adjacent to the tunnel. The major evolutionary burst of this family during the Middle Carboniferous is shown in Figure 4.

From the Fusulinellinae, a small group of genera (Figs. 2, 5) evolved near the end of the Middle Carboniferous and at the beginning of the Late Carboniferous, and formed the beginning of the family Schwagerinidae. These genera developed a honeycomblike structure in the layered outer wall (Fig. 3). During the early part of the Early Permian, several more-or-less independent lineages develop highly inflated shells, for example *Paraschwagerina, Acervoschwagerina, Robustoschwagerina, Pseudoschwagerina, Zellia, Eozellia, Sphaeroschwagerina,* and *Occidentoschwagerina.* Several inflated genera extended into the later part of the Early Permian and *Rugososchwagerina* extends into the lower part of the Upper Permian. It is possible that several of these genera, particularly *Sphaeroschwagerina, Zellia,* and *Robustoschwagerina,* may have been pelagic; however, most other fusulinaceans are considered to have been benthonic after their prolocular stage of development. The Schwagerinidae became extinct before the latest part of the Permian.

A second family that arose from the Ozawainellidae is the Schubertellidae (Figs. 2, 3, 6) which have a persistent but conservative Carboniferous history. During the Permian, however, the Schubertellidae started to expand, and one of these lineages, the *Boultonia,* eventually gave rise to a burst of Late Permian genera that extend to the end of the Permian. Although all the schubertellid genera are relatively small, they commonly become uncoiled and are closely associated with reefs and shallow lagoonal deposits in the later part of the Permian.

The third family that arose from the Ozawainellidae is the Staffellidae (Figs. 2, 3, 7). The Middle and Late Carboniferous portion of this family's range is marked by a few, small, conservative genera; however, in the later part of the Early Permian this family began a rapid evolutionary expansion and increased in average size. The greatest diversity for the family is in the early part of the Late Permian and only 1 or 2 genera occur in youngest Permian strata. This family is common in back reef and lagoonal sediments.

The family Verbeekinidae (Figs. 2, 3, 7) evolved in the later part of the Early Permian and rapidly evolved further into 5 or 6 generic lineages before becoming extinct. Some of these genera are large, robust to subspherical forms with complicated internal features. One of the early members of the verbeekinids gave rise to the beginnings of the Neoschwagerinidae which also evolved rapidly into many genera that are generally large and commonly occur in

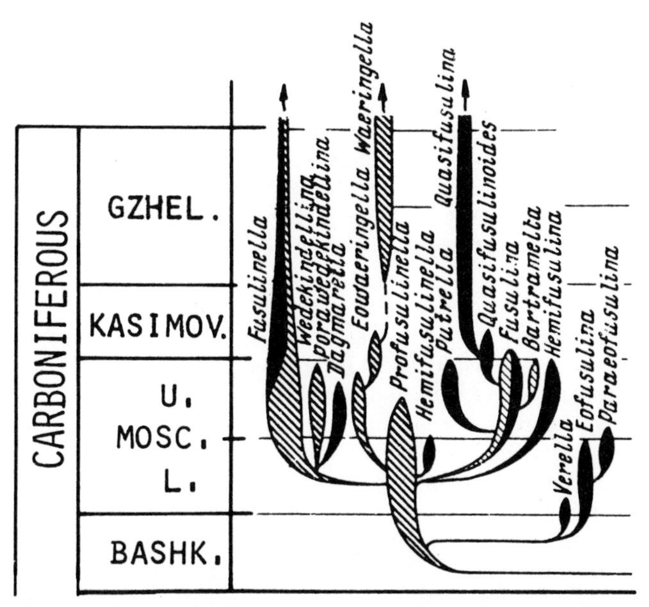

Figure 4. Phylogenetic history of the family Fusulinidae during the Middle and Late Carboniferous (modified from Rozovskaya, 1975). Black areas indicate genera or species complexes that are restricted to Tethyan and Uralian-Franklinian regions. Arrows indicate that genera range into the overlying Permian.

great numbers. Only *Yabeina* survived into the latest part of the Permian (Figs. 2, 7).

The Staffellidae, Verbeekinidae, and Neoschwagerinidae have a number of features in common, in addition to their common ancestry. They evolved rapidly in the later part of the Early Permian and early part of the Late Permian; only a very small number survived into the later part of the Late Permian. They also show an increase in size and strongly modified wall structure. The subfamily Boultoninae in the family Schubertellidae shows a similar increase in diversity in the early part of the Late Permian but a greater number of genera were successful in the latest part of the Permian before becoming extinct. Although the Schwagerinidae have a few new genera appearing in the Late Permian, most of the generic diversity in that family occurs in the early part of the Early Permian. The family Fusulinidae is most diverse in the Middle Carboniferous but does include a long-ranging lineage of several genera that extended into the Late Permian. The family Ozawainellidae had two peaks in diversity: the first, in the latest Early Carboniferous and early Middle Carboniferous; the second in the early part of the Late Permian. These phylogenetic patterns suggest that the Middle Carboniferous, the early part of the Early Permian, and the early part of the Late Permian are times of major diversification among the fusulinaceans while the Late Carboniferous and later parts of the Early Permian and the latest Permian are times of restriction of previously successful lineages and the gradual evolution of new modifications. Because these evolutionary patterns appear to be the type commonly associated with

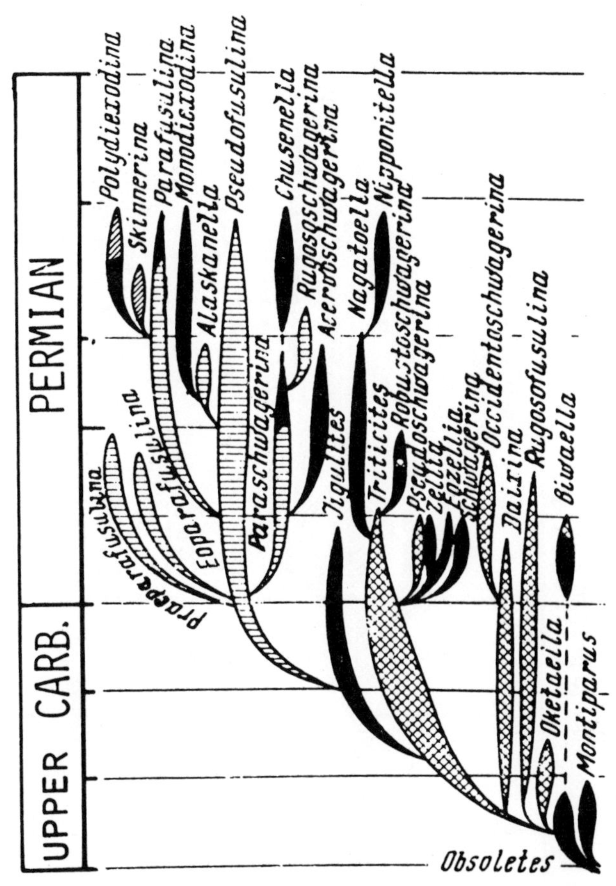

Figure 5. Phylogenetic history of the family Schwagerinidae during the Late Carboniferous and Permian (modified from Rozovskaya, 1975). Black areas indicate genera or species complexes that are restricted to the Tethyan and Uralian-Franklinian regions during the Late Carboniferous and early part of the Early Permian and restricted to the Tethyan Realm after that time. In this illustration *Schwagerina* as shown by Rozovskaya would be considered *Sphaeroschwagerina*, and *Pseudofusulina* as shown by Rozovskaya would be considered separable into *Schwagerina* (*sensu* Dunbar and Skinner) and *Pseudofusulina* of many authors.

changes in geographical distributions and the development of regional and endemic lineages, a summary of fusulinacean distribution is desirable.

GEOGRAPHIC DISTRIBUTION

Fusulinaceans are recorded from North America, South America, Europe, Asia, and northern Africa (Fig. 8). They are not known from Australia, Antarctica, or central and southern Africa (Ross, 1967), and considerable doubt exists that they occurred in these areas during the Permian. In general, fusulinaceans occur in normal marine limestones that are associated with bioherms and banks in which coral, algal, and echinodermal fragments are common. Some fusulinaceans occur in such numbers that they form the principle constituents of the limestones. Some elongate genera of Schwagerinidae in Permian strata are abundant in

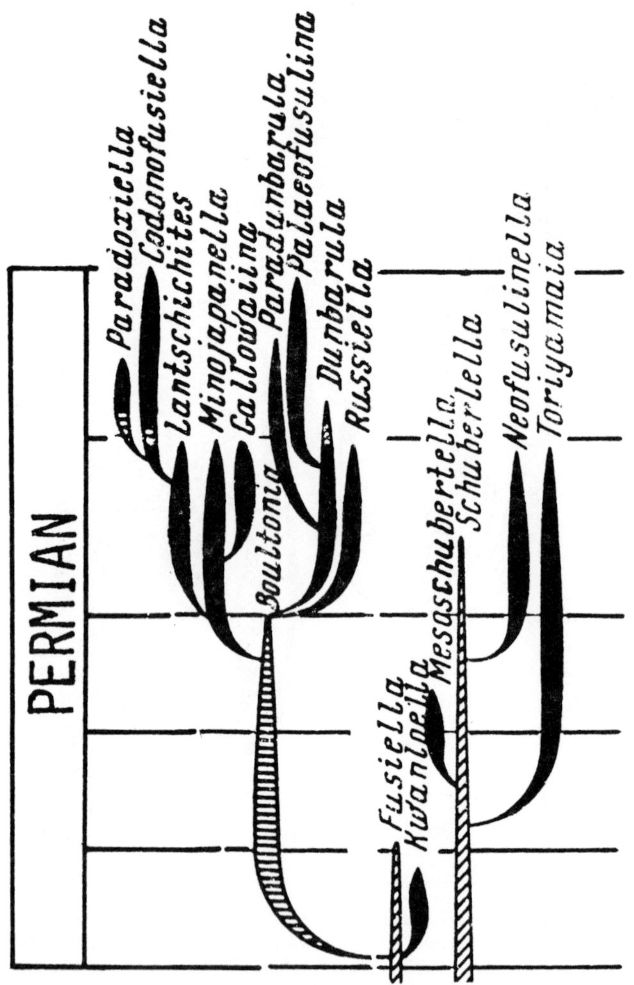

Figure 6. Phylogenetic history of the family Schubertellidae during the Permian (modified from Rozovskaya, 1975). Black areas indicate genera restricted to the Tethyan region or realm.

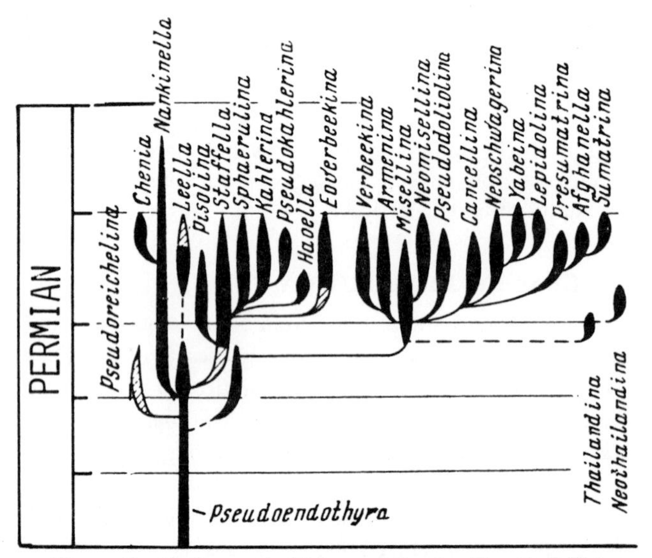

Figure 7. Phylogenetic history of the families Staffellidae, Verbeekinidae, and Neoschwagerinidae (modified from Rozovskaya, 1975). Black areas indicate genera restricted to the Tethyan region or realm.

well-sorted, shallow-water sandstones almost to the exclusion of other fossils. Reefy limestones in the early part of the Late Permian also commonly have great numbers of individuals.

Fusulinaceans are common in three major types of depositional environments. The first type is thin, algal-rich limestones on the cratonic shelves of the major Northern Hemisphere continents. This setting is usually associated with widespread, rapid transgressions and regressions of the shoreline across areas of extremely low relief during the Middle and Late Carboniferous and Early Permian. The number of ecological niches available on these shelves appears to have been high, and because the different shelves were more-or-less isolated from one another, endemic species abound and a few endemic genera also are present. The second type of depositional setting is the shelf edges which include reefs and lagoons as well as some deeper water carbonate environments between the reefs. Different species and even different genera inhabited each of the various shelf-edge environments. Both shelf and shelf-edge environments are particularly important because their stratigraphic successions are generally uncomplicated by major, later structural deformation. The third depositional setting is less thoroughly understood but includes thick carbonate deposits that may have abrupt lateral changes in facies into dark shales, greywackes, and ribbon chert. These depositional facies are associated with basaltic noncratonic igneous rocks and are probably closely related to former island arc and reef environments. The limestones are abundantly fossiliferous and normally include a wide variety of algae and corals in addition to many species and genera of fusulinaceans. Deposits of this type are usually strongly deformed by later structural events and their internal stratigraphy is complicated and generally difficult to work out in detail.

Carboniferous distributions of fusulinaceans are most readily treated in the three time increments, Early, Middle, and Late. The Early Carboniferous fusulinaceans include the Loeblichiidae and the early members of the Ozawainellidae. These were first extensively studied in Europe and later in parts of Asia, the Franklinian and Cordilleran Geosynclines of North America, and the central basins of North America (Lipina and Reitlinger, 1970; Mamet and Skipp, 1971). Several trends seem apparent. These early fusulinaceans are most common in the Tournaisian, Viséan, and early Namurian of the Russian Platform and Tethyan belt and also in the Uralian, Franklinian, and the northern part of the Cordilleran Geosynclines. Away from the Tethyan belt there is a general decrease in species and generic diversity and a decrease in the percentage of calcareous foraminiferids in the total foraminiferal fauna. Strata that form the type sections of the Mississippian in the Illinois basin have a reduced fauna of these early fusulinaceans (and also other foraminiferids and invertebrates) which has led to the suggestion that the Illinois basin may have been nearly landlocked or had special environmental conditions during the Early Carboniferous (Mamet and Skipp, 1971).

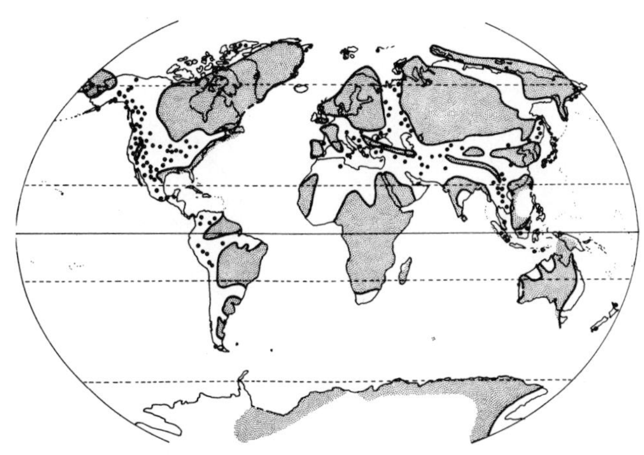

Figure 8. Distribution of principal fusulinacean localities as plotted on present distribution of continents.

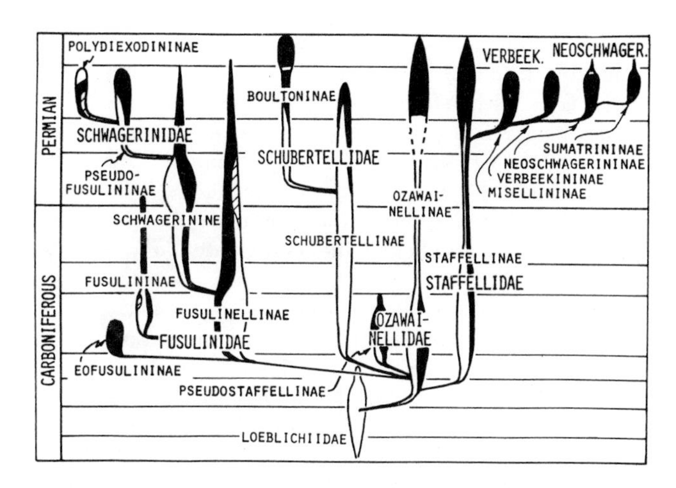

Figure 9. Phylogeny of the fusulinaceans showing the development of endemic genera and species complexes. Black areas indicate endemic taxa in the Uralian-Franklinian-Tethyan region during the Middle Carboniferous and early part of the Early Permian and in the Tethyan Realm during the remainder of the Permian.

During Middle Carboniferous time, the Fusulinidae (Figs. 4, 9) and Ozawainellidae (Fig. 9) begin to show considerable numbers of genera that are restricted in their geographic distributions and have phylogenetic histories that are nearly independent during a geologic stage or more. One of these paleogeographical regions included the Tethyan-Uralian-Franklinian geosynclinal belts and the northern part of what is now the Cordilleran belt of western North America; it contains a greater diversity of genera and species than other regions. In this region the endemic taxa include the Eofusulininae, 5 or 6 genera of Ozawainellinae, *Aljutovella* Rauser, *Taitzehoella* Sheng, *Dagmarella* Solovjeva, *Protriticites* Putrja, and *Hemifusulinella* Rumanjanceva in the Fusulinellidae, and most of the Fusulininae except *Beedeina* Galloway. Also several of the more widely distributed genera have different stratigraphic ranges in other regions implying they had irregular dispersals.

One of the other regions, the Midcontinent-southwestern North American region, demonstrates the difference in time of distribution and in dispersal of several more wide-ranging genera (Fig. 4). In these south-central and southwestern North American successions, *Profusulinella* Rauser and Beljaev occurs with little stratigraphic overlap with *Fusulinella* Moeller and in turn, *Fusulinella* occurs with little stratigraphic overlap with *Beedeina* Galloway. In the Tethyan-Uralian-Franklinian region each of these 3 genera appears in the same order but each continues into the range of the succeeding genus. *Fusulinella* (or descendents) even continues into the Late Carboniferous and Early Permian during which times stray introductions reappear briefly in the south-central and southwestern North American sections. During the later part of Middle Carboniferous time, *Beedeina* is common in the Midcontinent and southwestern North American region and is rare in the Tethyan-Uralian-Franklinian region where the related genus *Fusulina* is common. In the Midcontinent region *Fusulina* briefly appears near the base of the Late Carboniferous after the extinction of *Beedeina* and prior to the first appearance of Schwagerinidae in North America (Ross, 1973). In the Tethyan-Uralian-Franklinian region, the first Schwagerinidae, *Obsoletes* Kireeva and *Montiparus* Rozovskaya, are endemic (Fig. 5) and occur with the last *Fusulina* (Fig. 4) at about this equivalent stratigraphic horizon. The boundary between the Tethyan-Uralian-Franklinian and the Midcontinent-southwestern North American regions is not as sharply demarcated during the Middle Carboniferous as in the later part of the Permian, and is located in the eastern part of the North American Cordillera where incomplete faunal mixing extended from northern Alberta to southern Arizona. At times this broad area had extensive and continuous mixing as in the earliest or latest part of the Middle Carboniferous, and at other times the limits of mixing were narrower and tended to shift back and forth, as during the middle part of the Middle Carboniferous.

A few Middle Carboniferous fusulinaceans recorded from Brazil and parts of the Andean belt suggest these are most closely related to those in the Tethyan-Uralian-Franklinian region, possibly by way of north and northwest Africa. Additional information about these South American occurrences and faunas is needed.

Late Carboniferous fusulinaceans (Fig. 9) are less diverse, at least at the generic level, than are those from the Middle Carboniferous. In part this appears related to a widespread extinction of most species and many genera near the Middle-Late Carboniferous boundary that may relate to widespread withdrawal of epicontinental seas. These extinctions are possibly associated with major climatic and orographic changes at this time because this coincides with a major deformational phase of the Marathon-Ouachita orogenic belt in southern North America and with uplift of parts of Europe at the end of Westphalian and Moscovian deposition. Deposition during the Late Carboniferous shows an increase in shallow-marine clastic shelf sediments and an increase in topographic relief between shelves and basins. Biofacies suitable for most of the genera of Fusulinidae apparently were greatly reduced and new biofacies developed into which the Schwagerinidae were able to adapt and start their rapid evolution. Most of the Late Carboniferous genera of Schwagerinidae (Fig. 5) are long ranging and distributed in geographical patterns similar to the Fusulinidae of the Middle Carboniferous. *Triticites* Girty is the most common and the most widely distributed genus and shows endemism in a few species complexes. Other families in the Late Carboniferous, such as the Fusulinidae, Ozawainellidae, and Staffellidae (Figs. 4, 9), are significantly restricted in their taxonomic diversity and, in the case of the Fusulinidae, also in their geographic distribution.

Early Permian fusulinacean faunas, in terms of number of genera, number of species, and abundance of individuals, are dominated by the Schwagerinidae which evolved into many well-defined lineages at this time (Ross, 1967). Several of these genera, such as *Sphaeroschwagerina* Miklucho-Maclay, *Pseudoschwagerina* Dunbar and Skinner, *Zellia* Kahler and Kahler, and *Eozellia* Rozovskaya, are short ranging and found in the earlier part of the Epoch, and some others, such as *Robustoschwagerina* Miklucho-Maclay, *Parafusulina* Dunbar and Skinner, and *Acervoschwagerina* Hanzawa, first appear in the middle part of the Epoch. Most of these genera are common only in the Tethyan-Uralian-Franklinian region; most occurrences outside of that region are sporadic and brief incursions that did not survive for any appreciable length of time. In contrast, one major species lineage of *Pseudoschwagerina* was present for only the early part of its history in that region and then becomes confined to the Midcontinent-southwestern North American region.

Latest Carboniferous and early to middle Early Permian fusulinaceans are fairly widespread in the Andean belt of South America and, although not as completely known as those from North America and Eurasia, are apparently mostly closely related to those from the southwestern North American region and should be included with that region (Ross, 1967).

The later part of the Early Permian Epoch is marked by increasing endemism in species complexes in most genera of Schwagerinidae and in several other families also. Three regions are recognizable, a Tethyan region, a separate Uralian-Franklinian region, and a southwestern North American region. The separation of the Tethyan region from the Uralian-Franklinian region appears to have been sharp because after the early part of Artinskian time the fusulinaceans of the two regions become markedly different and faunal correlations between all three regions become more difficult as the amount of endemism increases.

In the Tethyan region, the genus *Pseudoendothyra* Mikhailov, a long ranging conservative genus from the Middle Carboniferous or earlier, began a remarkable evolutionary diversification in the later part of the Early Permian Epoch giving rise to a dozen genera of advanced Staffellidae. One of these gave rise to *Misellina* Schenck and Thompson which is the ancestral genus for both the Verbeekinidae and Neoschwagerinidae. The Verbeekinidae and Neoschwagerinidae evolved rapidly and became dominant forms

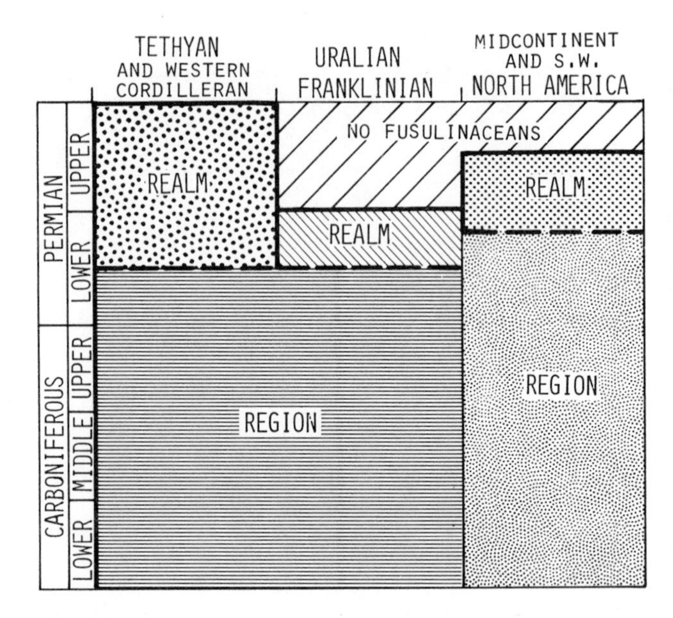

Figure 10. Development of fusulinacean provinciality during the Carboniferous and Permian. The differentiation of assemblages into regions during the Carboniferous and earlier part of the Permian is mainly at the species and species complex level and includes some genera. Differentiation during the later part of the Permian into realms is mainly at the generic and familial level.

in the Tethyan area so that by the beginning of Late Permian time the Tethyan fusulinacean faunas became taxonomically distinctive, and also difficult to correlate with other areas on the basis of any group of benthonic fossils. The Tethyan and other regions should be considered a separate realm beginning at this time (Figs. 7, 9, 10). In addition, the Schubertellidae (Fig. 6) also show a significant but smaller evolutionary burst and a few genera and several species complexes of the Schwagerinidae show increasing endemic distribution. Although the Verbeekinidae, the Schwagerinidae, and the Neoschwagerinidae die out at or very near the middle of the Late Permian, 1 genus of Staffellidae, *Nankingella* Lee, and 5 genera of boultoniin Schubertellidae survive into the latest part of the Permian (Figs. 6, 7, 9) in the Tethyan Realm.

Outside the Tethyan Realm fusulinaceans have a much different history in middle and late Artinskian and Late Permian times (Fig. 10). In the Uralian-Franklinian Realm the number and diversity of fusulinaceans decrease rapidly. The few surviving lineages of Fusulinidae and Staffellidae die out in the region, and *Pseudofusulina, Schwagerina,* and primitive *Parafusulina* continue until about the end of Early Permian time when they also die out. The Late Permian in the Uralian-Franklinian Realm lacks a fusulinacean fauna although locally small calcareous foraminiferids are reported in Upper Permian strata.

In the Midcontinent-southwestern North American region progressive reduction in the extent of epicontinental seas resulted in fusulinacean distribution being gradually reduced to the southwestern part of the region. In the later part of Early Permian time, species complexes of

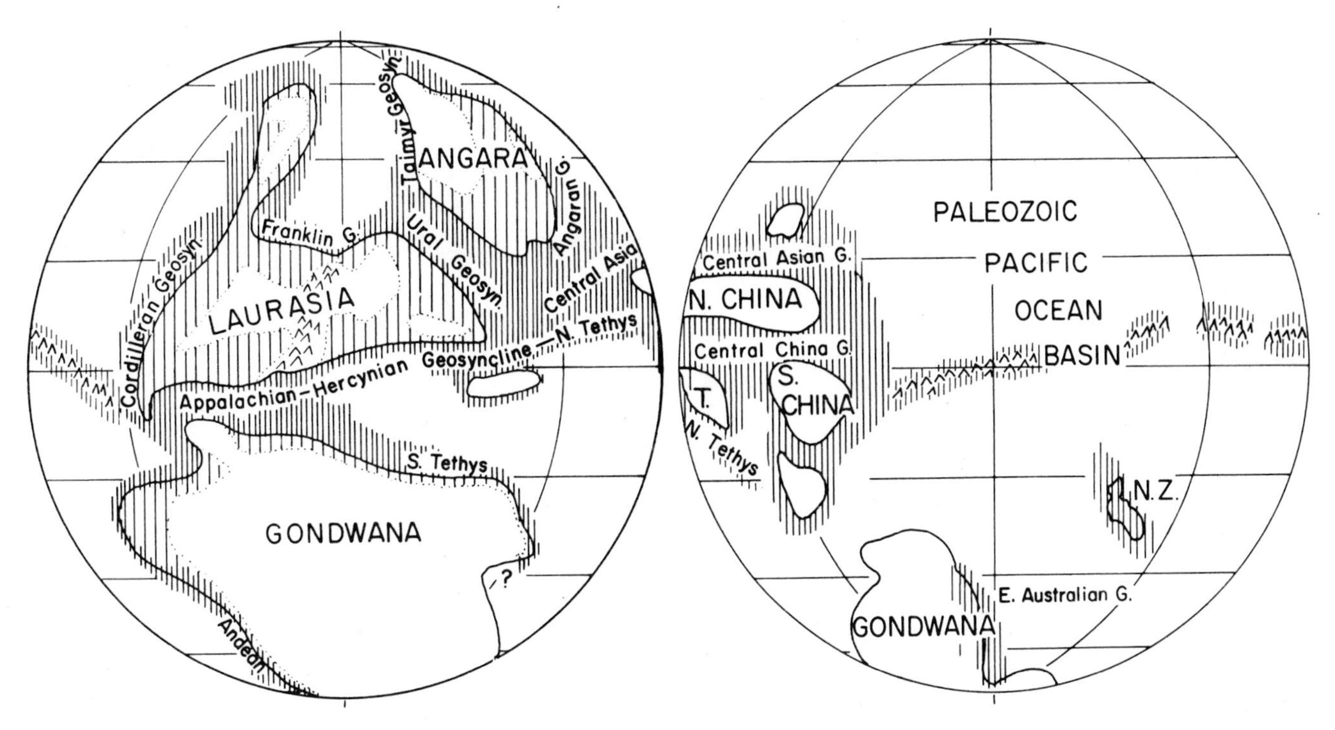

Figure 11. Early Carboniferous (Viséan) paleogeography (modified from Dietz and Holden, 1970). Nonmarine areas are stippled, epicontinental shelves are lightly ruled, miogeosynclines are heavily ruled, and eugeosynclines are shown with volcanoes.

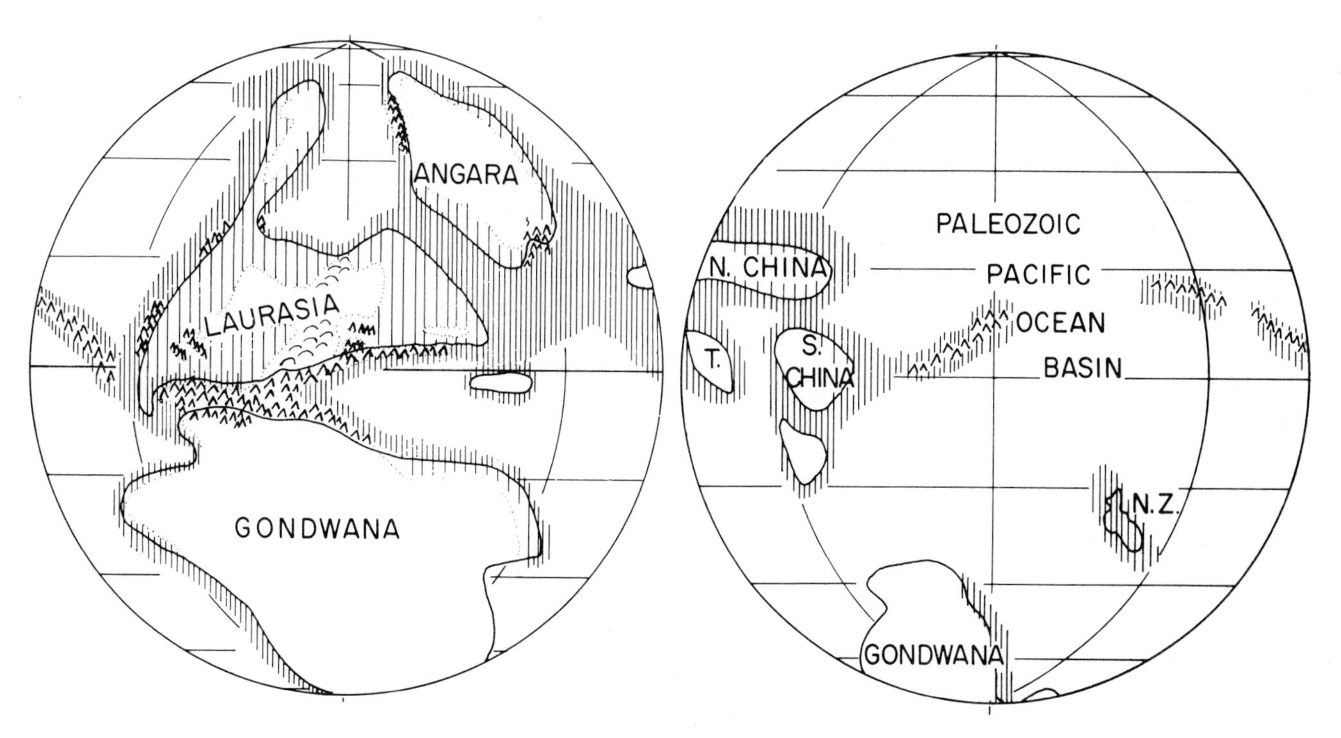

Figure 12. Middle Carboniferous (late Moscovian) paleogeography (modified from Dietz and Holden, 1970) (see Fig. 11 for explanation of symbols).

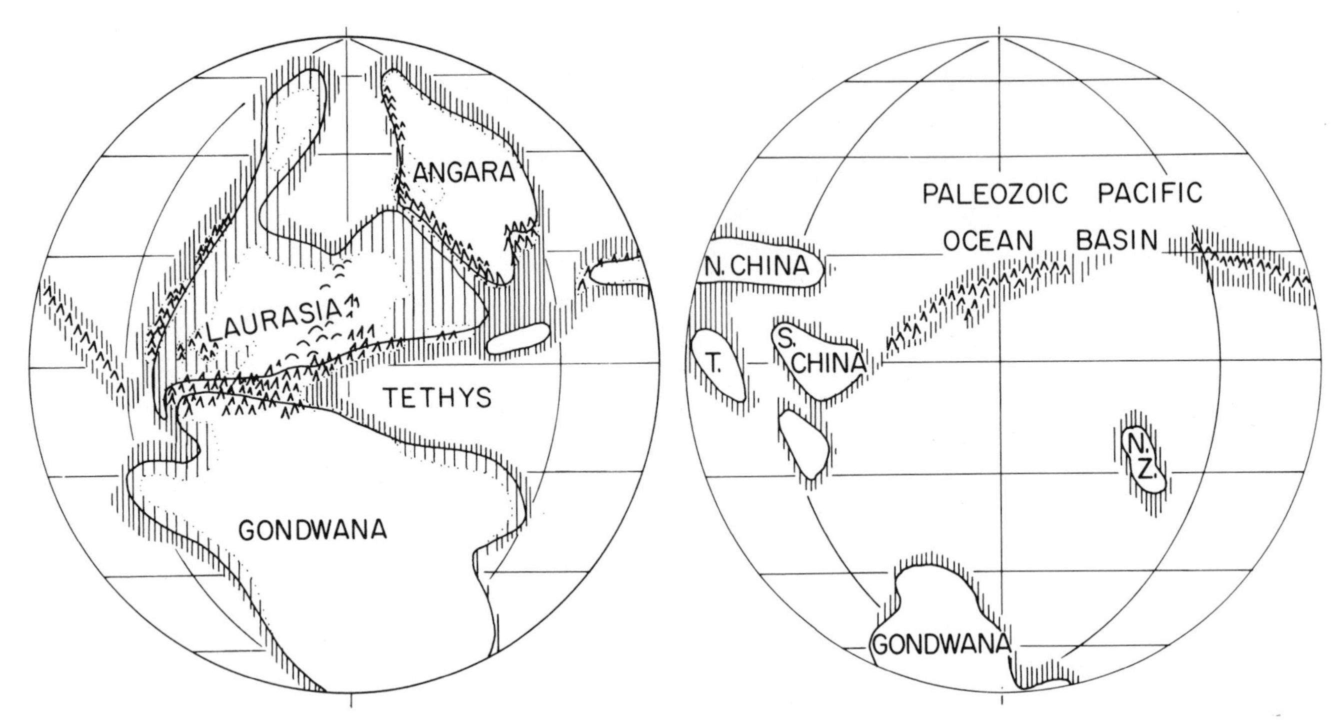

Figure 13. Early Permian (Asselian) paleogeography (modified from Dietz and Holden, 1970) (see Fig. 11 for explanation of symbols).

Schwagerina, primitive *Parafusulina*, and *Staffella* are the dominant fusulinaceans. Fusulinidae and early lineages of the Verbeekinidae and Neoschwagerinidae are absent. Representatives of *Schubertella* Staff and Wedekind, and *Boultonia* Lee appear in the region at different times (Fig. 6) and suggest separate successful dispersals during the Early Permian. Several other genera also show temporary dispersal into the region during the later part of the Early Permian, such as *Robustoschwagerina*. By the beginning of Late Permian time the fusulinacean genera were reduced to abundant advanced *Parafusulina*, rare *Skinnerina* Ross, and *Rauserella* Dunbar and Skinner. Shortly later, these genera are replaced by *Polydiexodina* Dunbar and Skinner (Fig. 5), a probable emigrant from the Tethyan fauna.

A few other genera appear for short intervals in the southwestern North American Realm during Late Permian time, such as *Paradoxiella*, *Codonofusiella*, *Yabeina*, and *Leella*, but they established no long lineages of species. In general, the southwestern North American Realm is identified by species complexes of *Parafusulina* and *Polydiexodina* which dominate the fusulinacean fauna and by the general lack of lineages of species and genera of Verbeekinidae or Neoschwagerinidae. By latest Permian time, fusulinaceans became extinct in this Realm.

Most of the Permian fusulinaceans described from South America are from the Early Permian and have close similarity with those from the southwestern North American Realm and show little indication of a direct connection with a Tethyan Realm or region.

DISCUSSION

The distribution of fusulinaceans during the Carboniferous and Permian shows several important trends (Fig. 10). Widespread and nearly cosmopolitan distributions of the Early Carboniferous give way to restricted distributions during the Middle Carboniferous in which infrequent dispersals take place of several of the more common genera. These dispersals were frequent enough, however, to be able to recognize enough faunal similarity between regions that correlations can be established at the stage, or even substage level, within the Epoch, although species, species complexes, and even some genera are endemic to particular regions and these endemic faunas form distinctive assemblages. During Late Carboniferous time, species and species complexes are usually endemic and many of the genera may also be endemic; however, the total number of genera is smaller and generic diversity is markedly lower. Dispersals between regions were probably less frequent than during the Middle Carboniferous. Near the end of Carboniferous and the beginning of Permian time, a few genera and species complexes became widely dispersed, perhaps at several different times within a relatively brief interval. These species groups established lineages in different regions that can be traced through several geologic stages. In the later part of Early Permian time, dispersals between different regions became less frequent and before the end of Early Permian time, three regions have strongly differentiated faunas (Fig. 10). By Late Permian time, fusulinaceans became extinct in one of these regions and the other two areas had distinctly different fusulinacean

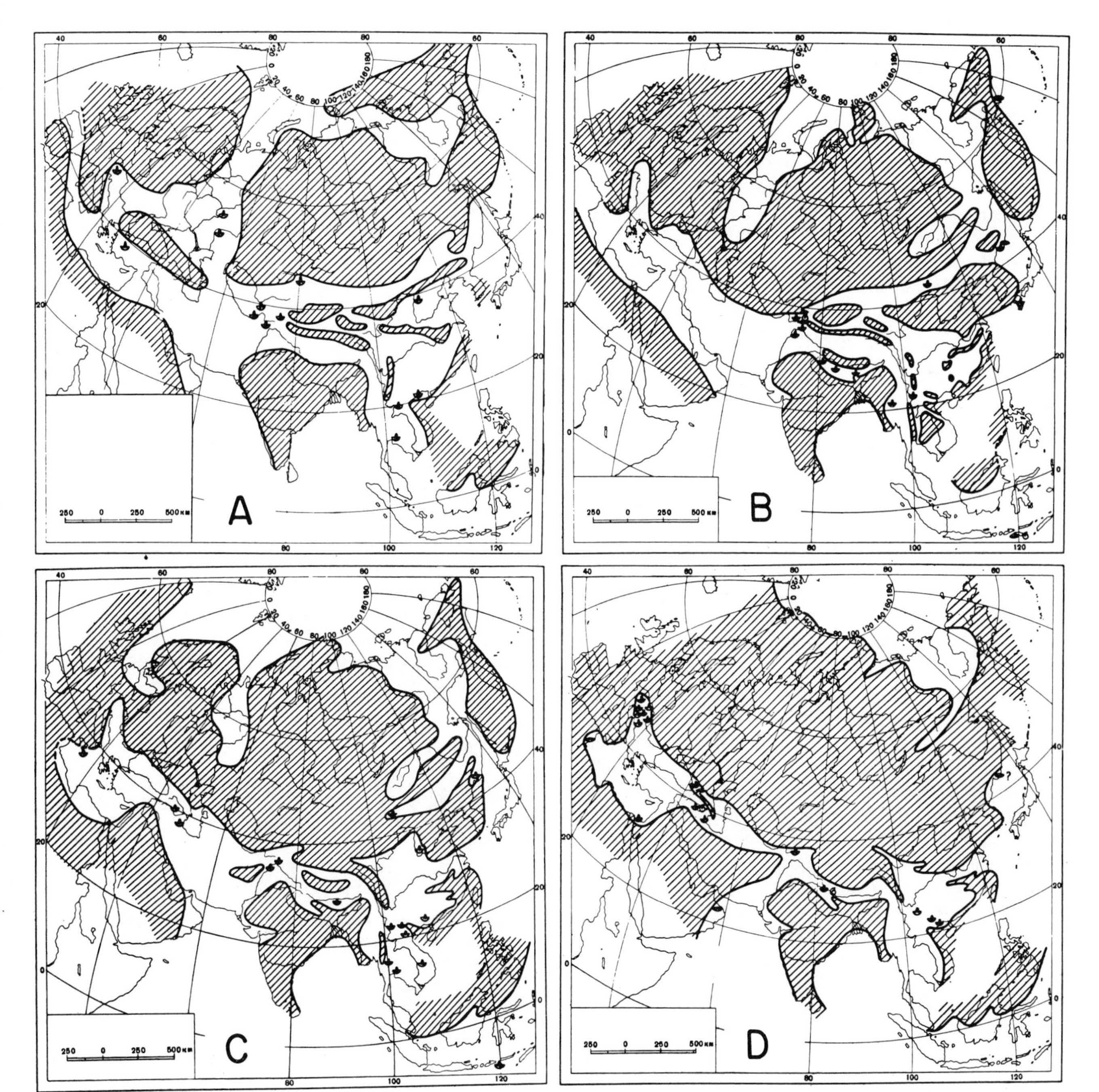

Figure 14. Four maps illustrating the successive changes in areas of marine deposition during the Permian as plotted using present geographic distribution of strata: (A) Asselian; (B) Darvasian (late Artinskian); (C) Murgabian; (D) Dzhulfian (from Grunt and Dmitrev, 1973).

faunas so that correlations of stages are difficult. These two areas take on the status of realms and form the beginning of the geographical provinciality that is prevalent through much of the Mesozoic.

PALEOGEOGRAPHIC RECONSTRUCTION

A review of possible geographic arrangements for the various epicontinental seas, miogeosynclines, and eugeosynclines during the Carboniferous and Permian suggests that a modified Dietz-Holden model (Dietz and Holden, 1970) may help to explain some of the apparent problems in fusulinacean phylogenetic history and distribution. Three time intervals are examined, Early Carboniferous (Viséan), Middle Carboniferous (Moscovian), and Early Permian (Asselian), on world reconstructions (Figs. 11, 12, 13).

The Early Carboniferous map (Fig. 11) depicts a marine connection near the paleoequator along the Ouachita-Appalachian-Hercynian Geosyncline and adjacent epicontinental shelves. Sea level was relatively high, mountain belts relatively low, and these features combined to produce fairly uniform climates in tropical and lower temperate latitudes. Foraminiferal faunas that are rich in calcareous species and algae have a diversity gradient toward predominantly arenaceous species at high latitudes suggesting cool water in high temperate and polar latitudes. Some areas had restricted water circulation indicated by low diversity of calcareous species, as in the central United States.

The later part of the Middle Carboniferous (late Moscovian) was again a time of relatively high sea level (Fig. 12) associated with fluctuations of sea level. Mountain building and the approach of northwestern Africa and northern South America to North America and Europe disrupted the marine connection along the Ouachita-Appalachian-Hercynian Geosyncline. Marine dispersal was along epicontinental shelves and geosynclines around the northern side of the Baltic and Canadian shield areas (i.e., the Russian Platform, and Uralian and Franklinian Geosynclines). A marine connection continued to exist at the southern end of the Uralian Geosyncline with a large remnant of the Hercynian Geosyncline which connected with an epicontinental sea that extended across parts of northern Africa into the Amazon Basin of South America at this time. Fusulinaceans are not known at this time from parts of Siberia or from southern South America, southern Africa, southern India, Australia, or Antarctica probably because these areas had cool temperatures as they have associated glacial deposits.

The third map (Figs. 13, 14A) shows the paleogeography during the early part (Asselian) of the Early Permian. Permian epicontinental seas reached their maximum distribution during this part of the Epoch. Mountain building and deformation of the Ouachita-Appalachian-Hercynian belt continued, and although mountains were high, marine embayments still extended onto many continents. The Uralian Geosyncline was very narrow but connected the Russian Platform with the Tethyan part of the Hercynian belt. A shallow-water marine connection with southwestern North America and western South America was by way of the Franklinian Geosyncline and northern Canadian shelf so that distance and latitude produced temperature gradients that reduced the dispersal of many faunas, including fusulinaceans.

After the Asselian (Fig. 14B), dispersal of genera becomes increasingly less frequent between the Tethyan region and the Uralian-Franklinian region and the southwestern North American region. By the later part of the Artinskian (Fig. 14B), marine connections between the Tethyan region and the Uralian Geosyncline ceased and from this time until the end of the Permian the Tethyan faunas become increasingly endemic and form a distinctive faunal realm. By the beginning of Late Permian (Fig. 14C), this northern Pangaea epicontinental sea lacks fusulinaceans and only a few genera of smaller Foraminiferida are present. The later part of the Late Permian (Dzhulfian) is a time of minor inundation of the epicontinental shelves (Fig. 14D) and those seas that did occupy a portion of these shelves had unusual environments of deposition that include considerable amounts of evaporites. Of the Early Permian geosynclines only the Tethyan Geosyncline remained a marine seaway in the later part of the Late Permian and even parts of it were uplifted.

Although a considerable amount is known about fusulinaceans and other shallow-water marine faunas of Carboniferous and Permian time, many questions remain unanswered about their distribution, dispersal, and phylogeny. There is little data from western South America and large parts of Asia where fusulinaceans are known to occur. New Zealand has at least two fusulinacean localities but a poorly understood geographic position in the late Paleozoic. Parts of southern Eurasia may have been separate sialic blocks that have been united at a much later time. Several problems are encountered in the Cordilleran and Andean structural belts of North and South America where fusulinacean faunas of relatively high diversity occur near faunas of low diversity or near glacial deposits. Some of these strata contain Tethyan Realm faunas that structurally lie against rocks containing contrasting non-Tethyan faunas. Because these strata are older than 150 m.y., and, therefore, older than our present sea floors, one may ask whether these belts are old tropical island arcs and other crustal fragments which have been added to the western side of the Americas during the Mesozoic and early Cenozoic? The late Paleozoic distribution of fusulinaceans suggests that type of history.

REFERENCES

Dietz, R. A., and Holden, J. C., 1970, Reconstruction of Pangaea: Breakup and dispersion of continents, Permian to Present: Jour. Geophys. Research, v. 75, no. 26, p. 4939-4956.

Grunt, T. A., and Dmitriev, B. J., 1973, Permiskie brakhipod'i Pamira: Akad. Nauk SSSR Paleont. Inst. Trudy, v. 136, 211 p.

Lipina, O. A., and Reitlinger, E. A., 1971, Stratigraphie zonale et paléozoogéographie du Carbonifère inférieur d'après les foraminifères: 6ᵉ Congr. Internat. Stratigr. Géol. Carbonifère, Comptes Rendus, v. 3, p. 1101-1112.

Mamet, B., and Skipp, Betty, 1971, Lower Carboniferous calcareous Foraminifera: Preliminary zonation and stratigraphic implications for the Mississippian of North America: 6ᵉ Congr. Internat. Stratigr. Géol. Carbonifère, Comptes Rendus, v. 3, p. 1120-1146.

Ross, C. A., 1967, Development of fusulinid (Foraminiferida) faunal realms: Jour. Paleontology, v. 41, p. 1341-1354.

——— 1973, Carboniferous Foraminiferida, *in* Hallam, A., ed., Atlas of palaeobiogeography: Amsterdam, Elsevier, p. 127-132.

——— 1974, Paleogeography and provinciality, *in* Ross, C. A., ed., Paleogeographic provinces and provinciality: Soc. Econ. Paleontologists and Mineralogists, Spec. Pub. 21, p. 1-17.

Rozovskaya, S. E., 1975, Sostav, sistema i filogeniya otryada fuzulinida: Akad. Nauk SSSR Paleont. Inst. Trudy, v. 149, p. 1-267.

Simpson, G. G., 1947, Holarctic mammalian faunas and continental relationships during the Cenozoic: Geol. Soc. America Bull., v. 58, p. 613-688.

Biological and Physical Factors in the Dispersal of Permo-Carboniferous Terrestrial Vertebrates

EVERETT C. OLSON, *Department of Biology, University of California, Los Angeles, California 90024*

ABSTRACT

The fossil record indicates that a rather abrupt dispersal of terrestrial tetrapods took place during the later Permian. Resultant faunas occur in Asia, India, South and East Africa, Europe, and South America. Three types of hypotheses have been advanced to explain these distributions: (1) that the record in its incompleteness is misleading; (2) that physical factors were at the base of the presumed dispersal; and (3) that biological factors were basic. These hypotheses are examined and evaluated. Elements of each are integrated into a tentative explanation of some of the phenomena, and a residue of questions not as yet answerable, is suggested.

INTRODUCTION

The patterns of distributions of Permo-Carboniferous and Triassic terrestrial vertebrates have been one of the "mysteries of the Permian" for many years. Discoveries of new fossil sites and growing acceptance of concepts of continental movements have revitalized interests in this area of biogeographic research. Central to the issues is the apparently rapid spread of terrestrial vertebrates from a restricted geographic distribution in the Late Carboniferous and Early Permian to an occupancy of all of the continents by Late Triassic. The major shift began early during the Late Permian.

Among those who have been especially active in the re-evaluations of the old concepts and development of new ones are C. Barry Cox, Pamela Robinson, A. R. Milner, and R. Panchen, all of London. Their work has been concerned primarily with the relation between faunal distributions and physical features of the continents. A second line of study has been concerned more with the ecological and physiological aspects of vertebrates in relation to their changing distributions, with special studies by E. C. Olson and R. T. Bakker. These two broad lines of investigation are not, of course, mutually exclusive and all such studies depend heavily upon theories of continental movements and the general principles of biogeography.

The various lines of investigation converge in several explanatory hypotheses relative to tetrapod distributions during the late Paleozoic and Triassic. This paper is concerned primarily with an analysis of these interrelated hypotheses. The hypotheses may be grouped into three general categories:

The hypothesis of insufficiency of the record. Under this concept the patterns of distributions as observed are considered to be in large part fortuitous, merely a function of the incompleteness and inadequacy of the record.

Physical hypotheses. These hypotheses suggest that physical barriers to migrations—water, topography, climate—were instrumental in the localization of faunas. Modifications with time permitted dispersals that were earlier impossible. In their pure form, these hypotheses do not consider biological changes of the organisms, or their population structure.

Biological hypotheses. Population structure, trophic structure, and (or) physiological properties were the primary limiting factors of dispersal, permitting it only as they changed with time. In the pure form, physical factors are considered insignificant.

That none of these can be considered mutually exclusive is self-evident, but separation makes it possible to point up the major aspects of the general problems of the biogeography of Permo-Carboniferous and Triassic tetrapods.

In this paper the distributional patterns of the Permo-Carboniferous will be examined first to set the stage for an analysis of the hypotheses erected to explain the distributions.

PATTERNS OF DISTRIBUTIONS

Attention in this section is directed primarily to strictly terrestrial tetrapods. Only where they have an important bearing on patterns of distribution will fishes and aquatic tetrapods be considered. Data from both sources can be potentially useful, but terrestrial organisms have supplied much of the information for several critical areas.

Stephanian-Sakmarian-Autunian

Milner and Panchen (1973) presented a succinct and penetrating analysis of the terrestrial vertebrate distributions of the Stephanian-Sakmarian-Autunian (Figs. 1, 2A). They emphasized the resemblance of samples from North America and Europe, and summarized this in a statement

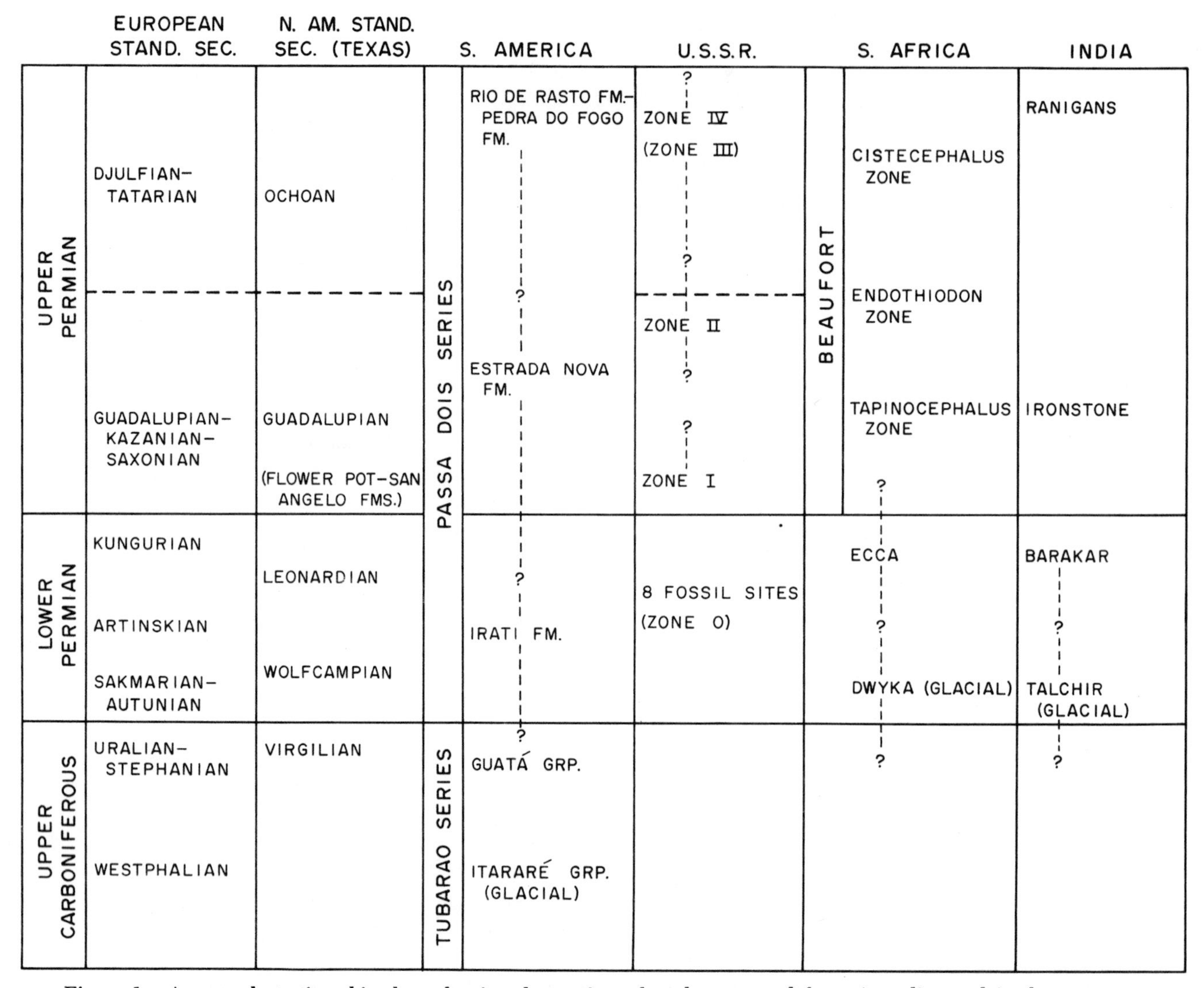

Figure 1. A general stratigraphic chart showing the sections, faunal zones, and formations discussed in the text.

to the effect that European collections essentially represent a small sample of the better known North American ones. Aquatic tetrapods and fishes, in general, show fewer resemblances between the two continents than do the strictly terrestrial vertebrates. Close affinities do, however, exist between some of the paleoniscoid fishes (Westoll, 1944), and some reptiles and amphibians (Romer, 1945). The principal barrier to intermigration and dispersals between Europe and North America appears to have been the Caledonian-Appalachian mountain chain, but details of the time of its effectiveness and just how it operated are not entirely clear.

The collecting localities and their concentration in North America and Europe, as well as their generally circum-equatorial distributions, are shown in Figure 2A (see Olson and Vaughn, 1970; Milner and Panchen, 1973). Also important, however, are the few occurrences elsewhere, especially in South America and South Africa (aquatic reptilian mesosaurs), India (labyrinthodont amphibians: Tewari, 1962; Verma, 1962; Wadia and Swinnerton, 1928; Woodward, 1905), and Siberia (amphibians: Efremov and Vjuschkov, 1955). These small occurrences are particularly important in the evaluation of each of the three types of hypotheses.

Leonardian-Early Guadalupian

Leonardian terrestrial vertebrates are known primarily from redbed deposits of Texas and Oklahoma (Fig. 1). Some aquatic vertebrates of this age occur in western Europe, and a few small sites, probably Leonardian, are present in the Cis-Uralian region of the Soviet Union. Important as the vertebrates of this time are phylogenetically and ecologically, they add very little to an understanding of Early Permian distributions, except perhaps in a negative fashion. Leonardian vertebrates are primarily a continuation of those found in the Late Carboniferous and earlier Permian, but they show some evolutionary modifications in response to climatic changes. The well-known

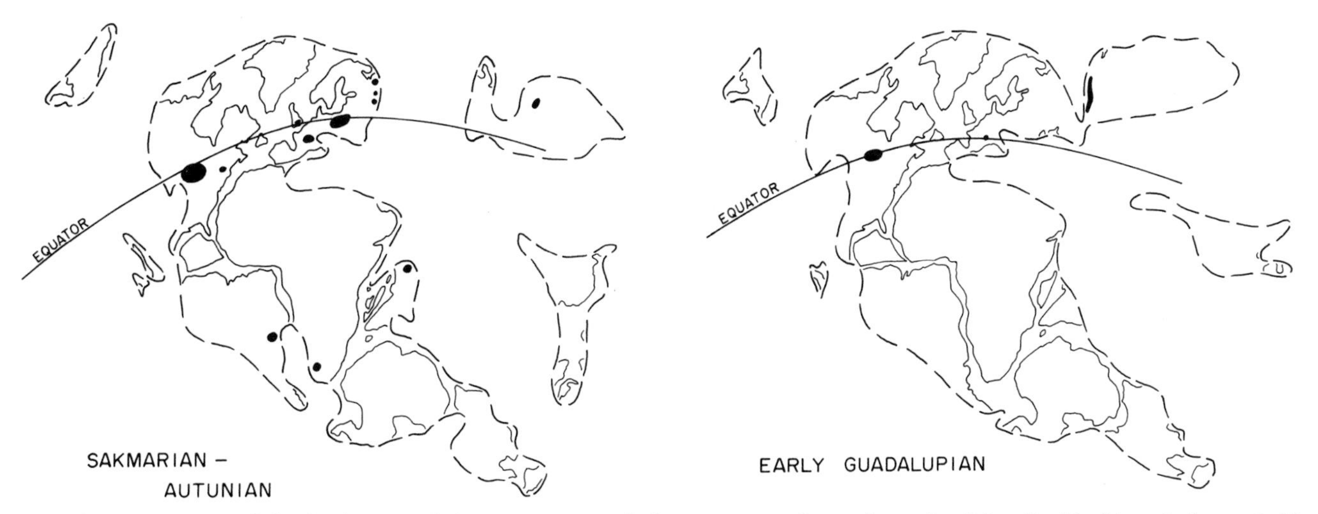

Figure 2. Maps of the land areas of the continents with the positions of vertebrate localities (in black) and the probable position of the paleoequator. The base maps have been modified after Schopf (1974) (on Winkel "triple" projection), and the position of the paleoequator has been derived from several of the sources cited in the reference list. Left (A): Early Permian (Sakmarian and Kungurian); Right (B): very early Late Permian (early Guadalupian).

vertebrates, however, furnish only meager indications of the emergence of the types of animals found in beds of early Guadalupian age. In retrospect, from the vantage point of the Guadalupian, it can be seen that a few species from the less well-known faunal complexes were actually forerunners of those seen in later times (Olson, 1975).

The map of the early Guadalupian (Fig. 2B) shows the few tetrapod-producing localities known from this time. Some remains have come from Texas and Oklahoma, 2 specimens are known from France (Sigogneau-Russell and Russell, 1974), and a moderate assemblage has come from the Cis-Uralian Soviet Union, with much the best preserved materials from the middle Cis-Uralian region in the vicinity of Ocher. The localities in the USSR belong in Zone I of Efremov, occurring in the basal part of the Kazanian deposits (see Efremov, 1954; Efremov and Vjuschkov, 1954; Olson, 1957, 1962; Chudinov, 1960, 1965). The North American and French sites appear to have been close to the Permian equator, but the Soviet sites extended to about 20° N latitude.

Materials from each of these early Guadalupian sites are limited. Those from Ocher, USSR, are well-preserved, while those from North America are largely fragmentary. Table 1 shows the distribution of families of reptiles in North America and the Soviet Union. Some lumping of North American families has been done. Without a critical analysis, the faunas appear to have been quite similar. Table 1 also includes an estimate of the confidence that can be placed in the supposed relationships, and this provides a less coherent picture. The Captorhinide, Caseidae, and Brithopodidae do have very similar representatives in North America and the Soviet Union, and the caseids from France are close to the others. Only the Brithopodids, however, belong to advanced reptilian groups, the other 2 families being well-known from the Leonardian. Opportunities for interchange of some types of animals evidently existed during the Leonardian and early Guadalupian to account for these close resemblances. The remaining familial resemblances are less certain, and, although they suggest that all of the reptiles may be part of a general

Table 1. Distributions of families of terrestrial amphibians and reptiles in the USA and USSR during the early Guadalupian. Notes in the last column give data on distributions not indicated in the areas compared. (P) present; (A) absent; (CL) confidence level. H, M, L confidence level with respect to familial assignments, based largely upon assignments of incomplete North American specimens.

Families	USA	USSR (Zone I)	CL	Notes
Amphibia				
Dissorophidae	P	A		Present, Zone II USSR
Reptilia				
Captorhinidae	P	P	H	Also Rep. of Niger
Varanopsidae	P	A		Present, S Africa
Sphenacodontidae	P	A		
Eotitanosuchidae (incl. Phthinosuchidae, Gorgonopsidae, in NA)	P	P	M	
Brithopodidae	P	P	M	
Caseidae	P	P	H	Also in France
Estemennosuchidae (incl. Driveriidae, Mastersonidae, Tappenosuridae, in NA)	P	P	L	
Venjukoviidae	P	P	L	

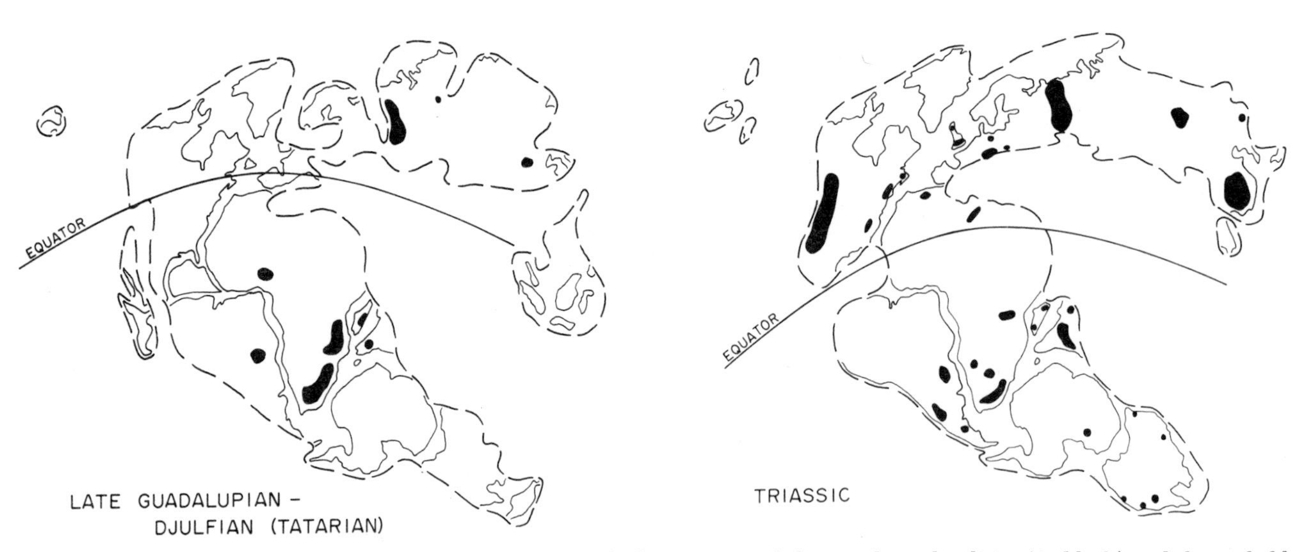

Figure 3. Maps of the land areas of the continents with the positions of the vertebrate localities (in black) and the probable position of the paleoequator. Data sources as for Figure 2. Left (A): Upper Permian (late Guadalupian and Djulfian or Tatarian); Right (B): Triassic, based primarily on distributions during the Late Triassic.

faunal complex, they are open to various interpretations and indicate at least considerable differential endemism.

Late Guadalupian and Djulfian (Tatarian)

The interval between the early and late Guadalupian (Fig. 1) was one in which the number of localities of terrestrial vertebrates increased enormously. The main intermediate stages are seen in the Soviet Union, Zone II, but many slightly younger sites are known both from the Soviet Union and from South and East Africa. The dramatic increase in number of localities is evident in a comparison of Figures 2A and 3A. Notable in the latter is the absence of localities in North America from which Lower Permian terrestrial vertebrates are best known. Much the same applies to western Europe, although a small site in England exists (not shown on the map because of the particular stage of the reconstruction).

Although the distances between Cis-Uralian USSR and South and East Africa are great and the kinds of deposits quite different (alluvial in Russia versus lacustrine in Africa), the faunal complexes of the two areas have much in common. As Cox (1974) has calculated, using data from Romer (1973), the similarity index is $C/C_1 \times 100 = 70$, based on suborders.[1] The similarity index for families is also quite high (66), but the wide discrepancy in the number of families, 22 in South Africa and 9 in the USSR, lowers the utility of comparisons. Genera within the common families are on the whole similar, even identical in one or two instances, so that intercommunication appears to have been established within the time interval represented.

Recent, as yet unpublished studies by M. Barberena and his colleagues in Brazil, have revealed the presence of tetrapods in the upper part of the Passa Dois Series in the Paraña Basin of Brazil. Dr. Barberena has kindly permitted me to cite these finds in this paper. *Endothiodon,* generically identical with the zone fossil of the *Endothiodon* zone of Africa, and a long snouted labyrinthodont, perhaps trematosaur-like, are currently known. The age appears to be Djulfian and, not unexpectedly, shows the presence of a South African-like complex in South America at this time. Also the Pedra do Fogo Formation of Northern Brazil (Price, 1948) now appears referable to the Upper Permian (M. Barberena and C. B. Cox, personal communs.).

Small collections of Late Permian age have also come from the Republic of Niger (Taquet, 1969), England, Asia, India, and Madagascar. By the end of the Permian, then, known localities span the areas of four continents. In general, the samples from the widely separated localities include similar types of tetrapods, but where large collections are available strong suggestions of endemism emerge from their comparisons.

Triassic

Figure 3B is a general map for the Triassic based largely upon distributions of tetrapods for the late part of the Period and intended primarily to show the continuation of trends initiated during the Permian. During the Triassic, a marked shift in the composition of terrestrial tetrapod faunas took place, with diminution of the therapsids and rapid increase of the sauropsids. By the end of the Period the latter were dominant. The distributions give evidence of a world-wide vertebrate fauna by the end of the Triassic. The composition of the faunal complexes from different areas indicate that the world fauna formed a coherent network of faunal assemblages. The development of this complex has been treated in detail by Cox (1973a, 1973b, 1973c), and by Robinson (1971).

[1] **C is the number of taxa in common, C_1 the number of taxa in the smaller of the two arrays. The index is moderately informative if the number of taxa in the two groups is similar and if the numbers are moderately large, say no less than 7 to 10.**

RELIABILITY OF THE RECORD

The first hypothesis, that of insufficiency of the record, must be explored before the others that explain distributions are considered. If it can be shown that the differences in distributions between the Lower and Upper Permian are, with little doubt, a function of a faulty record, then the other hypotheses are, of course, meaningless. Determination of the validity of the record requires evaluation of negative evidence, not a rewarding venture as a rule.

If there were no Lower Permian terrestrial deposits outside of the areas of Europe and North America, where the well-known faunas exist, reliance on data from these areas and the known Upper Permian distributions would at best produce very uncertain results. Although there has been a tendency to consider evidence from outside of these areas as secondary, two facts are important. First, epicontinental and geosynclinal marine deposits limit considerably the areas in which terrestrial vertebrates might have existed on the continents. The potential areas are shown in Figure 2A. Second, well-developed Upper Carboniferous and Lower Permian deposits occur over appreciable areas in Cis-Uralian USSR, in Siberia, and across Gondwanaland in Africa, South America, and India. These deposits, although they have few remains of tetrapods, give some hope of interpretation of distributions.

Vertebrate tetrapod remains from the Permo-Carboniferous sediments are preserved primarily in three types of deposits: (1) coal measures; (2) redbed facies; and (3) large, fresh-water lacustrine deposits. The redbed facies are a special type of red deposits which were for the most part formed in deltaic or quasi-deltaic situations in streams, lakes, and floodplains. Not all red sediments belong in this category, and some sediments so classified may have only a small increment of red materials.

The existence of tetrapod remains in each of these types of deposits in North America and Europe demonstrates that animals of this kind lived in, or near, the regions of deposition, or at least within the areas where erosion was producing the enclosing sediments. This evidence of vertebrate habitation cannot be extended automatically to include the areas of erosion and deposition of these types of sediments in which no vertebrate remains have been encountered. The frequency of tetrapod-bearing sites encountered in fossiliferous deposits of the three major types may provide clues, however, as to the likelihood of such habitation.

The frequency of sites in the three major deposit types where vertebrates are encountered differs sharply if taken over exposures of comparable dimensions. This fact can provide a crude estimate of the likelihood of finding a site in a given area of each type if it is assumed that the sources of tetrapods for each type of deposit were commensurate. The frequency of sites in coal measures is low, based on fossiliferous beds from middle and eastern North America and western Europe. Frequencies are considerably higher in redbed facies, judged particularly from the deposits in Texas and Oklahoma of the United States. They tend to be quite high in shallow, fresh-water lacustrine deposits, judged from such deposits in the Beaufort Beds of Africa, the Wellington (Lower Permian) deposits in central and northern Oklahoma, and parts of the Dunkard Series (Late Carboniferous and Lower Permian) in the eastern United States.

By use of this sort of a tool, crude as it is, it is possible to make some estimates of the likelihood of discovery of fossil tetrapods in the various facies that have not as yet yielded remains. In essence, this is more-or-less a formalization of the procedure often used in determining areas for prospecting for fossils in field work devoted to the exploration of new areas. If used with caution, it can be reversed to study the results of field explorations.

If we now look at the Permo-Carboniferous deposits outside of the Euramerican area from this point of view, an improved concept of the probable distributions of tetrapods at this time may emerge. Tables 2 and 3 summarize the types of sediments, the environments of deposition, and the fossil content of Late Carboniferous and Early Permian deposits pertinent to this problem.

Ectothermic tetrapods probably did not inhabit areas of active glaciation, such as those indicated by tillites in India, South Africa, and South America. They may have existed during times of deposition of the intercalated nonglacial sediments—coal measures, in a number of cases. These and other coal measures and coal measurelike deposits occur in the Upper Carboniferous of South America and in the Lower Permian of that continent and India. Chances of finding vertebrates in these coal-measure deposits are slight, even though tetrapods may have existed during their deposition. Experience indicates that such discoveries, if made, will be the result of "accidental" finds rather than purposeful exploration of the formations. Most likely to be found are semiaquatic or aquatic tetrapods, usually accompanied by fishes, if we may judge from coal measures elsewhere.

Higher beds in Brazil (Table 2), the Ironstone and Estrada Nova Formations, are not coal-measure deposits and approach somewhat more closely depositional characteristics of redbed facies, although red color is not prominent. In La Rioja Province of Argentina, the Patquia Formation (Stipanicic and Bonaparte, 1972) consists of red sandstones and layers of limestone. These beds were searched thoroughly for vertebrates by Romer and his party (Romer, 1966) with completely negative results. Although termed redbeds in some studies, the Patquia is more comparable to the totally barren orange and red sandstones of the large evaporite basins of the Upper Permian of North America. The frequency criterion of redbeds cannot be applied to them. The Ironstone and Estrada Nova of Brazil, however, fall close to this category and the apparent absence of terrestrial tetrapods is somewhat more favorable to an interpretation that none lived in or adjacent to the areas of deposition.

Remains of terrestrial tetrapods would not be expected throughout most of the Late Carboniferous and Early

Table 2. Permo-Carboniferous fresh-water deposits of South America (data from Mendes, 1967; Rocha-Campos, 1967; Romer, 1966; Stipanicic and Bonaparte, 1972).

Epochs	Formation	Description
BRAZIL		
Upper Permian	Estrada Nova Formation 100 to 1,000 m	Silt, sandstone, calcarenites, clay galls. Lacustrine, near-shore deposits. Pelecypods, crustaceans, fish, plants.
Lower Permian	Irati Formation 10 to 35 m	Black, pyrobituminous shale, nodular and lenticular dolomites, siltstones. Near-shore, shallow-basin lacustrine deposits, probably reductive conditions. Mesosaurs present, abundant in places. Fish, pelecypods, crustaceans, insects, plant impressions, and silicified wood.
Upper Carboniferous	Guata Formation 100 to 280 m	Crossbedded sandstone, siltstone, shale, and conglomerate. Coal measures in lower part. Lacustrine. Fish, plants.
	Itarare Formation	Glacial facies with some interbedded marine facies in middle section. Glacial facies with boulders in lower part.
ARGENTINA		
Permo-Triassic	Paganzo IV Paganzo III	Redbeds and vertebrates in the Triassic part of the section.
Lower Permian	Paganzo II Patquia Formation to 750 m	Red sandstones, some shales, with lenses and bands of limestone. No fossils of any type.
Upper Carboniferous	Paganzo I	Carbonaceous shales, pyrite, and some conglomerate at top. Glacial deposits, intercalated with sandstone and shale. Abundant plants in some horizons.

Permian deposits of the Soviet Union because they include primarily marine and evaporite facies. The coal measures of the Donetz Basin might be somewhat more favorable, but suffer from the low incidence of sites noted for coal measures in general. Along the eastern margins of the Cis-Uralian Soviet Union are lagoonal and terrestrial deposits of general lacustrine and redbed types. Here six sites of Early Permian vertebrates have been found (Efremov and Vjuschkov, 1955; Olson, 1957). The meager remains are of animals similar to those found in western Europe, suggesting that this area of the Soviet Union is best considered part of the Euramerican faunal complex. Fragmentary small amphibians are known from one or two sites in the area of Tungus, in Siberia, but almost nothing is known of the environment in which they lived, and the age is somewhat uncertain.

The Ecca Beds of South and East Africa were formed in subsiding basins (Hotton, 1967). The middle portion may have been deposited in fairly deep water, away from shore, but the sediments of the lower and upper parts would seem to indicate deposition under circumstances in which the inclusion of tetrapods, had any been present, would have been highly likely. None are found. Only in the underlying Dwyka, in the "White Band," are tetrapods present. *Mesosaurus*, a reptile, is identical to one of the genera found in the Irati Formation of Brazil.

The evidence of the Ecca Beds of Africa gives reason to assume that tetrapods did not exist in the basin of deposition or in areas adjacent to it. To a lesser degree this also may apply to the times and places of deposition of the Ironstone and Estrada Nova of Brazil, subject, of course, to additional exploration.

Tetrapods do occur in Lower Permian beds in Siberia, India, Africa, and South America. All of those known, however, are aquatic and this may be important. Aquatic mesosaurs are reptiles, but their phylogenetic source is extremely uncertain. The Indian fossils are labyrinthodont amphibians, very similar to genera well-known from Europe. In Europe and North America terrestrial tetrapods are found in associations with these types of amphibians, but only rarely associated with them in coal measures. This may have been the case for the Indian deposits, but the evidence is slight and any tetrapod distributions based on it may well be spurious. Critical to various hypotheses advanced to explain distributions, considered in what follows, is whether or not terrestrial tetrapods were present outside of the Euramerican area during the Late Carboniferous and Early Permian. Some of the later deposits of

Table 3. Permo-Carboniferous fresh-water deposits of South Africa, India, USSR.

Epochs	Formation	Description
SOUTH AFRICA (data partly from Hotton, 1967)		
Lower Permian	Ecca Formation 2,300 to 3,300 m	Upper part: sandstones, shales, grading into red and purple facies near top. Middle part: mainly dark blue, well-bedded shales. Lower part: sandstone and shales, in places, marly beds, coal. Pelecypods, fish, plants.
Upper Carboniferous	Dwyka Formation 840 m	At top, black, sapropelic shales forming "White Band" with *Mesosaurus*. Tillites.
INDIA (data primarily from Robinson, 1967)		
Upper Permian	Ironstone Formation 500 to ?1,500 m	Calcareous shales, fine green sandstones, coarse sediments in Satpura Region. Fluviatile to lacustrine.
Lower Permian	Barakar Formation to 1,700 m	Coal measures, some sandstone lenses. Lacustrine sediments with some fluviatile portions.
Upper Carboniferous	Talchir Formation to 275 m	Upper Carboniferous and Lower Permian glacial, mostly terrestrial but some marine (with marine invertebrates). Sparse flora.
USSR (data partly from Olson, 1962)		
Upper Permian	Kazanian	Redbed facies and evaporite. Many vertebrates, plants, insects.
	Ufimian	Redbed facies, no fossils.
Lower Permian	Kungurian 700 to 1,300 m	Largely marine and evaporite facies. Terrestrial to the east, some coal.
	Artinskian-Sakmarian to 3,500 m	Marine and evaporite facies.

India, South America, and South Africa are of types that have a high likelihood of preserving terrestrial tetrapods, but lack them. They *do* have some force in countering statements that such animals were present on these continents during the times of deposition. Negative evidence, however, remains a dangerous tool for conclusions, and additional work may well change this interpretation.

EXPLANATORY HYPOTHESES

Most of the recent hypotheses that offer explanations of the differences in distributions between the Late Carboniferous-Early Permian and Late Permian tetrapods have made the assumption that the record is a reasonable reflection of actual circumstances. Furthermore, they have tended to make the assumption that the Upper Permian tetrapods spread from a broad, geographic center, usually considered to be the Euramerican continent. Some older explanations envisaged a Gondwanaland center, but this has not been "popular" of late. Beyond these assumptions, the spread of tetrapods is visualized either as a replacement of then existing tetrapod occupants of the recipient areas or as occupancy of areas lacking tetrapods.

Rarely considered is the hypothesis that the Upper Permian distributions reflect origins of new types of tetrapods from a very widespread Early Permian faunal complex, as "vicariants" of a broad "track" in the sense of Croizat (Croizat and others, 1974). In the discussion of the proposed hypotheses that follows, it should be recognized that their reliance on a record has weak support and also that largely unexplored alternatives exist.

Physical Explanations

According to physical hypotheses, water, topography, and climate formed barriers to occupancy of specified areas and(or) to dispersals to them during the Early Permian. These barriers became ineffective during the Late Permian. Fundamental biological changes among the participating lineages were not involved in permitting the dispersals, and the biological potentials of the Early Permian tetrapods provided a sufficient base for radiations without basic modifications of either physiology or population structure.

Water Barriers. The role of water barriers has received the most attention. With the rise of confidence in continental mobility, and the development of the concept of a single large continent, Pangaea, during the late Paleozoic and early

Mesozoic, the older ideas of Permo-Carboniferous biogeography have been fundamentally altered. The stage for this change was set decades ago, but new ideas have been widely accepted only recently. The most challenging and provocative of the newer hypotheses is that proposed by Cox (1974). In essence, he has suggested that tetrapods originated and developed only on the Euramerican continent and that they did not penetrate to Gondwanaland and Asia until the Late Permian, when junctions of the formerly separated landmasses made this possible. The Late Carboniferous and Early Permian circum-equatorial distributions of tetrapods, as suggested by the record (Fig. 2A), are the result of geographic factors and not the physiological or populational aspects of the organisms. Cox presented his hypothesis with cautious qualifications, but the major points are those just noted.

Cox's hypothesis is a stimulating concept. What are its difficulties? First, of course, as noted by Cox, is the fact that the concept places faith in the data of the fossil record, a record seen in the preceding section to be somewhat shaky. We will ignore this difficulty for heuristic purposes. The second problem is that tetrapods *are* known from the Lower Permian of Siberia, only one or two to be sure, and from South America, South Africa, and India. Cox, likewise, was aware of this and made a note of it. Strictly, then, this hypothesis, if taken at full-face value, is false.

All of the tetrapods of the Lower Permian outside of the Euramerican continent, however, are aquatic and Cox implies in part of his statement that he is concerned with terrestrial tetrapods. It is among these that the increase in geographic range and in variety is most impressive. For them, as far as the record goes, the hypothesis could be valid. It is with the replacement of primitive synapsids (pelycosaurs) by more advanced ones (therapsids), from the Lower to Upper Permian, that most hypotheses are concerned.

A full meeting of Euramerica with Asia and Gondwanaland is a requisite for the expansions that this hypothesis envisages. Perhaps somewhat earlier partial contacts involving the areas with environments in which aquatic tetrapods and fishes but not terrestrial tetrapods thrived, made possible migrations of the former. A requisite of the hypothesis would seem to be that the origin and development of tetrapods was confined to the Euramerican continent, as Cox has suggested.

Viewed in its most favorable light, Cox's strictly physical hypothesis has merit in explaining the apparent radiations of the Late Permian. Many variants of the basic theme are possible, each assuming, in one way or another, that the source of Late Permian tetrapods was centered in the Euramerican continent. Currently, data are insufficient to give strong support to any one of the variants, but the evidence does seem to argue against the most radical departure from Euramerican origins, namely the concept that multiple origins of the Upper Permian tetrapods took place in several areas of the Earth.

Topographic Barriers. As the Euramerican, Asiatic, and Gondwanaland continents came into contact during the formation of Pangaea, mountain systems undoubtedly formed; we have good evidence for these in the Caledonian-Appalachian and Uralian systems. Without question such chains were important barriers to dispersals of terrestrial tetrapods. Milner and Panchen (1973) attempted to estimate the effects of the Caledonian-Appalachian chain during the Late Carboniferous and Early Permian, and Cox (1974) commented briefly on possible effects of mountain barriers in the formulation of his hypothesis of Late Permian dispersals. No very definitive analyses of long-range effects of topographic barriers during the Permo-Carboniferous have as yet been possible: partly, because of uncertainties on the exact times of the origins of the mountain ranges; partly, because of the difficulties in knowing the distributions of their elevations; and partly, because of the very limited areas that have yielded samples of terrestrial tetrapods. Clearly, strongly positive topographic structures have had local effects on the ecology of faunas, as discussed, for example, by Vaughn for Colorado and New Mexico (Vaughn, 1969). Presumably such effects were widespread and occurred in different guises in the vicinity of all mountain ranges. The lack of tetrapods in the Karroo Basin during the deposition of the Ecca Beds has been attributed to the interference of the Cedarberg Range with westerly oceanic winds to maintain cool climates until the beginning of Beaufort deposition (Hotton, 1967).

Such barriers did have local effects and surely must have been of great importance in channeling and filtering tetrapod dispersals. The increasing homogeneity of the tetrapod faunas of the Late Permian and the remarkably widespread distributions of similar faunal complexes during the Triassic indicate that over the longer periods of time topographic barriers did not produce a marked degree of continental endemism.

Climatic Barriers. Temperature and rainfall, both in amount and annual distribution, can be estimated from the characteristics of "climatically sensitive" rocks such as redbeds, coals, evaporites, and tillites, and from the relationships of different types of sedimentary beds (see, for example, Strakhov, 1967; Robinson, 1973; Olson, 1962). If it is assumed that the Permian tetrapods were basically ectothermic, the climates and climatological modifications could have been important in controlling their distributions. Most climatic interpretations have been closely related to physiology and are more conveniently considered later in that context.

In general, during the Late Carboniferous and very Early Permian, a considerable portion of the continent of Gondwanaland was glaciated; cool climates appear to have continued there after the end of glaciation. These cool climates have been suggested as the reason for the seeming lack of tetrapods, even in such deposits as coal measures that might occasionally carry them. The presence of some remains of amphibians and reptiles, however, indicates that ectothermic animals could exist in or near these areas of deposition. Except for times of deposition of

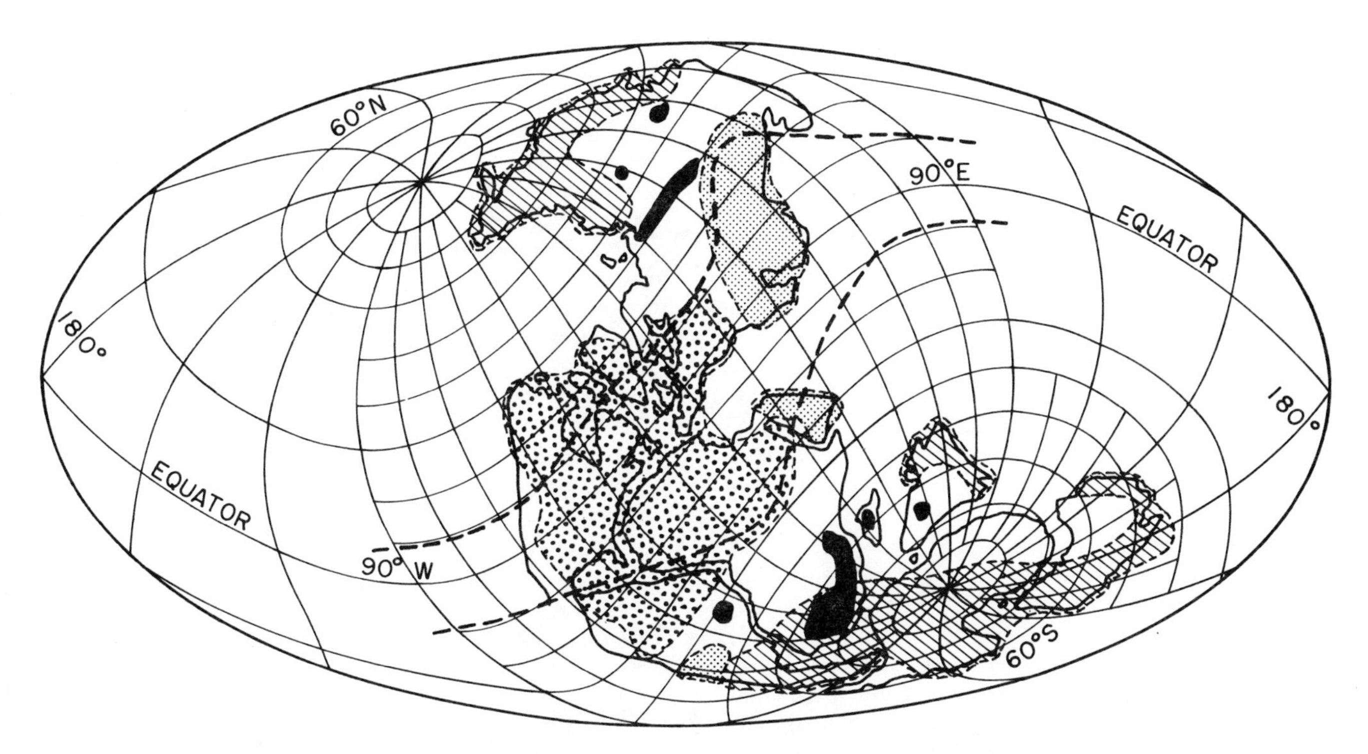

Figure 4. Redrawn and modified from Robinson's (1973) reconstruction of the climatic regions of the terrestrial areas of the world during the Upper Permian. Modified by utilization of different symbols and additions of localities in which terrestrial vertebrates are known to occur. The base is a Mollweide oblique projection. Crosshatch, middle and high latitude humid belt; heavy stipples, year-round dry belt; light stipples, sharply seasonal rainfall; solid black, tetrapod localities.

strictly glacial deposits, tillites, there seems no *a priori* reason to consider cool climates a bar to ectothermic tetrapods in the southern continent during, and immediately after, glacial times.

Except for the glacial episodes, the climates of the Permian have not been demonstrated to have been definitive with regard to terrestrial tetrapod distributions. The most extensive examination of climates has been by Robinson (1973), based, in part, on data from Strakhov (1967) and taking into account broad meteorological principles. Her map of distributions of major climatic types during the Upper Permian is shown in modified form in Figure 4. It is based on distributions of climatologically sensitive rocks and general interpretations of global atmospheric circulations. Plotted on this map are the known areas where terrestrial tetrapods are preserved in late Guadalupian and Djulfian rocks. It will be noted that, with one exception, the sites lie outside of the presumed arid zones, well away from the paleoequator and in, or adjacent to, areas with moderate temperatures and rainfall, either seasonal or throughout the year.

A major problem of distributions of terrestrial tetrapods that seems to emerge from the climatic distributions relates to the marked resemblances of the faunal complexes of the northern and southern parts of the massive continent during the Late Permian. Between the two areas vast regions of year-round dry climates appear to have existed. At least part of the time, lines of communication must have been developed, but when or where is not evident. The map and distributions show the state of the art of climatological reconstructions and interpretations of tetrapods of the Permian at the present time. The results are more provocative than definitive.

Biological Hypotheses

Biological hypotheses place primary emphasis upon modifications of organisms and their ecological interrelationships as requisites for the dispersal of terrestrial tetrapods during the Late Permian and Triassic. Fundamentally, there are two types of hypotheses, one involving physiological changes, and the other involving modifications of interspecies population structure and dynamics. Because physiological changes may depend upon populational features, the two are often merged into a synthetic hypothesis.

The most elaborate and provocative hypothesis was outlined in semipopular form by Bakker (1975a) and presented in a more detailed and conservative fashion in a technical publication (Bakker, 1975b). His hypothesis weaves together an intricate fabric involving population structure, especially the relative biomass and trophic structures of predators and prey; physiological properties, especially ectothermy and endothermy; morphology, posture, and function; and latitudinal zonation of the Earth relative to the distribution of climates. This all is cast in the framework of evolving bioenergetics. Simplified, the evolution of reptilian complexes from ectothermy to endothermy is seen as basic to the pattern and success of the apparent radiation. The complexities arise in the methods by which bioenergetics are determined, from many facets of the fossil record, from geological history, from extensive use of experimental work on modern analogues, and from analyses of the steps taken during the evolutionary transformations.

The intricate structure of this hypothesis and the evidence supporting it cannot be examined here with the care they deserve. The development of endothermy, seen as crucial in the semipopular account but less so in the more critical one, and changes in populational properties are at the heart of the hypothesis. I will try to examine the role of each separately, recognizing that Bakker's interpretation involves a thoroughly intimate relationship between the two.

First, we may ask, is endothermy a necessity for the dispersal and radiation that the record seems to show? This is apart from the question of whether endothermy actually arose among early therapsids. Robinson (1973) and Cox (1974) did not consider it necessary, and Robinson specifically states that ectothermy probably was characteristic of these therapsids. Upper Permian faunal complexes of the Soviet Union and South Africa include many therapsids for which endothermy has been postulated (see, for example, Brink, 1956; Olson, 1959; Ricqlés, 1974; Bakker, 1975a), but also include many tetrapods, both terrestrial and aquatic, that were quite surely ectotherms. Among these are small reptiles such as eosuchians and procolophons, large reptilian pareiasaurs, and large, labyrinthodont amphibians. Endothermy, thus, was not a necessary condition for existence in the areas where these faunal complexes have been found, even in very high latitudes (see Figs. 3A, 4). Endothermy could, of course, have enhanced the spread of active terrestrial tetrapods, perhaps shaped its course, but it does not appear to have been a requisite of this dispersal. Climates were not sufficiently severe, even at the very high latitudes, to inhibit the existence of either large or small ectotherms.

Interspecies population structure, especially related to trophic patterns, has been suggested as a basis for the dispersals of the Late Permian reptiles (see, for example, Olson, 1961, 1966, 1975) and is one aspect of Bakker's complex hypothesis (Bakker, 1975a, 1975b). The large predators of the early Permian fed, according to this concept, primarily upon water-based prey. The faunal complexes in which they existed thus were tightly tied to wet lowlands where abundant waters of lakes and streams were present. Only when this situation was altered by the development of an abundant source of terrestrial, herbivorous tetrapod prey could the extensive terrestrial dispersals take place.

The beginnings of the establishment of such a trophic system are evident during the Early Permian (Olson, 1975) among animals that were surely ectotherms. The change in trophic structure, it is proposed under this hypothesis, was not only a necessary prerequisite to the Late Permian dispersal, but was the principal key factor back of it and subsequent radiations. The virtual necessity of such trophic structure seems evident if the dispersal of the Late Permian terrestrial tetrapods is accepted, for dispersal under the limiting conditions of the supposed antecedent trophic structure appears highly improbable. As for all of the hypotheses we have discussed, reconstructions of the antecedent conditions are based on limited evidence, often from areas which seem to mirror major evolutionary events rather than show the central events themselves.

CONCLUSIONS

I have examined briefly the validity of the concept of a Late Permian adaptive dispersal of tetrapods. I have assumed that it did take place, and I have looked critically at several types of hypotheses that have been advanced to explain it. I would like to present my own interpretation of the one I consider most likely. My conclusions are based on some evidence which is definitive but on much, as we have seen, that is less than sufficient for sound interpretation.

The Late Permian radiation primarily involved therapsids; the extent to which their deployment was accompanied by other reptiles and amphibians is uncertain. Less active and less advanced tetrapods occur with the therapsids after the supposed dispersal, but whether these were derived from an invaded complex or were an integral part of the dispersing faunal complexes remains uncertain. The therapsid radiation appears to have stemmed from a broad center in the Euramerican continent. Traces of incipient therapsids are found in both North America and the Soviet Union during the very early Late Permian. No tetrapod-bearing beds of this age are found elsewhere. Unless we grant multiple origins of the therapsids, which is not impossible but is less plausible than a single origin, the Euramerican source seems fairly well established. The incipient therapsids probably did not come from elsewhere but developed from the preceding faunas of this broad area. It should be noted, however, that the precise transitional stages are not known.

Earlier, perhaps considerably earlier, aquatic tetrapods and fishes existed in parts of Gondwanaland and, it appears, in Asia. Their source remains uncertain. Probably the representatives came from a single source over a broad area. Although the mesosaurs are very distinct from any other known tetrapods, they are clearly reptilian. Because we know Late Carboniferous and Early Permian fossils best from North America and Europe, it is tempting to look at the Euramerican continent as the source. In that event it would appear that the aquatic tetrapods, perhaps accompanied by some primitive terrestrial reptiles and amphibians, crossed onto other continents no later than Late Carboniferous. This seems to me the most probable interpretation. The only known skeletal remains of Late Devonian tetrapods are from Greenland and, thereafter, a spotty record exists through the Lower and Upper Carboniferous in Euramerica. However, footprints have been reported from the Upper Devonian of Australia (Warren and Wakefield, 1972), and, although no tetrapods are otherwise known outside of Euramerica until Late Carboniferous (or Early Permian), this one find raises some interesting problems on distributions.

In any event, the migrations of the Late Permian terrestrial tetrapods were into continents that were to some extent populated by earlier tetrapods. With time, interactions certainly occurred. The resultant adaptive shaping of both the immigrants and the descendants of the tetrapods present earlier presumably is to be seen in the samples of tetrapods that we obtain from Upper Permian rocks.

Endothermy was not a prerequisite of the presumed radiation of the terrestrial tetrapods. Had it been developing it might well have been a factor in structuring the faunal complexes that we see. Evidence of incipient or fully developed endothermy in therapsids has come from several sources: bone histology (Ricqlès, 1974); morphological and functional interpretations (Brink, 1956); parallel development of basic structures that seem to cluster adaptively around incipient endothermy (Olson, 1959); bioenergetic interpretations (Bakker, 1975a, 1975b); and, of course, from the fact that derived mammals are endotherms. None of the evidence is conclusive, but it seems to me that the various lines are sufficient that the existence of at least partial endothermy is probable. If so, as Bakker has suggested, it may well have enhanced the dispersal, contributing to its rapidity and extent.

In summary, it appears to me that the evidence points to a Late Permian dispersal of terrestrial tetrapods from a Euramerican center following upon the attainment of a permissive trophic structure. This dispersal was accompanied by an adaptive radiation marked by increasing activity of the participating tetrapods, probably with the development of some degree of endothermy. The migrants entered areas already partially occupied by tetrapods. The well-matured community systems we observe in the Late Permian of Gondwanaland, England, the Soviet Union, and Asia, and in all of the continents during the Early and Middle Triassic, resulted from the interactions of the invaders and the tetrapods already present in these areas. This interpretation is based, in my opinion, on the best evidence of morphology, distributions, climatological analyses, and geology that we have at the present time. It can at best serve as a framework for raising questions and suggesting lines of study in order to support or refute it.

REFERENCES

Bakker, R. T., 1975a, Dinosaur renaissance: Sci. American, v. 232, p. 58-79.

——— 1975b, Experimental and fossil evidence for the evolution of tetrapod energetics, *in* Gates, D. M., and Schmerl, R. B., eds., Perspectives of biophysical ecology: New York, Springer-Verlag, p. 365-399.

Brink, A. S., 1956, Speculations on some advanced mammalian characteristics in the higher mammal-like reptiles: Palaeont. Africana, v. 4, p. 77-96.

Chudinov, P. K., 1960, Upper Permian therapsids of the Ezhovo locality: Paleont. Zhurn., v. 4, p. 81-94 (in Russian).

——— 1965, New facts about the fauna of the Upper Permian of the USSR: Jour. Geology, v. 73, p. 117-130.

Cox, C. B., 1973a, Triassic tetrapods, *in* Hallam, A., ed., Atlas of palaeobiogeography: Amsterdam, Elsevier, p. 213-233.

——— 1973b, Gondwanaland Triassic stratigraphy: Acad. Brasileira Ciências An., v. 45, p. 115-119.

——— 1973c, The distribution of terrestrial tetrapod families, *in* Tarling, D. H., and Runcorn, S. K., eds., Implications of continental drift to the Earth sciences, Vol. I: London, Academic Press, p. 369-371.

——— 1974, Vertebrate palaeodistributional patterns and continental drift: Jour. Biogeography, v. 1, p. 75-94.

Croizat, L., Nelson, G., and Rosen, D. E., 1974, Centers of origin and related concepts: Systematic Zoology, v. 23, p. 265-287.

Efremov, I. A., 1954, The fauna of terrestrial vertebrates in the Permian copper sandstones of the western Cis-Urals: Akad. Nauk SSSR Paleont. Inst. Trudy, v. 54, p. 1-416 (in Russian).

Efremov, I. A., and Vjushkov, B. P., 1955, Catalogue of localities of Permian and Triassic terrestrial vertebrates in the territories of the USSR: Akad. Nauk SSSR Paleont. Inst. Trudy, v. 56, p. 1-185 (in Russian).

Hotton, N., III, 1967, Stratigraphy and sedimentation in the Beaufort Series (Permian-Triassic), South Africa, *in* Teichert, C., and Yochelson, E. L., eds., Essays in paleontology and stratigraphy: Kansas Univ. Dept. Biol. Spec. Pub., v. 2, p. 390-428.

Mendes, J. C., 1967, The Passa Dois Group, *in* Bigarella, J. J., Becker, R. D., and Pinto, J. D., eds., Problems in Brazilian Gondwana geology: Brazilian Contr. Internat. Symposium Gondwana Stratigr. and Palaeont., Curitiba, p. 119-166.

Milner, A. R., and Panchen, A. L., 1973, Geographic variation in the tetrapod faunas of the Upper Carboniferous and Lower Permian, *in* Tarling, D. H., and Runcorn, S. K., eds., Implications of continental drift to the Earth sciences, Vol. 1: London, Academic Press, p. 353-367.

Olson, E. C., 1957, Catalogue of localities of Permian and Triassic terrestrial vertebrates of the territories of the USSR: Jour. Geology, v. 65, p. 196-226.

——— 1959, The evolution of mammalian characters: Evolution, v. 13, p. 344-353.

——— 1961, Food chains and the origin of mammals, *in* Internat. Colloquium on the evolution of lower and unspecialized mammals: Kon. Vlaamse Acad. Wetensch. Lett. Sch. Kunstens Belgie, Brussels 1961, pt. I, p. 97-116.

——— 1962, Late Permian terrestrial vertebrates, USA and USSR: Am. Philos. Soc. Trans., v. 52, pt. 2, p. 3-224.

——— 1966, Community evolution and the origin of mammals: Ecology, v. 47, p. 291-302.

——— 1975, Permo-Carboniferous paleoecology and morphotypic series: Am. Zoologist, v. 15, p. 371-389.

Olson, E. C., and Vaughn, P. P., 1970, The changes of terrestrial vertebrates and climates during the Permian of North America: Forma et Functio, v. 3, p. 113-138.

Price, L. I., 1948, A labyrinthodont amphibian from the Pedra de Fogo Formation, State of Maranhão: Brazil Div. Geol. and Mineral. Bull., v. 124, p. 7-32.

Ricqlès, A. de, 1974, Evolution of endothermy: Historical evidence: Evol. Theory, v. 1, p. 51-80.

Rocha-Campos, A. C., 1967, The Turbarão Group in the Brazilian portions of the Parana Basin, Pt. 1, *in* Bigarella, J. J., Becker, R. D., and Pinto J. D., eds., Problems in Brazilian Gondwana geology: Brazilian Contr. Internat. Symposium Gondwana Stratigr. and Palaeont., Curitiba, p. 27-106.

Robinson, P. L., 1969, The Indian Gondwana formations—a review, *in* Gondwana stratigraphy, I. U. G. S. Symposium, Paris 1967: UNESCO, p. 201-268.

——— 1971, A problem of faunal replacement in Permo-Triassic continents: Palaeontology, v. 14, p. 131-153.

——— 1973, Palaeoclimatology and continental drift, *in* Tarling, D. H., and Runcorn, S. K., eds., Implications of continental drift to the Earth sciences, Vol. 1: London, Academic Press, p. 451-476.

Romer, A. S., 1945, The Late Carboniferous vertebrate fauna of Kounovia (Bohemia) compared with that of the Texas red beds: Am. Jour. Sci., v. 243, p. 417-442.

——— 1966, The Chañares (Argentina) Triassic reptile fauna. I. Introduction: Harvard Univ. Mus. Comp. Zool., Breviora, v. 247, 14 p.

——— 1973, Permian reptiles, *in* Hallam, A., ed., Atlas of palaeobiogeography: Amsterdam, Elsevier, p. 150-167.

Schopf, T. J. M., 1974, Permo-Triassic extinctions: Relation to sea-floor spreading: Jour. Geology, v. 82, p. 129-143.

Sigogneau-Russell, D., and Russell, D. E., 1974, Étude du premier Caseide (Reptilia, Pelycosauria) d'Europe occidentale: Mus. Natl. d'Hist. Naturelle Bull., 3rd ser., no. 230, p. 145-205.

Stipanicic, P. N., and Bonaparte, J. F., 1972, Cuenca triasica de Ischigualasto-Villa Union, *in* Leanza, A. M., ed., Geologia Regional Argentina: Cordoba, Acad. Nac. Ciencias, p. 507-536.

Strakhov, N. H., 1967, Principles of lithogenesis, Vol. 1: New York, Consultants Bureau, 245 p.

Taquet, P., 1969, Première découverte en Afrique d'un reptile captorhinomorphe (Cotylosauria): Acad. Sci. Comptes Rendus, sér. D, v. 268, p. 779-781.

Tewari, A. P., 1962, A new species of *Archegosaurus* from the Lower Gondwana of Kashmir: India Geol. Survey Recs., v. 89, p. 427-434.

Vaughn, P. P., 1969, Early Permian vertebrates from southern New Mexico and their paleozoogeographic significance: Los Angeles County Mus. Contr. Sci., v. 166, 22 p.

Verma, K. K., 1962, *Chelydosaurus marahomensis* n. sp., a new fossil labyrinthodont from the Lower Gondwana near Maharom, Anaultuag District, Kashmir: Indian Minerals, v. 16, p. 180-182.

Wadia, D. N., and Swinnerton, W. E., 1928, *Actinodon risinensis* n. sp. in the Lower Gondwana of Vihi District, Kashmir: India Geol. Survey Recs., v. 61, p. 141-145.

Warren, J. W., and Wakefield, N. A., 1972, Trackways of tetrapod vertebrates from the Upper Devonian of Victoria, Australia: Nature, v. 238, p. 469-470.

Woodward, A. S., 1905, Permo-Carboniferous plants and vertebrates from Kashmir. II. Fishes and labyrinthodonts: India Geol. Survey Mem. Palaeont., v. 2, p. 10-13.

Westoll, T. S., 1944, The Haplolepidae, a new family of Late Carboniferous bony fishes: Am. Mus. Nat. History Bull., v. 83, p. 1-121.

Permian Positions of the Northern Hemisphere Continents as Determined from Marine Biotic Provinces

T. E. Yancey, *Department of Geology, Idaho State University, Pocatello, Idaho 83209*

ABSTRACT

A regular sequence of climatically-controlled Permian marine biotic provinces is present in North America and Europe. They are the Grandian (tropical), Cordilleran (temperate), and Boreal (boreal) in North America; the Tethyan (tropical), Mordvinian (temperate), and Boreal (boreal) in Europe. The Tethyan Province is redefined here for the Permian, and the Mordvinian Province is newly defined here. The distribution of provinces and the boundaries between them can be used effectively to determine paleolatitudes of the continental blocks they occupy.

The marine provinces of western North America and eastern Europe are continental margin provinces, occupying the margins of a single Euramerican continent during the Permian. In East Asia, only tropical and boreal biotas are present (belonging to the Tethyan and Boreal Provinces); these occur adjacent to each other without intervening temperate biotas, and indicate that the now adjacent Siberian and North China blocks were separated by at least 25° of latitude during the Permian, and later converged.

Many occurrences of Permian Tethyan biotas along the Pacific continental margins of Siberia, New Zealand, and North America are anomalous with respect to the provinces developed on the adjacent stable areas of the continents, and occur in areas of post-Permian accretion along the continents. The Tethyan Province was restricted to the Permian tropics of the Old World continents and did not occur on the American continents. The presence of Tethyan biotas in anomalous positions away from southwestern and southeastern Asia indicates dispersion and movement over thousands of kilometers from their Permian positions.

INTRODUCTION

Continental paleopositions are mostly determined from paleomagnetic data with added control coming from biotic distributions, sea floor anomaly patterns when available, and matching of continental margins, when possible. Biotic data have proven to be especially valuable in working out upper Paleozoic continental paleopositions. They are proportionately more significant for this purpose in the older portion of the Phanerozoic, as other data becomes more scarce. Fossil biotas can usually be well-dated, and are widely distributed; they can be recognized in many metamorphic terrains where paleomagnetic data are obliterated. One of the major advantages of using biotic data is their greater availability over other types of data, especially for crustal blocks in tectonically mobile belts.

Fossil biotas are especially valuable for determining Permian paleolatitudes, because of strong provincialism at that time. Anomalous biotic occurrences or the juxtaposition of two climatically remote provinces are readily explained by later tectonic movements, and the paleolatitude dissimilarity of the biotas will provide an estimate of the scale of the tectonic movement involved. For the Northern Hemisphere continents, the paleolatitude results obtained from Permian biotic analysis agree closely with the available Permian paleomagnetic data (McElhinny, 1973; Smith and others, 1973; Zijderveld and van der Voo, 1973), and both methods yield data suitable for paleotectonic interpretations. The two methods complement each other in paleolatitude determinations.

Permian biotas are strongly provincial, and these provinces occupied bands corresponding generally to latitudinal zones, which makes it possible to recognize paleolatitudes based on them [the latitudinal control of modern biotic provinces is apparent in the distribution of provinces shown by Hedgpeth (1957) and Ross (1974)]. In a traverse from equator to polar areas, there occurs a series of biotic provinces of both marine and land biotas, and in Permian mid- and low-latitudinal zones there is more than one province present within each zone. In a manner similar to modern provinces, boundaries of Permian provinces appear to be determined primarily by annual average temperatures and by geographic isolation. These controls result in a number of distinct provinces, but comparisons can be made between them because at any particular latitude the biotas generally have similar diversity and morphologically similar species, even if the species occurring there are unrelated.

Throughout the Permian, biotas belonging to warm and cold climatic extremes can be readily recognized among terrestrial plants (Chaloner and Lacey, 1973) and marine invertebrates (Stehli, 1964, 1971). Among marine invertebrates, the tropical and polar groups have been referred to as Tethyan and Boreal faunas, respectively. This is inadequate for paleolatitude analysis, and with our present level of knowledge of Permian biotas it is possible to recognize low-latitude (tropical), mid-latitude, and high-latitude (polar) biotas and provinces, rather than the very generalized warm (Tethyan) and cold (Boreal) subdivisions referred to in most of the present and past literature. Provinces are redefined here for Northern Hemisphere continents only, but a similar series of latitudinally restricted provinces should be recognizable on Southern Hemisphere continents as well.

A latitudinal series of Permian provinces is best seen along the western part of North America and in eastern Europe, on stable cratonal and miogeosynclinal areas within these two continents (see Figs. 1, 2). Within a latitudinal range of 55° in North America, there are three climatically produced marine biotic provinces, and within a similar range of latitude in eastern Europe there are also three easily distinguished marine biotic provinces. The tropical provinces, Grandian in North America and Tethyan in Europe, are more similar to each other than to the adjoining climatically different provinces. The same is true for the temperate Cordilleran (North America) and Mordvinian (Europe) Provinces.

The Permian temperate provinces occupy paleolatitude bands 20° to 25° wide, and the Boreal Province occupies a latitudinal band about as wide, although it may have been wider and included more area at the rotational pole beyond the cratonal areas. The tropical provinces had a minimum 10° width, but their full width is not present on either continent. By comparison with modern tropical provinces (Hedgpeth, 1957), the Permian tropical provinces probably did not extend more than 20° to 25° on either side of the equator. Within a latitudinal traverse from equator to pole, the Permian tropical and temperate provinces probably occupied bands of 20° to 25°, and the Boreal Province occupied a relatively wider band of up to 40°.

The sequence of Permian biotic provinces from low to high latitudes across both the North American and European continents demonstrates fairly conclusively that a threefold subdivision of Permian marine biotas is realistic, and the internal agreement between the two latitudinal traverses indicates that the character of the paleolatitude change within the biotas is well-determined. Similar climatic and biotic gradients should be determinable over other comparable segments of the Permian world.

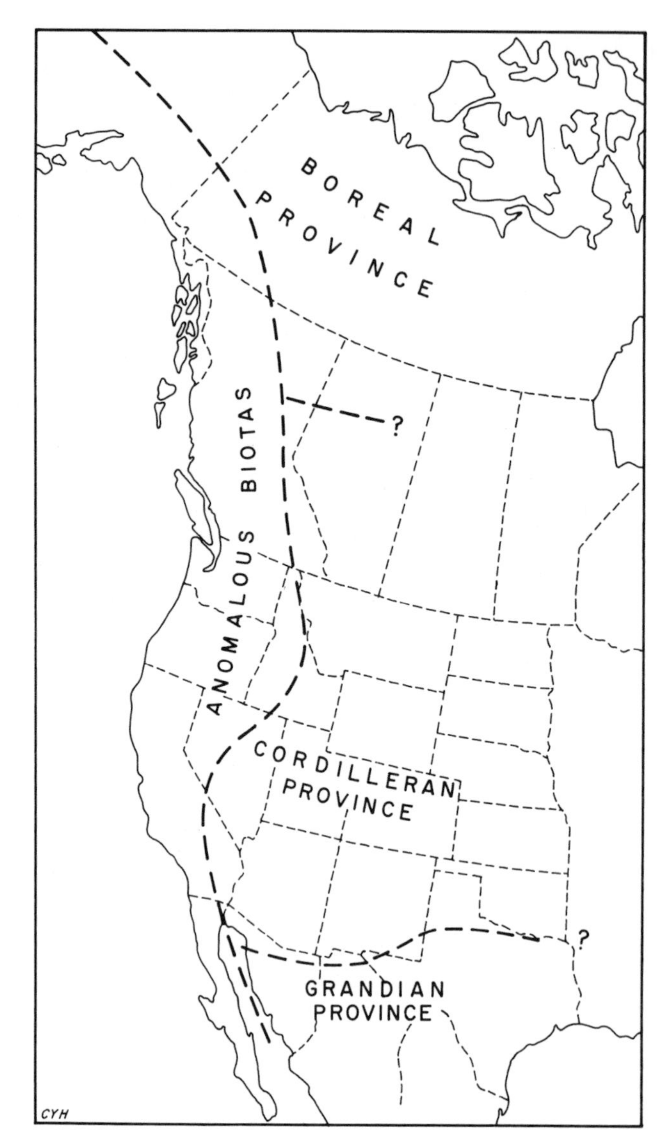

Figure 1. Permian biotic provinces in western North America (map adapted from Yancey, 1975).

NORTH AMERICAN PERMIAN BIOTIC PROVINCES

The correspondence of climatic zonation and provincialism is apparent in most parts of the Northern Hemisphere, but is best seen on the North American continent. Three geographically separate biotic provinces occur here (Yancey, 1975). These occupied low, middle, and high Permian latitudes, as determined by climatic indicators such as organic reefs and biotic gradients, and by paleomagnetic determinations on Permian rocks in these provinces (Smith and others, 1973). The North American continent stretched over a wide range of paleolatitudes, with climates ranging from tropical to polar.

The Grandian Province of Texas and northern Mexico is a tropical province as shown by the high biotic diversity, the abundance of highly ornamented species, and the common occurrence of large organic reefs. This Province is distinguished by the presence of lyttoniid and richthofeniacean brachiopods and common large fusulinacean species. It is distinct from the tropical Tethyan Province of southern Europe and Asia because it lacks waagenophyllid corals and verbeekinid fusulinids (apart from a very few occurrences of rare individuals).

The Cordilleran Province of the Rocky Mountains and Basin and Range area is a province of the Permian temperate or subtropical zone. This Province has a lower biotic diversity than the Grandian Province and lacks the richthofeniacean and (with one exception) lyttoniid families that are diagnostic of the Grandian. The Cordilleran Province is distinguished by the presence of the brachiopods *Squa-*

maria, Costellarina, and *Xestotrema,* and a few other brachiopod and molluscan genera. The brachiopod *Bathymyonia* occurs in the northern part of the Province and in the Boreal Province as well, and has a northern distribution.

The Boreal Province is circumpolar and occurs in the northernmost parts of the continents of North America, Europe, and Asia. It is distinguished by the presence of the brachiopod genera *Kuvelousia, Arctitreta, Camerisma* (*Callaiapsida*), and *Horridonia.* The last two genera are good indicators of the Province in North America, but in Europe these genera range outside the Boreal Province into the temperate Mordvinian Province.

The southern limit of distribution of *Horridonia* is difficult to determine. Cooper and Grant (1975) recognize King's *Horridonia texana* from west Texas, but base this taxon on 3 rather small valves that differ from typical *Horridonia* of the boreal region in both their cardinal process and size. Furthermore, Sutherland and Harlow (1973) recognize a small "*Horridonia*" in the Upper Carboniferous of New Mexico that they suggest is a new genus related to *Horridonia.* It is clear now that horridoniid brachiopods (though not necessarily *Horridonia s. s.*) occur rarely in low paleolatitudes, and in pre-Permian strata as well, but in North America typical large *Horridonia* is known only in the Permian Boreal Province where it is common. It has not been found in the Cordilleran Province, even in the northern part of the Province in central Alberta (see Logan and McGugan, 1968), and it remains an important but controversial means of recognizing the Boreal Province in North America.

The three North American provinces are distinct on the cratonal portions of North America (Fig. 1), but are not recognized within the tectonically complex Pacific margin of the continent (Yancey, 1975). Both tectonic (Silberling, 1973; Rogers and others, 1974) and biotic data suggest that the Pacific margin of North America was not part of the continent during the Permian. The extracratonal Pacific margin of North America contains an unusual mixture of Permian biotas (see Bostwick and Nestell, 1967; Monger and Ross, 1971) which are unlike biotas present in the cratonal provinces. Endemic fusulinid genera occur there, as well as verbeekinid fusulinids which are typical of the Tethyan Province of Asia and not of any North American cratonal province. These anomalous occurrences of Asian taxa and endemic genera all suggest that the crustal blocks of the continent were not part of North America during the Permian. The presence of Tethyan Province biotas at about 60° N latitude in British Columbia (Monger and Ross, 1971), 10° northward of Boreal Province biotas in Vancouver Island, British Columbia (Yole, 1963), indicates a minimum of 30° to 40° movement of these continental fragments since the Permian, in opposite directions. Boreal Province biotas were separated from tropical province biotas by perhaps as much as 3,000 km (1,800 miles) of temperate province biotas during the Permian, but they overlap Permian tropical biotas by at least 1,200 km (720 miles) in their present distribution along western North America.

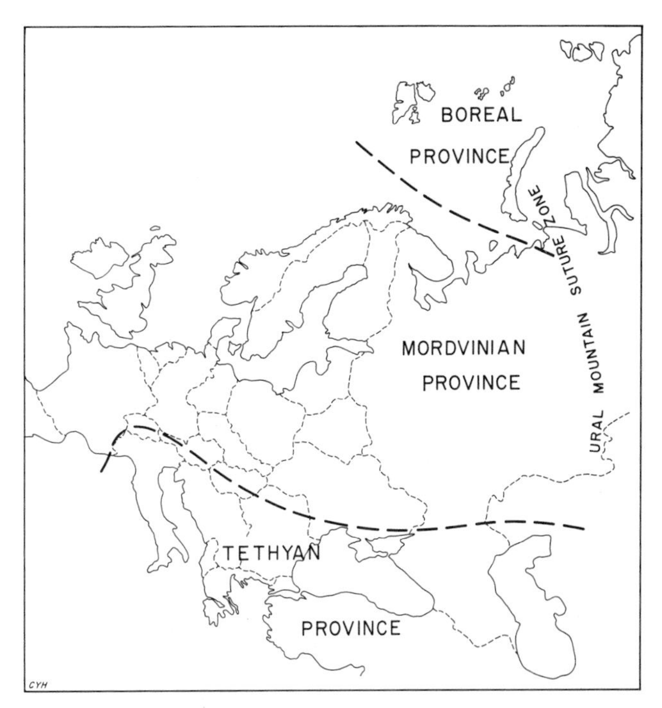

Figure 2. Permian biotic provinces in Europe (map based on a polar projection).

Alternative suggestions that the mixed distributions of Permian Boreal and Tethyan biotas along the Pacific margin of North America result from upwelling of colder waters adjacent to warmer inshore waters, or from occurrences of localized warmer or colder water currents, are improbable. Latitudinal overlap of adjacent modern provincial biotas along North America does not exceed 500 km (see Abbott, 1968, p. 34-37), and are normally much less. The latitudinal overlap of Permian tropical and boreal biotas is at least 1,200 km, whereas they should be latitudinally separated by 2,500 to 3,000 km. These distances of overlap are too large to explain as the result of ocean currents; they require tectonic causes.

EUROPEAN PERMIAN BIOTIC PROVINCES

A series of climatically produced marine biotic provinces similar to those of North America can be recognized in eastern Europe, west of the Ural Mountains (Fig. 2). Tropical marine biotas are present in the central and eastern Mediterranean and Black Sea regions. Polar marine biotas are present in the Svalbard Islands (Spitsbergen) and Novaya Zemlya. Between these two regions are temperate marine biotas of a third marine biotic province.

Tropical marine biotas of the Mediterranean and Black Sea regions are part of the well-known Permian Tethyan Province. This Province is characterized by the occurrence of verbeekinid foraminifera (Gobbett, 1967) and waagenophyllid corals (Minato and Kato, 1966), by many smaller foraminifera with heavy, thick-walled calcareous tests, especially the endothyraceans *Colaniella* and *Pachyphloia* and the miliolacean *Hemigordiopsis,* and by the unusual thick-walled bivalve *Tanchintongia.* Richthofeniacean and lyttoniid brachiopods are common in this Province, and, with few exceptions, are limited to the Tethyan Province of Eurasia and the Grandian Province of North America.

The Tethyan Province occupies an area from the central Mediterranean eastwards through the eastern Mediterranean and Black Sea areas to eastern and southeastern Asia, including China, southern Mongolia and Japan, exclusive of the Indian subcontinent. The western boundary extends from Tunisia to Sicily to the Carnic Alps (in southern Austria); the northern boundary extends from the Carnic Alps to the northern Black Sea to the Pamir Mountains and eastwards. Well-known Tethyan biotas occur in Sicily, the Carnic Alps, Yugoslavia, Greece, Cyprus, Turkey, the Caucasus Mountains, and other areas to the east. The southern boundary is not known, but occurs at least to the south of Tunisia and eastern Arabia.

The tropical nature of the Tethyan Province biotas is determined by several characteristic features. Many species have heavy, thick-walled calcareous shells of large size, and many species have very ornate sculpture on the shells. Tethyan biotas have very high taxonomic diversity, and include some highly modified endemic groups. All of these characteristics are associated in modern biotas with tropical climates. The highly modified lyttoniid and richthofeniacean brachiopods with heavy calcareous shells, coralloid form, and vesiculate shell spaces from the Sosio biotas of Sicily (Rudwick and Cowen, 1967) seem particularly tropical in affinities. There are no large organic reefs presently known within the Tethyan Province comparable to reefs of the Grandian Province in North America, but the composition of biotas and morphology of species is very similar. Tectonic complexities within the European part of the Tethyan Province makes paleomagnetic determinations of the Province's position unreliable, but it was located south of the stable European area which possibly reached down to 10° to 15° N of the paleoequator (Zijderveld and van der Voo, 1973).

Temperate Permian biotas of the Mordvinian Province, newly named here, occur over most of stable Europe, the Russian Platform, and the Ural Mountains. The Province is characterized by the endemic brachiopods *Dasyalosia, Paramarginifera, Tubaria,* and *Urushtenia* (which also occurs in the north Caucasus Mountains), and by moderate taxonomic diversity in the biotas, as well as by the absence of characteristic taxa of the Tethyan and Boreal Provinces. Furthermore, the distinctness of this Province from the Cordilleran Province of North America is shown by the occurrence of the fusulinid foraminifera *Sphaeroschwagerina, Quasifusulina, Occidentoschwagerina,* and *Zellia* (which also occur in the Tethyan Province, but not in the Cordilleran Province: see Ross, 1967). The Mordvinian and Cordilleran Provinces are temperate, and have similar biotic diversity. Although there are noticeable generic differences between their biotas, they differ primarily at the species level. The area occupied by the Mordvinian Province corresponds partly with the Ural-Kazakhstan nonmarine floral Province (Chaloner and Meyen, 1973), a temperate province intermediate between the Euramerican and Angaran floral Provinces of the Permian, and partly with the Euramerican floral Province in western Europe. These marine and nonmarine provincial biotas are temperate in character, which conforms with mid-latitude paleolatitudes determined from paleomagnetism (Smith and others, 1973; Zijderveld and van der Voo, 1973).

Some Russian paleontologists (Stepanov, 1973; Gorsky and Guseva, 1973) have recognized that Permian biotas of the Russian Platform are distinct from those of the Arctic and Caucasian-Himalayan regions. However, they have not defined the provincial character of the Russian Platform biotas or named the province. For this reason the name Mordvinian (after an ancient group of people of southern Russia) is introduced here.

The Mordvinian Province extends in an eastward band from the British Isles to the Ural Mountains. The southern boundary runs along the northern margin of the Tethyan Province, and the northern boundary runs to the south of Svalbard (Spitsbergen) and Novaya Zemlya. The Province includes stable Europe, the Russian Platform, the southern Urals, northern Urals, and Timan Mountains. The west European Zechstein biotas are included in this Province without reservation, although they are interbedded with strata deposited in high salinity waters, and are notably lacking in fusulinaceans. The occurrence of common brachiopods and echinoderms as well as small reefal accumulation in the British Isles (Pattison and others, 1973), indicates times of normal marine salinity for the Zechstein faunas.

Placement of the northern boundary of the Mordvinian Province in eastern Europe is relatively easy, but placement in western Europe is not as certain because of the scarcity of marine biotas of Permian age in that area. The northernmost temperate marine biotas of western Europe are in the Zechstein of Great Britian at about 55° N latitude, while the closest boreal marine biotas occur on Bjornøya (Bear Island) south of Svalbard (Spitsbergen) at 74° N. This gap can be reduced by including the Permian flora of the Oslo region (Høeg, 1935) at 60° N as a control point. The Oslo flora is a part of the Euramerican nonmarine floral Province. The Euramerican Province is considered to be a warm climate Province, and is definitely not of cold climatic origin (Chaloner and Lacey, 1973). This reduces the uncertainty for the western European end of the boundary line to a gap of 14° of latitude, between Oslo and Bjornøya.

Probable climatic differences are noticeable within the Province in the Urals region. The southern part of the Russian Platform and the southern Ural Mountains contain a taxonomically more diverse biota than the northern part of this area. In the south Urals, there are many small reefs within the Sakmarian and Artinskian strata (Nalivkin, 1973), indicating warm-water conditions (although not as warm as the tropical climates of the Tethyan Province). This is similar to the southern part of the Cordilleran Prov-

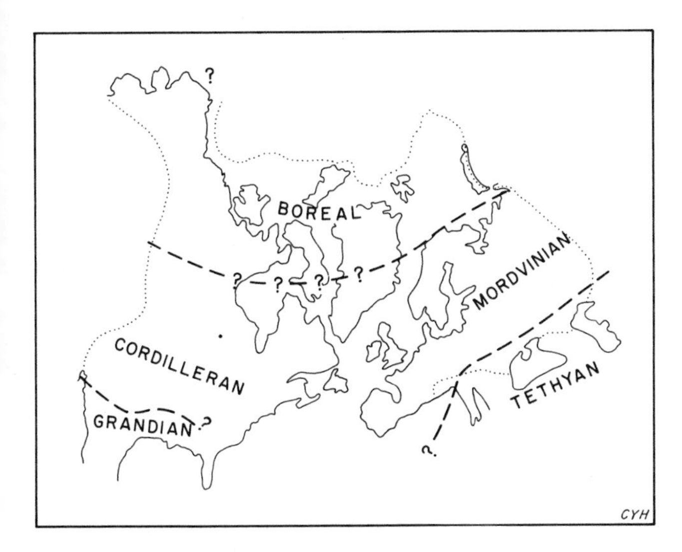

Figure 3. Permian biotic provinces on a reconstructed Euramerican continent. The dotted line marks the edges of the continental block during the Permian, except along the Arctic Ocean where it marks the edge of the modern continental shelf. The Arctic Ocean does not necessarily correspond to the Permian continental margin. Map modified from Smith and others (1973), with new interpretation of the limits of the continental blocks.

ince, which also contains small isolated reefs (Mudge and Yochelson, 1962). In the northern part of the Urals region the biotas are less diverse and contain a number of taxa with predominantly northern distribution that occur both in the Boreal Province and north temperate regions of Eurasia. These facts suggest the possibility that with further work two distinct subprovinces can be recognized within the temperate biotas of this region: a warm temperate one to the south, and a cold temperate one to the north.

The temperate Mordvinian Province contains the type section (or type area) of the Permian System, whereas the Permian standard section for North America is located within the tropical province of North America (in Texas). Difficulties in Permian correlation between North America and Europe are caused in part by this provincial difference in the standard sections.

Polar biotas of the present-day boreal areas of Eurasia are part of the well-known Permian Boreal Province. This Province is characterized by the occurrence of several genera that have a polar distribution; many of them are bipolar, being present at both the north and south poles (Waterhouse, 1967; Ustritskiy, 1973). The brachiopod genera *Arctitreta, Jakutoproductus*, and *Kuvelousia* are restricted to the Boreal Province, and the bipolar bivalve genera *Kolymia, Aphanaia*, and *Atomodesma* are probably restricted to the Boreal Province in the Northern Hemisphere part of their distribution. In Eurasia, the brachiopod genera *Camerisma* (*Callaiapsida*), *Septacamera, Horridonia*, and *Licharewia* occur in the Boreal Province and in the northern part of the temperate province, while in North America these genera are restricted to the Boreal Province.

The Permian Boreal Province has a modern circumpolar distribution, occurring in the arctic portions of the North American, European, and Asian continents. In Europe it includes Svalbard (Spitsbergen) and Novaya Zemlya, and most of Siberia in Asia (Ustritskiy, 1973). The southern boundary of the Province runs between Svalbard and Norway and between Novaya Zemlya and the northern Ural Mountains, and extends on into the southern parts of Siberia.

The polar nature of the Boreal Province biotas is indicated by the low taxonomic diversity of the biotas, and by the dominance of coarse shell ornament and the absence of fine, delicate ornament. Glaciomarine sediments have been reported interbedded with strata containing Boreal Province biotas in Novaya Zemlya, and extensively in eastern Siberia (Mikhaylov and others, 1970; Ustritskiy, 1973). The bipolar bivalve genera present in the Boreal Province have been found associated with glaciomarine deposits in both Australia and Siberia.

During the Permian, the European continental block was attached to North America, and was later split away by the rifting and opening of the North Atlantic Ocean (Smith and others, 1973). The North American biotic provinces were, therefore, those of the western side of the Euramerican continent, and the European biotic provinces were those of the eastern side of the Euramerican continent (Fig. 3); similar patterns of north-south provincialism occur along modern continental margins. These two sets of marine provinces were usually separated by large land areas in the middle of the continent (eastern North America and western Europe) that were only rarely covered by marine waters, and these incursions were usually of abnormal salinity.

TECTONIC IMPLICATIONS OF THE EUROPEAN PROVINCES

Biotic provinces in eastern Europe are arrayed in a south to north sequence of tropical to polar provinces that are similar to the North American continental margin provinces. This suggests that the eastern European area was a continental margin during much of the Permian. Lower Permian deposits contain common fusulinids and ammonoids that indicate normal marine waters and moderately deep water (on the continent). Conditions change to widespread nonmarine deposition in the Upper Permian. This change suggests that the joining of the Siberian block to the Euramerican block occurred in the Late Permian rather than the Late Carboniferous as conventionally portrayed on most continental plate reconstructions. The widespread occurrence of Permian boreal biotas on the Siberian block and the absence there of temperate biotas also suggests considerable separation between Euramerica and Siberia until the Late Permian. Paleomagnetic and tectonic data (Hamilton, 1970) do not provide clear evidence for the time of continental closure; a Late Permian closure fits best with biotic data.

PERMIAN BIOTIC PROVINCES IN EAST ASIA AND THEIR TECTONIC IMPLICATIONS

The biotas of southeast Asia, from Burma to Sumatra and from western China to Japan are part of the Tethyan marine biotic Province. This Province continues as far west as southern Europe and the Mediterranean region, and its biotic character is the same throughout this large area. In southeast Asia, the Tethyan marine biotic Province corresponds closely with the Cathaysian nonmarine floral Province. These two Provinces are the marine and nonmarine biotic provinces of the Permian tropical climatic zone in Asia, as shown by the diversity and specialization among the faunas (Stehli, 1973) and the diversity and large leaf size of the floras (Chaloner and Meyen, 1973). The northern boundary of the Tethyan Province in eastern Asia runs close to the northern border of China, and extends eastward to the north of Honshu Island in Japan. Along the northern border from Mongolia to Sikhote Alin (Vladivostok region) the boundary of the Cathaysian floral Province and the Tethyan marine Province do not exactly coincide, but the boundaries of both occur within the narrow zone of Mesozoic foldbelts between the North China stable block and the Siberian Platform stable block (Fig. 4).

Polar, cold climate Permian biotas occur on the Siberian Platform just north of the Tethyan tropical biotas of East Asia. These are part of the Boreal Province, which extends from Siberia to Spitsbergen, Greenland, and northernmost North America. The biotic character of the Province is similar throughout this large area. The Boreal Province corresponds closely with the Angaran nonmarine floral Province, and the two represent the marine and nonmarine provinces of the Permian northern cold climatic belt. Ustritskiy (1973) reports extensive glaciomarine sediments within the Boreal Province in the Verkhoyansk region and eastern Siberian block. The Province covers the entire Siberian block and the Kolyma block in the easternmost part of Siberia, but does not include the Koryak region just north of the Kamchatka Peninsula, located in the mobile belt along the Pacific Ocean.

The most conspicuous feature of the east Asian region is that there is no temperate biotic province, and there is no gradual trend from tropical to polar biotas comparable to the established paleoclimatic gradients in Europe and North America. The close positioning of the Siberian block with its boreal biotas beside the North China block with its tropical biotas, without an intervening temperate province, is a major biotic anomaly, and provides important evidence that these two crustal blocks were separated by large distances during the Permian. There has been a north-south closure between these blocks amounting to a minimum of 25°. The biotas indicate that the Siberian block has remained relatively close to the rotational pole, and that the North China block has moved a great distance northward away from the equatorial zone. The region containing the tropical Tethyan Province is composed of several crustal blocks that were probably separate during the Permian and since then have been joined (Stauffer, 1974). In addition to the movement of the North China block, the South China block, the Indochina block, and the southern Japanese Islands have all moved northward considerable distances since the Permian. The fundamental separation of eastern Asia into two (or more) distantly separated blocks during the late Paleozoic is suggested by the limited paleomagnetic data available (Kropotkin, 1971; McElhinny, 1973; McElhinny and others, 1974), and by climatic interpretation of palynofloras (Hart, 1974) and marine biotas (see Burrett, 1974; and herein). During the late Paleozoic the crustal blocks of southeast Asia were in a warm climatic belt in low latitudes, and were distantly separated from the cold climatic regions of Gondwanaland (Stauffer and Gobbett, 1972) and Angara (Hart, 1974; Kremp, 1974).

The boundary between the Boreal and Tethyan Provinces in East Asia is a narrow zone that runs close to the northern border of China, and extends eastward to the north of Honshu Island in Japan. In Mongolia-North China

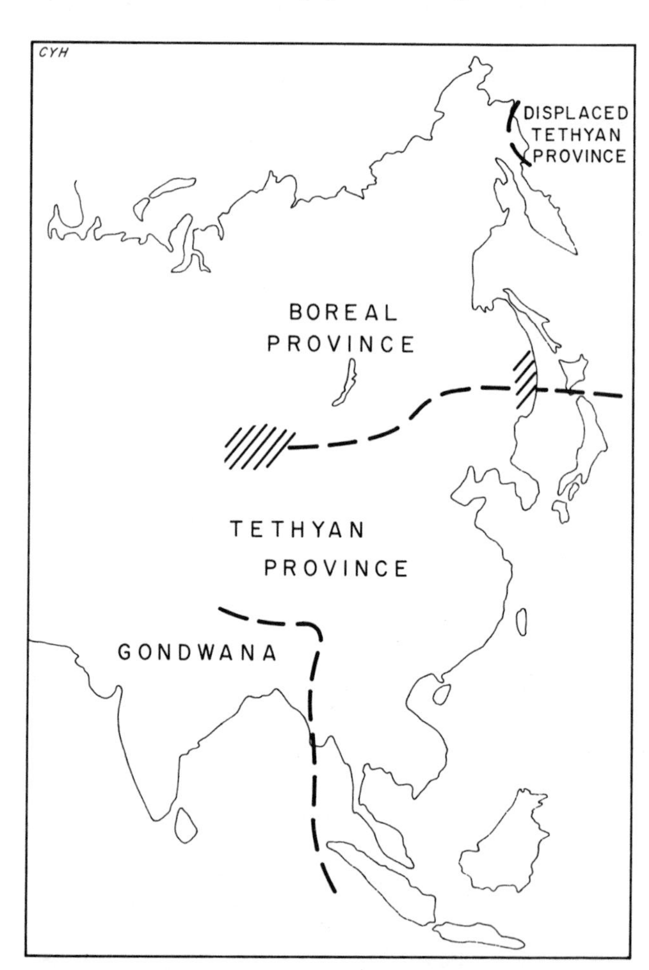

Figure 4. Permian provinces in East Asia. The diagonally hachured zones indicate areas of overlap between Tethyan and Boreal Province biotas. The Gondwana biotas of India were not in the Northern Hemisphere during the Permian and are not considered in this report. Map based on an azimuthal equal-area projection centered on 40° N, 90° E (Sinkiang, China).

this boundary occurs within Mesozoic foldbelts between stable blocks, and a tectonic mixing of rocks with cold non-marine floras of the Angara Province and warm marine biotas of the Tethyan Province results from tectonic displacement during collision and welding of the two blocks. In the Sikhote Alin region, there is also a latitudinal overlap between Angaran cold climate floras and Tethyan marine biotas, where Tethyan verbeekinid fusulinids occur in several parts of the region and Angaran floras occur at the south end of the region. The Sikhote Alin region is structurally very complex and lies within the Pacific margin zone of East Asia where deformation and tectonism have occurred repeatedly, perhaps partly as a result of continental accretion. Permian paleomagnetic results from near Vladivostok (McElhinny, 1973) indicate at least 15° of northward migration for one of the structural units of Sikhote Alin, and the occurrence of rich verbeekinid faunas 4° to 8° north of Vladivostok (Gobbett, 1967) indicates at least 20° to 25° northward movement for other structural units of Sikhote Alin.

The greatest biotic anomaly in East Asia is the occurrence of tropical Tethyan biotas in the Koryak district just north of the Kamchatka Peninsula, at a latitude of more than 60° N, and within the Pacific margin mobile belt. This isolated occurrence far beyond any other Tethyan biota is within 500 km of extensive Permian glaciomarine deposits on the the Kolyma block to the west. This anomaly cannot be explained from Permian paleoclimatic patterns alone, and the rocks containing the biota must have moved northward since the Permian, over a minimum distance of 30° to 35° latitude.

The Koryak anomaly provides compelling evidence of continental accretion within the Pacific margin areas since the Permian. Long distance movement of Permian landmasses within the Pacific Ocean Basin is also suggested in several places on other continents: there are isolated occurrences of distinctive Tethyan biotas in northern New Zealand, British Columbia in Canada and the adjoining Pacific margin of the United States, and in southernmost Mexico (Fig. 5). The Siberian, New Zealand, and British Columbia biotas are highly anomalous with respect to adjacent Permian biotas, and indicate thousands of kilometers of movement that terminate with accretion onto distant continents.

The New Zealand occurrence of Tethyan biotas at 35° S indicates a minimum of 5° to 10° southward movement. Permian biotas in the remainder of New Zealand are cold-water biotas, and these contrasting biotas must have converged over at least 25° of latitude to reach their present positions. The western North American occurrences in the belt from California to northern British Columbia are spread from 35° to 60° N latitude, and various units must have moved at least from 5° to 30° northward. The Mexican occurrences near 15° N are within acceptable latitudinal limits, but the biotas are anomalous in their dissimilarity to other New World Permian biotas. Because Tethyan biotas in the New World occur only within tectonically unstable areas, it is concluded that all occurrences of

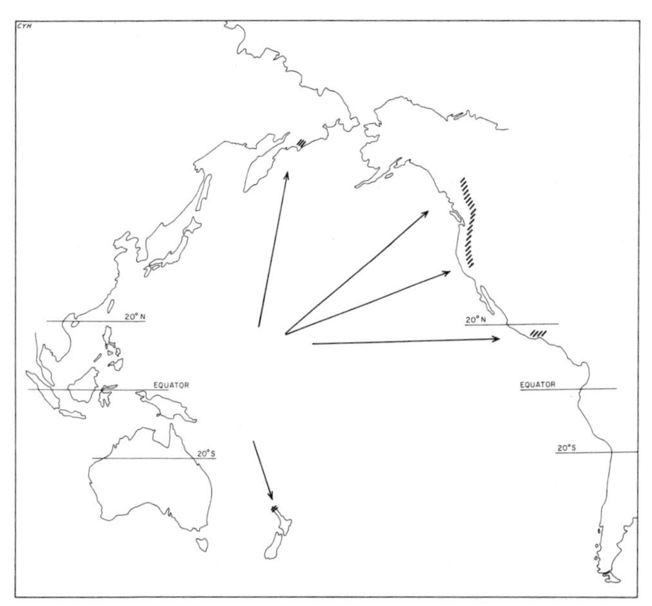

Figure 5. Anomalous occurrences of Tethyan biotas (diagonal hachures) in the circum-Pacific area, and their idealized directions of movement from a postulated western Pacific source area. The 20° N and S latitudes mark the probable limits of the tropical zone and Tethyan Province during the Permian. Locations of Tethyan biotas taken from Gobbett (1967, 1973), Bostwick and Nestell (1967), and Monger and Ross (1971). Map based on a transverse mercator projection with central meridian at 85° W.

Tethyan biotas in the New World are due to later accretion of crustal units onto the continents. The great scatter of Permian Tethyan biotas, especially within the circum-Pacific area, is perhaps due to the dispersion of a tropical Permian archipelago within the Pacific Ocean.

The distribution of rocks containing Permian Tethyan biotas within the Pacific margin areas places some limits on the type of sea-floor spreading that can be postulated for the Pacific Ocean in post-Permian times. The very minor occurrences of Permian Tethyan biotas on the southern continents (one occurrence in New Zealand: Gobbett, 1973) and their abundance in parts of North America and Asia indicates that the major post-Permian spreading axis (assuming a simple case with a single major spreading axis) must have been located to the south of the tropical zone in the early Mesozoic Pacific Ocean. This would mean a position on or south of 25° S latitude. The lack of Permian Tethyan biotas of Asiatic affinities in South America further indicates an absence of southeastern movement across the Pacific Ocean, and a predominant movement of crust to the north and northeast. An early Mesozoic axis of spreading aligned NW-SE would satisfactorily produce this pattern.

Larson and Chase (1972) have outlined the probable patterns of late Mesozoic spreading in the western Pacific Ocean, and these can be related well with postulated earlier patterns of sea-floor spreading. The axis of spreading continues to be aligned approximately NW-SE, and predominant movement of sea-floor crust is to the northeast and northwest. Accretion of Permian rock masses onto the Pacific margin of Asia probably resulted from movement of the Kula plate, while accretion of Permian rock masses onto the Pacific margin of North America is almost certainly related to movement of the Farallon plate. Other plates, since completely disappeared, may have been present in the western part of the Pacific Ocean Basin and may have been involved in westward movement of Permian rock masses.

ACKNOWLEDGEMENTS

I am grateful for the help and stimulation provided by P. H. Stauffer and C. S. Hutchison of the Geology Dept., University of Malaya, and for the critical comments of P. H. Stauffer on the first drafts of this study.

REFERENCES

Abbott, R. T., 1968, Seashells of North America: New York, Golden Press, 280 p.

Bostwick, D. A., and Nestell, M. K., 1967, Permian Tethyan fusulinid faunas of the northwestern United States, *in* Adams, C. G., and Ager, D. V., eds., Aspects of Tethyan biogeography: Systematics Assoc. Pub. no. 7, p. 93-102.

Burrett, C. F., 1974, Plate tectonics and the fusion of Asia: Earth and Planetary Sci. Letters, v. 21, p. 181-189.

Chaloner, W. G., and Lacey, W. S., 1973, The distribution of late Paleozoic floras, *in* Hughes, N. F., ed., Organisms and continents through time: Spec. Papers Palaeontology no. 12, p. 271-289.

Chaloner, W. G., and Meyen, S. V., 1973, Carboniferous and Permian floras of the northern continents, *in* Hallam, A., ed., Atlas of palaeobiogeography: Amsterdam, Elsevier, p. 169-186.

Cooper, G. A., and Grant, R. E., 1975, Permian brachiopods of west Texas, Pt. 3: Smithsonian Contr. Paleobiology no. 19, p. 795-1921.

Gobbett, D. J., 1967, Palaeozoogeography of the Verbeekinidae (Permian Foraminifera), *in* Adams, C. G., and Ager, D. V., eds., Aspects of Tethyan biogeography: Systematics Assoc. Pub. no. 7, p. 77-91.

——— 1973, Permian Fusulinacea, *in* Hallam, A., ed., Atlas of palaeobiogeography: Amsterdam, Elsevier, p. 151-158.

Gorsky, V. P., and Guseva, E. A., 1973, The Kungurian and Ufimian stages of the PriUrals, U.S.S.R., *in* Logan, A., and Hills, L. V., eds., The Permian and Triassic Systems and their mutual boundary: Canadian Soc. Petroleum Geologists Mem. 2, p. 168-172.

Hamilton, W., 1970, The Uralides and the motion of the Russian and Siberian platforms: Geol. Soc. America Bull., v. 81, p. 2553-2576.

Hart, G. F., 1974, Permian palynofloras and their bearing on continental drift, *in* Ross, C. A., ed., Paleogeographic provinces and provinciality: Soc. Econ. Paleontologists and Mineralogists Spec. Pub. 21, p. 148-164.

Hedgpeth, J. W., 1957, Marine biogeography, *in* Hedgpeth, J. W., ed., Treatise on marine ecology and paleoecology. Vol. 1. Ecology: Geol. Soc. America Mem. 67, p. 359-382.

Høeg, O. A., 1935, The lower Permian flora of the Oslo region: Norsk Geol. Tidsskr., v. 16, p. 1-43.

Kremp, G. O. W., 1974, A re-evaluation of global plantgeographic provinces of the late Paleozoic: Rev. Palaeobotany and Palynology, v. 17, p. 113-132.

Kropotkin, P. N., 1971, Eurasia as a composite continent: Tectonophysics, v. 12, p. 261-266.

Larson, R. L., and Chase, C. G., 1972, Late Mesozoic evolution of the western Pacific Ocean: Geol. Soc. America Bull., v. 83, p. 3627-3644.

Logan, A., and McGugan, A., 1968, Biostratigraphy and faunas of the Permian Ishbel Group, Canadian Rocky Mountains: Jour. Paleontology, v. 42, p. 1123-1139.

McElhinny, M. W., 1973, Palaeomagnetic results from Eurasia, *in* Tarling, D. H., and Runcorn, S. K., eds., Implications of continental drift to the Earth sciences, Vol. 1: London, Academic Press, p. 77-85.

McElhinny, M. W., Haile, N. S., and Crawford, A. R., 1974, Palaeomagnetic evidence shows Malay Peninsula was not a part of Gondwanaland: Nature, v. 252, p. 641-645.

Mikhaylov, Y. A., Ustritskiy, V. I., Cheryak, G. Y., and Yavshits, G. P., 1970, The upper Permian glaciomarine sediments of the northwest U.S.S.R.: Akad. Nauk SSSR Doklady, v. 190, p. 100-102 (in Russian).

Minato, M., and Kato, M., 1965, Waagenophyllidae: Hokkaido University, Jour. Fac. Sci., Geol. and Min., v. 12, p. 1-241.

Monger, J. W. H., and Ross, C. A., 1971, Distribution of fusulinaceans in the western Canadian Cordillera: Canadian Jour. Earth Sci., v. 8, p. 259-278.

Mudge, M. R., and Yochelson, E. L., 1962, Stratigraphy and paleontology of the uppermost Pennsylvanian and lowermost Permian rocks in Kansas: U. S. Geol. Survey Prof. Paper 323, 213 p.

Nalivkin, D. V., 1973, Geology of the U.S.S.R.: Edinburgh, Oliver and Boyd, 855 p. (English translation by N. Rast from the 1962 Russian publication.)

Pattison, J., Smith, D. B., and Warrington, G., 1973, A review of late Permian and early Triassic biostratigraphy in the British Isles, *in* Logan, A., and Hills, L. V., eds., The Permian and Triassic Systems and their mutual boundary: Canadian Soc. Petroleum Geologists Mem. 2, p. 220-260.

Rogers, J. J. W., and others, 1974, Paleozoic and lower Mesozoic volcanism and continental growth in the western United States: Geol. Soc. America Bull., v. 85, p. 1913-1924.

Ross, C. A., 1967, Development of fusulinid (Foraminiferida) faunal realms: Jour. Paleontology, v. 41, p. 1341-1354.

——— 1974, Paleogeography and provinciality, *in* Ross, C. A., ed., Paleogeographic provinces and provinciality: Soc. Econ. Paleontologists and Mineralogists Spec. Pub. 21, p. 1-17.

Rudwick, J. J. S., and Cowen, R., 1967, The functional morphology of some aberrant strophomenide brachiopods from the Permian of Sicily: Soc. Paleontológica Italiana Boll., v. 6, p. 113-176.

Silberling, N. J., 1973, Geologic events during Permian-Triassic time along the Pacific margin of the United States, *in* Logan, A., and Hills, L. V., eds., The Permian and Triassic Systems and their mutual boundary: Canadian Soc. Petroleum Geologists Mem. 2, p. 345-362.

Smith, A. G., Briden, J. C., and Drewry, G. E., 1973, Phanerozoic world maps, *in* Hughes, N. F., ed., Organisms and continents through time: Spec. Papers Palaeontology no. 12, p. 1-42.

Stauffer, P. H., 1974, Malaya and southeast Asia in the pattern of continental drift: Geol. Soc. Malaysia Bull., v. 7, p. 89-138.

Stauffer, P. H., and Gobbett, D. J., 1972, Southeast Asia a part of Gondwanaland?: Nature, v. 240, p. 139-140.

Stehli, F. G., 1964, Permian zoogeography and its bearing on climate, *in* Nairn, A. E. M., ed., Problems in palaeoclimatology: London, Wiley-Interscience, p. 537-549.

——— 1971, Tethyan and Boreal Permian faunas and their significance, *in* Dutro, J. T., Jr., ed., Paleozoic perspectives: A paleontological tribute to G. Arthur Cooper: Smithsonian Contr. Paleobiology no. 3, p. 337-345.

——— 1973, Permian brachiopods, *in* Hallam, A., ed., Atlas of palaeobiogeography: Amsterdam, Elsevier, p. 143-149.

Stepanov, D. L., 1973, The Permian System in the U.S.S.R., *in* Logan, A., and Hills, L. V., eds., The Permian and Triassic Systems and their mutual boundary: Canadian Soc. Petroleum Geologists Mem. 2, p. 120-136.

Sutherland, P. K., and Harlow, F. H., 1973, Pennsylvanian brachiopods and biostratigraphy in southern Sangre de Cristo Mountains, New Mexico: New Mexico Bur. Mines and Mineral Resources Mem. 27, 173 p.

Ustritskiy, V. I., 1973, Permian climate, *in* Logan, A., and Hills, L. V., eds., The Permian and Triassic Systems and their mutual boundary: Canadian Soc. Petroleum Geologists Mem. 2, p. 733-744.

Waterhouse, J. B., 1967, Cool-water faunas from the Permian of the Canadian Arctic: Nature, v. 216, p. 47-49.

Yancey, T. E., 1975, Permian marine biotic provinces in North America: Jour. Paleontology, v. 49, p. 758-766.

Yole, R. W., 1963, An Early Permian fauna from Vancouver Island, British Columbia: Canadian Petroleum Geology Bull., v. 11, p. 138-149.

Zijderveld, J. D. A., and van der Voo, R., 1973, Palaeomagnetism in the Mediterranean area, *in* Tarling, D. H., and Runcorn, S. K., eds., Implications of continental drift to the Earth sciences, Vol. 1: London, Academic Press, p. 133-161.

The Role of Fossil Communities in the Biostratigraphic Record and in Evolution

J. B. Waterhouse, *Department of Geology, University of Queensland, Brisbane, Australia*

ABSTRACT

Species-based fossil communities are likely to have played a critical role in the evolution of life because they contributed to the stability of species through various homeostats. Under stress conditions, generally caused by episodic climatic change, fossil communities broke down during enforced mass migration which allowed the development of new communities from a mixture of immigrants, persistent species, and new species that developed as isolated subpopulations. The fossil record thus shows a succession of biozones, each of which incorporates contemporaneous fossil communities and may be correlated around the world as substages. The boundaries between biozones are generally abrupt, not because of an imperfect record, but because of abrupt climatic changes reflecting changes in energy output by the sun, or in the rotation of the Earth, leading to rapid geographic shifts of biomes and provinces. This theory, called causal biostratigraphy, provides a natural explanation of the stratigraphic record without recourse to special pleading for gaps, and is in good accord with genetic and ecologic evidence.

INTRODUCTION

When the theory of evolution was enunciated by Darwin and Wallace, not for the first time, but at last in a respectable and well-documented form, insight into the evolutionary truth was hampered by ignorance about several major aspects. Of these, we are now, of course, well aware of the role played by genetics in offering substantial clarification of the theory. A second aspect is concerned with ecology. Both Darwin and Wallace were familiar with the variation of life across the globe and the importance of environmental parameters, but it is doubtful whether Darwin or his colleagues appreciated the importance of the way that species are associated in communities, and the relevance this had to species evolution and success. Darwin (1859, p. 286; *in* Darwin and Seward, 1903, p. 102) was thus able to write with considerable misapprehension on the question of the success of European species introduced into the foreign realm of New Zealand: "A multitude of British forms . . . would exterminate the natives." To Darwin, such species were about to outcompete, conquer, and replace the indigenous forms. We now know, from experience, that this prediction was false. Only man, with considerable input of energy and various undesired side effects, is able to sustain such intruders, especially amongst plant life. Indeed, this very arrival of immigrants, now under the hand of man, could have suggested to Darwin a parallel with climatically induced migration that successfully introduced new species in the past. But other than a few successful species, and apart from man's intervention, the local communities, rather than individual species, drive out the European intruders. In other words, communities appear to substantially restrain the spread of species and even evolution of species and will often inhibit the entry of any new species into the community.

We may add a second lesson from nature that Darwin would have had no opportunity to assess—that communities which are closely interlocked and adapted to the environment may be disrupted with damage to species by massive physical change. The building of the Aswan Dam in Egypt is a man-made effect comparable locally to a major climatic change or tectonic disruption. Over an interval of only a few years the dam has led to changes in flooding and salinity and to erosion of the coast. It has led to the increase of *Bilharzia,* malaria and trachoma, or their carriers, has affected life and distribution of snails, has led to diminished blooms of phytoplankton and zooplankton, and has substantially reduced marine fish (Collier and others, 1973, p. 4-11). Thus, it may be inferred that communities which protect constituent species may be disrupted by massive physical change and this in turn will affect the distribution and numbers of individuals in species. Moreover, sustained disruption of the environment and communities adopted to that environment, will facilitate the entry of new species.

Finally, there must be reservation about Darwin's views on time, the fossil record, and Earth history. Darwin set aside most of the fossil evidence on evolution with the proposal that it was massively incomplete. But there were polemic rather than scientific reasons for this attitude because he insisted on gradualistic evolution which most fossils did not substantiate. But the fossil record can no longer be set aside as woefully incomplete. More than 100 years of study demand instead that the gradualistic concept be reassessed. Furthermore, paleontological evidence strengthens the proposal that biotic associations have played a critical role in the evolution of life, and, therefore, in biostratigraphy.

FOSSIL COMMUNITIES, THEIR DEFINITION AND REALITY

A biotic community is broadly defined as "any assemblage of populations in a prescribed area of physical habitat" (Odum, 1959), but the prescription for delineating fossil communities lacks hard and fast rules. Perhaps the most clearly defined fossil communities would be the Upper Silurian ones recognized in the Welsh Borderland by Ziegler and others (1968). Each of these were said to contain several distinct species in constant proportions. Moreover, the communities were recurrent, helping to confirm their reality. It appears, from published claims, that a rigid and typological approach could be applied to these examples, though I have never myself known fossil communities to display so little variation. Indeed, Boucot (1975) has now reinterpreted the Silurian fossil communities of the Welsh Borderland as belonging to different benthic assemblage zones, and has shown that there is variation in the proportions of constituent species. Fossil communities that may better stand the test of time have to permit variation and flexibility, if they are to reflect the reality of which they represent the preserved and examined segment. Firstly, fossil communities must be based on species, a point stressed by Waterhouse (1973a), who suggested that "generic communities" should be discriminated. The easiest way, and certainly the most amenable to simple statistical analysis, is to focus attention on the species that either has the highest numbers or the largest biomass. Paleontologists have generally concentrated on individual numbers, but it is relatively simple to estimate biomass volumetrically by using plasticine models. Of course, these primary species may or may not be dominant, for the extent to which primary species affect other species is difficult to assess in many instances, although it is clear that species which build up substrate and contribute to shelter, such as corals, must have had a dominant role. As an alternative to preponderance in numbers or biomass, some authorities use key and discriminant species to distinguish communities, but this procedure may be more open to misinterpretation and it is perhaps more biostratigraphic than paleoecologic in technique.

The Reality or Otherwise of Fossil Communities

After Clements (1916) had conceived of an organismic complex of rigidly discrete plant communities in each of which the individual species were integrated into virtually a super organism, he almost inevitably provoked a starkly alternative concept, the individualistic model, which recognized extremely wide variation and no interplay between primary and associated species (Gleason, 1939). The ensuing debate has resulted, to some degree, in abandonment of the extreme positions and in some measure of agreement that primary species do influence associated species at least in terms of shelter and other parameters, and that there is some degree of reality to communities. I would base a cautious acceptance of communities on these reasons.

(1) In the marine world today, from personal observation, sand-dwelling communities quite clearly and sharply differ in terms of species and numbers from those of a gravel or rocky substrate. Similarly, in moving up a mountain side or out into a desert one sees quite abrupt and often total changes in floral association.

(2) Paleontologists working on a world scale find that associations of fossils are frequently recurrent, and of finite variation with often only a handful of species that oscillate in predominance, and a host of less important species. For the Permian of New Zealand, Waterhouse (1973a) found that several hundred species occurred in some fifty recurrent communities and subcommunities (defined in terms of numerical preponderance of species). In analyzing biosociological units, it is necessary to avoid the facile trap of considering too small a region and too short an interval of time. It is striking that Gleason (1939), who opposed the biologic reality of communities, considered only a segment of the United States severely disturbed by man; one may also point to marine biologists who analyse a relatively small area. The compilation of mid-Paleozoic fossil brachiopod communities and benthic associations by Boucot (1975) strongly suggests that there is more than just anthromorphic bias. But our analyses need to become more sophisticated to cope with associations which share a handful of species which vary in numerical or biomass preponderance, and it is clearly advisable to consider all fossil forms and not just the predominant phylum which so often has been brachiopods.

Communal Hierarchy

Terminology for communal associations is both complicated and unsettled, not surprisingly, because it is desirable to stress territorial or biosociological aspects or a mixture of both (Waterhouse, 1976b). Boucot (1975) assembled mid-Paleozoic brachiopod groups from different facies into major benthic assemblages that were in the main depth-correlated, and these in turn were grouped into faunal provinces. The benthic assemblage finds some counterpart with the "supercommunity" of Waterhouse (1973a) which embraces several communities that shared significant species. Each species was found to be preponderant in one set of communities, and occupied a significant part of other contemporaneous communities. In this latter concept and observation, the biosociological role is clearly important, and the type of community varied both with substrate and with depth. It is not clear whether the supercommunity should be equated with one or with several benthic assemblages. The long-time paleontological stress on fossil provinces and realms of biogeographic significance is well-known, and Williams (1973) has used Ordovician brachiopod genera and Jell (1974) has used Cambrian trilobite genera to analyse provincial implications, as well as perhaps depth, through Q-mode cluster analysis. For Permian marine life, Waterhouse and Bonham-Carter (1975) delimited marine groupings (which they considered to be fossil biomes on the basis of distribution and association of all known occurrences of Permian brachiopod families) with t-tests for association with other significant groups such as rugose corals and Fusulinacea. From distribution, associated rock type, oxygen-isotope data on temperature, and paleomagnetic evidence, the three major groupings were assigned to polar, temperate, and tropical biomes.

MAJOR PARAMETERS AFFECTING COMMUNITIES AND COMMUNAL ASSOCIATIONS

Early and simplistic attempts to relate the distribution and composition of fossil communities to variations of one parameter, such as substrate, salinity, or depth, have been replaced by a more realistic and cautious appraisal which attaches importance to a wide and varying range of parameters that interact and act directly to varying degree in different places on different species. As an initial and restricted generalization, it appears that within the marine world and within limits imposed by climate (and, therefore, to some degree, depth), species of brachiopods varied numerically according to their limits of distribution, the type of associated species, the substrate, and their mode of attachment. Amongst Permian Brachiopoda, for example, members of the Spiriferida with a delthyrial pedicle, or spinose Chonetacea and Strophalosiidae, were relatively independent of substrate, whereas the Productacea in particular appear to have evolved towards specialization for substrate (Waterhouse, 1973a). Of course, the role of unpreserved parameters should not be forgotten and some of these, notably plants and food supply, may have contributed substantially to the entity, reality, and homeostasis of fossil communities and may have been sensitive, perhaps more than the shellfish, to chemical and temperature regimes.

THE COMMUNITY IN BIOSTRATIGRAPHY

Although various geochemical and geophysical techniques are rich in promise for stratigraphic correlations, the burden of correlation, especially for Paleozoic rocks, still falls on fossils, because these appear to be much more precise and also far more resistant to change, whereas the other calibrations have proved highly sensitive to relatively slight but unpredictable initial and subsequent differences and changes in chemistry, temperature, pressure, and magnetic fields which often have to be subjectively "allowed for" or "adjusted." The apparent precision of numerical values is all too often spurious. Fossil collections, therefore, are critical for correlation and for timing the history of the Earth. These collections are generally samples of fossil communities affected by post-mortem transport, post-burial loss, and incomplete preservation to varying degrees that can be statistically assessed. For delineating biozones, which provide the key and basic unit for correlation (rather than stages which are broad, imprecise, and made up of several biozones), the paleontologist assembles correlative and adjacent fossil communities (Waterhouse, 1976b). Collections from a single community are normally assigned to a single biozone where sampling or preservation is substandard. If each different community were each assigned to a different biozone, we would end up with a confused plexus of repeated zones of no temporal value; the zones would be simply correlating parameters such as substrate or depth. Instead, the biostratigrapher strives to correlate fossil communities, as represented by collections from a single or associated suite of contemporaneous benthic assemblages, or from a single supercommunity. In that way, the biostratigrapher will arrive at the association of contemporaneous fossil communities that inhabited various depth zones and various substrates, and at other communal assemblages within a region, usually within a province. Specially distinctive communities may be distinguished as subzones. *If* sampling, or preservation, or study is inadequate, of course, only the data from one community may be available, and in that case only a subzone is known. Then correlation potential and overall knowledge is severely restricted unless we can be sure of comparing it with communities of exactly the same environment.

For example, I am currently studying the Permian of northwest Nepal which contains a series of fossil communities that may be assembled in two zones, the *Lamnimargus himalayenensis* Zone, followed by the *Echinalosia kalikotei* Zone. Between the two biozones are thin quartzites with a single bivalve fossil community rich in the burrowing bivalve *Pyramus*. No comparable substrate, or fossils, are found in either the underlying or succeeding zone. Because of the uniqueness of the substrate, we have a well-defined subzone, that, in the absence of outside evidence, belongs either to the preceding or succeeding zone, or to an intervening zone. Present evidence does not allow a decision to be made, and the fauna is best treated as *Pyramus* subzone and fossil community, zone not certain.

Of course, these suggestions are emphatically *not* understood by the present International Stratigraphic Commission (Hedberg, 1976). The stratigraphic code ignores paleoecology and as a result is confused about the role of fossil communities in biostratigraphy. This in turn has led to a confused and very restricted model for biozones. Fortunately, the practice of correlation has far outstripped the formal codification and philosophy, to the degree that it *works*, and so is used by oil and other companies for industrial as well as scientific purposes.

SPECIES AND FOSSIL COMMUNITIES IN THE STRATIGRAPHIC RECORD

In any extended stratigraphic sequence of one area, ranging through at least half a period, the fossil record will reveal a series of abrupt changes to both species and

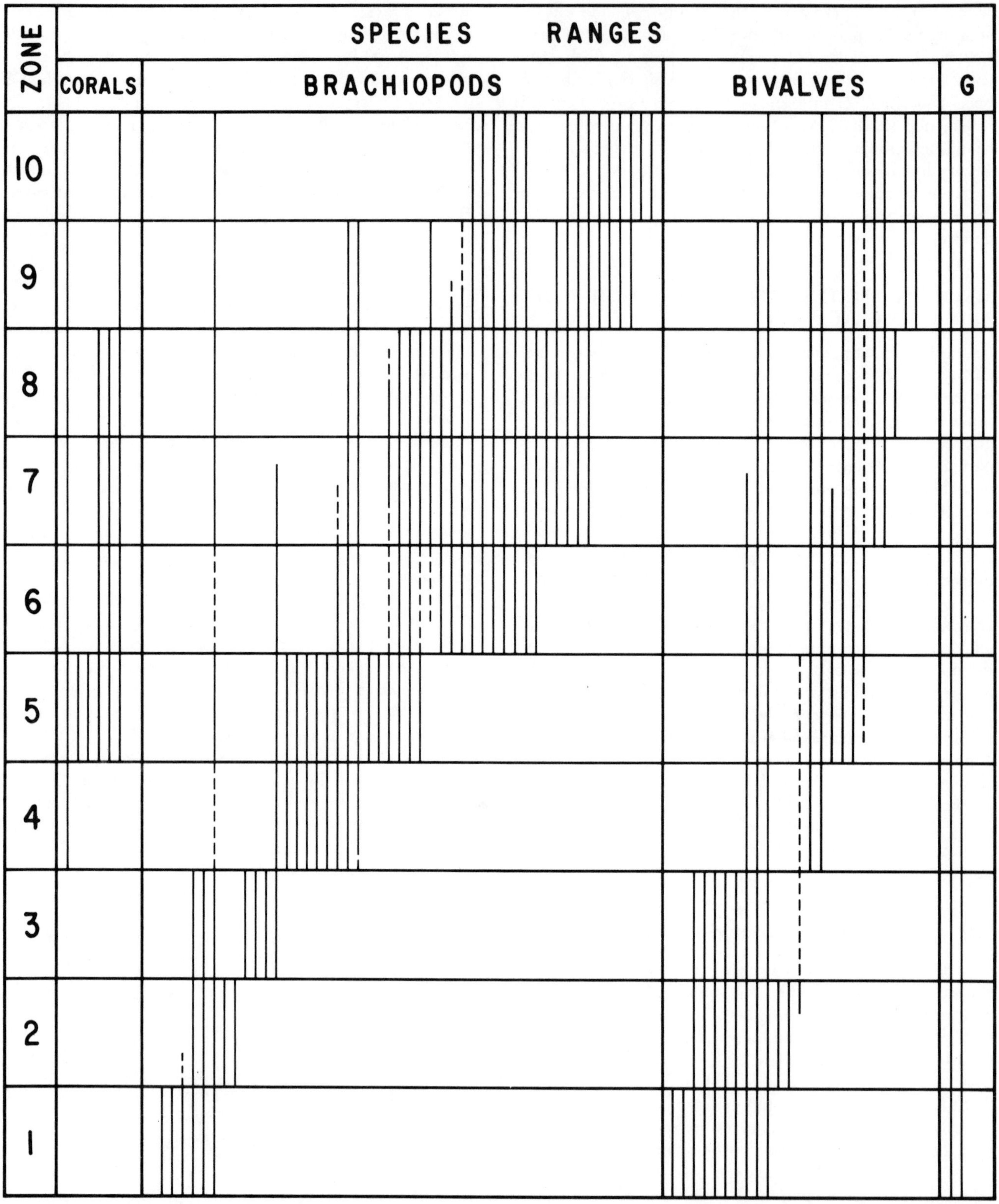

ZONE
SPECIES RANGES
CORALS
BRACHIOPODS
BIVALVES
G
10
9
8
7
6
5
4
3
2
1

Figure 1. Range chart for species (one per line) in the Permian of Tasmania (redrawn from Clarke and Banks, 1975, Fig. 33.2). Corals includes Anthozoa, Bryozoa, and sundry; (G) gastropods.

fossil communities (Fig. 1). Certainly this is true of the Carboniferous to Triassic Periods, with which I am most familiar, and a survey of literature suggests that the pattern is applicable throughout the history of life. The pattern is graphed by Waterhouse (1976a, Figs. 6 to 8) for the Permian of the Yukon Territory, New Zealand, and Afghanistan, the latter from data presented by Termier and others (1974). Recently, Clarke and Banks (1975) have presented much the same picture for the Permian of Tasmania although the changes are less striking perhaps because of the polar station Tasmania occupied during the Permian Period (Fig. 1). Plotting of the 1,000 species of a paleotropical biome being described by Cooper and Grant (e.g., 1976) from the Glass Mountains, would show very marked changes. The fossil record substantiates these observations on the evolution and succession of fossil species (Fig. 2):

(1) Most (but not all) species enter a succession fairly abruptly, usually close to the level at which underlying species disappear with almost comparable abruptness. The change coincides with a lithologic change only in some instances, but lithologic change is usually accompanied by change in species.

(2) Most species do not show morphological clines other than in random directions. The proportions of morphotypes may change a little up the column, but again not regularly or continuously unless there is obviously a gradual concomitant change in substrate, which suggests an ecocline. In this instance, the concoction of a cline from several discrete populations separated by long barren intervals may be discounted because such populations could equally well be linked by step-wise (ascensive) change.

(3) Successor species in persistent genera have not always evolved from the preceding species of the same genera in the region. They may be more closely allied to other species from different regions.

(4) Within the limits of present study, fossil communities changed just as abruptly, and as completely, as fossil species. This, of course, is to be expected, because the nature of communities depends on the nature of species of which they are composed. Some work on New Zealand Permian species confirms this (Waterhouse, 1973a), and a lengthy study of Permian communities in Nepal adds better confirmation. Unfortunately, the data require so much space and tabulation that it is impossible to set it down within the prescribed limits of the present paper.

(5) Within the limits of correlation, the changes appear to have been worldwide and synchronous. This is where a number of contemporaneous fossil communities and constituent species should be examined, because use of only one phylum—especially temperature-sensitive fossils, or study of only one or a few communities must frustrate adequate correlation.

Attention to aspects of the first two points has been focused by Eldredge and Gould (1972), who stated that they could find few examples in the fossil record of gradual morphologic change or what they called "phyletic gradualism." Instead, the fossil record appeared to show lengthy intervals of stability, punctuated by brief intervals of change, as especially well-documented for Pleistocene snails of Bermuda (Gould, 1969). To Eldredge and Gould (1972) the answer lay in peripheral speciation through isolation, especially on islands remote from the main gene pool, as enunciated by Mayr (1942) and indeed realized by Darwin (*in* Darwin, 1887, p. 159) and Wallace (*in* Darwin and Seward, 1903, p. 294). For it is clear that exceptional circumstances are required to preserve and proliferate any genetic change, no matter how advantageous, against the almost overwhelming tendency inherit in the gene pool to stabilize populations, reduce heteromorphy, and frustrate speciation. This strong thrust towards species stability is further reinforced by the conservative tendencies of the communal framework. Under these constraints, any new species, whether newly evolved or a migrant, must find it extremely difficult to penetrate an existent community. In spite of the reality of inter- and intra-specific competition for space, food, shelter, and light within an established community, there is also likely to be a high degree of interdependence, including adaptation or at least tolerance for cohabitation, and making viable, if not the best, use of local resources. Any new entry has to be very fortunate, or extremely vigorous, to

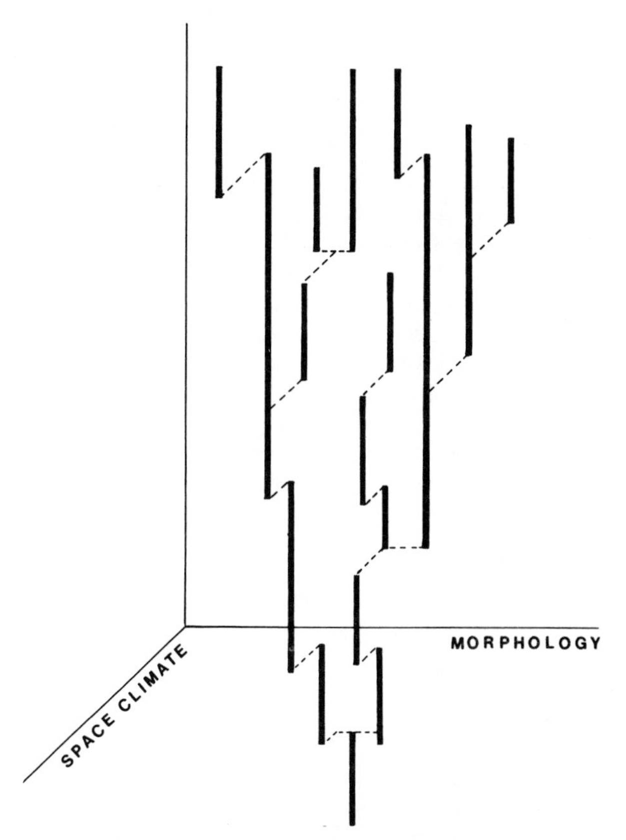

Figure 2. Scalariform or dendritic model for species evolution suggesting lengthy intervals of stability during which species and communities change little, punctuated by brief, climatic caused episodes of change.

penetrate established communities. It happens, as we know from observation and from the fossil record, but it is relatively rare. We may thus deduce that for most of the fossil record, for the duration of the biozones that have been recognized, evolution was relatively ineffectual because gene pools and community homeostasis resisted change. Species do not show clines, and communities did not change substantially, for intervals lasting for some 0.5 to 3 m. y. There are, of course, exceptions, which include the occasional superfit new species and the occasional successful migrant, sometimes replacing an underfit form. Slowly proceeding transgressions and regressions may have given rise to open and new situations suited for speciation and demise. Continental convergence and divergence, as summarized by Hallam (1974), also seriously affected species and communities to disrupt the normal pattern, although the speed at which plate movement occurred was clearly too slow, and the movements too erratic to explain the pattern of synchronous and abrupt changes to species and communities all over the world.

CAUSAL BIOSTRATIGRAPHY

For evolutionists of the 19th Century, the fossil record of more-or-less synchronous episodes of change with intervals of stability was to be dismissed as due to faulty preservation or nondeposition. To Darwin (1859, p. 412) "The noble science of geology loses glory from the extreme imperfection of the record." But these assertions are less convincing after another century of study. Synchronous world-wide faunal changes cannot be all due to unconformities, which should have been independent of changes of life, and the mystical gaps in the record would have been plugged, at least in many instances, if they were truly random. Darwin was obsessed with gradualism in biology, partly because this would conform with the fixest Earth model of steady up-down movement of oceans and continents called "uniformitarianism." As expressed by Hutton and Lyell, this would permit extremely slow and gradual changes in geology, with permanent oceans and continents, and certainly no plate tectonics (or even a fault scarp). In particular, the geological aphorism "no vestige of a beginning, no prospect of an end" was applied with relish to species. Yet from that day to this, biologists have been puzzled by the problem of "how discontinuity of groups is introduced into the biological continuum" (Huxley, 1940, p. 2). Biologists have dismissed the paleontological discontinuities as irrelevant to the biologic discontinuities, instead of perceiving that the two could be related, if not one and the same phenomenon.

Climatic Change and Mass Migration

It is not so easy to deny the fossil record today, other than on the basis of our long taught precepts based on the need for gradualism and the falsehood of catastrophism, but the reasons for episodic and apparently world-wide change are not obvious. We cannot appeal to changes in substrate or depth, or salinity, or to peripheral evolution, or to continental drift, for such physical changes are also of local extent, and most would have operated either continuously or erratically. Nor is there any obvious biologic cause because species show no clines for so much of their history. The most likely cause lies in climatic change. A glacial episode must have affected much of the life of the world, in the first place by causing wholesale migration of species towards the equator, in other words, shifting the geographic position of the biomes. We know this occurred during the Pleistocene episodes, and the analyses of family distributions of Permian brachiopods by Waterhouse and Bonham-Carter (1975) shows that the same massive shifts of faunas occurred during the Permian glacial intervals, with some confirmation from oxygen-isotope temperature analyses and diversity.

Permian Model

The model obtained for the Permian Period (Fig. 3), based on computer and diversity analyses, strongly supports the concept of a rapid, episodic shift of the biomes, and therefore all biota, towards and away from the poles concomitant with fluctuating temperatures that lead to a series of glaciations and interglacial stages and substages. The picture is supported by more detailed study. During warm episodes, glacial deposits are rare or absent. Indeed, Loughnan (1975) established that subtropical forest and soil-profiles developed during interglacials in the Hunter Valley, Australia, so that temperatures must have fluctuated from polar to subtropical. Tropical life expanded, and Fusulinacea especially plus various corals and brachiopods with tropical attributes, extended into high latitudes. During cold intervals, often ascertained by the recognition of glacial deposits in correlative beds, the tropical life-forms retreated, coral reefs diminished, and cold-water species and genera in particular entered waters of low paleolatitudes. These observations are summed from numerous examples set out in Waterhouse and Bonham-Carter (1975) and Waterhouse (1976a).

Krassilov (1974, 1975) has also pointed out that the succession of fossil ecosystems has been controlled mainly by climatic cycles. Particular attention has been paid to the Late Cretaceous floras of the Northern Hemisphere, which showed successive phytoclimatic changes, as well as some influence from orogenic and volcanic activity.

It is perhaps ironic that Darwin himself (1859, p. 250) suggested that some of the abrupt appearances and disappearances of fossils might be due to "a large amount of migration during faunal and other changes." Exactly, and if he (and especially Huxley) had followed this line, and been less concerned with attacking the fossil record for its failure to show gradualistic evolution, it might have been possible to graft a natural theory of evolution convincingly onto the paleontological observations made by Cuvier, Sedgwick, and Agassiz, and other professional paleontologists of those times.

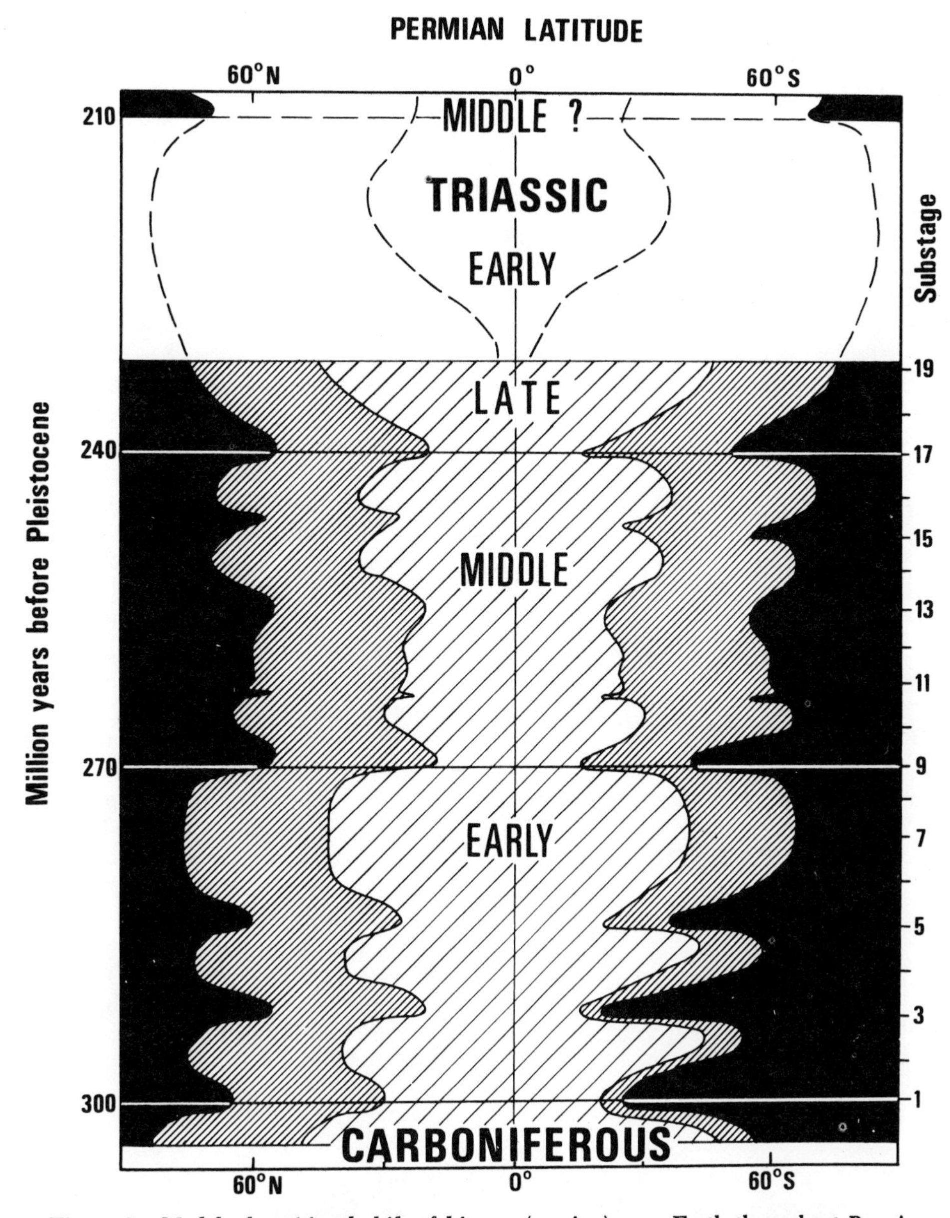

Figure 3. Model of positional shift of biomes (marine) over Earth throughout Permian Period, with north and south poles (N, S), and equator (0°). Time progresses upwards, zone (=substage) by zone. The model is based as far as possible on fossil distributions, assessed by Q-mode cluster analysis (Waterhouse and Bonham-Carter, 1975), with stratigraphic, climatic, and radiometric data summarized in part by Waterhouse (1976a).

Results of Climatic Change

When climates changed, it appears that cool times commenced fairly abruptly. It is easy to delineate the effects of glaciation on the marine environment: the depth of water changed and perhaps the substrate; the salinity and chemical content of the water and substrate may have been affected. It is probable that the food base was especially affected. Quite apart from the effect on individual species, entire communities must have been directly and indirectly affected to a drastic degree, to the extent that communal homeostasis broke down because of the change if not loss of food, the emigration of at least some species, the immigration of others, and the death of yet others. It is during these times of climatic change leading to community breakdown and species migration that there were the best opportunities for survival of mutants and new speciation. Certainly, "peripheral isolates" and subpopulations could have sprung up widely under such confused conditions.

At a simple level, it would seem unlikely that entire communities could have migrated *in toto*. It is to be expected that species would show different rates of migration, as reflected by differing longevity of larval stages, so that new communities would consist of a mixture of some that stayed unmoved (or were able to expand their domain during the time of change), of various newly evolved forms, and of some migrants into the area. The communities of a comparable station and more-or-less similar environmental parameters would have lost some species by both outward migration and death. However, this simple postulation may require qualification. It would appear that the tropical forms of life could at least survive, though not expand, under normal climatic change, and so were under less necessity for migratory mobility, whereas temperate and especially polar life forms would have such selective pressure in favor of mobility that ultimately all such life forms would become either relatively mobile, or relatively tolerant of temperature change. It does seem that this has indeed occurred. The great life crisis at the end of the Permian Period removed much life peculiar to the paleotropics rather than the poles (Waterhouse, 1973b), and amongst various life forms today the Bivalvia are clearly more mobile and conceivably more eurythermal than Brachiopoda which they have to some degree replaced.

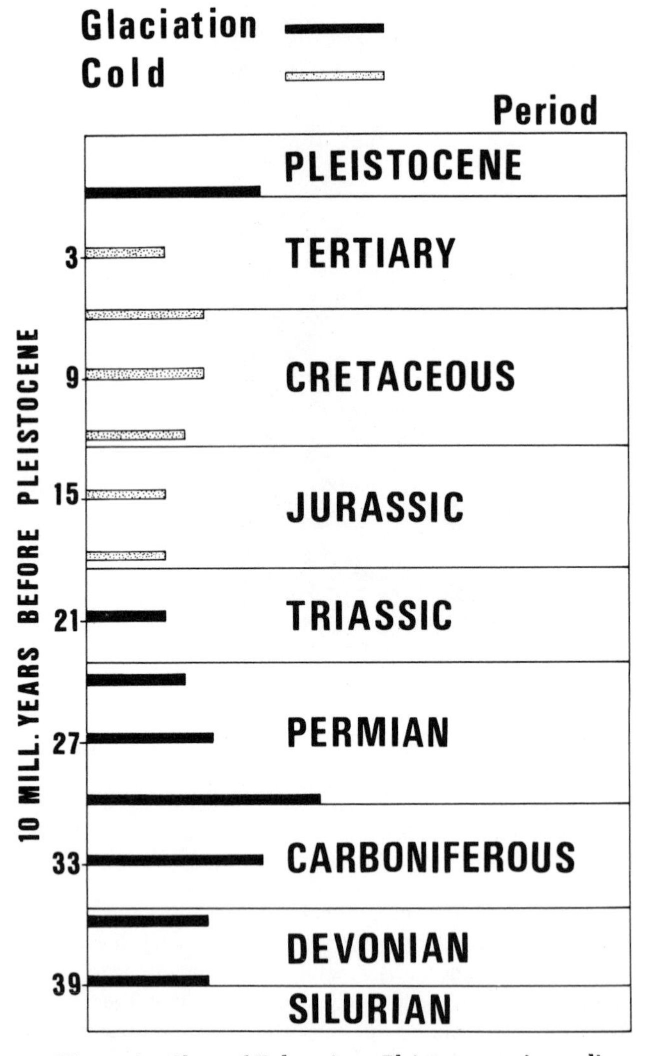

Figure 4. Chart of Paleozoic to Pleistocene major coolings (stippled bars) or glaciations (black bars) summarizing radiometric evidence largely set out by Hulston and McCabe (1972, Appendix 2), with some emendations to incorporate modern work; glacial episodes as in part discussed by Mahlzahn (1957) and Waterhouse (1974); and oxygen-isotope low temperature values based on oxygen-isotope data largely summarised from Bowen (1966).

But these cautions do not substantially alter the main thesis: life was still susceptible to climatically induced change even during the Mesozoic and Tertiary Eras. It is now becoming established that the Earth has endured substantial climatic oscillations throughout its history. Glaciations have affected the globe from the Late Ordovician to the Late Permian at what appear to be very rhythmic and regular cycles of 30 m. y. for the major episodes (Fig. 4). Earlier glaciations are less well-dated, but could have fitted into the rhythm. The Mesozoic record is different in kind, but suggests a similar rhythmicity of cooling for the Upper Jurassic and Cretaceous based on oxygen-isotope temperatures. There is a diamictite of Anisian age in New Zealand which would fall into the 30 m. y. pattern, though its glacial origin has not yet been confirmed, and the Early Jurassic cooling is poorly based, although there have been reports of it based partly on analyses of plants. The mid-Tertiary cooling, based on biotic changes and oxygen-isotope studies, may prove to be synchronous with tillite recently found in Ross Sea cores. In addition, minor subcycles may be recognised, and there was clearly a pattern of greater duration with major glaciations at intervals of 300 m. y. The causes may lie in variation in output of energy by the sun (Öpik, 1967), or in the obliquity of the Earth's orbit as discussed by Williams (1972).

CONCLUSIONS

The model outlined above differs substantially from the Darwinian and neo-Darwinian theory of evolution, and indeed falls closer to the pattern of development of life recognised from the fossil record by Cuvier and Agassiz. Yet it is flexible enough to recognize cases of gradual evolution with ecoclines and temporal clines, and certainly incorporates present understanding of genetic theory. In such a theory of evolution some of our philosophical distinctions must become blurred. Since time is to be measured by episodic changes to species and communities caused by climatic change, the changes have ecologic as well as morphologic reality. Moreover, successive intervals in time are marked off by successive spatial changes of biota, for time is, by my definition, change in space, or successive space, without any need to appeal to "flowage" of Newton or the nonsensical fourth dimension. The most satisfying point of all is the measure of reality to be restored to the paleontological record which has been under fire ever since the days of Darwin and Huxley because of its alleged arbitrary and inconstant or unpredictable nature, and because of the supposed relative rather than absolute nature of its calibrations. Now, biostratigraphy becomes as absolute, as sensitive, and as cumulative as other forms of sun-based time. The day and the year are essentially the same in kind as the substage (which incorporates correlative biozones), with a rhythmicity and periodicity that although, undoubtedly, likely to be complex, may prove just as amenable to analysis and prediction.

REFERENCES

Boucot, A., 1975, Evolution and extinction rate controls: Amsterdam, Elsevier, 427 p.

Bowen, R., 1966, Paleotemperature analysis: Amsterdam, Elsevier, 265 p.

Clarke, M. J., and Banks, M. R., 1975, The stratigraphy of the lower (Permo-Carboniferous) parts of the Parmeener Super-Group, Tasmania, *in* Campbell, K. S. W., ed., Gondwana geology: Canberra, Australian Natl. Univ. Press, p. 453-467.

Clements, F. E., 1916, Plant succession, an analysis of the development of vegetation: Carnegie Inst. Washington Pub. 242, 512 p.

Collier, B. D., Cox, G. W., Johnson, A. W., and Miller P. C., 1973, Dynamic ecology: Englewood Cliffs, N. J., Prentice-Hall, 563 p.

Cooper, G. A., and Grant, R. E., 1976, Permian brachiopods of west Texas, Pts. 1, 2: Smithsonian Contr. Paleobiology no. 21, p. 1923-2607.

Darwin, C., 1859, On the origin of species by means of natural selection. (Page references to 1950 reprint of First Edition, Watts and Co., London)

———— 1887, *in* Darwin, F., ed., Life and letters of Charles Darwin, Vol. 3: London, Murray, p. 159.

———— 1903, *in* Darwin, F., and Seward, A. C., eds., More letters of Charles Darwin, Vol. 1: London, Murray, p. 102.

Eldredge, N. and Gould, S. J., 1972, Punctuated equilibria: An alternative to phyletic gradualism, *in* Schopf, T. J., ed., Models in paleobiology: San Francisco, Freeman, p. 82-115.

Gould, S. J., 1969, An evolutionary microcosm: Pleistocene and Recent history of the land-snail *P.* (*Poecilozonites*) in Bermuda: Harvard Univ. Mus. Comparative Zoology Bull., v. 138, p. 407-532.

Gleason, H. A., 1939, The individualistic concept of the plant association: Am. Midland Naturalist, v. 21, p. 92-110.

Hallam, A., 1974, Changing patterns of provinciality and diversity of fossil animals in relation to plate tectonics: Jour. Biogeography, v. 1, p. 213-225.

Hedberg, H. D., 1976, International stratigraphic guide: New York, Wiley, 200 p.

Hulston, J. R., and McCabe, W. J., 1972, New Zealand potassium-argon age list 1: New Zealand Jour. Geology and Geophyics, v. 15, p. 406-432.

Huxley, J. S., ed., 1940, The new systematics: Oxford, The Clarendon Press, 583 p.

Jell, P. A., 1974, Faunal provinces and possible planetary reconstruction of the Middle Cambrian: Jour. Geology, v. 82, p. 319-350.

Krassilov, V. A., 1974, Causal biostratigraphy: Lethaia, v. 7, p. 173-179.

———— 1975, Climatic changes in eastern Asia as indicated by fossil floras. II. Late Cretaceous and Danian: Palaeogeography, Palaeoclimatology, Palaeoecology, v. 17, p. 157-172.

Loughnan, F. C., 1975, Correlatives of the Greta Coal Measures in the Hunter Valley and Gurredah Basin, New South Wales: Geol. Soc. Australia Jour., v. 22, p. 248-253.

Mahlzahn, E., 1957, Devonisches glazial im stat Piavi, Brasilien, ein neuer beitrag zur eiszeit des Devons: Geol. Jahrb. Beih., v. 25, p. 1-31.

Mayr, E., 1942, Systematics and the origin of species: New York, Columbia Univ. Press, 334 p.

Odum, E. P., 1959, Fundamentals of ecology: Philadelphia, W. B. Saunders, 546 p.

Öpik, E. J., 1967, Climatic changes, *in* Runcorn, K., ed., International dictionary of geophysics, Vol. 1: Oxford, Pergamon Press, p. 179-193.

Termier, G., Termier, H., de Lapparent, A. F., and Marin, P., 1974, Monographie du Permo-Carbonifère de Wardak (Afghanistan central): Lyons, Fac. Sci. Lab. Geol., Doc., Hors. sér. 2, p. 1-167.

Wallace, A. R., 1903, *in* Darwin, F., and Seward, A. C., eds., More letters of Charles Darwin, Vol. 1: London, Murray, p. 294.

Waterhouse, J. B., 1973a, Communal hierarchy and significance of environmental parameters for brachiopods: The New Zealand Permian model: Royal Ont. Mus. Life Sci. Contr., v. 92, p. 1-49.

———— 1973b, The Permian-Triassic boundary in New Zealand and New Caledonia and its relationship to world climatic changes and extinction of Permian life, *in* Logan, A., and Hills, L. V., eds., The Permian and Triassic Systems and their mutual boundary: Canadian Soc. Petroleum Geology Mem. 2, p. 445-464.

———— 1974, Upper Paleozoic Era: Encyclopaedia Britannica, 15th Edition, p. 921-930.

———— 1976a, World correlations for marine Permian faunas: Queensland Univ. Dept. Geol. Papers, v. 7, xvii + 232 p.

———— 1976b, The significance of ecostratigraphy and need for biostratigraphic hierarchy in stratigraphic nomenclature: Lethaia, v. 9, p. 317-325.

Waterhouse, J. B., and Bonham-Carter, G., 1975, Global distribution and character of Permian biomes based on brachiopod assemblages: Canadian Jour. Earth Sci., v. 12, p. 1085-1146.

Williams, A., 1973, Distribution of brachiopod assemblages in relation to Ordovician palaeogeography, *in* Hughes, N. F., ed., Organisms and continents through time: Spec. Papers Palaeontology no. 12, p. 242-269.

Williams, G. E., 1972, Geological evidence relating to the origin and secular rotation of the solar system: Modern Geology, v. 3, p. 165-181.

Ziegler, A. M., Cocks, L. R. M., and McKerrow, S. A., 1968, The Llandovery transgression of the Welsh Borderland: Palaeontology, v. 11, p. 736-782.

Permian Ectoprocts in Space and Time

June R. P. Ross, *Department of Biology, Western Washington University, Bellingham, Washington 98225*

ABSTRACT

Ectoprocts are represented in the Permian by about 95 genera belonging to 26 families of the orders Cryptostomata, Cystoporata, Trepostomata, and Ctenostomata. There is a gradual increase in generic diversity from 55 genera in the Early Permian to 65 genera in the Late Permian prior to widespread extinction near the Permo-Triassic boundary. The patterns of geographic distribution of genera indicate an increase in provincialism during the Permian. Generic faunal assemblages permit recognition of several distinct regions: a cool-water Tasman Geosyncline in eastern Australia; a subtropical southern Tethyan Sea which included central-eastern Afghanistan, parts of northern Pakistan, Thailand, Malaya, western Australia, and Timor; a tropical central Tethyan Sea which includes Transcaucasus, Darvas, Pamir, Tibet, Mongolia, and southwest China; a subtropical to tropical northern Tethyan Sea which included Japan, the Maritime Territory and Khabarovsk region of USSR, and possibly parts of northeast Siberia; and a temperate Uralian Sea, Russian Platform, Franklinian Sea, and Zechstein Sea. Some families and many genera have geographical distributions that are restricted to the Tethyan Seas.

INTRODUCTION

The patterns of phyletic evolution in one or more groups of organisms are used as aids by many biogeographers to establish biogeographic patterns from which centers of origin and routes of dispersal are reconstructed. Cracraft (1974) noted that biogeographic relationships can be examined and established based on evolutionary position and phylogenetic relationship. Two assumptions are usually made in this kind of analysis: first, phylogenetic closeness suggests close biogeographic proximity; second, primitive species will most likely appear near the center of origin of a group and more advanced species will tend to be peripheral in distribution. Such biogeographic relations which are dynamic and have constantly shifting geographic range boundaries in living populations are difficult to demonstrate in the fossil record because of the imperfection of that record. Another factor in the establishment and maintenance of biogeographic patterns is climate. If world climatic patterns at a particular time have marked temperature gradients, such as at the present, then an additional step in the analysis needs to be included to account for diversity gradients before attempting a reconstruction of world biogeographic patterns.

The various methods and concepts of establishing phylogenetic relationships as discussed by Anderson (1974), Ashlock (1974), and Cracraft (1974) show how widely divergent interpretations of biogeographic patterns may arise, particularly where analysis is based on only one individual taxonomic group that comprises only a small part of the total fauna. Biologists who work with a limited spatial distribution of organisms and a geologically short time scale seek explanations for taxonomic diversity of different species in associated communities and rely strongly on methods and concepts of community ecology (commonly called geographic ecology) as reviewed by Peet (1974). When wider spatial distribution and much longer spans of time are dominant parameters, the historical aspect provides another important dimension that may require considerable modification of concepts and methods.

The fossil record, upon which analysis of taxonomic diversity ultimately is based, is an incomplete record of the once living communities of organisms. This record may result in various parts of biotic communities being superimposed or mixed depending on various factors of fossilization such as conditions of burial and preservation. Considerable evidence for endemism and cosmopolitanism may be partially or totally lost during the fossilization process. Modification of the biota of communities during fossilization may lead to elimination of soft-bodied organisms, to changes in the site of deposition, to selective preservation of hard parts as a result of diagenetic processes, and to changes in or uneven rates of deposition of sediments enclosing the fauna and(or) flora. These and other features of the geologic processes contribute toward an imprecise recording of the original biotic composition and environmental site. The historical biogeographer has these considerations, as well as others, to include in his assessment of patterns of fossil distribution. Holland (1971) has examined many of these in his defining of faunal provinces during the Silurian.

In addition, historical biogeographers rely strongly on the concept of uniformitarianism. In many instances they must assume that the physical, chemical, and biologic

Table 1. Classification of families and genera of Permian ectoprocts.

Phylum Ectoprocta Nitsche 1869

Class Stenolaemata Borg 1926

Order Cryptostomata Vine 1883

Family Phylloporinidae Ulrich 1890
Genera: *Bashkirella* Nikiforova 1939; *Chainodictyon* Foerste 1887; *Rhombocladia* Rogers 1900.

Family Fenestellidae King 1849
Genera: *Fenestella* Lonsdale 1839; *Archimedes* Owen 1838; *Diploporaria* Nickles and Bassler 1900; *Fenestellata* Gregorio 1930; *Hemitrypa* Phillips 1841; *Hinganotrypa* Romanchuk and Kiseleva 1968; *Levifenestella* Miller 1961; *Lyrocladia* Shulga-Nesterenko 1931; *Minilya* Crockford 1944; *Penniretepora* D'Orbigny 1849 (=*Pinnatopora* Vine 1883); *Pseudounitrypa* Nekhoroshev 1926; *Ptiloporella* Hall 1885; *Ptylopora* McCoy 1844; *Wjatkella* Morozova 1970.

Family Septoporidae Morozova 1962
Genera: *Septopora* Prout 1859 (= *Silvaseptopora* Chronic 1953); *Synocladia* King 1849.

Family Fenestraliidae Morozova 1963
Genera: *Parafenestralia* Morozova 1963; *Triznella* Morozova 1963.

Family Polyporidae Vine 1883
Genera: *Polypora* McCoy 1845; *Anastomopora* Simpson 1897; *Kingopora* Morozova 1960 (=*Phyllopora* King 1849); *Lyropora* Hall 1857; *Matheropora* Bassler 1953; *Reteporidra* Nickles and Bassler 1900; *Thanniscus* King 1849.

Family Acanthocladiidae Zittel 1880
Genera: *Acanthocladia* King 1849; *Kalvariella* Morozova 1970.

Family Septatoporidae Engel 1975
Genus: *Septatopora* Engel 1975.

Family Rhabdomesidae Vine 1884
Genera: *Rhabdomeson* Young and Young 1875; *Ascopora* Trautschold 1876; *Callocladia* Girty 1911; *Megacanthopora* Moore 1929; *Neorhombopora* Shishova 1964; *Nicklesopora* Bassler 1952; *Pamirella* Gorjunova 1975; *Rhombopora* Meek 1872; *Saffordotaxis* Bassler 1952; *Streblocladia* Crockford 1944; *Syringoclemis* Girty 1911.

Family Hyphasmoporidae Vine 1885
Genera: *Hyphasmopora* Etheridge 1875; *Ogbinopora* Shishova 1965; *Streblascopora* Bassler 1952; *Streblotrypa* Vine 1885; *Streblotrypella* Nikiforova 1948.

Family Nikiforovellidae Gorjunova 1975
Genera: *Nikiforovella* Nekhoroshev 1948; *Clausotrypa* Bassler 1929; *Maychella* Morozova 1970; *Pinegopora* Shishova 1965.

Family Girtyoporidae Morozova 1966
Genera: *Girtyopora* Morozova 1966; *Girtyoporina* Morozova 1966; *Hayasakapora* Sakagami 1960; *Tavayzopora* Kiseleva 1969.

Family Timanodictyidae Morozova 1966
Genera: *Timanodictya* Nikiforova 1938; *Timanotrypa* Nikiforova 1938.

Order Cystoporata Astrova 1964

Family Fistuliporidae Ulrich 1882
Genera: *Fistulipora* McCoy 1850 (= *Dybowskiella* Waagen and Wentzel 1886; = *Cyclotrypa* Moore and Dudley 1944 *non* Ulrich; = *Triphyllotrypa* Moore and Dudley 1944); *Cyclotrypa* Ulrich 1896; *Eridopora* Ulrich 1882; *Fistulocladia* Bassler 1929; *Fistulotrypa* Bassler 1929; *Metelipora* Trizna 1950; ?*Fistuliramus* Astrova 1960.

Family Hexagonellidae Crockford 1947
Genera: *Hexagonella* Waagen and Wentzel 1886; *Coscinotrypa* Hall and Simpson 1887; *Evactinopora* Meek and Worthen 1865, *Evactinostella* Crockford 1957; *Fistulamina* Crockford 1947; *Meekopora* Ulrich 1889; *Meekoporella* Moore and Dudley 1944; *Prismopora* Hall 1883.

Family Sulcoreteporidae Bassler 1935
Genus: *Sulcoretepora* D'Orbigny 1849.

Family Goniocladiidae Nikiforova 1938
Genera: *Goniocladia* Etheridge 1876; *Ramipora* Toula 1875; *Ramiporella* Shulga-Nesterenko 1933; *Ramiporidra* Nikiforova 1938.

Family Etherellidae Crockford 1957
Genera: *Etherella* Crockford 1957; *Liguloclema* Crockford 1957.

Family Actinotrypidae Simpson 1895
Genera: *Actinotrypella* Gorjunova 1972; *Epiactinotrypa* Kiseleva 1973.

Order Trepostomata

Family Anisotrypidae Dunaeva and Morozova 1967
Genus: *Anisotrypella* Morozova 1967.

Family Eridotrypellidae Morozova 1960
Genera: *Hinganella* Romanchuk 1967; *Neoeridotrypella* Morozova 1970; *Permopora* Romanchuk 1967.

Family Stenoporidae Waagen and Wentzel 1886
Genera: *Stenopora* Lonsdale 1844 (= *Geinitzella* of some authors); *Arcticopora* Fritz 1961; *Rhombotrypella* Nikiforova 1933; *Stenodiscus* Crockford 1944; *Tabulipora* Young 1883.

Family Dyscritellidae Dunaeva and Morozova 1967
Genera: *Dyscritella* Girty 1911; *Dyscritellina* Morozova 1967.

Family Ulrichotrypellidae Romanchuk 1968
Genera: *Ulrichotrypella* Romanchuk 1967; *Ulrichotrypa* Bassler 1929; *Primorella* Romanchuk and Kiseleva 1968.

Table 1. (Cont.)

Family Araxoporidae Morozova 1970
  Genera: *Araxopora* Morozova 1965; *Paraleioclema* Morozova 1961 (= *Leioclema* of some authors); *Permoleioclema* Romanchuk 1966.
Families *Incertae Sedis*
  Genus: *Coeloclemis* Girty 1911.
  Genus: *Pseudobatostomella* Morozova 1960 (=*Batostomella* of some authors).

Class Gymnolaemata Allman 1856

Order Ctenostomata Busk 1852
  Suborder Stolonifera Ehlers 1876
    Family Vinellidae Ulrich and Bassler 1904
      Genus: *Condranema* Bassler 1952.
    Family Ascodictyidae Miller 1889
      Genera: *Ascodictyon* Nicholson and Etheridge 1887; *Bascomella* Morningstar 1922.

processes now at work on the Earth's surface were the same in past geologic ages. In this study I assume that Permian ectoprocts, all of which belong to extinct families, were normal marine, sessile, colonial organisms that could be dispersed by larval stages similar to those in existing ectoproct species. I also assume that morphologic convergence can be recognized as phylogenetically distinct so that independent lineages having similar types of structures resulting from adaptations to similar environmental conditions are separable.

The dispersal patterns of ectoprocts in Recent marine waters are affected markedly by the patterns of surface oceanic currents. Working with a model of sea-floor spreading and plate tectonics, it is apparent that the size, shape, and geographic position of ocean basins have changed with time; consequently the dynamics, particularly of the circulation patterns, of the ocean basins have changed also. Such changes would have caused changes in the levels of nutrient supplies, temperature stratification of the oceans, and areas of upwelling. These changes in turn would modify the abundance of organisms and the number of species in various marine communities. These modified conditions probably led to greater species diversity along particular margins of certain ocean basins similar to variations in species diversity that are seen in today's oceans. These, then, are some of the major considerations that need to be examined when attempting to establish distribution patterns for Permian ectoprocts.

CLASSIFICATION

Table 1 gives the classification of families and genera of Permian ectoprocts used in this report. Comprehensive systematics of many families distributed in Europe and Asia are found in Morozova (1970) and Gorjunova (1975) in which a number of relatively recently erected taxonomic groups are reviewed or discussed.

PERMIAN ECTOPROCT FAUNAS

The distribution of Permian ectoproct genera and families will be examined first and then these distributional patterns will be assessed in relation to the present interpretation of the Permian history of the Earth using the current hypotheses of sea-floor spreading and plate tectonics. Although faunal data are very incomplete for a number of regions of the world, it is possible to identify faunal patterns at different times in the Permian and also times of provincialism.

Figure 1 shows a correlation of the Permian units discussed in this study in a number of important regions in different parts of the world. The incompleteness of information on ectoproct faunas for many parts of these Permian sequences makes a world-wide set of stage nomenclature, such as Asselian, Sakmarian, Artinskian, Kazanian, and Dzhulfian (Djulfian), difficult to use. For example, biostratigraphers have great difficulty correlating Kazanian faunas of the Russian Platform with those of the Tethyan region, and many Soviet scientists working with Tethyan faunas use the Guadalupian of southwestern North America as their standard reference succession for the lower part of the Upper Permian. As a result two series of the Permian —the Lower and Upper Permian—have been used as the time-stratigraphic divisions in which the faunas have been assigned and analyzed. The Permian Period extended from about 280 m.y. to about 225 m.y. ago with the boundary of the Lower Permian-Upper Permian being at about 240 m. y. (Smith, 1964).

Invertebrate faunas of the Permian have received increasing examination and analysis as various biostratigraphic investigations have attempted to define the lower and upper boundaries of this late Paleozoic system (Furnish, 1966, 1973; Kummel and Teichert, 1966, 1973; Nakazawa, 1974; Shcherbovich, 1969; Spinosa and others, 1975; Toriyama, 1973; Tozer, 1969, 1971). Considerable data on the world distribution of some invertebrate and vertebrate faunas and also floras have already resulted in syntheses of paleobiogeographic histories for the Permian. Hill (1958) presented such a synthesis of Sakmarian (Early Permian) corals and more recently Rowett (1972, 1975) investigated provinciality in Early Permian coral faunas. Ross (1967) examined fusulinid faunal realms and their development. Other studies are available on vertebrates (Romer, 1968; Keast and others, 1972) and plants (Hart, 1969; Meyen, 1970; Plumstead, 1973; Hart, 1974), as

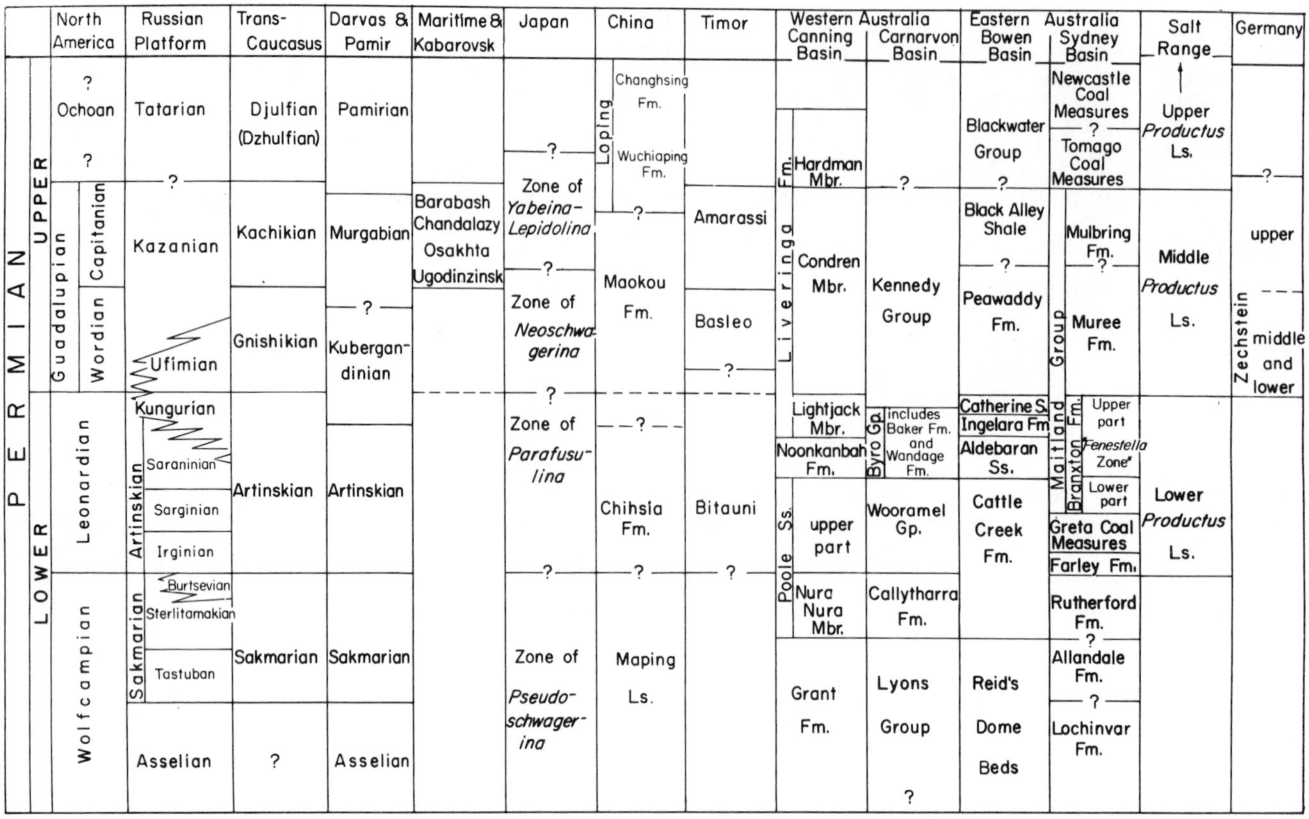

	North America	Russian Platform	Trans-Caucasus	Darvas & Pamir	Maritime & Kabarovsk	Japan	China	Timor	Western Australia Canning Basin	Western Australia Carnarvon Basin	Eastern Australia Bowen Basin	Eastern Australia Sydney Basin	Salt Range	Germany
PERMIAN UPPER	? Ochoan ?	Tatarian ?	Djulfian (Dzhulfian)	Pamirian		?	Loping: Changhsing Fm.; Wuchiaping Fm.		Liveringa Fm.: Hardman Mbr.	?	Blackwater Group ?	Newcastle Coal Measures ? Tomago Coal Measures	Upper *Productus* Ls.	?
Guadalupian: Capitanian		Kazanian	Kachikian	Murgabian	Barabash, Chandalazy, Osakhta, Ugodinzinsk	Zone of *Yabeina-Lepidolina* ?	? Maokou Fm.	Amarassi	Liveringa Group: Condren Mbr.	Kennedy Group	Black Alley Shale ? Peawaddy Fm.	Mulbring Fm. ?	Middle *Productus* Ls.	Zechstein: upper
Guadalupian: Wordian		Ufimian	Gnishikian	? Kubergandinian		Zone of *Neoschwagerina*		Basleo ?				Group: Muree Fm.		Zechstein: middle and lower
LOWER Leonardian		Kungurian; Artinskian: Saraninian, Sarginian, Irginian	Artinskian	Artinskian		? Zone of *Parafusulina*	– – ? – – Chihsia Fm.	Bitauni	Lightjack Mbr.; Noonkanbah Fm.; Poole Ss. upper part	Byro Gp. includes Baker Fm. and Wandage Fm.; Wooramel Gp.	Catherine S.; Ingelara Fm; Aldebaran Ss.; Cattle Creek Fm.	Maitland Gp. Branxton Fm.: Upper part, "*Fenestella* Zone", Lower part; Greta Coal Measures; Farley Fm.	Lower *Productus* Ls.	
Wolfcampian		Sakmarian: Burtsevian, Sterlitamakian, Tastubian	Sakmarian	Sakmarian		? Zone of *Pseudo-schwagerina*	? Maping Ls.	?	Poole Ss.: Nura Nura Mbr.; Grant Fm.	Callytharra Fm.; Lyons Group ?	Reid's Dome Beds	Rutherford Fm. ? Allandale Fm. ? Lochinvar Fm.		
		Asselian	?	Asselian										

Figure 1. Permian correlation chart for various parts of the world mentioned in text. Not all columns show the complete stratigraphic succession in a region but the relative position of ectoproct-bearing units is indicated. Data for the Australian successions are modified slightly from Dickins (1970).

well as general summaries that include discussion of the significance of Permian paleogeography (Ross, 1974, 1976).

Knowledge of Permian ectoprocts is now reaching the stage that we can undertake examination of biogeographic patterns for this group of organisms. Details of phyletic evolution within many of the 26 ectoproct families that existed in the Permian Period still remain unelucidated (Figs. 2, 3) because, as with other invertebrate groups, our knowledge is more complete for certain regions and sequences and marked by large gaps in other areas where ectoproct faunas are known to be present but have not been studied (Fig. 4). Soviet scientists are continuing to extensively document faunas of the Russian Platform, Ural Mountain region, Transcaucasus, Pamir and Darvas region, Maritime Territory, and Khabarovsk and Far East regions. Chinese investigators are examining faunas from parts of central and southern China and Tibet. Japanese bryozoologists have analyzed major faunas of Japan and also Thailand and Malaya. Documentation of faunas from western Australian and eastern Australian basins has provided considerable data. Early studies on the fauna from the famous Permian localities on Timor included many ectoprocts. Sketchy and incomplete information is available for the Permian ectoprocts of the Salt Range, Pakistan, and adjoining areas of India and Kashmir, and also for the Carnic Alps. In the mid-1960's ectoprocts of the Zechstein of Germany were re-examined. Some information is on hand for England and northeast Greenland and Spitsbergen. Data on the Permian faunas of North America and South America remain very incomplete.

Russian Platform and Adjacent Regions

During the Early Permian, ectoprocts occurred in a number of different ecological settings on the Russian Platform (Figs. 1, 4). They formed a major component of reefs dominated by hydractinian corals and adjacent shelf facies. Fenestellids are particularly abundant in these reef facies (Nikiforova, 1936, 1938, 1939; Novikova, 1937; Trizna, 1950; Shulga-Nesterenko, 1952; Elias and Condra, 1957). Asselian and Sakmarian strata are characterized by an abundance both in numbers of colonies and of species of the fenestrate cryptostomes, *Fenestella, Penniretepora, Reteporidra,* and *Polypora;* the rhabdomesid cryptostomes *Ascopora* and *Nicklesopora;* and the stenoporid trepostome *Rhombotrypella* as well as the cryptostome *Timanodictya.* The cystoporate *Fistulipora* is common in the fauna. A listing of described genera from Sakmarian strata indicates only 16, but this does not provide a true indication of the richness of the fauna. In Sakmarian reefs and adjacent shelf facies, the ectoproct assemblages consist of the cystoporates *Hexagonella, Metelipora,* and *Ramipora;* the cryp-

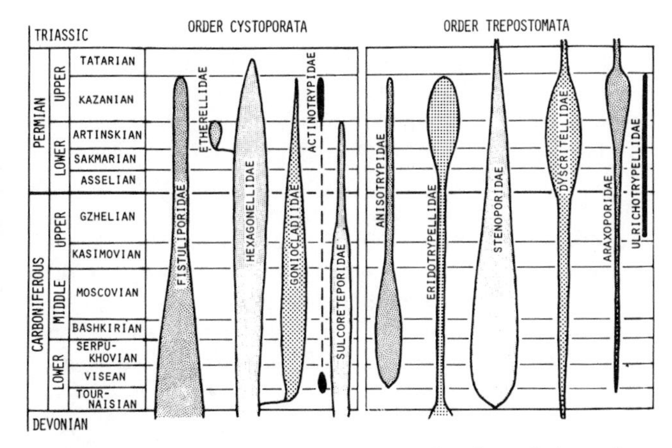

Figure 2. Distribution of ectoproct families of the orders Cystoporata and Trepostomata in the Carboniferous, Permian, and Early Triassic.

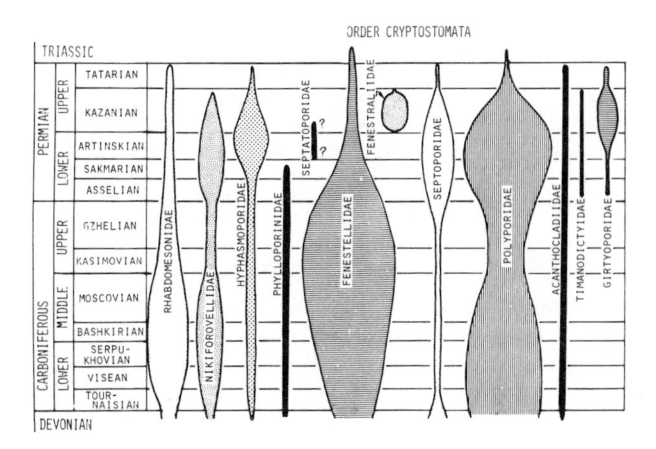

Figure 3. Distribution of ectoproct families of the order Cryptostomata in the Carboniferous, Permian, and Early Triassic.

tostomes *Clausotrypa, Septopora, Acanthocladia, Lyrocladia, Archimedes,* and *Streblotrypa;* and the phylloporinids *Chainodictyon* and *Bashkirella.* During early Artinskian (Irginian) time, ectoprocts were abundant in reef limestones and bedded limestone facies and included about 18 genera (Trizna and Klautsan, 1961). Although there are fewer Artinskian than Sakmarian species, only a small percentage of the Sakmarian species persisted into the Artinskian so that the majority of species are new. A large number of the new species belong to *Fenestella* and *Polypora. Reteporidra* and *Lyrocladia* are less abundant than in the underlying Sakmarian stage. *Rhombotrypella, Penniretepora, Clausotrypa,* and *Hexagonella* flourished. *Ascopora* and *Nicklesopora,* which were abundant and characteristic of the Sakmarian, are far less dominant. Likewise *Timanodictya,* which was especially characteristic of the early Sakmarian of the central Urals and of Bashkiria, is rare. Goniocladiids, such as *Gonocladia,* although present during the Sakmarian became far more significant in the early and middle part of Artinskian time. New genera, such as *Ptylopora* and *Ptiloporella,* characterized the early Artinskian facies.

In middle Artinskian (Sarginian) time, the great variety of lithofacies of reef and between-reef environments was highly suited for ectoprocts which are represented by about 13 genera. A large number of new species appear representing 50 percent of the ectoproct assemblage; 25 percent of the species extend upwards from beds of Sakmarian age and the other 25 percent continue on from beds of early Artinskian (Irginian) age. During the middle Artinskian, the trepostomes *Pseudobatostomella* and *Stenopora,* the fistuliporid cystoporate *Eridopora,* and some rhabdomesid cryptostomes became proportionally much more significant in these ectoproct assemblages. Species of *Fenestella* and *Polypora* remained significant.

In the later part of Artinskian (Saraninian) time, a marked change in the ectoproct species composition reflects changes in the Artinskian basin due to tectonic movements and subsidence. The ectoprocts are reduced in generic representation, there being about 11 genera at the beginning of this late part of the Artinskian. *Fenestella, Polypora, Rhombotrypella,* and *Pseudobatostomella* have new species assemblages and are well-represented. *Paraleioclema* and *Tabulipora* are additions to the assemblages. *Streblotrypa* is abundant. *Clausotrypa* and *Hexagonella* disappear in the early part of the late Artinskian, apparently not surviving the changing basinal conditions.

In early Kazanian time, the fluctuating diversity of ectoproct assemblages reflects different ecological conditions relating possibly to changes in salinity, turbidity, and depth in the shallow Kazanian Sea which extended from Tatarian ASSR northward and westward through Kirov Oblast and Mari and Bashkir ASSR to Arkangel Oblast. In the southern and shallowest part of the basin in the region of Tatarian ASSR, the ectoprocts are represented by 11 genera including the trepostomes *Dyscritella, Rhombotrypella,* and *Tabulipora,* and the cryptostomes *Pinegopora, Parafenestralia,* and *Triznella,* the latter two being rare. In the restricted southern part of the basin the species diversity is low. *Pseudobatostomella,* the fenestellid cryptostome *Wjatkella,* and the hyphasmoporid cryptostome *Streblascopora,* are abundant here and also are well-represented in the central and northern parts of the basin. *Fenestella* and *Thamniscus* also are common in the southern part. The central part of the basin has a greater diversity of genera and species with 10 of the 11 genera from the southern part of the basin as well as 14 other genera. *Thamniscus* is absent from the central part of the basin but reappears in the northern part. The trepostomes *Dyscritella* and *Rhombotrypella,* and the cryptostomes *Fenestella* and *Parafenestralia* are abundant. Nine of the genera from the central part of the basin, the trepostome *Neoeridotrypella,* the cystoporates *Cyclotrypa* and *Fistulipora,* and the cryptostomes *Polypora, Rhombopora, Reteporidra, Kingopora, Timanodictya,* and *Girtyopora,* extend into the northern region. The genera present in the southern part of the basin are also present in the northern region and, in addition, 6 other genera appear. The additional genera are the trepostomes *Ulrichotrypella* and *Anisotrypella,* the cystoporates *Goniocladia* and *Meekopora,* and the cryptostomes *Timanotrypa* and *Girtyoporina.* Ectoprocts are not present in higher units of the Upper Permian on the Russian Platform and adjacent regions.

Northwestern Arctic Regions and Adjacent Areas

Ectoproct faunas from the arctic region including Novaya Zemlya, Spitsbergen, and northeast Greenland have received little study (Figs. 1, 4). Faunas in other areas such as central-east Greenland and the Canadian arctic islands remain unidentified. Nikiforova (1936) reported from Novaya Zemlya a sparse fauna of *Fenestella, Ramipora, Hyphasmopora,* and *Pseudobatostomella*? from the Lower Permian Shales; from Spitsbergen the 2 genera *Synocladia* and *Polypora* were found in beds above the *Cyathophyllum* Limestone. Ross and Ross (1962) reported *Rhombotrypella, Stenopora, Tabulipora, Polypora, Fenestella,* and *Timanodictya* from the Sakmarian of northeast Greenland.

From the Upper Permian, probably Kazanian, at Green Harbor, Vestspitsbergen, Nikiforova (1936) noted *Fenestella, Polypora, Ptylopora, Septopora,* and *Ramipora* in shales below the siliceous *Productus*-bearing beds. Malecki (1968) described a fauna from the Tokrössoya Beds of Vestspitsbergen which had *Stenopora, Tabulipora,* and *Polypora.* These ectoprocts are possibly early Kazanian in age.

Western Europe

In western Europe (Baltic region, Poland, Germany, and England), ectoprocts are absent in the Lower Permian. In the Upper Permian (Figs. 1, 4) they are represented mostly by fenestrate cryptostomes. The faunas occur in the Zechstein or equivalent units such as the Magnesian Limestone of England. They have low generic and specific diversity, and the widespread occurrence of genera throughout these areas suggests similar ecological conditions and possibly interchange of faunas between the areas. The polyporid *Kingopora* occurs in all these areas as does *Fenestella.* The fauna from the Magnesian Limestone of northern England has, in addition, *Synocladia* and *Pseudobatostomella*?. The small reefs in the Zechstein Limestone of Germany have a greater number of genera; in addition to *Kingopora,* there are the cryptostomes *Fenestella, Thamniscus, Acanthocladia, Penniretepora,* and *Synocladia,* the trepostome *Stenopora* and the hexagonellid cystoporate *Coscinotrypa* (Dreyer, 1961). To the north in the region of Lithuania and Kaliningrad, an additional genus of acanthocladiids, *Kalvariella,* is present.

Southern Europe

No recent studies have been undertaken and the ectoproct fauna of the Lower Permian (Asselian and possibly Sakmarian) *Fusulina* Limestone has only *Fenestella* reported (Johnsen, 1906). In the Upper Permian (Guadalupian) of the Sosio Beds of Sicily, Gregorio (1930) recorded the cryptostome genus *Fenestellata.*

South Central USSR

In the Transcaucasus region in Armenian SSR and Nakichevan ASSR, Morozova (1970) identified an Upper Permian fauna of Guadalupian and Djulfian age (Figs. 1, 4). The lower part of the Guadalupian, the Gnishikian, has 13 genera of which the cystoporates were sparse in numbers. The cystoporates are the fistuliporids *Fistulipora* and *Cyclotrypa,* the hexagonellid *Hexagonella,* and the sulcoreteporid *Sulcoretepora.* Only 2 trepostomes are present, *Paraleioclema* and *Araxopora.* The cryptostomes are represented by 5 different families including the genera *Fenestella, Septopora, Polypora, Reteporidra, Rhabomeson, Ogbinopora,* and *Steblascopora.* In the upper part of the Guadalupian, the Kachikian, only the trepostome *Araxopora* is present. The Djulfian has 1 cystoporate, *Fistulipora,* and 5 cryptostomes, *Synocladia, Polypora, Septopora, Streblotrypa,* and *Girtyoporina.*

In the region of Darvas and Pamir (Figs. 1, 4), Gorjunova (1975) described an extensive ectoproct fauna ranging in age from Asselian into Pamirian. The Lower Permian of the Darvas section as described by Leven (1967), is assigned to the Sakmarian but this assignment requires modification because rocks of Asselian, Sakmarian, and possibly also early Artinskian age are present; some of the "Zone of *Schwagerina*" (Leven, 1967) may go as high as the early Artinskian. In southwest Darvas, the Asselian and Sakmarian has a diverse fauna of 11 genera: the cystoporates *Fistulipora, Actinotrypella, Goniocladia, Ramiporidra,* and *Sulcoretepora;* the trepostomes *Rhombotrypella* and *Primorella;* and the cryptostomes *Fenestella, Polypora, Rhabdomeson,* and *Streblascopora.* The Artinskian strata have a small fauna of cystoporates, the fistuliporids *Fistulipora, Eridopora,* and *Cyclotrypa,* abundant specimens of the hexagonellid *Hexagonella,* and the cryptostome *Fenestella.* In central Pamir in the Artinskian, *Fistulipora, Hexagonella,* and *Eridopora* occur with the cryptostomes *Rhombopora, Fenestella, Polypora,* and *Septopora.* In southeast Pamir, the Sakmarian has an abundance of the cryptostomes *Nikiforovella, Fenestella, Polypora,* and *Wjatkella,* together with a few specimens of *Dyscritella.* In the Artinskian, a mixed ectoproct assemblage consists of the cystoporates *Fistulamina* and *Ramiporidra,* the trepostomes *Rhombotrypella* and *Dyscritella,* and abundant cryptostomes of the genera *Pamirella, Streblascopora, Fenestella, Wjatkella, Polypora,* and *Chainodictyon.*

In Upper Permian strata ectoprocts are sparse. In southwest Darvas, the cystoporate *Fistulipora* and the cryptostome *Septopora* are the sole representatives and occur only in the high Permian, the Pamirian. There are no reports of ectoproct faunas from the intervening Kubergandian and Murgabian. In central Pamir, the Kubergandian, the Murgabian, and Pamirian stages are not differentiated and the Upper Permian fauna has the cystoporates *Fistulipora* and *Eridopora,* and the cryptostome *Ogbinopora.* In southeast Pamir, the fauna in the Kubergandian has *Hexagonella* and the trepostome *Araxopora.* The highest Permian, the Pamirian, has *Fenestella.*

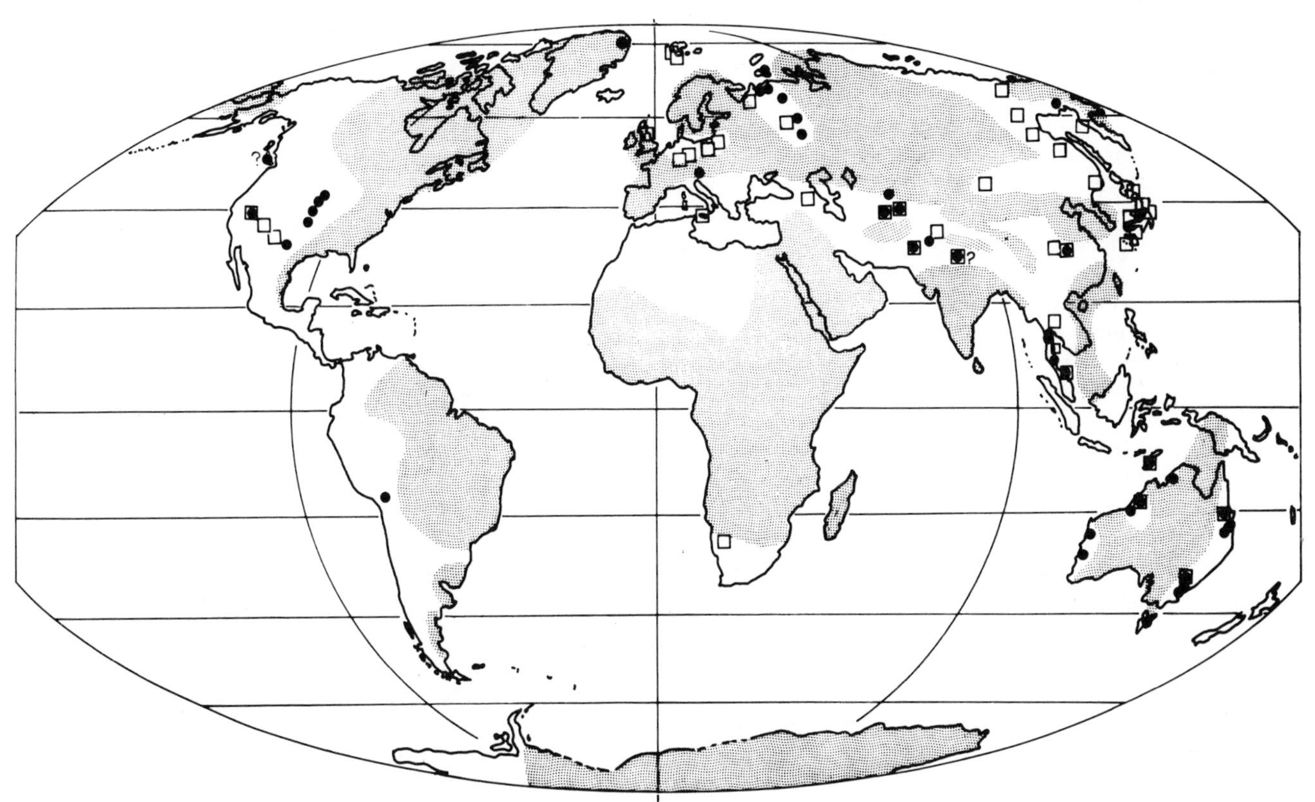

Figure 4. Plot of location of described Lower and Upper Permian ectoproct faunas using present geography. Stippling indicates inferred land areas during Kazanian time.

Afghanistan, Kashmir, Northern Pakistan and Northern India

The ectoproct fauna of the *Productus* Limestone (Figs. 1, 4) was extensively documented by Waagen and Pichl (1885) and Waagen and Wentzel (1886). Reed (1931) also reported on some ectoprocts. Although reference is frequently made to this fauna, particularly in comparisons of systematic studies, it has not been re-examined. The Lower Permian (Artinskian) Lower *Productus* Limestone has 10 genera: the cystoporates *Fistulipora* and *Hexagonella;* the trepostomes *Stenodiscus* and *Stenopora;* and the cryptostomes *Rhombopora, Fenestella, Polypora, Thamniscus, Acanthocladia,* and *Girtyopora.* The Upper Permian Upper *Productus* Limestone has the cystoporates *Fistulipora, Hexagonella,* and *Goniocladia,* and the cryptostomes *Synocladia, Rhombopora,* and *Polypora.*

In the central-eastern part of Afghanistan, abundant ectoprocts occur in strata ranging in age from Sakmarian into Kazanian (Termier and Termier, 1971). The Lower Permian faunas of Sakmarian and Artinskian age have numerous polyporids and fenestellids, such as *Polypora* (*Pustulopora*) Termier and Termier, and *Fenestella,* which range through the entire sequence, and *Polypora* (*Paucipora*) Termier and Termier, and *Minilya.* Other cryptostomes are *Rhombopora, Saffordotaxis, Septopora,* and *Rhabdomeson.* The cystoporates include *Cyclotrypa, Meekopora, Goniocladia,* and *Sulcoretepora.* The trepostomes have low representation with *Tabulipora, Dyscritella,* and *Rhombotrypella.* The Upper Permian (Kubergandinian) faunas have a number of genera which extend up from the Artinskian including *Rhombotrypella* and the 4 cystoporates noted above. The cryptostomes *Ascopora, Streblascopora, Septopora,* and *Acanthocladia* are new appearances in the Permian sequence. Higher in the Upper Permian (Murgabian), *Tabulipora* reappears and the cystoporates are represented by *Goniocladia, Hexagonella,* and *Coscinotrypa,* and the cryptostomes by *Streblascopora, Rhabdomeson, Thamniscus,* and *Reteporidra.*

Australia

Abundant Lower Permian ectoprocts (Figs. 1, 4) show great diversity in both western and eastern Australia and in the Northern Territory of Australia. In the Northern Territory in the Port Keats district of the Bonaparte Gulf Basin, a fauna considered to be Artinskian in age, has the hexagonellid cystoporates *Evactinostella, Fistulamina,* and *Hexagonella,* and the cryptostomes *Saffordotaxis, Streblascopora, Fenestella,* and *Polypora* (Crockford, 1943; Wass, 1967). To the southwest in the Fitzroy trough of the Canning Basin in the west Kimberley district (Crockford, 1957), the ectoprocts of Sakmarian age in the Nura Nura Member of the Poole Sandstone again have the hexagonellid cystoporates *Evactinostella* and *Hexagonella* as well as the fistuliporid cystoporate *Fistulipora,* the cryptostomes *Streblascopora, Fenestella,* and *Lyropora,* and the trepostomes *Dyscritella, Stenopora,* and *Paraleioclema.* In the Noonkanbah Formation, a rich fauna of Artinskian age contains all the genera from the lower Nura Nura Member except for the 2 distinctive taxa, *Lyropora* and *Paraleioclema*. In addition, a great diversity of cystoporates shows the expansion of this group as a significant part of the fauna. Newly appearing cystoporates are the hexagonellids *Prismopora* and *Fistulamina*, the fistuliporid *Eridopora*, the etherellids *Liguloclema* and *Etherella*, and the goniocladiids, *Goniocladia* and *Ramipora*. Additional cryptostomes are *Polypora*, *Synocladia*, *Minilya*, *Streblotrypa*, and *Acanthocladia*, the rhabdomesids *Saffordotaxis*, *Rhabdomeson*, *Megacanthopora*?, *Callocladia*?, and *Rhombocladia*, and the trepostomes have only 2 additional genera, *Stenodiscus* and *Tabulipora*.

In the Carnarvon Basin at the northern end of the Fitzroy Range, the Lower Permian has in its lowest unit, the Lyons Group (which is correlated with the Sakmarian), 2 genera, *Polypora* and *Stenopora* (Ross, 1963). The overlying Callytharra Formation has an amazingly rich ectoproct fauna of Sakmarian (Sterlitamak) age (Crockford, 1944, 1951). As in the Fitzroy trough, *Hexagonella*, *Fenestella*, and *Polypora* are present. The fauna is comprised of another hexagonellid cystoporate, *Evactinopora*, the goniocladiid cystoporate *Ramipora*, and an abundance of cryptostomes—*Minilya*, *Penniretepora*, *Lyropora*, *Septopora*, *Rhombopora*, *Streblotrypa*, *Rhombocladia*, and *Streblocladia*—and undescribed trepostomes. The lower part of the Barrabiddy Shale of the Wooramel Group has a limited fauna of Artinskian age of *Hexagonella*, *Stenopora*, and *Streblotrypa*. In the Cundelgo and Wandagee Formations of the Byro Group (upper part of the Artinskian), the cystoporates again are diverse with *Fistulipora*, *Hexagonella*, and *Ramipora*, and the cryptostomes include *Synocladia*, *Fenestella*, *Minilya*, *Polypora*, and *Streblotrypa*.

A very sparse ectoproct fauna has been reported in the Bonaparte Gulf Basin from a sequence which is considered to be in the lower part of Upper Permian (Etheridge, 1907; Crockford, 1943; Wass, 1967). The fauna has *Fistulipora*, *Rhombopora*, *Streblotrypa*, and *Ramipora*. To the southwest in the Fitzroy trough, Upper Permian strata of Tatarian age have *Stenodiscus* and *Dyscritella*.

In eastern Australia, the Bowen Basin and Springsure Shelf, Queensland, have an extensive fauna ranging in age from Sakmarian to Kazanian (Wass, 1968, 1969). Lower Permian strata of Sakmarian age have only *Fenestella* and *Polypora*. In rocks of Artinskian age, however, ectoprocts flourished. The cystoporates consist of diverse groups: the fistuliporid *Fistulipora*, the goniocladiids *Ramipora* and *Goniocladia*, and the etherellid *Liguloclema*. The trepostomes include *Stenopora*, *Stenodiscus*, and *Dyscritella*. The numerous cryptostomes are *Saffordotaxis*, *Rhombopora*, *Streblascopora*, *Diploporaria*, *Penniretepora*, *Polypora*, and *Minilya*. To the south in the Sydney Basin, New South Wales, the Allandale and Rutherford Formations (Sakmarian age) have a restricted fauna of *Stenopora* and *Dyscritella*. The *Fenestella* Shale and Ulladulla Mudstone (Artinskian age) show marked species diversity in the trepostome *Stenopora*, and other genera of the fauna are *Pseudobatostomella*?, *Rhombopora*, *Fenestella*, *Minilya*, and *Polypora*. Further south in Tasmania, the Berriedale Limestone (Artinskian) also has great species diversity in the trepostome *Stenopora* as well as having *Stenodiscus* and *Hermitrypa*?.

In the Upper Permian of the Bowen Basin, a fauna of Kazanian age is dominated by trepostomes, *Stenopora*, *Paraleioclema*?, and *Stenodiscus*, and cryptostomes *Saffordotaxis*, *Penniretepora*, *Ptylopora*, *Levifenestella*, *Fenestella*, *Polypora*, and *Septatopora*. In the Sydney Basin, a fauna of Kazanian age has *Stenopora*, *Ptylopora*, *Polypora*, and *Acanthocladia*. The Kazanian of Tasmania has *Fenestella*, *Polypora*, and *Acanthocladia*?.

Timor

Many comparisons are made with the rich Timor fauna (Figs. 1, 4) (Bassler, 1929) but presently only part of it is stratigraphically delimited so that specific ages are assignable to only a portion of the identified fauna. The Bitauni Beds of Artinskian age show less diversity of genera than many other Artinskian faunas. The only cystoporate is the cosmopolitan *Fistulipora*; the trepostomes *Ulrichotrypa* and *Hinganella* are restricted in occurrence, and the cryptostomes *Rhombopora*, *Streblascopora*, and *Fenestella* have a wide distribution. Two faunas of Kazanian age in the Basleo and Amarassi Beds display marked expansion in the number of genera. The Basleo Beds show diversification of cystoporates with the fistuliporids *Fistulipora*, *Eridopora*, *Fistulotrypa*, the goniocladiid *Goniocladia*, and the hexagonellid *Hexagonella*. *Ulrichotrypa*, *Streblascopora*, and *Fenestella* range up from the Artinskian. The overlying Amarassi Beds show a marked reduction in genera having only *Fistulipora*, *Fenestella*, *Stenopora*, and *Clausotrypa*. Higher in the Upper Permian (probably still Kazanian) only the cryptostome genera *Rhabdomeson* and *Streblotrypella* are present.

Thailand and Malaya

In the Lower Permian of Thailand, a fauna of late Sakmarian to middle Artinskian age (Figs. 1, 4) has the cystoporates *Fistulipora* and *Coscinotrypa*, the trepostomes *Dyscritella* and *Leioclema*?, and the cryptostomes *Fenestella*, *Polypora*, and *Thamniscus* (Sakagami, 1970, 1975). A rich middle to late Artinskian fauna is predominantly cystoporates and cryptostomes. The cystoporates include *Fistulipora*, the hexagonellid *Hexagonella*, the goniocladiids *Goniocladia* and *Ramipora*, the etherellid *Liguloclema*, and the sulcoreteporid *Sulcoretepora*. The cryptostomes are represented by a large diversity of *Fenestella*, *Polypora*, *Penniretepora*, *Acanthocladia*, *Rhabdomeson*, *Ascopora*, *Streblascopora*, *Streblotrypa*, *Steblotrypa*?, *Rhombopora*, *Ogbinopora*, and *Timanodictya*?. *Dyscritella* is the only trepostome.

To the south in Malaya, the fauna ranges higher in the Permian, extending from the upper part of the Lower Permian (Artinskian) into the Upper Permian (Guadalupian). Units that straddle the boundary of the Lower and Upper Permian have *Fistulipora, Polypora,* and *Cyclotrypa.* Higher units of Guadalupian age have a distinctive fauna of *Fenestella, Pseudobatostomella, Araxopora, Paraleioclema,* and *Clausotrypa.*

Japan

The Permian of Japan (Figs. 1, 4) occurs in a number of blocks and basins which apparently represent former inner- and outer-tectonic arcs that have been structurally deformed into the present arrangement of interlocking crustal pieces. The rich ectoproct faunas located on various blocks and basins range in age from Asselian (Zone of *Pseudoschwagerina*) to Pamirian (Zone of *Yabeina-Lepidolina*) (Sakagami, 1961, 1970). The ectoprocts of the Zone of *Pseudoschwagerina,* which range from the Asselian into the Sakmarian, occur in isolated beds in the western part of central and southern Honshu and have three groups of cystoporates—the fistuliporid *Fistulipora,* the hexagonellid *Coscinotrypa,* and the sulcoreteporid *Sulcoretepora.* Trepostomes are *Pseudobatostomella, Stenopora,* and *Tabulipora.* Many different cryptostomes are present: the fenestellids *Fenestella* and *Penniretepora;* the polyporids *Polypora, Anastomopora,* and *Thamniscus;* the girtyoporid *Hayasakapora;* and the hyphasmoporid *Streblascopora.*

Ectoprocts of the Zone of *Parafusulina,* correlated with the Artinskian, occur in the central-eastern part of Honshu. Many genera range up from the Zone of *Pseudoschwagerina* including *Fistulipora, Pseudobatostomella, Stenopora, Fenestella, Penniretepora, Hayasakapora,* and *Streblascopora.* The hexagonellid cystoporates are diverse with *Meekopora, Meekoporella,* and *Prismopora.* The trepostome *Paraleioclema*?, and the rhabdomesid cryptostomes *Rhombopora* and *Rhabdomeson,* are additions to this fauna.

Ectoprocts of the Zone of *Neoschwagerina,* correlated with the Wordian (Ross and Nassichuk, 1970) of the Guadalupian, occur in the eastern part of northern Honshu and eastern Shikoku. The genera *Fistulipora, Meekopora, Pseudobatostomella, Paraleioclema*?, *Fenestella, Penniretepora, Polypora, Rhabdomeson, Hayasakapora,* and *Streblascopora* range up from the Zone of *Parafusulina. Sulcoretepora, Tabulipora,* and *Thamniscus* reappear and are joined by the hexagonellid *Fistulamina,* the cryptostomes *Septopora* and *Saffordotaxis,* and the trepostome *Ulrichotrypella.* The highest ectoprocts, those of the Zone of *Yabeina-Lepidolina,* range in age from upper Guadalupian to possibly Djulfian. Faunas of this zone occur in northeastern Honshu, southwestern Honshu, and Kyushu. The cystoporates show marked diversity and include *Fistulipora, Meekopora, Prismopora,* and *Sulcoretepora* which extend into this highest fauna, as well as the hexagonellid *Coscinotrypa,* and the goniocladiids *Goniocladia* and *Ramipora. Fenestella, Penniretepora,* and *Polypora* are joined by the septoporids *Septopora* and *Synocladia.* The trepostomes are still limited in number to *Pseudobatostomella, Paraleioclema*?, *Coeloclemis*?, and *Dyscritella. Rhabdomeson, Hayasakapora,* and *Streblascopora* are present as well as the nikiforovellid cryptostome *Clausotrypa.*

China, Mongolia, and Tibet

Data continue to accumulate on the Permian faunas of central and southern China (Figs. 1, 4). In the Lower Permian Chihsia Limestone of western Chekiang, Loo (1958) reported *Fistulipora, Fenestella, Polypora,* and *Septopora. Fistulipora* was also found in this unit in western Hupeh (Yang, 1956). The Upper Permian Maokou Formation, correlated with the lower and middle Guadalupian, has *Fistulipora, Fenestella, Polypora, Acanthocladia, Stenopora,* and *Dyscritella* (Morozova, 1970). Higher in the sequence, the Loping Series, correlated with the upper Guadalupian and part of the Pamirian, contains a fauna with *Fistulipora, Polypora, Penniretepora, Septopora, Synocladia, Pseudobatostomella,* and *Paraleioclema.*

Faunas from Inner Mongolia from the Jisu Honguer Limestone, correlated with the Maokou Limestone of China, contain the cystoporates *Fistulipora, Fistulamina,* and *Hexagonella;* the cryptostomes *Fenestella, Polypora, Rhabdomeson, Streblascopora, Maychella, Girtyopora,* and *Girtyoporina;* and the trepostomes *Stenopora, Tabulipora, Dyscritella,* and *Paraleioclema* (Grabau, 1931; Morozova, 1970).

From the Qomolangma Feng region of Tibet, Yang and Hsia (1975) have described a fauna from the Upper Se-lung Formation which is assigned a Lower Permian age. No faunal data provide a precise age determination and it is possible that this fauna is Upper Permian in age. The ectoprocts consist of the fistuliporids, *Fistuliramus* and *Fistulotrypa;* the hexagonellid *Meekopora,* the trepostome *Stenopora,* and the cryptostomes *Fenestella, Polypora, Maychella, Streblotrypa,* and *Streblascopora.*

Maritime Territory and Khabarovsk Region, USSR

This structurally complex area has rich ectoproct faunas most of which are Upper Permian (Figs. 1, 4). The data on scattered Lower Permian faunas do not permit precise age determinations. Shishova (1960) described new species of the cryptostomes, *Fenestella, Lyrocladia, Hemitrypa,* and *Polypora,* from the Lower Permian Gutayskaya Suite of Zabaykal, Maritime Territory. From the Khabarovsk region, Romanchuk (1966) identified a fauna, apparently from the upper part of the Lower Permian, which had *Fistulipora, Fenestella, Paraleioclema,* and *Dyscritella.* Nikitina and others (1970) found *Stenopora, Dyscritella,* and *Paraleioclema* in rocks that may be Artinskian in age, but this unit could be younger or older. In the Osakhta Suite and associated rock units, a Late Permian fauna of Kazanian age has *Fenestella, Polypora, Coscinotrypa, Dyscritella, Paraleiclema,* and *Permoleioclema* (Romanchuk, 1966).

Of the 30 genera listed by Morozova (1970) from the upper part of the Zone of *Yabeina* in the Maritime Territory and Khabarovsk region, 18 are common to both areas. Genera common to both areas are the cystoporates *Fistulipora* and *Fistulamina,* the trepostomes *Dyscritella, Ulrichotrypella, Hinganella, Stenodiscus*?, *Tabulipora, Paraleioclema,* and *Permoleioclema,* and abundant cryptostomes—the rhabdomesids *Rhabdomeson* and *Streblasco-*

pora, the nikiforovellids *Maychella* and *Clausotrypa,* the fenestellid *Fenestella,* the polyporid *Polypora,* the septoporid *Septopora,* and the girtyoporids *Girtyoporina* and *Girtyopora.* Genera found only in the Maritime Territory are the hexagonellid cystoporates *Hexagonella* and *Meekopora,* the goniocladiid *Goniocladia,* the trepostomes *Stenodiscus* and *Pseudobatostomella,* and the cryptostomes *Penniretepora, Synocladia, Acanthocladia, Timanodictya,* and *Hayasakapora.* Genera occurring only in the Khabarovsk region are the hexagonellid *Coscinotrypa* and the trepostome *Permopora.*

In the southern Maritime Territory in Lower? and Upper Permian strata, ectoprocts are abundant and varied. Some 37 ectoproct genera have so far been recorded by Romanchuk (1966, 1967), Romanchuk and Kiseleva (1968), Kiseleva (1969, 1970), and Nikitina and others (1970) from four different rock suites, the Barabash, Chandalazy, Osakhta, and Ugodinzinsk Suites (Fig. 1). Ectoprocts are particularly abundant in the lower parts of the Barabash Suite and its approximate time equivalent the Chandalazy Suite. Most of the strata making up the four rock suites are correlated with the Guadalupian, however, some strata are questionably assigned an Artinskian age but may be younger or older. Nikitina and others (1970) identified four ectoproct assemblages from parts of these rock suites. The lowest assemblage with the trepostomes *Stenopora, Paraleioclema,* and *Dyscritella* was of questionable Artinskian age or possibly younger or older. The next higher assemblage of probable Kazanian age, but possibly of latest Artinskian age, has *Dyscritella* and *Arcticopora.* A higher assemblage, probably Kazanian in age, has a high diversity of cystoporates with the fistuliporid *Cyclotrypa,* the hexagonellid *Coscinotrypa,* the goniocladiid *Goniocladia,* and the etherellid *Liguloclema.* The cryptostome *Ogbinopora* and the trepostomes *Dyscritellina, Permoleioclema, Hinganella,* and *Ulrichotrypella* comprise the remainder of the fauna. The highest assemblage, of Kazanian age, has a distinctive fauna with the trepostome *Primorella* and two groups of cryptostomes, the fenestellid *Hinganotrypa,* and the girtyoporids *Girtyopora, Girtyoporina, Hayasakapora,* and *Tavayzopora.* In addition to these four ectoproct assemblages, Kiseleva (1973) found another assemblage of 3 distinctive cystoporates in rocks of Kazanian age in both the Barabash and Chandalazy Suites. The assemblage consists of *Prismopora, Etherella,* and *Epiactinotrypa.* From the Svetlankin Beds of the southern Maritime Territory of Upper Permian Djulfian age, 4 genera occur; they are the trepostome *Pseudobatostomella* and the cryptostomes *Streblascopora, Fenestella,* and *Girtyoporina* (Kiseleva and others, 1973).

Northeast USSR

From the Kolyma and Omolon massifs and adjacent regions, an Upper Permian fauna (Nekhoroshev, 1935, 1959; Morozova, 1970) occurs for which no precise correlation can be determined at this time (Fig. 4). It has a great abundance of species of *Fenestella,* a number of species of the fenestellid *Wjatkella,* and the nikiforovellid *Maychella.* Furthermore, the fauna includes the cystoporate *Fistulipora,* the trepostomes *Dyscritella* and *Primorella,* and the cryptostomes *Polypora, Synocladia,* and *Timanodictya.*

North America

Little analysis of ectoproct faunas has been undertaken since the studies by Girty (1908), Moore and Dudley (1944), and Elias and Condra (1957). The faunal lists reflect this lack of study. Data for this report has come from the studies noted above, and also from Branson (1948), Newton (1971), and Warner and Cuffey (1973). In the Lower Permian of the Midcontinent of the United States (Nebraska, Kansas, and Oklahoma), cryptostomes predominate and are represented in strata of Wolfcampian age by *Fenestella, Polypora, Septopora, Thamniscus, Streblotrypa, Rhombopora,* and *Syringoclemis* (Figs. 1, 4). The fistuliporid cystoporate *Cyclotrypa* occurs in the Wolfcampian of Nebraska and the hexagonellid cystoporate *Meekopora* in the Wolfcampian of Kansas. The ctenostomes, *Ascodictyon, Bascomella,* and *Condranema,* are also in the Wolfcampian of the Midcontinent region. *Bascomella* and also *Meekopora* range into the upper Guadalupian in Arizona. In the Wolfcampian of west Texas, both *Cyclotrypa* and *Meekopora* as well as an abundance of fenestrate ectoprocts, are present. These cystoporates are joined by *Meekoporella* and *Fistulipora* in the Leonardian and all 4 genera extend up into the Wordian (lower Guadalupian). Two other cystoporates also appear in the Guadalupian; they are *Goniocladia* and *Epiactinotrypa.* The cryptostomes have marked diversity with *Fenestella, Polypora, Thamniscus, Acanthocladia, Girtyopora,* and *Girtyoporina.* The trepostomes are *Pseudobatostomella*?, *Stenopora,* and *Paraleioclema*?. Higher in the upper Guadalupian (Capitanian), the fauna has the cystoporates *Fistulipora* and *Goniocladia,* the cryptostomes *Fenestella, Acanthocladia,* and *Girtyoporina,* and the trepostome *Paraleioclema.* In the southwest in Nevada, *Tabulipora* has been noted in rocks of Leonardian (Gilmour, 1962) and Guadalupian (Mayou, 1967) age.

Frequent comparison is made in systematic studies to an ectoproct assemblage of 11 genera described by Fritz (1932) from Strathcona Park, Vancouver Island, B.C., Canada. The fauna comes from strata which are complexly folded, faulted, and preserved in tilted fault blocks. Yole (1963), in examining a fauna from the Buttle Lake Formation, which is probably correlative with strata from which the fauna described by Fritz (1932) was obtained, considered the brachiopod and bryozoan assemblage to be most probably Early Permian in age. So far a more precise age determination has not been established. Fritz (1932) identified *Goniocladia, Ulrichotrypa, Rhombopora, Streblotrypa, Clausotrypa, Fenestella, Penniretepora, Polypora, Thamniscus*?, *Acanthocladia,* and *Phyllopora.* Yole's (1963) listing has all these genera except *Ulrichotrypa* (which he referred to *Stenopora*), *Streblotrypa,* and *Phyllopora* and, in addition, he included *Rhabdomeson.*

South America

So far ectoprocts have been reported only from the Lower Permian of southern Peru (Chronic, 1953) (Fig. 4).

The fauna occurs in the lower part of the Copacabana Group in the Zone of *Septopora* (= Zone of *Silvaseptopora*) and the Zone of *Triticites opimus*, both of Wolfcampian age. The fauna has the cystoporates *Meekopora* and *Goniocladia,* and the cryptostomes *Fenestella, Polypora, Septopora, Rhombopora,* and *Acanthocladia.*

Africa

Isolated occurrences of ectoprocts have been noted for the vast region of Africa including Egypt and Tunisia but no faunal data are published. In southwest Africa (Namibia) (Fig. 4), fenestellid fragments and the trepostome *Dyscritella* were found in the lower part of the Dwyka Tillite (Wass, 1972) and are considered to be Kazanian in age based on the associated pelecypod fauna.

DISTRIBUTION OF FAMILIES AND GENERA

Patterns of distributions of families and genera for different periods of time during the Permian are examined. The time periods are: (1) Asselian, Sakmarian, and early Artinskian (for brevity this is referred to as early Permian in this section of the report); (2) middle and late Artinskian; (3) Kazanian; and (4) Djulfian (= Pamirian). In describing geographic ranges of families and genera, they may be cosmopolitan; they may become dispersed or restricted to particular regions; or they may be endemic to certain regions. The regions (Fig. 5) which I have been able to identify based on ectoproct occurrences are: (1) Uralian Sea and Russian Platform (shelf); (2) Franklinian Sea and adjacent shelves including the Barents Shelf and shelf areas in northern North America; (3) Zechstein Sea of England, Germany, Poland, and southern Baltic region; (4) southern part of North America; (5) Andean Sea; (6) northern Tethyan Sea which includes Japan, Maritime Territory, Khabarovsk region, and possibly parts of the northeast USSR; (7) central Tethyan Sea which includes Transcaucasus, Darvas, Pamir, Tibet, Mongolia, and southwest China; (8) southern Tethyan Sea which includes central-eastern Afghanistan, the Salt Range of Pakistan, Thailand, Malaya, western Australia, and Timor; and (9) Tasman Geosyncline (eastern Australia).

Cystoporata

Permian Cystoporata are well-represented by several families and a considerable diversification of genera. Of the early Permian Fistuliporidae, *Fistulipora* is cosmopolitan and *Metelipora* is restricted to the Uralian Sea. During the Sakmarian, *Cyclotrypa* was present in the southern part of North America but in the Artinskian it appeared also in the Uralian Sea and Russian Platform, the northern and southern Tethyan Sea as well as the southern part of North America. During Artinskian and Kazanian times, 3 genera, *Fistulipora, Eridopora,* and *Fistulotrypa,* have also a Tethyan distribution. *Eridopora* is found also on the Russian Platform in the middle Artinskian, in the central and southern Tethys in the middle and late Artinskian, in the southern Tethys with *Fistulotrypa* in the Kazanian, then in the central Tethys in the late Kazanian and possibly Djulfian. *Fistulipora* occurs in all parts of the Tethyan Sea and also in the southern part of North America during late Artinskian time and continued on in the central and northern Tethyan Sea in Pamirian time. *Cyclotrypa* was cosmopolitan in the Kazanian.

In the Hexagonellidae, a distinctive family in the Permian, *Hexagonella* was cosmopolitan until Kazanian time when it became restricted to the central and southern Tethys and later to the central Tethys during the Djulfian. Three other genera are Tethyan in distribution in the Sakmarian and part of the Artinskian; *Coscinotrypa* is present both in the northern and southern Tethys, and *Evactinopora* and *Evactinostella* are endemic to the southern Tethys. *Coscinotrypa* also occurs in the Uralian Sea in the Lower Permian and in the Kazanian is restricted to the Zechstein Sea and northern Tethys. During the Artinskian, *Fistulamina* appeared in the central and southern Tethys, and in the Kazanian occurs in the central and northern Tethyan Seas, and *Prismopora* in the northern and southern Tethys. *Meekopora,* initially restricted to the southern part of North America and the Andean Sea in the early Permian, dispersed into the northern Tethys in Artinskian time, and thrived in the southern part of North America. This genus is present in strata of Wordian (Guadalupian) age in southern North America, in the lower Kazanian of the Russian Platform and Uralian Sea, and is also recorded from the Upper Permian of the northern Tethys. *Meekoporella,* another genus present in the northern Tethys and southern part of North America in the Artinskian, is restricted to the southern part of North America in the Kazanian.

The Sulcoreteporidae is represented by *Sulcoretepora,* a Tethyan genus occurring in the central and northern Tethyan Sea in the early Permian, the southern Tethyan Sea in the Artinskian, and the northern Tethyan Sea in the Kazanian where it survived well into the late Kazanian or even into the Djulfian.

The Goniocladiidae includes the cosmopolitan genus *Ramipora* which had a wide geographic distribution throughout the Permian (Sakmarian through Kazanian). *Goniocladia,* which is cosmopolitan in the early Permian, becomes restricted to the southern Tethyan Sea and Tasman Geosyncline in the Artinskian, becomes dispersed in Kazanian time, occurring in the southern and northern Tethyan Sea, the Russian Platform, and the southern part of North America, and lingers into the Djulfian of the northern Tethyan Sea. As far as known, *Ramiporidra* occurs in the early Permian and Artinskian. The Etherellidae is a Tethyan family occurring in the southern Tethys and Tasman Geocyncline in the Artinskian but becoming restricted to the northern Tethys in the Kazanian. *Etherella* appears to be endemic to the southern Tethys in the Artinskian and to the northern Tethys in the Kazanian. *Liguloclema* occurs in the southern Tethys and Tasman Geosyncline in the Artinskian and the northern Tethys in the Kazanian. The Actinotrypidae, which are limited in their occurrence, are found in the central Tethys in the

Sakmarian and Artinskian, and in the northern Tethys and southern part of North America in the Kazanian.

Cryptostomata

The large family Fenestellidae has the well-known cosmopolitan genus *Fenestella* which ranges through most of the Permian; only in Djulfian time is the genus restricted to the northern Tethys. During early Permian time *Penniretepora, Ptylopora, Ptiloporella, Diploporaria, Archimedes,* and *Lyrocladia* were part of the rich fauna of the Uralian Sea and Russian Platform. At this time, *Penniretepora* was also in the northern and southern Tethys, *Wjatkella* in the central and northern Tethys and the Tasman Geosyncline, and *Minilya* in the southern Tethys. In the Artinskian, *Penniretepora* maintained its wide distribution and dispersed into the Tasman Geosyncline. Although absent from the Uralian Sea during the Kazanian, *Penniretepora* is present in the Zechstein Sea, the Tethys, and Tasman Geosyncline. In the Artinskian, *Ptylopora* and *Ptiloporella* persist in the Uralian Sea together with *Wjatkella* which is also in the central and southern Tethys; *Diploporaria* appears in the northern Tethys and Tasman Geosyncline. In the Kazanian, *Ptylopora* occurs in the Franklinian Sea and Tasman Geosyncline, *Levifenestella* in the Tasman Geosyncline, and *Wjatkella* in the Uralian Sea, eastern part of the North American Cordillera, and northern regions of northeast Siberia.

The 2 genera of the Septoporidae, *Septopora* and *Synocladia,* have a wide distribution. *Septopora* was cosmopolitan in early and late Permian time and apparently became restricted during the Artinskian to the central Tethys. *Synocladia* was in the Uralian Sea in the early Permian, then in the southern Tethys in the Artinskian and the Zechstein, and in the central and northern Tethys in the Kazanian. The Septatoporidae, represented only by *Septatopora,* is a distinctive genus in the Kazanian of the Tasman Geosyncline. The Fenestraliidae, *Parafenestralia* and *Triznella,* are a specialized group in the Uralian Sea and Russian Platform in the Kazanian.

In the Polyporidae, only 1 genus, *Polypora,* is cosmopolitan through the Permian. *Anastomopora* is endemic to the northern Tethys and *Lyropora* is endemic to the southern Tethys in early Permian time. *Reteporidra* flourished in the Uralian Sea during the entire Permian and also extended its range into the central Tethys in the Kazanian. *Kingopora* is a distinctive element of the Uralian and the Zechstein Seas in the Kazanian. *Thamniscus* appeared in the early Permian in the southern and northern Tethys and the southern part of North America, continued in the southern Tethys in the Artinskian, and became cosmopolitan in the Kazanian. *Acanthocladia* of the Acanthocladiidae appears to parallel *Septopora* in its distribution through the Permian, being apparently cosmopolitan in the early Permian, then confined to the southern Tethys in the Artinskian, and again cosmopolitan in the Kazanian. The specialized *Kalvariella* is endemic to the Zechstein Sea during Kazanian time. The Timanodictyidae are represented by *Timanodictya* in the Uralian and Franklinian Seas in the early Permian, are found principally in the southern Tethys in the Artinskian, and in the Uralian Sea, southern part of North America, and northern Tethys in the Kazanian.

The Rhabdomesidae are well-represented in the Permian until the end of the Artinskian. In the early Permian, *Rhabdomeson, Ascopora, Nicklesopora,* and *Rhombopora* were common in the Uralian Sea. Of these, both *Rhabdomeson* and *Rhombopora* had wider distributions; *Rhabdomeson* was also in the Franklinian Sea and central Tethys in the early Permian, spread into the southern and northern Tethys in the Artinskian, and had an even wider distribution in all the Tethyan Seas in the Kazanian. *Rhombopora* was almost universal except for possibly the northern Tethys. *Syringoclemis* is endemic to the early Permian of the southern part of North America and *Streblocladia* is endemic to the southern Tethys in the early Permian. In the Artinskian, a number of rhabdomesid genera are restricted to the Tethys, namely, *Rhabdomeson, Ascopora, Pamirella Megacanthopora?,* and *Saffordotaxis.* The latter 2 genera are principally southern Tethys in distribution, although *Saffordotaxis* is also present in the Tasman Geosyncline. In Artinskian time, *Rhombopora* became cosmopolitan but *Nicklesopora* was still restricted to the Uralian Sea. *Callocladia?* occurs in the southern Tethys. In the Kazanian, the Tethyan genera are *Rhabdomeson, Rhombopora,* and *Saffordotaxis;* however, *Rhombopora* is also found in the Uralian Sea and *Saffordotaxis* in the Tasman Geosyncline.

The Hyphasmoporidae are mainly Tethyan in distribution except for the genus *Streblotrypa* and, at times, a few other genera. In the early Permian, *Hyphasmopora* occurs in the Franklinian Sea; *Streblotrypa* is widespread except for limited occurrence in the Tethys; and *Streblascopora* is found in central, southern, and northern Tethys. In the Artinskian, *Ogbinopora* and *Streblascopora* are widely distributed in the Tethys, the latter also occurring in the Tasman Geosyncline. In the Kazanian, the genera *Ogbinopora, Streblotrypa,* and *Streblotrypella* are Tethyan, and *Streblascopora* occurs in the Tethys and Uralian Sea.

The Nikiforovellidae were widely distributed in the Uralian Sea and Tethyan Seas. During the early Permian, *Nikiforovella* in the central Tethys and *Clausotrypa* in the Uralian Sea occurred in great abundance. *Clausotrypa* is also present in the Uralian Sea in the Artinskian and the southern and northern Tethys in the Kazanian. In the Kazanian, 2 additional nikiforovellid genera appear: *Maychella,* which is distributed in the central and northern Tethys, and *Pinegopora* which occurs in the Uralian Sea and Russian Platform.

The Girtyoporidae are represented in the northern Tethys during the early Permian by *Hayasakapora* which ranges through the Artinskian into the high Kazanian. *Girtyopora* appears in the Artinskian in the southern Tethys and by Kazanian time is widespread in the northern and central Tethys, the southern part of North America, and the Uralian Sea. *Tavayzopora* is endemic to the northern Tethys.

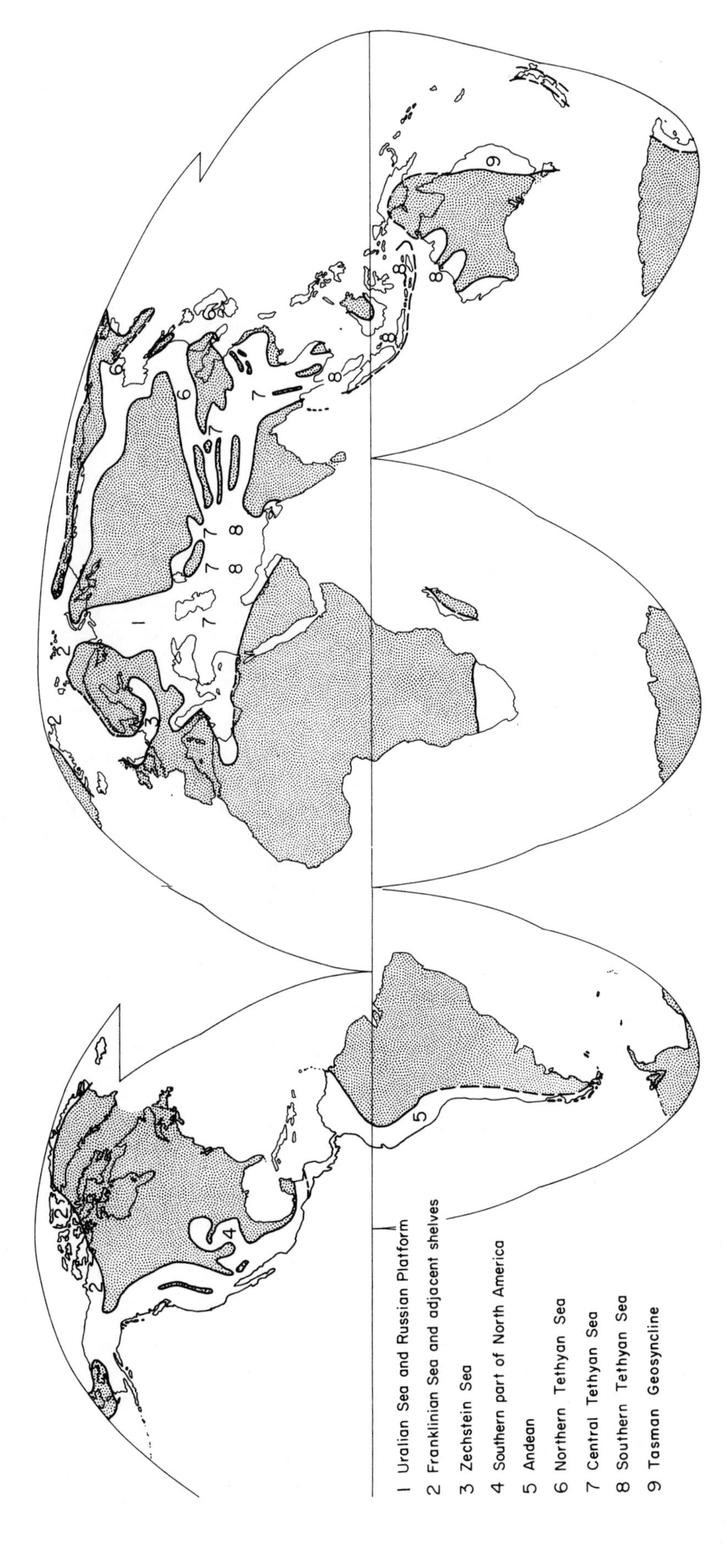

Figure 5. General distribution of Permian seaways and shallow-water continental shelves plotted on map of present geography.

The Phylloporinidae have little representation in the Permian in comparison to their abundance in older rocks of the lower and middle Paleozoic. *Chainodictyon* and *Bashkirella* both occur in the Uralian Sea and are present in the early Permian. *Chainodictyon* continues up into the Artinskian and is found in the Uralian Sea and central Tethys.

Trepostomata

Trepostome ectoprocts are far less abundant in the Permian than in other Paleozoic rocks and have a predominantly Tethyan distribution except for 3 or 4 cosmopolitan genera. The Anisotrypidae, represented by *Anisotrypella,* occurs in the Kazanian of the Uralian Sea. The Eridotrypellidae are a small group with 3 genera; *Hinganella* is endemic to the southern Tethys in the Artinskian and the northern Tethys in the Kazanian. In the Kazanian, *Neoeridotrypella* is endemic to the Uralian Sea and *Permopora* is endemic to the northern Tethys.

The widely known Stenoporidae have the cosmopolitan *Stenopora* and *Tabulipora* which range through the Permian except for the latest Permian. *Rhombotrypella* is also cosmopolitan in the early Permian but appears restricted to the Uralian Sea in the Kazanian. *Stenodiscus* occurs in the southern Tethys and the Tasman Geosyncline in the Artinskian; it occurs in the northern Tethys and Tasman Geosyncline during the Kazanian. *Arcticopora,* first described from the Triassic of the Arctic, occurs in the Kazanian in the northeastern Tethys.

The Dyscritellidae are mainly distributed in the Tethys. In the early Permian, *Dyscritella* is present in the central and southern Tethys and may also occur in the Uralian Sea. *Pseudobatostomella* is in the northern Tethys and Uralian Sea and may also be in the southern Tethys. During Artinskian time, this genus is widespread in the central, northern, and southern Tethys, and the Uralian and Zechstein Seas. Species that are questionably assigned to *Pseudobatostomella* would give the genus a nearly cosmopolitan distribution in both the Artinskian and Kazanian. Both *Dyscritella* and *Pseudobatostomella s. s.* are in the northern Tethys in the Kazanian and *Pseudobatostomella* ranges into the Djulfian in the northeastern Tethys.

In the Ulrichotrypellidae, *Primorella* is an early Permian genus in the central Tethys but reappears in the Kazanian in the northern Tethys and northeastern USSR. *Ulrichotrypa* is a cosmopolitan genus in the Artinskian and Kazanian. *Ulrichotrypella* is found in both the Uralian Sea and northern Tethys in the Kazanian.

The Araxoporidae are found mainly in the Tethys. *Paraleioclema* is present in the southern Tethys in the early Permian, in the northern Tethys during the Artinskian, and the central and northern Tethys during the Kazanian. *Permoleioclema* is in the northern Tethys in the Kazanian. Species questionably assigned to *Paraleioclema* are present in Uralian and Franklinian Seas in the Artinskian.

CONCLUSIONS

Analysis of the taxonomic and distributional data for Permian ectoprocts serves to help define patterns of paleobiogeographic distribution in marine invertebrates. Generic diversity of ectoprocts increases from 55 genera in the Asselian, Sakmarian, and early Artinskian and 53 genera in the middle and late Artinskian, to 65 genera in the Kazanian. This increase in diversity appears to be correlated with the development of provincial faunas that have a strong latitudinal pattern. Many genera, particularly cryptostomes, either have cosmopolitan distributions or occur in the well-studied southern Uralian fauna during the early part of the Early Permian but became restricted to the central or southern part of the Tethys during the later part of the Early Permian (Artinskian). In the succeeding Kazanian, many genera are widely dispersed in the Tethys and of these most are restricted to the Tethys. This pattern is followed by a rapid decrease in diversity during Djulfian time and into Early Triassic time.

Nine distinct regions (Figs. 5 to 7) are identified from faunal assemblages; these were interconnected at different times during the Permian.

Uralian Sea and Russian Platform. A great diversity of genera and families characterize this fauna during the Early Permian. Fenestellids, polyporids, rhabdomesids, and *Timanodictya* are abundant. Cystoporates (*Hexagonella, Ramipora,* and *Metelipora*) and representatives of several cryptostome families are present. By Artinskian time, fenestellids, polyporids, rhabdomesids, and the trepostome *Pseudobatostomella* are predominant and in late Artinskian, *Rhombotrypella* became a significant genus in the fauna. In Kazanian time (Late Permian), the rich fauna has fenestellids (*Fenestella* and *Wjatkella*), fenestraliids (*Parafenestralia* and *Triznella*), polyporids (*Polypora* and *Thamniscus*), the trepostomes *Pseudobatostomella, Dyscritella, Rhombotrypella, Neoeridotrypella,* and *Paraleioclema,* and the cryptostomes *Streblascopora, Rhombopora,* and *Timanodictya.*

Franklinian Sea and adjacent shelves. A sparse fauna of trepostomes, cryptostomes, and cystoporates is presently known from this region. Rare cystoporates, cryptostomes *Fenestella, Synocladia, Polypora,* and *Timanodictya,* and trepostomes *Stenopora* and *Tabulipora,* comprise the fauna during the Early Permian. The cystoporate *Ramipora* and cryptostomes *Fenestella* and *Polypora* range through the Early Permian into the Kazanian. *Ptylopora* apparently dispersed into this area from the Uralian Sea. The diversity of species of *Stenopora* parallels the diversity in this genus in the Tasman Geosyncline and suggests that these northern faunas were possibly part of a cooler water fauna than those in the northern part of the Uralian Sea although both areas had evaporite deposits at various times during the Early Permian.

Zechstein Sea. Considerable data on many different faunas and the geology are available for this region which was connected to the northwest with the Franklinian Sea. During Kazanian time, ectoprocts occurred in small patch

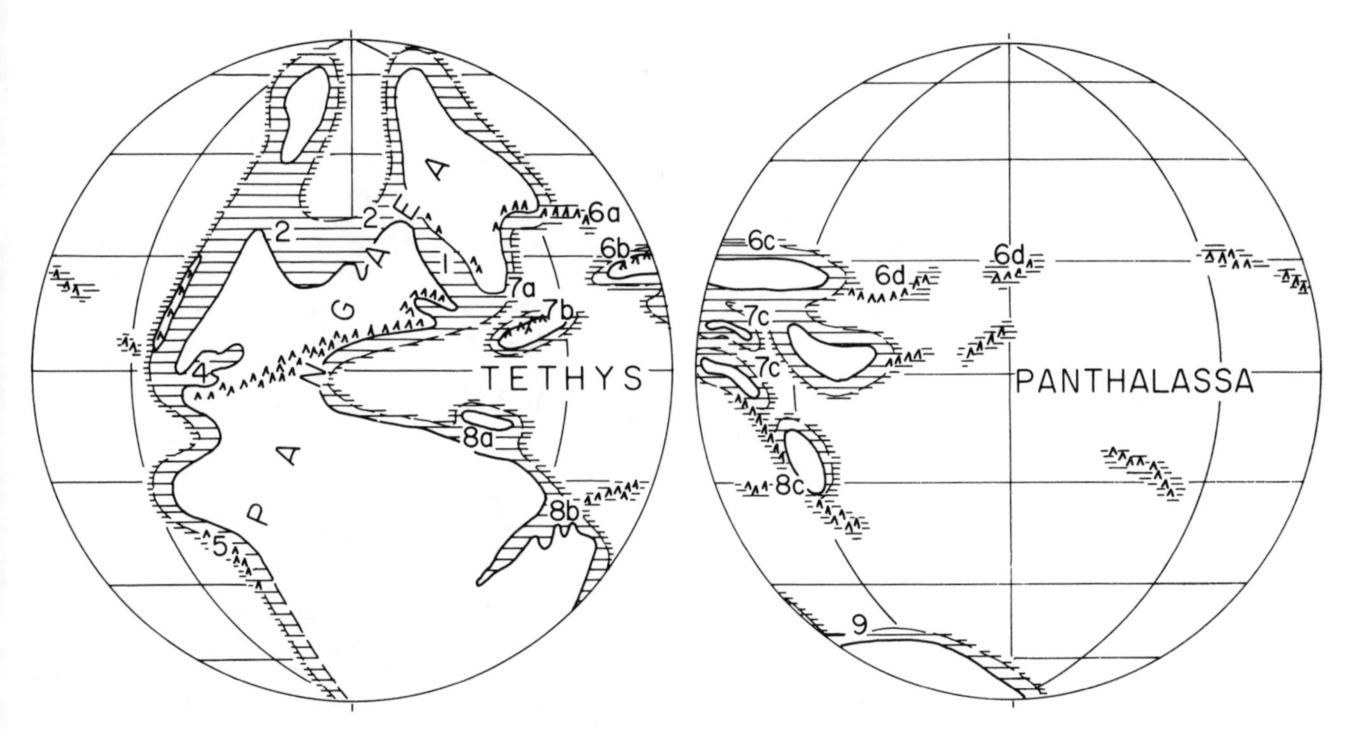

Figure 6. Paleogeographic reconstruction in the Artinskian based on map modified from Dietz and Holden (1970) and Smith and others (1973), with some data from Grunt and Dmitriev (1973). Numbers indicate regions referred to in text and letters refer to different parts of regions. (6a), (6b), (6c) Maritime Territory, Khabarovsk region and northeast USSR; (6d) Japan; (7a) Transcaucasus, Darvas, and Pamir; (7b) Tibet and Mongolia; (7c) southwest China; (8a) northern Afghanistan and Salt Range, Pakistan; (8b) western Australia; (8c) Malaya and Thailand.

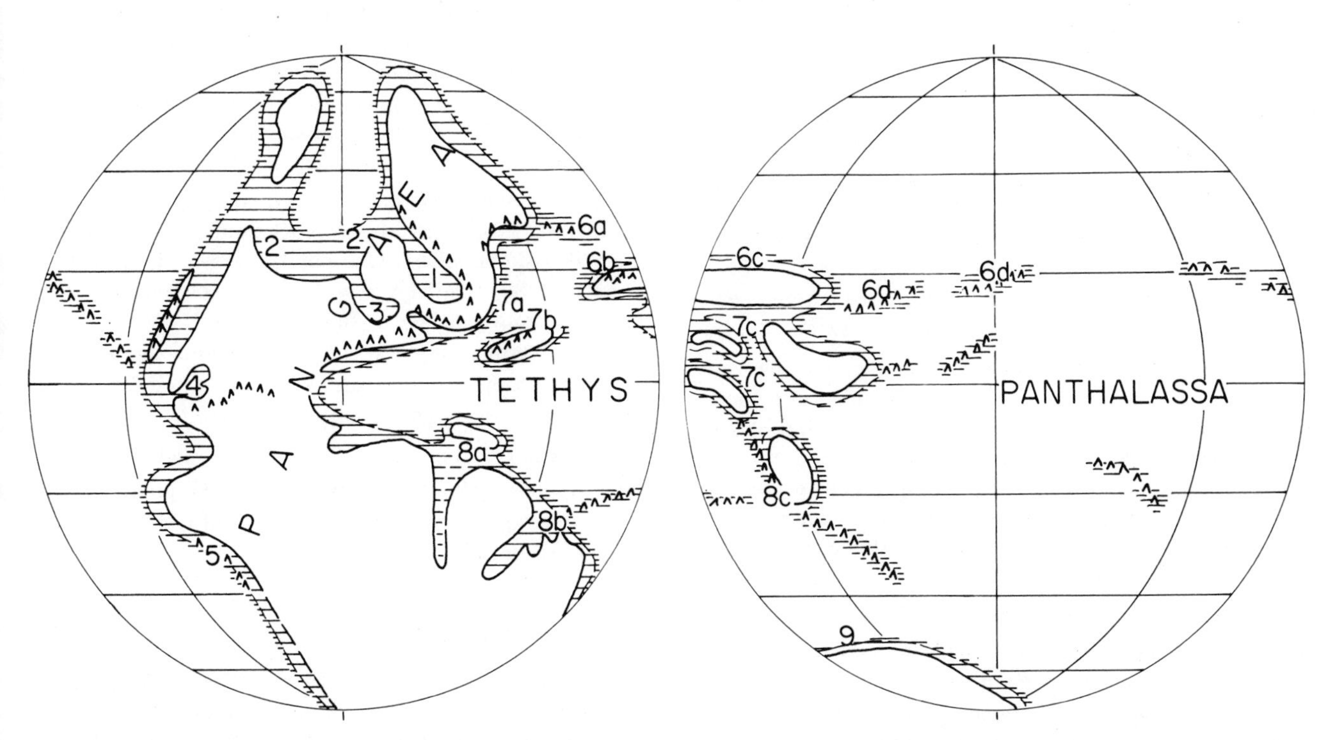

Figure 7. Paleogeographic reconstruction in the Kazanian based on map modified from Dietz and Holden (1970) and Smith and others (1973), with some data from Grunt and Dmitriev (1973). Numbers and letters are explained with Figure 6.

reefs laterally associated with evaporite deposits. This small fauna has limited diversity with predominantly fenestrate cryptostomes, such as *Kingopora, Fenestella, Thamniscus, Penniretepora, Synocladia,* and *Acanthocladia,* suggesting that some cryptostomes were tolerant of salinities that were higher or lower than normal.

Southern part of North America. Ectoproct faunas from the extensive Permian deposits of North America are reported only from the south-central and southwestern part of the region. In the Wolfcampian (Early Permian), fistuliporid and hexagonellid cystoporates and several groups of cryptostomes (*Fenestella, Polypora, Septopora, Thamniscus, Streblotrypa,* and *Rhombopora*) comprise the fauna. In the Leonardian (Early Permian), fistuliporid and hexagonellid cystoporates and fenestrate cryptostomes are abundant. In the Guadalupian (Late Permian), cystoporates and cryptostomes have increased diversity.

Andean Sea. During the Permian, the Andean Sea lay adjacent to the southern part of North America and the sparse ectoproct fauna indicates close connection between these two regions. In Early Permian, the fauna has cystoporates and cryptostomes that are similar to those in the southern part of North America.

Northern Tethyan Sea. The trepostomes *Pseudobatostomella, Stenopora,* and *Tabulipora,* and also some cryptostomes of the Uralian and Franklinian Seas are present during the Early Permian in this region. The girtyoporid cryptostome *Hayasakapora* is a distinct genus of the region. Cystoporates of Artinskian age diversified and this diversification continued through the Guadalupian (Late Permian) with an increased number of genera. Cryptostomes and trepostomes also became more abundant through the Permian and are represented by several distinctive genera, such as the trepostomes *Hinganella, Ulrichotrypella,* and *Permoleioclema,* and the cryptostomes *Maychella* (also found in the central Tethys), *Girtyopora,* and *Girtyoporina.*

The Upper Permian fauna of northeastern Siberia suggests connections with both the northern Tethys and the Uralian Sea because it has elements such as *Wjatkella,* which also is present in the Kazanian of the Uralian Sea, and *Maychella,* which also occurs in the northern and central Tethys.

Central Tethyan Sea. The Lower Permian has an abundance of cystoporates from several families and some cryptostomes including abundant *Nikiforovella* and a few trepostomes. In the Upper Permian, cystoporates of several families are still common, the cryptostomes show a gradual increase in diversity, and the trepostomes are sparse. The cryptostome *Ogbinopora* and the trepostome *Araxopora* are distinctive elements of the fauna.

Southern Tethyan Sea. The rich fauna of the Lower Permian of western Australia, Thailand, and Malaya has a high diversity of cystoporates and cryptostomes. In the lower part of the Upper Permian in western Australia, a reduced fauna of cystoporates, cryptostomes, and trepostomes may be an artifact of the geological record. The ectoproct data also suggest that western Australia was part of a southern Tethys subtropical and tropical biogeographic region (that extended also into Thailand and Malaya) based on the commonality of genera that are otherwise restricted to Tethyan faunas, an increased faunal diversity, and an abundance of biogenic carbonates. An abundance of different cystoporates persists in other parts of the southern Tethyan Sea in the Upper Permian. In Thailand and Malaya, cryptostomes show great diversity and trepostomes are rare. Connection with the central Tethyan Sea is established by the presence of such genera as *Ogbinopora* and *Maychella.*

Tasman Geosyncline. In the early part of the Early Permian, a sparse cryptostome fauna of *Fenestella* and *Polypora* in the Tasman Geosyncline suggests limited exchange or possibly no exchange with the southern Tethys of western Australia. During Artinskian time, the diversity of cystoporates, cryptostomes, and trepostomes increased markedly in the northern part of the Geosyncline (Bowen Basin and Springsure Shelf); this suggests continuing dispersal during the Artinskian of genera that were common in the Sakmarian of western Australia. The southern part of the Tasman Geosyncline during the Early Permian has a reduced cryptostome and trepostome fauna which may reflect cooler water conditions. This low generic diversity, combined with generally poorly developed carbonate facies in the Tasman Geosyncline, strongly suggest this was a cool-water region during the Permian. Because fenestellids and polyporids comprise a large proportion of the ectoproct fauna of this seaway and cystoporates are very sparse, it appears these groups were more tolerant of cool water than were many other ectoprocts. The lack of cystoporates may suggest that this group required certain environmental conditions, such as optimally warmer waters and higher salinities, that were lacking in the southern part of the Geosyncline. The Upper Permian has a small fauna of trepostomes and cryptostomes.

There remains no adequate explanation for the large number of extinctions of ectoproct families near the Permian-Triassic boundary. One point, however, is critical. By Djulfian time, relatively few localities contain ectoproct-bearing strata and those few occur in the central and northern Tethyan regions so that our knowledge of latest Permian ectoproct faunas is considerably more incomplete than for any other part of the Permian record.

More information about phylogenetic relations of ectoproct species, genera, and families would greatly aid in more precisely determining the patterns of distribution and times of dispersal.

ACKNOWLEDGMENTS

I thank E. Allen and E. Kaplan, Western Washington University, for aid in translating Chinese literature and J. Muller, Geological Survey of Canada, for discussion of localities on Vancouver Island.

REFERENCES

Anderson, S., 1974, Patterns of faunal evolution: Quart. Rev. Biol., v. 49, p. 311-332.

Ashlock, P. D., 1974, The uses of cladistics: Ann. Rev. Ecology and Systematics, v. 5, p. 81-99.

Bassler, R. S., 1929, The Permian Bryozoa of Timor: Paläontologie Timor, Leif. 16, Abh. 28, p. 37-90.

Branson, C. C., 1948, Bibliographic index of Permian invertebrates: Geol. Soc. America Mem. 26, 1049 p.

Chronic, J., 1953, Invertebrate paleontology (excepting fusulinids and corals), *in* Newell, N. D., Chronic, J., and Roberts, T. G., Upper Paleozoic of Peru: Geol. Soc. America Mem. 58, p. 43-165.

Cracraft, J., 1974, Continental drift and vertebrate distribution: Ann. Rev. Ecology and Systematics, v. 5, p. 215-261.

Crockford, J., 1943, Bryozoa from the Port Keats Bore, Northern Territory: Linnean Soc. New South Wales Proc., v. 68, p. 145-149.

——— 1944, Bryozoa from the Permian of western Australia. Pt. I. Cyclostomata and Cryptostomata from the North-West Basin and Kimberly District: Linnean Soc. New South Wales Proc., v. 69, p. 139-175.

——— 1951, The development of bryozoan faunas in the upper Palaeozoic of Australia: Linnean Soc. New South Wales Proc., v. 76, p. 105-122.

——— 1957, Permian Bryozoa from the Fitzroy Basin, Western Australia: Australia Bur. Mineral Resources, Geology and Geophysics Bull. no. 34, 134 p.

Dickins, J. M., 1970, Correlation and subdivision of the Permian of western and eastern Australia: Australia Bur. Mineral Resources, Geology and Geophysics Bull. no. 116, p. 17-27.

Dietz, R. S., and Holden, J. C., 1970, Reconstruction of Pangaea: Breakup and dispersion of continents, Permian to Present: Jour. Geophys. Research, v. 75, p. 4939-4956.

Dreyer, E., 1961, Die Bryozoen des mitteldeutschen Zechstein: Freiberg. Forschungsh., Paläontologie, v. 103, p. 5-51.

Elias, M. K., and Condra, G. E., 1957, *Fenestella* from the Permian of west Texas: Geol. Soc. America Mem. 70, 158 p.

Etheridge, R., 1907, Official contributions of the palaeontology of Western Australia. Vol. 19, Fossils of the Port Keats Bore, thirty miles north of Fossil Head, Treachery Bay: Suppl. Parliamentary Paper 55.

Fritz, M. A., 1932, Permian Bryozoa from Vancouver Island: Royal Soc. Canada Trans., ser. 3, v. 26, p. 93-109.

Furnish, W. M., 1966, Ammonoids of the Upper Permian *Cyclolobus*-Zone: Neues Jahrb. Geol. Paläont. Abh., v. 125, p. 265-296.

——— 1973, Permian stage names, *in* Logan, A., and Hills, L. V., eds., The Permian and Triassic Systems and their mutual boundary: Canadian Soc. Petroleum Geologists Mem. 2, p. 522-548.

Gilmour, E. H., 1962, A new species of *Tabulipora* from the Permian of Nevada: Jour. Paleontology, v. 36, p. 1019-1020.

Girty, G. H., 1908, The Guadalupian fauna: U. S. Geol. Survey Prof. Paper 58, 651 p.

Gorjunova, R. V., 1975, Permian Bryozoa of Pamir: Akad. Nauk SSSR Paleont. Inst. Trudy, v. 148, 127 p. (in Russian).

Grabau, A. W., 1931, The Permian of Mongolia, *in* Natural history of central Asia, Vol. 4: Am. Mus. Nat. History, 665 p.

Gregorio, A. de, 1930, Sul Permiana di Sicilia: Ann. Géol. Paléont., v. 52, p. 1-70.

Grunt, T. A., and Dmitriev, B. Ju., 1973, Permian brachiopods of Pamir: Akad. Nauk SSSR Paleont. Inst. Trudy, v. 136, 211 p. (in Russian).

Hart, G. F., 1969, Palynology of the Permian Period, *in* Tschudy, R. H., and Scott, R. A., eds., Aspects of palynology: New York, Wiley, p. 271-289.

——— 1974, Permian palynofloras and their bearing on continental drift, *in* Ross, C. A., ed., Paleogeographic provinces and provinciality: Soc. Econ. Paleontologists and Mineralogists Spec. Pub. 21, p. 148-164.

Hill, Dorothy, 1958, Sakmarian geography: Geol. Rundschau, v. 47, p. 590-629.

Holland, C. H., 1971, Silurian faunal provinces?, *in* Middlemiss, F. A., Rawson, P. F., and Newall, G., eds., Faunal provinces in space and time: Liverpool, Seel House Press, p. 61-76.

Johnsen, A., 1906, Bryozoen aus den karnischen Fusulinenkalk: Neues Jahrb. Mineral. Geol. Paläont., v. 2, p. 135-160.

Keast, A., Erk, F. C., and Glass, B., eds., 1972, Evolution, mammals and southern continents: Albany, N.Y., State Univ. New York Press, 543 p.

Kiseleva, A.V., 1969, New bryozoans of the family Girtyoporidae from the Upper Permian of the southern Maritime Territory: Paleont. Zhurn. no. 1, p. 90-94 (in Russian).

——— 1973, Some new and rare Cystoporata from the Upper Permian of the southern Maritime Territory: Paleont. Zhurn. no. 3, p. 65-70 (in Russian).

Kiseleva, A. V., Tashchi, S. M., and Vasil'ev, B. I., 1973, Age of the Svetlankin beds, southern Primorye region: Moskovskoe Obshchestvo Ispytatelei Prirody Bull., Otdel. Geol., v. 73, p. 24-34 (in Russian).

Kummel, B., and Teichert, C., 1966, Relations between the Permian and Triassic formations in the Salt Range and Trans-Indus ranges, West Pakistan: Neues Jahrb. Geol. Paläont. Abh., v. 125, p. 297-333.

——— 1973, The Permian-Triassic boundary beds in central Tethys, *in* Logan, A., and Hills, L. V., eds., The Permian and Triassic Systems and their mutual boundary: Canadian Soc. Petroleum Geologists Mem. 2, p. 17-34.

Leven, E. Ya. 1967, Stratigraphy and fusulinids of the Permian deposits of the Pamirs: Akad. Nauk SSSR Geol. Inst. Trudy, v. 167, p. 1-224 (in Russian).

Loo, L., 1958, Some bryozoans from the Chihsia Limestone of Hanchow, western Chekiang: Acta Paleontologica Sinica, v. 6, p. 293-304.

Malecki, J., 1968, Permian bryozoans from the Tokrössoya Beds, Sorkapp Land, Vestspitsbergen: Studia Geol. Polonica, v. 21, p. 7-32.

Mayou, T. V., 1967, Paleontology of the Permian Loray Formation in White Pine Co., Nevada: Brigham Young Univ., Research Studies Geology Series, v. 14, p. 101-122.

Meyen, S. V., 1970, Permian floras, *in* Vakhrameyev, V. A. and others, Paleozoic and Mesozoic floras and phytogeography of Eurasia: Akad. Nauk SSSR Geol. Inst. Trudy, v. 208, p. 111-157 (in Russian).

Moore, R. C., and Dudley, P. M., 1944, Cheilotrypid bryozoans from Pennsylvanian and Permian rocks of the Midcontinent region: Kansas Geol. Survey Bull. 52, p. 229-408.

Morozova, I. P., 1970, Bryozoa of the Late Permian deposits: Akad. Nauk SSSR Paleont. Inst. Trudy, v. 122, 347 p. (in Russian).

Nakazawa, K., 1974, On the Permian-Triassic boundary problem: Chigaku Zasshi Jour. Geography (Tokyo), v. 38, p. 1-24 (in Japanese, English abst.).

Nekhoroshev, V. P., 1934, Late Paleozoic Bryozoa of the Kolyma region: Akad. Nauk SSSR Kolyma Geol. Expedition 1929-1930, Soveta Izuch. Prirod., ser. Yakutskaya, no. 24, p. 65-79 (in Russian).

——— 1959, Bryozoa, *in* Kashirtsev, A. S., ed., Field atlas of Permian faunas of northeast U.S.S.R.: Yakutskiy Filial Sibirskoe Otdel., Moscow, p. 25-29 (in Russian).

Newton, G. B., 1971, Rhabdomesid bryozoans of the Wreford megacyclothem (Wolfcampian, Permian) of Nebraska, Kansas and Oklahoma: Kansas Univ. Paleont. Contr., Art. 56, 71 p.

Nikiforova, A. I., 1936, Some Lower Permian Bryozoa from Novaya Zemlya and Spitsbergen: Vses. Arkt. Inst. Trudy, v. 48, p. 113-141 (in Russian).

——— 1938, Stratigraphic distribution of Bryozoa in the Ishimbaevo reef limestones: Neft. Geologo-razved. Inst. Trudy, ser. A, no. 101, p. 76-89 (in Russian).

——— 1939, New species of upper Paleozoic bryozoans from the foothills border of Bashkiria (except the families Fenestellidae and Acanthocladiidae): Neft. Geologo-razved. Inst. Trudy, ser. A, no. 115, p. 70-101 (in Russian).

Nikitina, A. P., Kiseleva, A. V., and Burago, V. I., 1970, Scheme for biostratigraphic subdivision of the Upper Permian Barabash Suite in the southwest Maritime region: Akad. Nauk SSSR Doklady, v. 191, p. 187-189 (in Russian).

Novikova, E. N., 1937, Stratigraphic distribution of Bryozoa in the oil-bearing limestones of Ishimbaeva: Bashkirskyyu Neft. no. 6, p. 28-36 (in Russian).

Peet, R. K., 1974, The measurement of species diversity: Ann. Rev. Ecology and Systematics, v. 5, p. 285-307.

Plumstead, E. P., 1973, The late Paleozoic *Glossopteris* flora, *in* Hallam, A., ed., Atlas of palaeobiogeography: Amsterdam, Elsevier, p. 187-205.

Reed, F. R. C., 1931, New fossils from *Productus* limestones of the Salt Range, with notes on other species: Palaeontologica Indica, n. ser., v. 17, p. 1-56.

Romanchuk, T. V., 1966, New Permian bryozoans of the Khabarovsk region: Paleont. Zhurn., no. 2, p. 42-48 (in Russian).

——— 1967, New bryozoans of the order Trepostomata of the Upper Permian of the Khabarovsk Territory: Paleont. Zhurn., no. 2, p. 69-73 (in Russian).

Romanchuk, T. V., and Kiseleva, A. V., 1968, New Late Permian bryozoans of the Far East: Paleont. Zhurn., no. 4, p. 55-60 (in Russian).

Romer, A. S., 1968, Fossils and Gondwanaland: Am. Philos. Soc. Proc., v. 112, p. 335-343.

Ross, C. A., 1967, Development of fusulinid (Foraminiferida) faunal realms: Jour. Paleontology, v. 41, p. 1341-1354.

——— 1974, Paleogeography and provinciality, *in* Ross, C. A., ed., Paleogeographic provinces and provinciality: Soc. Econ. Paleontologists and Mineralogists Spec. Pub. 21, p. 1-17.

——— 1976, Introduction, *in* Ross, C. A., ed., Paleobiogeography: Stroudsburg, Pa., Dowden, Hutchinson and Ross, p. 1-11.

Ross, C. A., and Nassichuk, W. W., 1970, *Yabeina* and *Waagenoceras* from Atlin Horst area, northern British Columbia: Jour. Paleontology, v. 44, p. 779-781.

Ross, J. P., and Ross, C. A., 1962, Faunas and correlation of the late Paleozoic rocks of northeast Greenland. Pt. IV. Bryozoa: Medd. om Grønland, v. 167, no. 7, 65 p.

Ross, J. R. P., 1963, Lower Permian Bryozoa from Western Australia: Palaeontology, v. 6, p. 70-82.

Rowett, C. L., 1972, Paleogeography of Early Permian waagenphyllid and durhaminid corals: Pacific Geology, v. 4, p. 31-37.

——— 1975, Provinciality of late Paleozoic invertebrates of North and South America and a modified intercontinental reconstruction: Pacific Geology, v. 10, p. 79-94.

Sakagami, S., 1961, Japanese Permian Bryozoa: Palaeont. Soc. Japan Spec. Paper 7, 59 p.

——— 1970, On the Paleozoic Bryozoa of Japan and Thai-Malayan Districts: Jour. Paleontology, v. 44, p. 680-692.

——— 1975, Palaeozoic Bryozoa of Thailand and Malaya, *in* Toriyama, R. and others, The Carboniferous and Permian Systems in Thailand and Malaysia, Geology and Paleontology Southeast Asia, Vol. 15: Tokyo Univ. Press, p. 66-70.

Shcherbovich, S. F., 1969, Upper Gzhelian and Asselian fusulinids of the Caspian basin: Akad. Nauk SSSR Geol. Inst. Trudy, no. 176, 82 p. (in Russian).

Shishova, N. A., 1960, New Permian Bryozoa from northern Zabaykal: Paleont. Zhurn., no. 1, p. 73-83 (in Russian).

Shulga-Nesterenko, M. I., 1952, New Lower Permian Bryozoa from the area of the Urals: Akad. Nauk SSSR Paleont. Inst. Trudy, v. 37, 84 p. (in Russian).

Smith, A. G., Briden, J. C., and Drewry, G. E., 1973, Phanerozoic world maps, *in* Hughes, N. F., ed., Organisms and continents through time: Spec. Papers Palaeontology no. 12, p. 1-42.

Smith, D. B., 1964, The Permian Period, *in* Harland, W. B., Smith, A. Gilbert, and Wilcock, B., eds., The Phanerozoic time-scale. A symposium dedicated to Arthur Holmes: London, Geol. Soc. London, p. 211-220.

Spinosa, C., Furnish, W. M., and Glenister, B. F., 1975, The Xenodiscidae, Permian ceratitoid ammonoids: Jour. Paleontology, v. 49, p. 239-283.

Termier, H., and Termier, G., 1971, Bryozoaires du Paléozoique supérieur de l'Afghanistan: Lyons, Fac. Sci., Lab. Geol., Doc. no. 47, 52 p.

Tozer, E. T., 1969, Xenodiscacean ammonoids and their bearing on the discrimination of the Permo-Triassic boundary: Geol. Mag., v. 106, p. 348-361.

——— 1971, Triassic time and ammonoids: Problems and proposals: Canadian Jour. Earth Sci., v. 8, p. 989-1031.

Toriyama, R., 1973, Upper Permian fusulininan zones, *in* Logan, A., and Hills, L. V., eds., The Permian and Triassic Systems and their mutual boundary: Canadian Soc. Petroleum Geologists Mem. 2, p. 498-512.

Trizna, V. B., 1950, Characteristics of the reef and stratified facies of the central part of the Ufimian Plateau: Vses. Neft. Nauch.-issled. Geol.-razved. Inst. Trudy, no. 135, Microfauna SSSR, Sb. 3, p. 47-144 (in Russian).

Trizna, V. B., and Klautsan, R. A., 1961, Bryozoa of the Artinskian Stage of the Ufimian Plateau and their stratigraphic distribution in the Artinskian of the Ural region: Vses. Neft. Nauch.-issled. Geol.-razved. Inst. Trudy, no. 179, Microfauna SSSR, Sb. 13, p. 331-453 (in Russian).

Waagen, W., and Pichl, J., 1885, Salt Range fossils. Pt. 5. Bryozoa: Palaeontologica Indica, ser. 13, v. 1, p. 771-834.

Waagen, W., and Wentzel, J., 1886, Salt Range fossils. Pt. 6. Bryozoa, Echinodermata, Corals: Palaeontologica Indica, ser. 13, v. 1, p. 835-942.

Warner, D. J., and Cuffey, R. J., 1973, Fistuliporacean bryozoans of the Wreford megacyclothem (Lower Permian) of Kansas: Kansas Univ. Paleont. Contr., Paper 65, 24 p.

Wass, R. E., 1967, Permian Polyzoa from the Port Keats District, Northern Territory: Linnean Soc. New South Wales Proc., v. 92, p. 162-170.

——— 1968, Permian Polyzoa from the Bowen Basin: Australia Bur. Mineral Resources, Geology and Geophysics Bull. no. 90, 134 p.

——— 1969, Australian Permian polyzoan faunas: Distribution and implications, *in* Campbell, K. S. W., ed., Stratigraphy and palaeontology: Essays in honour of Dorothy Hill: Canberra, Australia Natl. Univ. Press, p. 236-245.

——— 1972, Permian Bryozoa from South Africa: Jour. Paleontology, v. 46, p. 871-873.

Yang, K., 1956, Some Cyclostomatous Bryozoa from the Permian rocks of western Hupeh: Acta Palaeontologica Sinica, v. 4, p. 169-174.

Yang, K., and Hsia, F., 1975, Bryozoan fossils from the Quomolangma Feng region, *in* A report of scientific investigations in the Quomolangma Feng region (Palaeontology, Fasc. 1): Academia Sinica, Nanking Inst. Geol. and Palaeontology, p. 39-70 (in Chinese).

Yole, R. W., 1963, An early Permian fauna from Vancouver Island, British Columbia: Canadian Petroleum Geology Bull., v. 11, p. 138-149.

Biogeographic Significance of Land Snails, Paleozoic to Recent

ALAN SOLEM, *Department of Zoology, Field Museum of Natural History, Chicago, Illinois 60605*

ABSTRACT

The known fossil records indicate that 3 of the 6 extant land-snail orders appeared suddenly in the late Paleozoic, and the other 3 appeared by the Cretaceous. Of the 34 family-level units of land snails known from Eocene or earlier deposits, 23 show no demonstrable change in distribution, 4 have shifted a few hundred miles, and only 7 have moved a significant distance. Of 15 family-level units that have only a mid-Tertiary to no fossil record, 6 have classic Gondwanaland or bicontinental disjunct distributions, also suggesting great antiquity. A summary of the few changes in distribution is given.

The Paleozoic land snails from North America and Europe include 1 family now essentially restricted to Polynesia, plus 2 families that could be derived from, but not be ancestral to, taxa that currently are restricted to the Pacific Islands. This strongly suggests that some of the land snails from this region are ancient relicts adapted to rare inter-island dispersal, and isolated on the Pacific plate since at least the mid-Mesozoic.

The stability of land-snail distribution through time is remarkable, and suggests that the discontinuities of land-snail distributions today may be explained by plate tectonic events in the Paleozoic and Mesozoic, rather than by Tertiary events that shaped the distribution of the land vertebrates.

INTRODUCTION

This biogeographic survey of the Paleozoic to Recent land snails is based in large part on a recent review of the Paleozoic nonmarine snails by Solem and Yochelson (in press). In brief summary, Solem and Yochelson demonstrated that: (1) 5 families and 3 orders of land snails appeared suddenly in the late Early Pennsylvanian to Early Permian; (2) all of the land-snail family units found in the Pennsylvanian are still extant today; (3) 2 of the Paleozoic genera, *Anthracopupa* and *Dendropupa,* lived in both North America and Europe; and (4) the land-snail superorder Stylommatophora, which generally has been considered to be derived from the mainly fresh and brackish water superorder Basommatophora, appeared in the Early Pennsylvanian, while the first fossil record of the Basommatophora is not until the uppermost Jurassic (Morrison Formation) or 175 m. y. later. Since only 6 orders of Gastropoda have colonized or differentiated on land, the presence of 3 of these (50 percent) in the Early Pennsylvanian is remarkable. The continuation of all 5 of the Pennsylvanian land-snail families until the present confirms the idea that the radiations of land snails are both ancient and relatively stable. The far earlier appearance of the supposedly derived Stylommatophora will require re-evaluation of the Basommatophora-Stylommatophora relationships.

The limited geographic range of Paleozoic land snails (only Europe and eastern North America have yielded records so far) is regrettable, as is the major time gap that exists in the record from the Early Permian to the Late Jurassic, or about 135 m. y. The European Late Jurassic records are followed by extensive Cretaceous records in both North America and Europe, then abundant Paleocene records from Europe, North America, and South America, plus a few from South Africa. Throughout the Tertiary, there are extensive records from North America (Henderson, 1935), Europe (Wenz, 1938; Zilch, 1959-1960), and South America (Parodiz, 1969). There is essentially no fossil land-snail record from southeast Asia, much of China (see Yen, 1943), Australia, and most of Africa. Despite these deficiencies, a review of the changing land-snail distributions through time, and a summary of the rather remarkable long term geographic stability shown by many families, do permit conclusions relevant to historical biogeography.

The exant land snails and slugs, snails that in the course of evolution have had the shell reduced to a plate-like remnant or completely lost, number about 25,000 species (Solem, in press). Classification of the 23 superfamilies of land mollusks into higher categories, plus the time and place of their first appearance in the fossil record, are given in Table 1. Two major groups of gastropods are not included, since they never colonized the land successfully. The subclass Opisthobranchia is marine, except for a handful of fresh-water species found in Indonesia, Micronesia, and Melanesia. The prosobranch order Neogastropoda (or Stenoglossa) contains active marine predators; only a few species of the family Buccinidae inhabit fresh waters of southeast Asia and Africa and 1 species of Marginellidae (*Rivomarginella* Brandt, 1968) has invaded fresh waters in Thailand. Both the other proso-

Table 1. Earliest record and location for land snail groups.

- Class Gastropoda
  - Subclass Prosobranchia
    - Order Diotocardia (= Archaeogastropoda)
      - Superfamily Neritacea
        - Family Helicinidae (late Paleozoic, North America)
    - Order Taenioglossa (= Mesogastropoda)
      - Superfamily Cyclophoracea
        - Family Cyclophoridae (*sensu* Wenz, 1938)
          - Subfamily Cyclophorinae (Late Cretaceous, Europe)
          - Subfamily Pupininae (Late Cretaceous, Europe)
          - Subfamily Diplommatininae (Late Jurassic, Europe)
          - Subfamily Craspedopominae (Paleocene, Europe)
          - Subfamily Cochlostomatinae (Paleocene, Europe)
      - Superfamily Littorinacea (Late Cretaceous, Europe)
      - Superfamily Rissoacea (Eocene, Europe)
  - Subclass Pulmonata
    - Superorder Systellommatophora (slugs, no record)
    - Superorder Basommatophora
      - Superfamily Ellobiacea (Late Jurassic, Europe)
    - Superorder Stylommatophora
      - Order Orthurethra
        - Superfamily Achatinellacea
          - Family Tornatellinidae (late Paleozoic, North America and Europe)
        - Superfamily Cionellacea
          - Family Cionellidae (Paleocene, Europe)
          - Family Amastridae (Recent, Hawaii)
        - Superfamily Pupillacea (late Paleozoic, North America)
          - Family Strobilopsidae (Eocene, Europe)
        - Superfamily Partulacea
          - Family Enidae (late Paleozoic, North America and Europe)
          - Family Partulidae (Recent, Polynesia and Micronesia)
      - Order Mesurethra
        - Superfamily Clausiliacea
          - Family Clausiliidae (Late Cretaceous, Europe)
          - Family Cerionidae (Miocene, West Indies)
        - Superfamily Strophocheilacea
          - Family Dorcasiidae (Paleocene, SW Africa)
          - Family Strophocheilidae (Paleocene, South America)
      - Order Sigmurethra
        - Suborder Holopodopes
          - Superfamily Achatinacea
            - Family Ferussaciidae (Eocene, Europe)
            - Family Subulinidae (Paleocene, Europe)
            - Family Megaspiridae (Late Cretaceous, Europe)
            - Family Achatinidae (Pleistocene, Africa)
          - Superfamily Streptaxacea (Late Cretaceous, Europe)
          - Superfamily Rhytidacea (Pliocene, New Zealand; Recent elsewhere)
          - Superfamily Acavacea (Pleistocene, Madagascar)
          - Superfamily Bulimulacea
            - Family Urocoptidae (Late Cretaceous, North America)
            - Family Bulimulidae (*s. l.*) (Eocene, South America)
        - Suborder Aulacopoda
          - Superfamily Arionacea
            - Family Endodontidae (Miocene, Bikini Atoll)
            - Family Charopidae (Miocene, Eniwetok Atoll)
            - Family Helicodiscidae (Pleistocene, North America)
            - Family Punctidae (Recent, many areas)
            - Family Discidae (late Paleozoic, North America)
          - Superfamily Succineacea (Paleocene, Europe)
          - Superfamily Limacacea (Paleocene, Europe and North America?)
            - Family Vitrinidae (unknown)
        - Suborder Holopoda
          - Superfamily Polygyracea
            - Family Sagdidae (Eocene, Wyoming)
            - Family Polygyridae (Late Cretaceous?, North America)
            - Family Corillidae (Recent, SE Asia, India)

Table 1. (Cont.)

Superfamily Oleacinacea
  Family Oleacinidae (Paleocene, Europe)
  Family Spiraxidae (Recent, West Indies, Central America)
Superfamily Camaenacea
  Family Camaenidae (Late Cretaceous, North America)
  Family Oreohelicidae (Late Cretaceous, North America)
  Family Ammonitellidae (Eocene?, North America)
Superfamily Helicacea
  Family Helminthoglyptidae (Cretaceous, North America)
  Family Bradybaenidae (Paleocene, Europe)
  Family Helicidae (Paleocene, Europe)

branch orders also are primarily marine. Of the Archaeogastropoda, only the Hydrocenidae and Helicinidae, which total more than 350 species, are terrestrial; several taxa of the Neritidae inhabitat fresh and brackish waters. Within the Mesogastropoda several groups (Hydrobiidae, Valvatidae, Viviparidae, Ampullariidae, "Thiaridae, *s. l.*", Bithynidae) are mainly to exclusively fresh-water inhabitants, and 3 lineages, containing approximately 3,850 species, have colonized the land.

The 3 superorders of the Pulmonata vary greatly in habitat preference and degree of diversity. The superorder Systellommatophora contains 3 families of slugs with a total of perhaps 250 to 300 species: the marine Oncidiidae, which is basically intertidal; the terrestrial and herbivorous Veronicellidae, which has a circum-tropical distribution roughly approximating the limits of palm trees; and the terrestrial and carnivorous slugs of the Rathouisiidae, which range from southeast China and Burma to northern Queensland. Members of the superorder Basommatophora are primarily found in fresh-water habitats. Exceptions are the few species of the Otinidae, Amphibolidae, and Stenacmidae, plus the rather speciose Siphonariacea, which are marine, and the Ellobiidae found in land (*Carychium* Müller, 1774; *Zospeum* Bourguignat, 1856; some species of *Pythia* Röding, 1798), brackish water, and marine habitats. There are probably less than 50 terrestrial species of Basommatophora, all belonging to the family Ellobiidae.

The vast majority of land snails, approximately 20,500 species, belong to the superorder Stylommatophora, which has no marine or fresh-water (some succineids are amphibious) taxa. The 3 orders of the Stylommatophora are not equally diverse, and are of uncertain relationship to each other. The order Orthurethra, with about 2,500 species, generally has been considered to be the most primitive. The order Mesurethra, with about 2,600 species, is intermediate in structural complexity, and the order Sigmurethra, with about 15,400 species, probably is the most advanced. It is quite probable that the Orthurethra and the Mesurethra represent parallel experiments. The Mesurethra might be ancestral to the Sigmurethra.

The above data on recent diversity are necessary background to discussing the time at which each taxon first appeared in the fossil record, and the documented changes in their geographic pattern, if any, that have occurred subsequently.

TIME AND PLACE OF APPEARANCE IN THE FOSSIL RECORD

The data on first occurrences are compiled from Wenz (1938) for the Prosobranchia and Zilch (1959-1960) for the Pulmonata. These sources have been modified by incorporating data for South American taxa from Parodiz (1969), for Cretaceous and Paleocene taxa of western Canada from Tozer (1956), and for Paleozoic taxa from Solem and Yochelson (in press). Where appropriate, new data and interpretations on the structure and probable affinities of Cretaceous, Paleocene, and Eocene taxa are included, based on materials in the United States National Museum (USNM), American Museum of Natural History (AMNH), Field Museum of Natural History (FMNH), Redpath Museum of McGill University, and Geological Survey of Canada. Data on family distributions are taken from the same basic sources, although extensively modified by changes recorded in Solem (1959) and many subsequent papers relating to family-level assignments of problematic taxa. The list of such modifications is far too extensive to be documented in this review.

Except for the superorder Systellommatophora, which contains only shell-less slugs and has no fossil record, all ordinal groups of land snails were present in the fossil record by the Late Cretaceous. Table 1 summarizes the superfamily and higher category classification, as well as indicating the time and place of first record. Where bicontinental records were *roughly* contemporaneous, both are listed. The Diotocardia, Orthurethra, and Sigmurethra were present in the Pennsylvanian and Permian, the Taenioglossa and Basommatophora in the Upper Jurassic, and the Mesurethra in the Late Cretaceous. The conventionally accepted position of the Basommatophora as a stem group to the various stylommatophoran land snails is not supported by this sequence in the fossil record, nor does the record support the equally conventional wisdom that the Mesurethra might be ancestral to the Sigmurethra. However, the time of first recorded appearance in the fossil record is not necessarily linked to the actual time of origin for a taxon. The first records of the Mesurethra (Cretaceous to Paleocene), for example, are of modern, fully differentiated families in Europe (Clausiliidae), South Africa (Dorcasiidae), and South America (Strophocheilidae). The same families still inhabit these areas today. We thus have no indication as to their place and time of origin. Indeed, 2 of the Paleozoic genera, *Anthracopupa* Whitfield, 1881, and a new genus (Solem and Yochelson, in press), are essentially typical members of the very distinctive modern families Tornatellinidae and Discidae, respectively. I cannot point to any fossil land snail that clearly is intermediate between family-level taxa, although 2 families, the Anadromidae from the Cretaceous to Eocene of Europe, and the Grangerellidae from the Paleocene of North America are of uncertain affinities. Many fossils are too fragmentary or simple in shell form to be assignable to family units, but most fossils, when the shell growth pattern and structure are analyzed, can be assigned to extant family (Paleozoic-Mesozoic) or even generic groups (early Tertiary).

The clustering of first appearances in North America and Europe indicates the location of fossil deposits, not necessarily of evolutionary origins. The size and location of the geologic community are also factors in discovery of fossils. The essential absence of records from elsewhere reflects the absence of appropriate terrestrial deposits from the ages concerned in the early record of land snails, possible lack of collecting effort, and probable disinterest by most invertebrate paleontologists in land deposits.

GEOGRAPHIC PATTERNS THROUGH TIME

Currently, there is no up-to-date review of land-snail biogeography on a world-wide basis. Within the space limitations of this review, only a sketchy outline can be presented. It is limited to a review of changing geographic patterns with respect to particular families or superfamilies. Brief comments are made on some general patterns that are evident, but no grand scheme of malacogeography is proposed. The sequence of classification in Table 1 is followed. Initially, comments are limited to a discussion of those taxa with great antiquity and(or) changing patterns of distribution through time.

Prosobranchia

The family Helicinidae is known first from the late Paleozoic of Illinois and Indiana (*Dawsonella* Bradley, 1874, late Early Pennsylvanian). One apparently primitive genus of the family, *Hendersonia* A. J. Wagner, 1905, lives today in southern Minnesota and Wisconsin, northern Iowa and Illinois, and in parts of Appalachia. Species consistent with belonging to *Hendersonia* have been described from the Paleocene of Wyoming and the Miocene of Oregon (Pilsbry, 1948, p. 1086-1087). The other early fossil of the Helicinidae is the genus *Dimorphoptychia* Sandberger, 1871, from the Paleocene "Calcaire de Rilly" of the Paris Basin. The type species is quite comparable to the modern *Calybium* L. Morlet, 1891, and *Heudeia* Crosse, 1885, from "Indo-China" and southeast China, respectively. The type species was discussed first by Pilsbry (1927-1935, p. 11). The species from Alberta that Tozer (1956, p. 45-50) referred to *Dimorphoptychia* have a greatly increased whorl count, a quite different body shape, and, on the basis of preliminary restudy, would seem to be referable to the pulmonate family Camaenidae, rather than the Helicinidae. The Helicinidae today are basically a tropical group with a discontinuous distribution. In the Old World, they range from India to southern Japan, Hawaii, Marquesas, and northern Australia. No helicinids live in the Nearctic, Africa (except one form on Mauritius), or in most of Australia. In the New World, they are most abundant in Middle America and the West Indies, with several Andean taxa, and a few reaching Florida and the southern states. One Pleistocene relict genus, *Hendersonia*, overlaps the Paleozoic distribution, and formerly was present in Wyoming and Oregon. The Helicinidae became extinct in Europe since the Paleocene, are today found nowhere near Oregon and Wyoming, but still are present in North America, their Paleozoic residence.

The Cyclophoracea is a complex of at least 5 families, whose limits and relationships are still controversial. Suffice it to record that they are most diverse in two centers: (1) West Indies, Middle America, and the Andes in the New World; and (2) India to Japan to Samoa and New Caledonia in the Old World. One group is sparsely represented in mainland Africa (Maizaniidae), and there are scattered records from the coast of East Africa and the Malagasy Islands. Only 1 family group, the Diplommatinidae (= Cochlostomatidae) is currently living in Europe, but most of the extant families have Cretaceous to Eocene representatives in Europe. There thus has been a clear distributional move away from Europe during the Tertiary.

The Littorinacea have European Late Cretaceous fossils (*Bauxia* Caziot, 1890) and sparse modern European representation (*Pomatias* Studer, 1789). Most modern genera are in Socotra, the Middle East, East and South Africa, and the Malagasy region. There is a closely related and highly diverse Miocene to Recent taxon in the West Indies, which has sparse Middle American and northern South American representation. There are no data on the actual point of origin or relationships between fossil and extant taxa, and the phyletic affinities of Old World (Pomatiasinae) and New World (Chondropominae, *s. l.*) subfamilies are unknown.

The Rissoacea land-snail taxa include the late Tertiary Assimineidae, which shows no change in modern distribution, and the Eocene to modern European Acmeidae.

To summarize the fossil to recent prosobranch distributions: (1) there has been a move away from Europe of the Helicinidae and Cyclophoracea; (2) there are no detectable changes in the distribution of the Rissoacea and Littorinacea; and (3) the North American Paleozoic helicinid taxon lies within the modern northern limits of the Helicinidae. Today the land prosobranchs are a basically tropical complex, far less well-represented in Europe than they were during the Eocene.

Pulmonata

The Ellobiidae, the only terrestrial family of the Basommatophora, contains mainly tropical mangrove and brackish water taxa. During the late Mesozoic and early Tertiary, there was a highly diverse European fauna. Many of these species have been referred to generic-level taxa that now are confined to the Oriental, Indonesian, and Melanesian regions. The truly land ellobiid, *Carychium,* has the same European distribution now as it had in the Mesozoic. It migrated at an unknown time into North and Middle America. *Carychium* was, and remains, a marginal terrestrial group, since several species are blind and confined to caves, and the others are mostly restricted to flood plains. The brackish water and marine taxa moved from Europe, while there is no documentable change in terrestrial distributions.

Within the Stylommatophora, several remarkable changes have occurred over the course of time but most families show a remarkable stability.

The Tornatellinidae are represented by the Paleozoic *Anthracopupa* Whitfield, 1881, in both North America (Illinois to Nova Scotia) and Europe (England to Vienna) and the little known Wyoming Late Cretaceous *Protornatellina isoclina* (White, 1895). Subsequently, they are known only as Recent or late Pleistocene taxa from Polynesia and Juan Fernandez. The few other Recent records probably are secondary introductions (often by modern commerce) into various tropical areas from the Polynesian center.

The Cionellacea conform to past distributions, as do most of the Pupillacea. There are a few reported moves in the Pupillacea, such as the genus *Negulus* O. Boettger, 1889, from the Eocene of Europe to East Africa today; *Ptychalaea* O. Boettger, 1889, from the Miocene of Europe to the Bonin Islands today; and *Microstele* O. Boettger, 1886, from the Miocene of Europe and China to India, Ceylon, and South Africa today. In view of the variability of pupilloid taxa these distributional changes need to be verified, but they do suggest that at least minor changes in distribution have occurred. In his classic discussion of Pupillacea distribution, Pilsbry (1927-1935, p. 139) pointed out that it ". . . is essentially a group of the northern continents. The data now at hand indicate Eurasia as the main area of evolution and radiation." In subsequent discussion, he pointed out (p. 140): "Few if any of the extinct Tertiary genera now known are generalized or synthetic types. The main evolution of the group seems to have been in the Paleocene and Cretaceous, and still remains to be recovered. The European Oligocene and Miocene pupillid fauna appears to be about as specialized and mature as the Recent, and contains many genera and subgenera still widely spread." This was followed (Pilsbry, 1927-1935, p. 141-169) by a regional analysis of the pupillid distributions.

One specialized family of the Pupillacea, the Strobilopsidae, does show a significant distributional move. Pilsbry (1927-1935, p. 4-19) reviewed its history and distribution, pointing out that there are about 20 western European Tertiary species, ranging in age from the late Eocene to Pliocene, with essentially modern subgeneric groups appearing in the Oligocene. No Recent European species are known. Strobilopsids today are confined to eastern China, Japan, Korea, the Philippines, eastern North America south to Guatemala, plus a few scattered South American localities. The earliest known North American fossil record is Pleistocene. Recent taxa are dually peripheral to the early Tertiary center of abundance (western Europe), where the family is now absent.

The superfamily Partulacea contains 2 families. The Recent family Partulidae is known only from the high volcanic islands of Polynesia (except Hawaii), Micronesia, and the fringes of Melanesia. The family Enidae, which today is basically Eurasian and African, has a few peculiar outliers in New Caledonia and the New Hebrides (*Draparnaudia* Montrouzier, 1859), and the monsoon areas of Western Australia to northern Queensland (*Amimopina* Iredale, 1933). The late Paleozoic genus *Dendropupa* Owen, 1859 (eastern Canada, France to Poland) is referred to the Enidae (Solem and Yochelson, in press). The Paleozoic Canadian record does not lie within extant distributional limits.

The order Mesurethra is of uncertain homogeneity and content. Until quite recently the Dorcasiidae and Strophocheilidae were lumped with such taxa as the Caryodidae, Acavidae (+ Clavatoridae), and Macrocyclidae as a single family, the Acavidae, with a disjunct Southern Hemisphere distribution (see Solem, 1969 for a distributional review). The relationships between the Dorcasiidae (Paleocene of southwest Africa) and Strophocheilidae (Paleocene of Argentina) are still uncertain. The Paleocene taxa are referable to exant genera and the families were as well-differentiated conchologically in the Paleocene as they are today.

The superfamily Clausiliacea is well-represented in the Upper Cretaceous of Europe. The extinct family Filholiidae from the middle Eocene to upper Oligocene of Europe is probably not a separate family. The family Clausiliidae still is Eurasian, with a secondary center of diversity in the Andes of South America, and a pair of species in the Greater Antilles. The approximately 2,300 species of Clausiliidae (*teste* F. E. Loosjes) make this one of the largest families of land mollusks. Other taxa, such as the West Indian Cerionidae with about 200 species, appear late in the fossil record and show no change in geographic range since their original appearance.

Because the shells of the Ferussaciidae and Subulinidae are relatively simple in shape and structure, many of the fossil "subulinids" seem dubious. For example, *Pseudocolumna teres* (Meek and Hayden, 1856), holotype USNM 2115, from the Eocene of Fort Union, western North Dakota and *P. vermicula* (Meek and Hayden, 1856) from the same Formation, holotype USNM 2113, can be interpreted as smooth-shelled bulimulids equivalent in shape and size to South American *Bostryx* (*Peronaeus*) rather than to any African subulinids. Available fossils of European subulinids do not suggest major changes in distribution. The large Achatinidae from Africa have only a very brief fossil record. In contrast, the family Megaspiridae, which is well-characterized by complex shell features, is known as Late Cretaceous to Oligocene fossils in Europe (*Palaeostoa* Andreae, 1884), plus Recent genera from

Brazil (*Megaspira* Jay, 1836; *Callionepion* Pilsbry and Vanatta, 1899), New Guinea (*Perrieria* Tapparone-Canefri, 1878), and Queensland (*Coelocion* Pilsbry, 1904). This represents a notable move in distribution that, if not for the European fossils, would automatically be interpreted as a "Gondwanaland relict distribution."

The superfamily Streptaxacea (see van Bruggen, 1967) is known from many Late Cretaceous fossils in Europe, and Miocene fossils from both Africa and Brazil. The European taxa became extinct in the Pliocene, except for a remnant Recent *Gibbulinella* Wenz, 1920, on the Canary Islands. Other species in that genus ranged from Late Cretaceous to late Eocene of mainland Europe. The Streptaxidae today are primarily African, South American, and southeast Asian with scattered adjacent records.

The superfamily Rhytidacea contains the New World Haplotrematidae, South American Systrophiidae, possibly the South American Macrocyclidae, plus the Rhytididae and its slug derivative, the South African Aperidae. The Rhytididae have a disjunct southern distribution—New Caledonia, parts of Melanesia, but primarily New Zealand, wetter portions of Australia, and South Africa. The only fossil is from the Pliocene of New Zealand, although several helicoid taxa from the Cretaceous and Eocene of western North America have been referred to the Macrocyclidae, almost certainly erroneously.

The Acavacea lack a significant fossil record. The Bulimulacea include several major taxa. The family Urocoptidae has records in the Late Cretaceous of Alberta and the Eocene of Wyoming (Tozer, 1956), although its current distribution is more southerly, with Arizona to Guatemala and the West Indies representing the main areas of diversity. The several families formerly clumped as the Bulimulidae mostly have a South American distribution and no extensive fossil record. The restricted Bulimulidae, in the sense of Zilch (1959-1960), appeared in the Eocene of Patagonia. One of the fossil species from that age belongs to the extant genus *Thaumastus* Albers, 1860 (Parodiz, 1969, p. 179-181). Extant Bulimulidae also are found in central and southwestern Australia, the northern tip of New Zealand to the Solomons and Fiji (*Bothriembryon* and *Placastylus*, *s. l.*), as well as South and Central America. Relating the North American Paleocene fossil family Grangerellidae to the Bulimulidae is uncertain. The Grangerellidae have some puzzling features that make me uncertain of its affinities, although it probably is not related to the Helicinidae as suggested by Russell (1931).

The suborder Aulacopoda contains 2 superfamilies, Limacacea and Succineacea, that appeared in the Paleocene and have nearly world-wide distributions, plus the superfamily Arionacea, with which I have been struggling for more than a decade.

Shells of the superfamily Limacacea generally are relatively featureless, simply coiled, and without expanded lip. It is very difficult to know if a fossil referred to an extant genus of zonitids or helicarionids is based on a juvenile of some other family, represents convergences in shell form, or actually is correctly assigned. Various Cretaceous to Eocene fossils from western North America have been referred to such genera as *Gastrodonta* Albers, 1850, and *Mesomphix* Rafinesque, 1819. Fossil species that now would be included in the latter genus, such as *Omphalina laminarum* Cockerell, 1906, and *Omphalina oreodontis* Cockerell and Henderson, 1912, might be helicoids (Pilsbry, 1946, p. 306), while the so-called *Gastrodonta* are based on young *Grangerella* Cockerell, 1915, such as *Gastrodonta evanstonensis sinclairi* Cockerell, 1912, which is a juvenile of *Grangerella megastoma* Cockerell, 1915 (based on examination of types in the AMNH; see also Pilsbry, 1946, p. 436-437). The earliest assured records for the Limacacea probably are *Grandipatula* Cossmann, 1889, *Archaegopsis* Wenz, 1914, and *Provitrina* Wenz, 1919, from the Paleocene of Europe. New World records probably only date from the mid-Tertiary. Of the extant families, the Zonitidae are basically holarctic; the Parmacellidae and Testacellidae are palearctic; the Helicarionidae (*s. l.*) occur from India through the Pacific Islands; the Urocyclidae are African; and the Limacidae (+ Milacidae) are basically European slugs that have been disseminated throughout the world by commerce in the last 100 years. The occurrence of such genera of zonitids as *Striatura* Morse, 1864, *Nesovitrea* Cooke, 1921, and *Godwinia* Sykes, 1900, in Hawaii (see Baker, 1941, p. 324-335) presumably is the result of accidental transportation by migrating birds, and does not affect the basic distribution pattern. The family Vitrinidae also is holarctic, with a rather extensive radiation on the wet mountains of East Africa. It also is represented in Hawaii (Baker, 1941, p. 321-322; Pilsbry, 1946, p. 501). Vitrinids are the only type of land snails, as contrasted with the marsh-dwelling Succineidae, that have been recorded on the feathers of birds with any frequency. Distribution of the Vitrinidae may be tied closely to bird migrations. Fossil taxa erroneously referred to *Vitrina*, such as the Cretaceous *V. obliqua* Meek and Hayden, 1857, are known to be umbilicated and thus are a helicoid or discid shell type (see Pilsbry, 1946, p. 501) and therefore not members of the family Vitrinidae. On the limited fossil record now available, and the poorly known phyletic relationships within and between the various limacacean families, I can identify no significant changes in distribution patterns. The time of origin for these families remains unknown, as do the true identity of most early fossils.

The shell-bearing families of the Arionacea have both complex shell structures and highly characteristic anatomical patterns. The family Discidae is holarctic. Its greatest diversity is in North America, where there are late Paleozoic, Late Cretaceous, and Paleocene records. The family Helicodiscidae has a disjunct distribution—North and Middle America, then Indonesia to northern Australia and the Solomon Islands, but no significant fossil record. The family Endodontidae is limited to Polynesia, the Lau Archipelago of Fiji, Palau; as a Miocene fossil it is known from Bikini Atoll. The Endodontidae probably were ancestral to the family Charopidae, which has a typical "Southern Relict" distribution—Australia, New Zealand,

New Caledonia, much of South Africa, and most of South America, with some taxa extending up into Middle America and even the western United States (Arizona to Idaho). New Zealand and South America apparently share at least 1 Recent genus (Solem, unpub. data). The exact intercontinental phyletic relationships of the arionacean families are very uncertain. Potentially, they offer highly important data on biogeography, since the Charopidae is one of the largest and most widely distributed land-snail families, and both the Discidae and the Endodontidae have fossil records.

The suborder Holopoda contains the most familiar and probably highly evolved land snails. It is thus rather surprising that their patterns of distribution are quite stable through time. The Polygyracea contains the families Sagdidae, Polygyridae, and Corillidae (= Plectopylidae). The Sagdidae today are mainly West Indian and Central American, but date from the Eocene of Wyoming (*Microphysula* Cockerell and Pilsbry, 1926). A few taxa today reach as far north as Vancouver Island (see Pilsbry, 1940, p. 979, 994). It is uncertain whether some Cretaceous to Eocene fossils from western North America [*Polygyra petrochlora* Cockerell, 1914; *P. parvula* (Whiteaves, 1885)] may be referred to the Polygyridae, but the presence of Oligocene species in the Rocky Mountain area and *Vespericola* Pilsbry, 1939, from the Miocene of Oregon (see Pilsbry, 1940, p. 892-893) establish the family within its modern distributional limits of North and Middle America by mid-Tertiary. The Indian and southeast Asian family Corillidae has no significant fossil record. Taxa from southwest Africa and China may be incorrectly included.

The family Oleacinidae extends today from Florida and the Gulf Coast south through Central America and the West Indies to Peru and Brazil. Some species live around the fringes of the Mediterranean, but the West Indies and Central America are the center of abundance. Fossil oleacinids are known from the Miocene of Florida [*Sigmataxis tampae* (Dall, 1915)] and the Paleocene of Europe (Pilsbry, 1946, p. 188).

The superfamily Camaenacea contains the families Camaenidae, Ammonitellidae, and Oreohelicidae. The Oreohelicidae range from southern Saskatchewan and British Columbia through the Rocky Mountain states to parts of northern Mexico. Records from Catalina Island, California, the Black Hills of South Dakota, and a loess record from eastern Iowa, define the east-west limits of Pleistocene to Recent distribution. Late Cretaceous and Paleocene fossils from Alberta, Canada; early Eocene fossils from New Mexico and Wyoming; and Miocene records from Oregon establish that the Oreohelicidae has occupied its basic range unchanged during the Tertiary. Similarly, the family Ammonitellidae is known with certainty from the lower Miocene of Oregon (*Ammonitella* Cooper, 1868; *Polygyrella* W. G. Binney, 1863) and probably the Eocene of Wyoming [*Glyptostoma spatiosa* (Meek and Hayden, 1861)], although Pilsbry (1940, p. 554) was uncertain as to the latter classification. Today the Ammonitellidae is a relict group inhabiting small portions of the western United States, all of which are peripheral to its Eocene record.

The family Camaenidae has a disjunct distribution today. One major center of diversity is the West Indies. Two genera, *Isomeria* Albers, 1850, and *Labyrinthus* Beck, 1837, inhabit Andean South America and the Caribbean border areas as far north as Costa Rica. The second center of camaenid diversity extends from eastern India and south China through the Solomon Islands, south into the northern three-quarters of Australia, with a few taxa extending as far north as southern Japan. One genus, *Ganesella* Blanford, 1863, is known from the Pliocene of Japan, but otherwise there are no Old World fossil records. In the New World, a subgenus, *Pleurodontites* Pilsbry, 1939, is known from the Miocene of Florida, and is related to the major radiation of the genus *Pleurodonte* Fischer von Waldheim, 1807, on Jamaica today.

Other supposedly camaenid fossils are controversial. Examination of several type specimens permits the following comments about them. The names applied below are those listed in the original descriptions. *Pleurodonte eohippina* Cockerell, 1915, holotype AMNH 22357, from the Eocene of Wyoming has the aspect of a helicinid genus, *Ceres* Gray, 1856, but is very poorly preserved. Pilsbry (1939, pp. 411-412) also rejected it as a camaenid. *Helix adipis* White, 1886, holotype USNM 20073, and *Helix evanstonensis* White, 1878, holotype USNM 12502, may be helminthoglyptids, but will require careful restudy before being assigned to a definite group. *Helix chriacorum* Cockerell, 1914, holotype AMNH 22365, possibly is a ribbed helicinid, but requires more detailed study. The *Dimorphoptychia* species that Tozer (1956, p. 45-50) described from Canada seem to show the shape, whorl count, and barrier pattern consistent with some of the West Indian camaenids, but detailed study of these was not possible.

The two most significant fossils are *Helix kanabensis* White, 1876, holotype USNM 8883, from the Laramie Group, Upper Kanab, Utah. The age is Upper Cretaceous (Maestrichtian). Pilsbry (1927-1935, p. 11-12) proposed a genus *Kanabohelix* for this species, stating that it seemed more helicoid than strobilopsid, but did not discuss it in detail. Analysis of its growth pattern, lip formation, whorl count, apertural deflection, and barrier position (Solem, in press) showed only slight differences from the Recent *Pleurodonte (Dentellaria) sinuosa* (Ferussac, 1850) from Jamaica. There are sufficient peculiarities of growth in this group to make convergence highly improbable, and *Kanabohelix kanabensis* (White, 1876) is a camaenid. Similarly, *Helix hesperarche* Cockerell, 1914, holotype AMNH 22362, from near Alpine, Texas (see Henderson, 1935, p. 134) and *Hodopoeus crassus* Pilsbry and Cockerell, 1945, from an unknown locality agree in observable details with the extant South American camaenid genus *Isomeria* Albers, 1850 (Solem, in press).

The presence of two undoubted camaenids in the Cretaceous to Eocene of North America, far to the north of present distributions, is significant in pointing to the area

Table 2. Geographic changes from origin to Recent for family-level taxa Eocene or older.

Time of appearance	"Sat tight"	"Shifted"	"Moved"
Late Paleozoic	Helicinidae Pupillacea Discidae	Enidae	Tornatellinidae
Jurassic	Ellobiidae		Diplommatininae
Cretaceous	Littorinacea Clausiliacea Polygyridae Oreohelicidae Helminthoglyptidae	Streptaxacea Urocoptidae	Megaspiridae Camaenidae Cyclophorinae Pupininae
Paleocene	Cochlostomatinae Dorcasiidae Strophocheilidae Subulinidae Succineacea Limacacea Oleacinacea Bradybaenidae Helicidae	Craspedopominae Bulimulidae (?)	
Eocene	Rissoacea Ferussaciidae Bulimulidae (?) Ammonitellidae		Strobilopsidae

and direction of origin for the camaenid land snails. The fact that the Cretaceous to Eocene fossils are equally differentiated as modern genera suggests that the origin of the group is at least Early Cretaceous, but gives no clue as to the place of origin. A Northern Hemisphere direction of derivation for the extant West Indian and South American taxa is indicated, rather than a possible Gondwanaland origin, as could be hypothesized in the absence of the North American fossils. Parodiz (1969) reported no camaenids from the Tertiary of South America.

The superfamily Helicacea contains 3 families, the Eurasian and North African Helicidae, the Eurasian Bradybaenidae, and the western North American, Central American, West Indian, and western South American Helminthoglyptidae (= Xanthonycidae). The Helicidae (*Loganiopharynx* Wenz, 1919) and Bradybaenidae (*Bradybaena* Beck, 1837, and *Coneulota* Pfeffer, 1929) date from the Paleocene, while the earliest Helminthoglyptidae (*Mesoglypterpes* Yen, 1952) are from the Cretaceous of Wyoming. Extant helminthoglyptids have a more western range, and the Epiphragmophorinae were characterized by Parodiz (1969, p. 187) as "a late Tertiary northern immigrant" into South America. Both the Bradybaenidae and Helicidae show no changes in distribution since their first appearance in the fossil record.

STABILITY AND CHANGE

Because the pre-Paleocene records of land snails are restricted to North America and Europe, and because even the Paleocene and Eocene records are rather limited in a geographic sense, it is not possible to present a sophisticated analysis of distributional patterns through time. Table 2 summarizes the geographic patterns for the family and superfamily groups discussed above for which Eocene or earlier fossil records exist. The three distributional categories used are somewhat arbitrary and obviously imprecise, since the total fossil record of a family may be based on only one or two localities. Thus the actual extent of the family range at time of first record is unknown. Nevertheless, the table does indicate whether a taxon (1) occupies today the same geographic area as it did in its first appearance in the fossil record ("Sat tight"); (2) now occurs at a distance of at least several hundred miles away from its first recorded appearance ("Shifted"); or (3) now occurs in an area that is several thousand miles removed from its first recorded locality ("Moved").

There are 34 family-level units of land snails known from Eocene or earlier fossils. They show an extraordinary degree of geographic stability. Twenty-three (68 percent) of these show *no* demonstrable change from the time of their first occurrence; 4 (12 percent) have shifted a few hundred miles; and 7 (20 percent) have moved a significant distance. Unquestionably, there were range expansions, contractions, and invasions of new continental areas that are not documented by the known fossil record.

An additional 15 family-level units of shell-bearing land snails have mid-Tertiary to Recent records. Some of these (Acavacea, Rhytididae, Charopidae) have a classic Gondwanaland distribution, while others (Helicodiscidae, Punctidae, Corillidae) are at least bicontinental and disjunct. Both of these situations also suggest great antiquity. The remaining land-snail families are known from only single major geographic areas.

Compared with the radical distributional changes shown by vertebrates, there is a true "snail's pace" to land

Table 3. Ordinal and age summary of distributional patterns, family-level taxa with pre-Eocene origin.

	"Sat tight"	"Shifted"	"Moved"
Paleozoic-Cretaceous			
Diotocardia	1		
Mesogastropoda	1		3
Basommatophora	1		
Orthurethra	1	1	1
Mesurethra	1		
Holopodopes		2	1
Aulacopoda	1		
Holopoda	3		1
Paleocene-Eocene			
Mesogastropoda	2	1	
Orthurethra	1		1
Mesurethra	2		
Holopodopes	3		
Aulacopoda	2		
Holopoda	4		
TOTALS	23	4	7

Table 4. Ordinal summary of family-level distribution patterns, pre-Eocene origin to Recent distribution.

	"Sat tight"	"Shifted"	"Moved"
Diotocardia	1		
Mesogastropoda	3	1	3
Basommatophora	1		
Orthurethra	2	1	2
Mesurethra	3		
Sigmurethra			
Holopodopes	3	2	1
Aulacopoda	3		
Holopoda	7		1
TOTALS	23	4	7

mollusk distributional changes. Even 3 of the 5 family units that appeared in the late Paleozoic are found in the same areas today. Two of these families, Discidae and Helicinidae, are far from being worldwide in range so that their stability is even more remarkable.

Data on a few of the more striking changes in distribution precede a brief summary of area changes. In general, many European Cyclophoracea fossils from the Cretaceous to Miocene compare well with extant taxa from southeast Asia and Indonesia (Cyclophorinae, Diplommatininae, and Pupininae, *sensu* Wenz, 1938); the Craspedopominae from the Paleocene of Europe have shifted to the Canaries, Azores, and Madeira; the Cochlostomatinae remain European in distribution since their first appearance in the Paleocene. There is thus a pattern of "withdrawal" from Europe, either shifting to offshore islands, or "moving" to southeast Asia. This situation is paralleled by the Helicinidae, which are known in Europe from an Eocene fossil, but today are found in the Old World only from southeast Asia through Polynesia.

The most striking moves are those shown by the Tornatellinidae, Megaspiridae, Camaenidae, and Strobilopsidae. Today the Tornatellinidae are restricted essentially to Polynesia and Juan Fernandez, although they were present in the late Paleozoic of eastern North America and Europe. The Megaspiridae existed in Europe from the Cretaceous to the Oligocene, but today are relict in Brazil, New Guinea, and Queensland. The Camaenidae were in Utah during the Cretaceous and in Florida in the Miocene; today camaenids are not found north of Costa Rica and Cuba in the New World. The Strobilopsidae were abundant in Europe from the Eocene to Pliocene, but today are absent from Europe and present on the fringes of eastern Asia, plus North and Central America.

The status of the Bulimulidae is uncertain. I do not know whether *Pseudocolumna* is a true bulimulid or not. If it is, then the Bulimulidae have shifted from North America; if not, then the Bulimulidae might be a true South American group with Gondwanaland relations to the Australian-Melanesian taxa. Determining whether the Grangerellidae should be classified with the Bulimulacea would be an important clue to the biogeographic status of this group.

In summary, several taxa have "withdrawn" from Europe since their first recorded appearances in the Paleozoic or Mesozoic. The Helicinidae, Diplommatininae, Cyclophorinae, and Pupininae moved from Europe to southeast Asia; the Streptaxacea shifted into Africa and southeast Asia; and the Megaspiridae went to New Guinea, Australia, and Brazil. The Eocene Strobilopsidae moved from Europe by the end of the Pliocene and colonized both eastern Asia and eastern North America; the Paleocene Craspedopominae retain a European fringing distribution today.

In the New World, the Urocoptidae shifted south, the Camaenidae moved south, and the Enidae vanished.

The most remarkable change is that of the Tornatellinidae which had an eastern North America to Austria distribution in the Early Pennsylvanian and Early Permian, was recorded from Wyoming in the Cretaceous, but now lives in Polynesia and Juan Fernandez.

Several actual or potential Gondwanaland families (Rhytididae, Acavacea) have no fossil record, or only mid-Tertiary records (Charopidae) from the northern fringes of their current distribution (Eniwetok, Marshall Islands). The Bulimulidae, known from the Eocene of Patagonia, may be another Gondwanaland family, but there may be North American Cretaceous to Eocene relatives.

PHYLETIC PATTERNS OF STABILITY

The patterns of stability and change, separated into time units, are summarized in Table 3. This compilation mixes taxa from the subfamily to superfamily level without discrimination, and thus is not amenable for statistical analysis. Table 4 lumps the time intervals. While the number of taxa is not large, the pattern is quite clear. In the higher pulmonates, the Mesurethra and the Sigmurethra, Recent distributions are virtually unchanged from those at time of origin; only the New World Camaenidae demonstrate a significant alteration in range since the Cretaceous. Only the Prosobranchia and Orthurethra have a fair portion of their early taxa showing a definite change in distribution.

PHYLETIC AND BIOGEOGRAPHIC IMPLICATIONS

The sudden appearance of half the extant orders of land snails in the late Paleozoic and the continuation of the Paleozoic land-snail families into the Recent means that: (1) the initial radiation of the land snails must have occurred prior to the Pennsylvanian; and (2) that the land snails were not noticeably affected by either the late Paleozoic vegetational shift to gymnosperms or the late Mesozoic rise to dominance of the angiosperms. Land snails presumably kept contentedly chewing away on dead plant matter. The appearance of the Stylommatophora in the Early Pennsylvanian, whereas the supposedly ancestral Basommatophora does not appear until the Jurassic-Cretaceous boundary, suggests that the Basommatophora might be totally independent of the Stylommatophora. Origins of both groups probably will have to be sought elsewhere among the Gastropoda.

It is equally important to identify phyletic "endpoints" and "generalized" taxa. Here I give a review based on the 5 late Paleozoic families and their phylogenetic positions. The Enidae (Solem, 1964) are the only group of the Orthurethra to obtain the advanced "sigmurethrous" excretory system. They thus probably are the most advanced group of the Orthurethra, but the enids were present in the late Paleozoic. In contrast, the Recent Partulidae are known only from the Pacific Islands, yet in anatomical structures they appear to be more generalized than the enids. The Partulidae could be ancestral to the enid pattern of structure, but have no fossil record. The Discidae, which are still Holarctic, are of uncertain phyletic position within the Arionacea. The Endodontidae are known to date only from the Miocene to Recent in Polynesia and part of Micronesia (Solem, 1976), and seem to be the most generalized group of the Sigmurethra, with a constellation of primitive features. The Charopidae, with their Gondwanaland distribution, have more advanced characters than the Endodontidae. Evidence is accumulating that would permit deriving the Charopidae from an endodontid ancestor (Solem, unpub. data), but not in the opposite direction. The Tornatellinidae contain a number of highly diverse subfamilies without intergrades (Cooke and Kondo, 1960, p. 40). No family of land snails ever has been suggested as being descended from the Tornatellinidae, and it stands as an isolated group of the Orthurethra, characterized by both shell and anatomy. The Pupillacea are still basically a holarctic group, and probably did not give rise to other groups. The Helicinidae are still the only significant land archaeogastropod group and did not give rise to other taxa.

We thus are faced with the fact that the land-snail families present from the Paleozoic are all without known derived families. They range from basal in a phyletic sense (Helicinidae, Tornatellinidae) to highly derived (Enidae); or they are members of a complex containing several more generalized taxa (Discidae). They cannot be considered to be the root stocks for the generally accepted more advanced taxa (Mesurethra, Holopodopes, Holopoda).

Space does not permit reviewing the full geographic implications, but one area requires special mention. The facts of Polynesia having (1) a collection of endemic families dating from the Paleozoic (Tornatellinidae), potentially ancestral to a Paleozoic taxon (Partulidae), and (2) the most generalized (Endodontidae) of a superfamily (Arionacea) with a Paleozoic representative (Discidae), is quite remarkable. Geologic data (Ladd and others, 1974) suggest that islands were present on the Pacific plate at least since the Cretaceous. While most, if not all, of the present islands are much younger, the persistence of isolated spots of land in the Pacific area since at least the Cretaceous is a minimum expectation. Rare but successful inter-island transport would suffice to maintain the fauna as a relict entity.

This is contrary to the view of Vagvolgyi (1976, p. 485) that the land snails of the Pacific Islands were derived "from the fauna of the nearest continent, in relatively recent times, through aerial immigration followed by local evolution." The demonstrated age of the Tornatellinidae and the ancestral position of the other groups relative to Paleozoic taxa suggest great age.

Particularly in view of the great distributional stability shown by most land-snail families, the endemic land-snail fauna of Polynesia can be viewed as a relict fauna, long isolated, and adapted for rare inter-island dispersal.

CONCLUSION

Initial radiation of land snails was prior to the Pennsylvanian, when 3 of 6 extant orders and 5 extant families appeared suddenly. Only a small minority of land snail family-level units can be shown to have undergone significant displacement in distribution since the time of their first appearance in the fossil record.

The extraordinary distributional stability and age of the land snails suggest that plate tectonic events of the Paleozoic and early Mesozoic will be the keys to an understanding of their biogeography, rather than the events of the Tertiary that determined the distribution of extant mammals.

ACKNOWLEDGMENTS

I am grateful to Ellis Yochelson for many stimulating discussions and much work on the Paleozoic taxa that led to our study, and to Daniel Axelrod and William Newman for several very helpful suggestions and comments. Help in manuscript preparation by Sharon Bacoyanis was essential.

REFERENCES

Baker, H. B., 1941, Zonitid snails from Pacific Islands, Pts. 3 and 4: Bernice P. Bishop Mus. Bull. 166, p. 205-370.

Cooke, C. Montague, Jr., and Kondo, Yoshio, 1960, Revision of Tornatellinidae and Achatinellidae (Gastropoda, Pulmonata): Bernice P. Bishop Mus. Bull. 221, p. 1-303.

Henderson, J., 1935, Fossil non-marine Mollusca of North America: Geol. Soc. America Spec. Paper 3, 313 p.

Ladd, H. S., Newman, W. A., and Sohl, N. F., 1974, Darwin Guyot, the Pacific's oldest atoll: Internat. Coral Reef Symposium, 2nd, Brisbane 1973, Proc., v. 2, p. 513-522.

Parodiz, J. J., 1969, The Tertiary non-marine Mollusca of South America: Carnegie Mus. Ann., v. 40, 242 p.

Pilsbry, H. A., 1927-1935, Manual of Conchology: Acad Nat. Sci. Philadelphia, Conchological Section, ser. 2, v. 28, 226 p.

——— 1939, Land Mollusca of North America (North of Mexico): Acad. Nat. Sci. Philadelphia, Mon. 3, v. 1, pt. 1, 573 p.

——— 1940, Land Mollusca of North America (North of Mexico): Acad. Nat. Sci. Philadelphia, Mon. 3, v. 2, pt. 1, p. 575-994.

——— 1946, Land Mollusca of North America (North of Mexico): Acad. Nat. Sci. Philadelphia, Mon. 3, v. 2, pt. 1, 520 p.

——— 1948, Land Mollusca of North America (North of Mexico): Acad. Nat. Sci. Philadelphia, Mon. 3, v. 2, pt. 2, p. 521-1113.

Russell, L. S., 1931, Early Tertiary Mollusca from Wyoming: Bull. Am. Paleontology, v. 18, 38 p.

Solem, A., 1959, Systematics and zoogeography of the land and freshwater Mollusca of the New Hebrides: Field Mus. Nat. Hist., Fieldiana, Zoology, v. 43, 359 p.

——— 1964, *Amimopina,* an Australian enid land snail: The Veliger, v. 6, p. 115-120.

——— 1969, Basic distribution of non-marine molluscs, *in* Symposium on Mollusca, Proc., Cochin 1968: Marine Biol. Assoc. India, Symp. ser. 3, pt. 1, p. 231-247.

——— 1976, Endodontoid land snails from Pacific Islands. Pt. 1. Family Endodontidae: Field Mus. Nat. Hist., Spec. Pub. Zoology, 508 p.

——— 1977, Classification of the land Mollusca, *in* Fretter, Vera, and Peake, John, eds., Pulmonates. Vol. 2, Systematics, evolution and ecology: New York, Academic Press (in press).

——— 1978, Cretaceous and early Tertiary camaenid land snails from western North America: Jour. Paleontology (in press).

Solem, A., and Yochelson, E. L., 1977, Affinities of the North American Paleozoic land snails: U. S. Geol. Survey Prof. Paper (in press).

Tozer, E. T., 1956, Uppermost Cretaceous and Paleocene non-marine molluscan faunas of western Alberta: Canada Geol. Survey Mem. 280, 125 p.

Vagvolgyi, J., 1976, Body size, aerial dispersal, and origin of the Pacific land snail fauna: Systematic Zoology, v. 24, p. 465-488.

van Bruggen, A. C., 1967, An introduction to the pulmonate family Streptaxidae: Jour. Conch., v. 26, p. 181-188.

Wenz, W., 1938, Prosobranchia: Handb. Paläozool., Gastropoda, Vol. 6, Pt. 1: Berlin, Borntraeger, 948 p.

Yen, T.-C., 1943, Review and summary of Tertiary and Quaternary non-marine mollusks of China: Acad. Nat. Sci. Philadelphia Proc., v. 95, p. 267-309.

Zilch, A., 1959-1960, Euthyneura: Handb. Paläozool., Gastropoda, Vol. 6, Pt. 2: Berlin, Borntraeger, 834 p.

Paleobiogeography of the Middle Jurassic Corals

Louise Beauvais, *Laboratoire de Paléontologie des Invertébrès, Université P. et. M. Curie, 4, place Jussieu, 75230 Paris-CEDEX 05, France*

ABSTRACT

No previous attempt has been made to study the biogeography of Middle Jurassic corals. The work discussed here is based on my own studies, supplemented by the published literature. The available data have unequal value because the taxonomy used by earlier authors is generally different from that of today. Moreover, the world distribution of Jurassic Madreporaria is still imperfectly known and the absence of ammonites in the coral-bearing formations makes their stratigraphy often imprecise. This work is thus a first attempt at coral paleobiogeography which subsequently will be completed and modified. In this attempt, I studied the distribution of the coral-bearing formations and assessed the genera and species stage-by-stage. The coral localities are plotted on maps that are synthesized from the best reconstructions of the oceanic plates. I have compared the madreporarian distribution with that of other groups such as the belemnites and ammonites. The paper concludes with a discussion of paleogeographic, paleoclimatologic, and stratigraphic events.

No coral is known in the Boreal Belemnite Realm. Coral genera have a cosmopolitan distribution but species are restricted to distinct basins. The world-wide distribution of the genera agrees with that of the bivalves and the flora; it confirms Hallam's (1971a, 1971b) opinions that a more equitable climate existed during the Jurassic than today and that no large, deep, oceanic barrier existed between eastern and western continents. The disappearance of corals from North America during the Bathonian and from northern Europe during the Callovian confirms that the North Atlantic began to open in the Bathonian-Callovian. From the stratigraphic point of view I found the following: 3 genera limited to the Dogger; 2 genera localized in the Bajocian; 48 genera that appeared during the Bathonian and 2 that appeared during the Callovian; 9 species that are characteristic of the Bajocian and 12 that are characteristic of the Bathonian. The species limited to the Callovian are endemic; consequently, I think that the planulae were not able to migrate during that stage.

INTRODUCTION

The work discussed here is based on my own studies of Middle Jurassic corals supplemented by a considerable bibliography. The data have very unequal value for several reasons: (1) taxonomy utilized by earlier authors is generally different from that used today; (2) the world distribution of the Jurassic Madreporaria is still imperfectly known; and (3) lack of ammonites in the coral-bearing formations makes their stratigraphy often imprecise. This work is a first attempt to study the biogeography of the Dogger corals; subsequently it will be modified.

The distribution of genera and species in the coral-bearing formations have been studied stage-by-stage. I will compare this distribution with that of some other groups such as the belemnites, the ammonites, the bivalves, and the foraminifera. The maps used to locate the coral formations are synthesized from Carey (1958), Dietz and Holden (1970), Dott and Batten (1971), Hallam (1971a), and Enay (1972). I have selected the reconstructions of oceanic plates that best agree with the zoological data.

I have used the stratigraphical scale established by the "Groupe français d'étude du Jurassique" in 1971. The *Sonninia sowerbyi* (Mill.) Zone is adopted for the lower limit of the Bajocian, and the *Parkinsonia parkinsoni* (Sow.) Zone for its upper boundary. However, it is possible that some corals or coral-bearing formations that were dated as Bajocian by earlier authors may belong to underlying zones such as the *concavum* or *murchisoniae* Zones. For example, at Crickley and Cheltenham the coral formations contain three or four superimposed reefs, the oldest belonging to the *murchisoniae* Zone of the Upper Liasic.

BAJOCIAN CORAL DISTRIBUTION

Bajocian madreporarian distribution is broad (Fig. 1), but the most numerous localities are situated in western Europe and on the western margin of the Mediterranean. It is also in this part of the world that the greatest number of species and genera are found. A similar pattern also characterizes the belemnites and ammonites; it may reflect better geologic investigation there than elsewhere. Nevertheless, it seems that conditions for reef growth were most favorable in Europe and from that region planulae could easily migrate along the platforms and settle in distant but geographically favorable areas without, however, developing true coral reefs. This appears to be the case for the Bolivian coral formation where only 4 genera and 6 species are known (*Montlivaltia, Stylina, Latomeandra, Enallocoenia*); for the Californian outcrops with the single genus *Latomeandra;* and for Wyoming where only *Actinastraea hyatti* is known.

Generally, except locally as in Normandy, Sarthe, Poitou, Portugal, and Mont d'Or Lyonnais, hermatypic genera dominate the Bajocian coral formations. It is thus impossible to distinguish during this stage, and even during the totality of the Dogger, a Tethyan zone with hermatypic corals, limited towards the north by an ahermatypic Madreporaria belt such as we can observe during the Upper Cretaceous (Beauvais and Beauvais, 1973). The predominance of nonreef-forming genera in some Middle Jurassic formations seems related to local ecologic events (influx of fresh water, silting up and so on) rather than to climatologic factors. However, it does not appear that great barrier reefs, similar to those present today in the Pacific Ocean, were growing during the Bajocian. Madreporaria were probably building patches in shallow, more-or-less agitated, epicontinental seas that suffered frequent tectonic movements which hindered the growth of reefs.

England

Lower Oolite Madreporaria formations (including the Bajocian and locally, the upper Aalenian) extended through Dorset, Somerset, and Gloucester. The main outcrops are those of Bridport, Burton, Horse Pools, Birdly, Leckhampton, Ravensgate, Seven Springs, Castle Carrey, Crickley, and Dundry. In Gloucestershire, Tomes (1878, 1882, 1884, 1885) indicates four corallian levels in the Lower Oolite, but in Dorset and Somerset, only the upper three are present. The English coral-bearing formations seem to constitute a group of bioherms separated by more-or-less wide passes: at Crickley, at Horse Pools, and near Gloucester, important reefy masses in which corals form the main mass of the rock occur. At Ravensgate Hill, small lenses with corals grade laterally to coralless marls. At Leckhampton, Madreporaria constitute patches which are scattered in marls containing numerous broken but no rolled corals.

The English Bajocian fauna is rich. I counted 36 genera and 63 species: 37 species are endemic and 10 are also found in the Paris Basin. The others have a wider, global distribution. The ahermatypic genera are: *Axosmilia, Bathmosmilia, Chomatoseris, Cyathophyllopsis, Discocyathus, Epismilia, Epistreptophyllum, Macgeopsis, Montlivaltia, Thecocyathus,* and *Thecoseris.* The hermatypic ones are: *Allocoeniopsis, Astraeofungia, Andemantastraea, Cladophyllia, Coenastraea, Confusastraea, Dimorpharaea, Dimorphastraea, Enallocoenia, Edwardsomeandra, Fungiasstraea, Isastracea, Latomeandra, Microsolena, Oroseris, Paraphyllogyra, Phyllogyra, Phylloseriopsis, Phylloseris, Stylina, Thamnasteria, Thamnoseris, Thecosmilia, Trigerastraea,* and *Vallimeandropsis.*

France

Normandy. Corals are abundant in the ferruginous oolite of Bayeux (Rioult, 1962) but they are only nonreef-building forms. The main outcrops are Bayeux, Croisille, St-Vigor, and Port-en-Bessin. The ahermatypic genera are: *Axosmila, Discocyathus, Montlivaltia,* and *Thecocyathus.* Only 1 genus, *Dimorpharaea,* is hermatypic, but it is a form which can easily survive in a turbid environment. The data furnished by corals confirms the theories of Bigot (1940a), Dangeard (1951a), and Rioult (1962) that the Bajocian of Normandy was an unstable period marked by sedimentary hiatuses and deposition of iron, glauconite, and phosphate.

Sarthe. In the Department of Sarthe (Alençon, Conlie, Hyères, Guéret) ahermatypic genera (*Chomatoseris, Epismiliopsis, Montlivaltia, Thecocyathus, Thecophyllia,* and *Trochocyathus*) are only found in the "oolite miliaire" facies which contains many gastropods. As in Normandy, the environment appears to have been variable and unstable. Among the 8 known species, 5 are endemic; the other 3 are found in England, Normandy, Switzerland, and in the vicinity of Mâcon.

Southern Paris Basin. In Vienne and Gartempe Valleys and in Poitou (at la Mothe-St-Heraye and St-Maixent), only ahermatypic genera are present (*Montlivaltia* and *Chomatoseris*). A little farther south on the southwestern edge of the Massif Central, Madreporaria are more abundant and true reefs with *Thamnasteria, Isastrea,* and *Cladophyllia* are reported by Glangeaud (1885) and Thevenin (1903) near Montron-aux-Brosses.

Luxembourg and the Ardennes. Coral-bearing formations in the Bar Valley, Longwy, and Charleville regions contain 8 hermatypic genera (*Coenastrea, Cyathophora, Confusastraea, Isastrea, Orbignycoenia, Stylina, Thamnasteria,* and *Trigerastraea*) and 9 species all known also in England, in the Boulonnais, in Normandy, Lorraine, Alsace, and in the Jura. The region thus appears as a place of transition between the English reef formations and those of the Jura and eastern France.

Lorraine and Adjacent Regions. Reef islets are found in the Department of Meurthe-et-Moselle, in the counties of Metz, Toul, Nancy, and Chaumont, on the "plateau de Langres" (at Noidant), and in the Department of Haute-Saône (Morey, Coulevon, Voncourt). This line of reefs parallels the crystalline massifs, part towards the Department of "la Nièvre" (at Coulanges), Mâcon (reefs of Milly, Flacé, Berzé-la-Ville), and the "Mont-d'Or Lyonnais" and the other part towards Belfort, Alsace (Bouxvilliers, Colmar), the French Jura (Salins, Nantua, Fort-St-André, Lons-le Saunier, "l'Ile Crémieux," Mont-Myon), and the Swiss Jura (Montmelon, Pichoux, Cornol, Roche-d'Or, Glovelier, Mythen, Ste-Croix). In all these regions, the fauna is abundant both in genera and species. Madreporaria

Figure 1. Bajocian coral distribution and paleogeography. Bajocian coral distribution (dots); approximate southern limit of the Bajocian Cylindroteuthididae Realm (x's: Stevens, 1973); approximate boreal limit of the Bajocian *Hibolites* Realm (Tethys) (dashed line: Stevens, 1973).

seem to have built islets surrounded with sediment similar to those developed during the same stage in England. Twenty-one genera are found in the Bajocian of Lorraine. The ahermatypic genera are: *Chomatoseris, Cyathophyllopsis, Montlivaltia,* and *Thecocyathus;* the hermatypic genera are: *Aggomorphastraea, Andemantastraea, Cladophyllia, Clausastraea, Coenastraea, Confusastraea, Dimorphastraea, Diplocoenia, Isastrea, Mesomorpha, Periseris, Phylloseris, Stylina, Stylosmilia, Thamnasteria,* and *Thecosmilia.* They include 28 species; among them only 8 are endemic and 7 are exclusively found in Europe. Thirteen others have a world-wide distribution either in the Bajocian or in the Bathonian and the Callovian. Hallam (1975) indicates that the Lorraine Bajocian reefs are represented by a series of low mounds that grew in shallow epicontinental waters. These patches are from 8 to 16 m high and their diameters are from 10 to 33 m. In these lenses may be observed numerous lamellar (*Isastraea, Thamnasteria*) or branching (*Stylosmilia*) Madreporaria concealing in their interstices a fine-grained bioclastic limestone. The fauna found in the inter-reef beds is rich and diversified; it includes bivalves, terebratulids, rhynchonellids, and other brachiopods, serpulids, crinoid ossicles, echinoid spines, gastropods, boring polychaetes, and very few coral fragments. This information suggests a littoral environment with little erosion.

The same facies and the same fauna is recognized by Bircher (1935) in Switzerland where the Bajocian fauna is very similar to that of the Mont-d'Or Lyonnais; it is a composite fauna with Swabian elements mixed with some Mediterranean forms. According to Bircher, the Bajocian is a littoral coral facies as contrasted with the Dauphinian cephalopod facies found in the French Alps. The latter lacks corals, echinoids, brachiopods, and gastropods. To the south of the French Alps, we are again on a platform: phaceloid hermatypic corals are found in the Department of Var (*Calamophylliopsis, Lochmaeosmilia*). Twenty genera are found in the Bajocian of Haute-Saône and Jura; they are the ahermatypic genera *Chomatoseris, Montlivaltia,* and *Thecocyathus,* and the hermatypic genera, *Cladophyllia, Collignonastraea, Columnocoenia, Confusastraea, Dimorpharaea, Isastraea, Latomaendra, Lochmaeosmilia, Meandraraea, Microphyllia, Periseris, Polyastropsis, Stylina, Stephanastraea, Stereocoenia, Thamnasteria,* and *Thecosmilia.* Among the 45 species known in this region, 14 are endemic, 13 are found in the adjacent districts (England, Alsace, Mâconnais), and 14 have a wide geographic and stratigraphic distribution.

In Mâconnais, the 14 known genera are hermatypic; they are also common to the Lorraine Basin and the Swiss Basin (*Aggomorphastraea, Andemantastraea, Cladophyllia, Clausastraea, Coenastraea, Collignonastraea, Dimorpharaea, Isastrea, Latomeandra, Microsolena, Periseris, Thamnasteria, Vallimeandra,* and *Vallimeandropsis*); among the 19 species representing these genera, only 4 are endemic. As in the case of the Ardennes, the Mâcon area seems to be a zone of passage between the English-Paris fauna and the Jura fauna.

The Mont-d'Or Lyonnais probably formed a shoal where unfavorable ecological conditions did not allow the growth of reef-building forms. In fact, only *Thecocyathus* and *Trochocyathus* are found there.

In Alsace, reefs seem less developed than in the Paris Basin; only 7 genera and 12 species are known. Three genera are ahermatypic: *Montlivaltia, Thecocyathus,* and *Thecocphyllia;* the others are hermatypic: *Confusastraea, Isastraea, Thamnasteria,* and *Thecosmilia.* Four species are endemic.

Only 3 species belonging to 3 genera are known in Württemberg and Swabia. One genus is ahermatypic (*Thecocyathus*), 2 species are endemic, and the third is found in the Bajocian of Alsace and Lorraine. A more direct opening towards northern seas with arrival of cool waters may explain the paucity of the coral fauna in the region.

There was communication between the Languedoc and Aquitain Basins; this communication is indicated by littoral deposits on the southern border of the Massif Central. Around Narbonne, the Madreporaria *Adelocoenia, Montlivaltia,* and *Chomatoseris* were collected.

Portugal

Only 1 ahermatypic genus (*Montlivaltia*) has been collected at Verride, Fatima, and Desgracias (Ruget-Perrot, 1961).

Morocco

Lenticular reef formations containing a rich hermatypic coral fauna are found in the eastern part of the "Moyen Atlas" at Rekkame and on the "Hauts-Plateaux" in the neighborhood of Bou-Rached, Dedbou, Matarka, and Tendrara (outcrops at Ternet, Kounif, Seheb-el-Kelb, Djebel Richa) (Choubert, 1938a, 1938b; du Dresnay, 1964, 1965; Lorenchet de Montjamont, 1964; Medioni, 1960, 1966; Beauvais, 1970). In these formations, 2 ahermatypic genera (*Bathmosmilia* and *Montlivaltia*) and 28 hermatypics are found (*Allocoeniopsis, Ampakabastraea, Astraraea, Coenastraea, Cladophyllia, Connectastraea, Dendraraea, Dimorpharaea, Dimorphastraea, Dimorphastraeopsis, Diplocoenia, Isastrea, Isastrocoenia, Kobya, Meandraraea, Melikerona, Mesomorpha, Microphyllia, Microsolena, Sakalavastraeopsis, Stereocoenia, Stylina, Stylohelia, Stylosmilia, Thamnasteria, Thamnoseris, Thecosmilia,* and *Trigerastraea*). Among the 45 species found, only 11 are endemic, consequently, Morocco seems to have had free communication with other parts of the world during the Bajocian.

North America

Some coral formations are known in the Pryor Mountains of Wyoming and from Montana; they include only 1 species, *Actinastraea hyatti* Wells. On the south side of Mont-Jura (California), a *Latimeandra* was described from the early Middle Jurassic.

South America

A Madreporaria outcrop containing 3 reef-building genera (*Enallohelia, Latomeandra,* and *Stylina*) and an ahermatypic one (*Montlivaltia*) is known from Caracoles, Bolivia. It was studied by Steinmann (1881) and includes 6 species widely distributed in England, the Paris Basin, Madagascar (Bathonian), and Tunisia (Callovian). Another outcrop is known in Chile at Manflas in the Cordillera of Copiapo from which Gerth (1926) mentions 2 species of *Isastrea* and 1 of *Thamnasteria.*

Madagascar and Africa

In Madagascar, some Bajocian Madreporaria limestones were reported by Besairie (1952) and Barrabé (1929), but no species or genera were listed and no collections are known. However, in Africa, coral formations studied by Thomas (1963) in Tanganyika begin in the Upper Bajocian and continued to grow into the Bathonian. Thus, it is possible that Madreporaria formations were developed in Madagascar as early as the Bajocian.

Summary and Conclusions

The Bajocian coral distribution does not permit recognition of faunal provinces similar to those determined from belemnites (Stevens, 1973) or ammonites (Arkell, 1956). However, no coral is found in the Boreal Realm associated with Cylindroteuthididae. This can be explained in the following way: the northern regions were probably not really cold (in North America Stevens found typical southern forms such as *Hibolites, Megateuthis,* and *Gastrobolus* mixed with the Cylindroteuthididae) but they were too cool, nevertheless, for the growth of coral reefs.

The 67 coral genera reported in the Bajocian are all cosmopolitan and the majority of them extend throughout the Mesozoic. Nevertheless, from 25 to 50 percent endemic species are found in some basins: Alsace-Swabia-Jura, Morocco, South and North America. The endemism can be explained by the living habits of Madreporaria: they are benthic and fixed and only the long-lived planulae are able to migrate for great distances. Concurring with Hallam (1971a, 1971b), I think that the absence of important physical barriers and deep oceanic areas in the Bajocian seas permitted the cosmopolitan distribution of coral genera.

The Bajocian climate was probably much more uniform than today: in Württemberg, a paleotemperature of 29°C is indicated for the lower Bajocian and temperatures from 14.2° to 18.1°C for the upper Bajocian. Temperatures from 19.1° to 20.3°C are indicated for the upper Bajocian of Greenland, and from 19.7° to 28.6°C for the same stage at Neuquen, Argentina.

The most abundant Bajocian genera are: *Isastrea, Thamnasteria, Stylina, Coenastraea, Cladophyllia, Confusastraea, Dimorpharaea, Thecosmilia, Chomatoseris, Montlivaltia,* and *Thecocythus. Polyastropsis* and *Cyathophyllopsis* seem to be characteristic of the Bajocian. Bajocian species in Europe and Morocco are: *Thamnoseris jurensis* (d'Orbigny), *Isastrea tenuistriata* M. Edw. and Haime, *Isastrea salinensis* Koby, *Latomeandra ramosa* (d'Orbigny), *Thecocyathus magnevillianus* (Mich.), *Confusastraea gregaria* (M'Coy), *Dimorpharaea defranciana* (Mich.), *Andemantastraea consobrina* (d'Orbigny), *Ampakabastraea crenulata* (d'Orbigny); and in the United States: *Actinastraea hyatti* (Wells).

BATHONIAN CORAL DISTRIBUTION

During the Bathonian (Fig. 2), corals disappear from the American continents. They are still widespread in Europe, however, and they appear in India and Iran. Some are known in Dobrudga (Barbulesci, unpub. data). In Madagascar, coral formations are at least as rich and as extensive as in Europe. The disappearance of reef-building corals from the Haut-Atlas in Morocco is easily explained by a local and temporary regression of the sea. The disappearance of corals from Wyoming and Montana may be due to the cooling mentioned by Stevens (1973). According to Stevens (1973), a Boreal Realm is more readily recognized in the Bathonian than in the Bajocian. The meridional limit of the Cylindroteuthididae which stands between 65° to 70°N in Europe shifts to 50°N along the western coast of North America. The retreat of reef corals from South America is more difficult to explain; it probably represents a hiatus in our knowledge.

Reef-building corals are well-developed in all Bathonian outcrops just as they were during the Bajocian, but the reefs never reach the dimensions of the present edifices in the Pacific Ocean.

England and France

Anglo-Paris Basin. Bathonian sedimentation is neritic and represents epicontinental shallow waters just as during the Bajocian. In England, despite regression of the sea from northern England and Scotland (Hallam, 1971b), there is no noticeable change in the location of coral reefs. The main outcrops are: Bath, Bradford Hill, Epwell, Banbury, Cheltenham, Rollright, Cirencester, Burford, Combe Down, Stonesfield, and Fairford. The English Bathonian possesses about the same facies as the Bajocian: lenticular reefs and coral beds in which corals transported from nearby reefs are broken and rounded. The numerous lateral changes of facies indicate the instability and agitation of the sea. The coral fauna is rich: 35 genera and 68 species are recorded. Only 4 genera are ahermatypic: *Acrosmilia, Chomatoseris, Montlivaltia,* and *Tricycloseris;* the others are hermatypic: *Allocoeniopsis, Adelocoenia, Andemantastraea, Bathycoenia, Cladophyllia, Comoseris, Confusastraea, Cryptocoenia, Cyathophora, Dendraraea, Dendrastraea, Edwardsomeandra, Enallhelia, Isastraea, Keriophyllia, Lochmeaosmilia, Meandraraea, Meandrophyllia, Mesomorpha, Microphyllia, Microsolena, Orbignycoenia, Pseudocoenia, Sakalavastraea, Stereocoenia, Stylina, Stylosmilia, Thamnasteria, Thamnocoenia, Thecosmilia,* and *Trigerastraea.* Twenty species are endemic in the Bathonian of England, and 7 in the Anglo-Paris Basin. Two of the Anglo-Parisian species also occur in the Bathonian of Poland and 4 in Madagascar and in India. The other species exist throughout the Dogger and have representatives in Morocco, Tunisia, Sinai, and elsewhere.

Normandy. In Normandy, as in England, frequent oscillations of the sea prevented reef growth. Numerous hardgrounds and facies changes can be observed in the Madreporaria formations. The main coral-bearing formations are located in the neighborhood of Caen, and at Blainville, Perrieres, Villedieu-les-Bailleuls, le Breuil, Courseuille, Luc, and Langrune. At Blainville (Calvados), the Bathonian reefs are represented by small patches scattered in a chalky, bedded mud. These lenses are from 8 or 10 to 20 or 25 m long and from 1 to 4 m high. They are built either by *Stromatopora* or by big, flat colonies of *Isastrea limitata* Lamouroux which may reach 0.50 m in diameter and which are superimposed atop one another. The lense pass laterally to a gravelly facies indicating strong currents (Dangeard, 1951b) and their upper surfaces are levelled, hardened, and bored. In Basse-Normandy, evidences of crossbedding and numerous changes of sediments indicate irregularities in the conditions of sedimentation (Bigot, 1940a). On the other hand, Mercier (1933) notes that at Perrieres the coral specimens are small, indicating a confined environment in a sheltered area. In general, strong currents accompanied the Norman Bathonian transgression but quiet conditions existed in sheltered gulfs. Rioult (1962) compared the limestones of Caen to a shore-marsh deposit in a tropical environment. He describes numerous similarities between the Caen limestone and the deposits on the low, marshy or lagoonal shores of Florida, Andros Island, or the Bermudas. During the Bathonian, Calvados would have been a broad, continental platform sinking slowly but sustaining various tectonic movements indicated by variations in sea level. The tropical nearshore marshes of the Bathonian were localized between the Paleozoic anticlines rising as rocks or shoals and thus were sheltered from the high sea. At Chailloué (Orne) (Bigot, 1926) the same facies are visible. The Bathonian sea transgressed upon the Armorican sandstones; in pre-Jurassic cavities of the submerged surface, are deposits of poorly bedded oolitic limestones containing a poor ahermatypic fauna (*Chomatoseris, Montlivaltia*). The small dimensions of the specimens are significant and characteristic of sheltered environments. Near Alençon, the "oolite miliaire" rich in gastropods, confirms the shallow-water sedimentation. Twenty-six coral genera are known in the Bathonian of Normandy. The ahermatypic genera are: *Axosmilia, Chomatoseris, Discocoenia, Genabacia, Montlivaltia,* and *Paramontlivaltia;* and the hermatypic genera: *Adelocoenia, Allocoeniopsis, Bathycoenia, Confusastraea, Cryptocoenia, Dactylocoenia, Dendrastraea, Dichotomosmilia, Euhelia, Isastraea, Lochmaeosmilia, Meandrophyllia, Microsolena, Semeloseris, Stephanocoenia, Stylohelia, Thamnasteria, Thamnoseris, Trigerastraea,* and *Vallimeandra.* These genera are represented by 30 species; 12 of them are endemic to Normandy and 7 are restricted to the Anglo-Paris Basin. A little toward the south at Domfront, Bourg-le-Roi, and Fresnay-sur-Sarthe, the upper Bathonian marls and the *Montlivaltia* limestones transgress on other levels. Near le Mans at Chemiré, a *Spongia* reef occurs in the *Montlivaltia* limestone. In the Department of Sarthe, the Bathonian seems to present the same characteristics as the Bajocian. The area seems to have been an unstable, intertidal zone with too much sedimentation to allow the growth of true reefs: only ahermatypic forms are known.

The Ardennes. In the Ardennes, numerous evidences of emergence and erosion indicate the rising of the massif. In this region, Madreporaria are more numerous on the anticlines which probably made shoals where corals could easily grow. It seems, however, that the instability of sea level did not allow the growth of true Madreporaria reefs; in fact, the genera are principally ahermatypic (*Chomatoseris, Genabacia, Epistreptophyllum, Montivaltia*) and only 2 are hermatypic (*Bathycoenia* and *Trigerastraea*). The same species are also found in the Anglo-Paris Basin, 2 are known in Madagascar, and 1 in the Sinai. The localities where these corals are found are: Les Vallées, Hirson, Eparcy, Leuze, and Rumigny. Many Madreporaria are also found in the Boulonnais at Hydrequent, le Wast, Boulogne, and Bléquenèque. The genera found are: *Cerathocoenia, Chomatoseris, Discocoenia, Genebacia,* and *Montlivaltia* (ahermatypic); and *Adelocoenia, Bathycoenia, Cladophyllia, Confusastraea, Cryptocoenia, Dendraraea, Isastraea, Meandrophyllia, Orbignycoenia, Scyphocoenia, Septastraea,* and *Thamnasteria* (hermatypic). In the Boulonnais, about 10 of the 28 species are endemic and 7 are found in the Anglo-Paris Basin. Almost all the coral genera of the Boulonnais and Ardennes are also found in England with the exceptions of *Epistreptophyllum, Cerathocoenia, Scyphocoenia,* and *Septastraea.* The similarity between the faunas agrees with the similarity of the facies on either side of the Channel; in both regions the lower Bathonian is poor and the middle Bathonian is rich and more developed. The opposite is observed to the east, in the Departments of Meuse and Moselle. There, Bathonian deposits are similar to those of Haute-Saône and the Jura; the lower Bathonian is well-developed and contains a rich fauna while the middle Bathonian has only scarce fossils or none. However, the coral genera and species of Lorraine are about the same as those of England and Normandy. The genera are: *Acrosmilia, Chomatoseris, Montlivaltia, Adelocoenia, Confusastraea, Stylina,* and *Thamnasteria.* The species are all common to the Anglo-Paris Basin and Switzerland. The main coral localities are Clapes (Moselle), Gorze, Auboué, Conflans, Jarny, and Thumerréville. At Vielley-St-Etiene (Meurthe-et-Moselle), the lower Bathonian reefs are a continuation of the upper Bajocian ones. Numerous corals are reported in the Bathonian of the Department of Haute-Marne (at Langres, Perrognay, south of Chaumont).

Alsace. In Alsace, 2 widespread genera and species (*Vallimeandropsis* and *Isastraea*) are reported.

Southern Paris Basin. In the southern part of the Paris Basin, beautiful Bathonian reefs are described

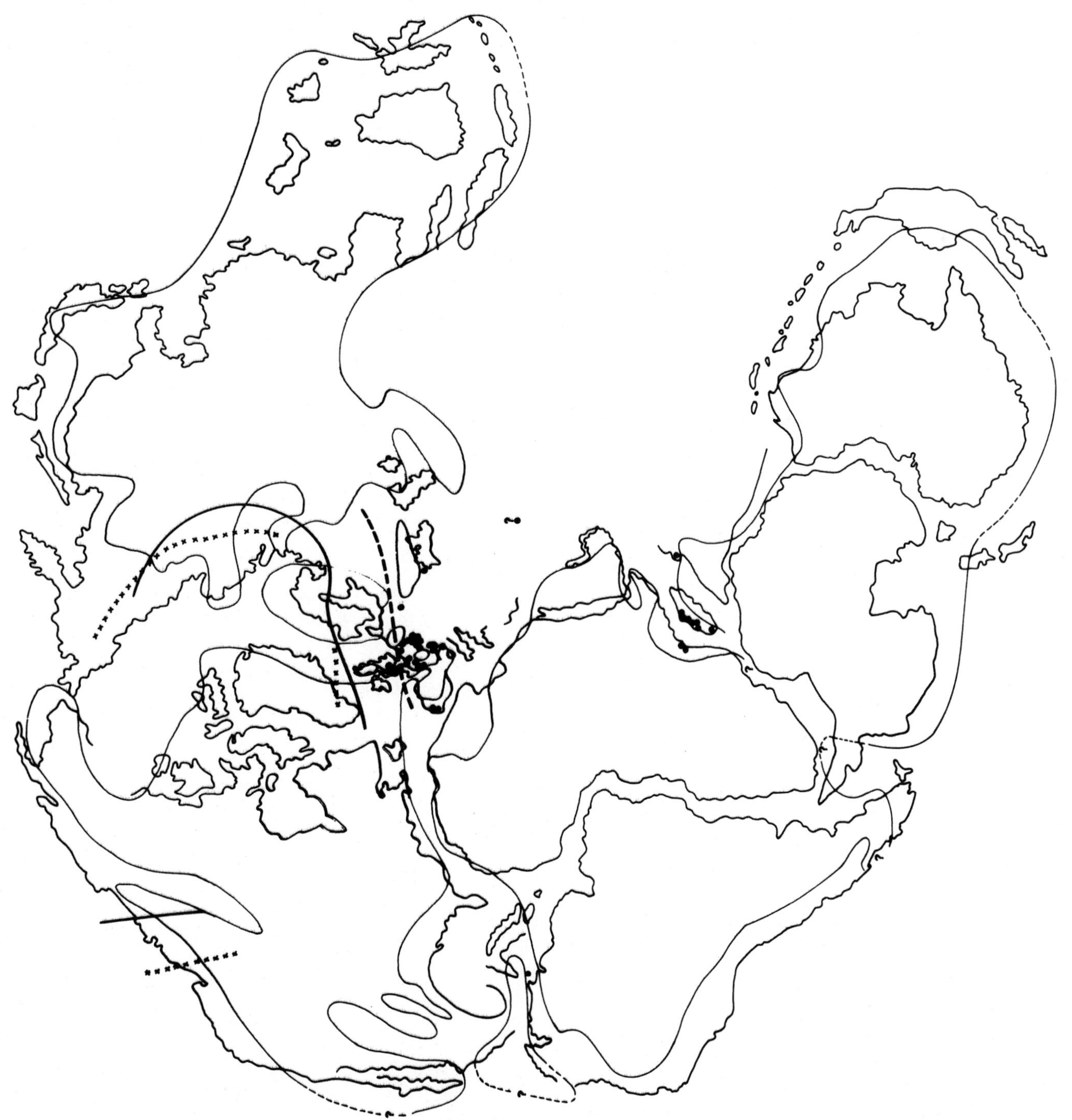

Figure 2. Bathonian coral distribution and paleogeography. Bathonian coral distribution (dots); approximate southern limit of the Boreal Realm (solid line: Hallam, 1971a); approximate southern limit of the Bathonian Cylindroteuthididae Realm (x's: Stevens, 1973); approximate boreal limit of the Bathonian *Hibolites* Realm (dashed line: Steven, 1973).

(Fischer, 1961, 1962) in the Creuse Valley (at Chasseneuil, St-Gaultier, Thenay) and in the Department of Vienne (at St-Savin-sur-Gartempe and Champigny). The coral fauna collected from the reef at St-Gaultier contains 30 genera of which 8 are nonreef formers. Of the 42 species, 16 are endemic. The ahermatypic genera from St-Gaultier are: *Acrosmilia, Axosmilia, Chomatoseris, Epistreptophyllum, Montlivaltia, Polymorphastraea, Protethmos,* and *Trochoplegma;* the hermatypic genera are: *Aggomorphastraea, Allocoenia, Allocoeniopsis, Ampakabastraea, Andemantastraea, Astraraea, Cladophyllia, Collignonastraea, Comophyllia, Cryptocoenia, Dendraraea, Dimorpharaea, Dimorphastraea, Edwardsomeandra, Isastraea, Lochmaeosmilia, Mesomorpha, Microsolena, Pseudocoenia, Stylina, Thamnasteria,* and *Trigerastraea.*

The Jura. In the vicinity of the Swiss and French Jura, data is limited: only 11 genera are known, 6 of them are hermatypic (*Adelocoenia, Confusastraea, Isastraea, Stylina, Stylohelia,* and *Thamnasteria*); the others are ahermatypic (*Chomatoseris, Discocyathus, Latiphylla, Montlivaltia*). True reefs do not characterize the Bathonian of the Jura. Among the 19 species described, 5 are endemic; the others have wide stratigraphic and geographic distribution (from the Bajocian to the Callovian). The main outcrops where Madreporaria are found in the Bathonian of the Jura are situated in Bas-Bugey (France), in Switzerland, and in the vicinity of Délémont, in the Vorbourg, at Movelier, Fringeli, Liesberg, Grellingen, Azuel, Mâle-Côte, and Soleute. Some corals are reported in the Glarnisch Alps at Oberlegi and a peculiar Bathonian facies with Madreporaria levels is found in the Romanic Alps between the Aar and Arve rivers, at la Raye, Boltingen, Simmerthal, Laitmaire, and Jaun. This facies is named "couches à Mytilus." It is brackish, even limnic, and contains previously worked coal deposits. The fossils are very numerous but their small size suggests a confined environment. The species of this facies have affinities with those of the Upper Jurassic. For this reason, the *Mytilus* beds were originally placed in the Malm. Recent studies both on corals and on other groups have put those beds unquestionably in the Bathonian-Callovian. Sixteen genera are described from the *Mytilus* beds. They are about the same as those recognized in other Bathonian outcrops, except for the genus *Ovalastraea* which is much more abundant in the Malm reefs but which is also found in the Bathonian of Madagascar. Ahermatypic genera are *Acrosmilia, Montlivaltia, Rhipidogyra,* and *Thecoseris;* hermatypic genera are *Adelocoenia, Baryphyllia, Coenastraea, Collignonastraea, Cryptocoenia, Diplocoenia, Ellipsocoenia, Latomeandra, Lochmaeosmilia, Microphyllia,* and *Stylina.* Of the 26 genera recognized, 16 are endemic.

Aquitain Basin. Some solitary corals (*Chomatoseris*) are recognized south of Charentes. No Madreporaria are found along the southeastern edge of the Massif Central which was probably partially emergent.

Provençal Basin. Numerous coral formations are found in the Provençal Basin (Lanquine, 1929; Koby, 1905). From the Koby collection, I identified the following ahermatypic genera: *Cerathocoenia, Chomatoseris, Macgeopsis,* and *Montlivaltia,* and the following hermatypic ones: *Adelocoenia, Allocoeniopsis, Andemantastraea, Dimorphastraea, Microsolena, Stylosmilia, Thamnasteria, Thecosmilia,* and *Trigerastraea.* Thirteen species represent these 12 genera, 5 of them are endemic. The others are found also in the Department of Sarthe at St-Gaultier, in Switzerland, and in Madagascar. Madreporaria localities occur near Toulon (Valette road), at Bandol, Valamy, St-Hubert, Forcalquieret, Puget-Ville, Roquefort-Clamarquier, between Gapeau and l'Argens, and in the neighborhood of Grasse and Grasse Roquevignon.

Portugal

The Bathonian seems to be rather rich in Madreporaria formations, but the corals are recrystallized and they cannot be identified. They are found at Obidos, Cesareda, Fatima, Desgracias, Cova de Fonte, and south of Barranco de Zambujal (Ruget-Perrot, 1962).

Poland

At Balin, near Cracow, 1 ahermatypic genus, *Montlivaltia,* and 5 hermatypic corals, *Dimorpharaea, Dimorphastraea, Dimorphastraeopsis, Thamnoseris,* and *Thecosmilia* were described by Reuss (1867) and Beauvais (1971). Of the 6 described species, 3 are endemic; the others are found in the Anglo-Paris Basin and in Morocco.

Iran

Corals are found in the "Qualeh Dokhtar Formation" on the eastern edge of the Shotori Chain (Flügel, 1966). Although they were formerly placed in the Upper Jurassic, they are now recognized as of Bathonian-Callovian age. This Iranian fauna possesses affinities with the Bathonian reefs of Cutch, India. The genera found are *Chomatoseris* (which indicates a Middle Jurassic age), *Epistreptophyllum, Metethmos, Trocharaea, Collignonastraea, Cyathophora, Dimorpharaea, Isastraea, Microsolena,* and *Stylina.* Among the 11 reported species, only 1 is endemic, 5 are found in the Bathonian of Cutch, 1 in the Callovian of Tunisia, and 4 survive into the Malm in Germany, the Crimea, and Azerbaidjan. According to Flügel (1966), the corals are found in beds that are part of a littoral, marly, more-or-less shelly, limestone. The Madreporaria are slightly cemented, and the sedimentary matrix which encloses them is rich in bioclastic material (fragments of bivalves, gastropods, and so forth).

Madagascar

Madreporaria are reported from almost all the western Bathonian outcrops. They are cited in numerous works (de Fromentel, 1873; Alloiteau, 1958; Collignon, 1959). Collignon (1959) described some formations in southern Madagascar which he identified as barriers, banks, and atolls. I am now studying the corals collected by him. Ninety-four species are currently known as a result of my work and that of Alloiteau (1958). Of these, 48 are endemic. The 46 genera recognized are almost all hermatypic except for *Chomatoseris, Macgeopsis, Montlivaltia,* and *Epistreptophyllum.* The following are known: *Adelocoenia, Aggomorphastraea, Allocoeniopsis, Ampakabastraea, Andemastastraea, Astraeofungia, Brachyseris, Cladophyllia, Collignonastraea, Collignonoseris, Columnocoenia, Connectastraea, Crateroseris, Cryptocoenia, Diplocoenia, Elasmofungia, Fungiastraea, Heliocoenia, Isastraea, Keriophyllia, Microphyllia, Mesomorpha, Microsolena, Morphastraea, Myriophyllia, Ogilviella, Ovalastraea, Parisastraea, Periseris, Plesiocoenia, Plesiostylina, Psammohelia, Pseudisastraea, Pseudocoenia, Pseudodiplocoenia, Sakalavastraea, Stylina, Thamnasteria, Trigerastraea, Valliculastraea, Vallimeandra,* and *Vallimeandropsis.* As with the Iranian corals, some of the Madagascar species survive into the Upper Jurassic. The Madagascar species also have affinities with the Indian forms. The main outcrops reported by Alloiteau (1958) are Andranomarivo, Ankadibé, Tongobory, Soaravikely, Vongoho, Ankazomiheva, Besavoa, Lazarina, Ampakabo, and Betioky.

India

The classic outcrops of Cutch provided Gregory (1900) with an abundant coral fauna which I have partially revised. Twenty-four genera are known. Seven of them are ahermatypic forms: *Metethmos, Montlivaltia, Placosmilia, Protethmos, Sematethmos, Tricycloseris,* and *Trocharaea.* Seventeen are hermatypic: *Adelocoenia, Ampakabastraea, Astraraea, Collignonastraea, Comoseris, Dimorpharaea, Dimorphastraea, Gregorycoenia, Isastraea, Kobya, Lochmaeosmilia, Melikerona, Microsolena, Stylina, Thamnasteria, Thamnoseris,* and *Vallimeandra.* Among the 48 species, 22 are endemic, 9 are also found at Mombasa (Kenya), some are found in Madagascar, and 14 have a world-wide distribution that includes Switzerland, England, and Morocco.

Africa

At Mombasa in Kenya and south of Amboui in Tanganyika, Thomas (1963) described a Madreporaria fauna of 10 genera and 11 species. The genera are *Metethmos* (ahermatypic) and *Calamophyllia, Connectastraea, Diplocoenia, Diplaraea, Kobya, Microphyllia, Pleurophyllia, Thamnasteria,* and *Thecosmilia* (hermatypic). Among the 11 species only 2 are endemic; the others are found in the Bathonian of Cutch.

Summary and Conclusions

Despite the disappearance of corals from America, there is generally an important increase in the number of genera and species during the Bathonian. One hundred and five genera are known, many of them new, including *Acrosmilia, Bathycoenia, Cerathocoenia, Genabacia, Placosmilia, Pleurophyllia, Paramontlivaltia, Throchoplegma, Tricycloseris, Trocharaea, Discocoenia, Allocoenia, Baryphyllia, Brachyseris, Crateroseris, Collignonoseris, Ellipsocoenia, Euhelia, Enallhelia, Elasmofungia, Heliocoenia, Gregorycoenia, Keriophyllia, Meandrophyllia, Morphastraea, Myriophyllia, Metethmos, Orbignycoenia, Ovalastraea, Ogilviella, Parisastraea, Psammohelia, Polymorphastraea, Protethmos, Pseudocoenia, Pseudiastraea, Pseudodiplocoenia, Plesiostylina, Plesiocoenia, Stephanocoenia, Semeloseris, Sematethmos, Septastraea, Scyphocoenia, Thamnocoenia, Dactylocoenia, Dendrastraea,* and *Dichotomosmilia.* Nearly all genera that appear during the Bathonian persist throughout the Mesozoic. The most common Bathonian genera are: *Chomatoseris, Montlivaltia, Allocoeniopsis, Adelocoenia, Cryptocoenia, Confusastraea, Collignonastraea, Isastrea, Lochmaeosmilia, Microsolena, Stylina, Thamnasteria,* and *Trigerastraea.* Some cosmopolitan species are confined to the Bathonian; they are consequently good stratigraphic species. These are *Epistreptophyllum articulatum* (de From. and Ferry), *Confusastraea cottaldina* (M. Edw. and Haime), *Montlivaltia cornutiformis* (Gregory), *Montlivaltia frustriformis* Gregory, *M. kachensis* Gregory, *M. obtusa* d'Orbigny, *M. sarthacensis* d'Orbigny, *Collignonastraea grossouvrei* Beauv., *C. jumarensis* (Gregory), *Stylina kachensis* Gregory, *Bathycoenia moneta* (d'Orbigny), and *Genabacia stellifera* (d' Archaic). Numerous other species have both wide geographic and stratigraphic distribution (Bajocian to Callovian). Thus, just as in the Bąjocian, no oceanic barriers hindered the free circulation of Bathonian coral planulae. Nevertheless, by means of endemic species we can distinguish a European basin and an Indian-Malagasy basin, but this distinction probably depends on the ecology of the Madreporaria. In fact, the cosmopolitan genera have ecological and climatic amplitudes that would have favored growth of coral reefs in both basins.

CALLOVIAN CORAL DISTRIBUTION

France

Coral distribution is more narrowly defined during the Callovian than during the Bathonian (Fig. 3). Madreporaria disappear from England, Boulonnais, Normandy, the Ardennes, and Iran, but reappear in South America. The disappearance of corals from the northern part of the Anglo-Paris Basin may be explained by the opening of the North Atlantic which introduced boreal influences. During the same interval, the boundary of the Boreal Belemnite Realm was displaced towards the south, and Cylindroteuthididae reach Portugal and the central United States (Stevens, 1973). However, paleotemperature data suggests that the opening of the North Atlantic did not really influence the climate of the boreal regions, but that it produced cold currents in which the Cylindroteuthididae could migrate and which concurrently prevented Madreporaria life. The cooling of the North American seas began during the Bathonian. During the Callovian, the Ardennes were rising (Bonte, 1941), creating a shoal which stopped the cold water and allowed the reefs on the eastern and southern sides of the Paris Basin to continue to develop. Consequently, some Madreporaria are still found in France in the Departments of Vienne and Indre, and in the southeastern part of the Paris Basin. At Roussac (Vienne), *Anabacia, Montlivaltia, Isastraea,* and *Epismilia* are reported; at Mont-St-Savin (Vienne), and at St-Germain (Indre), numerous hermatypic corals are found in balls or lamellar-shaped patches. At Etrochey (Cote-d'Or), also in the southern Paris Basin, a small patch of Callovian is known containing *Isastraea* and *Trochocyathus.*

In the Jura, as in Poitou, the lower Callovian is not present and the upper Callovian is extremely poor in corals. The rare occurrences where Madreporaria are cited are Chaumont, near St-Claude, and Ste-Croix (Switzerland). Three hermatypic genera are found (*Clausastraea, Microsolena,* and *Dimorphastraea*) represented by 3 endemic species. At St-Vallier-de-Thiey (Var), only 1 en-

Figure 3. Callovian coral distribution and paleogeography. Callovian coral distribution (dots); approximate southern limit of the Boreal Realm (solid line: Hallam, 1971a); approximate meridional limit of the Cylindroteuthididae Realm (x's: Stevens, 1973); approximate boreal limit of the *Hibolites* Realm (dashed line: Stevens, 1973).

demic and ahermatypic Madreporaria (*Axosmilia*) is reported and described by Koby (1905).

Portugal

In Portugal (at Quianios, Cap Mondego, Vedrogao, Cesareda, Fatima, Alvados), some small-sized hermatypic corals are known.

Tunisia

In southern Tunisia near the Libyan frontier, Callovian coral formations were described by Busson (1965). The Madreporaria (Beauvais, 1966a) include 20 hermatypic genera: *Adelocoenia, Aggomorphastraea, Ampakabastraea, Andemantastraea, Cladophyllia, Connectastraea, Cryptocoenia, Cyathopora, Dendrastraea, Diplocoenia, Isastraea, Melikerona, Mesomorpha, Meandrophyllia, Microphylliopsis, Pseudodiplocoenia, Stylina, Sakalavastraea, Thamnasteria,* and *Trigerastraea,* and 3 ahermatypic ones: *Bathmosmilia, Chomatoseris,* and *Paramontlivaltia.* The small size of the specimens points towards a sheltered environment. According to Freneix (1965), the bivalves associated with the corals are species which adapt easily to difficult conditions which might be found in essentially isolated, lagoonal environments, such as shallow, warm waters and repeated changes in salinity due to periodic influx of fresh water. Of the 33 species known from the Tunisian Callovian, only 8 are endemic; except for these, no other species are characteristic of the Callovian. The other species are found in the Bathonian; some even occur in the Bajocian.

India

From the Callovian Chari Beds southwest of Lodai, Gregory (1900) recorded several Madreporaria; they are: *Montlivaltia, Kobya, Connectastraea,* and *Stylina.* Of the 4 species, 1 is endemic, 2 are found in the Callovian of Tunisia and the other 2 are present in the Bathonian of Cutch and in the Bajocian of Morocco.

Madagascar

Numerous reef formations begun in the upper Bathonian, continued to grow without apparent interruption into the lower Callovian (Alloiteau, 1958). Although the genera and species are the same throughout, their number decreases through time. Fourteen genera are described by Alloiteau (1958). Among them 3 are ahermatypic: *Bathmosmilia, Macgeopsis,* and *Montlivaltia;* the hermatypic genera are: *Adelocoenia, Aggomorphastraea, Coenastraea, Collignonastraea, Dimorphastraea, Elysastraea, Isastraea, Lochmaeosmilia, Ovalastraeopsis, Thamnasteria,* and *Vallimeandra.* Of the 20 species identified, 10 are endemic; the others have a wide geographic range and stratigraphic distribution.

South America

Madreporaria are reported by Gerth (1926). In Bolivia (at Caracoles), the genera are: *Latimeandra, Isastraea, Stephanocoenia, Montlivaltia,* and *Stylophyllopsis;* and in Argentina (in the south of Mendoza and at Neuquen), they are *Latimeandra, Isastraea, Montlivaltia,* and *Convexastraea.*

Summary and Conclusions

The principal events of Callovian coral distribution are the diminution of the number of genera and the constriction of their geographical distribution. Only 39 genera are recorded in the Callovian, while 105 are found in the Bathonian, and 67 in the Bajocian. Extinction is not a factor, however, because genera which disappear during the Callovian reappear in the middle Oxfordian and some of them continue to exist into the Upper Cretaceous. Two genera first appear during the Callovian; they are *Microphylliopsis* and *Ovalastraeopsis. Adelocoenia, Thamnasteria,* and *Melikerona* are very common in the Callovian reef formations. During this stage we find no species that occur at more than a single locality. The species with an extensive world distribution were already found in the Bathonian and even in the Bajocian. Thus, it seems that the new species of the Callovian were the only endemic forms. Movement of the Mid-Atlantic Ridge in conjunction with volcanic eruptions probably changed the chemical composition of the sea water, and in particular its salinity, making it unfavorable to the planulae. Because of this, their migration through the oceans was hindered. Thus, the opening of the North Atlantic which introduced cold currents and the rising of the Mid-Atlantic Ridge seem to be the two essential events that explain Callovian faunal retrogression. The paleotemperature data we possess about this period indicates a subtropical climate in North America (31.7°C in British Columbia, 24.2°C in Alberta, 19.2° to 19.9°C in Montana, 16.9°C in Alaska, and 19.4°C in Greenland). These temperatures indicate agreement between the paleoclimatologic and paleomagnetic data and suggest for those localities a position from 15° to 20° on either side of the equator.

GENERAL CONCLUSIONS

The study of the distribution of Madreporaria genera and species provides no basis for differentiating coral provinces similar to the belemnite realms of Stevens (1973) or the ammonite realms of Arkell (1956). Nor is there differentiation between reef and nonreef zones similar to those of the Upper Cretaceous (Beauvais and Beauvais, 1974). However, no coral is reported in the Boreal Realm with Cylindroteuthididae, and Madreporaria always disappear from the areas where boreal belemnites appear.

The Middle Jurassic genera are cosmopolitan, but an average of 25 to 50 percent of the species are endemic. This endemism is restricted to some basins; during the Bajocian these include the Anglo-Paris Basin, the Alsace-Swabian-Jura Basin, the Morocco Basin, the Bolivian Basin, the North American basin, and the Indian-Malagasy basin (uniting together Tanganyika and Somalia); during the Bathonian-Callovian only the Tethyan and the Indian-Malagasy basins had endemic species.

The universal distribution of the coral genera agrees with the wide geographic distribution of some neritic bivalves and with the cosmopolitan flora. These facts led Burckhard (1930) and Hallam (1971a) to think that the Jurassic climate was poorly differentiated, allowing wide migrations through different latitudes, and that no deep oceanic barriers hindered these migrations. The world-wide distribution of Jurassic corals corroborates these opinions.

The disappearance of the Madreporaria from North America by the Bathonian and from Northern Europe by the Callovian, coincides with the displacement of the southern limit of the Boreal Belemnite Realm and tends to strengthen the theory that the North Atlantic opened during the Bathonian-Callovian.

From the stratigraphic point of view, 3 genera are found exclusively in the Dogger; they are *Chomatoseris, Genabacia,* and *Lochmaeosmilia.* Sixty-seven genera are known in the Bajocian and almost all of them are found throughout the Mesozoic. Only *Polyastropsis* and *Cyathophyllopsis* seem restricted to the Bajocian. During the Bathonian many genera evolved. One hundred and five genera are found in that level, all of which survive until the Upper Cretaceous. During the Callovian a clear diminution in the number of genera is observed (only 39 are found in the Callovian). However, the genera which seem to disappear during the Callovian may reappear in the Oxfordian and even later. Two genera appear in the Callovian: *Microphyllopsis* and *Ovalastraeopsis.* Some of the species while rigorously limited to a stage, have sufficiently wide geographic distribution to make them good stratigraphic fossils. In the Bajocian, these are *Thamnoseris jurensis* d'Orbigny, *Isastraea tenuistriata* M. Edw. and Haime, *I. salinensis* Koby, *Latomeandra ramosa* d'Orbigny, *Thecocyathus magnevillianus* (Mich.), *Confusastraea gregaria* (M'Coy), *Dimorphastraea defranciana* (Mich.), *Andemantastraea consobrina* (d'Orbigny), and *Ampakabastraea crenulata* (d'Orbigny). In the Bathonian, the species are *Epistreptophyllum articulatum* (de From. and Ferry), *Confusastraea cottaldina* (M. Edw. and Haime), *Montlivaltia cornutiformis* (Gregory), *M. frustriformis* Gregory, *M. kachensis* Gregory, *M. obtusa* d'Orbigny, *M. sarthacensis* d'Orbigny, *Collignonastraea grossouvrei* Beauv., *C. jumarensis* Gregory, *Stylina kachensis* Gregory, *Bathycoenia moneta* (d'Orbigny), and *Genabacia stellifera* (d'Archaic). In the Callovian, no species are both limited to the stage and have a wide enough geographical distribution to be considered characteristic of the stage. All widely distributed Callovian species were already present in the Bathonian or even in the Bajocian and all new species are restricted to the area where they appeared. It thus appears that during the Callovian, it was not possible for planulae to migrate. A change in the salinity of the oceans due to a pulsation of the Mid-Atlantic Ridge may explain this endemism and the diminution in the number of species. Unfortunately, no similar facts have been reported for other animal groups.

REFERENCES[1]

Alloiteau, J., 1958, Monographie des Madréporaires fossiles de Madagascar: Géol. Madagascar Ann., no 25, 218 p.

Arkell, W. J., 1930, A comparison between the Jurassic rocks of the Calvados coast and those of southern England: Geol. Assoc. London Proc., v. 41, p. 392-442.

——— 1956, Jurassic geology of the world: London, Oliver and Boyd, 806 p.

Barrabe, L., 1929, Contribution á l'étude stratigraphique et pétrographique de la partie médiane du pays Sakalawa (Madagascar): Soc. Géol. France Mém., n. sér. v. 5, p. 1-270.

Beauvais, L., 1965, Révision de quelques Madréporaires d'Angleterre de la collection Milne-Edwards: Soc. Géol. France Bull., sér. 7, v. 7, p. 871-875.

——— 1966a, Etude des Madréporaires jurassiques du Sahara tunisien: Ann. Paléont., v. 52, p. 113-152.

——— 1966b, Révision des Madréporaires du Dogger de la collection Koby: Eclog. Geol. Helv., v. 59, p. 989-1024.

——— 1967, Révision des Madréporaires du Dogger des collections A. d'Orbigny et H. Michelin: Soc. Géol. France Mém., n. sér., v. 46, p. 7-54.

——— 1970a, Madréporaires du Dogger: Etude des types. de Milne-Edwards et Haime: Ann. Paléont., v. 56, p. 39-74.

——— 1970b, Etude de quelques Polypiers bajociens du Maroc oriental: Notes et Mém. Serv. Géol. Maroc, no. 225, p. 39-50.

——— 1971, Essai de répartition stratigraphique des Madréporaires du Dogger: Acad. Sci. Comptes Rendus, sér. D, v. 272, p. 3256-3259.

——— 1972a, Contribution à l'étude de la faune bathonienne dans la vallée de la Creuse (Indre). Madréporaires: Ann. Paléont., v. 58, p. 35-87.

——— 1972b, Révision des Madréporaires du Dogger de Balin (Pologne), Collection Reuss: Naturhistorische Mus. Wien Ann., v. 76, p. 29-35.

——— 1975, Révision des types de Madréporaires décrits par Koby provenant des couches à Mytilus (Alpes Vaudoises): Fossil Cnidaria Newsletter no. 2, 4 p.

Beauvais, L., and Beauvais, M., 1974, Studies on the world distribution of the Upper Cretaceous corals, *in* Cameron, A.M. and others, eds., Biogeography: Internat. Coral Reef Symposium, 2nd, Brisbane 1973, Proc., v. 1, p. 475-494.

Beauvais, L., and Negus, P., 1975, The Fairford coral bed (English Bathonien), Gloustershire: Geol. Assoc. London Proc., v. 86, p. 183-199.

Benoist, E., 1900, Note pour servir à l'étude de la géologie du départment de l'Indre: Bathonien: Feuille Jeunes Natural., sér. 4, no. 131, p. 2-6.

Besaire, H., 1952, Carte géologique de Madagascar au 1/1000.000[e]: Serv. Géol. Maroc.

Bigot, A., 1926, Plate-forme littorale avec marmite du Bathonien de Chailloué (Orne): Acad. Sci. Comptes Rendus, sér. D, v. 183, p. 440-441.

——— 1927a, Les conditions de dépôt du Bathonien inférieur dans le Bessin et la région de Caen: Acad. Sci. Comptes Rendus, sér. D, v. 184, p. 1103-1106.

——— 1927b, Les conditions de dépôt du Bathonien supérieur dans le Bessin et la région de Caen: Acad. Sci. Comptes Rendus, sér. D, v. 184, p. 1149-1152.

——— 1934a, Les récifs bathoniens de Normandie: Acad. Sci. Comptes Rendus, sér. D, v. 199, p. 400-403.

——— 1934b, Les récifs bathoniens de Normandie: Soc. Géol. France Bull., sér. 5, v. 4, p. 697-736.

——— 1940a, Les surfaces d'usure et de remaniements dans le Jurassique de Basse Normandie: Soc. Géol. France Bull., sér. 5, v. 10, p. 165-176.

[1] References not cited in text provide additional documentation.

——— 1940b, Faunes néritiques de Jurassiques inférierur de la Normandie: Soc. Biogéographie Mém., v. 7, p. 31-40.

——— 1941, Jurassique inférieur de la Sarthe et du Maine-et-Loire: Soc. Géol. France Bull., sér. 5, v. 11, p. 227-240.

Bircher, W., 1935, Studien im oberen Bajocien der Ostschweiz (Glarner-und St. Galleralpen): Ph.D. thesis, Zürich Univ., Zürich, 179 p.

Bonte, A., 1941, Contribution à l'étude du Jurassique de la bordure septentrionale du Bassin de Paris: Serv. Carte Géol. France Bull., v. 42, 439 p.

——— 1960, Sur la composition du Bathonien dans la Nord et l'Est de la France: Soc. Géol. Nord Ann., v. 80, p. 161-167.

Boomgaard, W. H., 1948, Der Braunjura im Starzeltal: Neues Jahrb. Mineral. Geol. Paläont. Monatsh. Abt. B, no. 9-12, p. 319-333 (Jg. 1945-1948).

Burckhardt, C., 1930, Remarques sur le problème du climat jurassique, *in* Etude synthétique sur la Mésozoique méxicain, Pt. II: Soc. Paléont. Suisse Mém., v. 49, 280 p.

Busson, G., 1965, Sur les gisements fossilifères du Jurassiques moyen et supérieur du Sahara tunisien: Ann. Paléont., v. 51, p. 29-42.

Carey, S. W., 1958, The tectonic approach to continental drift, *in* Carey, S. W., convener, Continental drift: A symposium: Hobart, Geol. Dept. Univ. Tasmania, p. 177-355.

Chapuis, F., and Dewalque, G., 1854, Mémoire en réponse à la question suivante: Faire la description des Fossiles des terrains Secondaires de la Province de Luxembourg: Acad. Royale Sci. Belgique Mém., v. 25, 325 p.

Choubert, G., 1938a, Sur le Dogger du Haut-Atlas oriental: Acad. Sci. Comptes Rendus, sér. D, v. 206, p. 197-199.

——— 1938b, Le Dogger des Haut-Plateaux et de la moyenne Moulouya: Acad. Sci. Comptes Rendus, sér. D, v. 206, p. 265-267.

Collignon, M., 1959, Calcaires à Polypiers, récifs et atolls du Sud de Madagascar: Soc. Géol. France Bull., sér. 7, v. 1, p. 403-408.

Corroy, G., 1929, Le Bajocien supérieur et le Bathonien de Lorraine. Corrélation avec les régions voisines en particulier avec le Jura franc-comtois: Soc. Géol. France Bull., sér. 4, v. 29, p. 167-188.

——— 1932, Le Callovien de la bordure orientale du Bassin de Paris: Serv. Géol. France Mém., 337 p.

Crickmay, C. H., 1933, Mount-Jura investigation: Geol. Soc. America Bull., v. 44, p. 895-926.

Dangeard, L., 1930, Les récifs et galets d'Algues dans l'oolithe ferrugineuse de Normandie: Acad. Sci. Comptes Rendus, sér. D., v. 190, p. 66-68.

——— 1951a, Le Normandie; Géologie régionale de la France, No. 7: Actualités scientifiques et industrielles, no. 1140: Paris, Herman et Cie, 241 p.

——— 1951b, Les récifs du Bathonien de Blainville (Calvados): Soc. Linné Normandie Bull., sér. 9, v. 6, p. 16-18.

Dareste de la Chavanne, J., 1936, Zoanthaires, *in* Marzloff, Denise, Dareste de la Chavanne, J., and Moret, L., Etude sur la faune du Bajocien supérieur du Mont-d'Or lyonnais (Ciret); gastéropods, lamellibranches, brachiopodes, échinodermes, anthozoaires, spongiaires: Lyons, Fac. Sci., Lab. Géol., Doc. no. 28, p. 126-130.

Dietz, R. A., and Holden, J. C., 1970, Reconstruction of Pangaea: Breakup and dispersion of continents, Permian to Present: Jour. Geophys. Research, v. 75, p. 4939-4956.

Dott, R. H., and Batten, R. L., 1971, Evolution of the Earth: New York, McGraw-Hill, 649 p.

Douvillé, Fr., 1941, Jurassique à l'Ouest du Bassin de Paris; Première note préliminaire: Soc. Géol. France, Compte Rendu Somm. Séances, no. 16, p. 129-131.

Douvillé, H., 1916, Les terraine secondaires dans le Massif du Moghara à l'Est de l'Isthme de Suez: Paléontologie, Pts. 1, 2: Acad. Sci. Mém., sér. 2, v. 54, p. 1-184.

Dresnay, R. du, 1964, Stratigraphie du Jurassique moyen du Jbel Klakh (Haut-Atlas marocain oriental): Acad. Sci. Comptes Rendus, sér. D, v. 256, p. 2872-2874.

——— 1965, Relation entre "dalle des Hauts-Plateaux," "calcaire-corniche" et "marnes à Pholadomyes" dans la partie occidentale des Haut-Plateaux marocains: "l'oscillation vésulienne": Soc. Géol. France, Compte Rendu Somm. Séances, no. 7, p. 238-240.

Duncan, P. M., 1873, A monograph of the British fossil corals. Second series, Pt. III. Corals from the Oolitic strata: London, Palaeontographical Soc., v. 26, p. 1-24. (Supplement to the "Monograph of the British fossil corals" by M. Milne-Edwards and J. Haime).

Dutertre, A. P., 1920, Contribution à l'étude du Bathonien du Bas-Boulonnais: Soc. Geol. Nord Ann., v. 46, p. 157-169.

Enay, R., 1972, Paléobiogéographie des Ammonites du Jurassique terminal (Tithonique/Volgien/Portlandien *s. l.*) et mobilité continentale: Geobios, no. 5, p. 355-407.

Ferry, H.B.A.T. de, 1861, Note sur l'étage Bajocien des environs de Maçon: Soc. Linné. Normandie Mém., v. 12, p. 1-46.

Fischer, J. C., 1960, Observations stratigraphiques et tectoniques sur la Bathonien supérieur de l'Aisne: Soc. Géol. France Bull., sér. 7, v. 2, p. 895-905.

——— 1962, Sur les divisions du Dogger dans la vallée de la Creuse (Indre); Corrélations avec le sud-est du Bassin parisien: Soc. Géol. France Bull., sér. 7, v. 3, p. 588-598.

Flügel, E., 1966, Mitteljurassische Korallen vom Ostrand der Grossen Salzwüste (Shotori-Kette, Iran): Neues Jahrb. Geol. Paläont. Abh., v. 126, p. 46-89.

Freneix, S., 1965, Les bivalves du Jurassiques moyen et supérieur du Sahara tunisien (Arcacea, Pteriacea, Pectinacea, Ostreacea, Mytilacea): Ann. Paléont., v. 51, p. 49-113.

Fritz, P., 1965, O^{18}/O^{16}-Isotopenanalysen und palaeotemperaturbestimmungen an Belemniten aus dem Schwäb. Jura: Geol. Rundschau, v. 54, p. 261-269.

Fromentel, E. de, 1873, *in* Fischer, P., Sur le terrain jurassique de Madagascar: Acad. Sci. Comptes Rendus, sér. D, v. 76, p. 111-114.

Fromentel, E. de, and Ferry, H.B.A.T. de, 1865-1869, Paléontologie Française. Sér. I Terrains Jurassiques, Vol. 12. Zoophytes: Paris, Masson, 240 p.

Furon, R., 1958, Causes de la répartition des êtres vivants. Paléogéographie, biogéographie dynamique: Paris, Masson, 166 p.

Gardet, G., 1928, Position stratigraphique du calcaire à Polypiers de Villey-St. Etienne (Meurthe-et-Moselle): Soc. Géol. France Bull., sér. 4, v. 27, p. 437-441.

——— 1929, Le Bajocien supérieur et le Bathonien de Villey-St. Etienne (Meurthe-et-Moselle): Soc. Géol. France Bull., sér. 4, v. 29, p. 153-166.

——— 1943, Faciès à Polypiers du Bajocien supérieur (Dubisien) à l'Est de Toul (Meurthe-et-Moselle): Soc. Géol. France Bull., sér. 5, v. 13, p. 193-206.

——— 1945, Le Bathonien de la Lorraine: Serv. Carte Géol. France Bull., v. 45, 65 p.

——— 1945a, Lias et Bajocien du Sud de plateau de Langres (feuille de Langres au 80.000^{e}): Serv. Carte Géol. France Bull., v. 45, p. 33-50.

——— 1945b, Sur quelques points fossilifères du Bathonien et du Callovien des environs de Saint-Savin-sur-Gartempe (Vienne): Soc. Géol. France, Compte Rendu Somm. Séances, no. 16, p. 228-230.

——— 1949, Sur un niveau à Polypiers branchus de Bathonien (?) de l'Indre et de la Vienne: Soc. Géol. France, Compte Rendu Somm. Séances, no. 14-15, p. 330-331.

——— 1951, Sur la présence d'*Anabacia porpites* Smith dans les calcaires compacts à taches roses du Bathonien moyen de Haute Marne: Soc. Géol. France, Compte Rendu Somm. Séances, no. 9-10, p. 144-145.

Gerth, H., 1926, Anthozoa, *in* Jaworski, E., La fauna del Lias y Dogger de la Cordillera Argentina: Acad. Nac. Ciências Cordoba Actas, v. 9, p. 142.

Glangeaud, Ph., 1885, Le Jurassique à l'Ouest du Plateau Central: Serv. Carte Géol. France Bull., v. 8, 255 p.

Gordon, W. A., 1970, Biogeography of Jurassic Foraminifera: Geol. Soc. America Bull., v. 81, p. 1689-1703.

Gregory, J. W., 1900, Jurassic fauna of Cutch. The Corals: Palaeontologia Indica, ser. 9, v. 2, pt. 2, p. 1-195.

——— 1921, Fossil corals from British East Africa in the Rift Valleys and geology of East Africa: London, Seeley Serv., 479 p.

——— 1930, The fossil corals of Kenya colony collected by Miss McKinnon Wood: Glasgow Univ., Hunterian Mus., Geol. Dept., Mon. 4, p. 181-209.

——— 1938, Second collection of fossil corals from the Kenya coast lands made by Miss McKinnon Wood: Glasgow Univ., Hunterian Mus., Geol. Dept., Mon. 5, p. 90-97.

Grossouvre, A. de, 1887, Sur le système oolitique inférieur dans la partie occidentale du bassin de Paris: Soc. Géol. France Bull., sér. 3, v. 15, p. 513-538.

——— 1891, Sur le Callovien de l'Ouest de la France et sur sa fauna: Soc. Geol. France Bull., sér. 3, v. 19, p. 247-262.

Guebhard, O., 1904, Présentation de polypiers et d'un échinide nouveaux des Alpes-Maritimes: Soc. Géol. France Bull., sér. 4, v. 4, p. 355.

Hallam, A., 1969, Faunal realms and facies in the Jurassic: Palaeontology, v. 12, p. 1-18.

——— 1971a, Provinciality in Jurassic faunas in relation to facies and palaeogeography, *in* Middlemiss, F. A., Rawson, P. F., and Newall, G., eds., Faunal provinces in space and time: Liverpool, Seel House Press, p. 129-152.

——— 1971b, Mesozoic geology and opening of North Atlantic: Jour. Geol., v. 79, p. 129-157.

——— 1975, Coral patch reefs in the Bajocian (Middle Jurassic) of Lorraine: Geol. Mag., v. 112, p. 383-392.

Hoffman, K., 1952, Die Paläogeographie des Nordwestdeutschen Doggers: Erdöl und Kohle, no. 11, p. 696.

Huguet, J., and Lespinasse-Legrand, H., 1970, Preuves paléontologiques de l'existence du Dogger dans la partie nord-est de la nappe des Corbières orientales (Aude): Acad. Sci. Comptes Rendus, sér. D, v. 170, p. 279-282.

Imlay, R. W., 1956, Marine Jurassic exposed in Bighorn Basin, Pryor Mountains, and northern Bighorn Mountains, Wyoming and Montana: Am. Assoc. Petroleum Geologists Bull., v. 40, p. 562-599.

King, L. C., 1958, Basic palaeogeography of Gondwanaland during the late Palaeozoic and Mesozoic Eras: Geol. Soc. London Quart. Jour., v. 114, p. 47-77.

Koby, F., 1881-1890, Monographie des polypiers jurassiques de la Suisse: Soc. Paléont. Suisse Mém., pts. 1-9, v. 7-16, 582 p.

——— 1902, Sur les Polypiers jurassiques des environs de St.-Vallier-de-Thiey: Soc. Géol. France Bull., sér. 4, v. 2, p. 847-863.

——— 1907, Polypiers bathoniens de St. Gaultier: Soc. Paléont. Suisse Mém., v. 33, 61 p.

Koechlin, E., 1933, Ueber das Vorkommen von Bajocienkorallen im Kanton Baselland: Naturforsch. Gesell. Basel Verhandl., v. 43, p. 4-11.

Lanquine, A., 1929, Le Lias et le Jurassique des chaînes provençales. I. Le Lias et le Jurassique inférieur: Serv. Carte Géol. France Bull., v. 32, 385 p.

Lissajos, M., 1922, Etude sur la faune du Bathonien des environs de Mâcon: Lyons, Fac. Sci., Lab. Géol., Doc. no. 3, 281 p.

Lorenchet de Montjamont, M., 1964, Le Bajocien du Moyen-Atlas oriental (Maroc): Soc. Géol. France, Compte Rendu Somm. Séances, no. 3, p. 110-111.

Maubeuge, P., 1943, Sur l'extension des "calcaires à Polypiers de Husson" de Villey-St. Etienne vers Villey-le-Sec (Meurthe-et-Moselle) et sur leur attribution au Bajocien supérieur: Soc. Géol. France, Compte Rendu Somm. Séances, no. 4, p. 32-34.

——— 1952, Observations sur la stratigraphie du Bajocian supérieur et du Bathonien de la Haute-Marne et remarques sur le niveau stratigraphique du genre *Anabacia:* Soc. Sci. Nancy Bull., n. sér. v. 11, p. 41-47.

Médioni, R., 1960, Contribution à l'étude géologique des Hauts-Plateaux méridionaux marocains: Notes et Mém. Serv. Géol. Maroc, no. 149, p. 7-48.

——— 1966, Sur l'evolution des transgressions jurassiques sur un paléorelief à matérial hercynien dans la région de Debdov (Maroc oriental): Soc. Géol. France, Compte Rendu Somm. Séances, no. 9, p. 363-366.

Mercier, J., 1928, Etude sur le contact du Bathonien et du Callovien en Normandie et dans la Sarthe et sur l'équivalent du Cornbrash anglais: Soc. Linné. Normandie Bull., sér. 8, v. 1, p. 7-37.

——— 1933, Observations sur les dépôts bathoniens dans la zone des récifs en Normandie: Soc. Linné. Normandie Bull., sér. 8, v. 7, p. 87-91.

Meyer, G., 1888, Die Korallen des Doggers von Elsass-Lothringen: Geol. Spec. Karte Els.-Lothr. Abh., v. 4, 44 p.

Michelin, H., 1840-1847, Iconographie Zoophytologique: descriptions par localities et terrains des polypiers fossiles de France et pays environmants: Paris, P. Bertrand, 348 p.

Middlemiss, F. A., and Rawson, P. F., 1971, Faunal provinces in space and time: Some general considerations, *in* Middlemiss, F. A., Rawson, P. F., and Newall, G., eds., Faunal provinces in space and time: Liverpool, Seel House Press, p. 199-210.

Milne-Edwards, Henri and Haime, Jules, 1851, A monograph of the British fossil corals. Pt. 2. Corals from the Oolitic formations: London, Palaeontographical Soc., p. 72-145.

Mouterde, R., 1953, Etudes sur le Lias et le Bajocien des bordures nord et nord-est du Massif Central français: Serv. Carte Géol. France Bull., v. 50, 455 p.

Orbigny, A. d', 1849-1852, Prodrome de Paléontologie stratigraphique universelle: Paris, Masson, 3 vols.

Parent, H., 1935a, Observations sur le terrain Bathonien du Var, entre le Gapeau et l'Argens: Soc. Géol. France, Compte Rendu Somm. Séances, no. 6, p. 90-92.

——— 1935b, Observations sur le Bathonien du Var, entre Saint-Hubert et la Nord de Toulon: Soc. Géol. France, Compte Rendu Somm. Séances, no. 7, p. 100-102.

——— 1935c, Successions fossilifères comparées de Bathonien de Var et du Bassin de Paris: Soc. Géol. France, Compte Rendu Somm. Séances, no. 8, p. 119-121.

Pelletier, M., 1950, L'âge des calcaires à Entroques et des calcaires à Polypiers du Bajocien dans le Jura méridional: Soc. Géol. France, Compte Rendu Somm. Séances, no. 15-16, p. 293-295.

——— 1952a, Etude de quelques Polypiers bajociens du Jura méridional: Soc. Géol. France Bull., sér. 6, v. 1, p. 221-232 (1951).

——— 1952b, Le Bathonien du Bas-Bugey: Soc. Géol. France, Compte Rendu Somm. Séances, no. 15-16, p. 325-327.

——— 1954, Nouvelles observations sur le Bajocien supérieur du Jura méridional: Soc. Géol. France, Compte Rendu Somm. Séances, no. 11-12, p 245-247.

Piette, E., 1854-1855, Observations sur les étages inférieurs des terrains jurassiques dans les départements des Ardennes et de l'Aisne: Soc. Géol. France Bull., sér. 2, v. 12, p. 1083-1182.

Reuss, A. E., 1867, Die Bryozoen, Anthozoen und Spongiarien des Braunen Jura von Bilin bei Krakau: Akad. Wissensch. Wien, math.-natur. Kl., Denksch., v. 27, p. 1-26.

Rioult, M., 1962, Le calcaire de Caen, dépôt de rivage du Bathonien normand: Soc. Linné. Normandie Bull., sér. 10, v. 3, p. 119.

——— 1964, Le stratotype du Bajocien, *in* Colloque du Jurassique (Luxembourg, 1962): Inst. Grand-Ducal, Sect. Sci. Nat., Phys. Math., p. 239-258.

Ruget-Perrot, C., 1962, Etudes stratigraphiques sur le Dogger et le Malm inférieur du Portugal au Nord du Tage; bajocien, bathonien, callovien, lusitanien: Serv. Géol. Portugal Mém., n. sér., no. 7, 197 p.

Schmidtill, E., 1951, Korallenbänke im Dogger Gamma bei Thalmässing (Mfr.): Geol. Blätter Nordost-Bayern, v. 1, p. 49-57.

Stchépinsky, V., 1953, Le Bathonien moyen d'Arc-en-Barrois (Haute-Marne): Soc. Géol. France, Compte Rendu Somm. Séances, no. 3-4, p. 50-52.

Steinmann, G., 1881, Zur Kenntniss der Jura- und Kreideformation von Caracoles (Bolivia): Neues Jahrb. Mineral. Geol. Paläont., Beil.-Bd., v. 1, p. 239-301.

Stevens, G. R., 1963, Faunal realms in Jurassic and Cretaceous belemnites: Geol. Mag., v. 100, p. 481-497.

——— 1971, Relationship of isotopic temperatures and faunal realm of Jurassic-Cretaceous palaeogeography particularly of south-west Pacific: Royal Soc. New Zealand Jour., v. 1, p. 145-148.

——— 1973, Jurassic belemnites, *in* Hallam, A., ed., Atlas of palaeobiogeography: Amsterdam, Elsevier, p. 257-274.

Therquem, O., and Jourdy, E., 1871, Monographie de l'étage Bathonien dans le départment de la Moselle: Soc. Géol. France Mém., sér. 2, v. 9, 175 p.

Théobald, N., 1959, Variations de faciès du Bathonien du Jura franc-comtois entre Belfort et Besançon: Besançon Univ., Ann. Sci., sér. 2, Géol., no. 11, p. 15-26.

Théobald, N., and Maubeuge, P. L., 1949, Paléogéographie du Jurassique inférieur et moyen dans le nord-est de la France et le sud-ouest de l'Allemagne: Naturforsch. Gesell. Freiburg i. Br. Bericht, v. 39, p. 249-319.

Thevenin, A., 1903, Etude géologique de la bordure Sud-ouest du Massif Central: Serv. Carte Géol. France Bull., v. 14, 199 p.

Thiery, P., 1922, Sur la limite du Bathonien et du Bajocien en Lorraine: Acad. Sci. Comptes Rendus, sér. D, v. 174, p. 1243-1246.

——— 1922, Le Bajocien supérieur de Lorraine: Acad. Sci. Comptes Rendus, sér. D, v. 175, p. 38-41.

Thomas, H. D., 1934, Note préliminaire sur les Coraux de récif bathonien du Breuil: Soc. Géol. France Bull., sér. 5, v. 4, p. 726-736.

——— 1935a, On *Tricycloseris*, *Anabacia*, and some new genera of Hexacoralla: Geol. Mag., v. 72, p. 424-430.

——— 1935b, Jurassic corals and Hydrozoa, together with a redescription of *Astraea caryophylloides* Goldfuss: Geology and Palaeontology of British Somaliland (London), v. 3, p. 23-39.

——— 1963, Corals and the correlation of the Tanga limestone of Tanganyika: Overseas Geol. Mineral Resources, v. 9, p. 30-38.

Tomes, R. F., 1878, A list of the Madreporaria of Crickley Hill, Gloucestershire, with descriptions of some new species: Geol. Mag., ser. 2, v. 5, p. 297-305.

——— 1882, On the Madreporaria of the Inferior Oolite of the neighborhood of Cheltenham and Gloucester: Geol. Soc. London Quart. Jour., v. 38, p. 409-450.

——— 1883, On the Madreporaria of the Great Oolite of the counties of Gloucester and Oxford: Geol. Soc. London Quart. Jour., v. 39, p. 108-196.

——— 1884, A critical and descriptive list of the Oolitic Madreporaria of the Boulonnais: Geol. Soc. London Quart. Jour., v. 40, p. 698-723.

——— 1885, On some new or imperfectly known Madreporaria from the Great Oolite of the counties of Oxford, Gloucester, and Somerset: Geol. Soc. London Quart. Jour., v. 41, p. 170-190.

——— 1886, On some new or imperfectly known Madreporaria from the Inferior Oolite of Oxfordshire, Gloucestershire and Dorsetshire: Geol. Mag., ser. 3, v. 3, p. 1-23, 385-398, 443-452.

Wells, J. W., 1943, Palaeontology of the Harrar Province, Ethiopia. Pt. 3. Jurassic Anthozoa and Hydrozoa: Am. Mus. Nat. History Bull., v. 82, p. 31-54.

African Cretaceous Ostracodes and Their Relations to Surrounding Continents

KARL KRÖMMELBEIN, *Geologisch-Paläontologisches Institut und Museum, Universität Kiel, Kiel, West Germany*

ABSTRACT

African Ostracoda of Cretaceous age (nonmarine "Wealden" and marine species) have been investigated. Their regional, gondwanic occurrence is best understood if a Jurassic-Lower Cretaceous Pangaea reconstruction is assumed.

BIOGEOGRAPHY OF AFRICAN OSTRACODA

The African continent is surrounded by numerous Mesozoic sedimentary basins that extend partially onto the older basement of the continent. Cretaceous formations comprise a considerable portion of the Mesozoic sequences. They play an important role with respect to paleogeographic reconstructions because the Cretaceous is a crucial time period in Earth history: the final breakup of the assumed Gondwana supercontinent appears to have taken place essentially during this period. This is especially true for the separation of Africa from South America; the breakup and the drifting away of the remaining areas of Gondwana occurred earlier, i.e., during Permian and Triassic times.

The Cretaceous fossil faunas of the African continental margins offer a basis for deciphering hypothetical former land connections, whether Cretaceous or Permo-Triassic. Among the faunas, the ostracodes appear to be best suited for such paleogeographic studies, inasmuch as their species and genus diversity is great and they include nonmarine as well as marine forms.

Some of the African Mesozoic sedimentary basins contain a sequence of formations beginning with nonmarine strata ("Wealden facies" in a wide sense) and continuing with marine strata. This is true, for example, for the West African basins from the Ivory Coast to as far south as the Angola basins. Also, the basins along the eastern border of the continent contain Cretaceous nonmarine as well as marine strata, the ostracodes of which may contribute to knowledge about the general biogeographic situation.

For many years it has been known that the nonmarine ("West African Wealden") ostracodes of the Gabon and Congo coastal regions show a close similarity to those of the "Northeast Brazilian Wealden" of the Reconcavo-Tucano Basins. As early as 1966, Krömmelbein and Wenger were able to demonstrate that about 20 of the 40 species of the Gabon Basin also occur in the Bahia-Sergipe area of Brazil.

However, not all of the African "Wealden" basins show "Brazilian affinities" with respect to their formational sequences and their ostracode faunas. For instance, the "Wealden" of Ghana (borehole evidence only) has produced some species that appear more similar to those of the European Wealden than to species of the remaining equatorial African basins or to those of the Brazilian basins (Krömmelbein, 1968).

Unfortunately, practically nothing is known about the ostracodes of southern Africa basins in Angola and Namibia, or the Orange (Shelf) Basin.

It is not the purpose of this short paper to give a full account of the marine Cretaceous ostracodes of the African basins and their relations to "surrounding" areas, either nearby or more distant. Two examples only will be selected in order to demonstrate how meaningful Cretaceous ostracodes may turn out to be for paleobiogeographic reconstructions. Many rich ostracode faunas have been described in recent years in numerous papers by Grekoff, Neale, E. Miedi Himie Neufville, Dingle, Reyment, Bate, and Bayliss to mention only a few. I also have fairly good collections of marine Cretaceous ostracodes from many African basins which are as yet undescribed.

Among marine ostracodes, two groups of genera seem to deserve special attention: *Brachycythere,* with its many species, and the tribe Majungaellini, with the genera *Majungaella* and *Novocythere-Tickalaracythere* with their large-sized, richly ornamented species. The latter group has been selected to demonstrate some paleobiogeographic ties between Africa and the adjacent regions.

(1) The main species of *Majungaella* from various areas of Gondwana are shown in Figure 1, and the regional distribution of the Jurassic species is shown in Figure 2. The genus was first described from the Upper Jurassic to Lower Cretaceous of Madagascar (Grekoff, 1963). Most, and also the oldest, species occur in Tanzania (Bate, 1975) and range in age from Callovian to Albian. Some of the species are also known to occur in the Neocomian of South Africa (Dingle, 1969). Quite recently, species of *Majungaella* have been described from Western Australia (Gingin Chalk) by Neale (1975), from the Great Artesian Basin of central Australia by Krömmelbein (1975), and

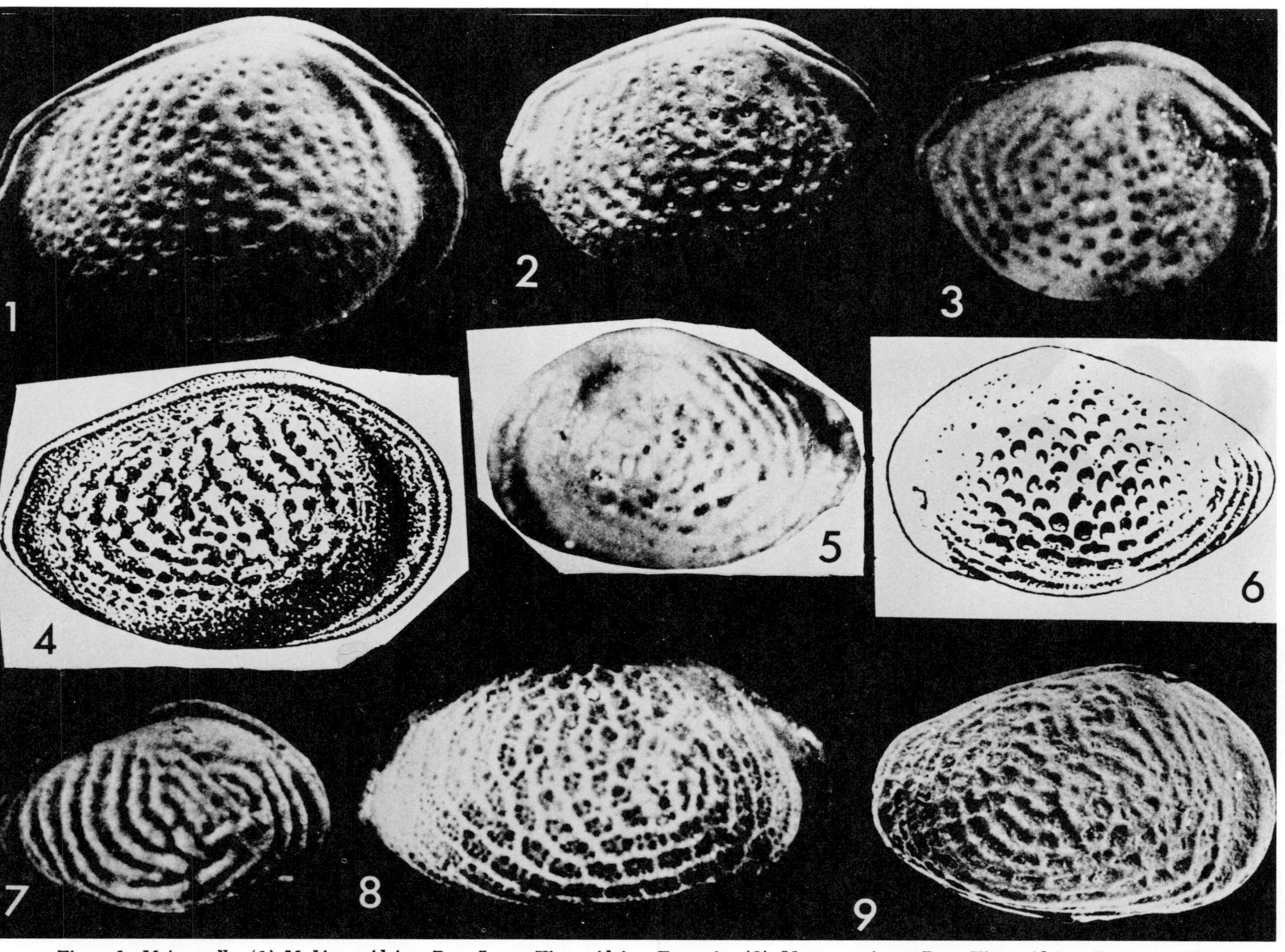

Figure 1. *Majungaella*. (1) *M. kimmeridgiana* Bate, Lower Kimmeridgian, Tanzania; (2) *M. praeperforata* Bate, Kimmeridgian, Tanzania; (3) *M. perforata* Grekoff, Portlandian, Madagascar; (4) *M. spitiensis* Jain and Mannikeri, Upper Jurassic, Himalayas (Spiti); (5) *M. nematis* Grekoff, Valanginian, Madagascar; (6) *M. uitenhagensis* (Dingle) = *M. nematis*, Neocomian, South Africa; (7) *M.*? sp. Oertli, Barremian, SE France; (8) *M. queenslandensis* Krö., Albian-Cenomanian, central Australia; (9) *M. verseyi* Neale, Santonian, W Australia.

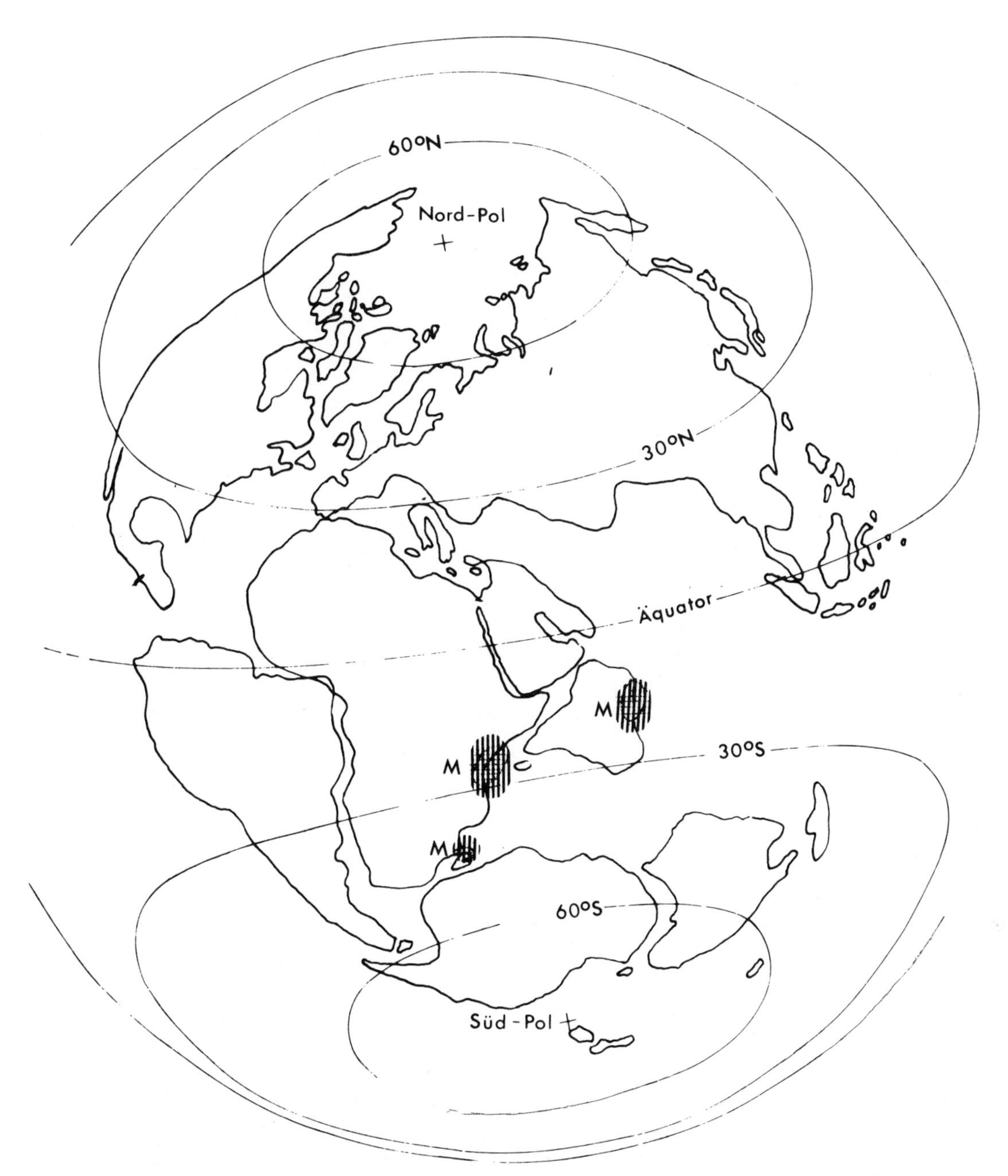

Figure 2. Jurassic paleogeography (reconstruction after Seyfert and Sirkin, 1973). (M) *Majungaella* spp., Madagascar, Tanzania, and Spiti area (see Fig. 1).

from the Spiti area of the Himalayas by Jain and Mannikeri (1975). Krömmelbein (1975) has earlier suggested that Oertli's (1963) "large *Neocythere*" from the Barremian of southeastern France might belong to *Majungaella;* however, this has still to be demonstrated. All of these occurrences are easily understood if a Jurassic-Lower Cretaceous Pangaea reconstruction of the continents is assumed, as shown in Figures 2 and 4.

(2) *Novocythere-Tickalaracythere*[1] probably has evolved from the *Majungaella* stock; the taxon ranges in age from Albian to Maastrichtian. The main species are shown in Figure 3, the regional distribution in Figure 4. Besides the Australian and East African occurrences, the genus has been shown to occur also in South America (Argentina and northeast Brazil). The occurrences of *Novocythere-Tickalaracythere* have posed more problems than those of *Majungaella,* because the genus has not yet been found in the Cretaceous of West Africa, a fact that is not easily understood. However, further research might reveal that the Majungaellina group, useful for stratigraphic and paleobiogeographic purposes, is more widespread than can presently be demonstrated.

[1] The 2 genera are most probably identical; if so, the correct name would be *Novocythere* (Malumian and others, 1972).

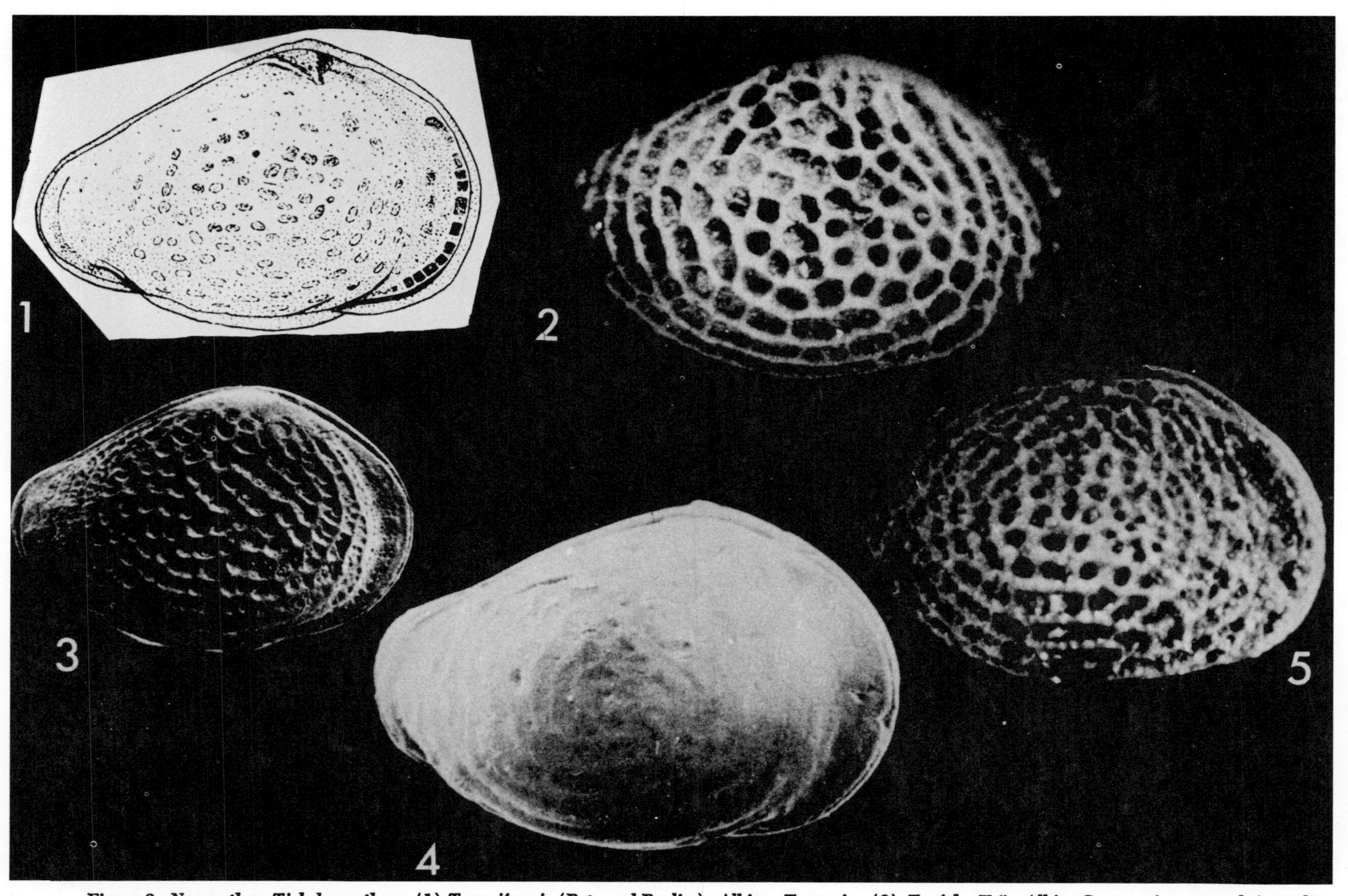

Figure 3. *Novocythere-Tickalaracythere.* (1) *T. pyriformis* (Bate and Bayliss), Albian, Tanzania; (2) *T. ticka* Krö., Albian-Cenomanian, central Australia; (3) *N. santacruciana* R. de Garcia, "Middle Cretaceous," Argentina; (4) *T. annula* (Bate), Santonian, W Australia; (5) *T.* n. sp. Krö., Campanian-Maastrichtian, NE Brazil.

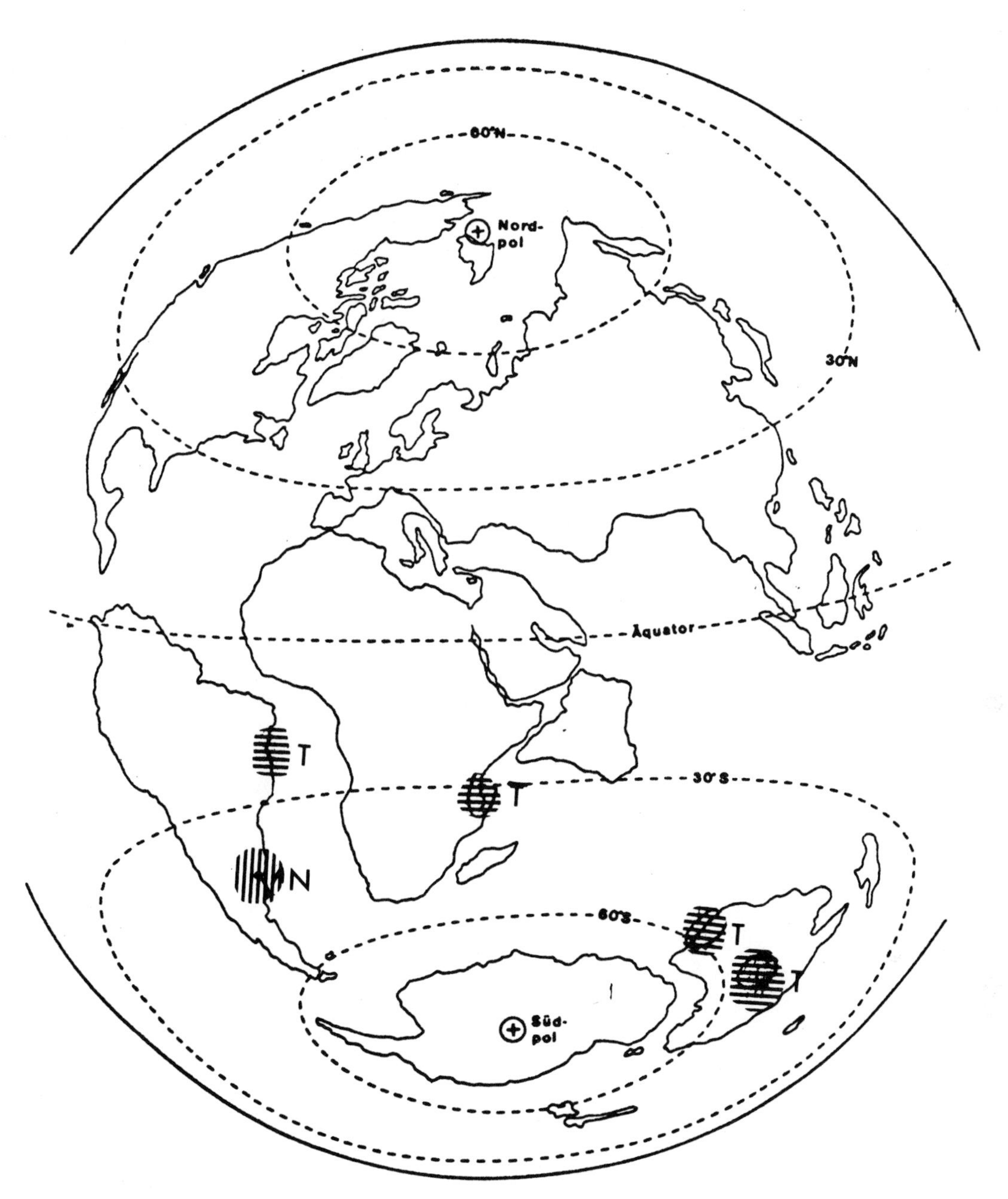

Figure 4. Cretaceous paleogeography (reconstruction after Seyfert and Sirkin, 1973). (N) *Novocythere;* (T) *Tickalaracythere* (see Fig. 3).

REFERENCES[2]

Bate, R. H., 1972, Upper Cretaceous Ostracoda from the Carnarvon Basin, Western Australia: Spec. Papers Palaeontology no. 10, 85 p.

——— 1975, Ostracods from Callovian to Tithonian sediments of Tanzania, East Africa: British Mus. (Nat. History) Bull., Geology, v. 26, p. 163-223.

Bate, R. H., and Bayliss, D. D., 1969, An outline account of the Cretaceous and Tertiary Foraminifera and of the Cretaceous ostracods of Tanzania: African Micropalaeontological Colloquium, 3rd, Cairo 1969, Proc., p. 113-164.

Dingle, R. V., 1969, Marine Neocomian Ostracoda from South Africa: Royal Soc. South Africa Trans., v. 38, p. 139-163.

[2] References not cited in text provide additional documentation.

Grekoff, N., 1963, Contribution à l'étude des Ostracodes du Mésozoique moyen (Bathonien-Valanginien) du Bassin de Majunga, Madagascar: Rev. Inst. Français Pétrole, v. 18, p. 1709-1783.

Jain, S. P., and Mannikeri, M. S., 1975, Ostracoda from the Spiti shales, Spiti Valley: Himalayan Geol., v. 5, p. 133-138.

Krömmelbein, K., 1968, The first non-marine Lower Cretaceous ostracods from Ghana, West Africa: Palaeontology, v. 11, p. 259-263.

——— 1975, Ostracoden aus der Kreide des Great Artesian Basin, Queensland, Australien: Senckenbergiana Lethaea, v. 55, p. 455-483.

——— 1976, Remarks on marine Cretaceous ostracods of gondwanic distribution: African Micropaleontological Colloquium, 5th, Addis Ababa 1972, Proc., p. 539-551.

Krömmelbein, K., and Wenger, R, 1966, Sur quélques analogies remarquables dans les microfaunes Crétacées du Gabon et du Brésil oriental (Bahia et Sergipe), *in* Reyre, D., ed., Bassins sédimentaires du littoral africain. Symposium, Pt. 1. Littoral Atlantique: Union Internat. Sci. Géol., Assoc. Serv. Géol. Africains (New Delhi 1964), p. 193-196.

Malumian, N., Masiuk, V., and Rossi de Garcia, E., 1972, Microfossiles del Cretacico Superior de la perforacion SC-1, Provincia de Santa Cruz, Argentina: Rev. Asoc. Geol. Argentina, v. 27, p. 265-272.

Neale, J. W., 1975, The ostracod fauna from the Santonian Chalk (Upper Cretaceous) of Gingin, Western Australia: Spec. Papers Palaeontology no. 16, 81 p.

Oertli, H. J., 1963, Faunes d'ostracodes du Mésozoique de France: Leiden, E. J. Brill, 57 p.

Seyfert, C. K., and Sirkin, L. A., 1973, Earth history and plate tectonics: New York, Harper and Row, 504 p.

Pre-Tertiary Phytogeography and Continental Drift—Some Apparent Discrepancies

CHARLES J. SMILEY, *Department of Geology, College of Mines, University of Idaho, Moscow, Idaho 83843*

ABSTRACT

Floral distribution patterns for later Paleozoic and Mesozoic times conform with no continental drift models that have been proposed to date. If plate tectonics is to be considered a synthesizing theory for interpretations of historical geology, certain modifications of existing models are required to incorporate the contradictory factors of historical phytogeography. The distribution of *Glossopteris* floras provides evidence that is neither for nor against the concept of continental drift, as such floras would not present a unified floral province in either event. Additionally, the geographic locations of ecotonal floras and floristic relationships seem to require positions of Australia and India near their present locations during later Paleozoic and Mesozoic times. Rich Mesozoic floral records across the Beringian region indicate a unified floral province for northeastern Eurasia and northwestern North America, and a land route for interchange of taxa in both directions through time. Mesozoic floras do not show the effects of a postulated oceanic separation of the Beringian region, yet such a separation is a requirement of current models. Continental drift models must take into account such obvious discrepancies in historical plant geography; otherwise, the theory of plate tectonics cannot be deemed the truly synthesizing principle of recent acclaim.

INTRODUCTION

Historical phytogeography concerns the distribution of plants in space and time. It involves not only the spacial distribution of taxa and floras of a particular age, but also the temporospacial changes in plant distribution that may occur through time and space. In paleobotany, both spacial and temporospacial distribution depend upon the adequacy of the fossil record, which is the only tangible evidence that a plant existed in a given area at a given time. Spacial distribution may be inferred by plotting on a map the known fossil occurrences of a taxon or of a distinctive association of taxa. Inferences on temporospacial distribution are dependent in large measure upon the adequacy of stratigraphic data to provide an accurate determination of the relative ages of fossil floras. Determining the route of floral interchange between areas is largely inferential, unless stratigraphically controlled and dated floral sequences are available in the intermediate area of interchange.

The spacial distribution of later Paleozoic floras that are characterized by the presence of *Glossopteris* has been plotted on maps depicting both continental drift and continental stability, to determine how well the data conform to widely used models of plate tectonics (Fig. 1). Plotted also are the locations of ecotonal floras—fossil assemblages containing plants that are characteristic of one floral province admixed with taxa that are characteristic of another.

The temporospacial distribution of later Mesozoic plants across the Beringian land region that now separates the Arctic and Pacific Oceans is also evaluated to determine how well these new data conform to the plate tectonics requirement of a Mesozoic oceanic separation at some position across the region. Stratigraphically controlled floral sequences are known from several areas extending from northeastern Eurasia to northwestern North America, which provide data for determining floral relationships throughout the Beringian region (Figs. 3, 4).

LATER PALEOZOIC

Glossopteris Distribution

The major floral provinces of later Paleozoic time are (1) the Gondwana Province of Southern Hemisphere continents and India, (2) the Euramerica Province of eastern North America and Europe, (3) the Angara Province of north-central Eurasia, and (4) the Cathaysia Province of eastern Eurasia and western North America.

It has long been assumed that the distinctive genera *Glossopteris* and *Gangamopteris* are confined to the Gondwana Province, and that continental drift models for later Paleozoic time will neatly unify this presently disjunct floral province within the limits of a postulated "Gondwanaland." However, Permian floras from Turkey (Wagner, 1962) eastward through the Kuznetzk Basin and Mongolia to the Pacific coast of Eurasia (Zimina, 1967), indicate an extensive distribution of *Glossopteris* and associated plants across the southern part of the "Laurasia" continent (Fig. 1, black areas). A disjunct distribution of *Glossopteris* and *Gangamopteris,* found in rocks on land areas that are widely separated at the present time, would appear to be the case whether continental drift occurred or not. The records of floras characterized by the presence of these genera are merely permissive of continental drift; they do not require it.

Ecotonal Floras

Mixed floras, containing characteristic plants from two or more of the major floral provinces, resemble ecotones that exist between contiguous floral communities or provinces. Because the distribution of the *Glossopteris* and *Gangamopteris* floras neither proves nor disproves continental drift, the locations of ecotonal floras, and their geographic relations to the provinces that have supplied the admixed taxa, are considered most critical to the present discussion.

An ecotonal flora reported from the New Guinea portion of the Australia-New Guinea "plate" contains a mixture of taxa from the Cathaysian Province of southeastern Eurasia and from the Australian *Glossopteris* flora (Gothan and Weyland, 1954). An expected geographic location of such an ecotonal flora coincides with the present proximity of the two regions. A wide oceanic separation that is required by plate tectonics theory creates what appears to be an impossible situation.

Distributed along the Himalaya Range are several areas containing records of *Glossopteris* floras in rocks commonly considered to have been deposited on "Laurasia" (Surange, 1966a, 1966b). These rocks are presumed in plate tectonics theory to have been later uplifted by a Cenozoic underthrusting of the Indian "plate." One of these Himalayan sites is Kashmir, where a sequence of Permian rocks contains *Glossopteris* floras in earlier deposits and Angara floras higher in the section. Only in its present location could Kashmir have been occupied first by a "Gondwana" flora from the south (India) and later by an Angara flora from the north (central Eurasia).

Glossopteris and associated taxa are found in the Kuznetzk region north of the Himalaya Range, where they are mixed with Angara taxa along the southern part of the Angara Province (Krishnan, 1954; Surange, 1966b; Zimina, 1967). *Glossopteris* and *Gangamopteris* have been reported from Permian deposits eastward to the Pacific coastal area of the Soviet Far East (Zimina, 1967). *Glossopteris* has been reported in Permian deposits of Thailand where it is associated with characteristic taxa of the Cathaysia Province (Kon'no, 1963), and it has also been reported as a relict in Triassic floras of the Tonkin area of Indo-China (Just, 1952).

The ecotonal admixture of Gondwana and Angara floral elements in the central Eurasian region north of India, the presence of *Glossopteris* floras along the Himalaya Range now bordering India on the north, the Gondwana to Angara floral replacement in the Permian sequence of Kashmir, and the presence of *Glossopteris* on the western border of the Cathaysia Province east of India, can best be explained by a Permian location of India in close proximity to the Eurasian landmass, rather than in high southern latitudes as currently shown on models of continental drift. India would then be in a position to serve as an upland source area for the outward dissemination of *Glossopteris* and associated plants into marginal areas of contiguous floral provinces (Angara on the north, Cathaysia on the east).

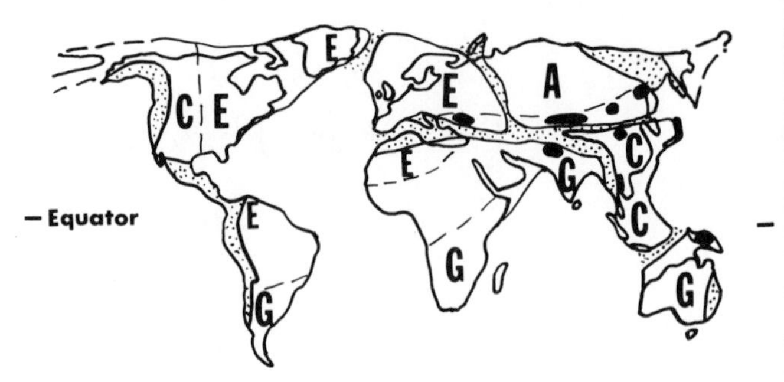

Figure 1. Map of later Paleozoic floral provinces and seaways (after Smiley, 1974). Floral provinces: (G) Gondwana; (C) Cathaysia; (A) Angara; (E) Euramerica. Stippled areas are locations of epicontinental seaways (after Kummel, 1970). Areas of known occurrences of *Glossopteris* include the Gondwana Province (G), and sites of ecotonal (mixed) floras shown by black dots. The distribution of *Glossopteris* floras thus is shown to be disjunct, whether plotted on world maps of the present or on Paleozoic maps depicting continental drift.

The Beringian Region

The recognition of a late Paleozoic land connection between eastern Eurasia and western North America is necessarily inferential because of the paucity of Permian floras in the Beringian region. However, Early Permian floras of western North America are so similar to those of the Cathaysia Province that the boundary between the Cathaysia and Euramerica Provinces is usually placed about midway across the North American continent (Fig. 1); and Late Permian floras of western North America closely resemble Eurasian Angara floras, as a reflection of the later Permian expansion of the Angara Province (White, 1912, 1929).

Plate tectonics theory requires a late Paleozoic separation of oceanic proportions across the present Beringian region. Application of the theory to this part of the world eliminates a Paleozoic proximity such as that which now exists between northeastern Eurasia and northwestern North America; yet, a close proximity seems required to explain the observed floral relationships of late Paleozoic time. On the other hand, application of the theory to the North Atlantic region would unite the Euramerica flora of eastern North America with the Euramerica flora of western Eurasia by bringing the two presently separated parts of the Euramerica Province into close geographic proximity. If plate tectonics theory is not applied, however, a moderate epeirogenic uplift of the shallow North Atlantic, east of Greenland, may be postulated. This would result in a North Atlantic route for floral interchange, without simultaneously eliminating a land route that is equally a requirement in the Beringian region.

Given equal needs for routes of floral interchange both eastward and westward from North America, an application of plate tectonics theory will eliminate the westward route that would be available otherwise. Alternatively, a slight uplift of the northern Atlantic area (without plate tectonics) will provide the avenues that are required for floral interchange in both directions.

LATER MESOZOIC

India-Australia

Jurassic and Early Cretaceous floras of India are more closely related generically to floras of Eurasia than to floras of Gondwanaland (Smiley, 1974). For example, 29 of 81 Jurassic genera (36 percent) occur on other Gondwana continents, in contrast to 48 genera (60 percent) that are present in Jurassic floras of Eurasia; only 5 of the genera having affinities with Gondwanaland floras are restricted to Gondwana areas. Vachrameev (1964) also considered the India Mesozoic flora to be a part of his Indo-European Province of Eurasia because of floral similarity.

A flora I described from Malaysia (1970) is generically identical to one from northern India that I observed in the laboratory of G. Pant at Allahabad (Fig. 2). The two floras are the same age (near the Jurassic-Cretaceous boundary) and species differences are slight. Such Mesozoic floral evidence also suggests a much closer proximity of India to Malaysia and to southern Eurasia than is permitted by current models of continental drift.

A recently described Early Cretaceous floral sequence near Melbourne in southern Australia (Fig. 2) has been interpreted to indicate a climate similar to that of the region at the present time (Douglas, 1969, 1973). These Australian floras also are similar to Early Cretaceous floral assemblages at present latitudes of 40° to 50°N, rather than resembling Early Cretaceous floras from latitudes of 60° to 70°N (Smiley, in prep.). Such evidence suggests an Early Cretaceous position of southern Australia near its present latitude (40°S), rather than at the 60° to 70°S latitude shown on continental drift models for Cretaceous time (Fig. 2, for example).

Beringian Region

Numerous later Mesozoic floras and floral sequences have been described across the Beringian region and in contiguous areas (Figs. 2 to 4). My research in northern Alaska has resulted in the richest of these, with precise stratigraphic control and marine faunal dating (Smiley, 1972a, 1972b). The Alaskan records thus represent the best reference data for the evaluation of spacial and temporospacial distribution of taxa across the Beringian region in later Mesozoic time (see Fig. 4). The Mesozoic floras extend from the Yukon River area (1 on Fig. 3) and more northern sites (2 on Fig. 3) on the Alaska landmass east of alternative plate boundaries, westward to the Kolyma River area (3 on Fig. 3) between the alternative plate boundaries, and farther west to the Lena River sites (4 and 5 on Fig. 3) on the Siberian landmass west of the two plate boundaries. The postulated boundaries are those proposed by Hamilton (1970) through the Bering Straits (B on

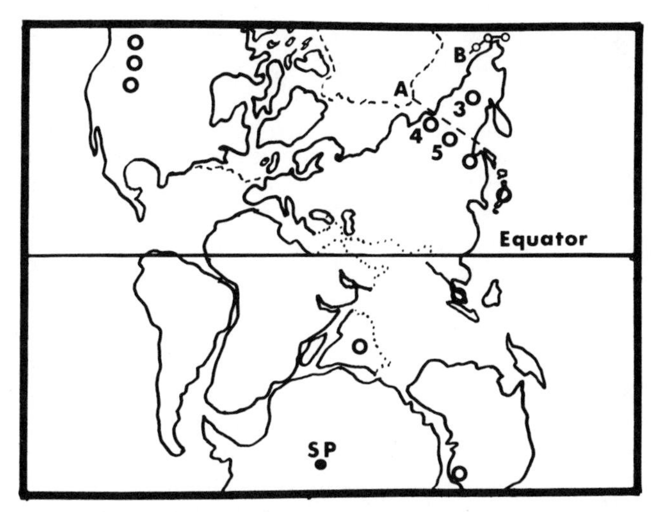

Figure 2. One of the current continental drift models for Cretaceous time (from Raven and Axelrod, 1974). Pertinent floral sites are indicated by circles. Numbered sites, and lines A and B, are the same as on Figure 3. The Cretaceous floras from Australia (present latitude 40°S) resemble ones from the latitude of Japan (presently 40°N). The floras from India and Malaysia are generically identical, although they are presumed on this model to have been historically separated and thus of different phylogenetic origins. The several medial Cretaceous floras of northwestern North America and northeastern Eurasia are so similar as to indicate a single unified floral province.

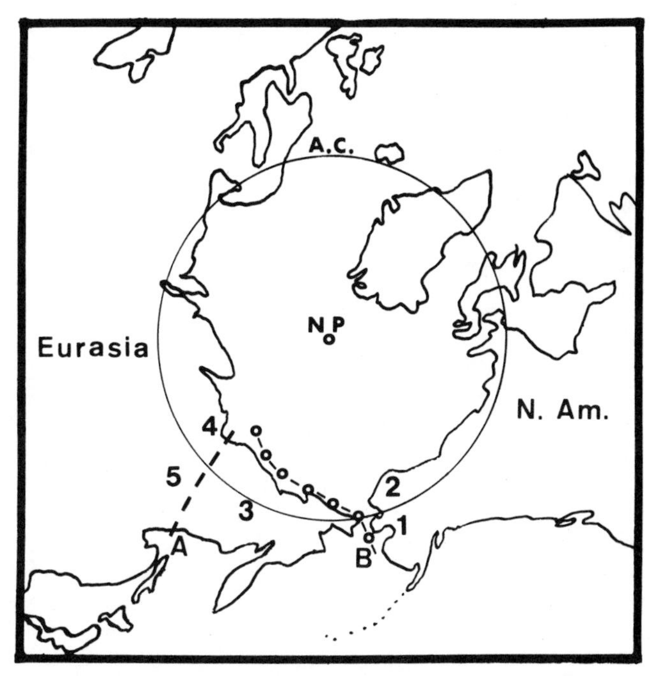

Figure 3. Map of present Holarctic region on North Polar projection. Mesozoic floral areas are indicated by numbers: (1) central Alaska (Yukon River area); (2) northern Alaska; (3) Kolyma River area; (4) Lower Lena River area; (5) Aldan River area. Alternative Cenozoic collision (plate) boundaries are A (Churkin, 1972) and B (Hamilton, 1970). Jurassic to Cretaceous floral records of the Beringian region indicate a unified floral province extending at least from areas 1 and 2 on the east to areas 4 and 5 on the west (see Fig. 4). (NP) present North Pole; (A.C.) present Arctic Circle.

Figs. 3, 4) and by Churkin (1972) in the area of the Verkhoyansk Mountains (A on Figs. 3, 4) east of the Lena River.

In the following analyses, the term "species identity" refers to the percentage of apparently identical species that are represented in two or more floras; the term "floral similarity" refers to the combined percentages of identical species and of similar forms that may be either the same or closely related species. The latter may include variants of a form that is characteristic of the floral province, or may include incomplete fossils of a regional form whose specific identity cannot be determined. The degree of floral similarity in deposits of the same age, and the temporo-spacial distribution of taxa across the region, should provide evidence for or against a presumed oceanic separation in Cretaceous time, and, consequently, the site of ultimate continental collision during Cenozoic time (Fig. 4, A? or B?).

Northern Alaska Area. The northern Alaska records are from 250 plant-bearing beds; each is stratigraphically plotted in local sections that range in thickness from 800 to 11,300 ft (250 to 3,400 m). More than 420 plant species are presently recognized from the approximately 10,000 good specimens collected. A stratigraphic sequence of 8 major floral types (floral zones) and 20 subzones has been recognized. In most of the sections, especially in the eastern part of the region, coal- and plant-bearing nonmarine units are interbedded with fossiliferous marine deposits. The Alaskan records thus are rich taxonomically and in abundance of good fossils from numerous horizons, and established floral zones can be traced laterally across the Cretaceous coastal plain for correlating with marine zonation (Smiley, 1972a, 1972b). The marine faunas indicate ages ranging from probable late Aptian to Maestrichtian, with upper Cenomanian and Coniacian stages apparently not represented.

Central Alaska Area. Re-evaluation of the taxonomy of the partly contemporaneous Yukon River (Shaktolik) floras from south of the Brooks Range (Hollick, 1930), indicates a species identity of 81 percent of the Yukon species in northern Alaska floras. Floral similarity is about 95 percent. Because the central Alaska floras are in closer proximity to those of northern Alaska than are other Mesozoic floras of the Beringian region, the degree of taxonomic similarity between central and northern Alaska serves as a point of reference for inferring the effect of distance on floral similarity from one side of the region to the other.

Aldan River Area. Samylina (1963) reported on a sequence of later Mesozoic floras from rocks exposed along the Aldan River, a tributary of the upper Lena River system (5 on Figs. 2 to 4). The sequence ranges in age from the upper part of the Jurassic, through the Neocomian, to the lower part of the Aptian. Taxonomic comparisons with the records from northern Alaska are based on comparisons of published photographs of Aldan specimens with Alaskan fossils.

The Jurassic Jaskoyskaya flora contains 14 species of foliar organs of which 7 (50 percent) appear identical to Alaskan forms: *Cladophlebis aldanensis* Vachr., *Raphaelia diamensis* Sew., *Ginkgo huttoni* (Sternb.) Heer, *G. lepida* Heer, *Czekanowskia rigida* Heer, *Sphenobaiera ikorfatensis* f. *papillata* Samyl., and *Phoenicopsis taschkessiensis* Krass. Two other forms are morphologically very similar but apparently not identical: *Cladophlebis serrulata* Samyl. and *Raphaelia* aff. *diamensis* Sew. The Upper Jurassic flora of the Aldan River area, although small, shows an apparent 64 percent floral similarity with the medial Cretaceous floras of northern Alaska. Five Aldan species of Jurassic age (36 percent) appear to have no close equivalents in the northern Alaskan floras: *Equisetites asiaticus* Pryn., *Neocalamites* sp., *Ctenis intermedia* (Krysht. and Pryn.) Pryn., *C. latiloba* Krysht. and Pryn., and *Pagiophyllum Kryshtofovichii* Samyl.

The lower Neocomian Ustj-Tyrsky flora contains 52 species of foliar organs of which 27 (52 percent) appear identical to Alaskan forms: *Equisetites* aff. *naktongensis* Tat., *Cladophlebis argutula* f. *lobata* Samyl., *Coniopteris arctica* (Pryn.) Samyl., *C. hymenophylloides* (Brongn.) Sew., *C. nympharum* (Heer) Vachr., *C. setacea* (Pryn.) Vachr., *Gonatosorus ketovae* Vachr., *Onychiopsis elongata* (Geyl.) Yok., *Raphaelia diamensis* Sew., *Aldania auriculata* Samyl., *Jacutiella amurensis* (Novop.) Samyl., *Nilssonia acutiloba* (Heer) Pryn., *Taeniopteris* sp. A, *T. rhitidorachis* Krysht., *Baiera concinna* (Heer) Kaw., *B. polymorpha* Samyl., *Ginkgo adiantoides* (Ung.) Heer, *G. huttoni* (Sternb.) Heer, *G. lepida* Heer, *Czekanowskia rigida* Herr, *Sphenobaiera angustiloba* (Heer) Fl., *S. longifolia* (Heer) Fl., *S. pulchella* (Heer) Fl., *Phoenicopsis angustifolia* (Eichw.) Heer, *Cephalotaxus cretacea* Samyl., *Podozamites angustifolius* (Eichw.) Heer, and *Pseudolarix dorofeevii* Samyl. Fifteen other forms are morphologically very similar but apparently not identical: *Thallites* aff. *jimboi* (Krysht.) Harris, *Cladophlebis denticulata* (Brongn.) Font., *C.* ex. gr. *haiburnensis* (L. and H.) Brongn., *C. lenaensis* Vachr., *C. multinervis* Gol., *C. pseudolobifolia* Vachr., *Coniopteris gleichenoides* Samyl., *Sagenopteris* sp., *Ctenis nana* Samyl., *Heilungia amurensis* (Novop.) Pryn., *Nilssoniopteris ovalis* Samyl., *N.* (*Sibiriophyllum*) *californicum* (Font.) Samyl., *Pterophyllum burejense* Pryn., *Taeniopteris jimboana* Krysht., and *Parataxodium* aff. *wigginsii* Arnold and Lowther. Thus, the lower Neocomian flora of the Aldan River area shows an apparent 80 percent floral similarity with the medial Cretaceous floras of northern Alaska. Ten Aldan species of this age (20 percent) appear to have no close equivalents in the northern Alaska floras: *Equisetites asiaticus* Pryn., *E. rugosus* Samyl., *Cladophlebis* aff. *distans* (Heer) Yabe, *Hausmannia* sp., *Aldania vachrameevii* Samyl., *Pterophyllum* cf. *cunielobum* Pryn., *Tyrmia polynovii* (Novop.) Pryn., *T.* aff. *tyrmensis* Pryn., *Ginkgodium amgaensis* Samyl., and *Sphenobaiera uninervis* Samyl.

The upper Neocomian Cherepanovsky flora contains 38 species of foliar organs of which 21 (55 percent) appear identical to Alaska forms: *Coniopteris arctica*

(Pryn.) Samyl., *C. nympharum* (Heer) Vachr., *C. setacea* (Pryn.) Vachr., *C.* sp. cf. *Sphenopteris silapensis* Pryn., *Gonatosorus ketovae* Vachr., *Onychiopsis elongata* (Geyl.) Yok., *Raphaelia diamensis* Sew., *Jacutiella amurensis* (Novop.) Samyl., *Nilssonia acutiloba* (Heer) Pryn., *N. jacutica* Samyl., *Taeniopteris rhitidorachis* Krysht., *Baiera polymorpha* Samyl., *Ginkgo adiantoides* (Ung.) Heer, *G. huttoni* (Sternb.) Heer, *Czekanowskia rigida* Heer, *Sphenobaiera pulchella* (Heer) Fl., *S. longifolia* (Heer) Fl., *Phoenicopsis angustifolia* Heer, *P.* sp., *Podozamites angustifolius* (Eichw.) Heer, and *Pseudolarix dorofeevii* Samyl. Eight other forms are morphologically very similar but apparently not identical: *Thallites* aff. *jimboana* (Krysht.) Harris, *Cladophlebis lenaensis* Vachr., *C. multinervis* Gol., *C. sangarensis* Vachr., *Sagenopteris* sp., *Heilungia amurensis* (Novop.) Pryn., *Nilssoniopteris ovalis* Samyl., and *Parataxodium* aff. *wigginsii* Arnold and Lowther. The upper Neocomian flora of the Aldan River area shows an apparent 76 percent floral similarity with the medial Cretaceous floras of northern Alaska. Nine Aldan species of this age (24 percent) appear to have no close equivalents in northern Alaska: *Equisetites rugosus* Samyl., *Cladophlebis* aff. *distans* (Heer) Yabe, *Hausmannia* sp., *Ctenis burejensis* f. *typica* Pryn., *C. sulcicaulis* (Phill.) Ward, *Heilungia aldanensis* Samyl., *Tyrmia* aff. *tyrmensis* Pryn., *T.* sp., and *Ginkgodium amgaensis* Samyl.

The lower Aptian Peschaniki flora contains 11 species of foliar organs of which 8 (73 percent) appear identical to Alaskan forms: *Coniopteris nympharum* (Heer) Vachr., *C.* sp. cf. *Sphenopteris silapensis* Pryn., *Jacutiella amurensis* (Novop.), *Nilssonia jacutica* Samyl., *Baiera polymorpha* Samyl., *Ginkgo adiantoides* (Ung.) Heer, *Podozamites*

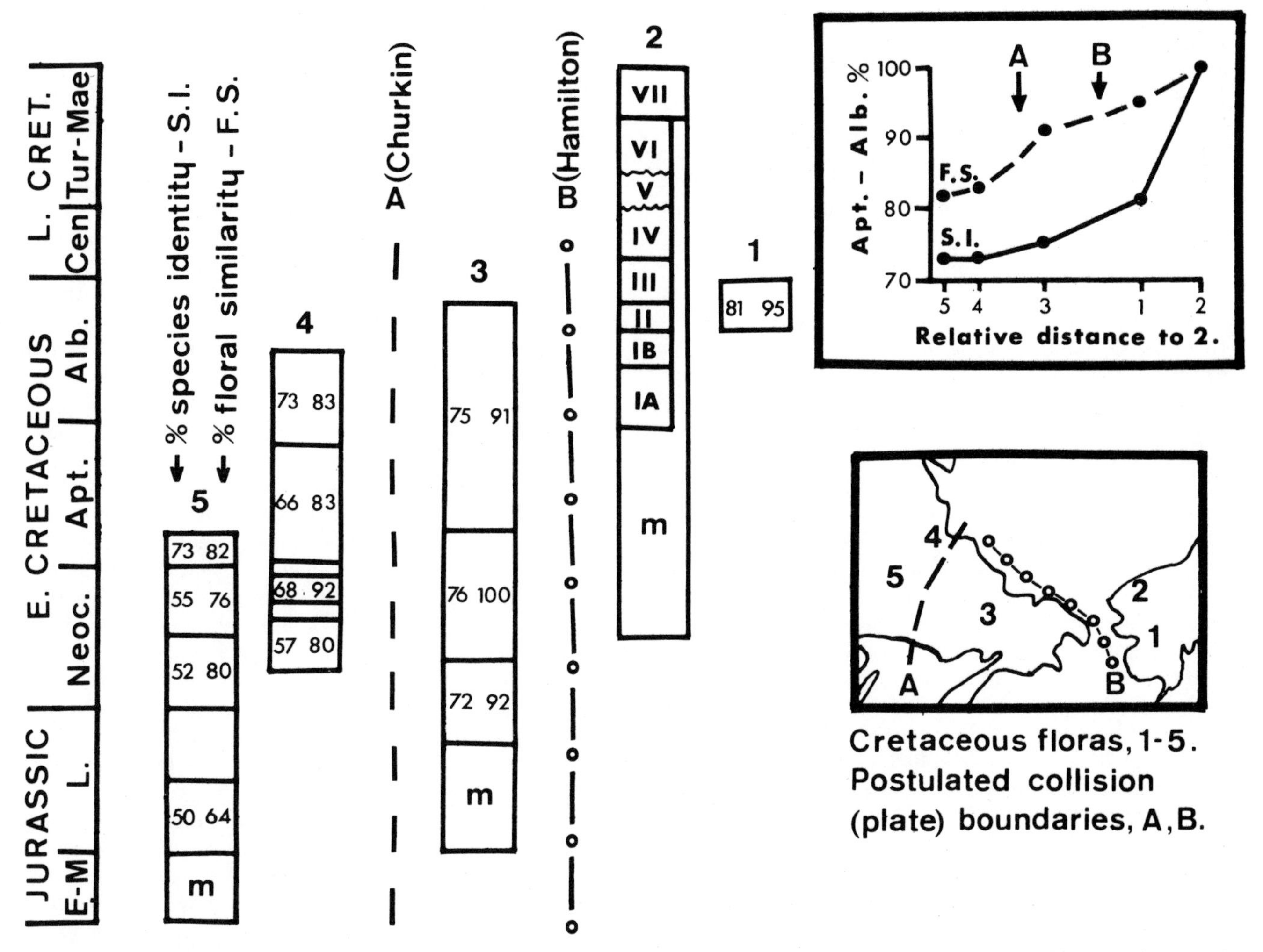

Figure 4. Columns of floral areas across the Beringian region showing Jurassic and Cretaceous floral sequences. Column and area numbers, and plate boundaries, as in Figure 3. Numbers at left without columns indicate percentages of species that are identical with ones in northern Alaska (Species Identity). Numbers at right within columns indicate percentages of species that are both identical and closely similar (Floral Similarity). Marine units are indicated by "m". Graph in upper right shows progressively increasing taxonomic relationships as a function of decreasing distance from northern Alaska, rather than as a function of historical isolation on two sides of broad oceanic separation.

angustifolius (Eichw.) Heer, and *Pseudolarix dorofeevii* Samyl. Another form referred to *Czekanowskia rigida* Heer has a questionable relationship with an Alaskan species. The small lower Aptian flora of the Aldan River area shows an apparent 82 percent floral similarity with the medial Cretaceous floras of northern Alaska. Two other Aldan species of this age (18 percent) appear to have no close equivalents in the northern Alaska floras: *Ctenis burejensis* f. *typica* Pryn. and *Sphenobaiera uninervis* Samyl.

The Aldan River floras show an apparent floral similarity with those of northern Alaska that increases from Late Jurassic to medial Cretaceous times, as the Aldan floras approach the age of the Alaskan sequence (Fig. 4): from 64 percent in the Jurassic, to 76 to 80 percent in Neocomian, to 82 percent in the lower Aptian. Also noted is a temporal increase in the percentages of apparently identical species in floras of the two areas: from 50 percent of the Aldan species in Upper Jurassic floras, to 52 to 55 percent in Neocomian floras, to 73 percent in Aptian floras. The percentage of lower Aptian species having apparently identical forms in northern Alaska (73 percent) is not significantly different from the percentage indicated for the Albian Yukon River flora of central Alaska (81 percent). This difference of less than 10 percent in apparent species identity seems to be directly related to distance from the northern Alaska area of reference (Fig. 4).

As the older Aldan floral sequence in Siberia approaches the age of the younger sequence in northern Alaska, such an observed increase in floral similarity between two areas of a more-or-less unified Beringian region could be expected. If, however, the eastern and western parts of the present Beringian region had been separated by oceanic distances, the noted degree of taxonomic similarity, and the noted temporal increase in similarity as age equivalency is approached, would not be expected for two historically isolated areas on different continents. Paleontologists usually interpret similar degrees of floral and faunal relationships as an indication of close proximity, or of interconnecting dispersal routes, when defending the concept of continental drift. But the same application in the Beringian region will lead to conclusions that are inconsistent with currently held interpretations of plate tectonics for this part of the world. A close proximity of northeastern Eurasia and northwestern North America would not have been established until probably late in Cenozoic time according to plate tectonics theory, whereas regional floral records require a close proximity of the two continents during the Mesozoic.

Lower Lena Area. Vasilevskaya and Pavlov (1963) reported on a sequence of later Mesozoic floras from the lower (northern) part of the Lena River system (4 on Figs. 2 to 4). The floral sequence ranges in age from lower Neocomian to middle Albian. The following taxonomic comparisons with floras from Northern Alaska are based on morphological comparisons of specimen photographs with Alaska fossils.

The lower Neocomian (Valanginian-Hauterivian) flora of the Lower Lena River area contains 30 species of foliar organs of which 17 (57 percent) appear identical to forms from northern Alaska: *Cladophlebis lenaensis* Vachr., *C.* cf. *novopokrovskii* Pryn., *C. tyrgyensis* Vass., *C. williamsonii* Brongn., *Coniopteris* cf. *arctica* Heer, *C. burejensis* (Zal.) Sew., *Aldania auriculata* Samyl., *Nilssonia* cf. *schaumbergensis* (Dunk.) Nath., *Nilssonia* sp. 2, *Nilssoniopteris ovalis* Samyl., *Baiera* sp., *Ginkgo huttoni* (Sternb.) Heer, *Sphenobaiera angustiloba* (Heer) Fl., *Pityophyllum staratchinii* (Heer) Nath., *Podozamites angustifolius* (Eichw.) Heer, *Taxocladus* sp., and *Leptostrobus limbatus* Vass. Seven other forms are morphologically very similar but apparently not identical: *Thallites* sp., *Equisetites burejensis* (Heer) Krysht., *Coniopteris kolymensis* Pryn., *Scleropteris* sp., *Ctenis tigyensis* Vass., *Nilssonia lobatidentata* Vass., and *Phoenicopsis angustissima* Pryn. The lower Neocomian flora of the Lower Lena River area shows an apparent 80 percent floral similarity with the medial Cretaceous floras of northern Alaska. Six Lower Lena species of this age (20 percent) appear to have no close equivalents in the northern Alaskan floras: *Adiantites* sp., *Jacutopteris lenaenis* Vass., *Thinnfeldia* sp., *Ctenis* sp., *Ginkgodium* (?) sp., and *Pseudotorellia nordenskioldii* (Nath.) Fl.

The upper Neocomian (Barremian) flora of the Lower Lena River area contains 25 species of foliar organs of which 17 (68 percent) appear identical to forms from northern Alaska: *Coniopteris onychioides* Vass. and K.-M., *Jacutiella amurensis* (Novop.) Samyl., *Nilssonia* sp., *Pterophyllum bulunense* Vass., *Ginkgo adiantoides* (Ung.) Heer, *G. digitata* (Brongn.) Heer, *G. huttoni* (Sternb.) Heer, *G. sibirica* Heer, *Czekanowskia rigida* Heer, *Sphenobaiera angustiloba* (Heer) Fl., *S. longifolia* (Heer) Fl., *S. pulchella* (Heer) Fl., *Phoenicopsis acutifolia* Vass., *Podozamites angustifolius* (Eichw.) Heer, *P. eichwaldii* Schimp., *P. lanceolatus* L. and H., and *Pityophyllum staratchinii* (Heer) Nath. Six other forms are morphologically very similar but apparently not identical: *Baiera* sp. 1, *Baiera* sp. 2, *Ginkgo angusticuneata* Vass., *G. obrutschewii* Sew., *Podozamites gramineus* Heer, and *P.* sp. The upper Neocomian flora of the Lower Lena River area shows an apparent 92 percent floral similarity with the medial Cretaceous floras of northern Alaska. Two Lower Lena species of this age (8 percent) appear to have no close equivalents in the northern Alaska floras: *Equisetites burejensis* (Heer) Krysht. and *Podozamites ovalifolius* Vass.

The Aptian flora of the Lower Lena River area contains 41 species of foliar organs of which 27 (67 percent) appear identical to Alaskan forms: *Equisetites burejensis* (Heer) Krysht., *Asplenium rigidum* Vass., *Cladophlebis huttoni* f. *minor* Vass., *C.* sp. 1, *C.* sp. 2, *Coniopteris nympharum* (Heer) Vachr., *C. onychioides* Vass. and K.-M., *C.* sp. 1, *Gleichenia lobata* Vachr., *Nilssonia comtula* Heer, *N. nipponensis* Yok., *N. orientalis* Heer, *Taeniopteris* cf. *arctica* Heer, *T.* sp. 2, *Ginkgo adiantoides* (Ung.) Heer, *G. huttoni* (Sternb.) Heer, *G.* cf. *huttoni*

(Sternb.) Heer, *Sphenobaiera angustiloba* (Heer) Fl., *S. longifolia* (Pomel) Fl., *Phoenicopsis acutifolia* Vass., *P. angustifolia* Heer, *P. speciosa* Heer, *Podozamites eichwaldii* Schimp., *P. gracilis* Vass., *P. latifolius* Heer, *P. reinii* Geyl., and *Pityophyllum nordenskioldii* (Heer) Nath. Seven other forms are morphologically very similar but apparently not identical: *Ruffordia* sp. 1, *Scleropteris ermolaevii* Vass., *Anomozamites angulatus* Heer, *Nilssonia gigantea* Krysht. and Pryn., *Otozamites* (?) sp., *Czekanowskia setacea* Heer, and *Taxocladus* cf. *sutchanensis* Pryn. The Aptian flora of the Lower Lena River area shows an apparent 83 percent floral similarity with the medial Cretaceous floras of northern Alaska. Seven Lower Lena species of this age (17 percent) appear to have no close equivalents in the northern Alaska floras: *Equisetites* sp. 2, *Adiantites gracilis* Vass., *Sphenopteris petiolipinnulata* Vass., *Anomozamites arcticus* Vass., *Baiera tripartita* Vass., *Podozamites striatus* Velen., and *Elatocladus* sp.

The Albian flora of the Lower Lena River area contains 41 species of foliar organs of which 30 (73 percent) appear identical to forms from northern Alaska: *Asplenium dicksonianum* Heer, *Cladophlebis* sp. 4, *Coniopteris nympharum* (Heer) Vachr., *C. onychioides* Vass. and K.-M., *C.* ex. gr. *saportana* Heer, *C. setacea* f. *compressa* Vass., *C. vachrameevii* Vass., *Onychiopsis elongata* (Geyl.) Yok., Polypodiaceae gen. and sp., *Sphenopteris ukinensis* Vass., *Nilssonia canadensis* Bell, *N. prinadae* Vachr., *Pterophyllum* (?) sp., *Taeniopteris* sp. 1, *Ginkgo adiantoides* (Ung.) Heer, *G. huttoni* (Sternb.) Heer, *G. lepida* Heer, *G. sibirica* Heer, *Sphenobaiera flabellata* Vass., *S. longifolia* (Pomel) Fl., *Phoenicopsis angustifolia* Heer, P. *speciosa* Heer, *Podozamites angustifolius* (Eichw.) Heer, *P. eichwaldii* Schimp., *Pityophyllum nordenskioldii* (Heer) Nath., *Cephalotaxopsis acuminata* Krysht. and Pryn., *C.* sp. 2, *C.* sp. 3, *C.* sp. 4, and *Pagiophyllum* sp. Four other forms are morphologically very similar but apparently not identical: *Czekanowskia rigida* Heer, *Podozamites gramineus* Heer, *Cyparissidium* sp., and *Elatocladus ketovae* Vass. The Albian flora of the Lower Lena River area shows an apparent 83 percent floral similarity with the medial Cretaceous floras of northern Alaska. Seven Lower Lena species of this age (17 percent) appear to have no close equivalents in the northern Alaska floras: *Equisetites* sp. 1, *Adiantites polymorphus* Vass., *Cladophlebis gluschinskii* Vass., *Anomozamites arcticus* Vass., *Ginkgo parvula*, *G. polaris* var. *pygmaea* Nath., and G. cf. *pusilla* Heer.

The Lower Lena River floras show a floral similarity with northern Alaska of about 80 to 90 percent during the Neocomian-Albian interval. This constantly high relationship implies a land connection between the two areas, rather than an oceanic barrier at either of the postulated plate boundaries. As in the Aldan River sequence, the proportion of similar species that appear to be represented in both areas increases as age equivalency is approached: from 57 percent of the Lower Lena lower Neocomian flora, to 67 to 68 percent of the upper Neocomian and Aptian floras, to 73 percent of the Albian species. Evidence from medial Cretaceous floras of the Aldan and Lower Lena Rivers areas, both of which are west of plate boundaries (thus west of an implied oceanic separation), indicates that about three-quarters of the Eurasian species occur also in northern Alaskan floras on the east of the postulated plate boundaries (or oceanic separation).

Kolyma River Area. Samylina (1964, 1967) reported on a sequence of later Mesozoic floras from the Zyrianka Coal Basin of the Kolyma River area of eastern Siberia (3 on Figs. 2 to 4). The sequence includes the Ozhoghinskaya flora of Late Jurassic-early Neocomian age, the Siliapskaya flora of middle Neocomian to early Aptian age, and the Buor-Kemiusskaya flora of middle Aptian to late (but not latest) Albian age. The following taxonomic comparisons with floras of northern Alaska are based on morphological comparisons of specimen photographs with Alaskan fossils.

The Jurassic-Neocomian (Ozhoghinskaya) flora of this area contains 25 species of foliar organs of which 18 (72 percent) appear identical to Alaskan forms: *Equisetites* sp. A, *Coniopteris burejensis* (Zal.) Sew., *C. onychioides* Vass. and K.-M., *C. setacea* Pryn., *C. silapensis* (Pryn.) Samyl., *Sphenopteris* sp., *Baiera polymorpha* Samyl., *Ginkgo* ex. gr. *lepida* Heer, *Sphenobaiera longifolia* (Pomel) Fl., S. *pulchella* (Heer) Fl., *Phoenicopsis* ex. gr. *angustifolia* Heer, *Podozamites angustifolius* (Eichw.) Heer, *P. eichwaldii* Schimp., *P. gracilis* Vass., *P. longifolius* Heer, *P. reinii* Geyl., *Pityophyllum* ex. gr. *nordenskioldii* (Heer) Nath., *P.* ex. gr. *staratchinii* (Heer) Nath. Five other forms are morphologically very similar but apparently not identical: *Onychiopsis* sp., *Nilssonia borealis* Samyl., *Ginkgo* ex. gr. *huttoni* (Sternb.) Heer, *Czekanowskia* ex. gr. *rigida* Heer, and *C.* ex. gr. *setacea* Heer. The Late Jurassic-early Neocomian flora of the Kolyma River area shows an apparent 92 percent floral similarity with the medial Cretaceous floras of northern Alaska. Two Kolyma species of this age (8 percent) appear to have no close equivalents in the northern Alaska floras: *Cladophlebis* ex. gr. *williamsonii* Brongn. and *Anomozamites arcticus* Vass.

The Neocomian-Aptian (Siliapskaya) flora contains 16 species of foliar organs of which 13 (81 percent) appear identical to Alaskan forms: *Asplenium dicksonianum* Heer, *Cladophlebis* ex. gr. *denticulata* (Brongn.) Font., *Coniopteris onychioides* Vass. and K.-M., *C. setacea* (Pryn.) Vachr., *Sphenopteris* sp., *Nilssonia grossinervis* Pryn., *Ginkgo* ex. gr. *lepida* Heer, *G. paradiantoides* Samyl., *Phoenicopsis* ex. gr. *angustifolia* Heer, *P.* (?) *magnum* Samyl., *Podozamites angustifolius* (Eichw.) Heer, *P. eichwaldii* Schimp., and *Pityophyllum* ex. gr. *nordenskioldii* (Heer) Nath. Three other forms are morphologically very similiar but apparently not identical: *Asplenium* sp., *Arctopteris* sp., and *Czekanowskia setacea* Heer. The small medial Neocomian-early Aptian flora of the Kolyma River area shows an apparent 100 percent floral similarity with the medial Cretaceous floras of northern Alaska. All of the Kolyma species of this age appear to have close equivalents in the northern Alaska floras.

The Aptian-Albian (Buor-Kemiusskaya) flora of the Kolyma River area contains 56 species of foliar organs of which 42 (75 percent) appear identical to Alaskan forms: *Mirella borealis* Samyl., *M.* sp., *Thallites* sp., *Acrostichopteris* sp., *Arctopteris kolymensis* Samyl., *A. rarinervis* Samyl., *Asplenium dicksonianum* Heer, *A. rigidum* Vass., *Cladophlebis argutula* (Heer) Font., *Coniopteris arctica* (Pryn.) Samyl., *C. bicrenata* Samyl., *C. gracillima* (Heer) Vass., *C.* aff. *maakiana* (Heer) Pryn., *C. nympharum* (Heer) Vachr., *C. onychioides* Vass. and K.-M., *C. saportana* (Heer) Vachr., *C.* sp., *Onychiopsis psilotoides* Vass. and K.-M., *Osmunda cretacea* Samyl., *O. denticulata* Samyl., *Osmundopsis efimoviae* Samyl., *Nilssonia comtula* Heer, *N.* aff. *grossinervis* Pryn., *N. magnifolia* Samyl., *N.* aff. *serotina* Heer, *N.* sp., *Nilssoniopteris prynadae* Samyl., *Ginkgo delicata* Samyl., *G. paradiantoides* Samyl., *G. pluripartita* (Schimp.) Heer, *G. singularis* Samyl., *Sphenobaiera biloba* Pryn., *S. flabellata* Vass., *Phoenicopsis* (?) *magnum* Samyl., *Podozamites angustifolius* (Eichw.) Heer, *P. eichwaldii* Schimp., *Pityophyllum nordenskioldii* (Heer) Nath., *Cephalotaxopsis borealis* Samyl., *C.* cf. *intermedia* Holl., *Cyparissidium gracile* (Heer) Heer, *Pagiophyllum triangulare* Pryn., and *Parataxodium* cf. *wigginsii* Arnold and Lowther. Nine other forms are morphologically very similar but apparently not identical: *Equisetites* sp. B, *Acrostichopteris* aff. *parvifolia* Font., *Asplenium popovii* Samyl., *Cladophlebis* aff. *lobifolia* (Phill.) Brongn., *C. pseudolobifolia* Vachr., *Glossophyllum magnum* Samyl., *Nilssonia schaumbergensis* (Dunk.) Nath., *Ginkgo polaris* Nath., and *Sciadopitys* sp. The Aptian-Albian flora of the Kolyma River area shows an apparent 91 percent floral similarity with the Cretaceous floras of northern Alaska. Five Kolyma species (9 percent) appear to have no close equivalents in the northern Alaskan floras: *Equisetites ramosus* Samyl., *Anomozamites* sp., *Baiera* cf. *ahnertii* Krysht., *Pseudotorellia pulchella* (Heer) Vass., and *Schizolepis cretaceus* Samyl.

The Kolyma River floras show a high and fairly constant (91 to 100 percent) floral similarity with northern Alaska from Late Jurassic through middle Albian time. Similarly, the incidence of apparently identical species also remained high and constant (72 to 75 percent) through this interval, indicating that three-quarters of the Kolyma species also are present in the Cretaceous floras of northern Alaska.

The very slight differences in apparent species identity between (1) medial Cretaceous floras of northern Alaska and those of the Aldan and Lower Lena areas (67 to 73 percent), (2) northern Alaska floras and those of the Kolyma River area (75 percent), and (3) northern Alaskan floras and those of the central Alaskan (Yukon River) area (81 percent) seem explicable on the basis of relative distance alone (Fig. 4). All areas seem to fall readily within a single unified floral province, with noted taxonomic differences reflecting differences in age, slight differences in latitude, geographic spacing of sampling sites, or differences in proximity to a cooler Arctic Sea or a warmer Pacific Ocean.

CONCLUSIONS

The geometry of recognized floral provinces does not always coincide with the global geometry that is postulated in current models of continental drift. The known distribution of leaves of the *Glossopteris* type, for example, shows a disjunct distribution pattern whether plotted on models of continental drift or on models of continental stability. Such known geographic distribution of the *Glossopteris* flora, therefore, can be considered only permissive of the theory of plate tectonics, as it argues neither for nor against it. The locations of many ecotonal floras do not conform with any of the drift models for Paleozoic time.

The evidence of floral interchange between Eurasia and western North America implies the presence of a land dispersal route across the Beringian region at least from later Paleozoic time. But plate tectonics theory requires an oceanic separation perhaps thousands of kilometers wide through this region prior to the Cenozoic. The close taxonomic relationship that exists among later Mesozoic floras of the Beringian region dictates against any late Mesozoic disruption of floral continuity from Alaska on the east to the Lena River area on the west. Thus, the evidence is contrary to any supposition of a prior separation and subsequent plate collision in the Beringian region, whether in the Verkhoyansk area or in the Bering Straits area.

Plate tectonics is primarily a geophysical theory, devived basically from geophysical data, originally applied to solve geophysical problems. Modern concepts of global tectonics originated from ocean floor data of the North Atlantic, and the evidence there and in the South Atlantic continues to provide the strongest support for these concepts. But too little attention has been given to the concommitant results, in the Beringian region, of continental movements related to the postulated history of the North Atlantic. When one applies the principle of cause and effect to interpretations of global history, it seems a geometric requirement for a presumed Beringian seaway to have progressively narrowed as the North Atlantic progressively widened during Cenozoic time. Either the applicability of plate tectonics theory is in question for this part of the world, or the time frame for inferred continental movements is in need of re-evaluation.

A continental drift model for late Paleozoic time, with India and Australia close to their present locations, would be acceptable from the viewpoint of floral relationships and distribution patterns. Models for late Paleozoic and Mesozoic times that exclude an oceanic separation in the Beringian region would be acceptable from the viewpoint of the rich floral records in that part of the world.

REFERENCES

Churkin, M., Jr., 1972, Western boundary of the North American continental plate in Asia: Geol. Soc. America Bull., v. 83, p. 1027-1036.

Douglas, J. G., 1969, The Mesozoic floras of Victoria, Pts. 1 and 2: Victoria Geol. Survey Mem. 28, 310 p.

——— 1973, The Mesozoic floras of Victoria, Pt. 3: Victoria Geol. Survey Mem. 29, 185 p.

Gothan, W., and Weyland, H., 1954, Lehrbuch der Palaobotanik: Berlin, Akademia Verlag, 316 p.

Hamilton, W., 1970, The Uralides and the motion of the Russian and Siberian platforms: Geol. Soc. America Bull., v. 81, p. 2553-2576.

Hollick, A., 1930, The Upper Cretaceous floras of Alaska: U. S. Geol. Survey Prof. Paper 159, 123 p.

Just, T. K., 1952, Fossil floras of the Southern Hemisphere and their phytogeographic significance: Am. Mus. Nat. History Bull., v. 99, p. 189-202.

Kon'no, E., 1963, Some Permian plants from Thailand: Japanese Jour. Geology and Geography, v. 34, p. 139-159.

Krishnan, M. S., 1954, History of the Gondwana era in relation to the distribution and development of flora: Lucknow, Birbal Sahni Inst. Palaeobotany, Sir Albert Charles Seward Memorial Lecture, p. 1-15.

Kummel, B., 1970, History of the Earth (2nd edition): San Francisco, W. H. Freeman, 707 p.

Raven, P. H., and Axelrod, D. I., 1974, Angiosperm biogeography and past continental movements: Ann. Missouri Botanical Garden, v. 61, p. 539-673.

Samylina, V. A., 1963, The Mesozoic flora of the lower course of the Aldan River: Akad. Nauk SSSR, Bot. Inst. Trudy, Paleobotanika, ser. 8, v. 4, p. 59-139 (in Russian).

——— 1964, Mesozoic flora of the area west of the Kolyma River (the Zyrianka Coal-Basin). Pt. 1. Equisitales, Filicales, Cycadales, Bennettitales: Akad. Nauk SSSR, Bot. Inst. Trudy, Paleobotanika, ser. 8, v. 5, p. 41-79 (in Russian).

——— 1967, Mesozoic flora of area west of Kolyma River (the Zyrianka Coal-Basin). Pt. 2. Ginkgos, conifers, general chapters: Akad. Nauk SSSR, Bot. Inst. Trudy, Paleobotanika, ser. 8, v. 6, p. 133-175 (in Russian).

Smiley, C. J., 1970, Later Mesozoic flora from Maran, Pahang, West Malaysia, Pts. 1 and 2: Geol. Soc. Malaysia Bull., v. 3, p. 77-113.

——— 1972a, Applicability of plant megafossil biostratigraphy to marine—non-marine correlations: An example from the Cretaceous of northern Alaska: Internat. Geol. Cong., 24th, Montreal 1972, Proc., Section 7, p. 413-421.

——— 1972b, Plant megafossil sequences, North Slope Cretaceous: Geoscience and Man, v. 4, p. 91-99.

——— 1974, Analysis of crustal relative stability from some late Paleozoic and Mesozoic floral records, *in* Kahle, Charles F., ed., Plate tectonics: Assessments and reassessments: Am. Assoc. Petroleum Geologists Mem. 23, p. 331-360.

Surange, K. R., 1966a, Indian fossil pteridophytes: New Delhi Council Sci. and Industrial Research, Botanical Mon. 4, 209 p.

——— 1966b, Distribution of *Glossopteris* flora in the Lower Gondwana formations of India, *in* Symposium on floristics and stratigraphy of Gondwanaland: Lucknow, Birbal Sahni Inst. Palaeobotany, p. 55-68.

Vakhrameev, V. A., 1964, Jurassic and Early Cretaceous floras of Eurasia and the paleofloristic provinces of this period: Akad. Nauk SSSR Doklady, v. 12, 263 p. (in Russian).

Vasilevskaya, N. D., and Pavlov, V. V., 1963, Stratigraphy and flora of Cretaceous deposits of the Lena-Oleneksk district, Lensk Coal Basin: Nauch.-issled. Inst. Geol. Arkt. Trudy, v. 128, 96 p. (in Russian).

Wagner, R. H., 1962, On a mixed Cathaysia and Gondwana flora from S. E. Anatolia (Turkey): Cong. Avanc. Études Stratigr. et Géol. Carbonifère, 4th, Heerlen 1958, Compte Rendu, v. 3, p. 745-752.

White, C. D., 1912, The characters of the fossil plant *Gigantopteris* Schenk and its occurrences in North America: U. S. Natl. Mus. Proc., v. 41, p. 493-516.

——— 1929, Flora of the Hermit Shale, Grand Canyon, Arizona: Carnegie Inst. Washington Pub. 405, 221 p.

Zimina, V. G., 1967, *Glossopteris* and *Gangamopteris* from the Permian deposits of the South Maritime Territory: Paleont. Jour., no. 2, p. 98-106.

Fossil Birds of Old Gondwanaland: A Comment on Drifting Continents and Their Passengers

PAT VICKERS RICH, *The National Museum of Victoria, Melbourne, Victoria 3000, and Monash University, Department of Earth Sciences, Clayton, Victoria 3168, Australia*

ABSTRACT

A detailed survey of avian fossils on Gondwana continents reveals a relatively rich record for Australia, but only a moderate representation for Africa, South America, and New Zealand. India's record is extremely poor. Although some of these continents have produced late Mesozoic or Paleogene taxa, most fossil birds are of Neogene vintage and in many cases restricted to the Pleistocene. This fossil record reflects long isolation of many gondwanic fragments as well as interchange between northern and southern continents but neither supports nor negates a once contiguous Gondwanaland; a richer and older avian record is required.

INTRODUCTION

In the early 1960's, studies of the Earth's magnetic properties began to convince large numbers of geologists and biologists alike that continents had not been locked forever in their present positions. One scientific group particularly affected by this information were biogeographers, most of whom had been content to move their animal and plant pawns about on a modern model of Earth geography. But since geophysics had now added another variable, mobile continents that needed inclusion in future calculations, most workers began considering how such changing geographies might have affected past dispersals of organisms.

Although several times alluded to in more general papers, the distribution of fossil birds on Gondwanaland has not been reviewed in detail to ascertain if this record tends to support or negate the clumping of Africa, India, South America, Australia, New Zealand, and Antarctica into one great landmass at some time in the past. This paper details the avian fossil record from those continents and attempts an evaluation of its biogeographic significance.

THE FOSSIL AVIAN RECORD ON GONDWANALAND CONTINENTS

Fossil avifaunas for two of the Gondwana continents, Africa and Australia, have been summarized recently (Rich, 1974, 1975a, 1975b). Maps from those papers are included here along with another summarizing avifaunas of South America (see Figs. 1 to 4). In general, most of the fossil taxa are predictable, based on modern avifaunas of these regions, but there are some surprises.

Africa

Africa's (Fig. 1) oldest bird fossils are Eocene and Oligocene in age and are exceedingly rare. *Gigantornis*, a probable pelecaniform (the avian order containing pelicans and cormorants), is known from central Africa (Nigeria). The remaining taxa of this age, however, are all from North Africa, mainly Egypt and Libya. *Psammornis*,[1] probably congeneric or very similar to *Struthio* (the ostrich), occurs in several North African localities. The Fayum area of Egypt has produced *Goliathia* (Ardeidae, herons), *Palaeoephippiorhynchus* (Ciconiidae, storks), as well as *Eremopezus* and *Stromeria* (closely related to the Malagasy Elephant Birds, Aepyornithidae), both thought to be ratites. The remainder of Africa's bird record is mid-Tertiary or younger in age, and comes mainly from North Africa (Algeria, Morocco, Tunisia, Egypt, and the Canary Islands); included are the ostrich (*Struthio*) and possible ostrich relatives (*Psammornis*), possible eggshell fragments of Elephant Bird (Aepyornithidae) from the Canary Islands (Rothe, 1964), a cormorant (*Phalacrocorax*), snake birds (*Anhinga*), a whale-billed stork (Balaenicipitidae), marabou storks (*Leptoptilos*), a jungle fowl (*Gallus*), barn owls (*Tyto*), a woodpecker (*Jynx*), and a thrush (*Luscinia*) (Rich, 1974). East African fossils from the mid-Tertiary localities of Ft. Ternan, Maboko, Songhor, and Rusinga include ostrich (Struthionidae), vulture (Accipitridae, Gypaetinae), herons (Ardeidae), game birds (Phasianidae), stone curlews (Burhinidae), possible hornbills (Bucerotidae), hawks (Accipitridae), possible cuckoos (Cuculidae), turacos (Musophagidae), storks (Ciconiidae), and flamingoes (Phoenicopteridae) (P. Rich, personal obs. 1973).

A few fragments of ducks and swans (Anatidae) and unidentified forms are known from the sub-Sahara in southwest Africa, but by far the most diverse assemblage has recently been recovered from the late Pliocene deposits of Langebaanweg, South Africa, including one

[1] *Psammornis* is based entirely on eggshells.

penguin, *Spheniscus predemersus,* closely related to the living *S. demersus* of southern and southwestern Africa (Hendey, 1976; Simpson, 1971c). Pleistocene birds are best represented in South Africa, Tanzania, and Madagascar, but small avifaunas have been reported from Algeria and Zambia.

The African Cenozoic avifaunas contain the same families, and even many of the same genera and species, that still occur there today. The only unusual elements are the Elephant Birds (Aepyornithidae), *Gigantornis,* and *Gallus.* The Elephant Birds were quite diverse in Madagascar during the Pleistocene (probably even before that despite the lack of a fossil record), and once probably ranged into North Africa as well as the neighboring Canary Islands. *Gigantornis,* a marine bird possibly related to the pelecaniforms, has no close *living* relatives in Africa, or for that matter anywhere. *Gallus,* present in mid-Tertiary deposits of North Africa, is now restricted to India, Ceylon, and southeast Asia.

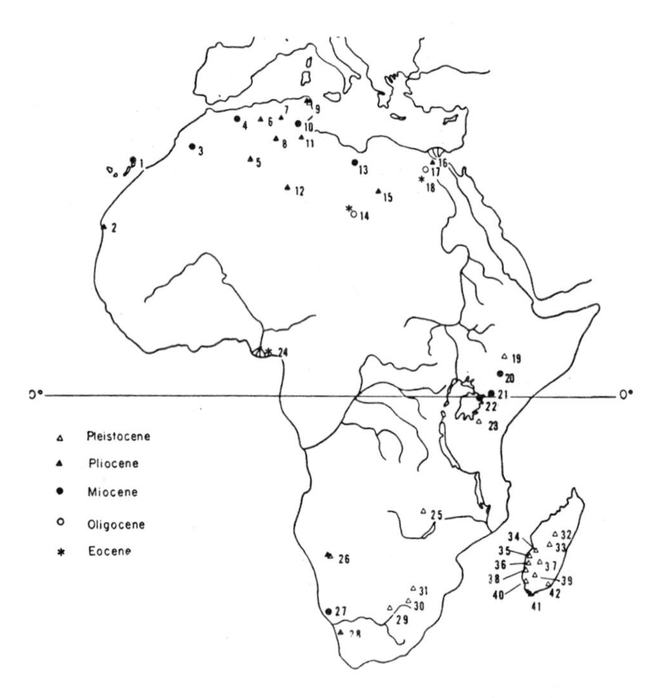

Figure 1. Localities producing fossil birds in Africa, the Canary Islands, and Madagascar: (1) Lanzarote, Canary Islands; (2) Cap Blanc (Cape Blanco); (3) Beni Mellal; (4) Oued el Hammam; (5) El Golea; (6) Wadi Djelfa; (7) Biskra; (8) Touggourt; (9) Hamada Damous; (10) Bled ed Douarah; (11) Southern Tunisia; (12) Iguidi; (13) Gabal Zelten; (14) Dor el Talha; (15) Giarabub; (16) Wadi Natrun; (17, 18) Fayum; (19) Omo; (20) Ngorora; (21) Ft. Ternan; (22) Rusinga Island; (23) Olduvai; (24) Ameki; (25) Broken Hill; (26) Etosha Pan; (27) Luederitz Bay; (28) Klein Zee; (29) Taungs; (30) Sterkfontein; (31) Kromdraai; (32) Ampasambazimba; (33) Antsirable; (34) Belo; (35) Morondava; (36) Lambohorana; (37) Ampoza; (38) Ambolisatra; (39) Taolambiby; (40) Itampolo; (41) Cape St. Marie; (42) Andrahomana (from Rich, 1974).

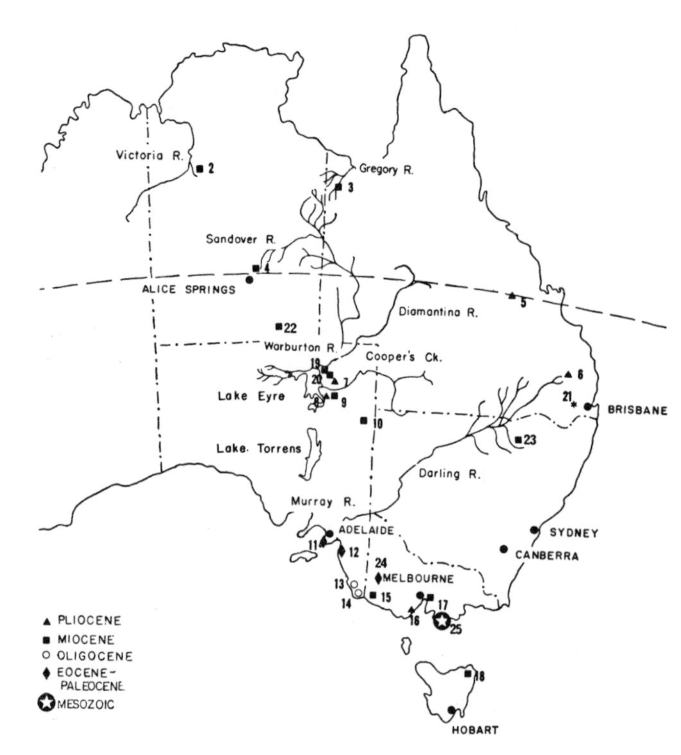

Figure 2. Tertiary and Mesozoic localities producing fossil birds in Papua-New Guinea and Australia: (1) Awe; (2) Bullock Creek; (3) Riversleigh; (4) Alcoota; (5) Peak Downs; (6) Chinchilla; (7) Lake Kanunka; (8, 9) Lake Palankarinna; (10) Eurinilla Creek and Lake Pinpa; (11) Port Noarlunga; (12) Christie's Beach; (13) Pritchard Brothers' Quarry; (14) Mt. Gambier; (15) Devil's Den; (16) Spring Creek (Minhamite); (17) Beaumaris; (18) Endurance Pit; (19) Lake Ngapakaldi; (20) Lake Pitikanta; (21) Redbank Plains; (22) Kangaroo Well; (23) Bugaldi (near Coonabarabran; (24) Redruth, Wannon; (25) Koonwarra (from Rich 1975a).

Australia

The early record of birds, although limited, extends to the Lower Cretaceous in Australia (Figs. 2, 3); this is the oldest record on any of the Gondwana continents. Feathers from the Koonwarra locality in southeastern Victoria (Fig. 4) are amongst the earliest occurrence of birds anywhere in the world (Talent and others, 1966; Waldman, 1970; Brodkorb, 1971b; Rich 1975a, 1975b, 1976). Unfortunately, despite the variety shown by the feathers, their identity even at the ordinal level, is uncertain; they simply record the presence of birds in Australia at a very early date.

The early Cenozoic, Paleogene, record of Australia like that of Africa, is pitifully poor. It consists of feather impressions from western Victoria and an unidentified bone, now lost, of supposed avian affinities (Cribb and others, 1960) from southeastern Queensland. The remaining record includes only penguins (Spheniscidae) from southeastern Australia, some of which were apparently closely related to New Zealand forms (Simpson, 1957, 1959, 1965, 1970a, 1970b, 1971b). One Eocene form, *Pachydyptes*

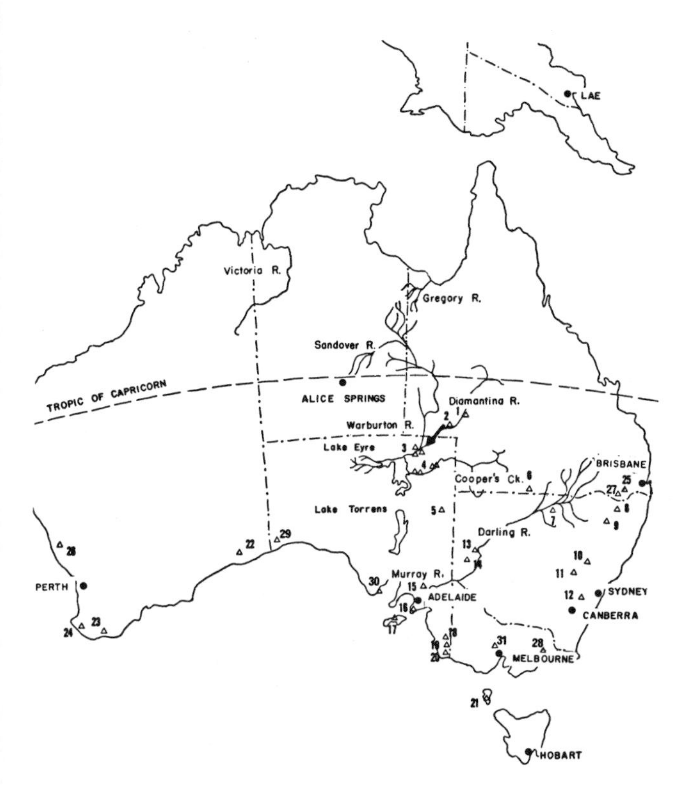

Figure 3. Quaternary localities producing fossil birds in Papua-New Guinea and Australia: (1) Diamantina River (no specific locality known); (2) Cassidy locality; (3) Warburton River localities; (4) Cooper's Creek localities; (5) Lake Callabonna; (6) Thorbindah; (7) Cuddie Springs; (8) Ashford Caves (Bone Cave); (9) Bingara; (10) Canadian Lead (Gulgong); (11) Walli and Wellington Caves; (12) Wombeyan Caves (Guineacor); (13) Lake Menindee; (14) Lake Tandou; (15) Baldina Creek; (16) Normanville (Salt Creek); (17) Kangaroo Island; (18) Henschke's Cave and Victoria Cave; (19) Penola; (20) Mt. Gambier; (21) King Island; (22) Madura Cave; (23) Scott River; (24) Mammoth Cave; (25) Darling Downs (Kings Creek, Warwick); (26) Drover's Cave; (27) Gore; (28) Buchan Caves; (29) Weeke's Cave; (30) Brother's Island, Pt. Lincoln; (31) Lancefield (from Rich 1975a).

Figure 4. Fossil feather from Early Cretaceous lake deposits in southeastern Victoria, Australia. Feather measures 20 mm length, 22 mm width across base (see Talent and others, 1966). Photograph by G. Wallis.

simpsoni (Jenkins, 1974), was distinctly larger than the largest living penguin, *Aptenodytes forsteri.*

Only in mid-Cenozoic or younger sediments are there abundant remains of birds; the Pleistocene cave deposits of southwestern, southern, and eastern Australia provide the best samples. From Miocene localities come a number of avian groups that are today present in Australia: ducks (Anatidae), stone plovers (Burhinidae), pelicans (Pelecanidae), cormorants (Phalacrocoracidae), rails (Rallidae), gulls (Laridae), owlet-nightjars (Aegothelidae), and emus (Dromaiidae) (see Rich, 1975a, 1975b for detailed summary). More unexpected elements are Mihirungs (Dromornithidae), an extinct group of large-to-enormous ground birds that were restricted to Australia and distantly related to emus. Also unexpected is a variety of flamigoes that were diverse in the mid-Cenozoic just as the Mihirungs were and became extinct during the last 2 m. y. Pleistocene sediments have produced additional families not previously recorded as fossils (see Rich, 1975a, 1975b) but have yielded no new elements that are absent from the living avifauna of Australia.

In summary, even though the Australian fossil bird record is the longest of all the Gondwana continents, a significant record of anything but marine penguins is not available until the Miocene. This avifauna is composed mainly of families still present in Australia, except for a rather diverse assemblage of flamingoes and Mihirungs, neither of which survived to the end of the Quaternary. The diversity of flamingoes and Mihirungs in Miocene faunas suggests a more lengthy history for them in Australia than is presently known.

South America

The earliest record of birds in South America (Fig. 5), a grebelike form placed in the order containing modern grebes (Podicipediformes) (see Table 1), comes from the Late Cretaceous (Maestrichtian) Quirquina Beds of Chile. Just as in Australia, early Cenozoic sediments have produced little—only the unstudied Paleocene Itaborai microfossils from Brazil and 2 species of primitive flamingo (*Telmabates*) from the Eocene of Argentina.

By the Oligocene, the record is enriched, but restricted geographically to Argentina. The first of the ground-dwelling carnivorous phororhacoids, nearly unique to South

Table 1. Summary of Cenozoic and late Mesozoic avifaunas from South America[1].

CRETACEOUS		
Chile	1[2]	**San Vicente Bay.** Quiriquina Beds, Cerro del Conejo, Vegas del Gualpen, southeast of San Vicente, Dept. Talcahuano, Late Cretaceous: *Neogaeornis wetzeli* (Podicipediformes, grebes).
PALEOCENE		
Brazil	2	**San Jose de Itaborai.** Marl fills of cavernous pre-Riochican Paleocene limestone: Aves, undetermined (small birds).
EOCENE		
Argentina	3	**Canadon Hondo.** Casamayoran Fauna, Casamayor Beds, Sarmiento Fm., near Paso Niemann, south of Rio Chico del Chubut, Terr. Chubut: *Telmabates antiquus* (Telmabatidae, extinct flamingo family).
	4	**Chubut Canadon.** Casamayoran Fauna, Casamayor Beds, Sarmiento Fm., south of Rio Chico del Chubut: *Telmabates howardae.*
OLIGOCENE		
Argentina	5	**Cabeza Blanca.** Deseadan Fauna, Deseado Fm., Terr. Chubut: *Andrewsornis abbotti* (Phorusrhacidae).
	6	**Golfo de San Jorge.** Deseadan Fauna, Deseado Fm., Terr. Santa Cruz: *Cruschedula revola* (Accipitridae, Buteoninae).
	7	**Rio Deseado.** Deseadan Fauna, Deseado Fm. (or middle Sarmiento Fm. or Group), southern Patagonia, Terr, Santa Cruz: *Cladornis pachypus* (Cladornithidae, Pelecaniformes); *Tiliornis senex* (Phoenicopteridae, flamingoes); *Ciconiopsis antarctica* (Ciconiidae, storks); *Teleornis impressus, Loxornis clivus* (Anatidae, Plectropterinae, ducks); *Climacarthrus incompletus* (Accipitridae, Buteoninae, hawks); *Aminornis excavatus* (Aramidae, limpkins); *Physornis fortis* (= *Aucornis*), *Riacama caliginea, Smiliornis penetrans, Pseudolarus guaraniticus* (Phorusrhacidae).
	8	**Mina Atala.** Atala Fm., Divisadero Largo Fm., Terr. Mendoza, Las Heras: *Cunampala simplex* (Cunampaiidae, Gruiformes).
MIOCENE		
Argentina	9	**Gulfode San Jorge, Rio Chubut, Rio Santa Cruz, San Julian, Trelew.** Patagonia Fm., early Miocene, Terr. Chubut or Santa Cruz: *Arthrodytes grandis, Palaeospheniscus gracilis, P. patagonicus, P. bergi, P. wimani, Paraptenodytes antarcticus, P. brodkorbi, Chubutodyptes biloculata* (Spheniscidae, penguins) (Simpson, 1972); *Argyrodyptes microtarsus* (Procellariidae, petrels and shearwaters).
	10	**Monte Leon and Lago Argentina.** Patagonia Fm., early Miocene, Terr. Santa Cruz: *Brontornis burmeisteri, Palaeociconia cristata, Phorusrhacos longissimus, Psilopterus australis* (Phorusrhacidae).
	11	**Corriguen kalk, Coy Inlet, Cueva, Kallik Aike, Karaiken, La Cueva, Lake Pueyrredon, Monte Observacion, Patagonia, Rio Sehuen, Tagua Quemada, and Take Harvey.** Santa Cruz Fm., middle Miocene: *Opisthodactylus patagonicus* (Opisthodactylidae, Rheiformes, primitive rhea)[3]; *Protibis cnemealis* (Threskiornithidae, ibises); *Eoneornis australis, Eurwloenia patagonicus* (Anatidae, Anatinae, ducks); *Thegornis musculosus, T. debilis* (Accipitridae, Circinae, kites); *Badiostes patagonicus* (Falconidae, falcons); *Anisolornis excavatus* (Cracidae, chachalacas); *Brontornis burmeisteri, Palaeociconia cristata, Phorusrhacos longissimus, Pseudolarus eocanenus, Psilopterus australis, P. communis, P. minutus, Lophiornis obliquus* (Phorusrhacidae).
Uruguay	12	**Arroyo Roman?.** Dept. Rio Negro (Miocene?): *Brontornis burmeisteri* (Phorusrhacidae).
Columbia	13	**Upper Magdalena Valley.** La Venta Fauna, La Venta Beds, Honda Group, Dept. Huila, along Villavieja-San Alfonso trail: *Hoazinoides magdalenae* (Opisthocomidae, hoatzin).
PLIOCENE		
Argentina	14	**Campo de Robilotte.** Southeast of Laguna Epecuen, partido de Adolfo Alsina, Prov. Entre Rios, El Brete, early Pliocene: *Onactornis pozzi* (Phorusrhacidae).
	15	**Rio Parana.** Entre Rios Series, Prov. Entre Rios, ravines of Rio Parana, early Pliocene: *Andalgalornis steulleti* (Phorusrhacidae).
	16	**?Valle de Santa Maria.** Andalgala or Corral Quemado Fm., Prov. Catamarca, early or middle Pliocene: *Phororhacus incertus* (= *Hermosiornis*) (Phorusrhacidae).
	17	**Andalgala and Chiquimil.** Andalgala Fm., Prov. Catamarca, middle Pliocene: *Andalgalornis ferox, Procariama simplex* (Phorusrhacidae).
	18	**Huayquerias.** Huayquerian Fauna, Prov. Mendoza, middle Pliocene: *Onactornis mendocinus* (Phorusrhacidae).

[1] If references are not cited, data in Tables 1 and 2 from Brodkorb 1963, 1964, 1967 or 1971a.
[2] Numbers refer to localities plotted in Figure 4. Only extinct genera are included in lists.
[3] Brodkorb (1963) listed this species as Eocene, but Pascual and Rivas (1971) place it in the Santa Cruz Fm., middle Miocene [see Simpson's discussion of this problem in Simpson (1972)].

Table 1. (Cont.)

	19	**Monte Hermoso.** Monte Hermosan Fauna, Monte Hermoso Fm., Prov. Buenos Aires, late Pliocene: *Tinamisornis intermedius, Cayetanornis parvulus, Querandiornis romani* (Tinamidae, tinamous); *Heterorhea dabbenei* (Rheidae, rheas); *Dryornis pampeanus* (Cathartidae, vultures); *Foetopterus ambiguus* (Accipitridae, hawks)[4]; *Prophororhacus australis* (Phorusrhacidae).
Bolivia	20	**Tarija.** Tarija Valley, Pliocene: *Vultur patruus* (Cathartidae, vultures).
PLEISTOCENE		
Argentina	21	**Anchorena.** Ensenadan, Pampas Fm., Prov. Buenos Aires, early-middle Pleistocene: *Rhea anchorenensis* (Rheidae, rheas).
	22	**Puerto de Olivos.** Ensenadan, Pampas Fm., Prov. Buenos Aires: *Pionus ensenadensis* (Psittacidae, parrots).
	23	**Arrecifes.** Pampas Fm., Prov. Buenos Aires: *Nothura paludosa* (Tinamidae, tinamous); *Pseudosterna pampeana* (Laridae, gulls and terns); *Euryonotus brachypterus, E. argentinus* (Rallidae, rails).
	24	**Arroyo Loberia.** Chapadmalal Fm., Prov. Buenos Aires, near mouth of Arroyo Loberia: *Prophororhacus rapax* (Phorusrhacidae).
	25	**La Plata.** Pampas Fm., Cuidad de la Plata, Prov. Buenos Aires: *Neochen debilis* (Anatidae, Plectropterinae, spur-winged geese).
	26	**Lujan.** Pampas Fm., Prov. Buenos Aires: *Phalacrocorax pampeanus* (Phalacrocoracidae, cormorants); *Lagopterus minutus* (Accipitnidae, Buteoininae, hawks); *Pseudosterna degener* (Laridae, terns and gulls).
	27	**Olivera.** Pampas Fm., Prov. Buenos Aires: *Pterocnemia fossilis* (Rheidae, rheas).
Brazil	28	**Lagoa Santa** (including Lapa da Escrivania nos. 5 and 11, Lapa dos Tatus). Cave deposits, Minas Gerais: *Prociconia lydekkeri* (Ciconiidae, storks); *Neochen pugil* (Anatidae, Plectropterinae, spur-winged geese).
Ecuador	29	**La Carolina, Santa Elena Peninsula.** Santa Elena asphalt sands, Rio Chico and Carolina Oil Co. camp: *Querquedula ambrechti* (Anatidae, ducks); 3 extinct species of Anatidae, 2 in the genus *Anas* (Campbell, 1976) (Anatidae, ducks); new genus of Cathartidae (vultures); new genus of Accipitridae (eagles), *Buteo hoffstetteri* (Accipitridae, hawks); *Oreopholus orcesi* (Charadriidae, plovers); a new genus of Scolopacidae (sandpipers); *Steganopus,* new species (Phalaropodidae, phalaropes); *Aratinga roosevelti*[5] (Psittacidae, parrots); plus other extant species (Campbell, 1976).
Peru	30	**Talara Tar Seeps.** 3 extinct species of Anatidae, 2 in the genus *Anas* (Campbell, 1976) (Anatidae, ducks); new genus of Cathartidae (vultures); new genus of Accipitridae (eagles); new genus of Scolopacidae (sandpipers); *Steganopus,* new species (Phalaropodidae, phalaropes); plus other extant species (Campbell, 1976).

[4] Brodkorb (1964) places *Foetopterus* in the Pleistocene, but Pascual and Rivas (1971) in a later review return it to the Montehermosan part of the Oligocene.
[5] May not represent an extinct species (see Campbell, 1976).

America (*Andrewsornis, Physornis, Riacama, Smiliornis,* and *Pseudolarus*), occur in the Oligocene Deseadean faunas along with hawks (*Cruschedula, Climacarthrus*), a presently extinct group of pelecaniforms (*Cladornis*), true flamingoes (*Tiliornis*), storks (*Ciconiopsis*), ducks (*Teleornis, Loxornis*), limpkins (*Aminornis*), and an extinct family, the Cunampaiidae (*Cunampala*), related to the cranes, rails, and phororhacoids.

Unlike the Oligocene avifaunas, those of the early Miocene include a large marine element in the Patagonian Formation. Four genera and several species of penguins (*Arthrodytes, Palaeospheniscus, Paraptenodytes,* and *Chubutodyptes*) are known from these sediments. Simpson (1972) suggested a close relationship between the Patagonian *Palaeospheniscus* and the New Zealand *Korora,* as well as possible relationship between *Perispheniscus* (= *Palaeospheniscus,* Patagonia) and *Duntroonornis* (New Zealand). Also known from the Patagonian sediments is a single petrel, *Argyrodyptes.*

Terrestrial Miocene sediments, known from Argentina, Uruguay, and Colombia, contain diverse phororhacoids (see Table 1), as well as the oldest record of the herbivorous, ground-dwelling rheas (*Opisthodactylus*), ibises (*Protibis*), ducks (*Eoneornis, Eurwloenia*), kites (*Thegornis*), falcons (*Badiostes*), chachalacas (*Anisolornis,* primarily a South and Central American group), as well as yet another South American endemic form, an hoatzin (*Hoazinoides*).

Pliocene avian fossils are known from Argentina and Bolivia, with only a condor (*Vultur*) coming from the latter country. The Argentinian record consists mainly of phororhacoids (*Onacroanis, Andalgalornis, Phororhacus, Procariama*), but included also is the first record of the South and Central American tinamous (*Tinamisornis, Cayetanornis, Querandiornis*), as well as rheas (*Heterorhea*), vultures (*Dryornis*), and hawks (*Foetopterus*).

Geographic representation of birds during the Pleistocene is much broader (Argentina, Brazil, Ecuador, and Peru) and included are many of the same families known from older rocks, as well as the first records for South America of parrots (*Aratinga, Pionus*), true rails (*Euryonotus*), gulls and terns (*Pseudosterna*), plovers (*Oreopholus*), sandpipers (*Scolopacidae*), phalaropes (*Steganopus*), and cormorants (*Phalacrocorax*).

Table 2. Fossil records of extinct avian genera on Gondwana continents exclusive of Africa, Australia, and South America.

INDIA	
Pliocene	Avian genera, all extinct, are restricted to the Pliocene and to one general geographic area, the Siwalik Hills. Named extinct forms are: *Struthio asiaticus* (Struthionidae, ostriches); *Pelecanus sivalensis* (Pelecanidae, pelicans); *Leptoptilos falconeri* (Ciconiidae, storks).
MAURITIUS	
Pleistocene	**From localities on Mauritius (mainly Mare aux Songes).** *Anhinga nana* (Anhingidae, darters or snake birds); *Butorides mauritianus* (Ardeidae, herons); *Podiceps gadowi* (Podicepididae, grebes); *Anas theodori, Sarkidornis mauritianus* (Anatidae, ducks); *Accipiter alphonsi* (Accipitridae, hawks); *Lophosittacus mauritanus* (Psittacidae, parrots); *Tyto sauzieri* (Tytonidae, barn owls); *Raphus cucullatus* (Raphidae, dodos). **From nearby Rodriquez Island** (extinct taxa). *Nycticorax megacephalus* (=*Botaurus lentiginosus*) (Ardeidae, herons); *Streptopelia rodericana* (Columbidae, pigeons); *Necropsittacus rodericanus* (Psittacidae, parrots); *Bubo leguati, Athene murivora* (Strigidae, typical owls); *Pexophaps solitaria,* perhaps *P. bourbonica* (Raphidae, solitares).
ANTARCTICA	
Mid-Tertiary	**Fildes Peninsula, King George Island.** Footprints of ducks (Anatidae), limpkins or herons (Aramidae or Ardeidae), and a ratite (V. Covacevich, personal commun. 1976).
Cenozoic	**Seymour Island.** Sediments of uncertain stratigraphic position; a small fauna of penguins with: *Anthropornis nordenskioeldi, A. grandis, Orthopteryx gigas, Eosphaeniscus gunnari, Delphinornis larsenii, Ichthyopteryx gracilis* (Simpson, 1971a, 1972).
NEW ZEALAND	
Late Eocene-Oligocene	**South Island, Dunedin.** *Palaeeudyptes antarcticus, P. marplesi* (Spheniscidae, penguins).
	South Island, Duntroon (north Otago). *Palaeeudyptes antarcticus, Archaeospheniscus lowei, A. lopdelli, Duntroonornis parvus, Platydyptes novaezealandiae, P. amiesi* (Spheniscidae, penguins); *Manu antiquus* (Diomedeidae, albatrosses).
	South Island, Hakataramea Valley (South Canterbury). *Platydyptes amiesi, Korora oliveri* (Spheniscidae, penguins).
	South Island, Oamaru. *Palaeeudyptes antarcticus, Platydyptes novaezealandiae, Pachydyptes ponderosus* (Spheniscidae, penguins).
	South Island, Waitaki Valley. *Platydyptes novaezealandiae* (Spheniscidae, penguins).
Miocene	**South Island, North Canterbury.** Odontopterygidae (bony-toothed birds) (Scarlett, 1972).
Pliocene	**South Island, Motunau (North Canterbury).** Greta Sandstone, Waitotaran Stage: *Palaeospheniscus novaezealandiae* (Spheniscidae, penguins); *Pseudodontornis stirtoni* (Pseudodontornithidae, bony-toothed birds).
Pleistocene	**North and South Islands.** *Anomalopteryx didiformis, A. oweni, Megalapteryx didinus, M. benhami, Pachyornis elephantopus, P. mappini, Euryapteryx curtus, E. geranoides, Emeus crassus* (Emeidae, moas); *Dinornis struthoides, D. torosus, D. novaezealandiae, D. giganteus* (Dinornithidae, moas) (after Cracraft, 1976); *Pseudapteryx gracilis, Apteryx australis, A. owenii, A. haasti* (Apterygidae, kiwis); *Cygnus sumnereusis* (also Chatham Islands), *Euryanus finschi, Cnemiornis calcitrans* (Cereopsinae), *C. septentrionalis, Pachyanus chathamica* (Anatidae, ducks) (Chatham Is.); *Harpagornis moorei* (Accipitridae, eagles); *Circus teauteensis* (Acciptridae, Circinae, harriers); *Pelecanus conspicillatus novaezealandiae* (Pelecanidae, pelican); *Capellirallus karamu, Gallirallus minor, Diaphorapteryx hawkinsi* (Chatham Is.), *Pyramida hodgeni, Nortornis mantelli, Nesophalaris prisca, Aptornis otidiformis, A. defossor* (Rallidae, rails); *Coenocorypha chathamica* (Scolopacidae, sandpipers) (Chatham Is.); *Megaegotheles novaezealandiae* (Aegothelidae, owlet-nightjars).

To sum up, the South American fossil avifauna contains a number of taxa completely or nearly restricted to this southern continent today, such as the rhea, tinamou, hoatzin, chachalachas, and limpkins. Fossils also include other South American endemic families that have not survived to the present, including several gruiforms, primarily in the family Phorusrhacidae, as well as a primitive flamingo (family Telmabatidae), and an extinct ratite family, the Opisthodactylidae. The remaining fossils, although perhaps generically unique to South America, represent families that have a much wider range worldwide.

New Zealand

New Zealand's earliest record of birds is late Eocene to Oligocene (see Table 2), a strictly marine record that consists only of penguins. The penguin fauna is closely related to that of South Australia of similar age.

The exact ages of many of the New Zealand penguins, have been recently questioned by Simpson (1970a, 1971b), who pointed out that the only reliable assignments are for *Pachydyptes ponderosus* to the latest Eocene, and *Archaeospheniscus lowei, A. lopdelli,* and *Duntroonornis parvus* to the middle or late Oligocene. The remaining species could have been derived from sediments ranging from latest Eocene to Miocene in age. The Mio-Pliocene record consists only of penguins and the large "bony toothed birds" in the genus *Pseudodontornis* (Brodkorb, 1971a, p. 174), also marine, whose close relatives (*Pseudodontornis longirostris* and *Osteodontornis*) are known from western North America and Brazil or Germany (see Brodkorb, 1963). It is the Quaternary avifaunas, particularly those of the sub-Recent, which hint at how rich the New Zealand Cenozoic avifaunas really were, and that the lack of early and mid-Tertiary vertebrate localities is the main reason for such low diversity. At least 13 extinct species of ground dwelling moas (Emeidae and Dinornithidae) are known (Cracraft, 1976), perhaps more, along with a variety of similarly extinct kiwis, swans, flightless geese (*Cnemiornis*) (thought to be closely related to the Australian Cape Barren Goose, *Cereopsis*), ducks, a giant eagle (*Harpagornis*), pelicans, a variety of rails, sandpipers, and a strange, possibly flightless owlet-nightjar (*Megaegotheles:* Rich and Scarlett, 1977 (see Williams, 1973, for overview of the sub-Recent record).

New Zealand's fossil record, particularly that of the last few thousand years, thus contains many birds unpredictable from today's surviving avifauna. That avifauna, however, is probably only a small part of the Cenozoic record for this island subcontinent, which will only be elucidated with the finding of additional, pre-Pleistocene terrestrial localities.

India

Three fossil birds, all extinct, are the pitifully inadequate record from India. All fossils come from Pliocene sediments in the Siwalik Hills of sub-Himalayan India and include an ostrich (*Struthio*), a pelican (*Pelecanus*), and a stork (*Leptoptilos*), which reflect a Eurasian rather than a Gondwanan influence. Certainly, by this time, however, India was geographically a part of the Northern Hemisphere continents under all current plate tectonic interpretations.

Madagascar and Mauritius

The avian fossil record for both Madagascar and Mauritius is limited to the Quaternary despite the fact that terrestrial conditions have existed much longer on them than the vertebrate evidence suggests. The Madagascar fossils have been summarized by Rich (1974).

Paleo-avifaunas from Mauritius include only 1 species in common with Madagascar, *Anhinga nana* (Anhingidae, snake bird). The remainder of its Pleistocene record of extinct taxa is composed of a heron, a grebe, ducks, a hawk, a parrot, an owl, and the unusual relative of the pigeon, the dodo (Raphidae) (see Table 2). Nearby Rodriquez Island produced a heron, a pigeon, a parrot, two owls, and the Solitaire (*Pezophaps soiltaria*), another relative of the pigeon, but in a separate family, the Raphidae.

These fossil records are of extremely short duration and give no insight to the paleogeography of Madagascar and Mauritius prior to the Pleistocene, by which time geographic arrangement was modern.

Antarctica

Penguin bones from Seymour Island in West Antarctica, whose precise stratigraphic placement is uncertain, represent 6 different species (see Table 2) and the majority of fossil birds known from Antarctica (Simpson, 1971a). Although the Miocene is the most frequent age assigned to the Seymour Island material, it could be as old as early Cenozoic. The South Shetland Islands have produced additional information: trackways assigned by Covacevich and Lamperein (1969, 1970) to three different families: rails (Rallidae), possible ducks (Anatidae), and either limpkins (Aramidae) or herons (Ardeidae).

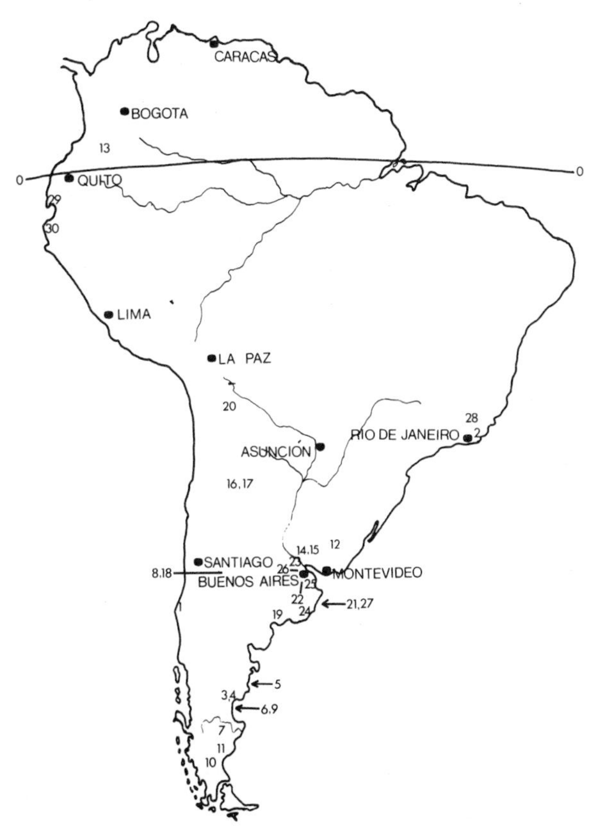

Figure 5. Mesozoic and Cenozoic localities producing birds in South America. *Cretaceous:* (1) San Vicente Bay. *Paleocene:* (2) San Jose de Itaborai. *Eocene:* (3) Canadon Hondo; (4) Chubut Canadon. *Oligocene:* (5) Cabeza Blanca; (6) Golfo de San Jorge; (7) Rio Deseado; (8) Mina Atala. *Miocene:* (9) Gulfo de San Jorge, Rio Chubut, Rio Santa Cruz, San Julian, Trelew; (10) Monte Leon, Lago Argentina; (11) Corriguen Kalk, Coy Inlet, Cueva, Kallik Aike, Karaiken, La Cueva, Lake Pueyrredon, Monte Observacion, Patagonia, Rio Sehuen, Tagua Quemada, Take Harvey; (12) Arroyo Roman?; (13) Upper Magdalena Valley. *Pliocene:* (14) Campo de Robilotte; (15) Rio Parana; (16) Valle de Santa Maria?; (17) Andalgala, Chiguimil; (18) Huayquerias; (19) Monte Hermoso; (20) Tarija. *Pleistocene:* (21) Anchorena; (22) Puerto de Olivos; (23) Arrecifes; (24) Arroyo Loberia; (25) La Plata; (26) Lujan; (27) Olivera; (28) Lagoa Santa; (29) La Carolina; and (30) Talara.

A large, broad-toed avian footprint from mid-Tertiary sediments on King George Island may represent one of the ground dwelling ratites (V. Covacevich, personal commun. 1976). The print lacks any large terminal claw impressions, which Covacevich believes would exclude phororhacoid affinities.

East Antarctica, the main cratonal landmass of this polar continent, unfortunately, has produced no avian fossils.

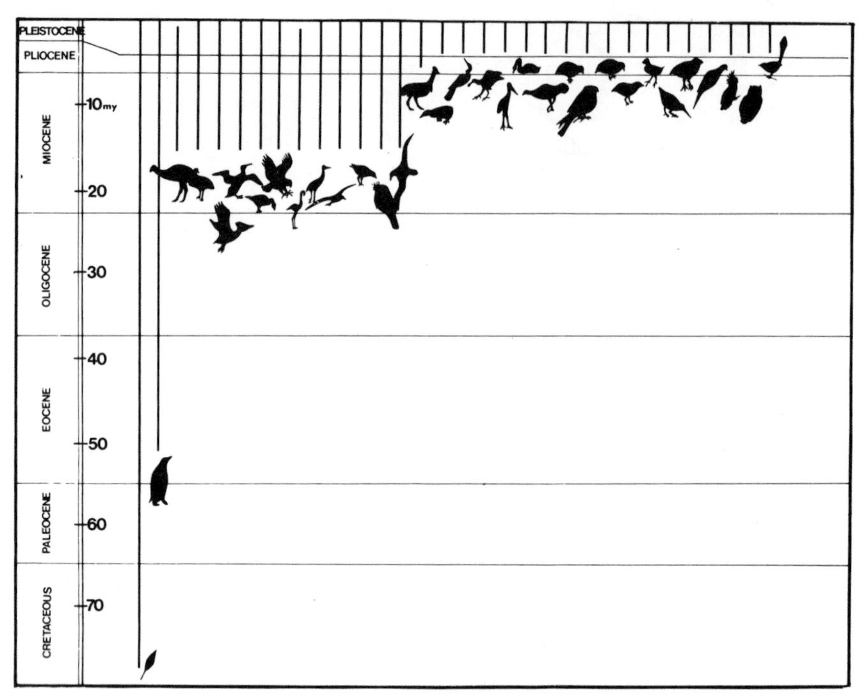

Figure 6. Geologic ranges of birds in Australia. Family names from left to right, excluding Cretaceous feather, include: Spheniscidae, Dromornithidae, Dromaiinae, Pelecanidae, Phalacrocoracidae, Anatidae, Accipitridae, Phoenicopteridae, Gruidae, Laridae, Burhinidae, Aegothelidae?, Diomedeidae, Casuariinae, Podicepedidae, Anhingidae, Threskiornithidae, Ciconiidae, *Cygnus*, Megapodiidae, Phasianidae, Falconidae, Turnicidae, Pedionomidae, Rallidae, Scolopacidae, Chionididae, Columbidae, Psittacidae, Tytonidae, Passeriformes.

SUMMARY AND CONCLUSIONS

The fossil record of birds from Gondwanaland, as is evident from the preceding summaries, is extremely sparse for the late Mesozoic and early Tertiary. In fact, South America has the only moderately diverse avifaunas known from terrestrial sediments older than mid-Tertiary. Although Australia has produced the oldest fossil bird (represented by early Cretaceous feathers) of any of the gondwanic fragments (see Fig. 6), only a few, mainly marine birds are known in Australia from rocks older than Miocene. Africa's record begins in the Eocene, with a marine form, and diverse assemblages are recorded only in sediments that are mid-Tertiary or younger. Most of these assemblages, as well as those from the Paleocene of South America, have not yet been adequately studied. New Zealand's record prior to the Pleistocene is mainly marine; that of Madagascar and Mauritius is entirely Quaternary. India has only a Pliocene record. Antarctica's fossil avifauna consists of a poorly-dated penguin assemblage and a few mid-Tertiary avian tracks, all from West Antarctica. East Antarctica, a cratonal mass that may have served as an important link between many gondwanic fragments, boasts no record at all.

Thus, at present, the avian fossil record can offer neither support for, nor criticism of, current Gondwanaland reconstructions. Well-studied, diverse avifaunas are not yet available for the late Mesozoic and early Cenozoic, a period when many of the gondwanic fragments may still have had connections (see Figs. 7, 8). But this is not to imply that the avian fossil record will forever be useless in helping to evaluate geophysical reconstructions. If assemblages ranging in age from 100 to 40 m. y. become better known, and a few are beginning to appear (e.g., the Itaborai avifauna), then valid faunal comparisons can be made for a time when a high degree of interchange may have been occurring. In comparing avifaunas living in the same areas today (Africa, South America, Australia, Antarctica), the biogeographer may be unable to resolve past similarities due to (1) the long isolation of these avifaunas after gondwanic breakup; (2) severe climatic conditions imposed on some by late Cenozoic cooling; and (3) the major interchange that has occurred during at least the latter half of the Cenozoic with the continents of the Northern Hemisphere.

What the fossil avifaunas (mid-Cenozoic or younger) of the southern continents do reflect is the long isolation of many of the fragments and a two-way faunal interchange with the northern continents rather than simply an invasion of the gondwanic continents by northern taxa.[2]

Long isolation of several of the gondwanic fragments is indicated by a mid-Tertiary or older record of several

[2] A major flaw in previous biogeographic thinking may be the assumption that most similarities of Southern and Northern Hemisphere avifaunas is a result of northern forms invading the southern landmasses, an assumption strengthened by Matthew's (1915) work on placental mammals. His observations for the placentals may not be as applicable for the birds; the fossil record is not sufficiently complete to allow resolution of this dilemma at present.

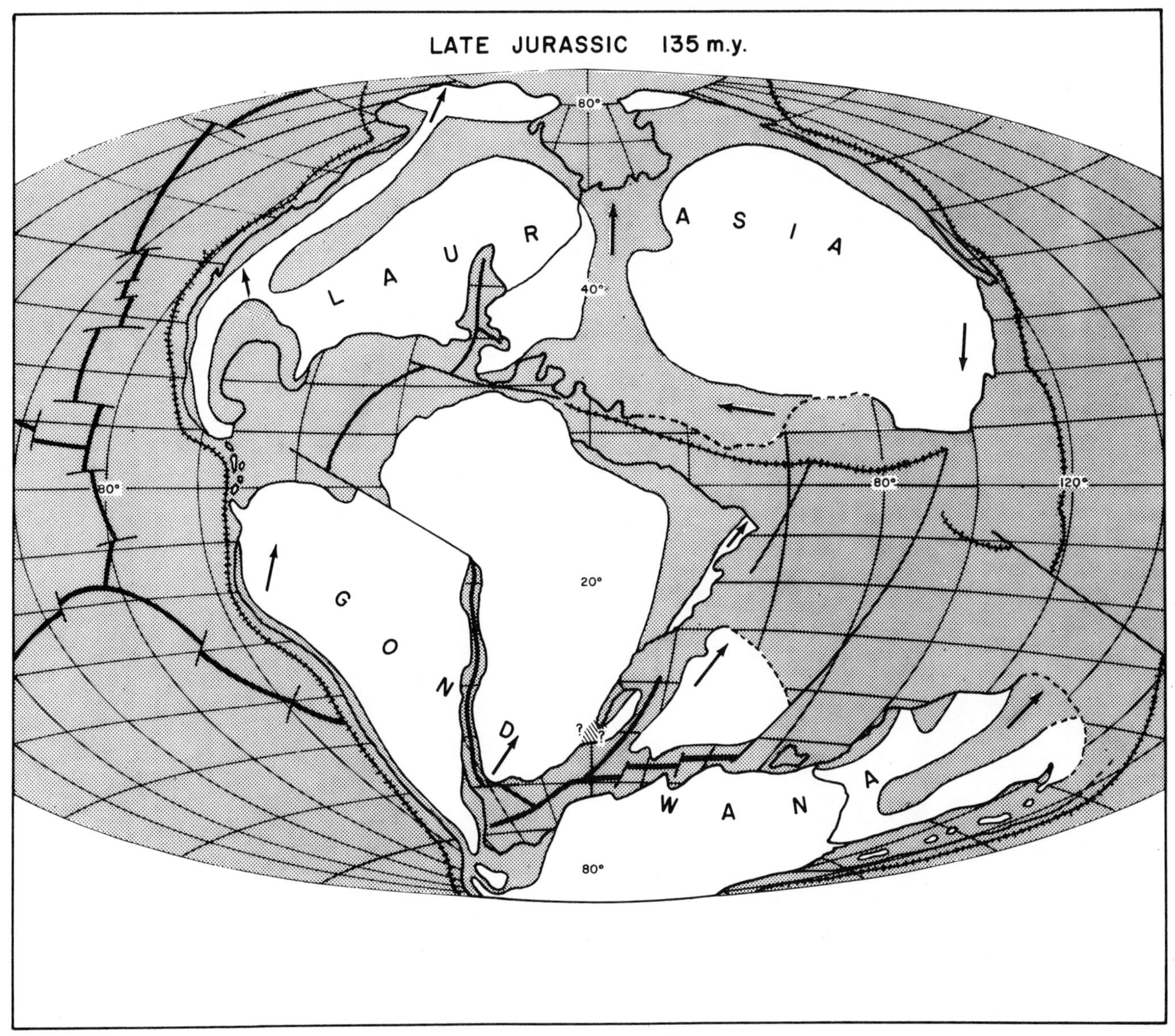

Figure 7. Continental reconstruction for the Late Jurassic. Map divides the Earth's surface into a number of plates that supposedly move with respect to one another over a period of time (arrows indicate direction). Boundaries between plates are threefold: ridges, where new volcanic material is added (thick lines); trenches, where one crustal plate slides beneath another (hatched lines); and transform faults, where two plates slide past one another laterally (thin lines). Epicontinental seas are also superimposed (from Rich, 1975a, 1975b, as modified from Tedford, 1974).

of the presently endemic taxa as well as appearance of several now extinct endemic taxa. In South America the impressive radiation of the phororhacoids (Phorusrhacidae) begins at least by the Oligocene, while relatives of the hoatzin (Hoazinoides), the rhea (Opisthodactylidae), and the chachalacas (*Anisolornis*) are known from the Miocene. Tinamous (Tinamidae) are first known from the Pliocene. In Australia the mid-Tertiary diversity of extinct Mihirungs (Dromornithidae) and flamingoes (Phoenicopteridae) hint at a much richer, yet undiscovered, earlier avifauna, while presently endemic groups such as the emus (*Dromaius*), cassowaries (*Casuarius*), owlet-nightjars (Aegothelidae), and plain wanderers (Pedionomidae: Rich, personal obs. 1976) had appeared no later than the Pliocene. In Africa, a close relative of the Malagasy Elephant Birds (Aepyornithidae) had appeared in the early Tertiary, while the ostrich (*Struthio*) and whale-billed storks (Balaenicipitidae) probably had appeared by mid-Tertiary.

Interchange of avifaunal elements is best exemplified by Africa, where dispersal probably occurred intermittantly throughout the Cenozoic. Ostrich fossils are known from both Africa and Europe (Brodkorb, 1963) as are turacos (Musophagidae), while a number of groups now restricted to Africa have European fossil records, e.g., the mousebirds (Coliidae) and secretary-birds (Sagittariidae) (Brod-

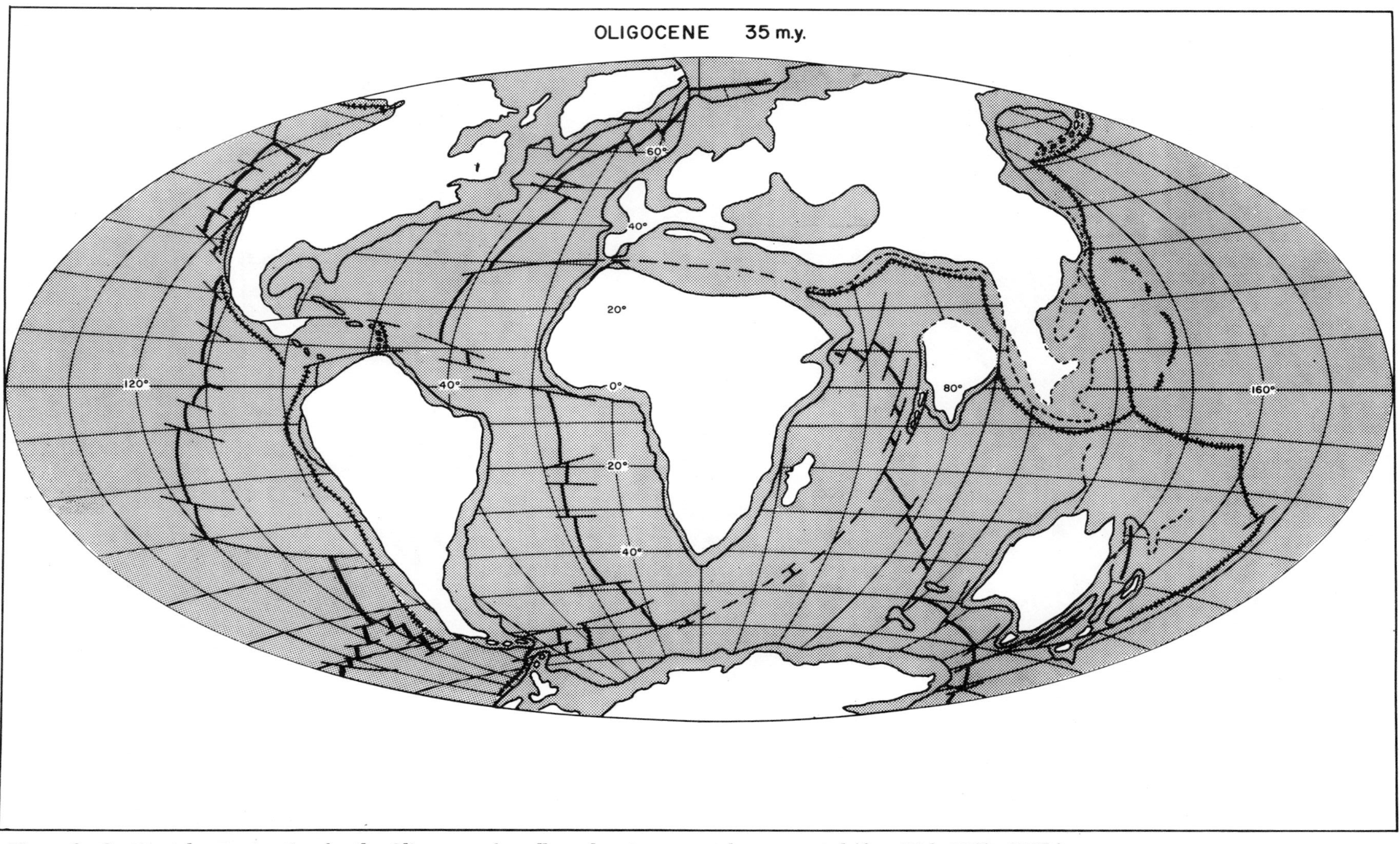

Figure 8. Continental reconstruction for the Oligocene, after all gondwanic segments have separated (from Rich, 1975a, 1975b).

korb, 1964, 1971a). Needless-to-say, with increased paleo-avifaunal diversity a great deal more could be learned about available dispersal routes, effects of isolation, latitudinal movement of gondwanic fragments with resulting climatic change, and the effects and timing of the collision with the Northern Hemisphere continents.

Marine avian data provides little information regarding continental arrangement. New Zealand and Seymour Island in West Antarctica share 2 genera of penguins in the Eocene-Oligocene, while Miocene rocks of Patagonia and New Zealand contain single genera thought to be closely related (*Korora*, New Zealand; *Palaeospheniscus*, Patagonia) (Simpson, 1972). With the realization that only a few localities produced the fossils, either a modern or an Eocene drift paleogeography could adequately explain such similarities. Simpson (1972, p. 35) noted that "All the fossil localities [containing early and mid-Tertiary penguins] are far within the latitudinal range of a single species (e.g. *Spheniscus magellanicus*) today, but there are also several Recent species, amongst them the large *Aptenodytes patagonicus*, that nearly reach the latitude of the southern but never reach anywhere near that of the northern fossil localities." Penguins distinctly larger than the Emperor Penguin, the largest living form, are known from probable Eocene deposits in West Antarctica and more northerly New Zealand, and southern Australia. It is impossible to verify at present, but perhaps this large size at seemingly anomalous northern latitudes may point to both Australia and New Zealand occupying more southerly and cool climes (thus applying Bergmann's Rule of size increase with latitude). What is needed to test this theory are more localities, both north and south, containing similar species whose sizes can be compared over several degrees of latitude.

ACKNOWLEDGMENTS

My thanks are due to Max Hecht who suggested that I write this paper, and to M. McKenna, R. Tedford, and W. Bock, who have influenced my thinking on this topic in times past, as well as to L. Frakes, T. Rich, G. Sanson, and I. Thornton whose discussion, comments, and research activities have been pertinent to development of this paper. M. L. Vickers skillfully typed the manuscript and P. Duce aided in graphic illustrations. The M. A. Ingram Trust kindly supported the research for this paper.

ADDENDUM

Gigantornis eaglesomei from the middle Eocene Ameki Formation of Nigeria is now thought to belong in the order of bony-toothed birds, Odontopterygiformes [Harrison, C. J. O., and Walker, C. A., 1976, A review of the bony-toothed birds (Odontopterygiformes) with descriptions of some new species: Tertiary Research Spec. Paper no. 2, Tertiary Research Group, London, 62 p.].

REFERENCES

Brodkorb, P., 1963, Catalogue of fossil birds. Pt. 1. Archaeopterygiformes through Ardeiformes: Florida State Mus. Bull., Bio. Sci., v. 7, p. 179-293.

——— 1964, Catalogue of fossil birds. Pt. 2. Anseriformes through Galliformes: Florida State Mus. Bull., Bio. Sci., v. 8, p. 195-335.

——— 1967, Catalogue of fossil birds. Pt. 3. Ralliformes, Ichthyornithiformes, Charadriiformes: Florida State Mus. Bull., Bio. Sci., v. 11, p. 99-220.

——— 1971a, Catalogue of fossil birds. Pt. 4. Columbiformes through Piciformes: Florida State Mus. Bull., Bio. Sci., v. 15, p. 163-266.

——— 1971b, Origin and evolution of birds, *in* Farner, D. S., and King, J. R., eds., Avian biology, Vol. 1: New York, Academic Press, p. 19-55.

Campbell, K. E., 1976, The late Pleistocene avifauna of La Carolina, southwestern Ecuador, *in* Olson, S. L., ed., Collected papers in avian paleontology honoring the 90th birthday of Alexander Wetmore: Smithsonian Contr. Paleobiology no. 27, p. 155-168.

Cracraft, J., 1976, The species of moas (Aves: Dinornithidae), *in* Olson, S. L., ed., Collected papers in avian paleontology honoring the 90th birthday of Alexander Wetmore: Smithsonian Contr. Paleobiology no. 27, p. 189-205.

Covacevich, C. V., and Lamperein, R. C., 1969, Nota sobre el hallazgo de icnitas fosiles de aves en Peninsula Fildes, Isla Rey Jorge, Shetland del Sur, Antártica: Inst. Anart. Chileno Bol. 4, p. 26-28.

——— 1970, Hallazgo de icnitas en Peninsula Fildes, Isla Rey Jorge, Archipielago Shetland del Sur, Antártica: Inst. Anart. Chileno, ser. Cient., v. 1, p. 55-74.

Cribb, H. G. S., McTaggert, N. R., and Staines, H. R. E., 1960, Sediments east of the Great Divide, *in* Hill, D., and Denmead, A. K., eds., The geology of Queensland: Geol. Soc. Australia Jour., v. 7, p. 345-355.

Hendey, Q. B., 1976, The Pliocene fossil occurrences in 'E' Quarry, Langebaanweg, South Africa: South African Mus. Ann., v. 69, p. 215-247.

Jenkins, R. J. F., 1974, A new giant penguin from the Eocene of Australia: Palaeontology, v. 17, p. 291-310.

Matthew, W. D., 1915, Climate and evolution: New York Acad. Sci. Ann., v. 24, p. 174-318.

Pascual, R., and Rivas, O. E. O., 1971, Evolucion de las comunidades de los vertebrados del Terciario Argentino. Los aspectos paleozoogeograficos y paleoclimaticos relacionados: Ameghiniana, v. 7, p. 372-412.

Rich, P. V., 1974, Significance of the Tertiary avifaunas from Africa (with emphasis on a mid to late Miocene avifauna from southern Tunisia): Egypt Geol. Survey Ann., v. 4, p. 167-210.

——— 1975a, Antarctic dispersal routes, wandering continents, and the origin of Australia's non-passeriform avifauna: Natl. Mus. Victoria Mem., v. 36, p. 63-126.

——— 1975b, Changing continental arrangements and the origin of Australia's non-passeriform continental avifauna: Emu, v. 75, p. 97-112.

——— 1976, The Cenozoic history of Australia's non-passeriform birds, *in* Frith, H. J., and Calaby, J. H., eds., The history of birds on the island continent Australia: Internat. Ornithological Congress, 16th, Canberra 1974, Proc. (Australian Acad. Sci.), p. 53-65.

Rich, P. V., and Scarlett, R. J., 1977, Another look at *Megaegotheles*, a large owlet-nightjar from New Zealand: Emu, v. 77, p. 1-8.

Rothe, P., 1964, Fossile Strausseneir auf Lanzarote: Natur. u. Mus., v. 94, p. 175-187.

Scarlett, R. J., 1972, Bone of presumed odontopterygian bird from the Miocene of New Zealand: New Zealand Jour. Geology and Geophysics, v. 15, p. 269-274.

Simpson, G. G., 1957, Australian fossil penguins, with remarks on penguin evolution and distribution: South Australian Mus. Recs., v. 13, p. 51-70.

——— 1959, A new fossil penguin from Australia: Royal Soc. Victoria Proc. v. 71, p. 113-119.

——— 1965, New record of a fossil penguin in Australia: Royal Soc. Victoria Proc., v. 79, p. 91-93.

——— 1970a, Ages of fossil penguins in New Zealand: Science, v. 168, p. 361-362.

——— 1970b, Miocene penguins from Victoria, Australia, and Chubut, Argentina: Natl. Mus. Victoria Mem., v. 31, p. 17-24.

——— 1971a, A review of fossil penguins from Seymour Island: Royal Soc. London Proc., ser. B, Biol. Sci., v. 178, p. 357-387.

——— 1971b, A review of pre-Pliocene penguins of New Zealand: Am. Mus. Nat. History Bull., v. 144, p. 319-378.

——— 1971c, Fossil penguin from the late Cenozoic of South Africa: Science, v. 171, p. 1144-1145.

——— 1972, Conspectus of Patagonian fossil penguins: Am. Mus. Novitates no. 2488, p. 1-37.

Talent, J. A., Duncan, P. M., and Handley, P. L., 1966, Early Cretaceous feathers from Victoria: Emu, v. 66, p. 81-86.

Tedford, R. H., 1974, Marsupials and the new paleogeography, *in* Ross, C. A., ed., Paleogeographic provinces and provinciality: Soc. Econ. Paleontologists and Mineralogists Spec. Pub. 21, p. 109-126.

Waldman, M., 1970, A third specimen of a lower Cretaceous feather from Victoria, Australia: Condor, v. 72, p. 377.

Williams, G. R., 1973, Birds, *in* Williams, G. R., ed., The natural history of New Zealand: Wellington, N.Z., A. H. and A. W. Reed Ltd., p. 304-333.

A Biogeographical Problem Involving Comparisons of Later Eocene Terrestrial Vertebrate Faunas of Western North America

JASON A. LILLEGRAVEN, *Department of Geology, The University of Wyoming, Laramie, Wyoming 82071*

ABSTRACT

Detailed taxonomic studies recently completed on later Eocene land vertebrate faunas from southern California allow comparisons with those of roughly equivalent age in the Rocky Mountain region. Although faunal assemblages of earliest Uintan (middle Eocene to early late Eocene) age show significant commonality at the species level, comparative faunal lists from younger Uintan rocks indicate much higher degrees of endemism. Geological data combined with an Eocene palinspastic reconstruction of the southwestern part of the United States and northwestern part of Mexico suggest that in Bridgerian (middle Eocene) time, faunal exchange between southern California and Wyoming was possible by way of a tropical to subtropical "gangplank of lowlands" that included coastal borderlands, the southern Mojave Desert, southern Nevada and(or) northwestern Arizona, the eastern slope of the erosionally-lowered Sevier orogenic belt, the borders of the northeastern Utah and northwestern Colorado lake system, and the Axial Basin Arch into south-central Wyoming. Climatic change (decrease in tropicality, dilation of temperate realms, and increase in aridity) temporally associated with epeirogenic uplift, increased mountain building, and volcanism in the late Eocene greatly restricted land vertebrate intracontinental dispersal and(or) made disjunct ranges among species that were previously continuous between coastal and western interior regions. Late Eocene climatic "deterioration" observed in western North America was apparently not completely controlled, however, by local changes in topography, but rather was a world-wide event.

INTRODUCTION

Eocene land vertebrate assemblages from North America are abundant and widespread. The record from the West Coast, however, is restricted to: (1) a possible earliest Eocene locality (described as latest Paleocene) from Punta Prieta, southern Baja California, Mexico; (2) a pair of localities, one probably of Bridgerian (roughly middle Eocene) age and the other younger (late Eocene or perhaps early Oligocene), from the Clarno Formation of west-central Oregon; and (3) a large number of middle through late Eocene localities from southern California. The southern California assemblages are of Uintan (latter half of the Eocene) age and currently are restricted to San Diego and Ventura Counties. The purpose of the present paper is to suggest the biogeographical implications of recognized similarities and differences between the faunas from the West Coast and the Rocky Mountains (especially northeastern Utah and western Wyoming) in the latter half of the Eocene as based upon recently completed detailed taxonomic studies.

Detailed locality-by-locality faunal lists of all currently recognized southern California Eocene land vertebrates (Golz and Lillegraven, 1977) are now available. Included in that paper are maps of the locality areas, the localities recognized to be involved in specific local faunas, a complete bibliography of original systematic paleontology of southern California Eocene vertebrates, and a correlation chart that attempts to summarize the temporal equivalencies of the following: (1) land vertebrate-bearing formations; (2) West Coast marine megafossil "stages"; (3) West Coast benthonic foraminiferal stages; (4) calcareous nannoplankton concurrent range zones; (5) correlative formations in the Uinta Basin of Utah; (6) North American land mammal "ages"; (7) approximate radiometric ages; and (8) "standard" European ages. A rather extensive discussion of the stratigraphic relationships between nonmarine vertebrate fossil-bearing rock units and the better known sequence in southern California is also included. Only a skeleton correlation chart is presented here (Fig. 1) that includes those terms necessary to the present discussion.

In most general terms, there are three major southern California outcrop areas known to include land vertebrates (Fig. 2a). These are: (1) the greater region of the city of San Diego (A in Fig. 2a) including the Ardath Shale, Scripps Formation, Friars Formation, Stadium Conglomerate, and Mission Valley Formation; (2) the northwestern corner of San Diego County (B in Fig. 2a) including the Santiago Formation; and (3) the southwestern part of Ventura County (C in Fig. 2a) including the Sespe Formation. The Sespe Formation includes rocks ranging in age from late Eocene through early Miocene; only the Eocene part of the Formation is considered here. Land vertebrate faunas from the San Diego area are known only from the Friars and Mission Valley Formations; only a minor number of specimens have been recovered from the Scripps Formation and Ardath Shale.

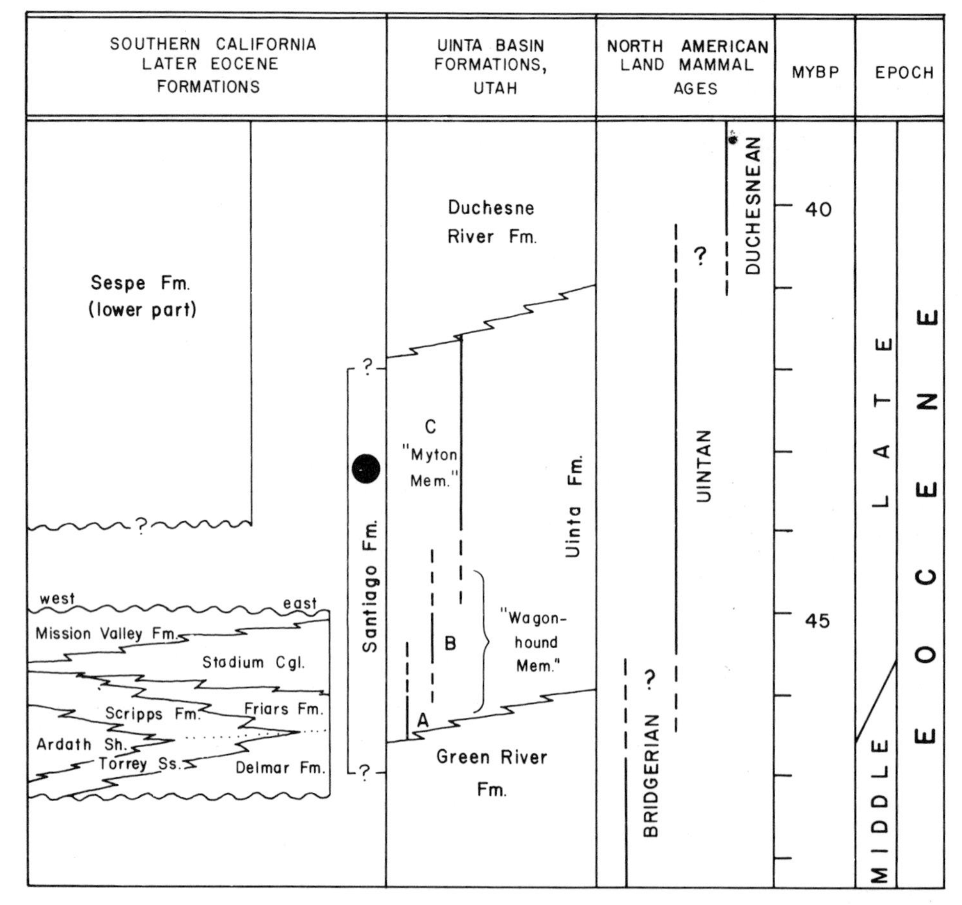

Figure 1. Correlation chart illustrating stratigraphic terminology used in text (simplified and slightly modified from Golz and Lillegraven, 1977). The black circle in the column for the Santiago Formation indicates the most likely temporal position of the land vertebrate faunas known from the unit. The terminology and facies relationships of the rock units from the greater San Diego area are from Kennedy and Moore (1971).

FAUNAL COMPARISONS—SOUTHERN CALIFORNIA VERSUS WESTERN INTERIOR

Tables 1 and 2 (from Golz and Lillegraven, 1977) were prepared from a comparison of the southern California mammalian lists with those of many localities from the North American Western Interior (especially Utah and Wyoming). In compiling the Tables, the southern California taxa (species level for the most part) were considered either markedly different from those of the Western Interior (and thus apparently endemic to the West Coast) or markedly similar with those of the Western Interior (and thus suggested fairly recent continuity of distributional ranges or dispersal). The lists were then subdivided by West Coast formational occurrences. Approximately contemporaneous formational occurrences were grouped in the Tables (e.g., Sespe + Santiago, "General San Diego Sections" = Friars + Mission Valley).

Although assignments of certain species (see Tables 1, 2) may be open to alternate personal interpretations, a basic pattern seems undeniable; the temporally earlier faunas as seen in the greater San Diego area show far less endemism than do the later faunas of the Santiago and Sespe Formations. It seems that long-distance intracontinental dispersal of land vertebrates during the middle part of the Eocene (Bridgerian or possibly into earliest Uintan) was far more common than later in the Eocene (most of the Uintan). The endemism of certain parts of the late Eocene West Coast fauna recognized by Black and Dawson (1966, p. 332) was apparently real for the Sespe and Santiago assemblages, but should not be generalized to include the earlier (San Diego) creatures. The similarity in the earliest Uintan West Coast assemblages with those of the Rockies extends even to physically tiny forms (certain primates, insectivores, and rodents) that might be expected to have been especially sensitive to vegetational and(or) edaphic barriers. As a miscellaneous point, it is important to realize that faunas of latest Bridgerian-earliest Uintan age from the Rockies are sparsely represented in existing collections. Thus, most comparisons of the ?earliest Uintan species from San Diego had to be made with Bridgerian or later Uintan kinds of Utah and Wyoming. The time differences involved, however, seem not to have been great.

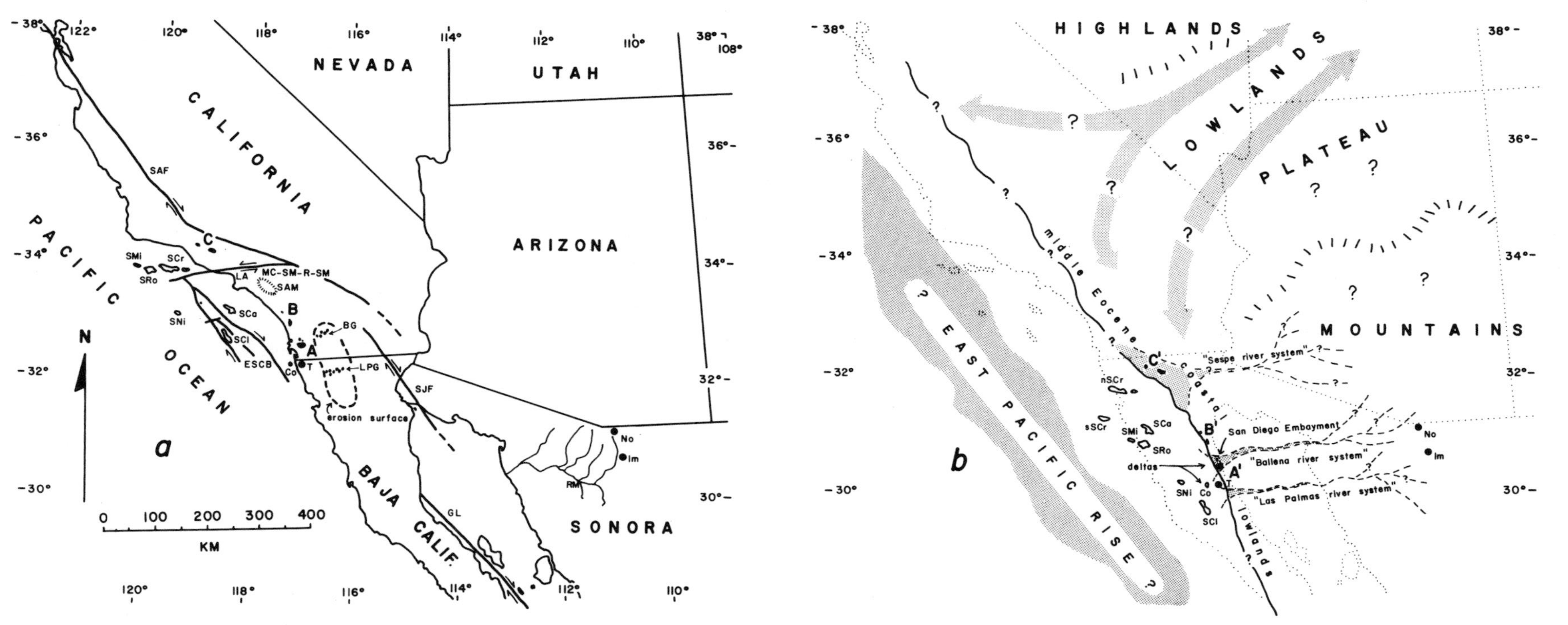

Figure 2. Comparison of present geography of southwestern North America: (a) with a palinspastic-paleogeographic reconstruction; (b) the same area in the middle Eocene at the time of deposition of the Friars or Mission Valley Formations. Figure 2b was constructed by combining information from Minch (1972), Howell and others (1974), Truex (1976), and various other sources discussed in this paper. The three major outcrop areas in southern California known to include land vertebrates (see text) are shown at A, B, and C on Figure 2a and at A′, B′, and C′ on Figure 2b. Although deposition of the Sespe Formation itself had not yet begun in the middle Eocene (the time represented by Figure 2b) abundant lower Eocene clastics in the area suggests the proximity of major fluviatile sources. A "ghost" outline of modern geographic boundaries and present latitudinal coordinates are superimposed on Figure 2b for spatial reference. The large stippled arrows on Figure 2b mark the general mid-Eocene land vertebrate dispersal routes suggested in the text. The "islands" indicated on Figure 2b were submerged to near bathyal depths during the middle Eocene. (BG) Ballena Gravel; (Co) Coronado Island; (ESCB) East Santa Cruz Basin Fault System; (GL) Guaymas Lineament; (Im) Imuris; (LA) Los Angeles; (LPG) Las Palmas Gravel; (MC-SM-R-SM) Malibu Coast-Santa Monica-Raymond-Sierra Madre Fault System; (No) Nogales; (nSCr) northern half of Santa Cruz Is.; (RM) Rio Magdalena drainage; (SAF) San Andreas Fault System; (SAM) Santa Ana Mountains; (SCa) Santa Catalina Is.; (SCI) San Clemente Is.; (SCr) Santa Cruz Is.; (SJF) San Jacinto Fault; (SMI) San Miguel Is.; (SNi) San Nicolas Is.; (SRo) Santa Rosa Is.; (sSCr) southern half of Santa Cruz Is.; (T) Tijuana.

ECOLOGICAL CONSIDERATIONS OF THE LATER EOCENE OF SOUTHERN CALIFORNIA

Strike-Slip Faulting and Paleolatitudes

The existing positions of known land vertebrate-bearing Eocene rocks in southern California are shown in Figure 2a. There is excellent evidence, however, that major post-Eocene strike-slip faunting near the areas has occurred; the construction of an Eocene palinspastic map (Fig. 2b) is required to provide a reasonable Eocene paleogeographic picture of southern California with respect to the remainder of western North America.

Minch (1972) mapped massive cobble deposits (Ballena and Las Palmas Gravels) to the east of San Diego and in northwestern Baja California. He determined that the cobbles represent middle to late Eocene river-bottom deposits formed by streams that flowed across early Tertiary erosion surfaces from east to west toward major marine deltas at San Diego (Stadium Conglomerate and other easterly equivalents within a massive conglomeratic lithosome) and south of Tijuana (Buenos Aires Formation), Mexico. The cobbles have distinctive mineralogical characteristics that allow determination of the source areas of the ancient rivers in the highlands to the east. Minch (1972) shows that the most likely provenience was in ancient volcanic provinces extending from south-central Arizona south to around Imuris (south of Nogales) in Sonora, Mexico. Channel deposits that he considers the eastern equivalents of the Ballena and Las Palmas Gravels are also present southwest of Imuris to the Gulf of California in the Rio Magdalena drainage (see Fig. 2a). The geographical matching of the Ballena-Las Palmas channels with the general highland source area requires a simple southeastward reconstructional movement of about 365 km parallel to the San Andreas-San Jacinto-Guaymas fault lineaments and the closure of the late Tertiary Basin and Range crustal dilation of northwestern Sonora. Thus, in the Eocene, San Diego was probably due west of the present area of Nogales (Fig. 2b) and has subsequently been carried about 365 km to the northwest.

Similarly, Howell and others (1974) recognized that massive cobble deposits on the present San Miguel and San Nicolas Islands and the submarine sandstone deposits on Santa Rosa Island (see Fig. 2a) are of the same mineralogical composition as those of the Stadium Conglomerate and Ballena channel deposits. They suggested that these rocks, now found on islands far to the northwest of the International Border, were adjacent to the greater San Diego area in the Eocene and represent parts of a bathyal alluvial fan deposit. Post-Eocene motion along right-lateral strike-slip faults of an East Santa Cruz Basin fault system carried the slivered fan 120 to 160 km northwest from its original position.

Truex (1976) suggested that part of the Santiago Formation as seen in the Santa Ana Mountains (Fig. 2a, southeast of Los Angeles) is time-correlative with the lower (Eocene) part of the Sespe Formation in the Santa Monica Mountains (northwest of Los Angeles). He furthermore pointed out that the depositional sequences of sedimentary rock in the two areas from the Upper Cretaceous into the Miocene shows great lithological similarities. He believed these similarities to be too great for coincidence and postulated that the two areas were nearly contiguous

Table 1. Mammalian species from Eocene rocks of southern California that were apparently endemic to the North American West Coast and are markedly distinct from those of the Rocky Mountain area.

From general San Diego sections
Cryptolestes vaughni
Batodonoides powayensis
Aethomylos simplicidens
Microsyops kratos
Ischyrotomus californicus
Ischyrotomus littoralis
Leptotomus caryophilus
Reithroparamys californicus
Metarhinus (?) *pater*
Merycobunodon littoralis
Leptoreodon major

From Sespe or Santiago Formations
Simidectes merriami
Craseops sylvestris
Chumashius balchi
Dyseolemur pacificus
Microparamys tricus
Microparamys sp. D
Leptotomus tapensis
Mytonomys burkei
Rapamys fricki
Presbymys lophatus
Eohaplomys serus
Eohaplomys matutinus
Eohaplomys tradux
Griphomys alecer
Griphomys toltecus
Hyaenodon vetus
Hyaenodon exiguus
Plesiomiacis progressus
Miacis (?) *hookwayi*
Tapocyon occidentalis
Amynodontopsis bodei
Tapochoerus egressus
Simimeryx hudsoni
Protoreodon pacificus
Eotylopus sp.
Protylopus pearsonensis
Protylopus stocki
Protylopus? robustus
Poebrodon californicus
Leptoreodon edwardsi
Leptoreodon leptolophus
Leptoreodon pusillus

From San Diego and Sespe or Santiago Formations
Sespedectes singularis
Proterixoides davisi
Namatomys fantasma
Simimys simplex
Amynodon reedi

at the pertinent time of sedimentary deposition. Separation later occurred by nearly 60 km of post-early Miocene left-lateral motion along the Malibu Coast-Santa Monica-Raymond-Sierra Madre fault system.

The determination of the paleolatitudinal position of southern California in the late Eocene is important in interpreting the local climatic conditions than prevalent. Peterson and Abbott (1977) state: "Paleomagnetic latitudes . . . indicate that in Paleocene-Early Eocene time the area [the USA-Mexico International Border] was located farther to the north at about 40°N well up into the westerlies," an interpretation that they based ". . . in part on illustrations from Dott and Batten (1976)" (probably Fig. 17.5). Numerous authors have stressed the importance of widespread right-lateral faulting in the evolution of the western North American Tertiary landscape. Beck (1976) recently reinforced this idea by analyzing paleomagnetic pole positions from the westernmost Cordillera that have been considered "discordant" with respect to pole positions recorded at localities closer to the "stable craton." He noted that in the discordant pole positions a remarkably consistent easterly declination was evident, associated with low inclinations; the pattern of "discordance" was clearly not random. Beck postulated that the most likely reason for these "discordant" pole positions is microcontinental movement involving right-lateral shear of the western Cordillera with respect to the craton. Minor microplate clockwise rotations may also have occurred. Such an interpretation is in basic agreement with the model of intermittent plate interactions proposed by Atwater (1970). Although Beck himself pointed out reservations with the theory, he emphasized (1976, p. 710) the adequacy of plate tectonic evidence for such a series of shear zones and stated: "It would seem . . . that enough evidence pointing to large-scale block movements and rotations in the Cordillera has accumulated to invalidate any regional tectonic synthesis that ignores the issue." The important point to be made here is that plate tectonic theory and geological evidence of major California Cenozoic right-lateral fault zones would suggest that southern California, in relation to mainland Mexico, moved northward during the Cenozoic, not southward. If the Eocene paleolatitude of the International Border south of San Diego, now at nearly 33°N, was 40°N as suggested by Peterson and Abbott (1977), all of North America would had to have moved significantly (10° latitude or more) southward since post-Eocene time. I know of no sound evidence for such an interpretation.

Paleoclimatic Evaluations in the Various Rock Units

Peterson and Abbott (1977) and Abbott and others (1976) have provided strong sedimentological evidence for the development of an ancient and deeply weathered (up to 13 m) lateritic (oxisol) soil zone in western San Diego County and northwestern Baja California. The paleosol is developed within Upper Cretaceous and older rock and underlies all known later Eocene fossiliferous deposits. It is interpreted to have formed during the Paleocene and early Eocene under hot, humid, fully tropical conditions similar to those of present equatorial regions.

Peterson and Abbott (1977) point out that the overlying later Eocene rocks, on the other hand, suggest the existence of quite difference climatic conditions. For example, there is only an immature development of clay minerals (although I have personally noticed that much of the clay making up the body of the northern exposures of the Mission Valley Formation was derived locally from late Eocene topographic highs capped by the above mentioned paleosol). Also, there are occasional cobbles that were broken in place after deposition; this they interpret as salt-weathering, a phenomenon characteristic of dry areas having an influx of soluble salt (as by tidal wash). Some degree of aridity is also indicated by the presence of scattered, generally thin, and unindurated Eocene caliche layers, especially in the northern and southern exposures of the upper measures of the Mission Valley Formation (see Pierce, 1974). My observations suggest that caliches are rare in the Friars Formation and Stadium Conglomerate. The caliches are interpreted as

Table 2. Mammalian species from Eocene deposits of southern California that are extremely similar to or conspecific with those of the Rocky Mountain area.

From general San Diego sections
Centetodon aztecus
cf. *Nyctitherium* sp.
Simidectes sp. cf. *S. medius*
Pelycodus sp. near *P. ralstoni*
Notharctus sp. near *N. robustior*
Omomys carteri
Hemiacodon sp. near *H. gracilis*
Washakius woodringi
Uintanius sp. near *U. ameghini*
Microsyops sp. cf. *M. annectens*
Uintasorex montezumicus
Microparamys sp. cf. *M. minutus*
Sciuravus powayensis
Pareumys sp. near *P. grangeri*
cf. *Harpagolestes* sp.
cf. *Uintatherium* sp.
Dilophodon leotanus
Protoreodon sp. cf. *P. parvus*
Leptoreodon sp. cf. *L. marshi*

From Sespe or Santiago Formations
Apatemys uintensis
Mytonolagus sp.
Ischyrotomus sp. near *I. compressidens*
Pareumys sp. near *P. milleri*
Teleodus californicus
Amynodon sp. cf. *A. intermedius*
Protoreodon pumilus

From San Diego and Sespe or Santiago Formations
Peratherium sp. cf. *P. knighti*
Nanodelphys californicus
Apatemys bellus
Protylopus sp. cf. *P. petersoni*

having developed in aggradational, heavily burrowed soil horizons forming on semiarid floodplains lateral to the river system entering the San Diego Embayment deltas (Peterson and Abbott, 1977). Analysis of the composition of the Ballena Gravels (of the westward-flowing Eocene rivers that drained to the San Diego delta) by Minch (1972) and Peterson and Abbott (1977) suggests the infrequent transport under high-energy conditions of a sedimentary load composed dominantly of cobble-sized particles. Transport was down a low-to-moderate gradient from the source area, 200 to 300 km eastward. The Ballena Gravels are remarkably free of smaller sized clastics. Interpretation practically requires a model of flashflood conditions of major magnitude developed from tropical storms dumping spectacular amounts of rain on the mountains to the east. Through most of the year, however, the river systems must have experienced only minor flow; the majority of the landscape probably showed semiarid conditions.

Perhaps the most reliable independent test of the paleoclimatic picture summarized above would be from paleobotanical evidence. Unfortunately, no megafloral or palynological studies have yet been published for any of the important land vertebrate-bearing formations of the West Coast. Pollen samples from the Mission Valley Formation have been prepared and will eventually be studied by Dr. Norman O. Frederiksen. A few leaf fossils have been found by my field parties in the Friars Formation, but the specimens have not been described. Plant megafossils are extremely rare in the San Diego late Eocene section in general. Two palynological studies have been made from the marine or perimarine parts of the section, but neither has been published. Lowe (1974) prepared an excellent undergraduate thesis on spore and pollen assemblages from the marine Ardath Shale. His analysis (1974, p. 59) suggests: "The coastal plain that drained into the Middle Eocene San Diego Embayment was characterized by a climate similar to that which presently occurs in areas vegetated by monsoonal forests tending toward tropical savanna woodland type plant assemblages, with little or no grass . . . The average annual temperature was probably in the vicinity of 20°C with less than six centigrade degrees fluctuation between the warmest and coolest months. Annual precipitation probably averaged 140 centimeters; falling mostly in the summer with a comparatively dry winter." Elsik and Boyer (in prep.) have studied palynomorphs from the Delmar and Torrey Formations that interdigitate laterally with the Ardath Formation (see Fig. 1). The Delmar and Torrey are middle Eocene in age and represent lagoonal and barrier bar or shoal deposits, respectively. The flora from both also suggests a subtropical to tropical environment.

As yet unpublished studies by Dr. John E. Fitch on at least 60 species representing at least 38 families of marine teleost fishes from the Ardath Shale also suggest subtropical to tropical marine waters. Fitch stated (personnal commun. 1976): "My best estimate of the S.D. Eocene is that it was considerably warmer than in the same area today—possibly equivalent to the area between Mazatlan and Acapulco (temperature-wise)."

Thus, although the sedimentological record of the Friars through Mission Valley section suggests some aridity, the nature of the older (middle Eocene) formations hints at considerable tropicality. Perhaps we are getting a glimpse of the transition in climate postulated by Peterson and others (1975) to have occurred sometime between the early and late Eocene in the southern California area. When palynological studies are completed for the Friars and Mission Valley Formations, the paleoecological picture should brighten considerably.

The Santiago Formation represents a combination of depositional environments including marine, lagoonal-estuarine, and very nearshore nonmarine-fluviatile. As suggested in Figure 1, its time of deposition was probably of considerable duration. I know of no published detailed paleoenvironmental studies of the Formation.

The Sespe Formation was also probably deposited for the most part near sea level, but it is dominantly fluviatile in origin. Although general statements on the nature of its deposition are common, I know of no published detailed paleoclimatological studies of the Sespe. Marine invertebrates in the area that are roughly temporal equivalents suggest, however, water conditions typical of subtropical to tropical areas (e.g., see Givens, 1974).

There is no geological or paleoenvironmental reason to believe that significant barriers to the dispersal of land vertebrates existed between the San Diego County and Ventura County areas during the later Eocene. Paleotopographic features certainly existed (and can be well seen in the greater San Diego area) but heavily vegetated coastal belts undoubtedly allowed broad corridors of dispersal (see Fig. 2b). Thus, we may consider the various southern California faunas as a biogeographical entity and investigate possible dispersal pathways to other areas. For more detailed treatments of paleogeographic relationships in the coastal areas of California during the Eocene, see Clarke and others (1975), Howell (1975), and Wilson and Link (1975).

POSSIBLE EOCENE DISPERSAL ROUTE—SOUTHERN CALIFORNIA-WESTERN INTERIOR

Paleogeography of the Corridor Between Southern California and Central Utah

The "Sevier orogenic belt" (of Armstrong, 1968) extended from southern Nevada and northwestern Arizona northeastward across west-central Utah and into eastern Idaho and western Wyoming (Fig. 3). It represents an area of major sediment accumulation during the late Precambrian through the Triassic that became complexly deformed and uplifted during the time interval from the Jurassic into the Late Cretaceous. It acted as the major source for the spectacular accumulation of sediments deposited along the western margins of the Cretaceous epicontinental seas. The rate of deformation of the arch decreased through the Paleocene. During most of the

Eocene the area was quiescent and its relief was rapidly reduced by erosion (Armstrong, 1968, p. 449). The arch became secondarily complicated by normal faulting as part of the Basin and Range Province during the latter half of the Tertiary.

Van Houten (1961, p. 616) presented a paleogeographic map that portrayed his impression of the western United States during the Paleocene and Eocene. The distribution of nonmarine sediments formed a strip that matches closely the southwestern borders of the Sevier orogenic belt as plotted by Armstrong (1968, p. 436). Van Houten (1956, p. 2819) stated: "In early Cenozoic time, and perhaps principally in the Eocene epoch, lake and swamp deposits consisting largely of calcareous mud and fine-grained sand, together with local floodplain deposits of sand and gravel derived from near-by uplands, accumulated in east-central and southern Nevada and adjacent Utah in a warm temperate to subtropical lowland. Apparently this broad area of aggradation marked the western limit of early Cenozoic accumulation that prevailed throughout the Rocky Mountain province on the east."

Volcanic activity was virtually nonexistent in the western United States south of Idaho during most of Eocene time (see Lipman and others, 1972, p. 221). Andesitic volcanics, however, began to accumulate during the latest Eocene in the general Great Basin area (McKee, 1971, p. 3499) and andesites-dacites plus rhyolites were erupted there in monumental fashion in the early Oligocene (McKee, 1971; Lipman and others, 1972). Tremendously extensive ignimbrites with some individual units involving as much as 500 mi^3 of flatulent ejecta spread across the southern half of Nevada and southwestern Utah (Cook, 1965).

Southern Arizona and northern Sonora would have been mountainous source areas for the major rivers that brought the masses of gravels onto the Eocene coastline of southern California and northern Baja California. The Eocene history of the part of the Colorado Plateau that includes northern Arizona and southeastern Utah is mainly conjectural due to extensive erosion. According to Hunt (1956, p. 63-64), however, it was probably a platform elevated well above the Uinta and San Juan Basins and may have contributed sediments to them. This part of the Plateau was low enough, however, to be receiving debris from more southerly mountains. The Eocene geography of southeastern California probably never will be known fully, as virtually all pre-Oligocene Tertiary sedimentary rocks have been removed by erosion. Neither Dibblee (1967, p. 117) nor Armstrong and Higgins (1973, p. 1097) show any Tertiary igneous activity in the Mojave or Colorado Desert areas prior to the Oligocene. The structure of the Sierra Nevada during the Eocene is also very imperfectly known. It was low enough, however, that sediments derived from highlands well to the east of its present axis were carried by streams westward over the summit of the range axis into western California. Axelrod (1950, p. 227) has also shown that considerable similarities existed between Paleogene floras of the Sierra Nevada and eastward into the State of Nevada. The similarities indicate that the Range could not have been a significant barrier to plant dispersal during that time. The greatest part of the uplift of the Sierra Nevada was surely in the later Tertiary (see Dalrymple, 1964).

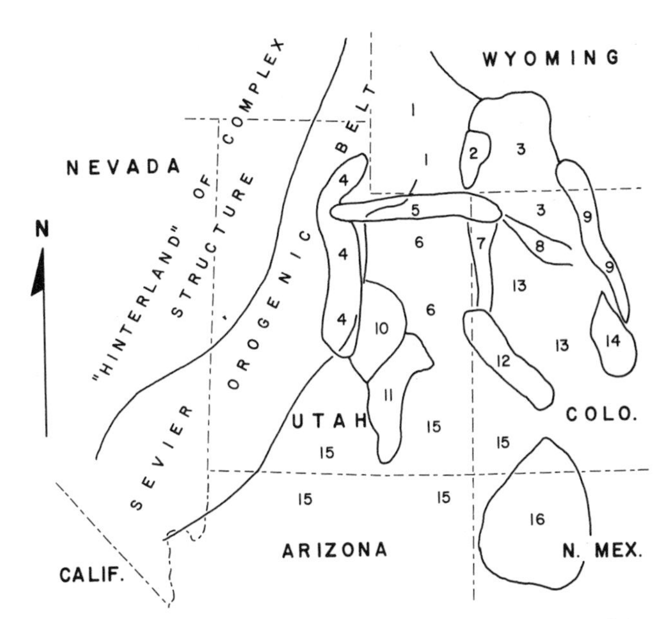

Figure 3. Index map showing major features mentioned in the text known to have been tectonically active during the Eocene. The basic map is simplified from Grose (1972) with the position of the Sevier orogenic belt from Armstrong (1968). (1) Green River Basin; (2) Rock Springs Uplift; (3) Washakie Basin; (4) Wasatch Plateau; (5) Uinta Uplift; (6) Uinta Basin; (7) Douglas Creek Arch; (8) Axial Basin Arch; (9) Sierra Madre-Park Range Uplift; (10) San Rafael Uplift; (11) Henry Mountain Basin; (12) Uncompahgre Uplift; (13) Piceance Basin; (14) Sawatch Uplift; (15) southern Colorado Plateau.

Paleogeography of the Region of Eastern Utah, Western Colorado, and Southwestern Wyoming

The Eocene geography of eastern Utah, western Colorado, and southwestern Wyoming can be interpreted much more precisely than the areas discussed above because of widespread basin-filling during that time. Even though the extensively preserved Eocene sediments of the region provide a rather detailed paleoecological picture, it helps little in explaining the reason for the decrease in similarity in land faunas between the West Coast and Rocky Mountains from the middle to the late Eocene. The reasons must lie outside the depositional areas and the extent of "paleopoetical interpretation" must remain high.

When evaluating the geography of the eastern half of Utah during the Eocene, it is critical to understand the relationships of four important sedimentary rock units. These are, in roughly decreasing age, the Wasatch (early to mid-Eocene), Green River (late early to early late Eocene), Uinta (early late to latest Eocene), and Duchesne River (latest Eocene to ?early Oligocene) Formations (for detailed stratigraphy, see Ryder and others,

1976). Cashion (1967, p. 8) provided a clear exposition of the temporal and environmental relationships of the earliest three of these units: "The Green River Formation can be visualized as a jagged-edged lens of lacustrine strata [of "Lake Uinta"] enveloped in a shell of fluvial strata. The upper part of the formation interfingers with fluvial beds of the Uinta Formation, and the lower part interfingers with fluvial beds of the Wasatch Formation. During part of geologic time the Uinta and Wasatch Formations probably formed a continuous fluvial sequence in the area peripheral to Lake Uinta." Unfortunately, much of this fluviatile border has subsequently been removed by esosion. It is clear, nevertheless, from Cashion's discussion that the Green River Formation is equivalent stratigraphically and temporally to at least the upper part of the Wasatch Formation and to the lower part of the Uinta Formation. The Duchesne River Formation is restricted to the northern end of the Uinta Basin and represents a southerly prograding wedge of clastic sediments being eroded from the rapidly uplifted Uinta Mountains in the latest Eocene (see Picard and Andersen, 1975).

The internal drainage of Lake Uinta was of enormous extent, covering at various times much of eastern Utah east of the ancestral Wasatch Range and south of the ancestral Uinta Mountains. It also extended, at least part of the time, across the Douglas Creek Arch just east of the Utah-Colorado border into the Piceance Basin of northwestern Colorado. The Henry Mountain Basin may have been sporadically inundated. Lake Uinta may also have been briefly continuous with Lake Gosiute in Wyoming by way of a gap across the Axial Basin Arch between the eastern extreme of the Uinta Mountains and the early uplifts of the main Rocky Mountains of north-central Colorado. During its maximum expanse, Lake Gosiute covered much of the southwestern quarter of Wyoming (Green River-Washakie Basins) and a small part of northwesternmost Colorado (see McDonald, 1972 for general discussion). Both Lake Uinta and Lake Gosiute varied greatly in size during different parts of their histories. There were also rather dramatic climatic changes locally during their existence. In no case could the lakes themselves be seriously considered as longstanding barriers to the dispersal of land vertebrates since there was probably plenty of well-vegetated land available around them from "Wasatch-" through "Uinta-time" (see reconstructive model in Ryder and others, 1976, p. 502). Surdam and Wolfbauer (1975) suggest that the shorelines of Lake Gosiute were extremely variable in distribution through time; ephemeral playa lake conditions were postulated. Although Ryder and others (1976, p. 504) also recognize considerable shoreline variation in Lake Uinta, they visualize the lake itself as being a perennial, more permanent feature. Any significant barriers to dispersal of land organisms between southern and northern Utah could only have been climatic.

Extent of Eocene Topographic Relief in the Western States

The extent of topographic relief in and surrounding the Eocene lake areas of Utah, Colorado, and Wyoming can be only indirectly detected by the extent of volcanic activity (see above), by structural clues (folding and faulting) in the erosionally denuded surrounding areas, and by interpretation of major clastic sediment transport directions as preserved within the sedimentary columns of the basins themselves. Superimposed upon these evidences can be the theoretical implications based upon major interactions of lithospheric plates.

Volcanic activity was extremely minor until late in the Eocene in the western United States south of Idaho (Lipman and others, 1972; McKee, 1971). It became important, however, in the late Eocene and spectacular during the Oligocene. The broad band (Sevier orogenic belt) from southern Nevada and northwestern Arizona northward through central Utah was structurally quiescent through most of the Eocene in comparison to its earlier history (Armstrong, 1968). The western part of the Sevier orogenic belt remained a topographically positive area throughout the Eocene and was an erosional source of sediments deposited in the surrounding basins. There is also extensive evidence for topographic highlands in areas to the south (San Rafael Uplift), east (Uncompaghre Uplift, Sawatch Uplift, and Park Ranges), and north (Uinta Mountains) of the Lake Uinta area. Cashion (1967), McDonald (1972), and Ryder and others (1976) provide excellent summaries of sediment source area directions before, during, and after the existence of Lake Uinta.

The Wasatch Formation in Utah is generally lithologically homogeneous throughout its extent as fluviatile mudstones, fine-grained sands, and restricted limestones; it appears that local bordering uplifts were not greatly elevated at the time. The Green River Formation shows indications of extensive deltaic deposits from major rivers interfingering with the lacustrine facies, although fluviatile deposits originally found around the margins of the lakes during maximum expansion have largely been removed by post-Eocene erosion. Thus, it is difficult to evaluate the extent of topographic development around the area during those times; water currents tended to distribute the materials from stream discharge rather evenly throughout the lake system. Nevertheless, major conglomeratic facies are unknown within the Green River Formation of Utah. The Uinta Formation, on the other hand, gives much more information about surrounding paleotopography. The Formation can be basically interpreted as the cycle of deposition during the major and final regression of Lake Uinta; the fluviatile elements progressively converged toward the center of the basin and the lacustrine deposits became more and more restricted. Also, the fluviatile sediments in several places show unequivocal evidence of significant uplift in nearby areas. For example, massive torrentially deposited conglomeratic deposits in the northwestern corner of the basin document a major uplift of the Wasatch and Uinta Mountains. Current-pattern studies also suggest the influx of large amounts of sands and silts from eastern and southern sources including the Park Range, Uncompahgre Uplift, and possibly the San Juan Uplift; an environment including generally greater topographic relief for "Uinta-" than for "Green River-time" is strongly suggested in the Utah-Colorado area. Still greater rejuvenation of uplift of the Uinta Mountains is clearly shown in the northern Uinta Basin through the study of the Duchesne River Formation (see Picard and Andersen, 1975); it shows southward transport of coarse debris flows down high-gradient streams from the Uintas. Pulses of uplift of that mountain range can be recognized from the latest Eocene into the early Oligocene.

According to Love (1960, p. 209), during the time of maximum lake expansion (early middle Eocene) in southwestern Wyoming, the western half of Wyoming subsided, or the eastern half was elevated, or both events occurred simultaneously. As a result, the eastward flowing drainage pattern established across the State in the Paleocene and early Eocene was interrupted and the major drainage was westward into the lakes (see Love and others, 1963, p. 204). This seems to have reversed again, however, during the late Eocene. "Before the beginning of Oligocene deposition major eastward drainage was followed by regional degradation, and a surface of moderately high relief was cut in eastern Wyoming" (Love, 1960, p. 209). Although one might postulate downwarp of the eastern half of Wyoming, there is probably more reason to suspect broadscale epeirogenic uplift in the western part of the State in the late Eocene.

Postulated Favored Route for Faunal Exchange

It seems most likely, from the summary provided above, that land faunal exchange during the Eocene between southern California and Wyoming, (see Fig. 2a) would have been first by way of coastal lowlands, then along a belt of comparative lowlands including what is now the southern Mojave Desert, across southern Nevada and(or) northwestern Arizona, up the eastern slope of the erosionally-lowered Sevier orogenic belt, along the borders of the eastern Utah and Colorado lakes, and finally across the Axial Basin Arch into south-central Wyoming. A multitude of other routes across central Nevada or central Arizona may also have been possible, but it seems clear that those areas were highlands under active erosion and dispersal would likely have been complexly "filtered" through intermontane basins of varying environments. The above-postulated route, on the other hand, seems to have been a comparative "gangplank of lowlands" with barriers to dispersal that were probably more climatic than topographic plus climatic.

CLIMATIC CHANGES, THEIR POSSIBLE CAUSES, AND THE DEVELOPMENT OF BARRIERS TO DISPERSAL

Eden Succumbs to Temperance

MacGinitie (1974, p. 40) stated: "The known early Eocene floras of North America from the Gulf coast to northwestern Washington, indicate a humid climate with annual rainfalls of approximately 50 inches, well distributed through the year." The flora of nearly the entire Eocene included many rain forest plant genera in common between Europe and western North America (Wolfe, 1975). Mammalian similarity between the two continents in the early Eocene (McKenna, 1975) is also pronounced at the generic level. Broad, warm, humid climatic belts made widespread dispersal of holarctic organisms possible during the earlier half of the Eocene. Climatic conditions in the Western Interior of the United States became less tropical (less equable and somewhat cooler) during the last one-third or more of the Eocene. In summarizing the late Eocene floral localities in the Rocky Mountains, Leopold and MacGinitie (1972, p. 162) showed that warm temperate aspects took over from the mid-Eocene maximum tropicality. The number of tropical plant taxa decreased markedly in the area in the late Eocene with a concomitant rise in temperate kinds (Leopold and MacGinitie, 1972, p. 166-167). The Rocky Mountain floras also show markedly increasing proportions of plants that grew in various circumstances of aridity in the late Eocene (Leopold and MacGinitie, 1972, p. 167; Axelrod, 1975, p. 291). The aridity probably had multiple causes, including such factors as mountainous uplift to the west and markedly increased ashfalls that would have caused edaphic aridity for plants attempting to grow in the powdery substrate (Leopold and MacGinitie, 1972, p. 174). It seems clear that the late Eocene climatic change, at least in the southern Rockies and eastern Great Basin area, was temporally associated with increased topographic relief and volcanism. In the words of Robinson (1972, p. 236): "For much of the Rocky Mountain region, then, a major erosional or nondepositional episode would appear to have taken place during the middle and, almost certainly, the Late Eocene . . . Development of very broad regional upwarps would seem to be the only explanation for these features." The vast quantities of Eocene volcanic ash debris reworked and deposited through fluviatile processes in the widespread White River Group of the eastern Rocky Mountain and High Plains regions during the Oligocene speaks forcefully of the intensity of Eocene volcanism and erosion in the highland source areas to the west.

Lithospheric Interactions and Late Eocene Uplift in Western North America

Were the temporal associations of climatic change and uplift purely coincidental, or were they in some way genetically related? One could argue that the latter is closer to the truth, especially when one takes into consideration supposedly time-correlative events in interactions between major plates of the Earth's crust. Current plate tectonic reconstructions of the evolution of western North America show the sea floor of the Farallon plate east of its spreading center (the East Pacific Rise) to have begun the process of subduction at a trench under the continental North American plate before the beginning of the Tertiary. McKee (1971, p. 3500) proposed that it was not until the later Eocene and Oligocene that the Farallon plate reached sufficient depths beneath the North American plate in the region of the Great Basin to initiate adequate melting and the establishment of andesitic to rhyolitic volcanism. Lipman and others (1972) suggested that the Farallon plate was subducted at an unusually shallow angle (ca. 20° relative to the horizontal) and that the subducted plate may have become imbricated.

Oxburgh and Turcotte (1976) proposed a thermal model for subduction zones (see Fig. 4), and showed a surface heat flow at the subduction trench that is somewhat lower than that of the oceanic plate before it begins its descent toward the Earth's mantle. The level of heat flow at the surface of the overriding plate above the Benioff Zone, however, is increased by more than a factor of three above that at the sea floor of the plate being subducted. Closely related to this in their model is the idea that crustal isotherms are markedly closer to the surface above the Benioff Zone than are their counterparts seaward from the subduction zone. For example, they postulated that the 800°C isotherm would be near 90 km in depth in the sea floor plate before subduction while it would be only about 30 km in depth below the overriding plate landward from the trench above the Benioff Zone. Although the detailed calculations have not yet been made for western North America, it seems only reasonable to assume that the largely granitic crust that is being underridden by basaltic oceanic crust (as was the case with the Farallon plate diving beneath the North American plate) would be differentially heated above that experienced seaward of the trench. The thicker column of heated mantle plus crust (see compacted isotherms of Fig. 4) should experience thermal expansion significantly greater than a column located outside the limits of the Benioff Zone. Broad uplift of the Earth's surface would then be expected above much of the expanse of the Benioff Zone. The expected extent of uplift, however, is not yet known. To test such an hypothesis for the late Eocene and Oligocene history of western North America is not an easy task. The timing of the initiation of extensive andesitic-rhyolitic volcanism in the Great Basin and the recognition, on erosional-depositional criteria, of a broad and general upwarp in the Rocky Mountain region does, however, fit with such an hypothesis. Huntoon (1974, p. 110) stated: "During the period between late Cretaceous and middle Eocene time, a period of roughly 40 million years, the Colorado Plateau and adjacent Basin and Range Province experienced regional uplift."

A Diversity of Evidence for Parallel Late Eocene Climatic Events Across the World

A decrease in tropicality and an increase in aridity would have been expected in association with major late Eocene continental upwarps, increased volcanic activity, and increased tectonism as seen in western North America. If one ignores events of the remainder of the world in the late Eocene, a rather "tidy" cause-and-effect picture of uplift, climatic change, and loss of facile "dispersibility" of land organisms can be painted for western North America. The painting gains another dimension of complexity, however, if one *does* consider what was happening elsewhere from the points of view of tectonic changes and climatology; the climatic changes seen in the Western Interior of North America were seemingly being paralleled by similar events around the world.

For example, Wolfe (1972) discussed three Eocene floras from north of the Gulf of Alaska (ca. 60°N latitude) that range in age from late middle (early Ravenian) to late Eocene (late Ravenian). The leaf-bearing beds interdigitate with marine mollusk- and foraminiferal-bearing strata and are thus well-dated. Also, the stratigraphic picture shows that the plants lived near the sea level of the time. A climatic transition from subtropical to temperate is indicated by the change from a "paratropical" (dominated by broad-leaved evergreen foliage with abundant drip tips) to a temperate flora (foliage of deciduous nature with families characteristic of temperate climes today). A similar late Eocene climatic "deterioration" is indicated by changes in the sea-level flora through the Puget Group of Washington State (Wolfe, 1971, p. 50). These two groups of floras are particularly important because ". . . the vegetational changes in either the Gulf of Alaska or main Puget sequence cannot be attributed to: (1) uplift; or (2) changes of successional nature due to volcanism; the Gulf of Alaska sequence lacks volcanic rocks and the main sequence in the type section of the Ravenian of the Puget Group . . . contains only a few intrusive volcanic rocks" (Wolfe, 1971, p. 50). Thus, local tectonic events in western North America cannot be directly applied to the changes in the floras of the northwestern Pacific in the late Eocene; a more general climatic event is being documented.

Savin and others (1975) summarized results of worldwide, low-latitude paleotemperature analyses based upon oxygen-isotope determinations from tests of open ocean benthonic and planktonic foraminifera recovered from cores from the Deep Sea Drilling Project. They too (1975, p. 1499), recognized a ". . . sharp temperature drop in late Eocene time followed by a more gradual lowering of temperature . . ." A particularly interesting point is that curves for surface and bottom temperatures parallel one another rather closely from the Late Cretaceous through the early Miocene. Savin and others (1975, p. 1506) stated: "The sympathetic variation of surface- and bottom-temperature curves suggest that during this time there was a change in capture of solar energy by the Earth as a whole. It is not possible to tell from our data whether this resulted from variations in output of solar energy, variations in the efficiency with which the Earth captured it, or both." Savin and others (1975, p. 1506-1507) cite several other, independent studies in which a late Eocene climatic deterioration is strongly suggested.

Frerichs (1970) noted that paleobathymetric data based upon taxa of microfaunal assemblages and(or) various geological evidences derived from several parts of the world (Trinidad, California, Barbados, Pyrenees region

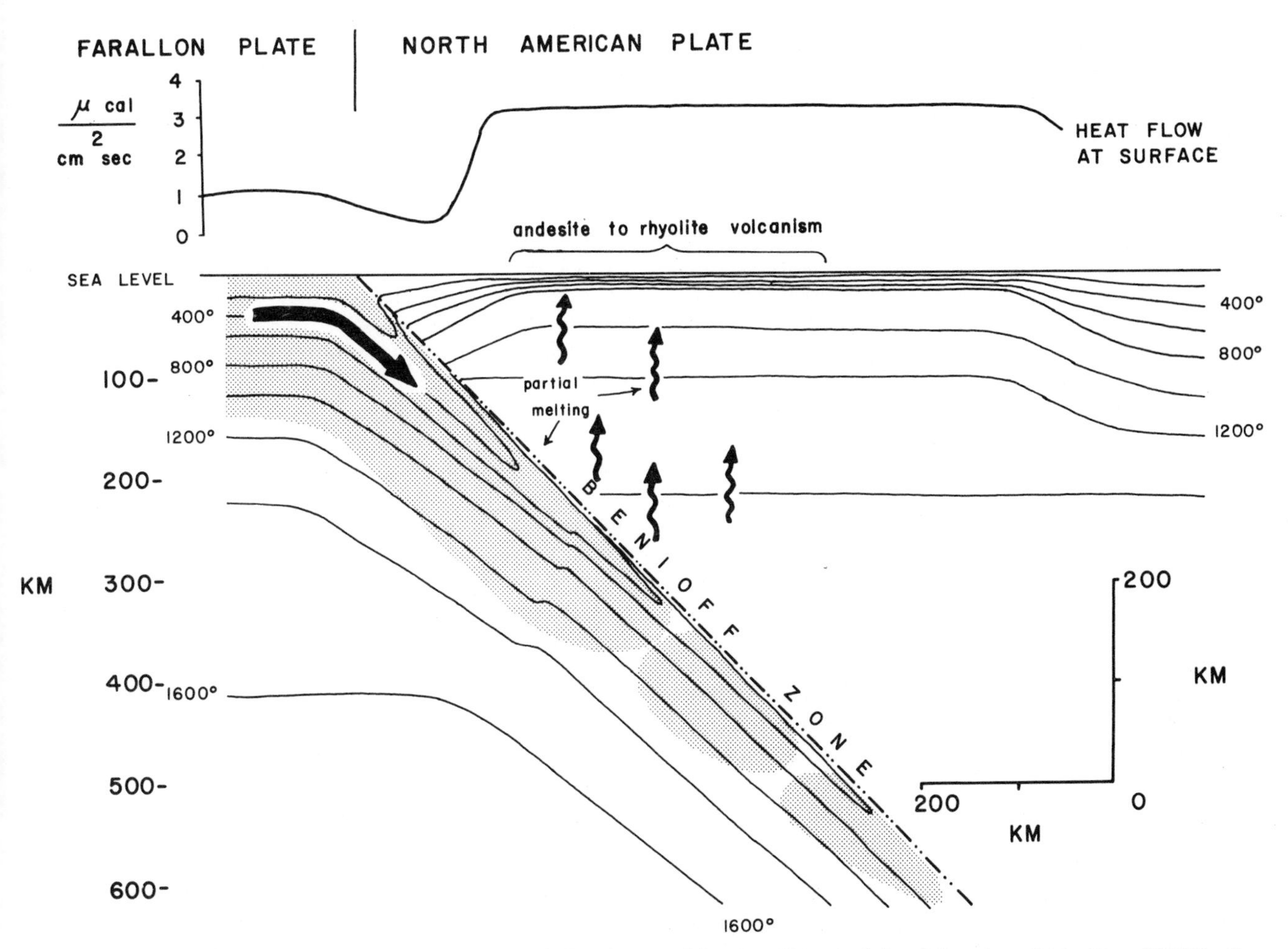

Figure 4. Modification of the thermal model of lithospheric subduction illustrated by Oxburgh and Turcotte (1976). Compaction of isotherms toward the Earth's surface above much of the area of subducted plate should result in a general uplift derived from increased thermal expansion of the crust. The idea is here applied to the hypothesized situation prevailing from the late Eocene through the Oligocene of the region of the North American West Coast, southern Basin and Range Province, and westernmost Rockies. The shaded area represents the crust of the oceanic Farallon plate. The heavy black arrow indicates the direction of subduction at the trench. Lipman and others (1972) suggested a significantly more shallow angle of subduction for the Farallon plate.

of Europe) indicate a general shallowing of marine waters in the late Eocene, culminating in the early Oligocene. He assumed these sample depth changes to be reflections of a major increase in world-wide tectonic activity with widespread uplifts in areas along the periphery of sea-floor spreading ridges (caused by increases in heat flow by way of convection cells from the interior of the Earth); and that the tectonic pulses should be causally related to world-wide climatic modifications. Frerichs and Shive (1971, p. 409) said: "We are suggesting that the correlation between magnetic discontinuities, changes in spreading rates and times of worldwide uplift is more than coincidental—that the same basic driving force, episodic in nature, is responsible for all these effects. At times when this force is active, sea floor spreading rates are accelerated and there is a period of worldwide uplift. When the force is relatively inactive sea floor spreading is either slow or completely stopped and there is a period of worldwide subsidence." Precisely opposite conclusions, however, were derived by Hays and Pitman (1973) who assumed that as sea-floor spreading rates increased, topographic elevation of the mid-oceanic ridges would likewise increase through thermal expansion; sea water would thus be displaced from the oceanic basins onto the continents as marine transgressions. The two theories appear irreconcilable at present. Although they did not mention the possibility, it would seem that the Frerichs and Shive hypothesis may require an episodically expanding diameter of the Earth, an idea strongly favored by Owen (1976) and much earlier by Bucher (1933).

Frerichs (1971) further noted that planktonic foraminifera, whose main controlling factor of dispersal is temperature, showed large-scale extinctions of genera during the late Eocene and early Oligocene following a major taxonomic radiation from the late Paleocene through the mid-Eocene. Extinction also occurred concurrently in coiled planktonic foraminifera in tropical areas (Tappan, 1971, p. 1097). Such a concept of widespread and general late Eocene extinction is in agreement with the completely independent work of Adams (1973) who noted a similar trend among most kinds of large foraminifera. He stated (p. 459): "By the end of the Upper Eocene a drastic (but not instantaneous) reduction in larger foraminiferal faunas has taken place over the whole world. Indeed, with the exception of a few species of *Nummulities, Operculina* and some peneroplids, most of the Palaeogene forms had become extinct, those remaining having enormous opportunities for expansion. Shelf habitats occupied by Eocene larger Foraminifera were largely empty, and the stage was set for a change of fauna." Also agreeing with Frerichs' (1970) and Frerichs' and Shive's (1971) idea of extensive late Eocene tectonic uplifts, Adams (1973, p. 459) stated: "There are few places where fossiliferous Upper Eocene beds are followed directly and conformably by fossiliferous Lower Oligocene sediments, and where they occur together there is usually a marked facies and faunal change at the boundary."

Frerichs (1971) noted that several morphological features among planktonic foraminifera restricted today to warm waters have developed repeatedly in the geologic past irrespective of taxonomic lines; they seem to have been somehow physiologically temperature related. Two such features include the origin of a keel on the test and the presence of accessory apertures in warm-temperate to tropical kinds; there are several other characters as well. Frerichs found that these features were developed in widespread groups during the Paleocene to mid-Eocene but that they largely disappeared during the late Eocene and early Oligocene. Similar patterns of Eocene diversification and Oligocene reduction in diversity were noted by Lipps (1970) for silicoflagellates and by Tappan (1967) for dinoflagellates and coccolithophorids. Prins (1971) noted that a wide variety of calcareous discoasters also became extinct at the end of the Eocene.

Jenkins (1974) postulated a northward penetration in the late Eocene and early Oligocene of cold-adapted penguins from the Antarctic regions to about a 45°S paleolatitude. A late Eocene cool interval is documented by land floras in eastern Asia (Wolfe, 1975, p. 270). Lillegraven (1972) showed that, at the familial taxonomic level, the late Eocene and early Oligocene was the time of the greatest modernization ever of the world-wide land mammal fauna; archaic kinds generally adapted to warmer climates gave way to modern varieties better able to cope with the temperate world of the late Cenozoic.

Refrigeration Began in the Late Eocene

Frakes and Kemp (1972) attempted to summarize the effects of continental positions (in terms of world-wide heat transfer via oceanic currents) on Eocene climates relative to Oligocene climates. They were working under the hypothesis that the widespread tropicality characteristic of the earlier Eocene continued until the Oligocene throughout most of the world. They stated (1972, p. 99): "The very warm and wet conditions which apparently characterized latitudes beyond 45° in the late Eocene were altered during the first half of the Oligocene to conditions more like the present." One point argued in the present paper is that widespread tendencies toward more temperate conditions developed during the late Eocene, perhaps 5 to 7 m. y. before the advent of the Oligocene.

Frakes and Kemp (1972) validly pointed out the importance of the development of the Circum-Antarctic Current to world climates. As stated by Kennett and others (1974, p. 144): "The Circum-Antarctic Current is of great oceanographic and climatic importance because it transports more than $200 \times 10^6 m^3$ of water per second, probably the largest volume transport of any ocean current. It also circulates completely around Antarctica, mixing waters of all oceans." Deep waters from Antarctic sources have profound cooling effects upon the climates of Northern Hemisphere oceanic basins today. The initial development of the current depended upon Australia moving far enough north away from Antarctica so that the South Tasman Rise would clear the northernmost part of Victoria Land. Analysis of cores from the Deep Sea Drilling Project suggests that at least shallow open ocean developed in this area in the late Eocene. Disruption of shallower bottom sediments by submarine current erosion did not occur, however, until the latest Eocene and erosion of the deeper basins did not begin until well into the Oligocene (Kennett and others, 1974; Pimm and Sclater, 1974). Thus, although mountain glaciers seem to have existed in Antarctica during the late Eocene (which probably had cooling effects upon the circum-Antarctic waters), the development of the deep, cold Circum-Antarctic Current characteristic of today appears to have developed too late in Earth history to be useful in explaining the extensive decrease in tropicality of the late Eocene of the world.

SUMMATION

Although considerable disagreement exists with respect to mechanisms of major tectonic-paleoclimatic events and the causes are as yet unknown, it seems nearly irrefutable that the late Eocene and early Oligocene was represented by a world-wide pulse of increased continentality, oceanic cooling, and a significant compression of tropical zones with dilation of temperate conditions. The time was marked by increased rates of extinctions and faunal replacements in many groups of organisms beyond that evident in the first two-thirds of the Eocene. Continental aridity increased in interior regions of North America and general world-wide climatic equability decreased. Even in the absence of sharp topographic barriers (i.e., mountain ranges, extensive canyonlands, and so on), one might have predicted *a priori* that climatic differentiation within a large continent during the late Eocene would have presented significant barriers (vegetational and edaphic) to the dispersal of land vertebrates from coastal to interior regions, beyond those present in the earlier Eocene.

This, indeed, seems to have been the case when one compares the faunal similarities from the various stratigraphic levels of the later Eocene of southern California with approximately contemporaneous faunas of northeastern Utah and Wyoming. Similarity is high between the Friars-Mission Valley faunas (earliest Uintan) and those of the Western Interior; extensive exchange apparently occurred through the earlier Eocene, at least into Bridgerian time. Similarity is low, however, between the Santiago-Sespe faunas (later Uintan and Duchesnean) and those of the Western Interior; post-Bridgerian exchange seems to have been markedly limited. The similarity between the mammalian faunas of the Friars-Mission Valley and Western Interior localities extends even to the very small species (e.g., certain rodents, insectivores, and primates) that might be expected to be more strictly controlled by edaphic barriers (Lillegraven, 1976, 1977, in prep.). Contrariwise, even the larger mammals (e.g., artiodactyls) are rather distinct between the Santiago-Sespe faunas and those of the Western Interior; artiodactyls are notably good cross-country travellers today and their dissimilarity between the two areas comes as a mild surprise. There is no doubt as to the phylogenetic affinities (at the generic level) between the Santiago-Sespe and Uinta Formation artiodactyls but the species distinctions are generally clear (see Golz, 1976), indicating that considerable time for the independent evolution of characters within geographically isolated groups had occurred. Likewise, members of the so-called Candelaria local fauna of late Uintan age from southwest Texas are taxonomically quite distinct from those of the Santiago-Sespe assemblages (see Wilson, 1971; Wilson and others, 1968; and Wood, 1974).

IS EVERYTHING SOLVED?

Many biogeographical mysteries remain concerning North American late Eocene land vertebrate faunas. For example, why have no fossil horses yet been recovered from southern California Eocene deposits? They abound in sediments of approximately correlative age in the Western Interior and have even been found in rocks of supposedly late Paleocene (or possibly earliest Eocene) age in the southern part of Baja California (Morris, 1968) and later Eocene of Oregon (West and others, in prep.). Another key problem not mentioned heretofore is the biogeographical relationship between Eocene faunas of southern California and the "Whistler Squat local fauna" of the Pruett Formation recently discovered in rocks of early Uintan age in west Texas. Unfortunately, not enough has been published on the Texan fauna to allow meaningful comparisons. The collection of more specimens is, however, eagerly awaited.

EPILOGUE

The above discourse has fluttered far and wide across subdisciplines of Earth Science; judgments of specialists in fields far outside those of my own competence have been variously accepted or challenged with abandon. Such a procedure, of course, is fraught with factual and conceptual pitfalls. It would have been much safer to remain entirely within the boundaries of information discovered by vertebrate paleontologists. That, however, is not fun nor is it particularly challenging intellectually. Drawing from the personal philosophy eloquently discussed by Dr. George Gaylord Simpson (1976), "compleat palaeontology" depends upon the analysis of information from a tremendous diversity of fields. Northing short of cosmology can be more eclectic than the study of paleobiogeography. Attempts at paleobiogeographic reconstructions, no matter how primitive or premature, are what in my opinion make systematic paleontology tolerable and, indeed, even worth doing.

ACKNOWLEDGMENTS

As yet unpublished information was generously provided for use in this review by the following individuals: Dr. John E. Fitch, State of California Department of Fish and Game, Dr. William C. Elsik, EXXON Company, and Drs. Patrick L. Abbott and Gary L. Peterson, Department of Geological Sciences, San Diego State University. Drs. Donald L. Blackstone, Jr., Edward R. Decker, and William E. Frerichs, and Mr. Jeffrey G. Eaton, Department of Geology, The University of Wyoming, variously provided technical assistance, valuable discussions, and critical review of the manuscript. My wife Bernice Ann Lillegraven, as usual, helped in a multitude of ways. The basic research on the San Diego land vertebrate assemblages was supported by a National Science Foundation Research Grant (BMS75-15285) administered through the Department of Zoology of San Diego State University and the Museum of Paleontology, University of California, Berkeley.

REFERENCES

Abbott, P. L., Minch, J. A., and Peterson, G. L., 1976, Pre-Eocene paleosol south of Tijuana, Baja California, Mexico: Jour. Sed. Petrology, v. 46, p. 355-361.

Adams, C. G., 1973, Some Tertiary Foraminifera, *in* Hallam, A., ed., Atlas of palaeobiogeography: Amsterdam, Elsevier, p. 453-471.

Armstrong, R. L., 1968, Sevier orogenic belt in Nevada and Utah: Geol. Soc. America Bull., v. 79, p. 429-458.

Armstrong, R. L., and Higgins, R. E., 1973, K-Ar dating of the beginning of Tertiary volcanism in the Mojave Desert, California: Geol. Soc. America Bull., v. 84, p. 1095-1100.

Atwater, T., 1970, Implications of plate tectonics for the Cenozoic tectonic evolution of western North America: Geol. Soc. America Bull., v. 81, p. 3513-3535.

Axelrod, D. I., 1950, Evolution of desert vegetation in western North America: Carnegie Inst. Washington Pub. 590, p. 215-306.

——— 1975, Evolution and biogeography of Madrean-Tethyan sclerophyll vegetation: Ann. Missouri Botanical Garden, v. 62, p. 280-334.

Beck, M. E., Jr., 1976, Discordant paleomagnetic pole positions as evidence of regional shear in the western Cordillera of North America: Am. Jour. Sci., v. 276, p. 694-712.

Black, C. C., and Dawson, M. R., 1966, A review of late Eocene mammalian faunas from North America: Am. Jour. Sci., v. 264, p. 321-349.

Bucher, W. H., 1933, The deformation of the Earth's crust; an inductive approach to the problems of diastrophism: Princeton, Princeton Univ. Press, 518 p.

Cashion, W. B., 1967, Geology and fuel resources of the Green River Formation, southeastern Uinta Basin, Utah and Colorado: U. S. Geol. Survey Prof. Paper 548, 48 p.

Clarke, S. H., Jr., Howell, D. C., and Nilson, T. H., 1975, Paleogene geography of California, *in* Weaver, D. W., Hornaday, G. R., and Tipton, A., eds., Paleogene symposium and selected technical papers: Am. Assoc. Petroleum Geologists-Soc. Econ. Paleontologists and Mineralogists-Soc. Econ. Geologists (Pacific Sections), p. 121-154.

Cook, E. F., 1965, Stratigraphy of Tertiary volcanic rocks in eastern Nevada: Nevada Bur. Mines Rept. 11, 61 p.

Dalrymple, G. B., 1964, Cenozoic chronology of the Sierra Nevada, California: California Univ. Pubs. Geol. Sci., v. 47, p. 1-41.

Dibblee, T. W., Jr., 1967, Areal geology of the western Mojave Desert, California: U. S. Geol. Survey Prof. Paper 522, 153 p.

Dott, R. H., Jr., and Batten, R. L., 1976. The evolution of the Earth (2nd edition): New York, McGraw-Hill, 504 p.

Elsik, W. C., and Boyer, J. E., in prep., Palynomorphs from the Middle Eocene of coastal southern California, U.S.A.

Frakes, L. A., and Kemp, E. M., 1972, Influence of continental positions on early Tertiary climates: Nature, v. 240, p. 97-100.

Frerichs, W. E., 1970, Paleobathymetry, paleotemperature, and tectonism: Geol. Soc. America Bull., v. 81, p. 3445-3452.

——— 1971, Evolution of planktonic Foraminifera and paleotemperatures: Jour. Paleontology, v. 45, p. 963-968.

Frerichs, W. E., and Shive, P. N., 1971, Tectonic implications of variations in sea floor spreading rates: Earth and Planetary Sci. Letters, v. 12, p. 406-410.

Givens, C. R., 1974, Eocene molluscan biostratigraphy of the Pine Mountain area, Ventura County, California: California Univ. Pubs. Geol. Sci., v. 109, p. 1-107.

Golz, D. J., 1976, Eocene Artiodactyla of southern California: Los Angeles County, Nat. History Mus. Sci. Bull., no. 26, 85 p.

Golz, D. J., and Lillegraven, J. A., 1977, Summary of known occurrences of terrestrial vertebrates from Eocene strata of southern California: Wyoming Univ. Contr. Geol., v. 15, p. 43-65.

Grose, L. T., 1972, Tectonics, *in* Mallory, W. M. and others, eds., Geologic atlas of the Rocky Mountain region: Denver, Rocky Mtn. Assoc. Geologists, p. 35-44.

Hays, J. D., and Pitman, W. C., III, 1973, Lithospheric plate motion, sea level changes and climatic and ecologic consequences: Nature, v. 246, p. 18-22.

Howell, D. G., 1975, Middle Eocene paleogeography of southern California, *in* Weaver, D. W., Hornaday, G. R., and Tipton, A., eds., Paleogene symposium and selected technical papers: Am. Assoc. Petroleum Geologists-Soc. Econ. Paleontologists and Mineralogists-Soc. Econ. Geologists (Pacific Sections), p. 272-293.

Howell, D. G., Stuart, C. J., Platt, J. P., and Hill, D. J., 1974, Possible strike-slip faulting in the southern California borderland: Geology, v. 2, p. 93-98.

Hunt, C. B., 1956, Cenozoic geology of the Colorado Plateau: U.S. Geol. Survey Prof. Paper 279, 99 p.

Huntoon, P. W., 1974, The post-Paleozoic structural geology of the eastern Grand Canyon, Arizona, *in* Breed, W. J., and Roat, E. C., eds., Geology of the Grand Canyon: Flagstaff, Mus. Northern Arizona, p. 82-115.

Jenkins, R. J. F., 1974, A new giant penguin from the Eocene of Australia: Palaeontology, v. 17, p. 291-310.

Kennedy, M. P., and Moore, G. W., 1971, Stratigraphic relations of Upper Cretaceous and Eocene formations, San Diego coastal area, California: Am. Assoc. Petroleum Geologists Bull., v. 55, p. 709-722.

Kennett, J. P. and others, 1974, Development of the Circum-Antarctic Current: Science, v. 186, p. 144-147.

Leopold, E. B., and MacGinitie, H. D., 1972, Development and affinities of Tertiary floras in the Rocky Mountains, *in* Graham, A., ed., Floristics and paleofloristics of Asia and eastern North America: Amsterdam, Elsevier, p. 147-200.

Lillegraven, J. A., 1972, Ordinal and familial diversity of Cenozoic mammals: Taxon, v. 21, p. 261-274.

——— 1976, Didelphids (Marsupialia) and *Uintasorex* (?Primates) from later Eocene sediments of San Diego County, California: San Diego Soc. Nat. History Trans., v. 18, p. 85-112.

——— 1977, Small rodents (Mammalia) from Eocene deposits of San Diego County, California: Am. Mus. Nat. History Bull., v. 158, p. 221-262.

——— in prep., Eocene primates from San Diego County, California.

Lipman, P. W., Prostka, H. J., and Christiansen, R. L., 1972, Cenozoic volcanism and plate-tectonic evolution of the western United States. I. Early and middle Cenozoic: Royal Soc. London Philos. Trans., ser. A, Math. and Phys. Sci., v. 271, p. 217-248.

Lipps, J. H., 1970, Ecology and evolution of silicoflagellates, *in* Ultra microplankton: North American Paleont. Conv., 1st, Chicago 1969, Proc., Pt. G, p. 965-993.

Love, J. D., 1960, Cenozoic sedimentation and crustal movement in Wyoming: Am. Jour. Sci., v. 258-A, p. 204-214.

Love, J. D., McGrew, P. O., and Thomas, H. D., 1963, Relationship of latest Cretaceous and Tertiary deposition and deformation to oil and gas in Wyoming, *in* Childs, O. E., and Beebe, B. W., eds., Backbone of the Americas: Am. Assoc. Petroleum Geologists Mem. 2, p. 196-208.

Lowe, G. D., 1974, Terrestrial paleoenvironmental implications of the Ardath flora (Ardath Shale, La Jolla Group), San Diego, California: senior thesis, San Diego State Univ. San Diego, 90 p.

MacGinitie, H. D. (with chapters by Leopold, E. B., and Rohrer, W. L.), 1974, An early middle Eocene flora from the Yellowstone-Absaroka volcanic province, northwestern Wind River basin, Wyoming: California Univ. Pubs. Geol. Sci., v. 108, p. 1-103.

McDonald, R. E., 1972, Eocene and Paleocene rocks of the southern and central basins, *in* Mallory, W. M. and others, eds., Geologic atlas of the Rocky Mountain region: Denver, Rocky Mtn. Assoc. Geologists, p. 243-256.

McKee, E. H., 1971, Tertiary igneous chronology of the Great Basin of western United States—implications for tectonic models: Geol. Soc. America Bull., v. 82, p. 3497-3502.

McKenna, M. C., 1975, Fossil mammals and early Eocene North Atlantic land continuity: Ann. Missouri Botanical Garden, v. 62, p. 335-353.

Minch, J. A., 1972, The late Mesozoic-early Tertiary framework of continental sedimentation, northern Peninsular Ranges, Baja California, Mexico: Ph.D. thesis, Univ. California, Riverside, 192 p.

Morris, W. J., 1968, A new early Tertiary perissodactyl, *Hyracotherium seekinsi,* from Baja California: Los Angeles County Mus. Cont. Sci., no. 151, 11 p.

Nilson, T. H., and Link, M. H., 1975, Stratigraphy, sedimentology and offset along the San Andreas fault of Eocene to lower Miocene strata of the northern Santa Lucia Range and the San Emigdio Mountains, Coast Ranges, central California, *in* Weaver, D. W., Hornaday, G. R., and Tipton, A., eds., Paleogene symposium and selected technical papers: Am. Assoc. Petroleum Geologists-Soc. Econ. Paleontologists and Mineralogists-Soc. Econ. Geologists (Pacific Sections), p. 367-400.

Oxburgh, E. R., and Turcotte, D. L., 1976, The physico-chemical behaviour of the descending lithosphere: Tectonophysics, v. 32, p. 107-128.

Owen, H. G., 1976, Continental displacement and expansion of the Earth during the Mesozoic and Cenozoic: Royal Soc. London Phil. Trans., ser. A, Math. and Phys. Sci., v. 281, p. 223-291.

Peterson, G. L., and Abbott, P. L., 1977, Sedimentological indications of an Eocene climatic change, southwestern California and northwestern Baja California, *in* Fairbridge, R., ed., Paleoclimatic indicators in sediments: Soc. Econ. Paleontologists and Mineralogists Spec. Pub. (in press).

Peterson, G. L., Pierce, S. E., and Abbott, P. L., 1975, Paleogene paleosols and paleoclimates, southwestern California, *in* Weaver, D. W., Hornaday, G. R., and Tipton, A., eds., Paleogene symposium and selected technical papers: Am. Assoc. Petroleum Geologists-Soc. Econ. Paleontologists and Mineralogists-Soc. Econ. Geologists (Pacific Sections), p. 401-408.

Picard, M. D., and Andersen, D. W., 1975, Paleocurrent analysis and orientation of sandstone bodies in the Duchesne River Formation (Eocene-Oligocene?), northern Uinta Basin, northeastern Utah: Utah Geol., v. 2, p. 1-15.

Pierce, S. E., 1974, Provenance and paleoclimatology of the Mission Valley Formation, San Diego County, California: M.S. thesis, San Diego State Univ., San Diego, 85 p.

Pimm, A. C., and Sclater, J. G., 1974, Early Tertiary hiatuses in the northeastern Indian Ocean: Nature, v. 252, p. 362-365.

Prins, B., 1971, Speculations on relations, evolution, and stratigraphic distribution of discoasters: Planktonic Conf., 2nd, Roma 1970, Proc., p. 1017-1037.

Robinson, P., 1972, Tertiary history, *in* Mallory, W. M. and others, eds., Geologic atlas of the Rocky Mountain region: Denver, Rocky Mtn. Assoc. Geologists, p. 233-242.

Ryder, R. T., Fouch, T. D., and Elison, J. H., 1976, Early Tertiary sedimentation in the western Uinta Basin, Utah: Geol. Soc. America Bull., v. 87, p. 496-512.

Savin, S. M., Douglas, R. G., and Stehli, F. G., 1975, Tertiary marine paleotemperatures: Geol. Soc. America Bull., v. 86, p. 1499-1510.

Simpson, G. G., 1976, The compleat palaeontologist?: Ann. Rev. Earth and Planetary Sci., v. 4, p. 1-13.

Surdam, R. C., and Wolfbauer, C. A., 1975, Green River Formation, Wyoming: A playa-lake complex: Geol. Soc. America Bull., v. 86, p. 335-345.

Tappan, H., 1967, Primary production, isotopes, extinctions and the atmosphere: Palaeogeography, Palaeoclimatology, Palaeoecology, v. 4, p. 187-210.

——— 1971, Microplankton, ecological succession and evolution, *in* Evolution of higher categories: North American Paleont. Conv., 1st, Chicago 1969, Proc., Pt. H, p. 1058-1103.

Truex, J. N., 1976, Santa Monica and Santa Ana Mountains—relation to Oligocene Santa Barbara Basin: Am. Assoc. Petroleum Geologists Bull., v. 60, p. 65-86.

Van Houten, F. B., 1956, Reconnaissance of Cenozoic sedimentary rocks of Nevada: Am. Assoc. Petroleum Geologists Bull., v. 40, p. 2801-2825.

——— 1961, Maps of Cenozoic depositional provinces, western United States: Am. Jour. Sci., v. 259, p. 612-621.

West, R. M. and others, in prep., Eocene chronology of North America, *in* Woodburne, M. O., ed., Vertebrate paleontology as a discipline in geochronology.

Wilson, J. A., 1971, Early Tertiary vertebrate faunas, Vieja Group, Trans-Pecos Texas: Agriochoeridae and Merycoidodontidae: Texas Memorial Mus. Bull., v. 18, 83 p.

Wilson, J. A., Twiss, P. C., DeFord, R. K., and Clabaugh, S. E., 1968, Stratigraphic succession, potassium-argon dates, and vertebrate faunas, Vieja Group, Rim Rock County, Trans-Pecos Texas: Am. Jour. Sci., v. 266, p. 590-604.

Wolfe, J. A., 1971, Tertiary climatic fluctuations and methods of analysis of Tertiary floras: Palaeogeography, Palaeoclimatology, Palaeoecology, v. 9, p. 27-57.

——— 1972, An interpretation of Alaskan Tertiary floras, *in* Graham, A., ed., Floristics and paleofloristics of Asia and eastern North America: Amsterdam, Elsevier, p. 201-233.

——— 1975, Some aspects of plant geography of the Northern Hemisphere during the late Cretaceous and Tertiary: Ann. Missouri Botanical Garden, v. 62, p. 264-279.

Wood, A. E., 1974, Early Tertiary vertebrate faunas, Vieja Group, Trans-Pecos Texas: Rodentia: Texas Memorial Mus. Bull. 21, 112 p.

Biogeographic Significance of the Late Mesozoic and Early Tertiary Molluscan Faunas of Seymour Island (Antarctic Peninsula) to the Final Breakup of Gondwanaland

WILLIAM J. ZINSMEISTER, *Institute of Polar Studies, The Ohio State University, Columbus, Ohio 43210*

ABSTRACT

The last fragmentation of Gondwanaland occurred during the Late Cretaceous and early Paleogene. During the initial part of this phase, the various continents (Antarctica, Australia, New Zealand, and South America) were separated by shallow seas. The similarities of the Late Cretaceous and early Paleogene molluscan faunas indicate that the region represented a single, broad, continuous faunal province. This shallow-water marine province which extended from West Antarctica into both southern South America and eastern Australia is here designated the Weddellian Province.

As the separation between the various southern continents increased, the amount of gene flow within the Weddellian Province decreased and eventually ceased. The gradual isolation of the faunas of each continent should be shown by progressive increase in local faunal provincialism in the Southern Hemisphere during the Paleogene. The Eocene molluscan fauna of Seymour Island on the northeastern side of the Antarctic Peninsula provides new data indicating that the increased faunal provincialism in the Southern Hemisphere during the Tertiary coincides with the final breakup and separation of Gondwanaland.

RESUMEN

La última fragmentación de Gondwanaland ocurrió durante El Cretácicio tardío y El Paleógeno temprano. Durante el principio de esta fase los varios continentes (Antártica, Australia, Nueva Zelanda, y la América del sur) fueron separados por mares vadosos. Las semejanzas de faunas molusco del Cretácicio tardío y El Paleógeno temprano indican que este región representó una provincia marina de mares vadosos que se extiende desde La Antártica del oeste penetrando la parte sur de Sudamérica y La Australia del este aquí se está designada como La Provincia Weddellian.

Mientras que la separacion entre los varios continentes del sur se aumento, la cantidad de comunicación faunistica dentro de la provincia se disminuó y por fin ceso. El aislamiento de la fauna de cada continente debe ser representado por el aumento progresivo en el provincialismo local de fauna en el hemisferio del sur durante El Paleógeno. La fauna molusco del Eoceno de la Isla de Seymour en el lado del nordeste de La Península Antártica provee nueva información indicando que el provincialismo aumentado de fauna es de Gondwanaland.

INTRODUCTION

During the 1974-1975 austral summer, a joint geological field party from the Institute of Polar Studies, The Ohio State University, and the Instituto Antártico Argentino visited Seymour Island (lat 64° 15′ S, long 56° 45′ W) on the northeast side of the Antarctic Peninsula in the Weddell Sea (Fig. 1). During the course of the season, a large collection of Cretaceous and Tertiary fossils was made. This material represents the first significant collection of fossils from that part of the Antarctic Peninsula since the Swedish South Polar Expedition of 1901-1903.

The Tertiary "Seymour Island Series," first described in detail by Andersson (1906), crops out on the northern end of the Island and forms a prominent flat-topped meseta (Fig. 2). The sequence consists of approximately 500 m of poorly consolidated sands and silts, with interbedded pebbly, fossiliferous sandstones. The sequence may be divided into two lithologic units. The lower unit (Fig. 2, section BB′) consists of a minimum of 105 m of immature, limonite-stained sandstones with locally abundant fossil wood and coalified plant debris; it occurs in fault splinters at the north tip of the Island at Cape Wiman and in Cross Valley (Elliot and others, 1975). The upper 350 m of the "Seymour Island Series," which crops out around the meseta (Fig. 2, section AA′), consists of loosely consolidated sand, silt, and clayey beds with local crossbedding, oscillation ripple marks, and both small and large scale cut-and-fill channels. The most striking feature of the sequence exposed around the meseta is the pebbly, fossiliferous shell banks which occur as discontinuous lenses throughout the lower 250 m of the section. The fossil material in these shell banks shows varying degrees of abrasion and fragmentation and is interpreted as representing accumulations in a high-energy, near-shore environment.

AGE OF THE SEYMOUR ISLAND SERIES

There has been considerable discussion about the age of the Tertiary sequence on Seymour Island. Wilckens

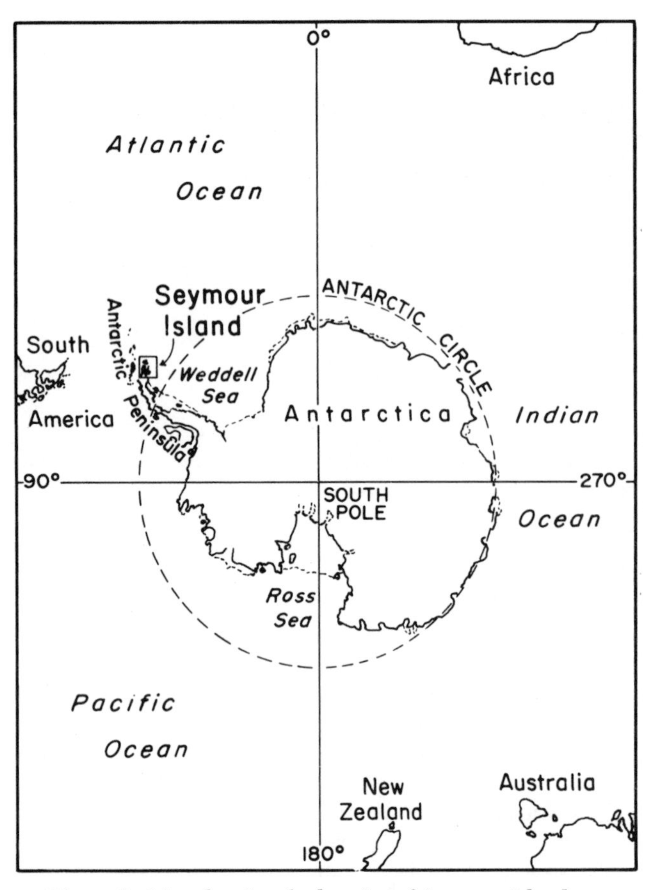

Figure 1. Map showing the location of Seymour Island.

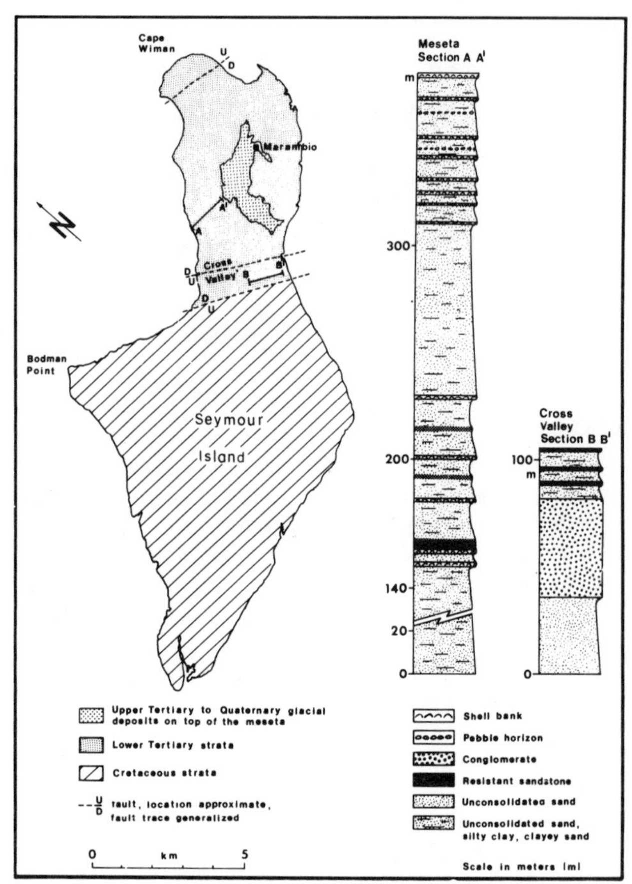

Figure 2. Geologic map and stratigraphic column of the Tertiary sequence on Seymour Island.

(1911) considered it to be Oligo-Miocene, Cranwell (1959, 1969) regarded it as Pâleocene, and Simpson (1971) assigned a late Eocene to possibly early Oligocene age. The differences of opinion stem in part from the lack of stratigraphic data concerning the relations of the various parts of the sampled sequence. Stratigraphic and faunal data collected during the 1974-1975 season indicate that the age of the sequence ranges from Paleocene to possibly early Oligocene with the early and middle Eocene missing.

Based on dinoflagellate data, Stephan Hall from Northern Illinois University (personal commun.) considers the plant-bearing strata exposed in Cross Valley to be Paleocene. It should be noted that the lithology of the sample Cranwell studied occurs only in the lower unit. The dinoflagellate assemblages together with Cranwell's palynological data suggests that the age of the lower 105 m of the "Seymour Island Series" is Paleocene.

The upper 350 m (Fig. 2) of the Tertiary exposed around the meseta is characterized by two distinct assemblages of molluscs. The diverse assemblage in the lower 250 m is dominated by *Cucullaea donaldi* (Sharman and Newton), *Antarctodarwinella nordenskjoldi* (Wilckens), and *Struthioptera* n. sp.; these occur together with a number of undescribed species of molluscs. An undescribed species of nautiloid of the genus *Aturia* similar to the Eocene *A. bruggeni* (Ihering) from Tierra del Fuego (Elliott and others, 1975), and a rich dinoflagellate flora (Hall, personal commun.) indicate a late Eocene age for the lower 250 m of the section. The molluscs which are so abundant in the lower part of the section are replaced in the uppermost 100 m by a new assemblage characterized by *Perissodonta laevis* (Wilckens), *Struthiolarella variabilis* Wilckens, and *Cyrtochetus bucciniformis* Wilckens. Almost all the penguin fossils occur in this part of the section. Although dinoflagellates are not abundant, Hall believes that the upper 100 m of the section may be as young as early Oligocene.

Wilckens' (1911) Oligo-Miocene age assignment of the "Seymour Island Series" was due to lack of precise chronostratigraphic knowledge of the Tertiary molluscan faunas of southern South America. He based his determination on comparisons of the Seymour Island fauna with the fauna of the "Patagonian Formation" of Argentina. At that time, the "Patagonian Formation" was considered to be Oligo-Miocene age. Although Wilckens' correlation of the "Seymour Island Series" with the "Patagonian Formation" was correct, his Oligo-Miocene age assignment was in error because of the limited chronostratigraphic knowledge of the Tertiary of the Southern Hemisphere at that time.

Biostratigraphic investigations have since demonstrated that the "Patagonian Formation" consists of a number of

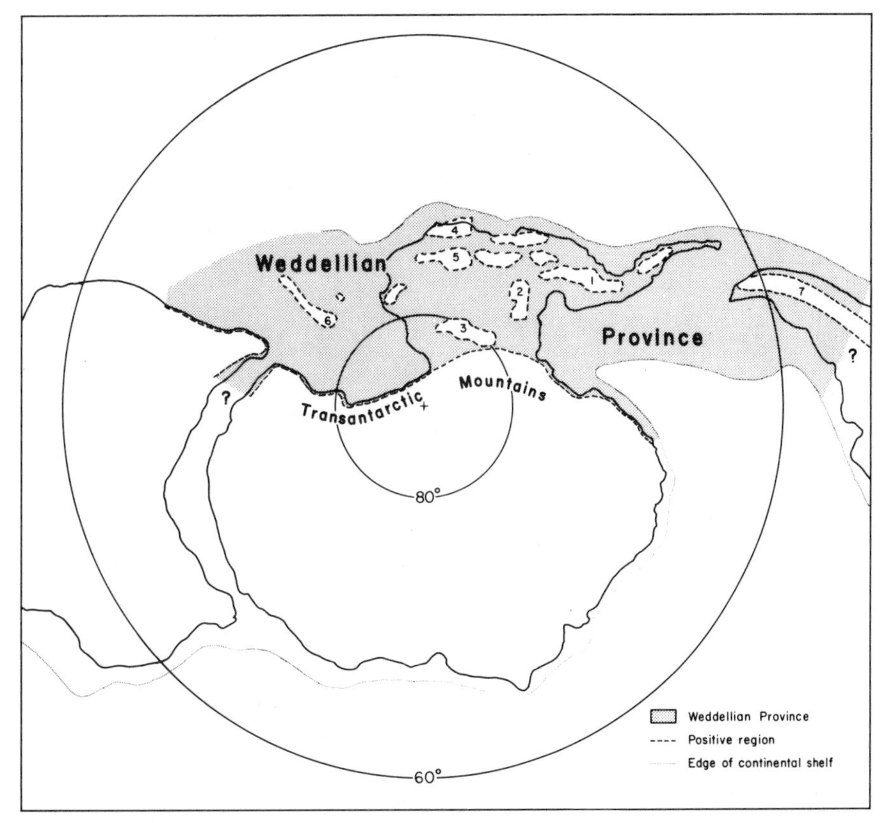

Figure 3. Paleogeography of the Weddellian Province (shaded area) during the Late Cretaceous (80 m. y.) before the final separation of New Zealand and Australia from Antarctica. Regions enclosed with dashed lines indicate probable land areas during the latest Cretaceous and earliest Tertiary: (1) Antarctic Peninsula; (2) Ellsworth Mountains; (3) Whitmore Mountains; (4) Thurston Island; (5) Marie Byrd Land; (6) New Zealand; (7) South America.

distinct, intricately interbedded marine and nonmarine units (Camacho, 1974), and that the age is older than previously thought. Camacho and Fernandez (1956) recorded the occurrence of several species of bivalves belonging to the cosmopolitan Eocene *Venericardia planicosta* group from the lower part of the "Patagonian Formation" near Comodoro Rivadavia in central Patagonia. The presence of this cosmopolitan Eocene bivalve genus suggests that the lower part of the "Patagonian Formation" is Eocene in age. In a later synthesis of the biostratigraphy of the marine "Patagonian Formation" Camacho (1974) considered the lower two-thirds (San Julian Formation and the overlying "Estrata con *Monophoraster* y *Venericor*") to be Eocene and the uppermost marine unit (Monte Leon Formation) to be upper Oligocene.

THE WEDDELLIAN PROVINCE

During the final fragmentation of Gondwanaland in the Late Cretaceous, the various continents (Antarctica, Australia, New Zealand, and South America) were separated by shallow seas extending from the southern part of South America along West Antarctica to eastern Australia. The similarity of the molluscan assemblages of the Southern Hemisphere during the Late Cretaceous-early Tertiary has been discussed by a number of investigators (Wilckens, 1911, 1922; Marshall, 1926; Spath, 1953; Howarth, 1966; Henderson, 1970; Zinsmeister, 1976). Because of the overall uniformity of the faunas, that region is considered to represent a single broad biogeographic province (Fig. 3). The name Weddellian Province is proposed for this shallow-water, cool-temperate marine province along the southern circum-Pacific that existed from the latest Cretaceous through the Eocene.

Kauffman (1973) proposed the Austral Province, based on distribution of Cretaceous bivalves, for the broad region in the southern part of the Indo-Pacific region during the Cretaceous. He included within the Austral Province, the southern tip of South America, the eastern side of India, and all of Australia, New Zealand, and Antarctica. He further sub-divided the Australian-New Zealand region into two distinct subprovinces. Because of the continued movement of the southern continents during the Cretaceous, the Austral Province had become fragmented into a number of distinct faunal provinces by the latest Cretaceous and earliest Tertiary. The Weddellian Province is considered to be one of those faunal provinces and one which had become distinguishable during the latest Cretaceous. Although study of the upper Eocene faunas has just begun, the presence of a number of undescribed genera indicates that the southern circum-Pacific region had become a center of endemism during the early Tertiary. This increase in the endemicity is most likely associated with the continued decrease in temperature and isolation of that region during the early Tertiary.

Geologic data accumulated during the last decade from West Antarctica and the Transantarctic Mountains allows a fairly detailed reconstruction of the paleogeography of the southern circum-Pacific during the Late Cretaceous (Fig. 3). The southern margin of the Weddellian Province in Antarctica coincides with the Transantarctic Mountains front. This major mountain belt, extending from north Victoria Land to the Weddell Sea, has been a positive highland since at least the mid-Mesozoic. The northern limit of the Province coincides with the edge of the continental shelf. The molluscan faunas of southern Argentina and central Chile (Steinmann and Wilckens, 1908) indicate that the Weddellian Province extended well into southern South America. Available data from Australasia suggests that the Province also included the southeastern edge of Australia and southern New Zealand (Singleton, 1943; Finlay and Marwick, 1937). Presuming that Antarctica has remained approximately in its present position since the Late Cretaceous, the northern limits of the Weddellian Province in South America and Australia were probably temperature controlled.

During the latter half of the Mesozoic, West Antarctica was the site of considerable magmatic and orogenic activity (Elliot, 1975). Uplift associated with the orogenesis produced land areas scattered throughout the Province, in the vicinity of Ellsworth and Marie Byrd Lands, Ellsworth and Whitmore Mountains, Thurston Island, and along the length of the Antarctic Peninsula.

LATE CRETACEOUS-EARLY PALEOGENE GEOLOGIC HISTORY OF THE WEDDELLIAN PROVINCE

The final phase of fragmentation of Gondwanaland occurred during the latest Cretaceous and earliest Tertiary. The sequence of final breakup and northward drift of New Zealand and Australia can be inferred from the magnetic lineations mapped by Weissel and Hayes (1972) and Hayes and Ringis (1973). Paleomagnetic data (Creer, 1973) indicates that the Antarctic portion of Gondwanaland had reached a position close to its present location by the Late Cretaceous. Separation of New Zealand from Antarctica and Australia commenced approximately 75 m.y. ago.[1] The eastward drift component associated with the opening of the Tasman Sea ceased about 56.5 m.y. ago while the northward drift of New Zealand from Antarctica has continued up to the present. The shallow-marine conditions that existed between New Zealand and the rest of Gondwanaland during the Cretaceous are believed to have ceased by about the end of the lower Paleocene.

The magnetic anomalies in the Southern Ocean indicates that active spreading between Australia and Antarctica began about 56.5 m.y. ago and has continued up to the present. Extensive transform faulting west of Tasmania accompanied the separation of Australia and Antarctica. This faulting has largely destroyed the magnetic record south of Tasmania. Although deep-marine conditions began to develop in the Southern Ocean by the late Paleocene, scattered highlands and shallow seas persisted between Australia and Antarctica from north Victoria Land northward along the Tasman Rise to southeastern Australia. This narrow corridor between Australia and Antarctica existed until about the late Eocene or early Oligocene. The development of the deep-flowing Circum-Antarctic Current south of the South Tasman Rise about 38 m.y. ago (Kennett and others, 1975) indicates that the last shallow-marine connection between Antarctica and Australia had disappeared by at least the middle Oligocene.

On the other hand, the tectonic history of southern South America and the opening of Drake Passage is not well understood. In a review of the origin of the Scotia Arc, Dalziel and Elliot (1971) suggested that the southern Andean Cordillera and the Antarctic Peninsula formed a continuous north-south mountain chain at the end of the Cretaceous. During the early Tertiary, the initial formation of the Scotia Arc began with the eastward bending and disruption of the Andean-West Antarctic Cordillera. The timing and rate of disruption is not known. Dalziel and Elliot (1971) believe that shallow-marine and scattered highland areas may have existed in the region of Drake Passage up to the end of the Oligocene or early Miocene. Magnetic data (Barker, 1972) indicate that deep-sea conditions between South America and the Antarctic Peninsula began to develop about 20 m. y. ago.

The development of local provincialism in the Southern Hemisphere tends to coincide with the final breakup of Gondwanaland. The geologic history of the following important groups of Southern Hemisphere molluscs illustrates the trend towards isolation and local provincialism.

The gastropod family Struthiolariidae has an extremely wide range and long geologic record. The earliest member of the family [*Conchothyra parasitica* (Hutton)] occurs in the late Senonian Mata Series of New Zealand. The adult form of this bizarre genus is completely enveloped by a thick heavy callus. During the latest Cretaceous, the Struthiolariidae underwent a period of diversification and by the Paleocene, the family had increased to 3 genera (*Conchothyra, Perissodonta, Monalaria*) representing two distinct stocks (*Struthiolaria* and *Struthiolarella*) (Zinsmeister, in prep.). Both of these stocks occur in the lower Paleocene (Teurian) Mitchell's Point facies of the Wangaloan Formation. The Tertiary species of *Conchothyra* is characterized by a greatly reduced callus. The genus *Perissodonta* (=? *Struthiolarella*) makes its appearance in the Wangaloan; it is small and ornamented by both axial and spiral ribs. The style of ornamentation is very similar to the sculpture of the later members of the *Struthio-*

[1] The age assignment for the magnetic anomaly data is based on Watkins' (1976) polarity scale.

Figure 4. Changes in the paleogeography of the Weddellian Province (stippled area) during the latest Cretaceous and Early Tertiary.

larella stocks and *Perissodonta* may represent the earliest member of that stock. *Monalaria,* the third genus which occurs in the Wangaloan, is small and relatively high spired. The moderately well-developed callus is restricted to the region around the aperture. Marwick (1960) has suggested that the later Tertiary struthiolariids of New Zealand may have been derived from *Monalaria.* After the Paleocene, the *Struthiolaria* stock is restricted to New Zealand and Australia, whereas the *Struthiolarella* stock occurs throughout southern South America and Antarctica. Members of the two stocks although abundant and widespread in their respective regions never occur together after the lower Paleocene.

The bivalve family Lahilliidae is widespread in the Southern Hemisphere from the Cretaceous to the Miocene. This bivalve appears to be related to the Cardiidae, but lacks the anterior lateral teeth and radial sculpture. The genus is known from the Late Cretaceous of New Zealand, South America, and Seymour Island. The typical Cretaceous *Lahillia* is moderately large and lacks a pallial sinus. The surface ornamentation consists only of fine concentric growth increments. In a monograph on the Wangaloan fauna, Finlay and Marwick (1937) proposed a new subgenus *Lahilleona* for the Lower Paleocene *Lahillia* of New Zealand. The criterion for the recognition of this new subgenus is the presence of a moderately deep pallial sinus. They considered *Lahilleona* to be unique to New Zealand. The Lahilliidae are not known to occur after the lower Paleocene in New Zealand. A *Lahillia* similar to specimens of the Wangaloan fauna was described by Singleton (1943) from the lower Paleocene Pebble Point locality of southeastern Australia. The poor preservation of the Pebble Point material makes it impossible to determine if the Australian species belongs to Finlay and Marwick's subgenus *Lahilleona.* As in the case of New Zealand, *Lahillia* is not known to occur after the Paleocene in Australia. Well-preserved material of *L. larseni* (Sharman and Newton) which I collected from the upper Eocene of Seymour Island shows the presence of a pallial sinus similar to that of *Lahilleona.* This indicates that the Seymour Island species is closely related to and probably derived from the New Zealand *Lahilleona.* The family Lahilliidae continues to be widespread in the Tertiary of southern South America until it became extinct by the end of the Miocene.

The third group of Southern Hemisphere molluscs is the large aporrhaid gastropod *Struthioptera.* This genus has a relatively long geologic record (Late Cretaceous through the lower Paleocene) in New Zealand. The earliest species, *Struthioptera haastana* (Wilckens), occurs in the upper Senonian Lower Amuri Group of the South Island of New Zealand (Wilckens, 1922). This species is characterized by a broad blunt wing with a short posterior spike. The general trend in later members of the lineage is toward an increase in the overall size of the shell, and a marked increase in the size of the wing and the length of the posterior spike. *Struthioptera osiris* (Finlay and Marwick) from the Wangaloan was the only known Tertiary representative of the *Struthioptera* lineage until a large number of individuals of a new species were collected from the upper Eocene of Seymour Island (Zinsmeister, 1977). The Seymour Island species is not as thick shelled as *S. osiris,* and the wing is more developed with an exceptionally long posterior spike. The occurrence of *Struthioptera* on Seymour Island extends the geologic range of the genus to at least the late Eocene.

All three of these groups of molluscs are known only from the Southern Hemisphere and make their first appearance during the Late Cretaceous. *Struthioptera* and the struthiolariids are known only from the Late Cretaceous of New Zealand. The genus *Lahillia* is abundant in the Late Cretaceous of southern South America, Antarctica, and New Zealand, but is not known until the lower Paleocene in Australia. The absence of a particular group in any of the regions is probably due to the incomplete knowledge of the fossil record or unfavorable facies.

New Zealand is the only region in the Southern Hemisphere where the paleontological record is fairly well known for the Late Cretaceous and early Tertiary. All three groups are known from the Cretaceous through to the lower Paleocene of New Zealand. While *Lahillia* and *Struthioptera* remain evolutionarily convervative and show little in the way of diversification, the Struthiolariidae underwent a marked period of diversification during the latest Cretaceous. By the early Paleocene, the struthiolariids had evolved into two distinct stocks. Both of these stocks occur in the Paleocene of New Zealand. After the lower Paleocene, *Lahillia, Struthioptera,* and the *Struthiolarella* branch of the Struthiolariidae disappear in New Zealand, but remain a common element of the later Tertiary faunas of Antarctica and southern South America.

The disappearance of these molluscs coincides with the final separation and isolation of New Zealand from the rest of Gondwanaland (Fig. 4b). The fossil record in New Zealand for the late Paleocene through the middle Eocene is not well documented because of unfavorable facies. By the late Eocene (Bartonian) the appearance of a number of warm-water Indo-West Pacific molluscan groups (Cypraeidae, Mitridae, Harpidae, and Conidae) indicates that conditions around New Zealand were becoming warmer (Fleming, 1962). This warming is most likely the consequence of the northward drift of New Zealand during the Paleocene and Eocene.

Paleontological data documenting the changes in the lower Tertiary molluscan faunas from southeastern Australia are meager due to unfavorable facies and exposures. Singleton (1943) described several new species of molluscs from the lower Paleocene Pebble Point Formation of Victoria. He considered the fauna of the Pebble Point to be "Lower Eocene or possible Paleocene" and closely related to the Wangaloan of New Zealand based on the presence of the two bivalves *Lahillia* and *Cucullaea.* Later foraminiferal studies (McGowran, 1965) indicate that the Pebble Point is of lower Paleocene age and the approximate equivalent of the Wangaloan of New Zealand. The absence of a fossil record in southeastern Australia during the late Paleocene and early Eocene makes it impossible to determine the time of final faunal separation between Australia and Antarctica; an approximate date can be determined, however, from geophysical data.

Recent data indicate that deep-sea conditions were initiated between Australia and Antarctica about 56.5 m.y. ago (Weissel and Hayes, 1972; Hayes and Ringis, 1973). Although deep-sea conditions began to develop in the Southern Ocean in the Paleocene, shallow seas persisted between Australia and Antarctica from north Victoria Land to southeastern Australia along the South Tasman Rise until the late Eocene. The development of the deep Circum-Antarctic Current about 38 m.y. ago (Kennett and others, 1975) indicates that the inferred shallow-marine connection along the South Tasman Rise disappeared by the middle Oligocene. The late Eocene and early Oligocene faunas of Victoria, as in the case of New Zealand, are characterized by the appearance of a number of warm Indo-West Pacific species (Fig. 4c).

In the eastern sector of the Weddellian Province, the faunal relations between southern South America and Seymour Island are not well understood. Although the two regions are geographically close, the early Cenozoic geologic history is difficult to evaluate. A continuous cordillera is believed to have joined southern South America and the Antarctic Peninsula in Late Cretaceous time (Dalziel and Elliot, 1971). Because of the complexity of the geologic history of the Scotia Arc region, the evolution of the paleogeography of this region between the Late Cretaceous and the early Miocene is not known.

Many of the genera (*Perissodonta, Struthioptera, Antarctodarwinella,* and a number of undescribed genera) which are abundant on Seymour Island are absent in equivalent-aged faunas from South America. These apparent faunal differences may be due either to the incomplete knowledge of the fossil record of Tierra del Fuego or to ecological factors related to changes in latitude. Although the faunas of Tierra del Fuego have not been studied since the early part of the century, the absence of many Seymour Island elements in South America when shallow-water conditions existed suggests that environmental factors due to changes in latitude are the primary cause of the observed faunal dissimilarities.

Complete isolation of the Weddellian Province from New Zealand, Australia, and southern South America along West Antarctica, is believed to have occurred prior to the late Eocene. The occurrence of *Struthiolarella* aff. *S. variabilis* Wilckens in a glacial erratic at Cape Crozier in the Ross Sea (Hertlein, 1969) and on Seymour Island, but nowhere else, suggests that the Weddellian Province was restricted to West Antarctica by the late Eocene. The absence of a fossil record in Antarctica after the early Oligocene makes it impossible to determine the history of the Weddellian Province after that time.

CONCLUSION

In the final stages of the breakup of Gondwanaland during the Late Cretaceous and early Tertiary, the various continents (Antarctica, New Zealand, Australia, and South America) were separated by shallow seas. These seas varied from narrow embayments to broad continental shelves. The overall similarity of the molluscan faunas from southern South America to southeastern Australia during the Late Cretaceous indicates that the region represented a single broad faunal province, along the southern circum-Pacific. The name Weddellian is proposed for this cool temperate province that existed from the latest Cretaceous through the late Eocene. As the separation between the continents increased, the amount of faunal communication within the province decreased and eventually ceased. The decrease in gene flow should be shown by progressive increase in local provincialism. Examination of several important groups of Southern Hemisphere molluscs indicate that the isolation and the development of local provincialism coincides with the final breakup of Gondwanaland.

ACKNOWLEDGMENTS

I would like to thank David Elliot, Director of the Institute of Polar Studies at The Ohio State University for his many suggestions and advice. This work was supported by National Science Foundation Grant OPP 74-21509 to The Ohio State University and the Institute of Polar Studies. This work on Seymour Island would not have been possible without the generous support of the Instituto Antártico Argentino; in particular it is a pleasure to acknowledge the continuing assistance of Dr. N. Fourcade.

REFERENCES

Andersson, J. G., 1906, On the geology of Graham Land: Uppsala Univ. Geol. Inst. Bull., v. 7, p. 19-71.

Barker, P. F., 1972, Magnetic lineations in the Scotia Sea, *in* Adie, R. J., ed., Antarctic geology and geophysics: Oslo, Universitetsforlaget, p. 17-26.

Camacho, H. H., 1974, Biostratigrafia de las formaciones marinas del Eoceno y Oligoceno de La Patágonia: Acad. Nac. Ciencias Exactas Fisicas Naturales Anal. (Buenos Aires), v. 26, p. 39-57.

Camacho, H. H., and Fernández, J. A., 1956, La transgresión en la costa atlántica entre Comodoro Rivadavia y el curso inferior del río Chubut: Asoc. Geol. Argentina Rev., v. 11, p. 23-45.

Cranwell, L. M., 1959, Fossil pollen from Seymour Island, Antarctica: Nature, v. 184, p. 1782-1785.

——— 1969, Antarctic and circum-Antarctic palynological contributions: Antarctic Jour. (U. S.), v. 4, p. 197-198.

Creer, K. M., 1973, A discussion of paleomagnetic poles on the map of Pangaea for epochs in the Phanerozoic, *in* Tarling, D. H., and Runcorn, S. K., eds., Implications of continental drift to the Earth sciences, Vol. 1: London, Academic Press, p. 47-76.

Dalziel, I. W. D., and Elliot, D. H., 1971, Evolution of the Scotia Arc and Antarctic margin, *in* Stehli, F. G., and Nairn, A. E. M., eds., The ocean basins and margins. Vol. 1. The South Atlantic: New York, Plenum Press, p. 171-245.

Elliot, D. H., 1975, Tectonics of Antarctica: A review: Am. Jour. Sci., v. 275-A, p. 45-106.

Elliot, D. H. and others, 1975, Geological investigations on Seymour Island, Antarctic Peninsula: Antarctic Jour. (U. S.), v. 10, p. 182-186.

Finlay, H. S., and Marwick, J., 1937, The Wangaloan and associated molluscan faunas of Kaitargata-Green Island subdivision: New Zealand Geol. Survey Palaeont. Bull. 15, p. 1-140.

Fleming, C. A., 1962, New Zealand biogeography—a palaeontologists' approach: Tuatara, v. 10, p. 53-108.

Hayes, D. E., and Ringis, J., 1973, Sea floor spreading in the Tasman Sea: Nature, v. 243, p. 454-458.

Henderson, R. A., 1970, Ammonoidea from the Mata Series (Santonian-Maastrichtian) of New Zealand: Spec. Papers Palaeontology no. 6, p. 1-82.

Hertlein, L. G., 1969, Fossiliferous boulders of early Tertiary age from Ross Island, Antarctica: Antarctic Jour. (U. S.), v. 4, p. 199-201.

Howarth, M. K., 1966, Ammonites from the Upper Cretaceous of the James Ross Island Group: British Antarctic Survey Bull., v. 10, p. 55-69.

Kauffman, E. G., 1973, Cretaceous Bivalvia, *in* Hallam, A., ed., Atlas of palaeobiogeography: Amsterdam, Elsevier, p. 353-383.

Kennett, J. P. and others, 1975, Cenozoic paleoceanography in the southwest Pacific Ocean, Antarctic glaciation, and the development of the Circum-Antarctic Current, *in* White, Stan M., ed., Initial Reports of the Deep Sea Drilling Project, Vol. 29: Washington D. C., U. S. Govt. Printing Office, p. 1155-1169.

McGowran, B., 1965, Two Paleocene foraminiferal faunas from the Wangerrip Group, Pebble Point coastal section, western Victoria: Royal Soc. Victoria Proc., v. 79, p. 9-74.

Marshall, P., 1926, The Upper Cretaceous ammonites of New Zealand: New Zealand Inst. Trans., v. 56, p. 129-210.

Marwick, J., 1960, Early Tertiary Mollusca from Otaio Gorge, South Canterbury, New Zealand: New Zealand Geol. Survey Palaeont. Bull. 39, p. 1-59.

Simpson, G. G., 1971, Review of fossil penguins from Seymour Island: Royal Soc. London Trans., ser. B, Biol. Sci., v. 178, p. 357-387.

Singleton, F. A., 1943, An Eocene molluscan fauna from Victoria: Royal Soc. Victoria Proc., v. 55, p. 267-281.

Spath, L. F., 1953, The Upper Cretaceous cephalopod fauna from Graham Land: Falkland Islands Dependencies Survey Sci. Rept., no. 3, p. 1-60.

Steinmann, G., and Wilckens, O., 1908, Kreide- und Tertiar fossilien aus den Magellanslandera: Arkiv Zoologi, v. 4, p. 1-118.

Watkins, N. D., 1976, Polarity subcommission sets up some guidelines: Geotimes, v. 21, no. 4, p. 18-20.

Weissel, J. K., and Hayes, D. E., 1972, Magnetic anomalies in the southeast Indian Ocean, *in* Antarctic oceanology. II. The Australian-New Zealand sector: Antarctic Research Ser., no. 19, p. 165-196.

Wilckens, O., 1911, Die mollusken der antarktischen Tertiar formation: Wissenschaftliche Ergebnisse der Schwedischen Sudpolarexpedition, 1901-1903, v. 3, p. 1-62.

——— 1922, The Upper Cretaceous gastropods of New Zealand: New Zealand Geol. Survey Palaeont. Bull. 9, p. 1-42.

——— 1924, *Lahillia* and some other fossils from the Upper Senonian of New Zealand: New Zealand Inst. Trans., v. 55, p. 539-544.

Zinsmeister, W. J., 1976, A new genus and species of the gastropod family Struthiolariidae *Antarctodarwinella ellioti*, from Seymour Island, Antarctica: Ohio Jour. Sci., v. 76, p. 111-114.

——— 1977, Note on a new occurrence of the Southern Hemisphere aporrhaid gastropod *Struthioptera* Finlay and Marwick on Seymour Island: Jour. Paleontology, v. 51, p. 399-404.

——— in prep., Revision of the gastropod family Struthiolariidae from Seymour Island, Antarctica.

Pinniped Biogeography

CHARLES A. REPENNING, *U.S. Geological Survey, Menlo Park, California 94025*
CLAYTON E. RAY, *Department of Paleobiology, Smithsonian Institution, Washington, D. C. 20560*
DAN GRIGORESCU, *Faculty of Geology and Geography, University of Bucharest, Bucharest, Romania*

ABSTRACT

The pinnipeds originated in Neogene temperate waters of the North Atlantic and North Pacific Oceans. Two superfamilies have been erected to recognize the distinctiveness of the 2 pinniped groups that resulted from this dual origin: the Otarioidea, including the sea lions and the walrus that originated in the North Pacific; and the Phocoidea, including the seals that originated in the North Atlantic. The distribution of both groups is and has been strongly controlled by climatic conditions, and both groups are subdivided into family group taxa that adjusted in different ways to climatic cooling. The phocine seals adapted to the cooling of the North Atlantic, beginning about 15 m. y. ago, and expanded northward, eventually into the Arctic, while the monachine seals maintained their temperate to tropical adaptations and retreated southward from latitudes at least as far north as Massachusetts and The Netherlands.

The adaptive pattern of the otarioids in the North Pacific is more complicated because their diversification was greater and entailed repeated variations. An ancestral family is recognized that derived, more than 22 m. y. ago, from ursid arctoid land carnivores and that lived in warming North Pacific seas. This family disappeared by way of evolution into 3 more advanced families. Oldest of these were the desmatophocids which lived along all shores of the North Pacific during the period of warmest seas, but like the ancestral enaliarctids, they never left this oceanic basin. The odobenids, the walrus family, evolved from the enaliarctids about 14 m. y. ago in mid-northern latitudes and warm-temperate seas and have lived almost exclusively in cooling seas. By about 8 m. y. ago they were the most diverse otarioids of the North Pacific and had split into 2 subfamilies: one remaining in temperate and cooler waters to the north, the other dispersing southward and adapting to subtropical to tropical waters.

The tropical odobenid subfamily is that from which the modern walrus was derived; between 6 and 8 m. y. ago it dispersed through the Central American Seaway into Caribbean waters to join the tropical monachine seals of that area. With closure of the Central American Seaway and consequent flooding of the Gulf Stream with the warm waters of the Atlantic equatorial currents, the odobenines and some monachines moved northward with the warming Gulf Stream and are known in the Cape Hatteras area about 5 m. y. ago. Like the phocine seals of that time, the odobenines then adapted to cooler waters, while the monachines were again depressed southward with further overall cooling of the North Atlantic.

In the North Pacific, the temperate to cool-water subfamily of the odobenids thrived for several million years in progressively cooling seas but eventually became extinct about 4 m. y. ago. At this time the North Pacific had cooled to its approximate present temperatures. These odobenids never left this oceanic basin. Less than 1 m. y. ago the modern walrus, having adapted to arctic waters in the Atlantic, returned to the North Pacific by way of the Arctic Ocean.

The temperate-water otariids of the North Pacific, the modern sea lions and fur seals, evolved from the last of the ancestral enaliarctids at least 11 m. y. ago. Their evolution was slow and unidirectional until about 6 m. y. ago when the earliest evidence of the lineage leading to the subarctic Alaskan fur seal is known. The rest of the otariids remained temperate-water animals; subtropical waters represented a major environmental barrier to them. However, about 5 m. y. ago the closing of the Central American Seaway permitted the Central American Pacific to cool enough for those otariids not already adapting to colder waters to disperse southward, and their first record in the Southern Hemisphere is of about this age in Peru. These otariids were ancestral to the living fur seal genus *Arctocephalus*, which has since dispersed entirely around Antarctica.

Those Atlantic monachine seals of the Caribbean-Pacific fauna that were left in the Pacific with the closure of the Seaway also dispersed southward; they are first known in the Southern Hemisphere about 5 m. y. ago, and have also encircled Antarctica.

In the North Pacific about 3 m. y. ago, the remaining ancestral otariids first evolved into forms that could be called sea lions. This appears to have happened on both sides of the North Pacific and from that time on the diversification of the sea lions was explosive in comparison

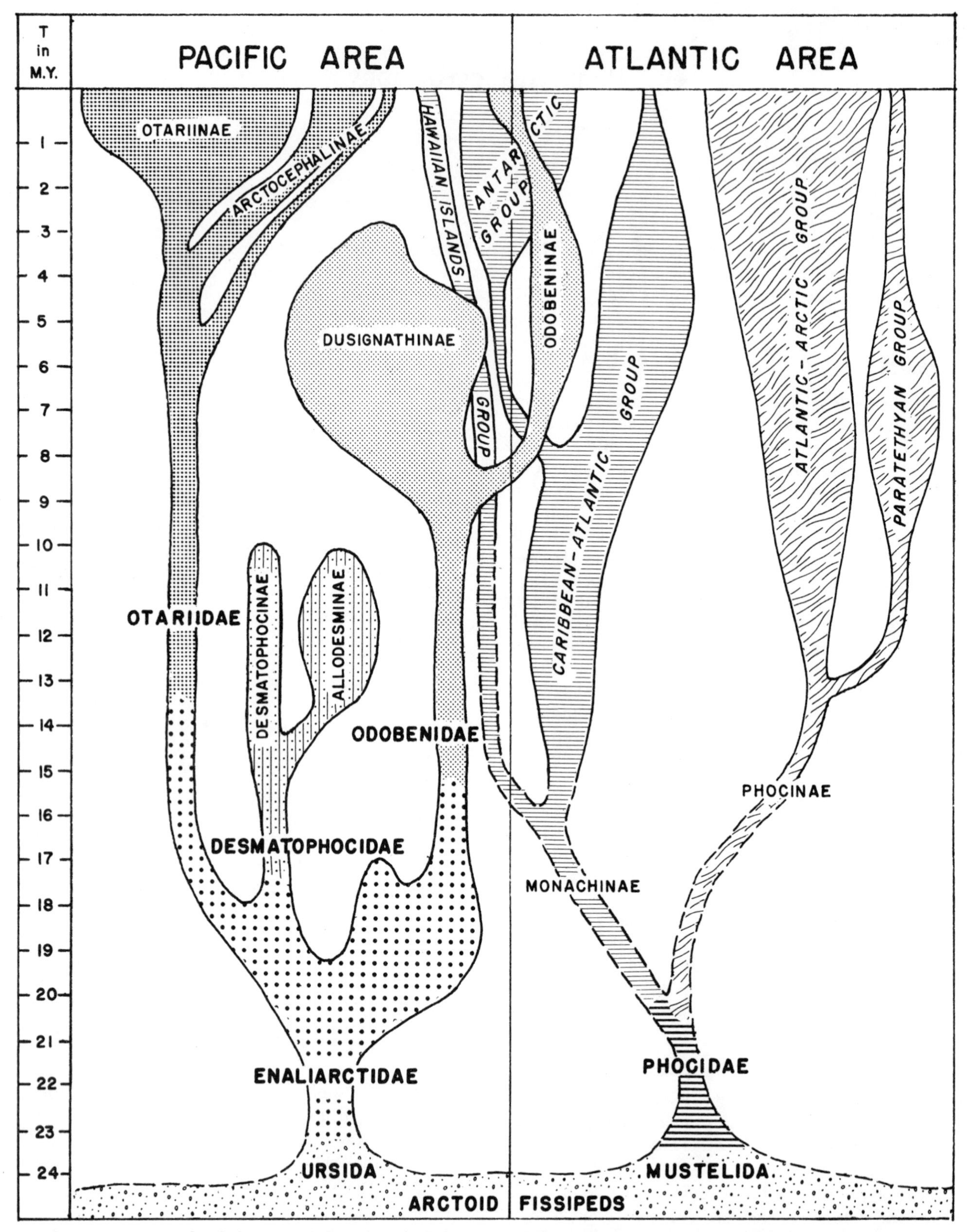

Figure 1. Generalized diagram of the inferred phylogeny of the pinnipeds showing the Pacific origin of the Otarioidea, the Atlantic origin of the Phocoidea, and the time of development of each of their families and subfamilies. Dashed lines indicate no fossil record and the width of the diagramed lineages is intended to reflect relative taxonomic diversity. The subfamilies of the Desmatophocidae are not discussed separately in the text.

to the preceding 10 m. y. of otariid evolution. At some time, probably in the early Pleistocene, the sea lions also dispersed into the South Pacific but, owing to the recency of their arrival in the Southern Hemisphere, they have spread only to the Australian region and to both sides of southern South America.

About 13 m. y. ago, one primitive form of the North Atlantic phocoid fauna moved from the North Sea area into Paratethys. It there evolved into a variety of endemic forms, most of which have become extinct but one of which evolved into the subgenus *Pusa,* still living as a relict in the Caspian Sea remnant of Paratethys. From there it spread to the North Atlantic and North Pacific by way of the Arctic Ocean about 3 m. y. ago.

CLASSIFICATION AND ORIGINS

The pinnipeds, or walruses, sea lions, and seals, are derived from land carnivores, or fissipeds. There are 2 major superfamilies of pinnipeds (see Fig. 1). The most useful character for their differentiation is their hind limbs; the otarioids can flex their hind limbs beneath them for use in terrestrial locomotion and the phocoids cannot. On land, the phocoids wriggle on their belly.

The otarioids are further subdivided into 2 living families, the sea lions and the walruses (Fig. 1). Sea lions, Family Otariidae, have small external ears; walruses, Family Odobenidae, do not. Two extinct families are known; the archaic Desmatophocidae and the ancestral Enaliarctidae. The oldest known enaliarctids are about 22 m. y. old and they share so many features with related fissiped carnivores that it is difficult to imagine the origin of the Otarioidea as being very much older.

The phocoids are all included in 1 family, the Phocidae, that is subdivided into 2 subfamilies, the Phocinae or northern seals, and the Monachinae or tropical and antarctic seals (Fig. 1). The oldest known phocids are about 15 m. y. old. Both monachine (Ray, 1976a) and phocine (True, 1906) seals are known at this time in the western North Atlantic although only phocine seals are known from somewhat younger deposits in the eastern Atlantic (Van Beneden, 1877). Primitive seals of about 13 m. y. age (Toula, 1897) are known from the Paratethys sea which extended eastward from Vienna to beyond the Caspian Sea. Paratethys is noteworthy because of the variety of seals that evolved together in this isolated sea.

The earliest known seals appear distinctly phocid (Ray, 1976b), showing little evidence of the nature of their fissiped ancestors not shown by living phocids. The much greater antiquity of phocoid origins implied by this modern look and the already established subfamilial diversity in the oldest known North Atlantic fossil phocid fauna suggest considerable unknown earlier history and consequent uncertainty about their ancestry and place of origin.

Regardless of their unknown point of entry into the sea, the phocoids lived and diversified mostly along both shores of the North Atlantic. Although the fossiliferous localities are much fewer in the Southern Hemisphere, as far as is now known the phocids did not enter the South Pacific until about 5 m. y. ago (Sacaco, Peru: R. J. Hoffstetter, personal commun. 1973), and the earliest records in the South Atlantic are probably 1 m. y. younger (Frenguelli, 1922; Hendey and Repenning, 1972).

Similarly, although their known history begins 22 m. y. ago and they are found in the North Pacific, the otarioids did not enter the South Pacific until about 5 m. y. ago (Sacaco, Peru: R. J. Hoffstetter, personal commun. 1973). The odobenids appear to have entered the Caribbean at an earlier date but their earliest North Atlantic record is about 4.5 to 5 m. y. old.

In view of the cold-water preferences of most living pinnipeds, one is inclined to feel that their ancestors also had this preference and that the equatorial region might have been an impassable barrier to dispersal. Although it does seem to have been a barrier, it does not seem to have been so great a one in the past as it is today, for associated invertebrate fossils indicate that known fossil pinnipeds lived in warmer waters than do most of their modern counterparts. It is clear that the degree of cold-water preference exhibited by most modern pinnipeds has increased with passing time.

The fossil record of the otarioids indicates an origin out of the ancestral family Enaliarctidae (Repenning and Tedford, 1977); the Enaliarctidae rather clearly have been shown to derive, more than 22 m. y. ago, from ursid arctoids (Mitchell and Tedford, 1973), terrestrial carnivores that are better known in the Old World than in the New. Hence, it is possible that the otarioid pinnipeds evolved in the western North Pacific over 22 m. y. ago, although the most extensive record is from the eastern Pacific along the shores of the United States.

The much less adequate record of fossil phocoids does not provide so clear an understanding of their origin. Following Mivart (1885) and Kellogg (1922), Savage (1957) has suggested a possible origin from a form similar to the fossil genus *Potamotherium,* for many years believed to be an otter. Similarities of the seals to the otters have been noted by many workers since that time. Because of several primitive features, assignment of *Potamotherium* to the lutrine mustelids is difficult except on superficial and adaptive similarities. However, Tedford (1976) has shown that it clearly was a primitive mustelid arctoid, not greatly, but nevertheless distinctly, different from the contemporary ursid arctoids of about 22 m. y. ago, and he has pointed out additional phocoid similarities. Hence, it would seem likely that the differentiation that led to the phocoids and otarioids probably took place on land in their fissiped ancestors. *Potamotherium*-like mustelid arctoids are also better known in the Palearctic than in the Nearctic. It is possible, therefore, that the phocoids evolved in the eastern Atlantic, although the most extensive record is from the western Atlantic along the shores of the United States.

No fossil pinnipeds are known from the area of the Indian Ocean. In fact, except for antarctic waters, this Ocean is not inhabited by modern pinnipeds. While equatorial temperatures dominate this oceanic basin and could be called upon as inferred cause for the lack of pinnipeds, it also seems probable that the Tethyan connection to the Indian Ocean was greatly reduced or closed by the joining of Africa and Asia before the unknown primitive phocoids developed great vagility, particularly so if they first entered the sea in the North Atlantic and were slow to enter the warmer Tethys.

OTARIOID BIOGEOGRAPHY

The 22 m. y. of otarioid history consists of four events of approximately equal duration, each successively replayed by each of the 4 recognized families. Each event may be described briefly as a sequence showing appearance of a family, adaptation, diversification and dispersal, and disappearance with replacement by the next family. The otariids of today, the sea lions and fur seals, are in their diversification and dispersal stage. All 4 families appear to have been amphi-North Pacific in distribution, but only the odobenids and the otariids were able to disperse beyond this center of otarioid evolution.

The Enaliarctids

The earliest enaliarctids are known from California and Oregon. Material in the National Museum of Natural History in preparation and under study by Ray and Repenning indicates that between 22 and perhaps 16 m. y. ago there existed an appreciable variety of enaliarctids. The enaliarctids appear to have been closely tied to the shore, much as is the living sea otter, although they were obviously better adapted to deep diving than are the sea otters (Repenning, 1976b). One undescribed specimen from Japan (Y. Hasegawa, personal commun. 1974) indicates that the family was also present in the western North Pacific. It is presumed, therefore, that they were distributed entirely around the shores of the North Pacific although their northernmost known occurrence was about 45°N latitude. They are known to have lived in warm- to cool-temperate seas.

The Desmatophocids

The archaic desmatophocids appear to have evolved out of the ancestral enaliarctids possibly 17 m. y. ago (undescribed specimen, National Museum of Natural History) although their earliest published records are 14 to 16 m. y. old. They are known from California, Oregon, Alaska (specimen questionably desmatophocid), and from several specimens in Japan. Although they were more fully adapted to pelagic marine life than were the enaliarctids (Repenning, 1976b), they also are not known in areas other than the North Pacific basin. They survived, in diminishing abundance, until about 9 m. y. ago.

The Odobenids

The oldest odobenids are about 14 m. y. old; by about 10 m. y. ago they were the most abundant and successful otarioids in the North Pacific. Excluding living otarioid genera, there are more named odobenid genera than all other otarioid genera together; including living genera, the named genera of odobenids equal those of the named otariids. There is, of course, only 1 living odobenid, the walrus *Odobenus,* but from 10 to 5 m. y. ago there was a remarkable variety. This great variety included both pelagic feeders that lived much as do modern sea lions, and bottom feeders that lived as does the modern walrus. The pelagic feeders included some of the largest pinnipeds the world has known. The diverse odobenids of 10 to 5 m. y. ago have been divided into 2 subfamilies by Repenning and Tedford (1977). One, leading to modern walrus, was subtropical to tropical in habit in the North Pacific. The second, including all pelagic genera as well as some that appear to have been bottom feeders, was warm to cold temperate in habit in the North Pacific and is known from northern Mexico, California, and Oregon. This second subfamily, known as the dusignathine odobenids, never left the North Pacific and became extinct about 4 m. y. ago.

Ancestral representatives of the lineage leading to the modern walrus, representing the subfamily group known as the odobenine odobenids, are known in the North Pacific only from Baja California. Two localities are now south of the Tropic of Cancer and a third is just north of it; palinspastic consideration of Pacific plate movements indicates that all were deposited south of the Tropic of Cancer. The dusignathine odobenids and other otarioids have never been found this far south in contexts older than 5 m. y. by which time the North Pacific Ocean had cooled to approximately modern temperatures (Addicott, 1970). Age control on the Baja California localities of the odobenine odobenids is weak at present but they appear to be considerably younger than 9 m. y. and at least somewhat older than 5 m. y., probably at least 7 m. y. old (see discussion in Repenning and Tedford, 1977). The next younger record of the odobenine odobenids is from the North Atlantic in the lower part of the Yorktown Formation of North Carolina of approximately 4.5 m. y. age (Hazel, 1977, Fig. 5; correlative with Neogene Calcareous Nannofossil Zone NN12 and Neogene Planktonic Foraminiferal Zone N19). There is no younger record of the odobenid subfamily leading to modern walrus in the North Pacific basin except for late Pleistocene records.

It seems inescapable that the tropically adapted odobenine odobenids entered the Caribbean from the Pacific via the Central American Seaway, in the vicinity of Costa Rica, Panama, and Colombia, and from there, as a part of the Caribbean pinniped fauna, dispersed northward into the North Atlantic. There are no fossil records of walrus in the Atlantic south of Florida.

On both shores of the North Atlantic the further record of the evolution of odobenine odobenids into modern walrus is rather well recorded, and the earliest record of *Odobenus* in the North Pacific, probably less than 300,000 yr ago (Repenning and Tedford, 1977), is from Alaska indicating its return through the Arctic Ocean.

The Otariids

The earliest otariid is between 10 and 12 m. y. old. The transition into this first otariid from the last of the ancestral enaliarctids represents an extremely minor change. The taxonomic distinction can be defended on the basis of slight morphologic differences in taxa that are separated by a hiatus of possibly 4 m. y. of no known fossil record. Nevertheless, this slight change marks the end of 6 m. y. of evolution of the enaliarctid otarioids and the beginning of 12 m. y. of evolution of the sea lions and fur seals, Family Otariidae, the most diverse otarioids of today.

Evolution of the otariids for the first half of their existence continued to follow the pattern of rather slight change and of single direction toward the living genus *Arctocephalus*. The first suggestion of diversification seems to be recorded in fossils of about 6 m.y. ago that appear to show the beginning of the lineage leading to the Alaskan fur seal, *Callorhinus*. The first suggestions of a separate sea lion lineage are known from rocks in California and Japan that are around 3 m. y. old, after which diversification and evolution appear to be explosive in comparison to the previous 9 m. y. of otariid existence. In the round numbers that we are forced to deal with, this diversification took place shortly after the extinction of the last North Pacific odobenid. The modern otariid fauna consists of 7 genera and 14 species.

The earliest otariids from the Southern Hemisphere are perhaps 5 m. y. old, judging from very preliminary estimates of their age (Hoffstetter, 1968). Two equatorial crossings are indicated: first, the fur seals, presumably at least 5 m. y. ago, on the basis of the supposed age of the Peruvian record, followed, less than 3 m. y. ago, by the sea lions. Modern distribution appears to be compatible with this assumption, for the sea lions have succeeded in occupying only Australia, New Zealand, and both sides of southern South America while the fur seals are now circum-Antarctic in distribution. Lice endemic on living sea lions and fur seals indicate a similar history (Kim and others, 1975).

It is to be noted here that the otariids did not follow the odobenids into the North Atlantic by way of the Central American Seaway even though they must have passed along the shores of Costa Rica, Panama, and Colombia to reach the Southern Hemisphere.

PHOCOID BIOGEOGRAPHY

The 15 m. y. of known phocoid history cannot as yet be broken down into convenient historical events to simplify their discussion. The oldest phocoid fossils complete enough for interpretation are clearly divisible into monachine and phocine categories (Ray, 1976a). Although these most ancient of known seals are recognizable in the classification system constructed around the living seals, they show many primitive features that make inescapable the conclusion that the 2 subfamilies derive from one common phocoid ancestor as noted by Hendey and Repenning (1972).

Only 2 lineages, the modern subfamilies, appear to derive from this protophocoid and their known histories appear to have no temporal, geographic, or ecologic distinction for nearly 10 m. y., although later records certainly suggest that the modern ecologic distinctiveness of the 2 subfamilies must have been evolving during this time. Following this period, the monachine seals invaded the tropical and southern temperate to antarctic regions of the Atlantic and Pacific while the phocine seals shifted northward into the temperate and arctic Atlantic and crossed over to the North Pacific by way of the Arctic Ocean. Today the ranges of the 2 subfamilies are quite distinct except that the monachine elephant seal has reinvaded (or has not been displaced from) the North Pacific phocine sea.

North Atlantic

Phocine seals, *Leptophoca lenis* (True, 1906), and monachine seals, *Monotherium? wymani* (Ray, 1976a), are first known together from the Calvert Formation or its equivalent in Maryland and Virginia, along the mid-latitude Atlantic coast of the United States. These localities are of mid-temperate aspect and contain planktonic foraminifers indicative of zone N8 of Blow (1969) which is believed to be about 15 m. y. old. Although by no means known only by an isolated specimen, these seals are nevertheless not common, and only 2 species are recognized. *Leptophoca lenis*, or a closely related species, is also represented in the eastern Atlantic, in Belgian deposits believed to be of somewhat lesser age (Ray, 1976b); this Belgian form was described as *Prophoca proxima* (Van Beneden, 1877). No monachine seals have been recognized in the eastern North Atlantic at this earliest stage of known seal history.

The following period of time from about 14 to perhaps 5 m. y. ago is not as well-represented in collections of fossil seals from the western Atlantic. In this part of the Belgian record the monachine seals first appear, represented by 2 or 3 species, and 2 phocine species presumably are also present. A few remains in the western Atlantic suggest the probable presence of at least 3 species related to these.

Beginning about 5 m. y. ago the complexion of the North Atlantic seal population appears to change on either side of the Atlantic. Many more fossils are known, and the number of species increases slightly. At least 4 phocine species and at least 1 monachine are known from Belgium in the eastern Atlantic; 2 monachine species and 2 phocine species are known from North Carolina. These are minimal numbers of species; more have been named but sexual dimorphism in one case and synonymies not yet proven seem to indicate fewer taxa (Ray, 1978). A close generic and specific similarity is present on both sides of the Atlantic and Ray (1978) has suggested an amphi-North Atlantic distribution pattern. Nevertheless, the phocine seals appear more abundant in the European Atlantic record and the monachine seals appear more abundant in the American Atlantic. The most abundant record of western Atlantic phocoids is from the Yorktown Formation near Cape Hatteras in North Carolina where monachine remains outnumber phocine remains about 10 to 1 (Ray, 1978). In the collection of contemporaneous seals from Belgium there are about three phocine specimens for each monachine specimen. Both collections contain abundant specimens, but this results from the activities of man. The Belgian material derives from dredging and excavation in the vicinity of Antwerp; the North Carolina material results from phosphate mining. Thus, no significance can be attributed to the unusual abundance of seal remains. However, the relative abundance of monachine remains in North Carolina suggests that a center of monachine diversification existed in the western Atlantic.

Hazel (1971) has indicated that oceanic temperatures along the southern Atlantic seaboard appear to begin warming with the deposition of the Yorktown Formation, reaching warm-temperate conditions in the Cape Hatteras area about 4.5 m. y. ago and continuing to subtropical conditions by about 2 m. y. ago (Croatan Formation). During Yorktown deposition subtropical waters moved northward to within 100 miles (161 km) south of Cape Hatteras (Duplin Formation, Hazel, 1978). The Yorktown Formation contains the first records of odobenids believed to have been introduced to the Caribbean from the Pacific. Thus, the relative abundance of monachine remains in the Yorktown Formation, together with the first record of odobenids derived from the south, plus the evidence of a warming sea, all suggest a possible Caribbean-Gulf of Mexico center of monachine diversification.

The warming, beginning between 4 and 5 m. y. ago along the southeastern Atlantic coast of the United States, is not characteristic of the North Atlantic as a whole and about 3 m. y. ago, when conditions at Cape Hatteras were becoming subtropical, glaciation and the probable formation of the Labrador Current as a significant mass are judged to have begun by Berggren and Hollister (1974, p. 176). These authors suggest, as have others such as Hopkins (1972a), that the closure of the Central American Seaway would have deflected the Atlantic equatorial currents into the Gulf Stream thus greatly increasing its warming influence south of Cape Hatteras.

Certainly, with the cooling of the North Atlantic in areas other than the southeastern coast of the United States, the monachines retreated southward and the phocines adapted to more boreal temperatures. That this may have begun before the time of deposition of the Yorktown Formation is suggested by the appearance of abundant monachine seals in this warm-temperate deposit (Ray, 1978), by the appearance of monachine seals in the South Pacific apparently about this time, and also by records in the southern South Atlantic in Argentina and South Africa about 4 m. y. ago (Hendey and Repenning, 1972). The amphi-North Atlantic phocine fauna seems to have persisted despite evidence of a more significant Labrador Current, suggesting that the phocine seals were adapting to cooler waters at least 3 m. y. ago. All of these earliest records appear to be approximately contemporaneous with the Yorktown Formation of the Cape Hatteras region, and the adaptations that permitted this dispersal to other oceans must have begun evolving earlier. Some of the dispersals themselves must also have begun at least somewhat earlier.

In summary, the phocid history of the North Atlantic shows clearly the origin and diversification of both phocine and monachine seals in this oceanic basin. In addition it offers suggestive evidence that the modern temperature preferences of these two groups had begun to develop by 5 m. y. ago; that the major center of monachine diversification may have been in the warmer latitudes of the western Atlantic; that the phocines progressively favored cooler waters in adjustment to the cooling Atlantic; and that they have maintained an amphi-North Atlantic population at least for the last 15 m. y.

The greatest weakness in the North Atlantic record is that there is, at present, a satisfactory record of phocid faunas only at one latitude on each side of this Ocean; in the vicinity of Cape Hatteras on the west and of the English Channel on the east. This seriously handicaps attempts to infer latitudinal differences in faunas. Nevertheless, it is evident that the earliest known phocine and monachine seals were temperate-water animals and perhaps this also is demonstrated by the nature of the early seals of Paratethys, an ancient sea that extended from Vienna on the west to east of the Caspian Sea and the precursor of the modern Black and Caspian Seas.

Paratethys

One of the earliest descriptions of a fossil seal, by Eichwald (1853), was of specimens from Paratethys, an isolated northern sea at times confluent with Tethys, which was characterized by flora and fauna of tropical aspect. This early description of a fossil seal has greatly influenced subsequent interpretation of phocoid origins and affinities. Some have assumed that the seals somehow originated in tropical Tethys. On the other hand, others have charged the windmill and suggested that part of Paratethys was arctic because of the cold-water adaptations of living forms. Associated invertebrate fossils, particularly in South Dobrogea, Romania (Grigorescu, 1976), indicate that the Sarmatian seals of Paratethys lived in warm-temperate waters and thus the cold-water adaptation of modern seals derived from these must have evolved with global cooling, as it did with the North Atlantic phocines, and as suggested by Chapskii (1970).

The oldest Paratethyan phocid of sufficient record for interpretation is *"Phoca" vindobonensis* (Toula, 1897). This species has variously been interpreted as a primitive phocine seal (Ray, 1976b) or suggested as a form possibly ancestral to both phocine and monachine seals (Hendey and Repenning, 1972). It shows considerable similarity to the 15-m. y.-old *Leptophoca lenis* of the western North Atlantic and could be considered as a species of this genus although it is somewhat younger in age, being around 13 m. y. old. No one has suggested that it is a monachine seal. A few other phocoid fragments, as *Miophoca vetusta*

(Zapfe, 1937), have been described from Paratethys in deposits older than *"Phoca" vindobonensis*, but these are too inadequately known to be judged as being either phocine or monachine, although Chapskii (1955) has noted similarities between *Miophoca vetusta* and the living *Halichoerus*.

"Phoca" vindobonensis is from Vienna in the western basin of Paratethys and seems to be slightly younger than the isolation of Paratethys from Tethys. Morphologically and chronologically it appears to be the likely ancestor of some or all of the seals which evolved in isolation in Paratethys 1 to 3 m. y. later. Although other marine mammals are known, there are no phocine seals from the contemporary Serravallian and Tortonian seas of Mediterranean Tethys to the south. All fossil and living seals from the Mediterranean are clearly monachine. There is, therefore, some question of the geographic origin of *"Phoca" vindobonensis*, and there exists the strong possibility that the phocine seals may have entered Paratethys from the north, from the North Sea and across northwestern Germany. The similarities of *"Phoca" vindobonensis* to the somewhat older *Leptophoca lenis* of western North Atlantic and to contemporaneous *Leptophoca proxima* (Ray, 1978) of the Rhine Delta seem to suggest such an origin.

From 14 to at least 10 m. y. ago, phocine seals evolved in isolation in Paratethys, which had at most only restricted and very temporary connections with Tethys and which was formed by several basins, at times partially separated by larger and smaller islands (some formed by bryozoan reefs). Possibly favored by this discontinuity of basins and isolation from Tethys, several Paratethyan seals evolved interesting specializations best indicated by their femora (Grigorescu, 1976). Least specialized and most abundant, and most like the older *"Phoca" vindobonensis*, was *Phoca pontica*. Recent discoveries in South Dobrogea, Romania, of the dentition and temporal bones of this species show close similarities to the living Caspian seal, *Phoca caspica*, strengthening earlier suggestions of the ancestral position of *Phoca pontica* (Chiriac and Grigorescu, 1975). *Phoca caspica*, *Phoca hispida* of the Arctic Ocean, and *Phoca sibirica* of Lake Baikal make up the living species of the subgenus *Pusa*. Paleogeographic considerations make it seem most reasonable that *P. sibirica* gained access to Lake Baikal from the Arctic Ocean. This possibility is strengthened by the propensity of seals of the subgenus *Pusa* to move up fresh-water streams to inland bodies of fresh water to become endemic, a habit also seen in some other members of the genus *Phoca* and possibly of great antiquity as it may have typified the entry of *"Phoca" vindobonensis* into Paratethys from the Rhine Delta and Antwerp some 13 m. y. ago.

Many Paratethyan seals show an interesting mixture of phocine and monachine characters that is attributed to their antiquity and the relative recency of their derivation from a common phocoid ancestor. Presumed monachine features in the limb bones of *Monotherium maeoticum* have caused several authors, beginning with Trouessart (1897), to consider it a monachine seal. Unless an unknown monachine ancestor also swam up the Rhine River from Belgium some 13 m. y. ago or was somehow derived from a Tethyan monachine even earlier, no possible monachine ancestor is known, particularly in Paratethys. The monachine characters of *M. maeoticum* may have no more taxonomic significance than do the phocine characters Hendey and Repenning (1972) found in the monachine *Prionodelphis capensis* from South Africa.

Southern Oceans

Phocoid remains, for that matter also otarioid remains, from the fossil record of the Southern Hemisphere are almost unknown. Nevertheless, monachine seals, as well as otariids, are now circum-Antarctic in distribution and are present in the South Pacific and South Atlantic. The history of pinnipeds in the Southern Hemisphere is largely undiscovered; only one pre-Pleistocene seal has been satisfactorily described from the Southern Hemisphere, *Prionodelphis capensis* from South Africa (Hendey and Repenning, 1972). A few records are known which suggest that both groups of pinnipeds had entered the Southern Hemisphere at least 5 m. y. ago, but nearly all known fossil records are late Pleistocene or younger in age, less than 1 m. y. old.

The most significant record in the South Pacific is that of Hoffstetter (1968) who mentioned that pinnipeds had been found in a locality near Sacaco, Peru. Subsequent information (R. J. Hoffstetter, personal commun. 1973) indicates that these fossils are probably late Miocene or early Pliocene in age, approximately 5 m. y. old, and that they include otariids and phocoids. The second fossil record of note in the South Pacific is that of Fleming (1968) who reports a mandibular ramus of *Ommatophoca* from the early Pleistocene of New Zealand. King (1973) has subsequently identified this specimen as being *Ommatophoca rossi*, the living Ross seal of Antarctica.

In the South Atlantic, Frenguelli (1922) described a few isolated teeth from Pliocene deposits of Argentina as a squalodont cetacean, *Prionodelphis rovereti*. Subsequent discovery of a mandibular fragment prompted Cabrera (1926) to state that it was a pinniped. In 1972, Hendey and Repenning described *Prionodelphis capensis* from South Africa; the deposits are believed to be 4 to 5 m. y. old based upon an extended correlation of the associated terrestrial fauna. Subsequently, a few humeri at least assignable to the same genus have been found in the Yorktown Formation of North Carolina (Ray, 1978), thus strengthening the South African correlation and suggesting the most interesting concept of 1 genus of seal known from South Africa, Argentina, and Atlantic North America.

Prionodelphis is the best known fossil phocid in the world. The material includes nearly complete skulls as well as most postcranial material (Hendey, 1976). The describers of *P. capensis* concluded that it was ancestral to some of the antarctic monachine seals but that it was already, at this early date, too specialized to be ancestral to *Lobodon* and *Hydrurga* in the modern antarctic fauna

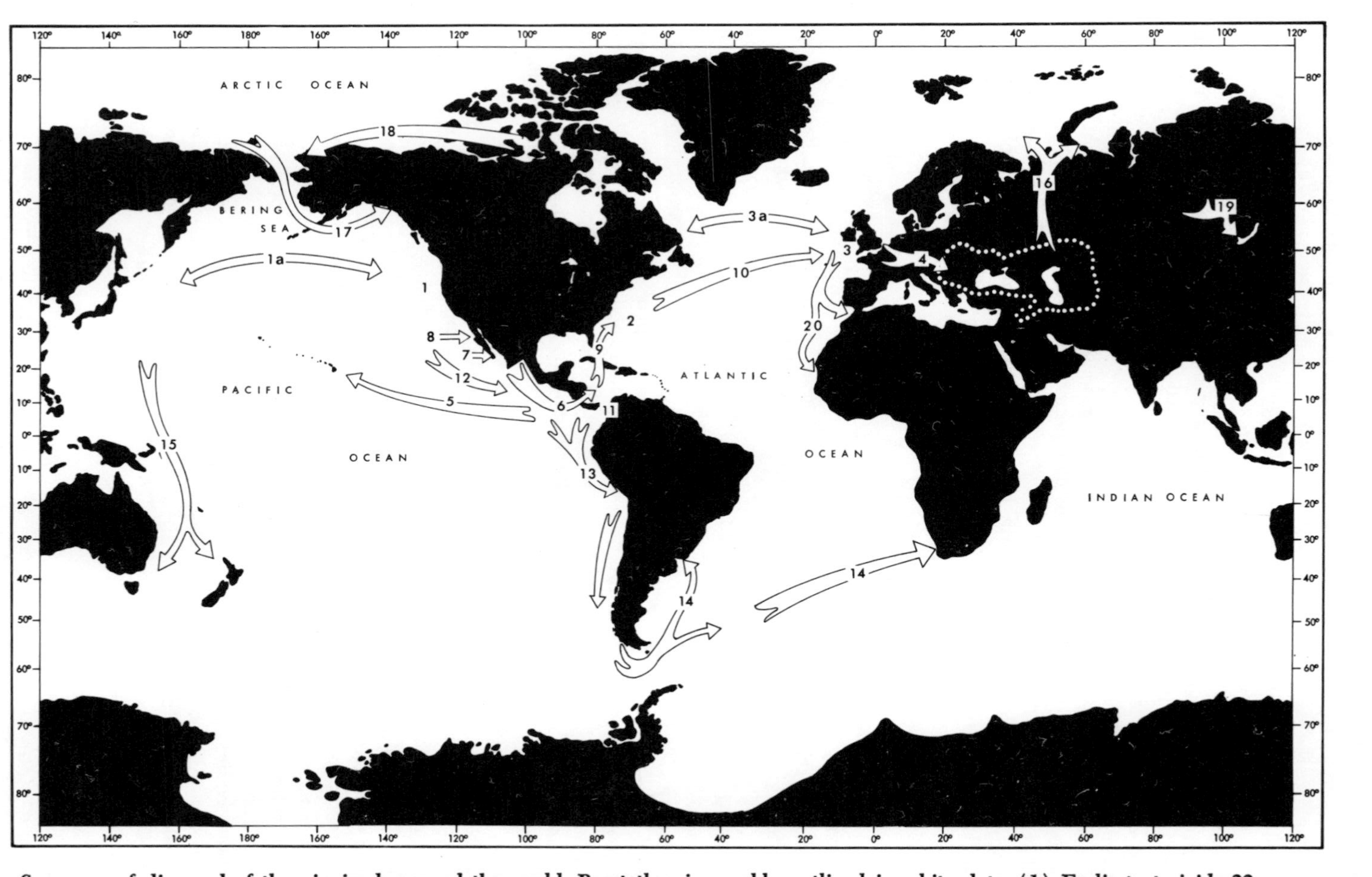

Figure 2. Summary of dispersal of the pinnipeds around the world. Paratethys is roughly outlined in white dots. (1) Earliest otarioids 22 m. y. ago; earliest odobenids 14 m. y. ago; earliest otariids 10 m. y. ago; odobenids became extinct 4 m. y. ago. (1a) Otarioids amphi-North Pacific at least during the last 18 m. y. (2) Earliest monachines and phocines 15 m. y. ago; earliest odobenids 5 m. y. ago. (3) Earliest phocines 13 m. y. ago; earliest monachines 10 m. y. ago; earliest odobenids 4 m. y. ago. (3a) Phocine and monachine seals amphi-North Atlantic at least from 10 to 5 m. y. ago; phocines and odobenids amphi-North Atlantic at least 4 m. y. ago to present. (4) Earliest phocine to Paratethys about 13 m. y. ago. (5) Caribbean-Pacific monachines follow equatorial currents to Hawaiian Islands during warmest time of North Pacific possibly 15 m. y. ago. (6) Odobenids enter Caribbean fauna possibly 8 m. y. ago. (7) Most southerly fossil odobenid, 7 to 8 m. y. ago. (8) Most southerly fossil otariid, 7 to 8 m. y. ago. (9) Caribbean monachines and odobenids move northward with warming Gulf Stream about 5 m. y. ago. (10) Odobenids disperse around North Atlantic about 4 m. y. ago. (11) Central American Seaway becomes impassable about 5 m. y. ago. (12), (13) Fur seals (otariids) move southward with cooling of Central American Pacific about 5 m. y. ago; sea lions (otariids) follow less than 3 m. y. ago. (13) With cooling of Central American Pacific, Pacific monachines cross the equator and move southward along the Peru Current with the fur seals about 5 m. y. ago or at an earlier date; sea lions (otariids) follow less than 3 m. y. ago. (14) Monachines enter South Atlantic (Argentina and South Africa) about 4.5 m. y. ago; fur seals follow later; sea lions still later. Fur seals and monachines are now circum-antarctic; sea lions are now present only on Atlantic and Pacific sides of southern South America. (15) Sea lions disperse to Australia-New Zealand region no more than 3 m. y. ago and possibly from western North Pacific. (16) *Pusa* enters Arctic Ocean from Paratethys 3 m. y. ago. (17) *Pusa* enters North Pacific perhaps 2.5 m. y. ago. (18) Odobenids (as *Odobenus*) return to North Pacific perhaps 600,000 yr ago. (19) *Pusa* enters Lake Baikal from lakes at the foot of the Central Siberian ice sheet about 300,000 yr ago. (20) Monachines invade the Mediterranean, probably from an eastern Atlantic source, both before and after the Messinian desiccation 5 to 6 m. y. ago and also enter the Black Sea and extend down the North African coast.

and could best be considered as a possible ancestor to *Ommatophoca* and *Leptonychotes* of Antarctica. They thus concluded that other unknown monachine seals must have been present in the Southern Hemisphere. Significantly, although the South African locality has produced abundant phocoid remains, no remains of otarioids have been found. Other marine vertebrates are abundant at the locality, and the invertebrates indicate a temperate environment with rocky coastlines nearby (Kensley, 1972). The locality is now well within the breeding range of otariid fur seals.

For both phocoid and otarioid paleobiogeographic history, the Southern Hemisphere is potentially the most productive area of discovery.

DISCUSSION

Both otarioids and phocoids originated in the north temperate oceans and adapted progressively to polar or tropical climates. From the North Pacific the otarioids moved southward; the odobenids entered the Atlantic, but the otariids did not. From the North Atlantic the phocoids moved both south and north; the monachine seals entered the South Pacific and the South Atlantic, and the phocine seals entered the Arctic Ocean. They were followed later by the North Atlantic odobenids, who returned to the North Pacific via the Arctic. Some phocoids moved through rivers to inland seas. The antarctic monachines adapted to tropical climates only to adapt then to antarctic environments. In the temperate southern seas, with no landmasses to block the way, both groups of pinnipeds have circled the world (Fig. 2).

Modern distribution of the pinnipeds clearly shows the maximum known specialization in temperature preference. Most tropical of the modern pinnipeds are the monachine species *Monachus monachus,* known from the Black Sea, throughout the Mediterranean, and down the west African coast to 20°N latitude; *Monachus tropicalis,* formerly present throughout the Caribbean and, of all seals, the record holder for tropical habitat; and *Monachus schauinslandi,* from the Hawaiian Island chain. Most polar in habit are, probably, the phocine *Phoca (Pusa) hispida,* the ringed seal of the Arctic Ocean, and the monachine *Leptonychotes weddelli,* the Weddell seal of the continental margins of Antarctica. Other monachines extend into the range of *Leptonychotes* in the Southern Hemisphere but tend to adjust to the position of the ice front, except for *Mirounga leonina* which ranges from Antarctica northward possibly to 35°S latitude. Other phocines extend into the range of *Phoca hispida* but tend to adjust to the position of the ice front (or at least cold water) and the temperature tolerant monachine, *Mirounga angustirostris,* ranges from Mexico northward nearly to 50°N latitude. The otariids distribute themselves in mid-latitudes; the odobenids are present in the Arctic Ocean, northern Atlantic, and are more northerly than the otariids in the Pacific although their southern limits are determined today by the activities of man.

From their temperate origins the change of climatic preferences has been marked in the pinnipeds. Climates have also changed markedly since the origins of the pinnipeds and interpretations are more complex because of this. For this reason, the discussion of pinniped biogeography will begin with the tropical zone, possibly the most stable climatically, if not in area.

The Equator

Evidence is nonexistent regarding pinnipeds in the equatorial area except that otariids live on the Galapagos Islands; this is undoubtedly because the Peru Current chills these Islands to a tolerable condition. Possibly the most significant bit of available evidence indicating when the pinnipeds crossed the equatorial barrier is the informal statement of R. J. Hoffstetter (personal commun. 1973) that both otarioids and phocoids have been found in deposits near Sacaco, Peru, tentatively dated as latest Miocene or early Pliocene, or about 5 m. y. old. Almost as significant is the discovery of a monachine seal apparently ancestral to *Leptonychotes* in temperate deposits of South Africa of perhaps 4 to 5 m. y. ago.

In the North Pacific area the period of time about 15 m. y. ago was marked by very warm waters which progressively cooled to modern temperatures by about 5 m. y. ago (Addicott, 1970). A comparable picture has been described in the South Pacific (Devereaux, 1967). Except for the complication of a strengthening Gulf Stream about 5 m. y. ago, the North Atlantic presents a comparable picture (Berggren and Hollister, 1974). From this picture of cooling oceans and from the age of the Sacaco, Peru, fossils, it seems probable that the otariids were unable to cross the equator until the Panamic Pacific was sufficiently cooled. Modern otariids are known farther south along the Mexican coast of Baja California than are any of their fossil records. Furthermore, it seems reasonable to infer that an evolving monachine fauna inhabited the Caribbean-Pacific faunal province on either side of the Central American Seaway as long ago as 15 m. y., and did not vacate this province until the North Atlantic equatorial currents were diverted into the Gulf Stream and the Pacific cooled.

The Mediterranean

In the Mediterranean there is no evidence of any phocine seal, despite a good record of other marine mammals. The only seals known in this area are monachine. All phocine seals loosely attributed to Tethys are from Paratethys, and available evidence indicates that they entered Paratethys from the north, rather than from Mediterranean Tethys to the south. Modern *Monachus* has invaded the area of Paratethys in the Black Sea, but there is no fossil Paratethyan seal which resembles *Monachus* and it is clear that *Monachus* came from the Mediterranean to the Black Sea by way of the Bosporus in Quaternary time. A Tethyan origin of the phocoids seems less likely than has been thought in the past.

The Caribbean

There are few fossil pinnipeds known from Florida (all monachine or odobenine), and none is known from the Caribbean. Nevertheless, modern *Monachus tropicalis* had a range extending from Florida and the Bahamas through the Gulf of Mexico and the Caribbean. The preponderance of monachine fossil remains near Cape Hatteras along the Atlantic Coast of North America, in deposits about 5 m. y. old that show the earliest trends toward warming waters along the southeastern coast of the United States, contrasts with the phocine preponderance in coeval faunas of Belgium, just as this local warming trend contrasts with the overall cooling of the North Atlantic Ocean. Thus, it seems reasonable to infer that there existed a monachine population in the Caribbean-Gulf of Mexico area, already adapted to warmer waters and depressed southward by the cooling Atlantic, but able to return northward with the warming of the Gulf Stream; just as it seems reasonable to infer that Pacific odobenids immigrating into the Caribbean remained a part of this tropical fauna until the warming of the southeastern coast of the United States enabled them to move northward. The oldest Atlantic odobenids, which earlier are known only from the Pacific coast of Mexico, are first found with the monachine seals in these same 5-m. y.-old deposits near Cape Hatteras.

This unknown Caribbean pinniped fauna may have been in existence for a long time. The Hawaiian monk seal, *Monachus schauinslandi*, is characterized by some of the most primitive features known in all phocid seals, living and fossil (Repenning and Ray, 1976), and, as will be discussed, must once have been a part of this unknown ancient Caribbean fauna. In several respects the Hawaiian monk seal appears to be more primitive than the oldest known monachine, *Monotherium*? *wymani*, of about 15 m. y. ago. From this, it appears likely that the differentiation that led to its relict survival in the Hawaiian Islands may have begun between the Caribbean and the Pacific Provinces more than 15 m. y. ago.

The Central American Seaway

Perpetual debate exists as to when this Seaway closed, and the debate persists because the inferred dates of closure are based upon different sorts of evidence. Obviously, it was closed to the east-west dispersal of seals and mollusks, uncomfortable in shallow-water tropical seas, before it was open to the free north-south dispersal of terrestrial life. Judging from coastal paleotemperatures near Cape Hatteras and southward, it would seem that by 5 m. y. ago the Central American Seaway had closed enough to begin diverting the Atlantic equatorial currents from their westward route into the Pacific to a northward course into the Gulf Stream; the shallowing Seaway must have warmed proportionately and the Central American Pacific must have cooled proportionately. The Seaway must have been completely closed to Atlantic currents by the time of the subtropical deposition of the Croatan Formation in the Cape Hatteras area (Hazel, 1971, personal commun. 1976) some 2 m. y. ago, at which time South American mammals were clearly well-distributed in North America.

In the eastern Pacific the otariids passed the Central American Seaway en route to Sacaco, Peru, sometime around 5 m. y. ago, and they did not pass into the Caribbean. Nor were they likely to have entered the Central American Pacific until the Atlantic equatorial currents were diverted and the Panamanian Pacific was allowed to cool, between 5 (Yorktown) and 2 (Croatan) m. y. ago. Odobenine odobenids, of tropical habit and of 7 or 8 m. y. age, are known from Baja California and their next known descendants, quite a bit more advanced toward living walrus, are known from the 5-m. y.-old Yorktown Formation near Cape Hatteras and elsewhere on the Atlantic Coast. Sometime between 5 and perhaps 8 m. y. ago these odobenids must have passed into the Caribbean through the Central American Seaway and there evolved more modern features. Then, presumably following the diversion of the equatorial currents northward into the Gulf Stream, they dispersed northward to North Carolina, about 5 m. y. ago, thence to Belgium in the eastern North Atlantic.

It would seem that monachine seals must have passed through the Central American Seaway, westward into the Pacific more than 5 m. y. ago. The living elephant seal, *Mirounga*, is present in both North and South Pacific. Its ancestry is uncertain, but Ray (1978) notes similarities to both *Monachus* and *Mirounga* in the 5-m. y.-old *Callophoca* of both western and eastern North Atlantic. The ability of this genus to participate in the amphi-North Atlantic fossil pinniped fauna may foreshadow the even wider temperature tolerance of the living *Mirounga*, in both North and South Pacific. Furthermore, Ray (1978) suggests that *Callophoca ambigua* is possibly the male of *Callophoca obscura*, also known on both sides of the Atlantic, because its single distinctive feature appears to be larger size. This also seems to correlate with living *Mirounga*, which is notable among phocid seals for its sexual dimorphism. In addition, this would indicate that *Callophoca* was the only monachine of amphi-North Atlantic distribution that was younger than 5 m. y. Other than these suggestions of similarity in morphology, sexual dimorphism, and temperature tolerance, there is no evidence of the ancestry of the living elephant seal.

The Hawaiian Islands

King and Harrison (1961), as well as other authors, indicate that the Hawaiian monk seal, *Monachus schauinslandi*, is more similar to the Caribbean *M. tropicalis* than it is to the Mediterranean *M. monachus*. Logically, it would seem to have entered the Pacific through the Central American Seaway and to have followed the North Equatorial Current to the Hawaiian Islands. The absence of *Monachus* in the eastern North Pacific fossil record (almost entirely known only north of 25°N latitude), and its restriction to southern localities in the Pleistocene of the western North Atlantic indicates that the genus is intolerant of cooler water. Its primitive nature and warm-water preference suggest that it crossed the Pacific to

Hawaii sufficiently far back in time that the fullest effect of the Atlantic equatorial currents passing between North and South America maintained near Caribbean temperatures en route and that extremely primitive features were still present. One is inclined also to suspect that it might have dispersed to Hawaii during the period of warmest water in the North Pacific, some 15 m. y. ago.

The lack of any significantly old fossil seals in the Caribbean and Gulf of Mexico makes any estimate of the antiquity of this inferred center of monachine and odobenine evolution quite conjectural. However, the retention of some features in the Hawaiian monk seal which are more primitive than those of the living Caribbean monk seal or, in fact, than those of the 15-m. y.-old *Monotherium*? *wymani* (Repenning and Ray, 1976), would seem to suggest a very early separation from the Caribbean monachine population. It seems more likely that the ancestors of *Monachus schauinslandi* had left the Caribbean-Pacific monachine population by 15 m. y. ago than that *Monotherium*? *wymani* of Virginia had no contact with the Caribbean population.

Current estimates of the age of the Hawaiian Island chain indicate that Midway Island is about 18 m. y. old, at which time it was also closer to the equator (Dalrymple and others, 1973). The seal may well have occupied the western islands first and moved eastward as new ones developed.

Paratethys and the *Pusa* Problem

Mention has been made of the endemic Paratethyan seals of 10 to 13 m. y. ago. The similarity of some of these seals, especially *Phoca pontica* and *Phoca pannonica*, to the living species of the subgenus *Pusa* has led a number of authors to suggest that this ancient seal was ancestral to *Pusa*. Recent discovery of much additional material of *Phoca pontica* greatly strengthens this interpretation (Grigorescu, 1976). The distribution of the 3 living species of the subgenus is difficult to explain biogeographically: *Phoca* (*Pusa*) *caspica* lives in the Caspian Sea, a remnant of Paratethys, and is a further argument for the ancestral position of *Phoca pontica; Phoca* (*Pusa*) *sibirica* lives in isolated Lake Baikal of central Asia; *Phoca* (*Pusa*) *hispida* lives in the Arctic Ocean and extends into the North Pacific and North Atlantic, including the Baltic Sea and lakes Saimaa and Ladoga.

Except for the presumably ancestral *Phoca pontica*, and *Phoca pannonica*, and a single radius from Alaska described in Barnes and Mitchell (1975), there is no fossil record of the subgenus *Pusa* upon which reconstruction of its dispersal can be based. The Alaskan radius is from the North Pacific, and as will be discussed, cannot be older than the opening of the Bering Strait connecting the Pacific and the Arctic Oceans about 3.5 m. y. ago.

At some time earlier than this Alaskan fossil, *Pusa* must have gained access to the Arctic Ocean from Paratethys. Paleogeographically, the most likely time was about 3 m. y. ago when the present Caspian Sea was greatly enlarged, extending along the western flank of the Ural Mountains northward beyond the latitude of Moscow to approximately 58°N latitude. At this time the Arctic Ocean had transgressed southward across northern Russia well below the Arctic Circle to about 61°N latitude. According to Grossheim and Khain (1967, Sheet 20), these two aquatic extensions were within 300 miles (483 km) of each other at the western foot of the Urals and were separated by lowlands supporting a spruce forest along the present course of the Kama River just north of the city of Perm. According to these authors, continental climatic deterioration had progressed to the point that the high mountains then separating the Caspian and Black Seas were glaciated. It would seem that *Pusa* could have left the Caspian Paratethys most easily at this time and could also have become adapted to rather cold climates before doing so. The Arctic Ocean probably was either ice-free or only moderately ice-bound 3 m. y. ago (Herman, 1970). If this was the origin of Arctic *Pusa*, the fossil radius from Alaska is probably less than 3 m. y. old. Diatoms from a sample 600 ft (183 m) higher in the section correlate with Schrader's DSDP-NPD Zones VII to V according to John A. Barron (personal commun. 1976) and are believed to be between 1.85 to 2.5 m. y. old.

During the middle Pleistocene ice advance, about 200,000 yr ago, a continental ice sheet filled the Western Siberian Plain and Grossheim and Khain (1967, Sheets 25 and 26) indicate that very large lakes existed along its advancing margins in the region of the Yenisey River. Should arctic *Pusa* have been depressed southward in these lakes by the advancing ice sheet, their entry into Lake Baikal would have been almost unavoidable even though they would have had to swim up the then abbreviated Yenisey River a distance considerably greater than that which presumably permitted passage from the enlarged Caspian to the extended Arctic Ocean.

Arctic Ocean

The Arctic Ocean was a part of the rest of the world long before it was joined with the North Pacific; the land bridge between Alaska and Siberia is common knowledge. According to Hopkins (1972b) "The Pacific and Arctic Oceans became connected . . . during late Miocene time . . . The seaway was disrupted and a . . . land bridge reestablished about 5 million years ago . . . Continuing crustal warping brought the Bering and Chukchi Seas into existence . . . about 3.5 m. y. ago."

The 3.5-m. y.-origin of the Bering and Chukchi Seas is based upon the earliest Pacific boreal mollusks found in the Tjörnes Beds of Iceland (Einarsson and others, 1967). The late Miocene connection of the Pacific with the Arctic Ocean was based upon the appearance of North Pacific mollusks in the western North Atlantic, primarily *Mya arenaria* in the Yorktown Formation, formerly considered to be of late Miocene age (MacNeil, 1965; Durham and MacNeil, 1967). Current evidence indicates that the type locality of the Yorktown Formation (where *Mya arenaria* is found) includes a fauna that belongs to the ostracode assemblage zone of *Orionina vaughani* (J. E.

Hazel, personal commun. 1971). This zone is correlated with Calcareous Nannofossil Zones NN13 to NN15 and is between 4.5 and 3.0 m. y. old (Hazel, 1977); it is younger than the seals in this Formation near Cape Hatteras. Thus, the earliest occurrence of North Pacific *Mya* in the western Atlantic, formerly believed to be of late Miocene age, may well represent the same Pacific mollusk invasion of the North Atlantic that is recorded in the Tjörnes Beds of Iceland, and probably was about 3.5 m. y. ago. Earlier thoughts, that the odobenids entered the Atlantic from the Pacific by way of the Arctic Ocean, were based upon the food of living walrus (including *Mya*) and the Yorktown association in the Atlantic; they now are clearly incorrect. The odobenids arrived in the Atlantic much earlier than did the North Pacific boreal mollusks and did so through the Central American Seaway, rather than following *Mya* through the Arctic Ocean.

Barnes and Mitchell (1975) discuss the earliest North Pacific records of phocoid seals. Their Specimen I is probably from the Port Orford Formation of Cape Blanco, Oregon, and is now believed to be early Pleistocene, less than 2 m. y. old (see discussion in Repenning, 1976a). Their Specimens II, III, and IV were considered to be late Pleistocene. They also report *Phoca* sp. from Humboldt County, California, and the reference they give in support of its age indicates that it is correlative to the Moonstone Beach fauna of this area and is less than 2 m. y. old (see discussion in Repenning, 1976a).

Their last-mentioned specimen was USNM 23876, a phocine radius collected by George Plafker 5,000 ft (1,524 m) below the top of the Yakataga Formation in the Malaspina District, Alaska. Available information on this specimen, mentioned earlier in connection with the origin of the subgenus *Pusa*, indicates it is from the Pliocene part of the Formation and probably middle Pliocene in age; this is in the West Coast Provincial Megafossil Chronology in which the "Pliocene" is as much as 9 m. y. old. The radius appears assignable to the subgenus *Pusa* and, judging from the above discussion of the time of connection between the North Pacific and the Arctic Ocean, it could be as much as 3.5 m. y. old. If the subgenus *Pusa* does derive from *Phoca pontica* of Paratethys, as seems likely, the fossil from Alaska must be less than 3 m. y. old based upon optimum paleogeographic conditions for entry of *Pusa* into the Arctic Ocean, as discussed earlier.

Clearly, phocines of Atlantic derivation were in the North Pacific 2 m. y. ago and could have entered with the opening of the Bering Strait as much as 3.5 m. y. ago.

The Southern Hemisphere

Clearly, the otariids invaded the Southern Hemisphere through the South Pacific. Once the tropical barrier was crossed some 5 m. y. ago, the cold waters of the Peru Current provided an avenue for dispersal. South of South America and Africa they dispersed around Antarctica on the West Wind Drift although the sea lions, crossing the equator no more than 3 m. y. ago, have not completed the encircling of Antarctica. It is not known if the otariids also went south along the western (Asiatic) side of the South Pacific; the warm-water barrier would seem to be more formidable in this area. Specimens in Japan under study by Y. Hasegawa may provide some clues.

The monachine seals in the Caribbean area, once they entered the Pacific, also appear to have entered the Southern Hemisphere along the Peru Current. If the record from southern Peru does indicate the first arrival of pinnipeds in the South Pacific, it indicates the joint arrival of otariid fur seals and monachine phocids. This suggests that, except for those that moved to Hawaii, the Caribbean-Pacific monachines remained in this area of the Pacific until it cooled because of the deflection of the Atlantic equatorial currents away from passage into the Pacific. Once cooled, the warm-water monachines had little reason to remain in the Central American Pacific and the cool-water otariids had less reason to avoid it.

On the other hand, the presence of monachine seals in abundance in South Africa, and in much less abundance in Argentina, as much as 4 m. y. ago and the total lack of otariid fossils in the South Atlantic at this time suggest that the Peruvian record may not record the earliest entry of monachine seals into the South Pacific, or that they moved southward into the South Atlantic from the Caribbean. Southward dispersal into the South Atlantic is uncertain; it seems quite possible that *Prionodelphis* of the western North Atlantic also crossed Panama to the Pacific, moved southward along the Peru Current to the West Wind Drift, and thus arrived in Argentina and South Africa.

ACKNOWLEDGMENTS

We wish to thank J. E. Hazel and D. M. Hopkins for information and discussion relative to the distribution of the pinnipeds. R. J. Hoffstetter and Y. Hasegawa have provided significant information about undescribed fossils from Peru and Japan. We thank D. M. Hopkins, J. G. Vedder, and F. H. Fay for thoughtful review of this paper. Jeffrey Lund prepared the illustrations, on funds provided through the Smithsonian Research Foundation.

REFERENCES

Addicott, W. O., 1970, Tertiary paleoclimatic trends in the San Joaquin Basin, California: U.S. Geol. Survey Prof. Paper 644-D, p. 1-19.

Barnes, L. G., and Mitchell, E. D., 1975, Late Cenozoic northeast Pacific Phocidae, *in* Ronald, K., and Mansfield, A. W., eds., Biology of the seal: Rapp. P.-v. Réun. Cons. Internat. Explor. Mer, v. 169, p. 34-42.

Berggren, W. A., and Hollister, C. D., 1974, Paleogeography, paleobiogeography and the history of circulation in the Atlantic Ocean, *in* Hay, W. W., ed., Studies in paleo-oceanography: Soc. Econ. Paleontologists and Mineralogists Spec. Pub. 20, p. 126-186.

Blow, W. H., 1969, Late middle Eocene to Recent planktonic foraminiferal biostratigraphy, *in* Bronnimann, P., and Renz, H. H., eds., Internat. Conf. Planktonic Microfossils, 1st, Geneva 1967, Proc., Vol. 1: Leiden, E. J. Brill, p. 199-421.

Cabrera, Á., 1926, Cetáceos fósiles del Museo de La Plata: Rev. Mus. La Plata, v. 29, p. 363-411.

Chapskii, K. K., 1955, An attempt at revision of the systematics and diagnoses of seals of the subfamily Phocinae: Akad. Nauk SSSR Zool. Inst. Trudy, no. 17, p. 160-199 (in Russian).

——— 1970, The concept of the arctic origin of pinnipeds and another solution to this problem, *in* Tolmachev, A. I., ed., The Arctic Ocean and its coastline in the Cenozoic: Leningrad, Hydrometeorological Publishers, p. 166-173 (in Russian).

Chiriac, M., and Grigorescu, D., 1975, Asupra focilor Bessarabiene din Dobrogea de Sud: Studii Cercet., Geol. Geofiz. Geogr., Geologie, v. 20, p. 89-110 (in Romanian, English abstract).

Dalrymple, G. B., Silver, E. A., and Jackson, E. D., 1973, Origin of the Hawaiian Islands: Am. Scientist, v. 61, p. 294-308.

Devereaux, I., 1967, Oxygen isotope paleotemperature measurements on New Zealand Tertiary fossils: New Zealand Jour. Sci., v. 10, p. 988-1011.

Durham, J. W., and MacNeil, F. S., 1967, Cenozoic migrations of marine invertebrates through the Bering Strait region, *in* Hopkins, D. M., ed., The Bering Land Bridge: Stanford, Calif., Stanford Univ. Press, p. 326-349.

Eichwald, C. T., von, 1853, Lethaea Rossica, ou Paléontologie de la Russie, Vol. 3: Stuttgart, E. Schweizerbart'sche, 533 p.

Einarsson, T., Hopkins, D. M., and Doell, R. R., 1967, The stratigraphy of Tjörnes, northern Iceland, and the history of the Bering Land Bridge, *in* Hopkins, D. M., ed., The Bering Land Bridge: Stanford, Calif., Stanford Univ. Press, p. 312-325.

Fleming, C. A., 1968, New Zealand fossil seals, *in* Notes from the New Zealand Geological Survey 5: New Zealand Jour. Geology and Geophysics, v. 11, p. 1184-1187.

Frenguelli, J., 1922, *Prionodelphis Rovereti.* Un representante de la familia "Squalodontidae" en el paranense superior de Entre Rios: Acad. Nac. Ciênc. Cordoba Bol., v. 25, p. 491-500.

Grigorescu, Dan, 1976, On the Paratethyan seals: Systematic Zoology, v. 25, p. 407-419.

Grossheim, V. A., and Khain, V. E., 1967, Paleogene, Neogene, and Quaternary, Vol. IV, *in* Vinogradov, A. P., ed., Atlas of the lithological-paleogeographical maps of the USSR: Moscow, Ministry of Geol. of the USSR and Academy of Sciences of the USSR (in Russian and English).

Hazel, J. E., 1971, Paleoclimatology of the Yorktown Formation (upper Miocene and lower Pliocene) of Virginia and North Carolina, *in* Oertli, H. J., ed., Paléoécologie d' Ostracodes: Centr. Recherche Pau-SNPA Bull., v. 5 (suppl.), p. 361-375.

——— 1978, Age and correlation of the Yorktown (Pliocene) and Croatan Formations (Pliocene and Pleistocene) at the Lee Creek, North Carolina, open pit mine: Smithsonian Contr. Paleobiology (in press).

Hendey, Q. B., 1976, The Pliocene fossil occurrences in 'E' Quarry, Langebaanweg, South Africa: South African Mus. Ann., v. 69, p. 215-247.

Hendey, Q. B., and Repenning, C. A., 1972, A Pliocene phocid from South Africa: South African Mus. Ann., v. 59, p. 71-98.

Herman, Yvonne, 1970, Arctic paleo-oceanography in the late Cenozoic time: Science, v. 169, p. 474-477.

Hoffstetter, R. J., 1968, Un gisement de vertébrès tertiaires à Sacaco (Sud-Pérou), témoin néogène d'une migration de faunes australes au long de la côte occidentale sud-américaine: Acad. Sci. Paris Comptes Rendus, sér. D, v. 267, p. 1273-1276.

Hopkins, D. M., 1972a, Changes in oceanic circulation and late Cenozoic cold climates: Internat. Geol. Cong., 24th, Montreal 1972, Abstracts, p. 370.

——— 1972b, The paleogeography and climatic history of Beringia during late Cenozoic time: Inter-Nord, no. 12, p. 121-150.

Kellogg, Remington, 1922, Pinnipeds from Miocene and Pleistocene deposits of California: California Univ. Pubs. Geol. Sci., v. 13, p. 23-132.

Kensley, Brian, 1972, Pliocene marine invertebrates from Langebaanweg, Cape Province: South African Mus. Ann., v. 60, p. 173-190.

Kim, K. C., Repenning, C. A., and Morejohn, G. V., 1975, Specific antiquity of the sucking lice and evolution of otariid seals, *in* Ronald, Keith, and Mansfield, A. W., eds., Biology of the seal: Rapp. P.-v. Réun. Cons. Internat. Explor. Mer, v. 169, p. 544-549.

King, J. E., 1973, Pleistocene Ross seal (*Ommatophoca rossi*) from New Zealand: New Zealand Jour. Marine and Freshwater Research, v. 7, p. 391-397.

King, J. E., and Harrison, R. J., 1961, Some notes on the Hawaiian monk seal: Pacific Science, v. 15, p. 282-293.

MacNeil, F. S., 1965, Evolution and distribution of the genus *Mya,* and Tertiary migrations of Mollusca: U. S. Geol. Survey Prof. Paper 483-G, p. G1-G51.

Mitchell, E. D., and Tedford, R. H., 1973, The Enaliarctinae, a new group of extinct aquatic Carnivora and a consideration of the origin of the Otariidae: Am. Mus. Nat. History Bull., v. 151, p. 201-284.

Mivart, St. G., 1885, Notes on the Pinnipedia: Zool. Soc. London Proc., p. 484-501.

Ray, C. E., 1976a, *Phoca wymani* and other Tertiary seals (Mammalia: Phocidae) described from the eastern seaboard of North America: Smithsonian Contr. Paleobiology no. 28, p. 1-36.

——— 1976b, Geography of phocid evolution: Systematic Zoology, v. 25, p. 391-406.

——— 1978, Seals and walruses (Mammalia: Pinnipedia) of the Yorktown Formation of Virginia and North Carolina: Smithsonian Contr. Paleobiology (in press).

Repenning, C. A., 1976a, *Enhydra* and *Enhydriodon* from the Pacific Coast of North America: U.S. Geol. Survey Jour. Research, v. 4, p. 305-315.

——— 1976b, Adaptive evolution of sea lions and walruses: Systematic Zoology, v. 25, p. 375-390.

Repenning, C. A., and Ray, C. E., 1977, The origin of the Hawaiian monk seal: Biol. Soc. Washington Proc., v. 89, p. 667-688.

Repenning, C. A., and Tedford, R. H., 1977, Otarioid seals of the Neogene: U.S. Geol. Survey Prof. Paper 992, p. 1-93.

Savage, R. J. G., 1957, The anatomy of *Potamotherium,* an Oligocene lutrine: Zool. Soc. London Proc., v. 129, p. 151-244.

Tedford, R. H., 1976, The relationship of the pinnipeds to other carnivores (Mammalia): Systematic Zoology, v. 25, p. 363-374.

Toula, Franz, 1897, *Phoca vindobonensis* n. sp., von Nussdorf in Wien: Beitr. Paläont. Geol. Öster.-Ungarns Orients, v. 11, p. 47-70.

Trouessart, E.-L., 1897, Catalogus Mammalium tam viventium quam fossilium, Nova editio: Berlin, R. Friedländer und Sohn, 664 p.

True, F. W., 1906, Description of a new genus and species of fossil seal from the Miocene of Maryland: U.S. Natl. Mus. Proc., v. 30, p. 835-840.

Van Beneden, P. J., 1877, Description des ossements fossils des environs d'Anvers: Mus. Royal Hist. Nat. Belgique Ann., v. 1, p. 1-88.

Zapfe, Helmuth, 1937, Ein bemerkenswerter Phocidenfund aus dem Torton des Wiener-Beckens: Zool.-Bot. Gesell. Wien Verhandl., v. 86-87, p. 271-276.

Fossil Beetles and the Late Cenozoic History of the Tundra Environment

JOHN V. MATTHEWS, JR., *Terrain Sciences Division, Geological Survey of Canada, Ottawa, Ontario K1A 0E8*

ABSTRACT

Recent paleoecological research has shown that the tundra ecosystem has a long history, during which its character at times differed greatly from that of present tundra. Most of the pertinent data come from study of plant macrofossils and pollen, but fossil beetles, another powerful source of paleoecological information, are now also being used to clarify tundra and taiga history.

Fossils of both beetles and plants show that the arctic (lowland) tundra of the Northern Hemisphere originated in late Tertiary time. A site where this fact is clearly illustrated is Meighen Island in the northern part of the Canadian Arctic Archipelago. There, late Miocene or early Pliocene Beaufort Formation sediments contain fossils indicative of a forest-tundra environment, but one differing from that of the present by its greater floral and faunal diversity.

Arctic tundra achieved holarctic, biome proportions by early Pleistocene time; however, at that time the tundra environment of Alaska and eastern Siberia was more xeric than at present. This characteristic, and its concomitant biotic facies, is so dominant during the late Pleistocene that tundra of that time is more appropriately termed "arctic-steppe." Typical of arctic-steppe regions were an abundance of grasses and sage, a diverse community of grazing ungulate herbivores, and grassland beetles.

Interglacial tundra in North America resembled that of the present, except that in one or more of the interglacials, larch, not spruce, formed the North American treeline, and during most interglacials the areal extent of tundra was less than now. Moreover, if some of the large Pleistocene mammals that are now extinct survived in interglacial tundra regions, then despite the apparent similarity of present and interglacial tundra, the two were different and our present tundra environment must be considered as a unique ecosystem.

INTRODUCTION

Tundra regions, often considered by North Americans as the inhospitable domain of the Inuit, polar bear, and dwarfed plants, lately have become the scene of unprecedented industrial activity in our search for new sources of crude oil and natural gas. Such endeavors raise the specter of environmental degradation since tundra ecosystems, while not as fragile as some, (e.g., tropical rainforest) are, because of the presence of permafrost, predisposed to self-perpetuating change once disturbance has occurred. Most people would classify such change as "damage," and efforts to prevent or control it have been associated in recent years with a spate of tundra-related studies. Ranging from floral and faunal inventories to ecosystem modeling (Bliss and others, 1973), many of these carry the implicit assumption that the tundra ecosystem of today is a model of past tundra environments. In fact, paleoecological research has shown this is not the case (Matthews, 1976a). The arctic (lowland) tundra ecosystem has a history stretching back into the late Tertiary, but at that time the tundra flora and fauna were more diverse than that of present. But the greatest deviation from present conditions occurred as little as 14,000 yr ago when large areas of unglaciated northern Asia and North America (Beringia) had a cold steppelike environment (Matthews, 1976a). These findings, and the way that they relate to the history of the tundra ecosystem, are the subject of this paper. The discussion, relying primarily on paleobotanical data, also makes use of a relatively new and potentially valuable type of paleoecological evidence—fossil insects.

Tundra—a Definition

Tundra is primarily a physiognomic term in that it refers to regions too cold for the growth of trees rather than to particular communities of plants. Many of the plant taxa and most of the community types which occur in tundra areas can be found as well in more temperate areas (Bliss, 1975). Two major types of tundra may be distinguished, namely arctic tundra which owes its existence to the cold climate of high latitude areas [and(or) proximity to cold-water bodies] and alpine tundra where the absence of trees is largely a function of altitude. Aside from the fact that they all lack trees, arctic tundra sites are vegetationally diverse. On small scale transects one may encounter numerous plant microcommunities, reflecting the vagaries of permafrost and drainage, while on a larger scale there exist climatically controlled vegetation zones (Young, 1971; Yurtsev, 1972). The warmest of these zones, abutting with the northern boreal forest (taiga) or in some places intermingled with it to form a forest-tundra ecotone, is characterized by dominance of shrubby species. Colder

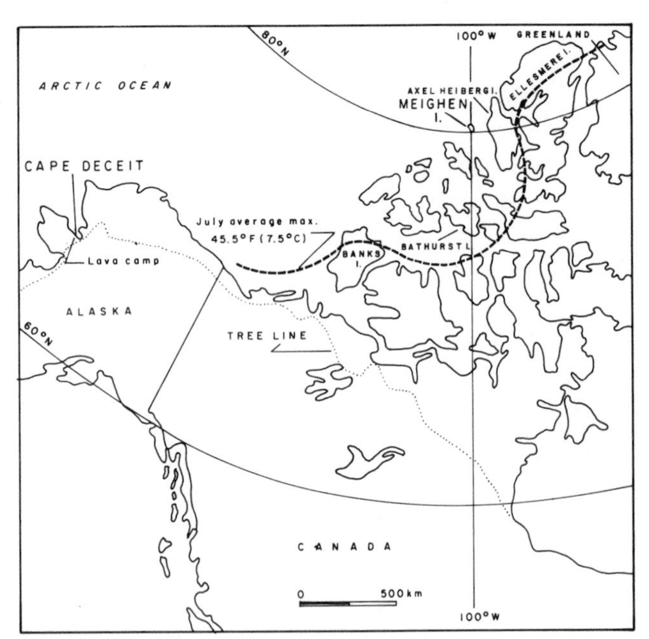

Figure 1. Localities mentioned in the text. Note the northward bulge of the July isotherm in the eastern part of the Arctic Archipelago.

tundra zones are herb dominated while the coldest portions of the Arctic fall within a polar desert zone where fellfield communities prevail. Meighen Island (Fig. 1), discussed later in conjunction with Tertiary tundra history, is within this most rigorous of tundra climatic zones.

It is more difficult to establish tundra zones on the basis of insects. Merriam's life zones are applicable as long as it is understood that some sites which are clearly beyond the limit of trees nevertheless have an insect fauna with strong Hudsonian (northern boreal or taiga) affinities (Mason, 1956, 1965; Munroe, 1956). Within tundra regions of North America, diversity of Coleoptera east of Hudson Bay is low (W. J. Brown, unpub. Ms), and beetles become rare compared to other insect orders in the more northern tundra floral zones. Their northern limit corresponds approximately with the 45.5°F (7.5°C) July average maximum isotherm (Fig. 1).

FOSSIL BEETLES

As Paleoecological Indicators

Most Quaternary age organic sediments of terrestrial origin contain insect fossils, usually pronota, heads, and elytra of beetles (Coleoptera) (Fig. 2). Brightly colored fragments of beetles have been noted in limnic sediments and peats for many years, but were considered mere curiosities until C. H. Lindroth (1948), and then G. R. Coope, in a host of subsequent studies (see Coope, 1970 for a review), showed that contrary to the prevailing opinion of most coleopterists, such fossils could be identified to the specific level. When this was attempted it became obvious that virtually all of the identifiable beetle fragments from middle and late Pleistocene sediments represented extant species (Coope, 1970), and recently (Matthews, 1974a) this has been shown to be true of early Pleistocene sediments as well. Even late Tertiary sediments contain a few fossils of living species (Gersdorf, 1969; Matthews, 1970, 1976b).

These findings set the stage for use of beetle fossils in paleoecological research, for if one knows the present distribution and habitat requirements of the species included in the fossil assemblage and assumes that the habitat requirements of these species have not changed with time, then it becomes possible to draw paleoecological inferences by comparing the composition of the fossil assemblages with the existing fauna at the site. Because the taxonomic diversity of Coleoptera assemblages is usually greater than that of pollen or plant macrofossil assemblages, beetle fossils provide a more detailed picture of the former environment. Important as well is the fact that beetles are more mobile than plants and can respond more quickly to climatic change (Coope, 1975).

To date some of the most incisive paleoecological studies using fossil Coleoptera have been carried out in England by G. R. Coope and his associates. The approach has been used in other parts of Europe, more recently in Siberia, and since 1968 several studies have also been conducted in North America (see Matthews, 1976c for a list). It will be some time, however, before the precision of these North American efforts approaches those of western Europe since one important prerequisite for fossil beetle research—a detailed taxonomic and ecological knowledge of the existing fauna—is inadequately fulfilled on this continent. The ground beetle (Carabidae) fauna of North America, especially northern areas, is relatively

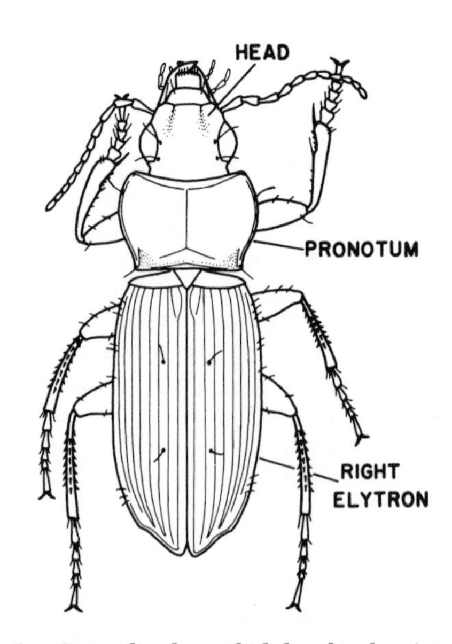

Figure 2. Generalized carabid beetle showing the parts which are usually found in Quaternary and northern Tertiary sediments.

well-known and this is fortunate because many of the beetle fossils are from carabids, which, unlike some other Coleoptera, are distributed in conformity to macroenvironmental conditions. Thus, ground beetle fossils provide the clearest insight to past regional environments.

Tertiary Beetle Fossils from the Arctic

Most paleontologists probably associate Tertiary insect fossils with either spectacular amber inclusions (Bachofen-Echt, 1949), or the delicate impressions of wings and flattened bodies which are occasionally found in certain types of terrestrial sediments (e.g., Gersdorf, 1969; Smiley and others, 1975). By contrast, all of the Tertiary beetles discussed below are, like their Quaternary counterparts, the actual, unaltered chitinous fragments of the original insect. Heads, pronota, and elytra dominate (Figs. 2, 3), and many of them preserve remnants of the original hairy covering or colors. They are much better preserved than most Tertiary insect fossils, but of course do not match those entrapped in amber. On the other hand, amber fossils are rare and represent only one type of environment, while the Tertiary fossils mentioned here are abundant (hundreds to thousands/kg of sediment) and portray a broad spectrum of paleoenvironments. Some specimens from autochthonous peats are partially articulated, revealing legs, antennae, wings, and genitalia.

As indicated above virtually all of the Quaternary beetle fossils studied to date are within the range of variation of existing species. In contrast, careful study of the Tertiary fossils from Alaska and northern Canada has shown that a good many of them are only closely related to existing species (Matthews, 1976b). Consequently, it cannot be assumed that the fossils have exactly the same ecological requirements as the living species, and the paleoecological value of the fossils is seemingly diminished. But this is not so as long as comparisons are based on assemblages of taxa rather than individual fossils, for while the requirements of an extinct species may differ from those of its extant relative, it is unlikely that the ecological implications of an entire assemblage will be much different from a comparable assemblage of extant species.

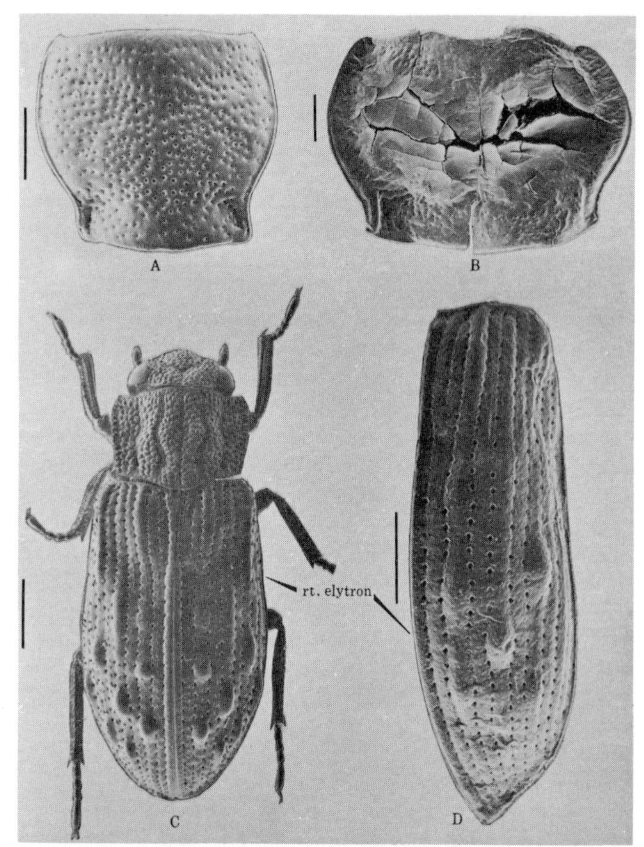

Figure 3. (A) *Diacheila* cf. *D. polita* Fald., pronotum, Beaufort Formation (late Tertiary), Meighan Is., NWT, GSC-46918. *D. polita*, the extant species to which the fossil is closely related, is usually found in tundra areas. (B) *Blethisa* cf. *B. multipunctata* L., pronotum, Beaufort Formation (late Tertiary), Meighan Is., NWT, GSC-48234. The specimen shows the cracked and curled cuticle which is characteristic of the most poorly preserved fossils from Meighen Island. (C) *Helophorus tuberculatus* Gyll., modern specimen for comparison with fossil *H. meighenensis*. (D) *Helophorus meighenensis* Matth., rt. elytron, Beaufort Formation (late Tertiary), Meighen Is., NWT, GSC-43314 (Holotype). The fossil illustrates the poorly developed elytral tubercles by which *H. meighenensis* is distinguished from *H. tuberculatus*. Its excellent state of preservation is indicated by comparison with the right elytron of the recent specimen. Scale bar = 0.5 mm.

TUNDRA HISTORY

Tertiary Beginnings

Some type of alpine tundra has probably existed since the evolution of land plants wherever and whenever mountains were tall enough to include a nival zone. Lowland tundra may have occurred near the margins of Paleozoic continental glaciers, but arctic tundra similar to that of the present is undoubtedly a Cenozoic phenomenon —arising in the Southern Hemisphere perhaps as early as the Oligocene when glaciers first formed in Antarctica (Savin and others, 1976), and in the Northern Hemisphere during the Miocene.

Several workers have speculated on the origin of Northern Hemisphere tundra (Dorf, 1960; Hoffman and Taber, 1968), but I believe the best indirect evidence comes from the paleobotanical work of J. A. Wolfe (1972; Wolfe and others, 1966). By careful study of foliar physiognomy and floral composition of Alaskan plant assemblages, he has formulated a picture of late Tertiary climatic change in Alaska while at the same time establishing a series of time-stratigraphic units (Wolfe and others, 1966) which have proved valuable for correlation of fossil floras in Siberia and Arctic Canada.

The Miocene Seldovian Stage, Wolfe's oldest Alaskan time-stratigraphic unit, is marked by floras dominated by (1) broad-leaved deciduous species (some of them Asian genera) south of 65°N and (2) by coniferous species (*Pinus, Picea, Tsuga, Abies, Larix*) as far north as 70°. Thus if arctic or lowland tundra existed at all in North

America during Seldovian time, it must have been restricted to areas of higher latitude than northernmost Alaska. Reflecting deteriorating Tertiary climate, floras of the next younger Homerian Stage are dominated by conifers and in some areas by shrubs, sedges, and grasses (Wolfe, 1972). The Homerian ended approximately 8 m.y. ago (Triplehorn and others, 1977) and compared to its coniferous forests, those of the subsequent Clamgulchian Stage are even more depauperate. Lava Camp, a Clamgulchian site on the Seward Peninsula in western Alaska (Hopkins and others, 1971), is dated by the K-Ar method at 5.7 m.y. and has yielded, in addition to plant macrofossils and pollen, a small assemblage of beetle fossils. The plants suggest a rich coniferous forest dominated by spruce and pine, but including several other conifer genera as well as tree birch and several shrubs not presently found in taiga regions. The beetles are chiefly forest taxa but a few appear to be close relatives of existing obligate tundra species. Another Clamgulchian flora from a site only slightly further north of Lava Camp contains a more depauperate, taiga-like coniferous flora, making it entirely probable that the Clamgulchian was a time when the northern limit of trees first occurred within Alaska. At this time most of the Canadian Arctic Archipelago was probably clothed in tundra.

Tertiary Tundra at Meighen Island

From Wolfe's Alaskan work we may infer that arctic tundra existed by the late Miocene, yet none of his floral assemblages actually represent such conditions. Instead the best direct evidence for late Tertiary tundra comes from sediments of the Beaufort Formation on Meighen Island in the northern part of the Canadian Arctic Archipelago (Fig. 1).

Beaufort Formation sediments, consisting mostly of sands and gravels with occasional interbedded peaty zones or marine clays, outcrop on the western parts of all the western islands of the Arctic Archipelago and as outliers on some of the more eastern islands (Balkwill and Bustin, 1975; Tozer and Thorsteinsson, 1964). The organic facies have yielded excellently preserved plant macrofossils (e.g., Hills and others, 1974; Hills and Matthews, 1974) and fossil insects (Matthews, 1974b, 1976b). On the basis of plant macrofossils, Hills (Hills and others, 1974; Hills and Bustin, 1976) suggests that the age of the entire Beaufort Formation is Seldovian and Homerian, while I believe it possible that some of the northern sites, like Meighen Island, may be as young as the Clamgulchian. Certainly the entire Formation is at least of early Pliocene age, and it was probably deposited at a time when the western islands of the Archipelago were linked into one rather large landmass.

Meighen Island apparently represents only the upper fraction of a thick clastic wedge of Beaufort and older sediments. Exposed above sea level on the Island are several hundred meters of sand and interbedded marine clay. The sands, of alluvial origin, contain organic horizons that have yielded fossil leaves so well preserved that they can be dried in a press like living plants and beetle fossils better preserved than those from most Quaternary sediments (e.g., Fig. 3). Wood is abundant, but not to the degree as at southern Beaufort exposures (e.g., Banks Island, Fig. 1) and the size of the tree stumps on Meighan Island is considerably smaller than those from southern Beaufort localities (Matthews, 1976b). This, and the fact that many of the fossil leaves and seeds represent shrubs common in open boreal or tundra sites (e.g., *Myrica*, dwarf birch, *Empetrum*, *Vaccinium*), suggests that deposition took place in a thinly forested or forest-tundra environment.

The beetle fossils, far outnumbering those of plants, confirm this conclusion. More than 66 genera representing a minimum of 87 species have been discovered to date, and many of the samples collected during the period of two short visits to Meighan Island still remain to be examined. Nearly all of the species to which the fossils are tentatively compared (Matthews, 1976b) now reside in taiga areas near treeline, but a few are most closely related to living species that are more or less restricted to tundra sites. Moreover, the carabid portion of the assemblage is dominated, like Pleistocene tundra assemblages (Matthews 1974a, 1975), by fossils of the subgenus *Cryobius*, and fossils of bark beetles (Scolytidae)—forest inhabitants—are very rare. Thus, the overall implication of the beetle and plant fossils is that Meighen Island was within a region of forest-tundra during deposition of its Beaufort sediments. Only sites in the central part of the Island have been sampled, so while forest-tundra existed there, true arctic tundra could have existed several tens of kilometers north, near the present northern tip of the Island. At the same time tundra most probably occurred on Axel Heiberg Island, northern Ellesmere Island, and northern Greenland; however, if the distribution of Miocene mean annual temperature isotherms bulged to the north in the eastern Arctic Archipelago, as they do today (Savile, 1961), forests, contemporaneous with Meighen Island tundra, may well have existed on Axel Heiberg Island and other sites north of Meighen Island (see Hills and Bustin, 1976).

Both plant and insect assemblages from Meighen Island differ from those expected of contemporary North American forest-tundra areas. For example, instead of a treeline comprised of white and black spruce with occasional outliers of poplar as at present in northern Canada and Alaska, Miocene treeline communities on Meighen Island contained spruce, pine, fir, larch, and even cedar. In this respect, Meighen Island forest-tundra was more like forests of southern James Bay or the Ontario-Quebec clay belt (Rowe, 1972) than contemporary forest-tundra. Distinctions of diversity are not confined to the plant assemblages, since several of the beetle genera from Meighen Island are presently absent in northern boreal regions. Some of them occur much to the south of present treeline, a few are confined to either western or eastern parts of the continent, one is common in the Palearctic region but absent from North America, and one of the leafhopper fossils represents a tribe restricted to the Himalayan region.

QUATERNARY TUNDRA

Early Pleistocene

In its initial late Tertiary stages, arctic tundra regions were undoubtedly disjunct, i.e., restricted to the northern parts of Asia and North America, but by the beginning of the Quaternary they probably had reached holarctic, biome proportions. This was true especially during the cold phases of the Pleistocene when lower sea levels caused Siberia and Alaska to be joined via the Bering Land Bridge, forming a huge, largely unglaciated realm known as Beringia.

The tundra ecosystem of Beringia differed greatly at times from that of the present, and one of the places where this is documented is at Cape Deceit in western Alaska (Matthews, 1974a). The Cape Deceit site (Fig. 1) is located in a climatically sensitive position, just beyond treeline, but the prime reason for its significance is that it reveals sediments of both early and late Pleistocene age.

The early Pleistocene sediments at Cape Deceit have pollen and plant macrofossils indicative of a tundra environment, albeit one somewhat different from that of the present. The percentage of spruce pollen is low as in most contemporary tundra surface spectra, but the amount of sage (*Artemisia*) pollen is slightly higher. Similarly, percentages of *Potentilla* are high and this trend is supported by an abundance of *Potentilla* macrofossils. Very few of the latter are from the hydrophytic species *Potentilla palustris* L., and the assemblage contains few *Carex* achenes as well. Both are relatively abundant in contemporary and Holocene tundra assemblages. The early Pleistocene insect assemblages at Cape Deceit are also distinctive. Attesting to treeless conditions, they lack obligate forest beetles yet include abundant fossils of two beetle species (*Tachinus apterus* Mäkl. and *Micralymma brevilinque* Schiødt.) which are rare to absent in contemporary tundra areas. The paleoecological significance of the latter species has been mentioned previously (Matthews, 1974a, 1975), and even though it is clear that *Micralymma brevilinque* is not a tundra species, interpretation of its fossils is complicated by the recent discovery of *M. brevilinque* or an undescribed species similar to it at a taiga site in the Yukon Territory (an example of the problem created by poor knowledge of our contemporary fauna). Some or all of the *Micralymma* fossils at Cape Deceit could conceivably represent this newly discovered species. Nevertheless, the genus *Micralymma* can still be considered as very rare at tundra and taiga sites, a contrast with its relative abundance in some of the Cape Deceit assemblages. And to date it has not been collected at poorly drained sites. Thus, it appears, that both insect and plant evidence from Cape Deceit indicate an early Pleistocene tundra more xeric than that of the present.

A similar conclusion has been drawn from paleobotanical study of sediments at an early Pleistocene site in eastern Siberia (Giterman, 1975). I have studied a small assemblage of insect fossils from this same site (Matthews, 1974c) and a detailed study is nearing completion (S. Kiselev, personal commun. 1975). My efforts, though very cursory, revealed the presence of tundra insects and abundant fossils of the pill-beetle *Morychus*, which is rare today but well represented in some late Pleistocene Alaskan-Yukon assemblages. It is unlikely that such great contrasts in relative abundance of fossils (*Morychus, Micralymma*) and their living counterparts are due only to some type of selective concentration of the fossils; they probably reveal something instead about the local and regional environment. In this regard it is of interest that *Morychus* individuals, like those of most species which are unexpectedly abundant in Pleistocene fossil assemblages from Alaska and the Yukon Territory, are beetles of open, scantily vegetated, dry sites.

Late Pleistocene

Fossils from sediments of presumed middle Pleistocene age at Cape Deceit also imply relatively dry, scantily vegetated tundra, containing a mixture of mesic to xeric adapted plants and beetles. In the assemblage *Micralymma* accounts for more than 14 percent of all the beetle individuals compared to only 8 percent in the early Pleistocene sediments.

The same type of environment is implied by Illinoian fossils at Cape Deceit. However, pollen spectra of the same age at a site not far from Cape Deceit (Colinvaux, 1964) suggest treeless conditions that contrast so much with present tundra in terms of continentality of climate, vegetation, and substrate conditions as to make the term "tundra" a misnomer. Instead terms such as "steppe-tundra," "periglacial steppe," and "arctic-steppe" are used (Matthews, 1976a). Paleontological indicators of arctic-steppe conditions, such as high percentages of *Artemisia* and grass pollen, abundant ungulate fossils, and presence of fossils of grassland insects, are most pronounced in sediments correlative with the last, Wisconsinan, glaciation. Ager (1975) shows that during the late Wisconsinan much of unglaciated interior Alaska possessed an arctic-steppe environment, but around 14,000 yr ago the herbaceous character of the vegetation changed as shrub birches became more abundant. Wisconsin age Cape Deceit samples postdate the shrub invasion, yet contain fossils indicative of persisting steppelike conditions. For example, the pollen spectra contain higher percentages of grass and *Artemisia* than earlier Pleistocene samples; *Potentilla* fruits and seeds of the grass, *Poa*, are abundant; fossils of the beetle *Micralymma* are present but rare compared to those of *Morychus;* and those of the weevil, *Lepidophorus lineaticollis* Kirby, a typical beetle of xeric tundra and forest sites, are abundant. Other late Pleistocene sites in Alaska and Yukon also contain abundant fossils of *Morychus* and (or) *Lepidophorus*, and most of them as well, like late Wisconsinan sediments at Cape Deceit, have a few fossils of the ground beetle *Harpalus amputatus* Say, a true grassland species.

Interglacial Tundra

All of the Quaternary evidence discussed above refers to Beringia during the cold periods of the Pleistocene. What was the character of tundra environments during interglacials? First, we know that Northern Hemisphere tundra was not continuous across the Bering Sea region, since sea level was as high or higher than it is today. Moreover, much of the area that was treeless during the cold phases of the Pleistocene was forested during interglacials, restricting arctic tundra to its present limits. In fact, it was at times less extensive than now since treeline in Alaska and northern Canada was west and north of its modern position during parts of several interglacials. At one of these times in the Cape Deceit area (Matthews, 1974a) and during an interglacial on Banks Island in the Arctic Archipelago (Kuc, 1974) (Fig. 1) treeline sites were characterized by larch (*Larix*) rather than spruce. Why this was so is not clear, for *Larix* does not form treeline in North America today.

During the last, Sangamon, interglacial, tundra existed at Cape Deceit, but spruce treeline was nearer than it is today. Unlike fossil assemblages of full glacial age, *Carex* (sedge) seeds are abundant in Cape Deceit Sangamon sediments and most of the insect fossils represent taxa now living at the site. *Micralymma* and *Morychus* fossils are absent and those of the weevil *Lepidophorus lineaticollis* are much less abundant than in full glacial assemblages. Sediments presumably representing the last interglaciation on Bathurst Island, Arctic Archipelago (Fig. 1), show that tundra persisted there throughout the interglacial, not however without some northward biotic shifts, notably of Coleoptera which do not occur on the Island today (Blake, 1974).

DISCUSSION

Plant and beetle fossils from Meighen Island represent a lowland forest-tundra ecosystem of late Tertiary age. The exact age of the ecosystem is not known. It is older than 8 m.y. if the plant assemblages are correlated with the Alaskan Homerian stage (Hills and others, 1974; Hills and Bustin, 1976), or it may be Clamgulchian and younger than 5.7 m.y. according to the stage of evolution of one of the beetle fossils (Matthews, 1976b, 1976c). Neverthelesss, it undoubtedly predates the middle Pliocene (3 to 4 m.y.).

Yurtsev (1972) has suggested that a frozen Arctic Ocean was critical to the origin of northern tundra. Was the Arctic Ocean perennially frozen when tundra first formed on Meighen Island? According to Clark (1971) freezing of the Arctic Ocean occurred earlier than 3. 5 m.y. ago,[1] but I doubt that it happened much before that date because erratics indicative of circum-arctic glaciers do not appear in levels of marine cores older than about 3 m.y. (Berggren and Van Couvering, 1974). The two phenomena —ocean freezing and glaciation—are not necessarily related in a causitive manner, but it seems unlikely that the former would occur significantly before the latter. Thus, if Meighen Island tundra is of Homerian age, it almost certainly predates a perennially frozen Arctic Ocean, and this may be true even if it is only as old at the early Pliocene. Late Tertiary tundra in the Arctic Islands probably formed when the climate was cold enough for a seasonally frozen Arctic Ocean, much like the present-day Bering Sea which itself is surrounded by coastal tundra.

The first appearance of tundra, even if it was only coastal, marks an important climatic threshold in the downward spiral of Tertiary temperatures. Therefore, it is important that this event be placed in context with other global events that are thought to have had climatic effects. Assuming the Meighen Island sediments to be approximately 6 m.y. old (a compromise age), then the formation of tundra there predated: glaciation of Iceland (Einarsson and others, 1967); breakup of the western islands of the Canadian Arctic Archipelago (L. V. Hills, personal commun. 1975); and closing of the Panamanian Isthmus (Berggren and Hollister, 1974). It was probably contemporaneous with: a magnetic pole position near that of the present (Symons, 1969); the existence of a Miocene seaway or a narrow land connection across the Bering Platform (Nelson and others, 1974); and desiccation of the Mediterranean Basin (associated with oscillatory sea-level changes) (Berggren and Van Couvering, 1974). It undoubtedly postdated: continent-wide glaciation of Antarctica; decoupling of high-latitude and low-latitude Tertiary temperature trends; and the formation of a well-defined thermocline in the Atlantic and Pacific Oceans (Savin and others, 1975). The importance of these phenomena to climatic change in the Arctic should become apparent as we gain more knowledge of Tertiary environments in the Arctic, and it can be safely predicted that the study of both plant and insect fossils will play an important role in this effort.

The late Tertiary tundra ecosystem of Meighen Island was taxonomically more diverse than that of today. Early Pleistocene tundra also differed from that of the present, but not to the same degree as late Pleistocene tundra, which is more properly termed arctic-steppe or periglacial steppe. The unique character of arctic-steppe was due to exposure of the Bering Platform and large areas of the Arctic continental shelf during glacial periods of lower sea level. The degree of exposure during any one glaciation was likely mediated by local or regional tectonism (Hopkins and others, 1974), thus explaining why early Pleistocene tundra was not as steppelike as that of the late Pleistocene. But the distinction may also be due in part to our poor knowledge of early Pleistocene environments.

Similarly, our understanding of interglacial tundra conditions is also meager. The data available portray conditions like those of the present, but it should not be forgotten that one of the chief distinguishing characteristics of arctic-steppe was a diverse assemblage of grazing herbivores and their associated predators (Guthrie, 1968). The extent to which a few of these mammals, now extinct, persisted in interglacial tundra areas is a measure of how interglacial tundra ecosystems could have differed from

[1] An earlier date ($<$ 4.5 m.y. ago) (Steuerwald and others, 1968) has apparently been discarded as a result of a new interpretation of the magnetic stratigraphy of one of the Arctic Ocean cores.

those of the present. Since it is possible that some of these mammals did survive in interglacial tundra areas, there is a strong possibility that the contemporary tundra ecosystem has no historical counterpart, that is to say, it is unique.

REFERENCES

Ager, T. A., 1975, Late Quaternary environmental history of the Tanana Valley, Alaska: Inst. Polar Studies, Rept. 54, 117 p.

Bachofen-Echt, A., 1949, Der Bernstein und Seine Einschlusse: Wien, Springer-Verlag, 204 p.

Balkwill, H. R., and Bustin, R. M., 1975, Stratigraphic and structural studies, central Ellesmere Island and eastern Axel Heiberg Island, District of Franklin: Canada Geol. Survey, Paper 75-1, pt. A, p. 513-517.

Berggren, W. A., and Hollister, C. D., 1974, Paleogeography, paleobiogeography and the history of circulation in the Atlantic Ocean, *in* Hay, W. W., ed., Studies in paleo-oceanography: Soc. Economic Paleontologists and Mineralogists Spec. Pub. 20, p. 126-186.

Berggren, W. A., and Van Couvering, J. A., 1974, The late Neogene: Palaeogeography, Palaeoclimatology, Palaeoecology, v. 16, p. 1-216.

Blake, W., Jr., 1974, Studies of glacial history in Arctic Canada. II. Interglacial peat deposits on Bathurst Island: Canadian Jour. Earth Sci., v. 11, p. 1025-1042.

Bliss, L. C., 1975, Tundra grasslands, herblands, and shrublands and the role of herbivores: Geoscience and Man, v. 10, p. 51-79.

Bliss, L. C., and others, 1973, Arctic tundra ecosystems: Ann. Rev. Ecology and Systematics, v. 4, p. 359-399.

Clark, D. L., 1971, Arctic Ocean ice cover and its late Cenozoic history: Geol. Soc. America Bull., v. 82, p. 3313-3324.

Colinvaux, P. A., 1964, The environment of the Bering Land Bridge: Ecological Mon., v. 34, p. 297-329.

Coope, G. R., 1970, Interpretation of Quaternary insect fossils: Ann. Rev. Entomology, v. 15, p. 97-120.

——— 1975, Climatic fluctuations in northwest Europe since the last interglacial, indicated by fossil assemblages of Coleoptera, *in* Wright, A. E., and Moseley, F., eds., Ice ages: Ancient and modern: Liverpool, Seel House Press, p. 153-168.

Dorf, E., 1960, Climatic changes of the past and present: Am. Sci., v. 48, p. 341-364.

Einarsson, T., Hopkins, D. M., and Doell, R. R., 1967, The stratigraphy of Tjörnes, northern Iceland, and the history of the Bering Land Bridge, *in* Hopkins, D. M., ed., The Bering Land Bridge: Stanford, Calif., Stanford Univ. Press, p. 312-325.

Gersdorf, E., 1969, Käfer (Coleoptera) aus dem Jungtertiar Norddeutschlands: Geol. Jahrb., v. 87, p. 295-332.

Giterman, R. E., 1975, Palynologische charakteristiic der unterpleistozanan Ablagerungen vom unterlauf der Kolyma: Quatär Paläontologie, v. 1, p. 7-11.

Guthrie, R. D., 1968, Paleoecology of the large mammal community in interior Alaska during the late Pleistocene: Am. Midland Naturalist, v. 79, p. 346-363.

Hills, L. V., and Matthews, J. V., Jr., 1974, A preliminary list of fossil plants from the Beaufort Formation, Meighen Island, District of Franklin: Canada Geol. Survey Paper 74-1, pt. B, p. 224-226.

Hills, L. V., Klovan, J. E., and Sweet, A. R., 1974, *Juglans eocinerea* n. sp. Beaufort Formation (Tertiary), southwestern Banks Island, Arctic Canada: Canadian Jour. Botany, v. 52, p. 65-90.

Hills, L. V., and Bustin, R. M., 1976, *Picea banksii* Hills and Ogilvie from Axel Heiberg Island, Arctic Canada: Canada Geol. Survey Paper 76-1, pt. B, p. 61-63.

Hoffmann, R. S., and Taber, R. D., 1968, Origin and history of holarctic tundra ecosystems, with special reference to their vertebrate faunas, *in* Wright, H. E., Jr., and Osburn, W. H., eds., Arctic and alpine environments, INQUA Cong., 7th, Denver 1965, Proc., v. 10: Bloomington, Indiana Univ. Press, p. 143-170.

Hopkins, D. M., Matthews, J. V., Jr., Wolfe, J. A., and Silberman, M. L., 1971, A Pliocene flora and insect fauna from the Bering Strait region: Palaeogeography, Palaeoclimatology, Palaeoecology, v. 9, p. 211-231.

Hopkins, D. M., Rowland, R. W., Echols, R. E., and Valentine, P. C., 1974, An Anvilian (early Pleistocene) marine fauna from western Seward Peninsula, Alaska: Quaternary Research, v. 4, p. 441-470.

Kuc, M., 1974, The interglacial flora of Worth Point, western Banks Island: Canada Geol. Survey Paper 74-1, pt. B, p. 227-231.

Lindroth, G. H., 1948, Interglacial insect remains from Sweden: Sveriges Geol. Undersökning Årsb., ser. C, no. 42, 29 p.

Mason, W. R. M., 1956, Distributional problems in Alaska: Internat. Cong. Entomology, 10th, Montreal 1956, Proc., v. 1, p. 703-710.

——— 1965, Ecological peculiarities of the Canadian north: Arctic Circular, v. 16, p. 15-17.

Matthews, J. V., Jr., 1970, Two new species of *Micropeplus* from the Pliocene of western Alaska with remarks on the evolution of Micropeplinae (Coleoptera: Staphylinidae): Canadian Jour. Zoology, v. 48, p. 779-788.

——— 1974a, Quaternary environments at Cape Deceit (Seward Peninsula, Alaska): Evolution of a tundra ecosystem: Geol. Soc. America Bull., v. 85, p. 1353-1384.

——— 1974b, A preliminary list of insect fossils from the Beaufort Formation, Meighen Island, District of Franklin: Canada Geol. Survey Paper 74-1, pt. A, p. 203-206.

——— 1974c, Fossil insects from the early Pleistocene Olyor suite (Chukochya River: Kolymian Lowland, U.S.S.R.): Canada Geol. Survey Paper 74-1, pt. A, p. 207-211.

——— 1975, Insects and plant macrofossils from two Quaternary exposures in the Old Crow-Porcupine region, Yukon Territory, Canada: Arctic and Alpine Research, v. 7, p. 249-259.

——— 1976a, Arctic-steppe—an extinct biome: AMQUA, 4th biennial meeting, Tempe, Arizona 1976, Abs., p. 73-77.

——— 1976b, Insect fossils from the Beaufort Formation: Geological and biological significance: Canada Geol. Survey Paper 76-1, pt. B, p. 217-227.

——— 1976c, Evolution of the subgenus *Cyphelophorus* (genus *Helophorus*, Hydrophilidae, Coleoptera): Description of two new fossil species and discussion of *Helophorus tuberculatus* Gyll: Canadian Jour. Zoology, v. 54, p. 652-673.

Munroe, E., 1956, Canada as an environment for insect life: Canadian Entomologist, v. 88, p. 372-476.

Nelson, C. H., Hopkins, D. M., and Scholl, D. W., 1974, Cenozoic sedimentary and tectonic history of the Bering Sea, *in* Hood, D. W., ed., Oceanography of the Bering Sea: Fairbanks, Alaska Inst. Marine Science, p. 485-516.

Rowe, J. S., 1972, Forest regions of Canada: Canadian Forestry Service Dept. Environment Pub. no. 1300, 172 p.

Savile, D. B. O., 1961, The botany of the northwestern Queen Elizabeth Islands: Canadian Jour. Botany, v. 39, p. 909-942.

Savin, S. M., Douglas, R. G., and Stehli, F. G., 1975, Tertiary marine paleotemperatures: Geol. Soc. America Bull., v. 86, p. 1499-1510.

Smiley, C. J., Gray, J., and Huggins, L. M., 1975, Preservation of Miocene fossils in unoxidized lake deposits, Clarkia, Idaho: Jour. Paleontology, v. 49, p. 833-844.

Steuerwald, B. A., Clark, D. L., and Andrew, J. A., 1968, Magnetic stratigraphy and faunal patterns in Arctic Ocean sediments: Earth and Planetary Sci. Letters, v. 5, p. 79-85.

Symons, D. T. A., 1969, Paleomagnetism of the late Miocene plateau basalts in the Cariboo region of British Columbia: Canada Geol. Survey Paper 69-43, 16 p.

Tozer, E. T., and Thorsteinsson, R., 1964, Western Queen Elizabeth Islands, Arctic Archipelago: Canada Geol. Survey Mem. 332, 242 p.

Triplehorn, D. M., Turner, D. L., and Naeser, C. W., 1977, K-Ar and fission-track dating of ash partings in Tertiary coals from the Kenai Peninsula, Alaska: A radiometric age for the Homerian-Clamgulchian Stage boundary: Geol. Soc. America Bull. (in press).

Wolfe, J. A., 1972, An interpretation of Alaskan Tertiary floras, *in* Graham, A., ed., Floristics and paleofloristics of Asia and eastern North America: Amsterdam, Elsevier, p. 201-233.

Wolfe, J. A., Hopkins, D. M., and Leopold, E. B., 1966, Tertiary stratigraphy and paleontology of the Cook Inlet region, Alaska: U. S. Geol. Survey Prof. Paper 398-A, 29 p.

Young, S. B., 1971, The vascular flora of St. Lawrence Island with special reference to floristic zonation in the Arctic regions: Harvard Univ., Contr. Gray Herbarium, v. 201, p. 11-115.

Yurtsev, B. A., 1972, Phytogeography of northeastern Asia and the problem of transberingian floristic interrelations, *in* Graham, A., ed., Floristics and paleofloristics of Asia and eastern North America: Amsterdam, Elsevier, p. 19-54.

In Search of Lost Oceans: A Paradox in Discovery

Richard H. Benson, *Department of Paleobiology, Smithsonian Institution, Washington, D.C. 20560*

ABSTRACT

Some paleontology is still in the early discovery stage. Few paleontologists have been able to examine fossils from the ocean floor. Fossil evidence of past oceans comes not only from the abyss, however. As crustal plates move and change the positions of continents, oceans have been destroyed. Also deep ocean waters undergo radical changes, especially temperature changes.

Today a formidable biological barrier exists at the interface of a two-layered ocean. In low latitudes, this barrier separates the deeper, cold psychrospheric benthic faunas from the shallower, warm thermospheric faunas. In times past, with warmer latitudinal World Ocean circulation dominant among the deeper water masses, the thermal barrier may not have existed. The deep ocean faunas of the Late Cretaceous and early Paleogene were not the same as now. The destruction of the Tethys Ocean brought about the loss of an important thermospheric water-mass generator. The formation of the Southern and Atlantic Oceans provided a significant generation of deep, cold waters, a meridional channel for their distribution, and the present psychrospheric dominance of ocean basins for the creation of a new deep fauna (about 40 m. y. ago). A study of the mechanics of currents in the deep ocean and the mechanics of skeletal reaction to thermally induced metabolic stress has led to a new view of evolution of both of these important systems.

INTRODUCTION

In one sense the results of the Deep Sea Drilling Project have been disappointing. Far fewer fossils representing past life on the ocean floor were found than were hoped for. The Ostracoda are an exception, however, and for this reason they have increased in importance for the study of deep ocean paleoenvironments. This report is a summation, with comments, of some of the more exciting results of about twelve years study of deep-sea ostracodes, both living and fossil, and of the search not only for oceanic life in the past, but also for lost oceans. Finding fossils of oceans older than about 40 m. y. is difficult, especially if these oceans were very different from those of today.

We know most of the different kinds of deep-sea ostracodes, even if many of them have yet to be described formally. The fossil record for these rare animals is most complete over the last 75 m. y. although some have been described from strata in the ocean floor as old as Lower Cretaceous (Albian, 100 m. y.: Oertli, 1974). These ostracodes have become architecturally distinct and genetically isolated from their shallow-sea counterparts. Their species seem to be remarkably long-lived and geographically widespread, and their assemblages are relatively small. Considering their remoteness in character and accessibility, they are becoming moderately well understood.

THE OSTRACODE

The marine ostracodes that leave fossil remains are tiny crustaceans, which are epibenthic in ecologic habit having no known planktonic larvae. They are heavily armored in a bivalved carapace (examples shown in Fig. 1), and have sexual reproduction. The oldest ones appeared first in the Cambrian and they become very important in the lower Paleozoic. None have ever evolved into anything more important, so far as we know. Their longevity as a group and their diversity of form is attributed to their ability to encapsulate themselves in a safe, often seemingly ponderous, external calcitic armor, which is fashioned according to the needs of a broad scope of environmental difference. Yet the animal is still able to retain enough metabolic energy after its considerable construction labors to remain very active and reproduce. Reproduction sometimes requires elaborate morphologic compensation in the adult stage. Producing an adequate population in the deep sea must be a slow deliberate process. The male sperm is about the size of that of an elephant, which says something about the ostracode's genetic complexity and limited fecundity, if not its potential size.

We will return to the ostracodes later, but to understand the ones of the deep sea, or the ocean, it is necessary first to look in a special way at oceans.

THE OCEANS

Most of us are more familiar with oceans than with ostracodes; however, oceans are as vast as ostracodes are small, and views of oceans differ considerably. To a petrologist, an ocean is that part of the Earth's crust primarily composed of sheared magnetically zoned basalt plates; to a sedimentologist, it is where traction loads turn into turbidity currents, sands into oozes; to a geophysicist, it is where rock strata become horizons; to a biologist, it is where the second greatest ecologic barrier begins. For paleontologists, especially the planktonic specialists, the sedimentary history of the oceans represents the ultimate reference for geologic time for the last 135 m. y. For a student of ostracodes, the oceans provide an opportunity to study the effects of extreme demands on the ability of the animal to build its skeletal structure. It is from this varying skeletal evidence that one can infer vast and unexpected change in the earlier history of the ocean.

Of the lessons learned during the conference on historical biogeography, one that impresses me most is the difference in point of view of those who study paleogeography from the historical record found on continents from those who study it from the ocean floor. Like the blind Hindus, we all see the elephant differently according to our opportunities and experience. And like Plato's slave in the fable of the shadows on the cave wall, it is not easy to convince others of what we see. Therefore, to reconstruct oceans of the past from ostracodes is, to say the least, an uncertain task.

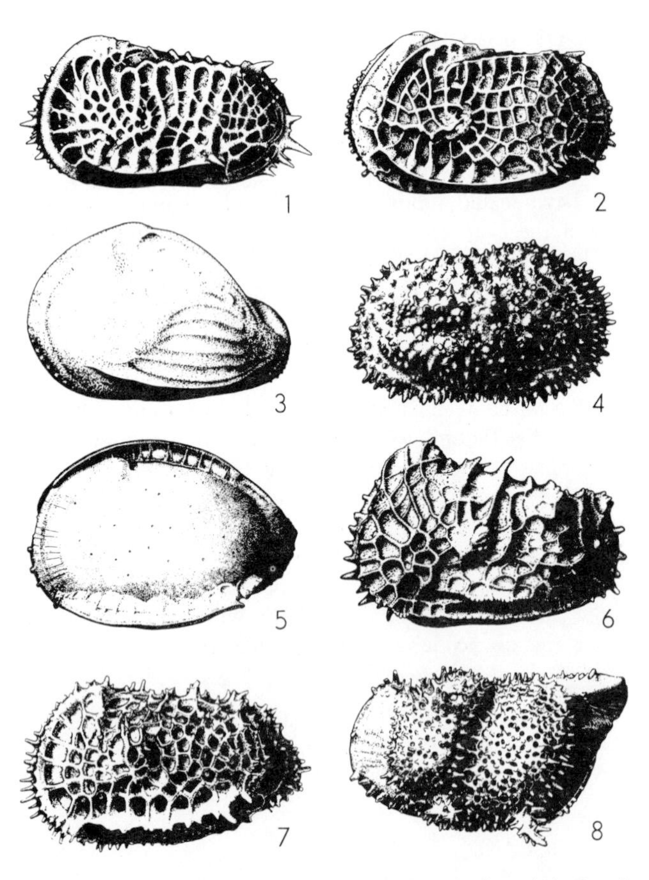

Figure 1. Typical psychrospheric ostracodes. (1) *Poseidonamicus major;* (2) *Bradleya dictyon;* (3) *Quasibuntonia sulcifera;* (4) *Henryhowella asperrima;* (5) *Brachycythere? mucronalatum;* (6) *Abyssocythere casca;* (7) *Agrenocythere spinosa;* (8) *Bythoceratina scaberrima.*

WHAT IS AN OCEAN?

One can no longer be assured that one's colleagues will assume any degree of permanency in world geography. Furthermore, it has become more and more difficult to shock one's colleagues with a discovery of anomalous paleobiogeographic distribution. This deprives us of the ability to refute a given paleobiogeographic model which is a significant analytical strategy for those of us dealing with "minor sorts of evidence." Recently, I was told that to attempt to find evidence for the Tethys Ocean was to deal with a "nonproblem." The argument of my critic (an oceanographer of some considerable respect) was that the existence of an east-west or latitudinal gap in Pangaea before the formation of the Atlantic was the inevitable consequence of geometric balance between moving plates and continental masses, and therefore, *sine qua non.* For him, the World Ocean was the consequence of an equitable redistribution of crustal mass. I was simply a modern Captain James Cook in quest of *mare incognita.* The task was rather mundane labor, not argument.

On the other hand, fortunately, there are some continental geologists who are dubious of concepts of a mobile Earth and of oceanographic evidence of lateral plate movement. In Europe these are the followers of Rutten (1970), Auboin (1965), Glangeaud (1962), and van Bemmelen (1969), who have either claimed isostatically raised or lowered landmasses in the area of present deep-sea or ocean basins of the Mediterranean or the Atlantic, or decried the fact that any conclusive evidence could be presented for the prior existence of deep-basin or oceanic conditions within continental Europe. It is they who force the enthusiast to give reasons and to demonstrate cause, no insignificant contribution to any dialectic.

I prefer to construct argument as a scientific strategy, because evidence without context is as meaningless as numbers without a structural reference. And as one whose experience comes more from the modern seas and oceans than from land, I will argue, with Anton Bruun (1957), Sanders (1968, 1969), and Hessler and Sanders (1967), that oceans have become biologically distinct from seas, because the general two-layer thermal structure of oceanic waters has created a formidable depth barrier to benthonic organisms. The fossil evidence of the lower of these two layers (psychrosphere; the upper is called the thermosphere) is *pro forma* evidence of prior oceanic conditions in low latitudes, because the psychrosphere cannot form outside of polar regions. Horizontal water-mass movement in general, especially to fill basins, is more important in this context than vertical mixing. From this spatial relation emerges one of the few structural reference systems on which a world paleobiogeography can be built from paleobiological evidence.

Furthermore, I will argue that psychrospheric conditions are not permanent in oceans throughout geologic time, but are the product of a World Ocean thermal economy whose budget intake or deficit varies according to whether circum-global circulation is dominantly latitudinal

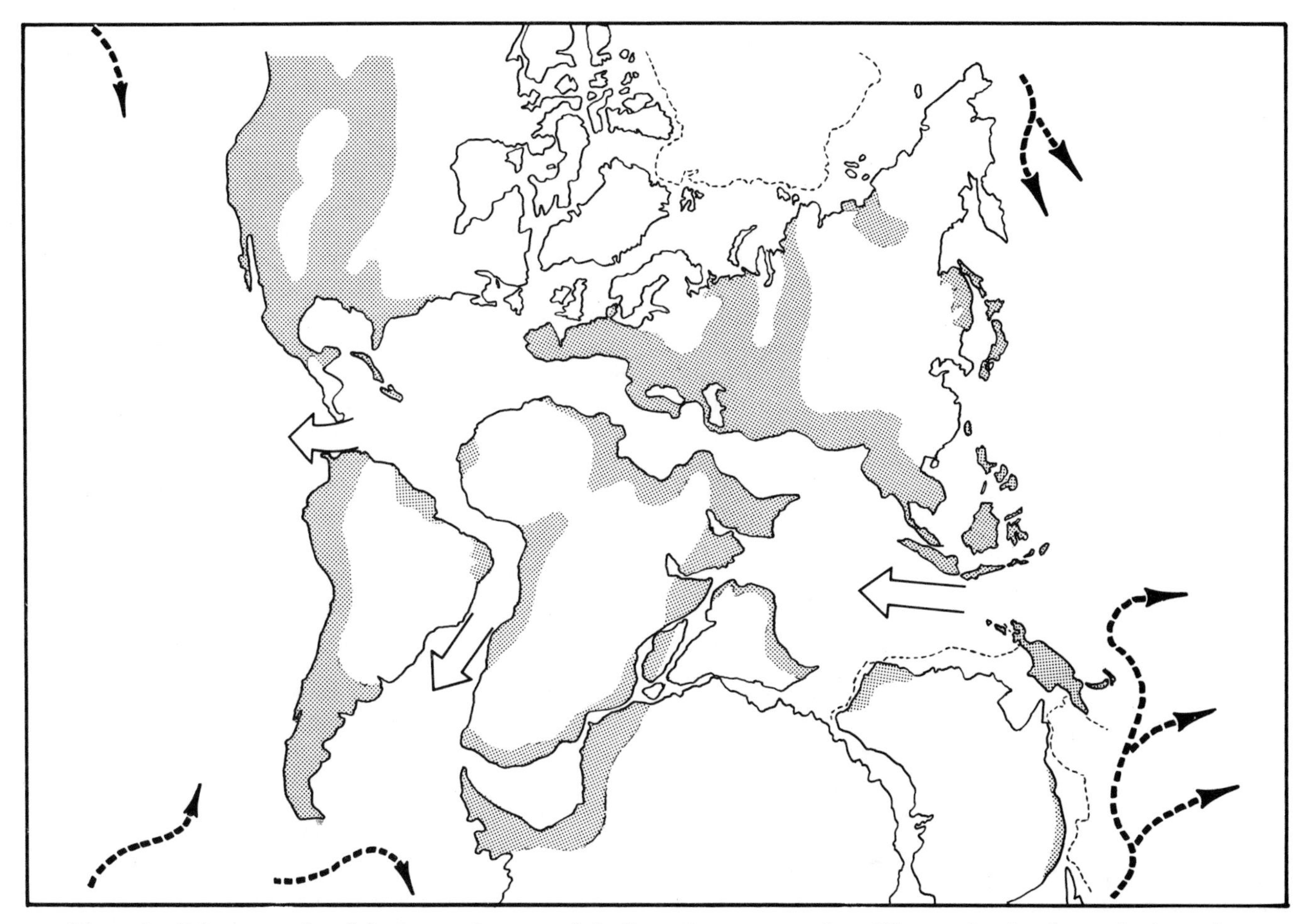

Figure 2. Paleogeography of the seas and oceans of the Late Cretaceous at about 75 m. y. showing the predominant thermospheric influence of Tethys (white arrows) and the lesser psychrospheric presence and influence (black arrows) of the polar regions.

or meridional. A dominance of continentality, or restriction to circulation of deep currents in either low latitudes or near the poles, is responsible for the changing character of oceans. Thermospheric oceans are the product of the channeling of the dominant deep arterial current systems to flow with the latitudes, especially at low latitudes (Fig. 2); psychrospheric oceans are the product of the channeling of these currents to flow with the meridians, especially from high latitudes (Fig. 3). Linked with the temperature difference between these two current systems is a chemical change leading to greater metabolic precipitation of carbonates at depth during times of thermospheric oceans, and to less metabolic precipitation during times of psychrospheric oceans. Unfortunately, the recognition of fossil evidence of past thermospheric ocean systems, as distinct from shallower warm seas, presents more problems to the paleobiogeographer than does the fossil record of a psychrospheric ocean. Morphologic adaptation is not as extreme for those organisms that invade warm depths, and segregation of these faunas is not as complete. Fortunately, there are other avenues of inquiry such as comparison to known shallow assemblages, relative diversity, tracing of evolutionary lineages, and finding depth-related adaptive structures that are independent of temperature (i.e., blindness).

The existence of the modern psychrospheric World Ocean is self-evident. All I can claim to have discovered about it is that its ostracode fauna is distinct, and that it began about 40 m. y. ago. Perhaps I could claim to have discovered (by reconstructed adaptive structural analogy) the pre-existing thermospheric ocean, which I suggest began about 140 m. y. ago (going back to Early Jurassic). I claim that it is possible to trace former psychrospheric invasions into dominantly continental regions. Outside of the crustal followers, not many workers have claimed that they could recognize evidence for the connection of deep-water filled basins that represent a former oceanic system from bits and pieces of the geologic record in tectonically disrupted regions.

As we will see, there have been at least four important oceanic events in the evolution of the World Ocean over the last 135 m. y.: the formation of the one-ocean Atlantic; the destruction of the thermospheric Tethys; and the creation of first a southern and then a northern generated psychrosphere.

The origin of the present Mediterranean Sea is conceptually interesting as a consequence of all of these events, plus the influence of continentalization. It is a major unresolved area of plate movement. It is not simply a subduction relic of the Tethys Ocean-Sea. Today the Mediterranean Sea is truly a deep sea, not an ocean, paleontologically speaking. One cannot determine its present depth from its fauna. Bathymetrically, it may be a small ocean; petrologically, it is both a sea and an ocean. Few psychrospheric animals have existed in its great depths for the past 2 m. y., but they did for several million years before that. It is a microcosm of former thermospheric ocean systems. Today the Mediterranean contributes dense thermospheric water to the Atlantic to about the same extent that the Labrador Current contributes cold, dense water. The thermospheric effects on the ostracode faunas of Mediterranean waters impinging on the Atlantic slope of France at about 1,200 m can now be demonstrated (Peypouquet, 1975).

MOTION IN THE OCEAN

Until a few years ago, global ocean circulation was thought to be driven by a combination of surface wind friction and the very slow overturn of thermally driven meridian cells extending from the equator to the poles. From the observations of Stommel (1957; Stommel and Arons, 1960, 1972; see discussions in von Arx, 1962; Fofonoff, 1962), it was shown that the main thermocline provided a floor beneath the wind-driven thermospheric circulation and a ceiling above the psychrospheric circulation, therefore providing a two-layer ocean model. The main driving force for the lower deep current system is "geostrophic" or the inertial response to the rotation of the Earth of an equitable density distribution of bodies of water having constantly mappable properties. That is to say, that the Earth turns beneath dense (cooled or more saline) and descending water masses. These water masses are pressed against onrushing continents that tend to squeeze their mass along yielding pressure gradients to form currents under the thickened western margins of spinning, floating discs of thermospheric gyrals.

In spite of the omissions of the above oversimplification, the main features of the two-layer, geostrophic concept help to explain why strong meridional psychrospheric currents now exist along eastern continental slopes as they disperse the cold waters produced near Labrador (only since the Miocene) or the Antarctic (since the Paleogene). What would happen if the present margins of the "shoving" continents were suddenly aligned parallel instead of normal to the Earth's rotation? This would have been the case with the continental masses bordering the Tethys Ocean when it was the primary channel of World Ocean deep circulation. The strong flow of bottom currents would have been through a low-latitude, high-density producing, warm-water, east-west system (analogous with the present Mediterranean). This would have increased the volume of dense, warmer bottom waters entering the basins of the World Ocean (Pacific) by several orders of magnitude over that of the present Mediterranean. The great amount of circum-global, cold bottom water that flows through the Southern Ocean today was not available, although what was is not known, so that a world dominance of a thermospheric ocean seems likely.

Berggren and Hollister (1977) have presented a series of maps of Atlantic Ocean circulation from its origin to the present (see also Ramsay, 1973; Smith and others, 1973). Although I would place the event of cold, bottom-water flooding (the origin of the psychrosphere) as much as 10 m. y. later (40 not 50 m. y. ago), this disagreement is a matter of dating the most noticeable effect of change and not the change itself. I believe that the authors overlooked the multiple effects of the closing of the Tethys system, and did not consider sufficient European-African meridional plate movement that would have closed a larger Tethys Ocean. Furthermore, I suggest that Africa probably moved southwest rather than northwest from South America (as suggested by van Andel, personal commun. 1976) to give Tethys a larger western opening. However, the gyralic plots of current systems are compelling and I think, in general, correct. If we could alter these explanations to include two-layer water-mass flow systems and introduce the major biologic boundary, I believe that the model becomes more usable. Lastly, if we introduce the idea of a conflict for dominance between Tethyan thermosphere generation and Polar (primarily Antarctic) psychrosphere generation, the model becomes developmental in an historical sense. One needs, of course, to add other influences such as the opening of Southern Ocean circulation (Kennett and others, 1972) and the movement of Antarctica into a polar position to increase the effectiveness of this region as a heat sink.

In this model of the evolution of the Atlantic circulatory system, there are several possible events to be tested by faunal distributions: (1) the formation of surface gyrals (Jurassic to Early Cretaceous); (2) the first connection of the North and South Atlantic by effective thermospheric flow (Early Cretaceous); (3) the invasion of the North Atlantic by the psychrosphere from the south (late Paleogene); and (4) the formation of an inflowing Arctic psychrospheric source and the effective closure of Tethys. Some of these will be discussed after first adding a few more elements to a rather complex and dynamic biogeographic, ocean systems equation.

ON "LOSING" AN OCEAN

Oceans can be "lost" by two means: firstly, by a change in water-mass structure to eradicate the psychrospheric presence or influence, which will remove biotic analogues of the present oceans; or secondly, by destruction of the ocean basins themselves. In the latter case, as with Tethys, subducted ocean crust and colliding continents may have removed all but a few fragments of the old ocean floor. The discovery of psychrospheric ostracodes in old Tethyan deep basin deposits would not only indicate that the waters were deep, but that they were deep continuously to the

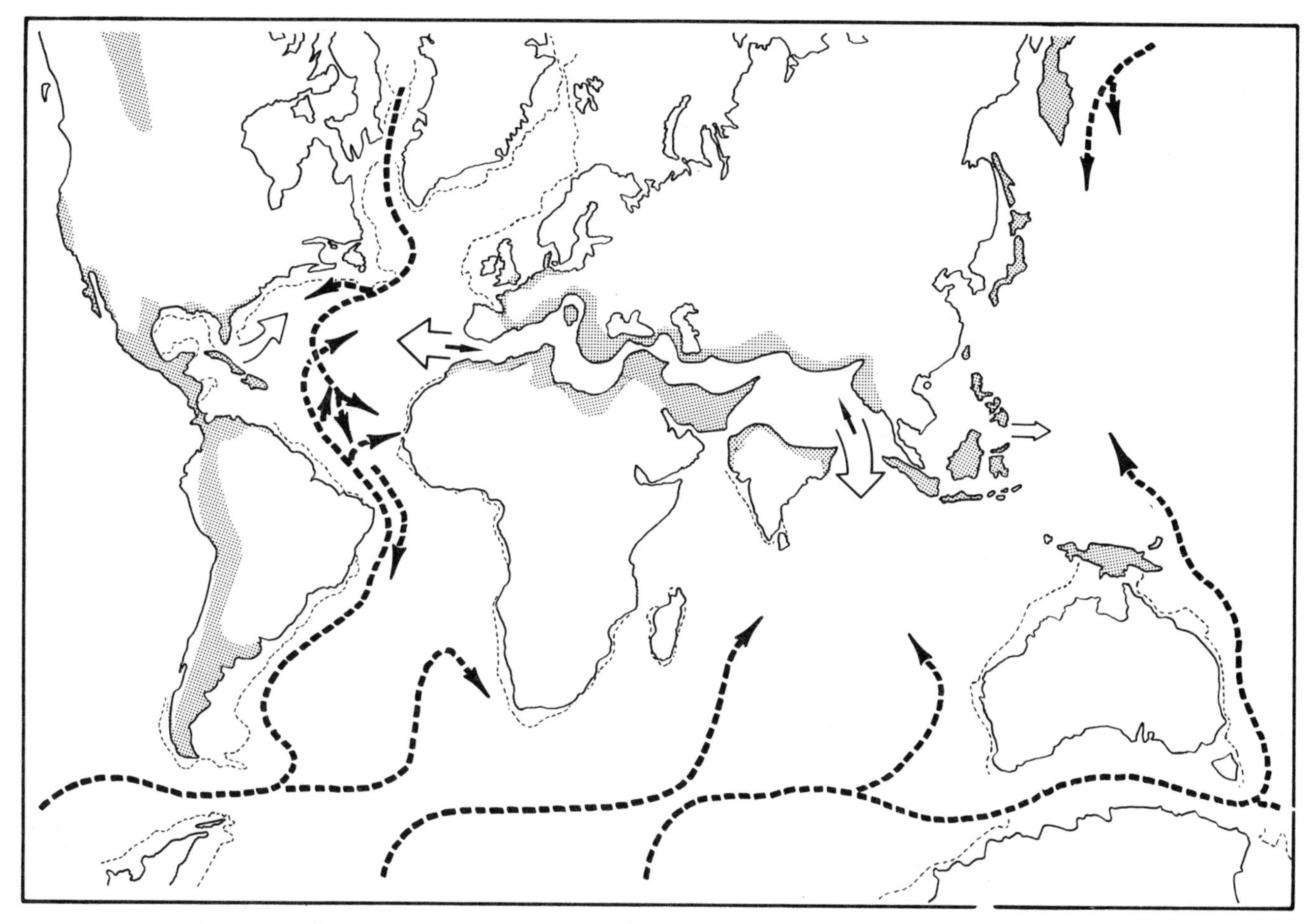

Figure 3. Paleogeography of the late Paleogene at about 30 m. y. showing the shrinkage of Tethys and the diminution of its thermospheric influence (white arrows) and the considerable increase of the psychrosphere (black arrows) with the possible junction of cold currents from both polar regions in the Atlantic (depending on whose dates are used for the beginning of the Arctic influence).

geographic source of psychrospheric water-mass generation. Too often those who would argue that "of course the pre-Alpine basins were deep" miss the significance of the presence or absence of connecting thresholds.

Yet the closer one comes to proving that Tethys was an ocean and not just a series of flooded epicontinental basins, the more one is confronted with the probable disappearance of the psychrospheric evidence for oceans or ocean connections. This is the paradox of the search for the most important lost ocean. If Tethys was a formidable thermospheric generator, and emptied a great volume of warm, dense water into the World Ocean, would it not have depressed any existing psychrosphere, thus forcing the psychrosphere out of a small but growing Atlantic, perhaps eliminating it altogether? This paradox exists in attempting to prove that the Mediterranean of Pleistocene times was paleontologically deep.

The search for "lost" oceans therefore becomes focused toward two regions that exercise potential threshold control over the deep water circulation of both Tethys and the Atlantic. It is necessary to turn to these regions in order to follow the progress of the conflict over control of the bottom waters of the World Ocean and the consequent depression of the psychrosphere-thermosphere biotic barrier. These regions are southwestern Europe where the last of Tethys dumped warm water into the Atlantic and the Rio Grande Rise in the southwestern Atlantic which was eventually overwhelmed by a building Antarctic psychrosphere. In the former, hopefully, some of the earlier psychrospheric faunas can be found. The latter is the site where we can hope to find the time of origin of the Atlantic as a full-fledged ocean, and to see what the pre-psychrospheric deep ocean fauna may have been like.

CHARACTERISTICS OF BENTHIC PSYCHROSPHERIC OSTRACODES

Ostracodes, like oceans, are steady state systems with certain mass property changes across critical boundaries and thresholds. Species composition is only one of these properties.

From a biological point of view, the interface between the thermosphere and the psychrosphere, like that of the psychrosphere and the bottom sediment, may be thought of as a high refractive transition zone. That is, energy and organisms tend to pass through this zone with varying degrees of difficulty, with the consequent alteration of adaptive structures. The alteration is maintained in its direction and increases in intensity with further penetration. If an organism chooses (or is selected for) existence near the interface, it must make special adaptive effort to remain morphologically and ecologically stable. Whether this is expressed as a structural response in morphology, in defense or food gathering mechanisms, or in diversity or population structure, it all can be viewed as reaction to change under stress. This stress can be seen as changes in skeletal design (mechanical stress in skeletons when mass is difficult to obtain), in low diversity, particular species dominance, and in specialized organ structures. The changes across these interfaces are great, but changes parallel to the interfaces tend to be small. Time, as an expression of biogeographic distance and the rate of metabolic or morphologic change, is of another scalar magnitude compared to that of organisms evolving in seas.

This conceptual model of unusual stresses and extended time frame is necessary to begin to understand the deep-sea ostracodes. There are few direct biologic analogues from lacustrine or shallow marine experience with which one can make comparison. On first view, the effect of great depths on the architecture of skeletons seems to be one of character exaggeration. In the past many ostracodes of the deep sea were considered to be particularly ornate. However, from a closer examination of "bizarre form," now as static skeletal structure, one is forced to the realization that selective inertia is operating differentially among subsets of reactant systems to reinforce old structure. This is quite a different viewpoint compared with the sense of progressive competition to create new structure in skeletal form typical of shallow-sea ostracodes. One wonders how many truly new kinds of organ systems can have evolved in the deep sea. Time, whatever time means in a place where seasons are perhaps the only short-term rhythms felt, is measured by a paleontologist in terms of "selective events." Time's arrow, that is the tendency toward greater entropy, seems reversed and then held in check in the deep sea. Concepts like gene pools and gene flow seem frozen as microscopic, bisexual species with astronomic populations and 10,000 mile (16,093 km) ranges change little over 10 or 20 m. y. time spans. Perhaps the key to understanding abyssal animals is both interface stress and a cessation of time in the periodic event sense.

If I have attempted to explain changes in ostracode skeletal morphology in terms of mechanical principles with different rates of reaction to stress, it is because other explanations, such as progressive, allopatric gene-linked alteration of discrete characters, have failed to make sense yet. I find that what has been called "ornamentation" can be explained logically as connected support structure, either active or vestigial, and that the evolutionary rates of these structures are different, depending upon which intrinsic organ system they are associated with. It is possible to conceive of mathematically describable structural sets within sets of such changing systems; each reacting at a different, decreasing rate to environmental stimuli; each hopefully linked to an adaptive chain within a coherent system of balanced stresses. As one portion of the system becomes strengthened under stress (by piezoelectric induced growth of calcite crystals along stress lines), another portion yields material. These structures become genetically fixed, of course, but the assumption of genetic control simply acknowledges the limits of historical continuity, rather than being the source of discovery of a new morphologic relationship of possible taxonomic importance.

I have written several descriptions of this point of view and of the different ostracode morphologies that result from adaptive stress (Benson, 1972a, 1974b, 1975b). It would be easy to repeat much of this, but I am afraid it would be of interest only to other ostracode workers. However, there are some interesting generalities that seem to be very useful and not complicated.

First, is the fact that although many shallow-water ostracodes are blind, all deep-sea ostracodes are blind. One can observe a decrease in the size of the ocular apparatus as shown in the eye tubercles with increasing depth, and the proportion of sighted species decreases until about 600 to 800 m (low latitudes) where eye tubercles disappear altogether (Benson, 1975b). A fossil assemblage with many sighted species succeeded by blind ones could indicate tectonic submergence (that much eustatic change would be unlikely).

Second, the average size of adults steadily increases with depth (Benson, 1975b; Morkhoven, 1972), as does their architectural sophistication (measured in terms of the disappearance of redundant mass) and although they may be of different species and more often different genera, the abyssal ostracodes will be twice the volume if not twice the length of shelf ostracodes.

Third, the amount of calcite and the wall thickness of the shell (exoskeleton) decreases proportionately with increasing depth. To maintain an equivalent strength some of the mass is moved outwards from the center of form of the skeleton in order to increase its reaction moment. This has the appearance of increased "ornamentation" (i.e., unexplained decoration formerly used for species recognition), but it can be demonstrated in almost all cases to be functionally related as singular connected structures to compensate for mechanical stress. This stress mostly originates with the adductor-closing muscles and is relieved at the shell margins or commissure. There are many morphologic designs resulting from this fundamental reaction series, but all can be traced into other solutions and ultimately reduced to about five general plans. Even the perfectly smooth ostracodes exhibit increased cantilevering and thinner shells with increased depth.

Fourth, the total number of deep-sea species globally is relatively small and equivalent to about that of one shallow-water province. This is rather interesting when one considers that the number of identifiable differences in shell form increases. The diversity of any one assemblage, from an adequate sample, is about one-half to one-third

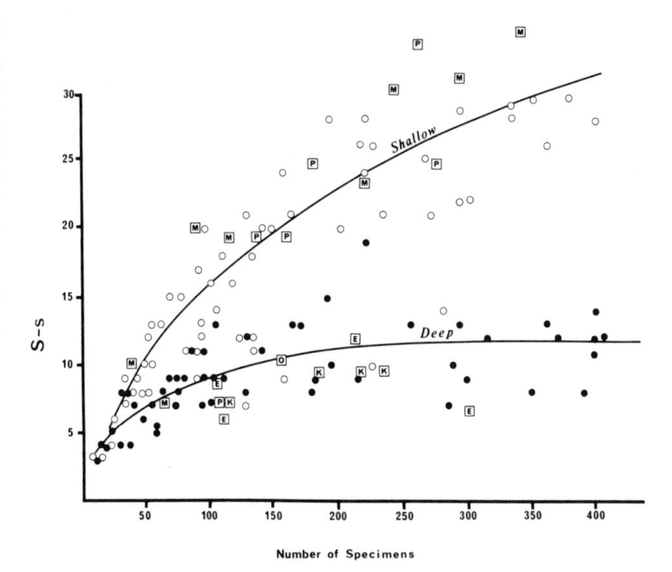

Figure 4. Simple species-diversity of ostracode assemblages found in individual Recent North Atlantic shallow-water samples (circles, less than 500 m); in world-ocean, deep-water samples (dots, more than 1000 m); and in similar older geological samples [squares, ages indicated: (P) Plio-Pleistocene (M) Miocene, (O) Oligocene, (E) Eocene, (K) Cretaceous] from the same regions. Plot is total species count (S) minus those represented by only one specimen (s) as potential contaminants versus sample size (after Benson, 1975a).

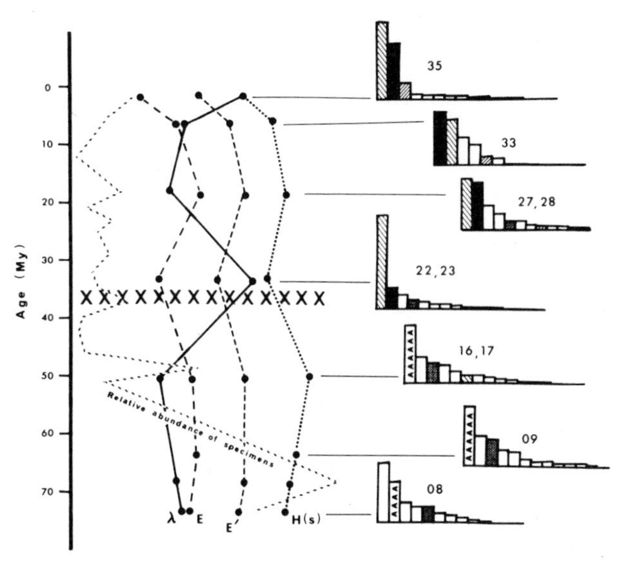

Figure 5. Changes in ostracode assemblage structure and specimen abundance through time in the South Atlantic (Rio Grande Rise region) (after Benson, 1975a), indicating the greatest change at about 40 m. y. (λ) Simpson's Index; (E) Sheldon's Equitability; (E') Buzas and Gibson's Equitability; H(s) Shannon-Weaver information function. Histograms are of species present with the identification numbers referring to composite foraminiferal zones with approximate ages as shown.

that of a shallow-water assemblage of the open shelf (Fig. 4). Low diversity among ostracode species indicates a high stress environment, but says nothing about time.

Fifth, the deep-sea ostracode taxa are long-lived and tend to be related to older rather than younger ostracode families or other higher groups. There are no hemicytherid ostracodes in the open deep sea. These latter ornate ostracodes are the dominant shallow-water forms of the Neogene. The modern deep-sea ostracode genera began about 40 m. y. ago, many by gradual evolution from the previous fauna, and have changed little since.

Lastly, although some psychrospheric ostracode genera come up into relatively shallow waters in high latitudes, they are, with a few exceptions, restricted to below 500 m in low latitudes. These exceptions, such as *Krithe* and *Henryhowella,* also change in size with depth; the respiratory area within the carapace increases as does the massiveness of the ornamentation. A similar psychrospheric assemblage is likely to be found in the central Atlantic as well as the northern Indian Ocean (Benson, 1974a) and to be equally useful for depth zonation.

THE PSYCHROSPHERE AND ITS ORIGIN

After the completion of the first circumnavigation of the *RV Glomar Challenger,* it was possible to examine the ostracode faunas (Benson, 1975a) of some 425 samples (50 cc) from 77 sites of the first 14 legs of cores of the Deep Sea Drilling Project. Over 80 species of 27 composite foraminiferal zones representing the last 75 m. y. of World Ocean floor history were examined and their distributions plotted. Special attention was given to the faunas found in cores of sites 15 through 22 of Leg 3 (later supplemented with cores of sites 356 and 357 of Leg 39) from the southwestern Atlantic. By cluster analysis (Dice Coefficient reflecting similarity by species in common) of assemblage structure, by tracing of the evolution of 3 exemplar genera, and by comparisons of species diversity (also with Recent analogues), species rank and predominance, and relative abundance, it was concluded that a single major change had taken place between an older Cretaceous to Eocene fauna and the Oligocene to Recent fauna (Fig. 5). The event of formation of the modern fauna took place, with relative suddenness, at about 40 m. y. or late Eocene. This was the beginning of significant generation of cold, bottom waters in the Antarctic to form the psychrosphere, whose faunal elements could be traced back into the Paleogene of the South Atlantic. Independently, others (Douglas and Savin, 1974; Shackleton and Kennett, 1974) were also concluding from other kinds of evidence that an important faunal change took place at this time. Kennett and Shackleton have recently proposed (Kennett, personal commun. 1976) that the conversion took less than 200,000 yr.

The opening of the Atlantic to form a north-south meridional corridor for deep-water circulation began to be effective probably in the Cretaceous. But the greater volume of dense water must have been Tethyan in origin and warm. It was not until Antarctica moved away from Australia into a polar position in late Eocene (Kennett and others, 1972) that circum-global, cold water circulation

(Southern Ocean) became sufficiently effective to counter an ever-shrinking Tethyan, low-latitude influence. Thus, the thermal dominance over dense water-mass formation shifted the major source of deep water-mass systems toward the poles to form general psychrospheric conditions. The suddenness of the change seems to be, simply, the result of overriding and passage of critical thresholds whose influence more than actual existence is known at present.

The major sill or threshold that controlled the invasion of the Antarctic psychrospheric waters into the Atlantic was and is in the area of the Rio Grande Rise in the southwestern Atlantic. A recent study of the ostracodes (Benson, 1977a) of two sites (356 and 357) of Leg 39 of the Deep Sea Drilling Project, involving 48 samples from cores of Maestrichtian to late Neogene, complemented the studies of the Leg 3 ostracodes (Benson, 1975a) to show that although the depths of the threshold had probably remained at oceanic depths (1,000 m or more), there had been a major change in the fauna in late Eocene to early Oligocene (especially well demonstrated at site 357). Examination of the phyletic lineages of several well-represented genera shows major architectural changes from more massive thermospheric forms to carbonate starved psychrospheric forms. These changes most dramatically appear in the increase in reticulation, which is one of several means of selectivity reinforcing shell-wall structure over an area with less mass. From these changes one can recognize many new taxa, but these identifications are based on what are believed to be significant structural changes.

The fauna of the Rio Grande Rise from the Oligocene to the present varies in abundance and diversity, but this variation is not great or well understood. In general, the modern fauna of the floor of most of the Atlantic is a direct reflection of this original composition with most of the major genera present. This fauna is the record of the Atlantic Ocean as we know it today with the possible addition of some new descendant species coming in with the northern Arctic invasion.

At present our knowledge of the effects of the formation of the Labrador Current on the psychrospheric ostracode fauna is very limited. *Echinocythereis* seems to be one of the newer immigrants, and possibly *Muellerina*, species of *Cytheropteron*, and a *Brachycythere*-like form. The North Atlantic remains a most interesting area, sealed off from the Arctic by the Wyville-Thompson Ridge and an almost sterile Norwegian Basin. Only a few Recent ostracode assemblages are known from the deep Arctic and these contain none of the typical World Ocean psychrospheric forms.

THE EXTINCTION OF TETHYS OCEAN

The Tethys Ocean was thermospheric for most of its 100 m. y. of existence, and for this reason its earlier existence as an ocean may be difficult to prove paleontologically. It is only because we can trace the origins of the modern psychrospheric fauna back into the Eocene, that we can recognize elements of deep ocean in the sediments of Priabonian age (late Eocene) in southern France, Italy,

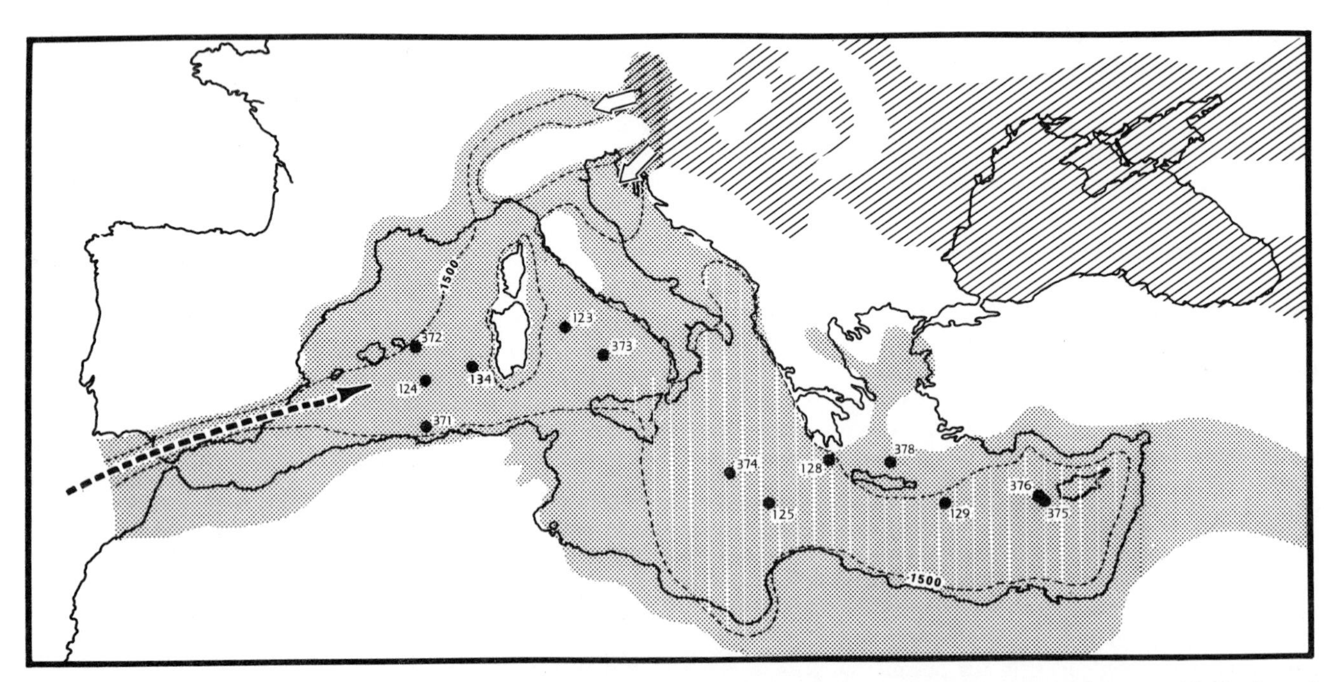

Figure 6. Paleogeography of Tethys (stippled) and Paratethys (cross-hatched) at about 12 m. y. without considering lateral movements in continental position. The dots represent the position of DSDP sites where ostracodes were found. Paratethyan influences are shown coming into the north, a shallow-sea link with the Indian Ocean to the east, and a deep-sea link with the Atlantic to the west. Psychrospheric conditions prevail in the western basins, while poor circulation and the first indications of the coming "salinity crisis" are seen in the eastern basins.

or Moravia (Pokorny, 1973). The presence of a few psychrospheric ostracodes in the middle Miocene of the floor of the present Mediterranean (Benson, 1977a) testifies to the presence of deep basins connected to the Atlantic during this time, but the Tethys Ocean was experiencing its last stages of devolution before total crisis (Fig. 6). The Mediterranean began its existence as an ocean with upper psychrospheric faunas during the Pliocene, but these too became extinct in the Pleistocene.

An initial Tethyan crisis came during the Oligocene when the northern passages of a broad Tethys seaway were blocked, and the Indian Ocean began to be severed from oceanic continuity with the west. Little is known of this stage because fossil evidence is poorly preserved.

The major faunal crisis began with gigantic ponding of continental waters to flood some of the old Tethys basins and form a "caspibrackish" Paratethys "lake-sea" fauna (Krstić, 1971; Sokač, 1967; Stancheva, 1963) that became characteristic of aquatic deposits over a broad area just north of the Caucasus-Himalayan region extending west past the Carpathian barriers. To the south, marine continuity was maintained through Asia Minor by a shallow seaway that emptied into large western basins now the site of the Mediterranean. It is not known if these basins are relics of the old oceanic margins or subducted blocks of the margins of the two colliding continental masses. The vertical displacements and lateral movements in the whole of the western region are among the most complex anywhere. It truly seems like a kind of giant's playground during a very short geologic interval. What appeared as mountains are now known to be slumped erratics.

The western Mediterranean basins stayed joined to the Atlantic through the Iberian Portal in Andalusia until late middle Miocene (Benson and Sylvester-Bradley, 1971; Benson, 1976), when the closure of this important threshold can be traced in the Betic region by ever-shallowing faunas into the ultimate throes of Tethyan extinction—the Messinian salinity crisis. A complete record of the death of the Tethys Ocean can be observed in the DSDP cores of Leg 42A from the Burdigalian and Serravallian, when relics of psychrospheric ostracodes and foraminifera occupied the western basins, to the Messinian, when the last of these marine benthic forms are replaced for a brief interval by a mixture of Paratethyan and lagoonal faunas (Benson, 1977b; Ramil Wright, personal commun.).

During this time, the eastern basins, restricted by a developing Tunisia-Sicily threshold, were largely thermospheric, although the deep faunas were generally very poor. With complete or almost complete closure of the Iberian Portal and a rapid lowering of world sea level, evaporation produced more than a million cubic kilometers of evaporite deposits in the Mediterranean intermixed with Paratethyan ostracode faunas (Benson, 1973a; Decima, 1964). It is calculated (Ryan, 1973) that the salinity of the World Ocean must have been reduced by as much as 6 parts per thousand during this time.

The end of the salinity crisis came rapidly with sudden flooding of the basins by deep marine waters (Ryan and others, 1973; Cita, 1973; Ruggieri, 1967; Benson, 1972b; Benson and Ruggieri, 1974). Uppermost psychrospheric ostracodes are typical of these earliest Pliocene deposits (Benson, 1973b). The rise in sea level or the opening of the newly formed threshold with the Atlantic must have come quickly although the presence of a giant waterfall at Gibraltar seems rather dramatic for some students of this event. Nevertheless, the Mediterranean Sea began with a strong oceanic influx into the western basins. The eastern basins felt some of this, but circulation remained relatively stagnant in the deeper basins with the result that many sapropelic layers were formed. The deepest cores of Leg 42B did not penetrate deposits of this earliest time, but they indicate that the Black Sea has probably changed little since Paratethyan times except in severity of restriction.

The events of the Pleistocene are less extreme. Restriction from the Atlantic cut off the psychrosphere to allow all of the Mediterranean to turn thermospheric. The deeper faunas became extinct, but were replaced by tolerant descending upper bathyal species. Near the shore lines many north European species replaced the last of the Tethyan refugees.

CONCLUSIONS

The purpose of this report is twofold. First, it is a commentary on ocean evolution from one who studies a particularly interesting benthic animal whose remains are found as microfossils in deep-sea sediments penetrated by DSDP coring, and sometimes in mountainous outcrops where oceans once stood. It is not a recitation of assemblage distributions. After twelve years of study, a viewpoint emerges that attempts to combine structural mechanics both from water-mass and skeletal change. Much of this will likely be proved inadequate or wrong. Yet it is a beginning, and fidelity to some procedural principles, such as connective causation or functional morphology, has been useful. The results have been geologically exciting and have outrun taxonomic description. Perhaps this is one demonstration that some oceanographic generalizations can be made from paleontological evidence, even before the more complex systematics of biologic systems are known. In any case, geology and geologists cannot wait, and the next researcher who is attracted to the same evidence will know where I went wrong.

Second, I have stressed the idea that one cannot hope to find oceans of the past as simply that part of the paleobiogeographic map between the point where one runs off the edges of continents and the map margins. Oceans evolve and have functional mechanics like the organic systems that inhabit them. Until we paleontologists attempt to know their changing morphology and physiology, as well as those of our favorite animals or plants, we cannot hope to ever find them once they have been lost.

ACKNOWLEDGMENTS

The author wishes to thank R. Cifelli and E. Kauffman for their reviews and thoughtful suggestions, Jane Gray and Art Boucot for their efforts to make me explain my ideas better, Laurie Brennan and Donna Casey for their assistance with the manuscript, and L. Isham for his help with preparation of some of the drawings.

REFERENCES

Arx, W. S. von, 1962, An introduction to physical oceanography: Reading, Mass., Addison-Wesley, 422 p.

Auboin, J., 1965, Geosynclines: New York, Elsevier, 335 p.

Bemmelen, R. W. van, 1969, Origin of the western Mediterranean Sea, *in* Symposium on the problems of oceanization in the western Mediterranean: Nederlands Geol. Mijnbouwk. Genoot. Verhandl., Geol. ser., v. 26, p. 13-52.

Benson, R. H., 1972a, The *Bradleya* problem, with descriptions of two new psychrospheric ostracode genera, *Agrenocythere* and *Poseidonamicus* (Ostracoda: Crustacea): Smithsonian Contr. Paleobiology no. 12, 138 p.

——— 1972b, Ostracodes as indicators of threshold depth in the Mediterranean during the Pliocene, *in* Stanley, D. J., ed., The Mediterranean Sea: Stroudsburg, Pa., Dowden, Hutchinson and Ross, p. 63-73.

——— 1973a, An ostracodal view of the Messinian salinity crisis, *in* Drooger, C. W., ed., Messinian events in the Mediterranean: Koninkl. Nederlandse Akad. v. Wetenschappen, Utrecht 1973: Amsterdam, North Holland Pub. Co., p. 235-242.

——— 1973b, Psychrospheric and continental Ostracoda from ancient sediments on the floor of the Mediterranean, *in* Kaneps, Ansis G., ed., Initial Reports of the Deep Sea Drilling Project, Vol. 13, Pt. 2: Washington, D. C., U. S. Govt. Printing Office, p. 1002-1008.

——— 1974a, Preliminary report on the ostracodes of Leg 24, *in* Musich, Lillian, ed., Initial Reports of the Deep Sea Drilling Project, Vol. 24: Washington, D. C., U. S. Govt. Printing Office, p. 1037-1043.

——— 1974b, The role of ornamentation in the design and function of the ostracode carapace: Geoscience and Man, v. 6, p. 47-57.

——— 1975a, The origin of the psychrosphere as recorded in changes of deep-sea ostracode assemblages: Lethaia, v. 8, p. 69-83.

——— 1975b, Morphologic stability in Ostracoda: Bull. Am. Paleontology, v. 65, p. 13-46.

——— 1976, Miocene deep-sea ostracodes of the Iberian Portal and the Balearic Basin: Marine Micropaleontology, v. 1, p. 249-262.

——— 1977a, The Cenozoic ostracode faunas of the Sao Paulo Plateau and the Rio Grande Rise, *in* Initial Reports of the Deep Sea Drilling Project, Vol. 39: Washington, D.C., U.S. Govt. Printing Office (in press).

——— 1977b, The paleoecology of the ostracodes of DSDP Leg 42A, *in* Initial Reports of the Deep Sea Drilling Project, Vol. 42: Washington, D.C., U.S. Govt. Printing Office (in press).

Benson, R. H., and Ruggieri, G., 1974, The end of the Miocene, a time of crisis in Tethys-Mediterranean history: Egypt Geol. Survey Ann., v. 4, p. 237-250.

Benson, R. H., and Sylvester-Bradley, P. C., 1971, Deep-sea ostracodes and the transformation of ocean to sea in the Tethys, *in* Oertli, H. J., ed., Paleoecologie d'Ostracodes: Centr. Recherche Pau-SNPA Bull., v. 5 (suppl.), p. 63-92.

Berggren, W. A., and Hollister, C. D., 1974, Paleogeography, paleobiogeography and the history of circulation in the Atlantic Ocean, *in* Hay, W. W., ed., Geological history of the ocean basins: Soc. Econ. Paleontologists and Mineralogists Spec. Pub. no. 20, p. 126-186.

Bruun, A. F., 1957, Deep sea and abyssal depths, *in* Hedgpeth, J. W., ed., Treatise on marine ecology and paleoecology, Vol. 1: Geol. Soc. America Mem. 67, p. 641-672.

Cita, M. B., 1973, Mediterranean evaporite: Paleontological arguments for a deep-basin model, *in* Drooger, C. W., ed., Messinian events in the Mediterranean: Koninkl. Nederlandse Akad. v. Wetenschappen, Utrecht 1973: Amsterdam, North Holland Pub. Co., p. 206-228.

Decima, A., 1964, Ostracodi del genere *Cyprideis* Jones del Neogene e del Quaternario italiani: Palaeontographica Italica, v. 57, p. 81-133.

Douglas, R. G., and Savin, S. M., 1974, Biogeography and bathymetry of Late Cretaceous-Cenozoic abyssal benthic foraminifera, *in* Berggren, W.A., ed., Paleogeography and paleobiogeography: Organisms and continents through time: Symposium Woods Hole, June 1974, pt. N, 7 p. (limited publication).

Fofonoff, N. P., 1962, Dynamics of ocean currents, *in* Hill, M. N., ed., The seas: New York, Wiley-Interscience, p. 323-396.

Glangeaud, L., 1962, Paleogeographie dynamique de la Mediterranee et de ses bordures, *in* Oceanographie Geologique et Geophysique de la Mediterranee Occidentale: Colloques Internat. CNRS, Villefranche, p. 125-165.

Hessler, R. R., and Sanders, H. L., 1967, Faunal diversity in the deep-sea: Deep-Sea Research, v. 14, p. 56-78.

Kennett, J. P. and others, 1972, Australian-Antarctic continental drift, paleocirculation changes and Oligocene deep-sea erosion: Nature, v. 239, p. 51-55.

Krstić, N. 1971, Ostracode biofacies in the Pannone, *in* Oertli, H. J., ed., Paleoecologie d'Ostracodes: Centr. Recherche Pau-SNPA Bull., v. 5 (suppl.), p. 391-397.

Morkhoven, F. P. C. M. van, 1972, Bathymetry of recent marine Ostracoda in the northwest Gulf of Mexico: Gulf Coast Assoc. Geol. Soc. Trans., v. 22, p. 241-252.

Oertli, H. J., 1974, Lower Cretaceous and Jurassic ostracods from DSDP Leg 27—a preliminary account, *in* Robinson, Paul T., ed., Shorebased paleontological studies, Initial Reports of the Deep Sea Drilling Project, Vol. 27: Washington, D. C., U. S. Govt. Printing Office, p. 947-965.

Peypouquet, J. P., 1975, Les variations des caracteres morphologiques internes chez les Ostracodes des genres *Krithe* et *Parakrithe*, relation possible aves la teneur en O_2 dissous dans l'eau: Inst. Geol. Bassin Aquitaine Bull., v. 17, p. 81-88.

Pokorny, V., 1973, *Abyssocythere*, a deep-sea ostracode in the Paleogene of Czechoslovakia: Acta Univ. Carolinae, Geol., v. 3, p. 243-252.

Ramsay, A. T. S., 1973, A history of organic siliceous sediments in oceans, *in* Hughes, N. F., ed., Organisms and continents through time: Spec. Papers Palaeontology no. 12, p. 199-234.

Ruggieri, G., 1967, The Miocene and later evolution of the Mediterranean Sea, *in* Adams, C. G., and Ager, D. V., eds., Aspects of Tethyan biogeography: London, Systematics Assoc. Pub. no. 7, p. 283-290.

Rutten, M. G., 1970, The geology of western Europe: Amsterdam, Elsevier, 520 p.

Ryan, W. B. F., 1973, Geodynamic implications of the Messinian salinity crisis, *in* Drooger, C. W., ed., Messinian events in the Mediterranean: Koninkl. Nederlandse Akad. v. Wetenschappen, Utrecht 1973: Amsterdam, North Holland Pub. Co., p. 26-43.

Ryan, W. B. F. and others, 1973, *in* Kaneps, Ansis G., ed., Initial Reports of the Deep Sea Drilling Project: Washington, D. C., U. S. Govt. Printing Office, Vol. 13, Pt. 2: p. 517-1447.

Sanders, H. L., 1968, Marine benthic diversity: A comparative study: Am. Naturalist, v. 102, p. 243-282.

———— 1969, Benthic marine diversity and the stability-time hypothesis, *in* Diversity and stability in ecological systems: Brookhaven Symposia in Biology, no. 22, p. 71-81.

Shackleton, N. J., and Kennett, J. P., 1974, Palaeotemperature history of the Cenozoic from oxygen isotope studies in D.S.-D.P. Leg 29, *in* Marine plankton and sediments and Third Planktonic Conference (chaired by Seibold, E.): Internat. Counc. Sci. Unions-UNESCO, Paris, p. 66.

Smith, A. G., Briden, J. C., and Drewry, G. E., 1973, Phanerozoic world maps, *in* Hughes, N. F., ed., Organisms and continents through time: Spec. Papers Palaeontology no. 12, p. 1-42.

Sokač, A., 1967, Pannonische und Pontische ostrakodenfauna des südwestlichen teiles des Pannonischen Beckens: Carpatho-Balkan Geol. Assoc. Cong., 8th, Belgrade 1967, Rept. Stratigr., p. 445-453.

Stancheva, M., 1963, Ostracoda from the Neogene in northwestern Bulgaria. II. Sarmatian Ostracoda: Trav. Geol. Bulgarie, Paléont., v. 5, p. 5-73.

Stommel, H., 1957, A survey of ocean current theory: Deep-Sea Research, v. 4, p. 149-184.

Stommel, H., and Arons, A. B., 1960, On the abyssal circulation of the world ocean. II. An idealized model of the circulation pattern and amplitude in oceanic basins: Deep-Sea Research, v. 6, p. 217-233.

———— 1972, On the abyssal circulation of the world ocean. V. The influence of bottom slope on the broadening of inertial boundary currents: Deep-Sea Research, v. 19, p. 707-718.

Dispersal of Pelagic Larvae and the Zoogeography of Tertiary Marine Benthic Gastropods

RUDOLF S. SCHELTEMA, *Department of Biology, Woods Hole Oceanographic Institution, Woods Hole, Massachusetts 02543*

ABSTRACT

The transport of planktonic veliger larvae by ocean currents is one way in which Recent warm-temperate and tropical gastropod species are dispersed thousands of miles along coastlines of continents and across ocean basins. Circulation in the contemporary tropical Atlantic Ocean makes possible dispersal between South America and West Africa both to westward on the North and South Equatorial Currents and to eastward on the Equatorial Undercurrent. Pelagic larvae of the gastropod family Architectonicidae are commonly found in the open sea; occurrence of teleplanic veligers of *Philippia krebsii* and *Architectonica nobilis* throughout the tropical Atlantic Ocean illustrate the long-distance dispersal of these 2 species. Planktonic larvae of the genera *Thais* and *Drupa* of the family Muricidae are also dispersed over very long distances.

Comparison of the well-preserved protoconchs on fossil architectonicids and muricids with those of Recent forms belonging to the same or very closely related genera, allow one to deduce the mode of development of species in past geological time. Not only can species having pelagic larvae be distinguished from those with a nonpelagic development, but also a rough estimate of the duration of the planktonic development can be made by a comparison of the size and number of whorls. The biological data on larval development and information on paleogeography and paleocirculation lead to the conclusion that transport of teleplanic veliger larvae has been a common means of gastropod dispersal in the warm-temperate and tropical Late Cretaceous and Tertiary Atlantic Ocean.

INTRODUCTION

There are three principal ways in which marine gastropods are dispersed. The first is active migration of the adult form. This method of dispersal has apparently accomplished the nearly circum-arctic distribution of the large prosobranch gastropods *Neptunea sutura* (Martyn) and *Neptunea communis* (Middendorf) described by Golikov (1963, Figs. 94, 97). As both species have a nonpelagic development they must have literally crawled around the arctic continental shelf since the Miocene. Migration thus restricted to shallow water, however, does not permit dispersal across such zoogeographical barriers as deep ocean basins. Second, gastropods may be dispersed by rafting, a means successful only among species able to attach to floating objects and to remain alive for long periods in the open sea. In cold waters at high latitudes, minute gastropods possibly can survive in the holdfasts of large drifting kelps detached from coastal regions during severe storms. Gastropod species too large to be rafted and lacking a pelagic development can attach their egg masses to debris that subsequently is set adrift. Without evidence, Marche-Marchad (1968) attributed the wide geographic distribution of large volutid gastropod species to the dispersal of their egg capsules on floating objects (genera *Cymba* and *Adelomelon*). The third, and probably most important means of dispersal among tropical and temperate gastropods, is the transport of veliger larvae by ocean currents, and it is this larval dispersal that I wish to consider in more detail, both in the present and in the geologic past.

LARVAL DISPERSAL OF CONTEMPORARY GASTROPOD SPECIES BY OCEAN CURRENTS

Until recently there has remained considerable uncertainty about the probability of long-distance larval dispersal. Ekman (1953) and Thorson (1961) minimized the role of larval transport over long distances because the duration of pelagic development of most shoal-water invertebrates seemed much too short. Darlington (1957, p. 213), when considering the distribution of shoal-water tropical invertebrates, thought that "it has not been settled whether or not some larvae may be carried across the Atlantic by ocean currents." Indeed, the compelling evidence that pelagic larvae are a means whereby benthic organisms can be dispersed over great distances along coastlines of continents and even across deep ocean basins has come only recently from the knowledge that larval stages occur regularly over temperate and tropical continental shelves, in shallow seas, and within major ocean current systems.

The evidence for larval transport comes largely from my own observations (Scheltema, 1964, 1966, 1968, 1971a, 1971b, 1972a). Seventy percent of all plankton tows made throughout the tropical and North Atlantic Ocean contain gastropod larvae that originated from shoal waters of the continental shelf. I have applied the term *teleplanic veliger*

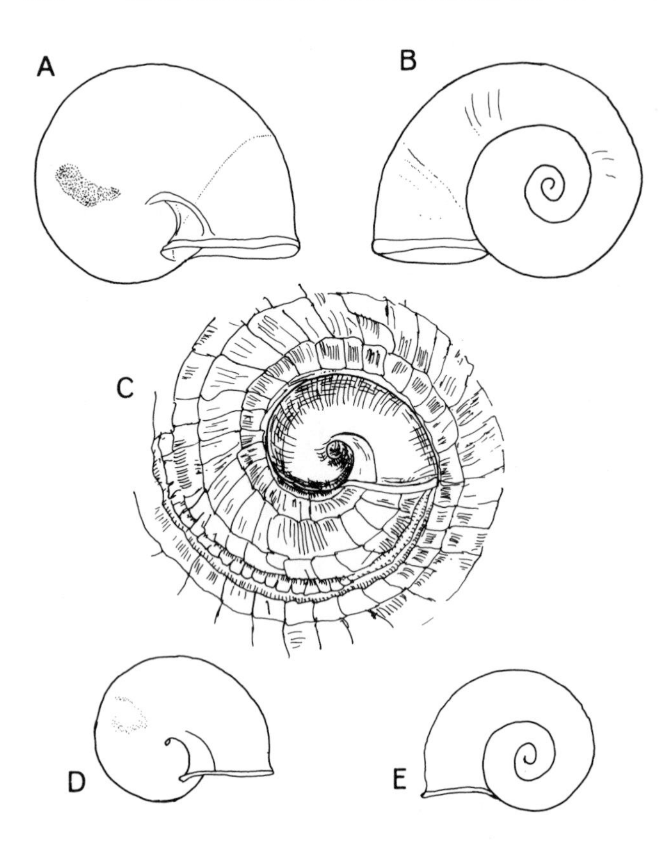

Figure 1. Larval shells and protoconch of gastropods belonging to the family Architectonicidae. (A) Apical view of *Philippia krebsii* (Mörch) veliger larva shell, longest dimension 1.4 mm; from plankton sample taken in the North Atlantic Drift, 39°26′N, 44°05′W. (B) Basal view of the same specimen as in A above. (C) Protoconch (larval shell) on apex of juvenile specimen of *Solarium trochleare* Sorgenfrei (redrawn from Sorgenfrei, 1958, Pl. 30, Fig. 101a), 0.93 mm in longest dimension; fossil from lower Miocene of Jutland, Denmark. Note striking similarity of protoconch on apex of this juvenile to the larval shells A,B,D and E. (D) Apical view of *Architectonica nobilis* (Röding) veliger larva shell, longest dimension 0.95 mm; from plankton sample taken in the North Equatorial Current, 10°48′N, 39°42′W. (E) Basal view of the same specimen as in D above. Note: Architectonicid larvae are hyperstrophic; the terminology of the adult shell is applied to the larval shell (see Robertson, 1963). All figures are similarly enlarged (drawn at X50).

larvae (from the Gr. *telep'lanos* meaning far-wandering) to veligers found in the open sea. Teleplanic larvae may be defined as those (1) that originate from shoal-water, continental-shelf benthos, (2) that are regularly found in the open sea, and (3) that have a pelagic development of long duration, usually many months, thereby serving as a means for long-distance dispersal.

Among the gastropods there are already 18 families known to be represented by at least some species with teleplanic veliger larvae. Many of these teleplanic veligers are modified morphologically and probably also physiologically for their long life in the open-sea plankton. Morphological adaptations that enhance their ability to remain afloat are long periostracal spines, reduction or complete lack of shell calcification, and unusually long velar lobes that are used in swimming. The larvae of some species as in the families Cymatiidae and Tonnidae have very large shells exceeding 0.5 cm in height.

Before considering the dispersal of some specific species of teleplanic veliger larvae, it is necessary to summarize the circulation of the equatorial Atlantic Ocean. For the present purpose only the region between 20°N and 20°S need be described. The North Equatorial Current is separated from the westwardly flowing South Equatorial Current by a weakly developed Equatorial Countercurrent flowing toward West Africa. The Countercurrent moves only a relatively small volume of water and apparently occurs seasonally (Schumacher, 1940, 1943; U.S. Hydrographic Office Monthly Pilot Charts). However, directly beneath the westwardly flowing South Equatorial Current a strong eastwardly moving Equatorial Undercurrent flows from South America toward the island of São Tomé off the West African coast (Voorhis, 1961; Metcalf and others, 1962; Khanaychenko and others, 1965). This warm, shallow current (20° to 27°C), whose core extends between 50 and 100 m depth, provides a means whereby tropical larvae can be dispersed from west to east across the tropical Atlantic (Scheltema, 1968, 1972a). Hence, transport from the eastern to the western tropical Atlantic is possible on the North and South Equatorial Currents, and conversely, from the western to eastern tropical Atlantic on the Equatorial Undercurrent.

Examples from 2 families of gastropods will suffice to illustrate long-distance dispersal. The Architectonicidae are found in tropical and warm-temperate waters throughout the world and in the Atlantic, with the exception of 1 species, are restricted to shallow depths of the continental shelf (i.e., not exceeding 200 m). Marche-Marchad (1969) found 10 species along the coast of West Africa and 7 of these are also found in the western Atlantic. Most species of Architectonicidae have a wide geographic range. Pelagic larval development insofar as known occurs in all architectonicid species, and indeed already 15 different kinds of veliger larvae have been found in the Atlantic open sea plankton samples (Robertson, *in litt.*).

Philippia krebsii is known in the western Atlantic from Cape Hatteras, Bermuda, Mexico, the Antilles, and tropical South America, and in the eastern Atlantic from the Canary and Cape Verde Islands. The species has also been recorded from the Ascension and St. Helena Islands in the central Atlantic. Larvae of *Philippia krebsii* (see Scheltema, 1971b, p. 296-297, for detailed description) are readily distinguished from other species of the family by their large size (1.3 to 1.4 mm) and the presence of a very distinct anal keel (Figs. 1A, 1B). The widespread dispersal of the veligers of this species is already well-known from previous accounts (Robertson, 1964, Fig. 11; Scheltema, 1968, Fig. 1, 1971b, Fig. 10) and is illustrated in a chart of the tropical Atlantic region in Figure 2.

Architectonica nobilis, its shell often illustrated because of its beauty, also has teleplanic veliger larvae commonly found in the open-ocean plankton. Adult populations occur

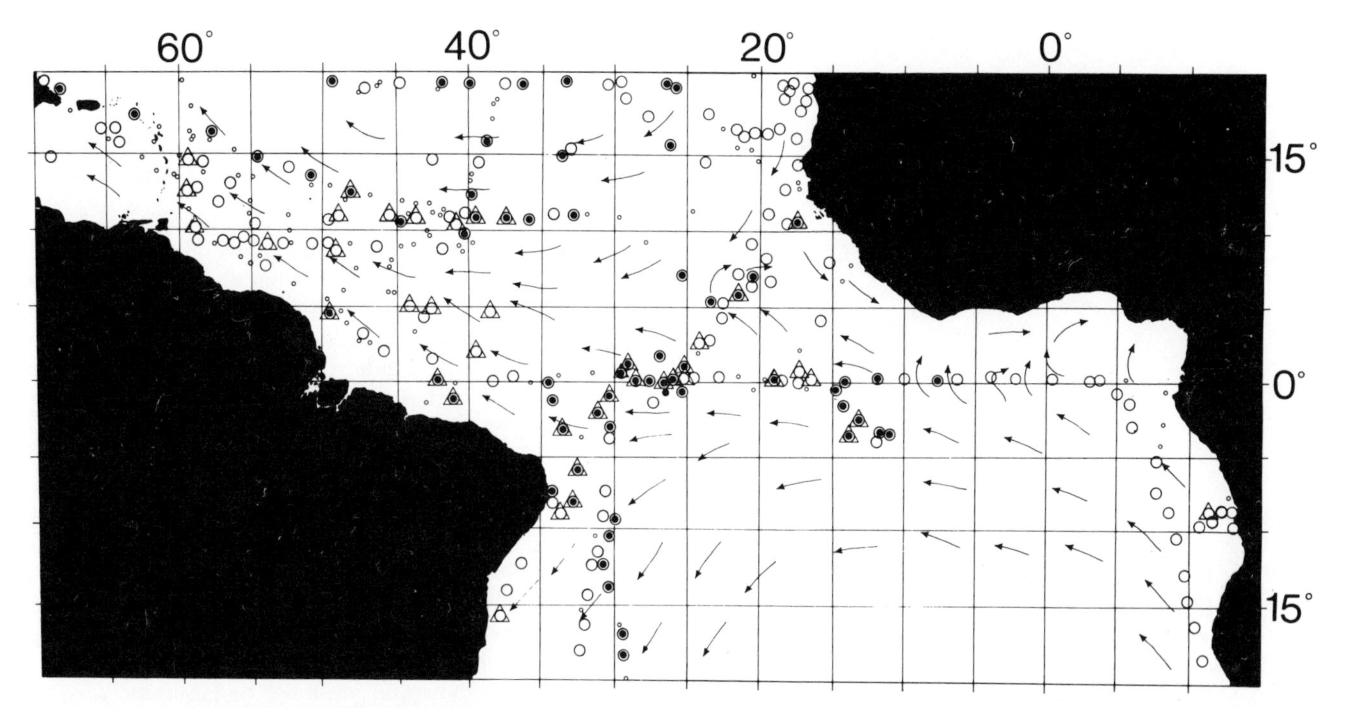

Figure 2. Geographical distribution in the tropical Atlantic Ocean of veliger larvae belonging to the gastropod family Architectonicidae. Filled circles are stations with *Philippia krebsii;* triangles, *Architectonica nobilis;* open circles, other species of Architectonicidae. Small circles indicate locations where larvae of Architectonicidae were absent. Arrows show surface currents. Inferred routes of dispersal in the upper 50 m are from east to west. Currents along the equator below 50 m depth flow in the opposite direction as those at the surface; passive dispersal of larvae at depths below 50 m will, therefore, be from the western to the eastern Atlantic Ocean (see Scheltema, 1968, 1972a).

in the western Atlantic from Cape Hatteras to Bahia, Brazil, and in the eastern Atlantic from Senegal to the Congo. The species has also been recorded from along the Pacific coast of Central and South America. Veliger larvae of *Architectonica nobilis* resemble those of *Philippia krebsii* but differ in being markedly smaller (ca. 0.9 mm) and in having a less distinct anal keel (Figs. ID, 1E). As in *Philippia krebsii,* the geographic distribution of *Architectonica nobilis* larvae is also widespread (Fig. 2) and extends across large areas of the tropical Atlantic Ocean.

Not only are *Philippia krebsii* and *Architectonica nobilis* larvae very widely dispersed by ocean currents, but other veligers belonging to the family Architectonicidae are also to be found throughout tropical waters (Fig. 2). In the region between 20°N and 20°S latitude that is being considered, approximately 70 percent of all samples contained Architectonicidae. I conclude from these data that larval transport has played a significant role in the wide geographic range of architectonicid species.

The second family exemplified here as having long-distance larval dispersal is the Muricidae, carnivorous forms whose members feed on bivalves, barnacles, and other gastropod species. The egg capsules of each muricid species are distinctive and usually small, about 5 to 8 mm in height. Unlike the Architectonicidae, many muricids lack a pelagic larval development and the young emerge from the egg capsules as minute juvenile benthic forms. However, in the genera *Thais* and *Drupa* some species have teleplanic larvae and the veligers emerge from the egg capsules as young swimming stages.

The larval shell of *Thais haemastoma* has been known since 1877 from an illustration by Craven who believed it to be an adult form and who consequently assigned to it the generic name *Sinusigera.* The latter name is retained at present to designate a type of veliger larva in which a typical projection separates the adapical from the basal portion of the outer lip. The larval shell of *Thais haemastoma* (Fig. 3A) is about 1.5 mm in height at maximum size before settlement. Details of the larvae are described elsewhere in the zoological literature (see Scheltema, 1971b).

The genus *Drupa* has a sinusigera larva that differs from *Thais haemastoma* in its amber to dark-brown color and elaborate ornamentation (Fig. 4A).

Larvae belonging to the Muricidae are found in the open waters of the tropical Atlantic Ocean (Fig. 5), and when encountered they often occur in large numbers. Identification of veligers belonging to the genera *Thais* and *Drupa* to precise species presents difficulties because of the great similarity of the pelagic larvae within each of these genera. Most of the larvae appear to be *Thais haemastoma* and *Drupa nodulosa,* species that both have wide geographic ranges. *Thais haemastoma* extends throughout the tropical eastern and western Atlantic (Clench, 1947), although sometimes under different specific or subspecific names (see Scheltema, 1971b, Fig. 9). *Drupa nodulosa* is found in the western Atlantic along the warm-temperate

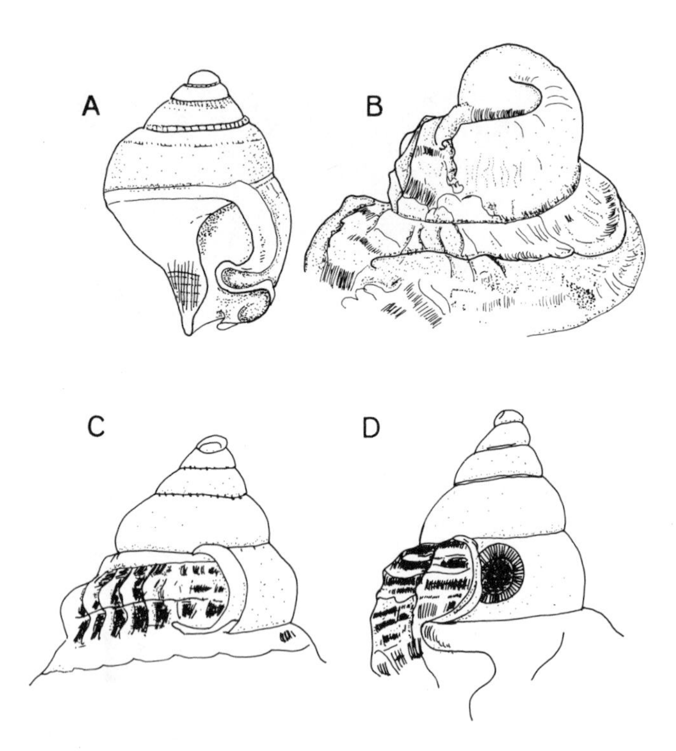

Figure 3. Larva and protoconchs of gastropods belonging to the family Muricidae. (A) Shell of pelagic veliger larva of *Thais haemastoma* (Linné) taken from plankton samples collected off the coast of West Africa (7°20'S, 7°23'E), height 1.5 mm. Note typical sinusigerous projection on outer lip. (B) Protoconch of the contemporary species *Thais emarginata* Deshayes from the central coast of California. This species has a nonpelagic development, and the protoconch differs markedly from the two with pelagic larval development (C,D) and from the larval shell (A). Note lack of a sinusigerous projection. (C) Protoconch of the contemporary Atlantic species *Thais haemastoma haysi* Clench from the coast of Texas. This species is known to have a pelagic larval development. The protoconch is slightly worn and shows the difficulty sometimes encountered when comparisons with larval shells are to be made (USNM 125561). (D) Protoconch of *Murex inornatus* Sorgenfrei from the lower Miocene of Jutland, Denmark. Comparison with the figure to left shows that this species had a pelagic development. Note sinusigerous projection (redrawn from Sorgenfrei, 1958, Fig. 13a). All figures are similarly enlarged (drawn at X50).

North American coast and in the Antilles, and in the eastern Atlantic along the west African coast from Senegal to Angola.

As in the Architectonicidae, larval transport appears to account for the wide geographic range of muricid species. There is an obvious difference, however, in the frequency with which the larvae of the 2 families are found in the open sea. Whereas 70 percent of all plankton tows between 20°N and 20°S latitude had Architectonicidae veligers, only 24 percent of these same samples contained muricid larvae. Differences in the frequency with which larvae occur in the samples reflect differences in the amount of long-distance dispersal found within the 2 families. The great geographic variation within species of the family Muricidae may be the result of a restricted genetic exchange between widely separated populations (see Scheltema, 1972b).

LARVAL DISPERSAL OF GASTROPOD SPECIES IN THE GEOLOGIC PAST BY OCEAN CURRENTS

To understand the geographical distribution of gastropod genera and species, one must know the method of their dispersal not only in the present but also in the geological past. How can one know which kind of larval development, pelagic or nonpelagic, a species had in remote geological time? Can one obtain indications that large-scale transport of teleplanic larvae has also occurred in the Late Cretaceous and throughout the Tertiary, times when the Atlantic Ocean first approached its present form?

Information on the kind of development in gastropod species is obtained from the characteristics of well-preserved protoconchs at the apex of juvenile or adult shells. Thorson (1950, p. 33) observed that "as a general rule, a clumsy, large apex points to a nonpelagic development, while a narrowly twisted apex often with delicate sculpture, points to a pelagic development." Other characteristics of the protoconch, however, must also be taken into account, and some of these were summarized recently by Shuto (1974, p. 243-245, Fig. 1). Thorson concluded that "only when we know the type of apex derived from a pelagic or a nonpelagic development within the individual genus, can reliable results be obtained"; but he also anticipated that by using the knowledge of contemporary species, valuable information could be found about larval development in fossil species of a genus.

Let us examine some specific examples from the families Architectonicidae and Muricidae. Comparison of the larval shells of the contemporary species *Philippia krebsii* and *Architectonica nobilis* with the protoconch of the lower Miocene species *Solarium trochleare* shows striking similarities between the fossil and modern forms (Fig. 1).

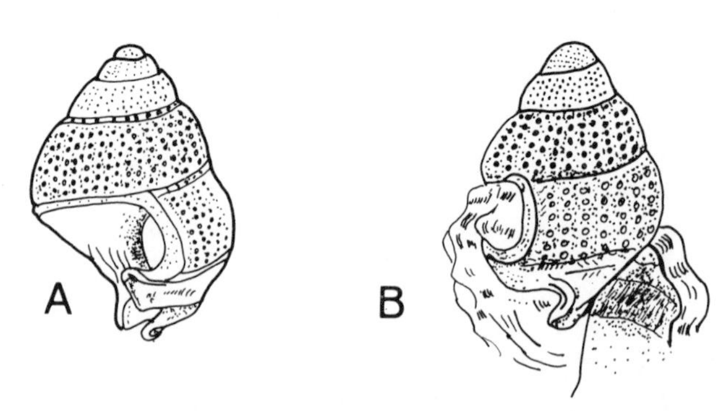

Figure 4. Larva and protoconch of gastropods belonging to the family Muricidae. (A) Larva of *Drupa* cf. *D. nodulosa* C. B. Adams; sculpture and ornamentation are very elaborate and the sinusigerous projection is conspicuous; color of shell is dark amber to brown. Larva taken off South America, 26°20'S, 46°05'W. (B) Protoconch of juvenile *Drupa nodulosa* from the collection of the Philadelphia Academy of Natural Sciences (ANSP 31389). The sinusigerous projection is still visible but the succeeding whorl would have concealed its basal portion as growth continued (see Figs. 3C, 3D).

Indeed, if the Miocene veliger swam under our lens, it could not be distinguished from a modern form! Details such as the anal keel and varix are identical in Miocene and Recent larval shells (compare A and D with C in Fig. 1), and it is quite clear from the protoconch that *Solarium trochleare* had a pelagic larva, as do all known modern forms of the family Architectonicidae. Turning to the Muricidae, a more complicated example emerges, since not all species among this family have a pelagic development. Compare first the protoconch of 2 modern species, one with a nonpelagic development and the other with a pelagic larval stage (Figs. 3B, 3C). The protoconch of *Thais emarginata* (Fig. 3B) has a large initial whorl and a shell that is low-spired and appears massive and "clumsy," whereas the protoconch of *Thais haemastoma* is high-spired and still shows the adapical edge of the sinusigerous projection better seen in its entirety on the larval shell taken from the plankton (Fig. 3A). A sinusigerous projection, never found in nonpelagic development, is a positive indication that a species has a pelagic larval stage. Compare the larva of *Thais haemastoma* (Fig. 3C) with that of *Murex inornatus* (Fig. 3D) from the lower Miocene of Jutland. There is a remarkable similarity between the 2 species and further comparison with the contemporary larvae (Fig. 3A), completely confirms that *Murex inornatus* had a pelagic larval development.

The length of the larval development of fossil species can also be very roughly estimated. The protoconch of a fossil form can be compared to a contemporary larva whose length of pelagic development is already known. Such comparisons can only be safely made, however, between very closely related genera or species. The similarity in size and number of whorls in *Murex inornatus* to those of *Thais haemastoma* suggests pelagic development of approximately equal duration (namely about two months). Comparison of fossil and recent Cymatiidae and Tonnidae suggest comparable lengths of pelagic larval development between the geologic past and the present (i.e., six or more months, see Scheltema, 1971b, Table I). The length of pelagic larval development in a gastropod species may vary and sometimes is dependent upon a settlement response (see Scheltema, 1961, 1967, p. 262-263; Robertson and others, 1970).

Evidence from contemporary gastropod species with teleplanic veligers allows us to conjecture about the effectiveness of larval dispersal in past geological time. Paleocirculation in the Late Cretaceous and early Tertiary had great similarities to contemporary Atlantic circulation; an anticyclonic gyre dominated the North Atlantic Ocean with a branch continuing eastward across Europe through the shallow epeiric Tethys Sea (Fig. 6). Westward moving flow from the Tethys Sea formed a current analogous to the contemporary North Equatorial Current. Upon reaching the western Atlantic, part of this North Equatorial Current was diverted through the passage separating the North and South American continents. Recent experimental tank studies of paleocirculation of Atlantic surface waters suggests that during the Late Cretaceous westwardly flowing

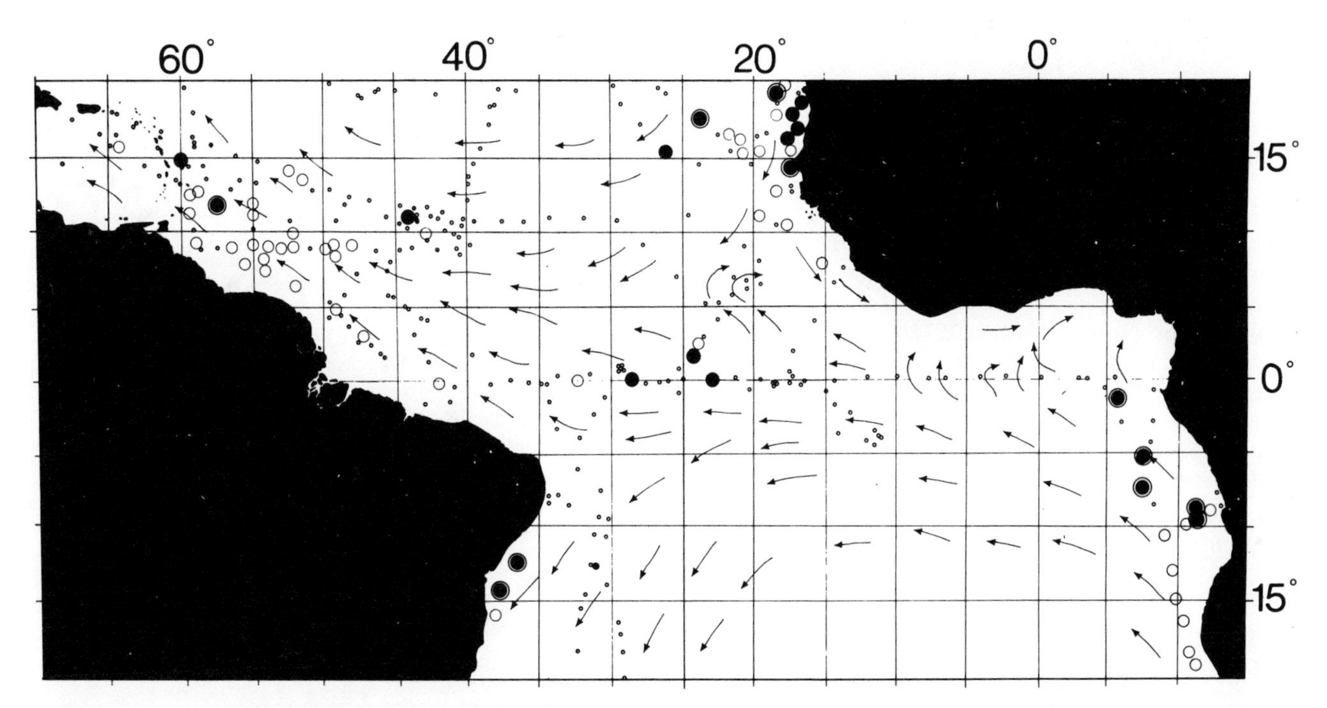

Figure 5. Geographical distribution in the tropical Atlantic Ocean of veliger larvae belonging to the gastropod family Muricidae. Open circles are stations where larvae of the genus *Thais* (mostly *Thais haemastoma*) were found; filled circles, the genus *Drupa* (mostly *Drupa nodulosa*). Small circles indicate locations where larvae of Muricidae were absent. Arrows show surface currents (refer to Fig. 2 for explanation).

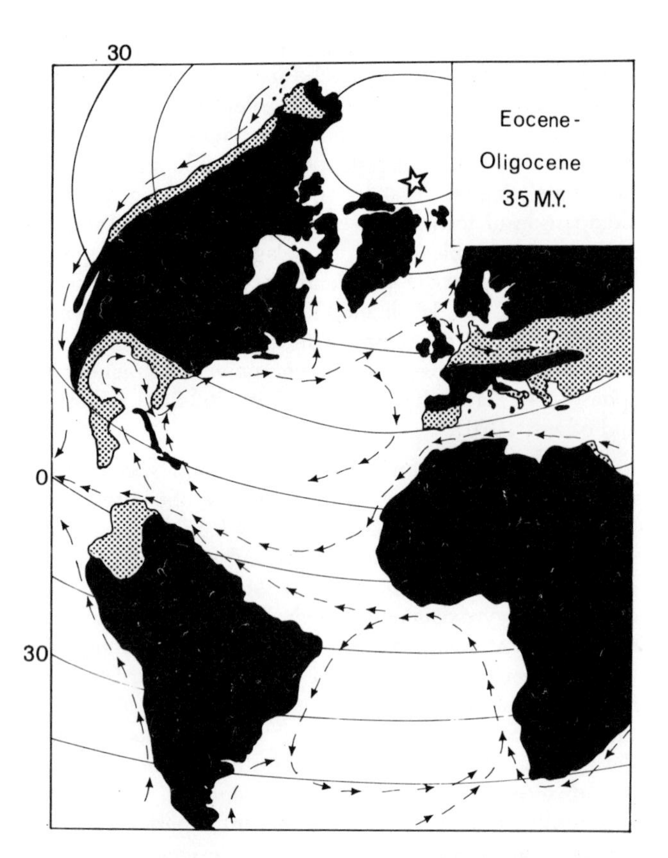

Figure 6. Surface circulation in the Eocene-Oligocene Atlantic Ocean as inferred by Berggren and Hollister (1974). The paleogeography of the continents is after Phillips and Forsyth (1972). Stippled areas are continental regions that were portions of epeiric seas during the epochs represented (Fell, 1967).

equatorial currents from western Tethys to the Caribbean had a velocity of 2 to 4 knots (Luyendyk and others, 1972, p. 2658). Allowing for the somewhat reduced size of the Late Cretaceous Atlantic (about 20 percent smaller), it can be computed that such a current speed would transport a passively drifting larva across the equatorial Atlantic in 28 to 56 days, less than half the time that would be required to cross the present Atlantic on the North Equatorial Current (see Scheltema, 1971b, Table II). Thus, during the Late Cretaceous and early Tertiary, currents were already favorable for transporting the larvae of Tethyan forms across the Atlantic to the West Indies and even to the eastern Pacific. The extension northward of a warm-temperate climate during the Cretaceous and early Cenozoic (see Berggren and Hollister, 1974, for review of this subject) must have made possible the survival of larvae of warm-water species into high latitudes and extended their range northward.

It is neither possible nor necessary to review here evidence for the introduction of Tethyan fauna to the Caribbean and eastern Pacific. That such a dispersal is possible for many gastropods by means of larval transport is the substance of the argument presented here. Just as the veliger larvae of *Philippia krebsii* and *Architectonica nobilis* are found dispersed on the North and Equatorial Atlantic Ocean today, so these same species must also have drifted in the Miocene Atlantic, as their fossil records appear to confirm (Dall, 1890; Woodring, 1928, 1959; Robertson, 1973).

ACKNOWLEDGMENTS

I wish to thank my friend and colleague Dr. Robert Robertson for assistance offered and information shared. From my wife Amélie, as always, I have received both help and encouragement. This research was supported by a grant from the National Science Foundation (No. OCE73-00439 A01). Contribution No. 3663 from the Woods Hole Oceanographic Institution.

REFERENCES

Berggren, W. A., and Hollister, C. D., 1974, Paleogeography, paleobiogeography, and the history of circulation in the Atlantic Ocean *in* Hay, W. W., ed., Studies in paleo-oceanography: Soc. Econ. Paleontologists and Mineralogists Spec. Pub. 20, p. 126-186.

Clench, W. J., 1947, The genera *Purpura* and *Thais* in the western Atlantic: Johnsonia, v. 2, p. 61-92.

Craven, A. E., 1877, Monographie du genre *Sinusigera* d'Orb.: Soc. Malacol. Belgique Ann., v. 12, p. 105-127.

Dall, W. H., 1890, Contributions to the Tertiary fauna of Florida: Wagner Free Inst. Sci. Trans., v. 3, pts. 1-6, 1654 p.

Darlington, P. J., 1957, Zoogeography: The geographical distribution of animals: New York, Wiley, 675 p.

Ekman, S., 1953, Zoogeography of the sea: London, Sidgwick and Jackson, 417 p.

Fell, H. B., 1967, Cretaceous and Tertiary surface currents of the oceans: Ann. Rev. Oceanogr. Marine Biol., v. 5, p. 317-341.

Golikov, A., 1963, Briukhonogie molluski roda *Neptunea* Bolten, Fauna SSSR, Molluski, Vol. 5, Pt. 1: Moskva, Akad. Nauk SSSR, 217 p.

Khanaychenko, H. K., Khlystov, N. Z., and Zhidov, V. G., 1965, The system of equatorial countercurrents in the Atlantic Ocean: Okeanologiya, v. 5, p. 222-229 (trans. Acad. Sci. USSR Oceanology, Scripta Technica, p. 24-32).

Luyendyk, B. P., Forsyth, D., and Phillips, J. D., 1972, Experimental approach to the paleocirculation of the oceanic surface waters: Geol. Soc. America Bull., v. 83, p. 2649-2664.

Marche-Marchad, I., 1968, Remarques sur le développement chez les *Cymba* Prosobranches Volutidés et l'hypothèse de leur origine sud américaine: Inst. Fond. Afrique Noire Bull. ser. A, v. 30, p. 1028-1037.

——— 1969, Les Architectonicidae (Gastropodes: Prosobranches) de la côte occidentale d'Afrique: Inst. Fond. Afrique Noire Bull., ser. A, v. 31, p. 461-486.

Metcalf, W. G., Voorhis, A. D., and Stalcup, M. C., 1962, The Atlantic Equatorial Undercurrent: Jour. Geophys. Research, v. 67, p. 2499-2508.

Phillips, J. D., and Forsyth, D., 1972, Plate tectonics, paleomagnetism, and the opening of the Atlantic: Geol. Soc. America Bull., v. 83, p. 1579-1600.

Robertson, R., 1963, The hyperstrophic larval shells of the Architectonicidae: American Malacol. Union Ann. Rept., 1963, p. 11-12.

——— 1964, Dispersal and wastage of larval *Philippia krebsii* (Gastropoda: Architectonicidae) in the North Atlantic: Acad. Nat. Sci. Philadalphia Proc., v. 116, p. 1-27.

——— 1973, On the fossil history and intrageneric relationships of *Philippia* (Gastropoda: Architectonicidae): Acad. Nat. Sci. Philadelphia Proc., v. 125, p. 37-46.

Robertson, R., Scheltema, R. S., and Adams, F. W., 1970, The feeding, larval dispersal, and metamorphosis of *Philippia* (Gastropoda: Architectonicidae): Pacific Sci., v. 24, p. 55-65.

Scheltema, R. S., 1961, Metamorphosis of the veliger larvae of *Nassarius obsoletus* (Gastropoda) in response to bottom sediment: Biol. Bull., v. 120, p. 92-109.

——— 1964, Origin and dispersal of invertebrate larvae in the North Atlantic: Am. Zoologist, v. 4, p. 299-300.

——— 1966, Evidence for trans-Atlantic transport of gastropod larvae belonging to the genus *Cymatium:* Deep-Sea Research, v. 13, p. 83-95.

——— 1967, The relationship of temperature to the larval development of *Nassarius obsoletus* (Gastropoda): Biol. Bull., v. 132, p. 253-265.

——— 1968, Dispersal of larvae by equatorial ocean currents and its importance to the zoogeography of shoal-water tropical species: Nature, v. 217, p. 1159-1162.

——— 1971a, The dispersal of the larvae of shoal-water benthic invertebrate species over long distances by ocean currents, *in* Crisp, D., ed., European Marine Biology Symposium, 4th, Bangor, North Wales 1969: Cambridge, Cambridge University Press, p. 7-28.

——— 1971b, Larval dispersal as a means of genetic exchange between geographically separated populations of shoal-water benthic marine gastropods: Biol. Bull., v. 140, p. 284-322.

——— 1972a, Eastward and westward dispersal across the tropical Atlantic Ocean of larvae belonging to the genus *Bursa* (Prosobranchia, Mesogastropoda, Bursidae): Internat. Rev. Gesamt. Hydrobiol., v. 57, p. 863-873.

——— 1972b, Dispersal of larvae as a means of genetic exchange between widely separated populations of shoal-water benthic invertebrate species, *in* Battaglia, B., ed., European Marine Biology Symposium, 5th, Venice 1970: Padua, Piccin Editore, p. 101-114.

Schumacher, A., 1940, Monatskarten der oberflächenströmungen im Nord Atlantischen Ozean (5° S bis 50° N): Ann. Hydrographie Maritimen Meteorologie, Berlin, v. 68, p. 109-123.

——— 1943, Monatskarten der oberflächenströmungen im äquatorialen und Südlichen Atlantischen Ozean: Ann. Hydrographie Maritimen Meteorologie, Berlin, v. 71, p. 209-219.

Shuto, T., 1974, Larval ecology of prosobranch gastropods and its bearing on biogeography and paleontology: Lethaia, v. 7, p. 239-256.

Sorgenfrei, T., 1958, Molluscan assemblages from the marine middle Miocene of South Jutland and their environments: Danmarks Geol. Undersøgelse, II Raekke, no. 79, 2 vols.

Thorson, G., 1950, Reproductive and larval ecology of marine bottom invertebrates: Biol. Rev., v. 25, p. 1-45.

——— 1961, Length of pelagic life in marine invertebrates as related to larval transport by ocean currents, *in* Sears, M., ed., Oceanography: Washington, D. C., Am. Assoc. Adv. Sci. Pub. 67, p. 455-475.

U. S. Hydrographic Office, (yearly), Pilot charts: Washington, D. C., U. S. Govt. Printing Office.

Voorhis, A. D., 1961, Evidence of an eastward equatorial undercurrent in the Atlantic from measurements of current shear: Nature, v. 191, p. 157-158.

Woodring, W. P., 1928, Contributions to the geology and palaeontology of the West Indies. Miocene mollusks from Bowden, Jamaica. Part II. Gastropods and discussion of results: Carnegie Inst. Washington Pub. 385, 564 p.

——— 1959, Geology and paleontology of Canal Zone and adjoining parts of Panama. Description of Tertiary mollusks (Gastropods: Vermetidae to Thaididae): U. S. Geol. Survey Prof. Paper 306, p. 147-228.

Californian Transition Zone: Significance of Short-Range Endemics

WILLIAM A. NEWMAN, *Scripps Institution of Oceanography, La Jolla, California 92093*

ABSTRACT

Existence of the Californian Transition Zone has been well-documented in previous studies of the latitudinal ranges of numerous molluscan species. The region consists of a steep thermal gradient and the overlap between the so-called Oregonian and Californian Faunal Provinces, centering on Point Conception (35°N). However, the significance of the so-called 4° short-range endemics occurring there has not been explored. It is the purpose of the present paper to evaluate them, utilizing the benthic barnacles. Of the 23 species involved, 4 are endemic: free-living *Arcoscalpellum californicum* and *Balanus aquila,* and obligate commensals *Armatobalanus nefrens* and *Trypetesa lateralis*. Of these, *Trypetesa* qualifies as a Transition Zone indicator since other species of the genus are found in comparable zones elsewhere in the North Pacific and Atlantic.

Ecological, biogeographical, and climatological considerations, of the past as well as for the present, indicate that 3 of the short-range endemic cirripeds of the Transition Zone are relicts of Tethys, and in keeping with the Age and Area Hypothesis, their average generic age is greater than that of members of either the Oregonian or Californian Provinces. Disharmony between overlapping provincial communities apparently results in incomplete utilization of the resources upon which the short-range endemics depend.

The Transition Zone is not restricted to the shore, but extends out into coastal, neritic, and oceanic waters. The present physical narrowness and biological characteristics of the Zone along the shore are the result of the northward movement of Point Conception and associated landmasses since the Pliocene, and of Pleistocene compression of the warm-temperate belt, placing the 14° to 15°C long-term mean sea temperature in the middle of the steepened thermal gradient.

INTRODUCTION

The existence of the Californian Transition Zone was empirically well-documented, first by Newell (1948) and more recently by Valentine (1966), both of whom utilized latitudinal distributional data on some 2,000 unspecified molluscan species. The Zone, a region of rapid faunal change extending over approximately 5° of latitude and centering on Point Conception (35°N), has been interpreted as the boundary between the so-called Oregonian and Californian Faunal Provinces. The cause of the faunal change has been attributed primarily to a marked steepening of the temperature gradient in the region of overlap (see Hedgpeth, 1957, for review).

A greater number of species is present within the Transition Zone than in the "pure" Provinces immediately to the north or south. In addition, there are a number of short-range endemics which further increases the number of species present there.

The present paper will utilize data on 23 relatively well-known, shallow-water barnacles (Appendix). The goal is to further elucidate the nature of the Transition Zone, especially with regard to the short-range endemics. The problem is approachable through present knowledge of their ecologies, combined with an historical view based on the substantial fossil record for these and related species. The attempt will be to demonstrate that, with the exception of 1 probably opportunistic species, the short-range endemics are biologically refugial species. Furthermore, they are geologically relicts of Tethys; that is, they are apparently remnants of a warm-temperate biota, surviving on resources made available because of spatial variation and biological disharmony between opposing provincial communities found in the Transition Zone.

It appears that the present temperature gradient centering on the Transition Zone became considerably sharpened, primarily by compression of the warm-temperate belt concomitant with and as a result of the Pleistocene. To a lesser extent, the northwesterly movement of the Pacific plate (of at least 300 km relative to latitude) steepened the gradient by bringing Point Conception and associated landmasses to their present position, where hydrographically, Point Conception coincides with a mean temperature of 14°C.

A comparable oceanic transition zone exists in the North Pacific (McGowan, 1974). It is suggested that its present discreteness is likewise due to inequitabilities in climate resulting from the Pleistocene.

From these considerations, it follows that the existence of endemic species in steep latitudinal gradients may generally require a more comprehensive explanation than simply physiological and morphological adaptation to physical rigors, as suggested by Vermeij (1972). A freeing of resources, due to the inability of the provincial species to fully utilize them, is likely also a necessary factor. When a latitudinal transition is relatively abrupt, such hiatuses in resource utilization would be exploitable by species belonging to neither community—in the present case, the Transition Zone endemics.

Finally, it is difficult to follow Jackson's (1974) criticism that marine biogeographers are unable to precisely define provincial boundaries, except those accompanying geographical barriers, for generally, the more subtle the gradient change, the more subtle the biotic expression of the boundary. While biological interactions likely sharpen boundaries, in the long run the existence of a precise boundary must be dependent upon and maintained by physical parameters (Hayden and Dolan, 1976).

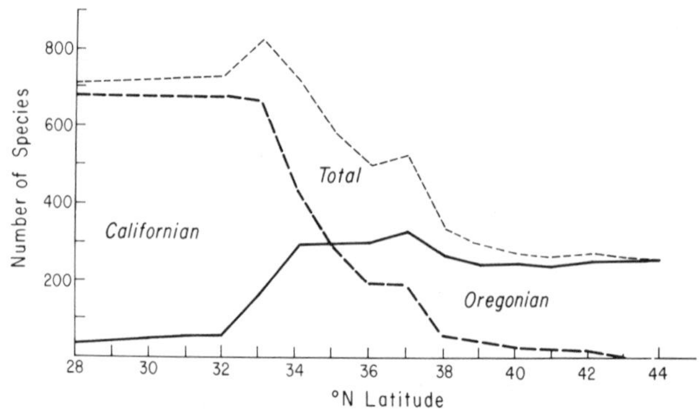

Figure 1. Number of species of benthic molluscs of the Oregonian and Californian Provinces, by latitude (from Newell, 1948). It can be observed that there are more species present in the Californian than in the Oregonian Province, as is generally the case when low- and high-latitude assemblages are compared. It can also be seen that there are more species present within the Transition Zone than in either Province. The "nodal point," where equal numbers of species from both Provinces occur, appears in the vicinity of Point Conception (35°N).

PATTERN

Distribution of Molluscan Species

A greater diversity of organisms inhabits shallow waters in low than in high latitudes. Consequently, one would expect to encounter more species in the Californian than in the Oregonian Faunal Province. This is corroborated by the data of Newell (1948) and Valentine (1966) (Fig. 1). As one progresses south into the Transition Zone between these two Provinces, species are added as others drop out. The change is rather abrupt. However, perhaps unexpectedly, the total number of species in the Transition Zone is greater than in the Provinces immediately to the north or south. The ultimate reasons for this are not immediately obvious, for it would seem that there could just as well be simply an increase to the level of the Californian Province. Or conceivably there could even be a reduction—that is, fewer species in the Transition Zone than on either side, similar to the situation observed in some estuaries when progressing from the sea into fresh water (Remane, 1934).

Two reasons are apparent for the large number of species in the Transition Zone. The first is that the ranges of species overlap; that is, many species of both the Californian and Oregonian Faunal Provinces range well into the Zone. While individuals may be less abundant, rare, or only occasionally present at the extremes of their ranges, the total number of species in the region of overlap is greater at any one time.

The second reason for the larger number of species in the Transition Zone than on either side, in addition to the piling up of overlapping ranges of the relatively long-range provincial species, is the presence of the short-range endemics. They contribute significantly to the total species count.

The density changes and altered relationships between prey, predators, competitors, symbionts, etc., probably disrupt patterns of resource utilization compared to the balances prevailing in the unmixed, more harmonious communities on either side. The disharmony created at the junction must be primarily responsible for the freeing of resources necessary for and utilized by the short-range endemics.

Distribution of Barnacle Species

Previous biogeographical studies including the Californian Transition Zone have emphasized statistical description and(or) verification of the Zone, and concern with this approach has been recently voiced by Jackson (1974). The 2,000 or so molluscan species utilized by Newell (1948) and by Valentine (1966) provided a significant quantitative data base on which to document the Zone, but relevant ecological parameters do not enter into such analyses. Furthermore, the quality of the data and biological aspects of the situation can neither be assessed nor readily discussed on a species to species basis.

The barnacles, on the other hand, are represented by relatively few and for the most part well-known species. Those present in the region of concern are listed in the Appendix, along with citations to the literature concerning them. The accompanying chart (Fig. 2), prepared for the most part from the literature, indicates which species might be found at a particular latitude along the coast of California. Without considering individual species or the ecological groups in which they have been arranged, it can readily be seen from the chart that the species one would expect to encounter in the vicinity of Cape Mendocino are, with one exception, different from those found around Punta Baja. Those unique to each extreme, for our purposes, can be considered members of the Oregonian and Californian Faunal Provinces, respectively.

The chart (Fig. 2) shows that the transition between one Province and the next is quite marked, even though northern and southern species drop out more-or-less gradually where the two biotas overlap. The region of overlap, centering on Point Conception (35°N), consequently consists of a mixture of Oregonian and Californian species. In addition, the region contains 4 short-range endemics.

The distributional pattern seen in the chart can be displayed graphically by simply plotting the number of species of each Province, and of the Transition Zone, by degrees of latitude (Fig. 3). From this plot it can readily be observed that there are more species in the Californian than the Oregonian Province, and that there are more species present in the Transition Zone than immediately adjacent in the Provinces. It will also be noted that the nodal point (the region where an equal number of species from both Provinces is present) centers in the region of Point Conception, as does the greatest number of short-range endemics. This is the same pattern displayed by the molluscs (Newell, 1948).

The data in the distributional chart can also be plotted as percent change per degree of latitude, by the end-point method, as was done for the molluscs by Newell (1948). This is accomplished simply by dividing the number of end-points by the number of species present per degree of latitude (Fig. 4). The result, the same as for the molluscs, may be initially surprising. Rather than a skewed bell, the curve is bimodal: the region of Point Conception, right in the middle of the Transition Zone, is a local minimum. A plausible explanation for this might be that change on either side of Point Conception is exaggerated by the short-range endemics, since they straddle the region and their end-points therefore fall on either side. This is partly the reason, but much the same result is achieved when the short-range endemics are excluded from the calculations.

If one ignores the species involved, and the species number curve (Fig. 3), and resorts to speculation to account for the bimodal situation, one could hypothesize that there are two classes of provincial species. There could be relatively stenotopic forms that drop out quickly when the physical and biological characteristics of the Transition Zone are encountered, species not only stenotopic but perhaps also influenced by the presence of the short-range endemics. The remainder then could be relatively eury-

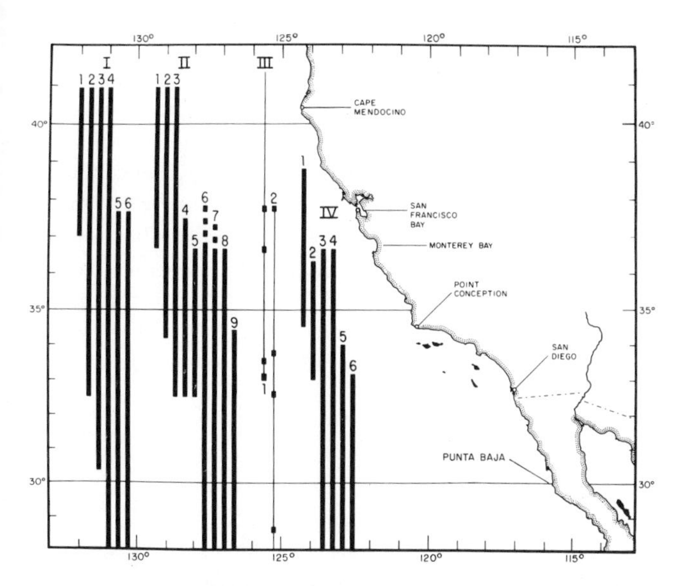

Figure 2. Latitudinal ranges of benthic cirripeds of the Oregonian and Californian Provinces (for sources, see Appendix). The 23 species present can be divided into four groups: I, Intertidal; II, Shallow water; III, Introduced (in bays and estuaries); and IV, Obligate commensals. Groups include: I—(1) *Semibalanus cariosus;* (2) *Chthamalus dalli;* (3) *Balanus glandula;* (4) *Pollicipes polymerus;* (5) *Chthamalus fissus;* (6) *Tetraclita rubescens.* II—(1) *Solidobalanus hesperius;* (2) *Balanus crenatus;* (3) *B. nubilus;* (4) *B. aquila;* (5) *Arcoscalpellum californicum;* (6) *Megabalanus californicus;* (7) *Balanus pacificus;* (8) *B. trigonus;* (9) *B. regalis.* III—(1) *Balanus improvisus;* (2) *B. amphitrite.* IV—(1) *Trypetesa lateralis;* (2) *Armatobalanus nefrens;* (3) *Conopea galeata;* (4) *Octolasmis californiana;* (5) *Membranobalanus orcutti;* (6) *Oxynaspis rossi.*

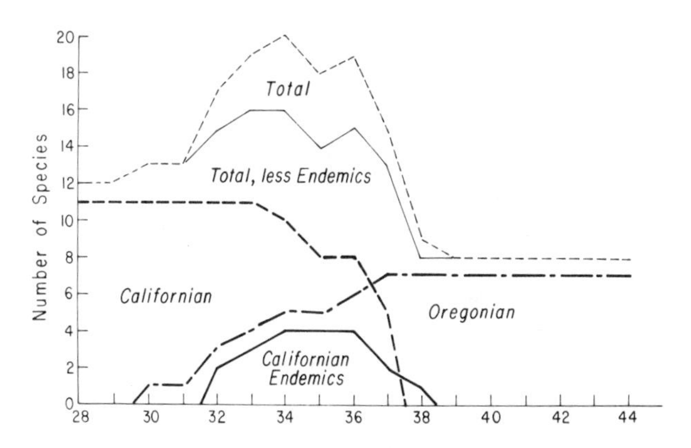

Figure 3. Number of species of benthic cirripeds of the Oregonian and Californian Provinces, plus short-range endemics, by latitude (from Fig. 2). As with the molluscs, it can be observed that there are more species present in the Californian than in the Oregonian Province and that there are more species present in the Transition Zone than in either Province, even when the short-range endemics are excluded.

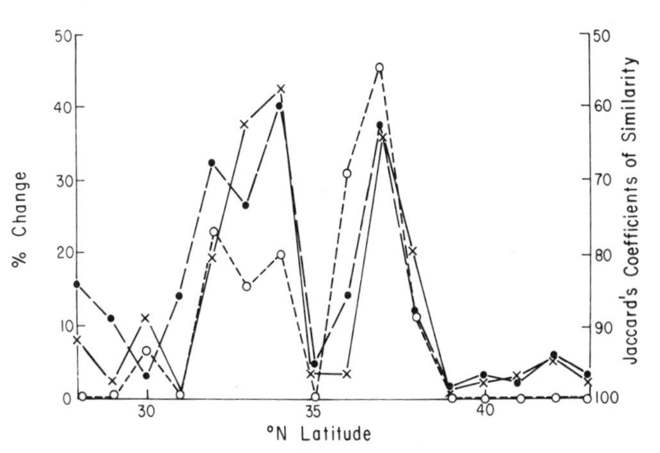

Figure 4. Comparison of percent change by degree of latitude for benthic molluscs (x's: Newell, 1948) and cirripeds (circles: Fig. 2), and again for the molluscs by Jaccard's Coefficients of Similarity (points: Valentine 1966). Note that the curve bracketing the Transition Zone is bimodal, there being very little change in the vicinity of Point Conception, the center of the Transition Zone (see text for discussion and explanation).

topic forms that are less susceptible to general properties of the Transition Zone and(or) the presence of the short-range endemics and therefore do not drop out until they encounter the full brunt of the opposing Provinces on either side of the nodal point.

Such interactions and responses may contribute to the bimodal nature of the plot. However, the primary explanation, at least as far as the barnacles are concerned, is simpler. The majority of species extend beyond the nodal point before terminating their ranges—4 out of 6 and 8 out of 10 for the Oregonian and Californian representatives, respectively. This fact for the most part accounts for the bimodality in the percent change curve (Fig. 4), and it seems likely that this is also the case for the molluscs, as indicated by the height of the nodal point above the abscissa in Figure 1.

Hydrography

Previous authors have considered the transition from Oregonian to Californian biotas, in the region centering on Point Conception, to be due to the change in hydrographic climate (Newell, 1948; Valentine, 1966). While biological interactions have generally been ignored, they must in many ways sharpen the faunal characteristics of the Zone. Nonetheless, the characteristics as presently observed must be due for the most part to temperature.

Valentine (1966) mentions that the isotherms converge on Point Conception, and indeed this headland separates the markedly different hydrographic climates of northern and southern California. The difference is not only in temperature, but in the intensity of onshore winds, concomitant surf conditions, the nature of substrata, and productivity. Also, the degree of seasonality, including day length and storm frequency, are likely factors of some importance. However, southern submergence, where high-latitude, shallow-water species occur in deeper water at the southern end of their ranges, and the appearance of cold-water species in isolated areas of upwelling well south of their normal ranges, are some of the field observations that indicate the importance of temperature, as do laboratory experiments on thermal tolerances, enzyme optima, and breeding cycles (Hutchins, 1947; Bullock, 1955). In addition, paleoecological work, such as that of Addicott (1966) based primarily on molluscan fossil evidence, has demonstrated that provincial boundaries shifted latitudinally with rapid climatic changes during the Pleistocene (see Enright, 1976, for theoretical considerations). Therefore, for present purposes, a brief look at the temperature pattern should suffice.

A chart of sea-surface isotherms does not readily illustrate that there is a marked change in temperature in the vicinity of Point Conception. Seasonal charts only make matters more confusing. Therefore, to demonstrate steepening of the gradient, data have been pooled from a number of sources (Coast and Geodetic Survey, 1952; Wyllie and Lynn, 1971; M. Robinson, personal commun.), for nine localities, ranging between approximately 45° and 25°N, and the mean values are plotted on Figure 5. It can

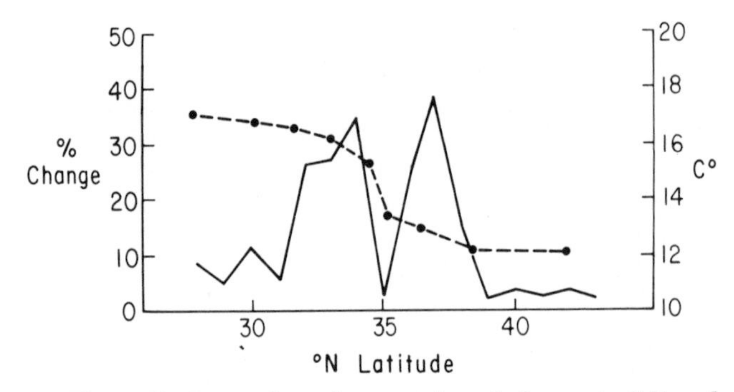

Figure 5. Comparison of percent faunal change (solid) and long-term temperature means (dashed) by latitude. Note the mean temperature in the vicinity of Point Conception is in the 14° to 15° C range and the marked steepening of the curve between the Oregonian from the Californian Faunal Provinces (sources for the temperature data are cited in the text).

be observed that the mean temperature rises from 12°C rather gradually as one progresses south from 45°N, and then abruptly as one passes into and through the Transition Zone, to around 16°C, before leveling off again to a gradual increase upon leaving the Zone further south. The region of rapid rise of 4°C over 5° latitude centers on Point Conception (35°N) where the mean temperature is approximately 14°C.

The alignment of the 14°C mean with the nodal point of the Transition Zone and Point Conception will be returned to shortly. For the present, it probably need only be noted that the two different hydrographic climates of northern and southern California are indeed separated by a marked temperature change extending over approximately 5° of latitude, the same latitudinal extent of many of the short-range endemics. As Ekman (1953) noted, the transition from subtropical to cold-temperate on the west coast of America is so sudden that there is practically no room for a warm-temperate fauna. We will return to this when considering the significance of the short-range endemics.

PLANKTON

So far we have been concerned with shallow-water benthic animals, the majority of which have planktonic larvae [*Trypetesa lateralis* is the only exception among the shallow-water barnacles (Tomlinson, 1960)]. While net transport of water along the coast is south, via the California Current, near-shore eddies, gyres, and counter currents can at times transport considerable amounts of warm water and the associated larvae north, as indicated by drift bottles (Crowe and Schwartzlose, 1972) and geostrophic interpretations (Wyllie, 1966). Conversely, the temperature of water coming south can vary considerably over time, and, along with local conditions such as upwelling, waters south of Point Conception can at times be unseasonably cool.

Variations in current direction and temperature presumably account for the occasional appearance of established adults of southern forms such as *Megabalanus californicus* as far north as Humboldt Bay (Zullo, 1968) and *Balanus pacificus* off San Francisco, and on other occasions the appearance of such northern species as *Chthamalus dalli* and *Balanus improvisus* as far south as San Diego and San Onofre, respectively. Occasional detection of adults north or south of their normal ranges indicates that their larvae must commonly be transported well beyond the ranges of parent populations, unlike the situation on the continental margin of the east coast of North America where it is inferred that the northern limit of certain southern species is limited by larval dispersal by currents rather than by survival of adults (Cutler, 1975). We can now ask, what is the biogeographic pattern of planktonic species inhabiting the same waters as many of the larvae of the benthic organisms of the California Current?

Data on 11 coastal and neritic copepods (A. Fleminger, personal commun.) illustrates the same basic pattern: that is, the nodal point for species ranging north and south from the Californian and Oregonian Provinces falls in the vicinity of Point Conception (Fig. 6). It seems reasonable to suspect that many other members of the plankton of the California Current are distributed in a comparable manner; that is, the Transition Zone for coastal and neritic forms is similar in geographical extent to that of the benthic organisms. This removes substratum and current *per se* as important considerations. The most obvious remaining physical parameter still seems to be temperature.

To carry the inquiry one step further, we can ask what is the oceanic situation? By good fortune, it has been nicely worked out by a number of Scripps scientists cited in and including McGowan and Williams (1973), and championed by McGowan (1974). A transition zone between approximately 38° and 43°N separates the so-called Subarctic and Central Water Masses of the North Pacific. It extends in a narrow band, for the most part as the North Pacific Current, across the ocean from south-central Japan to northern California and Oregon. In the east the current separates north and south, the southern limb forming the eastern boundary or California Current. The physical and chemical attributes of the water masses involved, as well as the biological setting, are beautifully illustrated by Reid and others (1976) and Reid and others (1977). The temperate–cold-temperate Subarctic Water Mass and subtropical Central Water Mass contain relatively distinct planktonic communities, as might be expected. In addition, as might not be so readily expected, the transition zone between these two faunal assemblages is inhabited by a characteristic assemblage of latitudinally short-range endemics.

Plankton workers have an advantage over benthic workers in that it is relatively easier to obtain quantitative estimates of abundance. McGowan (1974) has put this to advantage in plotting the distribution of some mixed macrozooplankton species, across the transition zone along the 155°W meridian (Fig. 7). Since the ordinal values are dependent upon the relative abundance of the species present along the latitudinal gradient, such plots are not directly comparable to the plots of number of species used here. However, Subarctic species abundances decline as they are replaced by the Central and Warm Water Cosmopolites in the transition zone. In addition, the short-range endemics can be observed centering on 42°N. While the values are derived in a different way, the pattern is the same as that given for benthic molluscs and barnacles, and for the coastal and neritic copepods (Figs. 1, 3, 6). While we don't have the numbers necessary to document the relative abundance of most benthic forms, it is common knowledge that individuals become less abundant as coastal populations approach the extremes of their ranges.

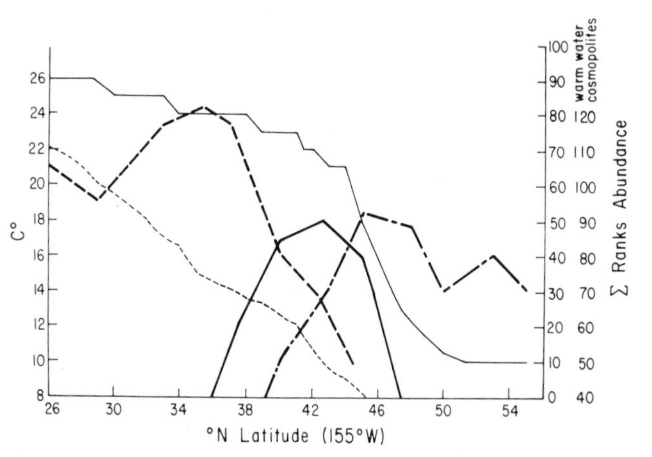

Figure 7. Summed ranks of abundance for some Subarctic (5 spp.: right), Transition (7 spp.: middle), and Warm Water (10 spp.: left) mixed zooplankton by latitude along 155°W longitude (from McGowan, 1971, *in* 1974), and near surface temperatures by latitude for August and September of 1964 (solid) and January of 1966 (fine dashed) (from McGowan and Williams, 1973). Note that the greatest abundance of Transition Zone species centers on 42°N, where the "mean" temperature is in the 14° to 15°C range (cf. temperature curve in Fig. 5). Note scale adjustment on right, for Warm Water Cosmopolites.

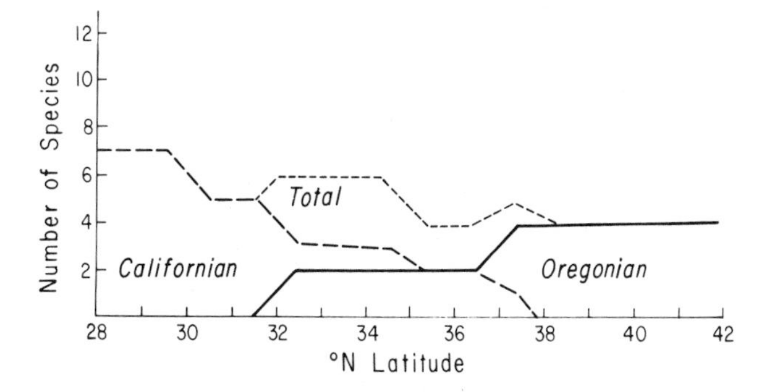

Figure 6. Number of species of coastal and neritic copepods, by latitude (Fleminger, personal commun., also cf. Fleminger, 1975). Similarities in pattern with molluscs and cirripeds are striking (cf. Figs. 1, 3).

McGowan and Williams (1973) present two meridional temperature sections along 155°W, but the pattern is complex. While it is clear that there is some compression and skewing of the isotherms in the vicinity of 42°N, it is difficult to see if there is a steepening of the gradient in this region. Therefore, I have taken the near-surface values from their plots, and entered them on Figure 7. The August-September curve is perhaps the most illuminating since it shows the full gradient with an abrupt change occurring between Subarctic and Central Water Masses. Both curves effectively bracket the transition zone. As far as these data go for near-surface conditions, it appears that the transition zone species, at their center of abundance, experience temperatures ranging between 8° and 22°C, that is, temperatures equidistant from 15°C. Indeed, if we look across the 14° to 15° ordinates, the intercepts of both temperature curves bracket the latitudinal ranges of the transition zone species. It will be recalled that the long-range mean temperature at the nodal point of the Californian Transition Zone was also in the vicinity of 14°C (Fig. 5). That such markedly different settings should appear this similar in temperature must be for the most part coincidental. However, as McGowan (1974) has noted, these conditions have prevailed over considerable stretches of geologic time. They must in some way be related to the perpetuation of the short-range endemics.

BARNACLES

Ecological Groups and Short-Range Endemics

The benthic cirripeds found in California can be divided into four ecological groups: (I) intertidal, (II) shallow water, (III) introduced, in bays and estuaries, and (IV) obligate commensals (Fig. 2). These groupings, while partially a matter of convenience, have natural attributes that prevent their being wholly arbitrary, even though southern submergence displayed by some, especially those in the lower intertidal or very shallow water, causes problems. For example, *Balanus aquila* is generally intertidal north and subtidal south of Point Conception.

Group I contains the only highly eurytopic form, *Pollicipes polymerus*, ranging through both the Californian and Oregonian Provinces. All other species in this and the remaining groups terminate one end of their distribution in, or reside wholly within, the Transition Zone.

All species falling into Group I are represented by individuals or aggregations of individuals that are community dominants persisting over long periods of time. These species also physically interact with each other in various combinations as adults, along both horizontal and vertical gradients. *Chthamalus* generally occurs higher in the intertidal than *Balanus*, *Semibalanus*, and *Tetraclita*. The northern forms, *Chthamalus dalli* and *Semibalanus cariosus* are replaced, in good part by *Chthamalus fissus* and *Tetraclita rubescens*, respectively, in the south. *Balanus glandula*, occurring for the most part between the aforementioned species in vertical distribution, ranges further south than *S. cariosus* and is eventually replaced in part by *Balanus regalis*, as are *B. nubilus* and *B. aquila*, and further south by the Panamic species, *B. inexpectatus*.

Members of Group II, from shallow water, follow a similar pattern, with *Solidobalanus hesperius* and *B. crenatus* being replaced by *Megabalanus californicus*, *B. pacificus*, and *B. trigonus* in the northern part of the Transition Zone. However, such species, with the possible exception of *B. nubilus* and *B. crenatus* in the northern parts of their ranges, are relatively opportunistic. That is, individuals neither dominate nor form long-standing constituents of the subtidal community.

The first two short-range endemics, *B. aquila* and *Arcoscalpellum californicum* belong to this group. They differ from the two short-range endemics in Group IV in that they are not obligate commensals. *Balanus aquila* occurs with *B. nubilus* in the northern part of its range, but both undergo submergence south of Point Conception where they are replaced in the intertidal by *B. regalis*, a close relative of *B. aquila*. *Balanus regalis* ranges south, at least to the southern limits of Baja California (Fig. 8).

Group III contains 2 bay or estuarine species. Both have been introduced to California, probably by ships (Newman, 1967; Carlton and Zullo, 1969). The first, *Balanus improvisus* from the temperate and warm-temperate Atlantic, ranges into the Transition Zone from the north, and has been found as far south as San Onofre in artificial situations. The second, *B. amphitrite*, probably introduced from the Indo-West Pacific, does likewise but from the south. Their apparent "recognition" of the Transition Zone could be used to bolster arguments that the temperature gradient is in good part responsible for maintaining the boundary between the two Provinces and in part it must be. However, *B. improvisus* may occur in warmer water on the Atlantic coast than it does on the Pacific, and this considerably weakens such a conclusion. Furthermore, while having fully marine physiological capabilities, it is generally restricted to estuarine situations, much more so than *B. amphitrite*. Southern California is more arid than northern California and estuarine conditions are either very seasonal or nonexistent south of Point Conception. It may be for this reason that *B. improvisus* has failed to become established on a regular basis, south of Point Conception.

On the other hand, the situation concerning the northern limit of *B. amphitrite* is clearly temperature related for it requires temperatures of approximately 20°C in order to breed (Graham and Gay, 1945). Indeed, its larvae must prefer to settle where the water is warm, for the species was first found in San Francisco Bay on rocks bathed by sea water being discharged from a power plant, and its present distribution is centered around Aquatic Park, Berkeley and Lake Merrit and the Oakland Estuary where suitable substrate and seasonal temperature conditions are met (Newman, 1967). In southern California, it is the most common barnacle in harbors. Its larvae settle and undergo metamorphosis on buoys along the outer coast, but the young barnacles rapidly disappear.

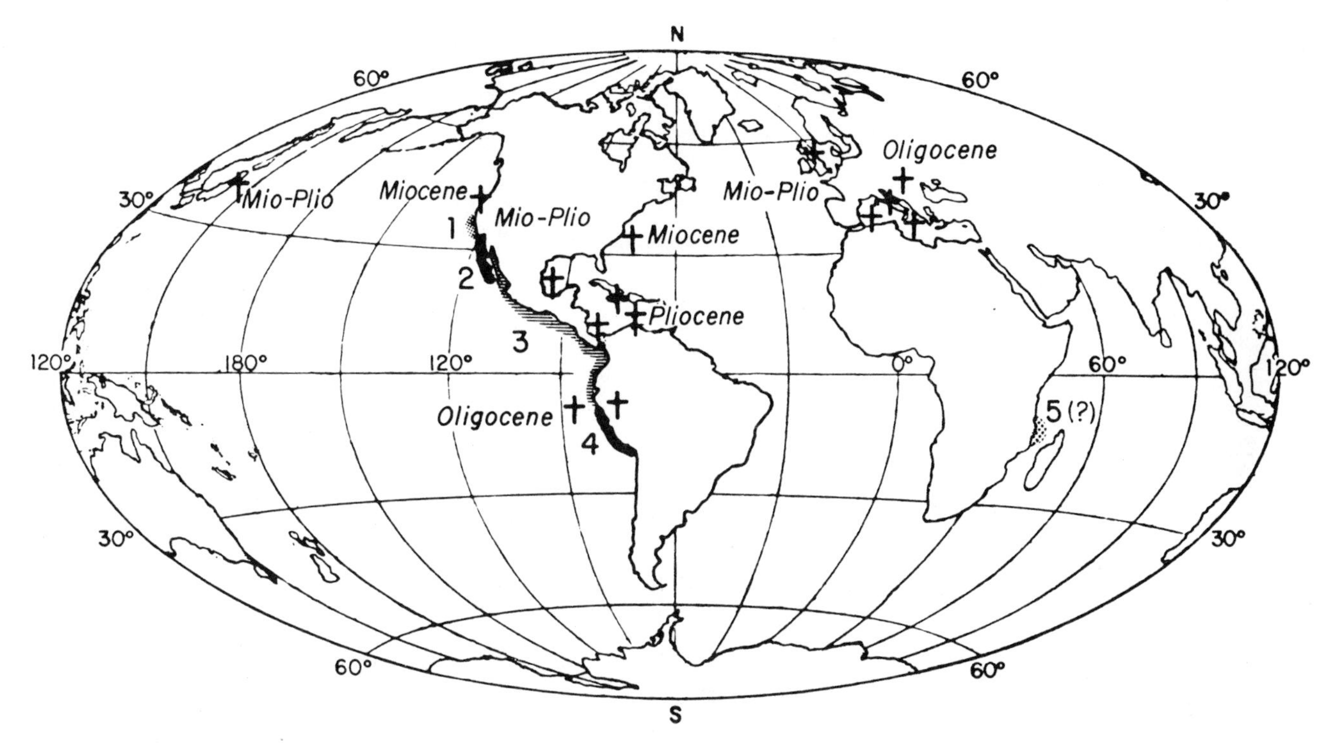

Figure 8. Distribution of extant and extinct members of the "*concavus*" group of *Balanus*: (1) *B. aquila;* (2) *B. regalis;* (3) *B. glyptopoma* (subspecies *eyerdami* and *panamensis*); (4) *B.* sp. (probably *B. concavus regalis* Pilsbry of Kolosváry, 1942); (5) *B. indicus* (membership in this group conjectural); and (†) extinct taxa (see Newman and Ross, 1976). It will be noted that this once widely distributed group is presently restricted to the eastern Pacific, and that the most northern member, *B. aquila,* is a short-range endemic in the Californian Transition Zone.

Members of Group IV are obligate commensals and as such they differ from all the aforegoing in being fully divorced as adults from interspecific interactions with other barnacles. They are afforded varying degrees of protection by their hosts, and are limited by host distribution. All are tropical or have tropical representatives, and 2 are short-range endemics.

Conopea galeata on gorgonians, *Membranobalanus orcutti* on sponges, *Oxynaspis rossi* on antipatharians, and *Octalasmis californiana* inhabiting the gill chambers of lobsters and brachyuran crabs, are wide-ranging species. While they are the sole representatives of their groups known from the eastern Pacific, numerous congeners are found in all other tropical regions; that is, they are members of viable, species-rich groups occurring elsewhere in the world.

Of the 2 short-range endemics, *Armatobalanus* (*Armatobalanus*) *nefrens* occurs on the hydrocorals *Allopora californica* and *Errinopora pourtalesii,* but most commonly on the former which is itself a short-range endemic. It is the sole representative of the subgenus in the eastern Pacific, but it has numerous consubgeners in other tropical seas (Fig. 9). The second, *Trypetesa,* is among the most interesting; there are but 4 species in the monogeneric family to which it belongs; all are effectively short-range endemics and all burrow in the interior of gastropod shells inhabited by hermit crabs (Fig. 10). While hermit crabs range widely in temperate and tropical regions, *T. lateralis* is restricted to the northern part of the Californian Transition Zone (Tomlinson, 1953).

Ekman (1953) pointed out that because of the exclusive occurrence of endemics, a greater value must be attached to them than to nonendemics in characterization of a region, and that the higher the taxonomic level, the potentially more important the endemic. He also noted the importance of bringing evidence from the various disciplines having a paleo-prefix into biogeographic analyses. It behooves us then to take a closer look at the 4 short-range endemics of the Californian Transition Zone.

The Significance of Short-Range Endemics

Opportunists and Relicts. As noted above, the Transition Zone is a region of steep environmental gradients as compared to the "pure" Provinces to the north and south. The most prominent physical parameter is the steepness of the temperature gradient, and it is presumably this factor that is primarily responsible for attenuation of the Oregonian and Californian biotas. Some species ranging into the Transition Zone not only change relationships with their traditional food sources, predators, competitors, or parasites, but they are confronted with a new variety of these. Furthermore, with the greater number of species in the Transition Zone, members of one Province not only encounter a changing array of species interactions, but a potentially greater variety of them. Like many islands (Udvardy, 1969), the Transition Zone has an unbalanced

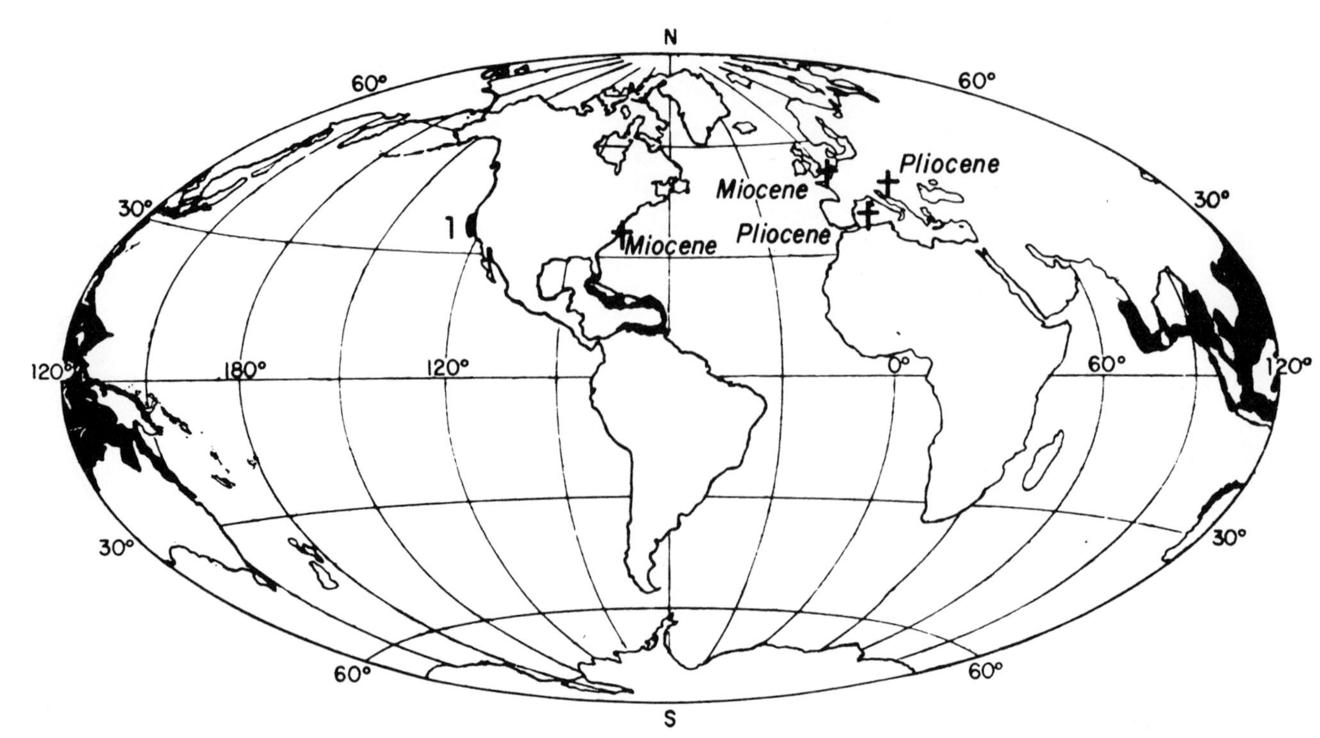

Figure 9. Distribution of extant and extinct members of *Armatobalanus* (*Armatobalanus*). It will be noted that this once widely distributed group is presently restricted for the most part to the Indo-West Pacific and Caribbean. The sole representative in the eastern Pacific, *Balanus nefrens* (1), is a short-range endemic of the Californian Transition Zone. It is an obligate commensal of hydrocoral, primarily *Allopora californica* which is also reputed to be a short-range endemic of this region.

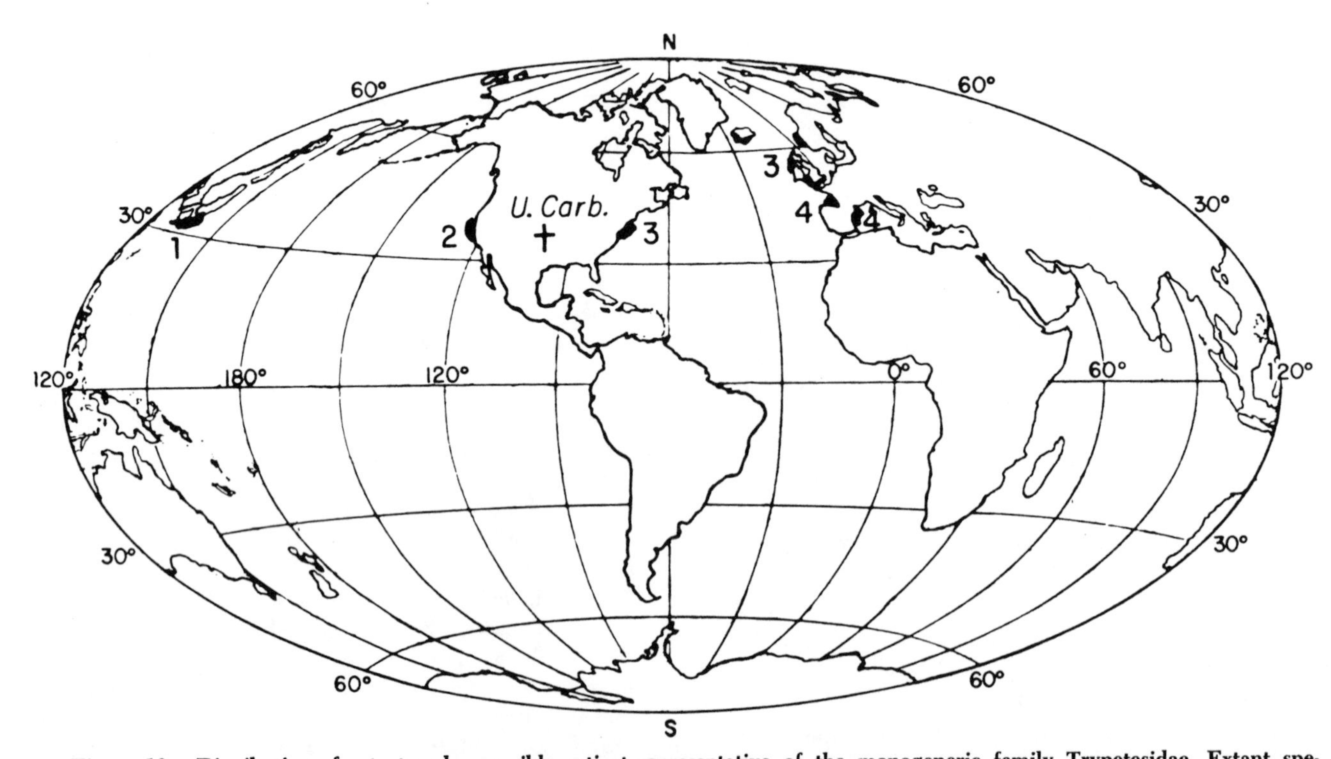

Figure 10. Distribution of extant and a possibly extinct representative of the monogeneric family Trypetesidae. Extant species are known only to occur burrowed in the interior of gastropod shells inhabited by hermit crabs: (1) *Trypetesa habei;* (2) *T. lateralis;* (3) *T. lampas;* (4) *T. nassarioides;* (5) *T. caveata* (Upper Carboniferous). *Trypetesa* is apparently a subtropical (1, 4) and warm-temperate (2, 3) transition zone indicator in the Northern Hemisphere (see text for discussion).

fauna, but unlike islands it has an exceptionally long rather than short faunal list. In this sense then, the Transition Zone is physically and biologically disharmonious as compared to the "pure" Provinces to the north and south. How then do the short-range endemics unique to the Transition Zone enter into the picture? To explore this satisfactorily, it will be necessary to consider the paleogeographic as well as the present situation for these and some additional species.

Of the 4 short-range endemic barnacle species of the Californian Transition Zone, 2 are free-living (*Balanus aquila* and *Arcoscalpellum californicum*) and 2 are obligate commensals (*Armatobalanus nefrens* and *Trypetesa lateralis*). The free-living forms are actually sedentary, but they commonly occur on a variety of inanimate substrates and are in a position to interact with other barnacle species. The obligate commensals, on the other hand, are found in association with certain organisms and no place else, and their interactions, at least as adults, do not involve other barnacles.

Balanus aquila is among the largest living barnacles. It reaches its greatest size and abundance in relatively shallow water and low intertidal north of Point Conception where it is (or was, prior to extensive collecting for physiological research) commonly found with the northern *Balanus nubilus*, a species comparable in size. In such situations, the two must compete, if in no other way than that space occupied by one cannot be occupied by the other.

Balanus nubilus ranges into deeper water than *B. aquila* north of Point Conception and neither has been found intertidally south where *B. aquila* too undergoes southern submergence. Southern deep-water (less than 30 m) specimens of both species are rarely encountered and not only are the ones I've examined relatively small but the shells have been badly eroded by sponge borings.

Both of these species are replaced intertidally south of Point Conception by *Balanus regalis*, a southern species closely related to *B. aquila*. From at least Point Conception south, *B. regalis* is a common surf zone barnacle of moderate size, but further south in Baja California it rivals *B. aquila* in stature. Thus, it appears that *B. aquila* is "confined," so to speak, by 2 comparable species with overlapping distributions, which, through failure to fully exploit available resources in the region of overlap, allow *B. aquila* to persist there.

It seems likely, as Zullo (1964) suggests, that the abundant Mio-Pliocene form of central-southern California and Baja California, *B. gregarius*, was either the progenitor or is a senior synonym of *B. aquila*. It would follow that the restriction of the latter to the Transition Zone was a Quaternary event. These species belong to the "*concavus* group" of *Balanus* (Zullo, 1966; Newman and Ross, 1976), and *B. aquila* is presently the most northern representative in the eastern Pacific. Further south, *Balanus regalis* is replaced by at least 2 other species. However, the *concavus* group, now nearly if not completely restricted to the eastern Pacific, has been on the wane, beginning in the Miocene in the tropical Atlantic (Fig. 8). The eastern Pacific extant representatives are all that remain of this once widely distributed Tethyan group, and there must be something about the suboptimal conditions for reef corals in the tropical eastern Pacific that has allowed these species to survive (cf. Dana, 1975). The impact of the Pleistocene will be explored later.

Much the same situation exists with *Balanus pacificus*, which is considered a member of the *concavus* group by some workers (Henry and McLaughlin, 1975). It is mentioned here, partly because of its ecology in the Transition Zone, but also to emphasize the relict nature of much of the tropical eastern Pacific fauna. *Balanus pacificus* differs from *B. aquila* in being opportunistic ("fugitive" of Hurley, 1973), at least at the northern end of its range, in the Transition Zone. However, it ranges south in the tropical eastern Pacific to Ecuador. It was sufficiently abundant in the tropical Atlantic to be recognized in the Miocene of Virginia (Ross, 1964). Thus, like the true concavoids, the history of *B. pacificus* illustrates the relict nature of the tropical eastern Pacific biota.

The second free-living, short-range endemic is the lepadomorph *Arcoscalpellum californicum*, and, as is generally the case with members of this Cretaceous genus, it is a deep-water species. Its latitudinal range along this coast is not well-documented, but if *A. osseum* is a junior synonym as I believe, then the species ranges in deep water from at least Fort Bragg, California to Isla Angel de la Guardia, Mexico (M. K. Wicksten, personal commun.). However, in the Transition Zone it extends up into relatively shallow water (less than 30 m or so).

While it is true that lepadomorphs have been for the most part replaced by balanomorphs in shallow water, a process that began in the Cretaceous (Newman and others, 1969), *A. californicum* has not been observed in association with other barnacles, so its appearance in the Transition Zone in shallow water does not seem to be contingent upon barnacle-to-barnacle interactions. The explanation then may be the lack of, or lower level of, interaction with other species, perhaps predators. In any event, its presence is more likely due to the biological disharmony of the region and concomitant incomplete utilization of resources, rather than some feature of the physical environment, that has proved favorable for it. That is, like *B. pacificus* and unlike *B. aquila*, *A. californicum* is a Transition Zone opportunist, rather than a relict.

We can now take up the obligate commensals. The first, *Armatobalanus* (*Armatobalanus*) *nefrens*, occurs on stylasterine hydrocorals, primarily *Allopora californica* which is itself a short-range endemic. It is the sole representative of the subgenus in the eastern Pacific, being only distantly related to *Armatobalanus* (*Hexacreusia*) *durhami*, a Panamic species occurring on scleractinian coral. Actually the closest relatives of *A. nefrens* are found in the Indo-West Pacific and the West Indies, and many are free-living rather than obligate commensals (Zullo, 1963).

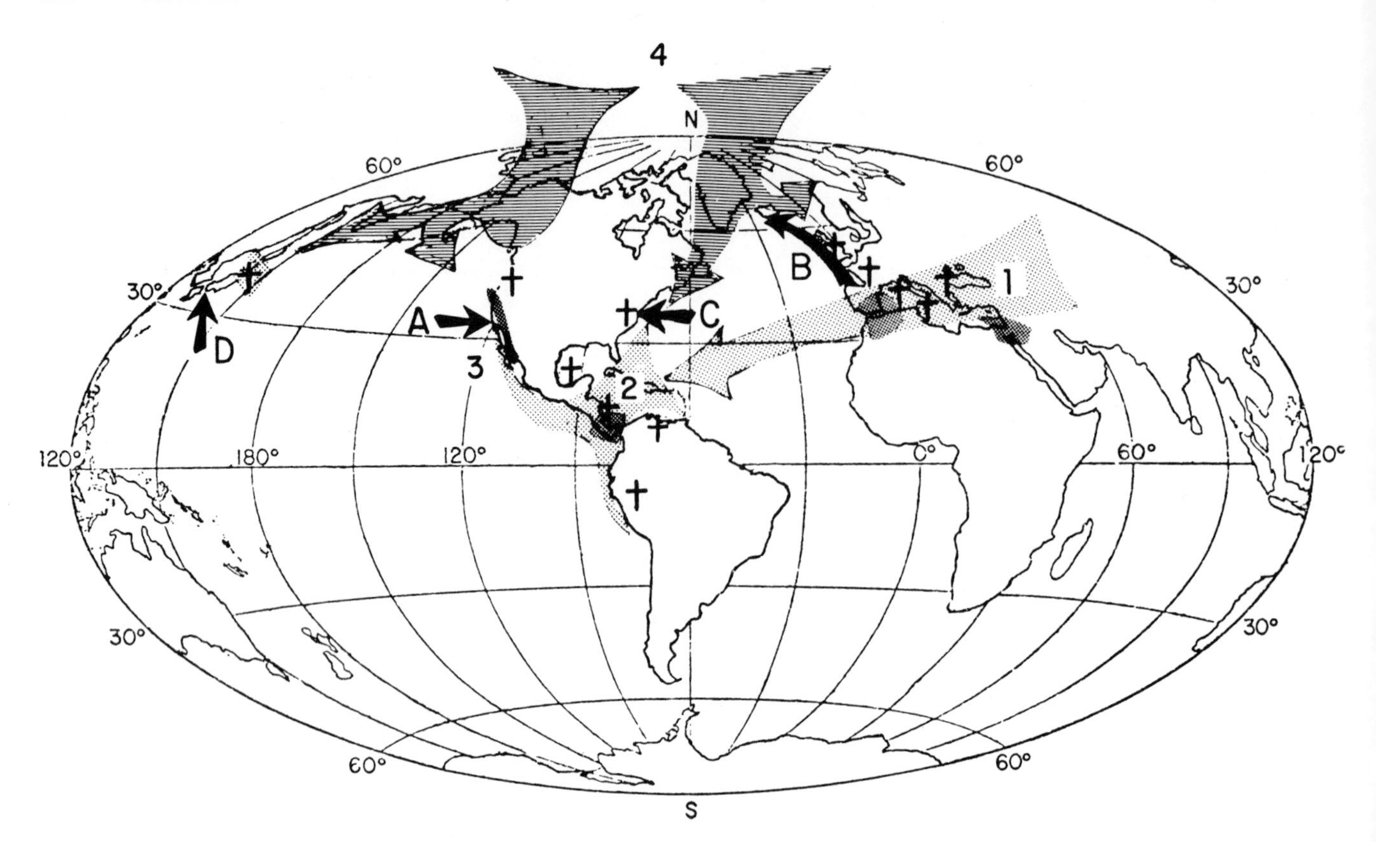

Figure 11. History of short-range endemics. Of the four among the barnacles of the Californian Transition Zone (A), one (*Arcoscalpellum californicum*) is an opportunist from deep water. The remaining three are descendents of an ancient Tethyan fauna, for the most part Oligo-Miocene in age, that once girdled the warm seas of the Earth (cf. Figs. 8, 9, 10). Deterioration of this fauna was well underway in the eastern Atlantic by the Miocene, and the ancestors of two of the short-range endemics (*concavus* group of *Balanus*, and *Armatobalanus*) disappeared there in the Pliocene, concomitant with separation of the Red Sea from the Mediterranean (1). Surviving members of the *concavus* group disappeared from the western Atlantic in the Plio-Pleistocene, concomitant with closure of the Panamic Isthmus (2). The third, *Trypetesa*, survived in both Oceans where it is represented today by two short-range endemics in the eastern Atlantic (B), one of the same in the western Atlantic (C), and two in the Pacific (D, A).

Beginning in the Pliocene, Point Conception and associated landmasses, upon which the Transition Zone centers (A), began its northward journey of approximately 3 cm/yr, relative to latitude (3). The Pleistocene followed (4) and had two relevant effects which are still apparent today: the first was the separation of many cold-temperate populations between the North Pacific and Atlantic, and some temperate populations on either side of both Oceans; and the second, severe compression and partial extinction of warm-temperate populations, some survivors of which persist as short-range endemics of the Californian Transition Zone.

The tropical American biota, from which *A. nefrens* descended, decreased markedly in diversity and otherwise changed in character during and since the Miocene (Newell, 1971). In the tropical Atlantic, the true coral barnacles (Pyrgomatidae) appeared in the Miocene (Newman and Ladd, 1974). In the tropical eastern Pacific the hermatypic corals apparently all but completely died out late in the Pliocene, later to be replaced by a few highly eurytopic Indo-Pacific hermatypes (Dana, 1975), but true coral barnacles did not return with them (Ross and Newman, 1973). *Armatobalanus nefrens* is, therefore, a relict, apparently having survived because of its association with stylasterine coral, and in this sense it is refugial (Fig. 9). Its principal host is limited to the Transition Zone; this indicates that the Transition Zone itself is in a sense a refugium.

The second obligate commensal and short-range endemic is *Trypetesa lateralis*, which burrows into the interior of gastropod shells inhabited by hermit crabs (Tomlinson, 1953). There are 3 other species in the genus: *T. habei* in the temperate-subtropical transition zone in Japan, *T. lampas* in the cold-temperate to temperate transition on the east coast of the United States and in northern Europe, and *T. nassarioides* (Tomlinson, 1969). The southern limit of *T. lampas* in Europe overlaps with the northern limit of *T. nassarioides,* the latter apparently ranging into the Mediterranean.

I consider all 4 to be short-range endemics (Table 1, Fig. 10). However, the distribution of *T. lampas* in the northeast Atlantic, in being wide-ranging latitudinally, looks peculiar. But we are dealing for the most part with temperature, not latitude, and in this situation the isotherms are canted northwestward from the British Isles to Iceland by the Gulf Stream, so that the winter mean minimum of approximately 5°C is the same for the Iceland

Table 1.

A. Californian Short-Range Endemics		
Free-living		
Balanus aquila	32°30′ to 37°45′	In a position to compete for space with other barnacles such as *B. nubilus* and *B. crenatus* from the north, *B. glandula* (complete overlap) and *B. regalis* from the south. Undergoes marked southern submergence south of Pt. Conception. Tethyan relict.
Arcoscalpellum californicum	24°45′ to 37°00′	Opportunist from deep-water to approximately 450 m, apparently not in competition with other barnacles. Known from depths less than 40 m between 32°12′ and 32°24′ south of Pt. Conception.
Obligate Commensals		
Balanus nefrens	33° to 36°25′	On hydrocoral, primarily *Allopora californica* which is also a short-range endemic. Not in competition with other barnacles, at least as an adult. Tethyan relict.
Trypetesa lateralis	35° to 38°55′	In gastropod shells inhabited by hermit crabs. Not in competition with other barnacles, but perhaps with other symbionts of the association. Unlike all other barnacles under consideration here, lacks planktonic larva. Possibly a Pleistocene temperate relict but more likely Tethyan, as two of the aforegoing. Other species of the genus also transition zone indicators, but elsewhere in the world.

B. Distribution of *Trypetesa* (see Fig. 10)

T. habei (Japan)	*T. lateralis* (California)	*T. lampas* (W Atlantic)	*T. lampas* and *T. nassarioides* (E Atlantic)
31°30′ to 33°40′, subtropical—warm-temperate transition zone, undergoes southern submergence.	35° to 38°55′, temperate, in northern half of Transition Zone.	?40° to ?42°, subtropical—cold-temperate transition zone.	Former, Iceland to Roscoff, temperate to cold-temperate transition zone; later, Roscoff into Mediterranean, temperate-subtropical transition zone.

population as for that on the east coast of the United States, even though it is some 20° of latitude further north.

The genus *Trypetesa* qualifies as a Transition Zone indicator (Fig. 10), as do the short-range endemics in general. What is really remarkable about *Trypetesa* is that it only occurs in, and marks, more-or-less comparable transition zones on both sides of the Atlantic and the Pacific.[1] Unlike the previously discussed short-range endemics, its restriction to the Transition Zone cannot be readily explained. It must not be due to interactions with other barnacles as in *B. aquila,* nor to the looseness of species packing as in *Arcoscalpellum,* nor to being an obligate commensal of a short-range host as with *Armatobalanus.* The resources required by it are perhaps available because of the disharmony between other symbionts of the gastropod-hermit crab association within the Transition Zone.

Trypetesa is, both in its morphology and symbiotic association, a very specialized acrothoracican. Unfortunately, its fossil record is uncertain. Suffice it to say for the moment that being restricted to the Northern Hemisphere at higher latitudes than other Northern Hemisphere acrothoracicans in the eastern Pacific and North Atlantic suggests that it might be a remnant of an ancient cold-temperate biota. It seems likely, though, that *Trypetesa* is a Tethyan relict since the Japanese and southern European species are clearly warm-temperate subtropical forms.

Age and Area, and Average Generic Age. We are fortunate in having a fossil record for the barnacles under consideration, and(or) for their allies, for otherwise one would have to depend primarily on the world-wide distribution of *Trypetesa* for clues to the relictual nature of 3 of the wholly Transition Zone species. Since the fossil record is substantial, the Age and Area Hypothesis of Willis (1922) may be applied with some interest. This hypothesis holds, in part, that groups spread from their point or center of origin and the area occupied by a species is therefore proportional to its age (Udvardy, 1969; Valentine, 1973). In order to side-step arguments over the theoretical aspects surrounding *spread* and *origin,* the hypothesis can be commuted to a statement of fact or principle: in general, older members of groups *range* further from the center of *distribution* than younger members. In this sense, the center of distribution is the region of optimal conditions, and the older taxa are generally eurytopic dominants (Newell, 1971). It follows that average generic age (AGA) will increase as one progresses from the center to the periphery of distribution. For the marine

[1] Two papers have appeared since the present article went to press, each reporting a new species of the family Trypetesidae from the southwest end of Madagascar (Turquier, 1976; Turquier and Carton, 1976). While these are noteworthy in being the first members of this relict family known from the Southern Hemisphere, their discovery is particularly relevant to the present report since they have been found inhabiting the hydrographically complex transition zone between the cold temperate and subtropical waters of the western Indian Ocean.

All previously recognized members of the family Trypetesidae were known from the Northern Hemisphere and, except for a Mediterranean population, all reside in transition zones between central and high latitude water masses, as is the case with the new representatives from the southern end of Madagascar. We can ask whether the new representatives are found there because of the complex hydrographic climate alone, or in addition, because Madagascar is sufficiently isolated to act as a refugium. If primarily the former, as it is apparently elsewhere, it would now be reasonable to ask if representatives are to be found where comparable transition zone waters touch on the shores of South Africa, and perhaps Australia, and the east coast of South America.

situation, the AGA pattern has been admirably demonstrated for corals (Stehli and Wells, 1971), and to some extent, for coral barnacles (Newman and others, 1976).

In the sense used here, the principle apparently holds for marine groups such as contemporary reef corals that became biotope dominants in the Cenozoic. Surviving members of Mesozoic or early Cenozoic communities likely would not demonstrate the principle since they are no longer widely distributed as populations. On the contrary, what remains of them is more apt to be short-range endemic species. As Udvardy (1969) and others have pointed out, the more endemics an area contains, the older the fauna of that area, if the endemics are relicts. It follows then that the short-range endemics of the California Transition Zone should be of greater AGA than members of the Oregonian and Californian Provinces, if they are indeed relict. Does average generic age support this expectation?

In the present study we are concerned with 14 genera representing 2 orders of Cirripedia: the Acrothoracica and Thoracica, appearing in the fossil record in the Carboniferous and Silurian, respectively. The Acrothoracica burrow almost exclusively in limestone or the calcareous hard parts of other organisms (Newman, 1974) and *Trypetesa* is the only extant form occurring in gastropod shells inhabited by hermit crabs (Tomlinson, 1969). Seilacher (1969) has questioned Tomlinson's (1963) identification of *Trypetesa* occurring in the exterior of Paleozoic myalinid shells, on the grounds that the burrows were made by two different animals. Indeed, the determination does seem unlikely since the habitat is also wrong and *Trypetesa* is otherwise the most advanced member of the order. Therefore, the age of *Trypetesa* can only be inferred. Hermit crabs appear in the Jurassic (Glaessner, 1969) and the present distribution of *Trypetesa* in the world indicates that it was Tethyan, so an inferred age somewhere between the Jurassic and the Cretaceous would be quite reasonable. However, to be on the safe side an Eocene age will also be used in the computation (Table 2).

Of the Thoracica, the Lepadomorpha are older than the Balanomorpha: Silurian, and Upper Cretaceous, respectively. Consequently, the average generic age of a biota high in lepadomorphs will almost invariably be greater than when they are not present. *Pollicipes* is not a potential problem in this regard since it ranges through both the Californian and Oregonian Provinces. It could be legitimately removed, therefore, from further consideration. But, as will be seen, the predicted relationship is achieved without doing so. However, 2 Californian lepadomorphs are obligate commensals (*Oxynaspis and Octolasmis*). Therefore the AGA of the obligate commensals is computed separately.

This leaves us with *Arcoscalpellum*, a short-range endemic, at least in the Transition Zone, where it appears to be an opportunist from deep water rather than a relict of an ancient warm-water biota in the same sense as the others. It is old, as being a lepadomorph would predict. Thus, as with the obligate commensals, we should make the computations with and without it to see what difference it makes.

The 2 remaining short-range endemics are *Armatobalanus nefrens* and *Balanus aquila.* The first is an obligate commensal of certain hydrocorals. It is the youngest of the 4 short-range endemics (lower Miocene), but it is an archaeobalanid which is an old group having roots in the Eocene (Newman and Ross, 1976). *Balanus s. s.*, the second free-living form, is also relatively old (lower Oligocene).

As can be seen from Table 2(A), the AGA of the 4 short-range endemics is, at best, approximately 30 m.y. greater than that of the Oregonian and Californian free-living genera. Including obligate commensals with the latter does not alter the situation. Then there is *Arcoscalpellum;* not a short-range endemic in the same sense as the others. Removing it from the calculation reduces the age difference by an order of magnitude, but the AGA is still greater for the short-range endemics than for the provincialites. If there are objections to the ages used here, one can compare ranks; the relationship remains the same. Whatever the approach, AGA supports the view that the short-range endemics of the Transition Zone are relict members of an essentially extinct biota.

From the distributional data it appears that the biota was a warm-temperate Tethyan one, and that restriction and extinction of most members was for the most part a Mio-Pliocene event on a world-wide basis, and Pleistocene locally. The question then is, how did the short-range endemics come to reside in the Transition Zone?

THE TRANSITION ZONE—COMPRESSED WARM-TEMPERATE AND(OR) A REFUGIUM?

The average generic age of the 3 Tethyan short-range Transition Zone endemics is greater than that of members of either the Oregonian or Californian Provinces. Clearly then, the endemics are the remnants of an old fauna. The world distribution of extant species closely related to the 3 indicates that they are not only old but that they are relicts, and the fossil record illustrates the course of relicturalization of 2 of them (*B. aquila* and *A. nefrens,* but not *Trypetesa*). Two questions remain: (1) are these 3 species surviving members of warm-water communities of a province that has been compressed latitudinally virtually to extinction, or (2) are they fugitive species that have accumulated in the refuge formed by the Transition Zone?

Ekman (1953) noted that the transition from subtropical to cold-temperate on the West Coast of North America is so sudden that there is practically no room for a warm-temperate fauna. This is because of the hydrophysiographic situation, waters above Point Conception being markedly colder on the average than waters below, where the coast cuts away to the east. However, the present configuration is not a static one, nor has it been through time. Recent studies have shown that lands west of the San Andreas and associated faults, attached to the

Table 2.

A

		yr B.P. x 10^6			Rank
Oregonian					
Chthamalus	Lower Pliocene	5			2
Semibalanus	Upper Miocene	7			3
Hesperibalanus	Lower Eocene	51			8
Balanus	Lower Oligocene	36			5
Pollicipes	Lower Eocene	51			8
		$\overline{X}$ = 30			5.2
Californian					
Chthamalus	Lower Pliocene	5			2
Tetraclita	Upper Miocene	7			3
Megabalanus	Lower Oligocene	36			5
Balanus	Lower Oligocene	36			5
Pollicipes	Lower Eocene	51			8
		$\overline{X}$ = 27			4.6
California Obligate Commensals					
Oxynaspis	Eocene	45			7
Octolasmis	Upper Eocene	40			6
Membranobalanus	Upper Pliocene	3			1
Conopea	Lower Miocene	20			4
		$\overline{X}$ = 27			4.5
Short-Range Endemics					
Trypetesa	Cretaceous (Eocene)	100	(45	45)	10
Arcoscalpellum	Upper Cretaceous	85	85		9
Armatobalanus	Lower Miocene	20	20	20	4
Balanus	Lower Oligocene	36	36	36	5
		$\overline{X}$ = 60.25	46.5	33.66	7.0

B

	yr B.P. x 10^6	Rank
Upper Pliocene	3	1
Lower Pliocene	5	2
Upper Miocene	7	3
Lower Miocene	20	4
Lower Oligocene	36	5
Upper Eocene	40	6
Eocene	45	7
Lower Eocene	57	8
Upper Cretaecous	85	9
Cretaceous	100	10

Pacific plate, are moving northwestward relative to the North American plate at a rate of about 6 cm per year (Atwater, 1970). This motion has apparently been in progress for the past 11 m.y., since subduction of the Farallon plate, and has led not only to the formation of the Gulf of California but to the northwesterly displacement of Point Conception and associated landmasses, some 600 km relative to the North American plate. But at the same time the North American plate has been in motion, and an analysis relative to latitude places Point Conception approximately 300 km north of its Pliocene position 11 m.y. ago (George Sharman, personal commun.).

The marine climatic regime was also dramatically altered over the same period of time. The Pliocene was decidedly warmer than today, and there were significant coral reefs in the vicinity of San Diego, which was therefore tropical rather than subtropical. There are no reefs until one reaches the vicinity of Panama today. During the Pleistocene the warm-temperate was compressed climatically by polar cooling, even more than it is today (Durham, 1950; Valentine, 1961; Addicott, 1966; McIntyre and others, 1976). Thus, a strong case can be made that the short-range endemics are remnants of an ancient, compressed, warm-temperate biota.

Yet the occurrence of *Trypetesa* in other transition zones in the Northern Hemisphere causes pause, since, from the aforegoing, it would follow that comparable compressions must have occurred there also. Have they? The answer is yes, since Pleistocene effects were generally felt throughout the temperate region although to differing degrees. As can be seen in the paleoclimatic chart of McIntyre and others (1976), both the Californian and Oceanic North Pacific transition zones likely had August temperatures in the 14° to 15°C range 18,000 yr ago, comparable to the long-term mean found at these localities today. As noted by Valentine (1966), since extratropical provincial chains owe their origin to comparable general climatic trends, there are resemblances among them. The present situation is a case in point.

Further evidence helps bring the compression into perspective. During warm periods, such extant cold-temperate species as *Semibalanus balanoides* and *Balanus crenatus* each had continuous populations across the Arctic, from the Pacific to the Atlantic (Zullo, 1966). Today these populations are disjunct. Distributions across the North Pacific were likewise affected; members of the warm-temperate *concavus* group became extinct in northern Japan (T. Yamaguchi, personal commun.) and Oregon, and the temperate—cold-temperate *Chthamalus dalli* population became disjunct.

It seems probable that the return of more equitable climates would not only reconnect the surviving cold-temperate populations but would greatly increase the northern latitudinal extent of the warm-temperate region and thereby blur the present Californian Transition Zone. Yet, it is doubtful that much of the ancient, and presently in good part refugial, warm-temperate fauna of the Transition Zone would once again form a significant portion of the warm-temperate biota. Extinction appears to have gone too far and only relatively unimportant members remain today, at least as far as the barnacles are concerned. The fate of the obligate commensals depends on the fate of their hosts and concomitant biological interactions. Only free-living *Balanus aquila* would seem to hold promise of ever again becoming a community or provincial dominant.

With a return to more equitable climates, blurring of the Transition Zone would likely occur in the open ocean as well as along the shore. While the surface circulation pattern should remain much the same, it would probably slow down with the decrease in temperature differential between the tropics and the poles and concomitant warming of the North Pacific. The North Pacific Current should then widen and become less distinguishable in temperature from the so-called Subarctic and Central Water Masses. If the short-range endemics presently occupying the Transition Zone there are remnants of a compressed warm-temperate biota, as is in good part the case along the shore, one would expect them to become more widely distributed latitudinally, especially towards the north, and thereby they would lose their status as short-range endemics, as would those along the shore.

The extent to which the Californian Transition Zone would blur with more equitable climates is problematical, since it is presently associated with major physiohydrographic features. Fate of the short-range endemics is also problematic, but only one (*B. aquila*) is likely capable of again being a long-range dominant. Assuming that physical features remained the same in configuration but not temperature, the northern terminus of the Californian Province would likely shift from Point Conception to as far north as Cape Mendocino. The Californian-Panamic, rather than the Californian-Oregonian Transition Zone would then center on Point Conception, since during the Pliocene it was at comparable latitudes.

ACKNOWLEDGMENTS

A chart of latitudinal ranges of cirripeds along the coast of California (comparable to Fig. 3, herein) was prepared many years ago for inclusion in one or the other of two handbooks on the invertebrates of California. I am grateful to the editors of these handbooks for not publishing the chart, for desire to eventually do so became the impetus for this paper. Dr. R. Rosenblatt alerted me to Newell's (1948) classic paper on faunal provinces of the California coast and K. Devonald pointed out the reciprocal relationship between Newell's percent change and Valentine's (1966) results utilizing Jaccard's Coefficient of Similarity. I would also like to acknowledge helpful discussions with Drs. A. Fleminger, D. Goodman, and J. McGowan, and to extend my appreciation to Fleminger and Goodman for reading the manuscript and making helpful suggestions. Ms. Gayle Kidder aided materially in the study of specimens from the coasts of the Americas and in preparation of the manuscript. Finally, I would like to thank Dr. A. Boucot for listening to an early synopsis of this work, when once on a visit here, and later for encouraging me to prepare this paper.

This study is for the most part a by-product of work on the systematics of Cirripedia, supported in part by the National Science Foundation (DEB75-17149).

REFERENCES

Addicott, W. O., 1966, Late Pleistocene marine paleoecology and zoogeography in central California: U. S. Geol. Survey Prof. Paper 523-C, p. 1-21.

Atwater, T., 1970, Implications of plate tectonics for the Cenozoic tectonic evolution of western North America: Geol. Soc. America Bull., v. 81, p. 3513-3536.

Barnes, H., and Reese, E. S., 1960, The behaviour of the stalked intertidal barnacle *Pollicipes polymerus* J. B. Sowerby, with special reference to its ecology and distribution: Jour. Animal Ecology, v. 29, p. 169-185.

Bullock, T. H., 1955, Compensation for temperature in the metabolism and activity of poikilotherms: Biol. Rev., v. 30, p. 311-342.

Carlton, J. T., and Zullo, V. A., 1969, Early records of the barnacle *Balanus improvisus* Darwin from the Pacific coast of North America: California Acad. Sci. Occas. Papers, no. 75, 6 p.

Coast and Geodetic Survey, 1952, Surface water temperatures at tide stations, Pacific Coast North and South America and Pacific Ocean Islands: Coast and Geodetic Survey Spec. Pub. 280, U. S. Dept. Commerce, 59 p.

Connell, J., 1970, A predator-prey system in the marine intertidal region. I. *Balanus glandula* and several predatory species of *Thais:* Ecological Mon., v. 40, p. 49-78.

Cornwall, I. E., 1936, On the nervous system of four British Columbian barnacles (one new species): Jour. Biol. Board Canada, v. 1, p. 469-475.

——— 1951, The barnacles of California: Wasmann Jour. Biol., v. 9, p. 311-346.

——— 1953, The central nervous system of barnacles (Cirripedia): Jour. Fish. Research Board Canada, v. 10, p. 76-84.

——— 1955, Canadian Pacific fauna. Vol. 10, Arthropoda. Pt. 10e, Cirripedia: Fish. Research Board Canada, p. 1-49.

——— 1960, Barnacle shell figures and repairs: Canadian Jour. Zoology, v. 38, p. 827-832.

——— 1962, The identification of barnacles, with further figures and notes: Canadian Jour. Zoology, v. 40, p. 621-629.

Crowe, F. J., and Schwartzlose, R. A., 1972, Release and recovery records of drift bottles in the California Current region 1955 through 1971: California Coop. Oceanic Fish. Inves. Atlas No. 16, State of California Marine Research Comm., 140 p.

Cutler, E. B., 1975, Zoogeographical barrier on the continental slope off Cape Lookout, North Carolina: Deep-Sea Research, v. 22, p. 893-901.

Dana, T. F., 1975, Development of contemporary eastern Pacific coral reefs: Marine Biol., v. 33, p. 355-374.

Darwin, C., 1854, A Monograph on the subclass Cirripedia with figures of all the species: Ray Soc. London, 684 p.

Dayton, P., 1971, Competition, disturbance, and community organization: The provision and subsequent utilization of space in a rocky intertidal community: Ecological Mon., v. 41, p. 351-389.

Durham, J. W., 1950, Cenozoic marine climates of the Pacific Coast: Geol. Soc. America Bull., v. 61, p. 1243-1264.

Ekman, S., 1953, Zoogeography of the sea: London, Sidgwick and Jackson, 417 p.

Enright, J. T., 1976, Climate and population regulation; the biogeographer's dilemma: Oecologia, v. 24, p. 295-310.

Fleminger, A., 1975, Geographical distribution and morphological divergence in American coastal-zone planktonic copepods of the genus *Labidocera, in* Cronin, L. E., ed., Estuarine Research, Vol. 1: New York, Academic Press, p. 392-419.

Glaessner, M. F., 1969, Decapoda, *in* Moore, R. C., ed., Treatise on invertebrate paleontology, Pt. R, Arthropoda 4, Vol. 2: Geol. Soc. America and Univ. of Kansas, p. R400-R533.

Gomez, E. D., 1975, Sex determination in *Balanus* (*Conopea*) *galeatus* (L.) (Cirripedia Thoracica): Crustaceana, v. 28, p. 105-107.

Graham, H. W., and Gay, H., 1945, Season of attachment and growth of sedentary marine organisms at Oakland, California: Ecology, v. 26, p. 375-386.

Hayden, B. P., and Dolan, R., 1976, Coastal marine fauna and marine climates of the Americas: Jour. Biogeography, v. 3, p. 71-81.

Hedgpeth, J. W., 1957, Marine biogeography, *in* Hedgpeth, J. W., ed., Treatise on marine ecology and paleoecology, Vol. 1: Geol. Soc. of America Mem. 67, p. 359-382.

Henry, D. P., 1942, Studies on the sessile Cirripedia of the Pacific coast of North America: Univ. Washington Pub. Oceanogr., v. 4, p. 95-134.

——— 1960, Thoracic Cirripedia of the Gulf of California: Univ. Washington Pub. Oceanogr., v. 4, p. 135-158.

Henry, D. P., and McLaughlin, P. A., 1975, The barnacles of the *Balanus amphitrite* complex (Cirripedia, Thoracica): Zool. Verhandl., v. 141, p. 1-254.

Hilgard, G. H., 1960, A study of reproduction in the intertidal barnacle *Mitella polymerus* in Monterey Bay, California: Biol. Bull., v. 119, p. 169-188.

Hurley, A. C., 1973, Fecundity of the acorn barnacle *Balanus pacificus* Pilsbry: A fugitive species: Limnology and Oceanography, v. 18, p. 386-393.

Hutchins, L. W., 1947, The bases for temperature zonation in geographical distribution: Ecological Mon., v. 17, p. 325-335.

Jackson, J. B. C., 1974, Biogeographic consequences of eurytopy and stenotopy among marine bivalves and their evolutionary significance: Am. Naturalist, v. 108, p. 541-560.

Kuhl, H., 1968, Die Beeinflussung der Metamorphose von *Balanus improvisus* Darwin durch Giftstoffe: Internat. Cong. Seawater Corrosion and Fouling, 2nd, Athens 1968, p. 1-8.

Lewis, C. A., 1975, Development of the gooseneck barnacle *Pollicipes polymerus* (Cirripedia: Lepadomorpha): Fertilization through settlement: Marine Biol., v. 32, p. 141-153.

McGowan, J. A., 1974, The nature of oceanic ecosystems, *in* Miller, C. B., ed., The biology of the oceanic Pacific: Corvallis, Oregon State Univ. Press, p. 9-28.

McGowan, J. A., and Williams, P. M., 1973, Oceanic habitat differences in the North Pacific: Jour. Exptl. Marine Biol. and Ecology, v. 12, p. 187-217.

McIntyre, T. C. and others, 1976, The surface of the ice-age Earth: Science, v. 191, p. 1131-1137.

McLaughlin, P. A., and Henry, D. P., 1972, Comparative morphology of complemental males in four species of *Balanus* (Cirripedia Thoracica): Crustaceana, v. 22, p. 13-30.

Newell, I. M., 1948, Marine molluscan provinces of western North America: A critique and a new analysis: Am. Philos. Soc. Proc., v. 92, p. 155-166.

Newell, N. D., 1971, An outline history of tropical organic reefs: Am. Mus. Novitates, no. 2465, p. 1-37.

Newman, W. A., 1960, *Octolasmis californiana* spec. nov., a pedunculate barnacle from the gills of the California Spiny Lobster: The Veliger, v. 3, p. 9-11.

——— 1967, On physiology and behaviour of estuarine barnacles, *in* Symposium on Crustacea, Proc., Erna Kulam 1965: Marine Biol. Assoc. India, Symp. ser. 2, pt. 3, p. 1038-1066.

——— 1972, An Oxynaspid (Cirripedia, Thoracica) from the eastern Pacific: Crustaceana, v. 23, p. 202-208.

——— 1974, Two new deep-sea Cirripedia (*Ascothoracica* and *Acrothoracica*) from the Atlantic: Jour. Marine Biol. Assoc. U. K., v. 54, p. 437-456.

——— 1975, Cirripedia, *in* Smith, R. I., and Carlton, J. T., eds., Intertidal invertebrates of the central California coast (3rd ed.): Berkeley, Univ. California Press, p. 259-260.

Newman, W. A., Jumars, P. A., and Ross, A., 1976, Diversity trends in coral-inhabiting barnacles (Cirripedia, Pyrgomatinae): Micronesica, v. 12, p. 69-82.

Newman, W. A., and Ladd, H. S., 1974, Origin of coral-inhabiting balanids (Cirripedia, Thoracica), *in* Contributions to the geology and paleobiology of the Caribbean and adjacent areas: Naturforsch. Gesell. Basel Verhandl., v. 84, p. 381-396.

Newman, W. A., and Ross, A., 1976, Revision of the balanomorph barnacles; including a catalog of the species: San Diego Soc. Nat. History Mem. 9, 108 p.

Newman, W. A., Zullo, V. A., and Withers, T. H., 1969, Cirripedia, *in* Moore, R. C., ed., Treatise on invertebrate paleontology, Pt. R, Arthropoda 4, Vol. 1: Geol. Soc. America and Univ. Kansas, p. R206-R-295.

Pilsbry, H. A., 1907, Cirripedia from the Pacific coast of North America: U. S. Bur. Fisheries Bull., v. 26, p. 193-204.

——— 1916, The sessile barnacles (Cirripedia) contained in the collections of the U. S. National Museum; including a monograph of the American species: U. S. Natl. Mus. Bull. 93, 366 p.

Reid, J. L., McGowan, J. A., Brinton, E., Fleminger, A., and Venrick, E. L., 1976, Ocean circulation and marine life: General Symposia, Joint Oceanographic Assembly, Edinburgh 1976, Abs., p. 5.

Reid, J. L., Brinton, E., Fleminger, A., Venrick, E. L., and McGowan, J. A., 1977, Ocean circulation and marine life, *in* Charnock, W., and Deacon, G., eds., General Symposia, Joint Oceanographic Assembly, Edinburgh 1976, Proc.: London, Plenum Press (in press).

Remane, A., 1934, Die Brackwasserfauna: Zool. Anz. Supp. 7, p. 34-74.

Rice, L., 1930, Peculiarities in the distribution of barnacles in communities and their probable causes: Puget Sound Biol. Station Pub., v. 7, p. 249-257.

Rogers, F. L., 1949, Three new subspecies of *Balanus amphitrite* from California: Jour. Entomology and Zoology, v. 41, p. 23-32.

Ross, A., 1962, Results of the Puritan-American Museum of Natural History expedition to western Mexico, 15. The littoral balanomorph Cirripedia: Am. Mus. Novitates, no. 2084, p. 1-44.

Ross, A., 1964, Cirripedia from the Yorktown Formation (Miocene) of Virginia: Jour. Paleontology, v. 38, p. 483-491.

Ross, A., Cerame-Vivas, M. J., and McCloskey, L. R., 1964, New barnacle records for the coast of North Carolina: Crustaceana, v. 7, p. 312-313.

Ross, A., and Newman, W. A., 1973, Revision of the coral-inhabiting barnacles (Cirripedia: Balanidae): San Diego Soc. Nat. History Trans., v. 17, p. 137-174.

Seilacher, A., 1969, Paleoecology of boring barnacles: Am. Zoologist, v. 9, p. 705-719.

Southward, A. J., and Southward, E. C., 1967, On the biology of an intertidal chthamalid (Crustacea, Cirripedia) from the Chuckchi Sea: Arctic, v. 20, p. 8-20.

Stehli, F. G., and Wells, J. W., 1971, Diversity and age patterns in hermatypic corals: Systematic Zoology, v. 20, p. 115-126.

Tomlinson, J. T., 1953, A burrowing barnacle of the genus *Trypetesa* (order Acrothoracica): Washington Acad. Sci. Jour., v. 43, p. 373-381.

——— 1960, Low hermit crab migration rates: The Veliger, v. 2, p. 61.

——— 1963, Acrothoracican barnacles in Paleozoic myalinids: Jour. Paleontology, v. 37, p. 164-166.

——— 1969, The burrowing barnacles (Cirripedia: order Acrothoracica): U. S. Natl. Mus. Bull. 296, p. 1-162.

Turquier, Yves, 1976, Étude de quelques Cirripèdes Acrothoraciques de Madagascar. II. Description de *Trypetesa spinulosa* n. sp.: Soc. Zool. France Bull., v. 101, p. 559-574.

Turquier, Yves, and Carton, Yves, 1976, Étude de quelques Cirripèdes Acrothoraciques de Madagascar. I. *Alcippoides asymetrica* nov. gen., nov. sp., et la famille des Trypetesidae: Arch. Zool. Exptl. Gènèrale, v. 117, p. 383-393.

Udvardy, M. D. F., 1969, Dynamic zoogeography: New York, Van Nostrand Reinhold, 445 p.

Utinomi, H., 1970, Studies on the cirripedian fauna of Japan. IX. Distributional survey of thoracic cirripeds in the southeastern part of the Japan Sea: Seto Marine Biol. Lab. Pub., v. 17, p. 339-372.

Valentine, J. W., 1961, Paleoecologic molluscan geography of the Californian Pleistocene: California Univ. Pubs. Geol. Sci., v. 34, p. 309-442.

——— 1966, Numerical analysis of marine molluscan ranges on the extratropical northeastern Pacific shelf: Limnology and Oceanography, v. 11, p. 198-211.

——— 1973, Evolutionary paleoecology of the marine biosphere: Englewood Cliffs, N. J., Prentice Hall, 511 p.

Vermeij, G. J., 1972, Endemism and environment: Some shore molluscs of the tropical Atlantic: Am. Naturalist, v. 106, p. 89-101.

Werner, W. E., 1967, The distribution and ecology of the barnacle *Balanus trigonus:* Marine Sci. Bull., v. 17, p. 64-84.

Willis, J. C., 1922, Age and area. A study in geographical distribution and origin of species: Cambridge, Cambridge Univ. Press, 259 p.

Wyllie, J. G., 1966, Geostrophic flow of the California Current at the surface and at 200 meters: California Coop. Oceanic Fish. Inves. Atlas No. 4, State of California Marine Research Comm., 288 p.

Wyllie, J. G., and Lynn, R. J., 1971, Distribution of temperature and salinity at 10 meters, 1960-1969, and mean temperature, salinity and oxygen at 150 meters, 1950-1968, in the California Current: California Coop. Oceanic Fish. Inves. Atlas No. 15, State of California Marine Research Comm., 188 p.

Zullo, V. A., 1963, A review of the subgenus *Armatobalanus* Hoek (Cirripedia: Thoracica) with the description of a new species from the California coast: Ann. Mag. Nat. History, ser. 13, v. 6, p. 587-594.

——— 1964, Re-evaluation of the late Cenozoic cirriped *Tamiosoma* Conrad: Biol. Bull., v. 127, p. 360.

——— 1966, Zoogeographic affinities of the Balanomorpha (Cirripedia: Thoracica) of the eastern Pacific, *in* Bowman, R. I., ed., The Galapagos: Berkeley, Univ. California Press, p. 139-144.

——— 1968, Extension of range for *Balanus tintinnabulum californicus* Pilsbry, 1916 (Cirripedia, Thoracica): California Acad. Sci. Occas. Papers, no. 70, 3 p.

——— 1969a, Thoracic Cirripedia of the San Diego Formation, San Diego County, California: Los Angeles County Mus. Contr. Sci., no. 159, p. 1-25.

——— 1969b, A late Pleistocene marine invertebrate fauna from Bandon, Oregon: California Acad. Sci. Proc., ser. 4, v. 36, p. 347-361.

APPENDIX

Species and Relevant Literature

Acrothoracica

Trypetesa lateralis Tomlinson 1953

Low intertidal pools, occurring burrowed in gastropod shells (particularly *Tegula*) inhabited by hermit crabs; a short-range endemic, from Point Arena to Point Conception. References: Tomlinson (1953, 1969).

Thoracica: Lepadomorpha

Pollicipes polymerus Sowerby 1833

Intertidal, with *Mytilus californianus* or in relatively pure stands; ranging south from British Columbia to Cape San Lucas, Baja California. References: Barnes and Reese (1960); Cornwall (1936, 1951, 1953); Hilgard (1960); Lewis (1975).

Arcoscalpellum californicum (Pilsbry) 1907

Occurring on rocks and attached organisms from 10 to 180 m; Fort Bragg, California south into Gulf of California but only known to range into shallow water from Monterey south to San Diego. References: Pilsbry (1907); Cornwall (1951).

Octolasmis californiana Newman 1960
Littoral, in gill chambers of large decapod crustaceans such as *Panulirus, Loxorhynchus,* and *Cancer;* Monterey to Panama. Reference: Newman (1960, 1975).

Oxynaspis rossi Newman 1972
In deep water from 183 m off Santa Catalina Island and 55 m off Baja California (22°55′N), on antipatharians. Reference: Newman (1972).

Thoracicia: Balanomorpha

Chthamalus dalli Pilsbry 1916
High intertidal; on rocks, pier pilings, and hard-shelled organisms; ranging from northern Japan and Alaska south to San Diego. References: Cornwall (1955); Dayton (1971); Newman (1975); Newman and Ross (1976); Pilsbry (1916); Rice (1930); Southward and Southward (1967); Utinomi (1970).

Chthamalus fissus Darwin 1854
High intertidal; in same habitats as *C. dalli,* but a southern species, ranging from San Francisco south to Baja California; *C. microtectus* (Cornwall 1955) from Monterey, is considered junior synonym of *C. fissus.* References: Connell (1970); Henry (1960); Newman and Ross (1976); Pilsbry (1916); Ross (1962).

Balanus glandula Darwin 1854
Common intertidally on rocks, pier pilings, and a variety of hard-shelled animals; bays and outer coast; ranging from the Aleutians south to San Quintin, Baja California. References: Cornwall (1955); Darwin (1854); Henry (1942); Newman and Ross (1976); Pilsbry (1916); Zullo (1969b).

Balanus crenatus Brugière 1789
From low intertidal to 182 m on various objects including seaweed; North Atlantic and Pacific; northern Japan and Alaska south to Santa Barbara, California. References: Cornwall (1955); Darwin (1854); Newman (1967); Newman and Ross (1976); Zullo (1969b).

Balanus trigonus Darwin 1854
From low intertidal to 90 m on a wide variety of substrates including rocks and hard-shelled invertebrates or embedded in sponges or corals; cosmopolitan in warm seas, ranging from Monterey to Peru in the eastern Pacific. References: Darwin (1854); Henry (1960); Newman and Ross (1976); Pilsbry (1916); Ross, Cerame-Vivas and McCloskey (1964); Werner (1967).

Balanus pacificus Pilsbry 1916
Low intertidal to 73 m; occasionally on rocks, usually on other organisms; Monterey Bay south to Baja California. References: Cornwall (1962); Darwin (1854); Henry (1960); Hurley (1973); Newman and Ross (1976); Pilsbry (1916); Ross (1962); Zullo (1969a).

Balanus amphitrite amphitrite Darwin 1854
Low intertidal to 18 m on rocks, shells, pier pilings, etc.; cosmopolitan in warm seas; introduced in the eastern Pacific where it ranges south in bays from San Francisco to Mexico; introduced into Salton Sea, California. References: Henry and McLaughlin (1975); Newman (1967); Newman and Ross (1976); Pilsbry (1916).

Balanus improvisus Darwin 1854
Low intertidal and subtidal, in estuaries on rocks, pilings and hard-shelled organisms, particularly tolerant of brackish waters. Introduced into the north Pacific; ranging south from the Columbia River to the Salinas River in California, occasionally in harbors south of Point Conception. References: Carlton and Zullo (1969); Darwin (1854); Kühl (1968); Newman (1967); Newman and Ross (1976).

Balanus regalis Pilsbry 1916
Low intertidal and subtidal, on rocks; Point Conception south to west Mexico. References: Henry (1960); Henry and McLaughlin (1975); Newman and Ross (1976); Pilsbry (1916); Ross (1962).

Balanus aquila Pilsbry 1907
Low intertidal to 18 m on rocks, pier pilings, abalone; a short-range endemic, San Francisco to San Diego, California. References: Cornwall (1960); Henry and McLaughlin (1975); Newman and Ross (1976); Pilsbry (1916); Zullo (1966).

Balanus nubilus Darwin 1854
Low intertidal to 90 m; on rocks, pier pilings, and hard-shelled organisms; from southern Alaska to La Jolla, California. References: Cornwall (1951); Darwin (1854); Newman and Ross (1976); Pilsbry (1916); Ross (1962); Zullo (1969a, 1969b).

Armatobalanus (Armatobalanus) nefrens (Zullo) 1963
Subtidal to 64 m; on and embedded in the hydrocorals *Allopora californica* Verrill and *Errinopora pourtalesii* (Dall); a short-range endemic, from Monterey south to the Channel Islands. References: Newman and Ross (1976); Ross and Newman (1973); Zullo (1963).

Conopea galeata (Linné) 1771
Subtidal to 90 m growing embedded in sea fans (gorgonians); Caribbean and eastern Pacific, Monterey to Central America. References: Cornwall (1951); Darwin (1854); Gomez (1975); McLaughlin and Henry (1972); Newman and Ross (1976); Pilsbry (1916); Ross (1962).

Megabalanus californicus (Pilsbry) 1916
Low intertidal to 9 m on rocks, pilings, buoys, mussels, and other hard-shelled organisms. Occasionally as far north as Humboldt and San Francisco Bays, but generally Monterey south to Guaymas, Gulf of California. References: Henry (1960); Newman and Ross (1976); Pilsbry (1916); Zullo (1968).

Membranobalanus orcutti (Pilsbry) 1907
Subtidal to 18 m in sponges; Point Conception, California, to Cape San Lucas, Baja California. Reference: Newman and Ross (1976); Pilsbry (1916).

Semibalanus cariosus (Pallas) 1788
Low intertidal, on rocks along exposed shores; Japan, Bering Sea, and Alaska south to San Francisco, California. References: Cornwall (1955); Darwin (1854); Newman and Ross (1976); Pilsbry (1916); Zullo (1969b).

Solidobalanus (Hesperibalanus) hesperius (Pilsbry) 1916
From 18 to 64 m, on hard-shelled organisms usually from soft bottoms; Bering Sea, Alaska, south to Monterey Bay, California. References: Cornwall (1955); Newman and Ross (1976).

Tetraclita rubescens Darwin 1854
Low intertidal, on rocks on exposed coasts, occasionally subtidal on hard-shelled organisms such as abalone; from north of Bodega Bay, California, south to Cape San Lucas, Baja California. References: Cornwall (1951); Darwin (1854); Henry (1960); Newman (1975); Pilsbry (1916); Ross (1962).

The Role of Circulation in the Parcelling and Dispersal of North Atlantic Planktonic Foraminifera

RICHARD CIFELLI, *Department of Paleobiology, National Museum of Natural History, Washington, D. C. 20560*

ABSTRACT

North Atlantic planktonic foraminifera are parcelled into cold-water (arctic and northern) and warm-water (subtropical and tropical) faunas. Faunal parcels refer to preferred, not restricted, habitats as species are dispersed widely by circulation. The major faunal change occurs at the margin of the North Atlantic Gyre where there is a turnover in species and an increase in diversity from the colder to warmer side of the Gyre. Because gyral circulation is clockwise, species are displaced to the north on the western side of the North Atlantic and to the south on the eastern side. Mixed associations of northern, subtropical, and tropical species occur at temperate latitudes on the western side of the Atlantic Ocean and at tropical latitudes on the eastern side. Faunal boundaries and species frequencies fluctuate seasonally but the faunas retain their identities throughout the year. The Mediterranean fauna is stocked from the eastern Atlantic and consists of a mixture of northern and subtropical species. Because of clockwise displacement on the eastern side of the ocean, tropical species have a difficult means of entry into the Mediterranean and are scarce.

An overview of the fossil record suggests that there were alternating expansions and contractions of warm gyral water in the past. Expansions seem to have occurred in the Late Cretaceous, Paleogene, and Miocene. Contractions are visible at the beginning and middle of the Cenozoic. Another contraction appears to have started at the beginning of the Pliocene.

INTRODUCTION

The open ocean is in a constant state of motion, with the water circulating horizontally (advectively) and vertically (convectively). In such a state one would expect, and does find, considerable mixing of the occupants in the epipelagic zone. This is especially true of the free-floating plankters, such as the foraminifera, that have ineffective means of self-propulsion. Mixing is an important aspect of plankton distribution that should not be minimized.

While the ocean is a kind of huge "mixing bowl," at the same time it is partitioned into several preferred faunal habitats referred to here as parcels. The term province is deliberately avoided because it implies static geographic limits. Plankton faunal boundaries are often diffuse, and even where they can be delineated, they change throughout the year in response to the temporal shifts in currents; plankton boundaries are never static.

Faunas occupying the parcels are more-or-less distinct, but species are dispersed widely by circulation. Limits of dispersal must depend on the adaptability of the species to changing conditions in the course of travel. In large parts of the oceans, faunal assemblages are composed of mixtures from more than one parcel. Since species are dispersed in a roughly circular pattern, faunal boundaries arch across the parallels of latitude. Therefore, while it is convenient to speak of a northern or high latitude fauna, this does not imply a latitudinal separation of that fauna. Latitude, actually, can be a misleading frame of reference for biogeographic analysis.

This paper will deal exclusively with North Atlantic and Mediterranean faunas since I have had little first-hand experience with any others. While it is unlikely that complete symmetry in faunal distribution exists among the oceans, the patterns in the North Atlantic should serve to illustrate the role of oceanic circulation in the parcelling of planktonic foraminiferal faunas.

CIRCULATORY FRAMEWORK OF THE NORTH ATLANTIC

There are several aspects of the circulatory framework that are common knowledge but are worth emphasizing because they bear directly on the discussion that follows:

(1) The greater part of the North Atlantic is dominated by a Gyre that moves in the fashion of a distorted rotating ring (Fig. 1). Most of the motion occurs around the margins of the Gyre, with very little motion at the core. Since rotation is clockwise, currents move from south to north on the western side of the Ocean and from north to south on the eastern side.

(2) There is a pronounced asymmetry to the Gyre. The western side is marked by the Gulf Stream, a strong set of currents. Everywhere else there are no clearly definable boundaries and currents are much weaker. On the southern side, the North Equatorial Current blends into the Sargasso Sea, the center of the Gyre.

(3) The Gulf Stream is more than just a moving body of water. It is also a thermal gradient that separates two distinct climatic regimes, the cold slope waters to the north, and the subtropical Sargasso Sea to the south. During a winter crossing of the Gulf Stream from north to south, the surface temperature may change from about 10° to 19°C. The Gulf Stream, which has an average width of about 130 miles (209 km) (Iselin, 1936), changes in position and intensity throughout the year. It also meanders and eddies, but always forms a distinct western boundary of the North Atlantic Gyre, a large subtropical water mass.

(4) The Gulf Stream can be traced eastward to about long 40° to 50°W where it splits into a northern and southern branch. Boundaries become less distinct and possibly a second Gyre is formed (Worthington, 1962). In any case, both branches are very diffuse with the northern branch (North Atlantic Current) spreading over much of the northeastern part of the ocean. The southern branch continues the clockwise direction of the North Atlantic Gyre, with most of the motion occurring along the Canaries Current, east of the Azores Islands.

(5) Associated with the bifurcation of the Gulf Stream is a deflection in ocean climate. Isotherms fan out to the north and to the south. North of lat 40°N, surface temperatures are higher on the eastern side of the North Atlantic than on the western side, while south of lat 40°N, they are lower (Fig. 2). There are no definable physical boundaries on the eastern side and currents are weak.

(6) The North Equatorial Current forms the lower part of the Gyre and is driven by the trade winds. Its intensity is dependent on the seasonal development of the trade winds. On the western side of the North Atlantic, the North Equatorial Current converges with the Guiana Current, a branch of the South Equatorial Current that crosses the equator. The Caribbean is fed by both North and South Atlantic water.

(7) There are two other features, not directly related to gyral circulation, that are important to plankton distribution. These are upwelling and undercurrents. Upwelling occurs off the coast of the greater part of West Africa. Because of upwelling, tropical waters along the eastern margin of the North Atlantic are relatively cool and nutrient rich.

The best known undercurrent is the Equatorial Undercurrent, which straddles the equator underneath the Equatorial Countercurrent. It moves in an easterly direction and is stronger than the overlying surface current. It is very shallow, less than 200 m and sometimes rises to the surface (Neumann, 1960). The other undercurrent flows poleward along the West African coast, following the zone of upwelling (Hughes and Barton, 1974).

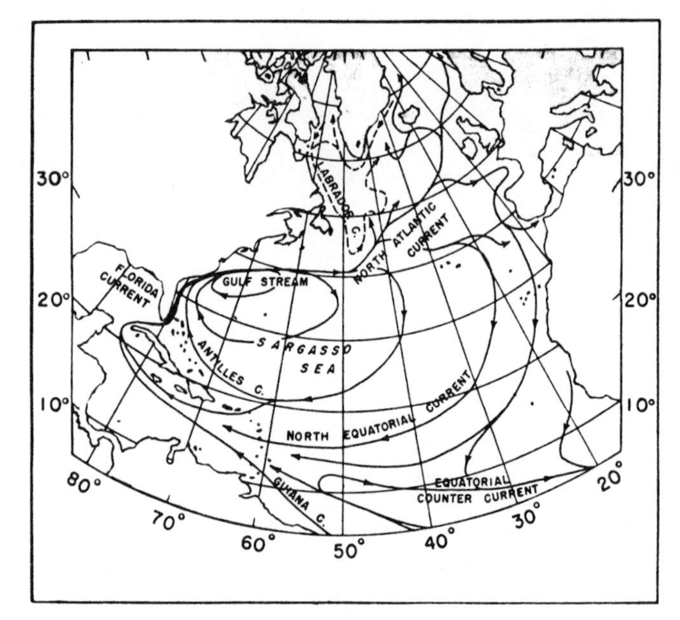

Figure 1. General circulation of the North Atlantic (after Sverdrup and others, 1946).

FAUNAL PARCELLING

Since planktonic foraminifera are moved by ocean currents, species tend to be long ranging and overlapping in distribution. At the same time, there are tendencies toward mutual associations of species that make it useful to view the ocean as being parcelled into several faunas. Of course, exacting distinctions between the faunas are difficult to make because distributional habits of species are still incompletely known. But it is possible to distinguish, with reasonable confidence, between endemic forms and expatriates. Over 40 years ago, Schott (1935) observed *Globigerina bulloides* and *Globorotalia inflata* in tropical waters near the Cape Verde Islands and did not hesitate to regard those species as intruders from the north.

An important question is how the boundaries between the faunas should be specified. Phleger and others (1953), as a first approximation, used a latitudinal basis of separation of faunas (low, middle, and high latitude). At the same time, these authors made the important observation that low-latitude forms are displaced toward the north on the western side of the North Atlantic and high-latitude forms displaced to the south on the eastern side. This observation gave the clue to the importance of circulation in the parcelling and dispersal of planktonic foraminifera. Gyral circulation generates rotary patterns of dispersal that cut across the parallels of latitude. Moreover, associated with the gyral circulation is a pronounced asymmetry in ocean climate, as indicated by the distribution of surface isotherms (Fig. 2). There is really no reason to expect a good correlation between plankton distribution and latitude, even if one assumes that species distributions are closely and directly regulated by temperature.

Various efforts have been made to use surface temperature to define boundaries of planktonic faunas. Bandy (1969), for example, grouped species in summer temperature steps of $< 2°C$, 2 to 5°C, 5 to 8°C, 9 to 18°C, and $> 18°C$. The species groups listed by Bandy represent reasonable natural faunal associations, but the temperature boundaries are not realistic. For example, in the North

Atlantic, summer surface temperatures of the slope waters exceed 18°C. Yet during the summer, as in the winter, these waters are occupied by species listed among Bandy's 2 to 5°C, 5 to 8°C, and 9 to 18°C faunas (e.g., *Globigerina bulloides, G. pachyderma, G. quinquelobaegelida,* and *Globorotalia inflata*). While faunas appear to be parcelled broadly according to ocean climate, i.e., warm versus cold water, absolute temperature boundaries of species are difficult to establish.

Phleger and others (1953) in their study of Atlantic cores found a poor correlation between species distribution and surface isotherms. In the living fauna, species show variable temperature relationships (Cifelli, 1971). It is possible that planktonic species, under certain circumstances, acclimate to temperature differences in the fashion of the benthic species *Ammonia beccarii* (Schnitker, 1974).

A more realistic view of faunal parcelling in the North Atlantic was given by Bé and Hamlin (1967), who arranged their faunas according to major climatic regions of the Ocean. They recognized arctic, subarctic, subtropical, and tropical faunas. Between the subarctic and subtropical faunas is a transition fauna. Importantly, this transition fauna is shown to spread in the eastern part of the North Atlantic, corresponding to the bifurcation of surface isotherms and splitting of the Gulf Stream. The southern limits of the transition fauna on the eastern side is placed off the Iberian Peninsula, but it is now known to continue along the African coast at least as far as the Cape Verde Islands (Cifelli and Beniér, 1976). On the western side of the Atlantic Ocean, northward displacement of species is indicated by a projection to the north of the tropical fauna.

This scheme of Bé and Hamlin (1967) begins to show the parallel between circulation and planktonic foraminiferal distribution. This parallel can be amplified as the circulation causes a mixing of species. It is not possible for faunas to exist in juxtaposition with no interaction. As viewed here the transition fauna is actually a mixed fauna composed of species from the North Atlantic Gyre and adjacent waters. Importantly, the composition of the mixed fauna changes along the course of the Gyre. The northward projection of the tropical fauna is also viewed as a mixture consisting of species from the Gyre and others displaced to the north by the clockwise movement of the circulation. Throughout the North Atlantic, species frequencies and especially numerical abundances, may vary appreciably over short distances.

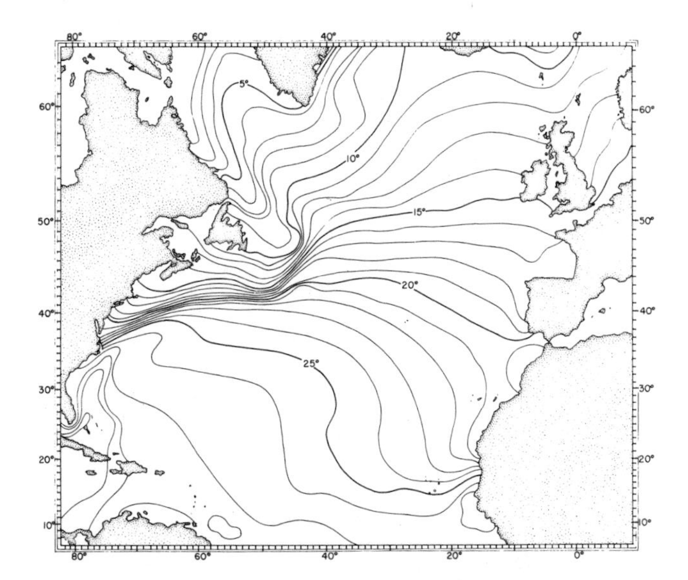

Figure 2. Surface isotherms in the North Atlantic for July (after Bé and Hamlin, 1967).

A view of parcelling in the North Atlantic is shown in Figure 3, where four faunas are recognized. There is an arctic fauna that occupies the arctic region largely between Labrador and Greenland; the limits of this fauna may be defined by the Subarctic Convergence (Ruddiman and Glover, 1975). A northern fauna is distributed to the west and north of the Gyre. A subtropical fauna is centered at the core of the Gyre. A tropical fauna occupies the equatorial waters but extends within the margins of the Gyre. Mixing between faunas (except the arctic fauna) is dictated by the clockwise movement of the Gyre, as will be discussed later.

Arctic fauna

Globigerina pachyderma (highly dominant)
Globigerina bulloides
Globigerina quinquelobaegelida

Northern fauna

Globigerina bulloides
Globigerina incompta
Globigerina quinquelobaegelida
Globorotalia inflata

Subtropical fauna

Globigerina rubescens
Globigerinella aequilateralis
Globigerinoides conglobatus
Globigerinoides ruber
Globigerinoides tenellus
Globigerinoides trilobus
Globorotalia crassaformis
Globorotalia hirsuta
Globorotalia truncatulinoides
Hastigerina pelagica
Orbulina universa

Tropical fauna

Candeina nitida
Globigerina dutertrei
Globigerina rubescens
Globigerinella aequilateralis
Globigerinoides ruber
Globigerinoides tenellus
Globigerinoides trilobus
Globorotalia crassaformis
Globorotalia hirsuta
Globorotalia menardii
Globorotalia tumida
Globorotalia ungulata
Hastigerina pelagica

Orbulina universa
Pulleniatina obliquiloculata
Sphaeroidinella dehiscens

Distinctions among these faunas are clearly of unequal importance. The relatively diverse tropical and subtropical faunas share many species in common and together stand in contrast to the species poor northern and arctic faunas. In a broad sense faunal parcelling is essentially twofold: warm water and cold water.

The boundary between the northern and subtropical faunas is well marked in the northwestern North Atlantic at the edge of the Gyre. Upon crossing the Gulf Stream from north to south there is a marked increase in diversity and a complete turnover in species. The zone of mixing is relatively narrow and contains not only northern and subtropical species but also tropical species displaced to the north by the clockwise movement of the Gyre. Most conspicuous of these expatriated tropical forms are *Globorotalia menardii* and *Pulleniatina obliquiloculata.*

It is difficult to establish boundary conditions between the subtropical and tropical faunas. A number of species seem equally represented in both gyral (Sargasso Sea) and equatorial waters, or there is an overlap in habitat. *Globigerinoides ruber* is a dominant species from the northern edge of the Gyre to the equator. *Globigerinella aequilateralis* is also an important species over a wide range of latitude but attenuates near the equator. *Globigerinoides conglobatus* and *Globorotalia truncatulinoides* occur mainly within the Gyre. *Globigerinoides trilobus* is mainly a tropical form but shows a secondarily good development in the eastern North Atlantic and Mediterranean.

Species definitely favoring the tropical habitat include: *Globorotalia menardii, G. tumida, G. ungulata, Candeina nitida, Sphaeroidinella dehiscens, Globigerina dutertrei,* and *Pulleniatina obliquiloculata.* Most of these species are scarce and seldom captured in plankton tows. The commonly occurring ones, especially *Globorotalia menardii, Pulleniatina obliquiloculata,* and *Globigerina dutertrei* are displaced to the north in the western North Atlantic and have been found in slope waters and Sargasso Sea adjacent to the Gulf Stream (Cifelli, 1965).

Diversity relationships in the western North Atlantic are affected by the northward displacement of tropical forms. Overall, the tropical fauna contains the highest number of species (15) of the four faunas recognized. However, diversities of individual assemblages captured in plankton tows are highest in the northern part of the Sargasso Sea. The reason for this is that assemblages in the northern Sargasso Sea consist of endemic subtropical species plus displaced migrants from the south and northern forms that filter across the Gulf Stream. On the other hand, exceptionally low diversities occur in the central to southern Sargasso Sea, near the core of the Gyre where neither northern or tropical forms effectively penetrate (Cifelli and Smith, 1974).

Tropical species are not displaced indefinitely along the Gyre. Somewhere east of long 65°W they virtually disappear. Very occasional representatives under exceptional circumstances may, of course, extend much farther, perhaps even completing a gyral circuit, but thus far they have gone unnoticed. The mixed zone east of about long 65°W consists of northern and subtropical mixtures.

East of long 50°W where the Gulf Stream splits into two branches and isotherms spread to the north and south (Figs. 1, 2) the zone of mixed northern and subtropical forms broadens appreciably. This is the area of the North Atlantic Current where faunal associations change rapidly over short distances. Completely northern assemblages are often found in sharp contact with mixed assemblages having appreciable numbers of subtropical species. These mixed associations, moreover, are not far removed from the arctic fauna which is found in the vicinity of the Grand Banks, at about long 40° to 50°W (Cifelli and Smith, 1970; Cifelli, 1973). The physical oceanography of this part of the ocean is still inadequately studied.

In Figure 3 the northern extension of the mixed zone is shown to occur somewhat below the British Isles. In most cases, that does represent the northern penetration of subtropical forms, but occasionally some are carried farther north.

On the eastern side of the North Atlantic the clockwise movement of the Gyre causes a displacement of northern forms to the south and a faunally mixed zone can be recognized down along the Iberian Peninsula and coast of Africa. Penetration of northern forms is remarkably deep, extending at least to lat 10°N (Cifelli and Beniér, 1976). The path along which the northern forms penetrate into tropical waters is actually a narrow one, close to the African coast. This is an area of upwelling where surface waters tend to be cooler than normal and nutrient rich. Upwelling must have an important regulatory effect on the depth of penetration of northern forms to the south.

As the northern forms penetrate into tropical waters they are found associated with tropical as well as subtropical species. Along the African coast, south of lat 20°N, the following species commonly occur together (Cifelli and Beniér, 1976):

Northern

Globigerina bulloides
Globigerina incompta
Globigerina quinquelobaegelida
Globorotalia inflata

Subtropical/Tropical

Globigerinoides ruber
Globigerinoides tenellus
Globigerinoides trilobus
Globorotalia crassaformis
Hastigerina pelagica
Orbulina universa

Tropical

Globigerina dutertrei
Globorotalia menardii
Pulleniatina obliquiloculata

Similar mixed associations are found on the western side of the North Atlantic near the Gulf Stream. One of the more important biogeographic consequences of gyral displacement is the occurrence of mixed associations of northern, subtropical, and tropical species at temperate latitudes on the western side of the North Atlantic and at tropical latitudes on the eastern side (Fig. 3). The ecologic factors favoring these associations are not understood at present.

The penetration of northern forms into tropical waters can be traced as far west as the Cape Verde Islands, close to where the North Equatorial Current is formed. From there on westward, the mixed zone consists of tropical and subtropical mixtures. However, occasional northern forms may survive the inimical conditions of the equatorial waters and complete the gyral circuit to the western side of the North Atlantic. *Globigerina bulloides* has been reported from Caribbean plankton tows and the most likely means of entry seems to be gyral circulation (Cifelli and Beniér, 1976).

SEASONAL VARIATION

Physical boundaries of North Atlantic surface waters change throughout the year. The Gulf Stream is known to oscillate in position, to meander, and to form detached eddies (Stommel, 1965), although the periodicity of movements has not been established yet. The North Equatorial Current is generated largely by the trade winds which migrate from north to south and vary in intensity seasonally (Iselin, 1936). African coastal upwelling is induced by winds from the continent and is much stronger during winter to summer than in the latter part of the year (Hughes and Barton, 1974). With such changes in physical conditions, it could not be expected that planktonic faunal boundaries would remain constant throughout the year. From a series of cruise tracks between Cape Cod and Bermuda, it was found that the faunal change from northern to subtropical was closest inshore during the fall and farthest offshore between winter and spring (Cifelli, 1962). In the area of the North Atlantic Current, boundary conditions appear to shift during the year (Cifelli and Smith, 1970). Along the West African coast the penetra-

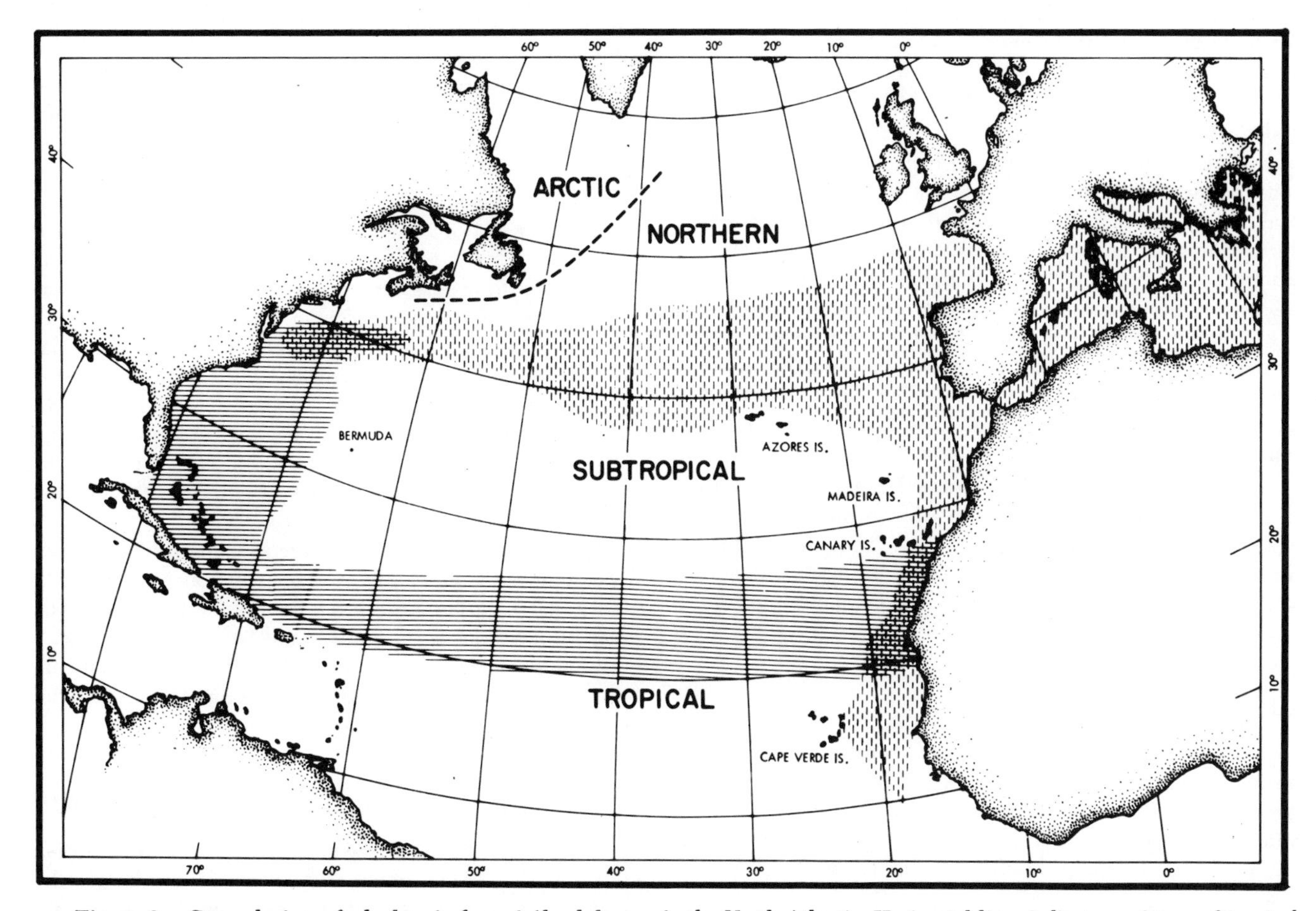

Figure 3. General view of planktonic foraminiferal faunas in the North Atlantic. Horizontal lines indicate a mixture of tropical and subtropical species. Vertical lines indicate a mixture of northern and subtropical species. Crossed lines indicate a mixture of all three kinds of species.

tion of northern forms appeared less during the fall, the time of minimum upwelling, than in the earlier parts of the year (Cifelli and Beniér, 1976).

Despite boundary changes, the faunal parcels retain their identities throughout the year. The northern fauna is not replaced by the arctic fauna during the winter nor is the subtropical fauna succeeded by the northern fauna. However, within parcels there is a seasonal succession of species, especially in the subtropical waters. Seasonal succession was first observed by Bé (1960) from monthly monitored stations near Bermuda. Bé found that *Globorotalia truncatulinoides* has a peak development in the northern Sargasso Sea during January and is dominant until about March when it declines. *Globorotalia hirsuta* is also strongly represented during the winter with a peak about February. Both of these species are poorly represented during most of the summer and fall months. *Globigerinoides ruber* and *Globigerinella aequilateralis* dominate during the summer months, but decline later in the year and are poorly represented during the winter. *Globigerinoides conglobatus* has a short developmental period extending from about October to January and is uncommon for most of the remainder of the year.

Peak development of *Globorotalia truncatulinoides* may occur later in the southern Sargasso Sea than in the northern Sargasso Sea. Cifelli and Smith (1974) observed an April peak for *G. truncatulinoides* in the southern Sargasso Sea, a time when that species had already declined in the northern Sargasso Sea.[1]

In the northern fauna, the principal seasonal change noted so far is that *Globigerina quinquelobaegelida* shows a peak development in the winter and is poorly represented during the summer. For the most part, however, the northern species do not show much evidence of seasonal cycles. Seasonality of tropical species has not been studied so far. Also, there is a lack of seasonal data from the eastern part of the Sargasso Sea. It is not known whether the cyclic development of species observed in the western part of the Sargasso Sea remains constant or changes along the course of the North Atlantic Gyre.

Glaçon and others (1971) studied seasonal variation of planktonic foraminifera in the Mediterranean from a monitored station in the Ligurian Sea. They found that *Globorotalia truncatulinoides* dominated during the fall and winter, and *Globigerina pachyderma* dominated in the spring.

THE MEDITERRANEAN

The Mediterranean fauna is, of course, completely stocked from the North Atlantic. While the opening through the Straits of Gibraltar is narrow, the flow of water is steady and insures a constant replenishment of species from the eastern edge of the Gyre. No living endemic species are known in the Mediterranean. The fauna consists of a mixture of northern and subtropical forms. Typically, this co-association of species is as follows (Cifelli, 1974):

Northern

Globigerina bulloides
Globigerina incompta
Globigerina quinquelobaegelida
Globorotalia inflata

Subtropical

Globigerinella aequilateralis
Globigerinoides ruber
Globigerinoides tenellus
Globigerinoides trilobus
Globorotalia truncatulinoides
Hastigerina pelagica
Orbulina universa

The full complement of northern species is represented. However, the subtropical species seem to be less well represented. *Globigerinoides conglobatus* and *Globorotalia crassaformis* are either rare or absent in the present Mediterranean. But these species may also be poorly represented along the eastern side of the North Atlantic. Meridional trends in species distributions are still inadequately established.

In the western part of the Mediterranean, the northern and subtropical forms occur in about equal numbers. East of the Straits of Sicily, the northern forms decrease in numbers; assemblages in the eastern part of the Mediterranean are almost wholly subtropical. *Globorotalia truncatulinoides* is also scarce in the eastern part of the Mediterranean.

Tropical forms are very scarce throughout the Mediterranean and they must follow a long, difficult route to reach it. The Straits of Gibraltar are situated at about lat 36°N, well above the northern limits of the tropical fauna in the eastern North Atlantic. Currents move from north to south on the eastern side of the North Atlantic, forming a barrier to the dispersal of tropical forms to the Mediterranean. The only apparent means of entry for the tropical forms would be a circuit around the Gyre.

Rare occurrences of *Globorotalia menardii* have been noticed recently in Mediterranean plankton tows (Cifelli, 1974). *Globigerina dutertrei* also occurs rarely. Of the other tropical species, *G. tumida* and *Pulleniatina obliquiloculata* have never been recorded from the Mediterranean. *Sphaeroidinella dehiscens* is known only from scattered occurrences from the Pliocene.

PARCELLING RELATIONSHIPS AMONG PLANKTON

Tesch (1946), in reviewing the pteropod distribution in the world oceans, noted that there are no sharply defined zoogeographic boundaries between lat 40°N and 36°S. However, between these latitudes pteropod species are not distributed indiscriminately but show some habitat

[1] In the Cifelli and Smith paper, there is an error in figure captions. Caption of text-figure 9 should be transposed with that of text-figure 11, and caption of text-figure 10 should be transposed with that of text-figure 12.

preferences. Tesch characterized the distributional habits of pteropods as follows: (A) species favoring equatorial waters; (B) species spreading beyond equatorial waters and populating all warmer areas; (C) species that are more subtropical than tropical, but not excluded from tropical waters; (D) a single species (*Limacina retroversa*) inhabiting temperate waters; and (E) a single species (*Limacina helicina*) adapted to polar water. The fact that species spread far beyond their preferred habitat was also stressed.

Here one can recognize a remarkably close parallel with the planktonic foraminifera. The pteropods are divided into cold- and warm-water parcels. Also, there are two cold-water parcels, each of which is characterized by very low diversity (single species). There are also two warm-water parcels that are relatively diverse and have much in common, like the foraminiferal subtropical and tropical parcels. Like foraminifera, pteropod species extend far beyond their preferred habitats and mix with species of other faunas, with the pattern of mixing dictated by oceanic circulation. Boundary relations cannot be compared directly, but the latitudes 40°N and 36°S given by Tesch (1946) suggest that the major pteropod boundaries occur at the margins of the Gyres, as is the case with the foraminifera. One would also expect pteropod displacements to parallel those of the foraminifera, but direct information is not available.

The parallels between parcelling of pteropods and foraminifera seem remarkably close considering that they are such different kinds of organisms. Pteropods are not only larger and more motile, but also more complex, with different feeding mechanisms and, one would guess, different food requirements. It would be interesting to see whether other planktonic organisms show parallel relationships. The habitats of the open ocean may be divided and utilized according to a consistent scheme, even though the ecologic reasons for the arrangement remain obscure.

FAUNAL PARCELLING AND HISTORICAL BIOGEOGRAPHY

Insofar as climatic differentiation existed in the past, it seems permissible to assume that planktonic foraminifera then, as today, were parcelled into cold- and warm-water faunas. Also, circulation may have been stronger or weaker in the past but there must always have been some clockwise motion of water in the Northern Hemisphere and counterclockwise motion in the Southern Hemisphere. Climatic differentiation should be associated with thermal gradients at the margins of the gyres where one would expect to find the greatest faunal change. Low diversity faunas should occur on the cold-water side of the gyre, high diversity faunas on the warm-water side. Importantly, gyral circulation would always result in mixing and displacement of species from their preferred habitats.

Viewed against this background, something puzzling emerges from the historic record of planktonic foraminifera. The record reveals two major radiations during the Cenozoic, one in the Paleogene, and another in the Neogene. Preceding each radiation, during the Danian and the Oligocene, there were severe reductions. Both reductions involved an elimination of subtropical and tropical forms, while the radiations involved a restoration of them (Cifelli, 1969). It is the reduced faunas of the Danian and Oligocene that are perplexing, because they display a lack of morphotypic differentiation and low taxonomic diversity. Some regionalism in species distribution exists, but there are no recognizable warm-water forms in these faunas, even in equatorial regions. Danian and Oligocene faunas, everywhere, are morphotypically analogous to modern cold-water faunas. Shackleton and Kennett (1975), from isotopic measurements, conclude that conditions in high latitudes during the Oligocene were freezing or near freezing. However, there still remains the larger question as to where the subtropics and tropics of the open ocean were at this time and during the Danian. The planktonic foraminiferal faunas give no evidence of them and suggest cool, climatically undifferentiated oceans. In contrast, Late Cretaceous and Paleogene faunas contain well-developed, warm-water morphotypes that are widely distributed and suggest greater expanses of warm water than exists today (Bandy, 1964).

The second radiation started at the beginning of the Miocene, and by middle Miocene there was a good development of tropical and subtropical forms. Moreover, these forms are widely distributed, and as in the Late Cretaceous and Paleogene, indicate a much greater expanse of warm water than exists today. The modern tropical species *Globorotalia menardii* occurs commonly in the Mediterranean from middle Miocene, when it first evolves, to the end of the Miocene. Assuming that habitat preferences of species do not change through time, *G. menardii* could migrate freely to the Mediterranean only if the tropical limits of the North Atlantic extended far beyond what they are now. Also, this species was recorded from late Miocene at lat 57°30′N, in the core at Deep Sea Drilling Site 116, from the Hatton-Rockall Basin, on the eastern side of the North Atlantic near the British Isles (Poore and Berggren, 1975). In that late Miocene core section, *G. menardii* occurs in association with the modern cold-water species *Globigerina bulloides* and *G. pachyderma*, along with modern warm-water, though not strictly tropical, species *Globigerinoides ruber*, *G. tribobus*, and *Orbulina universa*. This kind of mixed association is suggestive of waters near the margin of the Gyre. However, the northernmost Miocene limits of *G. menardii* and other warm-water forms in the North Atlantic are still not determined. Completely northern Miocene faunas may occur in the North Sea region, but those faunas are poorly known.

Beginning in the early Pliocene, the dispersal of *Globorotalia menardii* becomes restricted. This restriction seems to reflect a retraction of the exaggerated expanse of tropical-subtropical waters and a transition from Miocene to modern conditions. No longer does *G. menardii* migrate freely to the Mediterranean. After the Miocene, Mediterranean occurrences of this species are rare and scattered; nor do the other tropical species which evolve during the Pliocene become established in the Mediterranean.

Moreover, *Globorotalia menardii* does not appear in the Pliocene section of the Hatton-Rockall Basin core or the core taken from the Bay of Biscay (Berggren, 1972). It does occur in Pliocene cores taken from the vicinity of the Cape Verde Islands (Parker, 1973). Therefore, it would appear that northern limits of *G. menardii* on the eastern side of the North Atlantic fall south of the entrance to the Mediterranean during the Pliocene as they do today. Also, there is a clockwise displacement to the north on the western side but it was much greater during the early Pliocene than it is today (Fig. 4). Poore and Berggren (1974) have recorded *G. menardii* and other tropical species from the lower Pliocene section in a core taken from the Labrador Sea. Poore and Berggren note that the lower Pliocene section consists of a mixture of northern, subtropical, and tropical species—the kind of mixed assemblage found near the present Gulf Stream. Therefore, the western margin of the Gyre was still perhaps 8° north of where it is today. However, the change that occurred in the early Pliocene marked the beginning of a long-termed condition. Climate has fluctuated since the early Pliocene but never again has warm gyral water expanded sufficiently to allow migration of tropical species to the Mediterranean or the higher latitudes of the North Atlantic.

Just what effect tectonics may have had on the development of Cenozoic biogeography is difficult to judge from the standpoint of planktonic foraminifera. The fossil record suggests that planktonic foraminifera responded to some kind of recurring climatic phenomenon. Undirectional plate movements during the Cenozoic, therefore, merely introduce an additional complex variable to be dealt with. There appear to have been long-term cycles in the evolution of ocean climate and circulation. Expansions of warm gyral water are suggested in the Late Cretaceous, Paleogene, and Miocene; contractions at the beginning and middle of the Cenozoic. The contractions may have involved an elimination of warm gyral water. Another contraction seems to have started at the beginning of the Pliocene although there have been no signs since then of disintegration of the warm gyral water.

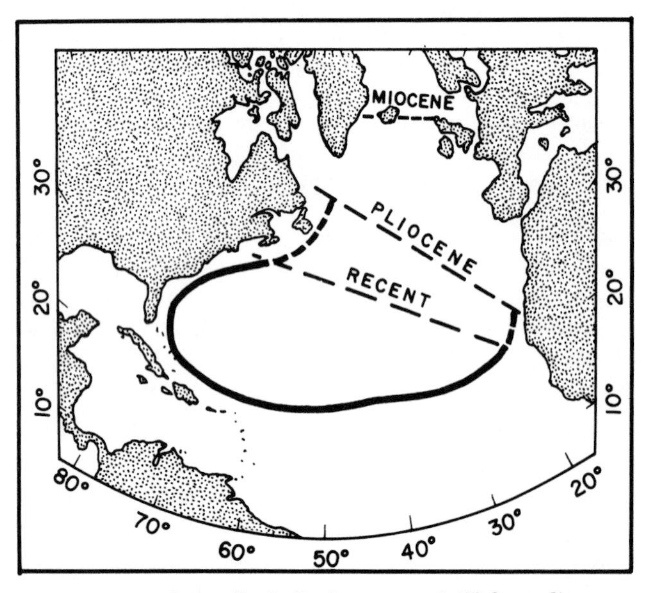

Figure 4. Latitudinal displacement of *Globorotalia menardii* in the late Cenozoic.

REFERENCES

Bandy, O. L., 1964, Cenozoic planktonic foraminiferal zonation: Micropaleontology, v. 10, p. 1-17.

——— 1969, Relationships of Neogene planktonic foraminifera to paleoceanography and correlation, *in* Brönnimann, P., and Renz., H. H., eds., Internat. Conf. Planktonic Microfossils, 1st, Geneva 1967, Proc., p. 68-81.

Bé, A. W. H., 1960, Ecology of Recent planktonic foraminifera. Pt. 2. Bathymetric and seasonal distributions in the Sargasso Sea off Bermuda: Micropaleontology, v. 6, p. 373-392.

Bé, A. W. H., and Hamlin, W. H., 1967, Ecology of Recent planktonic foraminifera in the North Atlantic during the summer of 1962: Micropaleontology, v. 17, p. 31-42.

Berggren, W. A., 1972, Cenozoic biostratigraphy and paleobiogeography of the North Atlantic, *in* Davies, Thomas A., ed., Initial Reports of the Deep Sea Drilling Project, Vol. 12: Washington D. C., U. S. Govt. Printing Office, p. 965-1001.

Cifelli, R., 1962, Some dynamic aspects of the distribution of planktonic foraminifera in the western North Atlantic: Jour. Marine Research, v. 20, p. 201-213.

——— 1965, Planktonic foraminifera from the western North Atlantic: Smithsonian Misc. Colln., v. 148, p. 1-36.

——— 1969, Radiation of Cenozoic planktonic foraminifera: Systematic Zoology, v. 18, p. 154-168.

——— 1971, On the temperature relationships of planktonic foraminifera: Jour. Foraminiferal Research, v. 1, p. 170-177.

——— 1973, Observations of *Globigerina pachyderma* (Ehrenberg) and *Globigerina imcompta* Cifelli from the North Atlantic: Jour. Foraminiferal Research, v. 3, p. 157-166.

——— 1974, Planktonic foraminifera from the Mediterranean and adjacent Atlantic waters (cruise 49 of the ATLANTIS II, 1969): Jour. Foraminiferal Research, v. 4, p. 171-183.

Cifelli, R., and Beniér, C., 1976, Planktonic foraminifera from near the West African coast and a consideration of faunal parcelling in the North Atlantic: Jour. Foraminiferal Research, v. 6, p. 258-273.

Cifelli, R., and Smith, R. K., 1970, Distribution of planktonic foraminifera in the vicinity of the North Atlantic Current: Smithsonian Contr. Paleobiology, no. 4, p. 1-5.

——— 1974, Distributional patterns of planktonic foraminifera in the western North Atlantic: Jour. Foraminiferal Research, v. 4, p. 112-125.

Glaçon, G., Verngaud Grazzini, C., and Sigal, M. J., 1971, Premiers résultats d'une série d'observations saisonières des Foraminifères du plancton Méditerraneen, *in* Farinacci, A., ed., Internat. Conf. Planktonic Microfossils, 2nd, Roma 1970, Proc., p. 611-648.

Hughes, P., and Barton, E. D., 1974, Stratification and water mass structure in the upwelling area off northwest Africa in April/May 1969: Deep-Sea Research, v. 21, p. 611-628.

Iselin, C. O'D., 1936, A Study of the circulation of the western North Atlantic: Papers Phys. Oceanogr. and Meteorology, v. 4, p. 1-101.

Neumann, G., 1960, Evidence for an equatorial undercurrent in the Atlantic Ocean: Deep-Sea Research, v. 6, p. 328-334.

Parker, F. L., 1973, Late Cenozoic biostratigraphy (planktonic foraminifera) of tropical deep-sea sections: Rev. Española Micropaleontologia, v. 5, p. 253-289.

Phleger, F. B., Parker, F. L., and Pierson, J. F., 1953, North Atlantic foraminifera, *in* Sediment cores from the North Atlantic Ocean: Swedish Deep-Sea Exped. Rept., v. 7, p. 1-122.

Poore, R. Z., and Berggren, W. A., 1974, Pliocene biostratigraphy of the Labrador Sea: Calcareous plankton: Jour. Foraminiferal Research, v. 4, p. 91-108.

——— 1975, Late Cenozoic planktonic foraminiferal biostratigraphy and paleoclimatology of Hatton-Rockall Basin: DSDP Site 116: Jour. Foraminiferal Research, v. 5, p. 270-293.

Ruddiman, W. F., and Glover, L. K., 1975, Subpolar North Atlantic circulation at 9300 yr. BP: Faunal evidence: Quaternary Research, v. 5, p. 361-389.

Shackleton, N. J., and Kennett, J. P., 1975, Paleotemperature history of the Cenozoic and the initiation of Antarctic glaciation: Oxygen and carbon isotope analyses in DSDP Sites 277, 279 and 281, *in* White, Stan M., ed., Initial Reports of the Deep Sea Drilling Project, Vol. 29: Washington, D. C., U. S. Govt. Printing Office, p. 743-755.

Schnitker, D., 1974, Ecotypic variation in *Ammonia beccarii* (Linné): Jour. Foraminiferal Research, v. 4, p. 216-223.

Schott, W., 1935, Die foraminiferen in den aequitorialen Teil des Atlantischen Ozeans: Deutsche Atlantischen Exped. Forsch. u. Vormess. "Meteor" 1925-1927, v. 3, pt. 3, p. 43-134.

Stommel, H., 1965, The Gulf Stream (2nd ed.): Berkeley, Univ. California Press, 248 p.

Sverdrup, H. U., Johnson, M. W., and Fleming, R. H., 1946, The oceans, their physics, chemistry and general biology: New York, Prentice Hall, 1060 p.

Tesch, J. J., 1946, The thecostomous pteropods. I. The Atlantic: Carlsberg Found. Oceanogr. Exped. 1928-1930: Dana Rept. no. 28, p. 1-82.

Worthington, L. V., 1962, Evidence for a two gyre circulation in the North Atlantic: Deep-Sea Research, v. 9, p. 51-57.

The Architectural Geography of Some Gastropods

GEERAT J. VERMEIJ, *Department of Zoology, University of Maryland, College Park, Maryland 20742*

ABSTRACT

A functional analysis of shell architecture in 5 circum-tropical families of low intertidal, open-surface gastropods from rocky shores reveals that predator resistance is generally greater in the tropical Indo-West Pacific and eastern Pacific than in the tropical Atlantic or in temperate regions. Contrary to earlier assertions, low spires confer predator resistance only when associated with small, elongate, and(or) dentate apertures. In groups where elongate or dentate apertures do not occur, Pacific and Indian Ocean representatives tend to be higher-spired than ecologically comparable tropical Atlantic or temperate zones species. In other groups, spires are lower among Indian and Pacific Ocean forms. The high incidence of antipredatory features among Panamic as compared to Caribbean gastropods is due less to the recent immigration of armored Indo-West Pacific species than to the retention and radiation of late Tertiary New World elements.

INTRODUCTION

In an earlier paper (Vermeij, 1974), I presented data on the incidence of certain architectural shell features in gastropod assemblages occurring in different parts of the marine tropics. These characteristics, which were interpreted as enhancing their possessors' predator resistance, include strong (usually spinose or nodose) external sculpture, teeth occluding entrance into the aperture, low spires (apical half-angle greater than 45°), and elongate apertures (aperture length : aperture width greater than 2.5). Strong sculpture not only strengthens the shell as a whole, but also renders it effectively larger and, therefore, more difficult to grasp or swallow. Low spires make it difficult for crabs and other predators to grasp and hold the shell, and reduce the mechanical advantage of the predator if, as is frequently attempted by crabs, the shell is broken from behind. Elongate and occluded apertures thwart attempts by predators to remove soft parts from the shell directly without breaking it. Teeth on the outer lip also increase the resistance of the lip to cracking.

Among open-surface, low intertidal gastropods from rocky shores, the incidence of antipredatory features has proved to be significantly higher in assemblages from the Indo-West Pacific and to a lesser extent the eastern Pacific than in ecologically comparable assemblages from either side of the tropical Atlantic (Vermeij, 1974). These curious biogeographical differences result not only from the occurrence in the low intertidal of well-armored groups restricted to the Indo-West Pacific or Panamic regions, but also from the higher proportion of such armored circum-tropical genera as *Conus* and *Cypraea* in Indian and Pacific Ocean faunas as compared with those of the Atlantic. Even beyond this, however, Indo-West Pacific and eastern Pacific representatives of circum-tropical families or genera are generally more heavily armored than are their ecological and taxonomic counterparts in the tropical Atlantic.

In this paper I provide data supporting these points and comment on the origin of the observed biogeographical pattern.

MATERIALS AND METHODS

During the past nine years I have collected shelled molluscs from a large number of tropical and temperate rocky shores. For each gastropod species at each site, up to six specimens were measured with Vernier calipers to obtain three parameters which together give a convenient description of overall shell form (for a detailed discussion on methods of measurement and derivations of parameters see Vermeij, 1973a, 1973b). These parameters are: (1) A/2, apical half-angle; (2) W, expansion rate of the whorls; and (3) S, ratio of aperture length : aperture width. Spire height (= apical half-angle) and expansion rate are determined from the point on the generating curve or aperture farthest from the axis of coiling. The ratio S is determined by dividing aperture length (distance from posterior end of aperture to tip of siphonal canal) by aperture width (distance from outer lip to the opposite edge of the smooth inductural deposits). Variation in sculptural intensity, erosion or encrustation of the shell in general, and the apical portion of the spire in particular, and variations in the extent of inductural deposition introduce considerable variance in the parameters even within populations; but the differences noted in the text between members of groups from different regions are significant at the 0.05 level by the Mann-Whitney U Test.

The various shell parameters may be qualitatively interpreted as follows: increasing A/2 values denote lower spires; higher W values usually indicate larger apertures,

Table 1. Architecture of some open-surface, low intertidal Thaididae from rocky shores.

Species	Locality	A/2[1]	S[1]	W[1]	Sc[1]	T[1]
Indo-West Pacific						
Morula granulata Duclos	Aimeliik, Palau	34.3	2.37	1.49	+	+
	Nosy Komba, Madagascar	36.5	2.66	1.62	+	+
	Nyali, Kenya	31.1	2.49	1.39	+	+
	Pago Bay, Guam	31.4	2.45	1.49	+	+
	The Fjord, Sinai	38.2	2.37	1.62	+	+
M. uva Roding	Pago Bay, Guam	36.3	2.54	1.54	+	+
	The Fjord, Sinai	34.6	2.33	1.52	+	+
M. triangulata Pease	Pago Bay, Guam	39.2	3.34	1.73	+	+
M. biconica Blainville	Pago Bay, Guam	34.3	2.93	1.77	+	+
M. cf. *biconica* Blainville	The Fjord, Sinai	36.9	2.46	1.49	+	+
M. fiscella Gmelin	Pago Bay, Guam	31.4	2.26	1.55	–	–
M. sp. (inshore)	Pago Bay, Guam	32.9	1.84	1.54	–	–
Drupa clathrata Lamarck	Pago Bay, Guam	55.7	1.54	3.44	+	+
D. rubusidaeus Roding	Pago Bay, Guam	54.5	1.67	2.08	+	+
D. grossularia Roding	Agat, Guam	63.8	1.14	1.90	+	+
D. morum Roding	Pago Bay, Guam	50.9	1.41	2.09	+	+
D. ricinus L.	Pago Bay, Guam	56.3	1.62	3.10	+	+
	Aimeliik, Palau	62.8	1.56	3.70	+	+
D. r. hadari Em. and Cern.	The Fjord, Sinai	54.2	1.66	1.95	+	+
Drupella elata Blainville	Agat, Guam	27.4	2.09	1.43	+	+
Thais armigera Lamarck	Aimeliik, Palau	42.8	1.61	1.52	+	–
	Pago Bay, Guam	44.2	1.72	1.37	+	–
T. tuberosa Roding	Pago Bay, Guam	45.6	1.67	1.76	+	–
T. intermedia Kiener	Pago Bay, Guam	29.8	1.55	1.43	+	–
Eastern Pacific						
T. melonis Duclos	Playa de Panama, Costa Rica	52.9	1.48	1.99	–	–
	Panama City, Panama	43.2	1.66	1.49	–	–
T. triangularis Blainville	Playa de Panama, Costa Rica	49.7	1.55	2.33	+	–
	Panama City, Panama	55.7	1.56	2.60	+	–
T. speciosa Val.	Playa de Panama, Costa Rica	57.8	1.41	1.69	+	–
T. biserialis Blainville	Punta Caldera, Costa Rica	42.1	1.65	2.01	–	–
	Panama City, Panama	32.3	1.63	1.59	–	–
Cymia tectum Wood	Panama City, Panama	40.5	2.10	1.71	+	+
Western Atlantic						
Thais rustica Lamarck	Discovery Bay, Jamaica	32.7	1.78	1.55	–	–
	Cahuita, Costa Rica	32.7	1.77	1.71	–	–
T. deltoidea Lamarck	Discovery Bay, Jamaica	42.4	1.70	1.54	+	–
	Cahuita, Costa Rica	38.5	1.69	1.67	+	–
'Morula' nodulosa C. B. Adams	Rio Bueno, Jamaica	24.5	2.35	1.31	–	+
Eastern Atlantic						
'M.' nodulosa C. B. Adams	Takoradi, Ghana	25.2	2.83	1.36	–	+
	Dakar, Senegal	25.5	2.23	1.35	–	+
Thais haemastoma Lamarck	Takoradi, Ghana	43.7	1.63	1.76	+	–
	Dixcove, Ghana	30.0	1.63	1.63	–	–
	Dakar, Senegal	33.0	1.63	1.75	–	–
	Ein Hat' Chelet, Israel	28.2	1.65	1.62	–	–
T. nodosa nodosa L.	Takoradi, Ghana	58.1	1.15	2.57	+	–
Southwest South America						
T. chocolata Duclos	Pucusana, Peru	36.9	1.75	1.73	–	–
North Atlantic						
Nucella lapillus L.	Plymouth, England	30.6	1.77	1.51	–	–
	Plymouth, England	35.1	1.71	1.65	–	–
	Plymouth, England	28.6	1.76	1.54	–	–
	Boothbay Harbor, Maine	28.3	1.70	1.50	–	–
	Boothbay Harbor, Maine	31.2	1.51	1.73	–	–

[1] A/2: Apical half-angle of shell in degrees; S: Ratio of aperture length : aperture width; W: Expansion rate of the whorls; Sc: Presence (+) or absence (–) of strong nodose or spinose external shell sculpture; T: Presence (+) or absence (–) of occluding teeth in the aperture.

although strong teeth may, in practice, reduce the opening to a narrow slit, as in *Drupa*. The change from a round or broadly ovate to a narrower, longitudinally elongate aperture, is reflected by an increase in the S ratio.

RESULTS

A careful inspection of Table 1 reveals that the Thaididae generally reflect the geographical patterns of shell architecture exhibited by gastropod assemblages as a whole. The genera *Drupa, Drupella,* and *Morula,* usually characterized by strong sculpture and well-developed teeth in the aperture, are limited to the Indo-West Pacific region. Species of *Morula* lacking apertural teeth large enough to occlude the aperture, such as a Guamanian species probably belonging to the *M. fusconigra* group, are restricted to inshore rocky habitats. Within the genus *Thais,* the combination of rather low spires and low W values render Indo-West Pacific species better armored than species in the tropical Atlantic. In the Panamic region, a low-spired, large-apertured, spinose species (*T. triangularis*) occurs side-by-side on sand-free rocks with heavily armored, smooth (*T. melonis*) and knobbed (*T. speciosa*) forms with thick, low-spired shells. The highly variable *T. biserialis,* which tends to possess a higher spire than other eastern Pacific *Thais,* is characteristically found on sand-scoured surfaces where other molluscs also show a reduction in armor (Vermeij, 1974).

The West African *T. nodosa nodosa* is morphologically like *T. triangularis* but attains larger sizes and has an even larger aperture. This species, characteristic of wave-exposed shores, co-occurs with the more wide-ranging *T. haemastoma,* which in spire height and development of knobs is extraordinarily variable. Cold-temperate Thaididae normally possess uninterrupted spiral ridges rather than axially distinct nodes or spines. While shell shape is highly variable within species, a combination of low spire and low W among temperate species is not achieved to the same degree that it is in tropical forms, nor are apertural teeth so large as to give the shell opening a slit-like appearance.

Experimental work (Ebling and others, 1964; Kitching and others, 1966; Kitching and Lockwood, 1974) on temperate Thaididae has demonstrated that specimens with large apertures and low spires found on exposed shores are more vulnerable to crab predation than the thicker, higher-spired, and smaller-apertured forms of more sheltered sites. This greater vulnerability of lower-spired snails to predation stems not only from the greater ease with which crabs can extract the soft parts from the intact shell, but also from the relatively larger surface area of the shell not supported from within by the walls of the preceding whorls. It is thus evident that low spires alone cannot impart immunity to predation; rather, the low spire must be accompanied by a small and(or) elongate aperture [low W and(or) high S].

Like the Thaididae, the Conidae (Table 2) and Fasciolariidae (Table 3) conform to the pattern of low spires and narrower apertures in Pacific and Indian Ocean species as compared to those from the tropical Atlantic. Especially among the species of *Conus* examined, a high spire is often associated with a wider aperture.

At first glance, the open-surface Trochidae (Table 4) appear to constitute an exception to the prevailing pattern of greater predator resistance in the Pacific and Indian Oceans. Indo-West Pacific species of *Trochus* and *Tectus,* contrary to expectation, are higher-spired than ecologically similar Caribbean species of *Tegula* and *Cittarium;* the two Panamic *Tegula* species examined are somewhat intermediate. Open-surface trochids, with the exception of the high intertidal *Monodonta punctulata* in Senegal, are absent on the West African shores I have visited.

All trochids have broadly ovate to nearly circular apertures, and none possess apertural teeth. In lower-spired trochids, W values are generally larger than among higher-spired forms, reflecting a larger aperture. In addition, the low-spired species tend to possess a well-marked umbilicus (e.g., the West Indian *Tegula excavata* and *Cittarium pica,* and the Red Sea lagoon-inhabiting *Trochus erythraeus*), while higher-spired species are either narrowly umbilicate or lack an umbilicus altogether.

Experiments in which various molluscivorous crabs in Guam were fed living *Trochus niloticus* and hermited Jamaican *C. pica,* indicate that the latter species achieves immunity from crab predation at a much larger size than the former species (Vermeij, 1976). This difference may result both from the presence in *C. pica* of an umbilicus, which mechanically weakens the shell, and from the much greater area of the body whorl not in contact with, or deriving added strength from, previous whorls. Thus, the lower spires of Caribbean and at least some temperate zone trochids (Table 4) may reflect less rather than greater predator resistance, as would have been predicted from a consideration of spire height alone.

With the exception of certain large and strongly knobbed species of *Cerithium* in the Indo-West Pacific (e.g., *C. nodulosum* and *C. echinatum*), rocky-shore cerithiids exhibit few adaptations obviously associated with predator resistance. There are no striking differences in form between small intertidal cerithiids from the various tropical oceans. Most of the smaller species are probably annuals, or in any case have short life-spans (Houbrick, 1974a), and, therefore, fall into the category of "weedy" species. In Guam, the largest available *C. columna* (34 mm in length) is still vulnerable to predation by the large (65 mm wide) crab, *Eriphia sebana.* As has already been suggested for plants (Cates, 1975; Cates and Orians, 1975; Levin, 1975), weedy species may exhibit less predator resistance, relying primarily on high fecundity and dispersability for survival. Thus, regional uniformity in shell architecture among small cerithiids is not surprising. The large Indo-Pacific species of *Cerithium* are either known or suspected to be perennials (M. Yamaguchi, personal commun.).

DISCUSSION

In general, the patterns of shell architecture seen in a number of circum-tropical families reflect the patterns of whole gastropod assemblages, although regional variation in, and interpretation of, spire height are more complex than originally believed (Vermeij, 1974). Low spires will be effective against crushing predators only if the prey's aperture is small (low W) or if the outer lip of a large aperture is beset with strong teeth. Predatory crabs of the families Xanthidae and Parthenopidae in Guam very often leave the outer lip of a prey snail intact if the shell is small enough for the spire to be broken off. At prey sizes too great for the spire to be damaged, the lip or base of the shell will generally be attacked. This behavior, which has also been observed in some temperate portunid and grapsid crabs (Ebling and others, 1964; Kitching and others, 1966; Kitching and Lockwood, 1974), is observable on prey ranging in shape from high-spired *Cerithium* through

Table 2. Architecture of some species of *Conus* from open rocky surfaces.

Species	Locality	A/2[1]	S[1]	W[1]
Indo-West Pacific				
Conus distans Hwass	Aimeliik, Palau	46.6	9.15	1.54
C. ebraeus L.	Nosy-Be, Madagascar	48.1	7.80	1.56
	Pago Bay, Guam	47.9	7.66	1.50
C. sponsalis Hwass	Pago Bay, Guam	46.3	7.00	1.33
	The Fjord, Sinai	46.3	7.35	1.31
C. miles L.	Agat, Guam	52.3	7.60	1.24
C. musicus Hwass	Nosy-Be, Madagascar	49.2	6.48	1.29
C. rattus Hwass	Pago Bay, Guam	46.4	7.27	1.26
C. imperialis L.	Pago Bay, Guam	55.1	8.48	1.60
C. vitulinus Hwass	Pago Bay, Guam	51.1	8.98	1.28
C. chaldaeus Roding	Pago Bay, Guam	48.7	6.41	1.36
C. lividus Hwass	The Fjord, Sinai	57.4	7.71	1.54
C. frigidus Hwass	Pago Bay, Guam	50.5	6.63	1.24
	Ras Muhamad, Sinai	48.3	7.50	1.22
Eastern Pacific				
C. nux Broderip	Playa de Panama, Costa Rica	49.8	6.03	1.45
C. princeps L.	Playa de Panama, Costa Rica	51.6	6.49	1.29
Western Atlantic				
C. regius Gmelin	Fort Point, Jamaica	41.0	7.19	1.25
C. mus Hwass	Rio Bueno, Jamaica	43.1	6.98	1.30
	Cahuita, Costa Rica	45.4	6.45	1.19
Eastern Atlantic				
C. mercator L.	Dakar, Senegal	42.2	6.69	1.41
C. ventricosus	Ein Hat' Chelet, Israel	42.8	6.02	1.42

[1] A/2: Apical half-angle of shell in degrees; S: Ratio of aperture length : aperture width; W: Expansion rate of the whorls.

Table 3. Architecture of some Fasciolariidae from open rocky surfaces.

Species	Locality	A/2[1]	S[1]	W[1]	Sc[1]
Indo-West Pacific					
Latirolagena smaragdula Lamarck	Aimeliik, Palau	29.8	2.13	1.61	–
Latirus barclayi Reeve	Pago Bay, Guam	30.2	2.42	1.44	+
L. polygonus Gmelin	Nosy Komba, Madagascar	29.3	2.96	1.45	+
L. polygonoides Kiener	Eilat, Israel	21.8	2.82	1.31	+
Eastern Pacific					
Opeatostoma pseudodon Burrow	Playa de Panama, C. R.	42.1	1.77	1.50	–
Leucozonia cerata Wood	Playa de Panama, C. R.	23.6	2.39	1.28	+
Western Atlantic					
L. leucozonalis Lamarck	Rio Bueno, Jamaica	24.4	2.21	1.37	–
L. nassa Gmelin	Cahuita, Costa Rica	26.6	2.17	1.33	–
L. ocellata Gmelin	Fort Point, Jamaica	26.3	2.07	1.37	–

[1] A/2: Apical half-angle of shell in degrees; S: Ratio of aperture length : aperture width; W: Expansion rate of the whorls; Sc: Presence (+) or absence (–) of strong nodose or spinose external shell sculpture.

moderately high-spired *Trochus* and *Thais* to low-spired *Conus* and *Drupa*. The advantage of a low spire in preventing apical fracture in relatively large prey would be offset in species with wide unarmed apertures by the disadvantage of providing crabs and other predators with an unrestricted opening from which to pick out soft parts or with a large surface of unsupported body whorl which may be peeled back to expose the flesh. The higher incidence of low spires among Indo-West Pacific and Panamic as compared to tropical Atlantic rocky-shore gastropods thus appears to be functionally tied to the higher incidence of elongate and dentate apertures. Groups which for various reasons cannot or do not produce apertural teeth or longitudinally elongate apertures (Trochidae, Turbinidae, Cerithiidae) exhibit higher spires in the Pacific and Indian Oceans than in other comparable tropical and temperate habitats.

The origins of the regional differences in tropical shell architecture, and the length of time over which these differences have endured, still remain obscure. Many well-armored genera have apparently always been restricted to the Indo-West Pacific region. These include the thaidids *Drupa, Drupella,* and *Morula* (Emerson and Cernahorsky, 1973); the subgenus *Cerithium s. s.,* which includes the large nodose species (Houbrick, 1974b); and the strombid genus *Lambis* (Abbott, 1961). Together with the stability of the Indo-West Pacific fauna during the Tertiary (Fell, 1967), this geographical restriction suggests that predator resistance has always been high among molluscs in the Pacific and Indian Oceans during much of the Cenozoic.

In the eastern Pacific, recent immigrations from the west have brought a modest number of Indo-West Pacific reef-associated fishes, scleractinian corals, reef molluscs, and coral-associated decapod crabs to the offshore islands and to a lesser extent the mainland of America (Emerson, 1967; Salvat and Ehrhardt, 1970; Porter, 1972; Garth, 1974; Dana, 1975). The Indo-West Pacific molluscan element in the Panamic fauna contains a high proportion of species with elongate or strongly dentate apertures (e.g., *Cypraea moneta, C. isabella, Conus ebraeus, Morula uva, Drupa morum, D. ricinus, Strigatella litterata, Nerita plicata*). In spite of this recent armored invasion, the generally high morphological predator resistance of eastern Pacific gastropods relative to those in the Atlantic is due largely to endemic groups and to species whose closest affinities lie with extant Caribbean forms. Examples include *Tegula, Jenneria, Anachis, Thais, Cymia,* and *Muricanthus.* The extinctions which decimated the Caribbean fauna of

Table 4. Architecture of some Trochidae from low intertidal, open rocky surfaces.

Species	Locality	A/2[1]	S[1]	W[1]	Sc[1]
Indo-West Pacific					
Trochus niloticus L.	Pago Bay, Guam	34.1	0.905	1.43	–
	Aimeliik, Palau	30.6	0.874	1.39	–
T. ochroleucus Gmelin	Pago Bay, Guam	33.1	0.954	1.35	–
T. virgatus Gmelin	The Fjord, Sinai	25.7	0.963	1.32	–
T. erythraeus Brocchi	Eilat, Israel	37.6	0.957	1.52	–
Tectus pyramis Born	Aimeliik, Palau	20.9	0.771	1.23	–
T. triserialis Lamarck	Koror, Palau	26.3	0.970	1.28	+
T. dentatus Forskal	The Fjord, Sinai	30.3	0.784	1.26	+
Eastern Pacific					
Tegula panamensis Phil.	Panama City, Panama	41.5	1.13	1.78	–
T. pellisserpentis Wood	Naos Island, Canal Zone	28.5	1.12	1.40	–
Western Atlantic					
T. viridula Gmelin	Cahuita, Costa Rica	37.1	0.965	1.57	–
T. excavata Lamarck	Cahuita, Costa Rica	38.4	0.863	1.53	–
Cittarium pica L.	Fort Point, Jamaica	35.5	0.885	1.99	–
	Cahuita, Costa Rica	42.0	1.00	1.99	–
Southwest South America					
Tegula atra Lesson	Montemar, Chile	41.8	0.817	1.67	–
	Pucusana, Peru	39.2	0.819	1.66	–
T. tridentata Pot. and Mich.	Montemar, Chile	36.6	1.07	1.59	–
Northeast Pacific					
T. funebralis A. Adams	La Jolla, California	36.0	1.08	1.64	–
T. aureotincta Forbes	La Jolla, California	33.7	1.07	1.50	–
T. eiseni Jordan	La Jolla, California	33.2	1.06	1.63	–
Northeast Atlantic					
Gibbula cineraria L.	Plymouth, England	47.0	1.14	1.75	–
G. umbilicalis da Costa	Plymouth, England	41.0	1.26	1.71	–

[1] A/2: Apical half-angle of shell in degrees; S: Ratio of aperture length : aperture width; W: Expansion rate of the whorls; Sc: Presence (+) or absence (–) of strong nodose or spinose external shell sculpture.

Table 5. Architecture of some intertidal species of *Cerithium* from open rocky surfaces.

Species	Locality	A/2[1]	S[1]	W[1]	Sc[1]
Indo-West Pacific					
Cerithium columna Sowerby	Ngerameayus, Palau	16.4	1.81	1.17	–
	Pago Bay, Guam	19.7	1.85	1.25	–
	Marset el Et, Sinai	16.8	1.78	1.19	–
C. sejunctum Iredale	Pago Bay, Guam	16.6	1.52	1.18	–
C. ravidum Philippi	Agat, Guam	17.4	1.76	1.21	–
C. nodulosum Brug.	Pago Bay, Guam	14.5	2.04	1.14	+
	Aimeliik, Palau	13.3	2.05	1.11	+
C. echinatum Lamarck	Nosy-Be, Madagascar	14.8	1.82	1.14	+
	The Fjord, Sinai	17.6	1.82	1.22	+
Clypeomorus morus Lamarck	Nosy-Be, Madagascar	17.4	1.67	1.20	–
	Pago Bay, Guam	20.4	1.71	1.31	–
	The Fjord, Sinai	18.7	1.65	1.24	–
C. sp.	Pago Bay, Guam	17.2	1.51	1.19	–
Eastern Pacific					
Cerithium menkei Carpenter	Panama City, Panama	20.0	1.62	1.24	–
C. stercusmuscarum Val.	Panama City, Panama	21.9	1.49	1.31	+
Western Atlantic					
C. literatum Born	Discovery Bay, Jamaica	19.3	1.72	1.20	–
C. eburneum Brug.	Discovery Bay, Jamaica	19.8	1.69	1.24	–
Eastern Atlantic					
C. atratum Born	Dixcove, Ghana	16.1	1.69	1.12	–

[1] A/2: Apical half-angle of shell in degrees; S: Ratio of aperture length : aperture width; W: Expansion rate of the whorls; Sc: Presence (+) or absence (–) of strong nodose or spinose external shell sculpture.

corals, molluscs, echinoids, and probably other groups during the late Tertiary (Woodring, 1966; Fell, 1967; Dana, 1975) apparently did not affect the eastern Pacific shore fauna as profoundly, though Panamic reefs were badly affected (Dana, 1975). Thus, a number of well-armored molluscs which before the closure of the Central American isthmus have been distributed throughout the tropical western Atlantic and eastern Pacific, have apparently been able to survive to the present only in tropical west America. If this is so, then the pre-Isthmian Caribbean may have supported a generally more predator-resistant molluscan fauna than it does today, though I regard it as unlikely that antipredatory adaptations would have evolved to the level now characterizing the Indo-West Pacific. These possibilities remain to be tested by careful analysis and interpretation of fossils.

ACKNOWLEDGMENTS

The kind cooperation of many marine laboratories and institutes enabled me to collect molluscs from many shores. These laboratories include the Marine Biological Association, Plymouth; University of Guam Marine Laboratory; Biological Laboratory and Micronesian Mariculture Demonstration Center, Koror, Palau; Smithsonian Tropical Research Institute, Panama Canal Zone; Organization of Tropical Studies, Costa Rica; Discovery Bay Marine Laboratory, Jamaica; Instituto del Mar del Peru, Callao; Estacion de Biologia Marina, Montemar, Chile; Hans Steinitz Memorial Marine Laboratory, Eilat, Israel; University of Ghana, Legon; IFAN, Dakar, Senegal; and ORSTOM, Nosy-Be, Malagassy Republic. Numerous persons at these institutes provided me with kind assistance. Among them I wish to thank in particular R. E. Dickinson, M. Yamaguchi, and L. G. Eldredge of Guam; A. J. and E. C. Southward of Plymouth; and E. Zipser and E. Dudley.

This work was supported by grants from the National Geographic Society, National Science Foundation (Oceanography Section), and the John Simon Guggenheim Memorial Foundation.

REFERENCES

Abbott, R. T., 1961, The genus *Lambis* in the Indo-Pacific: Indo-Pacific Mollusca, v. 1, p. 51-88.

Cates, R. G., 1975, The interface between slugs and wild ginger: Ecology, v. 56, p. 391-400.

Cates, R. G., and Orians, G. H., 1975, Successional status and palatability of plants to generalized herbivores: Ecology, v. 56, p. 410-418.

Dana, T. F., 1975, Development of contemporary eastern Pacific coral reefs: Marine Biol., v. 33, p. 355-374.

Ebling, F. J., Kitching, J. A., Muntz, L., and Taylor, C. M., 1964, The ecology of Lough Ine. XIII. Experimental observations of the destruction of *Mytilus edulis* and *Nucella lapillus* by crabs: Jour. Animal Ecology, v. 33; p. 73-83.

Emerson, W. K., 1967, Indo-Pacific faunal elements in the eastern Pacific with special reference to molluscs: Venus, v. 25, p. 85-93.

Emerson, W. K., and Cernahorsky, W. O., 1973, The genus *Drupa* in the Indo-Pacific: Indo-Pacific Mollusca, v. 1, p. 801-863.

Fell, H. B., 1967, Cretaceous and Tertiary surface currents of the oceans: Ann. Rev. Oceanogr. Marine Biol., v. 5, p. 317-341.

Garth, J. S., 1974, On the occurrence in the eastern tropical Pacific of Indo-West Pacific decapod crustaceans commensal with reef-building corals: Internat. Coral Reef Symposium, 2nd, Brisbane 1973, Proc., v. 1, p. 397-404.

Houbrick, J. R., 1974a, Growth studies on the genus *Cerithium* (Gastropoda: Prosobranchia) with notes on ecology and microhabitat: Nautilus, v. 88, p. 14-27.

——— 1974b, The genus *Cerithium* in the western Atlantic (Cerithiidae: Prosobranchia): Johnsonia, v. 5, p. 33-84.

Kitching, J. A., and Lockwood, J., 1974, Observations on shell form and its ecological significance in thaisid gastropods of the genus *Lepsiella* in New Zealand: Marine Biol., v. 28, p. 131-144.

Kitching, J. A., Muntz, L., and Ebling, F. J., 1966, The ecology of Lough Ine. XV. The ecological significance of shell and body forms in *Nucella:* Jour. Animal Ecology, v. 35, p. 113-126.

Levin, D. A., 1975, Pest pressure and recombination systems in plants: Am. Naturalist, v. 109, p. 437-451.

Porter, J. W., 1972, Ecology and species diversity of coral reefs on opposite sides of the Isthmus of Panama: Biol. Soc. Washington Bull., v. 2, p. 89-116.

Salvat, B., and Ehrhardt, J. P., 1970, Mollusques de l'île Clipperton: Mus. Natl. d'Hist. Naturelle Bull., ser. 2, Zoology, v. 42, p. 223-231.

Vermeij, G. J., 1973a, West Indian molluscan communities in the rocky intertidal zone: A morphological approach: Bull. Marine Sci., v. 23, p. 351-386.

——— 1973b, Molluscs in mangrove swamps: Physiognomy, diversity, and regional differences: Systematic Zoology, v. 22, p. 609-624.

——— 1974, Marine faunal dominance and molluscan shell form: Evolution, v. 28, p. 656-664.

——— 1976, Interoceanic differences in vulnerability of shelled prey to crab predation: Nature, v. 260, p. 135-136.

Woodring, W. P., 1966, The Panama land bridge as a sea barrier: Am. Philos. Soc. Proc., v. 110, p. 425-433.

The Roles of Plate Tectonics in Angiosperm History

DANIEL I. AXELROD, *Department of Botany, University of California, Davis, California 95616*

ABSTRACT

This paper summarizes the underlying roles that environmental changes due to plate tectonic events appear to have had in angiosperm history. Topics briefly considered include the area and environment of origin, early dispersal, opportunism, impoverishment, disjunct distribution, relict survival, and extermination.

INTRODUCTION

The times of major turnover in the plant world do not correspond with those in the animal world. They occurred earlier, in the Permo-Carboniferous transition and in the middle Cretaceous (e.g., Saporta, 1881; Clements, 1920; Seward, 1936). These revolutions, as well as lesser ones, largely occur at or near the peaks of the major transgressions (Axelrod, 1972c, 1974b) controlled chiefly by spreading on active ocean ridges and by other plate movements. These were the times when wholly new adaptive types were literally washed into the record (Fig. 1). The sites in which they underwent their earlier evolution probably were in more distant regions where greater environmental diversity would have favored their origin (Simpson, 1944, 1953; Stebbins, 1974; Axelrod, 1952). As seaways retreated, the taxa that represent these new adaptive types spread and diversified as increasingly more varied environments became available over the lowlands. The manner in which changing environments, controlled by plate movements, have shaped the course of angiosperm evolution is sketched here; more complete documentation and illustrations have been presented in earlier papers, as noted in the bibliography.

AREA OF ORIGIN

West Gondwanaland (South America-Africa-Madagascar) probably was the prime center of early angiosperm evolution. First, angiosperms are adapted, for the most part, to warm climates. Fully 80 percent of the 350 families are either restricted to, or find their optimum diversity in, tropical regions (Bews, 1927; Vester, 1940; Camp, 1947; Axelrod, 1952, 1960; Good, 1964). Second, their progressive time-space relations in the Cretaceous show that they were deploying into higher latitudes, implying an origin over the tropical regions (Axelrod, 1959). This is consistent with the later appearance of megafossils in higher latitudes, and also with the deployment of more complex pollen types into higher latitudes at later times. Third, West Gondwanaland was situated in the tropics and had an environment highly favorable for the origin of angiosperms and their early diversification. The area was characterized by broad regions of Precambrian crystalline rocks which would have provided a diversity of terrain and substrate for fragmentation of populations, and hence encouraged their early splitting into different adaptive types. A seasonally dry climate was undoubtedly present, as shown by the lithologic record in the lower middle latitudes where the oldest pollen types appear in Israel, tropical Africa, and South America, all then in equatorial regions. This environment was particularly suited for rapid evolution (Stebbins, 1952; Axelrod, 1967). In addition, the terrain of exceptionally hard crystalline rocks provided numerous arid edaphic sites which could aid in the early selection of alliances adapted not only to dry climate, but to different ways of living, as shown by the diverse, unique life forms that are restricted to dry regions (Axelrod, 1967, 1972b). In such an environment, early angiosperms would be well-removed from the lowland sites of sedimentation, and could, therefore, scarcely be expected to have contributed to a record there.

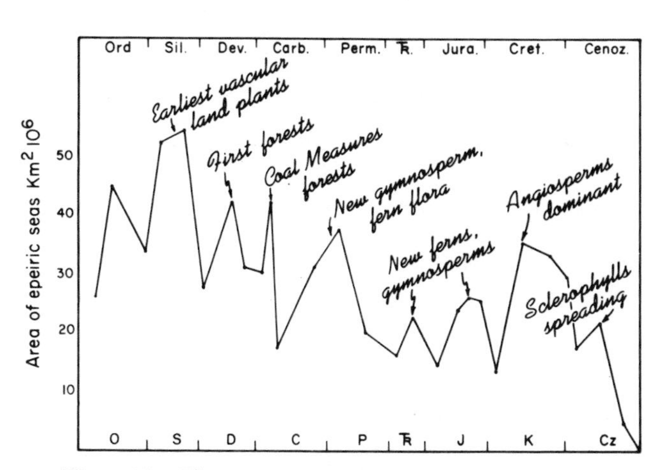

Figure 1. Floras representing major new taxa enter the record at times of maximum transgressions (simplified from Axelrod, 1974).

Although it has been asserted (Hughes, 1974) that early angiosperms will be found in the lowland record, this assumption is untenable in view of the conditions and circumstances that are involved in the origin of a new phylum (Simpson, 1944, 1953). Furthermore, since numerous rich Jurassic floras are already known from both low and high latitudes, one might well ask why early angiosperms (or their progenitors) have not already been discovered if, as Hughes (1974) and others suppose, they originated in well-watered lowland basins. In this regard, Krassilov (1975a, 1975b) suggests that angiosperms may have had diverse sources in the Caytoniales (Corystospermaceae, Peltaspermaceae), Bennettitales, and Gnetales. While some of these may well have been angiospermous in terms of their seeds, their leaf and stem structures seem far removed from angiosperms. To caption such plants proangiosperms, as well as the Dirhopalostacyaceae which bore *Nilssonia* foliage, seems unwise. However, the relation noted, that seed ferns and cycadophytes either had, or approached, the condition of angiospermy, gives support to the argument presented below.

AGE AND ANCESTRY

That angiosperm history extends well-down into pre-Cretaceous time seems highly probable. Even if we discount all records that have been asserted to represent pre-Cretaceous angiosperms (see Hughes, 1976), the group must still have had a long unrecorded history. Not a scrap of evidence has yet been found that bridges the gap between an undoubted angiosperm and the nearest possible ancestral group, the seed ferns. That the ancestral alliance may never be found seems highly probable in view of the nature of the factors involved in the evolution of a new phylum (Simpson, 1944), such as angiosperms. If the habit of angiospermy (enclosed seeds) was indeed an adaptation to drought stress in a tropical environment (Axelrod, 1970), then the phylum may well have diverged from a small, still unrecorded seed fern group (or groups?) that inhabited more remote sites on the Precambrian basement of West Gondwanaland, situated in areas well-removed from sites of sedimentation. In this regard, attention is directed to the work of Asama (1960, 1962, 1975) on the possible reduction series of seed-fern fronds into angiospermlike leaves. His theory finds further support in a current paper (Cheng and others, 1974) that illustrates Permian seed ferns which have pinnae with outlines and venation that simulate angiosperms, including such diverse taxa as *Cercis, Quercus* (Prinoidae, Dumoseae), *Magnolia,* and *Quiina.* In this connection, Melville (1975) reports that the relict families of Australia, Proteaceae and Epacridaceae, show vegetative (foliar) features that appear to relate them to Permian glossopterids.

The view (Doyle, 1969; Wolfe and others, 1975; Doyle and Hickey, 1976) that Early Cretaceous angiosperms in the Potomac flora represent, both in leaf- and pollen-type, the "first stages of their radiation" (Doyle and Hickey, 1976, p. 187) is not supported by the evidence. The oldest flora, the Patuxent, is no older than Aptian, yet by Albian time, only 5 to 6 m.y. later, angiosperms had already differentiated *globally* into very diverse floras that were adapted to distinctly different climates (Axelrod, 1970). Furthermore, while there is reason to assert that simple monosulcate pollen like that in the Patuxent flora is "primitive" (see Muller, 1970, Fig. 2), a more advanced structural type (i.e., tricolpate) occurs in rocks of Hauterivian-Barremian age in the USSR, in Barremian to Aptian rocks of Israel, and in the Aptian of Africa and Brazil (refs. in Raven and Axelrod, 1974, p. 558; Doyle and others, 1975; Doyle and Hickey, 1976). The fact that the pollen in those areas remains relatively undiverse until later, when the tricolpates appear and begin to differentiate rapidly in Laurasia, not only supports the idea of poleward migration of the angiosperms but that they invaded the outer tropical zone, as represented by the Potomac flora, much later. Clearly, the Potomac flora, when considered in relation to others of similar and probably somewhat greater age at lower latitudes, shows that the initial splitting and radiation had occurred much earlier, and in the tropics. This evidence indicates to me that the pollen taxa in the early Potomac sequence are *already relict,* and thus do not reveal the initial stage of angiosperm splitting. Furthermore, the tacit assumption of Doyle and Hickey (1976) that the "primitive" leaves and pollen reflect parallel evolution of structures in ancient angiosperm radiation is wholly untenable. Whereas the large leaves presumably were derived from plants living near-at-hand, the silt-size pollen grains obviously were transported in moving water by rivers from more distant sites to the area of deposition on the floodplain. The pollen was derived from wholly unknown plants. Furthermore, the great diversity of leaf type (size, margin, shape, venation) as compared with the low diversity of pollen, as illustrated by Doyle and Hickey (1976, Fig. 28), is not consistent with the concept of synchronous evolution of these plant structures. Clearly, there is no basis for associating the leaves and pollen in terms of the radiation of the angiosperms.

Rather than regarding the taxa of the Potomac sequence as ancestral, they more probably are members of an ancient, relict plexus surviving under moist, warm-temperate climate, having been shunted there by competition from more advanced plants living under different climates in the interior. This inference parallels the relation established years ago by Elias (1936) for the "Permian" flora in the Late Carboniferous. The same relation is implied also by the high stage of evolution indicated by plants in the Dakota (late Albian) as well as in the Cenomanian floras from the Woodbine, Tuscaloosa, Raritan, and Amboy Formations. All of them, which include hundreds of species representing scores of genera and family types, could scarcely have originated from the taxa in the Potomac flora in only 10 to 13 m.y.! Finally, it is evident that if angiosperms originated initially in response to drought stress (cf. Stebbins, 1974, p. 304), then the Potomac plants—which obviously lived under very moist rainforest conditions—must have adapted to that environment *after* the initial radiation of the group. The Potomac plants are best regarded as ancient angiosperms of unknown affinity that probably were removed from the "main line" of radiation, and were surviving in a favorable relict environment.

EARLY DISPERSAL

Angiosperms spread rapidly into the lowlands in the middle Cretaceous as climates moderated under advancing seas as the Laurasian and Gondwanan plates commenced to break up. Owing to their early dispersal, a common flora was shared by South America-Africa (and probably Madagascar-India) until the close of the Cretaceous. There are still numerous links across the region, including some 50-odd large pantropic families. Late Cretaceous dispersal southward across Antarctica into Australia-New Zealand explains much of the distribution of ancient families in the Southern Hemisphere (Wild, 1968), as in Cupressaceae, Podocarpaceae, Cunoniaceae, Proteaceae, Restioniaceae, and others which evidently had originated by then (see Disjunct Distribution). Isolation of Africa by the Late Cretaceous evidently accounts for the marked differences between the derivative taxa there and in the now widely separated austral lands, for instance in the unique genera *Braebium* and *Dilobea* in the proteads (Johnson and Briggs, 1975), and in other families. Closer relationships between genera and tribes of families now disjunct in South America-Australia reflect their continuity via Antarctica into the Eocene.

South America was widely separated from North America into the Eocene, but from then on supplied increasing numbers of taxa to it. However, dispersal northward from Africa into Eurasia was active in the middle to Late Cretaceous, and thence across to North America via a narrower, shallower Atlantic basin populated with many islands at middle latitudes. At this early date a number of ancient alliances, including forerunners of Annonales, Araliaceae, Arecaceae, Bombacaceae, Euphorbiaceae, Menispermaceae, and others, probably crossed into North America (Raven and Axelrod, 1974).

OPPORTUNISM

Scores of taxa had appeared by the middle of the Cretaceous (Cenomanian), and they rapidly increased in number and diversity. Many of them are extinct in terms of family and generic relations, and their affinities will always remain uncertain. Nonetheless, it is evident from their increasing diversity, and the rapidity with which they appeared, that they had undergone considerable radiation in well-drained sites removed from the wet (high water table), lowland basins of deposition where ferns, seed ferns, and gymnosperms had been contributing to the accumulating record.

New opportunities for adaptive radiation appeared as land areas were rifted and thence rafted to new positions during the later Cretaceous and Tertiary. These resulted from the development of new climates in response to local changes in terrain, as well as to the transport of lands into new climatic belts. Furthermore, isolated populations now diverged in a normal genetic sense, with selection for the newly developing conditions playing the guiding role in evolutionary trends.

South America and Africa (West Gondwanaland) shared a very similar flora at the end of the Cretaceous (Herngreen, 1974a, 1974b, 1975), and into the Paleocene. They were then separated by a relatively shallow ocean populated by numerous islands which have since been transported subsea in the widening, deepening basin. By Paleocene time, South America and Africa had moved about 600 km apart; by the close of the Eocene they were separated by about 1,300 km. There were then fewer islands linking them, the Atlantic basin had deepened, and South America was now closer to North America which in the Cretaceous had been isolated from it by fully 3,000 km of open ocean with only scattered islands.

Distinctness of the present floras of tropical South America and Africa is owed in part to the long time that evolution had progressed on West Gondwanaland following the appearance of numerous modern families in the Late Cretaceous (Muller, 1970, Fig. 5), and to spatial isolation on that large landmass. Probably in larger measure, however, it reflects progressive evolution following separation of Africa-South America, with the differentiation of unique species, genera, tribes, and subfamilies that link the American-African regions. Examples are provided by the Annonaceae, Flacourtiaceae, Moraceae, and others illustrated elsewhere (e.g., Axelrod, 1970, 1972a; Raven and Axelrod, 1974; Thorne, 1973; Moore, 1973). Also, there are closely related families that link the Old and New World, presumably because their ancestors dispersed between Africa and South America when these continents were much closer. Notable in this group are the related families Musaceae (*Musa*), Strelitziaceae (*Ravenala* in Madagascar, *Strelitzia* in southeast Africa, *Phenakospermum* in northern South America), and Heliconiaceae (with *Heliconia* in Central and South America).

Enrichment of the flora of tropical North America from a South American center increased following Eocene time as new islands enhanced opportunities for migration, and as the land bridge was gradually built up between them. Among the typically South American tropical families that gradually entered tropical Central America-Mexico were Begoniaceae, Bromeliaceae, Cyclanthaceae, Fabaceae-Mimosaceae, Marantaceae, Marcgravaceae, Myrsinaceae, Quiinaceae, and Viscacae (Raven and Axelrod, 1974, p. 627). As higher altitudes developed in response to volcanism and tectonism, taxa of montane vegetation zones, including the alpine paramo, shifted northward into Central America. Relatively fewer northern taxa (e.g., *Alnus, Juglans, Quercus*) penetrated into South America along the Andine axis, and these are confined to the relatively limited temperate montane habitats.

Continued buildup of the Cordilleran axis during the Quaternary provided a series of stepping stones for taxa of temperate requirements to disperse from western North

America southward into Fuegia. These fall into two major groups, those of moist temperate requirements and those of semiarid to subhumid requirements. Most of the migrants are herbaceous, self-fertilizers that invaded young, open areas (Raven, 1963). Interchange between the dry Argentine environment and the Chihuahuan region no doubt was favored by temporary way-stations that developed within the inner tropics at the peak of aridity in the middle Pliocene (Axelrod, 1948; Williams, 1975). This is consistent with the exchange of grazing mammals at this time, for they certainly did not crash through several thousand miles of rainforest to reach the plains or savannas. In addition, interchange probably occurred during periods of ice-age aridity. Probably at such a time the creosote bush (*Larrea*) invaded the North American desert, developing different polyploid populations as it spread northwesterly into California. However, most of the successful movement was from North America southward, probably because austral taxa were not adapted to the generally less equable northern climates, and especially to those of the cold glacial stages.

In similar manner, northward motion of the Australian plate established connections with the tropical lowlands of Malesia during the Miocene. The rich tropical flora of Malesia, including such families as Bignoniaceae, Sapotaceae, Melastomataceae, and Zingiberaceae, largely entered Australia in the Miocene and later (Raven and Axelrod, 1972). Subsequent uplift of the mountainous island chains through Malesia made possible the invasion of Australasia by many northern groups, such as Apiaceae-Apioideae, *Veronica, Epilobium,* and Asteraceae-Astereae, which in a number of instances have undergone spectacular evolutionary radiation in newly available subalpine and alpine habitats of New Zealand (Raven, 1973a).

That opportunities may appear as lands are rafted to new climates is exemplified in another way by the richly endemic (ca. 75 percent) flora of southwest Australia. During the early Tertiary the southern half of Australia was covered with an austral temperate rainforest, dominated by *Nothofagus, Podocarpus,* and evergreen dicots. The northern half was then in low-middle latitudes under the influence of a seasonal, subhumid climate suited to the origination of sclerophyllous taxa. Since climatic belts are largely stable, the sclerophyllous flora was displaced gradually southward as the continent moved north. As drier climate spread during the middle and later Tertiary, sclerophyllous vegetation enlarged its area in central and western Australia, probably reaching the south coast in the Pliocene. Spreading drought has more recently restricted it to its present areas in southwest and South Australia. Thus, a rich, diverse flora was literally swept into its present region from a much larger province, one now covered with desert and thorn scrub vegetation. This sweeping effect, which concentrated taxa of subhumid requirements into a smaller area, probably explains much of the high diversity of the sclerophyllous flora, and very equable climate (*M* 60-70) has favored their persistence.

Crocker and Wood (1947) emphasize that the sclerophyll vegetation of southwest Australia is exceedingly rich in taxa even though the area has been a peneplain since the Cretaceous. They attribute the richness of the flora to its occurrence on relict lateritic soils to which it is adapted. In this regard, it is highly probable that these soils originated under a climate suited to laterite formation, one with a distinct dry and wet season under warm temperature. The xerophyllous floras of southwest and South Australia probably were first isolated in the early drier phases of the Quaternary, and especially during each xerothermic period. This led to the emergence of related species in each area, and their isolation has been favored by the different substrates to which they are adapted (Wood, 1959). Further opportunism resulted from the fluctuating pluvial and dry climates which favored hybridization as populations shifted into the area of the present desert and thence retreated on several occasions. During the rapid expansion of the flora at times of relaxation of severe dry climate, the normal rates of differentiation must have been greatly exceeded for biotypes adapted to the new edapho-climatic conditions. This may well explain the high diversity of species in large genera, notably *Acacia, Eucalyptus, Grevillea, Goodenia, Hakea, Melalucea,* and others (see Crocker, 1959; Crocker and Wood, 1947, p. 118-119).

Additional evidence that the present Australian sclerophyllous genera developed in a warmer climate with summer rain is implied by the work of Specht and Rayson (1957). They determined the growth rhythm of the principal components of a heath community in South Australia by measuring the monthly dry-weight increment of terminal shoots of the larger sclerophyllous taxa. Distributed in the Australian sections of the families Proteaceae, Leguminosae, Epacridaceae, Rutaceae, and Casuarinaceae, the taxa inhabit infertile sands in a typically mediterranean-type climate with cool, wet winters and hot, dry summers. However, all the species show a period of maximum growth during summer, at a time of water stress. Growth is initiated only when mean temperature rises above 18°C and is inhibited when it falls below that level in autumn. In coastal Queensland, where there is a coincidence of summer rainfall and high air temperature, sclerophylls of the same genera but usually different species, have the *same* growth period as those in South Australia (Wood, 1959). In South Australia growth is therefore not initiated in winter, as in analogous sclerophyllous communities of California or the Mediterranean basin. The Australian taxa exhibit a growth rhythm markedly out of phase with the present climate, suggesting an origin in a warmer climate of wet summers. This implies that the present mediterranean-type climate is not ancient, but probably was initiated chiefly during the drier interpluvials.

The sclerophyllous flora at the southwest tip of Africa evidently has had a similar history. As Africa moved northward in the Late Cretaceous and Tertiary, the area of subhumid climate gradually shifted south. That it was covered with sclerophyllous vegetation by the late Eocene (37 m.y.) is implied by the small leaves representing the

Banke flora (Rennie, 1931). At that time sclerophylls probably covered much of the present area of the Namaqualand-Karroo-Kalahari deserts. As drier climate expanded in the Neogene, the sclerophyllous flora gradually shifted southward into the present Cape area where many taxa have survived under the mild oceanic climate. However, population shifts at times of moister climate during the Quaternary enabled the flora to return northward to the area of the present desert, a region where relict stands of sclerophyllous vegetation still persist in the higher hills above the Namaqualand and Karroo deserts. Those broad fluctuating population movements no doubt favored much hybridization, and hence the origin of many new species (or semispecies) in the large genera that now typify the area, such as *Erica, Phylica, Muraltia,* and others. They have survived in the Cape region under near-maritime climate, protected by the topographic wall of the Cape ranges which shelter them from the incursions of hot, dry desert conditions to the north. As in southern Australia, this flora also responds to a nonmediterranean climate. That is, the taxa enter their chief period of growth as the hot, dry, summer season commences, not in the moist winter period. This again suggests that the present mediterranean climate is quite young, and perhaps would largely be alleviated at times of higher rainfall, as in the pluvial periods.

Another related type of opportunism is suggested by the remarkable flora at the dry southern tip of Madagascar (Fig. 2). This is a thorn forest filled with taxa adapted variously to extreme drought, as exemplified by treelike succulents, leafless plants, and water-storage organs (succulent leaves, swollen trunks, or underground tubers). Derived from diverse families (see Koechlin, 1972, for photographs), they include many unique endemics, distributed in *Alluaudia* and *Didieria* (Didieriaceae, an endemic family), *Aloe* (Liliaceae), *Croton* (Euphorbiaceae:Crotonoideae), *Euphorbia* (Euphorbiaceae:Euphorbieae), *Folotsia* (Apocynaceae), *Moringia* (Moringiaceae), *Xerosicos* (Cucurbitaceae), and *Xerophyta* (Velloziaceae).

Current evidence regarding the controversy over the paleoposition of Madagascar now indicates that it was situated along the Somalia-Tanzania coast prior to the middle Cretaceous (90 m.y.), as discussed by Tarling and Kent (1976), P. J. Smith (1976a, 1976b), Kent (1972), Kent and others (1971), McElhinney and others (1976), and others they cite. Movement of Madagascar to its present position prior to the Late Cretaceous may explain the presence there of dinosaurs as well as the absence of primary fresh-water fish, for they had not yet originated. Although Madagascar was situated in the middle of the rainforest belt straddling the equator, the lee (west) side of the island was in a light rainshadow, and hence local savanna climates were present. Evolution of taxa adapted to drier sites was not only favored by the drier climate, but also by areas of very hard Precambrian crystalline rocks which regularly provide arid sites even in rainforest climates (Axelrod, 1972b). As Madagascar was shifting southward, selection was for drought-adapted alliances since the island was moving into the southern dry belt.

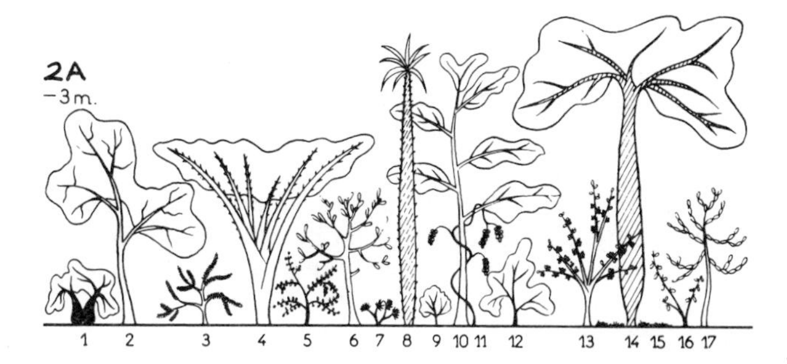

Figure 2. Adaptations to xeric edaphic and climatic environments by genera of diverse families in southern Madagascar, all contributing to a unique thorn scrub and open woodland. Most plants flower immediately after rain, and there is much convergence in plant form by different families. Adaptations include swollen trunks; spiny stems; very thin branches; leafless plants; drought deciduous plants; very small, often fleshy leaves; low fleshy parts; cactoid forms.

2a. Malagasy thicket of *Alluaudia comosa.* (1) *Commiphora monstruosa;* (2) *Cassia meridionalis;* (3) *Bauhinia grandidieri;* (4) *Alluaudia comosa;* (5) *Rhigozum madagascariensis;* (6) *Euphorbia leucodendron;* (7) *Xerophyta dasylirioides;* (8) *Pachypodium lamerei;* (9) *Blepharis calcitrapa;* (10) *Terminalia subserrata;* (11) *Xerosicyos danguyi;* (12) *Croton* sp.; (13) *Megistostegium perrieri;* (14) *Delonix adansonioides;* (15) *Selaginella nivea;* (16) *Alluaudiopsis fiherenensis;* (17) *Euphorbia oncoclada* (from Koechlin, 1972, with permission of W. Junk Publishers).

2b. Malagasy bush thicket of *Alluaudia procera.* (1) *Albizzia tulearensis;* (2) *Diospyros latispathulata;* (3) *Alluaudia procera;* (4) *Selaginella nivea;* (5) *Cyphostemma laza;* (6) *Pachypodium geayi;* (7) *Croton sp.;* (8) *Alluaudia dumosa;* (9) *Gyrocarpus americanus;* (10) *Aloe divaricata;* (11) *Euphorbia* sp.; (12) *Alluaudia humbertii;* (13) *Xerosicyos danguyi;* (14) *Maerua nuda* (from Koechlin, 1972, with permission of W. Junk Publishers).

Thus, selection was clearly for taxa that could withstand increasing drought, and they expanded into this widening environment to which they were largely preadapted. The origin of the unique arid taxa that are now there, including Didieriaceae, *Xerophyta* (Velloziaceae), and others (Fig. 2), may well have commenced in the middle Cretaceous. This inference is consistent with the disjunct occurrence of taxa that link tropical South America and Madagascar but are not found in Africa, including Monimiaceae (*s. s.*), Trigoniaceae, Winteraceae, Chloranthaceae, Elaeocarpaceae, *Weinmannia* (Cunoniaceae), the ceroxyloid and chamaedoreoid palms, and others (Dejardin and others, 1973).

The creation of new archipelagos, as marginal basins are formed during the fragmentation of plates, may lead to increased diversity. This is illustrated by the rampant speciation in figs (*Ficus*) in the Solomons and New Hebrides, islands formed during the Miocene fragmentation of the east Australian plate (Karig, 1971, 1972; Griffiths and Varne, 1972). The species are far more numerous and less distinct than those on New Caledonia, which is consistent with its continental origin, greater age, and rafting to isolation (Raven and Axelrod, 1972). In addition, the archipelagos provided stepping-stones for migration of a large Malesian element into the warm-temperate part of New Zealand. Dispersal into the region was aided by the Lord Howe and Norfolk Rises which were then comparatively large, emergent land areas that supported a very diverse flora.

Crustal movements seem largely responsible for the general trend to colder and drier climate during the Tertiary, and for the initiation of glaciation (Crowell and Frakes, 1970). The trend commenced in the middle Oligocene as Antarctica moved into higher polar latitudes. The Tethyan seaway was progressively restricted and then destroyed in the middle Miocene, the Arctic basin gradually closed, the Panamanian bridge was elevated, and the high American Cordillera and the Eurasian alpine system were rapidly built up. The change from a quasi-latitudinal circulation to one of strong meridional control as the Pacific, Atlantic, and Indian Oceans were isolated, accentuated the anticyclonic circulations which now increased in area, stability, and intensity. This brought spreading areas of progressively drier climate to lower-middle latitudes, and their areas increased further as the cordilleras were rapidly built up to form regional rain shadows in their lee.

As forests over temperate latitudes retreated under spreading drier and colder climates, prairie, grassland, steppe, and tundra expanded as new regional environments. These provided a wealth of opportunities for forest-border grasses and forbs to diversify in expanding open areas, as shown by the great diversity of Apiaceae, Asteraceae, Brassicaceae, Poaceae, Polemoniaceae, and other dominantly herbaceous families in temperate climates. In the warmer latitudes, savanna and thorn scrub environments spread as regions of seasonal rainfall increased in area and in intensity. This opened up abundant opportunities for scores of tropical and subtropical families that proliferated into numerous genera and species, as seen in the diversity in Mimosaceae, Caesalpiniaceae, Bombacaceae, Moraceae, Flacourtiaceae, Passifloraceae, Malvaceae, Euphorbiaceae, and others in the drier tropics and subtropics. Cactaceae also responded rapidly to new dry environments which range from coastal desert to alpine puna in Peru-Chile, as well as from tropic savanna into extremely dry thorn scrub and desert environments. Recent opportunism is exemplified by species and genera of various tribes of Cactaceae (e.g., Hylocereae, Notocacteae, Cereëae, Trichocereae) that occur in several or all of these environments (Buxbaum, 1969). In Africa, a comparable story is provided by the diversification of cactoid Euphoribeae, and of Aizoaceae, which typify its dry areas. Contrary to popular belief, these unique adaptive types did not originate in desert environments. They were preadapted to them, having originated earlier in dry sites within tropical savanna and thorn scrub environments (Axelrod, 1950, 1952, 1972b).

The gradual emergence of full mediterranean (dry summer) climate as the oceans further chilled also provided major opportunities for diversification. The California region is very rich in new taxa in the large families Asteraceae, Brassicaceae, Boraginaceae, Polygonaceae, Poaceae, Hydrophyllaceae, Onagraceae; and in the Mediterranean region there is a strong development of Apiaceae, Fabaceae, and Caryophyllaceae (Raven, 1973b). About half of all species in these areas are herbaceous. They are so new that they probably account for an increase of nearly one-third in the size of the entire flora during the Quaternary.

Diversification in the Himalayan region resulted from uplift as Peninsular India underrode the Asian plate in the Miocene and later. Proliferation of taxa in *Allium, Camellia, Epilobium, Lonicera, Magnolia, Rhododendron, Saxifraga,* and others (see Mani, 1974, p. 274) reflects the opportunism provided by wholly new montane environments. Many of the taxa show high polyploidy, especially at higher altitudes. During the Miocene and later, forests and savannas in western India-Pakistan retreated eastward as dry climate expanded, enabling invasion by elements of the subhumid to arid Mediterranean, Sahelian, and Iranian floras (Axelrod, 1974a). They are represented now by paired-species (or varieties) in the drier parts of India-Pakistan that provide links far to the west (Axelrod and Raven, 1972; Meusel and Schubert, 1971).

IMPOVERISHMENT

Rafting of a land area across latitude may lead to impoverishment, as when Peninsular India moved from southern, temperate latitudes through the inner, moist tropical latitudes to the northern, dry tropical belt (Axelrod, 1974a), from the Late Cretaceous (80 m.y.) into the Neogene (Sclater and Fisher, 1974; Molnar and Tapponier, 1975). Current evidence indicates that movement was not at a constant rate, but in a series of pulses (Johnson and others, 1976), two of which seem to have been especially critical. Following the Paleocene, the area of the Deccan Traps flora (near Nagpur, see Lakhanpal, 1970, for taxa) shifted rapidly north from near 15°S to 5°N, reaching it in the Eocene (53 m.y.). This rapid movement into the inner hot tropics evidently eliminated austral gymnosperms (*Araucaria,* Podocarpaceae), together with the leptodactylid frog (*Indobatrachus*), *Casuarina,* and others. They presumably could not adapt to the rapid shift to hotter climate, and escape into temperate regions was not possible because mountains of sufficient elevation were not available. The second pulse, from the late Oligocene (32 m.y.) into the Miocene, carried India rapidly northward another 15°, into the northern dry belt. This must have eliminated many inner tropical, moist rainforest taxa as the effects of drought spread, and as the global trend to aridity commenced to accentuate the dry conditions already present. The rapid shift into the dry belt may well explain the observation of Lakhanpal (1970, p. 687-688) that woody legumes and dipterocarps, which are rare in the older Tertiary floras of India, become abundant in the Neogene. Both families are especially well-represented in the dry Asian tropics today.

The flora of India is, therefore, impoverished, having only 1 small (2 sp.) endemic family (Hortoniaceae of Ceylon), far less than other tropical areas of the world. Mani (1974, p. 169) notes that East African and Malagasy taxa constitute a small but important part of the flora of Peninsular India, and they appear largely to represent older stocks. The Indian flora is made up chiefly of young migrants from humid tropical Asia in the late Paleogene, and thence from the drier savannas and more arid regions to the west as dry climates expanded during the Neogene and forest and savanna were confined to the moister east.

The climatic effects of plate movements in local and distant areas may combine to bring about major impoverishment of a continental flora. Compared with South America or Malesia, the tropical flora of Africa lacks large groups, or has only a poor representation of typically tropical alliances, notably palms, laurels, orchids, bamboos, as well as lianas and epiphytes (Richards, 1973). In addition, many species of tropical Africa have a wide distribution whereas numerous alliances in the moist American and Malesian tropics are highly endemic. Furthermore, many taxa are now disjunct between Madagascar, the nearby islands, and South America, and are not found in Africa, implying that they were once in Africa but were eliminated there. Among these are Monimiaceae, Trigoniaceae, Winteraceae, Chloranthaceae, Elaeocarpaceae, Bignoniaceae (Crescentieae), *Weinmannia* (Cunoniaceae), and the ceroxyloid and chamaedoreoid palms. By analogy, the relict endemics on Madagascar and neighboring islands, notably Didymeliaceae, Medusagynaceae, Sphaerosepalaceae, Didieriaceae, Sarcolaenaceae, and numerous genera and suprageneric taxa, may have once occurred on Africa. There is other indirect evidence for the extinction of at least some of these links on the African mainland: on Madagascar and nearby islands, groups such as ferns, palms, orchids, and bamboos are much better represented than on mainland Africa.

Impoverishment in Africa resulted from three major events that commenced in the Miocene: (1) the elevation of the East African rift belt rapidly increased bringing drier climate to the lee; (2) as the Tethyan seaway closed, latitudinal circulation was terminated, and each ocean basin now developed progressively stronger, more stable, and spreading anticyclonic systems; and (3) the Benguela Current commenced to bring colder water to the west coast. All three factors, acting more or less in concert, resulted in the spread of progressively drier climates over areas formerly covered with rainforest. This is inferred to have resulted in the extinction of rainforest taxa, as well as taxa that are represented now (often by related genera) on other continents, a view consistent with evidence that the present richest floristic areas in West Africa are those with least drought. In this regard, the Madagascar rainforest was less affected and has a more nearly balanced flora probably because it has a more oceanic climate and its wet montane area faces the Indian Ocean, and it is therefore less subject to drought. The Madagascar rainforest may well provide us with a glimpse of the probable nature of the Paleogene flora that covered Africa prior to the spread of aridity.

By contrast, the Amazonian rainforest was protected from the spreading drought along the west coast that resulted from the developing Humboldt Current. Commencing in the early Pliocene (6 to 8 m.y.), the Andean chain began to rise (Perch-Nielsen and others, 1975, p. 28), and its rapid uplift in the late Pliocene and Quaternary, as temperatures were lowered over Antarctica, effectively blocked the spread of the west coast aridity over Amazonia.

To the south, however, the *Nothofagus-Podocarpus*-evergreen dicot rainforest of southern Chile is highly impoverished when compared with the closely related forest of New Zealand at the same latitude (Godley, 1960). This reflects changes both in topography and regional climate following the Miocene, changes that are owed to plate movements (Axelrod and Raven, 1972). As the Andes began to rise, steppe and semidesert climates spread over the Argentine plain where the forest formerly dominated. The forest was restricted gradually to the wetter, windward slopes of the range. Then glaciation brought colder climate to the region. Periods of extreme drought followed in the north, and the forest was restricted southward to its present area. The rainforest was, therefore, trapped topographically and climatically, and much of it was decimated because it could neither escape nor adapt. Taxa formerly in Chile-Argentina that are now only in the Tasman region include *Podocarpus* (sect. Dacrycarpus, now in New Zealand, New Guinea), *Dacrydium* (*franklinii* group, now in Tasmania), *Phyllocladus* (now in New Zealand, Tasmania, New Guinea), and others (Florin, 1963; Couper, 1960). By contrast, the New Zealand flora remained in marine isolation and many of its taxa have persisted under mild, equable climate. Commencing in the Miocene, the flora was enriched by the southward spread of subtropical taxa as new archipelagos to the north (New Hebrides, Solomons) aided in their dispersal.

DISJUNCT DISTRIBUTIONS

Plate tectonics appears to provide a reasonable explanation for some of the troublesome problems of disjunct distribution.

A number of taxa that are disjunct between South America and Madagascar (e.g., Trigoniaceae, Winteraceae, the ceroxyloid and chamaedoreoid palms), or between South America and tropical southeast Asia (Chloranthaceae, Sabiaceae, Symplocaceae) are absent from Africa. This seems to be chiefly the result of extinction in Africa as aridity increased following Oligocene time, and as seasonal rainfall appeared along the west coast where earlier it had been more evenly distributed.

Numerous relicts on New Zealand, New Caledonia, and Fiji, as well as the small islands on Lord Howe Rise and Campbell Plateau, provide links with Queensland. Plants such as *Phormium, Coprosma,* and the palm *Rhapalostylis,* which occur in the Paleogene of New Zealand are still there, as well as on Norfolk Island. An endemic *Araucaria* is on Norfolk Island, and Lord Howe Island has an endemic palm and an endemic genus (*Howeria*) of the archaic austral hemipteran family Pelordiidae. The small island of New Caledonia is rich in archaic seed plants, including 40 endemic gymnosperms, several vesselless angiosperms (Winteraceae, *Amborella*), and other relict taxa that include members of the Monimiaceae, Escalloniaceae, Cunoniaceae, and Arecaceae. More advanced groups, notably Sympetalae, are poorly represented, which is consistent with New Caledonia's isolation since the Cretaceous (Griffiths and Varne, 1972). Its archaic land snails, some insects, and possibly galaxid fresh-water fishes, probably were derived from ancestors associated with the flora when the island was a part of Australia (Raven and Axelrod, 1972). The occurrence of terrestrial, giant horned turtles of the family Meoilaniidae in the early Tertiary of southern South America, and now on Lord Howe Island, and on Walpole Island southeast of New Caledonia in the Pleistocene, provides further evidence of earlier connections (Raven and Axelrod, 1972). Plants such as *Libocedrus (s. s.), Knightia,* and *Xeronema,* now restricted to New Zealand and New Caledonia, provide additional links consistent with the paleogeography reconstructed for the Late Cretaceous (e.g., Chase, 1971; Griffiths and Varne, 1972; Hays and Ringis, 1973).

Fiji is also an integral part of this picture. Some 23 percent of its 445 native genera reach their eastern limits in the Pacific on Fiji, including *Dacrydium, Acmopyle, Agathis, Kermadecia, Casuarina,* as well as the families Annonaceae, Cunoniaceae, and Epacridaceae. The monotypic family Balanopaceae links Queensland, New Caledonia and Fiji. Fiji also has the endemic family Degneriaceae, as well as land snakes and frogs. Numerous forest trees of Fiji have large seeds which are unlikely to have been dispersed over wide water gaps. It seems highly probable that all these lands were a part of the Australian plate into the Late Cretaceous. As they were rafted eastward various taxa were lost along the way owing to the different climates encountered and also to the reduction of lands by subsidence and erosion (Norfolk, Lord Howe, Walpole Islands), some of which have now wholly disappeared (Raven and Axelrod, 1972).

As suggested by J. D. Hooker (1853) more than a century ago, "The three great land areas in the southern hemisphere (New Zealand, Australia-Tasmania, temperate South America) show a botanical relationship . . . which is agreeable to the hypothesis of all being members of a once more extensive flora, which has been broken up by geological and climatic causes." These lands were closely tied together into the Eocene (Sclater and Fisher, 1974, Fig. 15), and then parted company in response to ocean-floor spreading.

The presence of related taxa (*Arbutus, Cercis, Cupressus, Helianthemum, Laurus-Umbellularia, Juniperus, Myrica, Platanus, Laurocerasus, Styrax*) in the sclerophyll vegetation of California and the Mediterranean region evidently reflects Paleogene links across lower middle latitudes of North America and across a shallower, narrower Atlantic populated with more numerous islands (Axelrod, 1975a, 1975b). Additional taxa (*Bumelia, Clethra, Persea, Pistacia, Quercus, Sageretia, Sapindus, Sabal*) linked these areas into the Miocene and Pliocene when they lived under a regime of summer rainfall and mild temperature. The links were reduced gradually as summer rain decreased over present areas of mediterranean climate and as colder climates developed on the east side of North America, eliminating the broad sclerophyll pathway that had extended across the southern United States-northern Mexico. Significantly, dry sites on the granite-gneiss domes in the Appalachian Piedmont still support unique taxa from the southwestern United States and northern Mexico (e.g., *Agave, Forestiera, Krameria, Nolina, Sageretia, Sedum, Yucca*), and attest to the formerly xerophyllous connections (McVaugh, 1943), as does sedimentological and microfossil evidence reviewed elsewhere (Axelrod, 1975a). This evidence is contrary to the view enunciated by Wolfe (1975, p. 276-277) that the Madrean-Tethyan links are of boreal derivation, an origin for which he provides no evidence and which is contrary to the nature of the temperate mixed forest that occupied that perhumid region and other factors (Axelrod, 1970, p. 309). Furthermore, we may well ask: (1) If the sclerophylls were derived from taxa that were regular members of a northern forest filled with deciduous hardwoods and conifers, how did they attain their adaptive features which ecologists agree are xerophytic? (2) If the sclerophylls were not in the subhumid Madrean-Tethyan region, what kinds of plants inhabited that area?

Extremely distinctive taxa of certain austral groups, such as Proteaceae (Grevilloideae-Brabeium: see Johnson and Briggs, 1963, 1975), Podocarpaceae (sect. Afrocarpus), Cupressaceae (*Widdringtonia*), Cunoniaceae, Restoniaceae, Rutaceae (Diomeae), Philesiaceae, Hydnoraceae, and others occur in South Africa today (Fig. 3). This implies that Africa separated first from South America-Antarctica-Australia, as noted earlier by Adamson (1948). The relation is consistent also with the phyletic relations of chironomid midges (Brundin, 1965, 1966), and many other groups (Keast, 1973; Cracraft, 1974, 1975; Johnson and Briggs, 1975).

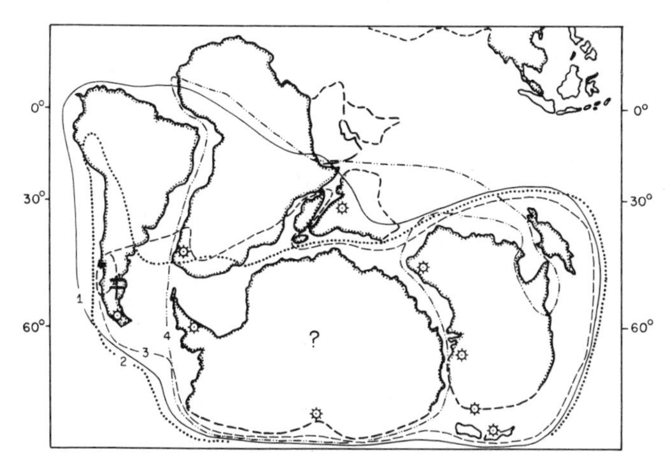

Figure 3. Distribution of 4 modern families typical of austral regions, assembled on a pre-drift world. (1) Proteaceae, with * marking some Eocene and Late Cretaceous fossil localities; (2) Restioniaceae; (3) Philesiaceae; (4) Aponogetonaceae, with ‡ marking a Late Cretaceous locality in Argentina, its sole known occurrence in South America.

RELICT SURVIVAL

Ancient taxa have been preserved by the break up of continental plates, and the subsequent transport of the fragments into marine isolation. There they have been removed from climatic stress, and also from competition with younger, more "aggressive" floras.

Accumulating information continues to clarify the complex history of the Mascarene area, including Madagascar, the Seychelles, and bordering islands. McKenzie and Sclater (1973) noted that no magnetic anomalies appear west of the Mascarene Plateau, on which the Seychelles now rest, implying that the entire Plateau is continental. Kutina (1975) postulates that between the mid-ocean ridge and the present coast of East Africa, there was a continuous land area with a Precambrian basement that disintegrated following the Late Cretaceous, when Madagascar lay along the Somalia-Tanzania coast (Smith and Hallam, 1970; Smith, 1976). The data are consistent with the derivation of the unbalanced mammal fauna by sweepstakes dispersal at various times during the Tertiary (Simpson, 1943). It also provides a route for Cretaceous dinosaurs to Madagascar-India-Australia by eastward migration from Africa prior to the Late Cretaceous.

As noted above, the flora of Madagascar includes many relicts that may be regarded as a sample (albeit modified) of the early Tertiary flora of tropical Africa, prior to its impoverishment. Among the notable endemic families are Didiereaceae and Sarcolaenaceae. Palms, bamboos, orchids, and woody ranalian taxa are abundant there, but poorly represented in Africa. Furthermore, many endemics characterize Madagascar as compared with Africa. The present Seychelles flora, with the relict family Medusagynaceae and many endemics, gives us a hint of the numerous unique taxa that probably were exterminated as erosion and regional subsidence reduced the larger Mascarene land area to the small, scattered, remaining islands.

Development of marginal basins by interaction of the Australian plate with the northwest-moving Pacific plate disrupted the east border of Australia (Griffiths and Varne, 1972). The Campbell Plateau (with New Zealand on it) separated from the Antarctic-Australian plate in the Late Cretaceous (80 m.y.), and by the Paleocene (65 m.y.), the south Tasman Sea was opening (Griffiths and Varne, 1972). During the rest of the Tertiary, the Tasman Sea continued to spread northward and eastward, transporting Lord Howe Rise and Norfolk Ridge (with New Caledonia) farther east. As the Fiji basin spread (Karig, 1971; Chase, 1971), Fiji was transported farther east. Endemic to New Caledonia are many archaic relicts, including Strasburgeriaceae; Degeneriaceae (with only 1 genus, *Degeneria*) is on Fiji only; and Balanopaceae (with only 1 enigmatic genus, *Balanops*) links New Caledonia with Queensland as do *Araucaria, Podocarpus,* Winteraceae, and others (Raven and Axelrod, 1972). New Caledonia and Fiji have preserved numerous relicts that may well be derivatives of the Eocene flora that covered Australia prior to plate fragmentation (see Smith, 1974).

Many relict Australian taxa survive in the mountains of Queensland, New Guinea, and New Caledonia which were built up in the Pliocene and more recently. Although much of northern Australia moved into tropical latitudes after the Eocene, uplift of mountains created climates similar to those which prevailed over the lowlands when Australia was farther south, making possible the survival under equable, temperate climate in tropical latitudes of otherwise largely temperate austral taxa (e.g., *Nothofagus, Agathis,* Winteraceae, and others).

As outlined by Takhtajan (1957), many woody annonalean taxa of diverse families have relict occurrences in the eastern Himalayas and southeast Asia at moderate elevations under moist, equable, warm-temperate climate. They appear to have been preserved from extinction by regional uplift during Miocene and later times as the Indian plate plunged under Asia. The new montane areas provided sites for survival as climates elsewhere became too severe for them.

In a related sense, the rich, woody flora of California was protected topographically from cold waves and dry spells of the Pleistocene by uplift of the Sierra Nevada-Peninsular Ranges and the Coast Ranges in the later Pliocene and Quaternary. In much the same way, elevation of the alpine system of southern Europe formed a protecting barrier for the sclerophyllous flora of the Mediterranean basin by blocking out the cold spells of the glacial ages. Also, the rich mixed deciduous hardwood forest of western Hupeh and nearby Szechuan, filled with relicts such as *Cathaya, Ginkgo, Metasequoia,* and others, is protected by mountains from the outbreaks of severe cold dry air from the Mongolian Plateau. As the dry air moves eastward in winter, it meets warm, moist air from the China Sea, and a thick fog-deck develops that damps the extremes of temperature.

Ancient taxa may be preserved on young volcanic islands that are built up marginally to the continents in response to plate movements. The laurel forest that persists on the volcanic, western Canary Islands is a remnant of a forest that inhabited the northwest African coast into the Miocene (Axelrod, 1975b). The laurel forest has relict taxa that find their nearest relatives in East Africa, in the mountains of tropical Africa, and in India. Guadalupe Island, off the coast of Baja California, has several relicts of earlier times, including the now-extinct sclerophyllous evergreen shrub *Hesperelaea* (Oleaceae), a monotype in the family. Relicts on Reunion and Mauritius appear to be survivors on these composite islands, which are sunken continental islands surmounted by younger volcanic piles (Axelrod, 1960, p. 283).

Juan Fernandez, which is not older than 4 m.y., has *Thrysopteris* which has been regarded generally as an ancient (Mesozoic) relict (see Skottsberg, 1956, p. 221). In addition, the endemic monotypic family Lactoridaceae (composed of *Lactoris* only) is there, as are unique genera in the palm (*Juania*), mint (*Cuminia*), and myrtle (*Nothomyrica*) families, as well as several striking tree-like genera of different tribes of Asteraceae (e.g., *Phoenicoseris, Robinsonia, Centurodendron*: see illustr. in Carlquist, 1965). Relict woody trees of Asteraceae occur also on St. Helena (e.g., *Melanodendron, Commidendron*), Ascension, and other oceanic islands.

In the Hawaiian Islands, the origin of new islands at the southeast end of the chain over a hot spot, and their eventual destruction at the northwest end, has doubtless added to the dynamism of evolution in the archipelago as a whole, and no doubt in other archipelagos of similar origin as well.

EXTERMINATION

A landmass may be rafted to a region that is climatically so unsuited to its flora that the flora is totally exterminated. Antarctica was joined to Australia up to about 50 m.y. ago, and they then shared a similar flora characterized by southern conifers and austral evergreen dicots of diverse families that ranged to southern Chile-Argentina as part of the Antarcto-Tertiary Geoflora. As Antarctica moved into progressively colder climate at higher latitudes, mesic evergreen dicot-podocarp forests gradually gave way to patchy forest, scrub, then to cushion plants, and finally these were overwhelmed by spreading ice in the Pliocene (5 m.y.).

Plate tectonics accounts also for the submersion of old sialic ridges such as Lord Howe Rise and Norfolk Ridge in the Tasman Sea, the Walvis Ridge in the south Atlantic (Perch-Nielsen and others, 1975), the Azores Plateau, and the Mascarene Plateau on which the Seychelles archipelago is perched. Floras that earlier occupied these areas have been entirely or largely (Seychelles, Lord Howe) lost by drowning. An indication of their uniqueness and diversity is apparent from the numerous endemics on the Seychelles, including the entire family Medusagynaceae, and the unique species of plants (*Araucaria*) and animals (terrestrial horned turtles) on small Norfolk and Lord Howe Islands, respectively (see Raven and Axelrod, 1972). Emphasis must be placed on the idea that old simatic rises that have since foundered also formed important paths for migration into the widening Indian, Atlantic, and south Pacific basins during the Cretaceous and Cenozoic. Subsidence of the Iceland-Faeroe Rise in the Miocene (Spuko and Perch-Nielsen, 1975) severed the broad pathway for forests between Europe and North America via Faeroes-Iceland-Greenland.

As islands of an oceanic archipelago are rafted away from hot spots, they are progressively eroded down to sea level and are thence transported subsea. That they earlier supported large, diverse floras is apparent from pollen studies of cores recovered from drilling at Midway Island (Leopold, in Ladd and others 1970), Eniwetok atoll (Leopold, 1969), and from the Ninetyeast Ridge (Kemp and Harris, 1975). Although widespread extinction must occur as the varied adaptive zones disappear, some taxa may survive by dispersing to adjacent young islands early in their history, when open areas are still present. Some unique taxa of oceanic islands may have attained considerable antiquity in this manner. They obviously could not have originated on the high islands where they now occur because most of the islands are quite young, less than 1 m. y. old (Axelrod, 1972a, p. 55-57).

If austral lands move only slightly farther north considerable extinction will result. The rich sclerophyllous floras in the mediterranean climates at the southwest tip of Africa and in southern Australia will totally disappear, as will the *Nothofagus-Podocarpus* evergreen forest of southeast Australia-Tasmania, and the relict temperate rainforest in the Knysna region of southeast Africa.

The fossil occurrence of austral taxa (Podocarpaceae, *Araucaria, Casuarina*, leptodactylid frogs) of temperate requirements in India, and of temperate rainforests (*Nothofagus, Laurelia, Araucaria*, Podocarpaceae) on Antarctica and Kerguelen Island, provide examples of a Viking funeral-ship type of dispersal (McKenna, 1973). The assemblages have been transported to areas well-removed from where they lived, and now occur in latitudes and under climates decidedly different than those that might be inferred for them on the basis of stable continents.

REFERENCES

Adamson, R. S., 1948, Some geographical aspects of the Cape flora: Royal Soc. South Africa Trans., v. 31, p. 437-464.

Asama, K., 1960, Evolution of the leaf forms through the ages explained by the successive retardation and neoteny: Tohoku Univ. Sci. Rept., 2nd ser., Geology, v. 4, p. 252-280.

——— 1962, Evolution of Shansi flora and origin of simple leaf: Tohoku Univ. Sci. Rept., 2nd ser., Geology, v. 5, p. 247-273.

——— 1975, Evolutionary biology in plants. IV. The origin of the angiosperms: Tokyo, Sanseido Co., 400 p.

Axelrod, D. I., 1948, Climate and evolution in western North America during middle Pliocene time: Evolution, v. 2, p. 127-144.

——— 1950, Evolution of desert vegetation in western North America: Carnegie Inst. Washington Pub. 590, p. 215-306.

——— 1952, A theory of angiosperm evolution: Evolution, v. 6, p. 29-60.

——— 1959, Poleward migration of the early angiosperm flora: Science, v. 130, p. 203-207.

——— 1960, The evolution of flowering plants, *in* Tax, S., ed., Evolution after Darwin. Vol. 1. The evolution of life: Chicago, Univ. Chicago Press, p. 227-307.

——— 1967, Drought, diastrophism and quantum evolution: Evolution, v. 21, p. 201-209.

——— 1970, Mesozoic paleogeography and early angiosperm history: Bot. Rev., v. 36, p. 277-319.

——— 1972a, Ocean-floor spreading in relation to ecosystematic problems, *in* Allen, R. T., and James, F. C., eds., A symposium on ecosystematics: Arkansas Univ. Mus. Occas. Paper no. 4, p. 15-76.

——— 1972b, Edaphic aridity as a factor in angiosperm evolution: Am. Naturalist, v. 106, p. 311-320.

——— 1972c, Revolutions in the plant world: Geol. Soc. America, Abs. with Programs, v. 4, p. 124.

——— 1974a, Plate tectonics in relation to the history of angiosperm vegetation in India: Lucknow, Birbal Sahni Inst. Palaeobotany Spec. Pub., v. 1, p. 5-18 (1971).

——— 1974b, Revolutions in the plant world: Geophytology, v. 4, p. 1-6.

——— 1975a, Plate tectonics and problems of angiosperm history: Mus. Natl. d'Hist. Naturelle Mém., n.s., ser. A., Zool., v. 88, p. 72-85 (1972).

——— 1975b, Evolution and biogeography of Madrean-Tethyan sclerophyll vegetation: Ann. Missouri Botanical Garden, v. 62, p. 280-334.

Axelrod, D. I., and Raven, P. H., 1972, Evolution and biogeography viewed from plate tectonic theory, *in* Behnke, J. A., ed., Challenging biological problems: Directions toward their solution: New York, Oxford Univ. Press, p. 218-236.

Bews, J. W., 1927, Studies in the ecological evolution of the angiosperms: New Phytologist Reprint 16, 134 p.

Brundin, L., 1965, On the real nature of transantarctic relationships: Evolution, v. 19, p. 496-505.

——— 1966, Transantarctic relationships and their significance, as evidenced by chironomid midges: Kgl. Svenska Vetenskapsakad. Handl., v. 11, p. 1-72.

Buxbaum, F., 1969, Die Entwicklingswege der Kakteen in Sudamerika, *in* Fittkau, J. and others, eds., Biogeography and ecology in South America, Monographiae Biologicae, Vol. 19: The Hague, W. Junk, p. 583-623.

Camp, W. H., 1947, Distribution patterns in modern plants and the problems of ancient dispersals: Ecological Mon., v. 17, p. 159-183.

Carlquist, S., 1965, Island life, a natural history of the islands of the world: Garden City, N.Y., Natural History Press, 451 p.

Chase, C. G., 1971, Tectonic history of the Fiji Plateau: Geol. Soc. America Bull., v. 82, p. 3087-3110.

Cheng, Lung-Hua and others, 1974, Chinese fossil plants, Vol. 1: Peking, Academia Sinica, 277 p. (in Chinese).

Clements, F. E., 1920, Plant succession: Carnegie Inst. Washington Pub. 242, 512 p.

Couper, R. A., 1960, Southern Hemisphere Mesozoic and Tertiary Podocarpaceae and Fagaceae and their palaeogeographic significance: Royal Soc. London Proc., ser. B, Biol. Sci., v. 152, p. 491-500.

Cracraft, J., 1974, Continental drift and vertebrate distribution: Ann. Rev. Ecology and Systematics, v. 5, p. 215-261.

——— 1975, Historical biogeography and Earth history: Perspectives for a future analysis: Ann. Missouri Botanical Garden, v. 62, p. 227-250.

Crocker, R. L., 1959, Past climatic fluctuations and their influence upon Australian vegetation, *in* Keast, A., Crocker, R. L., and Christian, C. S., eds., Biogeography and ecology in Australia, Monographiae Biologicae, Vol. 8: The Hague, W. Junk, p. 283-290.

Crocker, R. L., and Wood, J. G., 1947, Some historical influences on the development of the South Australian vegetation communities and their bearing on concepts and classification in ecology: Royal Soc. South Australia Trans., v. 71, p. 91-136.

Crowell, J. C., and Frakes, L. A., 1970, Phanerozoic glaciation and the causes of ice ages: Am. Jour. Sci., v. 268, p. 193-224.

Dejardin, J., Guillaumet, J.-L., and Mangenot, G., 1973, Contribution à la connaissance de l'élement non endémique de la flora malgache (végétaux vasculaires): Candollea, v. 28, p. 325-391.

Doyle, J. A., 1969, Cretaceous angiosperm pollen of the Atlantic Coastal Plain and its revolutionary significance: Jour. Arnold Arboretum, v. 50, p. 1-35.

Doyle, J. A., Van Campo, M., and Lugardon, B., 1975, Observations on exine structure of *Eucommiidites* and Lower Cretaceous angiosperm pollen: Pollen et Spores, v. 17, p. 429-486.

Doyle, J. A., and Hickey, L. J., 1976, Pollen and leaves from the mid-Cretaceous Potomac Group and their bearing on early angiosperm evolution, *in* Beck, C. B., ed., Origin and early evolution of angiosperms: New York, Columbia Univ. Press, p. 139-206.

Elias, M. K., 1936, Late Paleozoic plants of the midcontinent region as indicators of time and environment: Internat. Geol. Cong., 16th, Washington, D.C. 1933, Rept., v. 1, p. 691-700.

Embleton, B. J. J., and McElhinny, M. W., 1975, The paleoposition of Madagascar: Paleomagnetic evidence from the Isalo Group: Earth and Planetary Sci. Letters, v. 27, p. 329-341.

Florin, R., 1963, The distribution of conifer and taxad genera in time and space: Acta Horti Bergiani, v. 20, p. 121-312.

Gardner, C. A., 1959, The vegetation of Western Australia, *in* Keast, A., Crocker, R. L., and Christian, C. S., eds., Biogeography and ecology in Australia, Monographiae Biologicae, Vol. 8: The Hague, W. Junk, p. 274-282.

Godley, E. J., 1960, The botany of southern Chile in relation to New Zealand and the subantarctic: Royal Soc. London Proc., ser. B, Biol. Sci., v. 152, p. 457-475.

Good, Ronald, 1964, The geography of flowering plants (3rd ed.): New York, Wiley, 518 p.

Griffiths, J. R., and Varne, R., 1972, Evolution of the Tasman Sea, Macquarie Ridge and Alpine Fault: Nature Phys. Sci., v. 235, p. 83-87.

Hayes, D. E., and Ringis, J., 1973, Seafloor spreading in the Tasman Sea: Nature, v. 243, p. 454-458.

Herngreen, G. F. W., 1974a, Palynology of Albian-Cenomanian strata of borehole I-QS-1-MA, State of Maranhão, Brazil: Pollen et Spores, v. 15, p. 515-555.

——— 1974b, Middle Cretaceous palynomorphs from northeastern Brazil: Results of a palynological study of some boreholes and comparison with Africa and the Middle East: Sci. Géol. Bull. Strasbourg, v. 27, p. 101-116.

——— 1975, An upper Senonian pollen assemblage of borehole 3-PIA-10-AL State of Alagoas, Brazil: Pollen et Spores, v. 17, p. 93-140.

Hooker, J. D., 1853, Botany of the Antarctic Voyage of H. M. Discovery Ships "Erebus" and "Terror" in the years 1831-1843. Vol. 2. Flora Novae-Zelandiae, Pt. 1, Introductory Essay, 36 p.

Hughes, N. F., 1974, Angiosperm evolution and the superfluous upland origin hypothesis: Lucknow, Birbal Sahni Inst. Palaeobotany Spec. Pub., v. 1, p. 25-29.

——— 1976, Paleobiology of angiosperm origins: Problems of Mesozoic seed-plant evolution: Cambridge, Cambridge Univ. Press, 272 p.

Johnson, L. A. S., and Briggs, B. G., 1963, Evolution in the Proteaceae: Australian Jour. Botany, v. 11, p. 21-61.

——— 1975, On the Proteaceae—the distribution and classification of a southern family: Linnean Soc. Botany Jour., v. 70, p. 83-182.

Karig, D. E., 1971, Origin and development of marginal basins in the western Pacific: Jour. Geophys. Research, v. 76, p. 2542-2561.

——— 1972, Remnant arcs: Geol. Soc. America Bull., v. 83, p. 1057-1068.

Keast, A., 1973, Contemporary biotas and the separation sequence of the southern continents, *in* Tarling, D. H., and Runcorn, S. K., eds., Implications of continental drift to the Earth sciences, Vol. 1: London, Academic Press, p. 309-343.

Kemp, E. M., and Harris, W. K., 1975, The vegetation of Tertiary islands on the Ninetyeast Ridge: Nature, v. 258, p. 303-307.

Kent, P. E., 1972, Mesozoic history of the east coast of Africa: Nature, v. 238, p. 147-148.

Kent, P. E., Hunt, J. A., and Johnstone, D. W., 1971, The geology and geophysics of coastal Tanzania: Inst. Geol. Sci., Geophys. Paper 6, p. 1-101.

Koechlin, J., 1972, Flora and vegetation of Madagascar, *in* Battistina, R., and Richard-Vindard, G., eds., Biogeography and ecology in Madagascar, Monographiae Biologicae, Vol. 21: The Hague, W. Junk, p. 145-190.

Krassilov, V., 1975a, Dirhopalostachyaceae—a new family of proangiosperms and its bearing on the problem of angiosperm ancestry: Palaeontographica, Abt. B, v. 153, p. 100-110.

——— 1975b, Ancestors of the angiosperms, *in* Vorontsova, N. N., ed., Contemporary problems of evolution 4: Akad. Nauk SSSR, sect. General Biology, Far Eastern Sci. Center, Vladivostok, p. 76-106 (in Russian).

Kutina, J., 1975, Tectonic development and metallogeny of Madagascar with reference to the fracture pattern of the Indian Ocean: Geol. Soc. America Bull., v. 86, p. 582-592.

Lakhanpal, R. N., 1970, Tertiary floras of India and their bearing on the historical geology of the region: Taxon, v. 19, p. 675-694.

Ladd, H. S., Tracey, J. I., Jr., and Gross, G., 1970, Deep drilling on Midway Atoll: U.S. Geol. Survey Prof. Paper 680-A, p. A1-A22.

Leopold, E. B., 1969, Miocene pollen and spore flora of Eniwetok Atoll, Marshall Islands: U.S. Geol. Survey Prof. Paper 260-II, p. 1133-1185.

Mani, M. S., 1974, The vegetation and phytogeography of Assam-Burma, *in* Mani, M. S., ed., Ecology and biogeography in India, Monographiae Biologicae, Vol. 23: The Hague, W. Junk, p. 204-206.

McElhinny, M. W., Embleton, B. J. J., Daly, L., and Pozzi, J.-P. 1976, Paleomagnetic evidence for the location of Madagascar in Gondwanaland: Geology, v. 4, p. 455-457.

McKenna, M. C., 1973, Sweepstakes, filters, corridors, Noah's arks, and beached Viking funeral ships in paleogeography, *in* Tarling, D. H., and Runcorn, S. K., eds., Implications of continental drift to the Earth sciences, Vol. 1: London, Academic Press, p. 295-308.

McKenzie, D. P., and Sclater, J. G., 1973, The evolution of the Indian Ocean: Sci. American, v. 228, p. 62-74.

McVaugh, R., 1943, The vegetation of the granitic flat-rocks of the southeastern United States: Ecological Mon., v. 13, p. 119-166.

Melville, R., 1975, The distribution of Australian relict plants and its bearing on angiosperm evolution: Linnean Soc. Botany Jour., v. 71, p. 67-88.

Meusel, H. van, and Schubert, R., 1971, Beitrage zur pflanzengeographie des Westhimalajas. Pts. I-III: Flora, v. 160, p. 137-194, 373-432, 537-606.

Molnar, P., and Tapponnier, P., 1975, Cenozoic tectonics of Asia: Effects of a continental collision: Science, v. 189, p. 419-426.

Moore, H. J., 1973, Palms in the tropical forest ecosystems of Africa and South America, *in* Meggers, B. J., Ayensu, E. S., and Duckworth, W. D., eds., Tropical forest ecosystems in Africa and South America: A comparative review: Washington D. C., Smithsonian Inst. Press, p. 63-88.

Muller, J., 1970, Palynological evidence on the early differentation of angiosperms: Biol. Rev., v. 45, p. 417-450.

Perch-Nielsen, K. and others, 1975, Leg 39 examines facies changes in South Atlantic: Geotimes, v. 20, no. 3, p. 26-28.

Raven, P. H., 1963, Amphitropical relationships in the floras of North and South America: Quart. Rev. Biol., v. 38, p. 151-177.

——— 1973a, Evolution of subalpine and alpine plant groups in New Zealand: New Zealand Jour. Botany, v. 11, p. 177-200.

——— 1973b, The evolution of Mediterranean floras, *in* di Castri, F., and Mooney, H. A., eds., Mediterranean type ecosystems: Origin and structure, Ecological Studies, Vol. 7: New York, Springer-Verlag. p. 213-224.

Raven, P. H., and Axelrod, D. I., 1972, Plate tectonics and Australasian paleobiogeography: Science, v. 176, p. 1379-1386.

——— 1974, Angiosperm biogeography and past continental movements: Ann. Missouri Botanical Garden, v. 61, p. 539-673.

Rennie, J. V. L., 1931, Note on fossil leaves from the Banke clay: Royal Soc. South Africa Trans., v. 10, p. 251-253.

Richards, P., 1973, Africa, the "odd man out," *in* Meggers, B. J., Ayensu, E. S., and Duckworth, W. D., eds., Tropical forest ecosystems in Africa and South America: A comparative review: Washington, D. C., Smithsonian Inst. Press., p. 21-26.

Saporta, G., 1881, Die Pflanzenwelt vor dem Erscheinen des Menschen: Braunschweig, C. Vogt, 397 p.

Sclater, J. G., and Fisher, R. L., 1974, Evolution of the east-central Indian Ocean, with emphasis on the tectonic setting of the Ninetyeast Ridge: Geol. Soc. America Bull., v. 85, p. 683-702.

Seward, A. C., 1936, Plant life through the ages: London, Cambridge Univ. Press, 607 p.

Simpson, G. G., 1943, Mammals and the nature of continents: Am. Jour. Sci., v. 241, p. 1-31.

——— 1944, Tempo and mode in evolution: New York, Columbia Univ. Press, 237 p.

——— 1953, The major features of evolution: New York, Columbia Univ. Press, 434 p.

Skottsberg, C., 1956, Derivation of the flora and fauna of Juan Fernandez and Easter Island, *in* Skottsberg, C., ed., The natural history of Juan Fernandez and Easter Island, Vol. 1: Uppsala, Almquist and Wiksell, p. 193-438.

Smith, A. G., and Hallam, A., 1970, The fit of the southern continents: Nature, v. 225, p. 139-144.

Smith, J. M. B., 1974, Southern biogeography on the basis of continental drift: A review: Australian Mammal Soc. Jour., v. 1, p. 213-230.

Smith, P. J., 1976a, Madagascar issue settled: Nature, v. 259, p. 80.

——— 1976b, So Madagascar was to the north: Nature, v. 263, p. 729-730.

Specht, R. L., and Rayson, P., 1957, Dark Island heath (Ninety Mile Plain, South Australia). I. Definition of the ecosystem: Australian Jour. Botany, v. 5, p. 52-85.

Spuko, P. R., and Perch-Nielsen, K., 1975, Deep Sea Drilling Project: Geotimes, v. 20, p. 19-20.

Stebbins, G. L., 1952, Aridity as a stimulus to plant evolution: Am. Naturalist, v. 86, p. 33-44.

——— 1974, Flowering plants: Evolution above the species level: Cambridge, Belknap Press, 399 p.

Takhtajan, A., 1957, On the origin of the temperate flora of Eurasia: Bot. Zhurn., v. 42, p. 935-953 (in Russian, with English summary).

Tarling, D. H., and Kent, P. E., 1976, The Madagascar controversy still lives: Nature, v. 261, p. 304-305.

Thorne, R. F., 1973, Floristic relationships between tropical Africa and tropical America, *in* Meggers, B. J., Ayensu, E. S., and Duckworth, W. D., eds., Tropical forest ecosystems in Africa and South America: A comparative review: Washington, D. C., Smithsonian Inst. Press, p. 27-48.

Vester, H., 1940, Die Areale und arealtypen der Angiospermen-familien: Bot. Arch., v. 41, p. 203-275, 295-256, 530-577.

Wild, H., 1968, Phytogeography in south central Africa: Kirkia, v. 6, p. 197-222.

Williams, D. L., 1975, *Piptochaetium* (Gramineae) and associated taxa: Evidence for the Tertiary migration of plants between North and South America: Jour. Biogeography, v. 2, p. 75-85.

Wolfe, J. A., 1975, Some aspects of plant geography of the Northern Hemisphere during the Late Cretaceous and Tertiary: Ann. Missouri Botanical Garden, v. 62, p. 264-279.

Wolfe, J. A., Doyle, J. A., and Page, V. M., 1975, The bases of angiosperm phylogeny: Paleobotany: Ann. Missouri Botanical Garden, v. 62, p. 801-824.

Wood, J. G., 1959, The phytogeography of Australia (in relation to radiation of *Eucalyptus, Acacia,* etc.), *in* Keast, A., Crocker, R. L., and Christian, C. S., eds., Biogeography and ecology in Australia, Monographiae Biologicae, Vol. 8: The Hague, W. Junk, p. 291-302.

The Role of Biogeographic Provinces in Regulating Marine Faunal Diversity through Geologic Time

THOMAS J. M. SCHOPF, *Department of the Geophysical Sciences, University of Chicago, Chicago, Illinois, 60637*

ABSTRACT

Faunal diversity over geologic time may change according to the number of faunal provinces, as well as the size of individual provinces. The number of provinces is two to three times more important a factor in changing total diversity than is the effect of changing the size of existing provinces. A change in number of major faunal provinces is dependent upon alterations in major ocean currents which themselves are altered over the time scale that it takes major landmasses to move (millions of years for a major change). In contrast, the size of individual provinces can change significantly over a few thousand years owing to changes in sea level. Both the factors of endemism and change in size of individual provinces seem discernable in the pattern of extinctions and originations during the Permo-Triassic.

The number of major marine faunal provinces can be inferred from the pattern of oceanic circulation. Based on predicted patterns of ocean currents, there were approximately thirteen marine faunal provinces in the Silurian, fourteen in the Early Permian, eight in the latest Permian and eighteen today. The pattern of data on diversity over geologic time parallels these changes in endemism. These seemingly parallel patterns are in agreement with a steady-state faunal diversity which may be regulated by the number of faunal provinces which exist at any given time, owing to particular geographic configurations.

INTRODUCTION

This paper outlines a method of evaluating the effects of changing numbers of marine faunal provinces versus changing habitable area, on faunal diversity for geologically long periods of time. Thus, the paper addresses the question of the extent to which the increase in diversity from the Cambrian to the present time may be accounted for simply by changing the extent of endemism. Endemism is known to increase world faunal diversity (witness e.g., Australia), and so faunal diversity should be expected to change if endemism changes. Only after this established mechanism of control of faunal diversity is evaluated can one then test for any residual influence on observed diversity owing to a change in species packing or in preservation. The tentative conclusion is that changes in endemism may account for a large part of the observed changes in faunal diversity through geologic time.

The chief advantage of setting up a biogeographic hypothesis of a certain number of faunal provinces (independent of empirical data) is that it tells one what to expect. What one seeks to obtain is the null hypothesis of the expected number of faunal provinces with which one can then, with an open mind, compare the empirical data. This is especially important because differential erosion may selectively eliminate some faunal regions relative to others, thus leaving a highly biased rock record. In addition, faunal surveys of provinces based on world patterns of a single taxonomic group are subject to strong distortions if that group is unequally distributed in its regions of development, as indeed all groups are. The extreme alternative to predicting biogeographic patterns is to rely upon empirical summations of data that don't provide any *a priori* reason why one region should or should not be different than any other region.

ESTIMATING POSITION AND NUMBER OF FAUNAL PROVINCES

In the simplest of all geographical worlds, we can conceive of a single continent, as existed in Pangaea, and we can predict the ocean currents surrounding it, and its biogeographic provinces (Fig. 1). Given the laws of physics, a rotating stratified fluid like the ocean, and a zonal wind circulation resulting from the same causes as exist today, then the same *general* pattern of ocean currents will result. There will always be a westward flowing equatorial current, a mid-latitude "Gulf-Stream" circulation, and a highest latitude (say higher than 70°) clockwise flow associated with the polar high, as exists today in the Arctic Ocean (Ostenso, 1966). In such a world, we can predict eight major faunal provinces: northern and southern polar, eastern and western tropical, and four intermediate temperate provinces (one on each side of the continent, in both Northern and Southern Hemispheres). These provinces are related to major changes in water temperature owing to the patterns of ocean current circulation. In this sort of a world, there will always be a Gulf Stream, indeed two of them, one in each hemisphere; there will always be a westward-flowing equatorial current, and so on.

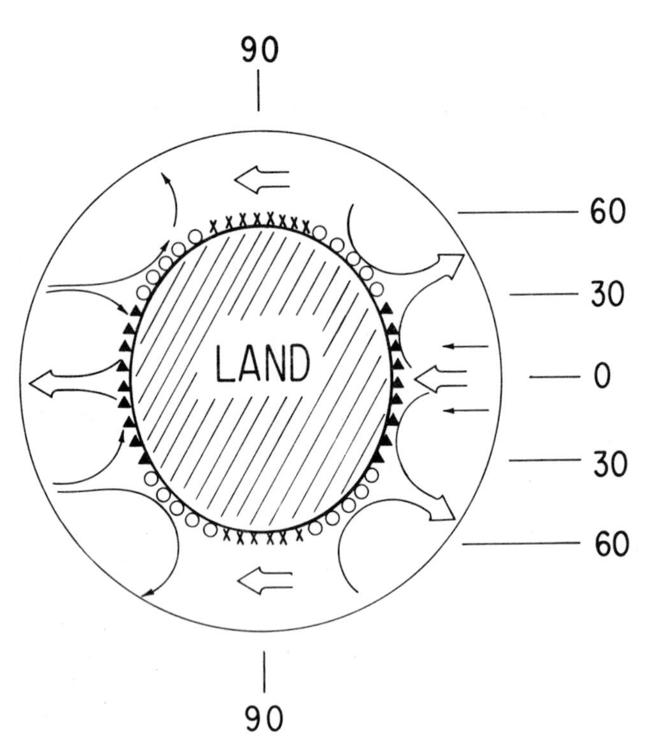

Figure 1. Hypothetical landmass surrounded by world ocean to illustrate current patterns and faunal provinces. Eight faunal provinces are indicated: two tropical (triangles), four temperate (circles), and two polar (x's).

Based on knowledge of the modern world, it also appears that a large body of water, even as newly separated as the Mediterranean, would constitute a separate province. And there seems to be one additional guideline. The tropical region of the Indian Ocean and western Pacific, extending from Somalia to the Solomons (about 15,000 km), seems to be divided into three faunal regions despite a "continuous" water connection. The reason for this biotic division may be that dispersal is sufficiently weak over those large distances so that gene flow is simply insufficient to maintain reproductive continuity. Thus, although larval transport is *possible* for great distances (Scheltema, 1971), it appears unlikely to be of *major* biogeographic importance for distances greater than 4,000 to 5,000 km.

This model is especially designed for the large portion (75 percent?) of Phanerozoic time when the extent of marine transgressions is less than, say 40 percent. Certain aspects of the model relating to continents as barriers may not apply as well to those times when epicontinental seas are said to have covered as much as 75 percent of continental regions, especially for narrow continents (say less than 5,000 km).

I believe the model is largely independent of the strength of the latitudinal temperature gradient. (For the alternative view that the steepness of the temperature gradient is of fundamental importance in regulating taxonomic diversity, see Valentine, 1968, 1973a, 1973b). Even in the absence of coastal geographic barriers, current systems alone are entirely sufficient to produce faunal provinces. Today, changes in the phytoplankton, zooplankton, and fish communities occur as one moves in the open Atlantic from the equator northward toward Bermuda, and then toward Greenland (Backus and others, 1965; John and Backus, 1976; Maynard, 1976). Species which are more-or-less confined to separate current systems maintain discrete populations, but not because changes in temperature per se present a fundamental physiological barrier to their distribution.

If one applies this general method of determining faunal provinces to the present world (Fig. 2), eighteen faunal provinces are predicted. These provinces are very similar to the major provinces of the present world, and are the minimum number likely to exist in a world of similar geographic separation. For comparison, I give (Fig. 3) the empirically derived modern faunal provinces for bryozoans, with twenty-three provinces (see Appendix). Approximately 30 to 50 percent of the species in each province are endemic to each province. Valentine (1973a, p. 356) cites thirty-one marine molluscan provinces of the world's continental shelves, but many of these are subdivisions of the major provinces shown for the bryozoa. In any event, certainly the major biogeographic provinces are indicated by considerations of the general oceanic circulation.

The differences between the anticipated and the real patterns are chiefly in the lack of recognition of small provinces of temperate latitudes and of oceanic islands, and in the exact configuration of tropical provinces where current systems are strongly controlled by local geography (e.g., see southeast Asia). The small "extra" temperate latitude provinces are chiefly geographically controlled (e.g., between Cape Hatteras and Cape Cod) and are characterized by steep temperature gradients. These *may* represent the degree of provincial addition which some

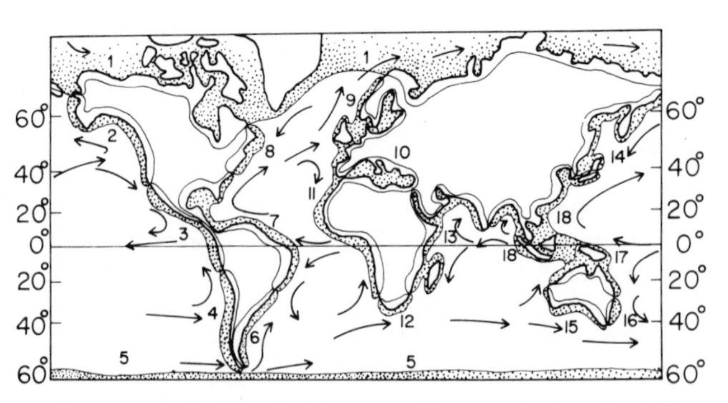

Figure 2. Predicted biogeographic provinces of modern shelf regions according to deductive model discussed in text. Note that in comparison to the "real" faunal provinces (Fig. 3), the predicted pattern omits the smaller oceanic islands such as Hawaii, and such geographically controlled provinces as the Virginian between Cape Cod and Cape Hatteras. The outlines of continents are generalized so as to be approximately commensurate with what might be expected for paleogeographic reconstructions.

believe is characteristic of a world with a steep thermal gradient as opposed to a shallow temperature gradient. Such provinces contribute very little to total diversity, and do not greatly disturb the major patterns.

If we apply the same method of prediction of faunal provinces to Pangaea, we will have eight provinces, the minimum number conceived. Empirical summaries of Pangaean brachiopod faunas of the Late Permian reveal tropical, temperate, and polar faunas in both the Northern and Southern Hemispheres (Waterhouse and Bonham-Carter, 1975). For the Early Permian, when the Siberian plate and the Angaran plate were set off as distinct continental regions, I estimate fourteen faunal provinces (geographic reconstructions as in Schopf, 1974, Figs. 3, 4).

During the Silurian, the continental regions of Gondwana, North China, South China, Baltica, Siberia, Malaysia, Tibetia, Kazkhstania, and Laurentia appear to have been distinct (McKerrow and Ziegler, 1972; Ziegler and others, 1977). Application of the principles outlined above yields a prediction of approximately thirteen faunal provinces, some of which, however, have very little rock remaining today. This estimate appears to be perhaps twice as high as is indicated in the empirical summaries (Boucot, 1974), although the empirical data are not plotted on drift reconstructions, thus making more difficult the evaluation of provincialism. It is of critical importance to have a firm understanding of mid-Paleozoic endemism set in mid-Paleozoic geography as a test for what we can call the biogeographic theory for faunal diversity.

ENDEMISM VERSUS AREA IN CONTROLLING DIVERSITY

Changes in diversity through time can be attributed to three basic causes: changes in habitable area, endemism, or the number of ways to make a living. Borrowing from modern ecology, we can predict the effects in diversity owing to changes in degree of habitable area versus changes in endemism.

For the present study, I will use (without affirming) the assumptions that (a) provinces are well-defined and bounded units and (b) species richness has a fixed limit, determined by the number of ways of making a living that relates to area. If these assumptions are relaxed, the essential picture of diversity relationships consequent on number of provinces would not be altered, but the over-all rate of increase in diversity from the past to the present would be increased. Over evolutionary time, an equilibrium should be reached between the rate of origination of new taxa and the rate of extinction of old taxa within a given faunal province much as Terborgh (1973) describes. The equilibrium number of species is strongly related to the size of a region being considered, and may very well apply even at continental size regions (Flessa, 1975; Rosenzweig, 1975). MacArthur and Wilson (1967, p. 8), for example, record the number of species of reptiles and amphibians against island size over five orders of magnitude, on a log-log plot. The equation of the line is:

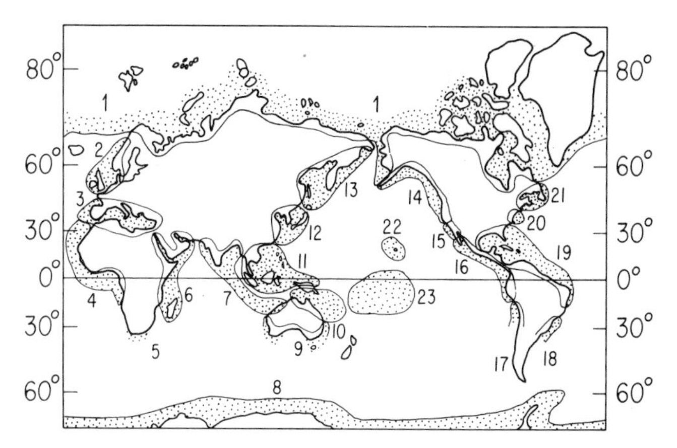

Figure 3. Bryozoan biogeographic provinces for continental shelves. Data given in Appendix.

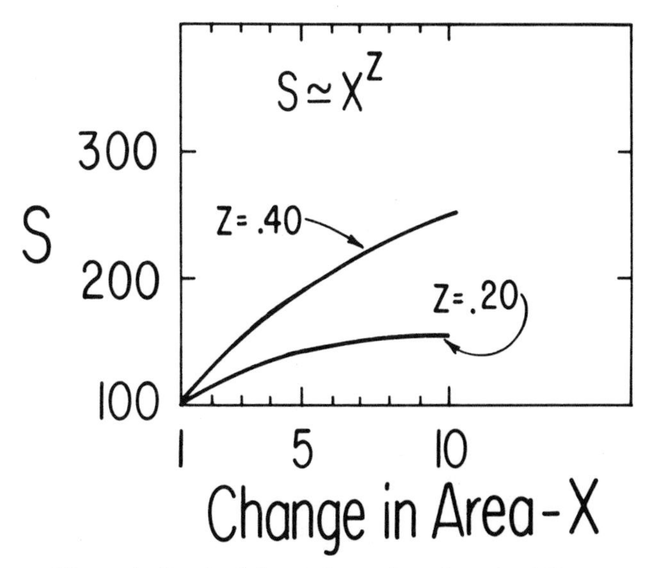

Figure 4. Graph of change in number of species (S) versus increasing habitable area (X). S is proportional to X to the z power where z is shown as 0.20 and as 0.40.

S (number of species) = c (intercept) × A (area) to the z power (the slope of the line).

Empirically z has been found to be close to 0.26 for species, and between 0.2 and 0.4 for species, genera, and families (Flessa, 1975). For the Permian families of marine invertebrates, z values are .27 to .30. We can also express this formula, $S = cA^z$, as:

The log of the number of species = z × the log of the area + the log of the constant (a straight line on the log-log plot, of the form $y = mx + b$).

Since the area of the shallow-marine environment is finite, let us investigate the changes in the diversity which would occur if one simply changed habitable area. I consider (Fig. 4) a faunal province initially with 100 taxa, and increase the area of the province to ten times the original area, and plot the effect on the number of taxa.

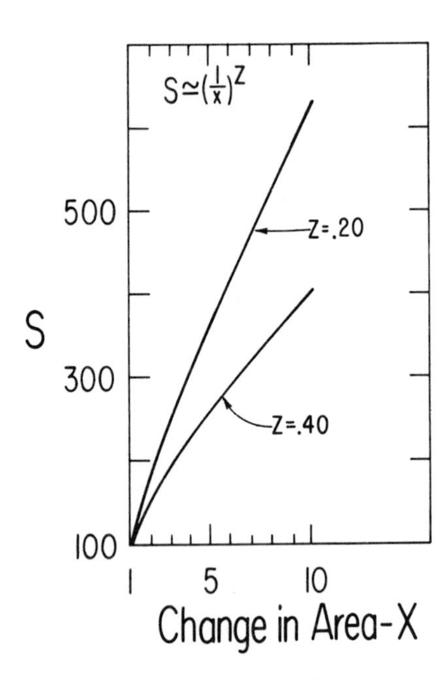

Figure 5. Graph of change in number of species (S) versus increasing numbers of islands (X). As a single island is subdivided into X smaller islands (each 1/X the size of the original), the number of species on each island is proportional to $(I/X)^Z$. The total number of species from X number of islands is what is shown for the two z values, assuming that evolutionary time has yielded different faunas on each island.

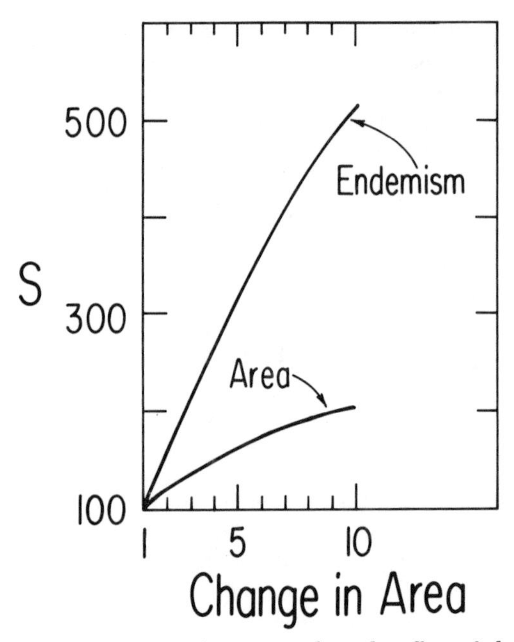

Figure 6. Summary diagram to show the effect of change in area and the change in degree of endemism (number of islands) on changes in total species (S) diversity. Lines are mean values for the range of values shown in Figures 2 and 3.

Here we see that the faunal diversity would increase by about a factor of 2, i.e., from 100 to 200, for a 10-fold increase in area. For many marine faunal provinces in existence, if each of them increased to a similar extent (say due to eustatic rise in sea level), then a 10-fold increase in number of square kilometers covered by the sea would only double the fauna, after speciation catches up with the area.

We can next compute how diversity would change by increasing endemism (Fig. 5). If an original area is cut into two regions, say by drifting apart, then the resulting fauna of each region is reduced by one-half to the 0.26 power; an original faunal province of 100 taxa would yield two regions, each able to accommodate 84 taxa. As evolution proceeds, and identity in taxa decreases, then a total fauna of 168 taxa could result whereas originally there were only 100 total taxa.

For a 10-fold reduction in area, i.e., ten provinces, each of them one-tenth the size of the original, the total fauna which could coexist becomes four to six times the original fauna (Fig. 5). The reverse situation of joining regions is also important. For two islands each with 84 taxa, or 168 total, joined into one island twice the size of the original island, only 100 total taxa can exist, and clearly a very significant extinction must occur if equilibrium is maintained (Flessa and Imbrie, 1973).

In general, if $S_1 = cA^Z$ and $S_2 = c\ (10A)^Z$, for a 10-fold increase in area, then $S_2/S_1 = 10^Z$ for increased area. In addition, if

$$S_3 = c^{10} \left(\frac{A}{10}\right)^Z$$

for a 10-fold subdivision in area, then $S_3/S_1 = 10^{1-Z}$, for subdividing area into provinces.

Thus, if $z < 0.5$, then the effect of adding provinces is greater than that of simply increasing area. As noted above, z values of about 0.3 are commonly observed for species, genera, and families of major taxa. This is in agreement with the idea that changes in endemism are more important than changes in area in regulating faunal diversity.

We can compare both processes on the same graph (Fig. 6). Changes in diversity owing to increased endemism are two to three times those owing simply to increase in area for each available habitat. Since 10-fold increases in habitable marine area are uncommon, I would conclude that changes in diversity owing only to changes in area may be modest, though recognizable. It seems quite clear that the easiest way to contribute to increased or decreased diversity is to increase or decrease endemism. The measure of endemism is simply the number of faunal provinces.

APPLICATION TO GEOLOGIC EXAMPLES

I believe that the effect of both changes in endemism and changes in area can be seen separately in the famous

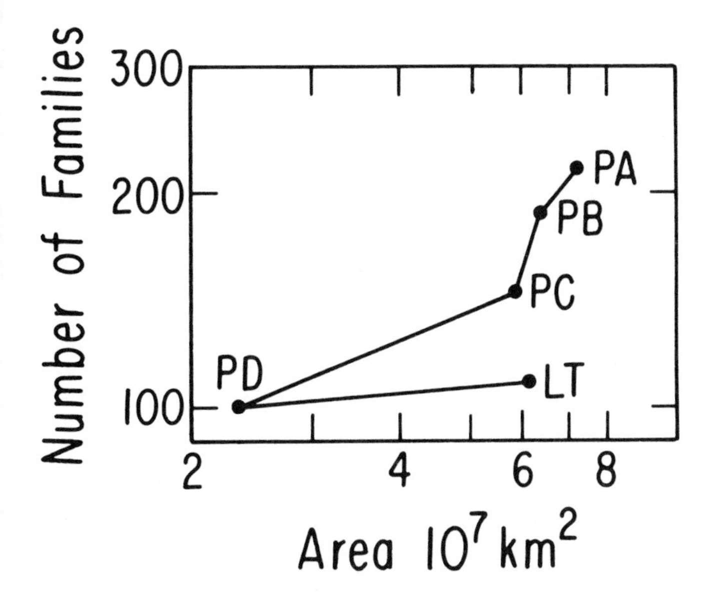

Figure 7. Plot on a log-log scale of number of Permian families of shallow-marine invertebrates versus area of Permian shallow-marine seas. PA is Early Permian, PB and PC are Middle Permian, and PD is Late Permian; LT is Early Triassic. Data on change in Permian diversity and area were computed by Schopf (1974), and are cited in Simberloff (1974, Table 1).

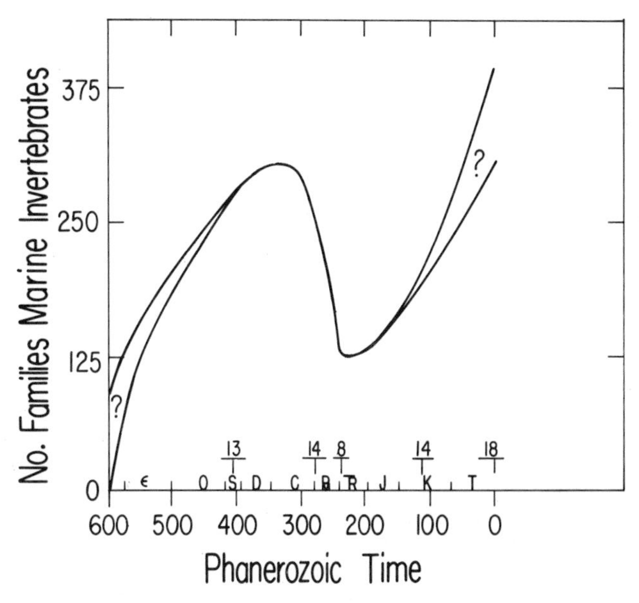

Figure 8. Plot of estimated number of families of marine invertebrates through the Phanerozoic (in millions of years before present). Letters at bottom are symbols for geologic periods; numbers above letters (18, 14, 8, 18) refer to estimated number of major faunal provinces for those periods of geologic time. Question mark for 600 m. y. refers to lack of knowledge of soft-bodied forms; question mark for Tertiary refers to unevaluated bias of more adequate fossil record which by itself will make it appear as though diversity has increased (Raup, 1972). Data for the Permian are tabulated by Schopf (1974). The Cretaceous appears to have approximately the same number of provinces as did the Lower Permian. Kauffman (1973) lists 17 provinces and subprovinces, which when examined by the method of this article leads to approximately 13 to 14 provinces.

Permo-Triassic extinctions. Figure 7 shows a log-log plot of Permian diversity versus Permian area. Note the steep slope going from lowermost Permian (PA) to Middle Permian (PB, then PC). These periods of time correspond to a changing paleogeography with the addition of the Siberian plate to Europe (and the formation of the Ural Mountains) and the addition of the Angaran plate to southern Asia (paleogeographic maps of these intervals given by Schopf, 1974, Figs. 3, 4). In applying the method outlined in the first section of this paper, I estimate that this paleogeographic change resulted in a reduction from fourteen to eight major faunal provinces. Thus, the main reduction in diversity from the Lower to Middle Permian is attributed to changes in the extent of endemism because of the steep slope in the species-area curve.

The subsequent changes in diversity have a much lower slope on the log-log plot going from the Middle Permian to latest Permian (PC to PD), and from the latest Permian into the Lower Triassic (PD to LT) (Fig. 7). There is no discernible major paleogeographic shift during these times. The major change in diversity is attributed to the effect of changing the size (via sea-level changes) of the provinces which did exist. There is no increase in endemism during the Lower Triassic, only a change in habitable area owing to a change in sea level of approximately 200 m (Forney, 1975).

Figure 8 shows a typical curve of taxonomic diversity through the Phanerozoic. In common with all such graphs, we see a rapid early Paleozoic rise, a mid-Paleozoic plateau of highly diverse faunas, and a significant reduction (perhaps about in half) at the end of the Paleozoic. There is an initially slow Triassic recovery, and then an increasing rise in diversity with present levels on the order of or higher than those of the mid-Paleozoic. I have indicated at the bottom of Figure 8 the number of faunal provinces estimated by the method previously outlined.

The post-Triassic rise in faunal diversity (Fig. 8) may be attributed to changes in increased endemism as continental masses fragmented from the Permo-Triassic Pangaea (Valentine, 1969; Valentine and Moores, 1970). I suggest that there has been an increase in the number of major faunal provinces from eight to approximately eighteen. Whereas only two of the provinces of the latest Permian-Early Triassic are equatorial, five or six of the major provinces inferred for the Silurian and for the Tertiary are tropical. Since modern sea level covers approximately as much shelf as was covered by the ocean during the latest Permian (Schopf, 1974; Forney, 1975), there is obviously no way to attribute the modern high diversity to changes in habitable area of individual faunal provinces. Changes in degree of endemism are certainly in the right direction to account for these changes in diversity.

CONCLUSION

Considerable empirical evidence suggests that there is an equilibrium number of species per unit area, i.e., an equilibrium diversity. In addition, the magnitude of the diversity is much more sensitive to the number of faunal regions than to the size of any given region. Thus, if there exists a way to determine the number of faunal regions at any given time, that number can be used as an index to total faunal diversity. Since the number of faunal provinces is regulated by geographic configurations, we may apply the aphorism that geologic history is the pacemaker of biologic diversity.

This paper advocates the deductive prediction of the number of faunal provinces based on ocean current systems and geography. At present, the correlation over geologic time between the number of inferred major faunal provinces and total estimated diversity is sufficiently great to warrant detailed testing.

ACKNOWLEDGMENTS

I thank my colleague, A. M. Ziegler, for the prepublication use of his Silurian reconstructions. For providing very useful reviews of an earlier draft of this paper, I thank K. Flessa, M. Rosenzweig, R. H. Whittaker, R. Osman, J. Terborgh, and A. M. Ziegler.

REFERENCES

Andosova, E., 1973, Moss animals (Bryozoa) Antarctic and sub-Antarctic: Soviet Antarctic Expeditions Bull. 87, p. 65-69 (in Russian).

Backus, R. H., Mead, G. W., Haedrich, R. L., and Ebeling, A. W., 1965, The mesopelagic fishes collected during Cruise 17 of the R/V Chain, with a method for analyzing faunal transects: Harvard Univ., Mus. Comparative Zoology Bull., v. 134, p. 139-157.

Boucot, A. J., 1974, Silurian and Devonian biogeography, *in* Ross, C. A., ed., Paleogeographic provinces and provinciality: Soc. Econ. Paleontologists and Mineralogists Spec. Pub. 21, p. 165-176.

Cook, P. L., 1968, Bryozoa (Polyzoa) from the coasts of tropical West Africa. Atlantide Report No. 10. Scientific results of the Danish Expedition to the coasts of tropical West Africa 1945-1946: Copenhagen, Danish Science Press, p. 115-262.

Flessa, K. W., 1975, Area, continental drift and mammalian diversity: Paleobiology, v. 1, p. 189-194.

Flessa, K. W., and Imbrie, J., 1973, Evolutionary pulsations: Evidence from Phanerozoic diversity patterns, *in* Tarling, D. H. and Runcorn, S. K., eds., Implications of continental drift for the Earth sciences, Vol. 1: London, Academic Press, p. 247-285.

Forney, G. G., 1975, Permo-Triassic sea-level change: Jour. Geology, v. 83, p. 773-779.

Gautier, Y. V., 1961, Récherches écologiques sur les Bryozoaires Chilostomes en Méditerranée occidentale: Thesis, Université d'Aix-Marseille, Aix-en-Provence, 434 p.

Hyman, L. H., 1959, The invertebrates: Smaller coelomate groups, Vol. 5: New York, McGraw-Hill, 783 p.

John, A. E., and Backus, R. H., 1976, On the mesopelagic fish faunas of slope water, Gulf Stream, and another Sargasso Sea: Deep-Sea Research, v. 23, p. 223- 234.

Kauffman, E. G., 1973, Cretaceous Bivalvia, *in* Hallam, A., ed., Atlas of palaeobiogeography: Amsterdam, Elsevier, p. 353-383.

Kluge, G. A., 1962, Bryozoa of the Northern Seas of USSR, *in* Guides to fauna of the USSR: Akad. Nauk SSSR, Zool. Inst., v. 8, p. 1-582 (in Russian).

MacArthur, R. H., and Wilson, E. O., 1967, The theory of island biogeography: Princeton, Princeton Univ. Press, 203 p.

MacGillivray, P. H., 1887, A catalogue of the marine polyzoa of Victoria: Royal Soc. Victoria Trans. Proc., v. 23, p. 187-224.

Marcus, E., 1921, Über die Verbreitung der Meeresbryozoen: Zool. Anzeiger, v. 53, p. 205-221.

Maturo, F. J. S., 1968, The distributional pattern of the Bryozoa of the east coast of the United States exclusive of New England: Atti Società Italiana Scienze Naturali Museo Civico Storia Naturale Milano, v. 108, p. 261-284.

Maynard, N. G., 1976, The relationship between diatoms in the surface sediments of the Atlantic Ocean and the biological and physical oceanography of overlying waters: Paleobiology, v. 2, p. 91-121.

McKerrow, S., and Ziegler, A. M., 1972, Paleozoic oceans: Nature, v. 240, p. 92-94.

Medioni, A., 1970, Les peuplements sessiles des fonds rocheux de la region de Banyuls-sur-Mer: Ascidies-Bryozoaires, Pt. 1: Vie et Milieu, ser. B, Oceanographie, v. 21, p. 591-656.

Moyano, G. H. I., 1968, Distribucion y profundidades de las especies exclusivamente Antarticas de Bryozoa Cheilostomata recolectadas por la Decimonovena Expedicion Antartica Chilena 1964-65: Soc. Biol. Concepcion Bol., v. 40, p. 113-123.

Nordgaard, O., 1918, Bryozoa from the Arctic regions: Tromsø Mus. Aarshefter, v. 40, p. 1-99 (1917).

——— 1923, Bryozoa, *in* Report of the scientific results of the Norwegian Expedition to Novaya Zemlya 1921, no. 17: Videnskaps. i Kristiania: Oslo, A. W. Brøggers Boktrykkeri, 19 p.

——— 1927, Bryozoa, *in* Gronlie, O. T., and Soot-Ryen, T., The Folden Fiord: Zoological, hydrographical and Quaternary geological observations made in the Folden Fiord during the summer of 1923: Tromsø Mus. Skrifter, v. 1, pt. 9, p. 1-10.

O'Donoghue, C. H., 1957, Some South African Bryozoa: Royal Soc. South Africa Trans., v. 25, p. 71-93.

O'Donoghue, C. H., and de Watteville, D., 1937, Notes on South African Bryozoa: Zool. Anzeiger, v. 117, p. 12-22.

——— 1944, Additional notes on Bryozoa from South Africa: Natal Museum Ann., v. 10, p. 407-432.

Okada, Y., and Mawatari, S., 1956, Distributional provinces of marine Bryozoa in the Indo-Pacific region: Pacific Sci. Congress, 8th, Quezon City 1949, Oceanography, Proc., v. 2, p. 391-402.

Osburn, R. C., 1955, The circumpolar distribution of Arctic-Alaskan Bryozoa, *in* Essays in the natural sciences in honor of Captain Allan Hancock: Los Angeles, Univ. Southern California Press, p. 29-38.

Ostenso, N. A., 1966, Arctic ocean, *in* Fairbridge, R. W., ed., Encyclopedia of oceanography: New York, Reinhold, p. 49-55.

Pastula, E. J., 1970, World atlas of coastal biological fouling. Pt. I. North America, South America, Iceland, and Greenland: Washington, D. C., Naval Oceanographic Office, 81 p. (informal report).

Raup, D. M., 1972, Taxonomic diversity during the Phanerozoic: Science, v. 177, p. 1065-1071.

Rogick, M. D., 1965, Bryozoa of the Antarctic, *in* Van Dye, P., and Van Mieghem, J., eds., Biogeography and ecology in Antarctica, Monographiae Biologicae, Vol. 15: The Hague, W. Junk, p. 401-413.

Rosenzweig, M. L., 1975, On continental steady states of species diversity, *in* Cody, M. L., and Diamond, J. M., eds., Ecology and evolution of communities: Cambridge, Mass., Belknap Press, p. 121-140.

Ryland, J. S., 1969, A nomenclatural index to "A history of the British Marine Polyzoa" by T. Hincks (1880): British Mus. (Nat. History) Bull., Zool., v. 17, p. 201-260.

Scheltema, R. S., 1971, Larval dispersal as a means of genetic exchange between geographically separated populations of shallow-water benthic marine gastropods: Bio. Bull., v. 140, p. 284-322.

Schopf, T. J. M., 1969a, Geographic and depth distribution of the phylum Ectoprocta from 200 to 6,000 meters: Am. Philos. Soc. Proc., v. 113, p. 464-474.

——— 1969b, Paleoecology of ectoprocts (bryozoans): Jour. Paleontology, v. 43, p. 234-244.

——— 1970, Taxonomic diversity gradients of ectoprocts and bivalves and their geologic implications: Geol. Soc. America Bull., v. 81, p. 3765-3768.

——— 1973, Ergonomics of polymorphism: Its relation to the colony as the unit of natural selection in species of the phylum Ectoprocta, *in* Boardman, R. S., Cheetham, A. H., and Oliver, W. A., eds., Animal colonies: Stroudsburg, Pa., Dowden, Hutchinson and Ross, p. 247-294.

——— 1974, Permo-Triassic extinctions: Relation to sea-floor spreading: Jour. Geology, v. 82, p. 129-143.

Simberloff, D. S., 1974, Permo-Triassic extinctions: Effects of area on biotic equilibrium: Jour. Geology, v. 82, p. 267-274.

Soule, D. F., and Soule, J. D., 1967, Faunal affinities of some Hawaiian Bryozoa (Ectoprocta): California Acad. Sci. Proc., v. 35, p. 265-272.

Spjeldnaes, N., 1963, Climatically induced faunal migrations: Examples from the littoral fauna of the late Pleistocene of Norway, *in* Nairn, A. E. M., ed., Problems in palaeoclimatology: London, Wiley-Interscience, p. 353-357.

Terborgh, J., 1973, On the notion of favorableness in plant ecology: Am. Naturalist, v. 107, p. 481-501.

Valentine, J. W., 1968, Climatic regulation of species diversification and extinction: Geol. Soc. American Bull, v. 79, p. 273-276.

——— 1969, Patterns of taxonomic and ecological structure of the shelf benthos during Phanerozoic time: Palaeontology, v. 12, p. 684-709.

——— 1973a, Evolutionary paleoecology of the marine biosphere: Englewood Cliffs, N. J., Prentice-Hall, 511 p.

——— 1973b, Plates and provinciality, a theoretical history of environmental discontinuities, *in* Hughes, N. F., ed., Organisms and continents through time: Spec. Papers Palaeontology no. 12, p. 79-92.

Valentine, J. W., and Moores, E. M., 1970, Plate-tectonic regulation of faunal diversity and sea level: A model: Nature, v. 228, p. 657-659.

Waterhouse, J. B., and Bonham-Carter, G. F., 1975, Global distribution and character of Permian biomes based on brachiopod assemblages: Canadian Jour. Earth Sci., v. 12, p. 1085-1146.

Williams, C. B., 1964, Patterns in the balance of nature: London, Academic Press, 324 p.

Ziegler, A. M. and others, 1977, Silurian continental distributions, paleogeography, climatology, and biogeography: Tectonophysics, v. 40, p. 13-50.

APPENDIX

Biogeography of Bryozoa

The purpose of this appendix is to provide documentation for Figure 3 and to summarize what is known of diversity gradients in Bryozoa.

Provinces. No world-wide review of bryozoan biogeography has appeared in more than a half century (Marcus, 1921). Accordingly, a more modern view of biogeographic provinces is given in Table 1 and Figure 3. These provinces refer only to the continental shelves. In water deeper than the shelf break (approximately 200 m), biogeographic patterns may be much different but bryozoans have been described from only a few of the potential biogeographic provinces; indeed, adequate data exists (Schopf, 1969a, Fig. 1) only for a determination of North Atlantic provinces.

Including duplicate records, the count of the number of species obtained so far is approximately 6,000 (Table 1). Okada and Mawatari (1956) indicate that of the 2,418 species in the Ethiopian, Indian, Malayan, China, Papuan, and Mexican Provinces, about a third (729) are endemic to some province. Adjacent provinces (Ethiopian versus Indian; Malayan versus Chinese; Malayan versus Papuan) share 30 percent of their species and have 70 percent of their species distinct. Thus, a province has about 30 percent endemics, shares 30 percent of its fauna with each of the two adjacent provinces, and has about 10 percent of its fauna which has been reported from elsewhere.

The percentage of endemics varies from region to region. In the Antarctic, 179 of 321, or 56 percent of species are endemics (Rogick, 1965). This high proportion of endemics is presumed to be because of Antarctica's geographic isolation. In the Virginian Faunal Province of the east coast of the United States, Maturo (1968) reports that of the 48 species found near its southern border, 19 species or 40 percent are endemics.

In general, we would appear to be approximately correct in assuming that about 30 percent of the total number of described species (6,000) are endemic to a given faunal province, or about 2,000 species are endemics, scattered among all provinces. Of the remaining 4,000 species, all are shared between at least two faunal provinces and so the number of species is at least less than half of that. It appears that most of these species are not shared by more than two other faunal provinces and so the number of records would not be reduced by much over one-third of 4,000.

Thus, we can estimate that there have been described on the order of 2,000 endemic species plus 1,600 shared species [between one-half (2,000) and one-third (1,333) reduction in other records], or a grand total of 3,600 living species of bryozoa. This figure agrees well with previous estimates, the bases for which, however, have not been presented.

The average number of species per genus does not seem to change from the tropics to the poles. The ratio is 2.45 in the tropics (ranging from 1.7 in the Ethiopian

Province to 4.0 in the Chinean Province: Okada and Mawatari, 1956). For the Pacific Arctic at Point Barrow, the ratio is 1.8 (Osburn, 1955); for the Northern Seas of the USSR it is 3.4 (Kluge, 1962); and for the endemic Antarctic fauna it is 2.6 (complete listing of Antarctic fauna given by Rogick, 1965).

If genera represent distinctive morphologic entities, and species represent variations on those themes, then the number of different variations per basic morphologic type is not latitude-dependent. Accordingly, the higher species diversity in the tropics must be owing to the opportunity for a greater number of morphologic types, for whatever reason.

As climatic zones changed during the Pleistocene, species ranges were disrupted and species isolated from one another. This has been recognized for various groups of animals trapped on the west coast of Florida, in the Gulf of California, to the north and south of Cape Cod, in the Mediterranean, and even in the Norwegian fjords (Nordgaard, 1918 p. 92-95; Spjeldnaes, 1963). A casual examination of coastal outlines around the world reveals that many similar *cul-de-sacs* also exist. The fauna of such places may be especially important in seeking the earliest stages of new species.

Nordgaard (1927, p. 9) discussing Bryozoa, argued that "circumtropic, circumboreal and probably also some circumpolar species have lived in the oceans of the world at least from Tertiary time, because we must go back to this epoch to find such an arrangement of mainland and oceans that a circumtropic or circumboreal distribution was made possible." This is considered merely an example of "a general law that old species have a great geographical distribution, and *vice versa,* an extensive spreading over the earth seems to involve a high age of the species" (Nordgaard, 1923, p. 16). As far as I am aware, this "general law" has never been tested by plotting distribution versus age of taxon for any group of Bryozoa. When that is done, one should also determine the degree of morphologic complexity of long-lived, widespread taxa relative to endemic, short-lived taxa. Present anecdotal data suggests that species presumed to be very widespread e.g., *Membranipora membrancea* (see Nordgaard, 1923, p. 16), *Bugula neritina,* and various cyclostomes once thought to be bipolar (see Hyman, 1959: 428-429), are all morphologically "simple" and "plastic," relative to other species to which they are related. This means that taxa which are presumed to be long-lived which are also widely distributed may simply be highly generalized species groups.

Diversity Gradients. Considering the world-wide data, bryozoan diversity is highest in the tropics, decreasing poleward (Schopf, 1970), as is typical of many higher taxa. Pacific diversities are two to three times higher than those of the Atlantic.

An additional way to examine patterns of evolution is to see how different depth zones in the ocean are utilized. Detailed records of 73 species collected from the intertidal to 40 m depth in the western Mediterranean French coast

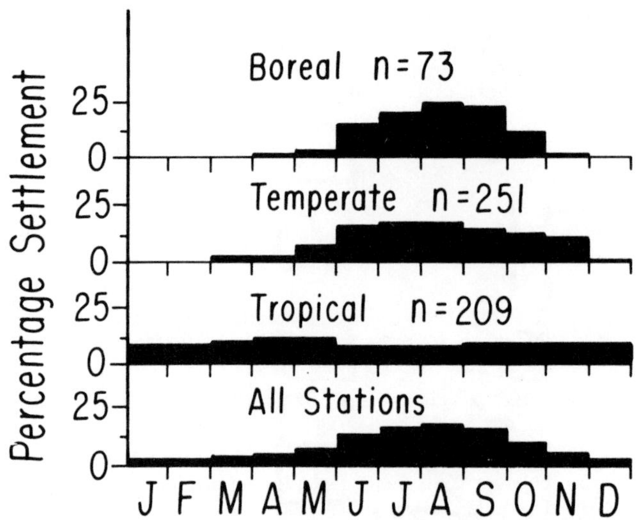

Figure 9. Time of year of maximum attachment of bryozoans in the Northern Hemisphere given as percentage of maximum attachment for each month, where n = number of months of maximum settlement for localities sampled. Graph for all stations normalizes collecting effort for the different regions. Tropical region from about 15°S to 30°N; temperate region from about 30°N to 45°N; and arctic region from about 45°N to 70°N (no samples poleward of that). All localities from both the Atlantic and Pacific coasts of the Americas in these regions were included. Note the increasing restriction of settlement to summer months in higher latitudes (unreduced data in Pastula, 1970).

(Banyuls-sur-Mer) reveals an increasing number of species from 0 to 5 m, 5 to 20 m, and 20 to 40 m (with 40 to 50 species: Medioni, 1970). Data of Gautier (1961, Fig. 67) for the whole of the western Mediterranean and of Schopf (1969b, Fig. 5) for off New England reveal an increasing number of species per station to depths of about 50 m (with 30 to 50 species), and then decreasing so that diversity at 150 m is about what it was at 5 m (15 to 20 species). A further decline of species per station is recorded as one goes into the deep sea (Schopf, 1969b; Androsova, 1973, Fig. 1). However, diversity (species per standard number of specimens sampled) may not decrease. We have no data on this.

A final biogeographic pattern that shows the way resources are partitioned is the season of year for breeding and larval settlement. The months of the year with maximum larval attachment of ten major groups of epifaunal organisms have been reported for each of 210 locations around the Americas, from Greenland to Chile to Point Barrow, Alaska (Pastula, 1970). Data for Bryozoa from these lists have been summarized and grouped into boreal, temperate, and tropical regions (Fig. 9). Settlement in the tropics is spread out over the year, with no month having more than 10 percent of the total maximum settlement. Settlement in the temperate region is concentrated in only six months (June through November). In the Arctic, settlement is yet further restricted, and is chiefly in only four months (June through September). This pattern of settlement corresponds to the time of year of phytoplankton production.

Table 1. Faunal provinces of bryozoans, approximate number of species, and sources of data. The most tentative and questionable of the estimates are in parentheses. Numbers are keyed to the map in Figure 3.

Province	Species	Source
1. Arctic	200	Various; see Schopf, 1970
2. Eastern North Atlantic Boreal	325	Ryland, 1969
3. Mediterranean	300	Gautier, 1961
4. Tropical West African	180	Cook, 1968
5. South African	175	O'Donoghue, 1957; O'Donoghue and Watteville, 1937, 1944
6. Ethiopian	100	Okada and Mawatari, 1956
7. Indian	135	Okada and Mawatari, 1956
8. Antarctic	320	Rogick, 1965; Moyano, 1968
9. South Australian	295	MacGillivray, 1887
10. Papauan	380	Okada and Mawatari, 1956
11. Malayan	640	Okada and Mawatari, 1956
12. Chinean	640	Okada and Mawatari, 1956
13. Western North Pacific Boreal	(200)	Schopf (estimated)
14. Eastern North Pacific Boreal	200	O'Donoghue (see Soule and Soule, 1967)
15. Southern Californian	220	Soule and Soule, 1967
16. Mexican	520	Okada and Mawatari, 1956
17. Chilean	(200)	Schopf (estimated)
18. Argentinean	(200)	Schopf (estimated)
19. Western Atlantic Tropical	280	Schopf, 1973
20. Virginian	120	Schopf (estimated)
21. Acadian	100	Schopf (estimated)
22. Hawaiian	80	Okada and Mawatari, 1956
23. Polynesian	90	Okada and Mawatari, 1956
	5900	

Crustose Coralline Algae as Microenvironmental Indicators for the Tertiary

WALTER H. ADEY, *Department of Paleobiology, Smithsonian Institution, Washington, D. C. 20560*

ABSTRACT

Crustose coralline algae are major framework contributors to Tertiary-Recent coral reefs and are builders of intertidal algal ridges in both the Caribbean and Indo-Pacific.

All Recent tropical coralline genera known are pantropic in distribution. Over 50 percent of the coralline species of the Caribbean and Indo-Pacific can be characterized as "species pairs" and some are indistinguishable. Based on studies in the Hawaiian Archipelago and in the Caribbean, it has been found that the ecology of most species pairs is quite similar in the two regions. At the generic and subfamily level, frequency of occurrence in the total coralline population relative to depth is also quite similar.

The Indo-Pacific and Caribbean have been effectively isolated from each other for from 3 to 20 m. y. Only 6 percent of the total hermatypic scleractinian coral genera are common to both oceans, and no coral species are unquestionably the same. The crustose corallines, however, are a conservative group, and at the subfamily, generic, and species levels they have changed relatively little in morphology and ecology during a similar period of isolation. It is concluded that crustose coralline algae are potentially excellent microenvironmental indicators for Tertiary bioherms.

INTRODUCTION

The first deep reef coring, in the central Pacific at Funafuti, demonstrated that crustose coralline algae ("Lithothamnion") can be dominant framebuilding organisms (Finckh, 1904; Gardiner, 1898). A similar picture was obtained for both Recent and Tertiary in the world's most northern atolls, Midway and Kure in the Hawaiian chain (Gross and others, 1969; Ladd and others, 1970). Intertidal algal ridges have also figured prominently in most of the marine geological treatments in the U.S. Geological Survey series of special publications treating the reefs of central Pacific islands and atolls, and more recently similar well-developed algal ridges have been described for the Caribbean (Adey and Burke, 1976).

Despite the volume of coralline carbonate in Tertiary-Recent biohermal structures (Wray, 1977, estimates "20-50% of the mass of modern reef accumulations"), little use has been made of these organisms in terms of paleoenvironmental, biogeographic, or stratigraphic interpretations. Johnson (1961) considered that crustose coralline species had time-stratigraphic value. However, his species characterizations, based in large part on cell measurements, have not generally been accepted, and there remains some question as to the statistical techniques that he employed (see e.g., Wray, 1977).

Littler (1971, 1973a, 1973b) measured the surface area coverage of organisms at several reef and shelf slope sites on Oahu, Hawaii. At the Waikiki fringing reef, he determined that the "crustose coralline algae cover 39% of the reef surface and exceed all other organisms as the major builders and consolidators of reef materials" while at 8 to 28 m "The deepwater crustose Corallinaceae (38% mean cover) overshadow all other calcareous organisms in terms of standing stock and also seem to have more biological influence than do other limestone producers." Without a modern taxonomic and systematic treatment for the Hawaiian Islands as background, Littler encountered persistent problems with identification at the species, generic, and even family levels. As a result, the summary treatment by Doty (1974) of the role of corallines in Pacific reef structures, based on the work of Littler, unfortunately contains little secure information that can be applied to a paleoecologic interpretation.

Adey and Boykins (in prep.) have completed a relatively extensive systematic and ecologic study of the crustose corallines of the Hawaiian Archipelago. Based on that study, an examination of a reef-flat collection from Kenya (see e.g., Isaac, 1971), the recent work on the corallines of Guam (Gordon and others, 1976), and studies in progress of the crustose corallines of the Caribbean, a brief interpretation of reef-associated crustose corallines of both the Indo-Pacific and Caribbean Sea is presented below.

Based on the considerable systematic and ecologic similarities between Hawaiian and Caribbean floras in comparison to scleractinian coral faunas, it is concluded that corallines are only relatively slowly evolving. The ecological specificity of many species and genera makes them excellent paleoecologic indicators for the late Tertiary.[1]

[1] Note that in the treatment to follow new species and genera in papers in preparation are treated as follows: *"Paragoniolithon typica."*

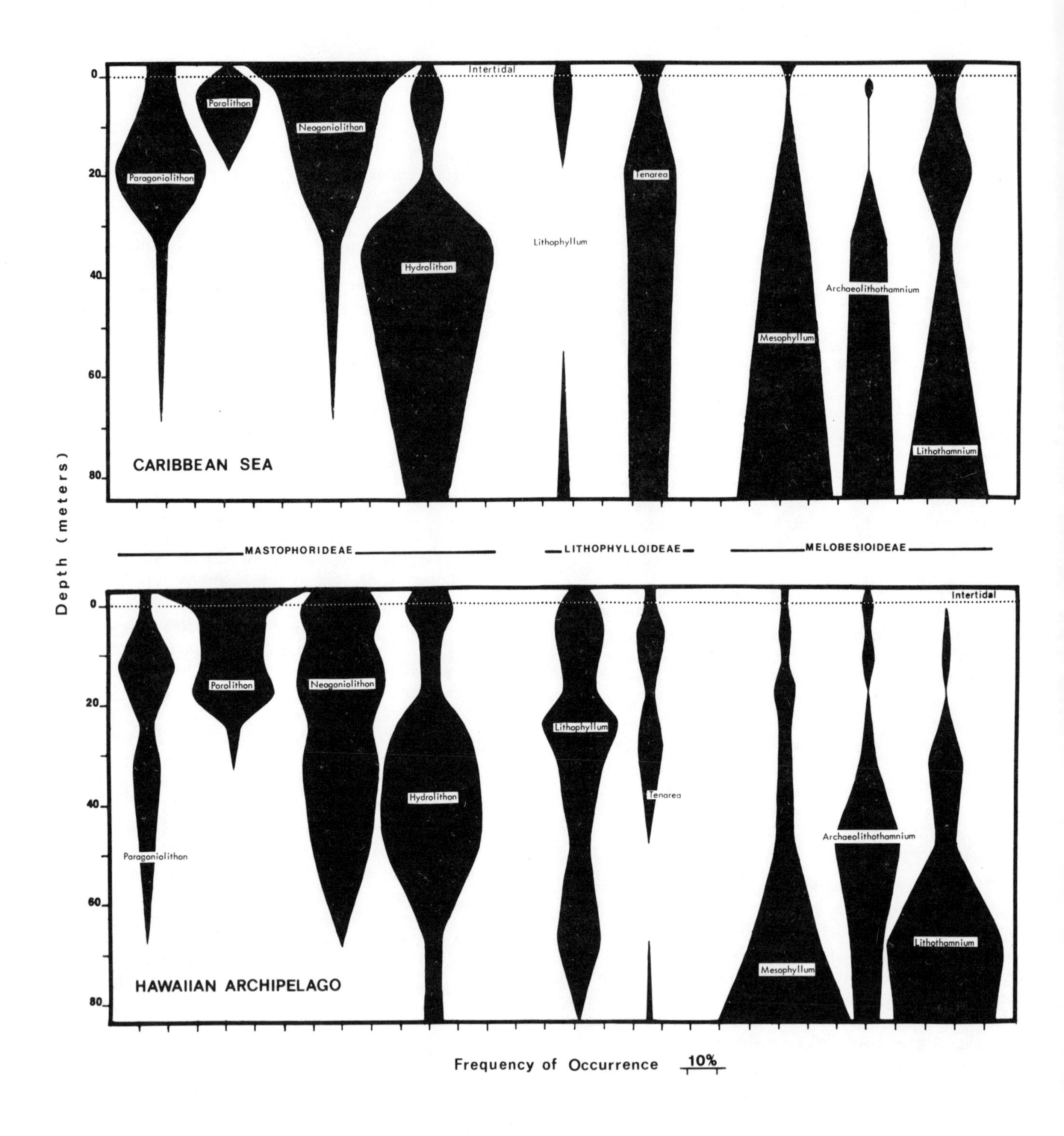

Figure 1. Comparison of distribution patterns of common crustose coralline species in idealized littoral zones in the Caribbean and Hawaiian Archipelago. For "species pairs" see Table 1.

Table 1. Species Relationships

Caribbean					Indo-Pacific
No obvious relationship	Similar	Very similar	Very similar	Similar	No obvious relationship
		Porolithon			
	P. antillarum	*P. pachydermum*	*P. onkodes*	*P. craespedium*	*P. gardineri*
		Paragoniolithon			
Pg. solubile	*Pg.* "*typica*"				*Pg.* "*conicum*"
		Neogoniolithon			
N. caribaeum *N.* "*ramusculum*" *N. mamillare* *N.* "*simplex*" *N.* "*moricatum*"	*N.* "*megacarpum*" *N.* "*clavynodum*" *N. strictum* *N.* "*lenormandi*" *N. accretum*			*N. fosliei* *N.* "*clavycymosum*" *N. frutescens* *N. rugulosum*	*N. rufus*
		Hydrolithon			
	H. børgesenii			*H. reinboldii*	*H. breviclavium* *H.* "*laeve*" *H.* "*megacystum*"
		Lithoporella			
		L. melobesioides	*L. melobesioides*		
		Lithophyllum			
	L. intermedium	*L. congestum*	*L. kotschyanum*		*L. pallescens* *L.* "*ganeopsis*" *L.* "*insipidus*" *L. mollucense* *L. punctatum*
		Tenarea			
T. bermudense	*T. prototypum*			*T. tessellata*	
		Archeolithothamnium			
	A. dimotum *A. episporum*			*A. erythraeum* *A.* "*episoredion*"	
		Lithothamnium			
L. ruptile	*L. occidentale*			*L. australe*	*L. pulchrum*
		Mesophyllum			
	M. erubescens *M.* "*friaselis*" *M.* "*levolaminum*"			*M. madagascariensis* *M. prolifer* *M. mesomorphum*	*M. purpurascens* *M.* "*syrphetodes*" *M. siamense* *M.* "*fluatum*"
		Total number of species			
8	3	13	3	13	3
			3		13

Listed Caribbean species—27

Listed Indo-Pacific species—32

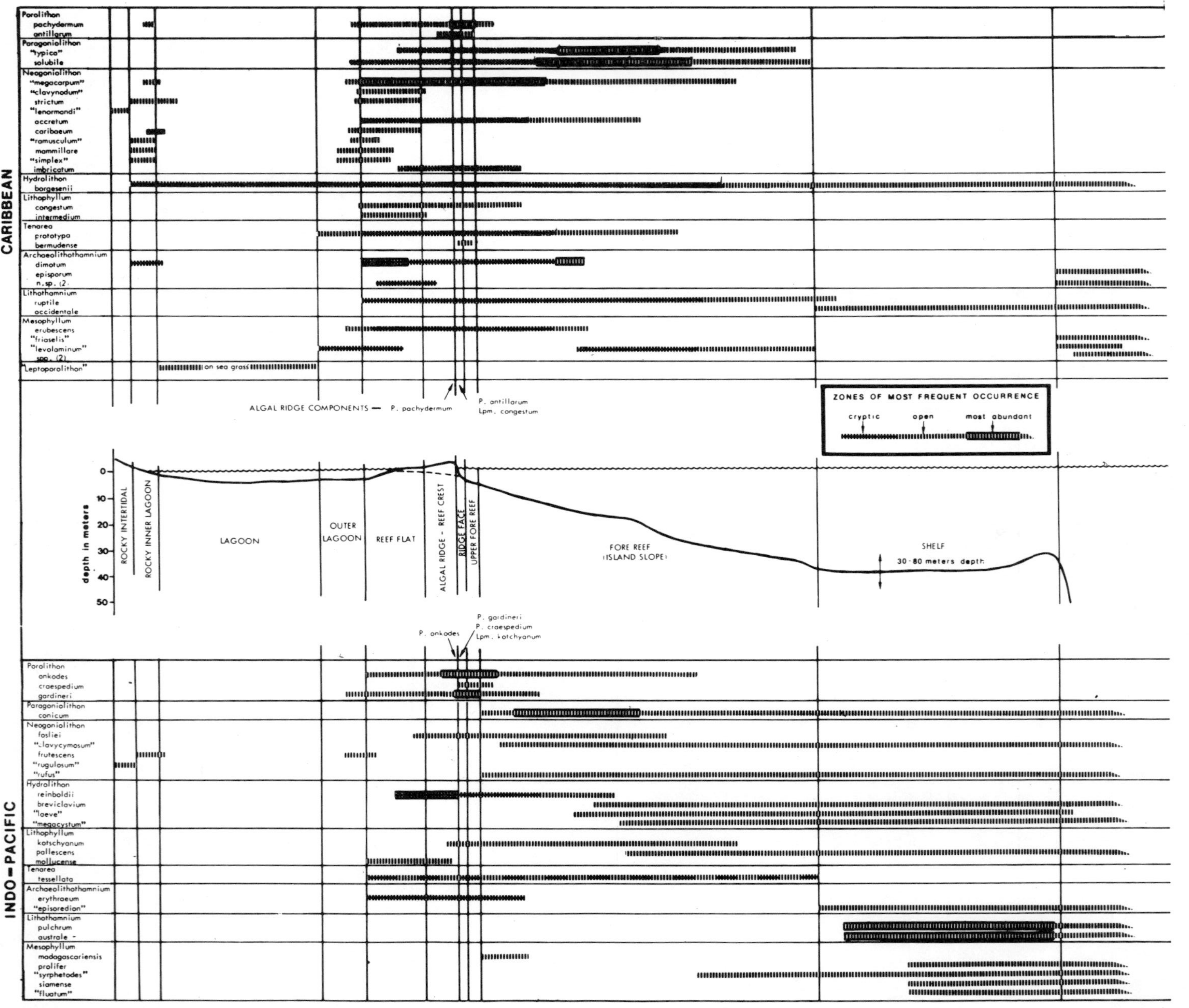

Figure 2. Comparison between Caribbean and Hawaiian Archipelago—generic and subfamily abundance with depth.

RESULTS

An idealized reef and shelf morphology and bathymetry, showing the present Caribbean and Indo-Pacific species most characteristic of major zones, is shown in Figure 1. In cases where an algal ridge is absent, the reef crest zone will have a coralline composition that is intermediate between reef flat and upper fore reef, with the branching species of *Lithophyllum* and *Porolithon* poorly represented or absent. At least in the Caribbean, the shelf is normally rich in coralline only if depths are greater than 40 m.

Common benthic crustose corallines found in this study in both the Caribbean and Indo-Pacific are listed in Table 1. About 55 to 65 percent of the species found in either ocean bear strong habit resemblances to those from the other and these have been termed "species pairs." In cases where hundreds of specimens of these species pairs have been examined from both oceans, it is seen that differences in vegetative morphology are consistent, however subtle. Generally, a size factor is involved, e.g., the "whorls" in *Tenarea tesselata* have a larger diameter and are wider than those in *Tenarea prototypum,* and the "branches" in *Porolithon craespedium* are considerably coarser than those in *Porolithon antillarum. Porolithon onkodes* and *Porolithon pachydermum,* on the other hand, and *Lithophyllum kotschyanum* and *Lithophyllum congeshum* are indistinguishable to me on the basis of external morphology. Whether they are to be kept as separate species awaits the results of anatomical and morphological statistical analysis still in progress on the Caribbean collection.

Most species pairs occupy quite similar ecological niches in both the Indo-Pacific and Caribbean. Of the 15 pairs recognized, only three show differences in distribution or microhabit. *Hydrolithon boergesenii* in the Caribbean is a eurybath covering nearly the entire depth range of crustose corallines, but its Pacific "pair," *Hydrolithon reinboldii,* is apparently restricted to depths of less than 20 m; the deeper zones are occupied by 3 species of the genus not known to occur in the Caribbean. The Pacific *Neogoniolithon "clavycymosum"* is known from a rather wide depth range, while its Caribbean pair, *Neogoniolithon "clavynodum"* appears to be a rare, subcryptic reef-flat species. *Mesophyllum "friaselis"* is a deep shelf-wall species of leafy habit, while its Pacific pair, *Mesophyllum prolifer,* is known only from deep bank sites. In the latter case, the apparent difference in ecology may not be real. I have no true shelf-wall collections from the Hawaiians and relatively few deep shelf dredgings from the Caribbean.

A frequency plot of Caribbean and Hawaiian genera as a function of depth is given in Figure 2. It is apparent from that figure that not only do most genera show a similar abundance-depth relationship in both oceans, but that even at the subfamily level the patterns are basically the same. *Lithophyllum* and *Porolithon* are more important at mid-depths in the Pacific than they are in the Caribbean, however, because of additional *Tenarea* and *Neogoniolithon* at mid-depths, the subfamily abundances remain about the same.

DISCUSSION

Considering the present state of the literature for Indo-Pacific corallines, it is difficult to compare the Hawaiian Archipelago with the region as a whole. Numerous Pacific corallines were received and described by Foslie at the turn of the century (see Littler, 1972), and several recent papers on Rongelap (Lee, 1967) and Guam (Gordon and others, 1976) also treat these plants. Mostly the collections are only reef-flat or shelf dredgings, and quantitative ecological data are lacking. However while species that are rare or lacking in Hawaii are apparently more common in the Indo-Pacific (e.g., *Porolithon craespedium* and *Lithophyllum mollucense*), the dominant elements of the coralline flora appear to be present in roughly the same proportions and in the same ecological niches. The total crustose coralline flora probably increases to about 50 species in the central Indo-Pacific.

It is generally accepted that the east Pacific and the Caribbean were connected by a Central American seaway until the late Pliocene (e.g., Woodring, 1966; Whitmore and others, 1965), with estimates of the time of separation ranging from 2 to 5 m. y. ago (see e.g., Berggren and Van Couvering, 1974). Because of the shallow-water isolation of the east Pacific and the complexities of ocean climate resulting from sporadic upwelling as well as the process of the closing the Isthmus, it is likely that Indo-Pacific and Caribbean coralline floras have been effectively separated for a period somewhat longer than 2 to 5 m. y. Too little is known about the corallines of the tropical east Pacific to determine the extent of their relationship to those of Hawaii.

Thus, while we can say that the pantropical coralline flora has little generic divergence, over 50 percent of the species being little changed from common ancestors, we do not know if the effective isolation has existed for as little as 3 m. y. or perhaps as much as 20 m. y.

Perhaps a more valid characteristic of interest to paleoecology is the rate of coralline evolution relative to that of the scleractinian hermatypic corals, companion framework builders in Neogene biohermal structures. Frost and Langenheim (1974) list 13 genera of Miocene hermatypic corals from Central America. At present, 1 of those is extinct, 4 live exclusively in the Indo-Pacific, and 4 in the Caribbean. Three genera are found in both regions. Of the total hermatypic genera of reef coral of both oceans, only about 6 percent occur in both, and no species are unquestionably the same. Some Caribbean and Indo-Pacific coral genera have ecological equivalents. However, a nearly equal number of important Pacific coral genera apparently do not have ecological equivalents (Wells, 1957).

On the other hand, while additional study is certainly needed, it appears that nearly all crustose coralline genera of the Indo-Pacific also occur in the Caribbean in similar ecologic situations. Furthermore, over 50 percent of the species have apparently not diverged significantly from their common ancestors, either morphologically or ecologically.

None of the coral or coralline species involved are biological species in the strict sense. Thus, the question of systematic comparability remains. On the other hand, in a paleoecologic context, the exact nature of the taxonomic units is of relatively little concern. Using the current systematic language for corallines and corals, the corallines are considerably more conservative than the hermatypic scleractinian corals.

CONCLUSIONS

Crustose coralline algae in the Indo-Pacific and Caribbean are characterized at the generic level by similar populations. Most genera occupy equivalent ecological situations in both areas, and over half of the species are "pairs," most of which occupy quite similar niches. This minimal evolutionary divergence has occurred over 3 to 20 m. y. of isolation and is considerably less than that to be found in the other major group of Tertiary bioherm builders, the scleractinian corals.

Many coralline species are very specific in their requirements for light and wave action, especially those typical of shallow water. Using coralline assemblages, while taking into account the mosaic of exposed and cryptic habitats within the reef framework, and the morphology of the species involved (extent and size of branches, thickness of crust, and so on), it should be possible to determine the depth of fossil coralline framework to within roughly ±0.5 m of mean, low water, to within ±4 m from 5 to 20 m, and to within ±15 m at depths greater than 20 m. In addition, island shore, reef-flat, reef-crest, upper fore-reef, and lower fore-reef zones, and shelf less than 40 m and shelf walls, each have characteristic assemblages which should be recognizable in the fossil record.

It would be desirable to undertake several modern paleoecologic studies of middle to late Neogene coralline reef assemblages. The major changes that occur in mid-Miocene, with the addition of *Porolithon* and *Neogoniolithon,* could then perhaps be evaluated and applied to Paleogene paleoecology. *Lithothamnium, Mesophyllum,* and crustose *Lithophyllum* were dominant in early Tertiary reefs. Yet, I have the impression that Paleogene distributions of corallines in reef systems are similar to those found at present, and that crustose corallines were largely cryptic organisms at that time. Since the genera *Mesophyllum* and *Lithothamnium* are dominantly antiboreal-antarctic and boreal-arctic, respectively, and *Lithophyllum* is temperate, I suggest that crustose corallines evolved in high latitudes during the Mesozoic, moved into cryptic situations in low latitudes during the early Tertiary, and eventually *Porolithon* and *Neogoniolithon* evolved as "sun forms" tolerant to bright light conditions.

REFERENCES

Adey, W., and Burke, R., 1976, Holocene bioherms of the eastern Caribbean: Geol. Soc. America Bull., v. 87, p. 95-109.

Adey, W., and Boykins, W. (in prep.), The crustose coralline algae of the Hawaiian Archipelago.

Berggren, W., and Van Couvering, J., 1974, The late Neogene: Palaeogeography, Palaeoclimatology, Palaeoecology, v. 16, p. 1-216.

Doty, M., 1974, Coral reef roles played by free-living algae: Intern. Coral Reef Symposium, 2nd, Brisbane 1973, Proc., v. 1, p. 27-33.

Finckh, A. E., 1904, Biology of the reef-forming organisms at Funafuti Atoll, Section 6, *in* The Atoll of Funafuti: Royal Soc. London, Rept. Coral Reef Committee of the Royal Society, p. 125-150.

Frost, S. H., and Langenheim, R., 1974, Cenozoic reef biofacies: Tertiary larger Foraminifera and scleractian corals from Chiapas, Mexico: De Kalb, Ill., Northern Illinois Univ. Press, 388 p.

Gardiner, J. S., 1898, The coral reefs of Funafuti, Rotuma and Fiji together with some notes on the structure and formation of coral reefs in general: Cambridge Philos. Soc. Proc., v. 9, p. 417-503.

Gordon, G. D., Masaki, T., and Akioka, H., 1976, Floristic and distributional account of the common crustose coralline algae on Guam: Micronesica, v. 12, p. 247-277.

Gross, M., Milliman, J., Tracey, J., and Ladd, H., 1969, Marine geology of Kure and Midway Atolls, Hawaii: A preliminary report: Pacific Sci., v. 23, p. 17-25.

Isaac, W. E., 1971, Marine botany of the Kenya coast. Pt. 5. A 3rd list of Kenya marine algae: East Africa Nat. History Soc. Natl. Mus. Jour., v. 28, p. 1-23.

Johnson, J. H., 1961, The use of calcareous algae in correlating Cenozoic deposits of the western Pacific area: Pacific Sci. Congr., 9th, Bangkok 1957, Proc., v. 12, p. 282-286.

Ladd, H., Tracey, J., and Gross, M., 1970, Deep drilling on Midway Atoll: U.S. Geol. Survey Prof. Paper 680-A, 22 p.

Lee, R. K. S., 1967, Taxonomy and distribution of the melobesioid algae on Rongelap Atoll, Marshall Islands: Canadian Jour. Bot., v. 45, p. 985-1001.

Littler, M., 1971, Standing stock measurements of crustose coralline algae and other saxicolous organisms: Jour. Exptl. Marine Biol. Ecology, v. 6, p. 91-99.

——— 1972, The crustose Corallinaceae: Ann. Rev. Oceanogr. Marine Biol. v. 10, p. 311-347.

——— 1973a, The population and community structure of Hawaiian fringing reef crustose Corallinaceae: Ann. Rev. Oceanogr. Marine Biol. v. 11 p. 103-120.

——— 1973b, The distribution, abundance and communities of deepwater Hawaiian crustose Corallinaceae: Pacific Sci., v. 27, p. 381-390.

Wells, J. W., 1957, Coral reefs, *in* Hedgpeth, Joel W., ed., Treatise on marine ecology and paleoecology, Vol. 1: Geol. Soc. America Mem. 67, p. 609-631.

Whitmore, F. C., and Stewart, R. H., 1965, Miocene mammals and Central American seaways: Science, v. 148, p. 180-185.

Woodring, W., 1966, The Panama land bridge as a sea barrier: Am. Philos. Soc. Proc., v. 110, p. 425-433.

Wray, J., 1977, Calcareous algae: Amsterdam, Elsevier, 185 p.

Epilogue: A Paleozoic Pangaea?

A. J. BOUCOT, *Department of Geology, Oregon State University, Corvallis 97331*
JANE GRAY, *Department of Biology, University of Oregon, Eugene 97403*

INTRODUCTION

Three major concepts of landmass division are currently used for the Paleozoic: fixist-modern geography; kaleidoscopic; pangaeic. Those unconvinced of the validity of the various plate tectonic concepts, or if convinced, uncertain which of the many available choices might be "correct," continue to plot their historical biogeographic data on a modern geographic base.

Those convinced of the concept's essential correctness tend to be either kaleidoscopists or pangaeists. Smith and others (1973) provide an adequate kaleidoscopic Paleozoic geography constructed from a mélange of paleomagnetic and geological data, the latter chiefly obtained from the present continents. Many paleontologists have used their reconstructions or variously modified them to better suit their own biogeographic information while retaining an essentially kaleidoscopic interpretation. Such kaleidoscopic reconstructions show a number of continental and subcontinental sized landmasses scattered about in accord with various geologic and paleontologic assumptions and data. The base map modifications, time interval by time interval, animal group by animal group, and for some of the plant groups, are in agreement with particular sets of data. One group of kaleidoscopists whirl the modern continents about, plus Peninsular India, Madagascar, and New Zealand, while maintaining their essential integrity. Another group fragment the modern continents in a manner which brings the biogeographically similar fragments for the particular Paleozoic time intervals close together. Inspection of these Paleozoic kaleidoscopic interpretations reveals that they are internally inconsistent for each time interval, depending on the organism used as well as on the individual investigator's inclination, and that they also require megamovements of a most rapid type *between* many time intervals (chiefly periods) but little or no movements *within* the same time intervals. In order that the various kaleidoscopic fragments remain in agreement with the data used by the various investigators, the megamovements required must coincide both with additional fragmentation of previously inferred landmasses, and with additional cementation of previously separate landmasses. The coincidence between these megamovements and additional fragmentation and cementation should concern the objective spectator of the Paleozoic scene.

The third possible major landmass concept—the pangaeic—has not been previously attempted for the entire Paleozoic, although many have considered a Cambrian through Permian gondwanic interpretation. Such a pangaeic reconstruction is provided here through a series of diagrams (Figs. 2 to 7) as a contrast with the far more complicated kaleidoscopic geographies of others. We consider the diagrams to be more-or-less cartoons rather than maps. No scale is provided for them, nor specific geographic coordinates. Their purpose is to suggest and explore possibilities rather than to "prove" anything. In the present state of uncertainty concerning the "correct" approach to reconstructing Paleozoic paleogeography we have made no attempt to be precise. We could have availed ourselves of the work of LePichon and others (1977) for many parts of the world, but this might have misled the reader into thinking that our pangaeic interpretation possesses greater precision than is actually the case.

PROBLEMS, LIMITATIONS, ANOMALIES

Although most geologists would agree with the assumption that there was a true ocean during the Paleozoic, there is no data on the abyssal geology and geophysics of the type critical to the various reconstructions of the post-Paleozoic. Van Andel (this volume) has pointed out how little assistance is provided by marine geologists and geophysicists to the biogeographer of the Paleozoic.[1] Nevertheless, the Paleozoic data are permissive of a variety of possibilities, depending on what assumptions are made, including, among others, a pangaeic possibility.

Many facets of Paleozoic lithofacies and biogeography indicate Paleozoic proximity of the circum-Mediterranean regions (see Berry and Boucot, 1973, for a typical Silurian

[1] We have intentionally disregarded paleomagnetic evidence for the following reasons, in addition to its limited extent: (1) conclusions derived from paleomagnetic evidence for the Paleozoic are far from unanimous; (2) some such conclusions are in complete disagreement with the biogeographic and stratigraphic evidence as we interpret them. For example, with regard to this last point, Irving (1977, Fig. 5a) postulates the presence of a vast equatorial ocean separating Laurasia on the north from Gondwana on the south during the Middle Devonian consistent with his interpretation of the paleomagnetic data. A hypothetical ocean in an equatorial position makes it difficult to explain the presence of Old World Realm Middle Devonian faunas in North Africa, and of Eastern Americas Realm faunas in northern South America. The only alternative to Irvings' reconstruction would be to deduce the presence of a huge continental equatorial block in the Middle Devonian, in the position of Irving's ocean, which could subsequently have been erased from the geologic record without trace.

argument). Similarly, Boucot (1969) has presented arguments based on mid-Paleozoic geology that view the Uralian region as merely a hingelike narrow sea, opening in a northerly direction but closed on the south in Kazakhstan, incised into the Eurasian landmass during the Paleozoic. The possible continuity of Africa and South America in the Paleozoic is too well known to require comment. North America, *vis-a-vis* South America and the Old World, does require more comment. Some have viewed the eastern edge of North America from eastern Newfoundland to Boston, and even including the Slate Belt of the Carolinas as well as the Paleozoic subsurface in northern Florida, as an Old World strip, segments of which became welded to North America at various times during the Paleozoic. There is no compelling evidence that such an Old World strip originally need have been far removed from North America. The relations of South America to North America during the Paleozoic are very poorly known and must be worked out only by very indirect means involving first South America with Africa, and then Africa with Eurasia. The relations of Eurasia with North America are fairly well known. However, the biogeographic data does suggest close affinities between North and South America for much of the Paleozoic.

Because such a limited body of physical and biological data exists for the Paleozoic it is essential that a variety of landmass and ocean current hypotheses be explored to find one which seems to satisfy the available data. The hypothesis which satisfies the most data is not necessarily "true." Contrariwise, any hypothesis which satisfies only one class of the available data, or which ignores an important class of data, cannot be considered satisfactory. Paleozoic paleogeographies, pangaeic or otherwise, assembled without consideration for *all* time intervals, run the risk that the reconstructions employed for adjoining time intervals will be so inconsistent as to make the inferred paleogeographies wholly unrealistic.[2] While internal consistency need not lead to "truth" in paleogeography and historical biogeography, obvious inconsistencies are unlikely to reflect anything but parochialism.

The success or failure of any Paleozoic biogeographic hypothesis will be determined by how satisfactorily it explains the following, chiefly biogeographic, anomalies in addition to the well-known physical data:

(1) A decrease in level of provincialism from Early to Late Ordovician; a more cosmopolitan Early Silurian; a post-Early Silurian increase in level of provincialism culminating in the late Early Devonian; a rapid decline to conditions of relatively high cosmopolitanism in the Late Devonian; a slow increase in level of provincialism culminating in the high provincialism of the Early and Middle Permian.

(2) The presence of warm temperate to tropical faunas along the western limits of South America from the Ordovician through Early Permian (Fig. 1). Subsequent discussion of this anomaly shows that the warm-cool boundary extending north-south in the Andean region remained fairly fixed from the Cambrian through the Early Permian. However, the east-west boundary separating warm northern units from cool southern units progressively migrates from a Cambrian position north of a line from Oaxaca to Florida to a position well south of and roughly parallel to the Amazon in the Early Permian.

(3) The presence in southeastern Kazakhstan during the Cambrian, Ordovician, Silurian, and Devonian of a biogeographically mixed group of taxa that is not found together elsewhere. In addition, many taxa first appear in Middle Ordovician time in southeastern Kazakhstan, but not elsewhere until the late Llandovery when they appear almost instantaneously throughout the North Silurian Realm.

(4) The presence in Manchuria and the adjacent Soviet Union north of the Amur during the Early Devonian of many Eastern Americas Realm taxa. The presence in the Tasman geosyncline and in New Zealand during the Early Devonian of some Eastern Americas Realm taxa.

(5) The presence of a North Atlantic Region in the Late Silurian surrounded to the east, north, and west by the Uralian-Cordilleran Region, and to the south by the Malvinokaffric Realm.

(6) The presence of a highly endemic Eastern Americas Realm in eastern North America, northern South America, and the west coast of South America, particularly during the Early Devonian, while the rest of the world was divided between Southern Hemisphere Malvinokaffric Realm and Old World Realm faunas.

(7) A far greater extent of Atlantic Realm and Malvinokaffric Realm into the Mediterranean region during later Ordovician and Silurian time than in the Devonian, a phenomenon that matches the southern migration in South America of the east-west Malvinokaffric Realm—extra-Malvinokaffric units boundary during the same time interval.

(8) A similar history of Cambrian through Silurian regression followed by Devonian transgression for much of the Malvinokaffric-Gondwanic regions in the southern half of South America, the Falkland Islands, Antarctica, and the southern half of Africa.

(9) During the Permian-Carboniferous the presence of a Midcontinent-Andean biogeographic unit distinct from the coeval Gondwana Realm, and other warm water, carbonate-rich Northern Hemisphere biogeographic units.

[2] For example, McKerrow and Cocks (1976) consider biogeographic evidence supporting the existence of a proto-Atlantic (Iapetus) Ocean. They suggest that an inferred Cambrian-Middle Devonian closure of Iapetus coincides with a steady increase of cosmopolitanism of marine benthos on either side of the separate regions. They terminate their tale with shared brackish or possibly fresh-water vertebrate faunas during the Early Devonian. The high level of Silurian cosmopolitanism following a high level of pre-Middle Ordovician provincialism fits their concept. But they fail to mention the globally high level of Early Devonian marine provincialism that included the Iapetus region and equalled the provincialism of any pre-Middle Ordovician interval, although they refer to a publication dealing exclusively with Early Devonian marine provincialism.

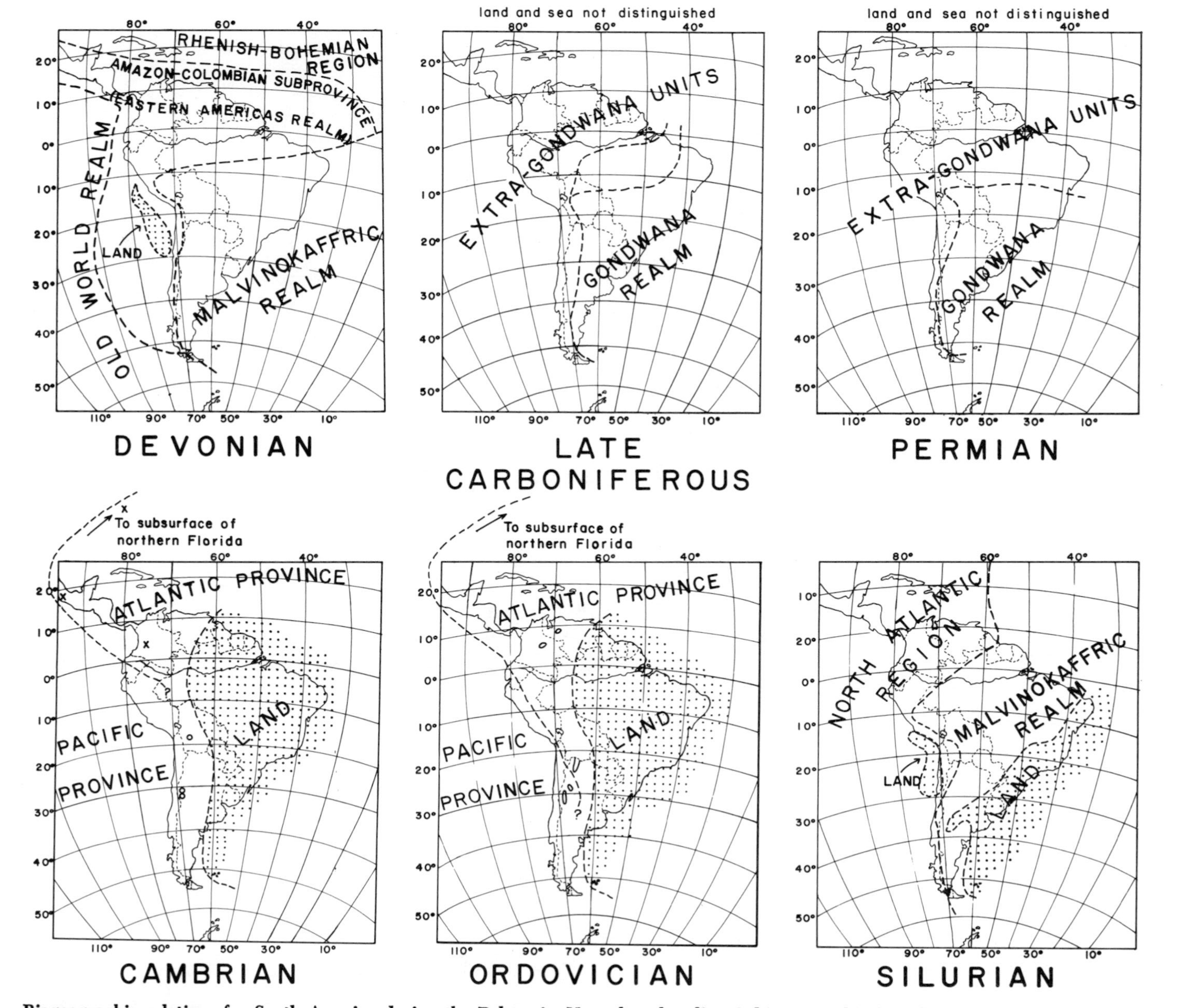

Figure 1. Biogeographic relations for South America during the Paleozoic. Note that the climatic-biogeographic boundary that parallels the central Andes remains stable throughout the Paleozoic.

(10) The presence of a unipolar set of noncarbonate sediments and appropriate organisms from the Cambrian into the Devonian, and of a bipolar set in the Permian and possibly in the Carboniferous.

In our opinion, most of these constraints are most readily solved by a Paleozoic pangaeic geography.

The South American Biogeographic Anomaly

The apparent continuity of a single north-south climatic boundary for South America throughout the Paleozoic plus the southward shift of the east-west boundary is one of the anomalies mentioned above. The data presented below and on Figure 1, suggest that a major biogeographic-climatic boundary between warm-cool areas has roughly paralleled the central and southern Andes throughout the Paleozoic, or at least from the Ordovician through the Permian. They also suggest that this boundary progressively shifted south with time from Central America and Florida in the early Paleozoic to well south of the Amazon in the latest Paleozoic. Both circumstances are explained by a steady northward movement of a pangaeic supercontinent, and both are inconsistent with previous non-pangaeic continental drift reconstructions.

Figure 1 (see Appendix for data sources) outlines the boundary between Gondwana—extra-Gondwana fauna, flora, and rocks, based on the distinction between noncarbonate rocks associated with tillites and carbonate rocks without tillites, and the climatically equivalent biogeographic units of earlier geologic periods extending back to the Cambrian. The Gondwana Realm is interpreted on a variety of evidence to represent a cool, and at times a glacial, climate. The Malvinokaffric Realm of the preceding Devonian and Silurian can be interpreted in the same way (see Boucot, 1975, for a discussion of some of the evidence), as can the Atlantic Realm of the Ordovician and Cambrian.

Attempts to synthesize the historical biogeography of the South American Paleozoic suffer from a paucity of data, compared with Eurasia, North America, and northern Africa. Because the best known intervals are the Permian-Carboniferous and the Devonian, the data will be presented in reverse order. Pre-Devonian fossils are essentially absent from Brazil, except for some Silurian, and pre-Silurian is absent from both Uruguay and Argentina east of the Pre-Cordillera.

(1) For the Late Carboniferous and the Permian, the climatic boundary between the Gondwana—extra-Gondwana units parallels the axis of the central and southern Andes, swinging to the east-northeast in the Lake Titicaca region.[3] Lack of data precludes consideration of the equivalent Malvinokaffric-Eastern Americas Realms boundary of the Early Carboniferous, although it may be presumed to conform generally to the Malvinokaffric-Eastern Americas Realms boundary for the Devonian and the Gondwana—extra-Gondwana units boundary of the Late Carboniferous-Permian (Fig. 1).

[3] The two boundary possibilities (Fig. 1) indicated for the Late Carboniferous in Brazil are based on uncertainties in the age of the subsurface rock (John C. Crowell, personal commun. 1978).

(2) The Malvinokaffric-Eastern Americas Realms boundary of the Devonian is situated in about the same position in the southern and central Andes, as the Gondwana—extra-Gondwana units boundary of the Late Carboniferous-Permian, although it extends north into northern Peru (evidence for the Devonian is lacking in northernmost Peru and Ecuador) before swinging east-northeast across Brazil. The east-northeast boundary between the Malvinokaffric-Eastern Americas Realms is thus north of the Gondwana—extra-Gondwana boundary of the Permian, and possibly of the Late Carboniferous.

(3) In the Silurian, the east-northeast boundary between the Malvinokaffric Realm—North Atlantic Region lies still further to the north of the boundary between the climatically equivalent units of the Devonian (based primarily on data on the north flank of the Amazon Basin and from the Merida Andes south of Lake Maracaibo), although the north-south boundary remains in essentially the same position that it occupied in the later Paleozoic.

(4) During the Ordovician, the north-south boundary, now separating the Atlantic Realm, a unit poor in carbonate rocks, from the Pacific Realm, a unit rich in limestone, generally continues to parallel the boundary of the Silurian to Permian. Minor east-west deviations in the Andean region are on a scale mainly of interest to the local specialist. The Ordovician boundary is extended into the southern Andean region, although there is no data for any of the Paleozoic south of the Pre-Cordillera de San Juan in central Argentina. Atlantic Realm data are also extended north of Peru because there is no evidence for the presence of Pacific Realm faunas in northern South America (pre-Silurian fossils are unknown south of the Merida Andes or on the Brazilian Shield). The Atlantic-Pacific Realms boundary is swung to the east-northeast because of data points in the subsurface of Florida and in Oaxaca, which are all that is available at this time.

(5) For the Cambrian, the best that can be said is that the interpretation shown in Figure 1 is consistent with those deduced for the post-Cambrian intervals.

The synthesis can be extended to the east to determine whether the data from Africa, the Near East, and southern Asia fit. They do in large part. Fossiliferous Cambrian, Ordovician, and Silurian are absent between the central Sahara and the Cape Mountain System. The Cambrian and Ordovician of North Africa and the Mediterranean region are of Atlantic type. The Malvinokaffric—extra-Malvinokaffric boundary for the Silurian lies in North Africa (see Boucot, 1975, Fig. 41), and Malvinokaffric-type Devonian faunas in Africa do not occur any farther north than Ghana (if there)—well to the south of the known Malvinokaffric-type Silurian of North Africa (Boucot, 1975, Fig. 2). Gondwana-type Permian-Carboniferous is unknown in North Africa although dominant elsewhere. In the Near

East no positive evidence exists for Gondwana-type biota or rocks, nor for Malvinokaffric faunas. In the western Himalayan region there is excellent evidence showing that Gondwana rocks, flora, and fauna overlie extra-Malvinokaffric Silurian and Devonian as well as extra-Atlantic Realm Ordovician and Cambrian. This evidence indicates that the Gondwana-type cold or cool climate transgressed over the limestone-rich Himalayan area earlier occupied by warm-water marine faunas that included reefs—a significant though less amazing change than the possible long-term fixity of the north-south boundary in the Andean region.

The alternative to such an appealingly simple conclusion is to divide the various outcrop areas in the central and southern Andes and sweep them away towards ones of their own biogeographic kind elsewhere. If this is done the Andean region will require a veritable kaleidoscopic behavior since "pieces" of the Malvinokaffric, Atlantic, and Gondwana Realms are not always located above or below their own biogeographic kind. For example, in the Pre-Cordillera de San Juan, Pacific Realm Ordovician rocks rich in limestone are overlain by Malvinokaffric Realm Silurian and Devonian with both Gondwana and extra-Gondwana type Permian-Carboniferous present over the latter from place to place. The extra-Gondwana Permian limestones of Bolivia and Peru occur above Malvinokaffric Silurian and Devonian rocks in many places.

A choice must be made between using climatic changes or geographic changes involving significant crustal drift, to explain specific biogeographic anomalies. We prefer to explain them by rapid changes in climate rather than the alternative appeal to rapid, kaleidoscopic movement of pieces of subcontinental sized real estate back and forth over great distances. The to-and-fro movements which would be required for the central and southern Andes during the Paleozoic seem unreasonable in view of what is known of Cenozoic plate movements.

Most attempts to interpret Paleozoic biogeography, including that of South America, align warm-cold boundaries in a latitudinal manner. The pangaeic reconstruction, however, aligns the Andean warm-cold boundaries in a north-south manner. This interpretation calls to mind a current system of Gulf Stream or Humboldt type. Such a current system doubtless would have had the capability of bringing portions of more than one continent into reproductive communication, and could provide an explanation for the extent of the Midcontinent-Andean biogeographic unit in the Carboniferous to Early Permian, and the Eastern Americas Realm in the Devonian.

ASSUMPTIONS

In attempting a pangaeic reconstruction we have made a number of assumptions:

(1) We assume that biogeographically similar terrestrial floras and biogeographically similar marine faunas of the same age were in more-or-less continuous reproductive communication.

(2) We assume that the presence of similar marine faunas across a large region, such as the North American Platform, indicates effective surface currents sweeping across rather than around the Platform to maintain reproductive communication.

(3) We assume that cool or cold regions are characterized by floras and faunas of relatively low taxonomic diversity (from the specific level on up) as compared with those inhabiting warmer regions.

(4) We assume that similar faunas (Cambrian faunas of Siberia, western Alaska, and Antarctica, for example) occurring on opposite sides of Pangaea can be explained in terms of teleplanic larvae rather than in terms of appropriately spaced, circum-pangaeic archipelagoes. We also assume that mixed biogeographic marine assemblages of one sort or another (those encountered during much of the early Paleozoic in southeastern Kazakhstan, for example), can be explained in terms of currents supplying teleplanic larvae from the requisite biogeographic units.

(5) We assume that shallow-water platform mudstone suite rocks lacking limestone, dolomite, evaporites, redbeds, and carbonate reefs more likely represent cool or cold regions, than do those in which such rock types are present.

(6) We assume that barriers to easy reproductive communication across portions of Pangaea will induce provincialism of one sort or another. Barriers such as land, excessively shallow regions, reef complexes, bodies of hypersaline water, and so on, will impede free marine circulation and reproductive communication across the supercontinent. Topographic barriers or oceans may also act as barriers for land plants. Changing climatic gradients are barriers to marine and continental organisms.

LANDMASS PLACEMENT, LANDMASS MOVEMENT, PANGAEIC CLIMATES, AND OCEAN CIRCULATION

During the Cambrian, we have centered our Pangaea (Fig. 2) more-or-less on the South Pole. With the supercontinent so placed a good bit of South America, Africa, the Mediterranean region, and the circum-North Atlantic region is in a high south latitude location. Peripheral areas occur at intermediate to low southern latitudes. During the Cambrian, these are considered to have been warm temperate to tropical.

Following the Cambrian, Pangaea is set on a slow northward movement throughout the Paleozoic which progressively shifts the landmass into higher northern latitudes. By the Permian-Carboniferous (Fig. 7), Siberia and the present arctic parts of the New World are in intermediate to high, but not highest, northern latitudes, while Australia, the "trailing edge" is in intermediate to high, but not highest, southern latitudes.

Various biogeographic units of the early Paleozoic, as detailed in the period by period summary to follow, find easily recognized counterparts in the biogeographic units of the late Paleozoic. For example, the Eastern Americas

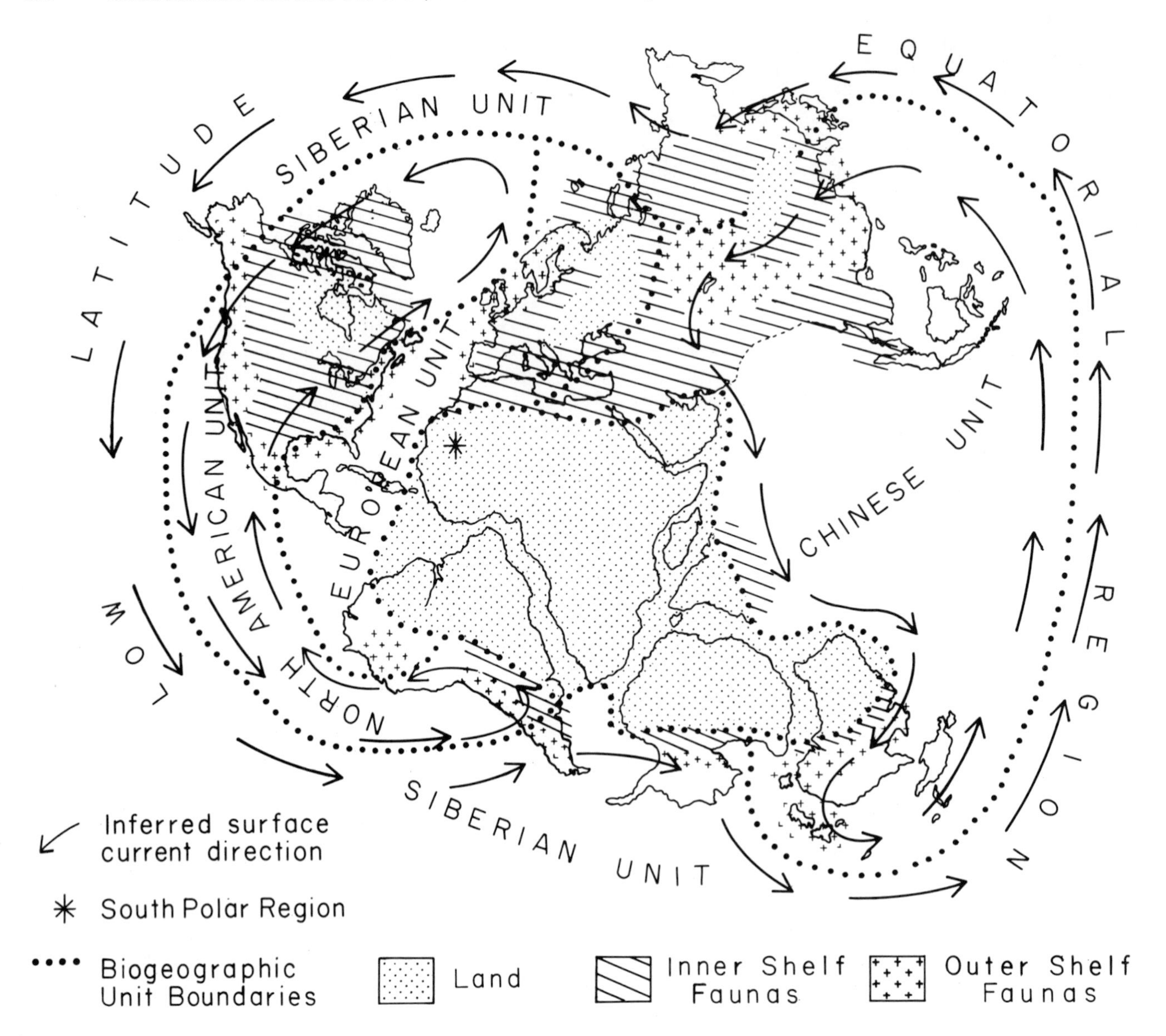

Figure 2. Pangaeic reconstruction for Middle and Late Cambrian time.

Realm of the Devonian, may be readily compared to the Carboniferous Midcontinent-Andean Unit of Ross (1973), both isolated to a certain degree on the western part of Pangaea by one mechanism or another. The Atlantic (European) unit of the Cambrian is a climatic precursor of the Atlantic (European) unit of the Ordovician, the successor Malvinokaffric Realm of the Silurian-Devonian, and the still later Gondwana Realm of the Permian-Carboniferous. These units stay in a south polar position throughout the pangaeic shift. The various physical evidences of Southern Hemisphere continental glaciation in the latest Ordovician and the Late Carboniferous-Early Permian, and the Permian glacial marine deposits on the northern rim of the Soviet Union (Ustritsky, 1973) fit with this geographic disposition, and faunas and floras of the south polar units can be categorized as cool to cold. Concurrently, the carbonate-rich floral and faunal biogeographic units in lower latitudes range from warm temperate to tropical throughout the Paleozoic.[4]

Accompanying the changing geographic position of the supercontinent is an everchanging surface current pattern in the regions deduced to have been truly oceanic as well as in those continental areas deduced to have been covered by epicontinental seas (Figs. 2 to 7). From the postulated geography it is possible to conceive of two major high-latitude gyres in the Southern Hemisphere

[4] Additional climatic insights might be provided by the location of marine evaporite deposits and the position of mountain ranges based on orogenic belts of specific ages. The absence of evidence for significant evaporite deposition during major parts of the Paleozoic, such as the Early Silurian and the latter half of the Late Devonian, require modifications of climate and (or) marine circulation. The positions of orogenic belts require assumptions about the duration of climatically significant topography. We present one example (Fig. 7) where an orogenic belt was used to explain Permian-Carboniferous vertebrate distribution.

which separated the extra-Malvinokaffric faunas into eastern and western biogeographic units early in the Paleozoic. To the north, at lower latitudes, a westerly moving surface-water current complex will have influenced a third, more cosmopolitan fauna. As the Paleozoic progresses, this low latitude current complex will continue to provide cosmopolitan faunas during times of shallow-water movement across Pangaea. Such a cosmopolitan biogeographic unit is a Paleozoic analog of the post-Paleozoic Tethyan biogeographic and climatic units. However, when the surface water currents are interrupted by a northern gyre with a boundary established adjacent to the pangaeic mass in a north-south direction, very provincial eastern and western low-latitude faunas will be established to the north of the cosmopolitan equatorial unit. In the later Paleozoic, potential for the presence of two southern biogeographic units on either side of a southern gyre, two northern biogeographic units on either side of a northern gyre, with an equatorial unit separating them, and cold southern and northern units, give rise to a complex biogeography.

When much of the pangaeic landmass was covered by epicontinental seas with few internal land or shoal obstructions, throughgoing circulation would have been conducive to shallow-water cosmopolitanism. Such conditions are encountered in the Early Silurian and the Late Devonian to Early Carboniferous. During times of significant regression, such as the later Ordovician, cosmopolitanism may have been induced by currents of the narrow shelf shoreline region. Important obstructions on the pangaeic landmass that interrupted the transcontinental circulation would have led to provincialism. Provincialism could have developed either internally or in various parts of the landmass. For example, during the Late Silurian, the North Atlantic Region may have been internally located on the northern half of the pangaeic landmass with the Malvinokaffric Realm to the south and the Uralian-Cordilleran Region on the other sides (either North Atlantic or Uralian-Cordilleran faunas are present on the western edge of South America). During the Early Devonian, the Eastern Americas Realm may have been cut off from the Old World Realm by a land barrier-induced current or by a water temperature barrier with the Malvinokaffric Realm occurring to the south. Teleplanic taxa of the Eastern Americas Realm could have been swept across the northern side of the Southern Hemisphere gyre to be mixed with Old World Realm elements in southeastern Kazakhstan and Manchuria at the same time that teleplanic elements from the Tasman Region were following the same route to southeastern Kazakhstan. A similar story can be inferred for the Cambrian and Ordovician of Kazakhstan.

ECOLOGIC VERSUS BIOGEOGRAPHIC DISTINCTIONS

There are significant, depth-correlated biological and lithological factors that may result in some scrambling of ecological and biogeographic distinctions for various time intervals of the Paleozoic.

At the lithologic level, it is important to recognize that a shelf-normal transect in a limestone-rich region extends seaward into a deeper, outer shelf-margin area where limestone and dolomites, as well as carbonate reefs, are uncommon or absent, and depth rather than biogeographically correlated.

Deeper water shelf-margin faunas commonly include rare endemic taxa that provide insights into their biogeographic affinities, *if* enough material is collected to include the rare taxa, despite the presence of many abundant cosmopolitan taxa that cannot be used to discriminate between warm-water and cold-water faunas. From the Cambrian to the Permian there tend to be a very high percentage of cosmopolitan genera in the outer shelf—shelf-margin benthos as compared to the inner shelf regions. In these communities, the more common taxa are cosmopolitan, whereas the rarer taxa tend to be endemic and allied to the endemic taxa occurring in the adjacent shallow-water areas of the shelf. Thus, the more off-shore faunas (those of Palmer's "outer detrital belt" and Boucot's Benthic Assemblages 4 through 6) globally have an overall cosmopolitan aspect that leads some into lumping them into a single biogeographic unit, rather than seeking the rarer endemic taxa that provide evidence for their affinities to nearby, nearshore biogeographic units.

Among the offshore faunas for which it is possible to scramble ecologic and biogeographic units, are the Cambrian agnostid-olenid trilobite faunas of the outer detrital belt; the Ordovician trinodid-triarthrid trilobite associations present in Benthic Assemblage 5 and 6 positions; the Middle Ordovician through Early Devonian *Skenidioides* and *Dicoelosia-Skenidioides* Community Group brachiopod faunas; the Middle and Late Devonian leiorhynchid brachiopod biofacies faunas now assigned to *Camarotoechia sensu strictu*, as well as the *Warrenella* dominated faunas; and the Permian bivalve *Atomodesma* fauna in the extra-Gondwana regions. Dutro (personal commun. 1977) suggests that the widespread *Atomodesma* may be a deeper water, shelf-margin fossil which is not too helpful for biogeographic unit discrimination, although very helpful as an environmental guide.

PANGAEIC SUMMARY BY PERIOD

Cambrian

Cambrian biogeography, paleogeography, lithofacies, and the broad biofacies changes encountered on the shelf transect from the shoreline into the bathyal region are ably summarized by Palmer (1973, 1974) within a non-pangaeic framework which does, however, unite the Gondwana landmasses into one unit. Palmer notes a biogeographic anomaly which arises from employing a non-pangaeic interpretation, i.e., the presence of Siberian affinity faunas in Antarctica far removed from those of Alaska and Siberia, but sandwiched between North American affinity faunas in west-central Argentina and Chinese affinity faunas in Australia and New Zealand. Palmer concludes that a non-pangaeic reconstruction provides only a partially satisfactory picture of total relations for the Cambrian. He also concludes (1974) that Cambrian data are inadequate for arriving at a satisfactory paleogeographic-biogeographic solution. The Cambrian data are easily reconciled with those of subsequent Paleozoic periods, however, *if* a pangaeic reconstruction is adopted for the entire Paleozoic, centered, as we have previously outlined, on the South Pole (Fig. 2).

Four biogeographic units (Palmer, 1973, 1974) of about realm to region rank (rankings high to low being realm, region, province, subprovince) are recognized: (1) the Siberian unit of low-latitude, equatorial regions; (2) the North American, and (3) the Chinese units of middle latitudes (two units due to the conclusion that a longitudinal land barrier extended through 180°, equator-to-pole-to-equator, and provided an effective circulation barrier); and (4) the European unit of high polar latitudes.

The European biogeographic unit (more-or-less the "Atlantic Region" of many authors) includes much of western Europe, the Mediterranean region, northwest Africa, probably subsurface Florida, central America (based on the Tremadoc age beds in Oaxaca), Colombia, and the Andean Ranges in northern Argentina (Salta and Jujuy Provinces) and adjacent southern Bolivia. The rarity of carbonate rocks throughout significant parts of this vast region during Middle and Late Cambrian times suggests a cold climate and the presence of a south pole in a central position somewhere in the land area including most of Africa, southern Arabia, Peninsular India, Madagascar, and most of eastern South America. Since much of this area was characterized by carbonate rocks during the Early Cambrian (Palmer, 1973, 1974) one may infer a global cooling during the Early-Middle Cambrian transition possibly comparable to that occurring in the later Cenozoic.

In such a reconstruction, the cold, pole-centered European biogeographic unit is bordered at intermediate and low latitudes by the other units, all of which have abundant carbonate rocks wherever waters are encountered shallower than those characteristic of the outer detrital belt. These rocks may be thought of as representing shallow-water, mid- to inner-shelf depth, and tropical to subtropical and warm temperate conditions conducive to the production and preservation of large volumes of biogenic carbonate. The Chinese biogeographic unit extends from Mongolia, Manchuria, and Korea through central and southern Asia into the Himalayan region (including the Salt Range), and thence to Australia and New Zealand. The North American unit extends from North America (exclusive of western Alaska) to the Pre-Cordillera de San Juan in west-central Argentina. The Siberian unit includes Siberia, northwest Alaska, Antarctica, and possibly the southern tip of South America where Cambrian rocks are not presently recognized; it is based on the close affinities of the Antarctic faunas with those of northeastern Alaska (Palmer, 1973, 1974).

The Siberian biogeographic unit is a problem for it is difficult to think of a current pattern combined with a shallow-water seaway *across* the pangaeic continent capable of explaining its anomalous faunal distribution. However, an equatorial and westerly directed current *encircling* Pangaea (Fig. 2) could account for its fauna, if it carried teleplanic larvae. If Siberia, western Alaska, and Antarctica were placed in lower latitudes than the North American and Chinese biogeographic units and the centrally located, polar-positioned European unit, a surface current would provide larval communication for the low-latitude areas, as well as means of isolating the Chinese and North American units from each other. In such a pangaeic reconstruction, two major high-latitude gyres would be set up in the Southern Hemisphere to the south of the low-latitude currents; areas of contact between these gyres would be expected to be areas of biogeographic mixing. Such mixtures occur in Turkey and elsewhere in the Middle East, where faunas include European, Chinese, and possibly North American taxa (Palmer, 1973, 1974). A second, somewhat stable center of mixing, involving faunas of Cambrian through Devonian ages, and including Siberian, Antarctic, and Alaskan taxa, is found in Kazakhstan (note Lake Balkhash as a reference point on Figs. 2 to 7) (Palmer, 1973, 1974).

A pangaeic reconstruction (Fig. 2) also provides a means of reconciling the cosmopolitan nature of Palmer's (1973, 1974) outer detrital belt faunas, which probably correspond to about Benthic Assemblage 4 through 6 (an outer shelf and upper continental slope position) since it is easy to infer interconnections between them. One may infer as well an important outer shelf—upper continental slope separating North America from adjacent northwest Africa and the platform part of Europe, and a broad marine incursion into central Asia from the east.

Ordovician

Ordovician marine biogeography fits a pangaeic interpretation (Fig. 3) by assuming that significant northward movement of the supercontinent placed the Siberian-American unit in a more northerly position than the corresponding Siberian unit of the Cambrian. The distinction between the Cambrian and Ordovician pangaeic positions is made clear by noting the shift in position of the "South Polar Region" (Figs. 2, 3).

The four biogeographic units of the Cambrian (Siberian, North American, Chinese, and European) can be only partly homologized with the many presently recognized units of the Ordovician. In part this complex situation results from the fact that during the Ordovician a variety of organisms have been used independently in biogeographic reconstructions, and as a result the number of units recognized and their areal extent are often widely different. Graptolite biogeography, for example, employs only two major units, conodont biogeography employs three major units (one of the conodont units has no conodonts where one of the graptolite units is represented), while trilobite-based and brachiopod-based biogeographic units include at least four units. Unfortunately, the various biogeographic boundaries for these major groups (graptolites, conodonts, shelly) do *not* invariably correspond to

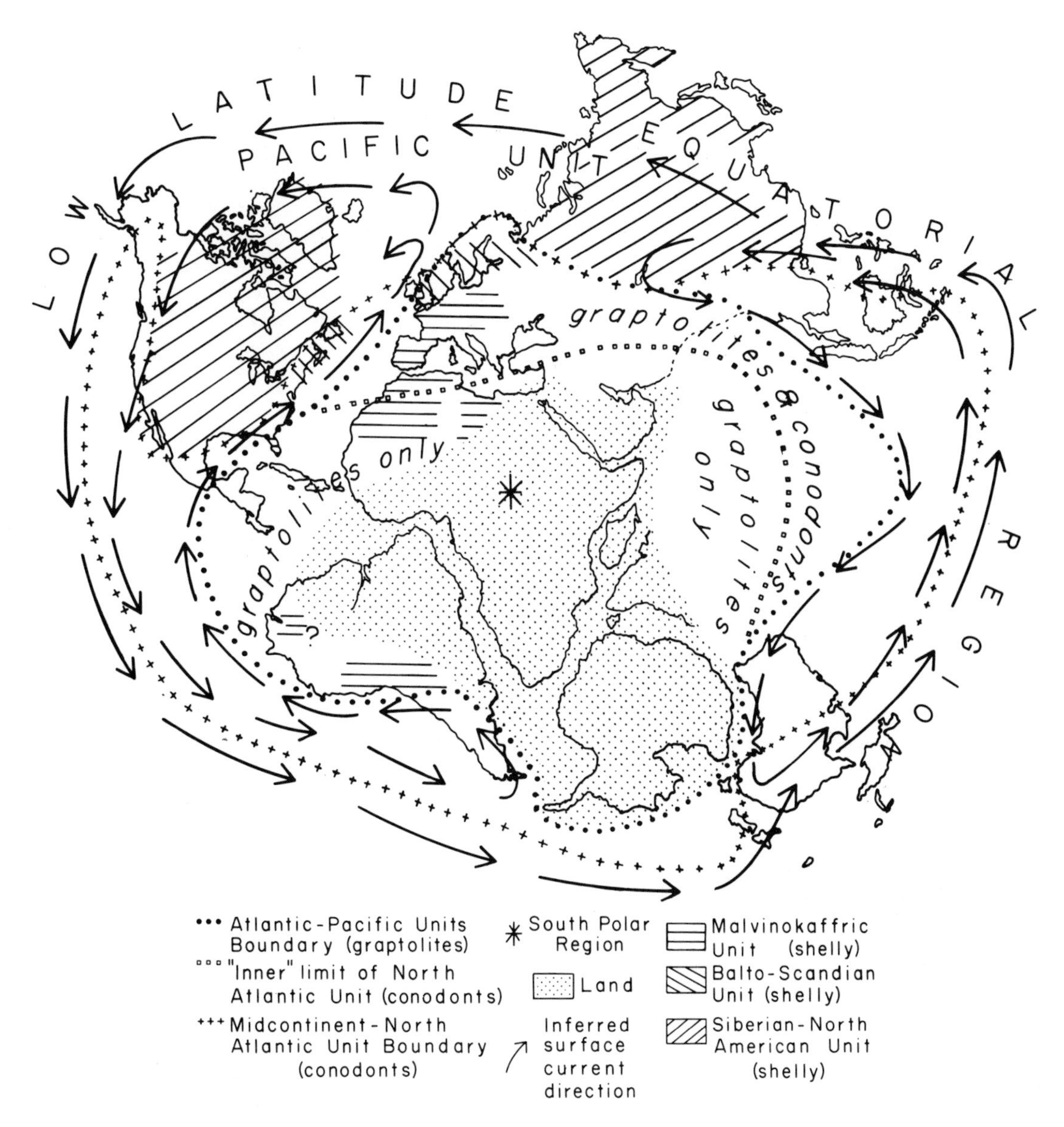

Figure 3. Pangaeic reconstruction for the Ordovician.

each other (see Fig. 3). For example, the "Inner" limit of the North Atlantic unit (conodonts, small squares on Fig. 3) lies well within the Atlantic unit (graptolites, filled circles on Fig. 3) in the area extending from Spain and northwest Africa to the east. There is also the possibility that some of the Ordovician biogeographic units may have partly intermingled biofacies differences due to environmental factors (community units) with those biofacies differences due to reproductive isolation (biogeographic units).

The European unit of the Cambrian is areally comparable to the Malvinokaffric unit of Figure 3, which corresponds to the Atlantic of some authors and the Tethyan of others. The Siberian unit of the Cambrian is comparable to the Siberian-North American unit.[5] The North American unit of the Cambrian is probably comparable to the Balto-Scandian unit. There is no Ordovician unit strictly comparable to the Chinese unit of the Cambrian, although such a unit will probably be split out in the future when the benthos of that region is given more detailed consideration. The Atlantic unit of Figure 3 (the region enclosed by filled circles) is based on graptolitic fauna (see Berry, this volume), and corresponds largely to the Malvinokaffric unit based on benthic organisms, and to a part of the Balto-Scandian unit. The Pacific unit of Figure 3 (region

[5] Kaljo and Klaaman (1973) assign the Ordovician corals to the Siberian-North American Unit; Chugaeva (1973) does the same with her Early Ordovician trilobites. Distinctive brachiopod faunas from Siberia and North America suggest, however, that separate Siberian and North American subunits may be practical within the Siberian-North American unit.

outside of filled circles) is also based on graptolites (see Berry, this volume), and corresponds to most of the extra-Malvinokaffric benthic units of the Ordovician.

The North Atlantic unit (Fig. 3), based purely on conodonts, overlaps part of the Malvinokaffric unit (in North Africa and Europe), and includes all of the Balto-Scandian unit (based on shelly benthos). The North Atlantic unit also includes such diverse areas as eastern Alaska, northern California and adjacent Nevada, the Pre-Cordillera de San Juan in Argentina, Malaysia, South Island of New Zealand, and possibly western Australia (Sweet and Bergström, 1974, Fig. 6). The conodont-based Midcontinent unit (boundary marked by crosses on Fig. 3) occurs elsewhere outside of the North Atlantic unit, particularly in much of North America, Europe, Asia, and eastern Australia plus parts of New Zealand.

The ultimate interpretation of the Ordovician conodonts of eastern Alaska (Sweet and Bergström, 1974) is critical. These conodonts have North Atlantic unit affinity. However, the conodonts of the North Atlantic unit may be a combination of endemic, platform conodonts (such as those occurring in the Balto-Scandian unit), *plus* a more cosmopolitan fauna of shelf margin to bathyal type which encircles the Midcontinent unit with its platform conodont faunas. If such is the case, then some of the areas from which North American unit conodonts are now recognized may turn out to represent deep water, with cosmopolitan assemblages plus as yet unrecognized endemic assemblages.

Throughout the Ordovician the south polar regions are in the Malvinokaffric unit. Carbonate-rich units peripheral to the Malvinokaffric represent tropical to warm temperate climates, chiefly at low latitudes. The conodont-based North Atlantic unit contains areas both rich in carbonate rocks and lacking in carbonate rocks. In northern Europe, carbonate-rich areas of the North Atlantic unit tend to occur near the boundary between the North Atlantic-Midcontinent units rather than near the Malvinokaffric unit, i.e. a temperature gradient existed from the polar Malvinokaffric unit to the low-latitude Midcontinent unit with the North Atlantic unit being intermediate. Evaporites are associated with the low-latitude Midcontinent unit rocks rather than with higher latitude units (North Atlantic and Malvinokaffric), and further support such a temperature gradient.

Ordovician biogeography is further complicated by the presence in the Malvinokaffric unit during the Ashgill (uppermost Ordovician) of widespread continental glaciation that left both physical (terrestrial glacial deposits and features, plus glacial-marine materials) and biological evidence (low diversity faunas that Sheehan, this volume, redefines as the "Hirnantian" fauna, plus the *Dalmanitina* faunas). These rocks and faunas are widespread in Malaysia, Africa, much of western, northern (Sheehan, this volume), central, and Mediterranean Europe, the central Andes of Peru, Bolivia, and west-central Argentina (Baldis and Blasco, 1975), and in the Provinces of Salta and Jujuy in northern Argentina. Uppermost Ordovician rocks of the Malvinokaffric unit are very poor in carbonate rock, and lack evaporites and redbeds as well as other geologic features of a warm climate. The record thus shows that the Malvinokaffric unit expanded during uppermost Ordovician time into areas where it was absent during most of the pre-Ashgill Ordovician as well as in the Silurian. We prefer to interpret this geographically expanded Malvinokaffric unit as due to more extensive cold or cool waters. Alternatively, a shift in the position of the pangaeic landmass, first to the south at the beginning of the Ashgill, then north again at the end of the Ashgill, would explain the cool and cold water rocks and faunas of the uppermost Ordovician, both preceded and followed by evidences of increased warmth.

Silurian

In comparison with the Ordovician, the Silurian is marked by biogeographic and climatic changes, and by some biological and climatic similarities. The Malvinokaffric unit (definitely a Realm in the Silurian) continued into the Silurian in about the same general regions that it occupied from the Middle Cambrian. In the Silurian its area diminished considerably, and in comparison with the Ordovician, the boundary of the unit shifted to the east and south (compare Figs. 1, 3, 4). There is evidence for cold- or at least cool-water faunas and rocks in part of North Africa, much of South America (except for the western portion of the Andean region, and the Merida Andes of Venezuela), and South Africa. There is also evidence favoring Malvinokaffric-type rocks and faunas in both the Ordovician and Silurian in the subsurface of Florida (see Laufeld, this volume). Boucot (1975) has summarized the benthic shelly evidence for endemism in the Malvinokaffric Realm, and Jaeger (1976) has analyzed the graptolitic information.

In the Early Silurian, regions peripheral to the Malvinokaffric Realm comprise a single biogeographic unit with cosmopolitan graptolites, conodonts, and benthic shelly fossils, in sharp contrast to the many units of the Ordovician. There are Early Silurian shelly and planktonic faunas, all of North Silurian Realm type, known from eastern Australia, the western Andes in Bolivia and northwesternmost Argentina, Venezuela, North America, most of Eurasia, the Istanbul region, and the western Himalayas. Antarctica, where Silurian fossils have yet to be found, is the only gap in these North Silurian Realm faunas encircling the cooler Malvinokaffric Realm. The low level of biogeographic differentiation in the warmer regions to the north of the Malvinokaffric Realm can be explained by the presence of teleplanic larvae in many animal groups that were spread widely in the surface currents circling westerly about Pangaea, as well as in the shallow transcontinental, epicontinental seas.

In the Late Silurian (Fig. 4), the cosmopolitan North Silurian Realm is subdivided into several units of region level (North Atlantic, Uralian-Cordilleran, Mongolo-Okhotsk) which herald further changes into the highly provincial Early Devonian (Fig. 5). Late Silurian biogeographic differentiation may be due to a somewhat changed

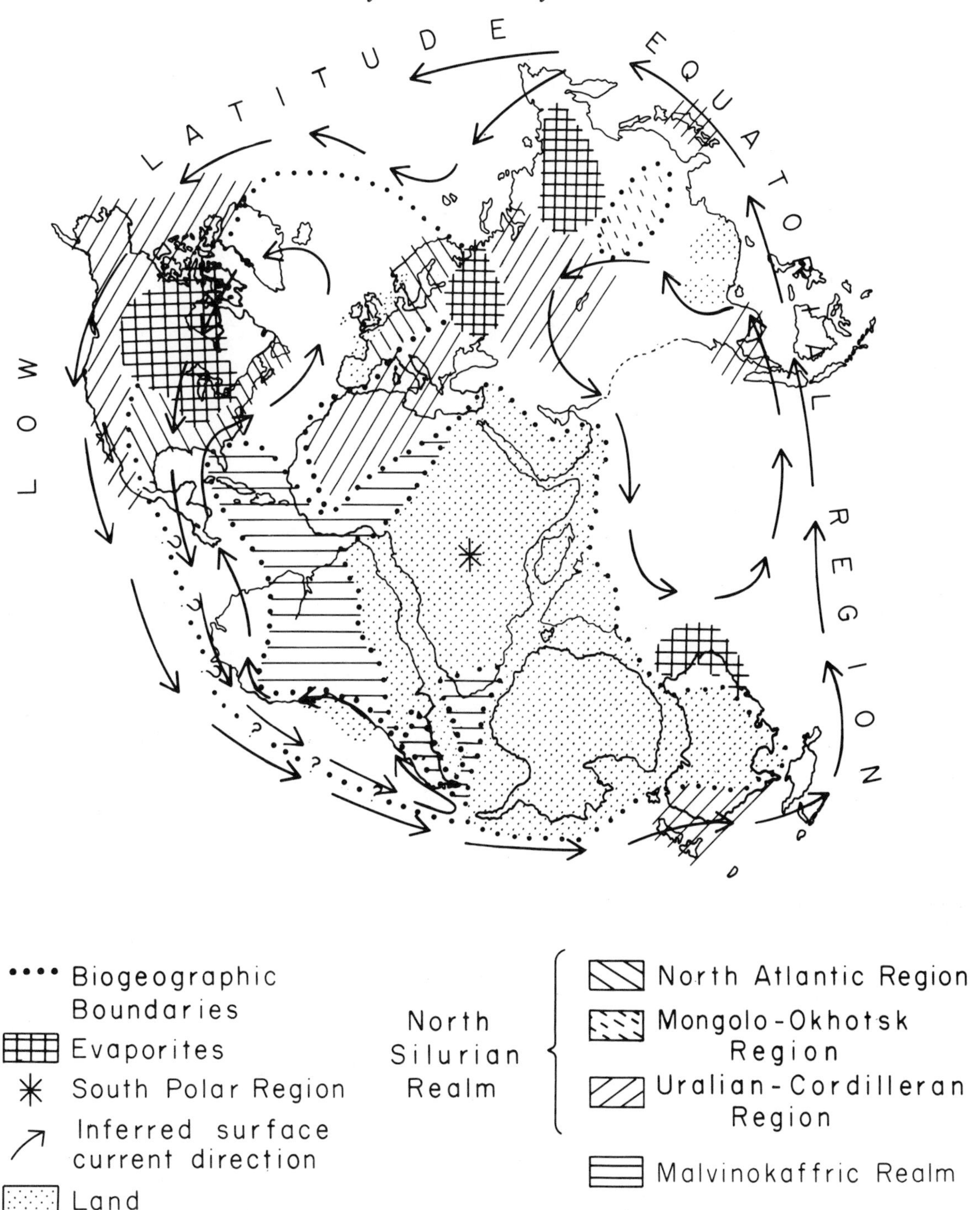

Figure 4. Pangaeic reconstruction for the Late Silurian.

shallow current circulation on the pangaeic landmass correlated with the presence of numerous reefs and widespread evaporites which suggest a lowered global climatic gradient. Widespread evaporites are unknown in the Early Silurian and Ashgill, but are found in the pre-Ashgillian Late and Middle Ordovician, in the same general regions as in the Late Silurian. Toward the end of the Silurian the gradual uplift of a number of low-lying land barriers on the pangaeic landmass further interfered with and complicated the current patterns on its surface. The hypersaline water masses probably also acted as at least partial barriers to panmixia across the supercontinent. The possibility of local hyposaline water masses, created by riverine input from the more extensive landmasses, may have additionally complicated the situation.

In comparison with the high climatic gradients of the Ashgill that correlate with high-latitude continental glaciation and low-latitude tropical conditions, climatic gradients

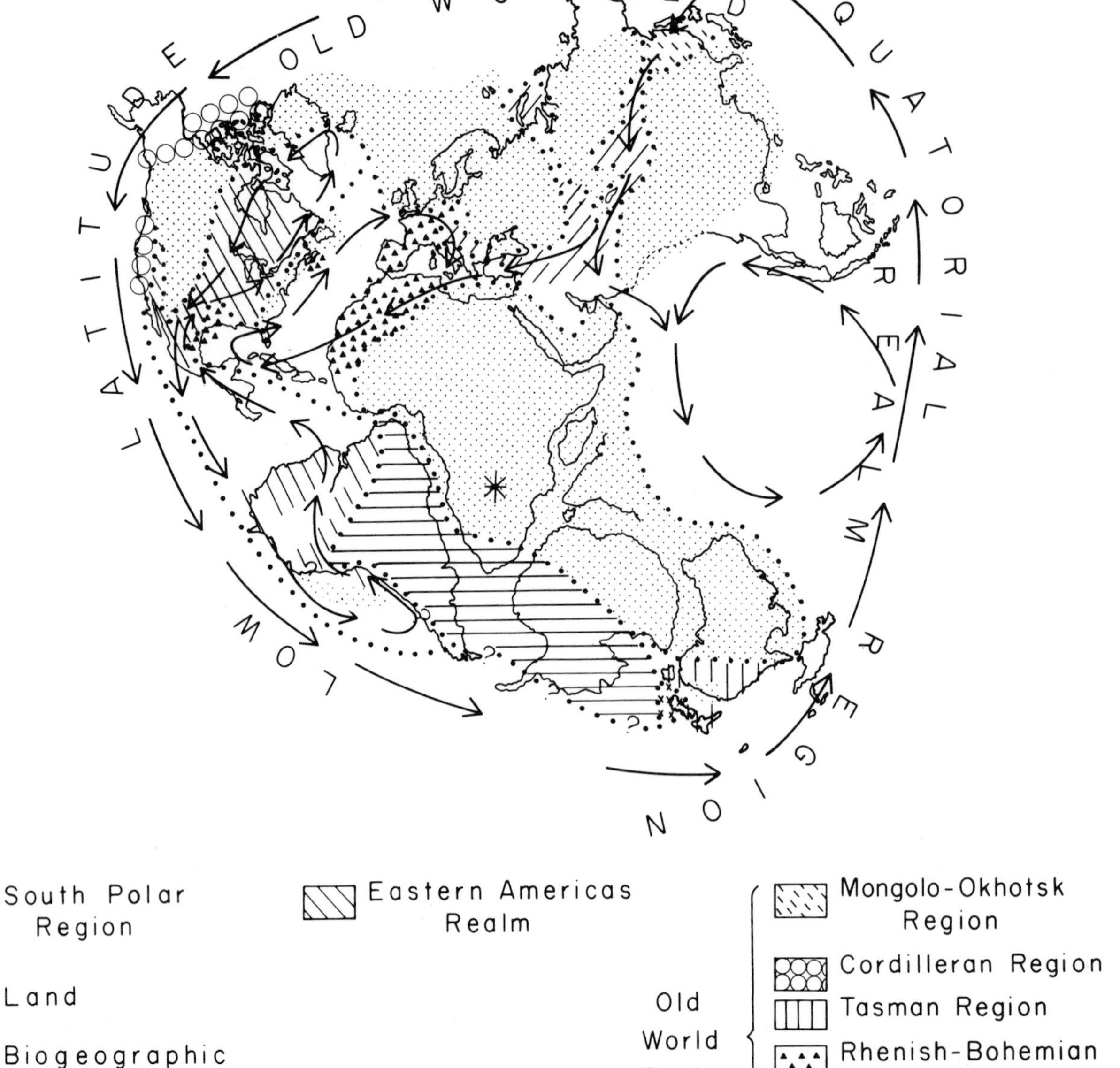

Figure 5. Pangaeic reconstruction for the later Early Devonian.

of the Late Silurian would appear to be lower, related to merely cool conditions at high latitudes, and grading into tropical conditions marked by the presence of evaporites at low latitudes.

Devonian

The moderately provincial Late Silurian grades into the highly provincial Early Devonian. The latter, in turn, grades into the very cosmopolitan Late Devonian. The Malvinokaffric Realm of the Silurian continues as the Malvinokaffric Realm of the pre-Late Devonian. The North Silurian Realm of the Late Silurian, with its three region-level units is replaced by two units of realm rank, the Eastern Americas Realm, and the Old World Realm. The Eastern Americas Realm corresponds in large part to the North Atlantic Region of the Late Silurian. The Old World Realm generally corresponds areally to the combined Uralian-Cordilleran and Mongolo-Okhotsk Regions of the Late Silurian, but now includes a total of six regions (Fig. 5) with the eastern portion of the former

North Atlantic Region now become the Rhenish-Bohemian Region. During the extreme cosmopolitanism of the Late Devonian none of the biogeographic units, including the stable Malvinokaffric Realm of the lower Paleozoic, is recognizable.

Devonian paleogeography and biogeography differs in two additional ways from that of the Silurian. First, there are a number of large land areas well away from the Malvinokaffric Realm for the first time (Fig. 5). These areas include the regions of Old Red Sandstone sedimentation in North America, Europe, Asia, and Australia. Second, the position of the boundary between the Malvinokaffric Realm and the combined Old World-Eastern Americas Realms in South America and Africa has shifted hundreds of kilometers to the south (on a modern base; see Fig. 1). This shift continues the movement in the same direction that began in the Cambrian.

Isolation of the Malvinokaffric Realm during the Early and Middle Devonian may have been caused by temperature gradient alone. Isolation of the Old World Realm from the Eastern Americas Realm, both carbonate-rich tropical to warm temperate areas, might have been a result of a combined current system and distribution of land area and hypersaline water bodies which served to separate all but the most teleplanic-eurytopic larvae. That the Rhenish-Bohemian Region was very effectively shielded from larval communication with the Eastern Americas Realm is shown by the lack of those Eastern Americas Realm taxa known in Kazakhstan, the Tasman Region, and Manchuria. The various biogeographic subdivisions of the Old World Realm may be local units isolated by distance alone and the absence of numerous teleplanic species.

A mixture of Old World and Eastern Americas Realms benthos (brachiopods and trilobites particularly) in Kazakhstan, on the other hand, may reflect in large part interacting surface currents and teleplanic larvae. Mixtures of some of the same taxa in the Tasman Region and in Manchuria (see Boucot, 1975) may be due to the same phenomena.

Edwards (1973) points out the floral distinctiveness of the Malvinokaffric Realm as contrasted with the remainder of the Devonian world; her conclusions are in agreement with the data provided by the shallow-water marine record. However, her failure to recognize additional nonmarine phytogeographic units elsewhere in the Early and Middle Devonian probably reflects deficiency in information rather than the true situation on land.

Early Carboniferous

Late Devonian cosmopolitaniety is matched by an almost equally cosmopolitan Early Carboniferous, although biogeographic units are recognizable during the latter interval.

On land, there is abundant plant evidence for which significant phytogeographic studies have been possible. However, the Early Carboniferous is a time of only modest biogeographic differentiation (Fig. 6). Chaloner and Lacey (1973) and Chaloner and Meyen (1973) summarize plant megafossil data indicating the presence of an Angaran flora in northern Asia, bounded by a Kazakhstan flora (not shown on Fig. 6) in a narrow belt on the south, in turn bounded by the very widespread *Lepidodendropsis* flora known from most other regions of the world except for the Gondwana[6] areas of Antarctica, South America, and Africa, where definitive floral evidence is lacking for the Early Carboniferous.

In the shallow seas, Mamet and Skipp (1978) recognize a broad division, the Taimyr-Alaskan unit, that includes much of the region characterized on land by the Angaran flora (northern Asia). To the south, they recognize a somewhat warmer Kuznetz-North American unit, bounded to the south by a low-latitude and warmer Tethyan unit (Fig. 6). The latter is bounded by the Gondwana region. The lack of a biogeographic unit between their Tethyan and Gondwana regions in the Southern Hemisphere corresponding to the Kuznetz-North American unit of the Northern Hemisphere may indicate a lack of information or a real difference in number of units that characterize the Northern and Southern Hemispheres.

A Gondwana shallow-marine fauna has not yet been differentiated for the Early Carboniferous. There may have been such a low level of climatic differentiation, a possibility suggested by the cosmopolitan nature of the shelly faunas of Australia during the interval (Runnegar and Campbell, 1976), that there was no really distinctive Early Carboniferous Gondwana fauna. The more widespread epicontinental seas of the Famennian-Early Carboniferous, in contrast to the earlier Devonian and later Late Carboniferous-Permian, may have been the predisposing factor limiting climatic differentiation.

An additional clue to a lowered climatic gradient in the Early Carboniferous and Late Devonian is the distribution of *Tropidoleptus*. The Devonian brachiopod *Tropidoleptus* and goniatites are common prior to the Early Carboniferous in subtropical to tropical regions (both the Eastern Americas and Old World Realms). They are absent in the older Devonian beds of the Malvinokaffric Realm, but appear in that Realm immediately prior to the change to nonmarine beds that terminates the Malvinokaffric Devonian sequence. Their appearance in Bolivia and some adjacent regions may be interpreted as evidence of climatic amelioration well before the end of the Devonian but continuing up into the Early Carboniferous. A change in climatic

[6] The late Paleozoic biogeographic term "Gondwana" is conceptually very similar to the Ordovician, Silurian, Devonian term "Malvinokaffric" as well as the Ordovician and Cambrian term "European" (="Atlantic," "Tethyan," "Mediterranean," and "Paleotethyan"). All four (Gondwana, Mediterranean, Malvinokaffric, European) refer to high latitude, cool-to-cold regions of the Southern Hemisphere in which carbonate rocks, redbeds, evaporites, evidences of lateritic weathering, and so on are lacking. In all four, the floras (where present) and marine faunas have many of the characters of modern high latitude biotas, such as growth rings in woods, low diversity benthic communities, absence of many of the higher taxa (phyla, classes, orders), low total numbers of species and so on. There is no evidence, however, that the Gondwana taxa are lineal descendants of earlier species living in the same region. The major terminal extinctions of the Paleozoic affected these high-latitude southern regions as well as the warmer, contemporary low-latitude regions. Many of the high-latitude southern region taxa are most reasonably derived, following each major terminal extinction, from nearby contemporary lower latitude, warm region taxa.

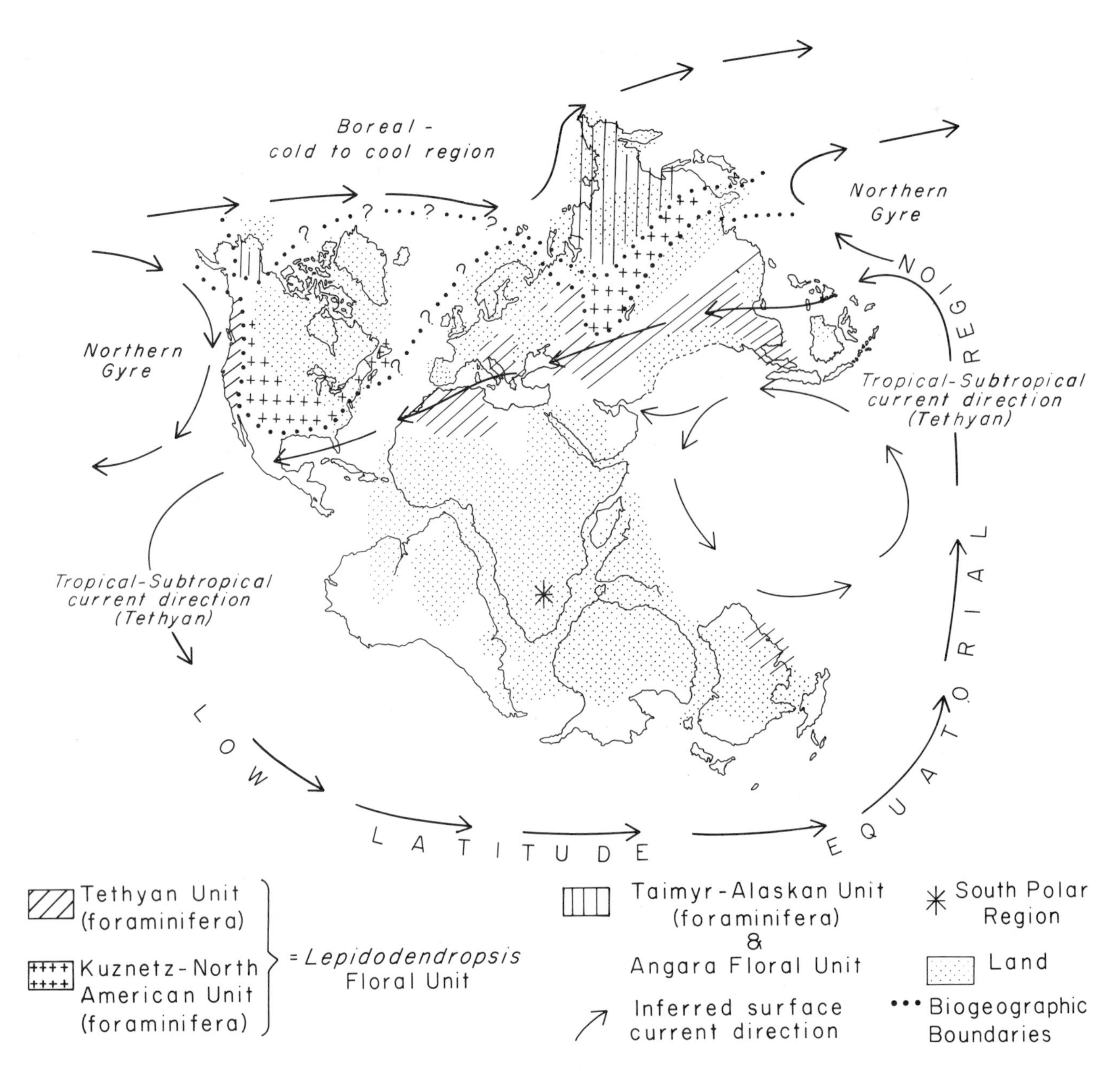

Figure 6. Pangaeic reconstruction for the Early Carboniferous.

gradient, opposite in direction, is indicated by the presence of *Dalmanitina*-type faunas in the Ashgill of the Pre-Cordillera de San Juan, west-central Argentina.

Late Carboniferous-Early Permian

The Late Carboniferous-Early Permian is a time of relatively high provincialism affecting both terrestrial and marine environments on a global scale (Fig. 7). Floral and faunal units of the Early Carboniferous nevertheless are recognizable in part, in addition to a variety of biogeographic units unique to this time interval. The Taimyr-Alaskan marine unit of the Early Carboniferous, as well as the Angara floral unit, persist. During the Permian, the Taimyr-Alaskan becomes the Boreal Realm (or Boreal Province of Yancey, this volume). The Tethyan unit also maintains its continuity during the Late Carboniferous-Early Permian (Tethyan unit plus a Midcontinent-Andean unit). The Kuznetz-North American unit is difficult to recognize with certainty, although the Cordilleran subunit may be its continuation in part.

In the Southern Hemisphere, the Gondwana Realm of the Late Carboniferous-Early Permian is characterized by a very distinctive fauna and flora (the *Glossopteris* flora) that follows the Early Carboniferous cosmopolitan shelly faunas of the marine environment. The rapid onset of Southern Hemisphere continental glaciation may have been

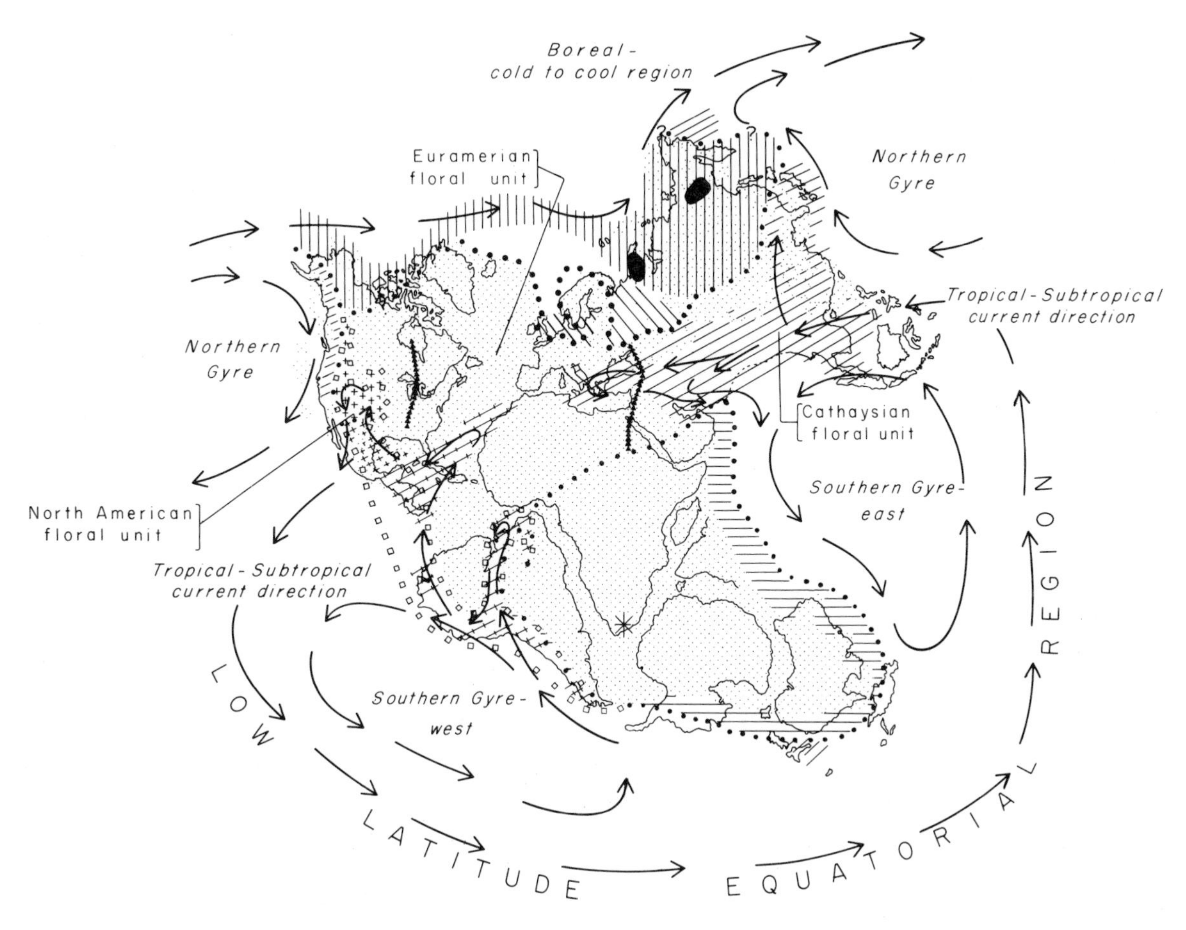

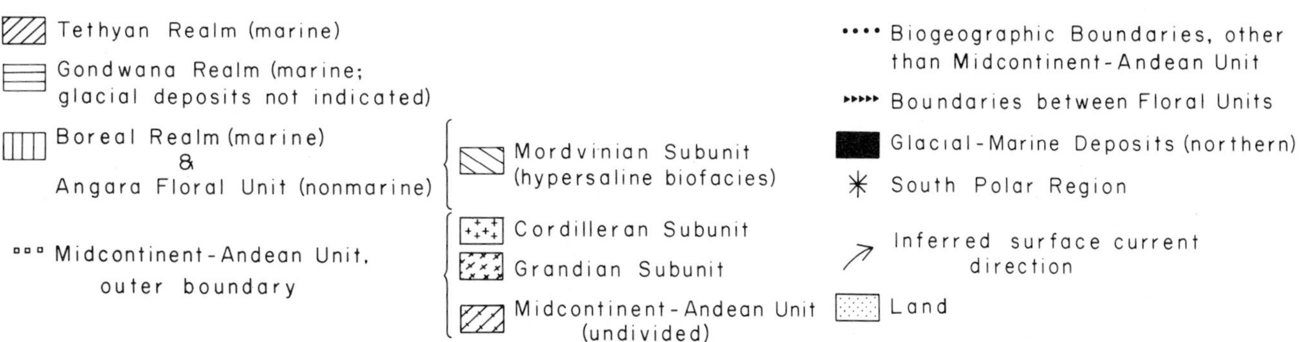

Figure 7. Pangaeic reconstruction for the interval Late Carboniferous-Early Permian.

controlled in part by the steady northward movement of Pangaea. The high level of climatic differentiation was responsible, in turn, for much of the differentiation of the Late Carboniferous-Early Permian fauna and flora. The presence in the Gondwana Realm of tillites, glaciated pavements, and similar evidences of glaciation (see Frakes, and others, 1975, and references therein), together with the available oxygen isotope temperature measurements, indicate clearly that the region was cool to cold.

In the Northern Hemisphere, the Angara floral unit also appears to have been occupied by a cool to cold flora (Chaloner and Meyen, 1973). It occurs in the same region as putative glacial-marine sediments of Permian age (Ustritsky, 1973) and is associated with the Arctic Permian Marine Fauna of Dutro and Saldukas (1973) based largely on brachiopods. The Arctic Permian Marine Fauna equates with the Boreal Province of Yancey (this volume). In the Northern Hemisphere, Yancey (this volume) differentiates a second marine unit, the Mordvinian Province, that may be viewed as a largely hypersaline biofacies of the Boreal unit, rather than a distinctive biogeographic unit. Yancey's Mordvinian unit appears to correspond to the largely hypersaline Zechstein and Perm seas with their restricted faunas of Arctic Permian derivation (Dutro and Saldukas, 1973). The Zechstein and Perm seas were closed to the Tethys to the south, except during the earlier Permian;

they were also separated from each other by a large land area (Dutro and Saldukas, 1973, Figs. 1, 2).

Sandwiched between the cool-to-cold northern Boreal-Angara and the southern Gondwana units are a variety of marine and nonmarine biogeographic units. The marine units are sufficiently rich in carbonate rocks, reefs, and evaporites to be considered warm temperate to tropical. The nonmarine redbeds locally occurring side-by-side with them may be similarly construed.

With the aid of fusuline foraminifera, Ross (1973) has distinguished a marine Tethyan unit (the Mediterranean region, southern Asia, parts of northeastern Asia, part of New Zealand, the coastal fringe of western North America) and a Midcontinent-Andean unit (much of middle latitude North America to the east of the Tethyan fringe adjacent to the present Pacific, the Andean regions of South America from Fuegia north) for the Late Carboniferous. Gobbett (1973) has recognized similar biogeographic units for the Permian. The Durhaminid and Cyathaxonid coral Provinces of Rowett (1975) for the Late Carboniferous and Early Permian largely parallel the fusuline-based units of Ross (1973) and Gobbett (1973).

For the North American Permian, Yancey (this volume) has subdivided the Midcontinent-Andean unit of Ross (1973) into a more northern Cordilleran unit, bounded on the south by a Grandian unit that includes reefy faunas as well as level-bottom faunas. R. E. Grant (oral commun. 1977) points out that some of the brachiopod faunas of the central Andes in Peru and Bolivia have much in common with those of the Grandian unit, which indicates the essential biogeographic unity of the fusuline and brachiopod faunas of the Midcontinent-Andean unit. The prominence of reefs in the Grandian unit suggests more fully tropical conditions than for other portions of the Midcontinent-Andean unit.

In western North America (California to southern Alaska), the present distribution of Permian fossils is anomalous, with Midcontinent-Andean fusuline faunas sandwiched between Tethyan faunas. All three belts (Tethyan, Midcontinent-Andean, Tethyan) occur to the west of Midcontinent-Andean faunas found on the North American Platform. Jones and others (1972) explain this in a manner consistent with the local geology by moving the two outer, seaward belts (Tethyan, Midcontinent-Andean) far enough to the south to join the Midcontinent-Andean of the North American Platform and its adjacent western Tethyan belt, both of the latter in place. Their reconstruction leaves a Pacific rim of Tethyan faunas margined to the east by Midcontinent-Andean faunas from southern Alaska to well south of California. The southern limits of this strip might even have extended as far south as the Andean region. This reconstruction is consistent with what is presently known of pre-Permian-Carboniferous biogeography (see Boucot and Potter, 1977, for summary of data).

On land, the intermediate, warm, plant-based phytogeographic unit that occurs between the Angara and Gondwana floral units is broken into three subunits, the Cathaysian, the Euramerian, and the North American (Fig. 7). The occurrence of these three units, suggests a degree of reproductive isolation that is at least permissively consistent with the biogeography of the contemporary nonmarine vertebrates (Panchen, 1973; Romer, 1973; Olson, this volume).

The high provincialism of the Early Permian appears to persist in the marine environment until near the end of the Period. On land, in about the mid-Permian, phytogeography suggests more cosmopolitan conditions similar to those of the Triassic, for which only a tripartite, latitudinal division has been recognized (see Florin, 1963, for examples drawn from the gymnosperms).

Post-Permian

Post-Permian biogeography begins with the well-known pangaeic reconstruction. We now suggest that this pangaeic reconstruction may have been a constant feature from the later Precambrian through the Permian, rather than one arrived at late in the Paleozoic. We also suggest that the position of this Pangaea kept shifting north from at least the Middle Cambrian to the Permian. Such a shift would account for the various physical and biogeographic data herein described, particularly when combined with intervals of marked change in global climatic gradients. Our interpretation views the various ophiolite suite rocks of the Paleozoic, such as those in the Urals, Appalachians, and western Cordillera of North America, as indicating mantle-derived materials of local significance rather than as evidence of long distance transport of originally far removed plates.

ACKNOWLEDGMENTS

We are very grateful to Dr. John C. Crowell, University of California, Santa Barbara for helpful comments made on an early version of the manuscript. We are also indebted to Dr. J. Thomas Dutro, U.S. Geological Survey, Washington, D.C., for his help and suggestions. Needless to say, neither can be held responsible for any of our errors or misconceptions.

REFERENCES

Amos, A. J., 1964, A review of the marine Carboniferous stratigraphy of Argentina: Internat. Geol. Cong., 22nd, India 1964, Pt. IX, Proc. Section 9, p. 54-72.

Amos, A. J., and Rocha-Campos, A. C., 1971, A review of South American Gondwana geology 1967-1969; I.U.G.S., Comm. Stratigr., Subcomm. Gondwana Stratigr. Palaeontol., Gondwana Symposium, Proc. and Papers, p. 1-13 (1970).

Baldis, B. A., and Blasco, G., 1975, Primeros trilobites Ashgillianos del Ordovicico Sudamericano: Congreso Argentino de Paleontologia y Bioestratigrafia, Tucuman 1975, Actas I, v. 1, p. 33-48.

Berry, W. B. N., 1978, Graptolite biogeography: A biogeography of certain lower Paleozoic plankton, *in* Gray, J., and Boucot, A. J., eds., Historical biogeography, plate tectonics, and the changing environment: Corvallis, Oregon State Univ. Press, p. 105-115.

Berry, W. B. N., and Boucot, A. J., eds., 1973, Correlation of the African Silurian rocks: Geol. Soc. America Spec. Paper 147, 83 p.

Boucot, A. J., 1969, The Soviet Silurian: Recent impressions: Geol. Soc. American Bull., v. 80, p. 1155-1162.

——— 1975, Evolution and extinction rate controls: Amsterdam, Elsevier, 427 p.

Boucot, A. J., and Potter, A. W., 1977, Middle Devonian orogeny and biogeographical relations in areas along the North American Pacific rim, *in* Murphy, M., Berry, W. B. N., and Sandberg, C., eds., Western North America: Devonian: Univ. California-Riverside Campus Mus. Contrib. no. 4, p. 210-219.

Chaloner, W. G., and Lacey, W. S., 1973, The distribution of late Paleozoic floras, *in* Hughes, N. F., ed., Organisms and continents through time: Spec. Papers Palaeontology no. 12, p. 271-290.

Chaloner, W. G., and Meyen, S. V., 1973, Carboniferous and Permian floras of the northern continents, *in* Hallam, A., ed., Atlas of palaeobiogeography: Amsterdam, Elsevier, p. 169-186.

Chugaeva, M. N., 1973, Biogeography of the uppermost Lower Ordovician, *in* Keller, B. M., ed., Biostratigraphy of the lower part of the Ordovician in the north-east of the USSR and biogeography of the uppermost Lower Ordovician: Akad. Nauk SSSR Trudy, no. 213, p. 237-280 (in Russian).

Douglass, R. C., and Nestell, M., 1976, Late Paleozoic Foraminifera from southern Chile: U. S. Geol. Survey Prof. Paper 858, 47 p.

Dutro, J. T., and Saldukas, R. B., 1973, Permian paleogeography of the Arctic: U. S. Geol. Survey Jour. Research, v. 1, p. 501-507.

Edwards, D., 1973, Devonian floras, *in* Hallam, A., ed., Atlas of palaeobiogeography: Amsterdam, Elsevier, p. 105-115.

Florin, R., 1963, The distribution of conifer and taxad genera in time and space: Acta Hortii Bergiani, v. 20, p. 122-326.

Frakes, L. A., Amos, A. J., and Crowell, J. C., 1969, Origin and stratigraphy of late Paleozoic diamictites in Argentina and Bolivia, *in* Gondwana stratigraphy, I.U.G.S. Symposium, Buenos Aires 1967: UNESCO, Paris 1969, p. 821-843.

Frakes, L. A., Kemp, E. M., and Crowell, J. C., 1975, Late Paleozoic glaciation: Part VI, Asia: Geol. Soc. America Bull., v. 86, p. 454-464.

Gobbett, D. J., 1973, Permian Fusulinacea, *in* Hallam, A., ed., Atlas of palaeobiogeography: Amsterdam, Elsevier, p. 151-158.

Harrington, H. J., 1962, Paleogeographic development of South America: Am. Assoc. Petroleum Geologists Bull., v. 46, p. 1773-1814.

Harrington, H. J., and Leanza, A. F., 1957, Ordovician trilobites of Argentina: Kansas Univ. Dept. Geology Spec. Pub. 1, 276 p.

Irving, E., 1977, Drift of the major continental blocks since the Devonian: Nature, v. 270, p. 304-309.

Jaeger, H., 1976, Das Silur und Unterdevon vom thuringischen Typ in Sardinien und seine regionalgeologische Bedeutung: Nova Acta Leopoldina, n. f., v. 45, p. 263-299.

Jones, D. L., Irwin, W. P., and Ovenshine, A. T., 1972, Southeastern Alaska—a displaced continental fragment?: U. S. Geol. Survey Prof. Paper 800-B, p. B211-B217.

Kaljo, D., and Klaaman, E., 1973, Ordovician and Silurian corals, *in* Hallam, A., ed., Atlas of palaeobiogeography: Amsterdam, Elsevier, p. 37-46.

Laufeld, S., 1978, Biogeography of Ordovician, Silurian, and Devonian chitinozoans, *in* Gray, J., and Boucot, A. J., eds., Historical biogeography, plate tectonics, and the changing environment: Corvallis, Oregon State Univ. Press, p. 75-90.

LePichon, X., Sibuet, J.-C., and Francheteau, J., 1977. The fit of the continents around the North Atlantic Ocean: Tectonophysics, v. 38, p. 169-209.

Levy, R., and Nullo, F., 1975, Braquiopodos Ordovicicos de Poron-Trehue, Bloque de San Rafael: Congreso Argentino de Paleontologia y Bioestratigrafia, Tucuman 1975, Actas I, v. 1, p. 23-32.

Mamet, B., and Skipp, B., 1978, Lower Carboniferous Foraminifera—paleogeographical implications: Cong. Internat. Stratigr. Géol. Carbonifère (in press).

McKerrow, W. S., and Cocks, L. R. M., 1976, Progressive faunal migration across the Iapetus Ocean: Nature, v. 263, p. 304-306.

Minato, M., and Tazawa, J., 1977, Fossils of the Huentelauquén Formation at the locality F, Coquimbo Province, Chile, *in* Ishikawa, T., and Aguirre, L., eds., Comparative studies on the geology of the circum-Pacific orogenic belt in Japan and Chile: Japan Soc. for the Promotion of Science, 1st Rept., p. 95-117.

Olson, E. C., 1978, Biological and physical factors in the dispersal of Permo-Carboniferous terrestrial vertebrates, *in* Gray, J., and Boucot, A. J., eds., Historical biogeography, plate tectonics, and the changing environment: Corvallis, Oregon State Univ. Press, p. 227-238.

Palmer, A. R., 1973, Cambrian trilobites, *in* Hallam, A., ed., Atlas of palaeobiogeography: Amsterdam, Elsevier, p. 3-12.

——— 1974, Search for the Cambrian world: American Scientist, v. 62, p. 216-224.

Panchen, A. L., 1973, Carboniferous tetrapods, *in* Hallam, A., ed., Atlas of palaeobiogeography: Amsterdam, Elsevier, p. 117-126.

Robison, R. A., and Pantojo-Alor, J., 1968, Tremadocian trilobites from the Nochixtlan region, Oaxaca, Mexico: Jour. Paleontology, v. 42, p. 767-800.

Rohr, D. M., 1978, Geographic distribution of the Ordovician gastropod *Maclurites*, *in* Gray, J., and Boucot, A. J., eds., Historical biogeography, plate tectonics, and the changing environment: Corvallis, Oregon State Univ. Press, p. 45-52.

Romer, A. S., 1973, Permian reptiles, *in* Hallam, A., ed., Atlas of palaeobiogeography: Amsterdam, Elsevier, p. 150-167.

Ross, C. A., 1973, Carboniferous foraminiferida, *in* Hallam, A., ed., Atlas of palaeobiogeography: Amsterdam, Elsevier, p. 127-132.

Rowett, C. L., 1975, Provinciality of late Paleozoic invertebrates of North and South America and a modified intercontinental reconstruction: Pacific Geology, v. 10, p. 79-94.

Runnegar, B., and Campbell, K. S. W., 1976, Late Paleozoic faunas of Australia: Earth-Sci. Rev., v. 12, p. 235-258.

Sheehan, P. M., 1978, Swedish Late Ordovician marine benthic assemblages and their bearing on brachiopod zoogeography, *in* Gray, J., and Boucot, A. J., eds., Historical biogeography, plate tectonics, and the changing environment: Corvallis, Oregon State Univ. Press, p. 61-73.

Smith, A. G., Briden, J. C., and Drewry, G. E., 1973, Phanerozoic world maps, *in* Hughes, N. F., ed., Organisms and continents through time: Spec. Papers Palaeontology no. 12, p. 1-42.

Smith, P. J., 1976, Does the key fit the locks of the past?: Nature, v. 260, p. 746-747.

Sweet, W. C., and Bergström, S. M., 1974, Provincialism exhibited by Ordovician conodont faunas, *in* Ross, C. A., ed., Paleogeographic provinces and provinciality: Soc. Econ. Paleontologists and Mineralogists Spec. Pub. 21, p. 189-202.

Ustritsky, V. I., 1973, Permian climate, *in* Logan, A., and Hills, L. V., eds., The Permian and Triassic Systems and their mutual boundary: Canadian Soc. Petroleum Geologists Mem. 2, p. 733-744.

van Andel, T. H., 1978, An eclectic overview of plate tectonics, paleogeography and paleoceanography, *in* Gray, J., and Boucot, A. J., eds., Historical biogeography, plate tectonics, and the changing environment: Corvallis, Oregon State Univ. Press, p. 9-25.

Whittington, H. B., 1953, A new Ordovician trilobite from Florida: Harvard Univ. Mus. Comp. Zool., Breviora, no. 17, p. 1-6.

Yancey, T. E., 1978, Permian positions of the Northern Hemisphere continents as determined from marine biotic provinces, *in* Gray, J., and Boucot, A. J., eds., Historical biogeography, plate tectonics, and the changing environment: Corvallis, Oregon State Univ. Press, p. 239-247.

APPENDIX

Data Sources for the Biogeographic Boundaries of Figure 1

Cambrian

Palmer (1974).

Ordovician

Berry (1978); Rohr (1978); Whittington (1953); Robison and Pantoja-Alor (1968). Trilobites and brachiopods have not yet been effectively employed for biogeographic purposes in South America. There is a monographic treatment of Ordovician trilobites (Harrington and Leanza, 1957). But most of the faunas occurring in Argentina, Bolivia, and Peru belong to the relatively deep-water agnostid-olenid biofacies (a complex of communities in about Benthic Assemblage 5 to 6 position) of the Early Ordovician or the succeeding trinodid-triarthrid biofacies of the Middle and Late Ordovician. The bulk of these trilobites are cosmopolitan genera that obscure the rarer, provincial taxa whose biogeographical affinities would be predicted to adhere to the pattern of the graptolites. The shelly faunas of Bolivia and adjacent northern Argentina (Salta and Jujuy Provinces), Peru, and Venezuela are predicted to be of "Tethyan" Realm type, i.e. "European" or "Atlantic," and similar to those described earlier by many authors from central Europe and North Africa. Those from the Pre-Cordillera de San Juan, on the other hand, are predicted to be similar to ones occurring in the Siberian-North American Realm and possibly other carbonate-rich areas such as the Tasman Region and the Baltic area. The presence of "Tethyan"- or "Hirnantian"-type trilobites in the Pre-Cordillera de San Juan in beds of Ashgill age (Baldis and Blasco, 1975) suggests that the cold Hirnantian-type biogeographic unit, essentially the Ashgillian precursor of the Malvinokaffric Realm, expanded to the west in the uppermost Ordovician over the previously warm-water extra-Malvinokaffric unit present in that region. Levy and Nullo (1975) summarize information about Middle Ordovician brachiopod and trilobite faunas in the Pre-Cordillera de San Juan, which indicates that they are compatible with North American unit faunas.

Silurian

Boucot (1975).

Devonian

Boucot (1975).

Carboniferous and Permian

Amos and Rocha-Campos (1970); Minato and Tazawa (1977); Frakes, Amos, and Crowell (1967); Harrington (1962); Douglass and Nestell (1976); Amos (1964).

Appendix

Participants in the 37th Annual Biology Colloquium included: (left to right, bottom row), Charles J. Smiley, Daniel Axelrod, Joel Hedgpeth, Everett C. Olson, William A. Oliver, Jr., Peter Telford; (middle row), Arthur Boucot, Alan Solem, Rudolf Scheltema, William A. Newman, June R. P. Ross, Curt Teichert, William Clemens, Allen Pedder; (back row), William Zinsmeister, Allan Ormiston, Jane Gray, Richard H. Benson, William B. N. Berry, Charles Ross, Earl G. Kauffman, and Ida Thompson.

The 37th Annual Biology Colloquium of Oregon State University at Corvallis was held April 23-24, 1976. The purpose of the Colloquium was to further explore the potentialities of biogeography, past and present. Twenty-one speakers presented contributions. The present volume consists of written contributions based largely on those oral presentations as well as contributions from scientists unable to participate in the program. The Editors, and the Organizing Committee of the 37th Annual Biology Colloquium wish to thank the participants for their ready cooperation in making both the two-day program held at Corvallis, and this volume, a realitiy. Special thanks are due Dean Robert Krauss of the College of Science and his staff for their help in preparing the indexes to this volume.

Standing Committee for the Biology Colloquium: Arthur J. Boucot, Chairman—Geology, Kenneth L. Chambers—Botany, Jane Gray—Biology, Jack D. Lattin—Entomology, William G. Pearcy—Oceanography, John A. Ruben—Zoology, Robert M. Storm—Zoology, Jorn Thiede—Oceanography, Henry Van Dyke—General Science

Sponsors: College of Science, School of Oceanography, Agricultural Experiment Station, School of Agriculture, School of Forestry, School of Pharmacy, School of Home Economics, Graduate Research Council, Sigma Xi, Phi Kappa Phi

Colloquium Speakers:

Daniel I. Axelrod, Department of Botany, University of California, Davis, California 95616

Richard H. Benson, Department of Paleobiology, U.S. National Museum, Washington, D.C. 20560

William B. N. Berry, Department of Paleontology, University of California, Berkeley, California 97405

William Clemens, Department of Paleontology, University of California, Berkeley, California 97405

Joel W. Hedgpeth, 5660 Montecito Avenue, Santa Rosa, California 95404

Erle G. Kauffman, Department of Paleobiology, U.S. National Museum, Washington, D.C. 20560

William A. Newman, Scripps Institution of Oceanography, La Jolla, California 92093

William A. Oliver, Jr., U.S. Geological Survey, Room E-501, U.S. National Museum, Washington, D.C. 20244

Everett C. Olson, Department of Biology, University of California, Los Angeles, California 90024

Allen Ormiston, Amoco Production Company, Tulsa, Oklahoma

Charles A. Ross, Department of Geology, Western Washington University, Bellingham, Washington 98225

June R. P. Ross, Department of Biology, Western Washington University, Bellingham, Washington 98225

Rudolf S. Scheltema, Woods Hole Oceanographic Institution, Woods Hole, Massachusetts 02543

Charles J. Smiley, Department of Geology, University of Idaho, Moscow, Idaho 83843

Alan Solem, Field Museum of Natural History, Chicago, Illinois 60605

Curt Teichert, Department of Geology, University of Rochester, Rochester, New York 14627

Peter G. Telford, Ontario Division of Mines, Queen's Block, Toronto, Ontario M7A 1X3

Ida Thompson, Department of Geological and Geophysical Sciences, Princeton University, Princeton, New Jersey 08540

Tjeerd H. van Andel, Department of Geology, Stanford University, Stanford, California 94305

William J. Zinsmeister, Institute of Polar Studies, The Ohio State University, 125 South Oval Drive, Columbus, Ohio 43210

TAXONOMIC INDEX

Abacocrinus, 121
Abathocrinus, 122
Abies, 373
Abiesgraptus, 111
Abyssocrinus, 122
Abyssocythere, 380
Acacia, 438
Acanthochitina, 78, 79
Acanthocladia, 260, 263-270, 274
Acanthocladia?, 266
Acanthocrinus, 122, 123
Acanthophyllum, 141, 143, 145
Acanthopyge, 151, 165, 166
Acanthopyge?, 151, 165
?Acanthopyge, 156
Acanthospirifer, 195
"Acanthospirifer", 173, 195
Acaste, 157, 159-162
?Acaste, 158
Acaste (Calmonia), 159
Acaste (Pennaia), 157
Acastella, 193
Acastella?, 159
Acastoides, 159
?Acastoides, 147-149, 151, 153, 157, 158
Accipiter, 326
Acervoschwagerina, 218, 219, 222
Acinophyllum, 143
Acrophyllum, 143
Acmophyllum?, 141
Acmopyle, 442
Acrosmilia, 293-297
Acrospirifer, 173
Acrostichopteris, 318
Actinastraea, 290, 292, 293
Actinomena, 66-68, 73
Actinotrypella, 260, 264
Adelocoenia, 292-294, 296, 297, 299
Adelomelon, 391
Adiantites, 316, 317
Adrenia, 173
Aemulophyllum, 143
Aesopomum, 173
Aethomylos, 336
Afghanella, 220
Agathis, 442, 443
Agave, 442
Aggomorphastraea, 292, 295, 296, 299
Agrenocythere, 380
Aknisophyllum, 142
Alaiophyllum, 144
Alakanella, 219
Alatiformia, 173
Albizzia, 439
Aldania, 314, 316
Alisocrinus, 119
Aljutovella, 221
Allanicytidium, 94, 102, 103
Allium, 440
Allocoenia, 295, 297
Allocoeniopsis, 290, 292-297
Allopora, 405-407, 409, 415
Allotropiophyllum?, 141
Alluaudia, 439
Alluaudiopsis, 439
Alnus, 437
Aloe, 439
Alpenachitina, 79, 85
Alsopocrinus, 123
Amarsupiocrinus, 129
Amauropsis, 38
Ambocoelia, 173
Amborella, 442
Ambothyris, 173
Amimopina, 281
Aminornis, 324, 325
Amissopecten, 173, 196
Ammonia, 419
Ammonitella, 283
Amoenospirifer, 173
Ampakabastraea, 292, 293, 295-297, 299, 300
Amphigenia, 173, 195
"Amphistrophia", 173, 195
"Amplexus", 145
Amsdenina, 173
Amsdenoides, 141
Amygdalotheca, 101, 103
Amynodon, 336, 337
Amynodontopsis, 336
Anabacia, 297
Anachis, 431
Anas, 325, 326
Anastomopora, 260, 267, 270
Anastrophia, 173
Anathyris, 173
Anatrypa-Desquamatia, 173, 195
Anchiopella, 158, 160
Ancillotoechia, 173
Ancylostrophia, 173
Ancyrochitina, 77-80, 82, 84, 85
"Ancyrochitina", 79, 84
Ancyrocrinus, 124
Ancyrodelloides, 204, 205, 213
Andalgalornis, 324, 325
Andemantastraea, 290, 292, 293, 295, 296, 299, 300
Andinacaste, 148, 151, 153, 156, 157, 165
Andinopyge, 160
Andrewsornis, 324, 325
Angochitina, 78-80, 83, 85
Anhinga, 321, 326, 327
Anisolornis, 324, 325, 329
Anisophyllum, 141
Anisopleurella?, 67
Anisotechnophorus, 28, 30, 33
Anisotrypella, 260, 264, 272
Anitiferocystis, 101, 103
Anomalocystites, 102, 103
Anomalopteryx, 326
Anomozamites, 317, 318
Anoplia, 173, 195
"Anoplia", 173, 195
Anoplotheca, 173
Anoptambonites, 69
Antarctodarwinella, 350, 354
Anthracopupa, 277, 280
Anthropornis, 326
Antifopsis, 94, 102
Antihomocrinus, 124
Antispirifer, 173
Aorocrinus, 122
Apatemys, 337
Aphanaia, 243
Aphanomena, 66, 67, 73
Aphoidophyllum, 143
Aphroidophyllum, 142-144
Aphyllum, 142
Apoptopegma, 28, 30, 33
Aptenodytes, 323, 331
Apteryx, 326
Aptornis, 326
Arachnophyllum, 141
Aratinga, 325
Araucaria, 440, 442-444
Araxopora, 261, 264, 267, 274
Arbizustrophia, 196
Arbutus, 442
Archaegopsis, 282
"Archaeocrinus", 121
Archaeospheniscus, 326
Archeolithothamnium, 460-462
Archimedes, 260, 263, 270
Architectonica, 391-394, 396
Arcoscalpellum, 399, 401, 404, 407-411, 414
Arcticopora, 260, 272
Arctitreta 241, 243
Arctocephalus, 357, 361
Arctopteris, 317, 318
Arcualla, 177
Arduspirifer, 173
Areostrophia, 173
"Argentopyge", 163
Argutastraea, 144, 145
Argyrodyptes, 324, 325
Armatobalanus, 399, 401, 405-411, 415
Armatobalanus (Armatobalanus), 405-407, 415
Armatobalanus (Hexacreusia), 407
Armenina, 220
Artemisia, 375
Arthrodytes, 324, 325
Ascodictyon, 261, 268
Ascopora, 260, 262, 263, 265, 266, 270
Aspidocrinus, 124, 129
Asplenium, 316-318
Asterobillingsa, 142-144
Asteropyge, 160
Astraea, 37, 38
Astraeofungia, 290, 296
Astraraea, 292, 295, 297
Astutorhyncha, 173, 195
"Astutorhyncha," 173, 195
Ateleocystites, 93, 97, 101, 103
"Ateleocystites", 102
"Ateleocystites(?)", 102
Athene, 326
Athyrhynchus, 173
Athyris, 173
Athyris?, 66, 73, 173
Atomodesma, 243, 471
Atrypa, 73, 172, 173, 187-189
Atrypopsis, 195
Atrypina, 173
Aturia, 350
Aucornis, 324
Aulacella, 195
Aulacophyllum, 142-144
Aulograptus, 107
Australirhynchia, 173
Australocoelia, 173, 196
Australocystis, 97, 102, 103
Australops, 162
Australosutura, 164, 165
"Australosutura," 151, 156, 164, 167
Axosmilia, 290, 294-296, 299
Azygograptus, 107

Bactrocrinites, 122
Badiostes, 324, 325
Baiera, 314-318
Bainella, 148, 151, 156, 158, 159, 162
Bainella (Bainella), 156, 158
Bainella (Belenops), 151, 156, 159
Balanocystites, 93, 101-103
Balanops, 443
Balanus, 399, 401, 403-412, 415
Barbaestrophia, 173
Barrandeophyllum, 142
Bartramelta, 219
Baryphyllia, 296, 297
Bascomella, 261, 268
Bashkirella, 260, 263, 272
Basidechenella?, 164
Bathericrinus, 121, 124
Bathmosmilia, 290, 292, 299
Bathycoenia, 293, 294, 297, 300
Bathymyonia 241
Bathyrhyncha, 173
Batodonoides, 336
Batostomella, 261
Baturria, 173, 195
Bauhinia, 439
Bauxia, 280
Beachia, 174
Beedeina, 221
Belenops, 151, 156, 159
Belodella, 203
Bethanyphyllum, 144
Biconostrophia, 174
Biernatium, 196
Bifida, 174
Bilharzia, 249
Bilobia, 67
Bisinocoelia, 174

Biwaella, 219
Blepharis, 439
Blethisa, 373
Blothrophyllum, 143
Bohemiaecystis, 101, 103
Bohemicocrinus, 122
Bojodouvillina, 174
Bojothyris, 174
Bolivianaspis, 162
Boliviproetus, 151, 156, 163, 164, 167
Bostryx (*Peronaeus*), 281
Botaurus, 326
Bothriembryon, 282
Botryocrinus, 121, 123, 124
Boucotia, 174, 195
"*Boucotia*", 174, 195
Bouleia, 148, 151, 156, 158, 161
Boultonia, 218, 219, 224
Bowenelasma, 143
Brachycythere, 305, 380, 386
Brachycythere?, 380
Brachyprion, 174
Brachyseris, 296, 297
Brachyspirifer, 174
Brachyzyga, 174
Bradleya, 380
Bradybaena, 284
Braebium, 437
Branikia, 174
Bransonia, 28-30, 32-35
Breviphrentis, 142, 143
"*Breviphyllum*", 145
Briantelasma, 142, 143
Brockocystis, 119, 120
Brontornis, 324
Browneella, 174
Bryograptus, 107
Bubo, 326
Buchanathyris, 174
Bucanophyllum, 143
Bugula, 456
Bumelia, 442
Burmeisteria, 151
Bursachitina, 78, 79, 84
"*Buschophyllum*", 144
Buteo, 325
Butorides, 326
Bythoceratina, 380

Calamophyllia, 297
Calamophylliopsis, 292
Calapoecia, 55
Calceocrinus, 119-121
Callaiapsida, 241, 243
Callicalyptella, 174
Calliocrinus, 120, 123, 124
Callionepion, 282
Callocladia, 260
Callocladia?, 266, 270
Callocystites, 121
Callophoca, 366
Callorhinus, 361
Calmonia, 148, 151, 153, 156, 158-160
Calpiocrinus, 121, 122
Colybium, 280
Calymene, 162, 166
"*Calymene*", 166
"*Camarella*", 174, 195
Camarium, 174
Camarocrinus, 126
Camarotoechia, 196, 471
Camellia, 440
Camerella, 66, 73
Camerisma (*Callaiapsida*), 241, 243
Camurophyllum, 142
Cancellina, 220
Cancer, 415
Candeina, 419, 420
Candispirifer, 174
Capellirallus, 326
Capnophyllum, 141
Cardiograptus, 107
Carex, 375, 376
Carinagypa, 174
Carinatina, 174
Carpocrinus, 121
Carychium, 279, 281
Caryocrinites, 120-123
Cassia, 439
Casuarina, 440, 442, 444
Casuarius, 329
Catenipora, 55
Cathaya, 443
Cavanophyllum, 143
Cayetanornis, 325
Centetodon, 337
Centronella, 174
Centurodendron, 444
Cephalotaxopsis, 317, 318
Cephalotaxus, 314
Cerathocoenia, 294, 296, 297
Ceratocystis, 92, 94, 100, 102, 103
Ceratopea, 37, 39-43, 45, 47
Ceratophyllum, 143
Ceraurinus, 54
Cercis, 436, 442
Cereopsis, 327
Ceres, 283
Cerithium, 429-432
Chacaltaya, 157
Chacomurus, 151, 153, 156, 162
Chainodictyon, 260, 263, 264, 272
Charactophyllum, 145
Charionella, 174
Charionoides, 174
Chauvelicystis, 101, 103
Chiarumanipyge, 151, 156, 162
Chinianocarpos, 92, 94, 101, 103
Chlamydophyllum, 142, 144
Chomatoseris, 290, 292-297, 299, 300
Chonetes, 174
"*Chonetes*", 174, 187, 195, 196
Chonophyllum 145
"*Chonophyllum*", 142, 145
Chonostrophia, 174
Chonostrophiella, 174
Christiania, 66, 69, 73
Chthamalus, 401, 403, 404, 411, 412, 415
Chubutodyptes, 324, 325
Chumashius, 336
Chusenella, 219
Cicerocrinus, 122
Ciconiopsis, 324, 325
Cigara, 102
Cimicinella, 174
Cingulodermis, 174
Circus, 326
Cittarium, 37, 38, 429, 431
Cladionophyllum, 143
Cladophlebis, 314-318
Cladophylla, 290, 292-296, 299
Cladornis, 324, 325
Clarkeia, 164
Clathrochitina, 84
Clausastraea, 292, 297
Clausotrypa, 260, 263, 266-268, 270
Clethra, 442
Clidochirus, 119, 120, 122
Cliftonia, 66-68, 73
Climacartus, 324, 325
Climacograptus, 107
Clonocrinus, 122
Clonograptus, 107
Clorinda, 62, 68, 174
Clorindina, 174
Closterocrinus, 120
Cloudella, 174
Cloudothyris, 174
Clypeomorus, 432
Cnemiornis, 326, 327
Codonofusiella, 220, 224
Coelocion, 282
Coeloclemis, 261
Coeloclemis?, 267
Coelocystis, 121
Coelospira, 174
Coelospirina, 174
Coenastraea, 290, 292, 293, 296, 299
Coenocorypha, 326
Colaniella, 242
Collignonastraea, 292, 295-297, 299, 300
Collignonoseris, 296, 297
Colossendeis, 4-6
Columnaria, 144
"*Columnaria*", 144
Columnocoenia, 292, 296
Commidendron, 444
Commiphora, 439, 444
Comophyllia, 295
Comoseris, 293, 297
Compressiphyllum, 143
Conchothyra, 352
Concinnispirifer, 174
Condramena, 261, 268
Coneulota, 284
Confusastraea, 290, 292-294, 296, 297, 300
Coniopteris, 314-318
Connectastraea, 292, 296, 297, 299
Conochitina, 78, 81, 83, 85
Conopea, 401, 405, 411, 415
Conus, 427, 430, 431
Convexastraea, 299
Coolinia, 67
Coprosma, 442
Cordania, 164
Cordatomyonia, 174, 195
Coronocrinus, 124, 129
Cortezorthis, 175
Corvinopugnax, 175
Corycephalus, 160
Corymbograptus, 107
Coscinotrypa, 260, 264-268
Costellarina, 241
Costellirostra, 175
Costellispirifer, 175
Costisorthis, 175
Costispirifer, 175
Cothurnocystis, 101, 103, 104
Craseops, 336
Craterophyllum, 141
Crateroseris, 296, 297
Crinistrophia, 175
Crotalocrinites, 120-122
Croton, 439
Crurithyris, 175
Cruschedula, 324, 325
Cryobius, 374
Cryphaeoides, 151, 156, 161
Cryphaeus, 159, 160, 161, 162
?Cryphaeus, 159
Cryphaeus (?), 160
Cryptatrypa, 175
Cryptocoenia, 293-297, 299
Cryptolestes, 336
Cryptonella, 175, 195
"*Cryptonella*", 175, 195
Cryptothyrella, 66, 67, 73
"*Cryptothyrella*", 67
Ctenis, 314-316
Ctenocrinus, 122, 123
Cucullaea, 350, 354
Cumberlandina, 175
Cuminia, 444
Cunampala, 324, 325
Cupressus, 442
"*Cupularostrum*", 175, 195
Curuyella, 151, 156, 159
Cyathactis, 142
Cyathactis?, 141, 142
Cyathochitina, 78, 79, 83
Cyathocrinites, 121, 124
Cyathocylindrium, 143
Cyathophora, 290, 293, 296, 299
Cyathophyllopsis, 290, 292, 293, 300
Cyathophyllum, 143
Cyathophyllum, 264
Cybelurus, 57
Cycladigera, 175
Cyclospira, 73
Cyclospira?, 66, 69
Cyclotrypa, 260, 263-265, 267-269
Cydimia, 175
Cygnus, 326
Cylindrophyllum, 143, 144
Cyliocrinus, 121
Cymatopegma, 27, 28, 30, 33
Cymba, 391
Cymia, 428, 431
Cymostrophia, 175
Cyparissidium, 317, 318
Cyphaspis, 157
?Cyphaspis, 148
Cyphostemma, 439
Cypraea, 427, 431
Cyrtina, 172, 175, 187-189
Cyrtinaella, 175
Cyrtinopsis, 196
Cyrtochetus, 350
Cyrtoniscus, 175
Cystiphylloides, 142-145
Cystiphyloides?, 143, 145
Cystiphyllum, 141, 142
Cytheropteron, 386

Cyttarocrinus, 124
Czekanowskia, 314-317

Dacrydium, 441, 442
Dactylocoenia, 294, 297
Dagmarella, 219, 221
Dairina, 219
Dalejina, 175, 187, 195
Dalejocystis, 102
Dalejodiscus, 175
Dalmanella, 67-69
Dalmanella?, 67
Dalmania, 160
Dalmanites, 151, 157, 158-163
"Dalmanites", 151, 153, 156, 163
?Dalmanites, 158-160, 162-164
Dalmanites (Acastella?), 159
Dalmanites (Anchiopella), 158
?Dalmanites (Anchiopella), 158
Dalmanites (Corycephalus), 160
?Dalmanites (Mesembria), 160
Dalmanitina, 63, 67, 474, 477
Dalmanitoides, 156, 163
?Dalmanitoides, 162
Dasylalosia, 242
Davidsoniatrypa, 175
Dawsonella, 280
Dawsonelloides, 175
Dazhucrinus, 121
Decalopoda, 5
Dechenella, 148, 156, 164
Dechenella?, 151
?Dechenella, 164
Decoropugnax, 195
"Decoropugnax", 175, 195
Dedzetina, 69
Degeneria, 443
Delonix, 439
Delphinornis, 326,
Deltacephalaspis, 151, 156, 158, 159
Deltacephalaspis (Deltacephalaspis), 151, 156, 158
Deltacephalaspis (Prestalia), 151, 156, 158
Delthyris, 67, 175
"Delthyris", 175, 195
Denayphyllum, 141, 142
Dendraraea, 292, 293, 295
Dendrastraea, 293, 294, 297, 299
Dendrocrinus, 119, 121
Dendropupa, 277, 281
Dendrostella, 143-145
Dentellaria, 283
Dentilasma, 141
Depasophyllum, 144
Dereimsia, 160
Desmidocrinus, 121, 122
Desmochitina, 78, 81, 84
Desquamatia, 173, 195
Diacheila, 373
Diaphorapteryx, 326
Dicellograptus, 63, 65-70
Dichotomosmilia, 294, 297
Dichozygopleura, 175
Dicoelosia, 66, 67, 73, 175, 471
Dicranograptus, 65
Dictenocrinus, 121, 124
Dictyonella, 66, 73, 175
Dictyonema, 107
Didieria, 439
Didymograptus, 107, 110
Digonophyllum, 143-145
Digonus, 156
Dilobea, 437
Dilophodon, 337
Dimerocrinites, 120, 121
Dimorpharaea, 290, 293, 295-297
Dimorphastraea, 290, 292, 295-297, 299, 300
Dimorphastraeopsis, 292, 296
Dimorphoptychia, 280, 283
Dinapophysia, 175
Dinophyllum, 142
Dinornis, 326
Diospyros, 439
Diplaraea, 297
Dipleura, 151, 153, 156, 162
Diplochone, 142, 143, 145
Diplocoenia, 292, 296, 297, 299
Diplograptus, 65, 68, 69
Diploporaria, 260, 266, 270
Discocoenia, 294, 297
Discocyathus, 290, 296
Discomyorthis, 175
Disphyllum, 143-145
"Disphyllum", 143
Disphyllum?, 145
Ditoecholasma, 141
Diversophyllum, 144
Dodecolopoda, 5
Dohmophyllum, 142, 143, 145
Dohmophyllum?, 144
Dolatocrinus, 123
Dolerorthis, 66, 175
Draborthis, 67
Draborthis?, 68
Drabovia, 67, 68, 70
Draparnaudia, 281
Dromaius, 329
Drupa, 391, 393-395, 428, 429, 431
Drupella, 428, 429, 431
Dryornis, 325
Dubaria, 175, 195
Dunarula, 220
Duncanella, 141
Duncanella?, 142
Duntroonornis, 325, 326
Duryeella, 175
Dybowskiella, 260
Dyscritella, 260, 263-269, 272
Dyscritellina, 260, 268
Dyseolemur, 336
Dyticospirifer, 175

Eatonia, 175
Eccentricosta, 122, 126
Echinalosia, 251
Echinocythereis, 386
Echinolichas, 166
Edaphophyllum, 143
Edriocrinus, 124, 129
Edwardsomeandra, 290, 293, 295
Eisenackitina, 78, 79, 82, 84
Elasmofungia, 296, 297
Elatocladus, 317
Ellipsocoenia, 296, 297
Elysastraea, 299
Elytha, 175
Elythyna, 176
Embolophyllum, 141-143
Emeus, 326
Empetrum, 374
Enallhelia, 293, 297
Enallocoenia, 290
Endothiodon, 230
Enoploura, 101, 103
Entelophylloides, 141
Entelophyllum, 141, 142
Eodevonaria, 176
Eofusulina, 219
Eoglossinotoechia, 176
Eognathodus, 195, 201, 204, 206, 208, 211, 213
Eohaplomys, 336
Eoischyrina, 28, 30, 33
Eoneornis, 324, 325
Eoparafusulina, 219
Eophacops, 148, 157
Eoplectodonta, 66, 69, 73
"Eoplectodonta", 66, 73
Eoplectodonta?, 67
Eoplectodonta (Kozlowskites), 69
Eopteria, 28, 30, 32, 33, 35
Eoreticularia, 176
Eoschuchertella, 176, 187-189, 195
Eosphaeniscus, 326
Eospirifer, 176
Eospirigerina, 66, 71, 73
Eostropheodonta, 67
Eotylopus, 336
Eoverbeekina, 220
Eowaeringella, 219
Eozellia, 218, 219, 222
Epiactinotrypa, 260, 268
Epilobium, 438, 440
Epismilia, 290, 297
Epismiliopsis, 290
Epistreptophyllum, 290, 294-297, 300
Epitomyonia, 66, 73
Equisetites, 314-318
Eremochitina, 78, 80
Eremopezus, 321
Erica, 439
Eridophyllum, 143, 144
Eridopora, 260, 263, 264, 266, 269
Eriphia, 429
Errinopora, 405, 415
Etherella, 260, 266, 268, 269
Etymothyris, 176
Eucalyptocrinites, 120-124
Eucalyptus, 438
Eucharitina, 176
Euchasma, 28, 30, 33
Euchasmella, 28, 30, 33
Euhelia, 294
Eunaticina, 38
Euphorbia, 439
Eurekaspirifer, 176
Eurwloenia, 324, 325
Euryanus, 326
Euryapteryx, 326
Eurydesma, 172
Euryonotus, 325
Euryspirifer, 176
Eurythyris, 176
Eutaxocrinus, 123
Evactinopora, 260, 266, 269
Evactinostella, 260, 265, 269
Evenkaspis, 53
Exilifrons, 143

Falsatrypa, 176
Fascicostella, 176, 195
Fasciphyllum, 142, 143, 145
Fascizetina, 195
Felinotoechia, 176
Fenestella, 260, 262-270, 272, 274
Fenestellata, 260, 264
Fenestraspis, 149, 151, 153, 156, 162, 163
?Fenestraspis, 151
Fibulistrophia, 176
Ficus, 440
Fimbrispirifer, 176, 196
"Fimbrispirifer", 176, 196
Fistulamina, 260, 264-267, 269
Fistulipora, 260, 262-269
Fistuliramus, 267
?Fistuliramus, 260
Fistulocladia, 260
Fistulotrypa, 260, 266, 267, 269
Fletcherina, 142
Foetopterus, 325
Foliomena, 69, 70
Folotsia, 439
Forestiera, 442
Francovichia, 151, 153, 156, 157, 162, 166
Franklinella, 196
"Franklinella", 176, 196
Frechastraea, 145
Fulcriphoria, 176
Fungiastraea, 290, 296
Fusiella, 220
Fusulina, 219, 221, 264
Fusulinella, 219, 221

Galliaecystis, 100, 103
Galliarallus, 326
Gallowiina, 220
Gallus, 321, 322
Gamonedaspis, 151, 156, 162, 163
?Gamonedaspis, 151
Ganesella, 283
Gangamopteris, 311, 312
Gastrobolus, 293
Gastrocrinus, 123
Gastrodonta, 282
Gazacrinus, 121, 122
Geinitzella, 260
Genabacia, 294, 297, 300
Gennaeocrinus, 124
Gibbula, 431
Gibbulinella, 282
Gigantornis, 321, 322, 331
Ginkgo, 314-318, 443
Ginkgodium, 314, 315
Giraldiella, 67
Girtyopora, 260, 263, 265, 267, 268, 270, 274
Girtyoporina, 260, 264, 267, 268, 274
Girvanella, 47
Gissocrinus, 121, 123, 124
Gladiostrophia, 176
Glansicystis, 121
Gleichenia, 316
Glidochirus, 124
Globigerina, 418, 419, 421-423
Globigerinella, 419, 420, 422

Globigerinoides, 419, 420, 422, 423
Globithyris, 176
Globorotalia, 418-420, 422-424
Glossinotoechia, 176
Glossinulina, 176
Glossinulus, 176
Glossoleptaena, 176
Glossophyllum, 143, 318
Glossopteris, 172, 311, 312, 318, 478
Glyptograptus, 107
Glyptostoma, 283
Glyptorthis, 66, 68, 69, 73
Godwinia, 282
Goliathia, 321
Gonatosorus, 314, 315
Gondylocrinus, 123
Goniocladia, 260, 263-269
Goniopteris, 314
Goodenia, 438
Gorgostrophia, 176
Grandipatula, 282
Grangerella, 282
Grayina, 176
Gregorycoenia, 297
Grevillea, 438
Grewgiphyllum, 143
Griphomys, 336
Grypophyllum, 143-145
"*Grypophyllum*", 143
Gurievskiella, 142
Gypidula, 176, 187, 188
Gypidulina, 176
Gyrocarpus, 439

Hadrophyllum, 143, 144
Hakea, 438
Halichoerus, 363
Hallia, 143, 144
Hallicystis, 121
Hanusatrypa, 176
Haoella, 220
Hapalocrinus, 121, 122
Harpagolestes, 337
Harpagornis, 326, 327
Harpalus, 375
Harpides, 57
Hastigerina, 419, 420, 422
Hausmannia, 162, 314, 315
Hayasakapora, 260, 267, 268, 270, 274
Hebetoechia, 176
Hedeina (*Macropleura*), 176, 196
Hedstroemophyllum, 142
Heilungia, 314, 315
Helianthemum, 442
Heliconia, 437
Heliocoenia, 296, 297
Heliophyllum, 143, 144
Helix, 283
Helophorus, 373
Hemiacodon, 337
Hemifusulina, 219
Hemifusulinella, 219, 221
Hemigordiopsis, 242
Hemitrypa, 260, 267
Hendersonia, 280
Henryhowella, 380, 385
Heraultipegma, 27, 28, 30, 33
Hercochitina, 78, 79
Hermitrypa?, 266
Hermosiornis, 324
Herpetocrinus, 122
Hesperelaea, 444
Hesperibalanus, 411, 416
Hesperorthis, 67
Heterophrentis, 142-144
Heterorhea, 325
Heterorthina?, 69
Heudeia, 280
Hexacreusia, 407
Hexacrinus, 122
Hexagonaria, 143-145
Hexagonaria?, 143
Hexagonella, 260, 262-266, 268, 269, 272
Hibolites, 291, 293, 295, 297
Hindella, 67
Hinganella, 260, 266-268, 272, 274
Hinganotrypa, 260, 268
Hipparionyx, 176, 196
"*Hipparionyx*", 176, 196
Hippocardia, 30, 32, 34, 35
Hirnantia, 64, 67-70
Hoazinoides, 324, 325
Hodopoeus, 283
Hoegisphaera, 78, 81, 85
Holmophyllum, 141, 142
Holorhynchus, 64, 66, 68, 73
Homalonotus, 151, 153, 156, 160, 162
?=*Homalonotus* (*Calymene*), 162
?*Homalonotus* (*Schizopyge*), 160
Homalophyllum, 143
Homeospira, 196
Homocrinus, 121
"*Homocrinus*", 124
Horderleyella, 67
Horridonia, 241, 243
"*Horridonia*", 241
Howellella, 172, 176, 187-189, 196
Howeria, 442
Howittia, 176
Hunanophrentis, 145
Hyaenodon, 336
Hyattidina, 66, 73
Hydrolithon, 460-463
Hydrurga, 363
Hyolithus, 67
Hyphasmopora, 260, 264, 270
Hysterolites, 176
"*Hysterolites*", 177, 196
Hystricurus, 57

Iberirhynchia, 177
Ichthyopteryx, 326
Icriodus, 201-211, 213
Icriodus-Polygnathus, 203
Icthyocrinus, 120, 121, 123, 124
Indobatrachus, 440
Innuitella, 177
Iowaphyllum, 144, 145
Iridistrophia, 177, 196
Isalaux, 55, 56
Isastracea, 290, 296
Isastrea, 292-297, 299, 300
Isastrocoenia, 292
Ischyrina, 30, 32, 34, 35
Ischyrotomus, 336, 337
Isomeria, 283
Isopoma, 177
Isorthis (*Arcualla*), 177
Isorthis (*Isorthis*), 177, 196
Isorthis (*Protocortezorthis*), 177
Isorthis (*Tyersella*), 177
Ivanothyris, 177

Jacutiella, 314-316
Jacutopteris, 316
Jakutoproductus, 242
"*Janius*", 177, 196
Jenneria, 431
Juania, 444
Juglans, 437
Juniperus, 442
Jynx, 321

Kahlerina, 220
Kalochitina, 78, 79
Kalvariella, 260, 264, 270
Kanabohelix, 283
Katunia, 177, 196
Kayserella, 177, 196
Keriophyllia, 293, 296, 297
Kermadecia, 442
"*Ketophyllum*", 142
Kimopegma, 27, 28, 30, 32
Kingopora, 260, 263, 264, 270, 274
Kinnella, 67
Kioelasma, 143
Kirkocystis, 102, 103
Kjerulfina?, 67
Knightia, 442
Kobeha, 142
Kobya, 292, 297, 299
Kodonophyllum, 142
Kolymia, 243
Korora, 325, 326, 331
Kozlowiaphyllum, 142-145
Kozlowskiaspis, 151, 153, 157, 158, 160
Kozlowskiaspis (*Kozlowskiaspis*), 151, 156, 160
Kozlowskiaspis (*Romaniella*), 151, 156, 160
Kozlowskites, 69
Krameria, 442
Kransia, 177
Krithe, 385
Kullervo, 67
Kunthia, 143, 144
Kunthia?, 144
Kuvelousia, 241, 243
Kwanloella, 220
Kymocystis, 142
Kymatothyris, 177

Labyrinthus, 283
Lactoris, 444
Lagenochitina, 78, 80
Lagopterus, 325
Lagynocystis, 101, 103
Lahilleona, 353
Lahillia, 353, 354
Lambis, 431
Lamnimargus, 251
Lamprophyllum, 141, 142
Lampterocrinus, 120-122
Lanceomyonia, 177
Lantschichites, 220
Larix, 373, 376
Larrea, 438
Lasiocrinus, 124
Latiphylla, 296
Latirolagena, 430
Latirus, 430
Latomeandra, 290, 292, 293, 296, 299, 300
Latonotoechia, 177
Laurelia, 444
Laurocerasus, 442
Laurus, 442
Leangella, 66-69, 73
Lecanocrinus, 121-123
Leella, 220, 224
Legrandella, 166
Leioclema, 261
Leioclema?, 266
"*Leiorhynchus*", 177, 196
Lekanophyllum, 143, 144
Lenzia, 181
Leonaspis, 149, 153, 156, 165
Lepanocrinus, 121
Lepidocyclus, 55
Lepidodendropsis, 477
Lepidoleptaena, 177
Lepidolina, 220, 267
Lepidophorus, 375, 376
Leptaena, 66-69, 73, 177, 187-189, 196
Leptaena-Leptagonia, 177, 187-189, 196
Leptaenisca, 177, 196
Leptaenopoma, 66-68, 73
Leptaenopyxis, 177
Leptagonia, 177, 187-189, 196
Leptathyris, 177
Leptestia, 67
Leptestiina, 69
Leptocoelia, 177, 196
Leptocoelina, 177
Leptodonta, 177
Leptonychotes, 365
Leptophoca, 361-363
"*Leptoporolithon*", 462
Leptoptilos, 321, 326, 327
Leptoreodon, 336, 337
Leptospira, 177
Leptostrobus, 316
"*Leptostrophia*", 177, 196
Leptotomus, 336
Leucozonia, 430
Levenea, 177
Levifenestella, 260, 266, 270
Libocedrus, 442
Licharewia, 243
Lichas, 165
Lievinella, 177
Liguloclema, 260, 266, 268, 269
Limacina, 423
Lingula, 62
Linguopugnoides, 177
Linochitina, 78, 79, 82, 85
Liospira, 39
Lissatrypa, 177
Lissopleura, 177
Lithocrinus, 121
Lithophyllum, 460-464
Lithoporella, 461
Lithothamnium, 460-462, 464
Llanoella, 177, 195
Lobodon, 363
Loboplasma, 142

Lobopyge, 165, 166
Lochmaeosmilia, 292-297, 299, 300
Loganiopharynx, 284
Lonicera, 440
Loomberaphyllum, 143
Lophiornis, 324
Lophosittacus, 326
Loxorhynchus, 415
Loxornis, 324, 325
Loyalophyllum, 145
Luscinia, 321
Lycokystiphyllum?, 142
Lyrielasma, 142
?Lyriocrinus, 120
Lyrocladia, 260, 263, 267, 270
Lyropora, 260, 265, 266, 270
Lysocystites, 120, 121, 129

Macgeea, 145
Macgeopsis, 290, 296, 299
Machaeraria, 177
Mackenziephyllum, 144
Maclurea, 40, 45
Maclurina, 45
Maclurites, 38-41, 43, 45-48, 55
Macropleura, 176, 196
Macrostylocrinus, 119-121, 124
Macrourus, 64
Maerua, 439
Magnolia, 436, 440
Maikottia, 142
Majungaella, 305-307
Malurostrophia, 177
Malvinella, 148-151, 156, 160, 161
Mansuyphyllum, 143
Manu, 326
Maoristrophia, 178
Margachitina, 78, 79, 83, 85
Markitoechia, 178
Marsupiocrinus, 120-123, 129
Marsupiocrinus (*Amarsupiocrinus*), 129
Martinophyllum, 142, 143
Mastigocrinus, 121
Matheropora, 260
Maurotarion, 151, 156, 165
Maychella, 260, 267, 268, 270, 274
Mazaphyllum, 142
"*McLearnites*", 178, 196
Meandraraea, 292, 293
Meandrophyllia, 293, 294, 297
Meekopora, 260, 264, 265, 267-269
Meekoporella, 260, 267
Megabalanus, 401, 403, 404, 411, 415
Megacanthopora, 260
Megacanthopora?, 266, 270, 277
Megaegotheles, 326, 327
Megakozlowskiella, 178, 196
Megalapteryx, 326
Meganterella, 178
Meganteris, 178
Megasalopina, 178
Megaspira, 282
Megastrophia, 178
Megateuthis, 293
Megistocrinus, 124
Megistostegium, 439
Melalucea, 438
Melanodendron, 444
Melikerona, 292, 297, 299
Membranipora, 456
Membranobalanus, 401, 405, 411, 416
Mendathyris, 178
Merista, 178
Meristella, 67, 178, 196
Meristella?, 73
Meristella-Meristina, 178, 196
Meristina, 178, 196
Merycobunodon, 336
Mesembria, 160
"*Mesodouvillina*", 178, 187, 188, 196
Mesoglypterpes, 284
Mesomorpha, 292, 293, 295, 296, 299
Mesomphix, 282
"*Mesopholidostrophia*", 178, 196
Mesophyllum, 143, 145, 460-463, 464
Mesosaurus, 172, 232, 233
Mesochubertella, 220
Metacryphaeus, 148, 149, 151, 153, 156, 158-162, 166
Metaplasia, 149, 178
Metarhinus (?), 336
Metasequoia, 443
Metelipora, 260, 262, 269, 272
Metethmos, 296, 297
Metrionaxon, 142
Metriophyllum, 142, 144, 145
Miacis (?), 336
Micidus, 178
Micralymma, 375, 376
Microcyclus, 143
"*Microcyclus*", 144
Microparamys, 336, 337
Microphyllia, 292, 293, 296, 297
Microphylliopsis, 299, 300
Microphysula, 283
Microplasma, 141, 142, 144, 145
Microplasma?, 141, 143
Microsolena, 290, 292-297
Microstele, 281
Microsyops, 336, 337
Mictophyllum, 145
"*Mictophyllum*", 145
Migmatophyllum, 142
Minilya, 260, 264, 266, 270
Miniprokopia, 196
Minojapanella, 220
Minussiella, 143, 144
Miophoca, 362, 363
Mirella, 318
Mirounga, 365, 366
Misellina, 220, 222
Mitrocystella, 93, 94, 101, 103
Mitrocystella?, 101
Mitrocystites, 93, 94, 101, 103
"*Mitrocystites?*", 94
Mitrocystites?, 101, 102
Molongella, 178
Molongia, 178
Monachus, 365-367
Monadotoechia, 178
Monalaria, 352, 353
Monodechenella, 164
Monodiexodina, 219
Monodonta, 429
Monograptus, 94, 111, 126, 127
Monomerella, 55
Monophoraster, 351
Monorakos, 53-55, 57
Monotherium, 363
Monotherium?, 361, 366, 367
Montiparus, 219, 221
Montlivaltia, 290, 292-297, 299, 300
Moravophyllum, 143-145
"*Moravophyllum*", 144
Moravophyllum?, 143
Moringia, 439
Morphastraea, 296, 297
Morula, 428, 429, 431
'*Morula*', 428
Morychus, 375, 376
Mucophyllum, 142
Mucronaspis, 68
Mucrospirifer, 178
Muellerina, 386
Multispirifer, 178
Muraltia, 439
Murex, 394, 395
Muricanthus, 431
Muriferella, 178
Musa, 437
Mutationella, 178
Mya, 367, 368
Myelodactylus, 120-123
Myocaris, 30, 32
Myona, 27, 28, 30, 33
Myona?, 31
Myrica, 374, 442
Myriophillia, 296, 297
Mystrocephala, 164
Mystrophora, 178, 187
Mytilus, 296, 414
Mytonolagus, 337
Mytonomys, 336

Nadiastrophia, 178
Nagatoella, 219
Najadospirifer, 178
"*Nalivkinella*", 142-144
Nalivkinella?, 144
Namatomys, 336
Nankingella, 222
Nanodelphys, 337
Nanothyris, 178
Nanshanophyllum, 142
Necropsittacus, 326
Negulus, 281
Nemagraptus, 65-69
Neobrachyelasma, 142
Neocalamites, 314
Neocalmonia, 147
Neochen, 325
Neocolumnaria, 144, 145
Neocythere, 307
Neoeridotrypella, 260, 263, 272
Neofusulinella, 220
Neogaeornis, 324
Neogoniolithon, 460-462, 463, 464
Neomisellina, 220
Neomphyma, 142
Neorhombopora, 260
Neoschwagerina, 220, 267
Neostringophyllum, 142-145
Neostringophyllum?, 144
Neothailandina, 220
Neptunea, 391
Nerita, 431
Nesophalaris, 326
Nesovitrea, 282
Nevadacystis, 101, 103
Nevadaphyllum, 143
Niajuphyllum, 142
Nicklesopora, 260, 262, 263, 270
Nicolella, 66, 67, 73
Nikiforovella, 260, 264, 270, 274
Nileus, 57
Nilssonia, 314-318, 436
Nilssoniopteris, 314, 315, 318
Nilssoniopteris (*Sibiriophyllum*), 314
Niobella, 57
Nipponitella, 219
Nolina, 442
Nortornis, 326
Notanoplia, 178
Notharctus, 337
Nothofagus, 438, 441, 443, 444
Nothomyrica, 444
Nothura, 325
Notiochonetes, 149
Notoconchidium, 178
Notoleptaena, 178
Notoparmella, 178
Novocythere, 305, 307-309
Novocythere-Tickalaracythere, 305, 307, 308
Nucella, 428
Nucleospira, 172, 178, 187, 188
Nummulites, 344
Nyaya, 57
Nycticorax, 326
Nyctitherium, 337
Nyctocrinus, 122
Nymphon, 4
Nymphorhynchia, 178

Obsoletes, 219, 221
Obturamentella, 178
Occidentoschwagerina, 218, 219, 242
Octolasmis, 401, 405, 410, 411, 415
Odobenus, 360, 364
Odontochile, 156, 162, 163
Odontophyllum, 144
"*Odontophyllum*", 142
Odontopleura, 165
Oepikila, 27, 28, 31, 33
Ogbinopora, 260, 264, 266, 268, 270, 274
Ogilviella, 178, 296, 297
Oketaella, 219
Oligophyllum?, 141
Oligoptycherhynchus, 178
Oliveria, 141
Ommatophoca, 363, 365
Omomys, 337
Omphalina, 282
Omphalocirrus, 38
Onacroanis, 325
Onactornis, 324
Oncograptus, 107
Onniella, 67, 68
Onychiopsis, 314, 315, 317, 318
Opeatostoma, 430
Operculina, 344
Opisthodactylus, 324, 325
Orbignycoenia, 290, 293, 294, 297
Orbulina, 419, 420, 422, 423

Oreopholus, 325
Orionina, 367
Oriskania, 178
Oroseris, 290
Orthis, 67, 73
"Orthis", 66, 69, 73
Orthocyathus, 143
Orthopteryx, 326
Orthostrophonella, 178
Orthostrophia, 178
Osculocystis, 120
Osmunda, 318
Osmundopsis, 318
Osteodontornis, 326
Otarion, 151, 153, 156, 165
Otarion (Maurotarion), 151, 156, 165
Otarion, (Otarion), 151, 156
Otozamites (?), 317
Ovalastraea, 296, 297
Ovalastraeopsis, 299, 300
Oxoplecia, 66, 67, 73
Oxynaspis, 401, 405, 410, 411, 415
Ozarkodina, 201, 213

Pachyanus, 326
Pachydyptes, 322, 326
Pachyornis, 326
Pachyphloia, 242
Pachyphyllum, 145
Pachypodium, 439
Pacificocoelia, 179, 196
Pagiophyllum, 314, 317, 318
Palaeeudyptes, 326
Palaeociconia, 324
Palaeocyathus, 141-143
Palaeoephippiorhynchus, 321
Palaeofusulina, 220
Palaeospheniscus, 324-326
Palaeostoa, 281
Palaestrophomena?, 68
Pallenopsis, 6
Palmatolepis, 203
Pamirella, 260, 264, 270
Panderodus, 203
Pandorinellina, 201, 213
Panulirus, 415
Papiliophyllum, 142
Papillostrophia, 179
Parabouleia, 148, 151, 156, 161
Paracalmonia, 150, 151, 156, 158
Parachonetes, 179
Paradoxiella, 220, 224
Paradunbarula, 220
Paraeofusulinia, 219
Parafenestralia, 260, 263, 270, 272
Parafusulina, 219, 222, 224, 267
Paraglossograptus, 107
Paragoniolithon, 459, 460-462
Paraleioclema, 261, 263-268, 272
Paraleioclema?, 266-268
Paramaclurites, 45
Paramarginifera, 242
Paramontlivaltia, 294, 297, 299
Paranacystis, 92-94, 102, 103
Parapholidostrophia, 179
Paraphyllogyra, 290
Paraptenodytes, 324, 325
Parapugnax, 196
Paraschwagerina, 218, 219
Paraspirifer, 179
Parastrophina, 66, 73
Parastrophinella, 66, 73
Parataxodium, 314, 315, 318
Parawedekindellina, 219
Pareumys, 337
Parisastraea, 296, 297
Parkinsonia, 289
Parmorthina, 196
Paucipora, 265
Paulocrinus, 121
Pauropegma, 28, 31, 33
Pedavis, 194, 204-206, 213
Pegmarhynchia, 179
Pelecanus, 326, 327
Peleicostella, 179
Pelekysgnathus, 204, 207, 209, 213
Peltocystis, 101, 103
Pelycodus, 337
Peneckiella, 143, 145
"Peneckiella", 143
Pennaia, 148, 151, 156-159
?Pennaia, 160
Penniretepora, 260, 262-264, 266-268, 270, 274
Pentacolossendeis, 5
Pentagonia, 179
Pentamerella, 179
Pentamerus, 54, 55
Peratherium, 337
Periechocrinus, 120-122
Peripaedium, 143, 144
Periseris, 292, 296
Perispheniscus, 325
Perissodonta, 350, 352-354
Permoleioclema, 261, 267, 268, 272, 274
Permopora, 260, 268, 272
Peronaeus, 281
Persea, 442
Perrieria, 282
Petalocrinus, 117, 120, 125-127
Petraia, 142
Petraia?, 141
Pezophaps, 326, 327
Phacellophyllum, 145
Phacopina, 148, 149, 151, 153, 156-158
?Phacopina, 151, 158
Phacopina (Phacopina), 157
Phacops, 148, 149, 151, 153, 156, 157-162, 166
?Phacops, 158, 160
Phacops (Cryphaeus), 159
?Phacops (Phacopina), 158
Phacops (Viaphacops), 148, 149, 151, 153, 156, 162, 166
Phalacrocorax, 321, 325
Phaulactis, 141, 142
Phaulactis?, 141
Phenakospermum, 437
Philippia, 391-394, 396
Phillipsastraea, 145
Phillopora, 260, 268
Phimocrinus, 124
Phoca, 363, 365, 367, 368
"Phoca", 362, 363
"Phoenicitoechia", 179, 196
Phoenicopsis, 314-318
Phoenicoseris, 444
"Pholidostrophia", 179, 196
Phormium, 442
Phororhacus, 324, 325
Phorusrhacos, 324
Phragmophora, 196
"Phragmophora", 179, 196
Phragmostrophia, 179
Phylica, 439
Phyllocladus, 441
Phyllocystis, 93, 100, 103, 104
Phyllograptus, 107
Phyllogyra, 290
Phylloseriopsis, 290
Phylloseris, 290, 292
Phymatophyllum, 143, 144
Physornis, 324, 325
Picea, 373
Pilophyllum?, 141
Pinegopora, 260, 263, 270
Pinguispirifer, 179
Pinnatopora, 260
Pinnocaris, 27, 28, 31-35
Pinus, 373
Pinyonastraea, 143
Pionus, 325
Pisocrinus, 120-123
Pisolina, 220
Pistacia, 442
Pityophyllum, 316-318
Placastylus, 282
Placocystella, 102, 103
"Placocystis", 102
Placocystites, 102, 103
"Placocystites", 94
Placosmilia, 297
Planetophyllum, 144
Planicardinia, 179
Plasmophyllum, 142
Platanus, 442
Platyceras, 157
Platydyptes, 326
Platyorthis, 179, 196
Platystrophia, 66, 67, 73
Plebejochonetes, 179
Plectatrypa, 73
Plectochitina, 77-79, 82, 85
Plectodonta, 179
Plectorhynchella, 179
Plectospira, 179
Plectothyrella, 67
Pleiopleurina, 179
Plesiocoenia, 296, 297
Plesiomiacis, 336
Plesiostylina, 296, 297
Plethorhyncha, 179
Pleurodonte, 283
Pleurodonte (Dentellaria), 283
Pleurodontites, 283
Pleurograptus, 65, 68-70
Pleuropegma, 27, 28, 31, 33
Pleurophyllia, 297
Plicanoplia, 179
Plicanoplites, 179
Plicocyrtina, 179, 196
Plicodevonaria, 179
Plicoplasia, 179
Plicostropheodonta, 179
Poa, 375
Podiceps, 326
Podocarpus, 438, 441, 443, 444
Podolella, 179
Podozamites, 314-318
Poebrodon, 336
Pollicipes, 401, 404, 410, 411, 414
Polyastropsis, 292, 293, 300
Polydiexodina, 219, 224
Polygnathus, 201-204, 207, 209, 213
Polygyra, 283
Polygyrella, 283
Polymorphastraea, 295, 297
Polypeltes, 121
Polypora, 260, 262-270, 272, 274
Polypora (Paucipora), 265
Polypora (Pustulopora), 265
Pomatias, 280
Pondia, 42
Porolithon, 460-464
Poseidonamicus, 380
Potamotherium, 359
Potentilla, 375
Pradoia, 179
Praegnantenia, 179
Praeparafusulina, 219
Presbymys, 336
Prestalia, 151, 156, 158
Presumatrina, 220
Primorella, 260, 264, 268, 272
Prionodelphis, 363, 368
Prionothyris, 179
Prismatophyllum, 143, 144
Prismopora, 260, 266-269
Proboloides, 158, 161
Probolops, 148, 151, 153, 156, 161, 162
Procariama, 324, 325
Procerulina, 179
Prociconia, 325
Productella, 179
Productus, 264, 265
Proetus, 157, 164
Proetus?, 164
Profusulinella, 219, 221
Prohexacrinites, 122
Prohexagonaria, 141, 142
Prokopia, 179, 196
Prophoca, 361
Prophororhacus, 325
Prorensselaeria, 179
Proreticularia, 179
Proschizophoria, 179
Protathyris, 179
Protaxocrinus, 119, 121, 122
Proterixoides, 336
Protethmos, 295, 297
Protibis, 324, 325
Protochonetes, 196
"Protochonetes", 180, 196
Protocortezorthis, 177
Protoleptostrophia, 180
Protoreodon, 336, 337
Proterixoides, 336
Protornatellina, 281
Protriticites, 221
Protylopus, 336, 337
Provitrina, 282
Psammohelia, 296, 297
Psammornis, 321
Pseudamplexus, 142, 145
Pseudapteryx, 326

Pseudisastraea, 296, 297
Pseudobatostomella, 261, 263, 264, 267, 268, 272, 274
Pseudobatostomella?, 264, 266, 268
Pseudoblothrophyllum, 142
Pseudocamarophoria, 196
Pseudoclathrochitina, 78, 80
Pseudocoenia, 293, 295-297
Pseudocolumna, 281, 285
Pseudocryptophyllum?, 141
Pseudodiplocoenia, 296, 297, 299
Pseudodoliolina, 220
Pseudodontornis, 326
Pseudoendothyra, 220, 222
Pseudoeuchasma, 28, 31, 33
Pseudofusulina, 218, 219
Pseudogrypophyllum, 142
Pseudokahlerina, 220
Pseudolarix, 314-316
Pseudolarus, 324
Pseudomicroplasma, 142
"Pseudomicroplasma", 145
Pseudomphyma, 142
Pseudoparazyga, 180
Pseudoreichelina, 220
Pseudoschwagerina, 218, 219, 222, 267
Pseudosterna, 325
Pseudotechnophorus, 28, 31, 33
Pseudotorellia, 316, 318
Pseudotryplasma, 145
Pseudounitrypa, 260
Psilopterus, 324
Psydracophyllum, 144
Pterochitina, 78, 81, 83, 85
Pterocnemia, 325
Pterophyllum, 314, 316, 317
Ptiloporella, 260, 263, 270
Ptychalaea, 281
Ptychoglyptus, 66, 69, 73
Ptychopegma, 28, 31, 33
Ptychophyllum, 142, 145
"Ptychophyllum", 145
Ptychophyllum?, 141
Ptychopleurella, 66, 73, 180
Ptylopora, 260, 263, 264, 266, 270, 272
Pugnax, 196
"Pugnax", 180, 196
Pulleniatina, 420, 422
Punctatrypa, 180
Pusa, 359, 363-365, 367, 368
Pustulopora, 265
Putrella, 219
Pycnactis, 141
Pycnosaccus, 122, 123
Pycnostylus?, 142
Pyramida, 326
Pyramus, 251
Pythia, 279

Quadrifarius, 180
Quadrikentron, 180
Quadrithyrina, 180
Quadrithyris, 180
Quasibuntonia, 380
Quasifusulina, 219, 242
Quasifusulinoides, 219
Quasimartinia, 180
Querandiornis, 325
Quercus, 436, 437, 442
Querquedula, 325
Quiina, 436

Radiastraea, 142-144
Radiomena, 180
Ramachitina, 85
Ramipora, 260, 262, 264, 266, 267, 269, 272
Ramiporella, 260
Ramiporidra, 260, 264, 269
Rapamys, 336
Raphaelia, 314, 315
Raphus, 326
Rauserella, 224
Ravenala, 437
Redstonea, 144
Reeftonia, 196
"Reeftonia", 180, 196
Reithroparamys, 336
Remipyga, 54, 55
Rensselaeria, 180
Rensselaerina, 180
Resserella, 68, 180, 196
Reteporidra, 260, 262-265, 270
Reticulariopsis, 182, 196
Reticulatrypa, 180
Reticulocarpos, 101, 103
Retzia, 196
Rhabdochitina, 78, 80
Rhabdomeson, 260, 264-268, 270
Rhapalostylis, 442
Rhea, 325
Rhenocystis, 102, 103
Rhenorensselaeria, 180
Rhenostrophia, 180
Rhenothyris, 180, 190
Rhigozum, 439
Rhipidogyra, 296
Rhizophylloides, 142
Rhizophyllum, 141-143
Rhododendron, 440
Rhombocladia, 260, 266
Rhombopora, 260, 263-270, 272, 274
Rhombotrypella, 260, 262-265, 272
Rhynchotreta, 196
"Rhynchotreta", 180, 196
Rhynchospirina, 196
"Rhynchospirina", 180, 196
Riacama, 324, 325
Ribeiria, 28, 31-35
Ribeirina, 28, 31, 33
Rivomarginella, 277
Robinsonia, 444
Robustoschwagerina, 218, 219, 222, 224
Romaniella, 151, 156, 160
Ruffordia, 317
Rugofusulina, 219
Rugoleptaena, 180
Rugososchwagerina, 218, 219
Russiella, 220

Sabal, 442
Saffordotaxis, 260, 265-267, 270
Sagenopteris, 314, 315
Sageretia, 442
Sakalavastraea, 293, 296, 299
Sakalavastraeopsis, 292
Salairophyllum, 143
"Salairophyllum?", 142
Salopina, 180, 196
Salopina (*crassiformis type*), 180
Sapindus, 442
Sarkidornis, 326
Saxifraga, 440
Scalez, 39
Scaphiocoelia, 147, 153, 157-163, 165
Scenophyllum, 143
Schellwienella, 196
Schizocystis, 121
Schizograptus, 107
Schizolepis, 318
Schizophoria, 180, 187, 188
Schizoproetoides, 164
Schizopyge, 160
Schizoramma?, 67
Schizostylus, 148, 151, 156, 159
Schizostylus (*Curuyella*), 151, 156, 159
?Schizostylus (*Curuyella*), 159
Schizostylus (*Schizostylus*), 151, 156, 159
Schubertella, 220, 224
Schwagerina, 219, 222, 224, 264
Sciadopitys, 318
Sciuravus, 337
Scleropteris, 316, 317
Scolopacidae, 325
Scotiacystis, 101, 103
Scotiella, 148, 157
Scyphocoenia, 294, 297
Scyphocrinites, 117, 122, 127
"Scyphocrinites", 119, 126, 127
Scyphophyllum, 142
Sedum, 442
Selaginella, 439
Sematethmos, 297
Semeloseris, 294, 297
Semibalanus, 401, 404, 411, 412, 416
Septacamera, 243
Septachonetes, 180
Septalaria, 196
"Septalaria", 180, 196
Septastraea, 294, 297
Septathyris, 180
Septatopora, 260, 266
Septatrypa, 180
Septopora, 260, 263-265, 267-270, 274
Sericoidea, 68, 69
Sespedectes, 336
Shaleria, 180
Sibiriophyllum, 314
Sibirispira, 196
"Sibirispira", 180, 196
Sicorhyncha, 180
Sieberella, 180
Sigmataxis, 283
Silvaseptopora, 260, 269
Simidectes, 336, 337
Simimys, 336
Simimeryx, 336
Sinospongophyllum, 143, 144
"Sinospongophyllum", 142
Sinusigera, 393
Siphonochitina, 78, 81
Siphonocrinus, 120-122
Siphonophrentis, 142-145
"Siphonophrentis", 143
Skenidioides, 180, 196, 471
Skenidioides?, 68
Skenidium, 180, 196
Skiagraptus, 107
Skinnerina, 219, 224
Skoliophyllum, 143
Smiliornis, 324, 325
Smithiphyllum, 145
"Smithiphyllum", 145
Sociophyllum, 143-145
Solarium, 392, 394, 395
Solidobalanus, 401, 404, 416
Solidobalanus (*Hesperibalanus*), 416
Sonninia, 289
Sowerbyella, 66, 67
"Sowerbyella", 66, 73
Sowerbyella?, 69
Spathognathodus, 122, 201-208, 210, 211, 213
Spermacystis, 102
Sphaerirhynchia, 196
"Sphaerirhynchia", 180, 196
Sphaerochitina, 77-79, 82, 85
"Sphaerochitina", 79, 85
Sphaeroidinella, 420, 422
Sphaeroschwagerina, 218, 219, 222, 242
Sphaerotocrinus, 122
Sphaerulina, 220
Sphenobaiera, 314-318
Spheniscus, 322, 331
Sphenopteris, 315, 317
Spinachitina, 78, 80
Spinatrypa, 181
Spinatrypina, 181
Spinella, 181
Spinolasma, 142
Spinoplasia, 181
Spinulicosta (*Lenzia*), 181
"Spirifer", 196
Spirigerina, 66, 73, 181
?Spirigerina, 66, 71, 73
Spirigerina (*Eospirigerina*), 66, 73
?Spirigerina (*Eospirigerina*), 66, 71
Spirinella, 181
Spirocrinus, 121
Spongia, 294
Spongonaria, 143
Spongophylloides, 141, 142
Spongophyllum, 142-145
"Spongophyllum", 144
Spurispirifer, 181
Spyridiocrinus, 123
Squamaria, 240
Staffella, 220, 224
Stathmoelasma?, 142
Stauria, 142
Steganopus, 325
Stegerhynchus, 181, 196
Stelidocrinus, 121
Stenodiscus, 260, 264, 266, 268, 272
Stenodiscus?, 267
Stenopora, 260, 263-268, 272, 274
Stephanastraea, 292
Stephanocoenia, 294, 297, 299
Stephanocrinus, 120, 122, 129
Stereocoenia, 292, 293
Stereolasma, 143, 144
Stereoxylodes, 141, 142
Stewartophyllum, 144
"Stewartophyllum", 143
Straelenia, 181

Straelina, 196
Streblascopora, 260, 263-268, 270, 272
Streblocladia, 260, 266, 270
Streblotrypa, 260, 263, 264, 266-268, 270, 274
Streblotrypa?, 266
Streblotrypella, 260, 270
Strelitzia, 437
Streptocrinus, 121
Streptopelia, 326
Striatura, 282
Strigatella, 431
Striispirifer, 181
Stringocephalus, 136
Stringophyllum, 144
Stringophyllum?, 144
Strixella, 181
Stromatopora, 294
Stromeria, 321
"Strophochonetes", 181, 196
Strophodonta, 196
"Strophodonta", 181, 196
Strophomena, 66, 73
"Strophonella", 181
Struthio, 321, 326, 327, 329
Struthiolarella, 350, 352-354
Struthiolaria, 352, 353
Struthioptera, 350, 353, 354
Struveina, 181
Sturtella, 181
Stylina, 290, 292-294, 296, 297, 299, 300
Stylocrinus, 122
Stylohelia, 292, 294, 296
Stylophyllopsis, 299
Stylopleura, 142
Stylopleura?, 142
Stylosmilia, 292, 293, 296
Styrax, 442
Subcuspidella, 181
Sudatea, 145
Sulcoretepora, 260, 264-267, 269
Sumatrina, 220
Sutherlandinia, 141
Synaptophyllum, 143
Synchirocrinus, 122
Syndetocrinus, 121, 122
Synocladia, 260, 264-268, 270, 272, 274
?Synphoria, 163
Syringaxon, 141-145
Syringoclemis, 260, 268, 270

Tabulipora, 260, 263-268, 272, 274
Tabulophyllum, 143-145
Tachinus, 375
Tadschikia, 196
Taemostrophia, 181
Taeniopteris, 314-317
Taimyrophyllum, 142-145
Taitzehoella, 221
Taleoleptaena, 196
Tanchintongia, 242
Tanuchitina, 78, 81
Tapochoerus, 336
Tapocyon, 336
Tarijactinoides, 151, 156, 161, 162
?Tarijactinoides, 151
Tastaria, 181
Tavayzopora, 260, 268, 270
Taxocladus, 316, 317
Technophorus, 28, 29, 31-35
Tectus, 429
Tegula, 414, 429, 431
Teichertina, 181
Teichostrophia, 181
Teiichispira, 37, 39-43, 45
Teleodus, 337
Teleornis, 324, 325
Telmabates, 323, 324
Temnocrinus, 121
Temnophyllum, 143-145
Tenarea, 460-463
Tenellodermis, 181
Tentaculites?, 67
Tenuicostella, 181
Terminalia, 439
Tetraclita, 401, 404, 411, 416
Tetradontella, 67
Tetragraptus, 107
Tetratomia, 181
Thailandina, 220
Thais, 391, 393-395, 428, 429, 431
Thalamocrinus, 120-122
Thallites, 314-316, 318
Thamnasteria, 290, 292-297, 299
Thamniscus, 260, 263-268, 270, 272, 274
Thamnocoenia, 293, 297
Thamnoseris, 290, 292-294, 297
Thaumastus, 282
Thecocyathus, 290, 292, 293, 300
Thecophyllia, 290, 292
Thecoseris, 290, 296
Thecosmilia, 290, 292, 293, 296, 297
Thegornis, 324, 325
Thelecalymene, 166
Thenarocrinus, 121
Thinnfeldia, 316
Thliborhynchia, 181, 196
"Thliborhynchia", 181, 196
Thoralicystis, 101, 103
Thrysopteris, 444
Thylacocrinus, 122, 123
Tibagya, 160
Tichalaracythere, 305, 307-309
Tigulites, 219
Tiliornis, 324, 325
Timanodictya, 260, 262-264, 268, 270, 272
Timanodictya?, 266
Timanotrypa, 260, 264
Tinamisornis, 325
Titanomena, 67
Tolmachovia, 28, 29, 32, 33, 35
Tomheganella, 181
Tonkinaria, 142
Toquimaella, 181
Toriyamaia, 220
Tortophyllum, 144
Totia, 196
Trapezophyllum, 145
Trematospira, 181, 196
Triarthus, 57
Triathyris, 181
Tricycloseris, 293, 297
Trigerastraea, 290, 292, 293, 295-297, 299
Trigonirhynchia, 181
Trimerus, 156
Triphyllotrypa, 260
Triticites, 219, 222, 269
Triznella, 260, 263, 270, 272
Trocharaea, 296, 297
Trochocyathus, 290, 292, 297
Trochoplegma, 296, 297
Trochus, 429, 431
Tromatopora, 294
Troosticrinus, 121, 122, 129
Tropidoleptus, 150, 181, 477
Trypetesa, 339, 401, 402, 405-411, 414
Tryplasma, 141, 142, 145
Tsuga, 373
Tubaria. 242
Tubulostrophia, 181
Turbo, 37, 38
Tyersella, 177
Typhloniscus, 148, 151, 156, 161
Tyrmia, 314, 315
Tyto, 321, 326

Uintanius, 337
Uintasorex, 337
Uintatherium, 337
Ulrichotrypa, 260, 266, 268, 272
Ulrichotrypella, 260, 264, 267, 268, 272, 274
Umbellularia, 442
Uncinulus, 196
"Uncinulus", 181, 196
Undispirifer, 196
Unguliproetus, 164
Urochitina, 78, 79, 83, 84
Urushtenia, 242
Utaratuia, 143, 144

Vaccinium, 374
Vagrania, 182, 196
Valliculastraea, 296
Vallimeandra, 294, 296, 297, 299
Vallimeandropsis, 290, 292, 294, 296
Vandercammenina, 182, 196
Vellamo, 67
Velostrophia, 182
Venericardia, 351
Venericor, 351
Vepresiphyllum, 142, 143
Verbeekina, 220
Verella, 219
Veronica, 438
Vespericola, 283
Viaphacops, 148, 149, 151, 153, 156, 162, 166
Victoriacystis, 94, 102, 103
Vitrina, 282
Vogesina, 148-151, 156, 157, 162, 164
Vogesina (Vogesina), 151, 157, 162
Vultur, 325

Waeringella, 219
Wanwanella, 28, 32, 33
Wanwania, 27, 28, 32, 33
Wanwanoidea, 28, 32, 33
Warrenella, 182, 196, 471
Warrenella-Reticulariopsis, 182, 196
Washakius, 337
Watsonella, 27, 28, 32, 33
Wedekindellina, 219
Weinbergina, 166
Weinmannia, 440, 441
Werneckeella, 182
Widdringtonia, 442
Wilsonicrinus, 121
Wintunastraea, 142
Wintunastraea?, 142
Wjatkella, 260, 263, 264, 268, 270, 272, 274

Xana, 182
Xenocyathellus, 144
Xenomartinia, 182
Xeronema, 442
Xerophyta, 439
Xerosicyos, 439
Xestrotrema, 241
Xyphelasma, 142
Xystriphyllum, 142-145

Yabeina, 218, 220, 224, 267
Yabeina-Lepidolina, 267
Yassia, 142
Yucca, 442

Zaphrentis, 143
Zdmir, 182
Zellia, 218, 219, 222, 242
Zelophyllia?, 141, 142
Zelophyllum, 142
Zelophyllum?, 141
Zlichopyramis, 196
Zlichorhynchus, 182
Zonophyllum, 142-144
"Zonophyllum", 145
Zophocrinus, 121, 122
Zospeum, 279

AUTHOR INDEX

Abbott, P. L., 337, 345, 347
Abbott, P. L., and others, 337, 345
Abbott, R. T., 241, 246, 431, 432
Abdusselamoglu, S., 89, 90
Achab, A., 86, 90
Adams, C. G., 344, 345
Adamson, R. S., 442, 444
Addicott, W. O., 360, 365, 368, 402, 411, 412
Adey, W., 459, 464
Agassiz, L., 254, 256
Ager, T. A., 375, 377
Ahlfeld, F., 154, 157, 160-163, 165, 166
Alberti, G. K. B., 154
Alloiteau, J., 296, 297, 299, 300
Amos, A. J., 155, 165, 480, 482
Amos, A. J., and others, 155, 165
Amsden, T. W., 139, 141, 207, 208, 211
Amsden, T. W., and others, 207, 208, 211
Andersen, D. W., 340, 341, 347
Anderson, S., 259, 275
Andersson, J. G., 349, 355
Androsova, E., 454, 456
Antevs, E., v
Arkell, W. J., 293, 299, 300
Armstrong, R. L., 338, 339, 345
Arons, A. B., 382, 389
Arx, W. S., von, 382, 388
Asama, K., 436, 444
Ashlock, P. D., 259, 275
Asselberghs, E., 196
Atkinson, K., 86
Atwater, T., 12, 24, 337, 345, 411, 412
Auboin, J., 380, 388
Aubouin, J. V., 2, 6
Axelrod, D. I., 313, 319, 339, 341, 345, 435-446

Bachmann, A., 86, 90
Bachofen-Echt, A., 373, 377
Backus, R. H., 450, 454
Backus, R. H., and others, 450, 454
Badham, J. P. N., 23, 24
Baker, H. B., 282, 287
Bakker, R. T., 227, 235-237
Balashov, Z. G., and others, 53, 58
Balashova, E. A., 53, 58
Baldis, B. A. J., 155, 158, 161, 165, 166, 474, 480, 482
Balkwill, H. R., 374, 377
Ball, I. R., 2, 6
Bandy, O. L., 15, 24, 418, 423, 424
Banks, M. R., 43, 46, 47, 200, 252, 253, 257
Barberena, M., 230
Barbulesci, 293
Barker, P. F., 352, 355
Barnes, C. R., 202, 203, 205, 210, 211
Barnes, C. R., and others, 202, 205, 211
Barnes, H., 412, 414
Barnes, L. G., 367, 368
Barnett, S. G., 40, 43, 202, 205, 207, 210, 211
Barnett, S. G., and others, 205, 207, 211
Barrabé, L., 293, 300
Barrande, J., 94, 99, 102, 165, 196
Barron, J. A., 367
Barton, E. D., 418, 421, 425
Bassler, R. S., 94, 99, 101, 102, 119, 128, 266, 275
Bate, R. H., 305, 309
Bather, F. A., 94, 99, 101, 102
Batten, R. L., 289, 301, 337, 346
Bayliss, D. D., 305, 309
Bé, A. W. H., 422, 424
Beauvais, L., 289, 292, 296, 299, 300
Beck, M. E., 337, 345
Beju, D., 86, 90
Bemmelen, R. W., van, 380, 388
Bengtson, S., 38, 43
Benier, C., 419, 422, 424
Benoist, E., 300
Benoit, A., 86, 90
Benson, R. H., 70, 71, 384-388
Berdan, J. M., 54
Berger, W. H., 13, 21, 24
Berggren, W. A., 15, 18, 24, 362, 365, 368, 376, 377, 382, 388, 396, 423-425, 463, 464
Bergström, J., 63, 67-69, 71, 86, 90
Bergström, J., and others, 86, 90
Bergström, S. M., 48, 86, 474, 482
Bernouilli, D., 23, 24
Berry, W. B. N., 48, 49, 62, 70, 71, 98, 99, 106-108, 110-114, 119, 126, 128, 139, 465, 473, 474, 480-482
Besaire, H., 293, 300
Beuf, S., and others, 62, 70, 71, 110, 112, 113
Bews, J. W., 435, 445
Bigot, A., 290, 293, 300
Billings, E., 41, 43, 50, 99
Binnekamp, J. G., 196
Bircher, W., 292, 301
Bird, J. M., 61, 71, 97, 99
Bischoff, G., 211
Black, C. C., 334, 346
Blackadar, R. G., 49
Blake, D. A. W., 50
Blake, W., Jr., 376, 377
Blasco, G., 474, 482
Bliss, L. C., 371, 377
Bliss, L. C., and others, 371, 377
Blow, W. H., 361, 368
Boekel, N. M., van, 86, 89, 90
Boersma, K. T., 211
Bogdanov, N. A., 53, 58
Bonaparte, J. F., 232, 238
Boneham, R. F., 86, 90
Bonham-Carter, G. F., 251, 254, 257, 451, 455
Bonte, A., 297, 301
Boomgaard, W. H., 301
Bostwick, D. A., 241, 245, 246
Boucek, B., 106-108, 110, 113
Bouché, P. M., 86, 90
Boucot, A. J., vi, 47-49, 56, 58, 61, 62, 64, 68, 70-72, 76, 78, 86, 98, 99, 106, 112, 113, 119, 124, 126, 128, 134, 139, 140, 148, 149, 151, 152, 155, 169, 186, 190-200, 202, 203, 207, 208, 210-212, 250, 257, 451, 454, 465, 466, 468, 474, 477, 480-482
Boucot, A. J., and others, 56, 58, 112, 113, 119, 124, 128, 169, 190-195, 197
Bourque, P. A., 195, 197
Bowen, R., 256, 257
Bowen, Z. P., 197
Boyer, J. E., 338, 346
Boykins, W., 459, 464
Bradshaw, J. S., 112, 113
Bramlette, M. N., 19, 22, 24
Branisa, L., 125, 128, 147, 154, 155, 157-166
Branson, C. C., 268, 275
Breimer, A., 125, 128
Brenchley, P. J., and others, 86, 90
Briden, J. C., 23, 24, 56, 59, 91, 99
Briden, J. C., and others, 91, 99
Bridge, J., 40, 43
Briggs, B. G., 437, 442, 445
Briggs, J. C., 3, 6
Brink, A. S., 155, 236, 237
Brodkorb, P., 322, 324-326, 329, 331
Broecker, W. S., 21, 24
Brower, J. C., 118, 119, 128
Brown, W. J., 372
Brück, P. M., 86, 90
Brück, P. M., and others, 86, 90
Brundin, L., 442, 445
Brunton, C. H. C., 196, 197
Brunton, C. H. C., and others, 196, 197
Bruun, A F., 380, 388
Bucher, W H., 343, 346
Bullard, E., and others, 11, 24
Bullock, T. H., 402, 412
Bulman, O. M. B., 106, 107, 113
Bultynck, P., 201, 203, 205, 207, 211
Burckhardt, C., 300, 301
Burke, R., 459, 466
Burrett, C. F., 23, 24, 58, 59, 61, 71, 244, 246
Burskyi, A. Z., 57, 59
Busson, G., 299, 301
Bustin, R. M., 374, 376, 377
Butts, C., 46, 48, 51, 52
Buxbaum, F., 440, 445

Cabrera, A., 363, 369
Camacho, H. H., 351, 355
Camp, W. H., 435, 445
Campau, D. E., 88, 90
Campbell, K. E., 325, 327, 331
Campbell, K. S. W., 149, 150, 155, 170, 193, 197, 477, 481
Carey, S. W., 96, 99, 289, 301
Carlquist, S., 444, 445
Carls, P., 195, 197, 205, 211, 213
Carls, P., and others, 197
Carlton, J. T., 404, 412, 415
Carter, C., 78, 86, 90
Carton, Y., 409, 414
Case, E. C., 52
Cashion, W. B., 339, 346
Caster, K. E., 91, 94, 97-99, 101, 102, 147, 151, 155, 162, 164
Cates, R. G., 429, 432
Cathcart, S. H., 54, 59
Cernahorsky, W. O., 431, 432
Chaiffetz, M. S., 77, 86, 90
Chaloner, W. G., 239, 242, 244, 246, 477, 479, 481
Chapskii, K. K., 362, 363, 369
Chapuis, F., 301
Chase, C. G., 245, 246, 442, 443, 445
Chase, T. E., and others, 16, 24
Chatterton, B. D. E., 195-197, 203, 211
Chauris, L., 87, 90
Chauvel, J.-J., 86, 87, 90, 94, 99, 101, 102
Chauvel, J.-J., and others, 86, 90
Chave, K. E., and others, 21, 24
Cheetham, A. H., 132, 140, 170, 197
Cheng, L.-H., and others, 436, 445
China, Geological Society of, and Institute of Geology of the Chinese National Academy of Science, 108, 113
Chiriac, M., 363, 369
Chlebowski, R., 86, 90
Chlupáč, I., 197
Choubert, G., 292, 301
Christie, R. L., 42, 48
Chronic, J., 268, 275
Chudinov, P. K., 227, 229, 237
Chugaeva, M. N., 53, 57, 59, 473, 481
Chugaeva, M. N., and others, 53, 59
Churkin, M., Jr., 53, 54, 139, 140, 313, 314, 319
Cifelli, R., 418-424
Cita, M. B., 387, 388
Clark, D. L., 376, 377
Clark, T. H., 50
Clarke, J. M., 147, 150, 151, 155, 157-163, 166, 197
Clarke, M. J., 252, 253, 257
Clarke, S. H., and others, 338, 346
Cleaves, A. B., 52
Clement, C., 102
Clements, F. E., 250, 257, 435, 445
Clench, W. J., 393, 396
Coast and Geodetic Survey, 402, 412
Cocks, L. R. M., 64, 68, 71, 209-211, 466, 481
Cohen, R., 242, 246
Colinvaux, P. A., 375, 377
Collier, B. D., 249, 257
Collier, B. D., and others, 249, 257
Collignon, M., 296, 301
Collinson, C., 86, 90
Combaz, A., 86, 90
Comte, P., 197
Condra, G. E., 262, 268, 275
Connell, J., 412, 415
Cook, E. F., 339, 346
Cook, H. E., 70, 71

Cook, P. L., 454, 457
Cooke, C. M., Jr., 286, 287
Coope, G. R., 372, 377
Cooper, B. J., 206, 211
Cooper, G. A., 56, 59, 61, 70-72, 136, 140, 196, 197, 241, 246, 253, 257
Cooper, R. A., 106, 108, 113
Copeland, M. J., 39, 43
Copper, P. A., 148, 155
Corliss, J. B., 20, 24
Cornwall, H. R., 51
Cornwall, I. E., 412, 414-416
Correia, M., 86, 90
Corroy, G., 301
Costa, Norma M. van Boekel, da, 86, 89, 90
Couper, R. A., 441, 445
Cousminer, H. L., 86, 90
Covacevich, V., 326, 327, 331
Cowen, R., 242
Cox, A., 10, 12, 24
Cox, C. B., 227, 230, 234, 236, 237
Cracraft, J., 259, 275, 326, 331, 442, 445
Cramer, F. H., 79, 86, 87, 90
Cramer, F. H., and others, 87, 90
Cranwell, L. M., 350, 355
Craven, A. E., 393, 396
Creer, K. M., 23, 24, 352, 355
Cribb, H. G. S., and others, 322, 331
Crickmay, C H., 140, 143, 144, 197, 301
Crocker, R. L., 438, 445
Crockford, J., 266, 275
Croizat, L., and others, 233, 237
Crowe, F. J., 402, 413
Crowell, J. C., 440, 445, 468
Cuffey, R. J., 268, 276
Culberson, C., 21, 24
Cumming, L. M., 50
Cutler, E. B., 403, 413
Cuvier, G., 254, 256

Dall, W. H., 396
Dalman, J. W., 67, 71
Dalrymple, G. B., 339, 346, 367, 369
Dalrymple, G. B., and others, 367, 369
Dalziel, I. W. D., 352, 355
Dana, T. F., 407, 408, 413, 431, 432
Dânet, N., 86, 90
Dangeard, L., 290, 301
Dareste de la Chavanne, J., 301
Darlington, P. J., 105, 113, 391, 396
Darton, N. H., 52
Darwin, C., 249, 253, 254, 256, 257, 413, 415, 416
Davis, G. A., 139, 140
Davis, W. E., Jr., 202, 203, 211
Davoren, P. J., 149, 150, 155, 170, 193, 197
Dawson, M. R., 334, 346
Dayton, P., 413, 415
Dean, B., 2, 3, 6
Dean, W. I., 61, 62, 71
Decima, A., 387, 388
Deflandre, G., 87, 90
Dehm, R., 99, 102
Dejardin, J., and others, 440, 445
Delo, D. M., 155, 157, 159, 161, 163
Derby, J., 40
Derstler, K., 91-94, 97-104
Destombes, J., 107, 110, 114
Deunff, J., 87, 88, 89, 90
Deunff, J., and others, 87, 90
Devereaux, I., 365, 369
DeVries, T. J., 102
Dewalque, G., 301
Dewey, J. F., 15, 24, 61, 71, 97, 99
Dewey, J. F., and others, 15, 24
Dibblee, T. W., Jr., 339, 346
Dicevitchius, E., 87, 90
Dickins, J. M., 262, 275, 276
Dietz, R. S., 15, 17, 18, 24, 153, 155, 216, 225, 226, 273, 275, 289, 301
Diez de Cramer, M. d. C., 79, 87
Dingle, R. V., 305, 309
Dinkelman, M. G., 15, 24
Dmitriev, B. J., 225, 226, 273, 275
Dolan, R., 400, 413
Dorf, E., 373, 377
Doty, M., 459, 464
Dott, R. H., Jr., 289, 301, 337, 346
Doubinger, J., 87, 89, 90
Douglas, J. G., 313, 319
Douglas, R. C., 481, 482
Douglas, R. G., 385, 388
Douvillé, Fr., 301
Douvillé, H., 301
Downie, C., 86, 88, 90
Doyle, J. A., 436, 445
Doyle, J. A., and others, 436, 445
Dresnay, R. du, 292, 301
Dreyer, E., 264, 275
Drot, J., 197
Druce, E. E., 202, 203, 211
Dubatolova, J. A., 117, 128
Dubertret, L., 50
Dudley, P. M., 268, 275
Dunbar, C. O., 52
Dunbar, M. J., 2, 5, 6
Duncan, P. M., 301
Dunn, D. L., 87, 90
Durham, J. W., 367, 369, 411, 413
Dutertre, A. P., 301
Du Toit, A. L., 12, 24
Dutro, J. T., 51, 53, 54, 56, 59, 61, 72, 471, 479, 480, 481

Ebling, F. J., and others, 429, 430, 432
Echols, D. J., 87, 90
Edwards, D., 477, 481
Efremov, I. A., 228, 232, 237
Ehlers, G. M., 140, 144
Ehrhardt, J. P., 431, 433
Eichwald, C .T., von, 362, 369
Einarsson, T., and others, 367, 369, 376, 377
Eisenack, A., 87, 90
Ekman, S., 2, 4, 6, 391, 396, 402, 405, 410, 413
Eldredge, N., 147, 155, 157-162, 165, 166, 253, 257
Elias, M. K., 262, 268, 275, 436, 445
Elliott, D. H., 349, 350, 352, 354, 355
Elliot, D. H., and others, 349, 350, 355
Elsik, W. C., 338, 346
Emerson, W. K., 431, 432
Emmons, E., 45
Enay, R., 289, 301
Engel, A. E. J., 23, 24
Enright, J .T., 402, 413
Erdtmann, B. D., 106, 114
Etheridge, R., 266, 275

Fager, E. W., 106, 114
Fagerstrom, J. A., 155, 166
Fåhraeus, L. E., 201-203, 205, 207, 209-211
Fairbridge, R. W., 22, 24
Fell, H. B., 396, 431, 432
Feton, C. L., 52, 140, 145
Fenton, M. A., 140, 145
Fernández, J. A., 351, 355
Ferrigno, K. F., 202, 212
Ferruglio, E., 155, 157
Ferry, H. B. A. T. de, 297, 301
Finckh, A. E., 459, 464
Fink, R. P., 87, 90
Finlay, H. S., 352, 353, 355
Fischer, J. C., 295, 301
Fisher, R. L., 13, 24, 440, 442, 446
Fisher, R. L., and others, 13, 24
Fitch, J. E., 338
Fix, M., 39
Fleming, C. A., 354, 355, 363, 369
Fleminger, A., 403, 413
Flessa, K. W., 451, 452, 454
Flood, P. G., 197
Florin, R., 441, 445, 480, 481
Flower, R. H., 41
Flügel, E., 296, 301
Fofonoff, N. P., 382, 388
Forbes, E., 1, 2, 6
Forney, G. G., 453, 454
Forsyth, D., 15, 24, 25, 396
Frakes, L. A., 344, 346, 440, 445, 479, 481, 482
Frakes, L. A., and others, 479, 481, 482
Francheteau, J., 13, 25
Frederickson, E. A., 55, 59
Frederiksen, N. O., 338
Freneix, S., 298, 299, 301
Frenguelli, J., 359, 363, 369
Frerichs, W. E., 342, 343, 346
Fritz, M. A., 268, 275
Fritz, P., 301
Fromentel, E. de, 296, 297, 301
Frost, S. H., 463, 464
Fry, W. G., 4, 6
Fuchs, A., 197
Furnish, W. M., 261, 275
Furon, R., 301

Gabrielse, H., 50, 52
Gabrielse. H., and others, 50
Gandl, J., 211, 213
Garcia-Alcalde, J. L., 197
Gardet, G., 301
Gardiner, J. S., 459, 464
Gardner, C. A., 445
Garth, J. S., 431, 432
Gass, I. G., and others, 10, 24
Gautier, Y. V., 454-457
Gay, H., 404, 413
George, W., 105, 114
Gersdorf, E., 372, 373, 377
Gerth, H., 293, 299, 301
Gigout, M., 99, 100
Gilbert-Tomlinson, J., 41, 43
Gill, E. D., 94, 99, 102, 197
Gilmour, E. H., 268, 275
Girty, G. H., 268, 275
Giterman, R. E., 375, 377
Givens, C. R., 338, 346
Glaçon, G., and others, 422, 424
Glaessner, M. F., 410, 413
Glangeaud, L., 380, 388
Glangeaud, Ph., 290, 302
Gleason, H. A., 250, 257
Glenister, B. F., 203, 212
Glimberg, C. F., 69, 71
Glover, L. K., 419, 425
Gobbett, D. J., 241, 244-247, 480, 481
Godley, E. J., 441, 445
Godwin-Austen, R., 1, 6
Goldstein, R. F., 87, 90
Goldstein, R. F., and others, 87, 90
Golikov, A., 391, 396
Golz, D. J., 333, 334, 345, 346
Gomez, E. D., 413, 415
Good, R., 435, 445
Gordon, G. D., and others, 459, 463, 464
Gordon, W. A., 302
Gorjunova, R. V., 261, 264, 275
Gorsky, V. P., 242, 246
Gothan, W., 312, 319
Gould, S. J., 253, 257
Grabau, A. W., 267, 275
Graham, H. W., 404, 413
Graindor, M.-J., and others, 87
Grant, R. E., 241, 246, 253, 257, 480
Gratsianova, R. T., 195, 198
Gray, J., and others, 76, 87
Gregorio, A., de, 264, 275
Gregory, J. W., 297, 299, 302
Grekoff, N., 305, 309
Griffiths, J., 61, 71
Griffiths, J. R., 440, 442, 443, 445
Grignani, D., 87
Grigorescu, D., 362, 363, 367, 369
Grose, L. T., 339, 346
Gross, M., and others, 459, 464
Grossheim, V. A., 367, 369
Grossouvre, A., de, 302
Groth, J., 155, 160, 162
Grunt, T. A., 225, 226, 273, 275
Guebhard, O., 302
Gupta, V. J., 122, 128
Guseva, E. A., 242, 246
Guthrie, R. D., 376, 377

Haas, W., 155
Hadding, A., 66, 69, 71
Haime, J., 302
Hall, J., 52, 99, 165, 166
Hall, S., 350
Hallam, A., vi, 15, 22, 24, 25, 254, 257, 289, 292, 293, 295, 298, 300, 302, 443, 446
Halls, C., 23, 24
Hamada, T., 150, 155
Hamilton, W., 56-59, 243, 246, 313, 319
Hamlin, W. H., 424
Hancock, N. J., and others, 70, 71
Harlow, F. H., 241, 247
Harper, C. W., Jr., 195, 196, 198
Harper, C. W., and others, 198
Harrington, H. J., 55, 59, 106-108, 110, 114, 148, 155, 166, 481, 482
Harrington, H. J., and others, 55, 59

Harris, W. J., 106, 107, 108, 114
Harris, W. K., 444, 446
Harrison, C. J. O., 331
Harrison, R. J., 366, 369
Hart, G. F., 244, 246, 261, 275
Hart, R. A., 20, 24
Hartt, C. F., 155, 160
Hasegawa, Y., 360, 368
Havlíček, V., 61, 62, 69, 71, 196, 198
Hayden, B. P., 400, 413
Hayes, D. E., 352, 354, 355, 442, 445
Hays, J. D., 22-24, 343, 346
Hazel, J. E., 132, 140, 170, 197, 360, 362, 366, 368, 369
Heath, G. R., 21, 24
Hedberg, H. D., 251, 257
Heddabout, C., 126, 128
Hede, J. E., 69, 71, 76
Hedgpeth, J. W., viii, ix, 1-6, 239, 240, 246, 399, 413
Henderson, J., 277, 283, 287
Henderson, R. A., 351, 355
Hendey, Q. B., 322, 331, 359, 361-363, 369
Henningsmoen, G., 69, 71
Henry, D. P., 407, 413, 415, 416
Henry, J.-L., 87, 90
Herdman, W. A., 1, 6
Herman, Y., 367, 369
Herngreen, G. F. W., 437, 445
Hertlein, L. G., 354, 355
Hessler, R. R., 3, 6, 7, 380, 388
Hickey, L. J., 436, 445
Higgins, R. E., 339, 345
Hilgard, G. H., 413, 414
Hill, D., 261, 275
Hills, L. V., 374, 376, 377
Hills, L. V., and others, 374, 376, 377
Hisinger, W., 67
Høeg, O. A., 242, 246
Hoek, H., 156, 163
Hoffman, K., 302
Hoffmann, R. S., 373, 377
Hoffstetter, R. J., 359, 361, 363, 365, 369
Holden, J. C., 15, 17, 18, 24, 153, 155, 216, 225, 226, 273, 275, 289, 301
Holland, C. H., 259, 275
Hollard., H., 126, 128
Hollick, A., 314, 319
Hollister, C. D., 15, 18, 24, 362, 365, 368, 376, 377, 382, 388, 396
Holtedahl, O., 40, 43, 51, 52
Hooker, J. D., 442, 445
Hopkins, D. M., 362, 367, 369, 374, 376, 377
Hopkins, D. M., and others, 374, 376, 377
Hotton, N., III, 232, 233, 237
Houbrick, J. R., 429, 431, 433
House, M. R., vii, ix, 203, 212
Howarth, M. K., 351, 355
Howell, D. G., 336, 338, 346
Howell, D. G., and others, 336, 346
Hsia, F., 267, 276
Hsü, K. J., and others, 20, 24
Huffman, G. G., 52
Hughes, C. P., 27, 29, 32, 36, 41, 43, 53, 55, 56, 58, 59, 61, 72
Hughes, N. F., vi, 436, 445
Hughes, P., 418, 421, 425
Hughes, T., 139, 140
Huguet, J., 302
Hulston, J. R., 256, 257
Hunt, C. B., 339, 346
Huntoon, P. W., 342, 346
Hurley, A. C., 407, 413, 415
Hurley, P. M., 23, 24
Hutchins, L. W., 402, 413
Hutt, J. E., and others, 106, 114
Hutton, J., 254
Huxley, J. S., 254, 256, 257
Hyman, L. H., 454, 456

Imbrie, J., 452, 454
Imlay, R. W., 302
Ingham, J. K., 57, 59, 91, 96, 100
Irving, E., 12, 24, 208, 212, 465, 481
Isaac, W. E., 459, 464
Isaacs, J. D., 3, 6
Isaacson, P. E., 96, 98-100, 147, 149, 150, 155, 166
Isaacson, P. E., and others, 98, 100
Isberg, O., 66, 68, 71
Iselin, C. O.'D., 418, 421, 424
Ivanov, A. N., 48

Jaanusson, V., 48, 51, 61, 62, 66-69, 71, 72, 107, 114
Jackson, D. E., 106-108, 114, 170, 199
Jackson, D. E., and others, 108, 114
Jackson, J. B. C., 400, 413
Jaeger, H., 107, 114, 127, 128, 474, 481
Jaekel, O., 100
Jahnke, H., 198
Jain, S. P., 307, 309
Jansonius, J., 87, 88, 90
Jardiné, S., 88, 90
Jefferies, R .P. S., 94, 100-102, 104
Jekhowsky, B. de, 89, 90
Jell, P. A., 27, 36, 91, 92, 96, 100, 250, 257
Jenkins, R. J. F., 323, 331, 344, 346
Jenkins, W. A. M., 78, 88, 90
Jenkyns, H. C., 23, 24
Jeppsson, L., 76, 88
Jodru, R. L., 88, 90
John, A. E., 450, 454
Johnsen, A., 264, 275
Johnson, D. B., 201, 204-206, 212
Johnson, D. G., 194, 198
Johnson, J. G., 71, 119, 124, 128, 134, 140-144, 149, 155, 169, 171, 190, 192, 195-198, 203, 205, 207-209, 211, 212
Johnson, J. G., and others, 196, 198
Johnson, J. H., 43, 46-48, 71, 459, 464
Johnson, L. A. S., 437, 440, 442, 445
Johnson, L. A. S., and others, 440
Johnson, M. E., 120, 128
Jones, C. R., 40, 43
Jones, D. L., and others, 480, 481
Jourdy, E., 303
Jumars, P. A., 3, 6
Just, T. K., 312, 319

Kaljo, D., 88, 90, 107, 114, 473, 481
Karig, D. E., 440, 443, 446
Kato, M., 241, 246
Katzer, F., 155, 157, 160-163
Kauffman, A. E., 88, 90
Kauffman, E. G., 147, 155, 351, 355, 453, 454
Kayser, E., 49, 155, 162
Keast, A., 261, 275, 442, 446
Keast, A., and others, 261, 275
Keble, R. A., 107, 114
Kegel, W., 198
Keller, B. M., and others, 51
Kellogg, R., 359, 369
Kelm, D. L., 23, 24
Kemp, E. M., 344, 346, 444, 446
Kendall, G. W., 198
Kennedy, M. P., 334, 346
Kennett, J. P., 15, 17, 18, 20, 24, 25, 344, 346, 352, 354, 355, 382, 385, 388, 423
Kennett, J. P., ad others, 15, 17, 20, 24, 344, 346, 352, 354, 355, 382, 385, 388
Kensley, B., 365, 369
Kent, P. E., 439, 446, 447
Kent, P. E., and others, 439, 446
Kerr, J. W., 49, 198, 208, 212
Khain, V. E., 367, 369
Khanaychencko, H. K., and others, 392, 396
Kim, K. C., and others, 361, 369
Kindle, C. H., 61, 70, 71
King, J. E., 363, 366, 369
King, L. C., 302
King, P. B., 52
Kinne, O., 70, 72
Kiselev, S., 375
Kiseleva, A. V., 268, 275, 276
Kiseleva, A. V., and others, 268, 275
Kitching, J. A., 429, 430, 433
Kitching, J. A., and others, 429, 430, 433
Klaaman, E., 473, 481
Klapper, G., 173-182, 186, 194, 195, 198, 201, 203-210, 212
Klapper, G., and others, 173-182, 186, 195, 198, 201, 205, 206, 212
Klautsan, R. A., 263, 276
Kleinhampl, F. J., 51
Kline, J. K., 77, 89, 90
Klootwijk, C. T., 91, 96, 100
Kluge, G. A., 454, 456
Knight, J. B., 43, 46-48
Knight, J. B., and others, 46-48
Knod, R., 155, 157, 160
Kobayashi, T., 27, 36, 50, 51, 110, 114, 150, 155
Koby, F., 296, 299, 302
Koechlin, E., 302
Koechlin, J., 439, 446
Kolosváry, G. von, 405
Kondo, P., 286, 287
Koninck, L. G., de, 99
Kon'no, E., 312, 319
Kozlowski, R., 155, 157, 159, 160-163, 165, 166, 198
Kramarenko, N. N., 53, 59
Krans, T. F., 200
Krassilov, V. A., 254, 257, 436, 446
Kremp, G. O. W., 244, 246
Krishnan, M. S., 312, 319
Krömmelbein, K., 305, 307, 309, 310
Kropotkin, P. N., 244, 246
Krstić, N., 387, 388
Kuc, M., 376, 377
Kuhl, H., 413, 415
Kummel, B., 261, 275, 312, 319
Kutina, J., 443, 446

Lacey, W. S., 239, 242, 246, 477, 481
Ladd, H. S., 286, 287, 408, 413, 444, 446, 459, 464
Ladd, H. S., and others, 286, 444, 446, 459, 464
Lake, P., 155, 158, 160, 161, 163
Lakhanpal, R. N., 440, 441, 446
Lambe, L. M., 140, 143, 144
Lamperein, R. C., 327, 331
Lane, H. R., 54, 206, 212
Lane, N. G., 124, 128
Lange, F. W., 79, 88, 90
Langenheim, R., 463, 464
Langenstrassen, F., 198
Lanquine, L., 296, 302
Larson, R. L., 245, 246
Laufeld, S., 76-78, 88, 90, 474, 481
Laufeld, S., and others, 88, 90
Leanza, A. F., 107, 108, 110, 114, 481, 482
Lee, C. K., 108, 114
Lee, R. K. S., 463, 464
Lefort, J.-P., 88, 90
Legault, J. A., 88, 90
Legrand, P., 107, 114
Leinen, M., 19, 21, 24
LeMâitre, D., 126, 128, 195, 198
Lemon, R. R. H., 49
Lenz, A. C., 106, 114, 186, 190, 198, 199
Leopold, E. B., 341, 346, 444, 446
LePichon, X., and others, 465, 481
Lespérance, P. J., 64, 68, 70, 72
Lespinasse-Legrand, H., 302
Lesueur, C. A., 45, 49
Leven, E. Y., 264, 275
Levin, D. A., 429, 433
Levin, H. L., 87, 90
Levy, R., 481, 482
Lewis, C. A., 413, 414
Lillegraven, J. A., 333, 334, 344-346
Lindroth, G. H., 372, 377
Lindström, G., 67, 71, 72
Lindström, M., 67, 68, 71, 72, 88, 90, 201, 212
Link, M. H., 338, 346
Linsley, R. M., 38, 44
Lipina, O. A., 220, 226
Lipman, P. W., and others, 339, 340, 342, 343, 346
Lipps, J. H., 344, 346
Lissajos, M., 302
Lister, T. R., 88, 90
Lister, T. R., and others, 88, 90
Littler, M., 459, 463, 464
Lockwood, J., 429, 430, 433
Loeblich, A. R., Jr., 22, 25
Logan, A., 241, 246
Longstaff, J., 51
Loo, L., 267, 275
Lorenchet de Montjamont, M., 292, 302
Loughnan, F. C., 254, 257
Love, J. D., 341, 346
Love, J. D., and others, 341, 346
Lowe, G. D., 338, 346
Lowenstam, H., 125
Ludvigsen, R., 54, 59, 198, 199
Luyendyk, B. P., 15, 17, 18, 24, 396

Luyendyk, B. P., and others, 15, 17, 18, 24, 396
Lyell, C., 254
Lynn, R. J., 402, 414

MacArthur, R. H., 451, 454
MacGillivray, P. H., 454, 457
MacGinitie, H. D., 341, 346
MacKenzie, W. S., and others, 140, 144
MacNeil, F. S., 367, 369
MacQueen, R. W., 50
Magloire, L., 88, 89, 90
Mahlzahn, E., 256, 257
Maillieux, E., 199
Ma'or, H., 40, 43
Maksimova, Z. A., 53-55, 59, 155
Malecki, J., 264, 275
Malfait, B. T., 15, 24
Malone, E. J., 199
Malumian, N., and others, 307, 310
Mamet, B., 220, 226, 477, 481
Mani, M. S., 440, 441, 446
Mannikeri, M. S., 307, 309
Männil, R. M., 63-65, 72, 88, 90
Männil, R. F., and others, 88, 90
Mantovani, M. P., 87
Marche-Marchad, I., 391, 392, 396
Marcus, E., 454, 455
Marek, L., 38, 43
Marincovich, L., 38
Marr, J. E., 100, 102
Marshall, P., 351, 355
Martin, F., 88, 90
Martin, F., and others, 88, 90
Martna, J., 66, 72
Marwick, J., 352, 353, 355
Mason, D., 207, 212
Mason, W. R. M., 372, 377
Massa, D., 69, 71
Masters, W. R., 86, 90
Matthew, W. D., 2, 6, 328, 331
Matthews, J. V., Jr., 371-377
Matthews, S. C., 27, 36
Maturo, F. J. S., 454, 455
Maubeuge, P. L., 302, 303
Mawatari, S., 455-457
Maynard, N. G., 450, 454
Mayou, T. V., 268, 275
Mayr, E., 253, 257
McCabe, W. J., 256, 257
McCammon, H., 140, 144
McDonald, R. E., 340, 346
McElhinny, M. W., 12, 23, 24, 56, 58, 59, 239, 244-246, 439, 445, 446
McElhinny, M. W., and others, 246, 439, 446
McGowan, J. A., 106, 114, 399, 403, 404, 413
McGowran, B., 354, 355
McGregor, D. C., 195, 199
McGugan, A., 241, 246
McIntyre, T. C., and others, 411, 413
McKee, E. H., 339-341, 346
McKeena, M. C., 341, 346, 444, 446
McKenzie, D. P., 443, 446
McKerrow, W. S., 61, 72, 119, 128, 209-211, 451, 454, 466, 481
McLaren, D. J., 140, 143-145
McLaren, D. J., and others, 140, 144, 145
McLaughlin, P. A., 407, 413, 415
McLean, R. A., 141, 142
McVaugh, R., 442, 446
Medioni, A., 454, 456
Médioni, R., 292, 302
Meek, F. B., 100
Melville, R., 436, 446
Mendes, J. C., 232, 237
Méndez-Alzola, R., 155, 157
Menzies, R. J., and others, 70, 72
Mercier, J., 294, 302
Merriam, C. W., 52, 140-145, 199
Merrill, G. K., 202, 212
Metcalf, W. G., and others, 392, 396
Meusel, H. van, 440, 446
Meyen, S. V., 242, 244, 246, 261, 275, 477, 479, 481
Meyer, G., 302
Meyerhoff, A. A., 12, 24
Michelin, H., 302
Middlemiss, F. A., vi, 199, 302
Middlemiss, F. A., and others, vi, 199
Mikhailova, N. F., 109, 114
Mikhaylov, Y. A., and others, 243, 246
Mikulic, D., 125
Miller, A. K., 50
Miller, M. A., 77, 88, 90
Miller, T. H., 87, 88, 90
Milne-Edwards, H., 2, 6, 302
Milner, A. R., 227, 234, 237
Minato, M., 241, 246, 481, 482
Minch, J. A., 335, 336, 338, 346
Minster, J. B., and others, 10, 11, 24
Mironova, M. G., 51
Missarzhevsky, V. V., 27, 36
Mitchell, E. D., 367, 368
Mittmeyer, H. G., 199
Mivart, St. G., 359, 369
Miyagkova, E. I., 48
Mizens, L. I., 196
Moberly, R., 15, 24
Moebius, K., 2, 3, 6
Molnar, P., 12, 24, 440, 446
Monger, J. W. H., 241, 245, 246
Monsen, A., 106, 107, 114
Moodey, M. W., 119, 128
Moore, G. W., 334, 346
Moore, H. J., 437, 446
Moore, R. C., 118, 128, 268, 275
Moore, T. C., Jr., and others, 19, 24
Moores, E. M., 22, 25, 453, 455
Morgan, D. H., 88, 90
Morgan, W. J., 11, 24
Morkhoven, F. P. C. M. van, 384, 388
Morozova, I. P., 261, 264, 267, 268, 275
Morris, W. J., 345, 346
Moskalenko, T. A., 206, 207, 209, 212
Mouterde, R., 302
Moy, R. L., 86, 90
Moyano, G. H. I., 454, 457
Mu, A. T., 106, 108, 114
Mu, A. T., and others, 108, 114
Mu, En-Chih, 50, 121, 126, 128
Mu, En-zhi, and others, 50
Mudge, M. R., 243, 246
Müller, H., 155, 164
Muller, J., 436, 437, 446
Munroe, E., 372, 377
Murphy, M. A., 204, 205, 208, 212

Nakazawa, K., 261, 275
Nalivkin, D. V., 242, 246
Nassichuck, W. W., 267, 276
Neale, J. W., 305, 310
Negus, P., 300
Nekhoroshev, V. P., 268, 275
Nelson, C. H., and others, 376, 377
Nelson, S. J., 50
Nestell, M. K., 241, 245, 246, 481, 482
Neufville, E. H., 305
Neuman, B., 67, 72
Neuman, R. B., 61, 66, 67, 70, 72
Neumann, G., 418, 425
Neville, R. W., 88, 90
Newell, I. M., 399-402, 412, 413
Newell, N. D., 70, 72, 399, 408, 409, 413
Newman, W. A., 404, 405, 407, 408, 410, 413-416
Newman, W. A., and others, 407, 410, 413
Newport, R. L., 89, 90
Newton, G. B., 268, 276
Nicoll, R. S., 122, 128
Niebuhr, W. W., II, 198
Nikiforova, A. I., 262, 264, 276
Nikiforova, O. I., 51, 120, 128, 199
Nikitin, I. E., 109, 114
Nikitina, A. P., 267, 268, 276
Nikitina, A. P., and others, 267, 268, 276
Nilson, T. H., 338, 346
Nilsson, R., 69, 86, 88, 90
Nordgaard, O., 454, 456
Norford, B. S., 49, 50, 120, 126, 128
Norin, E., 50
Novikova, E. N., 262, 276
Nullo, F., 481, 482

Obut, A. M., 89, 90, 107, 108, 114
Oder, C. R. L., 42, 43
O'Donoghue, C. H., 454, 457
Odum, E. P., 250, 257
Oehlert, D. P., 199
Oertli, H. J., 307, 310, 379, 388
Okada, Y., 455-457
Oldenburg, D. W., 13, 24
Olin, E., 68, 72
Oliver, W. A., Jr., 54, 58, 112-114, 131, 133-136, 139-145, 151, 155, 191-194, 199
Oliver, W. A., Jr., and others, 140, 142
Ollerenshaw, N. C., 50
Olson, E. C., 227-229, 232-234, 236, 237, 480, 481
Opdyke, N. D., 56, 58, 59
Öpik, E. J., 256, 257
Oradovskaya, M. N., 51, 53, 57, 59
Orbigny, A. d', 302
Orians, G. H., 429, 432
Ormiston, A. R., 54, 55, 150, 155, 193, 195, 199
Orr, R. W., 203, 212
Ortmann, A. E., 2, 6
Osborne, F. F., 107, 114
Osburn, R. C., 454, 456
Ostenso, N. A., 449, 454
Oxburgh, E. R., 343, 347
Owen, H. G., 96, 343, 347

Packard, E. L., v
Packham, G. H., 199
Palmer, A. R., 471, 472, 481, 482
Panchen, A. L., 227, 234, 237, 480, 481
Pant, G., 313
Pantojo-Alor, J., 481, 482
Parent, H., 302
Paris, F., 87, 89, 90
Parker, F. L., 424, 425
Parker, R. L., 13, 24
Parodiz, J. J., 277, 279, 282, 284, 287
Parsley, R. L., 100, 101
Pascual, R., 324, 325, 331
Pastula, E. J., 454, 456
Pattison, J., and others, 242, 246
Patrulius, D., and others, 89, 90
Paul, C. R. C., 124, 128
Pautot, G., and others, 20, 25
Pavlov, V. V., 316, 319
Pedder, A. E. H., 140-145, 186, 194, 195, 199, 203, 207, 212
Pedder, A. E. H., and others, 203, 212
Peel, J. S., 40, 43, 48, 75
Peet, R. K., 259, 276
Pelletier, M., 302
Perch-Nielsen, K., 441, 444, 446
Perch-Nielsen, K., and others, 441, 444, 446
Perry, D. G., 186, 187, 190, 199
Perry, D. G., and others, 199
Petersen, C. G. J., 3, 6
Peterson, G. L., 337, 338, 347
Peterson, G. L., and others, 338, 347
Petri, S., 88, 90, 155, 160
Peypouquet, J. P., 382, 388
Phelan, T., 136, 140
Philip, G. M., 186, 199, 201-207, 212
Phillips, J. D., 15, 24, 25, 396
Phleger, F. B., and others, 425
Pichl, J., 265, 276
Pichler, R., 89, 90
Picard, M. D., 340, 341, 347
Pickett, J. W., 207
Pielou, E. C., 148, 155
Pierce, S. E., 337, 347
Piette, E., 302
Pilsbry, H. A., 280-283, 413-416
Pimm, A. C., 344, 347
Pitman, W. C., III, 15, 22-25, 343, 346
Plafker, G., 368
Plodowski, G., 198
Plumstead, E. P., 261, 276
Pojeta, J., Jr., 27, 34, 36
Pojeta, J., Jr., and others, 27, 36
Pokorný, V., 387, 388
Pollack, J. M., 55, 59
Poole, F. G., 139, 141
Poore, R. Z., 423-425
Pope, J. K., 94, 100, 102
Porter, J. W., 431, 433
Potter, A. W., 61, 72, 480, 481
Poumot, C., 86, 89, 90
Predtechenskij, N. N., 51, 120, 128
Pribyl, A., 106, 113
Price, D., 64, 68, 71
Price, J. W., Sr., 99, 102
Price, L. I., 230, 237
Prins, B., 347
Prokop, R. J., 94, 100-102

Racheboeuf, P. R., 197
Răileanu, G., and others, 89, 90
Ramsay, A. T. S., 382, 388
Ramsbottom, W. H. C., 122, 128
Rathbun, R., 155, 160
Raup, D. M., 453, 454
Rauscher, R., 89, 90
Raven, P. H., 313, 319, 436-438, 440-443, 445, 446
Rawson, P. F., 302
Ray, C. E., 359, 361-363, 366, 367, 369
Raymond, P. E., 52
Rayson, P., 438, 446
Reed, F. R. C., 100, 102, 155, 158-161, 165, 265, 276
Reese, E. S., 412, 414
Regnéll, G., 66, 72, 100, 102, 119, 128
Rehmer, J., 196, 197
Reid, J. L., and others, 403, 413
Reitlinger, E. A., 220, 226
Remane, A., 400, 414
Renaud, A., 199
Rennie, J. V. L., 94, 100, 102, 156, 158-161, 439, 446
Renouf, J. T., 199
Repenning, C. A., 359, 360, 362, 363, 366-369
Reuss, A. E., 296, 302
Rexroad, C. B., 122, 128
Reyment, R., 305
Rhodes, F. H. T., 89, 90
Rice, L., 414, 415
Rich, P. V., 321-323, 327, 329-331
Richards, H. G., 58, 59
Richards, P., 441, 446
Richter, E., 147, 156, 164
Richter, R., 147, 156, 164
Rickards, R. B., and others, 106, 114
Ricqlés, A., de, 236, 237
Ringis, J., 352, 354, 442, 445
Rioult, M., 290, 294, 302
Rivas, O. E. O., 324, 325, 331
Robardet, M., and others, 89, 90
Robertson, R., 392, 395-397
Robertson, R., and others, 397
Robinson, M., 402
Robinson, P., 341, 347
Robinson, P. L., 227, 230, 233-237
Robinson, W. I., 52
Robison, R. A., 481, 482
Rocha-Campos, A. C., 232, 237, 480, 482
Rogers, F. L., 414
Rogers, J. J. W., and others, 241, 246
Rogick, M. D., 454-457
Rohr, D. M., 39, 51, 481, 482
Romanchuk, T. V., 267, 268, 276
Romer, A. S., 228, 230-232, 238, 261, 276, 480, 481
Rosen, D. E., 2, 6
Rosenweig, M. L., 451, 455
Ross, A., 405, 407, 408, 410, 413-416
Ross, A., and others, 414, 415
Ross, C. A., vi, 215, 219, 221, 222, 226, 239, 241, 242, 245, 246, 261, 262, 266, 267, 276, 470, 480, 481
Ross, J. R. P., 264, 276
Ross, R. J., Jr., 27, 28, 32, 34-36, 40, 43, 53-59, 61, 72, 91, 96, 100, 106-108, 119, 128
Rothe, P., 321, 331
Rowe, J. S., 374, 377
Rowett, C. L., 261, 276, 480, 481
Rowley, R. R., 119, 128
Rozman, K. S., 56, 59
Rozovskaya, S. E., 217, 219, 220, 226
Ruddiman, W. F., 419, 425
Rudwick, J. J. S., 242
Ruedemann, R., 108, 114
Ruget-Perrot, C., 292, 296, 303
Ruggieri, G., 387, 388
Runcorn, S. K., 15, 23, 25
Runnegar, B., 27, 34, 36, 38, 43, 477, 481
Runnegar, B., and others, 38, 43
Russell, D. E., 229, 239
Russell, F. S., 106, 114
Russell, L. S., 282, 287
Rutten, M. G., 380, 388
Ryan, W. B. F., 387, 388
Ryan, W. B. F., and others, 387, 389
Ryder, R. T., and others, 339, 347
Ryland, J. S., 455, 457
Rzhonsnitskaya, M. A., 196, 199
Saidji, M., 89
Sainsbury, C. L., and others, 51, 53-55, 59
Sakagami, S., 266, 267, 276
Saldukas, R. B., 479, 481
Salfeld, H., 156, 157
Salter, J. W., 47, 49, 50, 156, 161
Salvat, B., 431, 433
Samylina, V. A., 314, 317, 319
Sandberg, C. A., 202, 203, 212
Sandberg, C. A., and others, 212
Sanders, H. L., 3, 7, 380, 388, 389
Sando, J. W., 171, 199
Sando, J. W., and others, 171, 199
Sannemann, D., 211
Saporta, G., 435, 446
Sars, G. O., 4, 7
Savage, N. M., 186, 190-194, 196, 199, 200, 203
Savage, N. M., and others, 191-194
Savage, R. J. G., 359, 369
Savile, D. B. O., 374, 377
Savin, S. M., 70, 72, 342, 347, 373, 376, 377, 385, 388
Savin, S. M., and others, 70, 72, 342, 347, 373, 376, 377
Scarlett, R. J., 326, 327, 331
Scheltema, R. S., 391-397, 450, 455
Schmarda, L. K., 1, 2, 7
Schmid, M. E., 86, 90
Schmidt, F., 53, 59
Schmidt, K. P., 1, 2, 7
Schmidtill, E., 303
Schnitker, D., 418, 425
Schopf, T. J. M., 229, 238, 451, 453, 455-457
Schott, W., 418, 421, 425
Schubert, R., 440, 446
Schuchert, C., 52, 56, 59, 61, 70, 72, 100, 102, 122, 128
Schuchert, C., and others, 128
Schultz, G., 89, 90
Schumacher, A., 392, 397
Schwalb, H., 86, 90
Schwarz, E. H. L., 156, 158
Schwartzlose, R. A., 402, 413
Sclater, J. G., 11, 13-15, 25, 344, 440, 442, 443, 446
Sclater, J. G., and others, 11, 13-15, 25
Sclater, P. L., 2, 7
Scofield, W. H., 49, 51, 52
Scott, A. J., 86, 90
Scrutton, C. T., 141, 143
Scupin, H., 200
Seddon, G., 194, 200, 202, 203, 212
Sedgwick, A., 254
Seilacher, A., 410, 414
Seward, A. C., 249, 253, 435, 446
Seyfert, C. K., 307, 309, 310
Shackleton, N. J., 18, 20, 25, 385, 389, 423, 425
Shand, S. J., 156, 158
Sharman, G., 411
Shcherbovich, S. F., 261, 276
Sheehan, P. M., 64, 67-72, 127, 128, 474, 481
Shishova, N. A., 267, 276
Shive, P. N., 343, 346
Shulga-Nesterenko, M. I., 262, 276
Shuto, T., 394, 397
Sigogneau-Russell, D., 229, 238
Silberling, N. J., 241, 246
Simberloff, D. S., 453, 455
Simpson, G. G., 105, 106, 114, 170, 200, 215, 226, 322, 324-327, 331, 332, 345, 347, 350, 355, 435, 436, 443, 446
Simpson, G. G., and others, 105, 106, 114
Singleton, F. A., 352-355
Sirkin, L. A., 307, 309, 310
Skevington, D., 71, 72, 106, 107, 110-112, 115
Skipp, B., 202, 203, 212, 220, 226, 477, 481
Skoglund, R., 68, 72
Skottsberg, C., 444, 446
Sloss, L. L., 127, 128
Smiley, C. J., 312-314, 319, 373, 377, 379
Smiley, C. J., and others, 373, 377
Smith, A. G., 25, 27-29, 33, 36, 61, 62, 72, 92, 96-98, 100, 113, 115, 207, 208, 211, 212, 239, 240, 242, 246, 273, 276, 382, 389, 443, 446, 465, 481
Smith, A. G., and others, 27-29, 33, 36, 61, 62, 72, 92, 96-98, 100, 113, 115, 207, 208, 211, 212, 273, 276, 382, 389, 465, 481
Smith, D. B., 261, 276
Smith, G. A., and others, 152, 154, 156
Smith, J. M. B., 443, 446
Smith, L., 161, 165
Smith, P. J., 439, 446, 481
Smith, R. E., 195
Smith, R. K., 420, 421, 422, 424
Smith, S., 141, 144, 145
Snigireva, M. P., 207, 212
Sobolevskaya, R. F., 107, 108, 114
Sokač, A., 387, 389
Sokolova, M. N., 3, 7
Solem, A., 277, 279-281, 283, 286, 287
Solle, G., 200
Sommer, F. W., 89, 90
Sorauf, J. E., 141, 145
Sorgenfrei, T., 392, 394, 397
Soule, D. F., 455, 457
Soule, J. D., 455, 457
Southward, A. J., 414, 415
Southward, E. C., 414, 415
Spath, L. F., 351, 355
Specht, R. L., 438, 446
Spielman, J. R., 89, 90
Spinosa, C., and others, 261, 276
Spitz, A., 38, 43
Spjeldnaes, N., 61, 62, 70, 72, 106, 107, 115, 455, 456
Springer, F., 122, 126, 128
Sprinkle, J., 92, 94, 100, 102
Spuko, P. R., 444, 446
Stancheva, M., 387, 389
Staplin, F. L., 77, 89, 90
Starke, J. M., Jr., 52
Stauffer, C. R., 89, 90
Stauffer, P. H., 244, 246, 247
Stchépinsky, V., 303
Stebbins, G. L., 435, 436, 446
Stehli, F. G., 239, 244, 247, 410, 414
Steidtmann, E., 54, 59
Steinmann, G., 156, 163, 293, 303, 352, 355
Stelck, C. R., 141, 144
Stepanov, D. L., 242, 247
Steuerwald, B. A., and others, 376, 377
Stevens, G. R., 291, 293, 295, 297-299, 303
Stevens, N. C., 49
Stewart, G. A., 205, 212
Stewart, J. H., 139, 141
Stewart, R. H., 463, 464
Stipanicic, P. N., 232, 238
Stommel, H., 382, 389, 421, 425
Strakhov, N. H., 234, 235, 238
Strimple, H. L., 128
Strusz, D. L., 200
Strusz, D. L., and others, 200
Struve, W., 55, 156-161, 200
Stukalina, G. A., 118, 128
Stumm, E. C., 140, 141, 143-145
Suarez-Soruco, R., 156, 157, 162
Surange, K. R., 312, 319
Surdam, R. C., 340, 347
Sutherland, P. K., 141, 241, 247
Sverdrup, H. E., and others, 418, 425
Swartz, F. M., 156, 157, 160, 163, 165, 166
Swartz, C. K., 141
Sweet, W. C., 194, 200-202, 204, 205, 212, 474, 482
Sweet, W. C., and others, 201, 212
Swinnerton, W. E., 228, 238
Sylvester-Bradley, P. C., 387, 388
Symons, D. T. A., 376, 378
Szaniawski, H., 86, 90

Taber, R. D., 373, 377
Tait, R. V., 70, 72
Takhtajan, A., 443, 446
Talent, J. A., 191, 196, 198, 200, 322, 323, 332
Talent, J. A., and others, 322, 323, 332
Talwani, M., 15, 25
Tamain, G., 107, 115
Tappan, H., 22, 25, 343, 344, 347
Tapponnier, P., 440, 446
Taquet, P., 230, 238
Tarling, D. H., 15, 23, 25, 439, 447
Taugourdeau, P., 86, 89, 90
Taylor, H. E., 70, 71
Tazawa, J., 481, 482
Tedford, R. H., 329, 332, 359, 360, 369

Teichert, C., 49, 261, 275
Telford, P. G., 191-194, 200-208, 210, 212
Terborgh, J., 451, 455
Termier, G., 253, 257, 265, 276
Termier, G., and others, 253, 257
Termier, H., 265, 276
Ters, M., 87, 90
Tesch, J. J., 422, 423, 425
Tewari, A. P., 228, 238
Theobald, N., 303
Therquem, O., 303
Thevenin, A., 290, 303
Thiede, J., 13, 25
Thiery, P., 303
Thomas, D. E., 108, 114, 115
Thomas, H. D., 293, 297, 303
Thomas, I., 156, 162, 163
Thoral, M., 100
Thorne, R. F., 437, 447
Thorslund, P., 66, 67, 71, 72
Thorson, G., 1, 3, 7, 391, 394, 397
Thorsteinsson, R., 49, 374, 378
Tillman, C. G., 196, 200
Tjernvik, T., 107, 115
Tomes, R. F., 290, 303
Tomlinson, J. T., 402, 405, 408, 410, 414
Toriyama, R., 261, 276
Toula, F., 359, 369
Tozer, E. T., 261, 276, 279, 280, 282, 283, 287, 374, 378
Trettin, H. P., 40, 43, 49
Triplehorn, D. M., and others, 374, 378
Tripp, R. P., 156, 165
Trizna, V. B., 262, 263, 276
Troedsson, G. T., 50, 55, 59
Trouessart, E. L., 363, 369
True, F. W., 359, 361, 369
Trueman, E. R., 34, 36
Truex, J. N., 335, 347
Turcotte, D. L., 343, 347
Turner, J. C. M., 107, 108, 110, 115
Turquier, Y., 409, 414
Tyler, J. H., 141, 144
Tynni, R., 89, 90
Tzaj, D. T., 109, 115

Ubaghs, G., 91, 92, 94, 97, 100-104, 126, 128
Udvardy, M. D. F., 405, 409, 410, 414
Ulrich, A., 156, 160, 161
Ulrich, E. O., 39, 43, 49, 51, 52
Umnova, N. I., 89, 90
U. S. Hydrographic Office, Washington, D. C., Pilot Charts, 392, 397
Urban, J. B., 77, 89, 90
Ustritsky, V. I., 243, 244, 247, 470, 479, 482
Utinomi, H., 414, 415
Uyeno, T. T., 195, 199, 207, 212

Vagvolgyi, J., 286, 287
Vail, P. R., 21, 22
Vakhrameev, V. A., 313, 319
Valentine, J. W., 22, 25, 61, 62, 72, 202, 212, 399-402, 409, 411, 412, 414, 450, 453, 455
van Andel, Tj. H., vii, 12, 13, 17, 21, 22, 25, 382, 465, 482
van Andel, Tj. H., and others, 17, 21, 25
van Beneden, P. J., 359, 369
van Bruggen, A. C., 282, 287
van Couvering, J. A., 376, 377, 463, 464
van der Voo, R., 239, 242
van Houten, F. B., 339, 347
Vaněk, J., 69, 71, 155, 159, 160, 162, 166
Vandercammen, A., 200
Varne, R., 440, 442, 443, 445
Vasilevskaya, N. D., 316, 319
Vaughn, P. P., 228, 234, 237, 238
Veevers, J. J., 15, 25
Verma, K. K., 228, 238
Vermeij, G. J., 400, 414, 427, 429, 430, 433
Vester, H., 435, 447
Vjushkov, P. B., 228, 229, 232, 237
Vogt, C., 2, 7
von Bitter, P. H., 202, 203, 207, 210, 212
Voorhis, A. D., 392, 397
Vostokova, V. A., 45, 51

Waagen, W., 265, 276
Wadia, D. N., 228, 238
Wagner, R. H., 311, 319
Wakefield, N. A., 237, 238
Walcott, C. D., 52
Waldman, M., 322, 332
Walker, C. A., 331
Wallace, A. R., 105, 115, 249, 253, 257
Walliser, O. H., 212
Walmsley, V. G., 196, 200
Walmsley, V. G., and others, 200
Warburg, E., 66, 72
Warner, D. J., 268, 276
Warren, J. W., 237, 238
Warren, P. S., 141, 144
Warren, P. T., 106, 111, 115
Wass, R. E., 266, 269, 276
Waterhouse, J. B., 27, 36, 243, 247, 250, 251, 253-257, 451, 455
Watkins, N. D., 352, 355
Watkins, R. M., 62, 72, 111, 115
Watkins, R. M., and others, 62, 72
Watteville, D. de, 454, 457
Webby, B., 52
Webster, G. D., 119, 122, 128
Wegener, A., 12, 25
Weisbord, N. E., 147, 156
Weissel, J. K., 352, 354, 355
Wells, J. W., 141, 143, 303, 410, 414, 463, 464
Wenger, R., 305, 310
Wentzel, J., 265, 276
Wens, W., 277, 279, 285, 287
Werner, W. E., 414, 415
West, R. M., and others, 347
Westbroek, Peter, 197, 200
Westoll, T. S., 228, 238
Wetherby, A. G., 100
Weyl, R., 15, 25
Weyland, H., 312, 319
White, C. D., 312, 319
Whitfield, R. P., 52
Whitmore, F. C., 463, 464
Whittington, H. B., 27, 29, 32, 36, 41, 43, 48, 49, 53-56, 58, 59, 61, 72, 110, 115, 482
Wicksten, M. K., 407
Wilckens, O., 349-351, 353, 355
Wild, H., 437, 447
Willard, B., 52
Williams, A., 61, 62, 69, 73, 170, 200, 250, 256, 257
Williams, C. B., 455
Williams, D. L., 438, 447
Williams, G. E., 169, 200, 257
Williams, G. R., 327, 332
Williams, H., 50
Williams, P. M., 403, 404, 413
Willis, J. C., 409, 414
Wilson, A. E., 46, 47, 49, 50, 100, 101
Wilson, E. O., 451, 454
Wilson, J. A., 338, 345, 347,
Wilson, J. A., and others, 345, 347
Wilson, J. T., 23, 25, 61, 73, 139, 141
Wiman, C., 71, 73
Winterer, E. L., 12, 15, 21, 25
Wise, O. A., Jr., 39, 43
Wise, O. A., Jr., and others, 39, 43
Wisnes, T. S., 40, 43
Wittekindt, H., 203, 212
Wolfart, R., 148-151, 156, 157, 159, 160-162, 165
Wolfbauer, C. A., 340, 347
Wolfe, J. A., 341, 342, 344, 347, 373, 374, 378, 436, 442, 447
Wolfe, J. A., and others, 378, 436, 447
Wood, A. E., 345, 347
Wood, G. D., 89, 90
Wood, J. G., 438, 445, 447
Woodring, W. P., 39, 43, 396, 397, 432, 433, 463, 464
Woodring, W. P., and others, 39, 43
Woodward, A. S., 228, 238
Woodward, H., 100, 102
Woodward, S. P., 105, 115
Worthington, L. V., 418, 425
Wray, J., 459, 464
Wright, A. D., 67, 71, 73
Wright, A. J., 200
Wright, R. P., 89, 90
Wright, R., 387
Wu, Y.-Y., 121, 126, 128
Wyllie, J. G., 402, 414

Yamaguchi, M., 429
Yamaguchi, T., 412
Yancey, T. E., 127, 240, 241, 247, 479, 480, 482
Yang, K., 267, 276
Yapaudjian, L., 88, 90
Yeltysheva, R. S., 117, 128
Yen, T.-C., 277, 287
Yochelson, E. L., 38-40, 43, 45, 47, 49, 51, 52, 243, 245, 277, 279-281, 287
Yole, R. W., 241, 247, 268, 276
Young, S. B., 371, 378
Youngquist, W., 50
Yü, W., 50
Yurtsev, B. A., 371, 376, 378

Zapfe, H., 369
Ziegler, A. M., 61, 62, 68, 72, 73, 119, 126, 128, 250, 257, 451, 454, 455
Ziegler, A. M., and others, 68, 73, 128, 250, 257, 451, 455
Ziegler, P. A., 139, 141
Ziegler, W., 201, 203, 205-207, 209, 210, 212
Zijderveld, J. D. A., 239, 242, 247
Zilch, A., 277, 279, 282
Zima, M. B., 109, 115
Zimina, V. G., 311, 312, 319
Zinsmeister, W. J., 351, 353, 355
Zullo, V. A., 402-404, 407, 412, 414-416

BIOGEOGRAPHIC, ECOLOGIC INDEX

Acadian Province, 457
African Subprovince, 150
Agnostid-Olenid biofacies, 482
Agnostid-Olenid trilobite faunas, 471
Alpine tundra, 371, 373
Amazon-Colombian Subprovince, 119, 192, 467
American-Siberian Province, 57
Andean Province, 147, 150, 153, 154
Angara floral unit, 477-480
Angaran Province, 242, 244, 245, 311, 312
Antarctic Province, 457
Appalachian chitinozoan-facies, 79
Appalachian Province, 149, 150, 152-154, 190, 191, 193
Appohimchi Subprovince, 119, 125, 135, 144, 191, 192
Arctic Canada-Siberia Province, 56
Arctic Permian Marine Fauna, 479
Arctic Province, 457
Arctic-steppe, 371, 376
"Arctic-steppe", 375
Arctic tundra, 371, 373-376
Argentinean Province, 457
"Atlantic", 477, 482
Atlantic Realm, 466, 468, 469
Atlantic Province, 467
Atlantic Region, 105, 107-112, 472
Atlantic unit, 470, 473
Atomodesma fauna, 471
Austral Province, 351

Baltic chitinozoan-facies, 79
Baltic Province, 79
Balto-Europe craton, 92, 93, 95-99
Balto-Scandian unit, 473, 474
Benthic Assemblage, 62, 64, 65, 76, 186
Benthic Assemblages 1 through 5, 62
Benthic Assemblage 2, 196
Benthic Assemblage 2 to 5, 64, 66
Benthic Assemblage 3, 64, 68, 186
Benthic Asemblage 3 to 4, 47
Benthic Assemblage 4, 64, 67, 70
Benthic Asemblages 4 and 5, 186, 196
Benthic Assemblages 4 through 6, 471, 472
Benthic Assemblage 5, 64
Benthic Assemblage 5 to 6, 68, 471, 482
Benthic Assemblage 6, 62, 64, 68-70, 98
Benthic Assemblage analysis, 61, 63
Benthic assemblage zones, 250
Benthic Marine Life Zone 3, 76, 77
Bohemian Communities, 190
Bohemian Province, 62, 69
Boreal Province, 151, 239-244, 478, 479
Boreal-Angara unit, 480
Boreal Belemnite Realm, 289, 297, 300
Boreal Realm, 295, 298, 299, 478, 479
Brazilian Province, 147, 150, 154

Californian Province, 399-404, 410
Californian Transition Zone, 399, 400, 404-408, 410, 412
Canadian-Siberian biogeographical unit, 193
Caribbean Province, 366
Cathaysian floral unit, 479
Cathaysian Province, 244, 311, 312
Cathaysian subunit, 480
Chilean Province, 457
China Province, 455-457
Chinese unit, 470, 472, 473
Clorinda "Community", 62, 68
Cordilleran biogeographic unit, 169, 190, 192, 193, 480
Cordilleran Province, 239-243
Cordilleran Region, 119, 191, 192, 476
Cordilleran Subprovince, 191
Cordilleran subunit, 479
Cordilleran-Uralian Subprovince, 192
Cyathaxonid Province, 480
Cylindroteuthididae Realm, 291, 295, 298

Detrital carbonate lithotope, 63, 65-69
Dicoelosia-Skenidioides Community Group, 471
Durhaminid Province, 480

Eastern Australian "Province", 194
Eastern North America Province, 191
Eastern North America "Province", 191
Eastern Americas Realm, 119, 124, 125, 131, 133-135, 137, 138, 144, 169, 191, 192, 465-469, 471, 476, 477
Eastern North Atlantic Boreal Province, 457
Eastern North Pacific Boreal Province, 457
Ethiopian Province, 455, 457
Euramerian floral unit, 479
Euramerian subunit, 480
Euramerican Province, 242, 311, 312
Europe "Province", 193
European Province, 201, 208-211
"European", 477, 482
European Craton, 91
European Province, 208-210
European unit, 470, 472, 473

Foliomena Community, 69, 70
Forest-tundra, 371, 374
Forest-tundra ecosystem, 376

Gangamopteris flora, 312
Glossopteris flora, 311, 312, 318, 478
Gondwana, 15, 17, 91, 92, 95-99, 147, 153, 154, 169, 192, 244, 305, 313, 321-323, 326, 451, 465, 471, 477
"Gondwana", 477
Gondwana floral unit, 480
Gondwana Province, 244, 311, 312
Gondwana Realm, 466-470, 478, 479
Gondwana unit, 480
Gondwanaland, 61, 134, 137, 233, 234, 236, 237, 244, 277, 282, 284, 285, 313, 328, 349, 351, 352, 354
Grandian Province, 239, 240, 242, 243
Grandian subunit, 479
Grandian unit, 480
Graptolite shale pelagic community, 65
Graptolitic facies, 106
Graptolitic shale lithotope, 64, 68, 69

Hawaiian Province, 457
Hibolites Realm, 291, 295, 298
Hirnantia Community, 64, 68-70
Hirnantian-type biogeographic unit, 482
"Hirnantian" fauna, 474
Holorhynchus Community, 64, 68
Homalonotid Community, 149

Iberian-Sahara chitinozoan-facies, 69
Icriodus-Polygnathus Community, 203
Indian Province, 455
Indo-European Province, 313
Interglacial tundra, 371, 375-377
Interglacial tundra ecosystems, 376

Kazakhstan flora, 477
Kolyma-Alaska Provinces, 56
Kuznetz-North American unit, 477, 478

Laurasia, 15, 17, 311, 312, 436, 465
Leiorhynchid brachiopod biofacies, 471
Lepidodendropsis floral unit, 477, 478
Lingula "Community", 62

Malayan Province, 455, 457
Malvinokaffric Province, 112, 113, 147, 194
Malvinokaffric Realm, 48, 119, 125, 127, 131, 147-154, 157, 162, 164, 192, 194, 196, 211, 466-471, 474-477, 482
Malvinokaffric unit, 190, 191, 473, 474
"Mediterranean", 477
Mediterranean Province, 61, 62, 67-70, 457
Mexican Province, 455, 457
Michigan Basin-Hudson Bay Province, 135-137

Midcontinent-Andean unit, 466, 469, 470, 478-480
Midcontinent-North Atlantic unit, 473
Midcontinent-southwestern North American region, 221, 222
Midcontinent unit, 474
Mongolo-Okhotsk Region, 119, 192, 193, 474-476
Monorakid Subprovince, 53, 56, 58
Mordvinian Province, 239-243, 479
Mucronaspis Community, 68
Mud-clay lithotope, 64, 65, 68, 69

Nevadan Subprovince, 191, 192
New Zealand biogeographic unit, 190
New Zealand Region, 119, 476
Noncarbonate siltstone lithotope, 64, 67, 68
North American Craton, 91-93, 95-99
North American floral unit, 479
North American Province, 61, 62
North American Realm, 215, 224
North American subunit, 480
North American unit, 470, 472, 473, 482
"North Atlantic Area", 78
North Atlantic Region, 119, 466-468, 471, 474-477
North Atlantic unit, 473, 474
North European Province, 61-63, 66, 68-70
North Silurian Realm, 119, 466, 474-476

Old World Province, 191-194
Old World Realm, 119, 123-125, 131, 133-135, 137-139, 152, 154, 191-194, 465-467, 471, 476, 477
Old World unit, 190
Oregonian Province, 399-404, 410

Pacific Province, 201, 208, 211, 366, 467
Pacific Realm, 468, 469
Pacific Region, 105, 107-113
Pacific unit, 473
"Paleotethyan", 477
Palmatolepis Community, 203
Pangaea, 9, 15, 17, 23, 226, 233, 305, 307, 380, 449, 451, 453, 465, 469-472, 474, 479, 480
Panthalassa, 15, 17, 23
Papuan Province, 455, 457
Paratethys, 359, 362-365, 367, 368, 386, 387, 395
Pelagic Community, 69
Pelagic Graptolite-dominated Community, 68
Periglacial steppe, 376
"Periglacial steppe", 375
Phacops-Metacryphaeus Community, 149
Polynesian Province, 457
Province IV, 56
Pyramus fossil community, 251

Reef, 64-66, 68, 73, 76, 77, 242, 243, 262-264, 289, 290, 292-297, 299
Reef lithotope, 63
Remopleuridid Province, 56
Rhenish-Bohemian biogeographic unit, 190, 191
Rhenish-Bohemian Region, 117, 119, 124, 125, 127, 192, 193, 467, 476, 477
Rhenish-Bohemian Subprovince, 193
Rhenish Communities, 190
Rhenish Communities Complex, 153

Sahara Province, 79
Siberian and North American subunits, 473
Siberian unit, 470, 472, 473
Siberian-American unit, 472
Siberian-North American Realm, 482
Siberian-North American unit, 473
Siltstone lithotope, 65
Skenidioides Community Group, 471
South African Province, 154, 457
South African-Falklands-Antarctic Province, 150
South African-Malvinan Province, 147, 150

South American Province, 150
South Australian Province, 457
Southern area, 78, 79
Southern California Province, 457
Starfish Community, 98
"Steppe-tundra", 375

Taimyr-Alaskan unit, 477, 478
Tasman biogeographic unit, 190
Tasman Region, 119, 192, 194, 471, 476, 477, 482
Tasman Subprovince, 191, 194
"Tethyan", 477, 482
Tethyan Province, 239-245
Tethyan Realm, 215, 217, 219-222, 224, 226, 479
Tethyan Relict, 409
Tethyan unit, 471, 473, 477, 478, 480
Tethys, 15, 17, 21, 23, 216, 217, 269, 270, 272, 274, 291, 360, 362, 363, 365, 379-383, 386, 387, 395, 396, 399, 479
Toquima-Table Head "realm", 57
Transition Zone, 401-403, 409
Trilobite-dominated Communities, 64, 68
Trinodid-Triarthrid biofacies, 471, 482
Tropical West African Province, 457
Tundra ecosystem, 371, 375-377
Tundra zones, 372

Uralian biogeographic unit, 190, 193
Uralian Region, 119, 192, 193, 476
Uralian Subprovince, 193
Uralian-Cordilleran Region, 119, 466, 471, 474-476
Uralian-Franklinian Realm, 215, 222
Uralian-Franklinian region, 219, 221
Uralian-Mongolo-Okhotsk Region, 119
Uralian-Tasman Region, 124
Uralian-Tasman-Cordilleran Regions, 125

Virginian Province, 450, 455, 457
Vogesina-Metacryphaeus Community, 149

Warrenella dominated faunas, 471
Weddellian Province, 349, 351-354
West Gondwanaland, 435-437
Western North America "Province", 191, 192
Western Atlantic Tropical Province, 457
Western North Pacific Boreal Province, 457

Yangzte Valley Subprovince, 119

ERRATA

Addition to references, p. 155:

Hall, J., and Clarke, J. M., 1888, Descriptions of the trilobites and other Crustacea of the Oriskany, Upper Helderberg, Hamilton, Portage, Chemung, and Catskill Groups: New York Geol. Survey, Paleontology 7, pt. 64, 236 p.

Addition to references, p. 413:

Kolosváry, G. von, 1942, Studien an Cirripedien. I. Material aus dem Hamburger Museum. II. Material aus dem Sammlungen der ungarischen Universitat: Zool. Anz., v. 137, p. 138-140.